Thermodynamics of Materials with Memory

Giovambattista Amendola • Mauro Fabrizio
John Murrough Golden

Thermodynamics of Materials with Memory

Theory and Applications

Second Edition

Springer

Giovambattista Amendola
Dipartimento di Matematica
University of Pisa
Pisa, Italy

Mauro Fabrizio
Dipartimento di Matematica
University of Bologna
Bologna, Italy

John Murrough Golden
Grangegorman Campus
Technological University - Dublin
Dublin, Ireland

ISBN 978-3-030-80533-3 ISBN 978-3-030-80534-0 (eBook)
https://doi.org/10.1007/978-3-030-80534-0

Mathematics Subject Classification: 74A15, 74D05, 30E20, 35J10, 35K05

This Springer imprint is published by the registered company Springer Nature Switzerland AG
The registered company address is: Gewerbestrasse 11, 6330 Cham, Switzerland

Dedicated to Adele, Alessandra, and Marie

Preface to Second Edition

The major motivations for writing a second edition of this book were to include important results from a series of papers published since 2011, to present a fairly detailed discussion of thermoelectromagnetism, in particular, free energies in dielectric materials, and also to explore some features of fractional derivative models of materials with memory.

The first of these aims has resulted in the merger of Chaps. 15 and 16 of the first edition, with the omission, for the sake of simplicity, of certain interesting though somewhat peripheral topics, and the presentation of new material in Chaps. 17–21 and 23 of the second edition. The second aim has given rise to Chaps. 6 and 22 of the new edition and the omission of the discussion of nonsimple electromagnetism from Eq. (3.7.20) to before (3.7.27) in the first edition.

A few minor corrections, noticed since the publication of the original edition, have also been made, together with slight changes required in earlier chapters by the merger of Chaps. 15 and 16.

The authors would like to thank Dipartimento di Matematica—Università di Pisa (Italy), Dipartimento di Matematica—Università di Bologna (Italy), Technological University—Dublin (Ireland), respectively, for support during the writing of the second edition. We also thank Gennaro Amendola once again for his invaluable advice on matters related to LaTeX.

Pisa, Italy
Bologna, Italy
Dublin, Ireland

Giovambattista Amendola
Mauro Fabrizio
John Murrough Golden

Preface to First Edition

This book arose out of a conversation that took place in a bookshop in Berkeley, California, almost a decade ago. The original motivation was to provide a text on continuum thermodynamics that would allow a systematic derivation and discussion of free-energy functionals for materials with memory, including in particular explicit expressions for the minimum and related free energies, which were being developed at the time.

Progress was very slow, due to other commitments. The vision of what the book would explore broadened considerably over the years, in particular to include minimal states and a new single-integral free-energy functional that explicitly depends on the minimal state. Also, it was decided to include a detailed description of an alternative approach to the analysis of the integrodifferential equations describing the evolution of viscoelastic materials under varying loads, using minimal states and free-energy functionals depending on the minimal state. This is a novel approach to a well-known topic.

Our desire was to make the work as self-contained as possible, so chapters dealing with the general theory of continuum mechanics were included, with sections devoted to classical materials, specifically elastic bodies and fluids without explicit memory-dependence. These provided essential background to the more general and modern developments relating to materials with memory.

It was furthermore felt that certain other topics had not been covered previously in book form and should be included, in particular control theory and the Saint-Venant and inverse problems, as well as some discussion of nonsimple behavior, for materials with memory.

The book is divided into four parts. The mathematical presentation in the first three parts is largely accessible not only to applied mathematicians but also to mathematically oriented engineers and scientists. However, a higher standard is required for some of the chapters in the final part.

The authors wish to thank S. Chirita, A. Lorenzi, M.G. Naso, and V. Pata for their aid in writing Chaps. 25 (Naso), 26 (Chirita), 28 (Pata), and 29 (Lorenzi). One of the authors gratefully acknowledges support for research travel from the Dublin Institute of Technology during the period of preparation of this work. All of us express our thanks to Gennaro Amendola for his very useful advice and help on certain deeper aspects of LaTeX.

Pisa, Italy
Bologna, Italy
Dublin, Ireland
February 2011

Giovambattista Amendola
Mauro Fabrizio
John Murrough Golden

Contents

Part IV The Dynamical Equations for Materials with Memory

Introduction

In this work, we consider materials the constitutive equations of which contain a dependence upon the past history of kinetic variables. In particular, we deal with the constraints imposed upon these constitutive equations by the laws of thermodynamics. Such materials are often referred to as *materials with memory* or with hereditary effects.

The study of materials with memory arises from the pioneering articles of Boltzmann [38, 39] and Volterra [316–318], in which they sought an extension of the concept of an elastic material. The key assumption of the theory was that the stress at a time t depends upon the history of the deformation up to t. The hypothesis that the remote history has less influence than the recent history is already implicit in their work. This assumption, later termed *the fading memory principle* by Coleman and Noll [73], is imposed by means of a constitutive equation for the stress, of integral type, which in the linear case involves a suitable kernel (relaxation function) that is a positive, monotonic, decreasing function.

In the classical approach to materials with memory, the state is identified with the history of variables carrying information about the input processes. We show in this book how Noll's definition of state [277] is more convenient for application to such materials. Indeed, Noll takes the material response as the basis for the definition of state: if an arbitrary continuation of different given histories leads to the same response of the material, then the given histories are equivalent and the state is represented as the class of all such equivalent histories. We refer to this class as the *minimal state*.

The concept of a minimal state is developed and applied in [176] to the case of linear viscoelasticity with scalar relaxation functions given by a sum of exponentials. A subsequent paper [90] presents a treatment in three dimensions and in the more general context of thermodynamically compatible (tensor-valued) relaxation functions, taking into account weak regularity of histories and processes.

A generalization of minimal states to materials under nonisothermal conditions is discussed in Sect. 7.4 of the present book. A functional $\mathbf{I}^t$ is introduced, given by (7.4.2) with the crucial property expressed by (7.4.3). This quantity characterizes the minimal state. Special cases of it are used in a variety of contexts in later chapters.

It is closely related to the response of the material after time t, where the input variable is null for a finite period after this time on the material element (i.e., a "small" neighborhood of a fixed and arbitrary point of the body) under consideration. This characterization of the state is an interesting alternative to the usual one based on knowledge of the deformation history.

It seems more appropriate to refer to materials with states characterized in this way as *materials with relaxation* rather than materials with fading memory.

For the usual definition of state, a fading-memory property of the response functional [67] is required, as opposed to the case in which the minimal state is adopted, where indeed the relaxation property of the response functional suffices. Obviously, whenever the stress-response functional is such that knowledge of the minimal state turns out to be equivalent to knowledge of the past history, the property of relaxation of the stress response implies fading memory of the related functional. In this sense, the class of materials with relaxation is larger than the one described by constitutive equations with fading memory.

A significant advantage of the response-based definition of state relates to the physical features of the state itself. Indeed, the "future stress" $\mathbf{I}^t(\tau)$ can be determined through measurements and does not require knowledge of the past history at all.

For materials with memory, there are in general many different functional forms with the required properties for a free energy. Some of these are functions of the minimal state, while others do not have this property (see, e.g., [90]).

In Part III, these functional forms are explored for different categories of materials with memory. We note that for a material whose constitutive relation for the response functional has a linear memory term, all free energies associated with this material have memory terms that are quadratic functionals.

A new class of *single-integral-type* free energies, for certain categories of relaxation functions, is introduced in Sect. 10.1.3 as a quadratic form of the time derivative of the state variable $\mathbf{I}^t$ (see, e.g., [189, 190] for discussion and analysis of single-integral type free energies that are quadratic forms of histories). For exponentially decaying relaxation functions, it can be shown that the dissipation associated with such energies is bounded below by a time-decay coefficient multiplied by the purely memory-dependent part of the free energy. This property turns out to be crucial in the analysis of PDEs relating to linear viscoelastic materials, which is developed in Part IV.

An analogous property holds for a family of multiple-integral free-energy functionals that are the generalization of the previous single-integral-type free energy. We may refer to such a family as the *n-family*. For $n = 1$, one recovers the single-integral case.

In Chaps. 11–15, explicit forms of the minimum free energy are derived both in the general nonisothermal case and, more specifically, for viscoelastic solids, fluids, and rigid heat conductors. Different forms of relaxation functions are also considered. The minimum free energy is always a function of the minimal state. Indeed, an explicit formula is derived in Sect. 12.2 for this quantity as a quadratic functional of minimal-state variables related to $\mathbf{I}^t$.

Chapter 16 results from the merger of two chapters of the first edition and the omission of certain topics, as noted in the preface to the second edition. Relaxation functions consisting of sums of decaying exponentials are considered. A family of free energies, including the minimum, maximum, and intermediate forms, are given explicitly. All of these are functions of state. Some plots of these free energies are added in this edition. In the new Chaps. 17–21, certain results for free energies that are quadratic functionals, published after the appearance of the first edition, are discussed.

Chapter 6 in Part II and Chap. 22 in Part III are new chapters dealing with thermoelectromagnetism and in particular free energies for dielectric materials, while Chap. 23 is also new and seeks to provide an introduction to fractional derivative models of materials with memory.

In Part IV, we observe that the new approach outlined above and the new free energies, in both cases adapted to the theory of viscoelasticity, have interesting applications to the PDEs governing the motion of a suitable class of viscoelastic bodies. In particular, the use of the new free energies given by quadratic forms of the minimal state variables yields results relating to well-posedness and stability for the IBVP. This formulation allows for initial data belonging to broader functional spaces than those usually considered in the literature, which are based on histories.

Indeed, the response-based definition of state is useful for both the study of IBVP on the one hand and the evolution of linear viscoelastic systems on the other hand.

Furthermore, an application of semigroup theory to this class of materials is presented. Here, besides having the system of equations in a more general form than for the classical approach, results on asymptotic stability are again obtained for initial data belonging to a space broader than the one usually employed when states and histories are identified.

The book is divided into four parts, Part I dealing with the general principles of continuum mechanics and with elastic materials and classical fluids, which of course provide the foundation for developments in later chapters. A general treatment of continuum thermodynamics is presented in Part II.

In Part III, materials that are described by constitutive equations with linear memory terms are discussed in some detail. The specific cases included are viscoelastic solids and fluids, together with rigid heat conductors. Also, as noted earlier, the derivation of explicit forms of free energies is considered in depth. Part IV deals with the application of results and ideas from Part III to the equations of motion of linear viscoelastic materials.

Notation conventions are described at the beginning of Appendix A. Relevant mathematical topics are summarized in Appendices A–C.

Part I

Continuum Mechanics and Classical Materials

1

Introduction to Continuum Mechanics

1.1 Introduction

In this initial chapter, we introduce various fundamentals: description of deformation, definition and interpretation of the strain and stress tensors, balance laws, and general restrictions on constitutive equations. These provide the foundation for later developments.

A number of excellent, indeed hardly to be bettered, presentations of these basic topics exist in the literature, notably in [188, 205, 251, 313, 314]. Several formulations of standard arguments in this chapter and the next are based on those in [188, 251]. Other relevant texts are [281], the recent work [23], and the review [262].

An introduction to some notation and results relating to finite-dimensional vector spaces required in this and later chapters is given in Sect. A.2.

1.2 Kinematics

1.2.1 Continuous Bodies: Deformations—Strain Tensors

We will consider bodies the mass of which is distributed continuously. Moreover, a given body will occupy different regions at different times, but none of these regions will be intrinsically associated with the body. Thus, formally, a *continuous body* $\mathcal{B}$ is a set of material points $\mathbf{X}, \mathbf{Y}, \ldots$ endowed with a structure defined by a class Φ of one-to-one mappings $\varphi : \mathcal{B} \to \mathcal{E}$, where $\mathcal{E}$ is the three-dimensional Euclidean space, such that

(i) $\varphi(\mathcal{B})$ is a Kellogg regular region;*

---

* By a Kellogg regular region, we mean a domain of the Euclidean space $\mathcal{E}$ bounded by a union of a finite number of surfaces of class C^1. A more formal definition of a subbody is given in [32, 253, 278] (see also [2]).

G. Amendola et al., *Thermodynamics of Materials with Memory*,
https://doi.org/10.1007/978-3-030-80534-0_1

(ii) if $\varphi, \psi \in \Phi$, then the function $\lambda = \varphi \circ \psi^{-1} : \psi(\mathcal{B}) \to \varphi(\mathcal{B}) \in C^1(\psi(\mathcal{B}))$ is called a *deformation* (of class C^1) of $\mathcal{B}$ from $\psi(\mathcal{B})$ to $\varphi(\mathcal{B})$;
(iii) if $\varphi \in \Phi$ and $\lambda : \varphi(\mathcal{B}) \to \mathcal{E}$ is a deformation of class C^1, then the mapping $\lambda \circ \varphi$ is also in Φ.

The functions φ are referred to as localizations of $\mathcal{B}$, and they determine the possible configurations of the body in the space $\mathcal{E}$. A localization provides at any material point $\mathbf{X} \in \mathcal{B}$ the corresponding geometric point $\mathbf{x} = \varphi(\mathbf{X}) \in \mathcal{E}$.

The hypotheses (i)–(iii) introduce a unique structure of a differential variety on $\mathcal{B}$.[†]

The set Φ of all possible localizations of $\mathcal{B}$ allows us to locate $\mathcal{B}$ in $\mathcal{E}$, as well as to define the internal constraints of material systems. We consider as an example a rigid body for which the class Φ must be defined so that for each pair $\varphi_1, \varphi_2 \in \Phi$, we have

$$d(\varphi_1(\mathbf{X}),\ \varphi_1(\mathbf{Y})) = d(\varphi_2(\mathbf{X}),\ \varphi_2(\mathbf{Y}))$$

for all $\mathbf{X}, \mathbf{Y} \in \mathcal{B}$, where d is the metric of the Euclidean space $\mathcal{E}$.

Moreover, for any continuous body $\mathcal{B}$, it is possible to determine a class $\mathcal{S}$ of subbodies $A,\ B,\ C, \ldots$ of $\mathcal{B}$, characterized by the following properties:

(a) $\mathcal{B} \in \mathcal{S}$;
(b) any element $A \in \mathcal{S}$ is such that $\varphi(A)$ is a Kellogg regular region of $\mathcal{E}$, for any $\varphi \in \Phi$.[‡]

On the class $\mathcal{S}$ of subbodies, it is possible to define a measure that allows us to give a definition of the density and of the mass.

Definition 1.2.1. *The mass is a measure $M : \mathcal{S} \to \mathbb{R}^+$ absolutely continuous with respect to the ordinary volume measure; that is, for each $\varphi \in \Phi$, there is an integrable function $\hat{\rho}_\varphi : \varphi(\mathcal{B}) \to \mathbb{R}^+$, the density of mass, such that the mass relative to A is*

$$M(A) = \int_{\varphi(A)} \hat{\rho}_\varphi(\mathbf{x})\, dv,$$

for all $A \in \mathcal{S}$.

A motion of $\mathcal{B}$ with respect to a fixed observer O is a sufficiently regular function[§]

$$\tilde{\chi} : \mathcal{B} \times I \to \mathcal{E}, \tag{1.2.1}$$

[†] In other words, the body $\mathcal{B}$ does not identify itself with a particular configuration, but with the set of all possible configurations it can assume and hence with a differential variety.

[‡] The given definition for a subbody is independent of the chosen localization φ. In fact, if ψ is another localization, then the transformation $\lambda = \varphi \circ \psi^{-1} : \psi(\mathcal{B}) \to \varphi(\mathcal{B})$ possesses an inverse of class C^1. Therefore, if $\varphi(A)$ is a regular region, then $\psi(A)$ will be a regular region of $\mathcal{E}$.

[§] With respect to each context, the condition of being sufficiently regular may have various senses. For our purposes, the function χ is assumed to be twice continuously differentiable in the domain of existence.

where $I \subset \mathbb{R}$ is a time interval.

In what follows, we will identify the body $\mathcal{B}$ with one of its particular configurations, namely the reference configuration $\varphi_0(\mathcal{B})$ (see Fig. 1.1). Moreover, the function $\tilde{\chi}$ is such that for each $t \in I$, the new function $\tilde{\chi}_t : \varphi_0(\mathcal{B}) \to \varphi_t(\mathcal{B})$, which represents the deformation of the body $\mathcal{B}$ from $\varphi_0(\mathcal{B})$ to $\varphi_t(\mathcal{B})$, has an inverse, that is, there exists a function

$$\tilde{\chi}_t^{-1} : \varphi_t(\mathcal{B}) \to \varphi_0(\mathcal{B}). \tag{1.2.2}$$

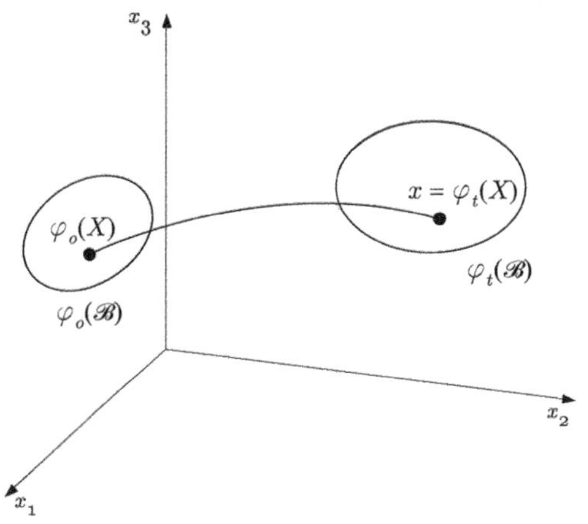

Fig. 1.1. The deformation of a body from $\varphi_0(\mathcal{B})$ to $\varphi_t(\mathcal{B})$

Hence $\tilde{\chi}_t$ is assumed to be one-to-one. This hypothesis expresses the requirement that the body does not penetrate itself. Thus, two distinct points of the configuration $\varphi_0(\mathcal{B})$ must be distinct in all other configurations.

It is possible to write the transformations (1.2.1) and (1.2.2) in the following forms:

$$\begin{aligned} \mathbf{x} &= \tilde{\chi}(\mathbf{X}, t), \\ \mathbf{X} &= \tilde{\chi}^{-1}(\mathbf{x}, t). \end{aligned} \tag{1.2.3}$$

The function defined by $(1.2.3)_1$ represents the position occupied by the particle $\mathbf{X}$ at the instant t, while relation $(1.2.3)_2$ locates the particle $\mathbf{X}$ that occupies the point $\mathbf{x}$ at the instant t. The variables $(\mathbf{X}, t)$ are the *Lagrangian* or *material coordinates*, while $(\mathbf{x}, t)$ are the *Eulerian* or *spatial coordinates*. The relations in (1.2.3) demonstrate that it is possible to express any physical quantity $\mathcal{F}$ in terms of material or spatial coordinates by

$$\tilde{\mathcal{F}}(\mathbf{X}, t) = \tilde{\mathcal{F}}(\tilde{\chi}^{-1}(\mathbf{x}, t),\ t) = \hat{\mathcal{F}}(\mathbf{x}, t). \tag{1.2.4}$$

Definition 1.2.2. *The Lagrangian description is the description of motion in terms of the variables* $(\mathbf{X}, t)$, *while the Eulerian description is that referring to the variables* $(\mathbf{x}, t)$.

As an example we consider the velocity of a particle $\mathbf{X}$ at the instant t, defined as

$$\tilde{\mathbf{v}}(\mathbf{X}, t) = \frac{\partial \tilde{\chi}}{\partial t}(\mathbf{X}, t);$$

on the basis of relation $(1.2.3)_2$, it is possible to express such a quantity in terms of the Eulerian variables as

$$\hat{\mathbf{v}}(\mathbf{x}, t) = \tilde{\mathbf{v}}(\tilde{\chi}^{-1}(\mathbf{x}, t),\ t). \tag{1.2.5}$$

Remark 1.2.3. The time derivative of a quantity $\mathcal{F}$ has different expressions, depending on the description. In fact, by direct differentiation with respect to t of (1.2.4), we obtain

$$\frac{\partial \tilde{\mathcal{F}}}{\partial t} = \frac{\partial \hat{\mathcal{F}}}{\partial t} + \nabla_{\mathbf{x}} \hat{\mathcal{F}} \cdot \mathbf{v}, \tag{1.2.6}$$

where $\nabla_{\mathbf{x}}$ is the spatial gradient operator. The partial derivative on the left is taken holding $\mathbf{X}$ fixed, while in that on the right, $\mathbf{x}$ is fixed.

The derivative $\frac{\partial \tilde{\mathcal{F}}}{\partial t}$ is the *material derivative* (or the total derivative), denoted by

$$\frac{d\hat{\mathcal{F}}}{dt} = \frac{\partial \tilde{\mathcal{F}}}{\partial t}. \tag{1.2.7}$$

If we choose as $\mathcal{F}$ the velocity $\mathbf{v}$, then, by virtue of (1.2.6), we have that the acceleration is given by

$$\mathbf{a} = \frac{\partial}{\partial t}\tilde{\mathbf{v}}(\mathbf{X}, t) = \frac{\partial \hat{\mathbf{v}}}{\partial t}(\mathbf{x}, t) + \nabla_{\mathbf{x}} \hat{\mathbf{v}}(\mathbf{x}, t)\, \mathbf{v}.$$

Definition 1.2.4. *The material gradient of deformation is the tensor*

$$\mathbf{F}(\mathbf{X}, t) = \nabla_{\mathbf{X}} \tilde{\chi}(\mathbf{X}, t), \qquad \text{that is, } F_{ij} = \frac{\partial \tilde{\chi}_i}{\partial X_j}, \tag{1.2.8}$$

where $\nabla_{\mathbf{X}}$ is the material gradient operator. The velocity gradient is the tensor

$$\mathbf{L}(\mathbf{X}, t) = \mathbf{L}(\tilde{\chi}(\mathbf{X}, t),\ t) = \nabla_{\mathbf{x}} \hat{\mathbf{v}}(\mathbf{x}, t). \tag{1.2.9}$$

Remark 1.2.5. If we set $\dot{\mathbf{F}} = \frac{\partial \mathbf{F}}{\partial t}$, then

$$\dot{\mathbf{F}} = \mathbf{L}\mathbf{F}. \tag{1.2.10}$$

In fact, we have

$$\dot{\mathbf{F}} = \nabla_{\mathbf{X}} \tilde{\mathbf{v}} = \nabla_{\mathbf{x}} \hat{\mathbf{v}} \nabla_{\mathbf{X}} \tilde{\chi}. \tag{1.2.11}$$

Remark 1.2.6. The requirement that the body does not penetrate itself is expressed by the assumption that

$$\det(\mathbf{F}) = \det(\nabla_{\mathbf{X}} \tilde{\chi}) \neq 0.$$

Furthermore, a deformation with $\det(\nabla_{\mathbf{X}} \tilde{\chi}) < 0$ cannot be reached by a continuous process of deformation starting from the reference configuration, that is, by a continuous one-parameter family $\tilde{\chi}_\sigma$ $(0 \le \sigma \le 1)$ of deformations with $\tilde{\chi}_0$ the identity,

$\tilde{\chi}_1 = \tilde{\chi}$, and $\det(\nabla_{\mathbf{X}}\tilde{\chi}_\sigma)$ never zero. Indeed, since $\det(\nabla_{\mathbf{X}}\tilde{\chi}_\sigma)$ is strictly positive at $\sigma = 0$, it must be strictly positive for all σ. Thus, we require that

$$\det \mathbf{F} > \mathbf{0}. \tag{1.2.12}$$

The above discussion motivates the following definition.

Definition 1.2.7. *By a deformation of $\mathcal{B}$, we mean a smooth one-to-one mapping $\tilde{\chi}$, which maps $\mathcal{B}$ onto a closed region in $\mathcal{E}$ and satisfies (1.2.12). The vector*

$$\mathbf{u}(\mathbf{X}, t) = \tilde{\chi}(\mathbf{X}, t) - \mathbf{X}$$

represents the displacement of $\mathbf{X}$. *A deformation with* $\mathbf{F}$ *constant is called homogeneous.*

The geometric significance of the tensor $\mathbf{F}$ becomes clear on observing that

$$\tilde{\chi}(\mathbf{X}', t) - \tilde{\chi}(\mathbf{X}, t) = \nabla_{\mathbf{X}}\tilde{\chi}(\mathbf{X}, t)(\mathbf{X}' - \mathbf{X}) + \mathbf{o}\left(|\mathbf{X}' - \mathbf{X}|\right),$$

for all $\mathbf{X}'$ in a neighborhood of $\mathbf{X}$, so that we can write

$$d\mathbf{x} = \mathbf{F}d\mathbf{X}. \tag{1.2.13}$$

Thus, the tensor $\mathbf{F}$ transforms the small quantity $d\mathbf{X}$ of the configuration $\varphi_0(\mathcal{B})$ into the small displacement $d\mathbf{x}$ of the configuration $\varphi_t(\mathcal{B})$ (see Fig. 1.2). Let

$$\mathbf{F} = \mathbf{R}\mathbf{U} = \mathbf{V}\mathbf{R} \tag{1.2.14}$$

be the polar decomposition of $\mathbf{F}$ at a given point, where $\mathbf{R}$ represents the rotation tensor, $\mathbf{U}$ is the right stretch tensor, and $\mathbf{V}$ is the left stretch tensor for the deformation $\tilde{\chi}$. Thus, $\mathbf{R}(P)$ measures the local rigid rotation of points near P, while $\mathbf{U}(P)$ and $\mathbf{V}(P)$ measure local stretching from P. The tensors $\mathbf{U}(P)$ and $\mathbf{V}(P)$ are symmetric. Since $\mathbf{U} = \sqrt{\mathbf{F}^T\mathbf{F}}$ and $\mathbf{V} = \sqrt{\mathbf{F}\mathbf{F}^T}$ involve the square roots of $\mathbf{F}^T\mathbf{F}$ and $\mathbf{F}\mathbf{F}^T$, their computation is often difficult. For this reason we introduce the *right and left Cauchy–Green strain tensors* $\mathbf{C}$ and $\mathbf{B}$, defined by

$$\mathbf{C} = \mathbf{U}^2 = \mathbf{F}^T\mathbf{F}, \quad \mathbf{B} = \mathbf{V}^2 = \mathbf{F}\mathbf{F}^T, \tag{1.2.15}$$

and note that

$$\mathbf{V} = \mathbf{R}\mathbf{U}\mathbf{R}^T, \qquad \mathbf{B} = \mathbf{R}\mathbf{C}\mathbf{R}^T.$$

In components, we have

$$C_{ij} = \sum_{m=1}^{3} \frac{\partial\tilde{\chi}_m}{\partial X_i}\frac{\partial\tilde{\chi}_m}{\partial X_j}, \quad B_{ij} = \sum_{m=1}^{3} \frac{\partial\tilde{\chi}_i}{\partial X_m}\frac{\partial\tilde{\chi}_j}{\partial X_m}.$$

Since $\mathbf{Cu} \cdot \mathbf{v} = \mathbf{Fu} \cdot \mathbf{Fv}$ for all $\mathbf{u}, \mathbf{v} \in V$ and $\mathbf{Cu} \cdot \mathbf{u} = \mathbf{Fu} \cdot \mathbf{Fu} > 0$ for all $\mathbf{u} \in V \backslash \{\mathbf{0}\}$, it follows that $\mathbf{C}$ is a symmetric and positive definite tensor (Sect. A.2.1).

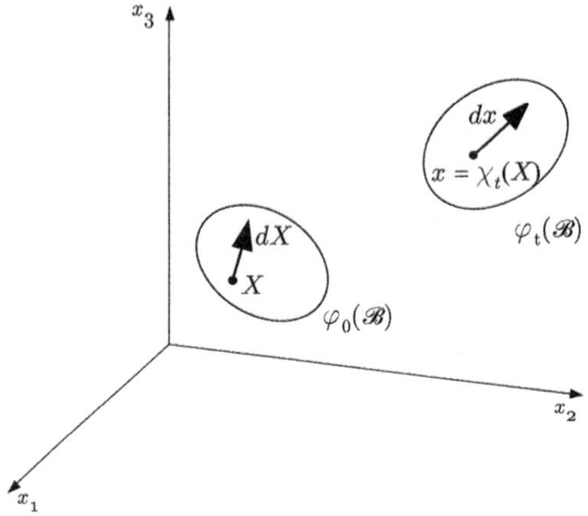

Fig. 1.2. The quantities $d\mathbf{X}$ and $d\mathbf{x}$ related by (1.2.13)

In view of the relation (1.2.12), it follows that $\mathbf{F}$ admits an inverse denoted by $\mathbf{F}^{-1}$, the *spatial gradient of deformation*, given by

$$\mathbf{F}^{-1} = \nabla_{\mathbf{x}} \mathbf{X}, \quad \text{or } F_{ij}^{-1} = \frac{\partial \tilde{\chi}_i^{-1}}{\partial x_j}.$$

With this we can introduce the *right and left Cauchy strain tensors*, $\mathbf{c}$ and $\mathbf{b}$, defined by

$$\mathbf{c} = \left(\mathbf{F}^{-1}\right)^T \mathbf{F}^{-1}, \quad \mathbf{b} = \mathbf{F}^{-1} \left(\mathbf{F}^{-1}\right)^T, \tag{1.2.16}$$

or, in components,

$$c_{ij} = \sum_{m=1}^{3} \frac{\partial \tilde{\chi}_m^{-1}}{\partial x_i} \frac{\partial \tilde{\chi}_m^{-1}}{\partial x_j}, \quad b_{ij} = \sum_{m=1}^{3} \frac{\partial \tilde{\chi}_i^{-1}}{\partial x_m} \frac{\partial \tilde{\chi}_j^{-1}}{\partial x_m}.$$

If $d\mathbf{X}$ and $\delta\mathbf{X}$ are two displacement elements related to the point $\mathbf{X}$ that at the instant t are transformed into two displacements $d\mathbf{x}$ and $\delta\mathbf{x}$, respectively, related to the point $\mathbf{x} = \tilde{\chi}(\mathbf{X}, t)$, so that

$$d\mathbf{x} = \mathbf{F} d\mathbf{X}, \quad \delta\mathbf{x} = \mathbf{F} \delta\mathbf{X}, \tag{1.2.17}$$

then

$$d\mathbf{x} \cdot \delta\mathbf{x} = d\mathbf{X} \cdot \mathbf{F}^T \mathbf{F} \delta\mathbf{X} = d\mathbf{X} \cdot \mathbf{C} \delta\mathbf{X}. \tag{1.2.18}$$

If the continuous body is rigid, then from the relation (1.2.18), we get necessarily $\mathbf{C} = \mathbf{1}$, the unit second-order tensor. When the body is not rigid, we can determine the elongation of the element $d\mathbf{X}$, associated with the tensor $\mathbf{C}$, by

$$|d\mathbf{x}|^2 = d\mathbf{x} \cdot d\mathbf{x} = d\mathbf{X} \cdot \mathbf{C} d\mathbf{X}, \tag{1.2.19}$$

so that the relative elongation is

$$|d\mathbf{x}|^2 - |d\mathbf{X}|^2 = 2\mathbf{E} d\mathbf{X} \cdot d\mathbf{X} = 2\mathbf{e} d\mathbf{x} \cdot d\mathbf{x},$$

where

$$\mathbf{E} = \frac{1}{2}(\mathbf{C} - \mathbf{1}) \quad \text{and} \quad \mathbf{e} = \frac{1}{2}(\mathbf{1} - \mathbf{c}) \tag{1.2.20}$$

are *Green's strain tensor* and *Almansi's strain tensor*, respectively. Obviously, for a rigid deformation of the body, we have $\mathbf{E} = \mathbf{0}$ and $\mathbf{e} = \mathbf{0}$. Thus, the tensor $\mathbf{E}$ appears as a measure of Lagrangian deformation, while the tensor $\mathbf{e}$ represents a measure of Eulerian deformation.

In terms of the displacement vector $\mathbf{u}(\mathbf{X}, t) = \tilde{\chi}(\mathbf{X}, t) - \mathbf{X}$ or $\mathbf{u}(\mathbf{x}, t) = \mathbf{x} - \tilde{\chi}^{-1}(\mathbf{x}, t)$, the gradients of deformation are

$$\mathbf{F} = \nabla_{\mathbf{X}} \mathbf{u} + \mathbf{1}, \qquad \mathbf{F}^{-1} = \mathbf{1} - \nabla_{\mathbf{x}} \mathbf{u},$$

and hence, from (1.2.20), the strain tensors are

$$\begin{aligned} \mathbf{E} &= \frac{1}{2}\left[\nabla_{\mathbf{X}} \mathbf{u} + (\nabla_{\mathbf{X}} \mathbf{u})^T + (\nabla_{\mathbf{X}} \mathbf{u})^T \nabla_{\mathbf{X}} \mathbf{u}\right], \\ \mathbf{e} &= \frac{1}{2}\left[\nabla_{\mathbf{x}} \mathbf{u} + (\nabla_{\mathbf{x}} \mathbf{u})^T - (\nabla_{\mathbf{x}} \mathbf{u})^T \nabla_{\mathbf{x}} \mathbf{u}\right]. \end{aligned} \tag{1.2.21}$$

The relations in (1.2.21) are known as the *strain–displacement (or geometrical) relations*.

Remark 1.2.8. (**Geometric Significance of the Strain Tensors**) The components E_{11}, E_{22}, and E_{33} of the strain tensor $\mathbf{E}$ characterize the relative elongations in the directions of $\mathbf{i}_1$, $\mathbf{i}_2$, and $\mathbf{i}_3$, respectively, while the components E_{ij}, with $i \neq j$, represent a measure of the variation of angles due to the process of deformation.

To see this, we first note that the relation (1.2.19) can be written in the form

$$\frac{|d\mathbf{x}|^2}{|d\mathbf{X}|^2} = \mathbf{N} \cdot \mathbf{C} \mathbf{N},$$

where $\mathbf{N} = \frac{d\mathbf{X}}{|d\mathbf{X}|}$. If we set $\Lambda_{(\mathbf{N})} = \frac{|d\mathbf{x}|}{|d\mathbf{X}|}$, then we have

$$\Lambda_{(\mathbf{N})} = (\mathbf{N} \cdot \mathbf{C} \mathbf{N})^{\frac{1}{2}} = \sqrt{\mathbf{N} \cdot (\mathbf{1} + 2\mathbf{E}) \mathbf{N}}.$$

We further introduce the unit elongation $E_{(\mathbf{N})}$ in the direction of unit vector $\mathbf{N}$, by

$$E_{(\mathbf{N})} = \Lambda_{(\mathbf{N})} - 1 = \frac{|d\mathbf{x}| - |d\mathbf{X}|}{|d\mathbf{X}|},$$

so that when $\mathbf{N} = \mathbf{i}_1$, for example, then

$$E_{(\mathbf{i}_1)} = \sqrt{1 + 2E_{11}} - 1,$$

and hence E_{11} appears as a measure for the elongation in the direction of $\mathbf{i}_1$.

Let us further consider the vectors $d\mathbf{X}_1 = dX_1\mathbf{i}_1$ and $d\mathbf{X}_2 = dX_2\mathbf{i}_2$, and let $d\mathbf{x}_1 = \mathbf{F}d\mathbf{X}_1$ and $d\mathbf{x}_2 = \mathbf{F}d\mathbf{X}_2$ be the corresponding vectors in the current configuration. Obviously, we have $d\mathbf{X}_1 \cdot d\mathbf{X}_2 = 0$, that is, the angle Θ_{12} between these vectors is $\frac{\pi}{2}$. On the other hand, the corresponding angle θ_{12} between the vectors $d\mathbf{x}_1$ and $d\mathbf{x}_2$ is given by

$$\cos\theta_{12} = \frac{d\mathbf{x}_1 \cdot d\mathbf{x}_2}{|d\mathbf{x}_1|\,|d\mathbf{x}_2|} = \frac{C_{12}}{\sqrt{C_{11}C_{22}}} = \frac{2E_{12}}{\sqrt{(1 + 2E_{11})(1 + 2E_{22})}},$$

and hence E_{12} appears as a measure of the variation of the angle Θ_{12} due to the deformation.

We now recall that given a tensor $\mathbf{S} \in \mathrm{Lin}(\mathbb{R}^3)$, the determinant of $\mathbf{S} - \lambda\mathbf{1}$ admits the representation (the Cayley–Hamilton theorem)

$$\det(\mathbf{S} - \lambda\mathbf{1}) = -\lambda^3 + I_1(\mathbf{S})\lambda^2 - I_2(\mathbf{S})\lambda + I_3(\mathbf{S})$$

for every $\lambda \in \mathbb{R}$, where

$$\begin{aligned} I_1(\mathbf{S}) &= tr\mathbf{S} = S_{11} + S_{22} + S_{33}, \\ I_2(\mathbf{S}) &= \frac{1}{2}\left[(tr\mathbf{S})^2 - tr\left(\mathbf{S}^2\right)\right], \\ I_3(\mathbf{S}) &= \det\mathbf{S}. \end{aligned} \tag{1.2.22}$$

We call $I_1(\mathbf{S})$, $I_2(\mathbf{S})$, and $I_3(\mathbf{S})$ the *principal invariants* of $\mathbf{S}$ and observe that they are invariant under changes of reference frames. We also note that any other invariant of $\mathbf{S}$ is a function of its principal invariants. When $\mathbf{S}$ is symmetric, the principal invariants are completely characterized by the spectrum $\{\lambda_1, \lambda_2, \lambda_3\}$ of $\mathbf{S}$. Indeed,

$$\begin{aligned} I_1(\mathbf{S}) &= \lambda_1 + \lambda_2 + \lambda_3, \\ I_2(\mathbf{S}) &= \lambda_1\lambda_2 + \lambda_2\lambda_3 + \lambda_3\lambda_1, \\ I_3(\mathbf{S}) &= \lambda_1\lambda_2\lambda_3. \end{aligned}$$

By substituting $\mathbf{S}$ by $\mathbf{C}$, $\mathbf{c}$, $\mathbf{E}$, or $\mathbf{e}$ in the above relations, we can determine expressions for the principal invariants of these tensors and relationships between them. Thus, from (1.2.20) and (1.2.22), we obtain

$$\begin{aligned}
&I_1(\mathbf{C}) = 3 + 2I_1(\mathbf{E}), \quad I_2(\mathbf{C}) = 3 + 4I_1(\mathbf{E}) + 4I_2(\mathbf{E}),\\
&I_3(\mathbf{C}) = 1 + 2I_1(\mathbf{E}) + 4I_2(\mathbf{E}) + 8I_3(\mathbf{E}),\\
&I_1(\mathbf{c}) = 3 - 2I_1(\mathbf{e}), \quad I_2(\mathbf{c}) = 3 - 4I_1(\mathbf{e}) + 4I_2(\mathbf{e}),\\
&I_3(\mathbf{c}) = 1 - 2I_1(\mathbf{e}) + 4I_2(\mathbf{e}) - 8I_3(\mathbf{e}).
\end{aligned}$$

Moreover, we observe that the relations (1.2.15), $(1.2.16)_1$, and $(1.2.22)_3$ give

$$I_3(\mathbf{C}) = (\det \mathbf{F})^2, \quad I_3(\mathbf{c}) = \frac{1}{(\det \mathbf{F})^2},$$

and hence

$$I_3(\mathbf{C})I_3(\mathbf{c}) = 1.$$

Definition 1.2.9. *The stretching* $\mathbf{D}$ *(or velocity of deformation) is*

$$\mathbf{D} = \frac{1}{2}\left(\mathbf{L} + \mathbf{L}^T\right) = \frac{1}{2}\left[\nabla_{\mathbf{x}}\hat{\mathbf{v}} + (\nabla_{\mathbf{x}}\hat{\mathbf{v}})^T\right], \tag{1.2.23}$$

where $\mathbf{L}$ *is defined by (1.2.9), while the spin* $\boldsymbol{\Omega}$ *is*

$$\boldsymbol{\Omega} = \frac{1}{2}\left(\mathbf{L} - \mathbf{L}^T\right) = \frac{1}{2}\left[\nabla_{\mathbf{x}}\hat{\mathbf{v}} - (\nabla_{\mathbf{x}}\hat{\mathbf{v}})^T\right]. \tag{1.2.24}$$

Thus, the stretching and the spin represent the symmetric and skew parts of the spatial gradient of velocity, respectively. Moreover, we have

$$\mathbf{L} = \mathbf{D} + \boldsymbol{\Omega}. \tag{1.2.25}$$

Note that

$$\begin{aligned}
\frac{d}{dt}|d\mathbf{x}|^2 &= \frac{d}{dt}(d\mathbf{x} \cdot d\mathbf{x}) = 2\frac{d}{dt}(d\mathbf{x}) \cdot d\mathbf{x}\\
&= 2\frac{d}{dt}(\mathbf{F}d\mathbf{X}) \cdot d\mathbf{x} = 2\frac{d}{dt}(\mathbf{F})d\mathbf{X} \cdot d\mathbf{x},
\end{aligned}$$

and hence, in view of relation (1.2.10),

$$\begin{aligned}
\frac{d}{dt}|d\mathbf{x}|^2 &= 2\mathbf{LF}d\mathbf{X} \cdot d\mathbf{x} = 2\mathbf{L}d\mathbf{x} \cdot d\mathbf{x} = 2d\mathbf{x} \cdot \mathbf{L}^T d\mathbf{x}\\
&= 2d\mathbf{x} \cdot \left(\frac{\mathbf{L} + \mathbf{L}^T}{2}\right) d\mathbf{x} = 2d\mathbf{x} \cdot \mathbf{D}d\mathbf{x}.
\end{aligned} \tag{1.2.26}$$

Thus, the stretching $\mathbf{D}$ is a measure of the variation per unit time of the arc $(d\mathbf{x})^2$. Therefore, when $\mathbf{D} = \mathbf{0}$, then there is no change in $|d\mathbf{x}|^2$ over time.

Theorem 1.2.10. *A necessary and sufficient condition for a motion to be locally rigid is* $\mathbf{D} = \mathbf{0}$.

Proof. From Taylor's formula, the velocity in a neighborhood of the point $\mathbf{x}_0$ is

$$\hat{\mathbf{v}}(\mathbf{x},t) = \hat{\mathbf{v}}(\mathbf{x}_0,t) + \nabla_{\mathbf{x}}\hat{\mathbf{v}}(\mathbf{x}_0,t)(\mathbf{x}-\mathbf{x}_0) + \mathbf{o}\left(|\mathbf{x}-\mathbf{x}_0|\right),$$

so that in view of the relation (1.2.25), we obtain

$$\hat{\mathbf{v}}(\mathbf{x},t) = \hat{\mathbf{v}}(\mathbf{x}_0,t) + \mathbf{D}(\mathbf{x}_0,t)(\mathbf{x}-\mathbf{x}_0) + \boldsymbol{\Omega}(\mathbf{x}_0,t)(\mathbf{x}-\mathbf{x}_0) + \mathbf{o}\left(|\mathbf{x}-\mathbf{x}_0|\right).$$

Since $\boldsymbol{\Omega}$ is a skew-symmetric tensor, it follows that it is possible to associate with it the vector $\omega = \Omega_{32}\mathbf{i}_1 + \Omega_{13}\mathbf{i}_2 + \Omega_{21}\mathbf{i}_3$, known as the *vorticity vector*, such that

$$\boldsymbol{\Omega}(\mathbf{x}-\mathbf{x}_0) = \omega \times (\mathbf{x}-\mathbf{x}_0). \tag{1.2.27}$$

Therefore, in a neighborhood of the point $\mathbf{x}_0$, neglecting terms of order higher than 1 in $|\mathbf{x}-\mathbf{x}_0|$, we have

$$\hat{\mathbf{v}}(\mathbf{x},t) = \hat{\mathbf{v}}(\mathbf{x}_0,t) + \omega \times (\mathbf{x}-\mathbf{x}_0) + \mathbf{D}(\mathbf{x}-\mathbf{x}_0). \tag{1.2.28}$$

Thus, when $\mathbf{D} = \mathbf{0}$, the velocity is a composition of a translation and a rotation, which is a rigid motion.

Conversely, when the motion is rigid, (1.2.28) implies that $\mathbf{D} = \mathbf{0}$. □

Remark 1.2.11. In general, as can be seen from (1.2.28), the motion is a superposed rigid motion on an instantaneous extension.

From (1.2.24) and (1.2.27), we have

$$\omega = \Omega_{32}\mathbf{i}_1 + \Omega_{13}\mathbf{i}_2 + \Omega_{21}\mathbf{i}_3 = \frac{1}{2}\left[\left(\frac{\partial v_3}{\partial x_2} - \frac{\partial v_2}{\partial x_3}\right)\mathbf{i}_1 + \left(\frac{\partial v_1}{\partial x_3} - \frac{\partial v_3}{\partial x_1}\right)\mathbf{i}_2 + \left(\frac{\partial v_2}{\partial x_1} - \frac{\partial v_1}{\partial x_2}\right)\mathbf{i}_3\right],$$

and hence

$$\omega = \frac{1}{2}(\nabla_{\mathbf{x}} \times \mathbf{v}).$$

If $\omega = \mathbf{0}$, then we say that the motion is *irrotational*, and the velocity field has no vortices. In this case there exists a scalar field such that $\mathbf{v} = \nabla_{\mathbf{x}}\varphi$, as stated by the following theorem.

Theorem 1.2.12. *Let D be a given simply connected volume in $\mathbb{R}^3$ and $\mathbf{v} : D \to \mathbb{R}^3$ a function of class $C^1(D)$ that satisfies*

$$\nabla_{\mathbf{x}} \times \mathbf{v} = \mathbf{0} \quad \text{in } D.$$

Then there exists a function $\varphi : D \to \mathbb{R}$ such that

$$\mathbf{v} = \nabla_{\mathbf{x}}\varphi \quad \text{in } D.$$

Proof. Let S be an arbitrary surface contained in D, and let C be its relative boundary curve. Under appropriate regularity assumptions upon S and C, we can use Stokes's formula

$$\int_S (\nabla_{\mathbf{x}} \times \mathbf{v}) \cdot \mathbf{n} d\sigma = \int_C \mathbf{v} \cdot d\mathbf{x} = \int_C v_1 dx_1 + v_2 dx_2 + v_3 dx_3,$$

where, since $\nabla_{\mathbf{x}} \times \mathbf{v} = \mathbf{0}$, the differential form $v_1 dx_1 + v_2 dx_2 + v_3 dx_3$ is a total differential. Therefore, there exists a potential φ such that

$$d\varphi = v_1 dx_1 + v_2 dx_2 + v_3 dx_3,$$

which yields $\mathbf{v} = \nabla_{\mathbf{x}} \varphi$. □

Let us consider the transformation between the configuration $\varphi_0(\mathcal{B})$ and the configuration $\varphi_t(\mathcal{B})$ given by

$$\tilde{\chi}_t : \varphi_0(\mathcal{B}) \to \varphi_t(\mathcal{B}).$$

The Jacobian of the transformation,

$$J(\mathbf{X}, t) = \det\left(\frac{\partial \tilde{\chi}_t}{\partial \mathbf{X}}\right) = \det \mathbf{F}(\mathbf{X}, t),$$

is a measure of volume change due to the deformation. If we denote by dv_0 a volume element in the configuration $\varphi_0(\mathcal{B})$ and by dv_t the corresponding volume element in the configuration $\varphi_t(\mathcal{B})$, then we have

$$dv_t = J dv_0. \tag{1.2.29}$$

Theorem 1.2.13. *The time derivative of the Jacobian is given by*

$$\frac{dJ}{dt} = \dot{J} = J \mathrm{div}_{\mathbf{x}} \mathbf{v}, \tag{1.2.30}$$

where $\mathrm{div}_{\mathbf{x}}$ *is the spatial divergence operator.*

Proof. Direct differentiation with respect to t of the relation $J = \det \mathbf{F}$ gives

$$\dot{J} = \frac{dJ}{dt} = \dot{\mathbf{F}} \cdot \mathbf{A},$$

where the tensor $\mathbf{A}$ has components $A_{hm} = J\left(\mathbf{F}^{-1}\right)_{mh}$. Therefore, using relation (1.2.10), we obtain

$$\dot{J} = L_{hk} F_{km} A_{hm} = J L_{hk} F_{km} \left(F^{-1}\right)_{mh} = J L_{hk} \delta_{kh} = J L_{hh},$$

which is relation (1.2.30). □

Remark 1.2.14. It is understood that these italic subscripts range over 1, 2, and 3. Moreover, we use the convention of summation over repeated subscripts, unless stated otherwise.

Definition 1.2.15. *A deformation* $\mathbf{x} = \tilde{\chi}(\mathbf{X}, t)$ *is isochoric (volume-preserving) if given any subbody* B *of* $\varphi_0(\mathcal{B})$, *we have* $\text{vol}(\tilde{\chi}(B)) = \text{vol}(B)$.

An immediate consequence of this definition is the following result.

Proposition 1.2.16. *A deformation is isochoric if and only if* $\det \mathbf{F} = 1$.

Remark 1.2.17. From relation (1.2.19), we deduce that

$$\frac{d}{dt}|d\mathbf{x}|^2 = d\mathbf{X} \cdot \dot{\mathbf{C}} d\mathbf{X},$$

so that comparing with relation (1.2.26) and using (1.2.17), we obtain

$$\dot{\mathbf{C}} = 2\mathbf{F}^T \mathbf{D} \mathbf{F}.$$

Moreover, $(1.2.20)_1$ gives

$$\dot{\mathbf{E}} = \mathbf{F}^T \mathbf{D} \mathbf{F}.$$

1.2.2 Small Deformations: The Saint-Venant Compatibility Conditions

We now study the behavior of the various kinematic fields when the displacement vector is of the form $\mathbf{u}_\epsilon = \varepsilon \mathbf{u}$, where ε is a parameter such that ε^p is negligible if $p \geq 2$, while $\mathbf{u}$ is a vector independent of ε. The theory corresponding to such small displacements is known as the *infinitesimal or linear theory of deformation*. In such a theory, we have

$$x_i = X_i + u_{\epsilon i},$$

and the partial derivatives of the displacement vector with respect to the spatial coordinates coincide with the partial derivatives of the same vector with respect to the material coordinates. In fact, we have, for example,

$$\frac{\partial u_{\epsilon i}}{\partial X_1} = \frac{\partial u_{\epsilon i}}{\partial x_j}\frac{\partial x_j}{\partial X_1} = \frac{\partial u_{\epsilon i}}{\partial x_j}\left(\delta_{1j} + \frac{\partial u_{\epsilon j}}{\partial X_1}\right) = \frac{\partial u_{\epsilon i}}{\partial x_1} + O\left(\varepsilon^2\right), \quad \text{etc.}$$

On the basis of relations of this type and from (1.2.21), we deduce that the Lagrangian and Eulerian strain tensors $\mathbf{E}$ and $\mathbf{e}$ coincide with the infinitesimal strain tensor $\boldsymbol{\varepsilon}$ defined by

$$\boldsymbol{\varepsilon} = \frac{1}{2}\left(\nabla \mathbf{u} + \nabla \mathbf{u}^T\right), \tag{1.2.31}$$

where $\nabla \mathbf{u} = \nabla_{\mathbf{x}} \mathbf{u} = \nabla_{\mathbf{X}} \mathbf{u}$. In component form, we have

$$\varepsilon_{ij} = \frac{1}{2}\left(u_{i,j} + u_{j,i}\right). \tag{1.2.32}$$

Theorem 1.2.18. (Saint-Venant's Compatibility Conditions) *The infinitesimal strain tensor* ε_{ij} *corresponding to a displacement vector field* $\mathbf{u}$ *of class* C^3 *satisfies the following compatibility equations:*

$$\begin{aligned}&\varepsilon_{ii,jj} + \varepsilon_{jj,ii} = 2\varepsilon_{ij,ij} \quad (i \neq j, \textit{ not summed}),\\ &\varepsilon_{ij,rr} + \varepsilon_{rr,ij} = \varepsilon_{jr,ir} + \varepsilon_{ir,jr} \quad (i \neq j \neq r \neq i, \textit{ not summed}).\end{aligned} \tag{1.2.33}$$

Moreover, if B_0 *is a simply connected region in* $\mathbb{R}^3$ *and* ε_{ij} *is a symmetric tensor of class* C^2 *defined on* B_0 *satisfying the conditions described by (1.2.33), then there exists a displacement vector field* **u** *such that its corresponding strain tensor calculated by means of relation (1.2.32) coincides with* ε_{ij}*. Such a displacement vector field is given by*

$$u_i(\mathbf{x}) = \int_{\mathbf{x}_0}^{\mathbf{x}} \left(\varepsilon_{ij} + \omega_{ij}^*\right) d\xi_j + \omega_{ij}^0\left(x_j - x_j^0\right) + u_i^0, \tag{1.2.34}$$

where

$$\omega_{ij}^*(\mathbf{x}) = \int_{\mathbf{x}_0}^{\mathbf{x}} \left(\varepsilon_{ik,j} - \varepsilon_{kj,i}\right) d\xi_k,$$

while $\omega_{ij}^0 = -\omega_{ji}^0$ *and* u_i^0 *are arbitrary constants. Also, the integrals are independent of the curve connecting the points* $\mathbf{x}_0$ *and* **x***.*

Proof. We first note that the relations in (1.2.33) are identically satisfied for ε_{ij} given by (1.2.32).

In order to prove the second part of the theorem, we introduce the skew-symmetric tensor

$$\omega_{ij} = \frac{1}{2}\left(u_{i,j} - u_{j,i}\right),$$

which, when coupled with (1.2.32), gives

$$u_{i,j} = \varepsilon_{ij} + \omega_{ij}.$$

Furthermore,

$$du_i = u_{i,j}dx_j = \left(\varepsilon_{ij} + \omega_{ij}\right)dx_j \tag{1.2.35}$$

is an exact differential in B_0 (that is, $u_{i,jk} = u_{i,kj}$) if and only if

$$\varepsilon_{ij,k} + \omega_{ij,k} = \varepsilon_{ik,j} + \omega_{ik,j}. \tag{1.2.36}$$

By a cyclic permutation of the indices i, j, and k in (1.2.36), we obtain

$$\varepsilon_{jk,i} + \omega_{jk,i} = \varepsilon_{ji,k} + \omega_{ji,k} \tag{1.2.37}$$

and

$$\varepsilon_{ki,j} + \omega_{ki,j} = \varepsilon_{kj,i} + \omega_{kj,i}. \tag{1.2.38}$$

If we now add (1.2.36) and (1.2.37) and from the result subtract (1.2.38), taking into account the relations $\varepsilon_{ij} = \varepsilon_{ji}$ and $\omega_{ij} = -\omega_{ji}$, then we obtain

$$\omega_{ij,k} = \varepsilon_{ik,j} - \varepsilon_{kj,i}.$$

Furthermore, $d\omega_{ij} = \omega_{ij,k}dx_k = \left(\varepsilon_{ik,j} - \varepsilon_{kj,i}\right)dx_k$ is an exact differential in B_0 (that is, $\omega_{ij,kl} = \omega_{ij,lk}$) if and only if

$$\varepsilon_{ik,jl} - \varepsilon_{kj,il} = \varepsilon_{il,jk} - \varepsilon_{lj,ik},$$

or

$$\varepsilon_{ij,kl} + \varepsilon_{kl,ij} - \varepsilon_{il,jk} - \varepsilon_{jk,il} = 0. \tag{1.2.39}$$

It is easy to verify that the relations in (1.2.39) are equivalent to those given by (1.2.33), and from these conditions, it follows that $d\omega_{ij} = \omega_{ij,k}dx_k = \left(\varepsilon_{ik,j} - \varepsilon_{kj,i}\right)dx_k$ is an exact differential, giving

$$\omega_{ij}(\mathbf{x}) = \int_{\mathbf{x}_0}^{\mathbf{x}} \left(\varepsilon_{ik,j} - \varepsilon_{kj,i}\right) d\xi_k + \omega_{ij}^0, \tag{1.2.40}$$

where $\omega_{ij}^0 = -\omega_{ji}^0$ are arbitrary constants. We note that the above integral is independent of the curve connecting the points $\mathbf{x}_0$ and $\mathbf{x}$.

At this stage we observe that the necessary and sufficient conditions for the integrability of the differential form (1.2.35) are satisfied and hence

$$u_i = \int_{\mathbf{x}_0}^{\mathbf{x}} \left(\varepsilon_{ij} + \omega_{ij}\right) d\xi_j + u_i^0, \tag{1.2.41}$$

where u_i^0 are arbitrary constants.

Finally, we substitute $\omega_{ij}(\cdot)$ given by (1.2.40) into (1.2.41) to obtain the relation (1.2.36). □

Remark 1.2.19. From the above analysis, we can deduce that $\varepsilon_{ij} = 0$ if and only if $\mathbf{u}$ is an infinitesimal rigid displacement $\mathbf{u}^*$, given by

$$u_i^* = a_i + e_{ijk}x_j b_k,$$

where $a_i = u_i^0 - \omega_{ij}^0 x_j^0$ and $b_i = e_{ijk}\omega_{jk}^0$ are arbitrary constants.

Remark 1.2.20. The relation (1.2.34) can be rewritten as

$$u_i = \int_{\mathbf{x}_0}^{\mathbf{x}} \left(\varepsilon_{ij} + \omega_{ij}^*\right) d\xi_j + u_i^*,$$

so that the displacement vector field is determined uniquely by ε_{ij} up to an infinitesimal rigid displacement.

1.2.3 Transformation of Areas and Volumes: Transport Theorems

We first discuss how the area and volume elements change as a result of a given deformation. To this end, let us consider the vectors $d\mathbf{X}_1 = dX_1\mathbf{i}_1$, $d\mathbf{X}_2 = dX_2\mathbf{i}_2$, and $d\mathbf{X}_3 = dX_3\mathbf{i}_3$, which, with the deformation $\mathbf{x} = \tilde{\chi}(\mathbf{X}, t)$, become $d\mathbf{x}_1 = \frac{\partial \mathbf{x}_1}{\partial X_1}dX_1$, $d\mathbf{x}_2 = \frac{\partial \mathbf{x}_2}{\partial X_2}dX_2$, and $d\mathbf{x}_3 = \frac{\partial \mathbf{x}_3}{\partial X_3}dX_3$, respectively. Let $d\mathbf{A}_3$ be the area vector associated with the rectangle determined by the vectors $d\mathbf{X}_1$ and $d\mathbf{X}_2$, and let $d\boldsymbol{\sigma}_3$ be the corresponding area vector associated with the parallelogram determined by the vectors $d\mathbf{x}_1$ and $d\mathbf{x}_2$, that is,

$$dA_3 = dX_1 \times dX_2, \quad d\sigma_3 = dx_1 \times dx_2.$$

Obviously, we have

$$d\sigma_3 = \frac{\partial \mathbf{x}_1}{\partial X_1} \times \frac{\partial \mathbf{x}_2}{\partial X_2} dA_3 = e_{ijk} \frac{\partial x_i}{\partial X_1} \frac{\partial x_j}{\partial X_2} \mathbf{i}_k dA_3. \tag{1.2.42}$$

Since

$$J = \det \mathbf{F} = e_{ijk} \frac{\partial x_i}{\partial X_1} \frac{\partial x_j}{\partial X_2} \frac{\partial x_k}{\partial X_3},$$

we can rewrite the relation (1.2.42) in the form

$$d\sigma_3 = J \frac{\partial X_3}{\partial x_j} \mathbf{i}_j dA_3. \tag{1.2.43}$$

A general area element $d\mathbf{A}$ will have components on all three axes. By an analogous procedure, one obtains

$$d\sigma_1 = J \frac{\partial X_1}{\partial x_j} \mathbf{i}_j dA_1, \quad d\sigma_2 = J \frac{\partial X_2}{\partial x_j} \mathbf{i}_j dA_2. \tag{1.2.44}$$

If we now set

$$d\mathbf{a} = d\sigma_1 + d\sigma_2 + d\sigma_3,$$

then, by (1.2.43) and (1.2.44),

$$d\mathbf{a} = J \frac{\partial X_k}{\partial x_j} dA_k \mathbf{i}_j. \tag{1.2.45}$$

Thus, putting

$$d\mathbf{a} = da_j \mathbf{i}_j,$$

it follows from (1.2.45) that

$$da_j = J \frac{\partial X_k}{\partial x_j} dA_k, \tag{1.2.46}$$

a relation that expresses the change of an area element due to the given deformation.

On the other hand, the volume element dv_t of the parallelepiped, determined by the vectors $d\mathbf{x}_1$, $d\mathbf{x}_2$, and $d\mathbf{x}_3$, is

$$dv_t = d\mathbf{x}_1 \times d\mathbf{x}_2 \cdot d\mathbf{x}_3 = J dX_1 dX_2 dX_3 = J dv_0. \tag{1.2.47}$$

It can be shown that for small deformations, in the limit of a linear theory, relation (1.2.47) gives

$$\frac{dv_t - dv_0}{dv_0} = \varepsilon_{ii} = \operatorname{tr} \boldsymbol{\varepsilon} = I_1(\boldsymbol{\varepsilon}),$$

so that $I_1(\boldsymbol{\varepsilon})$ represents the variation of volume per unit undeformed volume.

Let $\mathbf{x} = \tilde{\chi}(\mathbf{X}, t)$ be a motion of the body $\mathcal{B}$. For any subbody A of $\mathcal{B}$, we write $\varphi_t(A) = \tilde{\chi}(A, t)$ for the region of space occupied by A at time t. Then the volume of $\varphi_t(A)$ is

$$\text{vol}(\varphi_t(A)) = \int_{\varphi_t(A)} dv_t,$$

so that using a change of variables in this volume integral, we can write

$$\text{vol}(\varphi_t(A)) = \int_{\varphi_0(A)} J dv_0.$$

Thus, by virtue of relation (1.2.30), we have

$$\frac{d}{dt}[\text{vol}(\varphi_t(A))] = \int_{\varphi_0(A)} \dot{J} dv_0 = \int_{\varphi_t(A)} \dot{J} J^{-1} dv_t = \int_{\varphi_t(A)} \text{div}_{\mathbf{x}} \mathbf{v} dv_t.$$

This relation allows us to formulate the following results.

Theorem 1.2.21. (Transport of Volume) *For any subbody A of $\mathcal{B}$ and time t, denoting by $\mathbf{n}$ the outward unit normal vector on the boundary $\partial\varphi_t(A)$ of $\varphi_t(A)$, we have*

$$\frac{d}{dt}[\text{vol}(\varphi_t(A))] = \int_{\varphi_0(A)} \dot{J} dv_0 = \int_{\varphi_t(A)} \text{div}_{\mathbf{x}} \mathbf{v} dv_t = \int_{\partial\varphi_t(A)} \mathbf{v} \cdot \mathbf{n} da_t.$$

Since A is arbitrary, it follows from the third integral that $\text{div}_{\mathbf{x}}\mathbf{v}$ represents the rate of change of volume per unit volume in the current configuration.

Theorem 1.2.22. (Characterization of Isochoric Motions) *The following assertions are equivalent: (a) $\mathbf{x} = \tilde{\chi}(\mathbf{X}, t)$ is isochoric, (b) $\dot{J} = 0$, (c) $\text{div}_{\mathbf{x}}\mathbf{v} = 0$, and (d) $\int_{\partial\varphi_t(A)} \mathbf{v} \cdot \mathbf{n} da_t = 0$ for every subbody A and any time t.*

We can now establish the following general result.

Theorem 1.2.23. (Reynold's Transport Theorem) *Let $\mathcal{F}$ be a smooth spatial field, and assume that $\mathcal{F}$ is either scalar-valued or vector-valued. Then for any subbody A and time t, we have*

$$\begin{aligned} \frac{d}{dt} \int_{\varphi_t(A)} \hat{\mathcal{F}}(\mathbf{x}, t)\, dv_t &= \int_{\varphi_t(A)} \left[\frac{d}{dt} \hat{\mathcal{F}}(\mathbf{x}, t) + \hat{\mathcal{F}} \text{div}_{\mathbf{x}} \mathbf{v} \right] dv_t \\ &= \int_{\varphi_t(A)} \frac{\partial \hat{\mathcal{F}}}{\partial t}(\mathbf{x}, t)\, dv_t + \int_{\partial\varphi_t(A)} \hat{\mathcal{F}} \mathbf{v} \cdot \mathbf{n} da_t. \end{aligned} \tag{1.2.48}$$

Proof. For the transformation $\mathbf{x} = \tilde{\chi}(\mathbf{X}, t)$, since $dv_t = J dv_0$, we have

$$\int_{\varphi_t(A)} \hat{\mathcal{F}}(\mathbf{x}, t)\, dv_t = \int_{\varphi_0(A)} \tilde{\mathcal{F}}(\mathbf{X}, t) J(\mathbf{X}, t)\, dv_0,$$

and hence

$$\begin{aligned} \frac{d}{dt} \int_{\varphi_t(A)} \hat{\mathcal{F}}(\mathbf{x}, t)\, dv_t &= \int_{\varphi_0(A)} \frac{\partial}{\partial t} [\tilde{\mathcal{F}}(\mathbf{X}, t) J(\mathbf{X}, t)] dv_0 \\ &= \int_{\varphi_0(A)} \left[\frac{\partial}{\partial t} \tilde{\mathcal{F}}(\mathbf{X}, t) J(\mathbf{X}, t) + \tilde{\mathcal{F}}(\mathbf{X}, t) \frac{\partial}{\partial t} J(\mathbf{X}, t) \right] dv_0, \end{aligned}$$

so that using (1.2.7) and (1.2.30), we have

$$\begin{aligned}\frac{d}{dt}\int_{\varphi_t(A)} \hat{\mathfrak{F}}(\mathbf{x},t)\,dv_t &= \int_{\varphi_0(A)} \left[\frac{\partial}{\partial t}\tilde{\mathfrak{F}}(\mathbf{X},t) + \tilde{\mathfrak{F}}(\mathbf{X},t)\mathrm{div}_{\mathbf{x}}\mathbf{v}\right] J(\mathbf{X},t)\,dv_0 \\ &= \int_{\varphi_t(A)} \left[\frac{d}{dt}\hat{\mathfrak{F}}(\mathbf{x},t) + \hat{\mathfrak{F}}(\mathbf{x},t)\mathrm{div}_{\mathbf{x}}\mathbf{v}\right] dv_t,\end{aligned} \tag{1.2.49}$$

which is $(1.2.48)_1$. Relation $(1.2.48)_2$ follows from $(1.2.49)_1$ by using (1.2.6) and applying the divergence theorem. □

Remark 1.2.24. We note that

$$\int_{\varphi_t(A)} \frac{\partial \hat{\mathfrak{F}}}{\partial t}(\mathbf{x},t)\,dv_t = \left[\frac{d}{d\tau}\int_{\varphi_t(A)} \hat{\mathfrak{F}}(\mathbf{x},\tau)\,dv_t\right]_{\tau=t}.$$

Thus, $(1.2.48)_2$ asserts that the rate at which the integral of $\mathfrak{F}$ over $\varphi_t(A)$ is changing is equal to the rate computed as if $\varphi_t(A)$ were fixed in its current position plus the rate at which $\mathfrak{F}$ is carried out of this region across its boundary.

1.3 Principles of Continuum Mechanics

1.3.1 Principle of Conservation of Mass

Given a deformation $\mathbf{x} = \tilde{\chi}(\mathbf{X},t)$ of the body $\mathcal{B}$, we will write $\rho(\mathbf{x},t) = \rho_{\tilde{\chi}(\cdot,t)}(\mathbf{x})$ for the density at the position $\mathbf{x} \in \tilde{\chi}(\mathcal{B},t)$.

- **Principle of conservation of mass:** *The mass of any subbody A of $\mathcal{B}$ is conserved in time, so that we have*

$$\int_{\varphi_0(A)} \rho(\mathbf{X},0)\,dv_0 = \int_{\varphi_t(A)} \rho(\mathbf{x},t)\,dv_t. \tag{1.3.1}$$

In what follows, we will denote by $\rho_0(\mathbf{X})$ the reference mass density $\rho(\mathbf{X},0)$. Relation (1.3.1) expresses the principle of conservation of mass in integral form. We wish to establish a local form of this principle.

Theorem 1.3.1. *The local version of the principle of conservation of mass takes one of the following forms:*

$$\begin{aligned}\rho_0 &= \rho J, \\ \dot{\rho} + \rho\,\mathrm{div}_{\mathbf{x}}\mathbf{v} &= 0, \\ \frac{\partial \rho}{\partial t} + \mathrm{div}_{\mathbf{x}}(\rho\mathbf{v}) &= 0.\end{aligned} \tag{1.3.2}$$

Proof. If we change the variable of integration on the right-hand side of relation (1.3.1) from $\mathbf{x}$ to $\mathbf{X}$, we arrive at

$$\int_{\varphi_0(A)} \rho_0(\mathbf{X})\, dv_0 = \int_{\varphi_0(A)} \rho(\tilde{\chi}(\mathbf{X}, t),\ t) J dv_0,$$

so that

$$\int_{\varphi_0(A)} [\rho(\tilde{\chi}(\mathbf{X}, t),\ t)J - \rho_0(\mathbf{X})] dv_0 = 0, \tag{1.3.3}$$

for every subbody A of the body $\mathcal{B}$. We deduce from (1.3.3) the local form of the principle of conservation of mass expressed by $(1.3.2)_1$.

Furthermore, by differentiation of $(1.3.2)_1$ with respect to the time variable, we obtain

$$\rho \dot{J} + \dot{\rho} J = 0,$$

which with the aid of (1.2.30) yields $(1.3.2)_2$. Next, by (1.2.6), we have

$$\dot{\rho} = \frac{\partial \rho}{\partial t} + \mathbf{v} \cdot \nabla_{\mathbf{x}} \rho,$$

so that $(1.3.2)_2$, combined with this relation, implies $(1.3.2)_3$. □

Remark 1.3.2. The local form of the conservation of mass expressed by $(1.3.2)_1$ is referred to as the *continuity equation in Lagrangian form*, while $(1.3.2)_2$ is the *continuity equation in spatial form.*

By virtue of the above forms of the principle of conservation of mass, Reynold's transport theorem takes a simplified form.

Theorem 1.3.3. *Let $\mathcal{F}$ be a smooth spatial field, either scalar-valued or vector-valued. Then, for any subbody A of $\mathcal{B}$ and time t, we have*

$$\frac{d}{dt} \int_{\varphi_t(A)} \hat{\mathcal{F}}(\mathbf{x}, t) \rho(\mathbf{x}, t)\, dv_t = \int_{\varphi_t(A)} \frac{d}{dt} \left[\hat{\mathcal{F}}(\mathbf{x}, t)\right] \rho(\mathbf{x}, t)\, dv_t. \tag{1.3.4}$$

Thus, to differentiate the integral

$$\int_{\varphi_t(A)} \mathcal{F} \rho dv_t$$

with respect to time, we simply differentiate under the integral sign, treating the mass measure ρdv_t *as a constant.*

Proof. We replace $\mathcal{F}$ by $\mathcal{F}_\rho$ in Reynold's transport relation (1.2.48) and then use the form $(1.3.2)_2$ of the principle of conservation of mass to obtain (1.3.4). □

1.3.2 Momentum Balance Principles

Let $\mathbf{x} = \tilde{\chi}(\mathbf{X}, t)$ be a motion of the body $\mathcal{B}$, and let A be a subbody of $\mathcal{B}$. Then the linear momentum $\mathbf{Q}(A, t)$ and the angular momentum $\mathbf{K}_0(A, t)$ (about the origin) of A at time t are given by

$$\mathbf{Q}(A, t) = \int_{\varphi_t(A)} \mathbf{v}\rho dv_t \tag{1.3.5}$$

and

$$\mathbf{K}_0(A, t) = \int_{\varphi_t(A)} (\mathbf{x} - \mathbf{x}_0) \times \mathbf{v}\rho dv_t. \tag{1.3.6}$$

In view of the rule (1.3.4), we obtain, from (1.3.5) and (1.3.6),

$$\frac{d}{dt}\mathbf{Q}(A, t) = \int_{\varphi_t(A)} \dot{\mathbf{v}}\rho dv_t \tag{1.3.7}$$

and

$$\frac{d}{dt}\mathbf{K}_0(A, t) = \int_{\varphi_t(A)} (\mathbf{x} - \mathbf{x}_0) \times \dot{\mathbf{v}}\rho dv_t. \tag{1.3.8}$$

During a given motion, the mechanical interactions between parts of a body or between a body and its environment are described by forces. In what follows, we will be concerned with three types of force: (i) contact forces between parts of a body, (ii) contact forces exerted on the boundary of a body by its environment, and (iii) body forces exerted on the interior points of a body by the environment.

The environment can exert forces on interior points of $\mathcal{B}$, a classical example being the force field due to gravity. Such forces are determined by a prescribed vector field $\mathbf{b}$ on the trajectory $\mathcal{T}$ of the motion, so that $\mathbf{b}(\mathbf{x}, t)$ gives the force, per unit mass, exerted by the environment on $\mathbf{x}$ at time t. Thus, for any subbody A of $\mathcal{B}$, the integral

$$\int_{\varphi_t(A)} \mathbf{b}(\mathbf{x}, t)\rho dv_t$$

gives that part of the environmental force on A acting at a distance at time t (not due to contact).

Let us now consider the contact forces. To this end we use *Cauchy's hypothesis* concerning the form of the contact forces: *Assume the existence of a surface force density* $\mathbf{t} = \mathbf{t}(\mathbf{x}, t;\ \mathbf{n})$ *defined for every* $(\mathbf{x}, t)$ *in the trajectory* $\mathcal{T}$ *of the motion and for each unit vector* $\mathbf{n}$. To make this hypothesis more precise, we consider an oriented surface $\mathcal{S}$ in $\varphi_t(\mathcal{B})$ with positive unit normal $\mathbf{n}$ at $\mathbf{x}$. Then $\mathbf{t}(\mathbf{x}, t;\ \mathbf{n})$ represents the force, per unit area, exerted across $\mathcal{S}$ upon the material on the negative side of $\mathcal{S}$ by the material on the positive side. To determine the contact force between two subbodies A and C at time t, one integrates $\mathbf{t}$ over the surface of contact $\mathcal{S}_t = \varphi_t(A) \cap \varphi_t(C)$. Thus, denoting by $\mathbf{n}_x$ the outward unit normal to $\partial\varphi_t(A)$ at $\mathbf{x}$,

$$\int_{\mathcal{S}_t} \mathbf{t}(\mathbf{x}, t;\ \mathbf{n}_\mathbf{x}) da_t = \int_{\mathcal{S}_t} \mathbf{t}(\mathbf{n}) da_t \tag{1.3.9}$$

gives the force exerted on A by C at time t. Such a contact force depends on the intrinsic structure of the material and is therefore unknown in general.

For points on the boundary of $\varphi_t(\mathcal{B})$, $\mathbf{t}(\mathbf{x}, t;\ \mathbf{n})$, with $\mathbf{n}$ the outward unit normal to $\partial\varphi_t(\mathcal{B})$ at $\mathbf{x}$, gives the surface force, per unit area, applied to the body by the environment. This force is referred to as the *surface traction*, and it is usually known.

The above discussion motivates the following definition.

Definition 1.3.4. *By a system of forces for $\mathcal{B}$ during a motion (with trajectory $\mathcal{T}$), we mean a pair $(\mathbf{t}, \mathbf{b})$ of functions $\mathbf{t} : \mathcal{T} \times \mathcal{N} \to V$, $\mathbf{b} : \mathcal{T} \to V$, where $\mathcal{N}$ is the set of all unit vectors and V is the vector space $\mathbb{R}^3$, so that (i) $\mathbf{t}(\mathbf{x}, t;\ \mathbf{n})$ is a smooth function of $\mathbf{x}$ on $\varphi_t(\mathcal{B})$, for each $\mathbf{n} \in \mathcal{N}$ and $t \geq 0$ and (ii) $\mathbf{b}(\mathbf{x}, t)$ is a continuous function of $\mathbf{x}$ on $\varphi_t(\mathcal{B})$, for each $t \geq 0$. We refer to $\mathbf{t}$ as the surface force and $\mathbf{b}$ as the body force. The force $\mathbf{F}(A,\ t)$ and the moment $\mathbf{\Omega}_0(A,\ t)$ (about the origin) on a subbody A at time t are defined by*

$$\mathbf{F}(A,\ t) = \int_{\varphi_t(A)} \mathbf{b}\rho dv_t + \int_{\partial\varphi_t(A)} \mathbf{t}(\mathbf{n}) da_t \tag{1.3.10}$$

and

$$\mathbf{\Omega}_0(A,\ t) = \int_{\varphi_t(A)} (\mathbf{x} - \mathbf{x}_0) \times \mathbf{b}\rho dv_t + \int_{\partial\varphi_t(A)} (\mathbf{x} - \mathbf{x}_0) \times \mathbf{t}(\mathbf{n}) da_t. \tag{1.3.11}$$

- **The balance law of linear momentum:** *The time derivative of the linear momentum of every subbody A of $\mathcal{B}$ at time t is equal to the force $\mathbf{F}(A,\ t)$ acting on that subbody at time t, so that*

$$\frac{d}{dt}\mathbf{Q}(A,\ t) = \mathbf{F}(A,\ t). \tag{1.3.12}$$

- **The balance law of angular momentum:** *The time derivative of the angular momentum $\mathbf{K}_0(A,\ t)$ of every subbody A of the body at time t is equal to the moment $\mathbf{\Omega}_0(A,\ t)$ acting on that subbody at time t, that is,*

$$\frac{d}{dt}\mathbf{K}_0(A,\ t) = \mathbf{\Omega}_0(A,\ t). \tag{1.3.13}$$

Remark 1.3.5. We assume that there exists a *laboratory frame of reference* in which Newton's second law, (1.3.12), holds to a good approximation and refer to this and all frames of reference traveling at constant velocities relative to it as *inertial frames*. These are all connected by *Galilean transformations*, and Newton's second law applies equally in all of them.

In view of relations (1.3.7)–(1.3.11), the laws (1.3.12) and (1.3.13) of momentum balance can be written as follows:

$$\int_{\varphi_t(A)} \dot{\mathbf{v}}\rho dv_t = \int_{\varphi_t(A)} \mathbf{b}\rho dv_t + \int_{\partial\varphi_t(A)} \mathbf{t}(\mathbf{n}) da_t \tag{1.3.14}$$

and

$$\int_{\varphi_t(A)} (\mathbf{x} - \mathbf{x}_0) \times \dot{\mathbf{v}} \rho dv_t = \int_{\varphi_t(A)} (\mathbf{x} - \mathbf{x}_0) \times \mathbf{b} \rho dv_t + \int_{\partial\varphi_t(A)} (\mathbf{x} - \mathbf{x}_0) \times \mathbf{t}(\mathbf{n}) da_t. \tag{1.3.15}$$

Lemma 1.3.6. (Newton's Law of Action and Reaction) *For each* $\mathbf{x} \in \varphi_t(\mathcal{B})$ *and for each unit vector* $\mathbf{n} \in \mathcal{N}$, *it follows that*

$$\mathbf{t}(\mathbf{x}, t;\ \mathbf{n}) = -\mathbf{t}(\mathbf{x}, t;\ -\mathbf{n}), \tag{1.3.16}$$

for fixed time t.

Proof. Let us denote by R_δ a volume centered at $\mathbf{x}$, with rectangular sides. It has dimensions $\delta \times \delta \times \delta^2$, and $\mathbf{n}$ is the unit normal to the $\delta \times \delta$ faces. Let Σ_δ^+ be the face with the outward unit normal $\mathbf{n}$ and Σ_δ^- the face with the outward unit normal $-\mathbf{n}$. Furthermore, we set $\partial R_\delta = \Sigma_\delta^+ \cup \Sigma_\delta^- \cup \Sigma$. Obviously, we have

$$\mathrm{vol}(R_\delta) = \delta^4, \qquad \mathrm{Area}(\Sigma_\delta^+) = \mathrm{Area}(\Sigma_\delta^-) = \delta^2, \qquad \mathrm{Area}(\Sigma) = 4\delta^3. \tag{1.3.17}$$

We further note that R_δ is contained in the interior of $\varphi_t(\mathcal{B})$ for all sufficiently small δ, say $\delta \le \delta_0$.

We now apply (1.3.14) to the subbody A that occupies the region $\varphi_t(A) \equiv R_\delta$. Since $\mathbf{b}(\mathbf{x}, t)$, $\rho(\mathbf{x}, t)$, and $\dot{\mathbf{v}}(\mathbf{x}, t)$ are continuous in $\mathbf{x}$, it follows that the function $\mathbf{b}_*(\mathbf{x}, t)$ defined by $\mathbf{b}_* = \rho(\mathbf{b} - \dot{\mathbf{v}})$ is bounded on R_{δ_0} for t fixed, and hence

$$\kappa(t) = \sup_{\mathbf{x} \in R_{\delta_0}} |\mathbf{b}_*(\mathbf{x}, t)| < \infty. \tag{1.3.18}$$

For convenience, we fix the time t and suppress it as an argument in most of what follows.

From (1.3.14), we deduce

$$\left| \int_{\partial R_\delta} \mathbf{t}(\mathbf{n}) da_t \right| \le \kappa(t) \mathrm{vol}\,(R_\delta), \tag{1.3.19}$$

so that on the basis of relations (1.3.17) and (1.3.18), we obtain

$$\frac{1}{\delta^2} \int_{\partial R_\delta} \mathbf{t}(\mathbf{n}) da_t \to \mathbf{0} \text{ when } \delta \to 0. \tag{1.3.20}$$

But

$$\int_{\partial R_\delta} \mathbf{t}(\mathbf{n}) da_t = \int_{\Sigma_\delta^+} \mathbf{t}(+\mathbf{n}) da_t + \int_{\Sigma_\delta^-} \mathbf{t}(-\mathbf{n}) da_t + \int_{\Sigma} \mathbf{t}(\mathbf{n}) da_t. \tag{1.3.21}$$

Since $\mathbf{t}(\mathbf{x}; \mathbf{n})$ is continuous in $\mathbf{x}$ for each fixed $\mathbf{n} \in \mathcal{N}$, we have, using (1.3.17),

$$\frac{1}{\delta^2} \int_{\Sigma} \mathbf{t}(\mathbf{n}) da_t \to \mathbf{0} \text{ when } \delta \to 0, \tag{1.3.22}$$

$$\frac{1}{\delta^2} \int_{\Sigma_\delta^+} \mathbf{t}(+\mathbf{n}) da_t \to \mathbf{t}(\mathbf{x}; +\mathbf{n}), \qquad \frac{1}{\delta^2} \int_{\Sigma_\delta^-} \mathbf{t}(-\mathbf{n}) da_t \to \mathbf{t}(\mathbf{x}; -\mathbf{n}), \tag{1.3.23}$$

when $\delta \to 0$. Thus, relations (1.3.20)–(1.3.23) give

$$\mathbf{t}(\mathbf{x};+\mathbf{n}) + \mathbf{t}(\mathbf{x};-\mathbf{n}) = \mathbf{0},$$

which is (1.3.16). □

Theorem 1.3.7. **(Cauchy's Theorem for the Existence of Stress)** *Let* $(\mathbf{t},\mathbf{b})$ *be a system of forces for* $\mathcal{B}$ *during a motion. Then, a necessary and sufficient condition that the momentum balance laws be satisfied is that there exists a spatial tensor field* $\mathbf{T}$ *(the* Cauchy stress*) such that*

(i) for each unit vector $\mathbf{n} \in \mathcal{N}$,

$$\mathbf{t}(\mathbf{n}) = \mathbf{T}\mathbf{n}; \tag{1.3.24}$$

(ii) $\mathbf{T}$ *is symmetric;*
(iii) $\mathbf{T}$ *satisfies the equation of motion*

$$\rho\dot{\mathbf{v}} = \mathrm{div}_{\mathbf{x}}\mathbf{T} + \rho\mathbf{b}. \tag{1.3.25}$$

Proof. Necessity. Assume that the momentum balance laws (1.3.12) and (1.3.13) are satisfied. We first note that (1.3.16) holds. Furthermore, we choose $\mathbf{x}$ to belong to the interior of $\varphi_t(\mathcal{B})$ and let $\delta > 0$. Consider the tetrahedron $\mathcal{T}_\delta$ with the following properties: the faces of $\mathcal{T}_\delta$ are $\mathcal{S}_\delta$, $\mathcal{S}_{1\delta}$, $\mathcal{S}_{2\delta}$, and $\mathcal{S}_{3\delta}$, where $\mathbf{n}$ and $-\mathbf{i}_j$ are the outward unit normal vectors to $\partial\mathcal{T}_\delta$ on $\mathcal{S}_\delta$ and $\mathcal{S}_{j\delta}$, $j = 1, 2, 3$, respectively; the vertex opposite to $\mathcal{S}_\delta$ is $\mathbf{x}$; the distance from $\mathbf{x}$ to $\mathcal{S}_\delta$ is δ (Fig. 1.3). Clearly, $\mathcal{S}_\delta$ is contained in the interior of $\varphi_t(\mathcal{B})$ for all sufficiently small choices of δ, say $\delta \le \delta_0$. Thus, we can apply (1.3.14) to the subbody A that occupies the region $\mathcal{T}_\delta$ at time t, and since $\mathbf{b}_* = \rho(\mathbf{b} - \dot{\mathbf{v}})$ is bounded on $\mathcal{T}_{\delta_0}$, we can conclude that

$$\left|\int_{\partial\mathcal{T}_\delta} \mathbf{t}(\mathbf{n})da_t\right| \le \kappa_1(t)\mathrm{vol}(\mathcal{T}_\delta), \qquad \text{for all } \delta \le \delta_0, \tag{1.3.26}$$

where $\kappa_1(t)$ is finite and independent of δ.

Let $\mathcal{A}(\delta)$ denote the area of $\mathcal{S}_\delta$. Then $\mathrm{vol}(\mathcal{T}_\delta) = \frac{1}{3}\delta\mathcal{A}(\delta)$, and hence we can conclude from (1.3.26) that

$$\frac{1}{\mathcal{A}(\delta)}\int_{\partial\mathcal{T}_\delta} \mathbf{t}(\mathbf{n})da_t \to \mathbf{0} \quad \text{as } \delta \to 0. \tag{1.3.27}$$

But

$$\int_{\partial\mathcal{T}_\delta} \mathbf{t}(\mathbf{n})da_t = \int_{\mathcal{S}_\delta} \mathbf{t}(\mathbf{n})da_t + \sum_{j=1}^{3}\int_{\mathcal{S}_{j\delta}} \mathbf{t}(-\mathbf{i}_j)da_t, \tag{1.3.28}$$

and since $\mathbf{t}(\mathbf{x};\mathbf{n})$ is continuous in $\mathbf{x}$ for each fixed $\mathbf{n} = n_j\mathbf{i}_j \in \mathcal{N}$ and $\text{Area}\left(\mathcal{S}_{j\delta}\right) = \mathcal{A}(\delta)n_j$, we have

$$\frac{1}{\mathcal{A}(\delta)}\int_{\mathcal{S}_\delta} \mathbf{t}(\mathbf{n})da_t \to \mathbf{t}(\mathbf{x};\mathbf{n}) \quad \text{as } \delta \to 0, \tag{1.3.29}$$

$$\frac{1}{\mathcal{A}(\delta)}\int_{\mathcal{S}_{j\delta}} \mathbf{t}(-\mathbf{i}_j)da_t \to \mathbf{t}(\mathbf{x};-\mathbf{i}_j)n_j \quad \text{as } \delta \to 0 \quad (\text{not summed on } j).$$

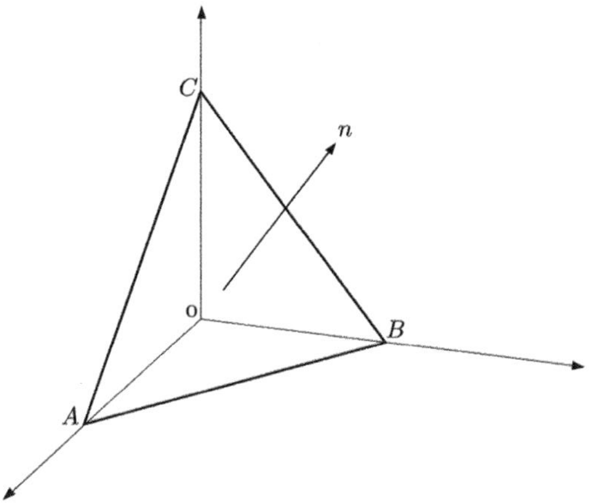

Fig. 1.3. The stress tetrahedron

Combining (1.3.27)–(1.3.29) with Newton's law of action and reaction (1.3.16), we conclude that

$$\mathbf{t}(\mathbf{x};\mathbf{n}) = \mathbf{t}(\mathbf{x};\mathbf{i}_j)n_j, \tag{1.3.30}$$

so $\mathbf{t}(\mathbf{x};\mathbf{n})$ is a linear function of the components of $\mathbf{n}$, or

$$\mathbf{t}(\mathbf{x};\mathbf{n}) = \mathbf{T}(\mathbf{x})\mathbf{n}. \tag{1.3.31}$$

We write this in components as

$$t_i(\mathbf{x};\ \mathbf{n}) = T_{ij}(\mathbf{x})n_j,$$

where $\mathbf{T}$, given by

$$\left(T_{ij}\right) = \begin{pmatrix} t_1\,(\mathbf{i}_1)\; t_1\,(\mathbf{i}_2)\; t_1\,(\mathbf{i}_3) \\ t_2\,(\mathbf{i}_1)\; t_2\,(\mathbf{i}_2)\; t_2\,(\mathbf{i}_3) \\ t_3\,(\mathbf{i}_1)\; t_3\,(\mathbf{i}_2)\; t_3\,(\mathbf{i}_3) \end{pmatrix},$$

is the *Cauchy stress tensor*.

Using (1.3.31), the balance of linear momentum (1.3.14) takes the form

$$\int_{\varphi_t(A)} \dot{\mathbf{v}}\rho dv_t = \int_{\varphi_t(A)} \mathbf{b}\rho dv_t + \int_{\partial\varphi_t(A)} \mathbf{T}\mathbf{n} da_t,$$

or equivalently, on applying the divergence theorem,

$$\int_{\varphi_t(A)} (\dot{\mathbf{v}}\rho - \mathbf{b}\rho - div_{\mathbf{x}}\mathbf{T})\, dv_t = \mathbf{0}.$$

This last relation can hold for every subbody A of the body and time t only if the equation of motion (1.3.25) is satisfied.

To complete the proof of necessity, we have only to establish the symmetry of the Cauchy stress $\mathbf{T}$. In fact, if we substitute (1.3.31) into (1.3.15), we obtain

$$\int_{\varphi_t(A)} (\mathbf{x} - \mathbf{x}_0) \times (\dot{\mathbf{v}}\rho - \mathbf{b}\rho)\, dv_t - \int_{\partial\varphi_t(A)} (\mathbf{x} - \mathbf{x}_0) \times \; (\mathbf{T}\mathbf{n})\, da_t = \mathbf{0},$$

or, using the divergence theorem,

$$\int_{\varphi_t(A)} (\mathbf{x} - \mathbf{x}_0) \times (\dot{\mathbf{v}}\rho - \mathbf{b}\rho - div_{\mathbf{x}}\mathbf{T})\, dv_t - \int_{\varphi_t(A)} e_{ijk}T_{jk}\mathbf{i}_i dv_t = \mathbf{0},$$

and hence, with the aid of the equation of motion (1.3.25), we deduce that

$$\int_{\varphi_t(A)} e_{ijk}T_{jk} dv_t = 0.$$

This relation can hold for every subbody A and time t only if

$$e_{ijk}T_{jk} = 0, \qquad \text{that is,} \quad T_{23} = T_{32}, \quad T_{31} = T_{13}, \quad T_{12} = T_{21}.$$

Sufficiency. Assume that there exists a symmetric spatial tensor field $\mathbf{T}$ consistent with the relations (1.3.24) and (1.3.25). Then it is an easy task to prove that the momentum balance laws (1.3.14) and (1.3.15) hold, and the proof is complete. □

Remark 1.3.8. Actually, one can see that the points (i) and (iii) are equivalent to balance of linear momentum, while, granted (1.3.14), the symmetry of the Cauchy stress $\mathbf{T}$ is equivalent to balance of angular momentum.

Definition 1.3.9. *If*

$$\mathbf{T}\mathbf{n} = \sigma\mathbf{n}, \qquad |\mathbf{n}| = 1,$$

then σ is a principal stress and $\mathbf{n}$ *is a principal direction, so that principal stresses and principal directions are eigenvalues and eigenvectors of* $\mathbf{T}$.

Since $\mathbf{T}$ is symmetric, it follows that there exist three mutually perpendicular principal directions and three corresponding principal stresses.

In general, the surface force $\mathbf{t} = \mathbf{T}\mathbf{n}$ can be decomposed into the sum

$$\mathbf{t} = T_0\mathbf{n} + \mathbf{t}_0,$$

where $T_0\mathbf{n}$ is the *normal force* and $\mathbf{t}_0$ is the *shearing force* perpendicular to $\mathbf{n}$. Obviously, we have

$$T_0 = \mathbf{n} \cdot \mathbf{T}\mathbf{n}, \qquad \mathbf{t}_0 = \mathbf{t} - (\mathbf{n} \cdot \mathbf{T}\mathbf{n})\mathbf{n} = (\mathbf{1} - \mathbf{n} \otimes \mathbf{n})\mathbf{T}\mathbf{n},$$

where $\otimes$ denotes the tensor product of two vectors.[¶] Clearly, $\mathbf{n}$ is a principal direction if and only if the corresponding shearing force vanishes. The normal component of the surface force is then a principal stress.

We now outline some important states of stress.

First consider a fluid at rest. It is incapable of exerting shearing forces, so that $\mathbf{Tn}$ is parallel to $\mathbf{n}$ for each unit vector $\mathbf{n}$, and hence every such unit vector is an eigenvector of $\mathbf{T}$. Thus, we have

$$\mathbf{T} = -p\mathbf{1},$$

where p is a scalar quantity referred to as the *pressure* of the fluid. The force per unit area on any surface in the fluid with unit normal $\mathbf{n}$ is $-p\mathbf{n}$.

Other states of stress are *pure tension* (or *compression*), where the tensile stress σ in the direction $\boldsymbol{\nu}$, with $|\boldsymbol{\nu}| = 1$, is defined by

$$\mathbf{T} = \sigma\,(\boldsymbol{\nu} \otimes \boldsymbol{\nu}),$$

and *pure shear* with shear stress τ relative to the direction pair $(\mathbf{k}, \mathbf{n})$, where $\mathbf{k}$ and $\mathbf{n}$ are orthogonal unit vectors, given by

$$\mathbf{T} = \tau(\mathbf{k} \otimes \mathbf{n} + \mathbf{n} \otimes \mathbf{k}).$$

1.3.3 Consequences of Momentum Balance Laws

Definition 1.3.10. *For every subbody A of a continuous body, we define the kinetic energy of A at time t by*

$$\frac{1}{2}\int_{\varphi_t(A)} \rho \mathbf{v}^2 dv_t;$$

we further define the stress power of A at time t by

$$\int_{\varphi_t(A)} \mathbf{T} \cdot \mathbf{D} dv_t, \tag{1.3.32}$$

where $\mathbf{D}$ *is the stretching, defined by (1.2.23).*

Theorem 1.3.11. *The power expended on any subbody A at time t by the surface and body forces is equal to the rate of change of kinetic energy plus the stress power, so that*

$$\int_{\varphi_t(A)} \rho \mathbf{b} \cdot \mathbf{v} dv_t + \int_{\partial\varphi_t(A)} \mathbf{t}(\mathbf{n}) \cdot \mathbf{v} da_t = \frac{d}{dt}\int_{\varphi_t(A)} \frac{1}{2}\rho \mathbf{v}^2 dv_t + \int_{\varphi_t(A)} \mathbf{T} \cdot \mathbf{D} dv_t. \tag{1.3.33}$$

Proof. Since $\mathbf{T}$ is symmetric, then with the aid of (1.2.23), we can write

$$\mathbf{T} \cdot \nabla_{\mathbf{x}}\mathbf{v} = T_{ij}\frac{\partial v_i}{\partial x_j} = T_{ji}\frac{\partial v_j}{\partial x_i} = T_{ij}\frac{\partial v_j}{\partial x_i} = T_{ij}\frac{1}{2}\left(\frac{\partial v_i}{\partial x_j} + \frac{\partial v_j}{\partial x_i}\right) = \mathbf{T} \cdot \mathbf{D}. \tag{1.3.34}$$

[¶] This is defined by the requirement that $(\mathbf{a} \otimes \mathbf{b})\mathbf{c} = (\mathbf{b} \cdot \mathbf{c})\mathbf{a}$, where $\mathbf{a}$ and $\mathbf{b}$ are given vectors and $\mathbf{c}$ is an arbitrary vector.

Furthermore, invoking (1.3.4), we have

$$\frac{d}{dt}\int_{\varphi_t(A)} \frac{1}{2}\rho \mathbf{v}^2 dv_t = \int_{\varphi_t(A)} \rho \mathbf{v}\cdot\dot{\mathbf{v}} dv_t. \tag{1.3.35}$$

Now, using (1.3.24) and the divergence theorem, we obtain

$$\begin{aligned}\int_{\varphi_t(A)} \rho \mathbf{b}\cdot\mathbf{v} dv_t + \int_{\partial\varphi_t(A)} \mathbf{t}(\mathbf{n})\cdot\mathbf{v} da_t &= \int_{\varphi_t(A)} \rho \mathbf{b}\cdot\mathbf{v} dv_t + \int_{\partial\varphi_t(A)} \mathbf{T}\mathbf{n}\cdot\mathbf{v} da_t \\ &= \int_{\varphi_t(A)} \rho \mathbf{b}\cdot\mathbf{v} dv_t + \int_{\partial\varphi_t(A)} \mathbf{n}\cdot\mathbf{T}^T\mathbf{v} da_t \\ &= \int_{\varphi_t(A)} [(\rho\mathbf{b} + div_{\mathbf{x}}\mathbf{T})\cdot\mathbf{v} + \mathbf{T}\cdot\nabla_{\mathbf{x}}\mathbf{v}] dv_t.\end{aligned} \tag{1.3.36}$$

Combining (1.3.34) with (1.3.36) and using the equation of motion (1.3.25), we obtain the relation (1.3.33). □

Definition 1.3.12. *Given a motion of the material, we refer to the list* $\{\mathbf{v}, \rho, \mathbf{T}\}$ *as a flow. The flow is steady if* $\varphi_t(\mathcal{B}) = \varphi_0(\mathcal{B})$ *for all t and*

$$\frac{\partial \mathbf{v}}{\partial t} = \mathbf{0}, \quad \frac{\partial \rho}{\partial t} = 0, \quad \frac{\partial \mathbf{T}}{\partial t} = \mathbf{0}. \tag{1.3.37}$$

In this case $\varphi_0(\mathcal{B})$ *is called the flow region. A flow is potential if the velocity is the gradient of a potential, that is, if there exists a function* ϕ *with the property that*

$$\mathbf{v} = \nabla_{\mathbf{x}}\phi. \tag{1.3.38}$$

Finally, a flow is irrotational if $\text{curl}_{\mathbf{x}}\mathbf{v} = \mathbf{0}$.

Theorem 1.3.13. (Bernoulli's Theorem) *Consider a flow* $\{\mathbf{v}, \rho, \mathbf{T}\}$, *where the stress tensor is given by a pressure* $-p\mathbf{1}$ *and the body force is conservative with potential energy* $\mathcal{V}$. *We have the following:*

(i) If the flow is potential, then

$$\nabla_{\mathbf{x}}\left(\frac{\partial \phi}{\partial t} + \frac{1}{2}\mathbf{v}^2 + \mathcal{V}\right) + \frac{1}{\rho}\nabla_{\mathbf{x}} p = \mathbf{0}. \tag{1.3.39}$$

(ii) If the flow is steady, then

$$\mathbf{v}\cdot\nabla_{\mathbf{x}}\left(\frac{1}{2}\mathbf{v}^2 + \mathcal{V}\right) + \frac{1}{\rho}\mathbf{v}\cdot\nabla_{\mathbf{x}} p = 0. \tag{1.3.40}$$

(iii) If the flow is steady and irrotational, then

$$\nabla_{\mathbf{x}}\left(\frac{1}{2}\mathbf{v}^2 + \mathcal{V}\right) + \frac{1}{\rho}\nabla_{\mathbf{x}} p = \mathbf{0}. \tag{1.3.41}$$

Proof. Since we have $\mathbf{T} = -p\mathbf{1}$, it follows that $\text{div}_{\mathbf{x}}\mathbf{T} = -\nabla_{\mathbf{x}} p$, and therefore the equation of motion (1.3.25) takes the form

$$\rho\dot{\mathbf{v}} = -\nabla_{\mathbf{x}} p + \rho\mathbf{b}. \tag{1.3.42}$$

Since the body force is conservative, we have $\mathbf{b} = -\nabla_{\mathbf{x}}\mathcal{V}$. Moreover, for a potential flow, we have, with the aid of (1.2.7),

$$\dot{\mathbf{v}} = \frac{\partial \mathbf{v}}{\partial t} + \frac{1}{2}\nabla_{\mathbf{x}}(\mathbf{v}^2) = \nabla_{\mathbf{x}}\left(\frac{\partial \phi}{\partial t} + \frac{1}{2}\mathbf{v}^2\right), \tag{1.3.43}$$

while for a steady flow,

$$\mathbf{v}\cdot\dot{\mathbf{v}} = \mathbf{v}\cdot\nabla_{\mathbf{x}}\left(\frac{1}{2}\mathbf{v}^2\right), \tag{1.3.44}$$

and for a steady irrotational flow,

$$\dot{\mathbf{v}} = \nabla_{\mathbf{x}}\left(\frac{1}{2}\mathbf{v}^2\right). \tag{1.3.45}$$

Relations (1.3.43)–(1.3.45), when combined with (1.3.42), yield the desired results (1.3.39)–(1.3.41). □

1.3.4 The Piola–Kirchhoff Stresses

The Cauchy stress tensor $\mathbf{T}$ measures the contact force per unit area in the deformed configuration, and it is convenient, especially for fluids whose current configuration is supposed known in advance. For many other problems of interest—especially those involving solids—it is convenient to work with a stress tensor that gives the force measured per unit area in the reference configuration. This is because in such problems the current configuration is not known in advance. To establish the form of this tensor, we have to formulate the momentum balance laws relative to the reference configuration $\varphi_0(\mathcal{B})$.

Note that by virtue of the mass balance law $(1.3.2)_1$, we can rewrite the momentum balance laws (1.3.14) and (1.3.15) in the following forms:

$$\int_{\varphi_0(A)} \dot{\mathbf{v}}\rho_0 dv_0 = \int_{\varphi_0(A)} \mathbf{b}\rho_0 dv_0 + \int_{\partial\varphi_0(A)} \mathbf{s}(\mathbf{N}) da_0 \tag{1.3.46}$$

and

$$\int_{\varphi_0(A)} (\mathbf{x}-\mathbf{x}_0)\times\dot{\mathbf{v}}\rho_0 dv_0 = \int_{\varphi_0(A)} (\mathbf{x}-\mathbf{x}_0)\times\mathbf{b}\rho_0 dv_0 + \int_{\partial\varphi_0(A)} (\mathbf{x}-\mathbf{x}_0)\times\mathbf{s}(\mathbf{N}) da_0, \tag{1.3.47}$$

where $\mathbf{s}(\mathbf{N})$ represents the force vector acting on the surface $\partial\varphi_t(A)$ but measured per unit area of the surface $\partial\varphi_0(A)$ in the reference configuration, the outward unit normal of which is denoted by $\mathbf{N}$. We have

$$\mathbf{s}(\mathbf{N})da_0 = \mathbf{t}(\mathbf{n})da_t,$$

so that using (1.2.46) and (1.3.30), where $da_i = n_i da_t$, $dA_j = N_j da_0$, with $\mathbf{n} = n_i \mathbf{i}_i$, $\mathbf{N} = N_i \mathbf{i}_i$, we obtain that

$$\mathbf{s}(\mathbf{N})da_0 = \mathbf{t}(\mathbf{i}_j)n_j da_t = \mathbf{t}(\mathbf{i}_j)da_j = \mathbf{t}(\mathbf{i}_j)J\frac{\partial X_i}{\partial x_j}dA_i = \mathbf{t}(\mathbf{i}_j)J\frac{\partial X_i}{\partial x_j}N_i da_0.$$

We can write this relation in the form

$$\mathbf{s}(\mathbf{N}) = \mathbf{S}(\mathbf{X})\mathbf{N}, \tag{1.3.48}$$

where $\mathbf{S}$ is a tensor given by

$$\mathbf{S} = J\mathbf{T}\left(\mathbf{F}^{-1}\right)^T, \tag{1.3.49}$$

known as the *first Piola–Kirchhoff stress*. In terms of components, this relation is

$$S_{ij} = JT_{ik}\frac{\partial X_j}{\partial x_k} \quad \Leftrightarrow \quad T_{ij} = \frac{1}{J}\frac{\partial x_j}{\partial X_k}S_{ik}.$$

Note that (1.3.48) is of the same form as (1.3.24). If we use (1.3.48) in the balance laws (1.3.46) and (1.3.47), then the following result is obtained.

Proposition 1.3.14. *The first Piola–Kirchhoff stress tensor satisfies the field equations*

$$\rho_0\ddot{\mathbf{x}} = \mathrm{Div}_{\mathbf{X}}\mathbf{S} + \rho_0\mathbf{b} \tag{1.3.50}$$

and

$$\mathbf{S}\mathbf{F}^T = \mathbf{F}\mathbf{S}^T. \tag{1.3.51}$$

Here, the operator $\mathrm{Div}_{\mathbf{X}}$ is evaluated with respect to the material point $\mathbf{X}$ in the reference configuration.

Remark 1.3.15. It is important to note that by (1.3.51), $\mathbf{S}$ generally is not symmetric. If we introduce the second Piola–Kirchhoff stress tensor $\widehat{\mathbf{S}}$, defined by

$$\widehat{\mathbf{S}} = \mathbf{F}^{-1}\mathbf{S}, \quad \text{or, in components, } \widehat{S}_{ij} = \frac{\partial X_i}{\partial x_k}S_{kj}, \tag{1.3.52}$$

then from (1.3.51), it follows that $\widehat{\mathbf{S}}$ is a symmetric tensor.

It is related to the Cauchy stress tensor by

$$\widehat{\mathbf{S}} = J\mathbf{F}^{-1}\mathbf{T}(\mathbf{F}^{-1})^{\mathsf{T}}. \tag{1.3.53}$$

We further have the following alternative version of the relation (1.3.33).

Theorem 1.3.16. (Theorem of Power Expended) *For every subbody A of the body, we have*

$$\int_{\varphi_0(A)}\rho_0\mathbf{b}\cdot\dot{\mathbf{x}}dv_0 + \int_{\partial\varphi_0(A)}\mathbf{s}(\mathbf{N})\cdot\dot{\mathbf{x}}da_0 = \frac{d}{dt}\int_{\varphi_0(A)}\frac{1}{2}\rho_0\dot{\mathbf{x}}^2dv_0 + \int_{\varphi_0(A)}\mathbf{S}\cdot\dot{\mathbf{F}}dv_0. \tag{1.3.54}$$

Proof. Let us take the inner product of the equation of motion (1.3.50) with $\dot{\mathbf{x}}$ and integrate over $\varphi_0(A)$. Using the divergence theorem and relation (1.2.11), we obtain (1.3.54). □

Note that using (A.2.4), we have

$$\mathbf{S}\cdot\dot{\mathbf{F}} = \mathrm{tr}(\mathbf{S}\dot{\mathbf{F}}^{\top}) = \mathrm{tr}(\widehat{\mathbf{S}}\dot{\mathbf{F}}^{\top}\mathbf{F}) = \mathrm{tr}(\mathbf{S}^{\top}\dot{\mathbf{F}}) = \mathrm{tr}(\widehat{\mathbf{S}}\mathbf{F}^{\top}\dot{\mathbf{F}}).$$

Thus, by virtue of (1.2.15) and (1.2.20),

$$\mathbf{S}\cdot\dot{\mathbf{F}} = \frac{1}{2}\mathrm{tr}(\widehat{\mathbf{S}}\dot{\mathbf{C}}) = \mathrm{tr}(\widehat{\mathbf{S}}\dot{\mathbf{E}}) = \widehat{\mathbf{S}}\cdot\dot{\mathbf{E}}.$$

1.4 Constitutive Equations

The mass and momentum balance principles apply to all bodies in nature and do not distinguish between different types of materials, in that they do not depend on the intrinsic structure of the material. Two different bodies of the same size and shape subjected to the same deformation will generally not have the same resulting stress distribution. For example, two thin wires of the same length and diameter, one of steel and one of copper, will require different forces to produce the same elongation. Therefore, the balance principles are insufficient to fully characterize behavior, and some additional hypotheses are required for a complete description of the behavior of a continuous body. Such supplementary hypotheses are known as *constitutive equations* and serve to distinguish different types of material behavior.

Constitutive equations also serve the purpose of providing a well-posed mathematical model for describing the deformation of a continuous body. In fact, supposing that the mass density of the body in the reference configuration is known and the body force field has been assigned, we have four differential equations (one is the continuity equation and the other three are the equations of motion) for the unknown set of functions defining, for example, the components of the displacement vector field, the mass density in the current configuration, and the components of the stress tensor. Clearly, the mathematical problem is underdetermined.

The possibility of dependence of constitutive quantities on not only the current values of field variables but also their past history is fundamental to the present work.

The importance of such memory properties in the study of the behavior of materials was first described by Cauchy in 1828 [60]. In this work, he observed that for solid bodies that are not quite elastic, "*les pressions ou tensions ne dépendent pas seulement du changement de form que le corps éprouve en passant de l'état naturel à un nouvel état, mais aussi des états intermédiaires et du temps pendant lequel le changement de form s'effectue*" [see [313] on page 56 (1960 edition)].

We now introduce the concept of objective tensors and the principle of material objectivity, which imposes constraints on the possible forms of constitutive equations. The remaining chapters of Part I and all of Parts II and III deal largely with properties of various specific constitutive equations, in most cases involving linear memory functionals, and of energy functionals associated with them.

1.4.1 Objectivity

Inertial frames are defined in Remark 1.3.5. However, we wish to consider more general frames of reference. Let $(\mathbf{x}, t)$ be the spatial coordinates in an inertial frame. Consider the frame of reference with coordinates $(\mathbf{x}', t')$ given by

$$\mathbf{x}' = \mathbf{Q}(t)\mathbf{x} + \mathbf{c}(t), \qquad t' = t + t_0, \tag{1.4.1}$$

where t_0 is a constant. The reference description coordinates $\mathbf{X}$ are unchanged. The quantity $\mathbf{Q}(t)$ is an orthogonal matrix, so that

$$\mathbf{Q}^{\top}(t)\mathbf{Q}(t) = \mathbf{1}.$$

Thus $\mathbf{Q}$ is a time-dependent rotation and $\mathbf{c}$ is a time-dependent translation. We take $\det \mathbf{Q} = 1$.

Relation (1.4.1) is a Euclidean transformation. In a Galilean transformation (Remark 1.3.5), $\mathbf{Q}$ is time-independent and $\mathbf{c}(t) = \mathbf{c}_0 + \mathbf{V}t$, the quantity $\mathbf{V}$ being the relative velocity of the origins of the two frames under consideration, while $\mathbf{c}_0$ is a fixed vector.

Tensor quantities transform in a well-defined manner under (1.4.1) for $\mathbf{Q}$ time-independent. A subset of these quantities have the same transformation properties even if $\mathbf{Q}$ is time-dependent. Tensors in this subset will be referred to as objective tensors. In particular, if ϕ is an objective scalar, $\mathbf{a}$ an objective vector, and $\mathbf{B}$ an objective second-order tensor, then

$$\begin{aligned} &\phi'(\mathbf{X}, t') = \phi(\mathbf{X}, t), \quad \mathbf{a}'(\mathbf{X}, t') = \mathbf{Q}(t)\mathbf{a}(\mathbf{X}, t), \\ &\mathbf{B}'(\mathbf{X}, t') = \mathbf{Q}(t)\mathbf{B}(\mathbf{X}, t)\mathbf{Q}^{\top}(t). \end{aligned} \tag{1.4.2}$$

Various physical quantities are assumed to be objective tensors. These assumptions are linked to the principle of material objectivity discussed in Sect. 1.4.2. Thermodynamic quantities introduced later such as the internal and free energies, the entropy, and the temperature are taken to be objective scalars, while the heat flux is assumed to be an objective vector. The Piola–Kirchhoff heat flux, defined analogously to the Piola–Kirchhoff stress tensor, is an objective scalar, by virtue of the device introduced in (1.4.5) below. The Cauchy stress tensor is assumed to transform as a second-order objective tensor under a change of observer, so that

$$\mathbf{T}' = \mathbf{Q}(t)\mathbf{T}\mathbf{Q}^{T}(t). \tag{1.4.3}$$

The particle velocity $\mathbf{v}$, given by (1.2.5), is not objective, nor is the kinetic energy density.

We note that the second-order tensor $\mathbf{F}$ is a transformation from the material to the spatial description. It acts like an objective vector in that under (1.4.1),

$$\mathbf{F}'(t) = \mathbf{Q}(t)\mathbf{F}(t). \tag{1.4.4}$$

Proposition 1.4.1. *The second Piola–Kirchhoff tensor* $\widehat{\mathbf{S}}$, *defined by* (1.3.52) *or* (1.3.53), *is an objective scalar; all its components have this property.*

Proof. This follows from (1.4.3), (1.4.4), and the observation that $J = \det \mathbf{F} = \det(\mathbf{QF})$ is an objective scalar. □

Observe that

$$\dot{\mathbf{F}}'(t) = \mathbf{Q}(t)\dot{\mathbf{F}}(t) + \boldsymbol{\Sigma}(t)\mathbf{F}'(t),$$

where $\boldsymbol{\Sigma}$ is the spin tensor, defined as

$$\boldsymbol{\Sigma}(t) = \dot{\mathbf{Q}}(t)\mathbf{Q}^{\mathsf{T}}(t).$$

Then by (1.2.10),

$$\mathbf{L}'(t) = \dot{\mathbf{F}}'(t)[\mathbf{F}'(t)]^{-1} = \mathbf{Q}(t)\mathbf{L}(t)\mathbf{Q}^{\mathsf{T}}(t) + \boldsymbol{\Sigma}(t).$$

We see from (1.4.4) that $\mathbf{C}$, given by $(1.2.15)_1$, and $\mathbf{E}$, given by $(1.2.20)_1$, are unaffected by the transformation (1.4.1). Thus, we have

$$\mathbf{C}' = \mathbf{C}, \qquad \mathbf{E}' = \mathbf{E},$$

and therefore, since, from $(1.2.15)_1$, $\mathbf{U} = \mathbf{C}^{1/2}$, we have

$$\mathbf{U}' = \mathbf{U}.$$

Thus, all these quantities are objective scalars, as shown for $\widehat{\mathbf{S}}$ in Proposition 1.4.1.

Note that if $\mathbf{a}$ is an objective vector and λ is an objective scalar, then by virtue of (1.4.4),

$$\mathbf{a}_s = \lambda\mathbf{F}^{-1}\mathbf{a} \tag{1.4.5}$$

is an objective scalar.

1.4.2 Principle of Material Objectivity

This principle [313], also termed the *principle of material frame indifference*, postulates that the intrinsic properties of a material, as expressed in its constitutive relations, do not depend on the observer frame. More recent discussions of the topic may be found in particular in [188, 195, 238].

For example, consider the simple case of a spring extended by an applied force [313]. Material frame indifference, in this case, is the statement that the spring constant is the same for all observers in all frames of reference given by (1.4.1).

Expressed more formally, it is the statement that the constitutive equations describing the response of a material must hold in all frames related by (1.4.1).

This principle is accepted as valid for most conditions, though breakdowns have been predicted, notably within the framework of rational extended thermodynamics [269]. An observation on page 258 of [195] is of interest in this context.

This principle is imposed by expressing constitutive relations in terms of objective tensors. For simple materials [195, 238, 313], these relations in general involve functionals of the history of $\mathbf{F}$ and thermodynamic variables that are taken to be objective scalars either by assumption or by construction as in (1.4.5).

Explicit dependence on time is excluded by indifference to the origin of time as expressed by $(1.4.1)_2$. Note, however, that if a material is aging ([167], for example), explicit time dependence can occur.

Explicit dependence on $\mathbf{X}$ will occur if the material is inhomogeneous. The property of inhomogeneity is generally assumed in the present work, though we will often omit explicit inclusion of the space dependence.

Thus, we write a general constitutive relation as

$$\begin{aligned} &\mathbf{A}(\mathbf{X},t) = \hat{\mathbf{A}}(\mathbf{F}^t(\mathbf{X},s),\ \mathbf{\Theta}^t(\mathbf{X},s); s \geq 0), \\ &\mathbf{F}^t(\mathbf{X},s) = \mathbf{F}(\mathbf{X},t-s), \qquad \mathbf{\Theta}^t(\mathbf{X},s) = \mathbf{\Theta}(\mathbf{X},t-s), \quad \forall s \geq 0, \end{aligned} \tag{1.4.6}$$

where $\mathbf{\Theta}$ represents a list of objective scalar variables and $\mathbf{A}$ is a scalar, vector, or higher order objective tensor. In the primed frame of reference, using only $(1.4.1)_1$, this becomes

$$\begin{aligned} \mathbf{A}'(\mathbf{X},t') &= \hat{\mathbf{A}}'(\mathbf{F}'(\mathbf{X},t-s),\ \mathbf{\Theta}'(\mathbf{X},t-s); s \geq 0) \\ &= \hat{\mathbf{A}}'(\mathbf{Q}\,(t-s)\mathbf{F}(\mathbf{X},t-s),\ \mathbf{\Theta}'(\mathbf{X},t-s); s \geq 0), \end{aligned}$$

where $(1.4.2)_1$ and (1.4.4) have been used. The principle of material objectivity can be stated as follows: the functional $\hat{\mathbf{A}}'$ is the same functional as $\hat{\mathbf{A}}$ for all frames of reference, or

$$\begin{aligned} &\hat{\mathbf{A}}'(\mathbf{Q}(t-s)\mathbf{F}(\mathbf{X},t-s), \mathbf{\Theta}'(\mathbf{X},t-s); s \geq 0) \\ &\qquad\qquad = \hat{\mathbf{A}}(\mathbf{Q}\,(t-s)\mathbf{F}(\mathbf{X},t-s),\ \mathbf{\Theta}'(\mathbf{X},t-s); s \geq 0), \end{aligned}$$

for all choices of the independent field variables. Thus, for $\hat{\mathbf{A}} = \phi,\ \mathbf{a}, \mathbf{B}$, transforming as specified by $(1.4.2)_2$, we have the conditions

$$\begin{aligned} &\hat{\phi}(\mathbf{Q}(t-s)\mathbf{F}^t(\mathbf{X},s),\ \mathbf{\Theta}^t(\mathbf{X},s); s \geq 0) = \hat{\phi}(\mathbf{F}^t(\mathbf{X},s),\ \mathbf{\Theta}^t(\mathbf{X},s); s \geq 0), \\ &\hat{\mathbf{a}}(\mathbf{Q}(t-s)\mathbf{F}^t(\mathbf{X},s),\ \mathbf{\Theta}^t(\mathbf{X},s); s \geq 0) = \mathbf{Q}(t)\hat{\mathbf{a}}(\mathbf{F}^t(\mathbf{X},s),\ \mathbf{\Theta}^t(\mathbf{X},s); s \geq 0), \\ &\hat{\mathbf{B}}(\mathbf{Q}(t-s)\mathbf{F}^t(\mathbf{X},s),\ \mathbf{\Theta}^t(\mathbf{X},s); s \geq 0) = \mathbf{Q}(t)\hat{\mathbf{B}}(\mathbf{F}^t(\mathbf{X},s),\ \mathbf{\Theta}^t(\mathbf{X},s); s \geq 0)\mathbf{Q}^\top(t), \end{aligned} \tag{1.4.7}$$

for all $\mathbf{F} \in \mathrm{Lin}^+$ and orthogonal tensors $\mathbf{Q}$, where the notation $(1.4.6)_{2,3}$ has been used.

The implications of the principle of material objectivity for the possible forms of constitutive equations are considerable [195, 238, 313], as shown by the following example.

Proposition 1.4.2. *Let* $\mathbf{F}(t-s) = \mathbf{R}(t-s)\mathbf{U}(t-s)$ *be the polar decomposition of* $\mathbf{F}$. *If an objective scalar obeys the principle of material objectivity* $(1.4.7)_1$, *then it can be expressed in terms of the current value and history of* $\mathbf{U}$ *and* $\mathbf{\Theta}$*; that is to say, it will not depend on* $\mathbf{R}$. *Since by* (1.2.15), *we have* $\mathbf{U} = \sqrt{\mathbf{C}}$, *it follows that the scalar is a function of the current value and a functional of the history of* $\mathbf{C} = \mathbf{F}^\top\mathbf{F}$.

Conversely, if it has this property, then $(1.4.7)_1$ *holds.*

Proof. The first assertion follows since we can always choose $\mathbf{Q}(t-s) = \mathbf{R}^\top(t-s)$ in $(1.4.7)_1$. The converse is immediate. □

Remark 1.4.3. Note that by virtue of (1.2.20), we can replace $\mathbf{C}(t-s)$ by the Green strain tensor $\mathbf{E}(t-s)$.

We must emphasize that Proposition 1.4.2 does not refer to the possible dependence on $\mathbf{F}$ of thermodynamic quantities in $\boldsymbol{\Theta}$ included to make them objective scalars, using (1.4.5).

Consider a particular class of bodies, the behavior of which depends on the history of the deformation gradient up to time t, $\mathbf{F}(\mathbf{X}, t-s) \forall s \in \mathbb{R}^+$. These materials are such that the stress tensor is given by the functional

$$\mathbf{T}(\mathbf{X}, t) = \hat{\mathbf{T}}(\mathbf{F}(\mathbf{X}, t),\ \mathbf{F}^t(\mathbf{X})), \tag{1.4.8}$$

where $\mathbf{F}(\mathbf{X}, t)$ is the current value and $\mathbf{F}^t(\mathbf{X}, s) = \mathbf{F}(\mathbf{X}, t-s)\ \forall s \in \mathbb{R}^{++}$ denotes the past history of $\mathbf{F}$. The functional $\hat{\mathbf{T}}$ may also depend on objective scalars as denoted by $\boldsymbol{\Theta}$ above.

The requirement of material frame indifference, as stated by $(1.4.7)_3$, yields that the functional $\hat{\mathbf{T}}$ must obey the relation (omitting $\mathbf{X}$)

$$\begin{aligned} &\hat{\mathbf{T}}(\mathbf{Q}(t)\mathbf{F}(t),\ \mathbf{Q}(t-s)\mathbf{F}(t-s)),\ s \in \mathbb{R}^{++}) \\ &\quad = \mathbf{Q}(t)\hat{\mathbf{T}}(\mathbf{F}(t),\ \mathbf{F}(t-s),\ s \in \mathbb{R}^{++})\mathbf{Q}^\top(t), \end{aligned} \tag{1.4.9}$$

for all $\mathbf{F} \in \mathrm{Lin}^+$ and orthogonal tensors $\mathbf{Q}$.

Proposition 1.4.4. *Property* (1.4.9) *is equivalent to the requirement that*

$$\hat{\mathbf{T}}(\mathbf{F}(t),\ \mathbf{F}(t-s)), s \in \mathbb{R}^{++}) = \mathbf{F}(t)\tilde{\mathbf{T}}(\mathbf{C}(t),\ \mathbf{C}(t-s),\ s \in \mathbb{R}^{++})\mathbf{F}^\top(t),$$

where $\mathbf{C}$ *is the right Cauchy–Green tensor, defined by* (1.2.15). *The dependence of* $\tilde{\mathbf{T}}$ *on* $\mathbf{C}$ *is not restricted by the property of material frame indifference. Note that from* (1.3.52), $\tilde{\mathbf{T}}$ *is related to the second Piola–Kirchhoff stress tensor by*

$$J\tilde{\mathbf{T}} = \widehat{\mathbf{S}}.$$

Proof. This follows immediately from Propositions 1.4.1 and 1.4.2. □

1.4.3 Fading Memory

We shall consider materials for which the property of fading memory holds. This property is expressible through the (Volterra) dissipation behavior of hereditary action [318], which states "*the modulus of the variation of the quantity* [given by (1.4.8)], *when* $\mathbf{F}^t$ *varies in any way ... in the interval* $(-\infty,\ t_1)$ *(with* $t_1 < t$*) can be made as small as we please by taking the interval* (t_1, t) *sufficiently large.*"‖

‖ In the Coleman and Noll theory [73], the fading memory property is given by the continuity of (1.4.8) with respect to the norm

$$\|\mathbf{F}^t\|^2 = \int_0^\infty h(s)|\mathbf{F}^t(s)|^2 ds,$$

where the map $h \in L^1(\mathbb{R}^+)$ is a suitable positive decreasing function.

A more precise definition of the property of fading memory at a material point $\mathbf{X} \in \mathcal{B}$ can be given by considering the set $\mathcal{D}$ of the histories that make up the domain of definition of the functional (1.4.8).

For this purpose, we suppose the set $\mathcal{D}$ has the following properties:

1. $\mathcal{D} = \mathrm{Lin} \times \mathcal{D}_r$, where $\mathcal{D}_r$ is a set of past histories that contains the space $L^\infty(\mathbb{R}^+)$.
2. The partly static history $\mathbf{F}^{t(\tau)}$, associated with $\mathbf{F}^t(\mathbf{X})$, is defined by

$$\mathbf{F}^{t(\tau)}(\mathbf{X}, s) := \begin{cases} \mathbf{F}(\mathbf{X}, t) & \forall s \in [0, \ \tau), \\ \mathbf{F}^t(\mathbf{X}, s - \tau) & \forall s \in [\tau, \ \infty), \end{cases} \tag{1.4.10}$$

where τ is the duration of the static part of the history. If $\mathbf{F}^t(\mathbf{X}) \in \mathcal{D}$, then $\mathbf{F}^{t(\tau)}$ belongs to $\mathcal{D}$.

Definition 1.4.5. *A viscoelastic material is characterized by the constitutive equation (1.4.8), where* $\mathbf{F}^t \in \mathcal{D}$*, and there exists a constitutive equation* $\mathbf{T}(\mathbf{X}, t) = \tilde{\mathbf{T}}(\mathbf{F}(\mathbf{X}, t))$ *of an elastic material such that*

$$\lim_{\tau \to \infty} \hat{\mathbf{T}}(\mathbf{F}^{t(\tau)}(\mathbf{X})) = \tilde{\mathbf{T}}(\mathbf{F}(\mathbf{X}, t)).$$

Moreover, $\hat{\mathbf{T}}(\mathbf{F}^{t(\tau)}(\mathbf{X})) - \tilde{\mathbf{T}}(\mathbf{F}(\mathbf{X}, t))$ *is a function of* τ*, which belongs to* $L^2(\mathbb{R}^+)$.

This definition includes an expression of the fading memory property. Consider its application to the simplest case, namely a linear constitutive relation defining a linear viscoelastic material. Such linear relations will be systematically derived and discussed in Part III. For a linear viscoelastic body, we have

$$\mathbf{T}(\mathbf{X}, t) = \mathbb{G}_0(\mathbf{X})\mathbf{E}(\mathbf{X}, t) + \int_0^\infty \mathbb{G}'(\mathbf{X}, s)\mathbf{E}^t(\mathbf{X}, s)ds, \tag{1.4.11}$$

where $\mathbf{E} \in \mathrm{Sym}$ is the strain tensor.** The infinitesimal approximation to this quantity, as given by (1.2.31), is generally, though not necessarily, used in this context. The quantities $\mathbb{G}_0$ and $\mathbb{G}'$ are fourth-order tensors in Lin(Sym). The domain $\mathcal{D}$ consists of the set of pairs $(\mathbf{E}(t), \ \mathbf{E}^t)$ such that $\mathbf{E}(t) \in \mathrm{Sym}$ and $\mathbb{G}'\mathbf{E}^t \in L^1(\mathbb{R}^+)$.

In the linear theory, $\mathcal{D}$ includes constant histories by property 1. It follows that the kernel $\mathbb{G}'$ belongs to $L^1(\mathbf{R}^+)$. Then if $\mathbb{G}' \in L^1(\mathbb{R}^+)$, we conclude that $\mathbb{G}'\mathbf{E}^{t(\tau)} \in L^1(\mathbb{R}^+)$, where $\mathbf{E}^{t(\tau)}$ is the partly static history associated with $\mathbf{E}^t$. Hence,

$$\lim_{\tau \to \infty} \hat{\mathbf{T}}(\mathbf{E}^{t(\tau)}) = \mathbb{G}_\infty \mathbf{E}(t), \qquad \mathbb{G}_\infty = \mathbb{G}_0 + \int_0^\infty \mathbb{G}'(s)ds. \tag{1.4.12}$$

We observe that (1.4.11) represents a viscoelastic material with the fading memory property, according to Definition 1.4.5, because the right-hand side of $(1.4.12)_1$ is the stress associated with an elastic material. For the same reason, the tensor $\mathbb{G}_\infty$ must be a positive definite tensor in the case of a solid, though it may vanish for a liquid. Thus, we have

$$\mathbb{G}_\infty \geq \mathbf{0}. \tag{1.4.13}$$

** This follows from the principle of material frame indifference as expressed through Proposition 1.4.4.

2

Materials with Constitutive Equations That Are Local in Time

2.1 Introduction

We now consider the constitutive equations relating to fluids and solids for which memory effects are negligible. This contrasts with subsequent chapters, which are devoted almost entirely to materials with memory. In fact, however, one example included in the discussion, namely viscous fluids, can be visualized as possessing very short-term memory, expressed by the presence of time derivatives of certain field quantities. This is consistent with the general correlation that will arise throughout the present work between memory effects and energy dissipation.

In what follows, we will consider three types of constitutive equations: (i) constraints on the possible deformations the body may undergo, (ii) assumptions concerning the form of the stress tensor, and (iii) constitutive equations relating the stress to the deformation. Each of these three types of constitutive equations is appropriate for a certain class of materials, and its validity is verified by experiment.

As an example of a constitutive equation of type (i) we give the constraint that only rigid motions are possible, a constraint that underlies rigid-body mechanics. Another example in this class is the assumption of incompressibility, in which only isochoric deformations are permissible. Such an assumption is realistic for liquids such as water under normal flow conditions. An example of a constitutive equation of type (ii) is the widely used assumption that the stress is a pressure, an assumption appropriate for most fluids when viscous effects are negligible. Finally, an example of a constitutive equation of type (iii) is Hooke's law relating the deformation of a body to the state of stress, which is appropriate for linear elastic materials.

2.2 Fluids: Ideal Fluids

In this section, the constitutive equations for various classes of fluids are discussed. To this end we note that the contact forces in such materials are most naturally considered in the Eulerian description of deformation (Definition 1.2.2), that is, they

G. Amendola et al., *Thermodynamics of Materials with Memory*,
https://doi.org/10.1007/978-3-030-80534-0_2

are regarded as depending on the variables $(\mathbf{x}, t)$. From a physical point of view, this means that the contact forces are determined by the kinematic properties of the fluid at the present time.

Definition 2.2.1. *By a dynamical process, we mean a pair* $(\mathbf{x}, \mathbf{T})$*, where* $\mathbf{x}$ *is a deformation in the sense of (1.2.3)*$_1$ *and* $\mathbf{T}$ *a symmetric tensor field on the trajectory of* $\mathbf{x}$*. The quantity* $\mathbf{T}(\mathbf{x}, t)$ *is a smooth function of* $\mathbf{x}$ *on* $\varphi_t(\mathcal{B})$*.*

Definition 2.2.2. *A dynamical process* $(\mathbf{x}, \mathbf{T})$ *is isochoric if* $(\mathbf{x}, t)$ *is an isochoric or volume-preserving deformation at each t. A material body is incompressible if each* $(\mathbf{x}, \mathbf{T})$ *is isochoric, so that for every subbody A of the body, we have*

$$\mathrm{vol}\,(\varphi_t(A)) = \mathrm{vol}\,(\varphi_0(A)), \qquad \text{for all } t. \tag{2.2.1}$$

The above relation states that every deformation preserves volume, and moreover, the volume of any subbody throughout the deformation must be the same as its volume in the reference configuration. In view of relations (1.2.29), (1.2.30), and (1.3.2)$_1$*, the condition of incompressibility (2.2.1) takes one of the following forms:*

$$J = \det \mathbf{F} = 1 \quad \text{or} \quad \mathrm{div}_{\mathbf{x}} \mathbf{v} = 0 \quad \text{or} \quad \rho = \rho_0.$$

Definition 2.2.3. *A dynamical process* $(\mathbf{x}, \mathbf{T})$ *is Eulerian if the Cauchy stress is a pressure, given by*

$$\mathbf{T} = -p\mathbf{1},$$

where p is a scalar field on the trajectory of $\mathbf{x}$*.*

Definition 2.2.4. *An ideal fluid is a material body that can support only isochoric Eulerian dynamical processes and whose density* ρ_0 *is constant.*

The tension on an arbitrary elementary surface $d\sigma$, with unit normal vector $\mathbf{n}$, is given by $\mathbf{t}(\mathbf{n}) = -p\mathbf{n}$, so that it is parallel to the normal (see Fig. 2.1). We observe that an ideal fluid is an incompressible material body for which the Cauchy stress is a pressure in every flow. Furthermore, the pressure is not determined uniquely by the deformation; there exists an infinite number of pressure fields corresponding to the same deformation. That such a property is physically reasonable can be inferred from the following example. Consider a ball composed of an ideal fluid under a time-independent uniform pressure p, and assume, for the moment, that all body forces are absent. Then the ball should remain in the same configuration for all time. Moreover, since the ball is incompressible, an increase or a decrease in pressure should not result in a deformation. Thus, the same deformation corresponds to all uniform pressure fields.

Summarizing, the equations of motion for an ideal fluid with density ρ_0 (see (1.3.25) and (1.3.2)$_2$) are

$$\rho_0 \dot{\mathbf{v}} = -\nabla_{\mathbf{x}} p + \rho_0 \mathbf{b}, \qquad \mathrm{div}_{\mathbf{x}}\, \mathbf{v} = 0. \tag{2.2.2}$$

We note that for a conservative body force with the potential energy $\mathcal{V}$, relation (2.2.2)$_1$ becomes

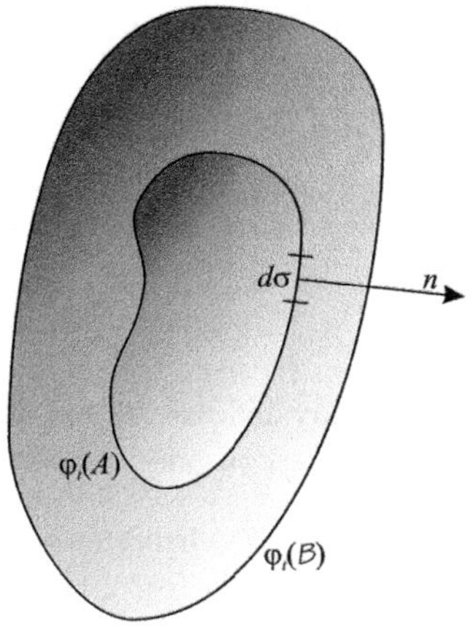

Fig. 2.1. The normal unit vector to an elementary surface element

$$\dot{\mathbf{v}} = -\nabla_{\mathbf{x}}\left(\frac{p}{\rho_0} + \mathcal{V}\right),$$

so that the acceleration is the gradient of a potential.

On the basis of Bernoulli's theorem (Theorem 1.3.13), we can formulate the following result.

Theorem 2.2.5. (**Bernoulli's Theorem for Ideal Fluids**) *Let* $\{\mathbf{v}, \rho_0, -p\mathbf{1}\}$ *be a flow of an ideal fluid under a conservative body force with potential energy* $\mathcal{V}$. *Then,*

(i) if the flow is potential ($\mathbf{v} = \nabla_{\mathbf{x}}\phi$), *then*

$$\nabla_{\mathbf{x}}\left(\frac{\partial \phi}{\partial t} + \frac{1}{2}\mathbf{v}^2 + \frac{p}{\rho_0} + \mathcal{V}\right) = \mathbf{0}. \tag{2.2.3}$$

(ii) if the flow is steady, then

$$\frac{d}{dt}\left(\frac{1}{2}\mathbf{v}^2 + \frac{p}{\rho_0} + \mathcal{V}\right) = 0. \tag{2.2.4}$$

(iii) if the flow is steady and irrotational, then

$$\frac{1}{2}\mathbf{v}^2 + \frac{p}{\rho_0} + \mathcal{V} = constant. \tag{2.2.5}$$

Proof. Since $\rho = \rho_0 = constant$, the relation (1.3.39) implies (2.2.3). Furthermore, if we set

$$\eta = \frac{1}{2}\mathbf{v}^2 + \frac{p}{\rho_0} + \mathcal{V},$$

then for a steady flow, $\frac{\partial \eta}{\partial t} = 0$, and moreover, by (1.3.40), we have $\mathbf{v} \cdot \nabla_{\mathbf{x}}\eta = 0$, so that

$$\frac{d\eta}{dt} = \frac{\partial \eta}{\partial t} + \mathbf{v} \cdot \nabla_x \eta = 0,$$

which is (2.2.4). Finally, for a steady and irrotational flow, (1.3.41) yields $\nabla_{\mathbf{x}}\eta = 0$, which, with $\frac{\partial \eta}{\partial t} = 0$, implies that η is a constant in space and time, or (2.2.5). □

Remark 2.2.6. With the aid of Bernoulli's theorem, we see that for a steady and irrotational flow under a conservative body force, Eqs. $(2.2.2)_1$ and $(2.2.2)_2$ reduce to

$$\mathrm{div}_{\mathbf{x}}\mathbf{v} = 0, \quad \mathrm{curl}_{\mathbf{x}}\mathbf{v} = \mathbf{0}, \quad \frac{1}{2}\mathbf{v}^2 + \frac{p}{\rho_0} + \mathcal{V} = constant. \tag{2.2.6}$$

In a steady motion the velocity is tangent to the boundary, so that (2.2.6) should be supplemented by the boundary condition

$$\mathbf{v} \cdot \mathbf{n} = 0 \quad \text{on} \quad \partial\varphi_0(\mathcal{B}).$$

Remark 2.2.7. For an unsteady flow, we have to solve the system of differential equations described by (2.2.2). Using (1.2.6), we can write this system in the form of *Euler's equations*

$$\frac{\partial \mathbf{v}}{\partial t} + (\nabla_{\mathbf{x}}\mathbf{v})\mathbf{v} = -\nabla_{\mathbf{x}}\left(\frac{p}{\rho_0}\right) + \mathbf{b},$$
$$\mathrm{div}_{\mathbf{x}}\mathbf{v} = 0,$$

which is a nonlinear differential system because of the presence of the term $(\nabla_{\mathbf{x}}\mathbf{v})\mathbf{v}$.

2.2.1 Elastic Fluids

In what follows, we consider a material body in which compressibility effects are not ignored and for which the pressure is completely specified by the deformation.

Definition 2.2.8. *An elastic fluid is a material body for which the constitutive class is defined by the smooth response function $\hat{p}\colon \mathbb{R}^+ \to \mathbb{R}$ giving the pressure when the mass density is known:*

$$p = \hat{p}(\rho). \tag{2.2.7}$$

For such a fluid, the constitutive class is the set of all Eulerian dynamical processes $(\mathbf{x}, -p\mathbf{1})$ (Definition 2.2.1) that obey the constitutive equation (2.2.7).

The basic equations for the flow $\{\mathbf{v}, \rho, -p\mathbf{1}\}$ of an elastic fluid are the equations of motion

$$\rho\dot{\mathbf{v}} = -\nabla_{\mathbf{x}}p + \rho\mathbf{b}, \tag{2.2.8}$$

conservation of mass

$$\frac{\partial \rho}{\partial t} + \mathrm{div}_{\mathbf{x}}(\rho \mathbf{v}) = 0, \tag{2.2.9}$$

and the constitutive equation (2.2.7).

We assume that $\hat{p}$ has a strictly positive derivative and define the functions $\kappa > 0$ and ε on $\mathbb{R}^+$ by

$$\kappa^2(\rho) = \frac{d\hat{p}(\rho)}{d\rho}, \quad \varepsilon(\rho) = \int_{\rho_*}^{\rho} \frac{\kappa^2(\xi)}{\xi} d\xi,$$

where ρ_* is an arbitrarily chosen value of the mass density. The function $\kappa(\rho)$ is called the speed of sound. By the chain rule, we obtain

$$\nabla_{\mathbf{x}}\varepsilon(\rho) = \frac{\kappa^2(\rho)}{\rho}\nabla_{\mathbf{x}}\rho = \frac{1}{\rho}\frac{d\hat{p}(\rho)}{d\rho}\nabla_{\mathbf{x}}\rho = \frac{1}{\rho}\nabla_{\mathbf{x}}p,$$

so that for a conservative body force with potential energy $\mathcal{V}$, the equation of motion (2.2.8) takes the form

$$\dot{\mathbf{v}} = -\nabla_{\mathbf{x}}[\varepsilon(\rho) + \mathcal{V}],$$

and hence the acceleration is the gradient of a potential. Thus, by Bernoulli's theorem, we get the following result.

Theorem 2.2.9. (Bernoulli's Theorem for Elastic Fluids) *Let $\{\mathbf{v}, \rho, -p\mathbf{1}\}$ be a flow for an elastic fluid under a conservative body force with potential energy $\mathcal{V}$. Then,*

(i) if the flow is potential, then

$$\nabla_{\mathbf{x}}\left[\frac{\partial \phi}{\partial t} + \frac{1}{2}\mathbf{v}^2 + \varepsilon(\rho) + \mathcal{V}\right] = \mathbf{0}.$$

(ii) if the flow is steady, then

$$\frac{d}{dt}\left[\frac{1}{2}\mathbf{v}^2 + \varepsilon(\rho) + \mathcal{V}\right] = 0.$$

(iii) if the flow is steady and irrotational, then

$$\frac{1}{2}\mathbf{v}^2 + \varepsilon(p) + \mathcal{V} = constant.$$

Remark 2.2.10. If we set

$$\alpha(\rho) = \frac{\kappa^2(\rho)}{\rho},$$

then the basic equations (2.2.7)–(2.2.9) lead to

$$\begin{aligned} \frac{\partial \mathbf{v}}{\partial t} + (\nabla_{\mathbf{x}}\mathbf{v})\mathbf{v} + \alpha(\rho)\nabla_{\mathbf{x}}\rho &= \mathbf{b}, \\ \frac{\partial \rho}{\partial t} + \mathrm{div}_{\mathbf{x}}(\rho\mathbf{v}) &= 0. \end{aligned} \tag{2.2.10}$$

The relations in (2.2.10) furnish a nonlinear differential system for ρ and $\mathbf{v}$. Concerning such a system we can establish the following result.

Proposition 2.2.11. *In a steady flow of an elastic fluid under vanishing body forces, we have*

$$\frac{d}{dt}(\rho v) = \rho\left(1 - m^2\right)\frac{dv}{dt}, \tag{2.2.11}$$

where $v = |\mathbf{v}|$ *is assumed different from zero and*

$$m = \frac{v}{\kappa(\rho)} \tag{2.2.12}$$

is the Mach number.

Proof. Since we have a steady flow, it follows that $\frac{\partial \rho}{\partial t} = 0$, and hence, from (1.2.6),

$$\dot{\rho} = \mathbf{v} \cdot \nabla_{\mathbf{x}}\rho.$$

Furthermore, with $\mathbf{b} = \mathbf{0}$, Eq. $(2.2.10)_1$ implies that

$$\mathbf{v} \cdot \dot{\mathbf{v}} = -\frac{\kappa^2(\rho)}{\rho}\mathbf{v} \cdot \nabla_{\mathbf{x}}\rho = -\frac{\kappa^2(\rho)}{\rho}\dot{\rho},$$

and therefore, with the aid of (2.2.12), we have

$$\mathbf{v} \cdot \frac{d}{dt}(\rho\mathbf{v}) = \mathbf{v} \cdot (\rho\dot{\mathbf{v}} + \dot{\rho}\mathbf{v}) = \rho(\mathbf{v} \cdot \dot{\mathbf{v}})\left(1 - m^2\right).$$

This last relation, when combined with the observation that

$$\mathbf{v} \cdot \dot{\mathbf{v}} = \frac{d}{dt}\left(\frac{1}{2}v^2\right) = v\dot{v}$$

and hence

$$\mathbf{v} \cdot \frac{d}{dt}(\rho\mathbf{v}) = v(\dot{\rho} + \rho\dot{v}) = v\frac{d}{dt}(\rho v),$$

proves (2.2.11).

It follows that for $m < 1$, the mass flow $\rho(\mathbf{x})v(\mathbf{x})$ increases, while for $m > 1$, the mass flow decreases. □

Proposition 2.2.11 motivates the following definition.

Definition 2.2.12. *A flow is subsonic, sonic, or supersonic at* $(\mathbf{x}, t)$ *according to whether* $m(\mathbf{x}, t)$ *is less than, equal to, or greater than 1, respectively.*

For a steady flow under a conservative body force with potential energy $\mathcal{V}$, *we conclude, from Bernoulli's theorem, that the basic equations characterizing such a flow are*

$$\mathrm{div}_{\mathbf{x}}(\rho\mathbf{v}) = 0, \quad \mathrm{curl}_{\mathbf{x}}\mathbf{v} = \mathbf{0}, \quad \frac{1}{2}\mathbf{v}^2 + \varepsilon(\rho) + \mathcal{V} = constant.$$

2.2.2 Newtonian Fluids: The Navier–Stokes Equations

The ideal and elastic fluids previously discussed never exhibit shearing stress, and therefore they are incapable of describing frictional force. Friction in fluids generally manifests itself through shearing forces that retard the relative motion of fluid particles. A measure of the relative motion of fluid particles is furnished by the velocity gradient.

Definition 2.2.13. *A Newtonian fluid is an incompressible material with constitutive equation*

$$\mathbf{T}(\mathbf{x},t) = -p(\mathbf{x},t)\mathbf{1} + 2\mu\mathbf{D}(\mathbf{x},t), \tag{2.2.13}$$

where $\mathbf{D}$ *is the stretching, and the scalar constant* μ *is known as the viscosity of the fluid. The term* $\mathbf{T}_0 = 2\mu\mathbf{D} = \mathbf{T} + p\mathbf{1}$ *is referred to as the extra stress. Since* $\mathrm{tr}\mathbf{D} = \mathrm{div}_{\mathbf{x}}\mathbf{v} = 0$, *we have that* $\mathbf{T}_0 = \mathbf{T} - \frac{1}{3}(tr\mathbf{T})\mathbf{1}$.

Remark 2.2.14. Since $\mathbf{D}$ vanishes for a material at rest, by (1.2.23), it follows from (2.2.13) that a Newtonian fluid at rest behaves like an ideal fluid.

The basic equations for a Newtonian fluid are

$$\rho\left[\frac{\partial\mathbf{v}}{\partial t} + (\nabla_{\mathbf{x}}\mathbf{v})\mathbf{v}\right] = \mathrm{div}_{\mathbf{x}}\mathbf{T} + \rho\mathbf{b}, \quad \mathbf{T} = -p\mathbf{1} + 2\mu\mathbf{D}, \quad \mathrm{div}_{\mathbf{x}}\mathbf{v} = 0. \tag{2.2.14}$$

We have $\rho = \rho_0$, because of $(1.3.2)_2$ and $(2.2.14)_3$. Also,

$$2\,\mathrm{div}_{\mathbf{x}}\mathbf{D} = \mathrm{div}_{\mathbf{x}}\left[\nabla_{\mathbf{x}}\mathbf{v} + (\nabla_{\mathbf{x}}\mathbf{v})^T\right] = \Delta\mathbf{v} + \nabla_{\mathbf{x}}\mathrm{div}_{\mathbf{x}}\mathbf{v} = \Delta\mathbf{v},$$

and Eqs. (2.2.14) reduce to

$$\frac{\partial\mathbf{v}}{\partial t} + (\nabla_{\mathbf{x}}\mathbf{v})\mathbf{v} = \nu\Delta\mathbf{v} - \nabla_{\mathbf{x}}p_0 + \mathbf{b}, \quad \mathrm{div}_{\mathbf{x}}\mathbf{v} = 0, \tag{2.2.15}$$

where

$$\nu = \frac{\mu}{\rho_0}, \quad p_0 = \frac{p}{\rho_0}.$$

The scalar constant ν is known as the *kinematic viscosity*, and Eqs. (2.2.15) are the *Navier–Stokes equations*. These constitute a nonlinear system of partial differential equations for the velocity $\mathbf{v}$ and the pressure p.

Suppose that the flow takes place in a region R. To the Navier–Stokes equations, we add the restriction that the fluid adheres, without slipping, to the boundary ∂R. For a stationary boundary, this means that $\mathbf{v} = \mathbf{0}$ on ∂R. If the boundary moves, then at each point on the boundary the fluid velocity must coincide with the velocity of the boundary.

Theorem 2.2.15. (Balance of Energy for a Viscous Fluid) *For any flow of a Newtonian fluid, we have*

$$\frac{d}{dt}\int_{\varphi_t(A)} \frac{1}{2}\rho_0 \mathbf{v}^2 dv_t + 2\mu \int_{\varphi_t(A)} |\mathbf{D}|^2 dv_t = \int_{\varphi_t(A)} \rho_0 \mathbf{b}\cdot \mathbf{v} dv_t + \int_{\partial\varphi_t(A)} \mathbf{Tn}\cdot \mathbf{v} da_t, \tag{2.2.16}$$

for every part A of the fluid.

Proof. The constitutive equation (2.2.13), together with the fact that $\mathbf{D}$ is traceless, gives

$$\mathbf{T}\cdot\mathbf{D} = -p\mathrm{tr}\mathbf{D} + 2\mu|\mathbf{D}|^2 = 2\mu|\mathbf{D}|^2,$$

and hence the theorem of power expended, expressed by relation (1.3.33), yields (2.2.16). □

Remark 2.2.16. The term

$$2\mu \int_{\varphi_t(A)} |\mathbf{D}|^2 dv_t$$

represents the rate at which the fluid in A dissipates energy. The energy equation (2.2.16) asserts that the total power expended on A must equal the rate of change of kinetic energy plus the rate of energy dissipation.

Corollary 2.2.17. *For a flow of a Newtonian fluid in a finite region $\mathcal{B}$ under the hypotheses of zero body force $\mathbf{b} = \mathbf{0}$ and $\mathbf{v} = \mathbf{0}$ on $\partial\varphi_t(\mathcal{B})$ at all times and $\mu > 0$, we have*

$$\frac{d}{dt}\int_{\varphi_t(\mathcal{B})} \frac{1}{2}\rho_0 \mathbf{v}^2 dv_t \leq 0,$$

so that the kinetic energy decreases with time.

It is useful to write the Navier–Stokes equations in dimensionless form. For convenience we will assume that $\mathbf{b} = \mathbf{0}$. Consider l a typical length (such as the diameter when a cylindrical body is considered) and v a typical velocity. Let us further identify points $\mathbf{x}$ with their position vectors from a given origin O and introduce the dimensionless position vector $\bar{\mathbf{x}} = \frac{\mathbf{x}}{l}$, the dimensionless time $\bar{t} = \frac{tv}{l}$, the dimensionless velocity $\bar{\mathbf{v}}(\bar{\mathbf{x}}, \bar{t}) = \frac{1}{v}\mathbf{v}(\mathbf{x}, t)$, and the dimensionless pressure $\bar{p}_0(\bar{\mathbf{x}}, \bar{t}] = \frac{1}{v^2} p_0(\mathbf{x}, t)$. Thus, we have

$$\nabla_{\bar{\mathbf{x}}}\bar{\mathbf{v}} = \frac{l}{v}\nabla_{\mathbf{x}}\mathbf{v}, \quad \frac{\partial\bar{\mathbf{v}}}{\partial\bar{t}} = \frac{l}{v^2}\frac{\partial\mathbf{v}}{\partial t}, \quad \nabla_{\bar{\mathbf{x}}}\bar{p}_0 = \frac{l}{v^2}\nabla_{\bar{\mathbf{x}}} p_0.$$

Hence, the Navier–Stokes equations (2.2.15) become

$$\begin{aligned} \frac{\partial\bar{\mathbf{v}}}{\partial\bar{t}} + (\nabla_{\bar{\mathbf{x}}}\bar{\mathbf{v}})\bar{\mathbf{v}} &= \frac{1}{Re}\bar{\Delta}\bar{\mathbf{v}} - \nabla_{\bar{\mathbf{x}}}\bar{p}_0, \\ \mathrm{div}_{\bar{\mathbf{x}}}\bar{\mathbf{v}} &= 0, \end{aligned} \tag{2.2.17}$$

where $\bar{\Delta}$ is the Laplacian in dimensionless coordinates and

$$\mathrm{Re} = \frac{lv}{\nu}$$

is a dimensionless quantity known as the *Reynolds number* of the flow. The Navier–Stokes equations in the dimensionless form (2.2.17) show that a solution of the Navier–Stokes equations with a given Reynolds number can be used to generate solutions that have different length and velocity scales but the same Reynolds number. This fact allows one to model a given flow situation in the laboratory by adjusting the length and velocity scales and the viscosity to give an experimentally tractable problem with the same Reynolds number of the effective flow.

Remark 2.2.18. Returning to the general Navier–Stokes equations, we note that if the flow is steady and if we neglect the nonlinear term $(\nabla_{\mathbf{x}}\mathbf{v})\mathbf{v}$, then the relations in (2.2.15) reduce to

$$\nu\Delta\mathbf{v} = \nabla_{\mathbf{x}} p_0 - \mathbf{b}, \qquad \operatorname{div}_{\mathbf{x}}\mathbf{v} = 0.$$

Solutions of this equation are called *Stokes flows* and are presumed to describe slow or creeping flows of Newtonian fluids.

For the compressible flow of a fluid, the constitutive equation (2.2.13) is replaced by

$$\mathbf{T}(\mathbf{x}, t) = -p(\mathbf{x}, t)\mathbf{1} + \lambda(\operatorname{div}_{\mathbf{x}}\mathbf{v})\mathbf{1} + 2\mu\mathbf{D}(x, t), \tag{2.2.18}$$

where λ and μ are the coefficients of viscosity. The basic equations (2.2.14) for such a flow become

$$\rho\frac{\partial\mathbf{v}}{\partial t} + \rho(\nabla_{\mathbf{x}}\mathbf{v})\mathbf{v} = \mu\Delta\mathbf{v} + (\lambda + \mu)\nabla_{\mathbf{x}}(\operatorname{div}_{\mathbf{x}}\mathbf{v}) - \nabla_{\mathbf{x}} p + \rho\mathbf{b},$$
$$\frac{\partial\rho}{\partial t} + \operatorname{div}_{\mathbf{x}}(\rho\mathbf{v}) = 0.$$

Now let us study the steady flow of a Newtonian fluid in a pipeline of cylindrical form having as cross-section a circle of radius R. Body forces are neglected. A reference frame is chosen such that the x_3-axis is parallel to the generators of the cylinder. We seek a solution of the following boundary value problem:

$$\nu\Delta\mathbf{v} - \nabla_{\mathbf{x}} p_0 = (\nabla_{\mathbf{x}}\mathbf{v})\mathbf{v}, \qquad \operatorname{div}_{\mathbf{x}}\mathbf{v} = 0, \tag{2.2.19}$$

in the cylinder $\mathcal{C} = \{\mathbf{x} \in \mathbb{R}^3 \colon x_1^2 + x_2^2 < R^2, x_3 \in \mathbb{R}\}$ with

$$\mathbf{v} = \mathbf{0} \quad \text{on} \quad \partial\mathcal{C},$$

which is the frequently adopted no-slip assumption, noted before Theorem 2.2.15, that no relative motion can take place between the viscous fluid and the solid cylinder.

Let us try a solution of the above boundary value problem in the form

$$\mathbf{v} = v(x_1, x_2)\mathbf{i}_3,$$

for which the relation $\operatorname{div}_{\mathbf{x}}\mathbf{v} = 0$ is identically satisfied and the right-hand side of $(2.2.19)_1$ vanishes. This last equation then gives

$$\frac{\partial p_0}{\partial x_\alpha} = 0 \ (\alpha = 1, 2),$$

$$\nu \Delta_0 v = \frac{\partial p_0}{\partial x_3},$$

where $\Delta_0 v = \frac{\partial^2 v}{\partial x_1^2} + \frac{\partial^2 v}{\partial x_2^2}$ is the Laplacian in two dimensions. Thus, from the first two equations, we conclude that $p_0 = p_0(x_3)$, while the third equation implies that $\frac{dp_0}{dx_3}$ is independent of x_3. Therefore, we deduce that

$$\frac{dp_0}{dx_3} = -m, \quad \nu \Delta_0 v = -m,$$

where m is an unknown constant. Hence,

$$p_0 = p_1 - m x_3,$$

where p_1 is a constant of integration. Moreover, the function v is the solution of the following boundary value problem:

$$\Delta_0 v = -\frac{m}{\nu} \text{ in } \Sigma, \quad v = 0 \quad \text{on} \quad \partial\Sigma,$$

where Σ denotes the cross-section $x_1^2 + x_2^2 < R^2$. We try a solution of this boundary value problem in the form

$$v = v(r), \quad r = \sqrt{x_1^2 + x_2^2},$$

so that

$$\frac{\partial v}{\partial x_\alpha} = \frac{x_\alpha}{r}\frac{dv}{dr} = \frac{x_\alpha}{r} v', \qquad \frac{\partial^2 v}{\partial x_\alpha \partial x_\beta} = \left(\frac{1}{r}\delta_{\alpha\beta} - \frac{x_\alpha x_\beta}{r^3}\right) v' + \frac{x_\alpha x_\beta}{r^2} v'',$$

and hence

$$\Delta_0 v = \frac{1}{r} v' + v''.$$

Thus, the function $v(r)$ satisfies the following differential equation:

$$(r v')' = -\frac{m}{\nu} r, \quad \text{with } v(R) = 0.$$

The general solution of this differential equation is

$$v(r) = -\frac{m}{4\nu} r^2 + C_1 \ln r + C_2,$$

where C_1 and C_2 are arbitrary constants. Because v is finite for $r = 0$, it follows that $C_1 = 0$, while the condition $v(R) = 0$ gives $C_2 = \frac{m}{4\nu} R^2$. Thus, the solution is

$$\mathbf{v} = \frac{m}{4\nu}\left(R^2 - r^2\right)\mathbf{i}_3, \quad p_0 = p_1 - m x_3.$$

The above motion is known as *Poiseuille flow*.

2.2.3 Uniqueness of Solutions

For the classical viscous flow problem, we assume that the following data are given: a bounded regular region R, a kinematic viscosity $\nu > 0$, a body force field $\mathbf{b}$ on $R\times[0, \infty)$, an initial velocity distribution $\mathbf{v}_0$ on R, and a boundary velocity distribution $\hat{\mathbf{v}}$ on $\partial R \times [0, \infty)$. The problem is to find a class C^2 velocity field $\mathbf{v}$ and a smooth pressure field p on $R \times [0, \infty)$ that satisfy the Navier–Stokes equations

$$\frac{\partial \mathbf{v}}{\partial t} + (\nabla_{\mathbf{x}}\mathbf{v})\mathbf{v} = \nu\Delta\mathbf{v} - \nabla_{\mathbf{x}}p + \mathbf{b}, \qquad \text{div}_{\mathbf{x}}\mathbf{v} = 0, \tag{2.2.20}$$

with the initial condition

$$\mathbf{v}(\mathbf{x}, 0) = \mathbf{v}_0(\mathbf{x}), \quad \text{for every } \mathbf{x} \in R, \tag{2.2.21}$$

and the boundary condition

$$\mathbf{v} = \hat{\mathbf{v}} \quad \text{on} \quad \partial R \times [0, \infty). \tag{2.2.22}$$

A pair $(\mathbf{v}, p)$ with these properties will be called a *solution* of the above initial boundary value problem. We now prove the following result.

Theorem 2.2.19. (Uniqueness of Solution) *Let* $(\mathbf{v}_1, p_1)$ *and* $(\mathbf{v}_2, p_2)$ *be solutions of the viscous flow problem corresponding to the same data. Then we have*

$$\mathbf{v}_1 = \mathbf{v}_2, \quad p_1 = p_2 + \alpha, \tag{2.2.23}$$

where α *does not depend on* $\mathbf{x}$.

Proof. By setting

$$\mathbf{u} = \mathbf{v}_1 - \mathbf{v}_2, \quad \alpha = p_1 - p_2,$$

we obtain, from relations (2.2.20)–(2.2.22),

$$\mathbf{u}(\mathbf{x}, 0) = \mathbf{0}, \quad \mathbf{u} = \mathbf{0} \quad \text{on} \quad \partial R \times [0, \infty), \quad \text{div}_{\mathbf{x}}\mathbf{u} = 0. \tag{2.2.24}$$

Moreover, by subtracting $(2.2.20)_1$ with $\mathbf{v} = \mathbf{v}_2$ and $p = p_2$ from $(2.2.20)_1$ with $\mathbf{v} = \mathbf{v}_1$ and $p = p_1$, we obtain

$$\frac{\partial \mathbf{u}}{\partial t} + (\nabla_{\mathbf{x}}\mathbf{v}_1)\mathbf{v}_1 - (\nabla_{\mathbf{x}}\mathbf{v}_2)\mathbf{v}_2 = \nu\Delta\mathbf{u} - \nabla_{\mathbf{x}}\alpha.$$

Since

$$(\nabla_{\mathbf{x}}\mathbf{v}_1)\mathbf{v}_1 = (\nabla_{\mathbf{x}}\mathbf{u})\mathbf{v}_1 + (\nabla_{\mathbf{x}}\mathbf{v}_2)\mathbf{v}_1,$$

we have

$$\frac{\partial \mathbf{u}}{\partial t} + (\nabla_{\mathbf{x}}\mathbf{u})\mathbf{v}_1 + (\nabla_{\mathbf{x}}\mathbf{v}_2)\mathbf{u} = \nu\Delta\mathbf{u} - \nabla_{\mathbf{x}}\alpha. \tag{2.2.25}$$

Note the following identities:

$$
\begin{aligned}
\mathbf{u}\cdot\Delta\mathbf{u} &= \operatorname{div}_{\mathbf{x}}\left[(\nabla_{\mathbf{x}}\mathbf{u})^T\mathbf{u}\right] - |\nabla_{\mathbf{x}}\mathbf{u}|^2,\\
\mathbf{u}\cdot(\nabla_{\mathbf{x}}\mathbf{u})\mathbf{v}_1 &= \mathbf{v}_1\cdot(\nabla_{\mathbf{x}}\mathbf{u})^T\mathbf{u} = \mathbf{v}_1\cdot\nabla_{\mathbf{x}}\left(\frac{1}{2}\mathbf{u}^2\right),\\
\mathbf{u}\cdot(\nabla_{\mathbf{x}}\mathbf{v}_2)\mathbf{u} &= \mathbf{u}\cdot\mathbf{D}_2\mathbf{u},
\end{aligned}
\tag{2.2.26}
$$

where (see (1.2.23))

$$
\mathbf{D}_2 = \frac{1}{2}\left[\nabla_{\mathbf{x}}\mathbf{v}_2 + (\nabla_{\mathbf{x}}\mathbf{v}_2)^{\top}\right]. \tag{2.2.27}
$$

In view of relation (2.2.26), we conclude from (2.2.25) that

$$
\frac{1}{2}\frac{\partial}{\partial t}\left(\mathbf{u}^2\right) + \mathbf{v}_1\cdot\nabla_{\mathbf{x}}\left(\frac{1}{2}\mathbf{u}^2\right) + \mathbf{u}\cdot\mathbf{D}_2\mathbf{u} = \nu\operatorname{div}_{\mathbf{x}}\left[(\nabla_{\mathbf{x}}\mathbf{u})^T\mathbf{u}\right] - \nu|\nabla_{\mathbf{x}}\mathbf{u}|^2 - \mathbf{u}\cdot\nabla_{\mathbf{x}}\alpha.
$$

If we integrate this relation over R, then from the divergence theorem and the boundary conditions $(2.2.24)_{1,2}$, together with $(2.2.20)_2$ (for $\mathbf{v}_1$), we conclude that

$$
\frac{1}{2}\frac{d}{dt}||\mathbf{u}||^2 + \int_R \mathbf{u}\cdot\mathbf{D}_2\mathbf{u}dv = -\nu\int_R |\nabla_{\mathbf{x}}\mathbf{u}|^2 dv \leq 0, \tag{2.2.28}
$$

where

$$
||\mathbf{u}||^2(t) = \int_R \mathbf{u}^2(\mathbf{x},t)dv.
$$

Since $\operatorname{div}_{\mathbf{x}}\mathbf{v}_2 = 0$ also, it follows that $\operatorname{tr}\mathbf{D}_2 = 0$, and thus the lowest eigenvalue of the symmetric tensor $\mathbf{D}_2(\mathbf{x},t)$ defined by (2.2.27) will be nonpositive. Let $-\gamma(\mathbf{x},t)$ denote this eigenvalue (with $\gamma \geq 0$), so that

$$
\mathbf{u}\cdot\mathbf{D}_2\mathbf{u} \geq -\gamma\mathbf{u}^2.
$$

Let us choose $\tau > 0$ and put

$$
\lambda = 2\sup_{(\mathbf{x},t)\in R\times[0,\tau]}\gamma(\mathbf{x},t).
$$

This quantity is finite by virtue of the C^2 property of $\mathbf{v}$. Then,

$$
\mathbf{u}\cdot\mathbf{D}_2\mathbf{u} \geq -\frac{\lambda}{2}\mathbf{u}^2,
$$

so that the relation (2.2.28) gives

$$
\frac{d}{dt}||\mathbf{u}||^2 - \lambda||\mathbf{u}||^2 \leq 0 \quad \text{on} \quad [0,\tau],
$$

or

$$
\frac{d}{dt}\left(||\mathbf{u}||^2 e^{-\lambda t}\right) \leq 0 \quad \text{on} \quad [0,\tau],
$$

and hence

$$
||\mathbf{u}||^2(\tau) \leq ||\mathbf{u}||^2(0)e^{\lambda\tau}.
$$

Since $||\mathbf{u}||^2(0) = 0$, the above relation implies that $||\mathbf{u}||^2(\tau) = 0$, and hence

$$
\mathbf{u}(\mathbf{x},\tau) = \mathbf{0} \quad \text{for every } \mathbf{x}\in R.
$$

Since τ was arbitrarily chosen, it follows that $\mathbf{u} \equiv \mathbf{0}$, and hence relation $(2.2.23)_1$ holds. Finally, (2.2.25) implies that $\nabla_{\mathbf{x}}\alpha = \mathbf{0}$, and the proof is complete. □

2.3 Elastic Solids

The force on an elastic spring depends only on the change in length of the spring, and it is independent of the past history of the length as well as the rate at which the length is changing with time. We have seen previously that the deformation gradient $\mathbf{F}$ measures local changes in distance. Thus, it seems natural to define an elastic body as one for which the constitutive equation prescribes the Cauchy stress $\mathbf{T}(\mathbf{x}, t)$ at $\mathbf{x} = \tilde{\chi}(\mathbf{X}, t)$ when the deformation gradient $\mathbf{F}$ is known, that is,

$$\mathbf{T}(\mathbf{x}, t) = \hat{\mathbf{T}}(\mathbf{F}(\mathbf{X}, t), \mathbf{X}).$$

We now proceed to make more precise the dependence of the Cauchy stress on the deformation gradient.

2.3.1 Finite Elasticity

We assume that the response of the material body is independent of the observer, so that by $(1.4.7)_3$ (see (1.4.3) and (1.4.4)), we have

$$\mathbf{Q}\hat{\mathbf{T}}(\mathbf{F})\mathbf{Q}^T = \hat{\mathbf{T}}(\mathbf{Q}\mathbf{F}), \tag{2.3.1}$$

for every tensor $\mathbf{F}$ with $\det \mathbf{F} > 0$ and every orthogonal tensor $\mathbf{Q}$ with $\det \mathbf{Q} = 1$.

The polar decomposition of $\mathbf{F}$ is given by (1.2.14). Let us choose $\mathbf{Q} = \mathbf{R}^{\mathrm{T}}$ such that (2.3.1) becomes

$$\mathbf{R}^{\mathrm{T}}\hat{\mathbf{T}}(\mathbf{F})\mathbf{R} = \hat{\mathbf{T}}(\mathbf{U}),$$

or

$$\hat{\mathbf{T}}(\mathbf{F}) = \mathbf{R}\hat{\mathbf{T}}(\mathbf{U})\mathbf{R}^{\mathrm{T}} = \mathbf{F}\mathbf{U}^{-1}\hat{\mathbf{T}}(\mathbf{U})\mathbf{U}^{-1}\mathbf{F}^{\mathrm{T}} = \mathbf{F}\tilde{\mathbf{T}}(\mathbf{C})\mathbf{F}^{\mathrm{T}},$$

by virtue of $(1.2.15)_1$, on putting $\mathbf{U} = \mathbf{C}^{1/2}$. This final form motivates the following definition.

Definition 2.3.1. *An elastic solid is a material body characterized by a constitutive equation of the form*

$$\mathbf{T}(\mathbf{x}, t) = \hat{\mathbf{T}}(\mathbf{F}(\mathbf{X}, t), \mathbf{X}), \tag{2.3.2}$$

where the response function $\hat{\mathbf{T}}$ *is completely determined by a tensor function* $\tilde{\mathbf{T}}(\mathbf{C}, \mathbf{X})$ *according to the formula*

$$\hat{\mathbf{T}}(\mathbf{F}, \mathbf{X}) = \mathbf{F}\tilde{\mathbf{T}}(\mathbf{C}, \mathbf{X})\mathbf{F}^T,$$

with $\mathbf{C} = \mathbf{U}^2 = \mathbf{F}^T\mathbf{F}$, *the right Cauchy–Green strain tensor corresponding to* $\mathbf{F}$.

The above definition emphasizes the importance of the strain tensors $\mathbf{U}$ and $\mathbf{C}$ for describing the deformation of an elastic solid.

Remark 2.3.2. The complete system of field equations for an elastic solid consists of the constitutive equation

$$\mathbf{T} = \mathbf{F}\tilde{\mathbf{T}}(\mathbf{C}, \mathbf{X})\mathbf{F}^T, \quad \mathbf{C} = \mathbf{F}^T\mathbf{F}, \tag{2.3.3}$$

the equation of motion

$$\rho\dot{\mathbf{v}} = \mathrm{div}_{\mathbf{x}}\mathbf{T} + \rho\mathbf{b}, \tag{2.3.4}$$

and the balance of mass

$$\rho \det \mathbf{F} = \rho_0, \tag{2.3.5}$$

where ρ_0 is the density in the reference configuration.

Definition 2.3.3. *An elastic material is homogeneous provided both* $\rho_0(\mathbf{X})$ *and* $\hat{\mathbf{T}}(\mathbf{F},\mathbf{X})$ *are independent of the material point. In other cases the elastic material is inhomogeneous.*

Definition 2.3.4. *A symmetry transformation at* $\mathbf{X}$ *is an orthogonal tensor* $\mathbf{Q}$ *with* $\det \mathbf{Q} = 1$ *such that*

$$\hat{\mathbf{T}}(\mathbf{F},\mathbf{X}) = \hat{\mathbf{T}}\,(\mathbf{F}\mathbf{Q},\mathbf{X}).$$

An elastic material is isotropic if every rotation is a symmetry transformation; otherwise, the elastic material is anisotropic.

Remark 2.3.5. Assume that the material at $\mathbf{X}$ is isotropic. Then the constitutive equation can be written in the form (for example, [188])

$$\mathbf{T} = \beta_0(\mathfrak{I}_{\mathbf{B}})\mathbf{1} + \beta_1(\mathfrak{I}_{\mathbf{B}})\mathbf{B} + \beta_2(\mathfrak{I}_{\mathbf{B}})\mathbf{B}^{-1}, \tag{2.3.6}$$

where $\mathbf{B} = \mathbf{F}\mathbf{F}^T$ is the left Cauchy–Green tensor and β_0, β_1, and β_2 are scalar functions of the principal invariants of $\mathbf{B}$ (see (1.2.22)), denoted by $\mathfrak{I}_{\mathbf{B}}$.

We further note that

$$\det \mathbf{C} = \det\left(\mathbf{F}^T\mathbf{F}\right) = (\det \mathbf{F})^2.$$

The second Piola–Kirchhoff stress tensor $\widehat{\mathbf{S}}$, defined by (1.3.52), is given by

$$\widehat{\mathbf{S}} = \det \mathbf{F}\tilde{\mathbf{T}}(\mathbf{C},\mathbf{X}) = \sqrt{\det \mathbf{C}}\tilde{\mathbf{T}}(\mathbf{C},\mathbf{X}). \tag{2.3.7}$$

As shown in general by Proposition 1.4.1, this quantity is an objective scalar. In terms of the first Piola–Kirchhoff stress tensor defined by (1.3.49), the constitutive equation (2.3.2) becomes

$$\mathbf{S} = \mathbf{F}\widehat{\mathbf{S}}(\mathbf{C},\mathbf{X}), \tag{2.3.8}$$

on using $(2.3.3)_1$ and (2.3.7).

In view of relations (1.3.50)–(1.3.51) with (2.3.8), we can rewrite the basic equations in (2.3.3)–(2.3.5) in the form

$$\mathbf{S} = \mathbf{F}\widehat{\mathbf{S}}(\mathbf{C},\mathbf{X}), \quad \mathbf{C} = \mathbf{F}^T\mathbf{F}, \quad \mathbf{F} = \nabla_{\mathbf{X}}\mathbf{x}, \quad \rho_0\ddot{\mathbf{x}} = \mathrm{Div}_{\mathbf{X}}\mathbf{S} + \rho_0\mathbf{b}. \tag{2.3.9}$$

Note that relation (1.3.51) follows automatically from the fact that $\widehat{\mathbf{S}}$ is a symmetric tensor. Moreover, since the density enters into (2.3.9) only through its reference value ρ_0, which is assumed known a priori, the balance of mass (2.3.5) need not be included in the list of field equations.

In what follows, we assume that the material body is identified with the regular region $\mathcal{B}$ occupied by the body in the reference configuration. All the fields in (2.3.9) are defined on $\mathcal{B} \times \mathbb{R}$, and the operator $\text{Div}_{\mathbf{X}}$ is with respect to the material point $\mathbf{X}$ in $\mathcal{B}$. In contrast, some of the fields in (2.3.3)–(2.3.5) are defined on the trajectory of $\mathbf{x}$, and moreover, the operator $\text{div}_{\mathbf{x}}$ is with respect to the position $\mathbf{x}$ in the current configuration. For this reason the formulation (2.3.9) is more convenient than that furnished by (2.3.3)–(2.3.5) in problems for which the trajectory is not known in advance.

The initial–boundary value problems of finite elasticity are obtained by including with (2.3.9) suitable initial and boundary conditions. As initial conditions one usually specifies the initial position and velocity

$$\mathbf{x}(\mathbf{X}, 0) = \mathbf{x}_0(\mathbf{X}), \quad \dot{\mathbf{x}}(\mathbf{X}, 0) = \mathbf{v}_0(\mathbf{X}),$$

where $\mathbf{x}_0(\mathbf{X})$ and $\mathbf{v}_0(\mathbf{X})$ are prescribed functions on $\mathcal{B}$. As boundary conditions one usually specifies

$$\mathbf{x}(\mathbf{X}, t) = \hat{\mathbf{x}}(\mathbf{X}, t) \text{ on } \Sigma_1 \times [0, \infty), \quad \mathbf{S}(\mathbf{X}, t)\mathbf{n} = \hat{\mathbf{s}}(\mathbf{X}, t) \text{ on } \Sigma_2 \times [0, \infty),$$

where $\hat{\mathbf{x}}$ and $\hat{\mathbf{s}}$ are prescribed vector fields on $\Sigma_1 \times [0, \infty)$ and $\Sigma_2 \times [0, \infty)$, respectively, and Σ_1 and Σ_2 are regular subsets of $\partial\mathcal{B}$ such that $\overline{\Sigma_1} \cup \Sigma_2 = \partial\mathcal{B}$ and $\Sigma_1 \cap \Sigma_2 = \emptyset$.

In the static theory, all fields are independent of time, and the underlying boundary value problem consists in finding a deformation $\mathbf{x} = \tilde{\chi}(\mathbf{X})$ that satisfies the field equations

$$\mathbf{S} = \mathbf{F}\widehat{\mathbf{S}}(\mathbf{C}, \mathbf{X}), \quad \mathbf{C} = \mathbf{F}^T\mathbf{F}, \quad \mathbf{F} = \nabla_{\mathbf{X}}\mathbf{x}, \quad \text{Div}_{\mathbf{X}}\mathbf{S} + \rho_0\mathbf{b} = \mathbf{0} \tag{2.3.10}$$

and the boundary conditions

$$\mathbf{x}(\mathbf{X}) = \hat{\mathbf{x}}(\mathbf{X}) \quad \text{on} \quad \Sigma_1, \quad \mathbf{S}(\mathbf{X})\mathbf{n} = \hat{\mathbf{s}}(\mathbf{X}) \quad \text{on} \quad \Sigma_2,$$

where again $\hat{\mathbf{x}}$ and $\hat{\mathbf{s}}$ are prescribed functions on Σ_1 and Σ_2, respectively.

When the traction is prescribed over the entire boundary, that is, when we have the boundary condition

$$\mathbf{S}\mathbf{n} = \hat{\mathbf{s}} \quad \text{on} \quad \partial\mathcal{B},$$

then an integration on $\mathcal{B}$ of (2.3.10) implies that

$$\int_{\mathcal{B}} \rho_0 \mathbf{b} dV + \int_{\partial\mathcal{B}} \hat{\mathbf{s}} dA = \mathbf{0}.$$

This relation is a particular case of (1.3.46), and it involves only the prescribed data. It furnishes a necessary condition for the existence of a solution. On the other hand, (1.3.47) yields

$$\int_{\mathcal{B}} \rho_0(\mathbf{x} - \mathbf{x}_0) \times \mathbf{b} dV + \int_{\partial\mathcal{B}} (\mathbf{x} - \mathbf{x}_0) \times \hat{\mathbf{s}} dA = \mathbf{0},$$

which, because of the presence of $\mathbf{x}$, is not a restriction on the data, but rather a compatibility condition automatically satisfied by any solution of the boundary value problem.

2.3.2 Hyperelastic Bodies

We now characterize a class of elastic materials for which the Piola–Kirchhoff stress is given by the derivative of a scalar function. To this end we consider a dynamical process $(\mathbf{x}, \mathbf{T})$ of the body $\mathcal{B}$ (Definition 2.2.1) corresponding to the body force $\mathbf{b}$. Then, given any subbody A of $\mathcal{B}$, the work on A during a time interval $[t_0, t_1]$ is given by

$$\begin{aligned}&\int_{t_0}^{t_1}\left\{\int_{\varphi_t(A)} \rho\mathbf{b}\cdot\mathbf{v}dv_t + \int_{\partial\varphi_t(A)} \mathbf{Tn}\cdot\mathbf{v}da_t\right\}dt \\ &\quad = \int_{t_0}^{t_1}\left\{\int_{\varphi_0(A)} \rho_0\mathbf{b}\cdot\mathbf{v}dv_0 + \int_{\partial\varphi_0(A)} \mathbf{SN}\cdot\mathbf{v}da_0\right\}dt.\end{aligned} \tag{2.3.11}$$

Definition 2.3.6. *The dynamical process* $(\mathbf{x}, \mathbf{T})$ *is closed during the interval* $[t_0, t_1]$ *if*

$$\mathbf{x}(\mathbf{X}, t_0) = \mathbf{x}(\mathbf{X}, t_1), \quad \dot{\mathbf{x}}(\mathbf{X}, t_0) = \dot{\mathbf{x}}(\mathbf{X}, t_1), \tag{2.3.12}$$

for all $\mathbf{X} \in \varphi_0(\mathcal{B})$.

For such a process, it follows from (1.2.8) and (2.3.12) that

$$\mathbf{F}(\mathbf{X}, t_0) = \mathbf{F}(\mathbf{X}, t_1), \qquad \dot{\mathbf{F}}(\mathbf{X}, t_0) = \dot{\mathbf{F}}(\mathbf{X}, t_1). \tag{2.3.13}$$

If we integrate (1.3.54) between t_0 and t_1 (recalling (1.3.48)) and use (2.3.12), then we deduce that for closed processes, (2.3.11) reduces to

$$\int_{t_0}^{t_1}\left\{\int_{\varphi_0(A)} \rho_0\mathbf{b}\cdot\mathbf{v}dv_0 + \int_{\partial\varphi_0(A)} \mathbf{SN}\cdot\mathbf{v}da_0\right\}dt = \int_{t_0}^{t_1}\int_{\varphi_0(A)} \mathbf{S}\cdot\dot{\mathbf{F}}dv_0dt.$$

Definition 2.3.7. *We say that the work is nonnegative in closed processes if given any subbody A and any time interval* $[t_0, t_1]$, *we have*

$$\int_{t_0}^{t_1}\int_{\varphi_0(A)} \mathbf{S}\cdot\dot{\mathbf{F}}dv_0dt \geq 0$$

for any process that is closed during $[t_0, t_1]$.

We can conclude, from the above definition, the following result.

Proposition 2.3.8. *The work is nonnegative in closed processes if and only if, given any* $\mathbf{X} \in \varphi_0(\mathcal{B})$ *and any time interval* $[t_0, t_1]$, *we have*

$$\int_{t_0}^{t_1} \mathbf{S}(\mathbf{X}, t)\cdot\dot{\mathbf{F}}(\mathbf{X}, t)dt \geq 0, \tag{2.3.14}$$

for any process that is closed during $[t_0, t_1]$.

Definition 2.3.9. *An elastic material is hyperelastic if there exists a scalar function* $\hat{e} : \text{Lin}^+ \times \varphi_0(\mathcal{B}) \to \mathbb{R}$ *such that the first Piola–Kirchhoff stress tensor* $\mathbf{S}$ *is the derivative of* $\hat{e}$ *with respect to* $\mathbf{F}$*:*

$$\mathbf{S}(\mathbf{X}, t) = \hat{\mathbf{S}}(\mathbf{F}, \mathbf{X}) = \mathbf{D}_{\mathbf{F}}\hat{e}(\mathbf{F}, \mathbf{X}) \quad \text{or} \quad S_{ij} = \frac{\partial \hat{e}}{\partial F_{ij}}, \tag{2.3.15}$$

where the derivative is with respect to $\mathbf{F}$*, holding* $\mathbf{X}$ *fixed. The scalar function* $\hat{e}(\mathbf{F}, \mathbf{X})$ *is known as the strain–energy density. It is assumed to be an objective scalar.*

Remark 2.3.10. Observe that material objectivity as expressed by $(1.4.7)_1$ gives that

$$\hat{e}(\mathbf{F}, \mathbf{X}) = \hat{e}(\mathbf{Q}(t)\mathbf{F}, \mathbf{X}),$$

for all $\mathbf{F} \in \text{Lin}^+$ and orthogonal $\mathbf{Q}$. It follows from Proposition 1.4.2 that

$$\hat{e}(\mathbf{F}, \mathbf{X}) = \tilde{e}(\mathbf{C}, \mathbf{X}),$$

where $\tilde{e}$ is also an objective scalar function and (2.3.15) yields (2.3.8). In fact, the second Piola–Kirchhoff stress tensor is given by

$$\widehat{\mathbf{S}}(\mathbf{C}, \mathbf{X}) = \frac{1}{2}\frac{\partial}{\partial \mathbf{C}}\tilde{e}(\mathbf{C}, \mathbf{X}).$$

We now give a characterization of the class of hyperelastic materials in terms of the work in closed processes.

Theorem 2.3.11. *An elastic material is hyperelastic if and only if the work is zero in closed processes.*

Proof. Suppose first that the body is hyperelastic so that there exists the strain–energy density $\hat{e}$ such that (2.3.15) holds true. Then, for a closed process during the time interval $[t_0, t_1]$, we have

$$\frac{d}{dt}\hat{e}(\mathbf{F}, \mathbf{X}) = D_{\mathbf{F}}\hat{e}(\mathbf{F}, \mathbf{X}) \cdot \dot{\mathbf{F}}(\mathbf{X}, t) = \hat{\mathbf{S}}(\mathbf{F}, \mathbf{X}) \cdot \dot{\mathbf{F}}(\mathbf{X}, t),$$

and since $\mathbf{F}(\mathbf{X}, t_0) = \mathbf{F}(\mathbf{X}, t_1)$ by virtue of $(2.3.13)_1$, we obtain

$$\begin{aligned}\int_{t_0}^{t_1} \hat{\mathbf{S}}(\mathbf{F}(\mathbf{X}, t), \mathbf{X}) \cdot \dot{\mathbf{F}}(\mathbf{X}, t) dt &= \int_{t_0}^{t_1} \frac{d}{dt}\hat{e}(\mathbf{F}(\mathbf{X}, t), \mathbf{X}) dt \\ &= \hat{e}(\mathbf{F}(\mathbf{X}, t_1), \mathbf{X}) - \hat{e}(\mathbf{F}(\mathbf{X}, t_0), \mathbf{X}) = 0.\end{aligned}$$

Thus, the work is zero in closed processes.

We assume now that the work is zero in closed processes, so that

$$\int_{t_0}^{t_1} \hat{\mathbf{S}}(\mathbf{F}) \cdot \dot{\mathbf{F}} dt = 0. \tag{2.3.16}$$

Relation (2.3.16) shows that the integral of $\hat{\mathbf{S}}$ over any piecewise smooth closed curve in Lin^+ vanishes. Since Lin^+ is an open and connected subset of the vector space Lin, a standard theorem in vector analysis tells us that $\hat{\mathbf{S}}$ is the derivative of a smooth scalar function $\hat{e}$ on Lin^+, which is the strain–energy density. Clearly, $\hat{\mathbf{S}}$ determines $\hat{e}$ only up to an arbitrary function of $\mathbf{X}$ alone. □

Theorem 2.3.12. *If the work is nonnegative in closed processes, then it is zero in such processes.*

Proof. We define

$$\mathbf{F}^*(t) = \mathbf{F}(t_0 + t_1 - t). \tag{2.3.17}$$

The quantity $\mathbf{F}^*$ represents the reversal in time of $\mathbf{F}$. In view of $(2.3.13)_1$ and (2.3.17), we have

$$\mathbf{F}^*(t_0) = \mathbf{F}(t_1) = \mathbf{F}(t_0) = \mathbf{F}^*(t_1),$$

and moreover,

$$\dot{\mathbf{F}}^*(t) = \frac{d}{dt}\mathbf{F}(t_0 + t_1 - t) = -\dot{\mathbf{F}}(t_0 + t_1 - t),$$

giving

$$\dot{\mathbf{F}}^*(t_0) = -\dot{\mathbf{F}}(t_1) = -\dot{\mathbf{F}}(t_0) = \dot{\mathbf{F}}^*(t_1).$$

Therefore, $\mathbf{F}^*(t)$ is a closed process, and hence the work is nonnegative on this process, so that

$$\begin{aligned} 0 \le \int_{t_0}^{t_1} \hat{\mathbf{S}}(\mathbf{F}^*) \cdot \dot{\mathbf{F}}^* dt &= -\int_{t_0}^{t_1} \hat{\mathbf{S}}(\mathbf{F}(t_0 + t_1 - t)) \cdot \dot{\mathbf{F}}(t_0 + t_1 - t) dt \\ &= -\int_{t_0}^{t_1} \hat{\mathbf{S}}(\mathbf{F}(\tau)) \cdot \dot{\mathbf{F}}(\tau)\, d\tau. \end{aligned}$$

Thus, we can conclude that for every $\mathbf{F}$ satisfying (2.3.13), we have

$$\int_{t_0}^{t_1} \hat{\mathbf{S}}(\mathbf{F}) \cdot \dot{\mathbf{F}} dt = 0. \tag{2.3.18}$$

□

It can be shown that the condition $(2.3.13)_2$ can be avoided without affecting the validity of the result (2.3.18) [188].

Combining Theorems 2.3.11 and 2.3.12, we see that the property that the work is nonnegative in closed processes implies that the material is hyperelastic.

Remark 2.3.13. Theorem 2.3.11 proves that the work is zero in closed processes for hyperelastic materials. Moreover, if we set $e(\mathbf{X}, t) = \hat{e}(\mathbf{F}(\mathbf{X}, t), \mathbf{X})$, then we have

$$\dot{e} = \mathbf{S} \cdot \dot{\mathbf{F}},$$

and so the theorem of power expended (1.3.54) leads to the following important corollary.

Theorem 2.3.14. (Balance of Energy for Hyperelastic Materials) *Each dynamical process for a hyperelastic body satisfies the energy equation*

$$\frac{d}{dt}\int_{\varphi_0(A)} \left(e + \frac{1}{2}\rho_0 \mathbf{v}^2\right) dv_0 = \int_{\varphi_0(A)} \rho_0 \mathbf{b} \cdot \mathbf{v} dv_0 + \int_{\partial\varphi_0(A)} \mathbf{Sn} \cdot \mathbf{v} da_0, \tag{2.3.19}$$

for each subbody A.

Remark 2.3.15. The term

$$\int_{\varphi_0(A)} e dv_0$$

represents the *strain energy* of the subbody A. The energy equation (2.3.19) asserts that the rate at which the total energy of A is changing must equal the power expended on A.

A direct consequence of the above balance of energy is the following result concerning the conservation of energy.

Proposition 2.3.16. (Conservation of Energy) *For a dynamical process in a hyperelastic finite body with body force* $\mathbf{b} = \mathbf{0}$ *and subject to the condition* $\mathbf{Sn} \cdot \mathbf{v} = \mathbf{0}$ *on the boundary of* $\mathcal{B}$, *the total energy is constant, that is,*

$$\int_{\varphi_0(\mathcal{B})} \left(e + \frac{1}{2}\rho_0 \mathbf{v}^2 \right) dv_0 = constant.$$

2.4 Linear Elasticity

Let us consider an elastic material described by the general constitutive equation

$$\mathbf{S} = \hat{\mathbf{S}}(\mathbf{F}). \tag{2.4.1}$$

We now consider the linearized theory appropriate to situations in which the displacement vector is small, in the sense described in Sect. 1.2.2. The crucial point is the linearization of the general constitutive equation (2.4.1) near $\mathbf{F} = \mathbf{1}$, and of great importance in this context is the elasticity tensor.

Definition 2.4.1. *The elasticity tensor* $\mathbb{C}$ *for the material point* $\mathbf{X}$ *is the derivative of the first Piola–Kirchhoff stress with respect to* $\mathbf{F}$ *at* $\mathbf{F} = \mathbf{1}$*:*

$$\mathbb{C} = D_{\mathbf{F}}\hat{\mathbf{S}}(\mathbf{1}), \quad \text{or} \quad C_{ijkl} = \frac{\partial S_{ij}}{\partial F_{kl}}(\mathbf{1}). \tag{2.4.2}$$

The derivation of the linearized form of the general constitutive equation (2.4.1) requires the following two fundamental assumptions:

- the displacement vector $\mathbf{u}$ is small;
- the residual stress vanishes, i.e.,

$$\hat{\mathbf{S}}(\mathbf{1}) = \mathbf{0}. \tag{2.4.3}$$

In order to derive this linearized form, we note that

$$\mathbf{x} = \mathbf{X} + \mathbf{u}, \quad \mathbf{F} = \mathbf{1} + \mathbf{H}, \quad \mathbf{H} = \nabla_{\mathbf{X}}\mathbf{u}, \tag{2.4.4}$$

and consider $\hat{\mathbf{S}}(\mathbf{F})$ as a function of $\mathbf{H}$. In view of the hypothesis that the displacement vector $\mathbf{u}$ is small, we have the infinitesimal theory of deformation and recall that the measures of deformation reduce to the infinitesimal strain tensor $\boldsymbol{\varepsilon}$ defined by (1.2.31) with $\nabla\mathbf{u} = \nabla_{\mathbf{X}}\mathbf{u} = \nabla_{\mathbf{x}}\mathbf{u}$. Moreover, the powers of $\mathbf{H} = \nabla_{\mathbf{X}}\mathbf{u} = \nabla_{\mathbf{x}}\mathbf{u} = \nabla\mathbf{u}$ greater than or equal to two are all negligible. In view of relations (1.3.49) and (2.4.3), it follows that

$$\hat{\mathbf{T}}(\mathbf{1}) = \mathbf{0}$$

and

$$\mathbf{S} = \hat{\mathbf{S}}(\mathbf{F}) = \hat{\mathbf{T}}(\mathbf{F}) + \mathbf{o}(\mathbf{H}) = \widehat{\mathbf{S}}(\mathbf{C}) + \mathbf{o}(\mathbf{H}). \tag{2.4.5}$$

Thus, the various stress tensors coincide at the limit of the linearized theory, and we can use $\mathbf{S}$ and $\mathbf{T}$ interchangeably. Moreover, on the basis of relations (2.4.1)–(2.4.4), we deduce the following form of the constitutive relation:

$$\hat{\mathbf{S}}(\mathbf{F}) = \mathbb{C}\boldsymbol{\varepsilon} + \mathbf{o}(\mathbf{H}), \tag{2.4.6}$$

as $\mathbf{H} \to \mathbf{0}$, where $\boldsymbol{\varepsilon}$ is the infinitesimal strain tensor.

The importance of the elasticity tensor now becomes apparent from the linearized constitutive relation (2.4.6). Let us outline some of its properties. In view of relations (1.3.49) and (2.4.2), we deduce that

$$\mathbb{C} = D_{\mathbf{F}}\hat{\mathbf{T}}(\mathbf{1}), \quad \text{or} \quad C_{ijkl} = \frac{\partial T_{ij}}{\partial F_{kl}}(\mathbf{1}),$$

so that

$$C_{ijkl} = C_{jikl}. \tag{2.4.7}$$

If the material at $\mathbf{X}$ is hyperelastic, then

$$\mathbb{C} = D_{\mathbf{F}}\hat{\mathbf{S}}(\mathbf{1}) = D^2_{\mathbf{F}}\hat{e}(\mathbf{1}),$$

and therefore we have the supplementary symmetry

$$C_{ijkl} = C_{klij}. \tag{2.4.8}$$

In view of the symmetry of ε in (2.4.6), we need to only consider C_{ijkl} with the further property

$$C_{ijkl} = C_{ijlk}. \tag{2.4.9}$$

Thus, $\mathbb{C} \in \mathrm{Lin(Sym)}$.

We have the following definitions (see Sect. A.2).

Definition 2.4.2. *We say that* $\mathbb{C}$ *is symmetric if*

$$\mathbf{H} \cdot \mathbb{C}\mathbf{G} = \mathbf{G} \cdot \mathbb{C}\mathbf{H}$$

for all tensors $\mathbf{H}$ *and* $\mathbf{G}$. *This is the case if (2.4.8) holds. Also,* $\mathbb{C}$ *is positive definite if*

$$\boldsymbol{\varepsilon} \cdot \mathbb{C}\boldsymbol{\varepsilon} > 0$$

for all symmetric tensors $\boldsymbol{\varepsilon} \neq \mathbf{0}$. *We call* $\mathbb{C}$ *strongly elliptic if*

$$\mathbf{A} \cdot \mathbb{C}\mathbf{A} > 0$$

whenever $\mathbf{A}$ *has the form* $\mathbf{A} = \mathbf{a} \otimes \mathbf{c}$, *with* $\mathbf{a} \neq \mathbf{0}$, $\mathbf{c} \neq \mathbf{0}$.

An important consequence of the linearized constitutive equation (2.4.6) and relation (2.3.6) is the following result.

Theorem 2.4.3. *Assume that the elastic material at* $\mathbf{X}$ *is isotropic. Then there exist scalars* μ *and* λ *such that*

$$\mathbb{C}\boldsymbol{\varepsilon} = 2\mu\boldsymbol{\varepsilon} + \lambda(tr\ \boldsymbol{\varepsilon})\ \mathbf{1},$$

for every symmetric tensor $\boldsymbol{\varepsilon}$. *The scalars* $\mu = \mu(\mathbf{X})$ *and* $\lambda = \lambda(\mathbf{X})$ *are the* Lamé moduli *at* $\mathbf{X}$.

Summarizing, we conclude that the basic equations of the linear theory of elasticity consist of the stress–strain relation

$$\mathbf{S} = \mathbb{C}\boldsymbol{\varepsilon}, \tag{2.4.10}$$

the strain–displacement relation (see (1.2.31))

$$\boldsymbol{\varepsilon} = \frac{1}{2}\left(\nabla\mathbf{u} + \nabla\mathbf{u}^T\right), \tag{2.4.11}$$

and the equation of motion

$$\rho_0\ddot{\mathbf{u}} = \mathrm{Div}\mathbf{S} + \rho_0\mathbf{b}, \tag{2.4.12}$$

where ∇ and Div are with respect to $\mathbf{X}$ or $\mathbf{x}$. Note that these equations are expressed in terms of the displacement $\mathbf{u}(\mathbf{X}, t) = \mathbf{x}(\mathbf{X}, t) - \mathbf{X}$, rather than the motion $\mathbf{x}$. Given $\mathbb{C}$, ρ_0, and $\mathbf{b}$, the system described by relations (2.4.10)–(2.4.12) is a linear system of partial differential equations for the fields $\mathbf{u}$, $\boldsymbol{\varepsilon}$, and $\mathbf{S}$.

When the body is isotropic, the constitutive equation (2.4.10) is replaced by

$$\mathbf{S} = 2\mu\boldsymbol{\varepsilon} + \lambda\ (\mathrm{tr}\ \boldsymbol{\varepsilon})\ \mathbf{1}. \tag{2.4.13}$$

Moreover, when the body is homogeneous, then ρ_0, μ, and λ are constants.

Sometimes it is convenient to have the stress–strain law (2.4.13) inverted to give $\boldsymbol{\varepsilon}$ as a function of $\mathbf{S}$. This inversion is easily accomplished upon noting that (2.4.13) gives

$$\mathrm{tr}\ \mathbf{S} = (3\lambda + 2\mu)\ \mathrm{tr}\ \boldsymbol{\varepsilon},$$

and hence we have

$$\boldsymbol{\varepsilon} = \frac{1}{2\mu}\left[\mathbf{S} - \frac{\lambda}{3\lambda + 2\mu}(\mathrm{tr}\ \mathbf{S})\mathbf{1}\right],$$

or

$$\boldsymbol{\varepsilon} = \frac{1}{E}[(1 + \nu)\mathbf{S} - \nu\ (\mathrm{tr}\ \mathbf{S})\ \mathbf{1}], \tag{2.4.14}$$

where

$$E = \frac{\mu(3\lambda + 2\mu)}{\lambda + \mu}, \qquad \nu = \frac{\lambda}{2(\lambda + \mu)}.$$

The modulus E is known as *Young's modulus*, while ν is *Poisson's ratio*.

Let us assume that the body is homogeneous and isotropic. Then, since

$$\mathrm{Div}\left(\nabla\mathbf{u} + \nabla\mathbf{u}^T\right) = \Delta\mathbf{u} + \nabla(\mathrm{Div}\,\mathbf{u}), \quad tr\,\boldsymbol{\varepsilon} = \mathrm{Div}\,\mathbf{u},$$

Eqs. (2.4.11)–(2.4.13) are easily combined to give the displacement equations of motion

$$\rho_0\ddot{\mathbf{u}} = \mu\Delta\mathbf{u} + (\lambda + \mu)\nabla(\mathrm{Div}\,\mathbf{u}) + \rho_0\mathbf{b}. \tag{2.4.15}$$

In the case of static theory, we have $\ddot{\mathbf{u}} = \mathbf{0}$, and the displacement equations of equilibrium are

$$\mu\Delta\mathbf{u} + (\lambda + \mu)\nabla(\mathrm{Div}\,\mathbf{u}) + \rho_0\mathbf{b} = \mathbf{0},$$

which hold approximately for slow deformations.

We now discuss some particular solutions of the equilibrium equations, in the absence of body force, for a homogeneous and isotropic body.

- *Pure shear*

Let us consider the following state of displacement:

$$u_1 = \gamma X_2, \quad u_2 = u_3 = 0.$$

The matrices for the corresponding $\boldsymbol{\varepsilon}$ and $\mathbf{S}$ are

$$(\boldsymbol{\varepsilon}) = \frac{1}{2}\begin{pmatrix} 0 & \gamma & 0 \\ \gamma & 0 & 0 \\ 0 & 0 & 0 \end{pmatrix}, \qquad (\mathbf{S}) = \begin{pmatrix} 0 & \tau & 0 \\ \tau & 0 & 0 \\ 0 & 0 & 0 \end{pmatrix}, \qquad \tau = \mu\gamma.$$

Thus, μ determines the response of the body in shear and for this reason is called the *shear modulus*.

- *Uniform compression or expansion*

The state of displacement

$$u_1 = \epsilon X_1, \quad u_2 = \epsilon X_2, \quad u_3 = \epsilon X_3$$

corresponds to

$$\boldsymbol{\varepsilon} = \epsilon\mathbf{1}, \quad \mathbf{S} = \pi\mathbf{1}, \quad \pi = 3\kappa\epsilon,$$

where

$$\kappa = \frac{2}{3}\mu + \lambda \tag{2.4.16}$$

is the *modulus of compression*.

- *Pure tension*

We consider the following state of stress:

$$(\mathbf{S}) = \begin{pmatrix} \sigma\ 0\ 0 \\ 0\ 0\ 0 \\ 0\ 0\ 0 \end{pmatrix},$$

which obviously satisfies the equilibrium equations. We want to find the corresponding state of displacement. Because we know the state of stress, it is convenient to use (2.4.14), which gives

$$(\boldsymbol{\varepsilon}) = \begin{pmatrix} \epsilon\ 0\ 0 \\ 0\ l\ 0 \\ 0\ 0\ l \end{pmatrix} \quad \text{with} \quad \epsilon = \frac{\sigma}{E}, \qquad l = -\nu\epsilon.$$

Furthermore, we note that this state of strain satisfies Saint-Venant's conditions of compatibility (1.2.33), and the corresponding state of displacement is given by

$$u_1 = \epsilon X_1, \quad u_2 = l X_2, \quad u_3 = l X_3.$$

Remark 2.4.4. Since an elastic solid should increase its length when pulled, decrease its volume when acted on by a pure pressure, and respond to a positive shearing strain by a positive shearing stress, one would expect that

$$E > 0, \quad \kappa > 0, \quad \mu > 0.$$

Also, a pure tensile stress should produce a contraction in the direction perpendicular to it and hence $\nu > 0$.

2.4.1 Linear Elastostatics

The basic system of field equations for the static behavior of an elastic material consists of the strain–displacement relation

$$\boldsymbol{\varepsilon} = \frac{1}{2}\left(\nabla \mathbf{u} + \nabla \mathbf{u}^T\right), \tag{2.4.17}$$

the stress–strain relations

$$\mathbf{S} = \mathbb{C}\boldsymbol{\varepsilon}, \tag{2.4.18}$$

and the equations of equilibrium

$$\text{Div}\mathbf{S} + \rho\mathbf{b} = \mathbf{0}. \tag{2.4.19}$$

The body is assumed to occupy a volume $B \in \mathbb{R}^3$ corresponding to the reference volume $\mathcal{B}$ but also closely approximating the current deformed shape.

Definition 2.4.5. *We call a list* $[\mathbf{u}, \boldsymbol{\varepsilon}, \mathbf{S}]$ *of fields that are smooth on B and satisfy (2.4.17)–(2.4.19) an elastic state corresponding to* **b**.

Let Σ_1 and Σ_2 denote complementary regular subsets of the boundary of B, so that $\overline{\Sigma}_1 \cup \Sigma_2 = \partial B$ and $\Sigma_1 \cap \Sigma_2 = \emptyset$. The *mixed problem of elastostatics* can be formulated as follows:

- *given*: B, Σ_1, Σ_2, an elasticity tensor $\mathbb{C}$ on B, a body force field $\mathbf{b}$ on B, surface displacements $\hat{\mathbf{u}}$ on Σ_1, and surface tractions $\hat{\mathbf{s}}$ on Σ_2;
- *find*: an elastic state $[\mathbf{u}, \boldsymbol{\varepsilon}, \mathbf{S}]$ that corresponds to $\mathbf{b}$ and satisfies the boundary conditions

$$\mathbf{u} = \hat{\mathbf{u}} \quad \text{on} \quad \Sigma_1, \qquad \mathbf{Sn} = \hat{\mathbf{s}} \quad \text{on} \quad \Sigma_2. \tag{2.4.20}$$

An elastic state with these properties will be called a *solution of the mixed problem of elastostatics*.

Three fundamental theorems of elastostatics are now proved.

Theorem 2.4.6. (Theorem of Work and Energy) *Let $[\mathbf{u}, \boldsymbol{\varepsilon}, \mathbf{S}]$ be an elastic state corresponding to the body force $\mathbf{b}$. Then,*

$$2\mathfrak{U}(\boldsymbol{\varepsilon}) = \int_B \rho \mathbf{b} \cdot \mathbf{u} dv + \int_{\partial B} \mathbf{Sn} \cdot \mathbf{u} da, \tag{2.4.21}$$

where $\mathfrak{U}(\boldsymbol{\varepsilon})$ is the strain energy of the body defined by

$$\mathfrak{U}(\boldsymbol{\varepsilon}) = \frac{1}{2} \int_B \boldsymbol{\varepsilon} \cdot \mathbb{C} \boldsymbol{\varepsilon} dv.$$

Proof. By the symmetry of $\mathbf{S}$ and the divergence theorem, we obtain

$$\int_{\partial B} \mathbf{Sn} \cdot \mathbf{u} da = \int_{\partial B} \mathbf{Su} \cdot \mathbf{n} da = \int_B \mathrm{Div}(\mathbf{Su}) dv = \int_B (\mathbf{u} \cdot \mathrm{Div}\mathbf{S} + \mathbf{S} \cdot \nabla \mathbf{u}) dv,$$

and further, from (2.4.17) and (2.4.18),

$$\mathbf{S} \cdot \nabla \mathbf{u} = \mathbf{S}^T \cdot \nabla \mathbf{u}^T = \mathbf{S} \cdot \left\{ \frac{1}{2} \left(\nabla \mathbf{u} + \nabla \mathbf{u}^T \right) \right\} = \mathbf{S} \cdot \boldsymbol{\varepsilon} = \boldsymbol{\varepsilon} \cdot \mathbb{C} \boldsymbol{\varepsilon}.$$

These relations, when combined with the equations of equilibrium (2.4.19), give (2.4.21). □

Theorem 2.4.7. (Uniqueness Theorem) *Assume that the elasticity tensor is positive definite. Let $[\mathbf{u}_1, \boldsymbol{\varepsilon}_1, \mathbf{S}_1]$ and $[\mathbf{u}_2, \boldsymbol{\varepsilon}_2, \mathbf{S}_2]$ be solutions of the same mixed problem of linear elastostatics. Then,*

$$\mathbf{u}_1 = \mathbf{u}_2 + \mathbf{u}^*, \quad \boldsymbol{\varepsilon}_1 = \boldsymbol{\varepsilon}_2, \quad \mathbf{S}_1 = \mathbf{S}_2,$$

where $\mathbf{u}^*$ is an infinitesimal rigid displacement of B.

Proof. Let

$$\mathbf{u} = \mathbf{u}_1 - \mathbf{u}_2, \quad \boldsymbol{\varepsilon} = \boldsymbol{\varepsilon}_1 - \boldsymbol{\varepsilon}_2, \quad \mathbf{S} = \mathbf{S}_1 - \mathbf{S}_2.$$

Then, $[\mathbf{u}, \boldsymbol{\varepsilon}, \mathbf{S}]$ is an elastic state that corresponds to a null body force and satisfies the following boundary conditions:

$$\mathbf{u} = \mathbf{0} \quad \text{on} \quad \Sigma_1, \qquad \mathbf{Sn} = \mathbf{0} \quad \text{on} \quad \Sigma_2,$$

which give

$$\mathbf{Sn} \cdot \mathbf{u} = 0 \quad \text{on} \quad \partial B.$$

In view of relation (2.4.21), we conclude that

$$\int_B \boldsymbol{\varepsilon} \cdot \mathbb{C}\boldsymbol{\varepsilon} dv = 0.$$

Since $\mathbb{C}$ is positive definite, this last relation can hold only if $\boldsymbol{\varepsilon} = \mathbf{0}$; this in turn implies that $\mathbf{S} = \mathbf{0}$ and $\mathbf{u} = \mathbf{u}^*$, where $\mathbf{u}^*$ is an infinitesimal rigid displacement of B. Thus, the proof is complete. □

Theorem 2.4.8. **(Betti's Reciprocal Theorem)** *Assume that the elasticity tensor is symmetric. Let* $[\mathbf{u}_1, \boldsymbol{\varepsilon}_1, \mathbf{S}_1]$ *and* $[\mathbf{u}_2, \boldsymbol{\varepsilon}_2, \mathbf{S}_2]$ *be elastic states in* B *corresponding to body force fields* $\mathbf{b}_1$ *and* $\mathbf{b}_2$*, respectively. Then,*

$$\int_B \rho \mathbf{b}_1 \cdot \mathbf{u}_2 dv + \int_{\partial B} \mathbf{S}_1 \mathbf{n} \cdot \mathbf{u}_2 da = \int_B \rho \mathbf{b}_2 \cdot \mathbf{u}_1 dv + \int_{\partial B} \mathbf{S}_2 \mathbf{n} \cdot \mathbf{u}_1 da. \tag{2.4.22}$$

Proof. In view of the equations of equilibrium (2.4.19) and the divergence theorem, we have

$$\int_B \rho \mathbf{b}_1 \cdot \mathbf{u}_2 dv + \int_{\partial B} \mathbf{S}_1 \mathbf{n} \cdot \mathbf{u}_2 da = \int_B \mathbf{S}_1 \cdot \nabla \mathbf{u}_2 dv. \tag{2.4.23}$$

Since the elasticity tensor is symmetric, we conclude from the strain–displacement relation that

$$\mathbf{S}_1 \cdot \nabla \mathbf{u}_2 = \mathbf{S}_1 \cdot \left[\frac{1}{2}\left(\nabla \mathbf{u}_2 + \nabla \mathbf{u}_2^T\right) \right] = \mathbf{S}_1 \cdot \boldsymbol{\varepsilon}_2,$$

so that relation (2.4.23) becomes

$$\int_B \rho \mathbf{b}_1 \cdot \mathbf{u}_2 dv + \int_{\partial B} \mathbf{S}_1 \mathbf{n} \cdot \mathbf{u}_2 da = \int_B \mathbf{S}_1 \cdot \boldsymbol{\varepsilon}_2 dv. \tag{2.4.24}$$

On the other hand, the symmetry of the elasticity tensor and the stress–strain relation give

$$\mathbf{S}_1 \cdot \boldsymbol{\varepsilon}_2 = \mathbb{C}\boldsymbol{\varepsilon}_1 \cdot \boldsymbol{\varepsilon}_2 = \boldsymbol{\varepsilon}_1 \cdot \mathbb{C}\boldsymbol{\varepsilon}_2 = \mathbf{S}_2 \cdot \boldsymbol{\varepsilon}_1. \tag{2.4.25}$$

Then, (2.4.24) and (2.4.25) give relation (2.4.22), and the proof is complete. □

When $\Sigma_1 = \partial B$ (that is, $\Sigma_2 = \emptyset$), the boundary condition (2.4.20) takes the form

$$\mathbf{u} = \hat{\mathbf{u}} \quad \text{on} \quad \partial B.$$

We refer to this as the *displacement problem*.

When $\Sigma_2 = \partial B$ (that is, $\Sigma_1 = \emptyset$), the boundary condition (2.4.19) takes the form

$$\mathbf{Sn} = \hat{\mathbf{s}} \quad \text{on} \quad \partial B,$$

and we have the *traction problem*.

Proposition 2.4.9. *A necessary condition that the traction problem has a solution is that*

$$\int_B \rho \mathbf{b} dv + \int_{\partial B} \hat{\mathbf{s}} da = \mathbf{0}, \quad \int_B \rho \mathbf{r} \times \mathbf{b} dv + \int_{\partial B} \mathbf{r} \times \hat{\mathbf{s}} da = \mathbf{0}, \tag{2.4.26}$$

where $\mathbf{r}$ *is the position vector, with respect to the origin* O*, of the point of evaluation of* $\mathbf{b}$ *and* $\hat{\mathbf{s}}$*.*

Proof. The relation (2.4.26) is a direct consequence of the equilibrium equations (2.4.19) and the divergence theorem. □

2.4.2 Saint-Venant's Problem

We consider a homogeneous and isotropic cylindrical bar with generators parallel to the x_3-axis. Let the end faces $\mathcal{S}_1$ and $\mathcal{S}_2$ be located at $x_3 = 0$ and $x_3 = l$, respectively, with the origin at the centroid of $\mathcal{S}_1$ and with the x_1- and x_2-axes coincident with the principal axes of inertia, so that

$$\int_{\mathcal{S}_1} x_1 da = \int_{\mathcal{S}_1} x_2 da = 0, \int_{\mathcal{S}_1} x_1 x_2 da = 0. \tag{2.4.27}$$

We assume that the bar is loaded only on the end faces, so that the lateral surface $\mathcal{L}$ is traction-free. Moreover, we assume that the body forces are zero.

Saint-Venant's problem consists in the determination of an equilibrium displacement field $\mathbf{u}$ on B (that is, a displacement field satisfying the basic equations of elastostatics with null body force), subjected to the requirements

$$\mathbf{S}\mathbf{n} = \mathbf{0} \quad \text{on} \quad \mathcal{L},$$

or equivalently, since $n_3 = 0$ on $\mathcal{L}$,

$$S_{\alpha\beta} n_\beta = 0, \quad \alpha = 1, 2, \qquad \text{on} \quad \mathcal{L}, \tag{2.4.28}$$

and

$$\mathbf{S}\mathbf{n} = \hat{\mathbf{s}}^{(\alpha)} \quad \text{on } \mathcal{S}_\alpha, \quad \alpha = 1, 2. \tag{2.4.29}$$

Necessary conditions for the existence of a solution to this problem are

$$\int_{\mathcal{S}_1} \hat{\mathbf{s}}^{(1)} da + \int_{\mathcal{S}_2} \hat{\mathbf{s}}^{(2)} da = \mathbf{0}, \quad \int_{\mathcal{S}_1} \mathbf{r} \times \hat{\mathbf{s}}^{(1)} da + \int_{\mathcal{S}_2} \mathbf{r} \times \hat{\mathbf{s}}^{(2)} da = \mathbf{0}. \tag{2.4.30}$$

These are an immediate consequence of (2.4.26).

Under suitable smoothness hypotheses on the given data, a solution of Saint-Venant's problem exists and it is uniquely determined.

In the *relaxed formulation of Saint-Venant's problem*, the local conditions (2.4.29) are replaced by the following global conditions:

$$\int_{\mathcal{S}_1} \mathbf{S}\mathbf{n} da = \mathbf{R}, \quad \int_{\mathcal{S}_1} \mathbf{r} \times \mathbf{S}\mathbf{n} da = \mathbf{M},$$

or equivalently,

$$\int_{\mathcal{S}_1} S_{3i} da = -R_i, \qquad \int_{\mathcal{S}_1} e_{ijk} x_j S_{3k} da = -M_i, \tag{2.4.31}$$

where **R** and **M** represent the resultant force and the resultant moment about O of the tractions acting on $\mathcal{S}_1$. We do not specify the loading on $\mathcal{S}_2$, since balance of forces and moments require that (2.4.30) be satisfied.

Definition 2.4.10. *By a solution of Saint-Venant's relaxed problem, we mean any equilibrium displacement field that satisfies the conditions (2.4.28) and (2.4.31).*

Remark 2.4.11. It is obvious that the relaxed statement of the problem fails to characterize the solution uniquely. However, we recall the so-called Saint-Venant's principle (e.g., [203]), which states in effect that any solution obeying (2.4.31) will not differ significantly from any other solution with the same property, except in the vicinity of $\mathcal{S}_1$. Thus, if this principle is accepted, it suffices to outline an appropriate representative solution of this class. The classification of the relaxed problem rests on various assumptions concerning the resultants **R** and **M**. We will exemplify this with some particular cases.

- **Extension problem**

The total force on the end $\mathcal{S}_1$ is equipollent to a force directed along the negative x_3-axis, of magnitude R_3 and null moment about O, that is,

$$\int_{\mathcal{S}_1} S_{3\alpha} da = 0, \quad \int_{\mathcal{S}_1} S_{33} da = -R_3, \quad \int_{\mathcal{S}_1} e_{ijk} x_j S_{3k} da = 0. \tag{2.4.32}$$

The *problem of extension* consists in the determination of an equilibrium displacement field **u** satisfying the boundary conditions (2.4.28) and (2.4.32). We seek a representative solution of this problem by assuming the following state of stress:

$$S_{11} = S_{22} = S_{12} = S_{23} = S_{31} = 0, \quad S_{33} = a_1, \tag{2.4.33}$$

with a_1 a constant. This state of stress satisfies the equilibrium equations and the lateral boundary condition (2.4.28). From relations (2.4.32) and (2.4.33), we obtain

$$a_1 = -\frac{R_3}{A},$$

where A is the area of $\mathcal{S}_1$. To find a solution of the extension problem, we need to find the displacement field corresponding to the state of stress (2.4.33). We use the stress–strain relation in the form expressed by (2.4.14) to obtain the state of strain

$$\varepsilon_{11} = \varepsilon_{22} = -\frac{\nu}{E} a_1, \quad \varepsilon_{33} = \frac{1}{E} a_1, \quad \varepsilon_{12} = \varepsilon_{23} = \varepsilon_{31} = 0.$$

This state of strain satisfies the compatibility conditions (1.2.33), and the corresponding displacement field is given (up to an arbitrary infinitesimal rigid displacement) by

$$u_1 = \frac{\nu R_3}{EA} x_1, \quad u_2 = \frac{\nu R_3}{EA} x_2, \quad u_3 = -\frac{R_3}{EA} x_3.$$

- **Bending of a beam**

The total force on the end $\mathcal{S}_1$ is equipollent to a moment of magnitude M_2 about the negative x_2-axis, so that

$$\int_{\mathcal{S}_1} S_{3i} da = 0, \quad \int_{\mathcal{S}_1} x_2 S_{33} da = \int_{\mathcal{S}_1} (x_1 S_{32} - x_2 S_{31}) da = 0, \tag{2.4.34}$$

and

$$\int_{\mathcal{S}_1} x_1 S_{33} da = -M_2. \tag{2.4.35}$$

The *problem of bending* involves determining an equilibrium displacement field $\mathbf{u}$ satisfying the boundary conditions (2.4.28), (2.4.34), and (2.4.35). We seek a representative solution of this problem by assuming the following state of stress:

$$S_{11} = S_{22} = S_{12} = S_{23} = S_{31} = 0, \quad S_{33} = a_2 x_1,$$

where a_2 is a constant. This field satisfies the lateral boundary conditions (2.4.28) and the conditions (2.4.34), by virtue of (2.4.27). Moreover, we have

$$\text{Div}\mathbf{S} = \mathbf{0}.$$

Furthermore, the condition (2.4.35) implies that

$$a_2 = -\frac{M_2}{I},$$

where

$$I = \int_{\mathcal{S}_1} x_1^2 da$$

represents the moment of inertia of $\mathcal{S}_1$ about the x_2-axis. To obtain a solution of the bending problem we need to only construct the displacement field corresponding to this state of stress. Using the stress–strain relation in the form (2.4.14), we see that

$$\begin{aligned} &\varepsilon_{23} = \varepsilon_{31} = \varepsilon_{12} = 0, \\ &\varepsilon_{11} = \varepsilon_{22} = \frac{\nu M_2 x_1}{EI}, \quad \varepsilon_{33} = -\frac{M_2 x_1}{EI}. \end{aligned}$$

The compatibility relations (1.2.33) are satisfied. Moreover, the corresponding displacement field is given by

$$u_1 = \frac{M_2}{2EI}\left[x_3^2 + \nu\left(x_1^2 - x_2^2\right)\right], \quad u_2 = \frac{M_2 \nu}{EI} x_1 x_2, \quad u_3 = -\frac{M_2}{EI} x_1 x_3.$$

- **Torsion of a cylinder**

The total force on the end $\mathcal{S}_1$ is equipollent to

$$\int_{\mathcal{S}_1} S_{3i} da = 0, \quad \int_{\mathcal{S}_1} x_\alpha S_{33} da = 0, \quad \int_{\mathcal{S}_1} (x_1 S_{32} - x_2 S_{31}) da = -M_3. \tag{2.4.36}$$

The *problem of torsion* involves determining an equilibrium displacement field $\mathbf{u}$ satisfying the boundary conditions (2.4.28) and (2.4.36). We seek such a solution by assuming the following state of displacement:

$$u_1 = -\tau x_2 x_3, \quad u_2 = \tau x_1 x_3, \quad u_3 = \tau \psi(x_1, x_2), \tag{2.4.37}$$

where τ is a constant and ψ is termed the *warping function*. The state of stress corresponding to the displacements given by (2.4.37) is

$$S_{\alpha\beta} = S_{33} = 0,$$
$$S_{31} = \tau\mu(\psi_{,1} - x_2), \quad S_{23} = \tau\mu(\psi_{,2} + x_1).$$

The equilibrium equations are satisfied if and only if

$$\psi_{,11} + \psi_{,22} = 0 \quad \text{in } \mathcal{S}_1, \tag{2.4.38}$$

while the lateral boundary conditions are equivalent to

$$\frac{\partial \psi}{\partial n} = x_2 n_1 - x_1 n_2 \quad \text{on} \quad \partial\mathcal{S}_1, \tag{2.4.39}$$

where $\frac{\partial \psi}{\partial n}$ is the normal derivative of ψ on $\partial\mathcal{S}_1$. The condition (2.4.36) implies

$$\tau D = -M_3, \tag{2.4.40}$$

where

$$D = \mu \int_{\mathcal{S}_1} \left(x_1^2 + x_2^2 + x_1\psi_{,2} - x_2\psi_{,1} \right) da$$

is the *torsional rigidity of the cross-section* $\mathcal{S}_1$. Thus, the torsion solution is given by (2.4.37) with τ determined by (2.4.40) and where ψ is the solution of the Neumann problem defined by (2.4.38) and (2.4.39).

Part II

Continuum Thermodynamics and Constitutive Equations of Mechanics and Electromagnetism

3

Principles of Thermodynamics

In this chapter, we discuss various fundamental concepts and results in continuum thermodynamics. Some examples are given in terms of the materials discussed in Part I, generalized to a nonisothermal context.

3.1 Heat Equation

Since the heat of a body is not conserved, we do not have a balance equation corresponding to those for mass and momentum, given by (1.3.1) and (1.3.12), respectively. However, there must be a balance between the net quantity of heat entering a body and the net heat absorbed by that body over a given time period.

We denote by $Q_A(t)$ the quantity of heat per unit time entering the subbody A of the body under consideration at time t by conduction or radiation. Also, let $H_A(t)$ be the heat absorbed by the subbody A per unit time at the moment t.

The heat flow density $c(\mathbf{x}, t) \in \mathbb{R}$ is assigned to each point $\mathbf{x}$ on the surface of the body. If $d\mathbf{a} = \mathbf{n}da$ is the area vector associated with a surface element on $\partial\varphi_t(A)$, the outward unit normal of which is $\mathbf{n}$, then $-cda$ is the amount of heat crossing da, going inward, per unit time. The quantity $c(\mathbf{x}, t)$ depends linearly on the orientation $\mathbf{n}$ of $d\mathbf{a}$; indeed, we have

$$c(\mathbf{x}, t) = \mathbf{q}(\mathbf{x}, t) \cdot \mathbf{n},$$

where $\mathbf{q} \in \mathbb{R}^3$ is known as the (Eulerian) heat flux vector. This is the analogue of Cauchy's theorem (1.3.24) for heat flow, where $\mathbf{q}$ corresponds to the Cauchy stress tensor. We can also define a heat flux vector $\mathbf{q}_L(\mathbf{X}, t)$ in the reference or Lagrangian configuration, corresponding to the first Piola–Kirchhoff stress tensor, given by (1.3.49). This quantity is given by

$$\mathbf{q}_L = J\mathbf{F}^{-1}\mathbf{q}. \tag{3.1.1}$$

It is an objective scalar, as defined by (1.4.7). The heat supply per unit time due to external sources, such as radiation, is denoted by $r(\mathbf{x}, t) \in \mathbb{R}$ or $\tilde{r}(\mathbf{X}, t)$ in the reference

G. Amendola et al., *Thermodynamics of Materials with Memory*,
https://doi.org/10.1007/978-3-030-80534-0_3

configuration. Then, for both configurations,

$$Q_A(t) = -\int_{\partial\varphi_t(A)} \mathbf{q}(\mathbf{x},t)\cdot\mathbf{n}da + \int_{\varphi_t(A)} \rho(\mathbf{x},t)r(\mathbf{x},t)dv_t$$
$$= -\int_{\partial\varphi_0(A)} \mathbf{q}_L(\mathbf{X},t)\cdot\mathbf{N}da_0 + \int_{\varphi_0(A)} \rho_0(\mathbf{X})\tilde{r}(\mathbf{X},t)dv_0,$$

where $\mathbf{N}$ is the outward unit normal to $\partial\varphi_0(A)$. If we denote by $h(\mathbf{x},t)(\tilde{h}(\mathbf{X},t)$ in the Lagrangian description) the heat absorbed by the body per unit mass in unit time (specific heat power), then

$$H_A(t) = \int_{\varphi_t(A)} \rho(\mathbf{x},t)h(\mathbf{x},t)dv_t = \int_{\varphi_0(A)} \rho_0(\mathbf{X})\tilde{h}(\mathbf{X},t)dv_0.$$

At any time t and for any subbody A of the body, the heat equation is expressed by

$$H_A(t) = Q_A(t).$$

The Eulerian form of this equation is given by

$$\int_{\varphi_t(A)} \rho(\mathbf{x},t)h(\mathbf{x},t)dv_t = -\int_{\partial\varphi_t(A)} \mathbf{q}(\mathbf{x},t)\cdot\mathbf{n}da + \int_{\varphi_t(A)} \rho(\mathbf{x},t)r(\mathbf{x},t)dv_t, \qquad (3.1.2)$$

from which, under hypotheses of continuity for the integrand functions and the arbitrariness of A, we can deduce the *Eulerian local form of the heat equation*

$$\rho(\mathbf{x},t)h(\mathbf{x},t) = -\mathrm{div}_{\mathbf{x}}\,\mathbf{q}(\mathbf{x},t) + \rho(\mathbf{x},t)r(\mathbf{x},t). \qquad (3.1.3)$$

The Lagrangian form of (3.1.2) is

$$\int_{\varphi_0(A)} \rho_0(\mathbf{X})\tilde{h}(\mathbf{X},t)dv_0 = -\int_{\partial\varphi_0(A)} \mathbf{q}_L(\mathbf{X},t)\cdot\mathbf{n}da_0 + \int_{\varphi_0(A)} \rho_0(\mathbf{X})\tilde{r}(\mathbf{X},t)dv_0,$$

with corresponding local form

$$\rho_0(\mathbf{X})h(\mathbf{X},t) = -\mathrm{Div}_{\mathbf{X}}\mathbf{q}_L(\mathbf{X},t) + \rho_0(\mathbf{X})\tilde{r}(\mathbf{X},r).$$

3.2 Definition of a Material as a Dynamical System

For a large class of continuum systems, it is possible to obtain a good description of their thermomechanical properties using the concepts of *state* σ and *thermomechanic process* P. For such systems, it is necessary to introduce the concepts of *absolute temperature* θ and *gradient of temperature* $\mathbf{g} = \nabla_{\mathbf{x}}\theta$. The state is an entity that depends on material properties, while the process for any material is a function of time for a duration $d_P \in \mathbb{R}^+$ of type $P : [0, d_P) \to \mathrm{Lin}(\mathbb{R}^3) \times \mathbb{R} \times \mathbb{R}^3$, defined as

$$P(t) = \left(\mathbf{L}(t), \dot{\theta}(t), \mathbf{g}(t)\right). \qquad (3.2.1)$$

Alternatively, there are advantages to using the Lagrangian description. In this case, $P : [0, d_P) \mapsto \mathrm{Sym} \times \mathbb{R} \times \mathbb{R}^3$ is given by

$$P(t) = \left(\dot{\mathbf{E}}(t), \dot{\theta}(t), \mathbf{g}_L(t)\right), \qquad \mathbf{g}_L = \nabla_{\mathbf{X}}\theta, \tag{3.2.2}$$

and is an objective scalar. In the present chapter, we use (3.2.1) for definiteness, though a choice closely related to (3.2.2) is adopted in Chap. 5.

Definition 3.2.1. *A material is simple if relative to a point* $\mathbf{X}$ *of the body* $\mathcal{B}$, *it is possible to define a state* σ *such that the constitutive equations for* $\mathbf{T}$, h, *and* $\mathbf{q}$ *are functions of* σ, P *and for which we have:*

(i) the space of states Σ *(that is, the set of possible states for the material body) is a metric space;*

(ii) any process $P : [0, d_P) \to \mathrm{Lin}(\mathbb{R}^3) \times \mathbb{R} \times \mathbb{R}^3$ *is a piecewise smooth function on* $[0, d_P)$ *and is in the space of processes* Π; *if* $P \in \Pi$, *then its restriction* $P_{[t_1,t_2)}$ *to the interval* $[t_1, t_2) \subset [0, d_P)$ *belongs to* Π. *The restriction* $P_{[0;t)}$ *is denoted by* P_t.
If $P_1, P_2 \in \Pi$, *then* $P_1 * P_2 \in \Pi$, *where*

$$P_1 * P_2(\tau) = \begin{cases} P_1(\tau), & \tau \in [0, d_{p_1}), \\ P_2(\tau - d_{p_1}), & \tau \in [d_{p_1}, d_{p_1} + d_{p_2}); \end{cases}$$

(iii) there exists a function $\hat{\varrho} : \Sigma \times \Pi \to \Sigma$, *the evolution function, which determines the final state* σ^f *when the initial state* σ^i *and the process* P *are known. Such a function has the semigroup property, that is, if* $(\sigma_1, \sigma_2) \in \Sigma$ *and* $P_1 \in \Pi$ *is such that* $\hat{\varrho}(\sigma_1, P_1) = \sigma_2$, *then for any* $P \in \Pi$,

$$\hat{\varrho}(\sigma_1, P_1 * P) = \hat{\varrho}(\hat{\varrho}(\sigma_1, P_1), P) = \hat{\varrho}(\sigma_2, P); \tag{3.2.3}$$

(iv) for any fixed $P \in \Pi$, *the function* $\hat{\varrho}(\cdot, P) : \Sigma \to \Sigma$ *is continuous.*

Remark 3.2.2. Referring in particular to (1.2.10), it is easy to prove that for thermoelastic materials, the function $\hat{\varrho}$ can be constructed as a general integral of the differential system

$$\frac{d\mathbf{F}}{dt} = \mathbf{L}\mathbf{F}, \qquad \frac{d\theta}{dt} = \dot{\theta}, \tag{3.2.4}$$

and moreover, it is also easy to prove that (3.2.4) implies that $\hat{\varrho}(\cdot, P) : \Sigma \to \Sigma$ *is continuous. In a similar manner, the function* $\hat{\varrho}$ *can be determined for viscous fluids.*

The constitutive variables for a material depend on σ_t and $P(t)$, that is,

$$\begin{aligned} \mathbf{T}(t) &= \hat{\mathbf{T}}(\sigma_t, P(t)), \\ h(t) &= \hat{h}(\sigma_t, P(t)), \\ \mathbf{q}(t) &= \hat{\mathbf{q}}(\sigma_t, P(t)), \end{aligned} \tag{3.2.5}$$

where $\sigma_t = \hat{\varrho}(\sigma, P_t)$, the quantity P_t being the restriction of P to the time interval $[0, t) \subset [0, d_P)$ (see Definition 3.2.1).

For *thermoelastic materials*, we have the form

$$\mathbf{T}(t) = \hat{\mathbf{T}}(\mathbf{F}(t), \theta(t)), \quad h(t) = \mathbf{A}(\mathbf{F}, \theta) \cdot \mathbf{L} + B(\mathbf{F}, \theta)\dot{\theta}, \ \mathbf{q}(t) = -\mathbf{K}(\mathbf{F}, \theta)\mathbf{g}, \tag{3.2.6}$$

where $\hat{\mathbf{T}}$, $\mathbf{A}$, B, and $\mathbf{K}$ are continuous functions of $(\mathbf{F}, \theta)$. For these constitutive equations, the state is the pair $\sigma = (\mathbf{F}, \theta)$, and the space of possible states Σ will be a subspace of the vector space $\mathrm{Lin}(\mathbb{R}^3) \times \mathbb{R}$.

We also consider viscous fluids for which the thermomechanical characteristics are expressed by means of the constitutive equations (see (2.2.18))

$$\begin{aligned} \mathbf{T}(t) &= -p(\rho, \theta)\mathbf{I} + \lambda(\rho, \theta)(\mathrm{div}_{\mathbf{x}}\, \mathbf{v})\, \mathbf{I} + 2\mu(\rho, \theta)\mathbf{D}, \\ h(t) &= a(\rho, \theta)\dot{\rho} + b(\rho, \theta)\dot{\theta}, \qquad \mathbf{q}(t) = -K(\rho, \theta)\mathbf{g}, \end{aligned} \tag{3.2.7}$$

where p, λ, μ, a, b, and K are continuous functions of (ρ, θ). Here, the state is defined by the pair $\sigma = (\rho, \theta)$. Recalling (1.2.9), (1.2.23), and $(1.3.2)_2$, we see that the quantities $\dot{\rho}$, $\mathrm{div}_{\mathbf{x}}\, \mathbf{v} = \mathrm{tr}\nabla_{\mathbf{x}}\, \mathbf{v}$, and $\mathbf{D}$ are expressible in terms of $\mathbf{L}$, so that the Cauchy stress and the heat power will be functions of type $(3.2.5)_{1,2}$.

A limiting special case of viscous fluids, namely ideal fluids, will be considered in order to derive the *absolute temperature scale*. We deal only with the case that the fields are independent of the position variable $\mathbf{x}$. For such fluids, (3.2.7) reduces to

$$\mathbf{T}(t) = -p(\rho, \theta)\mathbf{I}, \qquad h(t) = a(\rho, \theta)\dot{\rho} + b(\rho, \theta)\dot{\theta}, \qquad \mathbf{q}(t) = \mathbf{0}. \tag{3.2.8}$$

The first and the second laws of thermodynamics introduced in the next two sections are constraints on the constitutive equations (3.2.5) rather than new equations imposed on the field variables. In other words, the constitutive equations (3.2.5) cannot depend in an arbitrary way on the variables $(\sigma_t, P(t))$. It is well known that it is impossible to devise a machine or a piece of equipment that executes a perpetual motion of the first or the second kind. Such a machine could be realized if it were possible to have materials with general constitutive equations of type (3.2.5). The principles of thermodynamics allow us to establish which materials, as described by equations of type (3.2.5), are compatible with the physical world.

3.3 First Principle of Thermodynamics

This principle leads to the law of conservation of energy under conditions whereby mechanical energy can be transformed into heat and vice versa.

Definition 3.3.1. *A closed cycle is a pair $(\sigma, P) \in \Sigma \times \Pi$ such that $\hat{\varrho}(\sigma, P) = \sigma$.*

- **First principle of thermodynamics for simple materials** (on cycles): *In any closed cycle $(\sigma, P) \in \Sigma \times \Pi$, the sum of the heat absorbed by the body and the work done by internal forces is equal to zero, that is,*

$$\int_0^{d_P} \left[\hat{h}(\sigma_t, P(t)) + \frac{1}{\rho} \hat{\mathbf{T}}(\sigma_t, P(t)) \cdot \mathbf{D}(t) \right] dt = 0. \tag{3.3.1}$$

Before dealing with the general case, let us consider a thermoelastic material represented by the system of equations (3.2.6). The space of states Σ is a particular subspace of the space $\mathrm{Lin}(\mathbb{R}^3) \times \mathbb{R}$. With any closed cycle (σ, P), we can associate the corresponding closed curve c in Σ given by

$$\sigma_t = \hat{\varrho}(\sigma, P_t), \quad t \in [0, d_P).$$

The case in which c is a polygonal closed curve in Σ will be considered. Then, the integral in (3.3.1) takes the form

$$\int_0^{d_P} \left(\mathbf{A} \cdot \mathbf{L} + B\dot{\theta} + \frac{1}{\rho} \mathbf{T} \cdot \mathbf{D} \right) dt = 0,$$

which can be rewritten as a curvilinear integral in the space of states Σ in the form

$$\int_c \mathbf{A} \left(\mathbf{F}^{-1} \right)^T \cdot d\mathbf{F} + B d\theta + \frac{1}{\rho} \mathbf{T} \left(\mathbf{F}^{-1} \right)^T \cdot d\mathbf{F} = 0, \tag{3.3.2}$$

where (3.2.4) has been used. This is the integral on c of the following differential form:

$$\left(\mathbf{A} + \frac{1}{\rho} \mathbf{T} \right) \left(\mathbf{F}^{-1} \right)^T \cdot d\mathbf{F} + B d\theta. \tag{3.3.3}$$

On the basis of (3.3.2), one can assert that the differential form (3.3.3) is integrable. This means that under hypotheses of continuity for **A**, **T**, and B, there exists a function of state $e : \Sigma \to \mathbb{R}$, differentiable on Σ and such that

$$de = \left(\mathbf{A} + \frac{1}{\rho} \mathbf{T} \right) \left(\mathbf{F}^{-1} \right)^T \cdot d\mathbf{F} + B d\theta. \tag{3.3.4}$$

It is possible to obtain an analogous result by repeating the above procedure for an ideal or viscous fluid. These observations lead to the idea that it is always possible to prove the existence of a function of such a type for any simple material.

The disadvantage of (3.3.1) is that for materials with memory, which are our main concern in later chapters, closed cycles rarely occur. We now state, without proof, the first law for a general material, in a form that makes no reference to cycles.

- **The first principle of thermodynamics** (general form): *For any simple material, there exists a function of state $e : \Sigma \to \mathbb{R}$, known as the internal energy, such that for any pair of states $\sigma_1, \sigma_2 \in \Sigma$ and for any process P with $\hat{\varrho}(\sigma_1, P) = \sigma_2$, one has*

$$e(\sigma_2) - e(\sigma_1) = \int_0^{d_P} \left[\hat{h}(\sigma_t, P(t)) + \frac{1}{\rho} \hat{\mathbf{T}}(\sigma_t, P(t)) \cdot \mathbf{L}(t) \right] dt, \tag{3.3.5}$$

where $\sigma_t = \hat{\varrho}(\sigma_1, P_t)$.

Remark 3.3.2. The internal energy e is uniquely determined up to an additive constant, which can be fixed by assigning the value of the energy in the reference state of the material. We refer to e as *a thermodynamic potential.*

At all points of continuity of the function

$$\hat{h}(\sigma_t, P(t)) + \frac{1}{\rho}\hat{\mathbf{T}}(\sigma_t, P(t)) \cdot \mathbf{L}(t),$$

it follows from (3.3.5) that

$$\frac{de}{dt}(\sigma_t) = \hat{h}(\sigma_t, P(t)) + \frac{1}{\rho}\hat{\mathbf{T}}(\sigma_t, P(t)) \cdot \mathbf{L}(t). \tag{3.3.6}$$

Moreover, using the heat equation (3.1.3), we obtain the *energy equation*

$$\rho\frac{de}{dt}(\sigma_t) = -\mathrm{div}_{\mathbf{x}}\, \hat{\mathbf{q}}(\sigma_t, P(t)) + \hat{\mathbf{T}}(\sigma_t, P(t)) \cdot \mathbf{L}(t) + \rho r. \tag{3.3.7}$$

Remark 3.3.3. Note that (3.3.6) allows us to express the quantity h in terms of the new function of state e and the mechanical rate of work. Indeed, we can now regard e as a fundamental quantity, more convenient than $\hat{h}$ because it is a function of state.

In Chap. 5, we will use the Lagrangian form of (3.3.7), given by

$$\rho_0\dot{e}(\sigma_t) = -\mathrm{Div}_{\mathbf{X}}\hat{\mathbf{q}}_L\,(\sigma_t, P(t)) + \widehat{\mathbf{S}}(\sigma_t, P(t)) \cdot \dot{\mathbf{E}}(t) + \rho_0 r, \tag{3.3.8}$$

where $\widehat{\mathbf{S}}$ is the second Piola–Kirchhoff stress tensor, defined by (1.3.52), and $\mathbf{E}$ is the strain tensor given by $(1.2.21)_1$.

3.4 Second Principle of Thermodynamics

The first principle of thermodynamics allows the possibility of transformation of various types of energy from one into another, provided the total energy is conserved. The second principle of thermodynamics imposes some limits on such transformations of energy and asserts that not all types of energy have the same capability of transforming themselves into mechanical work. These two assertions are not in contradiction. In fact, for the first principle we can prove the existence of an energetic balance, while the second principle asserts that natural transformations, for a system in a cycle, are those that transform valuable energy into a less valuable form. In other words, this means that in such processes there is a loss of energy.

3.4.1 The Absolute Temperature Scale

Before discussing general materials, we briefly consider perfect fluids with field variables that are independent of the position variable $\mathbf{x}$, defined by the constitutive equations (3.2.8), in order to derive the absolute temperature scale and to motivate the general definition of entropy.

With the aid of (3.2.8), the heat equation (3.1.3) can be written as

$$\rho\left[a(\rho, \theta)\dot{\rho} + b(\rho, \theta)\dot{\theta}\right] = \rho r, \tag{3.4.1}$$

where θ in this relationship is some empirical temperature. For materials with dissipation, the entropy relationship will be an inequality, as we shall see below. However, for a perfect fluid, it will be an equality, which is assumed to be of the form

$$\rho\dot{\eta} = -div_{\mathbf{x}}\mathbf{j}_\eta + \rho s, \tag{3.4.2}$$

where $\mathbf{j}_\eta$ is a flux term describing entropy flow and s is an entropy source term. We can neglect the divergence term due to the assumption of field homogeneity, so that (3.4.2) becomes

$$\begin{gathered}\rho\left[c(\rho,\theta)\dot{\rho} + d(\rho,\theta)\dot{\theta}\right] = \rho s,\\ c(\rho,\theta) = \frac{\partial\eta(\rho,\theta)}{\partial\rho}, \qquad d(\rho,\theta) = \frac{\partial\eta(\rho,\theta)}{\partial\theta}.\end{gathered} \tag{3.4.3}$$

We do not regard this equation as independent of the heat equation (3.4.1) but rather as a nontrivial relation that can be derived from that equation, so that any solution of the entropy equation is also a solution of the heat equation. Thus, we suppose the existence of an integrating factor $R(\rho,\theta)$ such that

$$c(\rho,\theta)\dot{\rho} + d(\rho,\theta)\dot{\theta} - s = R(\rho,\theta)\left[a(\rho,\theta)\dot{\rho} + b(\rho,\theta)\dot{\theta} - r\right]$$

for any $\dot{\rho}$, $\dot{\theta}$, and r. It follows that

$$c = Ra, \qquad d = Rb, \qquad s = Rr. \tag{3.4.4}$$

We now show that the factor R does not depend on ρ and is a monotonic decreasing function of θ. In relation $(3.4.4)_3$, the quantities s and r are source terms and do not depend on the material properties.

Let us now consider a further special case that of an ideal fluid. This is a fluid with internal energy depending only on temperature and a pressure function $p(\rho,\theta)$ that is a monotonic increasing function of θ.

Now (3.3.6), with the aid of $(1.3.2)_2$, yields

$$h(\rho,\theta,P(t)) = e'(\theta)\dot{\theta} - \frac{1}{\rho^2}p(\rho,\theta)\dot{\rho},$$

which, combined with $(3.2.8)_2$, gives

$$a(\rho,\theta) = -\frac{1}{\rho^2}p(\rho,\theta), \qquad b(\rho,\theta) = b(\theta) = e'(\theta),$$

whence $(3.4.4)_{1,2}$ become

$$c(\rho,\theta) = -\frac{R(\theta)}{\rho^2}p(\rho,\theta), \qquad d(\rho,\theta) = d(\theta) = R(\theta)b(\theta). \tag{3.4.5}$$

From $(3.4.5)_2$ and $(3.4.3)_{2,3}$, we conclude that

$$\eta(\rho,\theta) = A(\rho) + B(\theta),$$

where

$$A'(\rho) = -\frac{R(\theta)}{\rho^2}p(\rho,\theta), \qquad B'(\theta) = d(\theta) = R(\theta)b(\theta).$$

Thus, $R(\theta)p(\rho,\theta)$ cannot depend on θ. Let θ' be a fixed temperature. Then,

$$R(\theta) = R(\theta')\frac{p(\rho,\theta')}{p(\rho,\theta)}.$$

We can always choose $R(\theta') > 0$. The quantities $p(\rho,\theta')$ and $p(\rho,\theta)$ both are positive, and the latter increases monotonically with θ. Thus, R is a positive decreasing function of the empirical temperature θ.

Recalling $(3.4.4)_3$ and the fact that the source functions s and r do not depend on the material under consideration, we deduce from $(3.4.4)_3$ that R has a universal character. Its inverse

$$T(\theta) = \frac{1}{R(\theta)}$$

is positive and increasing and can be chosen as a suitable measure of temperature. This function represents *the absolute temperature scale*.

Dropping the constancy assumption with respect to $\mathbf{x}$ on the fields, $(3.4.4)_3$, (3.1.3), and (3.4.2) give

$$R(\theta)(\rho h + \mathrm{div}_{\mathbf{x}}\,\mathbf{q}) = \rho\dot{\eta} + \mathrm{div}_{\mathbf{x}}\,\mathbf{j}_\eta,$$

which must hold for all choices of fields. Using the identity

$$R(\theta)\mathrm{div}_{\mathbf{x}}\,\mathbf{q} = \mathrm{div}_{\mathbf{x}}\,[R(\theta)\mathbf{q}] - \mathbf{q}\cdot\nabla R(\theta),$$

we deduce that

$$\mathrm{div}_{\mathbf{x}}\,\mathbf{j}_\eta = \mathrm{div}_{\mathbf{x}}(R\mathbf{q}), \qquad \rho\dot{\eta} = \rho R h - \mathbf{q}\cdot\nabla R(\theta). \tag{3.4.6}$$

From now on, we denote the absolute temperature $1/R$ by θ. The relations in (3.4.6) provide motivation for those now introduced, though the former were derived for very particular materials.

3.4.2 Entropy Action

We consider the function

$$s(\sigma,P) = \int_0^{d_P}\left[\frac{\hat{h}(\sigma_t,P(t))}{\theta(t)} + \frac{1}{\rho\theta^2(t)}\hat{\mathbf{q}}(\sigma_t,P(t))\cdot\mathbf{g}(t)\right]dt, \tag{3.4.7}$$

which will be referred to as the *entropy action*. By means of this functional we can introduce the following principle:

- **The second principle of thermodynamics for simple materials** (on cycles): *On any closed cycle* $(\sigma,P)\in\Sigma\times\Pi$, *the entropy action is such that*

$$\oint_0^{d_P}\left[\frac{\hat{h}(\sigma_t,P(t))}{\theta(t)} + \frac{1}{\rho\theta^2(t)}\hat{\mathbf{q}}(\sigma_t,P(t))\cdot\mathbf{g}(t)\right]dt \le 0. \tag{3.4.8}$$

Remark 3.4.1. For a homogeneous process ($\mathbf{g} = \mathbf{0}$), relation (3.4.8) takes the classical form

$$\oint_0^{d_P} \frac{\hat{h}(\sigma_t, P(t))}{\theta(t)} dt \leq 0.$$

Consider (3.4.8) for isothermal processes. If we set $\theta = \textit{constant}$ and use (3.3.1), then it follows that

$$-\frac{1}{\theta} \oint_0^{d_P} \frac{1}{\rho} \hat{\mathbf{T}}(\sigma_t, P(t)) \cdot \mathbf{D}(t) dt \leq 0, \tag{3.4.9}$$

which is a statement that the mechanical work is nonnegative in closed processes or the principle of dissipation of mechanical energy (see (1.3.33) or (2.3.14)in the material description).

For a general material, the second law is now stated, again without proof and making no reference to closed processes.

- **The second principle of thermodynamics (general form)**: *There exists a function of state $\eta : \Sigma \to \mathbb{R}$, known as the entropy, such that*

$$\eta(\sigma_2) - \eta(\sigma_1) \geq \oint_0^{d_P} \left(\frac{h}{\theta} + \frac{1}{\rho \theta^2} \mathbf{q} \cdot \mathbf{g} \right) dt = s(\sigma_1, P), \tag{3.4.10}$$

for all $(\sigma_1, \sigma_2) \in \Sigma$ and $P \in \Pi$ with $\hat{\varrho}(\sigma_1, P) = \sigma_2$.

Remark 3.4.2. Under appropriate hypotheses of regularity on the constitutive functionals, it follows from the inequality (3.4.10) that

$$\dot{\eta}(\sigma_t) \geq \frac{\hat{h}(\sigma_t, P(t))}{\theta(t)} + \frac{1}{\rho(t)\theta^2(t)} \hat{\mathbf{q}}(\sigma_t, P(t)) \cdot \mathbf{g}(t). \tag{3.4.11}$$

We note the connection between this inequality and the equality $(3.4.6)_2$. By means of the heat equation (3.1.3), we deduce from relation (3.4.11) the well-known Clausius–Duhem inequality

$$\rho \dot{\eta} \geq -\text{div}_{\mathbf{x}} \left(\frac{\mathbf{q}}{\theta} \right) + \frac{\rho}{\theta} r. \tag{3.4.12}$$

Again, we observe the connection between this relation and (3.4.2) combined with $(3.4.6)_1$.

Let us define the function of state known as the *Helmholtz free energy* by

$$\psi_H(\sigma_t) = e(\sigma_t) - \theta(t)\eta(\sigma_t). \tag{3.4.13}$$

Using (3.3.7), (3.4.12), and (3.4.13), one can show that

$$\dot{\psi}_H(\sigma_t) \leq \frac{1}{\rho(t)} \hat{\mathbf{T}}(\sigma_t, P(t)) \cdot \mathbf{L}(t) - \eta(\sigma_t)\dot{\theta}(t) - \frac{1}{\rho(t)\theta(t)} \hat{\mathbf{q}}(\sigma_t, P(t)) \cdot \mathbf{g}(t), \tag{3.4.14}$$

which is an alternative form of the Clausius–Duhem inequality. The free energy function of state is not uniquely determined in general, for a given material. Very general properties of this function are proved in the next chapter, including the result that the free energies associated with a given state of the material with memory form a convex set. This nonuniqueness also applies to entropy.

In Chap. 5, we will use the Lagrangian form of (3.4.12) and (3.4.14), given by

$$\rho_0\dot{\eta} \geq -\mathrm{Div}_{\mathbf{X}}\left(\frac{\mathbf{q}_L}{\theta}\right) + \frac{\rho_0}{\theta}r \tag{3.4.15}$$

and

$$\dot{\psi}_H(\sigma_t) \leq \frac{1}{\rho_0}\widehat{\mathbf{S}}(\sigma_t, P(t)) \cdot \dot{\mathbf{E}}(t) - \eta(\sigma_t)\dot{\theta}(t) - \frac{1}{\rho_0\theta(t)}\hat{\mathbf{q}}_L(\sigma_t, P(t)) \cdot \mathbf{g}_L(t), \tag{3.4.16}$$

where $\mathbf{g}_L$ is given by (3.1.1).

We now consider two examples: nonlinear elastic bodies and viscous liquids, already introduced in an isothermal context in Part I. Recalling Remark 3.3.3, it will be observed that we use the heat power h but immediately relate it to the internal energy. In the second example, h is not introduced; instead, e is used.

The principles of thermodynamics impose some restrictions on the constitutive functionals. They become automatically satisfied if the constitutive equations obey certain conditions. In what follows, we seek to determine such conditions for the constitutive equations describing elastic materials and viscous fluids.

3.5 Applications to Elastic Bodies

For an elastic body, we have $\sigma = (\mathbf{F}, \theta)$, while from (3.2.6) and (3.3.5), for any $(\sigma, P) \in \Sigma \times \Pi$, where $\hat{\varrho}(\sigma, P) = \sigma_1$,

$$e(\sigma_1) - e(\sigma) = \int_0^{d_P} \left[B(\mathbf{F}, \theta)\dot{\theta} + \mathbf{A}(\mathbf{F}, \theta) \cdot \mathbf{L} + \frac{1}{\rho(\mathbf{F})}\mathbf{T}(\mathbf{F}, \theta) \cdot \mathbf{L}\right] dt.$$

From (3.3.6) (indeed, more specifically recalling (3.3.4)), we have that $e = \tilde{e}(\mathbf{F}, \theta)$ obeys the relation

$$\frac{d\tilde{e}}{dt}(\mathbf{F}, \theta) = B(\mathbf{F}, \theta)\dot{\theta} + \mathbf{A}(\mathbf{F}, \theta) \cdot \mathbf{L} + \frac{1}{\rho(\mathbf{F})}\mathbf{T}(\mathbf{F}, \theta) \cdot \mathbf{L}.$$

The function (3.4.7) can be written as

$$s(\sigma, P) = \int_0^{d_P} \left\{\frac{1}{\theta}\left[-\frac{1}{\rho}\hat{\mathbf{T}}(\mathbf{F}, \theta) \cdot \mathbf{L} + \frac{\partial \tilde{e}}{\partial \mathbf{F}} \cdot \dot{\mathbf{F}} + \frac{\partial \tilde{e}}{\partial \theta}\dot{\theta}\right] + \frac{1}{\rho\theta^2}\hat{\mathbf{q}}(\mathbf{F}, \theta; \mathbf{g}) \cdot \mathbf{g}\right\} dt.$$

We now consider the integral on an arbitrary curve c in $\mathrm{Lin}(\mathbb{R}^3) \times \mathbb{R}$,

$$I(c) = \int_c \frac{1}{\theta}\left[B(\mathbf{F}, \theta)d\theta + \mathbf{A}(\mathbf{F}, \theta)\left(\mathbf{F}^T\right)^{-1} \cdot d\mathbf{F}\right]. \tag{3.5.1}$$

If c^+ is a closed curve in $\mathrm{Lin}(\mathbb{R}^3) \times \mathbb{R}$ and c^- is the same curve with opposite orientation, then

$$I(c^+) = -I(c^-). \tag{3.5.2}$$

Let $\sigma \in \Sigma$ be a finite element, and let c^+ be an oriented closed polygonal curve containing σ. Then for any $t > 0$, there is a parametrization $\tau \to (\mathbf{F}(\tau), \theta(\tau))$ of the curve c^+, defined on $[0, d_P)$, such that

$$(\mathbf{F}(d_P), \theta(d_P)) = (\mathbf{F}(0), \theta(0)) = \sigma.$$

Corresponding to this parametrization, there is a process P such that

$$\hat{\varrho}(\sigma, P_t) = (\mathbf{F}(t), \theta(t)), \quad \hat{\varrho}(\sigma, P) = \sigma.$$

Using this parametrization of c^+, we can rewrite (3.5.1) in the form

$$I(c^+) = \int_0^{d_p} \frac{1}{\theta(t)} \left[B(\mathbf{F}, \theta)\dot{\theta} + \mathbf{A}(\mathbf{F}, \theta) \left(\mathbf{F}^T\right)^{-1} \cdot \dot{\mathbf{F}} \right] dt = \int_0^{d_P} \frac{\hat{h}\left(\mathbf{F}, \theta, \dot{\mathbf{F}}, \dot{\theta}\right)}{\theta} dt,$$

by virtue of $(3.2.6)_2$. The second principle of thermodynamics (3.4.8) states that on any closed process P with duration d_P, we have

$$\int_0^{d_P} \frac{\hat{h}}{\theta} dt + \int_0^{d_P} \frac{1}{\rho\theta^2} \hat{\mathbf{q}} \cdot \mathbf{g} dt \leq 0. \tag{3.5.3}$$

Therefore, we obtain

$$I(c^+) \leq -\int_0^{d_P} \frac{1}{\rho\theta^2} \hat{\mathbf{q}} \cdot \mathbf{g} dt. \tag{3.5.4}$$

If we set $M(c^+) = \sup\{-\frac{1}{\rho\theta^2}\hat{\mathbf{q}}(\mathbf{F}, \theta, \mathbf{g}) \cdot \mathbf{g} \mid (\mathbf{F}, \theta) \in c^+\}$, then $M(c^+)$ is finite because c^+ is a compact subset of Σ and $\mathbf{q}$ is assumed continuous on Σ. Moreover, $\hat{\mathbf{q}}(\mathbf{F}, \theta, \mathbf{g}) \cdot \mathbf{g}$ is always nonpositive for an elastic body. In fact, if we consider the process P defined by

$$P(t) \equiv (\mathbf{0}, 0, \mathbf{g}) \quad \text{for all } t \in [0, d_P)$$

and suppose that the initial state is $\sigma_0 = (\mathbf{F}, \theta)$, where $\mathbf{g}$, $\mathbf{F}$, and θ are arbitrary given quantities, independent of t, then from (3.5.3), we have

$$\hat{\mathbf{q}}(\mathbf{F}, \theta, \mathbf{g}) \cdot \mathbf{g} \leq 0 \quad \text{for all } (\mathbf{F}, \theta, \mathbf{g}). \tag{3.5.5}$$

Therefore, from (3.5.4),

$$I(c^+) \leq d_P M(c^+),$$

and since this relation holds for any $d_P > 0$, we obtain that

$$I(c^+) \leq 0.$$

Let us now consider the curve c^- with opposite orientation with respect to c^+. Using a procedure similar with that leading to (3.5.5), we obtain

$$I(c^-) \leq 0,$$

and hence, by virtue of (3.5.2), on any closed curve c,

$$I(c) = 0.$$

Therefore, using well-known theorems concerning the integrability of differential forms, we deduce from (3.5.1) that there exists a differentiable function of state $\eta : \Sigma \to R$, such that

$$\dot{\eta}(\mathbf{F}, \theta) = \frac{1}{\theta(t)} \left[B(\mathbf{F}, \theta)\dot{\theta} + \mathbf{A}(\mathbf{F}, \theta) \left(\mathbf{F}^T\right)^{-1} \cdot \dot{\mathbf{F}} \right] = \frac{\hat{h}(\mathbf{F}, \theta, \dot{\mathbf{F}}, \dot{\theta})}{\theta}. \tag{3.5.6}$$

Using $(3.5.6)_2$, we can write the expression (3.4.7) for $s(\sigma, P)$ as

$$s(\sigma, P) = \eta(\hat{\rho}(\sigma, P)) - \eta(\sigma) + \int_0^{d_P} \frac{1}{\rho\theta^2} \hat{\mathbf{q}}(\sigma_t, P(t)) \cdot \mathbf{g} dt,$$

so that using the fact that $\mathbf{q} \cdot \mathbf{g} \leq 0$, we obtain

$$s(\sigma, P) \leq \eta(\hat{\rho}(\sigma, P)) - \eta(\sigma).$$

Also, from $(3.5.6)_2$, the property $\mathbf{q} \cdot \mathbf{g} \leq 0$, and (3.1.3), we recover the Clausius–Duhem inequality (3.4.12):

$$\rho\dot{\eta} \geq \frac{1}{\theta}\rho h + \frac{1}{\theta^2}\mathbf{q} \cdot \mathbf{g} = \frac{1}{\theta} \left(-\mathrm{div}_{\mathbf{x}}\, \mathbf{q} + \rho r + \frac{1}{\theta}\mathbf{q} \cdot \mathbf{g} \right) = -\mathrm{div}_{\mathbf{x}} \left(\frac{\mathbf{q}}{\theta} \right) + \frac{\rho}{\theta} r.$$

3.6 Thermodynamic Restrictions for Viscous Fluids

We consider here the compressible flow of a Newtonian fluid defined by the constitutive equations (3.2.7) and determine the restrictions imposed by the principles of thermodynamics on these equations. Combining (3.3.6) and (3.4.11), we obtain

$$\theta\dot{\eta} \geq \dot{e} - \frac{1}{\rho}\mathbf{T} \cdot \mathbf{L} + \frac{1}{\rho\theta}\mathbf{q} \cdot \mathbf{g}.$$

Recalling that for viscous fluids the state is given by the pair $\sigma = (\rho, \theta)$, it follows that the internal energy e and entropy η are functions of the following type:

$$e = \hat{e}(\rho, \theta), \quad \eta = \hat{\eta}(\rho, \theta).$$

As noted earlier for e, these quantities are referred to as thermodynamic potentials.

The function of state introduced in (3.4.13), namely the Helmholtz free energy ψ_H, is also a thermodynamic potential, which here assumes the form

$$\hat{\psi}_H(\rho, \theta) = \hat{e}(\rho, \theta) - \theta\hat{\eta}(\rho, \theta).$$

It must satisfy (3.4.14):

$$\dot{\psi}_H \leq -\eta\dot{\theta} + \frac{1}{\rho}\mathbf{T}\cdot\mathbf{L} - \frac{1}{\rho\theta}\mathbf{q}\cdot\mathbf{g}. \tag{3.6.1}$$

On the basis of the principles of thermodynamics, the inequality (3.6.1) must be verified by all processes $P(t) = \big(\mathbf{L}(t), \dot{\theta}(t), \mathbf{g}(t)\big)$ representing possible solutions of the equation of motion. Recalling (1.2.23), the symmetry of $\mathbf{T}$, and in particular its form given by $(3.2.7)_1$, we can write (3.6.1) in the form

$$\frac{\partial\psi_H}{\partial\rho}\dot{\rho} + \frac{\partial\psi_H}{\partial\theta}\dot{\theta} \leq -\eta\dot{\theta} + \frac{1}{\rho}\Big[-p\,\mathrm{div}_{\mathbf{x}}\,\mathbf{v} + \lambda\,(\mathrm{div}_{\mathbf{x}}\,\mathbf{v})^2 + 2\mu|\mathbf{D}|^2\Big] + \frac{1}{\rho\theta}K|\mathbf{g}|^2. \tag{3.6.2}$$

With the aid of the equation of continuity $(1.3.2)_2$, relation (3.6.2) gives

$$\left(\frac{\partial\psi_H}{\partial\rho} - \frac{p}{\rho^2}\right)\dot{\rho} + \left(\frac{\partial\psi_H}{\partial\theta} + \eta\right)\dot{\theta} \leq \frac{\lambda}{\rho^3}\dot{\rho}^2 + \frac{2\mu}{\rho}|\mathbf{D}|^2 + \frac{K}{\rho\theta}|\mathbf{g}|^2. \tag{3.6.3}$$

The quantities ψ_H, p, η, λ, μ, and K are functions of the state (ρ, θ). For a given state (ρ, θ), we consider a process P,

$$P(t) = (\mathbf{0}, \dot{\theta}, \mathbf{0}),$$

for some arbitrary $\dot{\theta}$. In this case, since $\mathbf{L} = \mathbf{0}$ yields $\mathbf{D} = \mathbf{0}$, it follows that $\mathrm{div}_{\mathbf{x}}\,\mathbf{v} = \mathrm{tr}\nabla_{\mathbf{x}}\,\mathbf{v} = 0$, and hence $\dot{\rho} = 0$ by virtue of (1.2.9), (1.2.23), and $(1.3.2)_2$, relation (3.6.3) reduces to

$$\left(\frac{\partial\psi_H}{\partial\theta} + \eta\right)\dot{\theta} \leq 0,$$

from which, in view of the fact that $\dot{\theta}$ is arbitrary, we deduce that

$$\eta(\rho, \theta) = -\frac{\partial\psi_H}{\partial\theta}(\rho, \theta). \tag{3.6.4}$$

Now consider a process of the type

$$P(t) = \left(\left(\frac{1}{3}div_{\mathbf{x}}\,\mathbf{v}\right)\mathbf{I}, 0, \mathbf{0}\right) = \left(-\frac{\dot{\rho}}{3\rho}\mathbf{I}, 0, \mathbf{0}\right). \tag{3.6.5}$$

Then (3.6.3) yields

$$\left(\frac{\partial\psi_H}{\partial\rho} - \frac{p}{\rho^2}\right)\dot{\rho} \leq \frac{1}{\rho^3}\left(\lambda + \frac{2}{3}\mu\right)\dot{\rho}^2,$$

where $\dot{\rho}$ is arbitrary. Dividing by the quantity $\dot{p}^2$ and using the fact that $\dot{\rho}$ is arbitrary, we see that this inequality holds if and only if

$$p = \rho^2\frac{\partial\psi_H}{\partial\rho}. \tag{3.6.6}$$

With these results, the initial inequality (3.6.3) reduces to

$$\frac{\lambda}{\rho^3}\dot{\rho}^2 + \frac{2\mu}{\rho}|\mathbf{D}|^2 + \frac{K}{\rho\theta}|\mathbf{g}|^2 \geq 0. \tag{3.6.7}$$

Taking into account the arbitrariness of $\dot{\rho}$, $\mathbf{D}$, and $\mathbf{g}$, we obtain, from (3.6.7), the following inequalities:

$$\frac{K}{\rho\theta}|\mathbf{g}|^2 \geq 0, \quad \frac{\lambda}{\rho^3}\dot{\rho}^2 + \frac{2\mu}{\rho}|\mathbf{D}|^2 \geq 0. \tag{3.6.8}$$

The first inequality in (3.6.8) furnishes the condition

$$K \geq 0. \tag{3.6.9}$$

If we first set $\mathbf{D} = \frac{1}{3}(\mathrm{div}_{\mathbf{x}}\, \mathbf{v})\,\mathbf{I}$ (see (3.6.5)) in the second inequality of (3.6.8), then

$$\left(\lambda + \frac{2}{3}\mu\right)(\mathrm{div}_{\mathbf{x}}\, \mathbf{v})^2 \geq 0,$$

from which we deduce that

$$3\lambda + 2\mu \geq 0. \tag{3.6.10}$$

If we now choose $\mathbf{D}$ arbitrarily but put $\mathrm{div}_{\mathbf{x}}\, \mathbf{v} = 0$ so that $\dot{\rho} = 0$, then the second inequality of (3.6.8) yields

$$\mu \geq 0. \tag{3.6.11}$$

Therefore, the restrictions imposed by the principles of thermodynamics on the constitutive equation for compressible flow of a Newtonian fluid defined by the constitutive equation (3.2.7) are those contained in relations (3.6.4), (3.6.6), (3.6.9), (3.6.10), and (3.6.11).

We have used as examples in this chapter the classical materials introduced in Part I. However, the main focus of later chapters will be on more general materials, namely those with memory. The principles of thermodynamics introduced here and their consequences in the form of the energy equation (3.3.7) or (3.3.8) and the Clausius–Duhem inequality (3.4.14) or (3.4.16) apply quite generally to simple materials, and indeed we will use them in Chap. 5. In fact, general constitutive equations for materials with memory are derived using a generalization of the kind of arguments developed in Sect. 3.6 to obtain (3.6.4) and (3.6.6).

The statements of the thermodynamic principles that we have presented in these sections apply to simple materials. In the next section we study the generalization of the laws of thermodynamics required for nonsimple materials.

3.7 Principles of Thermodynamics for Nonsimple Materials

Just as for a simple material, we suppose that a nonsimple material is formally characterized by a state σ, which is an element of the space Σ of the possible states of the body, and a process P, belonging to the space Π of the possible processes of the body, with a duration $d_P \in \mathbb{R}^+$, defined more generally than by (3.2.1). Furthermore, we assume that for this system, there exists a function $\hat{\rho} \in \Sigma \times \Pi \to \Sigma$ with the same properties as described in Definition 3.2.1. Finally, the behavior of the material is always described by the triplet $(\mathbf{T}, h, \mathbf{q})$, defined in (3.2.6).

The important difference between simple and nonsimple materials, from a thermodynamic point of view, emerges in the context of defining the mechanical power.

For simple materials, the internal mechanical power, denoted by $\mathcal{P}^i_m$, is always expressed by

$$\mathcal{P}^i_m = \mathbf{T} \cdot \mathbf{L}, \tag{3.7.1}$$

and therefore the first law assumes the form (3.3.1) or (3.3.7).

In the general case, this quantity is not necessarily defined by (3.7.1). Its expression can be derived by means of the balance equation of power.

For a mechanical system, denoting by $\tilde{\mathcal{P}}^i_m$ the internal mechanical power, this balance equation is written as

$$\frac{d}{dt}T(t) + \tilde{\mathcal{P}}^i_m(\sigma(t), P(t)) = \tilde{\mathcal{P}}^e_m, \tag{3.7.2}$$

where $\tilde{\mathcal{P}}^e_m$ denotes the external mechanical power, defined by the flux and the mechanical sources due to external forces, while T is the kinetic energy.

3.7.1 First Law of Thermodynamics

For nonsimple mechanical systems, the first law is stated by assuming the existence of a function of state $e(\sigma)$, the internal energy, such that

$$\rho\dot{e}(\sigma(t)) = \tilde{\mathcal{P}}^i_m(\sigma(t), P(t)) + \rho h(\sigma(t), P(t)), \tag{3.7.3}$$

where h denotes the internal thermal power per unit mass and ρ is the mass density.

The form (3.7.3) is the classical statement of the first law, given by R. Clausius in 1850, and can be applied also to electromagnetic, chemical, and biological systems, which are characterized by the triplet $(\Sigma, \Pi, \hat{\rho})$.

An alternative and more general formulation of the first law is now given.

• **First principle of thermodynamics** (on cycles). *For any cyclic process* (σ^i, P), *i.e.*, $\hat{\rho}(\sigma^i, P) = \sigma^i$, *the sum of the internal thermal power* ρh *and of other internal powers* $\tilde{\mathcal{P}}^i(\sigma, P)$ *satisfies*

$$\oint \left[\tilde{\mathcal{P}}^i(\sigma(t), P(t)) + \rho h(\sigma(t), P(t))\right] dt = 0, \tag{3.7.4}$$

where $\sigma(t) = \hat{\rho}(\sigma^i, P_t)$.

Referring to (3.3.4), we note that it is similarly possible to deduce from (3.7.4) the existence of an internal energy $e(\sigma(t))$ such that the equality (3.7.3) holds.

An example is now formulated for which, in the representation of the first law, we cannot use the expression (3.7.1) for the internal mechanical power.

Let us consider a thermoelastic plate, for which, following [116, 225, 226] (see also [3]), the stress tensor (in this context a vector) assumes the form

$$\mathbf{T}(\mathbf{x}, t) = -a\nabla\left[\nabla^2 u(\mathbf{x}, t)\right] + b\nabla\ddot{u}(\mathbf{x}, t) + c\nabla\theta(\mathbf{x}, t), \tag{3.7.5}$$

where u is the vertical displacement, while a, b, and c are suitable material constants. The equation of motion is given by

$$\rho\ddot{u}(\mathbf{x},t) = \nabla \cdot \mathbf{T}(\mathbf{x},t) + \rho f(\mathbf{x},t), \tag{3.7.6}$$

where f denotes the external forces. We multiply this by $\dot{u}$ to obtain

$$\frac{d}{dt}\left(\frac{1}{2}\rho\dot{u}^2\right) = \nabla \cdot (\mathbf{T}\dot{u}) - \mathbf{T} \cdot \nabla\dot{u} + \rho f\dot{u}. \tag{3.7.7}$$

To derive the contribution due to the internal and external powers in the balance equation (3.7.7), we must use the representation (3.7.5) for the stress $\mathbf{T}$. Then (3.7.7) becomes

$$\begin{aligned}\frac{1}{2}\frac{d}{dt}&\left[\rho\dot{u}^2 + a\left(\nabla^2 u\right)^2 + b\,(\nabla\dot{u})^2\right] + c\nabla\theta\cdot\nabla\dot{u}\\ &= \nabla\cdot\left\{\left[-a\nabla\left(\nabla^2 u\right) + b\nabla\ddot{u} + c\nabla\theta\right]\dot{u} + a\nabla^2 u\nabla\dot{u}\right\} + \rho f\dot{u}.\end{aligned} \tag{3.7.8}$$

On comparing with (3.7.2), it follows that the internal mechanical power is given by

$$\tilde{\mathcal{P}}^i_m = \frac{1}{2}\frac{d}{dt}\left[a\left(\nabla^2 u\right)^2 + b\,(\nabla\dot{u})^2\right] + c\nabla\theta\cdot\nabla\dot{u}, \tag{3.7.9}$$

which is completely different from the definition of $\mathcal{P}^i_m$ given in (3.7.1). Consequently, the local form (3.7.3) of the new formulation of the first law (3.7.4) is expressed by

$$\rho\dot{e} = \frac{1}{2}\frac{d}{dt}\left[a\left(\nabla^2 u\right)^2 + b\,(\nabla\dot{u})^2\right] + c\nabla\theta\cdot\nabla\dot{u} + \rho h. \tag{3.7.10}$$

Moreover, by virtue of (3.7.2) and putting

$$\mathbf{N}' = \left[a\nabla\left(\nabla^2 u\right) - b\nabla\ddot{u} - c\nabla\theta\right]\dot{u} - a\nabla^2 u\,\nabla\dot{u}, \tag{3.7.11}$$

we derive the expression for the external mechanical power

$$\tilde{\mathcal{P}}^e_m = -\nabla\cdot\mathbf{N}' + \rho f\dot{u}.$$

It is now possible to better understand the difference between the new formulation and the classical one, proposed by Truesdell and Noll [313] and Coleman [67] (also [311]) for simple materials, in which the first law relating to mechanical systems is expressed by (see (3.3.7))

$$\rho\dot{e} = \mathbf{T}\cdot\mathbf{L} - \nabla\cdot\mathbf{q} + \rho r. \tag{3.7.12}$$

This expression clearly agrees with the representation (3.7.3) when the internal power $\tilde{\mathcal{P}}^i_m$ has the form (3.7.1) and h, by virtue of the balance equation for heat (see (3.1.3)), is given by

$$\rho h = -\nabla\cdot\mathbf{q} + \rho r. \tag{3.7.13}$$

The representation (3.7.12) does not hold for nonsimple materials. In fact, already in the example of the plate, we have seen that (3.7.10), deduced from (3.7.4) with $\tilde{\mathcal{P}}^i$ defined by (3.7.9) and h given by (3.7.13), is not compatible with (3.7.12), where $\mathbf{T}$ is expressed by (3.7.5). Therefore, (3.7.12) must be modified.

Dunn and Serrin [97], but subsequently also Fabrizio and Morro [122], have proposed a modification of (3.7.12), by assuming

$$\rho\dot{e} = \mathbf{T}\cdot\mathbf{L} - \nabla\cdot\mathbf{N} - \nabla\cdot\mathbf{q} + \rho r, \tag{3.7.14}$$

where the vector $\mathbf{N}$ is a suitable extra flux.

For the example of the plate, we firstly observe that using (3.7.8) and (3.7.11), we can eliminate in (3.7.10) the expression for $\tilde{\mathcal{P}}^i_m$ given by (3.7.9), thus obtaining

$$\rho\dot{e} = -\frac{d}{dt}T + \rho f\dot{u} - \nabla\cdot\mathbf{N}' + \rho h; \tag{3.7.15}$$

hence, taking into account (3.7.7) and (3.7.13), we also have

$$\rho\dot{e} = \mathbf{T}\cdot\nabla\dot{u} - \nabla\cdot(\mathbf{T}\dot{u} + \mathbf{N}') - \nabla\cdot\mathbf{q} + \rho r, \tag{3.7.16}$$

that is, (3.7.14) is satisfied with $\mathbf{N}$ expressed by

$$\mathbf{N} \equiv (\mathbf{T}\dot{u} + \mathbf{N}') = -a\nabla^2 u\,\nabla\dot{u},$$

by virtue of (3.7.5) and again of (3.7.11).

An analogous problem arises when we consider a thermoelastic plate with memory characterized by the following constitutive equations:

$$\begin{aligned}
\mathbf{T}(\mathbf{x},t) &= -\nabla\left[\int_0^\infty C'(s)\nabla^2 u^t(\mathbf{x},s)ds - C_0\nabla^2 u(\mathbf{x},t)\right] + c\nabla\theta(\mathbf{x},t),\\
e(\mathbf{x},t) &= \tilde{e}(\theta(\mathbf{x},t), \nabla^2 u(\mathbf{x},t), \nabla^2 u^t(\mathbf{x},s)),\\
\mathbf{q}(\mathbf{x},t) &= -k_0\nabla\theta(\mathbf{x},t),
\end{aligned} \tag{3.7.17}$$

where the scalar kernel $C'(\cdot)$ is a suitable smooth function.

From the equation of motion (3.7.6), we have already derived the balance equation of the mechanical power (3.7.7), where, using $(3.7.17)_1$, we obtain

$$\begin{aligned}
-\mathbf{T}\cdot\nabla\dot{u} &= \left\{\nabla\left[\int_0^\infty C'(s)\nabla^2 u^t(s)ds - C_0\nabla^2 u\right] - c\nabla\theta\right\}\cdot\nabla\dot{u}\\
&= \nabla\cdot\left\{\left[\int_0^\infty C'(s)\nabla^2 u^t(s)ds - C_0\nabla^2 u - c\theta\right]\nabla\dot{u}\right\}\\
&\quad - \int_0^\infty C'(s)\nabla^2 u^t(s)ds\nabla^2\dot{u} + \frac{C_0}{2}\frac{d}{dt}\left(\nabla^2 u\right)^2 + c\theta\nabla^2\dot{u}.
\end{aligned}$$

Thus, (3.7.7) yields

$$\frac{d}{dt}\left(\frac{1}{2}\rho\dot{u}^2\right)+\int_0^\infty C'(s)\nabla^2u^t(s)ds\nabla^2\dot{u}-\frac{C_0}{2}\frac{d}{dt}\left(\nabla^2u\right)^2-c\theta\nabla^2\dot{u}$$
$$=\nabla\cdot\left(\left\{-\nabla\left[\int_0^\infty C'(s)\nabla^2u^t(s)ds-C_0\nabla^2u\right]+c\nabla\theta\right\}\dot{u}\right.$$
$$\left.+\left[\int_0^\infty C'(s)\nabla^2u^t(s)ds-C_0\nabla^2u-c\theta\right]\nabla\dot{u}\right)+\rho f\dot{u}.$$

Hence, it follows that

$$\tilde{\mathcal{P}}^i_m=\int_0^\infty C'(s)\nabla^2u^t(s)ds\nabla^2\dot{u}-\frac{C_0}{2}\frac{d}{dt}\left(\nabla^2u\right)^2-c\theta\nabla^2\dot{u}.$$

Using (3.7.3), we can give an expression for the thermal power as follows:

$$\rho h=\rho\dot{e}-\tilde{\mathcal{P}}^i_m=\rho\dot{e}-\int_0^\infty C'(s)\nabla^2u^t(s)ds\nabla^2\dot{u}+\frac{C_0}{2}\frac{d}{dt}\left(\nabla^2u\right)^2+c\theta\nabla^2\dot{u}.$$

To obtain the expression (3.7.14), we put

$$\begin{aligned}\mathbf{N}'=&\left\{\nabla\left[\int_0^\infty C'(s)\nabla^2u^t(s)ds-C_0\nabla^2u\right]-c\nabla\theta\right\}\dot{u}\\&-\left[\int_0^\infty C'(s)\nabla^2u^t(s)ds-C_0\nabla^2u-c\theta\right]\nabla\dot{u},\end{aligned}\tag{3.7.18}$$

instead of (3.7.11). Relations (3.7.15) and (3.7.16) follow, where by virtue of $(3.7.17)_1$ and (3.7.18), the quantity $\mathbf{N}$ is now given by

$$\mathbf{N}=\mathbf{T}\dot{u}+\mathbf{N}'=-\left[\int_0^\infty C'(s)\nabla^2u^t(s)ds-C_0\nabla^2u-c\theta\right]\nabla\dot{u}.$$

Other interesting examples, for which the classical representation of the power, used in the first law, does not coincide with that following from a correct balance between the internal and external powers, occur in electromagnetism, in the study of phase transitions, and also in other physical systems, such as those describing processes of phase separation, for example, by means of the Cahn–Hilliard equation [44]. This topic is discussed below.

A further interesting application relates to the thermodynamics of superconducting materials. It has been shown in [122] that for these materials, an extra flux must be introduced into the first law, which is proportional to the superconducting current.

The case of electromagnetism is discussed in Chap. 6.

As a final example of nonsimple materials, we consider the thermodynamics of phase transitions in a binary mixture, characterized by the Cahn–Hilliard equation

$$\dot{c}=\nabla\cdot\mathbf{J},\tag{3.7.19}$$

where $\mathbf{J}$ is the mass flux given by

$$\mathbf{J}=M(c)\nabla\mu(c).\tag{3.7.20}$$

Here c denotes the concentration of one of the two components, with a range of values in the interval $[0, 1]$. The quantity $M(c)$ is the *mobility*, representing the density with which the two phases are mixed, and is such that

$$M(0) = M(1) = 0,$$

while the *chemical potential* $\mu(c)$ is given by

$$\mu(c) = -\gamma\nabla^2 c + \theta G'(c) + \theta_0 F'(c), \qquad \gamma > 0. \tag{3.7.21}$$

Here, θ_0 is the transition temperature, while $G(c)$ and $F(c)$ are assigned the polynomial forms

$$G(c) = \beta\frac{c^2}{2}, \quad F(c) = \beta\left(\frac{c^4}{4} - \frac{c^2}{2}\right), \qquad \beta > 0,$$

corresponding to a second-order transition as described by the Cahn–Hilliard equation.

From (3.7.19) and (3.7.20), we have

$$\dot{c} = \nabla \cdot [M(c)\nabla\mu(c)], \tag{3.7.22}$$

whence, using (3.7.21), we obtain

$$\dot{c} = \nabla \cdot \left\{M(c)\nabla\left[\theta_0 F'(c) + \theta G'(c) - \gamma\nabla^2 c\right]\right\}.$$

This equation must be related to the heat equation, derived from the energy balance law

$$\dot{e} = \tilde{\mathcal{P}}^i_c + h, \tag{3.7.23}$$

where $\tilde{\mathcal{P}}^i_c$ is the internal power of the mixture. Multiplying (3.7.22) by the potential $\mu(c)$, we have the equation for the power

$$\dot{c}\mu(c) = \nabla \cdot [M(c)\mu(c)\nabla\mu(c)] - M(c)[\nabla\mu(c)]^2.$$

Using (3.7.21) again, we obtain

$$\theta_0\dot{F}(c) + \theta\dot{G}(c) + \frac{\gamma}{2}\frac{d}{dt}(\nabla c)^2 + M(c)[\nabla\mu(c)]^2 = \nabla \cdot [\gamma\dot{c}\nabla c + M(c)\mu(c)\nabla\mu(c)].$$

Thus, the power $\tilde{\mathcal{P}}^i_c$ is expressed by

$$\tilde{\mathcal{P}}^i_c = \theta_0\dot{F}(c) + \theta\dot{G}(c) + \frac{\gamma}{2}\frac{d}{dt}(\nabla c)^2 + M(c)\,[\nabla\mu(c)]^2.$$

By assuming for e the form

$$e = \theta_0 F(c) + \frac{\gamma}{2}(\nabla c)^2 + \tilde{e}(\theta),$$

Eq. (3.7.23) becomes

$$\tilde{e}_\theta\theta_t = \theta\dot{G}(c) + M(c)\,[\nabla\mu(c)]^2 - \nabla \cdot \mathbf{q} + r,$$

with the aid of (3.7.13).

3.7.2 Second Law of Thermodynamics

For simple materials, the second law is expressed by the Clausius–Duhem inequality, which assumes the form (3.4.11), i.e.,

$$\rho\dot{\eta} \geq -\nabla \cdot \frac{\mathbf{q}}{\theta} + \rho\frac{r}{\theta}, \tag{3.7.24}$$

where η denotes the entropy function and r the heat supply.

In 1967, I. Müller [268] proposed a new version of this inequality by replacing (3.7.24) with the more general expression

$$\rho\dot{\eta} \geq -\nabla \cdot \mathbf{\Phi} + \rho\frac{r}{\theta}, \tag{3.7.25}$$

where the vector $\mathbf{\Phi}$, the *entropy flux*, is characterized by a constitutive equation. If we assume for $\mathbf{\Phi}$ the form

$$\mathbf{\Phi} = \frac{\mathbf{q}}{\theta} + \mathbf{\Phi}',$$

where $\mathbf{\Phi}'$ denotes an *entropy extra flux*, (3.7.25) can be written as

$$\rho\dot{\eta} \geq -\nabla \cdot \frac{\mathbf{q}}{\theta} - \nabla \cdot \mathbf{\Phi}' + \rho\frac{r}{\theta}. \tag{3.7.26}$$

In order to better understand the difference between simple and nonsimple materials, let us derive the balance law of entropy power. If the heat equation (3.7.13) is divided by the absolute temperature θ, we obtain the identity

$$\rho\frac{h}{\theta} + \frac{1}{\theta^2}\mathbf{q} \cdot \nabla\theta = -\nabla \cdot \frac{\mathbf{q}}{\theta} + \rho\frac{r}{\theta}, \tag{3.7.27}$$

which is the desired relation. For a simple material, we can claim that the internal entropy power is defined by

$$\mathcal{P}^i_\eta := \rho\frac{h}{\theta} + \frac{1}{\theta^2}\mathbf{q} \cdot \nabla\theta, \tag{3.7.28}$$

while the external entropy power is given by

$$\mathcal{P}^e_\eta := -\nabla \cdot \frac{\mathbf{q}}{\theta} + \rho\frac{r}{\theta}.$$

Thus, by (3.7.27), we see that the formulation (3.7.24), relating to simple materials, reduces to the inequality

$$\rho\dot{\eta} \geq \mathcal{P}^i_\eta = \rho\frac{h}{\theta} + \frac{1}{\theta^2}\mathbf{q} \cdot \nabla\theta. \tag{3.7.29}$$

In general, for nonsimple materials, the internal entropy power is not given by (3.7.28). Using (3.7.27) again, the inequality (3.7.26) becomes

$$\rho\dot{\eta} \geq \rho\frac{h}{\theta} + \frac{1}{\theta^2}\mathbf{q} \cdot \nabla\theta - \nabla \cdot \mathbf{\Phi}'. \tag{3.7.30}$$

From the inequalities (3.7.24) and (3.7.26) (or (3.7.29) and (3.7.30)), we see that the second law assumes different expressions for simple and nonsimple materials. Following [75], we propose the second law in a unique general form as follows.

• **Second principle of thermodynamics** (on cycles). *For any cyclic process* (σ^i, P), *i.e.*, $\hat{\rho}(\sigma^i, P) = \sigma^i$, *the internal entropy power* $\tilde{\mathcal{P}}^i_\eta$ *satisfies the inequality*

$$\oint \tilde{\mathcal{P}}^i_\eta(\sigma(t), P(t))dt \leq 0, \tag{3.7.31}$$

where $\sigma(t) = \hat{\rho}(\sigma^i, P_t)$.

By a similar argument to that used in the context of (3.4.11), we can prove that from (3.7.31) there exists the entropy function $\eta(\sigma)$ such that

$$\rho\dot{\eta}(\sigma) \geq \tilde{\mathcal{P}}^i_\eta(\sigma, P), \tag{3.7.32}$$

so that we recover (3.7.30) on putting

$$\tilde{\mathcal{P}}^i_\eta = \rho\frac{h}{\theta} + \frac{1}{\theta^2}\mathbf{q}\cdot\nabla\theta - \nabla\cdot\boldsymbol{\Phi}'. \tag{3.7.33}$$

We prefer to state the second law by the inequality (3.7.31) or (3.7.32), since the representation (3.7.30) of the second law is ambiguous, because of the term $\nabla\cdot\boldsymbol{\Phi}'$.* In order to understand the natural expression for the second law by means of internal entropy power $\tilde{\mathcal{P}}^i_\eta$ used in (3.7.31) and (3.7.32), let us introduce the example considered in the work of Cimmelli and Frischmuth [66], in which they have studied the thermodynamics of the model proposed by Guyer and Krumhansl [193] for the heat flux, characterized by the constitutive law

$$\dot{\mathbf{q}} + \frac{1}{\tau_R}\mathbf{q} = -c_V\nabla\theta + \tau_N\left[\nabla^2\mathbf{q} + 2\nabla(\nabla\cdot\mathbf{q})\right], \quad \tau_R > 0,\ \tau_N > 0. \tag{3.7.34}$$

For this equation, when the coefficient c_V has the form

$$c_V = \frac{c_0}{\theta^2}, \quad c_0 > 0, \tag{3.7.35}$$

an entropy extra flux can be introduced.

To derive the balance equation (3.7.27), we consider the inner product of (3.7.34) by $\mathbf{q}$, taking account of (3.7.35); thus, we obtain the quantity $\frac{1}{\theta^2}\mathbf{q}\cdot\nabla\theta$, which allows us to rewrite (3.7.27) as

$$\begin{aligned}&\rho\frac{h}{\theta} - \frac{1}{2c_0}\frac{d}{dt}\mathbf{q}^2 - \frac{1}{c_0\tau_R}\mathbf{q}^2 - \frac{\tau_N}{c_0}\left[(\nabla\mathbf{q})^2 + 2(\nabla\cdot\mathbf{q})^2\right]\\ &\quad = -\nabla\cdot\frac{\mathbf{q}}{\theta} + \rho\frac{r}{\theta} - \frac{\tau_N}{c_0}\nabla\cdot[(\nabla\mathbf{q})\,\mathbf{q} + 2(\nabla\cdot\mathbf{q})\,\mathbf{q}.\end{aligned} \tag{3.7.36}$$

---

* In the literature, examples for which it is necessary to introduce an entropy extra flux $\boldsymbol{\Phi}'$ are very few. Only materials for which there is a nonlocal constitutive equation for the heat flux $\mathbf{q}$ have need of an extra flux in (3.7.26) or in (3.7.30). It has been shown (see [122, Chapter 11.2]) that for isothermal processes, the introduction of an energy extra flux in the first law is equivalent to the inclusion of an entropy extra flux in the second law. But this equivalence holds only for isothermal processes.

Hence, it follows that the expression of $\tilde{\mathfrak{P}}^i_\eta$ considered in (3.7.32) is given by

$$\tilde{\mathfrak{P}}^i_\eta = \rho\frac{h}{\theta} - \frac{1}{2c_0}\frac{d}{dt}\mathbf{q}^2 - \frac{1}{c_0\tau_R}\mathbf{q}^2 - \frac{\tau_N}{c_0}\left[(\nabla\mathbf{q})^2 + 2(\nabla\cdot\mathbf{q})^2\right], \tag{3.7.37}$$

while the external entropy power has the form

$$\tilde{\mathfrak{P}}^e_\eta = -\nabla\cdot\frac{\mathbf{q}}{\theta} + \rho\frac{r}{\theta} - \nabla\cdot\mathbf{\Phi}'_0, \tag{3.7.38}$$

where we have put

$$\mathbf{\Phi}'_0 = \frac{\tau_N}{c_0}[(\nabla\mathbf{q})\,\mathbf{q} + 2(\nabla\cdot\mathbf{q})\,\mathbf{q}]. \tag{3.7.39}$$

Finally, from (3.7.32), using (3.7.36)–(3.7.38), we have

$$\rho\dot{\eta} \geq \tilde{\mathfrak{P}}^i_\eta \equiv \tilde{\mathfrak{P}}^e_\eta = -\nabla\cdot\frac{\mathbf{q}}{\theta} + \rho\frac{r}{\theta} - \nabla\cdot\mathbf{\Phi}'_0,$$

that is, (3.7.30) is satisfied with the entropy extra flux (3.7.39).

However, the natural form of the second law is the representation (3.7.32) with the expression (3.7.37) for $\tilde{\mathfrak{P}}^i_\eta$.

Remark 3.7.1. Further examples of nonsimple materials are discussed in some detail in Chap. 21. These include second-gradient thermoviscoelastic fluids and heat flow in a rigid conductor, with nonlocal behavior and thermal memory. The rigid conductor problem is related to the material associated with (3.7.34). A general theory of nonsimple or nonlocal materials is also proposed. The main emphasis in that chapter is to introduce generalizations of some of the free energy functionals discussed in Chaps. 10 and 11 for simple materials.

4

Free Energies and the Dissipation Principle

4.1 Axiomatic Formulation of Thermodynamics

We present in this chapter an axiomatic formulation of thermodynamics in order to introduce free energies in a very general manner and to prove certain fundamental properties of these quantities. For most of the discussion, no underlying model is assumed, in contrast to the previous chapter. However, in the context of an equivalence relation between states, we ascribe a form to the work function consistent with the general nonisothermal theory introduced in Chap. 5.

The treatment presented here is based on [104, 110] and ultimately on Noll [277] and Coleman and Owen [75, 76], who present a general axiomatic formulation of thermodynamics in which the existence of the free energy and entropy is deduced from more fundamental considerations rather than assumed a priori.

We consider a simple material, as specified by Definition 3.2.1, for which a process is defined by a map $P : [0, d_P) \mapsto \mathcal{V}$, piecewise continuous on the time interval $[0, d_P)$, where $\mathcal{V}$ is a given finite-dimensional vector space. The set of all processes is Π, while Σ is the set of all states. Also, $\hat{\varrho}$ is the evolution function, which transforms the state σ_1 under the process P into $\sigma_2 = \hat{\varrho}(\sigma_1, P)$; it obeys (3.2.3).

Remark 4.1.1. Nonsimple materials can also be included in the general formulation, in particular by generalizing the definition of P as given by (3.2.1). Examples are given in Chap. 21.

We introduce a quantity $W : \Sigma \times \Pi \mapsto \mathbb{R}$, which will be referred to as the *work* or the *work function*, and $\mathbf{\Sigma} : \Sigma \times \Pi \to \mathcal{V}$, the *response function.*

The following important property of the work function is assumed. If $P_1, P_2 \in \Pi$, then

$$W(\sigma, P_1 * P_2) = W(\sigma, P_1) + W(\hat{\varrho}(\sigma, P_1), P_2). \tag{4.1.1}$$

We can now introduce an equivalence relation, denoted by $\mathcal{R}$, between states in the space state Σ by means of the following definition, in terms of the response function.

G. Amendola et al., *Thermodynamics of Materials with Memory*,
https://doi.org/10.1007/978-3-030-80534-0_4

Definition 4.1.2. *Two states $\sigma_1, \sigma_2 \in \Sigma$ are said to be equivalent if*

$$\boldsymbol{\Sigma}(\sigma_1, P) = \boldsymbol{\Sigma}(\sigma_2, P) \tag{4.1.2}$$

for all $P \in \Pi$.

If in the space state Σ of a simple material there are equivalent states, then it is possible to consider a new state space $\Sigma_{\mathcal{R}}$ for the given material element, namely the quotient of Σ on $\mathcal{R}$. The elements of this quotient space, denoted by $\sigma_{\mathcal{R}} \in \Sigma_{\mathcal{R}}$, will be called *minimal states*. However, for the given material element, it is not necessary to use the minimal-state space $\Sigma_{\mathcal{R}}$; in fact, any definition of state that characterizes the simple material can be correctly used.

We ascribe a particular form to $W(\sigma, P)$ consistent with that introduced in Chap. 5. Let $\boldsymbol{\Lambda}(t) \in \mathcal{V}$, and let $\dot{\boldsymbol{\Lambda}}(t)$ be its time derivative. We put

$$P(t) = \dot{\boldsymbol{\Lambda}}(t), \tag{4.1.3}$$

which actually corresponds to (3.2.2), for the definition of $\boldsymbol{\Lambda}$, closely related to that introduced in Chap. 5. We shall define a state σ as $(\boldsymbol{\Lambda}^t, \boldsymbol{\Lambda}(t))$ below in Chap. 5, for materials with memory, where $\boldsymbol{\Lambda}^t$ is the history $\boldsymbol{\Lambda}(t-s)$, $s \in \mathbb{R}^{++}$.

Let

$$W(\sigma, P) = \int_0^{d_P} \boldsymbol{\Sigma}(\sigma, P_t) \cdot \dot{\boldsymbol{\Lambda}}(t) dt. \tag{4.1.4}$$

With this choice of work function, the following result can be proved.

Proposition 4.1.3. *We have*

$$W(\sigma_1, P) = W(\sigma_2, P) \qquad \forall P \in \Pi \tag{4.1.5}$$

if and only if σ_1 and σ_2 are two equivalent states.

Proof. Relation (4.1.5) states that

$$\int_0^{d_P} \boldsymbol{\Sigma}(\sigma_1, P_t) \cdot \dot{\boldsymbol{\Lambda}}(t) dt = \int_0^{d_P} \boldsymbol{\Sigma}(\sigma_2, P_t) \cdot \dot{\boldsymbol{\Lambda}}(t) dt, \tag{4.1.6}$$

which is clearly true by (4.1.2) if the two states σ_1 and σ_2 are equivalent. Conversely, if (4.1.6) holds, we have

$$\int_0^{d_P} [\boldsymbol{\Sigma}(\sigma_1, P_t) - \boldsymbol{\Sigma}(\sigma_2, P_t)] \cdot \dot{\boldsymbol{\Lambda}}(t) dt = 0,$$

for all $P \in \Pi$ and therefore for arbitrary $\dot{\boldsymbol{\Lambda}}(t)$, so that

$$\boldsymbol{\Sigma}(\sigma_1, P_t) = \boldsymbol{\Sigma}(\sigma_2, P_t), \quad t \in [0, d_p).$$

In particular, putting $t = d_P$, we have $\boldsymbol{\Sigma}(\sigma_1, P) = \boldsymbol{\Sigma}(\sigma_2, P)$, which gives (4.1.2). □

Definition 4.1.4. *A pair (σ, P) is called a cyclic process if $\hat{\varrho}(\sigma, P) = \sigma$.*

This concept is of course related to that of closed processes, defined by (2.3.12). Cyclic processes (σ, P) for material elements of a body with fading memory are realized only by means of a periodic history σ and a process P, the duration of which is equal to a finite number of periods of the history.

The second law of thermodynamics can be expressed in the following form [75].

The Dissipation Principle. *For every cyclic process* (σ, P), *the work is such that*

$$W(\sigma, P) \geq 0. \tag{4.1.7}$$

Definition 4.1.5. *A set* $\mathcal{S} \subset \Sigma$ *is invariant under* $\hat{\varrho}$ *if for every* $\sigma_1 \in \mathcal{S}$ *and* $P \in \Pi$, *the state* $\sigma = \hat{\varrho}(\sigma_1, P)$ *is in* $\mathcal{S}$.

Definition 4.1.6. *A function* $\psi : \mathcal{S}_\psi \to \mathbb{R}^+$ *is a free energy if*

(i) *the domain* $\mathcal{S}_\psi$ *is invariant under* $\hat{\varrho}$;
(ii) *for any pair* $\sigma_1, \sigma_2 \in \mathcal{S}_\psi$ *and* $P \in \Pi$ *such that* $\hat{\varrho}(\sigma_1, P) = \sigma_2$, *we have*

$$\psi(\sigma_2) - \psi(\sigma_1) \leq W(\sigma_1, P). \tag{4.1.8}$$

An important inequality follows from the expression (4.1.4) for the work and from the inequality (4.1.8). Let σ_2 in the latter relation be $\sigma(t) = \hat{\varrho}(\sigma_1, P_t)$. We put $d_P = t$. Then, under the hypothesis that $\psi(\sigma(t))$ is differentiable with respect to t, we obtain from (4.1.4)

$$\dot{\psi}(\sigma(t)) \leq \mathbf{\Sigma}(\sigma, P_t) \cdot \dot{\mathbf{\Lambda}}(t). \tag{4.1.9}$$

The explicit form (4.1.4) for the work function will not be required again in this chapter.

It is well known (see, for example, [75, 90, 105]) that there may be many free energies for a material with fading memory. The family of all free energies that are possible for the material element under consideration is denoted by $\mathcal{F}$. If $\psi_1, \psi_2 \in \mathcal{F}$, then $\mathcal{S}_{\psi_1} \cap \mathcal{S}_{\psi_2}$ is invariant under $\hat{\varrho}$. It is assumed that $\mathcal{S}_{\psi_1} \cap \mathcal{S}_{\psi_2}$ is nonempty under $\hat{\varrho}$. This will be true in particular if $\mathcal{S}_{\psi_1} \subseteq \mathcal{S}_{\psi_2}$ or $\mathcal{S}_{\psi_2} \subseteq \mathcal{S}_{\psi_1}$. A fundamental property of this set is now proved [121, 266].

Proposition 4.1.7. *The set* $\mathcal{F}$ *is convex.*

Proof. Let $\psi_1, \psi_2 \in \mathcal{F}$, and consider

$$\psi = \alpha\psi_1 + (1 - \alpha)\psi_2, \qquad \alpha \in [0, 1]. \tag{4.1.10}$$

Its domain is $\mathcal{S}_{\psi_1} \cap \mathcal{S}_{\psi_2}$. It is nonnegative and obeys the inequality (4.1.8). Thus, it is a free energy of the material element. □

It is assumed that the set $\mathcal{F}$ is bounded, so it will have a minimum and a maximum element.

We now consider the problem of the existence of a minimum free energy and study its properties.

Let

$$\mathcal{W}(\sigma) = \{W(\sigma, P);\ P \in \Pi\} \tag{4.1.11}$$

be the set with elements given by the work done by starting from σ and applying any process $P \in \Pi$.

The Strong Dissipation Principle. *The set $\mathcal{W}(\sigma)$ is bounded below for all $\sigma \in \Sigma$. Furthermore, there is a unique state $\sigma^\dagger$, which we refer to as the zero state, with the property that*

$$\inf \mathcal{W}(\sigma^\dagger) = 0. \tag{4.1.12}$$

Definition 4.1.8. *A state $\sigma \in \Sigma$ is attainable from all of Σ if for any initial state σ^i, there exists a process $P \in \Pi$ such that $\hat{\varrho}(\sigma^i, P) = \sigma$. A simple material system is attainable if every state σ is attainable from every other state $\sigma' \in \Sigma$.*

Remark 4.1.9. Materials with fading memory are in general not attainable. Indeed, we see that a state $\sigma(t)$, defined by a history up to time t, is attainable from a history $\sigma_0(t_0)$, $t_0 < t$, only if the history up to time t_0 in $\sigma(t)$ is the same as the history in $\sigma_0(t_0)$.

The following theorem establishes the connection between the dissipation principle and the strong dissipation principle [104].

Theorem 4.1.10. *The dissipation principle follows from the strong dissipation principle.*

Proof. Firstly, for any pair $\sigma, \sigma' \in \Sigma$, we consider the quantity

$$M(\sigma, \sigma') := \inf\{W(\sigma, P);\ P \in \Pi,\ \hat{\varrho}(\sigma, P) = \sigma'\}, \tag{4.1.13}$$

which, by virtue of the strong dissipation principle, is bounded below. Thus, there exists M_σ such that

$$M(\sigma, \sigma') \geq M_\sigma > -\infty \quad \forall \sigma \in \Sigma.$$

If $(\sigma, P) \in \Sigma \times \Pi$ is a cycle, then so is the pair $(\sigma, P * P)$. Putting $P^n = P * P * \cdots * P$, we see that (σ, P^n) is a cycle for any integer n. Thus, taking into account (4.1.1) and (4.1.13), we have

$$W(\sigma, P^n) = nW(\sigma, P) \geq M(\sigma, \sigma) \quad \forall n \in \mathbb{N}.$$

Hence, it follows that $W(\sigma, P)$ is nonnegative, which proves (4.1.7), since otherwise, the condition $M(\sigma, \sigma) > -\infty$ would not be satisfied. Furthermore, $M(\sigma, \sigma)$ vanishes, since it is the infimum over a set of nonnegative numbers that contains zero because there exists the trivial process P_0, which satisfies $\hat{\varrho}(\sigma, P_0) = \sigma$ and gives $W(\sigma, P_0) = 0$. □

The dissipation principle therefore follows from the strong dissipation principle. We shall adopt the strong dissipation principle in the next section.

Let us introduce the set

$$\Phi := \left\{\phi : \Sigma \to \mathbb{R}^+; \phi(\sigma) \leq \psi(\sigma)\ \forall \psi \in \mathcal{F},\ \forall \sigma \in \mathcal{S}_\psi;\ \phi(\sigma^\dagger) = 0\right\} \tag{4.1.14}$$

and the function $\phi_M : \Sigma \to \mathbb{R}^+$ defined as

$$\phi_M(\sigma) = \sup\{\phi(\sigma); \phi \in \Phi\}. \tag{4.1.15}$$

The functional ϕ_M is the largest one with the property that it is less than or equal to every free energy for all states.

We also define $\tilde{\mathcal{W}}$ as

$$\tilde{\mathcal{W}}(\sigma) := \{W(\sigma, P) - \phi_M(\hat{\varrho}(\sigma, P));\ P \in \Pi\}. \tag{4.1.16}$$

4.2 Minimum and Maximum Free Energies

Definition 4.2.1. *A functional ψ_m is the minimum free energy if*

(i) ψ_m is a free energy with domain $\mathcal{S}_\psi = \Sigma$,
(ii) the zero state $\sigma^\dagger \in \Sigma$ is such that $\psi_m(\sigma^\dagger) = 0$, and
(iii) for any free energy $\psi : \mathcal{S}_\psi \to \mathbb{R}^+$ such that $\sigma^\dagger \in \mathcal{S}_\psi$ and $\psi(\sigma^\dagger) = 0$, we have

$$\psi(\sigma) \geq \psi_m(\sigma) \quad \forall \sigma \in \mathcal{S}_\psi. \tag{4.2.1}$$

Proposition 4.2.2. *The minimum free energy, if it exists, is unique.*

Proof. Suppose we have two free energies ψ_{m_1} and ψ_{m_2} with $\mathcal{S}_\psi = \Sigma$. For any $\sigma \in \Sigma$, from (4.2.1), we should have $\psi_{m_1}(\sigma) \geq \psi_{m_2}(\sigma)$ by considering $\psi_{m_2}(\sigma)$ as the minimum. Taking $\psi_{m_1}(\sigma)$ as the minimum, we have $\psi_{m_2}(\sigma) \geq \psi_{m_1}(\sigma)$. Hence, $\psi_{m_1}(\sigma) = \psi_{m_2}(\sigma)$ for any $\sigma \in \Sigma$. □

The following result characterizes the minimum free energy [104].

Theorem 4.2.3. *The functional*

$$\psi_m(\sigma) := -\inf \mathcal{W}(\sigma), \tag{4.2.2}$$

the quantity $\mathcal{W}$ being defined by (4.1.11), *is the minimum free energy.*

The proof is a corollary of the following lemmas.

Lemma 4.2.4. *For every state $\sigma \in \Sigma$,*

$$N(\sigma) := -\inf \mathcal{W}(\sigma) = \sup\{-W(\sigma, P);\ P \in \Pi\} \tag{4.2.3}$$

is finite, and moreover, the functional

$$\psi_m(\sigma) := N(\sigma) \tag{4.2.4}$$

is a free energy defined on Σ.

Proof. It follows immediately from the strong dissipation principle that $\{-W(\sigma, P);$ $P \in \Pi\}$ is bounded above for all $\sigma \in \Sigma$. Hence, $N(\sigma)$ is defined on Σ, and requirement *(i)* of definition 4.1.6 is satisfied. For requirement *(ii)*, let σ_1 and σ_2 be two states and P' a process such that $\hat{\varrho}(\sigma_1, P') = \sigma_2$. Then for any $P \in \Pi$, (4.2.3) yields

$$-W(\sigma_1, P' * P) \le N(\sigma_1). \tag{4.2.5}$$

We also have that for any $\varepsilon > 0$, there exists a process P_ε such that

$$-W(\sigma_2, P_\varepsilon) > N(\sigma_2) - \varepsilon. \tag{4.2.6}$$

Thus, from (4.2.4)–(4.2.6), we obtain

$$\begin{aligned}\psi_m(\sigma_2) - \psi_m(\sigma_1) - \varepsilon &\le -W(\sigma_2, P_\varepsilon) + W(\sigma_1, P' * P)\\ &= -W(\sigma_2, P_\varepsilon) + W(\sigma_1, P') + W(\sigma_2, P).\end{aligned}$$

The arbitrariness of P and ε gives the desired result (4.1.8) on putting $P = P_\varepsilon$. □

Lemma 4.2.5. *The free energy ψ_m, as defined by (4.2.4), is nonnegative,*

$$\psi_m(\sigma) \ge 0 \quad \forall \sigma \in \Sigma, \tag{4.2.7}$$

and $\psi_m(\sigma) = 0$ if and only if $\sigma = \sigma^\dagger$.

Proof. If (4.2.7) holds, the set $\{-W(\sigma, P);\ P \in \Pi\}$ contains the zero value, since for any zero process P_0, we have $W(\sigma, P_0) = 0$; therefore, the supremum of such a set is nonnegative.

Let

$$\psi_m(\sigma^\dagger) = \sup\left\{-W(\sigma^\dagger, P);\ P \in \Pi\right\} = \inf \mathcal{W}(\sigma^\dagger).$$

Then (4.1.12) gives that $\psi_m(\sigma^\dagger) = 0$. The converse follows by the uniqueness of $\sigma^\dagger$ obeying (4.1.12). □

Lemma 4.2.6. *Any free energy $\psi : \mathcal{S}_\psi \to \mathbb{R}^+$ with $\sigma^\dagger \in \mathcal{S}_\psi$ and $\psi(\sigma^\dagger) = 0$ satisfies*

$$\psi_m(\sigma) \le \psi(\sigma) \quad \forall \sigma \in \mathcal{S}_\psi. \tag{4.2.8}$$

Proof. Let $\sigma \in \mathcal{S}_\psi$. For every $\varepsilon > 0$, there exists a process P_ε such that

$$\psi_m(\sigma) < -W(\sigma, P_\varepsilon) + \varepsilon. \tag{4.2.9}$$

Putting $\sigma_\varepsilon = \hat{\varrho}(\sigma, P_\varepsilon)$, we obtain, for any free energy ψ,

$$\psi(\sigma_\varepsilon) - \psi(\sigma) \le W(\sigma, P_\varepsilon). \tag{4.2.10}$$

Use of (4.2.9) and (4.2.10) gives

$$\psi_m(\sigma) < -W(\sigma, P_\varepsilon) + \varepsilon \le -\psi(\sigma_\varepsilon) + \psi(\sigma) + \varepsilon < \psi(\sigma) + \varepsilon,$$

from which, ε being arbitrary, we obtain (4.2.8). □

Recalling the definitions of $\phi_M(\sigma)$ and $\tilde{\mathcal{W}}(\sigma)$, given by (4.1.15) and (4.1.16), we have the following result [110].

Remark 4.2.7. We shall assume that $\inf \tilde{\mathcal{W}}$, where $\tilde{\mathcal{W}}$ is defined by (4.1.16), exists for all $\sigma \in \Sigma$. This is not guaranteed by the strong dissipation principle since $\phi_M(\hat{\varrho}(\sigma, P))$ must also be restricted.

Theorem 4.2.8. *The functional*

$$\tilde{\psi}_m(\sigma) := -\inf \tilde{\mathcal{W}}(\sigma) \tag{4.2.11}$$

is a free energy such that $\tilde{\psi}_m(\sigma) = \psi_m(\sigma)\ \forall\sigma \in \Sigma$.

Proof. We can write (4.2.11) in the equivalent form

$$\tilde{\psi}_m(\sigma) = \sup\{-W(\sigma, P) + \phi_M(\hat{\varrho}(\sigma, P));\ P \in \Pi\}. \tag{4.2.12}$$

Since for the zero process P_0, we have $W(\sigma, P_0) = 0$, $\phi_M(\hat{\varrho}(\sigma, P_0))$ is nonnegative, so that $\tilde{\psi}_m(\sigma) \geq 0$. We also have $\psi_m(\sigma^\dagger) = 0$.

Given two states σ_1 and σ_2 and a process P' such that $\hat{\varrho}(\sigma_1, P') = \sigma_2$, from (4.2.12), we have

$$\tilde{\psi}_m(\sigma_1) \geq -W(\sigma_1, P' * P) + \phi_M(\hat{\varrho}(\sigma_1, P' * P)) \tag{4.2.13}$$

for any $P \in \Pi$; moreover, for any $\varepsilon > 0$, there exists a process P_ε such that

$$\tilde{\psi}_m(\sigma_2) - \varepsilon < -W(\sigma_2, P_\varepsilon) + \phi_M(\hat{\varrho}(\sigma_2, P_\varepsilon)). \tag{4.2.14}$$

Comparison of (4.2.13) and (4.2.14) yields

$$\begin{aligned}
&\tilde{\psi}_m(\sigma_2) - \tilde{\psi}_m(\sigma_1) - \varepsilon \\
&\qquad < -W(\sigma_2, P_\varepsilon) + \phi_M(\hat{\varrho}(\sigma_2, P_\varepsilon)) + W(\sigma_1, P' * P) - \phi_M(\hat{\varrho}(\sigma_1, P' * P)) \\
&\qquad = -W(\sigma_2, P_\varepsilon) + \phi_M(\hat{\varrho}(\sigma_2, P_\varepsilon)) + W(\sigma_1, P') + W(\sigma_2, P) - \phi_M(\hat{\varrho}(\sigma_2, P)),
\end{aligned}$$

by virtue of (4.1.1). Identifying P_ε and P, the latter being arbitrary, we deduce that

$$\tilde{\psi}_m(\sigma_2) - \tilde{\psi}_m(\sigma_1) \leq W(\sigma_1, P'),$$

which completes the proof that $\tilde{\psi}_m(\sigma)$ is a free energy.

We must show that $\psi_m(\sigma) = \tilde{\psi}_m(\sigma)$. The quantity ϕ_M being nonnegative, $W(\sigma, P) \geq W(\sigma, P) - \phi_M(\hat{\varrho}(\sigma, P))\ \forall P \in \Pi$, and hence $\inf \mathcal{W}(\sigma) \geq \inf \tilde{\mathcal{W}}(\sigma)$ or

$$\psi_m(\sigma) \leq \tilde{\psi}_m(\sigma) \quad \forall\sigma \in \Sigma. \tag{4.2.15}$$

Let $\sigma \in \mathcal{S}_\psi$. There exists a process P_ε, for every $\varepsilon > 0$, such that

$$\tilde{\psi}_m(\sigma) < -W(\sigma, P_\varepsilon) + \phi_M(\hat{\varrho}(\sigma, P_\varepsilon)) + \varepsilon, \tag{4.2.16}$$

and putting $\sigma_\varepsilon = \hat{\varrho}(\sigma, P_\varepsilon)$, for any free energy ψ, we have

$$\psi(\sigma_\varepsilon) - \psi(\sigma) \le W(\sigma, P_\varepsilon).$$

Comparison of these inequalities yields

$$\begin{aligned}\tilde{\psi}_m(\sigma) &< -W(\sigma, P_\varepsilon) + \phi_M(\sigma_\varepsilon) + \varepsilon \\ &\le -\psi(\sigma_\varepsilon) + \psi(\sigma) + \phi_M(\sigma_\varepsilon) + \varepsilon \\ &\le \psi(\sigma) + \varepsilon\end{aligned}$$

because by definition, $\phi_M(\sigma_\varepsilon) \le \psi(\sigma_\varepsilon)$. From the arbitrariness of ε, it follows that

$$\tilde{\psi}_m(\sigma) \le \psi(\sigma) \tag{4.2.17}$$

for all ψ and, in particular, for $\psi = \psi_m$. Thus, (4.2.15) and (4.2.17) yield $\tilde{\psi}_m(\sigma) = \psi_m(\sigma)\ \forall \sigma \in \Sigma$, which completes the proof. □

Remark 4.2.9. A similar result holds for any choice of $\phi \in \Phi$ defined by (4.1.14). On examining the last part of the proof of Theorem 4.2.8, we can in fact weaken considerably the constraint on the functions $\phi \in \Phi$ that they must be less than or equal to all free energies for all states.

Corollary 4.2.10. *If* $\exists\, \varepsilon_0 > 0$ *such that* $\varepsilon < \varepsilon_0$,

$$\phi_M(\hat{\varrho}(\sigma, P_\varepsilon)) \le \psi(\hat{\varrho}(\sigma, P_\varepsilon)) \quad \forall \psi \in \mathcal{F},\ \sigma \in \mathcal{F}_\psi, \tag{4.2.18}$$

where P_ε *is defined by (4.2.16), then Theorem 4.2.8 holds with* $\tilde{\mathcal{W}}(\sigma)$ *defined by (4.1.16), though now* ϕ_M *is constrained only by (4.2.18) rather than by (4.1.15).*

Thus, in fact, the property must hold only for the final states of processes in the vicinity of the optimal process.

Remark 4.2.11. It is always possible to represent the minimum free energy as a function of the minimal state σ_R. From the definition of $\mathcal{W}(\sigma)$, given by (4.1.11), and from the fact that $W(\sigma, P) = W(\sigma_R, P)$ for all $P \in \Pi$, which follows from Proposition 4.1.3, it is clear that

$$\inf \mathcal{W}(\sigma) = \inf \tilde{\mathcal{W}}(\sigma_R).$$

Thus, if $\sigma \in \sigma_R$, we have

$$\psi_m(\sigma) = \psi_m(\sigma_R). \tag{4.2.19}$$

Therefore, the minimum free energy is independent of the representation of state that we use.

Let us denote by Σ_σ the set of all $\sigma' \in \Sigma$ attainable from σ,

$$\Sigma_\sigma = \{\sigma' \in \Sigma;\ \exists\, P \in \Pi,\ \sigma' = \hat{\varrho}(\sigma, P)\}.$$

Let $\sigma_0, \sigma \in \Sigma$ be any pair of states such that $\sigma \in \Sigma_{\sigma_0}$. We introduce the set

$$N(\sigma_0, \sigma) = \{W(\sigma_0, P);\quad \forall \quad P \in \Pi,\ \hat{\varrho}(\sigma_0, P) = \sigma\}. \tag{4.2.20}$$

This set is bounded below by virtue of the strong dissipation principle.

Theorem 4.2.12. *For any fixed σ^i, the functional $\psi_M^{\sigma^i} : \Sigma_{\sigma^i} \to \mathbb{R}^+$, defined by*

$$\psi_M^{\sigma^i}(\sigma) = \inf N(\sigma^i; \sigma) + \psi_m(\sigma^i),$$

is a free energy, called a maximum free energy. For any free energy $\psi : \mathcal{S}_\psi \to \mathbb{R}^+$ such that $\mathcal{S}_\psi \supset \Sigma_{\sigma^i}$ and $\psi(\sigma^i) = \psi_m(\sigma^i)$, we have

$$\psi(\sigma) \leq \psi_M^{\sigma^i}(\sigma), \qquad \forall\, \sigma \in \Sigma_{\sigma^i}. \tag{4.2.21}$$

Proof. The functional $\psi_M^{\sigma^i}$ is clearly well defined on the set Σ_{σ^i}, which is invariant under $\hat{\varrho}$. Let $\sigma_1, \sigma_2 \in \Sigma_{\sigma^i}$ be a pair of states and $P \in \Pi$ a process such that $\sigma_2 = \hat{\varrho}(\sigma_1, P)$; then, for any $\varepsilon > 0$, there is a process P_1^ε such that

$$\hat{\varrho}(\sigma^i, P_1^\varepsilon) = \sigma_1, \quad \psi_M^{\sigma^i}(\sigma_1) > W(\sigma^i, P_1^\varepsilon) + \psi_m(\sigma^i) - \varepsilon \tag{4.2.22}$$

and

$$\psi_M^{\sigma^i}(\sigma_2) \leq W(\sigma^i, P_1^\varepsilon * P) + \psi_m(\sigma^i). \tag{4.2.23}$$

The inequalities (4.2.22) and (4.2.23) give

$$\psi_M^{\sigma^i}(\sigma_2) - \psi_M^{\sigma^i}(\sigma_1) \leq W(\sigma_1, P).$$

Let $\psi : \mathcal{S}_\psi \to \mathbb{R}^+$ be a free energy such that $\mathcal{S}_\psi \supset \Sigma_{\sigma^i}$ and $\psi(\sigma^i) = \psi_m(\sigma^i) = \psi_M^{\sigma^i}(\sigma^i)$. For any $\varepsilon > 0$, there exists a P_ε such that $\hat{\varrho}(\sigma^i, P_\varepsilon) = \sigma$ and $W(\sigma^i, P_\varepsilon) + \psi_m(\sigma^i) < \psi_M^{\sigma^i}(\sigma) + \varepsilon$. Moreover, ψ being a free energy, we have $\psi(\sigma) \leq W(\sigma^i, P_\varepsilon) + \psi(\sigma^i) = W(\sigma^i, P_\varepsilon) + \psi_m(\sigma^i)$, and hence (4.2.21) follows because of the arbitrariness of ε. □

Remark 4.2.13. Of course, for any $\sigma^i \in \Sigma$, we may obtain a different free energy. Moreover, for a fixed $\sigma^i \in \Sigma$, the definition of maximum free energy may depend on the definition of state. We can, however, construct a maximum free energy that is defined on the space of minimal states. In other words, if we consider the definition of minimal state, then (4.2.20) is replaced by

$$N(\sigma_{0R}, \sigma_R) = \{W(\sigma_{0R}, P);\ \forall\, P \in \Pi,\ \hat{\varrho}(\sigma_{0R}, P) = \sigma_R\}.$$

Since this set is generally larger than $N(\sigma_0, \sigma)$, if $\sigma_0 \in \sigma_{0R}$ and $\sigma \in \sigma_R$, the maximum free energy, defined on Σ_R as

$$\psi_M^{\sigma_R^i}(\sigma_R) = \inf N(\sigma_R^i, \sigma_R) + \psi_m(\sigma_R^i),$$

satisfies the following inequality:

$$\psi_M^{\sigma_R^i}(\sigma_R) \leq \psi_M^{\sigma^i}(\sigma),\ \sigma^i \in \sigma_R^i,\ \sigma \in \sigma_R.$$

Relation (4.2.21) will apply to any free energy $\psi(\sigma_R)$ defined on Σ_R, provided $\psi(\sigma_R^i) = \psi_m(\sigma_R^i)$.

5

Thermodynamics of Materials with Memory

We now apply thermodynamic principles to field theories with memory. For general nonlinear, nonisothermal theories, we assume that a free energy is given, this being the fundamental constitutive assumption. Applying a generalization of the approach of Coleman [67], Coleman and Mizel [71], and Gurtin and Pipkin [191], we derive the constitutive equations for the theory in Sect. 5.1. Also, fundamental properties of free energies are derived. Furthermore, some observations are made on the case of periodic histories and in relation to constraints on the nonuniqueness of free energies. In Sect. 5.2, an expression for the maximum recoverable work is given for general materials, together with an integral equation for the process yielding this maximum. Finally, in Sect. 5.3, we discuss how free energies can be constructed from combinations of simpler free energies.

In Part III, the entire emphasis is on determining suitable explicit forms of free energies for materials with memory. All these involve quadratic functionals of histories.

5.1 Derivation of the Constitutive Equations

Let us begin by stating the first and second laws of thermodynamics. The theory is developed in terms of the material description.

The internal energy per unit mass and the entropy per unit mass at $(\mathbf{X}, t)$, both scalar quantities, are denoted, respectively, by $e(\mathbf{X}, t)$ and $\eta(\mathbf{X}, t)$. The local absolute temperature is $\theta(\mathbf{X}, t) \in \mathbb{R}^+$. The Piola–Kirchhoff heat flux vector, defined by (3.1.1), is denoted by $\mathbf{q}(\mathbf{X}, t) \in \mathbb{R}^3$; the subscript L is dropped. We introduce a variable $\kappa(\mathbf{X}, t)$, which is the coldness, given by $1/\theta > 0$. All these quantities can also be expressed in terms of the current position $\mathbf{x}$ and time t. The quantity $\mathbf{g}_L \in \mathbb{R}^3$ is defined by (3.2.2). The subscript L will now be dropped. We also introduce $\mathbf{d} \in \mathbb{R}^3$ as the gradient of κ, so that

$$\mathbf{g} = \nabla_{\mathbf{X}}\theta, \qquad \mathbf{d} = \nabla_{\mathbf{X}}\kappa = -\frac{1}{\theta^2}\mathbf{g}, \tag{5.1.1}$$

G. Amendola et al., *Thermodynamics of Materials with Memory*,
https://doi.org/10.1007/978-3-030-80534-0_5

where the gradient $\nabla_{\mathbf{X}}$, here and below, is with respect to $\mathbf{X}$. The energy balance equation or the first law of thermodynamics has the form (see (3.3.8))

$$p - \rho\dot{e} - \mathrm{Div}_{\mathbf{X}}\mathbf{q} + \rho r = 0, \qquad p = \widehat{\mathbf{S}} \cdot \dot{\mathbf{E}}. \tag{5.1.2}$$

The quantity $\rho(\mathbf{X})$ (denoted by ρ_0 in Chap. 3) is the mass density in the material configuration, and r is the external radiation absorbed per unit time, per unit mass at $(\mathbf{X}, t)$. We write the second law of thermodynamics as

$$D = \dot{\eta} + \frac{1}{\rho}\mathrm{Div}_{\mathbf{X}}\mathbf{j}_\eta - \kappa r \geq 0, \tag{5.1.3}$$

where $\mathbf{j}_\eta$ is the entropy flux, which in the present work will be taken to have the equilibrium form (see (3.4.15))

$$\mathbf{j}_\eta = \kappa\mathbf{q}. \tag{5.1.4}$$

The quantity $D(\mathbf{X}, t)$ is the *total rate of entropy production* per unit mass.

The superimposed dot notation in (5.1.2), (5.1.3), and below indicates the material time derivative, i.e., holding $\mathbf{X}$ constant.

The quantities $\widehat{\mathbf{S}}$, $\mathbf{E}$, $\mathbf{q}$, and $\mathbf{g}$ are unaffected by a time-dependent coordinate transformation in $\mathbf{x}$. Thus, their components are objective scalars.

The Helmholtz free energy per unit mass, introduced in Sect. 3.4.2, is defined by

$$\psi_H = e - \theta\eta.$$

In terms of this quantity, we can write (5.1.2) as

$$\frac{p}{\rho} - \dot{\psi}_H - \eta\dot{\theta} - \frac{1}{\rho\theta}\mathbf{q} \cdot \mathbf{g} = \theta D. \tag{5.1.5}$$

The fact that D is nonnegative means that (5.1.5) implies the Clausius–Duhem inequality (3.4.16). Relation (5.1.5) is unsatisfactory for materials with memory because, as discussed in Chap. 4 for free energies, neither ψ_H nor η is in general uniquely defined. On the other hand, e is always uniquely defined. It is therefore advantageous to modify the above standard formulation. We introduce the following free energy:

$$\psi = \kappa e - \eta = \kappa\psi_H. \tag{5.1.6}$$

Instead of (5.1.5), we now have

$$\kappa\frac{p}{\rho} - \dot{\psi} + e\dot{\kappa} + \frac{1}{\rho}\mathbf{d} \cdot \mathbf{q} = D \geq 0. \tag{5.1.7}$$

Observe that in this formulation, the natural temperature variable is κ rather than θ.

In [159], an integral of $\mathbf{g}$ was adopted as an independent variable, in a modification of the approach of [191, 260]. It was pointed out in [159] that an alternative and arguably better procedure would be to adopt the quantity $\mathbf{m} \in \mathbb{R}^3$, defined by

$$\mathbf{m}(t) = \int_0^t \mathbf{q}(u)du, \qquad \dot{\mathbf{m}}(t) = \mathbf{q}(t), \tag{5.1.8}$$

as an independent variable. This choice has the advantage that $\mathbf{m}$ has an immediate physical interpretation in that $\mathbf{m} \cdot \mathbf{a}$ is the cumulative heat flow over the time interval $[0, t]$, through a unit area perpendicular to the unit vector $\mathbf{a}$ in the reference configuration. For the general development of the theory, we shall adopt $\mathbf{m}$ as an independent variable rather than the time integral of $\mathbf{g}$, though it is an easy matter to exchange their roles where required.

We shall be assuming that the free energy is a functional of the history of this quantity, namely $\mathbf{m}^t(s) = \mathbf{m}(t-s)$, $s > 0$, or specifically a functional of

$$\mathbf{m}(t) - \mathbf{m}^t(s) = \int_{t-s}^{t} \mathbf{q}(u)du, \tag{5.1.9}$$

with no separate dependence on $\mathbf{m}(t)$; such dependence cannot occur, because $\mathbf{m}(t)$ depends on the choice of the time origin. This is essentially the basis of the approach in [191], though in that reference, $\mathbf{g}$ is used in (5.1.8) rather than $\mathbf{q}$.

Modifying and extending the compact notation of Coleman [67], we introduce $\mathbf{\Lambda} : \mathbb{R} \mapsto \Gamma^+, \mathbf{\Sigma} : \mathbb{R} \mapsto \Gamma^+$, where

$$\begin{aligned}
\mathbf{\Lambda}(t) &= (\mathbf{E}(t), \kappa(t), \mathbf{m}(t)), \\
\mathbf{\Sigma}(t) &= \frac{1}{\rho}\left(\kappa\widehat{\mathbf{S}}(t), \rho e(t), \mathbf{d}(t)\right), \\
\Gamma^+ &= \mathrm{Sym} \times \mathbb{R}^+ \times \mathbb{R}^3.
\end{aligned} \tag{5.1.10}$$

We assume that $\mathbf{\Lambda}$ is continuously differentiable. In terms of this notation, $(5.1.7)_1$ can be written as

$$\dot{\psi} + D = \mathbf{\Sigma} \cdot \dot{\mathbf{\Lambda}} \tag{5.1.11}$$

with $\dot{\mathbf{\Lambda}} \in \Gamma$, where*

$$\Gamma = \mathrm{Sym} \times \mathbb{R} \times \mathbb{R}^3. \tag{5.1.12}$$

The dot product here denotes a scalar product in the space Γ (Appendix A). This relation is an expression of the first law and indicates that of the work done on the material element per unit time, some is stored ($\dot{\psi}$) and the rest is dissipated (D). The second law is imposed through the Clausius–Duhem inequality $(5.1.7)_2$ or

$$D = \mathbf{\Sigma} \cdot \dot{\mathbf{\Lambda}} - \dot{\psi} \geq 0, \tag{5.1.13}$$

where D is defined by (5.1.3) with (5.1.4).

Remark 5.1.1. Following (4.1.3), we choose $\dot{\mathbf{\Lambda}}(t)$ to be the thermodynamic process in later chapters. If the time integral of $\mathbf{g}$, rather than $\mathbf{q}$, is used in (5.1.8) and $\theta(t)$ replaces $\kappa(t)$, then this corresponds to (3.2.2).

---

* The general developments described in later chapters go through for Γ understood to be any finite-dimensional vector space, rather than just that defined by (5.1.12) and therefore may apply to a variety of other physical applications, for example, those involving electromagnetic fields.

Let $\mathbf{\Lambda}^t : \mathbb{R}^{++} \mapsto \Gamma^+$ be defined by

$$\mathbf{\Lambda}^t(s) = \mathbf{\Lambda}(t-s), \quad s \in \mathbb{R}^{++}. \tag{5.1.14}$$

We assume that these belong to a real Hilbert space $\mathcal{H}$ of functions with values in Γ^+, possessing a suitable inner product and fading memory norm, the latter denoted by $\|\cdot\|$ [67, 73]. A constitutive assumption is now made by requiring that the free energy ψ depends in a specified way on the history and current value of $\mathbf{\Lambda}$. We put

$$\psi(t) = \widetilde{\psi}\left(\mathbf{\Lambda}^t, \mathbf{\Lambda}(t)\right), \tag{5.1.15}$$

denoting that $\widetilde{\psi}$ is a functional of the history $\mathbf{\Lambda}^t$ and depends also on the current value $\mathbf{\Lambda}(t)$. Note that since all components of $\mathbf{\Lambda}^t, \mathbf{\Lambda}$ are objective scalars, we have automatically that ψ is an objective scalar.

To eliminate the arbitrariness of an additive constant, which is a feature of all physical energies, we impose the condition

$$\widetilde{\psi}\left(\mathbf{0}^\dagger, \mathbf{0}\right) = 0, \tag{5.1.16}$$

where $\mathbf{0}$ is the zero and $\mathbf{0}^\dagger$ is the zero history, both in Γ^+.

We now summarize the core argument of rational thermodynamics [67]. The free energy given by (5.1.15) is constrained to obey the second law (5.1.13) for arbitrary choices of $\mathbf{\Lambda}^t \in \mathcal{H}$. This yields constitutive equations and an expression for the rate of dissipation. The balance laws (1.3.2) and (1.3.50) with (1.3.52) (or, in the Eulerian description, (1.3.25)) and (5.1.2) must hold, and these involve elements of $\mathbf{\Lambda}^t$. This can always be arranged by suitable choices of body forces $\mathbf{b}$ in (1.3.50) and the external radiation r in (5.1.2).

An important aspect of this approach is that balance laws, which have universal application, are clearly separated from constitutive equations that apply to particular categories of materials.

Of course, once constitutive equations have been established from such general arguments, these relations, combined with the balance laws and suitable boundary and initial conditions, can be used to determine $\mathbf{\Lambda}^t$ for specified choices of $\mathbf{b}$ and r.

Assuming that $\widetilde{\psi}$ is differentiable with respect to $\mathbf{\Lambda}(t)$ and Fréchet differentiable with respect to $\mathbf{\Lambda}^t$ within $\mathcal{H}$ (fading memory principle [67, 73], generalized in [71]), we can apply the chain rule to obtain

$$\frac{d}{dt}\widetilde{\psi}\left(\mathbf{\Lambda}^t, \mathbf{\Lambda}(t)\right) = \partial_{\mathbf{\Lambda}}\widetilde{\psi}\left(\mathbf{\Lambda}^t, \mathbf{\Lambda}(t)\right) \cdot \dot{\mathbf{\Lambda}}(t) + \delta\widetilde{\psi}\left(\mathbf{\Lambda}^t, \mathbf{\Lambda}(t) \mid \frac{\partial}{\partial t}\mathbf{\Lambda}^t\right), \tag{5.1.17}$$

where $\partial_{\mathbf{\Lambda}}$ indicates the derivative of $\widetilde{\psi}$ with respect to the current value $\mathbf{\Lambda}(t)$ and $\delta\widetilde{\psi}$ is the Fréchet differential of $\widetilde{\psi}$ at $\mathbf{\Lambda}^t$ in the direction $\partial\mathbf{\Lambda}^t/\partial t$, where

$$\frac{\partial}{\partial t}\mathbf{\Lambda}^t(s) = -\frac{\partial}{\partial s}\mathbf{\Lambda}^t(s).$$

The functional $\delta\widetilde{\psi}$ is linear in $\partial\mathbf{\Lambda}^t/\partial t$. These derivatives with respect to field quantities are assumed to be continuous in their arguments. It follows from (5.1.17) combined with (5.1.13), by virtue of a standard argument [67], that

$$\Sigma(t) = \widetilde{\Sigma}\left(\Lambda^t, \Lambda(t)\right) = \partial_{\Lambda}\widetilde{\psi}\left(\Lambda^t, \Lambda(t)\right), \tag{5.1.18}$$

which are the constitutive equations of the material and are objective relationships; in fact, all components of both sides are objective scalars. Also, with the aid of (5.1.11),

$$D(t) = -\delta\widetilde{\psi}\left(\Lambda^t, \Lambda(t) \mid \frac{\partial}{\partial t}\Lambda^t\right). \tag{5.1.19}$$

Recalling that the free energy functional depends on $\mathbf{m}(t) - \mathbf{m}^t(s)$, with no separate dependence on $\mathbf{m}(t)$, we shall generally, except in Sect. 5.1.3, write (5.1.15) in the form

$$\psi(t) = \widetilde{\psi}_d\left(\Lambda_r^t, \Lambda_0(t)\right), \tag{5.1.20}$$

where

$$\Lambda_r^t(s) = \Lambda^t(s) - \Lambda(t), \qquad \Lambda_0(t) = (\mathbf{E}(t), \kappa(t), \mathbf{0}_3), \quad \Lambda_r^t,\ \Lambda_0 \in \Gamma, \tag{5.1.21}$$

and the quantity $\mathbf{0}_3$ indicates the zero in $\mathbb{R}^3$. Note that (5.1.17) can be written in terms of $\widetilde{\psi}_d$ as follows:

$$\frac{d}{dt}\widetilde{\psi}_d\left(\Lambda_r^t, \Lambda_0(t)\right) = \partial_{\Lambda}\widetilde{\psi}_d\left(\Lambda_r^t, \Lambda_0(t)\right) \cdot \dot{\Lambda} + \delta_r\widetilde{\psi}_d\left(\Lambda_r^t, \Lambda_0(t) \mid \frac{\partial}{\partial t}\Lambda_r^t\right),$$

where $\delta_r\widetilde{\psi}_d$ is the Fréchet differential of $\widetilde{\psi}_d$ at Λ_r^t in the direction $\partial\Lambda_r^t/\partial t$ with

$$\frac{\partial}{\partial t}\Lambda_r^t(s) = \frac{\partial}{\partial t}\Lambda^t(s) - \dot{\Lambda}(t).$$

The quantity $\partial_{\Lambda}\widetilde{\psi}_d\,(\Lambda_r^t, \Lambda_0(t)) \cdot \dot{\Lambda}$ will have zero in the $\mathbb{R}^3$ contribution. Using the linearity of $\delta_r\widetilde{\psi}_d$ with respect to $\partial\Lambda_r^t/\partial t$, we have that (5.1.18) can be written as follows:

$$\Sigma(t) = \partial_{\Lambda}\widetilde{\psi}_d\left(\Lambda_r^t, \Lambda_0(t)\right) - \delta_c\widetilde{\psi}_d\left(\Lambda_r^t, \Lambda_0(t)\right), \tag{5.1.22}$$

where the first term on the right is the derivative with respect to the second argument in $\widetilde{\psi}_d$, yielding zero for the $\mathbb{R}^3$ component, and $\delta_c\widetilde{\psi}_d \in \Gamma$ is defined by the relation

$$\delta_c\widetilde{\psi}_d\left(\Lambda_r^t, \Lambda_0(t)\right) \cdot \Lambda_c = \delta_r\widetilde{\psi}_d\left(\Lambda_r^t, \Lambda_0(t) \mid \Lambda_h(s)\right), \tag{5.1.23}$$

where, for arbitrary $\Lambda_c \in \Gamma^+$, Λ_h is a history such that

$$\Lambda_h(s) = \Lambda_c, \quad \forall\, s > 0.$$

For the quadratic free energy functional introduced in Chap. 7, it is not necessary to use (5.1.22). The simpler relation (5.1.18) can in fact be applied. Writing this out explicitly, we obtain

$$\frac{\kappa}{\rho}\widetilde{\mathbf{S}} = \partial_{\mathbf{E}}\widetilde{\psi}, \quad e = \partial_{\kappa}\widetilde{\psi}, \quad \frac{1}{\rho}\mathbf{d} = \partial_{\mathbf{m}}\widetilde{\psi}.$$

Relation (5.1.19) becomes

$$D(t) = -\delta_r \widetilde{\psi}\left(\mathbf{\Lambda}_r^t, \mathbf{\Lambda}_0(t) \mid \frac{\partial}{\partial t}\mathbf{\Lambda}^t\right).$$

We define the equilibrium free energy ϕ to be given by (5.1.15) for the static history $\mathbf{\Lambda}^t(s) = \mathbf{\Lambda}^\dagger(s) = \mathbf{\Lambda}(t)$, $s \in \mathbb{R}^+$. It follows from (5.1.20) that this quantity depends only on $\mathbf{\Lambda}_0$, so that

$$\phi(t) = \widetilde{\phi}(\mathbf{\Lambda}_0(t)). \tag{5.1.24}$$

It can be deduced from (5.1.13) that [67, 191]

$$\phi(t) \leq \psi(t), \quad \forall\, t \in \mathbb{R}, \tag{5.1.25}$$

giving that the equilibrium free energy is less than or equal to the free energy for an arbitrary history. From (5.1.16), we have $\widetilde{\phi}(\mathbf{0}) = 0$.

The notation $\phi(t)$ will be used in most cases instead of $\widetilde{\phi}(\mathbf{\Lambda}_0(t))$. The quantity ϕ is conventionally chosen to be nonnegative so that ψ has the same property.

We can write (5.1.20) as

$$\psi(t) = \widetilde{\phi}(\mathbf{\Lambda}_0(t)) + \widetilde{\psi}_h(\mathbf{\Lambda}_r^t, \mathbf{\Lambda}_0(t)) = \phi(t) + \psi_h(t), \tag{5.1.26}$$

where $\psi_h(t) > 0$ is the history-dependent part of the free energy. Note that by definition,

$$\widetilde{\psi}_h(\mathbf{0}, \mathbf{\Lambda}_0(t)) = 0. \tag{5.1.27}$$

It follows from (5.1.22) that the generalized stress can also be expressed as the sum of an equilibrium part and a history-dependent part:

$$\mathbf{\Sigma}(t) = \mathbf{\Sigma}_e(t) + \mathbf{\Sigma}_{rh}(t), \tag{5.1.28}$$

where

$$\begin{aligned} \mathbf{\Sigma}_e(t) &= \widetilde{\mathbf{\Sigma}}(\mathbf{\Lambda}_0(t)) = \frac{d}{d\mathbf{\Lambda}(t)}\widetilde{\phi}(\mathbf{\Lambda}_0(t)), \\ \mathbf{\Sigma}_{rh}(t) &= \widetilde{\mathbf{\Sigma}}_{rh}(\mathbf{\Lambda}_r^t, \mathbf{\Lambda}_0(t)) = \partial_{\mathbf{\Lambda}}\widetilde{\psi}_h\left(\mathbf{\Lambda}_r^t, \mathbf{\Lambda}_0(t)\right) - \delta_c\widetilde{\psi}_h\left(\mathbf{\Lambda}_r^t, \mathbf{\Lambda}_0(t)\right). \end{aligned} \tag{5.1.29}$$

The quantity $\widetilde{\mathbf{\Sigma}}_{rh}$ vanishes as $\mathbf{\Lambda}_r^t$ tends to the zero history, provided that

$$\lim_{\mathbf{\Lambda}^t \to \mathbf{\Lambda}^\dagger} \partial_{\mathbf{\Lambda}}\widetilde{\psi}_h(\mathbf{\Lambda}^t, \mathbf{\Lambda}(t)) = \partial_{\mathbf{\Lambda}} \lim_{\mathbf{\Lambda}^t \to \mathbf{\Lambda}^\dagger} \widetilde{\psi}_h(\mathbf{\Lambda}^t, \mathbf{\Lambda}(t)),$$

where $\mathbf{\Lambda}^\dagger$ is defined before (5.1.24). We see this by observing that the right-hand side is $\mathbf{\Sigma}_e(t)$, given by $(5.1.29)_1$, while the left-hand side is the static history limit of $\widetilde{\mathbf{\Sigma}}(t)$ by virtue of (5.1.18).

5.1.1 Required Properties of a Free Energy

Let us draw together for future reference the properties of a free energy, stated earlier, which will be used to determine whether a given functional is a free energy. These are not independent conditions, since they include both a statement of the second law and consequences of that law.

P1 We have

$$\frac{\partial \widetilde{\psi}(\mathbf{\Lambda}^t, \mathbf{\Lambda}(t))}{\partial \mathbf{\Lambda}(t)} = \frac{\partial \psi(t)}{\partial \mathbf{\Lambda}(t)} = \mathbf{\Sigma}(t) = \widetilde{\mathbf{\Sigma}}(\mathbf{\Lambda}^t, \mathbf{\Lambda}(t)), \tag{5.1.30}$$

which is (5.1.18).[†]

P2 Let $\mathbf{\Lambda}^{\dagger}$ be a static history equal to $\mathbf{\Lambda}(t)$ at the current and all past times. Then,

$$\widetilde{\psi}(\mathbf{\Lambda}^{\dagger}, \mathbf{\Lambda}(t)) = \widetilde{\phi}(\mathbf{\Lambda}_0(t)), \tag{5.1.31}$$

where $\widetilde{\phi}(\mathbf{\Lambda}_0(t))$ is the equilibrium free energy. This is, in fact, a definition of $\widetilde{\phi}$, as given by (5.1.24), included here for completeness.

P3 For any history and current value $(\mathbf{\Lambda}^t, \mathbf{\Lambda}(t))$,

$$\widetilde{\psi}(\mathbf{\Lambda}^t, \mathbf{\Lambda}(t)) \geq \widetilde{\phi}(\mathbf{\Lambda}_0(t)), \tag{5.1.32}$$

which is (5.1.25).

P4 Condition (5.1.11) holds, namely

$$\dot{\psi}(t) + D(t) = \mathbf{\Sigma}(t) \cdot \dot{\mathbf{\Lambda}}(t), \quad D(t) \geq 0, \tag{5.1.33}$$

where $D(t)$ is given by (5.1.19). The first relation is a statement of the first law, while the non-negativity of $D(t)$ is in effect the second law.

These are the Graffi conditions for a free energy [174, 175].

5.1.2 Periodic Histories for General Materials

Integrating (5.1.33) from a past time when the material was undisturbed and assuming the integrals exist, we obtain

$$\begin{aligned} \mathcal{D}(t) = W(t) - \psi(t) \geq 0, \qquad \mathcal{D}(t) = \int_{-\infty}^{t} D(u)du, \\ W(t) = \int_{-\infty}^{t} \mathbf{\Sigma}(u) \cdot \dot{\mathbf{\Lambda}}(u)du, \end{aligned} \tag{5.1.34}$$

where $W(t)$ is the work function (see (4.1.4)) and $\mathcal{D}(t)$ is the total dissipation up to time t.

In the present context, the state is defined as the history and current value[‡] $(\mathbf{\Lambda}^t, \mathbf{\Lambda}(t))$, and ψ is a state variable by virtue of (5.1.15). Indeed, this is a very general requirement, as we see from Definition 4.1.6.

Let the history be periodic with period T. We have, for all t,

$$\mathbf{\Lambda}(t + T) = \mathbf{\Lambda}(t). \tag{5.1.35}$$

[†] This applies only to simple materials, which are the focus of interest in all of the present work apart from Sect. 3.7 and Chap. 21. A generalization of (5.1.30) for certain nonsimple materials is given by (21.1.19).

[‡] In certain later contexts (for specific independent variables), we will also use what amounts to the convention $(\mathbf{\Lambda}(t), \mathbf{\Lambda}^t)$ to maintain consistency with some of the literature.

Differentiating with respect to t, we obtain

$$\dot{\mathbf{\Lambda}}(t+T) = \dot{\mathbf{\Lambda}}(t).$$

Once transient effects have faded away, all state functions will return to the same value after a time T has passed, and we have

$$\psi(t+T) = \psi(t), \qquad \dot{\psi}(t+T) = \dot{\psi}(t).$$

The quantity $\widetilde{\mathbf{\Sigma}}$, defined by (5.1.18), is also a function of state, so that

$$\mathbf{\Sigma}(t+T) = \mathbf{\Sigma}(t).$$

From (5.1.33), it follows that

$$D(t+T) = D(t).$$

Each period is generated by a cyclic process as given by Definition 4.1.4. Integrating (5.1.33) over $[t, t+T]$, we obtain

$$\int_t^{t+T} D(u)du = \int_t^{t+T} \mathbf{\Sigma}(u) \cdot \dot{\mathbf{\Lambda}}(u)du > 0 \tag{5.1.36}$$

for materials with dissipation, or

$$\mathcal{D}(t+T) - \mathcal{D}(t) = W(t+T) - W(t) > 0. \tag{5.1.37}$$

Now,

$$\frac{d}{dt}[W(t+T) - W(t)] = \mathbf{\Sigma}(t+T) \cdot \dot{\mathbf{\Lambda}}(t+T) - \mathbf{\Sigma}(t) \cdot \dot{\mathbf{\Lambda}}(t) = 0,$$

so that both sides of $(5.1.36)_1$ are independent of t. The infinite integral defining $\mathcal{D}$ and $W(t)$ in (5.1.34) therefore must diverge for periodic histories, since they consist of an infinite sum of integrals over a period, as in (5.1.36), each being equal to all the others.

Observe that if $\widetilde{\psi}$ depends only on $\mathbf{\Lambda}(t)$ and is independent of the history, then, using (5.1.30),

$$\begin{aligned}\int_t^{t+T} \mathbf{\Sigma}(u) \cdot \dot{\mathbf{\Lambda}}(u)du &= \int_{\mathbf{\Lambda}(t)}^{\mathbf{\Lambda}(t+T)} \frac{\partial}{\partial \mathbf{\Lambda}(u)} \widetilde{\psi}(\mathbf{\Lambda}(u)) \cdot d\mathbf{\Lambda}(u) \\ &= \widetilde{\psi}(\mathbf{\Lambda}(t+T)) - \widetilde{\psi}(\mathbf{\Lambda}(t)) = 0,\end{aligned}$$

so that the total dissipation must be zero, a result that is consistent with (5.1.19).

5.1.3 Constraints on the Nonuniqueness of the Free Energy

Let us consider in general terms the nature of the arbitrariness in ψ and the constraints imposed on this arbitrariness by the properties summarized in Sect. 5.1.1, which we will refer to as the thermodynamic constraints. It will be required also that the equilibrium free energy $\widetilde{\phi}(\mathbf{\Lambda}_0(t))$ contains no arbitrariness.

Proposition 5.1.2. *The most general form of the free energy* (5.1.15) *consistent with the thermodynamic constraints is*

$$\psi(t) = \widetilde{\phi}(\mathbf{\Lambda}_0(t)) + \widetilde{\psi}_f\left(\mathbf{\Lambda}^t, \mathbf{\Lambda}(t)\right) + \widetilde{\psi}_r\left(\mathbf{\Lambda}^t\right),$$

where $\widetilde{\phi}(\mathbf{\Lambda}_0(t))$ *is the equilibrium free energy, and* $\widetilde{\psi}_f$ *is a fixed, nonnegative functional, with no restriction on its* $\mathbf{\Lambda}(t)$ *dependence other than that* $\widetilde{\phi}+\widetilde{\psi}_f$ *yields a satisfactory form for* (5.1.30). *The quantity* $\widetilde{\psi}_r$, *which is independent of* $\mathbf{\Lambda}(t)$, *is arbitrary within thermodynamic constraints and in particular may be zero. These thermodynamic constraints give that* $\widetilde{\psi}_f$ *and* $\widetilde{\psi}_f + \widetilde{\psi}_r$ *must be nonnegative-valued functionals and that* $-\delta\widetilde{\psi}_f$ *and* $-\left(\delta\widetilde{\psi}_f + \delta\widetilde{\psi}_r\right)$ *also must be nonnegative, where, as before,* δ *indicates the Fréchet derivative specified in* (5.1.17). *Furthermore,* $\widetilde{\psi}_f$ *and* $\widetilde{\psi}_r$ *must vanish for the static history* $\mathbf{\Lambda}^t(s) = \mathbf{\Lambda}(t)$, $s \in \mathbb{R}^+$.

Proof. Let

$$\psi_1(t) = \Psi_1\left(\mathbf{\Lambda}^t(s), \mathbf{\Lambda}(t)\right)$$

be an alternative choice of free energy, obeying the thermodynamic constraints. Then P1 or (5.1.30) gives that

$$\partial_{\mathbf{\Lambda}}\left(\widetilde{\psi} - \widetilde{\psi}_1\right) = \mathbf{0},$$

so that

$$\widetilde{\psi}_1\left(\mathbf{\Lambda}^t, \mathbf{\Lambda}(t)\right) = \widetilde{\psi}\left(\mathbf{\Lambda}^t, \mathbf{\Lambda}(t)\right) + \widetilde{\psi}_r\left(\mathbf{\Lambda}^t\right),$$

where $\widetilde{\psi}_r$ is arbitrary within thermodynamic constraints on $\widetilde{\psi}_1$ and of course may be zero. We can always separate out $\widetilde{\phi}(\mathbf{\Lambda}_0(t))$ (by adding and subtracting this quantity), and the resulting memory terms $\widetilde{\psi}_f$ or $\widetilde{\psi}_f + \widetilde{\psi}_r$ must obey the thermodynamic constraints. Finally, since $\widetilde{\phi}(\mathbf{\Lambda}_0(t))$ is uniquely defined, $\widetilde{\psi}_f$ must vanish for the static history if $\widetilde{\psi}_r$ is zero, as noted in (5.1.27), and any choice of $\widetilde{\psi}_r$ must have the same property, which completes the proof. □

This result is a simpler version of a proposition given in [159]. The origin of the simplification is the use of ψ rather than ψ_h. It is also a general statement of a property noted below (see (7.1.16)) in connection with free energies that are quadratic in the memory terms.

The internal energy is unique. A list of other unique quantities involving the Helmholtz free energy ψ_h, entropy, and dissipation is given in [159].

5.2 The Maximum Recoverable Work for General Materials

In the next and subsequent chapters, we deal with the case of free energies given by quadratic functionals, leading to constitutive equations with linear memory terms. Before leaving the general formulation, it is of interest to note that results can be obtained, using functional differentiation, that correspond to key formulas derived later in the linear memory case. One example will be given, namely the formulas determining the maximum recoverable work. Others can be derived, with somewhat greater difficulty, notably formulas for the work function as a double integral over histories, the generalization of (7.5.7).[§] Let us define

$$\mathbf{V}(t) = \dot{\mathbf{\Lambda}}(t), \qquad \mathbf{V} \in \Gamma,$$

noting that

$$\mathbf{\Lambda}(t) = \mathbf{\Lambda}(-\infty) + \int_{\infty}^{t} \mathbf{V}(s)ds, \tag{5.2.1}$$

where $\mathbf{\Lambda}(-\infty)$ is a constant, which is taken to be zero. The integral is assumed to exist. Let

$$\mathbf{V}^t = \{\mathbf{V}(s);\ s \leq t\}, \tag{5.2.2}$$

which is a similar convention to that in (5.1.14). The sum of two such sets is understood to mean

$$\mathbf{V}_1^t + \mathbf{V}_2^t = \{\mathbf{V}_1(s) + \mathbf{V}_2(s);\ s \leq t\}.$$

Relations (5.2.1) and (5.2.2) allow us to write $\widetilde{\mathbf{\Sigma}}(\mathbf{\Lambda}^t, \mathbf{\Lambda}(t))$ in (5.1.18) compactly as

$$\mathbf{\Sigma}(t) = \widetilde{\mathbf{\Sigma}}(\mathbf{V}^t) = \mathbf{\Sigma}_e(t) + \widetilde{\mathbf{\Sigma}}_{rh}(\mathbf{V}^t),$$

in the notation of (5.1.28) and (5.1.29). The work done on the material up to time t is given by $(5.1.34)_3$ or

$$W(t) = \widetilde{W}(\mathbf{V}^t) = \int_{-\infty}^{t} \widetilde{\mathbf{\Sigma}}(\mathbf{V}^s) \cdot \mathbf{V}(s)ds,$$

where the integral is assumed to be finite. Using (5.1.29), the recoverable work from the state at time t may be written as

$$W_R(t) = \widetilde{W}_R(\mathbf{V}^t) = -\int_{t}^{\infty} \widetilde{\mathbf{\Sigma}}(\mathbf{V}^s) \cdot \mathbf{V}(s)ds = \phi(t) - \int_{t}^{\infty} \widetilde{\mathbf{\Sigma}}_{rh}(\mathbf{V}^s) \cdot \mathbf{V}(s)ds, \tag{5.2.3}$$

where $\mathbf{V}(s)$, $s \geq t$, is the process determining the evolution of the state after time t, chosen so that the integrals in (5.2.3) exist. We take $\phi(\infty)$ to be zero. Let us assume that $\widetilde{\mathbf{\Sigma}}_{rh}$ is continuously Fréchet differentiable in the fading memory Hilbert space $\mathcal{H}$. Let $\mathbf{\Delta} : \mathbb{R} \mapsto \Gamma$ belong to $\mathcal{H}$, and let $\mathbf{\Delta}^t$ be defined in the same way as $\mathbf{V}^t$ by (5.2.2). Then,

$$\widetilde{\mathbf{\Sigma}}_{rh}(\mathbf{V}^s + \mathbf{\Delta}^s) = \widetilde{\mathbf{\Sigma}}_{rh}(\mathbf{V}^s) + \delta\widetilde{\mathbf{\Sigma}}_{rh}(\mathbf{V}^s|\mathbf{\Delta}^s) + \mathbf{o}\|\mathbf{\Delta}^s\| \qquad \forall\ \mathbf{\Delta}^s \in \Gamma, \tag{5.2.4}$$

where $\delta\widetilde{\mathbf{\Sigma}}$ is the Fréchet differential, continuous in $\mathbf{V}^s$ and linear in $\mathbf{\Delta}^s$. Also, $\|\cdot\|$ is the norm in $\mathcal{H}$. By the Riesz representation theorem, we can write

$$\delta\widetilde{\mathbf{\Sigma}}_{rh}(\mathbf{V}^s|\mathbf{\Delta}^s) = \int_{-\infty}^{\infty} \mathbb{L}(s, u;\ \mathbf{V}^s)\mathbf{\Delta}(u)du, \tag{5.2.5}$$

where $\mathbb{L} \in \mathrm{Lin}(\Gamma)$ and

$$\mathbb{L}(s, u;\ \mathbf{V}^s) = \mathbf{0}, \quad s < u, \tag{5.2.6}$$

since values of $\mathbf{\Delta}(u)$, $u > s$, cannot contribute.

[§] J. M. Golden, unpublished notes.

Proposition 5.2.1. *The maximum recoverable work, which, by Theorem 4.2.3, is equal to the minimum free energy, can be expressed in the form*

$$W_{rm}(t) = \phi(t) + \frac{1}{2}\int_t^\infty ds \int_t^\infty du\, \mathbb{L}_S\,(u, s;\, \mathbf{V}_m)\, \mathbf{V}_m(u,t)\cdot \mathbf{V}_m(s,t)$$

with

$$\mathbb{L}_S\,(u, s;\, \mathbf{V}_m) = \begin{cases} \mathbb{L}(s, u;\, \mathbf{V}_m^s), & s > u, \\ \mathbb{L}^\top(u, s;\, \mathbf{V}_m^u), & s < u, \end{cases} \tag{5.2.7}$$

where $\mathbf{V}_m(\cdot, t)$ *is the solution of the equation*

$$\widetilde{\Sigma}_{rh}(\mathbf{V}^s) + \int_s^\infty \mathbb{L}^\top(u, s;\, \mathbf{V}^u)\, \mathbf{V}(u)du = \mathbf{0}, \qquad s \geq t. \tag{5.2.8}$$

Proof. We seek to maximize the recoverable work given by (5.2.3). Applying a variation

$$\mathbf{V}^s \to \mathbf{V}^s + \mathit{\Delta}^s, \qquad \mathit{\Delta}(u) = \mathbf{0} \text{ if } u < t,$$

we obtain the condition

$$\int_t^\infty \widetilde{\Sigma}_{rh}(\mathbf{V}^s)\cdot \mathit{\Delta}(s)ds + \int_t^\infty ds \int_t^s du\, \mathbb{L}(s, u;\, \mathbf{V}^s)\mathit{\Delta}(u)\cdot \mathbf{V}(s) = 0, \tag{5.2.9}$$

with the aid of (5.2.4)–(5.2.6). Noting the identities

$$\int_t^\infty ds \int_t^s du\, F(s,u) = \int_t^\infty du \int_u^\infty ds\, F(s,u) = \int_t^\infty ds \int_s^\infty du\, F(u,s), \tag{5.2.10}$$

relation (5.2.9) becomes, using (A.2.3),

$$\begin{aligned} &\int_t^\infty \widetilde{\Sigma}_{rh}(\mathbf{V}^s)\cdot \mathit{\Delta}(s)ds + \int_t^\infty ds \int_s^\infty du\, \mathbb{L}(u, s;\, \mathbf{V}^u)\mathit{\Delta}(s)\cdot \mathbf{V}(u) \\ &\quad = \int_t^\infty \widetilde{\Sigma}_{rh}(\mathbf{V}^s)\cdot \mathit{\Delta}(s)ds + \int_t^\infty ds \int_s^\infty du\, \mathbb{L}^\top(u, s;\, \mathbf{V}^u)\mathbf{V}(u)\cdot \mathit{\Delta}(s) = 0. \end{aligned}$$

The arbitrariness of $\mathit{\Delta}$ gives (5.2.8), the solution of which yields the optimal future process $\mathbf{V}_m(\cdot, t)$. Using (5.2.8) in (5.2.3), we obtain, by adding the leftmost and rightmost forms of (5.2.10), an expression for the maximum recoverable work or the minimum free energy

$$\begin{aligned} W_{rm}(t) &= \psi_m(t) \\ &= \phi(t) + \int_t^\infty ds \int_s^\infty du\, \mathbb{L}^\top(u, s;\, \mathbf{V}_m^u)\mathbf{V}_m(u,t)\cdot \mathbf{V}_m(s,t) \\ &= \phi(t) + \frac{1}{2}\int_t^\infty ds \int_t^\infty du\, \mathbb{L}_S(u, s; \mathbf{V}_m)\mathbf{V}_m(u,t)\cdot \mathbf{V}_m(s,t), \end{aligned} \tag{5.2.11}$$

where $\mathbb{L}_S$ is given by (5.2.7). □

Note that

$$\mathbb{L}_S^\top(u, s;\, \mathbf{V}_m) = \mathbb{L}_S(s, u;\, \mathbf{V}_m).$$

We observe that the form $(5.2.11)_2$ is a generalization of a result given in Sect. 7.5, while (5.2.8) is a generalized form of the Wiener–Hopf equation (11.2.26).

5.3 Generation of New Free Energies

If some free energies are known for a certain category of materials, for example, those with constitutive equations that have linear memory, we ask in this section, and answer affirmatively, whether it is possible to construct (for example, nonlinear) functions of the known quantities that are free energies relating to more general materials (for example, those with constitutive equations that have nonlinear memory). Note that the findings discussed here are quite different from those in Chap. 17, which deals only with quadratic functionals producing linear memory constitutive equations.

Let $\psi_1(t), \psi_2(t), \ldots, \psi_n(t)$ be a set of n free energies relating to a state $(\mathbf{\Lambda}^t, \mathbf{\Lambda}(t))$ in a given material, or perhaps in different materials, at time t. To allow for the latter possibility, we assign to each $\psi_i(t)$, $i = 1, 2, \ldots, n$, different constitutive equations

$$\mathbf{\Sigma}_i(t) = \widetilde{\mathbf{\Sigma}}_i(\mathbf{\Lambda}^t, \mathbf{\Lambda}(t))$$

and work functions

$$W_i(t) = \int_{-\infty}^{t} \mathbf{\Sigma}_i(s) \cdot \dot{\mathbf{\Lambda}}(s) ds,$$

where

$$\frac{\partial \psi_i(t)}{\partial \mathbf{\Lambda}(t)} = \mathbf{\Sigma}_i(t), \quad i = 1, 2, \ldots, n, \tag{5.3.1}$$

and, by virtue of (5.1.33),

$$\dot{\psi}_i(t) \leq \mathbf{\Sigma}_i(t) \cdot \dot{\mathbf{\Lambda}}(t), \quad i = 1, 2, \ldots, n.$$

If all these free energies belong to the same material, the dependent field variables $\mathbf{\Sigma}_i$ are all equal and the index i refers to different free energies of the same material.

Proposition 5.3.1. *The quantity*

$$\psi(t) = f(\psi_1(t), \psi_2(t), \ldots, \psi_n(t)) \tag{5.3.2}$$

is a free energy for the state $(\mathbf{\Lambda}^t, \mathbf{\Lambda}(t))$ *with the dependent field given by*

$$\mathbf{\Sigma}(t) = \sum_{i=1}^{n} \frac{\partial f}{\partial \psi_i(t)} \mathbf{\Sigma}_i(t), \tag{5.3.3}$$

provided that $f : (\mathbb{R}^+)^n \mapsto \mathbb{R}^+$ *has the properties*

$$\frac{\partial}{\partial y_i} f(y_1, y_2, \ldots, y_n) \geq 0, \quad i = 1, 2, \ldots, n, \tag{5.3.4}$$

and

$$f(0, 0, \ldots, 0) = 0. \tag{5.3.5}$$

Proof. We have

$$\dot{\psi}(t) = \sum_{i=1}^{n} \frac{\partial f}{\partial \psi_i(t)} \dot{\psi}_i(t) \le \sum_{i=1}^{n} \frac{\partial f}{\partial \psi_i(t)} \mathbf{\Sigma}_i(t) \cdot \dot{\mathbf{\Lambda}}(t) = \mathbf{\Sigma}(t) \cdot \dot{\mathbf{\Lambda}}(t),$$

where $\mathbf{\Sigma}$ is defined by (5.3.3). Thus, property 4 of a free energy, given by (5.1.33), holds. Also, by virtue of (5.3.1) and (5.3.3),

$$\frac{\partial \psi(t)}{\partial \mathbf{\Lambda}(t)} = \mathbf{\Sigma}(t), \tag{5.3.6}$$

which is property 1 as given by (5.1.30). If $\phi_i(t)$, $i = 1, 2, \ldots, n$, are the equilibrium free energies corresponding to $\psi_i(t)$, according to the prescription (5.1.31), then

$$\phi(t) = f(\phi_1(t), \phi_2(t), \ldots, \phi_n(t))$$

and property 3, i.e., (5.1.32), is obeyed by virtue of the assumptions (5.3.4).

If (5.1.16) is to hold for all free energies ψ and $\psi_i(t)$, $i = 1, 2, \ldots, n$, then we must have (5.3.5). □

This result can be used, for example, as follows: assume we have a nonlinear dependent field variable of the form (5.3.3), where f obeys (5.3.4) and the $\mathbf{\Sigma}_i$ are determined by (5.3.1). Then (5.3.2) immediately gives a free energy that generates $\mathbf{\Sigma}(t)$ through (5.3.6).

Taking f to be an analytic function of its arguments at the origin, we can write

$$\psi(t) = \sum_{i=1}^{n} \lambda_i \psi_i(t) + \text{higher powers}. \tag{5.3.7}$$

A constant term is excluded by (5.3.5). If we omit higher powers, taking ψ to be a linear combination of the ψ_i, it follows from (5.3.4) that

$$\lambda_i \ge 0, \quad i = 1, 2, \ldots, n. \tag{5.3.8}$$

If the free energies ψ_i, $i = 1, 2, \ldots, n$, relate to the same material, then (5.3.3) becomes

$$\mathbf{\Sigma}(t) = \kappa(t)\mathbf{\Sigma}_{sm}(t), \qquad \kappa(t) = \sum_{i=1}^{n} \frac{\partial f}{\partial \psi_i(t)},$$

$$\mathbf{\Sigma}_{sm}(t) = \mathbf{\Sigma}_1(t) = \mathbf{\Sigma}_2(t) = \cdots = \mathbf{\Sigma}_n(t).$$

Let higher powers be neglected in (5.3.7). If ψ is assumed to relate to the same material as the ψ_i, $i = 1, 2, \ldots, n$, then

$$\mathbf{\Sigma}_{sm}(t) = \mathbf{\Sigma}(t)$$

and

$$\sum_{i=1}^{n} \frac{\partial f}{\partial \psi_i(t)} = \sum_{i=1}^{n} \lambda_i = 1.$$

This, together with (5.3.8), amounts to convexity (see Proposition 4.1.7).

The next chapter is something of a diversion from the main flow of the discussion, to consider Thermoelectromagnetism.

6

Thermoelectromagnetism of Continuous Media

We now explore Continuum Electromagnetism in the context of thermodynamic principles.

6.1 Electromagnetism of Continuous Media

The classical theory of electromagnetism of continua, without memory effects, is first explored. The case of nonlocal materials is also discussed briefly.

6.1.1 Balance Laws in Electromagnetic Media

The theory of electromagnetism is characterized by a few basic principles which are now introduced. The discussion is based on the book [122].

Three of the relevant physical quantities are the *electric charge* q, the *electric field* $\mathbf{E}$ and the *magnetic induction* $\mathbf{B}$.

By assuming that charged particles are continuously distributed in a region $\Omega \subset \mathbb{R}^3$, there exists an *electric charge density* ϱ such that the charge q, contained in Ω, is given by the integral

$$q = \int_{\Omega} \varrho dv.$$

If electric charge is considered to be a primitive concept, the electric field $\mathbf{E} = \mathbf{E}(\mathbf{x}, t)$ is defined as the force acting on a unitary charge placed at a point $\mathbf{x} \in \mathbb{R}^3$ at time $t > 0$.

The magnetic induction $\mathbf{B} = \mathbf{B}(\mathbf{x}, t)$, at any $\mathbf{x} \in \mathbb{R}^3$ and $t > 0$, is the mechanical torque exerted on a unitary magnetic dipole located at the point $\mathbf{x}$.

Three other important physical quantities are the *electric current density* $\mathbf{j}$, the *electric displacement* $\mathbf{D}$ and the *magnetic field* $\mathbf{H}$. These are introduced by using certain laws, discussed below, which express the basic axioms of electromagnetism and relate all these fields $\mathbf{E}$, $\mathbf{B}$, $\mathbf{D}$, $\mathbf{H}$ and $\mathbf{j}$.

G. Amendola et al., *Thermodynamics of Materials with Memory*,
https://doi.org/10.1007/978-3-030-80534-0_6

Balance Law of Charge

The time variation of the charge in a domain Ω is equal to the charge entering or leaving Ω through its boundary $\partial\Omega$. This law is expressed mathematically by the relation

$$\frac{d}{dt}\int_{\Omega} \varrho dv = -\int_{\partial\Omega} \mathbf{j}\cdot\mathbf{n} da,$$

where we have introduced the current density vector $\mathbf{j}$, which expresses the charge which flows across $\partial\Omega$, per unit area and unit time, and $\mathbf{n}$ which denotes the unit outward normal to $\partial\Omega$, at a given point. Thus, using the divergence theorem, we have

$$\int_{\Omega}\left(\frac{\partial\varrho}{\partial t} + \nabla\cdot\mathbf{j}\right) dv = 0,$$

whence, by virtue of the arbitrariness of Ω and by assuming that the functions are C^1, we obtain the local form

$$\frac{\partial\varrho}{\partial t} + \nabla\cdot\mathbf{j} = 0, \tag{6.1.1}$$

which is known as the *continuity equation*.

The electric displacement vector $\mathbf{D}$ is introduced as follows.

Gauss's Law

The total charge inside any region $\Omega \subset \mathbb{R}^3$ is equal to the flux of the vector $\mathbf{D}$, the electric displacement, across the boundary $\partial\Omega$ of Ω, which can be expressed mathematically as

$$\int_{\partial\Omega} \mathbf{D}\cdot\mathbf{n} da = \int_{\Omega} \varrho dv.$$

The divergence theorem, the arbitrariness of Ω and the regularity of the functions give the local form

$$\nabla\cdot\mathbf{D} = \varrho. \tag{6.1.2}$$

Conservation of Magnetic Flux

The flux of the magnetic induction $\mathbf{B}$ through any closed surface $\partial\Omega$ vanishes, that is

$$\int_{\partial\Omega} \mathbf{B}\cdot\mathbf{n} da = 0,$$

from which it follows that

$$\nabla\cdot\mathbf{B} = 0. \tag{6.1.3}$$

Faraday's Law

For any surface S surrounded by a given closed curve ∂S, the electromagnetic fields $\mathbf{E}$ and $\mathbf{B}$ are related by

$$\int_{\partial S} \mathbf{E}\cdot d\mathbf{l} = -\int_{S} \frac{\partial\mathbf{B}}{\partial t}\cdot\mathbf{n} da.$$

Hence, by using Stokes's theorem, we obtain the local form of this law given by

$$\nabla \times \mathbf{E} = -\frac{\partial \mathbf{B}}{\partial t}. \tag{6.1.4}$$

Ampere–Maxwell Law

For any surface S surrounded by a given closed curve ∂S, the electromagnetic fields **H**, **D** and **j** are related by

$$\int_{\partial S} \mathbf{H} \cdot d\mathbf{l} = \int_S \left(\mathbf{j} + \frac{\partial \mathbf{D}}{\partial t} \right) \cdot \mathbf{n} da.$$

From this equation, by means of Stokes's theorem, it follows that

$$\nabla \times \mathbf{H} = \mathbf{j} + \frac{\partial \mathbf{D}}{\partial t}. \tag{6.1.5}$$

6.1.2 Constitutive Equations

In the previous section we have derived a system of differential equations, given by (6.1.1)–(6.1.5), known as *Maxwell's equations*. These equations are not all independent; in fact the continuity equation (6.1.1) can be derived from (6.1.2) and (6.1.5); also (6.1.3) follows from (6.1.4). Consider the following system

$$\begin{aligned} \frac{\partial \mathbf{B}}{\partial t} &= -\nabla \times \mathbf{E}, \\ \frac{\partial \mathbf{D}}{\partial t} &= \nabla \times \mathbf{H} - \mathbf{j}, \\ \nabla \cdot \mathbf{D} &= \varrho, \\ \nabla \cdot \mathbf{B} &= 0. \end{aligned} \tag{6.1.6}$$

Only the first and third equations are independent. Consequently, two of the five fields **E**, **B**, **H**, **D**, and **j** are arbitrary and the other three fields must be assigned in terms of the previous two.

Such connections are expressed by suitable constitutive equations, which are the relations that characterize the behavior of the materials under consideration.

A choice of constitutive equations is given by functions

$$\mathbf{D}(t) = \hat{\mathbf{D}}(\mathbf{E}(t), \mathbf{B}(t)), \qquad \mathbf{H}(t) = \hat{\mathbf{H}}(\mathbf{E}(t), \mathbf{B}(t)), \tag{6.1.7}$$

which describe a general dependence of $\mathbf{D}(t)$ and $\mathbf{H}(t)$ on $\mathbf{E}(t)$ and $\mathbf{B}(t)$.

The current density **j** is usually written as the sum of two terms

$$\mathbf{j} = \mathbf{J} + \mathbf{J}_f,$$

where

$$\mathbf{J}(t) = \hat{\mathbf{J}}(\mathbf{E}(t), \mathbf{B}(t))$$

is the part of the electric current generated by the action of the electromagnetic fields $\mathbf{E}$ and $\mathbf{B}$ in the material, while $\mathbf{J}_f$, known as the *forced current density*, does not depend on the electromagnetic fields inside the material but derives from external influences.

A particular form of (6.1.7) is given by the constitutive equations

$$\mathbf{D}(t) = \hat{\mathbf{D}}(\mathbf{E}(t)), \qquad \mathbf{H}(t) = \hat{\mathbf{H}}(\mathbf{B}(t)),$$

which describe the behavior of many types of material. For brevity, let us omit the explicit time dependence. The simplest model is that characterized by a linear dependence, given by

$$\mathbf{D} = \boldsymbol{\varepsilon}\mathbf{E}, \qquad \mathbf{H} = \boldsymbol{\mu}^{-1}\mathbf{B}, \tag{6.1.8}$$

where $\boldsymbol{\varepsilon}$ and $\boldsymbol{\mu}$ are second-order symmetric tensors.

If the media are homogeneous, $\boldsymbol{\varepsilon}$ and $\boldsymbol{\mu}$ are independent of the position $\mathbf{x}$; if the media are isotropic, then $\boldsymbol{\varepsilon}$ and $\boldsymbol{\mu}$ are proportional to the identity tensor and relations (6.1.8) become

$$\mathbf{D} = \varepsilon\mathbf{E}, \qquad \mathbf{H} = \frac{1}{\mu}\mathbf{B}, \tag{6.1.9}$$

where ε and μ are scalar functions. In particular, in free space we have

$$\mathbf{D} = \varepsilon_0\mathbf{E}, \qquad \mathbf{H} = \frac{1}{\mu_0}\mathbf{B},$$

where ε_0 and μ_0 are the *dielectric constant* and the *magnetic permeability* of the vacuum, respectively.

For the current density $\mathbf{J}$ in conducting materials we can consider the case where this quantity depends only on the electric field $\mathbf{E}$. A very simple constitutive equation is expressed by *Ohm's law*

$$\mathbf{J} = \sigma\mathbf{E}, \tag{6.1.10}$$

where σ is the *electrical conductivity* of the medium.

Two further vector fields, the *electric polarization* $\mathbf{P}$ and the *magnetization* $\mathbf{M}$, are introduced to describe the electromagnetic behavior of a material. These are defined by

$$\mathbf{P} = \mathbf{D} - \varepsilon_0\mathbf{E},$$
$$\mathbf{M} = \frac{1}{\mu_0}\mathbf{B} - \mathbf{H}.$$

If the media are isotropic, so that $\mathbf{D}$ and $\mathbf{B}$ satisfy the constitutive equations (6.1.9), the vectors $\mathbf{P}$ and $\mathbf{M}$ are given by

$$\mathbf{P} = \varepsilon_0\chi_e\mathbf{E},$$
$$\mathbf{M} = \chi_m\mathbf{H},$$

where we have introduced

$$\chi_e = \frac{\varepsilon}{\varepsilon_0} - 1, \qquad \chi_m = \frac{\mu}{\mu_0} - 1,$$

which are known as the *electric* and *magnetic susceptibilities*, respectively.

6.1.3 Boundary Conditions

When in a body there is a surface, or a line, of discontinuity for an electromagnetic field, Maxwell's equations allow us to evaluate the related jump conditions at the surface, or at the line, of discontinuity.

Let Γ, or γ, be a surface, or a line, dividing Ω into two regions Ω^+ and Ω^-; for the sake of simplicity, we suppose that Γ, and γ, are fixed in the electromagnetic body. We shall use the superscripts + and – for the limit values of a vector field, evaluated from Ω^+ and Ω^- at Γ, or at γ, respectively, and denote by $\mathbf{n}_\Gamma$ the unit normal vector to Γ oriented from Ω^- to Ω^+ and by $\mathbf{t}_\gamma$ the unit tangent vector to γ.

By using the integral form of the laws, which express the basic axioms of the electromagnetism, it is possible to derive the jumps of the electromagnetic fields at the surface, or at the line, which are now reproduced below.

Gauss's law yields

$$\mathbf{D}^+ \cdot \mathbf{n}_\Gamma - \mathbf{D}^- \cdot \mathbf{n}_\Gamma = \varrho_S,$$

where ϱ_S denotes the *surface charge density* on Γ. Therefore, the normal component of $\mathbf{D}$ is discontinuous across the surface with a jump given by the surface charge density ϱ_S; consequently, if $\varrho_S = 0$, the normal component of $\mathbf{D}$ is continuous.

From the conservation law of the magnetic flux it follows that the normal component of $\mathbf{B}$ is continuous across Γ, that is

$$\mathbf{B}^+ \cdot \mathbf{n}_\Gamma = \mathbf{B}^- \cdot \mathbf{n}_\Gamma.$$

Moreover, Faraday's law implies the continuity of the tangential component of $\mathbf{E}$ across a line γ, i.e.,

$$\mathbf{E}^+ \cdot \mathbf{t}_\gamma = \mathbf{E}^- \cdot \mathbf{t}_\gamma,$$

and the Ampere–Maxwell equation gives

$$\mathbf{H}^+ \cdot \mathbf{t}_\gamma - \mathbf{H}^- \cdot \mathbf{t}_\gamma = \mathbf{J}_S \cdot \mathbf{n},$$

where $\mathbf{J}_S$ denotes the *surface current density* and $\mathbf{n}$ is the unit vector orthogonal to $\mathbf{t}_\gamma$.

6.1.4 Balance of Energy and the First Law of Thermodynamics

A balance law of energy for electromagnetic media can be derived from Maxwell's equations. To this purpose we introduce the vector

$$\mathbf{S} = \mathbf{E} \times \mathbf{H}, \tag{6.1.11}$$

which is the instantaneous *Poynting's vector*, and consider the following identity

$$\nabla \cdot (\mathbf{E} \times \mathbf{H}) = \mathbf{H} \cdot \nabla \times \mathbf{E} - \mathbf{E} \cdot \nabla \times \mathbf{H}.$$

Using Maxwell's equations $(6.1.6)_1$ and $(6.1.6)_2$, we obtain

$$\nabla \cdot (\mathbf{E} \times \mathbf{H}) = -\mathbf{H} \cdot \frac{\partial \mathbf{B}}{\partial t} - \mathbf{E} \cdot \mathbf{j} - \mathbf{E} \cdot \frac{\partial \mathbf{D}}{\partial t}.$$

Hence, on integrating over Ω and using the divergence theorem, we get the following global form

$$\int_\Omega \left(\mathbf{H} \cdot \frac{\partial \mathbf{B}}{\partial t} + \mathbf{E} \cdot \frac{\partial \mathbf{D}}{\partial t} \right) dv + \int_{\partial\Omega} \mathbf{E} \times \mathbf{H} \cdot \mathbf{n} da = -\int_\Omega \mathbf{E} \cdot \mathbf{j} dv. \tag{6.1.12}$$

This formula is known as *Poynting's theorem.*

If we consider the particular case of free space, characterized by the values ε_0 and μ_0 for the dielectric constant and the magnetic permeability, respectively, the first integral of this last formula becomes

$$\int_\Omega \left(\mathbf{H} \cdot \frac{\partial \mathbf{B}}{\partial t} + \mathbf{E} \cdot \frac{\partial \mathbf{D}}{\partial t} \right) dv = \frac{d}{dt} \int_\Omega \left(\frac{\mu_0}{2} \mathbf{H}^2 + \frac{\varepsilon_0}{2} \mathbf{E}^2 \right) dv = \frac{d}{dt} \int_\Omega u dv,$$

where

$$u = \frac{1}{2} (\varepsilon_0 \mathbf{E}^2 + \mu_0 \mathbf{H}^2).$$

This is the *energy density of the electromagnetic field.*

An analogous result is obtained if the constitutive equations of **D** and **B** are expressed by (6.1.8), where $\boldsymbol{\varepsilon}$ and $\boldsymbol{\mu}$ are second-order symmetric tensors, that is by

$$\mathbf{D} = \boldsymbol{\varepsilon} \mathbf{E}, \qquad \mathbf{B} = \boldsymbol{\mu} \mathbf{H};$$

in this case we obtain

$$u = \frac{1}{2} (\mathbf{E} \cdot \boldsymbol{\varepsilon} \mathbf{E} + \mathbf{H} \cdot \boldsymbol{\mu} \mathbf{H}).$$

Moreover, the right hand side of (6.1.12) can be identified as π, the *power of the charge which flows with current density* **j**, that is

$$\pi = \int_\Omega \mathbf{E} \cdot \mathbf{j} dv.$$

Therefore, by using the definition (6.1.11) of Poynting's vector, (6.1.12) can be put in the following form

$$\frac{d}{dt} \int_\Omega u dv = -\int_{\partial\Omega} \mathbf{S} \cdot \mathbf{n} da - \int_\Omega \mathbf{E} \cdot \mathbf{j} dv,$$

which expresses the variation of the electromagnetic energy in terms of the flux of the Poynting vector across the boundary $\partial\Omega$ and of the dissipated power. For this reason the Poynting vector is identified as the energy flux vector of the electromagnetic field.

In order to consider thermal effects, we must state the laws of thermodynamics for electromagnetic materials. It is assumed that the thermoelectromagnetic body is rigid. Consequently, the mass density can be considered constant and assumed equal to one. Thus, the heat equation can be written in the following form

$$h = -\nabla \cdot \mathbf{q} + r, \tag{6.1.13}$$

where h is the rate at which heat is absorbed per unit volume, known as the *internal heat power*, while $\mathbf{q}$ is the heat flux and r is the heat supply.

A thermoelectromagnetic system may be considered as a dynamical system, for which the concepts of state and process are well defined. These concepts, which are introduced in Chap. 4, allow us to perform a thermodynamic analysis based on the properties of cycles.

We denote by Σ the space of states, which are characterized by the triplets $\sigma = (\mathbf{E}, \mathbf{H}, \theta)$, where θ is the absolute temperature and $\mathbf{E}$ and $\mathbf{H}$ are two vectors of $\mathcal{V}$, which is usually $\mathbb{R}^3$. A process is defined as follows.

Definition 6.1.1. *A thermoelectromagnetic process P of duration d_P is a piecewise continuous function $P : [0, d_P) \to \mathcal{V} \times \mathcal{V} \times \mathbb{R}$ given by*

$$P(t) = (\dot{\mathbf{E}}^P(t), \dot{\mathbf{H}}^P(t), \dot{\theta}^P(t)), \qquad \forall t \in [0, d_P).$$

We denote by Π the space of processes and by P_t the restriction of a process P to the time interval $[0, t)$. Given an initial state $\sigma_0 = (\mathbf{E}_0, \mathbf{H}_0, \theta_0)$ and a process P, the *transition function* is the map

$$\hat{\rho} : \Sigma \times \Pi \to \Sigma, \qquad \hat{\rho}(\sigma_0, P) = \sigma,$$

where the state $\sigma(t)$ is given by

$$\sigma(t) = (\mathbf{E}(t), \mathbf{H}(t), \theta(t)), \qquad \forall t \in [0, d_P).$$

In particular, if

$$\hat{\rho}(\sigma_0, P) = \sigma_0,$$

the pair (σ_0, P) is called a *cycle*.

For a rigid electromagnetic body the quantity

$$\begin{aligned}\mathcal{P}^i_{el}(\sigma(t), P(t)) = {} & \dot{\mathbf{D}}(\sigma(t), P(t)) \cdot \mathbf{E}(t) + \dot{\mathbf{B}}(\sigma(t), P(t)) \cdot \mathbf{H}(t) \\ & + \mathbf{J}(\sigma(t), P(t)) \cdot \mathbf{E}(t)\end{aligned}$$

can be identified as the internal power due to the electromagnetic field and therefore is referred to as the *internal electromagnetic power.*

Thus, the first law of thermodynamics for a rigid thermoelectromagnetic system can be formulated as follows.

First Law of Thermodynamics *There exists a state function* $e : \Sigma \to \mathbb{R}$, *called the internal energy, such that*

$$\dot{e} = h + \mathcal{P}^i_{el},$$

that is

$$\dot{e}(\sigma(t)) = h(\sigma(t), P(t)) + \dot{\mathbf{D}}(\sigma(t), P(t)) \cdot \mathbf{E}(t) + \dot{\mathbf{B}}(\sigma(t), P(t)) \cdot \mathbf{H}(t) \\ + \mathbf{J}(\sigma(t), P(t)) \cdot \mathbf{E}(t). \tag{6.1.14}$$

6.1.5 Second Law of Thermodynamics and the Clausius–Duhem Inequality

The second law of thermodynamics places restrictions on the set of the admissible processes. Its formulation requires the notion of reversibility of a process.

Second Law of Thermodynamics: *There exists a thermodynamic potential* $\eta : \Sigma \to \mathbb{R}$, *called the entropy, such that*

$$\dot{\eta}(\sigma(t)) \geq \frac{h(\sigma(t), P(t))}{\theta(t)} + \frac{1}{\theta^2(t)} \mathbf{q}(\sigma(t), P(t)) \cdot \nabla\theta(t). \tag{6.1.15}$$

Substituting the expression (6.1.14) for h, we obtain the *Clausius–Duhem inequality*

$$\theta\dot{\eta} \geq \dot{e} - \dot{\mathbf{D}} \cdot \mathbf{E} - \dot{\mathbf{B}} \cdot \mathbf{H} - \mathbf{J} \cdot \mathbf{E} + \frac{1}{\theta}\mathbf{q} \cdot \nabla\theta. \tag{6.1.16}$$

The Helmholtz free energy is defined by (3.4.13) in a mechanics context. We adopt the same definition here, putting

$$\psi = e - \theta\eta.$$

Using this quantity, the inequality (6.1.16) can be written in the following equivalent form

$$\dot{\psi} + \dot{\theta}\eta \leq \dot{\mathbf{D}} \cdot \mathbf{E} + \dot{\mathbf{B}} \cdot \mathbf{H} + \mathbf{J} \cdot \mathbf{E} - \frac{1}{\theta}\mathbf{q} \cdot \nabla\theta.$$

6.1.6 Thermodynamics of Nonlocal Materials

The expression (6.1.14) for the first law of thermodynamics, i.e.

$$\dot{e} = h + \dot{\mathbf{D}} \cdot \mathbf{E} + \dot{\mathbf{B}} \cdot \mathbf{H} + \mathbf{J} \cdot \mathbf{E}, \tag{6.1.17}$$

holds for a wide class of electromagnetic materials. Using the terminology introduced for mechanical systems, these are called *simple materials*. Classical or hereditary dielectrics and conductors obeying Ohm's law (6.1.10) belong to this class.

However, for other cases, there exist many physical phenomena that cannot be described by the model of simple materials. These are termed nonsimple or nonlocal materials , for which the form (6.1.17) of the first law is not appropriate. We note that this general topic is also delay with in Sect. 3.7 and Chap. 21.

One example of a nonsimple material is given by a dielectric with quadrupoles, which are characterized by the constitutive equations

$$\begin{aligned} \mathbf{D}(\mathbf{x}, t) &= \varepsilon_0 \mathbf{E}(\mathbf{x}, t) - \varepsilon_1 \Delta \mathbf{E}(\mathbf{x}, t) - \varepsilon_2 \nabla \left[\nabla \cdot \mathbf{E}(\mathbf{x}, t) \right], \\ \mathbf{B}(\mathbf{x}, t) &= \mu \mathbf{H}(\mathbf{x}, t), \\ \mathbf{J}(\mathbf{x}, t) &= \mathbf{0}, \end{aligned} \tag{6.1.18}$$

where ε_0, ε_1, ε_2 together with μ are positive characteristic constants.

Referring to Poynting's theorem and using Maxwell's equations $(6.1.6)_2$ and $(6.1.6)_1$, we have

$$\begin{aligned} \dot{\mathbf{D}} \cdot \mathbf{E} + \dot{\mathbf{B}} \cdot \mathbf{H} + \mathbf{J} \cdot \mathbf{E} &= \nabla \times \mathbf{H} \cdot \mathbf{E} - \nabla \times \mathbf{E} \cdot \mathbf{H} \\ &= -\nabla \cdot (\mathbf{E} \times \mathbf{H}). \end{aligned}$$

Hence, using the constitutive relations (6.1.18), we obtain

$$\begin{aligned} &\frac{1}{2} \frac{d}{dt} \left[\mu \mathbf{H}^2 + \varepsilon_0 \mathbf{E}^2 + \varepsilon_1 |\nabla \mathbf{E}|^2 + \varepsilon_2 (\nabla \cdot \mathbf{E})^2 \right] \\ &= -\nabla \cdot \left[\mathbf{E} \times \mathbf{H} - \varepsilon_1 \left(\nabla \dot{\mathbf{E}} \right) \mathbf{E} - \varepsilon_2 \left(\nabla \cdot \dot{\mathbf{E}} \right) \mathbf{E} \right]. \end{aligned}$$

Thus, the internal and external electromagnetic powers are given, respectively, by

$$\begin{aligned} \mathcal{P}^i_{el} &= \frac{1}{2} \frac{d}{dt} \left[\mu \mathbf{H}^2 + \varepsilon_0 \mathbf{E}^2 + \varepsilon_1 |\nabla \mathbf{E}|^2 + \varepsilon_2 (\nabla \cdot \mathbf{E})^2 \right], \\ \mathcal{P}^e_{el} &= -\nabla \cdot \left[\mathbf{E} \times \mathbf{H} - \varepsilon_1 \left(\nabla \dot{\mathbf{E}} \right) \mathbf{E} - \varepsilon_2 \left(\nabla \cdot \dot{\mathbf{E}} \right) \mathbf{E} \right]. \end{aligned}$$

Therefore, the first law of thermodynamics can be put in the general form

$$\dot{e} = h + \mathcal{P}^i_{el} = h + \frac{1}{2} \frac{d}{dt} \left[\mu \mathbf{H}^2 + \varepsilon_0 \mathbf{E}^2 + \varepsilon_1 |\nabla \mathbf{E}|^2 + \varepsilon_2 (\nabla \cdot \mathbf{E})^2 \right]. \tag{6.1.19}$$

An alternative approach is to modify the representation (6.1.17) by introducing an extra flux **N**. Thus, the first law is written in the form

$$\dot{e} = \dot{\mathbf{D}} \cdot \mathbf{E} + \dot{\mathbf{B}} \cdot \mathbf{H} + \mathbf{J} \cdot \mathbf{E} - \nabla \cdot \mathbf{N} - \nabla \cdot \mathbf{q} + r. \tag{6.1.20}$$

In the case of a dielectric with quadrupoles, the extra flux **N** is given by

$$\mathbf{N} = -\varepsilon_1 \left(\nabla \dot{\mathbf{E}} \right) \mathbf{E} - \varepsilon_2 \left(\nabla \cdot \dot{\mathbf{E}} \right) \mathbf{E}.$$

Such an example establishes that the formulations (6.1.17) and (6.1.20) of the first law are not general; they must be adapted to the particular material under consideration.

In contrast, relation (6.1.19) applies to all materials; it is the classical representation of the first law in the original framework of equilibrium thermodynamics.

6.1.7 Two Potentials Related to the Electromagnetic Fields

The Maxwell equation

$$\nabla \cdot \mathbf{B} = 0$$

establishes that **B** is a solenoidal field. Consequently, there exists a vector field **A**, the *magnetic potential*, such that

$$\mathbf{B} = \nabla \times \mathbf{A}. \tag{6.1.21}$$

By substituting (6.1.21) into Faraday's equation,

$$\nabla \times \mathbf{E} + \frac{\partial \mathbf{B}}{\partial t} = \mathbf{0},$$

it follows that

$$\nabla \times \left(\mathbf{E} + \frac{\partial \mathbf{A}}{\partial t} \right) = \mathbf{0}.$$

Therefore, there exists a scalar potential Φ, the *electric potential*, given by

$$\mathbf{E} = -\nabla \Phi - \frac{\partial \mathbf{A}}{\partial t}. \tag{6.1.22}$$

The magnetic and electric potentials, **A** and Φ, are not uniquely determined; different potentials can lead to the same fields **B** and **E**, satisfying (6.1.21) and (6.1.22). In fact, for any scalar function χ, the transformations

$$\mathbf{A} \to \mathbf{A}' = \mathbf{A} + \nabla \chi, \qquad \Phi \to \Phi' = \Phi - \frac{\partial \chi}{\partial t} \tag{6.1.23}$$

do not change the values of **B** and **E**. These are known as *gauge transformations*. Using (6.1.23), we obtain the following equality

$$\frac{\partial \Phi'}{\partial t} + \nabla \cdot \mathbf{A}' = \frac{\partial \Phi}{\partial t} - \frac{\partial^2 \chi}{\partial t^2} + \nabla \cdot \mathbf{A} + \Delta \chi.$$

Thus, if χ is a solution of the equation

$$\Delta \chi - \frac{\partial^2 \chi}{\partial t^2} = -\frac{\partial \Phi}{\partial t} - \nabla \cdot \mathbf{A},$$

then the potentials Φ' and $\mathbf{A}'$ satisfy the following equation

$$\frac{\partial \Phi'}{\partial t} + \nabla \cdot \mathbf{A}' = 0,$$

which is the *Lorentz condition*.

6.2 Electromagnetic Systems with Memory

There are very simple phenomena in electromagnetic materials that can be studied by considering these as materials with memory.

In fact, in the linear case, if we consider time-harmonic variation of the independent fields, the resultant dependent quantities exhibit dependence on frequency through multiplying coefficients; therefore, the constitutive relations are expressed by means of convolutions.

Some interesting aspects of frequency dependence (Kramers–Kronig relations, integral theorems) follow from Causality, which states that the response of the material at time t does not depend on the values assumed by causal parameters after this instant t.

6.2.1 Memory Effects Justified by Waves in Water

For an isotropic, linear dielectric which allows some conduction (an imperfect dielectric), the constitutive equations are given by (6.1.9) and (6.1.10), that is

$$\mathbf{D} = \epsilon \mathbf{E}, \qquad \mathbf{B} = \mu \mathbf{H}, \qquad \mathbf{J} = \sigma \mathbf{E}.$$

Thus, we can write Maxwell's equations (6.1.6) in the form

$$\begin{aligned} \nabla \times \mathbf{E} &= -\mu \dot{\mathbf{H}}, \qquad & \nabla \times \mathbf{H} &= \sigma \mathbf{E} + \epsilon \dot{\mathbf{E}}, \\ \nabla \cdot \mathbf{E} &= 0, & \nabla \cdot \mathbf{H} &= 0, \end{aligned}$$

where it is assumed that ϵ and μ do not depend on the space variable $\mathbf{x}$ and that $\varrho = 0$.

Moreover, assuming also that σ does not depend on $\mathbf{x}$, we consider plane waves, for which the fields $\mathbf{E}$ and $\mathbf{H}$ are expressed in terms of the exponential function $\exp[i(\mathbf{k} \cdot \mathbf{x} - \omega t)]$. For such plane wave forms, we obtain

$$\begin{aligned} \nabla \cdot \mathbf{E} &= i\mathbf{E} \cdot \mathbf{k}, \qquad & \nabla \cdot \mathbf{H} &= i\mathbf{H} \cdot \mathbf{k}, \\ \nabla \times \mathbf{E} &= i\mathbf{k} \times \mathbf{E}, & \nabla \times \mathbf{H} &= i\mathbf{k} \times \mathbf{H}. \end{aligned}$$

If these quantities are to satisfy Maxwell's equations, we must have

$$\begin{aligned} \mathbf{E} \cdot \mathbf{k} &= 0, \qquad & \mathbf{H} \cdot \mathbf{k} &= 0, \\ \mathbf{k} \times \mathbf{E} &= \mu\omega \mathbf{H}, & \mathbf{k} \times \mathbf{H} &= -\omega \left(\epsilon + i\frac{\sigma}{\omega}\right)\mathbf{E}. \end{aligned} \tag{6.2.1}$$

Taking the cross product with $\mathbf{k}$ of the relations $(6.2.1)_{3,4}$, we obtain

$$\mathbf{k} \times (\mathbf{k} \times \mathbf{E}) = -\mu\omega^2 \left(\epsilon + i\frac{\sigma}{\omega}\right)\mathbf{E}, \quad \mathbf{k} \times (\mathbf{k} \times \mathbf{H}) = -\mu\omega^2 \left(\epsilon + i\frac{\sigma}{\omega}\right)\mathbf{H}.$$

Since $\mathbf{E}$ and $\mathbf{H}$ are both orthogonal to $\mathbf{k}$, by virtue of $(6.2.1)_{1,2}$, it follows that

$$k^2 = \omega^2 \mu \hat{\epsilon}, \qquad \hat{\epsilon} = \epsilon + i\frac{\sigma}{\omega}.$$

Thus, homogeneous waves are characterized by a phase speed v_{ph} and an absorption coefficient α given by

$$v_{ph} = \frac{\omega}{\mathrm{Re}k} = \frac{1}{\mathrm{Re}\,\sqrt{\mu\hat{\epsilon}}}, \qquad \alpha = 2\,\mathrm{Im}k = 2\omega\mathrm{Im}\,\sqrt{\mu\hat{\epsilon}},$$

respectively, and by an index of refraction n of the form

$$n = \frac{c}{v_{ph}} = \mathrm{Re}\sqrt{\frac{\mu\hat{\epsilon}}{\mu_0\epsilon_0}}.$$

Using these relations, we obtain the following results for a relevant case of wave propagation.

For water, n and α change significantly with the frequency ω of the wave. When the frequency assumes very low values, we have $n \simeq 9$; such a value is due to a partial orientation of the dipole moments of the molecules of the water. Near 10^{10} Hz the curve of n decreases to a value in the infrared; with values of $\omega/2\pi$ in the range 3.8×10^{14} Hz to 7.9×10^{14} Hz, that is in the visible region, $n \simeq 1.34$ with a small variation. Moreover, for values greater than 6×10^{15} Hz, there is no significant data on the index of refraction; asymptotically, we assume that n approaches unity.

6.2.2 Some Simple Models to Study Material Behavior

6.2.2.1 Dielectrics

The Lorentz [240] description of the material behavior of a dielectric is based on the effect which bound electrons have on the polarization **P** and so on **D**.

Each of these bound electrons is considered as a harmonic oscillator tied by a spring, with zero equilibrium length, to the massive molecular core. Moreover, together with the force $-e\mathbf{E}$ we suppose that a damping force proportional to the velocity of the electron is applied on this. For speeds of the electron, supposed small with respect to the speed of the light in vacuo, we may neglect the effect of magnetic induction **B** on the force.

Thus, the equation of motion of the electron assumes the form

$$m(\ddot{\mathbf{x}} + g\dot{\mathbf{x}} + \omega_0^2\mathbf{x}) = -e\mathbf{E},$$

where $\mathbf{x}$ and m denote the average position and the mass of the electron, while g and ω_0^2 are the damping and spring forces per unit mass. Let

$$\mathbf{E}(t) = \mathbf{E}_\omega \exp(-i\omega t).$$

We look for solutions of the form

$$\mathbf{x}(t) = \mathbf{x}_\omega \exp(-i\omega t).$$

This yields an expression for $\mathbf{x}_\omega$, which in turn gives

$$\mathbf{p}_\omega = -e\mathbf{x}_\omega = \frac{e^2}{m(\omega_0^2 - \omega^2 - i\omega g)}\mathbf{E}_\omega$$

for the dipole moment of the electron.

If in the unit volume there are $\tilde{N}$ molecules, each of which has Z electrons, and if f_k electrons, such that

$$\sum_k f_k = Z,$$

have damping constant g_k and binding frequency ω_k, we obtain

$$\mathbf{P}_\omega = \frac{\tilde{N}e^2}{m}\sum_k \frac{f_k}{\omega_k^2 - \omega^2 - i\omega g_k}\mathbf{E}_\omega.$$

Hence, by introducing the inductive capacity κ, defined by the relation

$$\mathbf{P} = (\kappa - 1)\epsilon_0\mathbf{E},$$

we obtain

$$\kappa = 1 + \frac{\tilde{N}e^2}{m\epsilon_0}\sum_k \frac{f_k}{\omega_k^2 - \omega^2 - i\omega g_k}.$$

Thus, the permittivity of the dielectric displacement

$$\mathbf{D} = \epsilon_0\mathbf{E} + \mathbf{P}$$

assumes the following form

$$\epsilon = \epsilon_0\left(1 + \frac{\tilde{N}e^2}{m\epsilon_0}\sum_k \frac{f_k}{\omega_k^2 - \omega^2 - i\omega g_k}\right),$$

which varies with the frequency ω.

For gases the values of ϵ do not differ much from ϵ_0.

If there is a continuous distribution of oscillators and their damping g can be neglected, we obtain

$$\epsilon - \epsilon_0 = \frac{e^2}{m}\int_0^\infty \frac{f(x)}{x^2 - \omega^2}dx,$$

where f is the density of oscillator strength, which obeys the relation

$$N = \int_0^\infty f(x)dx,$$

where N is the number of oscillators or electrons per unit volume.

If there is only one oscillator, we have

$$\kappa(\omega) - 1 = \lambda\frac{1}{\omega_0^2 - \omega^2 - i\omega g_0}, \qquad \lambda = f_0\frac{\tilde{N}e^2}{m\epsilon_0}.$$

For this case, using the identity

$$\int_0^\infty \exp(-g_0 u)\sin\left(\sqrt{\omega_0^2 - g_0^2}\,u\right)\exp(i\omega u)du = \frac{\sqrt{\omega_0^2 - g_0^2}}{\omega_0^2 - 2i\omega g_0 - \omega^2}, \tag{6.2.2}$$

and putting

$$h(u) = \frac{\lambda}{\sqrt{\omega_0^2 - g_0^2}} \exp(-g_0 u)\sin\left(\sqrt{\omega_0^2 - g_0^2}\,u\right),$$

we can write

$$\kappa(\omega) - 1 = \int_0^\infty h(u)\exp(i\omega u)du.$$

It follows from (6.2.2) that

$$\mathbf{P}(t) = \epsilon_0 \int_0^\infty h(u)\mathbf{E}(t-u)du, \tag{6.2.3}$$

for $\mathbf{E}(t) \propto \exp(-i\omega t)$. It may be observed, on considering $\mathbf{E}(t)$ given by sums of exponentials of this kind that (6.2.3) has more general validity.

6.2.2.2 Magnetic Materials

Some materials have the property that, in presence of an external magnetic field, their bound or free electrons start to travel, thereby causing a magnetic field with a magnetic moment opposite to that of the external field.

These are called *magnetic materials* and are of the following types: *paramagnetic*, *ferromagnetic*, *antiferromagnetic* and *ferrimagnetic*.

Finally, there are *diamagnetic* materials, which have less applications.

The atoms or molecules of some substances present permanent magnetic moments, randomly oriented in the absence of an externally applied field. These magnetic dipoles become aligned with any external field, which is applied to the material. Thus a field in the direction of the applied field is induced: this phenomenon is *paramagnetism*.

The ratio

$$\chi = \frac{M}{H}, \tag{6.2.4}$$

where M is the induced magnetic moment and H is the applied field, is called the *magnetic susceptibility*. When an increase of the temperature occurs, the induced field decreases since the randomness of the dipoles increases. This behavior is governed by *Curie's law*, which has the following form

$$\frac{M}{H} = \frac{C}{\theta}, \tag{6.2.5}$$

where θ is the absolute temperature and C is the *Curie constant*.

In *ferromagnetic materials*, contrary to what occurs in paramagnetic materials, the magnetic moments interact strongly with of the surrounding atoms; such an action produces an alignment of the moments in a given localized region. Moreover, in the absence of an external magnetic field we can have a residual magnetic moment in the region. This alignment is justified by considering a fictitious internal field H_E, known as *Weiss's field*, which is assumed proportional to the magnetization, i.e.,

$$H_E = \lambda M, \tag{6.2.6}$$

where the constant λ is *Weiss's field constant*. Equation (6.2.5), substituting $H + H_E$ for H and taking account of (6.2.6), becomes

$$\frac{M}{H + \lambda M} = \frac{C}{\theta},$$

whence we obtain for the susceptibility (6.2.4) the following expression

$$\chi = \frac{C}{\theta - \lambda C}. \tag{6.2.7}$$

This Eq. (6.2.7), known as the *Curie–Weiss law*, well represents the susceptibility above the *Curie point temperature*, defined as $\theta = \lambda C$.

In *antiferromagnetic materials* the interaction of neighboring magnetic moments causes antiparallel orientations of the moments; thus, the moment of one atom cancels that of its neighbor. When the temperature increases above the Curie point, the thermal energy becomes greater than the interaction energy; thus, the materials become paramagnetic.

There exist some materials, known as *ferrimagnetic*, in which the parallel moments and the antiparallel ones have different strengths; thus, we have a net magnetization.

Ferrites are interesting examples of this kind of material; a simple model for the magnetization in ferrites can be derived in the following manner.

A spinning electron is characterized by a magnetic dipole moment $\mathbf{m}$ and angular momentum $\mathbf{l}$, which are connected by

$$\mathbf{m} = \gamma \mathbf{l}, \qquad \gamma = -\frac{e}{m}, \tag{6.2.8}$$

where γ is the gyromagnetic ratio, e the electron charge and m the mass. The time derivative of $\mathbf{l}$, which is determined by the torque due to the magnetic induction $\mathbf{B}$, is given by

$$\dot{\mathbf{l}} = \mathbf{m} \times \mathbf{B}.$$

Substitution from $(6.2.8)_1$ gives

$$\dot{\mathbf{m}} = \gamma \mathbf{m} \times \mathbf{B}. \tag{6.2.9}$$

If there are N electrons per unit volume, the magnetization is given by

$$\mathbf{M} = N\mathbf{m},$$

whence, by using (6.2.9) and the relation $\mathbf{B} = \mu_0 \mathbf{H}$, we obtain

$$\dot{\mathbf{M}} = \gamma \mu_0 \mathbf{M} \times \mathbf{H}. \tag{6.2.10}$$

This equation, if the magnetic field is $\mathbf{H} = \mathbf{H}_0 = constant$, taking into account the definition $(6.2.8)_2$, assumes the form

$$\dot{\mathbf{M}} = \frac{e}{m} \mu_0 \mathbf{H}_0 \times \mathbf{M}. \tag{6.2.11}$$

From (6.2.11), it follows that the magnetization $\mathbf{M}$ precesses about $\mathbf{H}_0$ at the precession frequency

$$\omega_c = \frac{e}{m} \mu_0 |\mathbf{H}_0|.$$

Let $\mathbf{e}_z$ be the unit vector along the z axis. We assume that, in addition to a constant magnetic field $\mathbf{H}_0 = H_0 \mathbf{e}_z$, there is a magnetic field $\mathbf{H}_1$ changing over time. Thus,

$$\mathbf{H} = H_0 \mathbf{e}_z + \mathbf{H}_1$$

and hence

$$\mathbf{M} = \mathbf{M}_0 + \mathbf{M}_1.$$

Moreover, by assuming for simplicity that $|\mathbf{H}_1| \ll |\mathbf{H}_0|$ and $|\mathbf{M}_1| \ll |\mathbf{M}_0|$, it follows from (6.2.10) that

$$\mathbf{M}_0 \times \mathbf{H}_0 = \mathbf{0}, \qquad \mathbf{M}_0 = M_0 \mathbf{e}_z$$

and hence

$$\dot{\mathbf{M}}_1 = -\frac{e}{m} \mu_0 \mathbf{e}_z \times (M_0 \mathbf{H}_1 - H_0 \mathbf{M}_1). \tag{6.2.12}$$

We observe that Eq. (6.2.12) is the linear approximation of (6.2.10) and that it is a constitutive equation of rate-type, $f(\mathbf{M}_1, \dot{\mathbf{M}}_1, \mathbf{H}_1) = 0$.

Inner multiplication of (6.2.10) by $\mathbf{M}$ gives that $\mathbf{M}^2$ is constant in time.

Landau and Lifshitz [227] observed that, if $\mathbf{M}^2$ is constant, $\dot{\mathbf{M}}$ can be considered as a combination of $\mathbf{M} \times \mathbf{H}$ and a damping term $\mathbf{M} \times (\mathbf{M} \times \mathbf{H})$, giving

$$\dot{\mathbf{M}} = \gamma_e \mathbf{M} \times \mathbf{H} - \frac{\lambda}{\mathbf{M}^2} \mathbf{M} \times (\mathbf{M} \times \mathbf{H}),$$

where $\gamma_e = \gamma \mu_0$ and λ is a parameter determining the level of damping. This equation allows for the tendency of $\mathbf{M}$ to align with $\mathbf{H}$ after a certain time.

6.2.2.3 Metals

The following simple model allows us to give a precise theory of the optical properties of metals.

In a metallic conductor there are positive ions, which are considered fixed in a space region; moreover, all the atoms have electrons, called *conduction electrons*, in the conduction band.

These electrons can be considered to be a gas.

If no electric field is applied, the force acting on any electron is on average zero, as is the total charge inside any volume element. Thus, the free electrons have thermal velocities, such that the net current is zero.

If an electric field is applied, the electrons develop velocities in the opposite direction to the field. The average of these is the *drift velocity* of the electron gas and constitutes the net current.

The conduction electrons are not entirely free since they engage with the defects in the crystalline lattice of the atoms of metal or with the distortions of the same lattice. Thus, a mean collision time τ is considered in order to study these effects; after a collision the electron velocity is taken to be random. Subsequently, the electrons, subject to the electric field $\mathbf{E}$, assume a velocity in the direction of this field; with the next collision, after a time τ, the velocity again becomes random. The drag force, due to these collisions, is determined by equating the change of momentum, $-m\dot{\mathbf{x}}$, to the force times τ.

The equation of motion of electrons subject to an electric field $\mathbf{E}$ and a drag force has the form

$$m\ddot{\mathbf{x}} + \frac{m}{\tau}\dot{\mathbf{x}} = -e\mathbf{E}, \tag{6.2.13}$$

where $\dot{\mathbf{x}}$ is the average velocity of the electrons. For an electric field $\mathbf{E}$ with sinusoidal behavior, we put

$$\mathbf{E}(t) = \mathbf{E}_\omega \exp(-i\omega t),$$

and look for solutions of the form

$$\mathbf{x} = \mathbf{x}_\omega \exp(-i\omega t). \tag{6.2.14}$$

From (6.2.14), it follows that

$$\ddot{\mathbf{x}} = -i\omega\dot{\mathbf{x}}. \tag{6.2.15}$$

Then, by substituting (6.2.15) into the differential equation (6.2.13), we can derive the following expression

$$\dot{\mathbf{x}} = -\frac{e}{m}\frac{\tau}{1 - i\omega\tau}\mathbf{E} \tag{6.2.16}$$

for the drift velocity. Therefore, the macroscopic current density $\mathbf{J}$, related to N electrons per unit volume, assumes the form

$$\mathbf{J} = -Ne\dot{\mathbf{x}} = \mathbf{J}_\omega \exp(-i\omega t), \qquad \mathbf{J}_\omega = \frac{Ne^2}{m}\frac{\tau}{1 - i\omega\tau}\mathbf{E}_\omega,$$

whence we obtain the following complex conductivity

$$\sigma = \frac{Ne^2}{m}\frac{\tau}{1 - i\omega\tau},$$

which is a function of the frequency ω.

The DC limit, that is when $\omega \to 0$, has the form

$$\sigma_{DC} = \frac{Ne^2}{m}\tau,$$

which is the *Drude conductivity*.

The collision time τ, for good conductors, such as copper, and at room temperature, is of the order of 10^{-14} s. For frequencies $\omega < 2 \times 10^{15}$ Hz, that is from *DC* to infrared, the conductivity is real and does not depend on the frequency. Finally, if $\omega \simeq 1/\tau$, or greater, the conductivity is imaginary and the behavior of the metal is comparable to that of a plasma.

6.2.2.4 The Ionosphere

The ionosphere of the Earth can be studied by considering it as a plasma or a diluted ionized gas. Thus, we take account only of the motions of free electrons while the positive ions, which are much heavier, can be considered as a fluid, which renders the ionosphere globally neutral.

Since we must assume a damped motion for the electrons because of their collisions, the model for metals can be used.

We can write (6.2.13) in the form

$$m\ddot{\mathbf{x}} + m\nu_c\dot{\mathbf{x}} = -e\mathbf{E}, \qquad \nu_c = \frac{1}{\tau}, \tag{6.2.17}$$

where $\mathbf{x}$ denotes the average displacement of an electron, $\dot{\mathbf{x}}$ is its velocity and we have introduced the *collision frequency* ν_c. Relation (6.2.15) follows from (6.2.14) as does

$$\dot{\mathbf{x}} = -i\omega\mathbf{x}.$$

Substituting these relations into (6.2.17), we derive the expressions

$$\mathbf{x} = \frac{e}{m}\frac{1}{\omega(\omega + i\nu_c)}\mathbf{E}, \qquad \dot{\mathbf{x}} = -\frac{e}{m}\frac{i}{\omega + i\nu_c}\mathbf{E} \tag{6.2.18}$$

for the displacement vector $\mathbf{x}$ and the velocity $\dot{\mathbf{x}}$. This relation for the velocity is equivalent to (6.2.16). Thus, we have

$$\mathbf{J} = -Ne\dot{\mathbf{x}} = \frac{Ne^2}{m}\frac{i}{\omega(1 + i\nu_c/\omega)}\mathbf{E}.$$

It follows that the conductivity

$$\sigma = \frac{Ne^2}{m}\frac{i}{\omega(1 + i\nu_c/\omega)}$$

is complex and depends on the frequency. Then, taking account of (6.2.18), the polarization $\mathbf{P}$ and the permittivity ϵ assume the following forms

$$\mathbf{P} = -Ne\mathbf{x} = -\frac{Ne^2}{m}\frac{1}{\omega(\omega + i\nu_c)}\mathbf{E}, \qquad \epsilon = \epsilon_0 - \frac{Ne^2}{m}\frac{1}{\omega(\omega + i\nu_c)}.$$

The permittivity ϵ can be written as

$$\epsilon = \epsilon_0 \left[1 - \frac{\omega_p^2}{\omega(\omega + i\nu_c)} \right], \qquad \omega_p = \sqrt{\frac{Ne^2}{m\epsilon_0}},$$

where we have introduced *the plasma frequency* ω_p, which is a measure of the electron density N in the plasma. Finally, the index of refraction is expressed by

$$n = \Re \sqrt{1 - \frac{\omega_p^2}{\omega(\omega + i\nu_c)}}.$$

Therefore, the wave is characterized by two frequencies, the collision frequency ν_c and the plasma frequency ω_p. Since normally $\nu_c \ll \omega_p$, three frequency domains can be considered:

- *Conductor domain*, if $\omega < \nu_c$. The permittivity is roughly pure imaginary and n is approximately equal to $\alpha/2\omega$. This result coincides with the *DC* limit of an ordinary conductor.
- *Evanescent domain*, if $\nu_c < \omega < \omega_p$. With these frequencies the ratio ϵ/ϵ_0 is real and negative; thus, the index of refraction is approximately zero and the wave amplitude decays as $\exp[-(\omega_p/c)z]$; therefore, we have evanescent waves.
- *Dielectric domain*, if $\omega > \omega_p$. For high values of the frequency, the ratio ϵ/ϵ_0 is roughly real and positive and

$$n \simeq \sqrt{1 - \left(\frac{\omega_p}{\omega}\right)^2}.$$

  The plasma becomes like a very transparent and dispersive dielectric with $n < 1$; thus, the phase velocity is greater than c.

6.2.3 The Clausius–Duhem Inequality and Its Consequences

We now discuss the thermodynamic theory of an electromagnetic field, introduced in the seventies [69, 70]; such a theory allows us to consider results derived in previous work and to introduce important arguments. An earlier discussion of thermodynamic concepts in the context of classical electromagnetic theory is given in Sects. 6.1.4 and 6.1.5.

The *Dissipation Principle*, applied to every set of fields satisfying the balance equations, states that the *rate of production of entropy* $R = R(P, t)$, for any subregion P of the body and any time t, must satisfy the following inequality

$$R(P,t) := \frac{d}{dt} \int_P \eta \, dv + \int_{\partial P} \frac{1}{\theta} \mathbf{q} \cdot \mathbf{n} \, da - \int_P \frac{r}{\theta} \, dv \geq 0.$$

Let us introduce a function γ, such that

$$R := \int_P \gamma \, dv.$$

In order to determine the form of this new function γ, we apply the divergence theorem to the second integral and use (6.1.13); it follows that (*cf.* (6.1.15))

$$R(P,t) = \int_P \frac{1}{\theta}\left(\theta\dot{\eta} - \frac{1}{\theta}\mathbf{q}\cdot\nabla\theta - h\right) dv. \tag{6.2.19}$$

The *free enthalpy density* is given by

$$\zeta = e - \theta\eta - \mathbf{D}\cdot\mathbf{E} - \mathbf{B}\cdot\mathbf{H}.$$

Taking the time derivative gives

$$\dot{\zeta} + \eta\dot{\theta} + \mathbf{D}\cdot\dot{\mathbf{E}} + \mathbf{B}\cdot\dot{\mathbf{H}} = \dot{e} - \theta\dot{\eta} - \dot{\mathbf{D}}\cdot\mathbf{E} - \dot{\mathbf{B}}\cdot\mathbf{H}. \tag{6.2.20}$$

Now, by substituting in (6.2.19) the expression for h derived from (6.1.17), we obtain

$$R(P,t) = \int_P \frac{1}{\theta}\left(\theta\dot{\eta} - \frac{1}{\theta}\mathbf{q}\cdot\nabla\theta - \dot{e} + \dot{\mathbf{D}}\cdot\mathbf{E} + \dot{\mathbf{B}}\cdot\mathbf{H} + \mathbf{J}\cdot\mathbf{E}\right) dv,$$

whence, on using (6.2.20), it follows that

$$\gamma = \frac{1}{\theta}\left[\mathbf{E}\cdot\mathbf{J} - \left(\dot{\zeta} + \eta\dot{\theta} + \mathbf{D}\cdot\dot{\mathbf{E}} + \mathbf{B}\cdot\dot{\mathbf{H}}\right)\right] - \frac{1}{\theta^2}\mathbf{q}\cdot\nabla\theta. \tag{6.2.21}$$

Moreover, the Dissipation Principle is equivalent to the Clausius–Duhem inequality

$$\gamma \geq 0, \qquad \forall \mathbf{x} \in \Omega, \quad \forall t \in \mathbb{R}, \tag{6.2.22}$$

for all fields which satisfy the balance equations. This may be shown by invoking the arbitrariness of the subregion P and the continuity of γ.

We now consider the particular case of an electromagnetic material with fading memory (see [69, 70]).

Putting $\Gamma = (\mathbf{E}, \mathbf{H}, \theta, \mathbf{g})$, where $\mathbf{g} = \nabla\theta$, we assume that all the physical quantities ζ, $\mathbf{D}$, $\mathbf{B}$, $\mathbf{J}$, η and $\mathbf{q}$ depend on the histories $\Gamma^t = (\mathbf{E}^t, \mathbf{H}^t, \theta^t, \mathbf{g}^t)$, each of which is an element of a suitable fading memory space with norm $\|\ \|$, such that $\|\Gamma^t\| < \infty$.

We separate the dependence on the present value $\Gamma(t)$ and on the restricted history $_r\Gamma^t$, defined by $_r\Gamma^t(s) = \Gamma(t-s)\ \forall s \in \mathbb{R}^{++}$. The quantities $\Gamma(t)$, $_r\Gamma^t \in \mathfrak{V}_{10}$, a 10-dimensional vector space.

By assuming appropriate smoothness for the constitutive functionals $\hat{\zeta}$, $\hat{\mathbf{D}}$, $\hat{\mathbf{B}}$, $\hat{\mathbf{J}}$, $\hat{\eta}$ and $\hat{\mathbf{q}}$, we have

$$\dot{\zeta} = \partial_\Gamma\hat{\zeta}(\Gamma^t)\cdot\dot{\Gamma}(t) + \delta\hat{\zeta}(\Gamma^t|_r\dot{\Gamma}^t),$$

where we have denoted by ∂_Γ the derivative of $\hat{\zeta}$ with respect to the current value $\Gamma(t)$ and by $\delta\hat{\zeta}$ the Fréchet differential of $\hat{\zeta}$ at the first argument in the direction of the second one. Note that

$$\partial_\Gamma \hat{\zeta}(\Gamma^t) = \partial_{\mathbf{E}} \hat{\zeta}(\Gamma^t) \cdot \dot{\mathbf{E}} + \partial_{\mathbf{H}} \hat{\zeta}(\Gamma^t) \cdot \dot{\mathbf{H}} + \partial_\theta \hat{\zeta}(\Gamma^t)\dot{\theta} + \partial_{\mathbf{g}} \hat{\zeta}(\Gamma^t) \cdot \dot{\mathbf{g}}.$$

Thus, the expression for γ can be put in the following form

$$\begin{aligned}\gamma = &-\frac{1}{\theta}\Big\{[\partial_{\mathbf{E}}\hat{\zeta}(\Gamma^t) + \hat{\mathbf{D}}(\Gamma^t)] \cdot \dot{\mathbf{E}} + [\partial_{\mathbf{H}}\hat{\zeta}(\Gamma^t) + \hat{\mathbf{B}}(\Gamma^t)] \cdot \dot{\mathbf{H}} \\ &+[\partial_\theta\hat{\zeta}(\Gamma^t) + \hat{\eta}(\Gamma^t)]\dot{\theta} + \partial_{\mathbf{g}}\hat{\zeta}(\Gamma^t) \cdot \dot{\mathbf{g}}\Big\} \\ &+\frac{1}{\theta}\left[\mathbf{E} \cdot \hat{\mathbf{J}}(\Gamma^t) - \frac{1}{\theta}\hat{\mathbf{q}}(\Gamma^t) \cdot \mathbf{g}\right] - \frac{1}{\theta}\delta\hat{\zeta}(\Gamma^t|{}_r\dot{\Gamma}^t).\end{aligned} \tag{6.2.23}$$

Let us now examine the inequality (6.2.22). We must allow free variation of the whole history and of its time derivative, that is of the present values $\Gamma(t)$ and $\dot{\Gamma}(t)$ and of the histories ${}_r\Gamma^t$ and ${}_r\dot{\Gamma}^t$, at any point. Thus, after substituting the constitutive functionals $\hat{\zeta}$, $\hat{\mathbf{D}}$, $\hat{\mathbf{B}}$, $\hat{\mathbf{J}}$, $\hat{\eta}$ and $\hat{\mathbf{q}}$ in the balance equations, we must study the functional differential equations so obtained.

Corresponding to each independent choice of ${}_r\Gamma^t$, $\Gamma(t)$ and $\dot{\Gamma}(t)$ for any point, we assume that there exists at least one admissible history Γ^t. In [69, 70] an analogous assertion is described as the *assumption of solvability*.

For any vector $\boldsymbol{\Phi} \in \mathfrak{V}_{10}$ and any history $\boldsymbol{\Psi} : \mathbb{R}^+ \to \mathfrak{V}_{10}$, there exist Γ^t and $\dot{\Gamma}(t)$ such that

$$\dot{\Gamma}(\mathbf{x}, t) = \boldsymbol{\Phi}, \qquad \Gamma^t(\mathbf{x}, s) = \Psi(s) \quad \forall s \in \mathbb{R}^{++},$$

that is they can be chosen arbitrarily, for any point $\mathbf{x}$ of the body and any time t. Thus, from the arbitrariness of $\boldsymbol{\Phi}$ and hence of $\dot{\Gamma}$ it follows that in (6.2.23) the first three quantities in brackets must be equal to zero; also

$$\partial_{\mathbf{g}}\hat{\zeta} = \mathbf{0},$$

while the remaining part is nonnegative.

Therefore, we can summarize these results in the following manner.

The Dissipation Principle holds if

(a) ζ together with $\mathbf{D}$, $\mathbf{B}$ and η are expressed as functionals of $\mathbf{E}^t$, $\mathbf{H}^t$, θ^t and ${}_r\mathbf{g}^t$,
(b) the functionals $\hat{\mathbf{D}}$, $\hat{\mathbf{B}}$ and $\hat{\eta}$ can be derived by means of the functional $\hat{\zeta}$ through the relations

$$\hat{\mathbf{D}} = -\partial_{\mathbf{E}}\hat{\zeta}, \qquad \hat{\mathbf{B}} = -\partial_{\mathbf{H}}\hat{\zeta}, \qquad \hat{\eta} = -\partial_\theta\hat{\zeta}, \tag{6.2.24}$$

(c) the following inequality

$$\theta\delta\hat{\zeta}(\Gamma^t|{}_r\dot{\Gamma}^t) \leq \theta\mathbf{E} \cdot \hat{\mathbf{J}}(\Gamma^t) - \mathbf{g} \cdot \hat{\mathbf{q}}(\Gamma^t)$$

holds for all $\Gamma^t \in C^1$ such that $||\Gamma^t|| < \infty$.

Using conditions (a) and (c), and putting

$$\mathcal{D} = -\delta\hat{\zeta}(\Gamma^t|_r\Gamma^t),$$

we see that the expression (6.2.23) for γ is reduced to

$$\theta\gamma = \mathbf{E}\cdot\hat{\mathbf{J}}(\Gamma^t) - \frac{1}{\theta}\mathbf{g}\cdot\hat{\mathbf{q}}(\Gamma^t) + \mathcal{D} \geq 0. \tag{6.2.25}$$

Thus, the necessary conditions (a), (b), and (c) are also sufficient to ensure that $\gamma(x,t) \geq 0$ for any time evolution, which is smooth at position $\mathbf{x}$ and time t and satisfies Maxwell's equations.

Moreover, from the Dissipation Principle, by assuming that $\hat{\mathbf{D}}$, $\hat{\mathbf{B}}$ and $\hat{\eta}$ are continuously differentiable with respect to the present values $\mathbf{E}(t)$, $\mathbf{H}(t)$ and $\theta(t)$ and by calculating in (6.2.24) the derivatives of $\hat{\mathbf{D}}$ and $\hat{\mathbf{B}}$ with respect to $\mathbf{H}$ and $\mathbf{E}$, respectively, and analogously the derivatives of $\hat{\mathbf{D}}$ and $\hat{\eta}$ with respect of θ and $\mathbf{E}$ and the derivatives of $\hat{\mathbf{B}}$ and $\hat{\eta}$ with respect of θ and $\mathbf{H}$, we obtain the following relations

$$\partial_{\mathbf{H}}\hat{\mathbf{D}} = \partial_{\mathbf{E}}\hat{\mathbf{B}}, \qquad \partial_{\theta}\hat{\mathbf{D}} = \partial_{\mathbf{E}}\hat{\eta}, \qquad \partial_{\theta}\hat{\mathbf{B}} = \partial_{\mathbf{H}}\hat{\eta}.$$

Also, we deduce the symmetry of the following quantities

$$\partial_{\mathbf{E}}\hat{\mathbf{D}} = [\partial_{\mathbf{E}}\hat{\mathbf{D}}]^T, \qquad \partial_{\mathbf{H}}\hat{\mathbf{B}} = [\partial_{\mathbf{H}}\hat{\mathbf{B}}]^T.$$

Note in particular that these symmetry relations hold without recourse to potentials, such as ζ, if $\hat{\mathbf{D}}$ and $\hat{\mathbf{B}}$ are linear.

Let us now examine the forms assumed by the inequality (6.2.22) and the expression (6.2.25) in three particular cases. To this purpose we recall that the current density $\mathbf{J}$ can depend on the temperature gradient $\mathbf{g}$.

Firstly, we consider the case when the temperature is uniform in the body. In such a situation we obtain $\mathbf{g} = \mathbf{0}$ and the following inequality

$$\mathcal{D} \geq -\mathbf{E}\cdot\mathbf{J}(\Gamma^t, \mathbf{0}).$$

If it is assumed further that the electric field $\mathbf{E} = \mathbf{0}$, or if the material is nonconducting and hence $\mathbf{J} = \mathbf{0}$, this inequality reduces to

$$\mathcal{D} \geq 0.$$

Finally, if a material does not conduct either heat or electricity, we have $\mathbf{q} = \mathbf{0}$ and $\mathbf{J} = \mathbf{0}$; thus, (6.2.25) becomes

$$\theta\gamma = \mathcal{D} \tag{6.2.26}$$

and (6.2.21), taking into account this last relation (6.2.26), assumes the following form

$$-\left[\mathbf{D}\cdot\dot{\mathbf{E}} + \mathbf{B}\cdot\dot{\mathbf{H}} + \eta\dot{\theta}\right] = \dot{\zeta} + \mathcal{D},$$

where, apart from the sign, the quantity $\mathbf{D}\cdot\dot{\mathbf{E}}+\mathbf{B}\cdot\dot{\mathbf{H}}+\eta\dot{\theta}$ is the *generalized power* and is expressed by the sum of two quantities: $\dot{\zeta}$, which is the *recoverable power*, since it is the time derivative of the free enthalpy ζ, and $\mathcal{D}$, which is the *non recoverable power*. This last quantity is nonnegative by virtue of (c).

Consequently, we can state that $\mathcal{D}$ expresses the measure of dissipation; therefore, in non-dissipative bodies the condition $\mathcal{D} = 0$ holds for every time evolution.

6.3 Thermodynamics of Simple Electromagnetic Materials

The laws of thermodynamics have an important role in constructing mathematical models, which describe the behavior of various types of electromagnetic materials. In fact, they allow us to state whether a constitutive equation is admissible for a given material and to establish the function spaces in which the related mathematical problems have their natural surroundings. Moreover, there is the possibility of deriving some important thermodynamic potentials, such as entropy, free energy and enthalpy, for these materials.

All fields which play a part in any given model must satisfy the balance equations, which in the present case are Maxwell's equations, and the assumed constitutive equations. Moreover, among the many constitutive relations that we may choose, only those which fulfil the second law of thermodynamics are admissible.

Therefore, the second law is an important tool, which allows us to select the physically admissible constitutive equations, while, from a mathematical point of view, it leads to the determination of some restrictions on the constitutive functionals, related to the inequalities which express this law.

In order to describe the behavior of an electromagnetic material, as in the models of other materials, it is useful to introduce the concepts of states and processes (see Chap. 4 and Sect. 6.1.4) and to consider cyclic and reversible processes. In this section, the second law is established for cyclic electromagnetic processes. This provides a principle of dissipation of electromagnetic energy.

6.3.1 Electromagnetic Materials

We now examine only exclusively electromagnetic phenomena; for the sake of simplicity, thermal effects will be neglected.

We denote by $\mathcal{B}$ a body which occupies the region $\mathcal{R}$ in a three-dimensional Euclidean space. Any point of $\mathcal{B}$ is identified by the position vector $\mathbf{x} \in \mathcal{R}$.

The electromagnetic field is given by the set of the vectors **E**, **H**, **D**, **B** and **J**, each of which is a function of **x** and of the time t. As previously, the dependence on **x** will not usually be shown explicitly, while the dependence on t will be specified; thus, we shall write $\mathbf{E}(t)$, $\mathbf{H}(t)$, etc.

Definition 6.3.1. *An electromagnetic process P of duration d_P, defined by*

$$P(t) = (\dot{\mathbf{E}}^P(t), \dot{\mathbf{H}}^P(t)), \qquad \forall t \in [0, d_P), \tag{6.3.1}$$

is a piecewise continuous function on $[0, d_P)$, with values in $\mathcal{V} \times \mathcal{V}$, $\dot{\mathbf{E}}^P$ and $\dot{\mathbf{H}}^P$ being the time derivatives of the electric field **E** *and the magnetic field* **H**, *at the fixed point* **x**.

The restriction of P to $[t_1, t_2) \subset [0, d_P)$ will be denoted by $P_{[t_1,t_2)}$; moreover, we shall write P_t instead of $P_{[0,t)}$.

Let P_1 and P_2 be two processes with durations d_{P_1} and d_{P_2}, respectively. Recalling Definition 3.2.1, we define the composition $P_1 * P_2$ of P_2 with P_1 by

$$(P_1 * P_2)(t) = \begin{cases} P_1(t), & t \in [0, d_{P_1}), \\ P_2(t - d_{P_1}), & t \in [d_{P_1}, d_{P_1} + d_{P_2}). \end{cases}$$

Materials, characterized by processes defined by (6.3.1), in which the time evolutions of $\mathbf{E}$ and $\mathbf{H}$ at the fixed point $\mathbf{x} \in \mathcal{B}$ are involved are *simple materials*.

There are other materials for which, at a fixed point, the process P depends by the time evolutions of $\mathbf{E}$ and $\mathbf{H}$ in the whole body $\mathcal{B}$ or in a neighborhood of the fixed point. These materials are *nonsimple* and the related phenomena are considered *nonlocal*; this occurs for example when the process depends by the time evolutions of the gradients $\nabla\mathbf{E}$ and $\nabla\mathbf{H}$.

We shall consider simple materials unless otherwise specified.

The behavior of an electromagnetic material depends on the process P and is expressed by the following function

$$R(t) = (\mathbf{D}(t), \mathbf{B}(t), \mathbf{J}(t)), \qquad \forall t \in [0, d_P),$$

such that $R : [0, d_P) \to \mathcal{V} \times \mathcal{V} \times \mathcal{V}$, in which $\mathbf{D}$ denotes the displacement vector, $\mathbf{B}$ the magnetic induction and $\mathbf{J}$ the current density.

We shall specify below that R is a compound function of t through the state $\sigma(t)$ and the process $P(t)$.

The set Π of electromagnetic processes P has the following properties:

(a) if $P \in \Pi$, for any $[t_1, t_2) \subset [0, d_P)$ the restriction $P_{[t_1,t_2)} \in \Pi$,
(b) if P_1 and $P_2 \in \Pi$, their composition $P_1 * P_2 \in \Pi$ also.

Definition 6.3.2. *A simple electromagnetic system is characterized by the set* $\{\Pi, \Sigma, \varrho, R\}$, *such that*

(1) Π is the set of processes,
(2) Σ, known as the state space, is the set of elements σ each of which is a state of the system,
(3) $\varrho : \Sigma \diamond \Pi \to \Sigma$, where $\Sigma \diamond \Pi \subset \Sigma \times \Pi$, is the evolution function, which transforms the state σ_1 under a process P into $\sigma_2 = \varrho(\sigma_1, P)$ and is such that

$$\varrho(\sigma, P_1 * P_2) = \varrho(\varrho(\sigma, P_1), P_2), \qquad \forall P_1, P_2 \in \Pi, \ \forall \sigma \in \Sigma,$$

(4) $R : \Sigma \times \Pi \to \mathcal{V} \times \mathcal{V} \times \mathcal{V}$ is the response function, which associates, to any initial state and any process, the response of the system at the end of the process.

This property can be expressed by

$$\mathbf{D}(t) = \hat{\mathbf{D}}(\sigma(t), P(t)), \ \mathbf{B}(t) = \hat{\mathbf{B}}(\sigma(t), P(t)), \ \mathbf{J}(t) = \hat{\mathbf{J}}(\sigma(t), P(t)). \qquad (6.3.2)$$

The function ϱ, for any state σ and process P, gives a *family of states** $\sigma(t) = \varrho(\sigma, P_t)$, $\forall t \in [0, d_P)$. A family of states is called a *cycle* if $\varrho(\sigma, P) = \sigma$. The time-domain of any process P and of the family of states will be denoted by $[0, d)$.

We now give the definition of equivalent states.

---

* In textbooks on thermodynamics the terms transformation (cf. [132]) or process (cf. [329]) are used.

Definition 6.3.3. *Two states σ_1 and σ_2 are* equivalent *if the following relations*

$$\hat{\mathbf{D}}(\sigma_1, P_\tau) = \hat{\mathbf{D}}(\sigma_2, P_\tau), \ \hat{\mathbf{B}}(\sigma_1, P_\tau) = \hat{\mathbf{B}}(\sigma_2, P_\tau), \ \hat{\mathbf{J}}(\sigma_1, P_\tau) = \hat{\mathbf{J}}(\sigma_2, P_\tau) \qquad (6.3.3)$$

are satisfied for any $\tau > 0$ and any process P of duration τ.

Such a definition allows us to introduce the concept of a *minimal state* as the equivalence class of the states which satisfy (6.3.3).

A non-conducting dielectric is an example of an electromagnetic system, for which the relations (6.3.2) assume the form

$$\mathbf{D}(t) = \hat{\mathbf{D}}(\mathbf{E}(t)), \qquad \mathbf{B}(t) = \hat{\mathbf{B}}(\mathbf{H}(t)), \qquad \mathbf{J}(t) = \mathbf{0}. \qquad (6.3.4)$$

Hence, it follows that the resulting state σ is formed by the pair $(\mathbf{E}, \mathbf{H})$; thus $\Sigma = \mathcal{V} \times \mathcal{V}$. Moreover, the transition function ϱ, given a state $\sigma_0 = (\mathbf{E}_0, \mathbf{H}_0)$ and a process $P = (\dot{\mathbf{E}}^P, \dot{\mathbf{H}}^P)$, assumes the form

$$\mathbf{E}(t) = \int_0^t \dot{\mathbf{E}}^P(\xi)d\xi + \mathbf{E}_0, \qquad \mathbf{H}(t) = \int_0^t \dot{\mathbf{H}}^P(\xi)d\xi + \mathbf{H}_0,$$

while the response function, independent of the process, is given by (6.3.4). This gives the fields $\mathbf{D}$, $\mathbf{B}$ and $\mathbf{J}$ which correspond to each pair $(\mathbf{E}, \mathbf{H})$ so that $R : \Sigma \to \mathcal{V} \times \mathcal{V} \times \mathcal{V}$.

In the particular case when $\hat{\mathbf{D}}$ and $\hat{\mathbf{B}}$ are assumed linear, we have

$$\mathbf{D} = \boldsymbol{\epsilon}\mathbf{E}, \qquad \mathbf{B} = \boldsymbol{\mu}\mathbf{H},$$

where $\boldsymbol{\epsilon}$ and $\boldsymbol{\mu}$ are tensors, which belong to the space Lin($\mathcal{V}$) and are usually assumed to be positive definite.

6.3.2 Materials with Fading Memory

We have already introduced the model of an electromagnetic system with fading memory (Sect. 6.2). A norm for the state space of this model has not yet been assigned. The influence function, which is associated with a norm, may not depend physically on the material properties, but its choice can greatly influence the theory (see [135, 136]).

We note that Definition 6.3.2 of a simple material does not imply any norm (for such a definition see [277], also).

For this reason, we shall now define two models which characterize dielectrics and conductors with memory, respectively, without introducing any norm in the state spaces.

6.3.2.1 Dielectrics with Memory

As already stated, any history can be expressed by means of the present value and the past history; thus, we shall denote the history of the electric field $\mathbf{E}$ by the pair

$\mathbf{E}^t = (\mathbf{E}(t), \mathbf{E}_r^t)$, where $\mathbf{E}_r^t(u) = \mathbf{E}^t(u)\ \forall u > 0$ and, analogously, the history of the magnetic field $\mathbf{H}$ by $\mathbf{H}^t = (\mathbf{H}(t), \mathbf{H}_r^t)^\dagger$ Given a pair of histories $(\mathbf{E}^t, \mathbf{H}^t)$, we can introduce two particular histories:

- the *static continuation* of duration $a \geq 0$ expressed by the pair of histories $({}_a\mathbf{E}^t,\ {}_a\mathbf{H}^t)$ thus defined

$$({}_a\mathbf{E}^t(\xi),\ {}_a\mathbf{H}^t(\xi)) = \begin{cases} (\mathbf{E}^t(\xi - a), \mathbf{H}^t(\xi - a)), & \xi > a, \\ (\mathbf{E}(t), \mathbf{H}(t)), & \xi \leq a, \end{cases}$$

- the *static relaxation* before a time $b \geq 0$ denoted by $({}^b\mathbf{E}^t,\ {}^b\mathbf{H}^t)$ and defined by

$$({}^b\mathbf{E}^t(\xi),\ {}^b\mathbf{H}^t(\xi)) = \begin{cases} (\mathbf{E}^t(\xi), \mathbf{H}^t(\xi)), \ 0 \leq \xi < b, \\ (\mathbf{E}^t(b), \mathbf{H}^t(b)), \ \xi \geq b. \end{cases}$$

In order to characterize dielectric materials, we now give the following definition.

Definition 6.3.4. *The constitutive equations of a dielectric with memory have the following form*

$$\mathbf{D}(t) = \hat{\mathbf{D}}(\mathbf{E}^t, \mathbf{H}^t), \qquad \mathbf{B}(t) = \hat{\mathbf{B}}(\mathbf{E}^t, \mathbf{H}^t), \qquad \mathbf{J} = \mathbf{0}. \tag{6.3.5}$$

The domain of the functions $\mathbf{E}(t)$, $\mathbf{H}(t)$, $\mathbf{E}_r^t$ *and* $\mathbf{H}_r^t$, *which express the pair of histories* $(\mathbf{E}^t, \mathbf{H}^t)$, *is* $\mathcal{D} = \mathcal{V} \times \mathcal{V} \times \mathcal{D}_r$, *where* $\mathcal{D}_r$ *denotes the set of pairs of past histories* $(\mathbf{E}_r^t, \mathbf{H}_r^t) : \mathbb{R}^{++} \times \mathbb{R}^{++} \to \mathcal{V} \times \mathcal{V}$ *such that*

1. $\mathcal{D}_r \supset L^\infty(\mathbb{R}^{++}) \times L^\infty(\mathbb{R}^{++})$;
2. *if* $(\mathbf{E}^t, \mathbf{H}^t) \in \mathcal{D}$, *then the static continuation* $({}_a\mathbf{E}^t,\ {}_a\mathbf{H}^t) \in \mathcal{D}\ \forall a \geq 0$;
3. *there exist two functions* $\tilde{\mathbf{D}}$ *and* $\tilde{\mathbf{B}}$ *on* $\mathcal{V} \times \mathcal{V}$ *such that*

$$\lim_{a\to\infty} \hat{\mathbf{D}}({}_a\mathbf{E}^t,\ {}_a\mathbf{H}^t) = \tilde{\mathbf{D}}(\mathbf{E}(t), \mathbf{H}(t)),\ \lim_{a\to\infty} \hat{\mathbf{B}}({}_a\mathbf{E}^t,\ {}_a\mathbf{H}^t) = \tilde{\mathbf{B}}(\mathbf{E}(t), \mathbf{H}(t)); \tag{6.3.6}$$

4. *if* $\mathbf{E}^t$, $\mathbf{H}^t \in \mathcal{D}$, *then* ${}^b\mathbf{E}^t$, ${}^b\mathbf{H}^t \in \mathcal{D}$, $\forall b \geq 0$, *and*

$$\lim_{b\to\infty} \hat{\mathbf{D}}({}^b\mathbf{E}^t,\ {}^b\mathbf{H}^t) = \hat{\mathbf{D}}(\mathbf{E}^t, \mathbf{H}^t),\ \lim_{b\to\infty} \hat{\mathbf{B}}({}^b\mathbf{E}^t,\ {}^b\mathbf{H}^t) = \hat{\mathbf{B}}(\mathbf{E}^t, \mathbf{H}^t).$$

We observe that a dielectric without memory have the functions $\tilde{\mathbf{D}}$ and $\tilde{\mathbf{B}}$, introduced in (6.3.6), as constitutive functions.

Moreover, from the form (6.3.5) of the constitutive equations and the related properties (6.3.6), it follows that the dielectrics with memory, which we are considering, are simple materials; the domain $\mathcal{D}$ of their constitutive equations coincides with the state space, that is $\Sigma = \mathcal{D} = \mathcal{V} \times \mathcal{V} \times \mathcal{D}_r$.

If we consider a linear model, the property of fading memory has some particular constraints on the memory kernels.

---

† The notation $\mathbf{E}_r$, $\mathbf{H}_r$ in this context is different from the usage in $(5.1.21)_1$ which is widely used in Part III.

In fact, for a linear choice of constitutive functionals (see [228, §§77–80], and [208, Chapter 7]), i.e.,

$$\mathbf{D}(t) = \boldsymbol{\varepsilon}^0 \mathbf{E}(t) + \int_0^\infty \boldsymbol{\varepsilon}'(\xi)\mathbf{E}^t(\xi)d\xi, \ \mathbf{B}(t) = \boldsymbol{\mu}^0 \mathbf{H}(t) + \int_0^\infty \boldsymbol{\mu}'(\xi)\mathbf{H}^t(\xi)d\xi, \tag{6.3.7}$$

where $\boldsymbol{\varepsilon}^0$, $\boldsymbol{\mu}^0 \in Sym$ and $\boldsymbol{\varepsilon}'$, $\boldsymbol{\mu}' : \mathbb{R}^+ \to Sym$, it follows, by assuming $\mathbf{E}^t$, $\mathbf{H}^t \in L^\infty(\mathbb{R}^+)$, and the memory kernels $\boldsymbol{\varepsilon}'$, $\boldsymbol{\mu}' \in L^1(\mathbb{R}^+)$. Moreover, if

$$\mathcal{D}_r = \left\{ (\mathbf{E}_r^t, \mathbf{H}_r^t) : \mathbb{R}^{++} \to \mathcal{V} \times \mathcal{V};\ \left| \int_0^\infty \boldsymbol{\varepsilon}'(s+a)\mathbf{E}_r^t(s)ds \right| \right.$$
$$\left. + \left| \int_0^\infty \boldsymbol{\mu}'(s+a)\mathbf{H}_r^t(s)ds \right| < +\infty,\ \forall a \geq 0 \right\}$$

and

$$\lim_{a\to\infty} \hat{\mathbf{D}}({}_a\mathbf{E}^t) = \boldsymbol{\varepsilon}_\infty \mathbf{E}(t), \qquad \lim_{a\to\infty} \hat{\mathbf{B}}({}_a\mathbf{H}^t) = \boldsymbol{\mu}_\infty \mathbf{H}(t),$$

with

$$\boldsymbol{\varepsilon}_\infty = \boldsymbol{\varepsilon}^0 + \int_0^\infty \boldsymbol{\varepsilon}'(s)ds, \qquad \boldsymbol{\mu}_\infty = \boldsymbol{\mu}^0 + \int_0^\infty \boldsymbol{\mu}'(s)ds,$$

properties 2 and 3 in (6.3.6) are satisfied.

Therefore, the functionals of a dielectric with memory, as $a \to \infty$, must become the response functions of a dielectric without memory; moreover, the equilibrium values $\boldsymbol{\varepsilon}_\infty$ and $\boldsymbol{\mu}_\infty$ are positive definite tensors. Property 4 shows that the pairs $(\mathbf{E}_r^t, \mathbf{H}_r^t)$ of the past histories of $\mathbf{E}$ and $\mathbf{H}$, which belong to $\mathcal{D}_r$, must satisfy the following relations

$$\lim_{b\to\infty} \int_b^\infty \boldsymbol{\varepsilon}'(s)ds\, \mathbf{E}_r^t(b) = \mathbf{0}, \qquad \lim_{b\to\infty} \int_b^\infty \boldsymbol{\mu}'(s)ds\, \mathbf{H}_r^t(b) = \mathbf{0}.$$

By considering the constitutive functionals (6.3.7), we see that the equivalence relation (6.3.3) between states can be expressed by an equivalence relation between the histories [90].

In fact, on using (6.3.7) again, we can show that two states $(\mathbf{E}_1^t, \mathbf{H}_1^t)$ and $(\mathbf{E}_2^t, \mathbf{H}_2^t)$ are equivalent in accordance to the definition of equivalence between states, expressed by (6.3.3), if the following equalities

$$\mathbf{E}_1(t) = \mathbf{E}_2(t), \qquad \mathbf{H}_1(t) = \mathbf{H}_2(t)$$

hold, together with

$$\int_0^\infty \boldsymbol{\varepsilon}'(s+a)\mathbf{E}_{r1}^t(s)\,ds = \int_0^\infty \boldsymbol{\varepsilon}'(s+a)\mathbf{E}_{r2}^t(s)\,ds, \quad \forall a \geq 0,$$

in terms of the past histories $\mathbf{E}_{r1}^t$, $\mathbf{E}_{r2}^t$ with the kernel $\boldsymbol{\varepsilon}'$, and the analogous equality

$$\int_0^\infty \boldsymbol{\mu}'(s+a)\mathbf{H}_{r1}^t(s)\,ds = \int_0^\infty \boldsymbol{\mu}'(s+a)\mathbf{H}_{r2}^t(s)\,ds, \quad \forall a \geq 0,$$

in terms of $\mathbf{H}_{r1}^t$, $\mathbf{H}_{r2}^t$ with the kernel $\boldsymbol{\mu}'$.

Thus, two past histories $(\mathbf{E}^t_{r1}, \mathbf{H}^t_{r1})$, $(\mathbf{E}^t_{r2}, \mathbf{H}^t_{r2}) \in \mathcal{D}_r$ are equivalent if their differences $\mathbf{E}^t_r = \mathbf{E}^t_{r1} - \mathbf{E}^t_{r2}$ and $\mathbf{H}^t_r = \mathbf{H}^t_{r1} - \mathbf{H}^t_{r2}$ are equivalent to the zero past history $(\mathbf{0}^\dagger_r, \mathbf{0}^\dagger_r)$, that is if they satisfy

$$\int_0^\infty \boldsymbol{\varepsilon}'(s+a)\mathbf{E}^t_r(s)\,ds = \mathbf{0}, \quad \int_0^\infty \boldsymbol{\mu}'(s+a)\mathbf{H}^t_r(s)\,ds = \mathbf{0}, \ \forall a \geq 0. \tag{6.3.8}$$

This means that, by denoting the set of past histories $\mathbf{E}^t_r$ and $\mathbf{H}^t_r$, which satisfy (6.3.8), by $\mathcal{D}_0$ and the quotient space by $\mathcal{D}_r/\mathcal{D}_0$, the minimal state is an element of $\Sigma^{(m)} := \mathcal{V} \times \mathcal{V} \times (\mathcal{D}_r/\mathcal{D}_0)$.

6.3.2.2 Conductors with Memory

Two examples of conductors, which have a complex-valued conductivity when the related fields have a harmonic variation in time, are metals and the ionized atmosphere (see [171, 186]), discussed in Sects. 6.2.2.3 and 6.2.2.4. For these materials we can construct a model of conductors with memory, by introducing the memory functional

$$\mathbf{J}(t) = \hat{\mathbf{J}}(\mathbf{E}^t), \tag{6.3.9}$$

expressing $\mathbf{J}$ in terms of $\mathbf{E}^t$, the history of the electric field $\mathbf{E}$, while both the electric displacement $\mathbf{D}$ and the magnetic induction $\mathbf{B}$ are expressed by functions of $\mathbf{E}$ and $\mathbf{H}$.

Thus, we give the following definition.

Definition 6.3.5. *The constitutive equations, which characterize a conductor with memory, have the form*

$$\mathbf{D}(t) = \hat{\mathbf{D}}(\mathbf{E}(t), \mathbf{H}(t)), \quad \mathbf{B}(t) = \hat{\mathbf{B}}(\mathbf{E}(t), \mathbf{H}(t)), \quad \mathbf{J}(t) = \hat{\mathbf{J}}(\mathbf{E}^t)$$

and are such that:

1. *the set $\mathcal{D}$, of histories $\mathbf{E}^t$ of the electric field $\mathbf{E}$, is the domain of the functional $\hat{\mathbf{J}}$ and satisfies the condition $\mathcal{D} \supset L^\infty(\mathbb{R}^+)$;*
2. *if $\mathbf{E}^t \in \mathcal{D}$, then the static continuation $_a\mathbf{E}^t \in \mathcal{D}$ for any $a \geq 0$ and there exists $\hat{\mathbf{J}}$, a function on $\mathcal{V}$, such that*
$$\lim_{a\to\infty} \hat{\mathbf{J}}(_a\mathbf{E}^t) = \hat{\mathbf{J}}(\mathbf{E}(t));$$
3. *if $\mathbf{E}^t \in \mathcal{D}$, then the static relaxation $^b\mathbf{E}^t \in \mathcal{D}$ for any $b \geq 0$ and*
$$\lim_{b\to\infty} \hat{\mathbf{J}}(^b\mathbf{E}^t) = \hat{\mathbf{J}}(\mathbf{E}^t).$$

We suppose that the functional (6.3.9) has the linear form

$$\mathbf{J}(t) = \int_0^\infty \sigma'(\xi)\mathbf{E}^t(\xi)d\xi,$$

where the function σ', which belongs to $L^1(\mathbb{R}^+)$ since $\mathcal{D} \supset L^\infty(\mathbb{R}^+)$, must satisfy

$$\lim_{a\to\infty} \int_0^\infty \sigma'(\xi + a)\mathbf{E}^t(\xi)d\xi = \mathbf{0} \qquad \forall \mathbf{E}^t \in \mathcal{D},$$

by virtue of property 2; moreover, the limit in property 2 gives

$$\lim_{a\to\infty} \hat{\mathbf{J}}({}_a\mathbf{E}^t) = \sigma_\infty \mathbf{E}(t), \qquad \sigma_\infty = \int_0^\infty \sigma'(\tau)\,d\tau.$$

We observe that materials with memory can be described by means of *integrated histories*, which are defined as follows

$$\grave{\mathbf{E}}^t(\xi) = \int_{t-\xi}^t \mathbf{E}(u)du = \int_0^\xi \mathbf{E}^t(s)ds \qquad \forall \xi \in \mathbb{R}^+,$$

whence

$$\frac{d}{d\xi}\grave{\mathbf{E}}^t(\xi) = \mathbf{E}^t(\xi), \qquad \grave{\mathbf{E}}^t(0) = \mathbf{0}.$$

If $\sigma' \in L^1(\mathbb{R}^+)$, then $\sigma'(\infty) = 0$ and, on integrating by parts, we obtain

$$\int_0^\infty \sigma'(\xi)\mathbf{E}^t(\xi)d\xi = -\int_0^\infty \sigma''(\xi)\grave{\mathbf{E}}^t(\xi)d\xi,$$

where it is assumed that $\sigma'' \in L^1(\mathbb{R}^+)$.

Therefore, when linear constitutive functionals are assumed, we can consider either dependence on $\mathbf{E}^t$ or on the related integrated history $\grave{\mathbf{E}}^t$; these are equivalent. This is not true when the functionals are nonlinear.

6.3.3 Thermodynamic Laws in Terms of Cycles

We shall now consider a thermodynamic theory constructed by means of the properties of cycles. For this purpose, the second law is introduced and will be considered as a principle of electromagnetic energy dissipation.

Let us consider a thermoelectromagnetic system, which occupies a region $\mathcal{R}$ with a smooth boundary $\partial\mathcal{R}$. It is assumed to be rigid, so that the mass density ρ depends only on position, even if thermal variations occur, and can be taken to be unity. It will be omitted.

The local form of the balance of energy has the following form

$$\dot{e} = -\nabla \cdot \mathbf{q} + r + \mathbf{E} \cdot \dot{\mathbf{D}} + \mathbf{H} \cdot \dot{\mathbf{B}} + \mathbf{E} \cdot \mathbf{J}, \tag{6.3.10}$$

where, as in Sect. 6.1.4, e is the internal energy, $\mathbf{q}$ denotes the heat flux and r is the heat source. The sum of the first two terms of this equality denotes the rate at which heat is absorbed per unit volume, that is (6.1.13).

The thermoelectromagnetic processes $\bar{P}$ now assumes the form

$$\bar{P}(t) = (\dot{\mathbf{E}}(t), \dot{\mathbf{H}}(t), \dot{\theta}(t), \mathbf{g}(t)), \quad \forall t \in [0, d_{\bar{P}}),$$

where the contributions of absolute temperature θ and the temperature gradient $\mathbf{g} = \nabla\theta$ are added. We shall denote by $\bar{\Pi}$ the space of these processes, by $\bar{\Sigma}$ the space of the thermoelectromagnetic states and by $\bar{\varrho}$ the map that gives the state $\bar{\sigma}$ obtained by applying a process $\bar{P} \in \bar{\Pi}$ to the state $\bar{\sigma}_0 \in \bar{\Sigma}$.

A closed path in the state space $\bar{\Sigma}$ is called *cycle*; therefore, a pair $(\bar{\sigma}, \bar{P}) \in \bar{\Sigma} \times \bar{\Pi}$ gives a cycle if $\bar{\varrho}(\bar{\sigma}, \bar{P}) = \bar{\sigma}$.

Finally, the time-domain of the process is always denoted by $[0, d)$.

Taking into account (6.3.10) and (6.1.13) and the fact that in a cycle we have

$$e(\bar{\sigma}(\mathbf{x}, d)) = e(\bar{\sigma}(\mathbf{x}, 0)),$$

at any $\mathbf{x} \in \mathcal{R}$, we make the following claim. *First Law of Thermodynamics*. The equality

$$\int_0^d \int_{\mathcal{R}} \left[h(t) + \dot{\mathbf{D}}(t) \cdot \mathbf{E}(t) + \dot{\mathbf{B}}(t) \cdot \mathbf{H}(t) + \mathbf{J}(t) \cdot \mathbf{E}(t)\right] dvdt = 0 \tag{6.3.11}$$

must hold for any cycle, the duration of which is denoted by d.

We must consider also the very important assumption that the inequality

$$\int_0^d \left[-\nabla \cdot \left(\frac{\mathbf{q}}{\theta}\right) + \frac{r}{\theta}\right] dt \leq 0 \tag{6.3.12}$$

must be satisfied for any cycle and at any point $\mathbf{x}$ of the body.

This applies where the mass density ρ is unity. If, however, the mass density ρ varies over time, a factor $1/\rho$ must be included in the integrand of (6.3.12).

We note that the assumption made on the inequality (6.3.12) is equivalent to assuming that the integral of the entropy production per unit volume, which is expressed by $\dot{\eta} + \nabla \cdot (\mathbf{q}/\theta) - r/\theta$, is nonnegative for any cycle.

Another version, more general than the one we have considered, must be used if the material is nonlocal. For such materials, the statement of thermodynamic laws may involve the whole body. We now state this following fundamental law of thermodynamics. *Second Law of Thermodynamics*. For any cycle with duration d we have

$$\int_0^d \int_{\mathcal{R}} \left[\frac{h(t)}{\theta(t)} + \frac{1}{\theta^2(t)}\mathbf{q}(t) \cdot \mathbf{g}(t)\right] dvdt \leq 0. \tag{6.3.13}$$

This integrand is equal to that in (6.3.12), as may be seen with the aid of (6.1.13). A space integration over the body has been included.

If thermal effects are not considered and we study only electromagnetic phenomena, the temperature θ is constant and thus $\mathbf{g} = \mathbf{0}$; we shall continue to consider a constant mass density ρ in the body. With these hypotheses the inequality (6.3.13) becomes

$$\int_0^d \int_{\mathcal{R}} h(t) dvdt \leq 0. \tag{6.3.14}$$

Hence, by using the first law (6.3.11), for these electromagnetic phenomena, taking account of (6.3.14), it follows that

$$\int_0^d \int_{\mathcal{R}} \left[\dot{\mathbf{D}}(t) \cdot \mathbf{E}(t) + \dot{\mathbf{B}}(t) \cdot \mathbf{H}(t) + \mathbf{J}(t) \cdot \mathbf{E}(t)\right] dvdt \geq 0. \tag{6.3.15}$$

This inequality, which must hold for every cycle of duration d, expresses the global form of the principle of electromagnetic energy dissipation.

For simple materials we assume that the inequality (6.3.15) holds for any sub-body $\mathcal{A} \subset \mathcal{R}$.

The continuity of the integrand in this last inequality and the arbitrariness of the sub-body $\mathcal{A} \subset \mathcal{R}$ allow us to derive the local form of this principle, expressed by the following inequality

$$\int_0^d \left[\dot{\mathbf{D}}(t) \cdot \mathbf{E}(t) + \dot{\mathbf{B}}(t)\dot{\mathbf{H}}(t) + \mathbf{J}(t) \cdot \dot{\mathbf{E}}(t)\right] dt \geq 0,$$

which must hold for any cycle of duration d.

We now leave the topic of Thermoelectromagnetism until Chap. 22 and return to our main line of discussion.

Part III

Free Energies for Materials with Linear Memory

7

A Linear Memory Model

We now address the problem of finding explicit forms for the free energy of materials with constitutive relations given by linear memory functionals. Such materials are referred to in this work as *linear memory materials*. As we will see, the equilibrium (or alternatively, the instantaneous) contribution, which is to say the portion of the constitutive equation without memory effects, need not be linear. If the part of the constitutive equation without memory is also linear, we use the description *a completely linear material*. A linear viscoelastic material is understood to be completely linear, while a viscoelastic material with linear memory need not have this property.

A general form for the free energy with a quadratic memory term is now determined, which is the basis for all developments in Part III. In particular, we show that the associated constitutive equations have linear memory functionals.

7.1 A Quadratic Model for Free Energies

The assumption is made that $\|\mathbf{\Lambda}^t - \mathbf{\Lambda}(t)\|$ is small. Since the norm has the fading memory property, this amounts to the assumption that $\mathbf{\Lambda}^t(s)$ does not vary greatly from $\mathbf{\Lambda}(t)$ for small values of s [73].

In [159], the theory of Crochet and Naghdi [78], incorporating a temperature-dependent timescale, was adapted to make more realistic the modeling of mechanical components of the equations. For simplicity, this refinement is omitted in the present treatment. A physical consequence of this is that the allowed variation in coldness must be restricted for materials that are very temperature sensitive. There will be associated constraints on the range of $\mathbf{m}$, defined by (5.1.8).

Remark 7.1.1. The theory developed here applies to solids for finite or infinitesimal strain and to liquids if strains and strain rates are infinitesimal.

G. Amendola et al., *Thermodynamics of Materials with Memory*,
https://doi.org/10.1007/978-3-030-80534-0_7

Let $\widetilde{\psi}_h$ in (5.1.26) be twice Fréchet differentiable at the zero relative history and let the quadratic term be expressible as an integral, so that we can make the approximation

$$\psi(t) = \widetilde{\phi}(\mathbf{\Lambda}_0(t)) + \frac{1}{2}\int_0^\infty \int_0^\infty \mathbf{\Lambda}_r^t(s) \cdot \mathbb{K}(s, u, \mathbf{\Lambda}(t))\mathbf{\Lambda}_r^t(u) ds du, \qquad \mathbf{\Lambda}_r^t \in \Gamma, \qquad \mathbb{K} \in \mathrm{Lin}(\Gamma), \tag{7.1.1}$$

in the notation of (A.2.2). It will be assumed that with a small error, $\mathbf{\Lambda}(t)$ in $\mathbb{K}$ can be replaced by a constant quantity $\mathbf{\Lambda}_c$. This dependence is henceforth omitted [161]. We write (7.1.1) as

$$\psi(t) = \widetilde{\phi}(\mathbf{\Lambda}_0(t)) + \frac{1}{2}\int_0^\infty \int_0^\infty \mathbf{\Lambda}_r^t(s) \cdot \mathbb{K}(s, u)\mathbf{\Lambda}_r^t(u) ds du. \tag{7.1.2}$$

The operator $\mathbb{K}$ is at least positive semidefinite in the sense that the function $\mathbb{K}(\cdot, \cdot)$ is such that the integral is nonnegative. This must be true in view of (5.1.25). It should be noted that (5.1.25) also excludes linear functionals in (7.1.1). It is assumed that the integral on (7.1.2) exists for $\mathbf{\Lambda}_r^t$ finite, though not necessarily zero, at infinity. It will therefore be assumed that as $s \to \infty$, the kernel $\mathbb{K}(s, u)$ goes to zero as

$$\mathbb{K}(s, u) \sim s^{-1-b}, \quad b > 0, \quad u \in \mathbb{R}^+, \tag{7.1.3}$$

or more strongly; similarly for the limit of large u at fixed s. With the aid of (A.2.3), we see that there is no loss of generality in taking

$$\mathbb{K}^\top(s, u) = \mathbb{K}(u, s), \tag{7.1.4}$$

where the transpose refers to $\mathrm{Lin}(\Gamma)$.

Let us define $\mathbb{L} \in \mathrm{Lin}(\Gamma)$ by

$$\mathbb{L}(s, u) :\equiv \int_s^\infty \int_u^\infty \mathbb{K}(t_1, t_2) dt_1 dt_2, \tag{7.1.5}$$

so that

$$\mathbb{L}_{12}(s, u) :\equiv \frac{\partial^2}{\partial s \partial u}\mathbb{L}(s, u) = \mathbb{K}(s, u), \qquad \mathbb{L}^\top(s, u) = \mathbb{L}(u, s), \tag{7.1.6}$$

and

$$\lim_{s\to\infty} \mathbb{L}(s, u) = \mathbf{0}, \qquad \lim_{s\to\infty} \frac{\partial}{\partial u}\mathbb{L}(s, u) = \mathbf{0}, \quad u \in \mathbb{R}^+, \tag{7.1.7}$$

with similar limits at large u holding for fixed s. Also, from (7.1.3),

$$\lim_{u\to\infty} \frac{\partial}{\partial u}\mathbb{L}(s, u) = \mathbf{0}, \quad s \in \mathbb{R}^+, \qquad \lim_{s\to\infty} \frac{\partial}{\partial s}\mathbb{L}(s, u) = \mathbf{0}, \quad u \in \mathbb{R}^+, \tag{7.1.8}$$

and

$$\lim_{u\to\infty} \mathbb{L}_{12}(s, u) = \lim_{s\to\infty} \mathbb{L}_{12}(s, u) = \mathbf{0}, \quad u \in \mathbb{R}^+.$$

Remark 7.1.2. We could include a constant term $\mathbb{L}_\infty$ on the right of (7.1.5), which would mean that the right of $(7.1.7)_1$ is replaced by $\mathbb{L}_\infty$. This is the conventional approach taken in [158] and later papers, where the constant $\mathbb{L}_\infty$ has an important physical role. However, we can take it to be zero, and certain simplifications result from doing so, specifically in relation to partial integrations. This means that our $\mathbb{L}(\cdot,\cdot)$ is in fact $\mathbb{L}_c(\cdot,\cdot)-\mathbb{L}_\infty$, where $\mathbb{L}_c(\cdot,\cdot)$ is the more conventional choice. We shall, when discussing specifically mechanical applications, revert to the earlier convention.

Of course, for fluids in shear flow and for heat flow, $\mathbb{L}_\infty$ is in fact zero.

We write (7.1.2) as

$$\psi(t) = \widetilde{\phi}(\mathbf{\Lambda}_0(t)) + \frac{1}{2}\int_0^\infty \int_0^\infty \mathbf{\Lambda}_r^t(s)\cdot \mathbb{L}_{12}(s,u)\mathbf{\Lambda}_r^t(u)dsdu. \qquad (7.1.9)$$

By partial integrations, one can put this in the form

$$\begin{aligned} \psi(t) &= \widetilde{\phi}(\mathbf{\Lambda}_0(t)) + \frac{1}{2}\int_0^\infty \int_0^\infty \dot{\mathbf{\Lambda}}^t(s)\cdot \mathbb{L}(s,u)\dot{\mathbf{\Lambda}}^t(u)dsdu, \\ \dot{\mathbf{\Lambda}}^t(u) &= \frac{\partial}{\partial t}\mathbf{\Lambda}^t(u) = -\frac{\partial}{\partial u}\mathbf{\Lambda}^t(u) = -\frac{\partial}{\partial u}\mathbf{\Lambda}_r^t(u). \end{aligned} \qquad (7.1.10)$$

The form $(7.1.10)_1$ is interesting in view of the definitions of $\mathbf{m}$ given by (5.1.8). Relations (7.1.9) and (7.1.10) are the basis of all developments in Part III.

Remark 7.1.3. Note that $\mathbb{L}$ and $\mathbb{L}_{12}$ must be nonnegative operators by virtue of (5.1.25), since the integrals in (7.1.9) and (7.1.10) must be nonnegative. This is the continuous equivalent of nonnegativity in a matrix, as discussed in Sect. A.2.1. Necessary conditions for this property can be established by considering histories that are zero everywhere except in a set of diminishing measure around a point s, giving that the "diagonal elements" are nonnegative or

$$\mathbb{L}(s,s) \geq \mathbf{0}, \qquad \mathbb{L}_{12}(s,s) \geq \mathbf{0}.$$

In fact, let us assume that $\mathbb{L}$ is infinitely differentiable in both arguments and consider histories that are differentiable to any required degree. We insert $\partial\mathbf{\Lambda}_r^t$, instead of $\mathbf{\Lambda}_r^t$, into (7.1.9), where ∂ indicates the first derivative of $\mathbf{\Lambda}_r^t(\cdot)$. By partial integration and the use of very localized histories, we can deduce that

$$\mathbb{L}_{1122}(s,s) \geq \mathbf{0}. \qquad (7.1.11)$$

Continuing in this way, we can deduce that

$$\mathbb{L}_{(n)}(s,s) \geq \mathbf{0}, \qquad n = 0,1,2,\ldots, \qquad (7.1.12)$$

where the subscript indicates differentiation with respect to both the first and second arguments, n times. For example, $\mathbb{L}_{(2)}(s,s)$ is the quantity on the left of (7.1.11).

The argument breaks down if $\mathbb{L}$ is not differentiable and $\mathbb{L}_{12}$ is unbounded. This is true for the Graffi–Volterra free energy, where $\mathbb{L}$ is given by (10.1.3). It also applies to the work function given by (7.5.3) in view of (7.5.6).

7.1.1 Constitutive Relations

We can determine the generalized stress within the present model from either (5.1.18) (see also P1 in Sect. 5.1.1) or (5.1.22). The differentiation of the occurrence of $\mathbf{\Lambda}(t)$ in $\mathbf{\Lambda}_r^t$ is equivalent to the functional differentiation defined by (5.1.23). Equation (5.1.18) is simpler to apply. One obtains

$$\begin{aligned}\mathbf{\Sigma}(t) &= \widetilde{\mathbf{\Sigma}}_e(\mathbf{\Lambda}_0(t)) + \int_0^\infty \mathbb{L}'(u)\mathbf{\Lambda}_r^t(u)du \\ &= \widetilde{\mathbf{\Sigma}}_0(\mathbf{\Lambda}(t)) + \int_0^\infty \mathbb{L}'(u)\mathbf{\Lambda}^t(u)du \\ &= \widetilde{\mathbf{\Sigma}}_e(\mathbf{\Lambda}_0(t)) + \int_0^\infty \mathbb{L}(u)\dot{\mathbf{\Lambda}}^t(u)du,\end{aligned} \tag{7.1.13}$$

where (7.1.4) and results from Sect. A.2.2 have been used; also, we have put

$$\begin{aligned}&\widetilde{\mathbf{\Sigma}}_e(\mathbf{\Lambda}_0(t)) = \mathbf{\Sigma}_e(t) = \frac{d\widetilde{\phi}(\mathbf{\Lambda}_0(t))}{d\mathbf{\Lambda}(t)}, \qquad \mathbb{L}(u) = \mathbb{L}(0,u), \\ &\widetilde{\mathbf{\Sigma}}_0(\mathbf{\Lambda}(t)) = \mathbf{\Sigma}_0(t) = \widetilde{\mathbf{\Sigma}}_e(\mathbf{\Lambda}_0(t)) + \mathbb{L}_0\mathbf{\Lambda}(t), \quad \mathbb{L}_0 = \mathbb{L}(0).\end{aligned} \tag{7.1.14}$$

The prime in $(7.1.13)_{1,2}$ indicates differentiation with respect to the argument. Relations (7.1.7) have been used. Also, from $(7.1.7)_1$ and (7.1.8), we have

$$\mathbb{L}(\infty) = \mathbf{0}, \qquad \mathbb{L}'(\infty) = \mathbf{0}. \tag{7.1.15}$$

It is interesting to note, in general terms, how a lack of uniqueness may arise in free energies given by quadratic functionals. If $\mathbb{L}$ is replaced by $\mathbb{L}^{(1)}$ with the same positivity property as $\mathbb{L}$ and

$$\begin{aligned}&\mathbb{L}^{(1)}(s,u) = \mathbb{L}(s,u) + \mathbb{L}^{(2)}(s,u), \\ &\mathbb{L}^{(2)}(s,0) = \mathbb{L}^{(2)}(0,u) = \mathbf{0}, \quad s,u \in \mathbb{R}^+,\end{aligned} \tag{7.1.16}$$

then using $\mathbb{L}^{(1)}$ in (7.1.9) also produces a satisfactory free energy, by virtue of $(7.1.14)_3$, since

$$\frac{1}{2}\int_0^\infty \int_0^\infty \mathbf{\Lambda}_r^t(s)\cdot \mathbb{L}_{12}^{(2)}(s,u)\mathbf{\Lambda}_r^t(u)dsdu$$

can be shown to be independent of $\mathbf{\Lambda}(t)$, using the derivatives of $(7.1.16)_{2,3}$. This is in effect the statement of Proposition 5.1.2 for a free energy given by a quadratic functional.

The notation (see (5.1.28))

$$\begin{aligned}&\mathbf{\Sigma}(t) = \mathbf{\Sigma}_e(t) + \mathbf{\Sigma}_{rh}(t) = \mathbf{\Sigma}_0(t) + \mathbf{\Sigma}_h(t), \\ &\mathbf{\Sigma}_{rh}(t) = \int_0^\infty \mathbb{L}'(u)\mathbf{\Lambda}_r^t(u)du = \int_0^\infty \mathbb{L}(u)\dot{\mathbf{\Lambda}}^t(u)du, \\ &\mathbf{\Sigma}_h(t) = \int_0^\infty \mathbb{L}'(u)\mathbf{\Lambda}^t(u)du,\end{aligned} \tag{7.1.17}$$

will be used later. It will be assumed that

$$\mathbb{L}(u) = \mathbb{L}^{\top}(u), \quad u \in \mathbb{R}^{+}. \tag{7.1.18}$$

Remark 7.1.4. Note that Σ_{rh} vanishes at the zero relative history. The quantity Λ_r^t does not belong to $L^1(\mathbb{R}^+) \cap L^2(\mathbb{R}^+)$, so that for convergence of the integral, we must have $\mathbb{L}'$ decaying sufficiently strongly at large times, which is a statement of the fading memory property in the linear memory context.

Remark 7.1.5. With the aid of $(5.1.21)_1$ and (7.1.17), relation (7.1.9) can be written as

$$\begin{aligned} \psi(t) &= S(t) + \frac{1}{2}\int_0^\infty \int_0^\infty \Lambda^t(s) \cdot \mathbb{L}_{12}(s,u)\Lambda^t(u)dsdu, \\ S(t) &= \phi(t) + \Lambda(t) \cdot [\Sigma(t) - \Sigma_0(t)] + \frac{1}{2}\Lambda(t) \cdot \mathbb{L}_0\Lambda(t), \end{aligned} \tag{7.1.19}$$

where $\mathbb{L}_0$ is defined by the last relation of (7.1.14). Observing that $\Sigma(t) - \Sigma_0(t)$ does not depend on $\Lambda(t)$, we have, with the aid of (7.1.14),

$$\frac{d}{d\Lambda(t)} S(t) = \Sigma(t), \tag{7.1.20}$$

which yields (5.1.30). A disadvantage of using this form, with histories rather than relative histories, is that $S(t)$ is not a nonnegative quantity.

Recalling $(5.1.10)_{1,2}$, we now write out $(7.1.13)_1$ in detail as follows:

$$\begin{aligned} \frac{\kappa(t)}{\rho}\widehat{\mathbf{S}}(t) &= \frac{\kappa(t)}{\rho}\widehat{\mathbf{S}}_e(t) + \int_0^\infty \mathbb{L}'_E(u)\mathbf{E}_r^t(u)du + \int_0^\infty \mathbb{L}'_\kappa(u)\kappa_r^t(u)du \\ &\quad + \int_0^\infty \mathbb{L}'_m(u)\mathbf{m}_r^t(u)du, \\ \mathbb{L}'_E &: \mathbb{R}^+ \mapsto \mathrm{Lin(Sym)}, \quad \mathbb{L}'_\kappa : \mathbb{R}^+ \mapsto \mathrm{Sym}, \quad \mathbb{L}'_m : \mathbb{R}^+ \mapsto \mathrm{Lin}(\mathbb{R}^3, \mathrm{Sym}). \end{aligned} \tag{7.1.21}$$

Also

$$\begin{aligned} e(t) &= \epsilon_e(t) + \int_0^\infty \Xi'_E(u) \cdot \mathbf{E}_r^t(u)du + \int_0^\infty \Xi'_\kappa(u)\kappa_r^t(u)du \\ &\quad + \int_0^\infty \Xi'_m(u) \cdot \mathbf{m}_r^t(u)du, \\ \Xi'_E &: \mathbb{R}^+ \mapsto \mathrm{Sym}, \quad \Xi'_\kappa : \mathbb{R}^+ \mapsto \mathbb{R}, \quad \Xi'_m : \mathbb{R}^+ \mapsto \mathbb{R}^3, \end{aligned} \tag{7.1.22}$$

and

$$\begin{aligned} \frac{1}{\rho}\mathbf{d}(t) &= \int_0^\infty \mathbf{V}'_E(u)\mathbf{E}_r^t(u)du + \int_0^\infty \mathbf{V}'_\kappa(u)\kappa_r^t(u)du + \int_0^\infty \mathbf{V}'_m(u)\mathbf{m}_r^t(u)du, \\ \mathbf{V}'_E &: \mathbb{R}^+ \mapsto \mathrm{Lin}(\mathrm{Sym}, \mathbb{R}^3), \quad \mathbf{V}'_\kappa : \mathbb{R}^+ \mapsto \mathbb{R}^3, \quad \mathbf{V}'_m : \mathbb{R}^+ \mapsto \mathrm{Lin}(\mathbb{R}^3). \end{aligned} \tag{7.1.23}$$

There is no equilibrium term in the heat flow equation (7.1.23). The quantities $\widehat{\mathbf{S}}_e$ and ϵ_e depend only on $\Lambda_0(t)$.

Note that by virtue of (7.1.18),

$$\begin{aligned} &\mathbb{L}_E'^{\top}(u) = \mathbb{L}_E'(u), \quad \mathbb{L}_\kappa'(u) = \Xi_E'(u), \quad \mathbb{L}_m'(u) = \mathbf{V}_E'^{\top}(u), \\ &\Xi_m'(u) = \mathbf{V}_\kappa'^{\top}(u), \quad \mathbf{V}_m'(u) = \mathbf{V}_m'^{\top}(u). \end{aligned} \tag{7.1.24}$$

Relaxation functions are generally found in experimental work [133] to be monotonically decaying, certainly for mechanical systems. However, this does not follow from any general thermodynamic principle, though special results consistent with such behavior can be obtained from reasoning based on thermodynamics ([104] and (7.2.16), (7.2.17) below). Several choices adopted in later chapters (sums or integrals of decaying exponentials) have this property, but we do not assume it in general. Indeed, in Chap. 16, cases of nonmonotonic behavior are considered.

7.1.2 Dissipation Rate

The rate-of-dissipation function D, given by (5.1.19), is obtained by differentiating the functional dependence of $\widetilde{\psi}$ on $\mathbf{\Lambda}^t$ with respect to t, i.e., ignoring $\mathbf{\Lambda}(t)$, which occurs in $\mathbf{\Lambda}_r^t$. This follows by elementary considerations, avoiding explicit functional differentiation, if we realize that differentiating $\mathbf{\Lambda}(t)$ gives the term $\mathbf{\Sigma} \cdot \dot{\mathbf{\Lambda}}$ in (5.1.11). Thus, D must emerge from the remaining differentiation with respect to t of $\psi(t)$. Let us differentiate (7.1.9), selecting only the functional dependence. In the term where the s-dependent left-hand side of the quadratic form is differentiated, we carry out a partial integration with respect to the u derivative on $\mathbb{L}$ and vice versa, finally obtaining, with the aid of $(7.1.10)_{2,3,4}$,

$$D(t) = -\frac{1}{2}\int_0^\infty \int_0^\infty \dot{\mathbf{\Lambda}}^t(s) \cdot [\mathbb{L}_1(s,u) + \mathbb{L}_2(s,u)]\dot{\mathbf{\Lambda}}^t(u)ds du. \tag{7.1.25}$$

We see that the existence of dissipation in the material is intrinsically connected with the constitutive dependence on the history of the independent variable. By partial integration, we can also write

$$D(t) = -\frac{1}{2}\int_0^\infty \int_0^\infty \mathbf{\Lambda}_r^t(s) \cdot [\mathbb{L}_{121}(s,u) + \mathbb{L}_{122}(s,u)]\mathbf{\Lambda}_r^t(u)ds du, \tag{7.1.26}$$

where $(7.1.10)_{2,3,4}$ have been used again.

7.1.3 Complete Material Characterization

The lack of uniqueness of the free energy of a material with memory is an intrinsic and unavoidable feature of the theory. This is clear from the very general framework developed in Chap. 4, in particular, the discussion relating to Proposition 4.1.7, the content of Proposition 5.1.2, and the point made in relation to (7.1.16). This arbitrariness raises the following question: what is the true physical free energy $\psi_p(t)$ of the material and what is its relationship to the various forms of $\psi(t)$ that will be proposed in later chapters?

The physical free energy in the context of materials with memory is perhaps best defined as the quantity obeying (5.1.34), where $W(t)$ and $\mathcal{D}(t)$ are assumed to have clear physical definitions. This is certainly true of $W(t)$. In an isothermal mechanical context, $\mathcal{D}(t)$ is the total heat produced per unit mass, up to time t. Also, we have total clarity for cyclic processes, which yield (5.1.37). It will be assumed that for a material with memory, the actual physical total dissipation $\mathcal{D}_p(t)$ and rate of dissipation $D_p(t)$ are in principle experimentally measurable quantities. Thus, from (5.1.34) and the value of $\mathcal{D}_p(t)$, we can determine the physical free energy $\psi_p(t)$.

All the free energies proposed in later chapters are determined by the relaxation function $\mathbb{L}(s)$, $s \in \mathbb{R}^+$, or special cases of this quantity. At least their memory-dependent parts are; we assume that the instantaneous terms are always fully specified. These provide either bounds on or approximations to $\psi_p(t)$. A fundamental question is now posed. Is it the case that ψ_p

1. is determined completely by $\mathbb{L}(s)$, $s \in \mathbb{R}^+$ (leaving aside the equilibrium term $\phi(t)$), so that the kernel of the quadratic form $\mathbb{L}(s, u)$, $s, u \in \mathbb{R}^+$, is fully determined by $\mathbb{L}(s)$, $s \in \mathbb{R}^+$; or
2. do we have to determine $\mathbb{L}(s, u)$ separately?

The first possibility is an expression of the intuitive idea that a material is completely characterized by its constitutive relationship (7.1.13), which is possibly supported by the fact that all forms of the free energy proposed to date depend only on $\mathbb{L}(s)$, $s \in \mathbb{R}^+$. This assumption is the basis of the theory developed in Sect. 16.10. The only experimental requirement for this approach to material characterization is the measurement of the relaxation function $\mathbb{L}(s)$, $s \in \mathbb{R}^+$, which is discussed below.

The second possibility requires that to completely characterize a material, we must determine $\mathbb{L}(s)$, $s \in \mathbb{R}^+$, by a series of measurements, as for the first option, but also $\mathbb{L}(s, u)$, $s, u \in \mathbb{R}^+$, by a separate program of measurement. In other words, the energy characteristics of the material are not contained in the constitutive equation, but must be determined separately.

By $\mathbb{L}(s, u)$ we mean of course the physical kernel $\mathbb{L}_p(s, u)$, and once this is determined experimentally, ψ_p is given uniquely.

In fact, referring to our earlier discussion of the measurability of dissipation, the most natural way to proceed is to determine by experiment $\mathbb{L}(s)$, $s \in \mathbb{R}^+$, and also the quantity $\mathbb{K}(s, u)$, $s, u \in \mathbb{R}^+$, given by*

$$\mathbb{K}(s, u) = \mathbb{L}_1(s, u) + \mathbb{L}_2(s, u), \tag{7.1.27}$$

which is the kernel of (7.1.25). Then there is the mathematical problem of determining $\mathbb{L}(s, u)$ from (7.1.27) and the boundary condition $(7.1.14)_3$ or

$$\mathbb{L}(0, u) = \mathbb{L}(u, 0) = \mathbb{L}(u), \tag{7.1.28}$$

which is discussed later.

It is interesting to consider how $\mathbb{K}(s, u)$, $s, u \in \mathbb{R}^+$, might be measured. We outline a method that, while probably not practical, at least shows that it is possible. Recall

* Not equal to the quantity used briefly at the beginning of Sect. 7.1.

how in principle, $\mathbb{L}(s)$, $s \in \mathbb{R}^+$, might be measured with the aid of $(7.1.13)_3$. We choose a history $\boldsymbol{\Lambda}^t(s) = \boldsymbol{\Lambda}(t-s)$ and measure the corresponding $\boldsymbol{\Sigma}(t)$ for a range of values of t. A standard example (possible only in approximation) is to take $\boldsymbol{\Lambda}(u)$ to be zero for $u < 0$, with a discontinuity at $u = 0$. After this time, it has a constant value $\boldsymbol{\Lambda}_1$. Thus, $\dot{\boldsymbol{\Lambda}}(u)$ is given by a singular delta function $\boldsymbol{\Lambda}_1\delta(u)$, so that $\dot{\boldsymbol{\Lambda}}^t(u) = \boldsymbol{\Lambda}_1\delta(t-u)$ and (7.1.13) yields that

$$\boldsymbol{\Sigma}(t) = \widetilde{\boldsymbol{\Sigma}}_e(\boldsymbol{\Lambda}_{10}) + \boldsymbol{\Lambda}_1\mathbb{L}(t),$$

where $\boldsymbol{\Lambda}_{10}$ is $\boldsymbol{\Lambda}_1$ with zero in the $\mathbb{R}^3$ contribution, as indicated by (5.1.21). Assuming, as noted earlier, that $\widetilde{\boldsymbol{\Sigma}}_e$ is known, it follows that detailed measurements over time of $\boldsymbol{\Sigma}(t)$ will provide numerical estimates of $\mathbb{L}(t)$. Using a similar history in (7.1.25) gives

$$D(t) = \boldsymbol{\Lambda}_1 \cdot \mathbb{K}(t,t)\boldsymbol{\Lambda}_1,$$

so that the measurements of $D(t)$ over a wide range of times yield the diagonal values $\mathbb{K}(t,t)$, provided suitably widely varied choices of $\boldsymbol{\Lambda}_1$ are also used.

Nondiagonal terms can be obtained by considering histories with two discontinuities, so that

$$\dot{\boldsymbol{\Lambda}}^t(u) = \boldsymbol{\Lambda}_1\delta(t_1-u) + \boldsymbol{\Lambda}_2\delta(t_2-u),$$

yielding

$$\begin{aligned} D(t) = \boldsymbol{\Lambda}_1 \cdot \mathbb{K}(t_1,t_1)\boldsymbol{\Lambda}_1 + \boldsymbol{\Lambda}_2 \cdot \mathbb{K}(t_2,t_2)\boldsymbol{\Lambda}_2 \\ + \boldsymbol{\Lambda}_1 \cdot \mathbb{K}(t_1,t_2)\boldsymbol{\Lambda}_2 + \boldsymbol{\Lambda}_2 \cdot \mathbb{K}(t_2,t_1)\boldsymbol{\Lambda}_1. \end{aligned}$$

The first two terms are diagonal and are already known from the experiments with one discontinuity. The other two terms are equal because

$$\mathbb{K}^\top(s,u)) = \mathbb{K}(u,s),$$

which follows from $(7.1.6)_2$. Thus, a suitable range of measurements of $D(t)$ for histories with two discontinuities yield the nondiagonal terms.

One can give an explicit solution of the first-order partial differential equation (7.1.27) for $\mathbb{L}(s,u)$, $s,u \in \mathbb{R}^+$, where $\mathbb{K}(s,u)$, $s,u \in \mathbb{R}^+$, is presumed to be known. Define new variables

$$x = s + u, \quad y = s - u, \tag{7.1.29}$$

in terms of which (7.1.27) becomes

$$\frac{\partial}{\partial x}\mathbb{L}_n(x,y) = \frac{1}{2}\mathbb{K}_n(x,y), \quad \mathbb{L}_n(x,y) = \mathbb{L}(s,u), \quad \mathbb{K}_n(x,y) = \mathbb{K}(s,u),$$

with general solution

$$\mathbb{L}_n(x,y) = \mathbb{L}_n(x_0,y) + \frac{1}{2}\int_{x_0}^{x} \mathbb{K}_n(x',y)dx', \tag{7.1.30}$$

where x_0 is an arbitrary nonnegative real quantity. Observe that (7.1.28) becomes

$$\mathbb{L}(u) = \mathbb{L}_n(u, u) = \mathbb{L}_n(u, -u), \quad u \in \mathbb{R}^+.$$

Then, choosing $x_0 = |y|$ in (7.1.30), we have

$$\mathbb{L}(s, u) = \mathbb{L}_n(s + u, s - u) = \mathbb{L}(|s - u|) + \frac{1}{2}\int_{|s-u|}^{s+u} \mathbb{K}_n(x', s - u)dx',$$

which is the solution obeying the boundary conditions (7.1.28).

These mathematical developments and in effect the choice between options 1 and 2 above are explored in more detail in Chap. 17, leading to option 2 as the required approach for uniquely determining the material properties. This is in spite of the fact that option 1 is the only one used to date.

7.1.4 Linear Equilibrium Response

A very important limiting case occurs when $\widetilde{\phi}$ in (7.1.10) has the form

$$\widetilde{\phi}(\boldsymbol{\Lambda}_0(t)) = \phi(t) = \frac{1}{2}\boldsymbol{\Lambda}_0(t) \cdot \mathbb{L}_e\boldsymbol{\Lambda}_0(t) = \frac{1}{2}\boldsymbol{\Lambda}(t) \cdot \mathbb{L}_e\boldsymbol{\Lambda}(t), \tag{7.1.31}$$

where $\mathbb{L}_e \in \mathrm{Lin}(\Gamma^+)$ is a nonnegative tensor, which can be taken to be a symmetric matrix acting on Sym × $\mathbb{R}$ and zero on $\mathbb{R}^3$. This yields for $\widetilde{\boldsymbol{\Sigma}}_e$ and $\widetilde{\boldsymbol{\Sigma}}_0$ in (7.1.13)

$$\begin{aligned}\widetilde{\boldsymbol{\Sigma}}_e(\boldsymbol{\Lambda}_0(t)) &= \mathbb{L}_e\boldsymbol{\Lambda}_0(t) = \mathbb{L}_e\boldsymbol{\Lambda}(t),\\ \widetilde{\boldsymbol{\Sigma}}_0(\boldsymbol{\Lambda}_0(t)) &= \mathbb{L}_e\boldsymbol{\Lambda}_0(t) + \mathbb{L}_0\boldsymbol{\Lambda}(t) = (\mathbb{L}_e + \mathbb{L}_0)\boldsymbol{\Lambda}(t),\end{aligned} \tag{7.1.32}$$

where (7.1.14) has been used. Then (7.1.13) becomes

$$\begin{aligned}\boldsymbol{\Sigma}(t) &= \mathbb{L}_e\boldsymbol{\Lambda}_0(t) + \int_0^\infty \mathbb{L}'(u)\boldsymbol{\Lambda}_r^t(u)du\\ &= \mathbb{L}_e\boldsymbol{\Lambda}_0(t) + \mathbb{L}_0\boldsymbol{\Lambda}(t) + \int_0^\infty \mathbb{L}'(u)\boldsymbol{\Lambda}^t(u)du\\ &= \mathbb{L}_e\boldsymbol{\Lambda}_0(t) + \int_0^\infty \mathbb{L}(u)\dot{\boldsymbol{\Lambda}}^t(u)du.\end{aligned} \tag{7.1.33}$$

The quantity S, given by $(7.1.19)_2$, now takes the form

$$S(t) = \boldsymbol{\Lambda}(t) \cdot \boldsymbol{\Sigma}(t) - \frac{1}{2}\boldsymbol{\Lambda}(t) \cdot (\mathbb{L}_0 + \mathbb{L}_e)\boldsymbol{\Lambda}(t), \tag{7.1.34}$$

as can be seen with the aid of (7.1.31) and (7.1.32).

In Part III, the formulas are valid mainly for general equilibrium response. However, the linear case is adopted in several contexts for definiteness.

If the conventional notation, in the sense of Remark 7.1.2, is adopted, then $\mathbb{L}_0$ must be replaced by $\mathbb{L}_0 - \mathbb{L}_\infty$, and (7.1.34) is replaced by

$$S(t) = \boldsymbol{\Lambda}(t) \cdot \boldsymbol{\Sigma}(t) - \frac{1}{2}\boldsymbol{\Lambda}(t) \cdot \mathbb{L}_0\boldsymbol{\Lambda}(t). \tag{7.1.35}$$

Linear equilibrium response may be applicable for large values of $\mathbf{\Lambda}(t)$ but will certainly be applicable for small values of this field variable, including small strains for which the relations of Sect. 1.2.2 apply. In this case, the Lagrangian and Eulerian descriptions coincide.

Remark 7.1.6. Referring to Remark 7.1.1, we note that the discussion of fluids in later sections will be confined to the case that the strain tensor in the extra stress (Definition 2.2.13) is assumed to be infinitesimal. In this limit, the tensor $\mathbf{D}$ in (2.2.13) can be replaced by $\dot{\mathbf{E}}$, and we can use the coordinates $(\mathbf{x}, t)$.

7.1.5 Time-Independent Eigenspaces

An important special case of materials with linear memory occurs when the eigenvectors of $\mathbb{L}(t)$ do not depend on t, so that we have

$$\mathbb{L}(t) = \sum_{k=1}^{m} L_k(t)\mathbb{B}_k, \tag{7.1.36}$$

where $\mathbb{B}_k = \mathbf{B}_k \otimes \mathbf{B}_k$, $k = 1, 2, \ldots, m$, are the orthonormal projectors on the m eigenspaces of $\mathbb{L}$, while $\{\mathbf{B}_k\}$ are their normalized eigenvectors, which constitute an orthonormal basis of Γ. This is the spectral representation of $\mathbb{L}(t)$ (see (A.2.10)). Under our assumption, the (real) eigenvectors $\mathbf{B}_k$ and projectors $\mathbb{B}_k$, $k = 1, 2, \ldots, m$, are independent of time.

From (7.1.18), the constant eigenvectors $\mathbf{B}_k$, in a particular basis, are the rows or columns of an orthogonal transformation $\mathbf{R}$. The time-independence allows us to diagonalize the constitutive equation (7.1.17). Let us put

$$\mathbf{\Lambda}^t(s) = \sum_{k=1}^{m} \Lambda_k^t(s)\mathbf{B}_k, \qquad \mathbf{\Sigma}_h(s) = \sum_{k=1}^{m} \Sigma_{hk}(s)\mathbf{B}_k,$$

where the scalar quantities Λ_k^t, Σ_{hk} are given by

$$\Lambda_k^t(s) = \mathbf{\Lambda}^t(s) \cdot \mathbf{B}_k, \qquad \Sigma_{hk}(s) = \mathbf{\Sigma}_h(s) \cdot \mathbf{B}_k.$$

Then $(7.1.17)_5$ and (7.1.36) yield

$$\Sigma_{hk}(t) = \int_0^\infty L_k'(u)\Lambda^t(u)du, \qquad k = 1, 2, \ldots, m,$$

which constitute a series of scalar relationships.

In the case of linear equilibrium response, as given by $(7.1.33)_2$, the quantity $\mathbb{L}_0 = \mathbb{L}(0)$ will be diagonalized by the same transformation as $\mathbb{L}(u)$, $u = 0$, under our assumption. We need also, however, to assume this property for $\mathbb{L}_e$.

The time-independent eigenvectors property implies restrictions on $\mathbb{L}(t)$, which will be illustrated by a simple example below.

Let us consider a mechanically and thermally isotropic viscoelastic solid subject to infinitesimal strain, with thermal expansion effects and isotropic heat flow properties for which the coldness gradient is determined only by the heat flow history.

The tensor $\widehat{\mathbf{S}}$ in (7.1.21) is replaced by the Cauchy stress tensor $\mathbf{T}$, and $\boldsymbol{\Lambda}$ by $\mathbf{E}$. The traceless parts of $\mathbf{T}$ and $\mathbf{E}$ are denoted by $\hat{\mathbf{T}}$, $\hat{\mathbf{E}}$, and their volumetric components by T, E. Thus, (7.1.21) becomes

$$\begin{aligned}\hat{\mathbf{T}}(t) &= G_{S\infty}\hat{\mathbf{E}}(t) + \int_0^\infty G'_S(u)\hat{\mathbf{E}}^t_r(u)du,\\ T(t) &= G_{B\infty}\left[E(t) + \beta\kappa_d(t)\right] + \int_0^\infty G'_B(u)E^t_r(u)du + \int_0^\infty G'_K(u)\kappa^t_r(u)du,\end{aligned}\tag{7.1.37}$$

where G'_S, G'_B, and G'_κ are scalar functions and $G_{S\infty}$, $G_{B\infty}$ are positive constants. Furthermore,

$$\kappa_d(t) = \kappa(t) - \kappa_0,\tag{7.1.38}$$

where the quantity κ_0 is the equilibrium coldness and β is related to the equilibrium coefficient of thermal expansion. We have absorbed the factor $\kappa(t)/\rho \approx \kappa_0/\rho$ into G'_S, G'_B. This is a natural completion of the linearization of the problem, though it hides an interesting prediction of the theory, namely that the relaxation function is proportional to $\rho\theta(t)$, which was initially derived in [159], using a formulation of thermodynamics based on the standard Helmholtz free energy ψ_H. However, it emerges very naturally from the formulation used here, based on ψ defined by (5.1.6).

Relation (7.1.22) takes the form

$$\epsilon(t) = \epsilon_1 E(t) + \epsilon_2\kappa_d(t) + \int_0^\infty G'_\kappa(u)E^t_r(u)du + \int_0^\infty \Xi'_\kappa(u)\kappa^t_r(u)du,\tag{7.1.39}$$

where Ξ'_κ is a scalar function and a consequence of (7.1.18) as expressed by $(7.1.24)_2$ has been used. Note that (7.1.18) further requires that

$$\epsilon_1 = G_{B\infty}\beta.\tag{7.1.40}$$

Finally, (7.1.23) becomes

$$\frac{1}{\rho}\mathbf{d}(t) = \int_0^\infty V'_m(u)\mathbf{m}^t_r(u)du = \int_0^\infty V_m(u)\mathbf{q}^t_r(u)du\tag{7.1.41}$$

in terms of the scalar function V_m, by virtue of (5.1.8).

We need to diagonalize the bulk or volumetric component of the mechanical equation $(7.1.37)_2$ and the relation for the internal energy (7.1.39) by means of a constant rotation matrix. This places restrictions on the form of the relaxation functions involved. We require that

$$\mathbf{R}\begin{pmatrix} G_{B\infty} & \epsilon_1\\ \epsilon_1 & \epsilon_2\end{pmatrix}\mathbf{R}^\top = \begin{pmatrix} G_1 & 0\\ 0 & G_2\end{pmatrix},$$

$$\mathbf{R}\begin{pmatrix} G'_B(u) & G'_\kappa(u)\\ G'_\kappa(u) & \Xi'_\kappa(u)\end{pmatrix}\mathbf{R}^\top = \begin{pmatrix} G'_1(u) & 0\\ 0 & G'_2(u)\end{pmatrix},$$

$$\mathbf{R} = \begin{pmatrix} c & s\\ -s & c\end{pmatrix},\quad c^2 + s^2 = 1,$$

where G_1, G_2, G_1', G_1' are the required eigenvalues. It follows that

$$\begin{pmatrix} G_{B\infty} & \epsilon_1 \\ \epsilon_1 & \epsilon_2 \end{pmatrix} = G_1 \begin{pmatrix} c^2 & cs \\ cs & s^2 \end{pmatrix} + G_2 \begin{pmatrix} s^2 & -cs \\ -cs & c^2 \end{pmatrix},$$

$$\begin{pmatrix} G_B'(u) & G_\kappa'(u) \\ G_\kappa'(u) & \Xi_\kappa'(u) \end{pmatrix} = G_1'(u) \begin{pmatrix} c^2 & cs \\ cs & s^2 \end{pmatrix} + G_2'(u) \begin{pmatrix} s^2 & -cs \\ -cs & c^2 \end{pmatrix}.$$

These relations yield that

$$\begin{aligned} (c^2 - s^2)\epsilon_1 &= cs(G_{B\infty} - \epsilon_2), \\ (c^2 - s^2)G_\kappa'(u) &= cs(G_B'(u) - \Xi_\kappa'(u)), \end{aligned} \tag{7.1.42}$$

by solving for (G_1, G_2), (G_1', G_2') in terms of $(G_{B\infty}, \epsilon_2)$, (G_B', Ξ_κ'), respectively. Thus (7.1.42) gives the constraints imposed by the requirement of time-independent eigenspaces. Specifically, we have

$$\frac{G_B'(u) - \Xi_\kappa'(u)}{G_K'(u)} = \frac{G_{B\infty} - \epsilon_2}{\epsilon_1} = \frac{c^2 - s^2}{cs}, \qquad u \in \mathbb{R}^+, \tag{7.1.43}$$

so that the ratio on the left is independent of u. The value of this ratio for any choice of u or the ratio on the right of $(7.1.43)_1$ gives the angle of rotation.

This is not a physical requirement but rather a condition that renders the factorization problem underlying the determination of the minimum and other free energies possible by the method described in Sect. 11.1.1 for scalar fields. However, it clearly represents a physical restriction on the material.

Relations (7.1.37)–(7.1.41) can be written in the form

$$\begin{aligned} \hat{\mathbf{T}}(t) &= G_{S\infty}\hat{\mathbf{E}}(t) + \int_0^\infty G_S'(u)\hat{\mathbf{E}}_r^t(u)du, \\ X_1(t) &= G_1 Y_1(t) + \int_0^\infty G_1'(u)Z_1^t(u)du, \\ X_2(t) &= G_2 Y_2(t) + \int_0^\infty G_2'(u)Z_2^t(u)du, \\ \frac{1}{\rho}\mathbf{d}(t) &= \int_0^\infty V_m(u)\mathbf{q}_r^t(u)du, \end{aligned}$$

where

$$\begin{pmatrix} X_1(t) \\ X_2(t) \end{pmatrix} = \mathbf{R} \begin{pmatrix} T(t) \\ \epsilon(t) \end{pmatrix},$$

$$\begin{pmatrix} Y_1(t) \\ Y_2(t) \end{pmatrix} = \mathbf{R} \begin{pmatrix} E(t) \\ \kappa_d(t) \end{pmatrix},$$

$$\begin{pmatrix} Z_1^t(t) \\ Z_2^t(t) \end{pmatrix} = \mathbf{R} \begin{pmatrix} E_r^t(u) \\ \kappa_r^t(u) \end{pmatrix},$$

which completes the diagonalization.

7.1.6 Short-Term Memory

There are well-known models, notably incompressible viscous liquids and Fourier heat flow, in which the rate of dissipation is given by a quadratic form in terms of the time derivatives of the field quantities, and there is no corresponding term relating to energy storage in (5.1.11). Thus, for incompressible viscous liquids, the rate of dissipation, which is equal to the rate of work done, is proportional to $|\dot{\mathbf{E}}|^2$ in the light of Remarks 2.2.16 and 7.1.6, while for Fourier heat flow, it is proportional to $|\dot{\mathbf{m}}|^2 = |\mathbf{q}|^2$ for small temperature variations (or $|\mathbf{d}|^2$ if $\mathbf{d}$ is chosen to be the independent variable). We can take account of such models in a general theory by adding a term

$$\dot{\mathbf{\Lambda}}(t) \cdot \mathbb{L}_d \dot{\mathbf{\Lambda}}(t)$$

to D, given by (7.1.25), and leave the free energy (7.1.9) or (7.1.10) unchanged.

However, rather than simply introducing such a term, we can extract it from the theory developed above, in the following way. Let us replace $\mathbb{L}$ by

$$\mathbb{L}_\lambda(s,u) = \mathbb{L}(s,u) + \lambda \mathbb{L}_d e^{-\lambda(s+u)},$$

where $\mathbb{L}_d \in \mathrm{Lin}(\Gamma)$ is a constant, and take the short memory limit $\lambda \to \infty$ in the rightmost term. By changing variables in the integrals from s, u to $v = \lambda s$ and $w = \lambda u$, we find that for large λ, the quantity ψ, given by (7.1.10), is unchanged, while D, given by (7.1.25), is replaced by

$$D + \dot{\mathbf{\Lambda}}(t) \cdot \mathbb{L}_d \dot{\mathbf{\Lambda}}(t).$$

Also, if $\mathbb{L}(u)$ in $(7.1.13)_3$, is replaced by

$$\mathbb{L}_\lambda(u) = \mathbb{L}(u) + \lambda \mathbb{L}_d e^{-\lambda u},$$

then for large λ, this relation becomes

$$\mathbf{\Sigma}(t) = \widetilde{\mathbf{\Sigma}}_e(\mathbf{\Lambda}_0(t)) + \int_0^\infty \mathbb{L}(u)\dot{\mathbf{\Lambda}}^t(u)du + \mathbb{L}_d \dot{\mathbf{\Lambda}}(t).$$

7.2 Constitutive Equations in the Frequency Domain

Let us now consider the frequency-domain representation of (7.1.13). In the notation of (C.1.3) and using (C.2.2), we write the transform of the causal function $\mathbb{L}'$ in (7.1.13) as

$$\mathbb{L}'_+(\omega) = \int_0^\infty \mathbb{L}'(s)e^{-i\omega s}\,ds = \mathbb{L}'_c(\omega) - i\mathbb{L}'_s(\omega). \tag{7.2.1}$$

Unless indicated otherwise, the frequency may be taken to be real. The notation $\mathbb{L}'_F$ will be reserved for a somewhat different use in (7.2.25) below. We have

$$\mathbb{L}'_c(\omega) = \frac{1}{2}[\mathbb{L}'_+(\omega) + \overline{\mathbb{L}'_+}(\omega)], \qquad \mathbb{L}'_s(\omega) = -\frac{1}{2i}[\mathbb{L}'_+(\omega) - \overline{\mathbb{L}'_+}(\omega)] \tag{7.2.2}$$

and

$$\overline{\mathbb{L}'_+(\omega)} = \mathbb{L}'_+(-\omega), \qquad \mathbb{L}'_c(-\omega) = \mathbb{L}'_c(\omega), \qquad \mathbb{L}'_s(-\omega) = -\mathbb{L}'_s(\omega). \tag{7.2.3}$$

As in Appendix C, we assume that

$$\mathbb{L}' \in L^1(\mathbb{R}^+) \cap L^2(\mathbb{R}^+),$$

giving

$$\mathbb{L}'_+ \in L^2(\mathbb{R}).$$

Indeed, recalling Remark 7.1.2, the stronger assumption

$$\mathbb{L} \in L^1(\mathbb{R}^+) \cap L^2(\mathbb{R}^+) \tag{7.2.4}$$

can and will be made.

7.2.1 Sinusoidal Histories for the General Theory

Fabrizio and Morro [118, 120] (also [121]) examine the implications of thermodynamics for the Fourier transform of the relaxation function derivative in the case of a linear viscoelastic solid. They deduce simple necessary and sufficient conditions for the validity of the second law applied to such materials. It follows from these considerations that the Fourier sine transform of the relaxation function derivative must be negative definite. This approach also provides a very simple demonstration of the symmetry and positivity of the instantaneous and equilibrium elasticity tensors. These and corresponding results for fluids and rigid heat conductors are given in Sects. 8.1.1, 8.10.1, and 9.2. In this subsection, we derive corresponding results for the general theory.

It is assumed that the independent quantity $\boldsymbol{\Lambda}$ has been varying sinusoidally for a sufficiently long time to allow transient effects to be neglected, so that the response $\boldsymbol{\Sigma}$ is also varying sinusoidally with the same period. Each period is generated by a cyclic process (Definition 4.1.4), as described for more general materials in Sect. 5.1.2.

Consider a history and current value $(\boldsymbol{\Lambda}^t, \boldsymbol{\Lambda}(t))$ defined by

$$\boldsymbol{\Lambda}(t) = \mathbf{C}e^{i\omega_0 t} + \overline{\mathbf{C}}e^{-i\omega_0 t}, \quad \boldsymbol{\Lambda}^t(s) = \boldsymbol{\Lambda}(t-s), \quad \omega_0 \in \mathbb{R}^{++}, \tag{7.2.5}$$

where $\mathbf{C} \in \Gamma$ is an amplitude and $\overline{\mathbf{C}}$ its complex conjugate. The quantity $\boldsymbol{\Lambda}^t_+$ has the form

$$\boldsymbol{\Lambda}^t_+(\omega) = \mathbf{C}\frac{e^{i\omega_0 t}}{i(\omega^- + \omega_0)} + \overline{\mathbf{C}}\frac{e^{-i\omega_0 t}}{i(\omega^- - \omega_0)},$$

using the notation of (C.2.10). The generalized stress, as given by $(7.1.17)_2$, becomes

$$\begin{aligned} \boldsymbol{\Sigma}(t) &= \widetilde{\boldsymbol{\Sigma}}_0(\boldsymbol{\Lambda}(t)) + \boldsymbol{\Sigma}_h(t), \\ \boldsymbol{\Sigma}_h(t) &= \int_0^\infty \mathbb{L}'(s)\boldsymbol{\Lambda}^t(s)\,ds = \mathbb{L}'_+(\omega_0)\mathbf{C}e^{i\omega_0 t} + \mathbb{L}'_+(-\omega_0)\overline{\mathbf{C}}e^{-i\omega_0 t}. \end{aligned} \tag{7.2.6}$$

Let us, for purposes of the following discussion, drop the assumption (7.1.18) that $\mathbb{L}$ is symmetric. Referring to (4.1.4), we see that the work done on the material over $[t_0, t]$ is given by

$$\begin{aligned} W(t_0, t) &= \int_{t_0}^{t} \mathbf{\Sigma}(s) \cdot \dot{\mathbf{\Lambda}}(s) ds \\ &= \phi(t) - \phi(t_0) + \int_{t_0}^{t} \mathbb{L}(0)\mathbf{\Lambda}(s) \cdot \dot{\mathbf{\Lambda}}(s) ds + \int_{t_0}^{t} \mathbf{\Sigma}_h(s) \cdot \dot{\mathbf{\Lambda}}(s) ds, \end{aligned} \tag{7.2.7}$$

where relations (7.1.14) have been used. Now, by splitting $\mathbb{L}_0$ into its symmetric and skew-symmetric parts and carrying out an integration by parts, we obtain

$$\begin{aligned} \int_{t_0}^{t} \mathbb{L}_0\mathbf{\Lambda}(s) \cdot \dot{\mathbf{\Lambda}}(s) ds &= \frac{1}{2}[\mathbb{L}_0\mathbf{\Lambda}(t) \cdot \mathbf{\Lambda}(t) - \mathbb{L}_0\mathbf{\Lambda}(t_0) \cdot \mathbf{\Lambda}(t_0)] \\ &\quad + \frac{1}{2}\int_{t_0}^{t} \left(\mathbb{L}_0 - \mathbb{L}_0^{\mathsf{T}}\right)\mathbf{\Lambda}(s) \cdot \dot{\mathbf{\Lambda}}(s) ds. \end{aligned}$$

Also

$$\begin{aligned} &\int_{t_0}^{t} \mathbf{\Sigma}_h(s) \cdot \dot{\mathbf{\Lambda}}(s) ds \\ &= \frac{1}{2}\left[\mathbf{C} \cdot \mathbb{L}'_+(\omega_0)\mathbf{C}(e^{2i\omega_0 t} - e^{2i\omega_0 t_0}) + \overline{\mathbf{C}} \cdot \mathbb{L}'_+(-\omega_0)\overline{\mathbf{C}}(e^{-2i\omega_0 t} - e^{-2i\omega_0 t_0})\right] \\ &\quad - i\omega_0(t - t_0)\left[\overline{\mathbf{C}} \cdot \mathbb{L}'_+(\omega_0)\mathbf{C} - \mathbf{C} \cdot \mathbb{L}'_+(-\omega_0)\overline{\mathbf{C}}\right]. \end{aligned} \tag{7.2.8}$$

Let us take $t = t_0 + T$, where $T = \frac{2\pi}{\omega_0}$. The term $\phi(t) - \phi(t_0)$ on the right of (7.2.7) vanishes, as do the first two terms on the right of (7.2.8). Then, we have

$$\begin{aligned} &\int_{t_0}^{t} \mathbb{L}(0)\mathbf{\Lambda}(s) \cdot \dot{\mathbf{\Lambda}}(s) ds + \int_{t_0}^{t} \mathbf{\Sigma}_h(s) \cdot \dot{\mathbf{\Lambda}}(s) ds \\ &= -i\frac{\omega_0 T}{2}\left[\overline{\mathbf{C}} \cdot (\mathbb{L}_0 - \mathbb{L}_0^{\mathsf{T}})\mathbf{C} - \mathbf{C} \cdot (\mathbb{L}_0 - \mathbb{L}_0^{\mathsf{T}})\overline{\mathbf{C}}\right. \\ &\qquad \left. + 2\overline{\mathbf{C}} \cdot \mathbb{L}'_+(\omega_0)\mathbf{C} - 2\mathbf{C} \cdot \mathbb{L}'_+(-\omega_0)\overline{\mathbf{C}}\right] \\ &= -i\omega_0 T\left\{\overline{\mathbf{C}} \cdot (\mathbb{L}_0 - \mathbb{L}_0^{\mathsf{T}})\mathbf{C} + \overline{\mathbf{C}} \cdot [\mathbb{L}'_+(\omega_0) - \mathbb{L}'^{*}_+(\omega_0)]\,\mathbf{C}\right\} \geq 0, \end{aligned} \tag{7.2.9}$$

where $\mathbb{L}'^{*}_+(\omega_0)$ is the Hermitian conjugate of $\mathbb{L}'_+(\omega_0)$, defined by (A.2.8) to be $\overline{\mathbb{L}'^{\mathsf{T}}_+}(\omega_0) = \mathbb{L}'^{\mathsf{T}}_+(-\omega_0)$. The final inequality is taken to be strict for $0 < \omega_0 < \infty$; the case of nondissipative materials is excluded. We can divide through by the positive quantity $\omega_0 T$. Taking the limit $\omega_0 \to \infty$ and using (C.2.13) gives

$$-i\overline{\mathbf{C}} \cdot (\mathbb{L}_0 - \mathbb{L}_0^{\mathsf{T}})\mathbf{C} = 2\mathbf{C}_r \cdot (\mathbb{L}_0 - \mathbb{L}_0^{\mathsf{T}})\mathbf{C}_i \geq 0, \tag{7.2.10}$$

where $\mathbf{C}_r$, $\mathbf{C}_i$ are the real and imaginary parts of the arbitrary complex quantity $\mathbf{C}$. Note that the inequality need not be strict in this limit. Relation (7.2.10) implies that

$$\mathbb{L}_0 = \mathbb{L}_0^{\top}, \tag{7.2.11}$$

which is a special case of (7.1.18). Using $(7.2.2)_2$ and $(7.2.3)_1$, we deduce from (A.2.12) that

$$\mathbb{L}_s'(\omega_0) < \mathbf{0}, \quad 0 < \omega_0 < \infty. \tag{7.2.12}$$

Let us now reinstate (7.1.18) for all times (and therefore all frequencies for Fourier-transformed quantities). Then (7.2.11) and (7.2.12) imply (7.2.9). Thus, we have the following important result.

Proposition 7.2.1. *The second law, as expressed by* (7.2.9), *implies* (7.2.11) *and* (7.2.12). *Conversely, if* (7.1.18) *is assumed,* (7.2.12) *implies* (7.2.9).

The symmetry of $\mathbb{L}'$, as expressed by (7.1.18), is assumed henceforth, unless otherwise stated.

The converse proved in Proposition 7.2.1 refers only to the particular history (7.2.5). The following, more general, result can also be proved.

Proposition 7.2.2. *Relation* (7.2.12) *implies that* $W(t_0, t)$*, taken over a cycle, is non-negative.*

Proof. Let us take the period of the cycle to be that corresponding to the frequency $\omega_0 > 0$. Any history with such cyclic behavior can be expressed by the Fourier series

$$\mathbf{\Lambda}(t) = \sum_{n=1}^{\infty} \left(\mathbf{C}_n e^{in\omega_0 t} + \overline{\mathbf{C}}_n e^{-in\omega_0 t} \right).$$

Relation (7.2.5) is of course a special case. The steps leading to the generalization of (7.2.9) can be carried through without difficulty. Using the symmetry of $\mathbb{L}'$, we obtain that

$$W\left(t_0, t_0 + \frac{2\pi}{\omega_0}\right) = -2\pi \sum_{n=1}^{\infty} \overline{\mathbf{C}}_n \cdot \mathbb{L}_s'(n\omega_0)\mathbf{C}_n.$$

The result follows immediately. □

7.2.2 Properties of $\mathbb{L}'$

We have, by $(C.1.3)_3$ and (7.2.12),

$$\mathbb{L}_s'(\omega) = \int_0^{\infty} \mathbb{L}'(s) \sin \omega s ds < \mathbf{0}.$$

Note that

$$\mathbb{L}_s'(\omega) = -\omega \mathbb{L}_c(\omega). \tag{7.2.13}$$

The sine inversion formula $(C.1.6)_2$ gives

$$\mathbb{L}'(s) = \frac{2}{\pi} \int_0^{\infty} \mathbb{L}_s'(\omega) \sin \omega s d\omega. \tag{7.2.14}$$

Integration on $(0, s)$ yields

$$\mathbb{L}(s) - \mathbb{L}_0 = \frac{2}{\pi}\int_0^\infty \frac{1-\cos\omega s}{\omega}\mathbb{L}'_s(\omega)d\omega \leq \mathbf{0}, \tag{7.2.15}$$

so that

$$\mathbb{L}(s) \leq \mathbb{L}_0. \tag{7.2.16}$$

We divide by s and take the limit $s \to 0$, which is assumed to exist. This yields

$$\mathbb{L}'(0) \leq \mathbf{0}. \tag{7.2.17}$$

Indeed, if

$$|\mathbb{L}'_s(\omega)| \sim |\omega|^{-p},\ p > 1,$$

for large, real $\pm\omega$, then on making a change of variable $u = \omega s$ in (7.2.14), we obtain that $\mathbb{L}'(0) = \mathbf{0}$, but from $(C.2.17)_2$, it is clear that this need not be true in general. Indeed, we shall assume that

$$\mathbb{L}'(0) < \mathbf{0}. \tag{7.2.18}$$

Since $\mathbb{L}'_s \in L^1(\mathbb{R}^+) \cap L^2(\mathbb{R}^+)$, then $\mathbb{L}'_s(\omega)/\omega \in L^1(\mathbb{R}^+)$. Taking the limit $s \to \infty$ in (7.2.15) and recalling that $\mathbb{L}(\infty)$ vanishes, we obtain, with the aid of the Riemann–Lebesgue lemma (C.2.13),

$$\mathbb{L}_0 = -\frac{2}{\pi}\int_0^\infty \frac{\mathbb{L}'_s(\omega)}{\omega}d\omega > 0. \tag{7.2.19}$$

We shall assume that

$$0 < \left|\int_0^\infty \mathbb{L}(s)ds\right| < \infty, \tag{7.2.20}$$

where the rightmost inequality follows from (7.2.4). Near the origin in the frequency domain,

$$\begin{aligned}\mathbb{L}'_+(\omega) &= \int_0^\infty \mathbb{L}'(s)ds - i\omega\int_0^\infty s\mathbb{L}'(s)ds + \mathbf{O}(\omega^2)\\ &= -\mathbb{L}_0 + i\omega\int_0^\infty \mathbb{L}(s)ds + \mathbf{O}(\omega^2),\end{aligned}$$

where $(7.1.15)_1$ has been used. Observe that from the definition of the cosine transform $(C.1.3)_4$ and (7.2.13), we obtain

$$\mathbb{L}'_c(0) = -\mathbb{L}_0, \qquad \frac{\mathbb{L}'_s(\omega)}{\omega} \underset{\omega\to 0}{\longrightarrow} -\int_0^\infty \mathbb{L}(s)ds. \tag{7.2.21}$$

The quantity $\mathbb{L}'_+$ is analytic on $\Omega^{(-)}$, by Proposition C.2.1. This property is extended by assumption to an open set containing Ω^-. It is further assumed, for simplicity, to be analytic at infinity, though this assumption must be dropped for materials with finite memory, as discussed in Chap. 15.

The quantity $\mathbb{L}'_s$ has singularities in both $\Omega^{(+)}$ and $\Omega^{(-)}$ that are symmetric under reflection in the origin, by $(7.2.3)_3$. They are also symmetric under reflection in the real axis, by virtue of $(7.2.2)_2$ and Remark B.1.2, so that they are mirror images of each other also with respect to the imaginary axis. The function $\mathbb{L}'_s$ goes to zero at the origin and must also be analytic on the real axis. It vanishes linearly at the origin by (7.2.20) and (7.2.21). By virtue of (7.2.12), (7.2.13), and the evenness of $\mathbb{L}_c$,

$$\mathbb{L}_c(\omega) > \mathbf{0}, \quad |\omega| < \infty.$$

A quantity central to our considerations in later chapters is defined by

$$\mathbb{H}(\omega) = -\omega\mathbb{L}'_s(\omega) = \omega^2\mathbb{L}_c(\omega) \geq \mathbf{0}, \quad \omega \in \mathbb{R}. \tag{7.2.22}$$

It is a nonnegative, even tensor function of the frequency, vanishing quadratically at the origin. It is a symmetric tensor in $\mathrm{Lin}(\mathbb{R}^m)$, where $m = 10$. The relation (see (C.2.16) and (C.2.17))

$$i \lim_{\omega\to\infty} \omega\mathbb{L}'_+(\omega) = \lim_{\omega\to\infty} \omega\mathbb{L}'_s(\omega) = \mathbb{L}'(0) \tag{7.2.23}$$

yields

$$\mathbb{L}'(0) = -\mathbb{H}(\infty) \equiv -\mathbb{H}_\infty. \tag{7.2.24}$$

The symmetric quantity $\mathbb{H}_\infty$ is positive definite by virtue of (7.2.18).

If $\mathbb{L}(s)$, $s \in \mathbb{R}^+$ is extended to the even function $\mathbb{L}(|s|)$ on $\mathbb{R}$, then $d\mathbb{L}(|s|)/ds$ is an odd function with Fourier transform given by

$$\mathbb{L}'_F(\omega) = -2i\mathbb{L}'_s(\omega), \tag{7.2.25}$$

which follows from $(C.1.6)_1$. We have, from (C.2.22),

$$\mathbb{L}'_+(\omega) = -\frac{1}{2\pi i}\int_{-\infty}^{\infty} \frac{\mathbb{L}'_+(\omega')}{\omega' - \omega} d\omega', \qquad \omega \in \Omega^{(-)}.$$

One can replace $\mathbb{L}'_+(\omega')$ by $\mathbb{L}'_+(\omega') + \mathbb{F}(\omega')$, where $\mathbb{F}$ is analytic in $\Omega^{(+)}$ and goes to zero as ω^{-1} for large ω, since closing the contour on $\Omega^{(+)}$ results in its contributing zero by Cauchy's theorem. In particular, we can put $\mathbb{F}(\omega') = \lambda\overline{\mathbb{L}'_+}(\omega')$, where λ is any complex number. Choosing $\lambda = -1$ gives

$$\mathbb{L}'_+(\omega) = \frac{1}{\pi}\int_{-\infty}^{\infty} \frac{\mathbb{L}'_s(\omega')}{\omega' - \omega} d\omega', \qquad \omega \in \Omega^{(-)},$$

by virtue of $(7.2.2)_2$. Referring to (7.2.1) and the Plemelj relation $(B.2.14)_2$, we deduce

$$\mathbb{L}'_c(\omega) = \frac{1}{\pi} P\int_{-\infty}^{\infty} \frac{\mathbb{L}'_s(\omega')}{\omega' - \omega} d\omega' = \frac{2}{\pi} P\int_0^{\infty} \frac{\omega'\mathbb{L}'_s(\omega')}{(\omega')^2 - \omega^2} d\omega', \qquad \omega \in \mathbb{R}, \tag{7.2.26}$$

where $(7.2.3)_3$ has been used. This expresses $\mathbb{L}'_c$ at nonnegative frequencies, in terms of $\mathbb{L}'_s$ for nonnegative frequencies at which physical measurements are possible. Choosing $\lambda = 1$ interchanges the roles of $\mathbb{L}'_c$, $\mathbb{L}'_s$. These are examples of "dispersion relations" ([167] and references therein).

7.2.3 Frequency-Domain Representation of the History

Consider the Fourier transform of $\mathbf{\Lambda}^t \in L^1(\mathbb{R}^+) \cap L^2(\mathbb{R}^+)$, given by (see (C.1.3))

$$\mathbf{\Lambda}^t_+(\omega) = \int_0^\infty \mathbf{\Lambda}^t(u)e^{-i\omega u}du.$$

The derivative of $\mathbf{\Lambda}^t_+$ with respect to t will be required. Assuming that $\mathbf{\Lambda}^t \in C^1(\mathbb{R}^+)$, we find, with the aid of (5.1.14) and an integration by parts, that

$$\frac{d}{dt}\mathbf{\Lambda}^t_+(\omega) = -i\omega\mathbf{\Lambda}^t_+(\omega) + \mathbf{\Lambda}(t). \tag{7.2.27}$$

By virtue of (C.2.16), (C.2.17),

$$i \lim_{\omega\to\infty} \omega\mathbf{\Lambda}^t_+(\omega) = \lim_{\omega\to\infty} \omega\mathbf{\Lambda}^t_s(\omega) = \mathbf{\Lambda}(t). \tag{7.2.28}$$

We deduce from (C.2.10) that the Fourier transform of $\mathbf{\Lambda}^t_r(s) = \mathbf{\Lambda}^t(s) - \mathbf{\Lambda}(t)$, $s \in \mathbb{R}$, is given by

$$\mathbf{\Lambda}^t_{r+}(\omega) = \mathbf{\Lambda}^t_+(\omega) - \frac{\mathbf{\Lambda}(t)}{i\omega^-}, \tag{7.2.29}$$

which, on noting (7.2.28), can be seen to behave as ω^{-2} at large frequencies. Also, with the aid of (7.2.27), we see that

$$\frac{d}{dt}\mathbf{\Lambda}^t_+(\omega) = -i\omega\mathbf{\Lambda}^t_{r+}(\omega), \qquad \frac{d}{dt}\mathbf{\Lambda}^t_{r+}(\omega) = -i\omega\mathbf{\Lambda}^t_{r+}(\omega) - \frac{\dot{\mathbf{\Lambda}}(t)}{i\omega^-}. \tag{7.2.30}$$

A continuation of the history $\mathbf{\Lambda}^t(u)$, $u \in \mathbb{R}^+$, is $\mathbf{\Lambda}^t(u)$, $u \in \mathbb{R}^{--}$. Its Fourier transform is given by

$$\mathbf{\Lambda}^t_-(\omega) = \int_{-\infty}^0 \mathbf{\Lambda}^t(u)e^{-i\omega u}du.$$

The Fourier transform over $\mathbb{R}$ is

$$\mathbf{\Lambda}^t_F(\omega) = \int_{-\infty}^\infty \mathbf{\Lambda}^t(u)e^{-i\omega u}du = \mathbf{\Lambda}^t_+(\omega) + \mathbf{\Lambda}^t_-(\omega).$$

Corresponding to (7.2.29), we have

$$\mathbf{\Lambda}^t_{r-}(\omega) = \mathbf{\Lambda}^t_-(\omega) + \frac{\mathbf{\Lambda}(t)}{i\omega^+}.$$

Thus

$$\mathbf{\Lambda}^t_{rF}(\omega) = \mathbf{\Lambda}^t_{r+}(\omega) + \mathbf{\Lambda}^t_{r-}(\omega) = \mathbf{\Lambda}^t_F(\omega) + i\mathbf{\Lambda}(t)\left(\frac{1}{\omega^-} - \frac{1}{\omega^+}\right). \tag{7.2.31}$$

The last term on the right is proportional to the singular delta function $\delta(\omega)$, by virtue of (C.2.19).

7.2.4 Constitutive Equations in Terms of Frequency-Domain Quantities

Applying Parseval's formula (C.3.1) to (7.1.13)$_1$ by extending the range of integration to $\mathbb{R}$ in the latter formula, one obtains

$$\boldsymbol{\Sigma}(t) = \boldsymbol{\Sigma}_e(t) + \frac{1}{2\pi}\int_{-\infty}^{\infty} \overline{\mathbb{L}'_+}(\omega)\boldsymbol{\Lambda}^t_{r+}(\omega)\,d\omega. \tag{7.2.32}$$

As above, if we replace $\overline{\mathbb{L}'_+}(\omega)$ by $\overline{\mathbb{L}'_+}(\omega) + \mathbb{F}(\omega)$, where $\mathbb{F}(\omega)$ is analytic on Ω^- and is a constant or zero at infinity, the relationship still holds. This follows by a simple application of Cauchy's theorem over Ω^-, using the behavior of $\boldsymbol{\Lambda}^t_{r+}$ at large ω. In particular, we have

$$\boldsymbol{\Sigma}(t) = \boldsymbol{\Sigma}_e(t) + \frac{1}{2\pi}\int_{-\infty}^{\infty} [\overline{\mathbb{L}'_+}(\omega) + \lambda\mathbb{L}'_+(\omega)]\boldsymbol{\Lambda}^t_{r+}(\omega)\,d\omega, \tag{7.2.33}$$

where λ is any complex constant. Choosing $\lambda = -1$ yields

$$\boldsymbol{\Sigma}(t) = \boldsymbol{\Sigma}_e(t) + \frac{i}{\pi}\int_{-\infty}^{\infty} \mathbb{L}'_s(\omega)\boldsymbol{\Lambda}^t_{r+}(\omega)\,d\omega. \tag{7.2.34}$$

In fact, (7.2.34) corresponds to taking the even extension of $\mathbb{L}$ to $\mathbb{R}$. This amounts to writing (7.1.13)$_1$ as

$$\boldsymbol{\Sigma}(t) = \boldsymbol{\Sigma}_e(t) + \int_{-\infty}^{\infty} \frac{d}{ds}\mathbb{L}(|s|)\boldsymbol{\Lambda}^t_r(s)\,ds, \tag{7.2.35}$$

where $\boldsymbol{\Lambda}^t_r$ is taken to be zero on $\mathbb{R}^-$. Parseval's formula applied to (7.2.35) gives that $\overline{\mathbb{L}'_F}$ replaces $\overline{\mathbb{L}'_+}$ in (7.2.32) and (7.2.34) follows from (7.2.25). Relation (7.2.34) was first given for viscoelastic materials in [104].

If we take $\lambda = 1$, then $i\mathbb{L}'_s(\omega)$ in (7.2.34) is replaced by $\mathbb{L}'_c(\omega)$. This is less convenient for future use because there is no relation for $\mathbb{L}'_c$ corresponding to (7.2.12), yielding (7.2.22), a quantity guaranteed to be nonnegative by thermodynamics, which allows an important factorization to be carried out. In fact, $\mathbb{L}'_c(\omega)$ is nonnegative in many standard models, but this is not guaranteed by thermodynamics. A similar drawback applies to other choices of λ.

7.3 The Form of the Generalized Relaxation Function

A material can be characterized by the singularity structure of $\mathbb{L}'_+$ on $\Omega^{(+)}$, as may be seen, at least in the case that $\mathbb{L}'_+$ is analytic at infinity, by evaluating

$$\begin{aligned} \mathbb{L}'(s) &= \frac{1}{2\pi}\int_{-\infty}^{\infty} \mathbb{L}'_+(\omega)e^{i\omega s}\,d\omega, \quad s \in \mathbb{R}^{++}, \\ \mathbb{L}'(s) &= \mathbf{0}, \quad s \in \mathbb{R}^{--}, \qquad \mathbb{L}'(0) = \mathbb{L}'(0^+), \end{aligned} \tag{7.3.1}$$

using (B.1.14).

The possible types of singularities are isolated singularities, discontinuities associated with branch cuts, and essential singularities (Sects. B.1.4, B.1.5). We take $\mathbb{L}$ to be a scalar quantity L for purposes of this discussion. In the context of assumptions made later, in Sect. 11.6, there is no loss of generality in doing this.

7.3.1 Isolated Singularities

Let us first consider the case that L'_+ has only isolated singularities. We confine the discussion to the case that the number of such singularities is finite, not only to avoid convergence issues but also in recognition of the fact that the determination of L'_+ for a given material is always, at least partially, an exercise in phenomenology, in effect, curve-fitting to data points. Clearly, the outcome of such a phenomenological exercise automatically produces a finite, rather than an infinite, number of such singularities. Predictions of molecular theories, such as those based on Brownian motion, may also provide guidance [133, 157].

Thus, L'_+ can be expressed as a sum of pole terms of varying orders, at different positions in $\Omega^{(+)}$. We have

$$\int_0^\infty e^{-\alpha s} e^{-i\omega s} ds = -\frac{i}{\omega - i\alpha}, \qquad \alpha \in \mathbb{R}^{++}, \tag{7.3.2}$$

and, by differentiation with respect to α,

$$\int_0^\infty s^n e^{-\alpha s} e^{-i\omega s} ds = \frac{(-i)^{n+1} n!}{(\omega - i\alpha)^{n+1}}, \qquad \alpha \in \mathbb{R}^{++}. \tag{7.3.3}$$

Thus, a simple pole at $i\alpha$ in the frequency domain corresponds to a decaying exponential $e^{-\alpha s}$, while a pole of order $n+1$ corresponds to $s^n e^{-\alpha s}$. A sum of poles of different orders at $i\alpha$ corresponds to a polynomial multiplying $e^{-\alpha s}$.

Consider the case that α is complex. The formulas (7.3.2) and (7.3.3) apply unchanged. However, $e^{-\alpha s}$ will now contain oscillating trigonometric functions. We adopt the terminology that $e^{-\alpha s}$ is a strictly decaying exponential if α is real and positive, and a decaying exponential if $\mathrm{Im}\alpha \neq 0$ and $\mathrm{Re}\alpha \in \mathbb{R}^{++}$.

The positions of the singularities are subject to the condition that the components of L' be real, namely (C.1.7), or for complex ω,

$$\overline{L'_+(\omega)} = L'_+(-\overline{\omega}), \tag{7.3.4}$$

which implies that the singularities of L'_+ in $\Omega^{(+)}$ must be symmetric with respect to reflection in the imaginary axis.

Note that if L'_+ is given by sums of isolated singularities, it can always be expressed as a ratio of polynomials, in other words, a rational function of ω.

If the singularities are simple poles on the imaginary axis, with negative imaginary coefficients, then we see from (7.3.2) that the resulting time-domain relaxation function derivatives are monotonically decreasing in magnitude as time increases. If the singularities are off the imaginary axis, so that trigonometric factors are generated in the time-domain representation, L' need not exhibit monotonic behavior.

7.3.2 Branch Cuts

When only branch-cut singularities exist, we obtain from (7.3.1)

$$L'(s) = \frac{1}{2\pi} \int_C d(\omega) e^{i\omega s} \, d\omega, \quad s \in \mathbb{R}^+.$$

The integral over C encloses each branch cut. Consider one of these, from a to b on $\Omega^{(+)}$. Then the contribution to L' is, from (B.1.11),

$$L'(s) = \frac{1}{2\pi} \int_a^b d(\omega) e^{i\omega s} \, d\omega, \quad s \in \mathbb{R}^+, \tag{7.3.5}$$

with

$$d(\omega) = L_+'^-(\omega) - L_+'^+(\omega), \tag{7.3.6}$$

where $L_+'^-(\omega)$ is the limiting value of $L'_+(\omega)$ approaching the cut segment on the side where the shrunken contour around the cut is going from a to b and $L_+'^+(\omega)$ is the limiting value approaching from the other side.

Indeed, we see from (7.3.5) that

$$L'_+(\omega) = -\frac{1}{2\pi i} \int_a^b \frac{d(\omega')d\omega'}{\omega' - \omega}, \quad \omega \in \Omega^{(-)}. \tag{7.3.7}$$

Having derived this form for $\omega \in \Omega^{(-)}$, we can extend it by analytic continuation to ω anywhere in the complex plane except on the cut. The Plemelj formulas (B.2.6) give

$$L_+'^\pm(\omega) = \mp\frac{1}{2}d(\omega) - \frac{1}{2\pi i} P \int_a^b \frac{d(\omega')d\omega'}{\omega' - \omega}, \quad \omega \in [a, b],$$

from which (7.3.6) follows immediately.

The condition (7.3.4) implies that branch cuts not on the imaginary axis must have mirror images of each other in that axis.

It is shown in Sect. 16.5 that branch cuts lead to a situation in which minimal states are singletons, the maximum free energy is the work function and the lower bound on the dissipation is the weakest possible estimate, namely zero. Thus, they have disadvantages from a theoretical point of view but are widely used in modeling experimental results [133].

If a branch cut is along the positive imaginary axis, then (7.3.5) becomes

$$L'(s) = \frac{1}{2\pi} \int_{a_0}^{b_0} d_c(\alpha) e^{-\alpha s} \, d\alpha, \quad s \in \mathbb{R}^+, \quad d_c(\alpha) = id(i\alpha), \tag{7.3.8}$$

where $a_0 = -ia$ and $b_0 = -ib$; the quantities a_0, b_0 are real and nonnegative. The function d is defined by (7.3.6). This important special case is referred to as the continuous-spectrum form [94]. An expression for the minimum free energy of such materials is derived in Chap. 14. Clearly, d_c must be a real quantity. In this case, (7.3.7) becomes

$$L'_+(\omega) = \frac{1}{2\pi} \int_{a_0}^{b_0} \frac{d_c(\alpha)d\alpha}{\alpha + i\omega}, \quad \omega \notin [a, b].$$

On the negative imaginary axis, at $\omega = i\beta, \beta \le 0$, and on the positive imaginary axis for $\beta < a_0$,

$$L'_+(i\beta) = \frac{1}{2\pi}\int_{a_0}^{b_0} \frac{d_c(\alpha)d\alpha}{\alpha - \beta}, \tag{7.3.9}$$

which is a real quantity. This also follows from (7.2.1). For $\beta \in [a_0, b_0]$,

$$L'^{\pm}_+(i\beta) = \pm\frac{i}{2}d_c(\beta) + \frac{1}{2\pi}P\int_{a_0}^{b_0} \frac{d_c(\alpha)d\alpha}{\alpha - \beta},$$

which are complex quantities, where the real part of each is a principal value integral but otherwise similar to that in (7.3.9); the imaginary parts are proportional to the cut discontinuity function.

If d_c is a nonnegative quantity on the cut, then L', given by (7.3.8), is a monotonically decreasing function.

Branch-cut singularities off the imaginary axis may yield oscillatory behavior in L', combined with relaxation behavior, due to trigonometric functions multiplying the exponentials.

7.3.3 Essential Singularities

Essential singularities at infinity of a certain kind are associated with finite memory, i.e., where $L'(s)$, or a term in this function, vanishes for $s > s_d > 0$, the quantity s_d being the duration of the memory [111]. The minimum free energies associated with simple examples of such materials are discussed in Chap. 15.

Essential singularities at finite points on $\Omega^{(+)}$ are the remaining possibility. It is difficult to imagine a choice of relaxation behavior that would generate such behavior in L'_+. Such singularities are excluded from consideration in this work.

With the exception of Chap. 15, we consider materials for which L'_+ is analytic at infinity, i.e., obeys (7.2.23) for the limit taken in any direction in the complex plane, and indeed (C.2.16) if higher derivatives of $L'(s)$ exist at the origin.

7.4 Minimal States in the Nonisothermal Case

In the classical approach to materials with memory, the state is identified with the history of the independent variables. Noll's characterization of state [277] is more convenient for application to such materials. Indeed, Noll takes the material response as the basis for the definition of state: if an arbitrary process, acting on different given histories up to time t, leads to the same response of the material after time t, then the given histories are equivalent and the state is represented as the class of all these equivalent histories. We shall refer to it as the *minimal state* [110].

The concept of a minimal state arises out of Definition 4.1.2 of equivalent states. A minimal state is in effect an equivalence class under this definition. The idea has been applied to completely linear materials in [31, 89, 90, 92, 110, 176]. In the present nonisothermal context, where the memory terms in the constitutive equations are linear, the definition of minimal state is now discussed. Two states $(\mathbf{\Lambda}_1^t, \mathbf{\Lambda}_1(t))$, $(\mathbf{\Lambda}_2^t, \mathbf{\Lambda}_2(t))$ are equivalent, or in the same minimal state, if

$$\begin{aligned} \boldsymbol{\Sigma}_1(t+s) &= \boldsymbol{\Sigma}_2(t+s), \\ \boldsymbol{\Lambda}_1(t+s) &= \boldsymbol{\Lambda}_2(t+s) + \boldsymbol{\Lambda}_c, \quad s \geq 0, \end{aligned} \tag{7.4.1}$$

where $\boldsymbol{\Sigma}_1$, $\boldsymbol{\Sigma}_2$ are the generalized stresses corresponding to the states $(\boldsymbol{\Lambda}_1^t, \boldsymbol{\Lambda}_1(t))$, $(\boldsymbol{\Lambda}_2^t, \boldsymbol{\Lambda}_2(t))$; we are using actual rather than relative histories. In the second relation of (7.4.1), $\boldsymbol{\Lambda}_c$ is a constant member of Γ^+. This is the condition that the process $\dot{\boldsymbol{\Lambda}}_i(t+s)$, $i = 1, 2, s \in \mathbb{R}^+$ (see Remark 5.1.1), acting on each is the same. Using $(7.1.13)_2$ and the definition of $\boldsymbol{\Sigma}_0$ in (7.1.14), relation $(7.4.1)_1$ becomes

$$\begin{aligned} \boldsymbol{\Sigma}_e(\boldsymbol{\Lambda}_{01}(t+s)) - \boldsymbol{\Sigma}_e(\boldsymbol{\Lambda}_{02}(t+s)) + \mathbb{L}(s)\boldsymbol{\Lambda}_c + \mathbf{I}^t(s, \boldsymbol{\Lambda}_d^t) &= \mathbf{0}, \\ \mathbf{I}^t(s, \boldsymbol{\Lambda}_d^t) = \int_0^\infty \mathbb{L}'(s+u)\boldsymbol{\Lambda}_d^t(u)du&, \\ \boldsymbol{\Lambda}_d^t(u) = \boldsymbol{\Lambda}_1^t(u) - \boldsymbol{\Lambda}_2^t(u)&, \end{aligned} \tag{7.4.2}$$

where $\boldsymbol{\Lambda}_{01}, \boldsymbol{\Lambda}_{02}$ are $\boldsymbol{\Lambda}_0$, given by $(5.1.21)_2$, for states 1, 2, respectively. Letting $s \to \infty$ and recalling (7.1.15), we obtain

$$\boldsymbol{\Sigma}_e(\boldsymbol{\Lambda}_{01}^t(\infty)) = \boldsymbol{\Sigma}_e(\boldsymbol{\Lambda}_{01}^t(\infty) - \boldsymbol{\Lambda}_{0c}),$$

where $\boldsymbol{\Lambda}_{0c}$ is $\boldsymbol{\Lambda}_c$ with the heat flow component equal to zero. It will be assumed that $\boldsymbol{\Sigma}_e$ is uniquely invertible, so that

$$\boldsymbol{\Lambda}_{0c} = \mathbf{0}.$$

This condition does not, however, constrain the cumulative heat flow variable $\mathbf{m}$. Thus, we conclude that the conditions for equivalence take the form

$$\begin{aligned} \boldsymbol{\Lambda}_{01}(t) &= \boldsymbol{\Lambda}_{02}(t), \\ \mathbf{I}^t(s, \boldsymbol{\Lambda}_d^t) &= \mathbf{I}^t(s, \boldsymbol{\Lambda}_1^t) - \mathbf{I}^t(s, \boldsymbol{\Lambda}_2^t) = \mathbf{0}, \quad s \geq 0. \end{aligned} \tag{7.4.3}$$

By applying the same process to each state, we conclude from $(7.4.3)_1$ that

$$\boldsymbol{\Lambda}_{01}(t+s) = \boldsymbol{\Lambda}_{02}(t+s), \quad s \geq 0. \tag{7.4.4}$$

A functional of $(\boldsymbol{\Lambda}^t, \boldsymbol{\Lambda}(t))$ or $(\boldsymbol{\Lambda}_r^t, \boldsymbol{\Lambda}(t))$ that yields the same value for all members of the same minimal state will be referred to as a function or functional of the minimal state or as a minimal-state variable.

Note that (cf. Proposition 4.1.3) from (7.4.1),

$$\begin{aligned} W_1(t+s) - W_1(t) &= \int_t^{t+s} \boldsymbol{\Sigma}_1(u)\dot{\boldsymbol{\Lambda}}_1(u)du \\ &= \int_t^{t+s} \boldsymbol{\Sigma}_2(u)\dot{\boldsymbol{\Lambda}}_2(u)du = W_2(t+s) - W_2(t). \end{aligned}$$

Conversely, the condition

$$W_1(t+s) - W_1(t) = W_2(t+s) - W_2(t), \qquad s \in \mathbb{R}^+,$$

which is satisfied for any process

$$\dot{\mathbf{\Lambda}}(u) = \dot{\mathbf{\Lambda}}_1(u) = \dot{\mathbf{\Lambda}}_2(u), \quad u \in [t, t+s],$$

implies that

$$\mathbf{\Sigma}_1(u) = \mathbf{\Sigma}_2(u), \quad u \in [t, t+s],$$

and the two states are equivalent.

For relaxation functions in the form of finite sums of (more than one) exponentials, there are typically many different histories in the same minimal state [160, 176]. For the case of integrals of exponentials, as in (7.3.5) and (7.3.8), this is not so: the sets of equivalent states have only one member [94, 160]. A generalization of these results is given in Sect. 16.5.

Let us assume for simplicity that $(7.4.3)_1$ and (7.4.4) are replaced by $\mathbf{\Lambda}_1(t+s) = \mathbf{\Lambda}_2(t+s)$, $s \geq 0$. This is true for all isothermal models, which are those mainly of interest in later chapters. It is then clear from (7.4.3) that the statement that $(\mathbf{\Lambda}_1^t, \mathbf{\Lambda}_1(t))$ is equivalent to $(\mathbf{\Lambda}_2^t, \mathbf{\Lambda}_2(t))$ implies that $(\mathbf{\Lambda}_d^t, \mathbf{\Lambda}_d(t))$ is equivalent to the zero history and current value. In particular, this means that $\mathbf{\Lambda}_d(t) = \mathbf{\Lambda}_1(t) - \mathbf{\Lambda}_2(t) = \mathbf{0}$. The converse is also true.

The following notation is introduced:

$$\langle \mathbf{\Lambda}_1^t, \mathbf{\Lambda}_2^t \rangle = \int_0^\infty \int_0^\infty \mathbf{\Lambda}_1^t(s_1) \cdot \mathbb{L}_{12}(s_1, s_2) \mathbf{\Lambda}_2^t(s_2) ds_1 ds_2. \tag{7.4.5}$$

We have

$$\langle \mathbf{\Lambda}_1^t, \mathbf{\Lambda}_2^t \rangle = \langle \mathbf{\Lambda}_2^t, \mathbf{\Lambda}_1^t \rangle$$

by virtue of (7.1.4) with $(7.1.6)_1$.

Let $(\mathbf{\Lambda}_1^t, \mathbf{\Lambda}(t))$, $(\mathbf{\Lambda}_2^t, \mathbf{\Lambda}(t))$ be any equivalent states. Then a free energy is a functional of the minimal state if

$$\psi(\mathbf{\Lambda}_1^t, \mathbf{\Lambda}(t)) = \psi(\mathbf{\Lambda}_2^t, \mathbf{\Lambda}(t)). \tag{7.4.6}$$

It is not necessary that a free energy have this property, though it holds for many of the free energies introduced in later chapters. Notably, this is true for the minimum free energy, as can be shown in great generality (Remark 4.2.11) but also in other cases. It follows from (7.1.19) and (7.4.6) that

$$\langle \mathbf{\Lambda}_1^t, \mathbf{\Lambda}_1^t \rangle = \langle \mathbf{\Lambda}_2^t, \mathbf{\Lambda}_2^t \rangle. \tag{7.4.7}$$

Note that if $\mathbf{\Lambda}_d^t$ is equivalent to the zero history, then from (7.4.6),

$$\psi(\mathbf{\Lambda}_d^t, \mathbf{0}) = 0, \tag{7.4.8}$$

or from (7.4.7),

$$\langle \mathbf{\Lambda}_d^t, \mathbf{\Lambda}_d^t \rangle = 0. \tag{7.4.9}$$

If the material has minimal states that are nonsingleton, then (7.4.9) implies that the quadratic functional with kernel $\mathbb{L}_{12}$ is positive semidefinite. If the material has singleton minimal states, then this quadratic functional is positive definite.

In general, one cannot deduce (7.4.9) from (7.4.3) or vice versa, so that (7.4.9) does not necessarily imply that the free energy is a functional of state. Also, (7.4.3) does not in general imply (7.4.7) or indeed (7.4.9). However, assume that

$$\langle \mathbf{\Lambda}^t, \mathbf{\Lambda}^t \rangle = (p^t, p^t), \tag{7.4.10}$$

where $(\cdot, \cdot)$ is the norm in some specified function space and $p^t = p^t(u, \mathbf{\Lambda}^t)$, $u \in U \subseteq \mathbb{R}$, is a linear functional of $\mathbf{\Lambda}^t$. Assume also that the requirement

$$p^t(u, \mathbf{\Lambda}_d^t) = p^t(u, \mathbf{\Lambda}_1^t) - p^t(u, \mathbf{\Lambda}_2^t) = 0 \tag{7.4.11}$$

is equivalent to $(7.4.3)_2$. Then we can deduce the following result. Relation $(7.4.3)_2$ implies (7.4.11), which implies (7.4.7) and also (7.4.9) or (7.4.8). Thus, it follows that free energies whose history-dependent part obeys (7.4.10) are functionals of the minimal state and also obey (7.4.9).

We will introduce in later chapters categories of free energies derivable from extremum requirements (for example, the minimum free energies), all of which meet condition (7.4.10).

If (7.4.7) is used in (7.4.9), we deduce that

$$\langle \mathbf{\Lambda}_1^t, \mathbf{\Lambda}_1^t \rangle = \langle \mathbf{\Lambda}_2^t, \mathbf{\Lambda}_2^t \rangle = \langle \mathbf{\Lambda}_1^t, \mathbf{\Lambda}_2^t \rangle. \tag{7.4.12}$$

These relations in turn imply (7.4.9).

7.5 Forms of the Work Function

We seek general expressions for the total work done on the material. Allowing for a disturbed initial state in $(7.2.7)_1$ with $t_0 = -\infty$ and using $(7.1.14)_2$ together with $(7.1.17)_{1,4}$, we have that

$$\begin{aligned}
\widetilde{W}(\mathbf{\Lambda}^t, \mathbf{\Lambda}(t)) = W(t) &= \phi(-\infty) + \int_{-\infty}^{t} \mathbf{\Sigma}(u) \cdot \dot{\mathbf{\Lambda}}(u) du \\
&= \phi(-\infty) + \int_{-\infty}^{t} \mathbf{\Sigma}_e(u) \cdot \dot{\mathbf{\Lambda}}(u) du + \int_{-\infty}^{t} \mathbf{\Sigma}_{rh}(u) \cdot \dot{\mathbf{\Lambda}}(u) du \\
&= \phi(t) + \int_{-\infty}^{t} \mathbf{\Sigma}_{rh}(u) \cdot \dot{\mathbf{\Lambda}}(u) du \\
&= \phi(t) + \int_{-\infty}^{t} \int_{0}^{\infty} \mathbb{L}(s) \dot{\mathbf{\Lambda}}(u-s) \cdot \dot{\mathbf{\Lambda}}(u) ds du \\
&= \phi(t) + \int_{-\infty}^{t} \int_{-\infty}^{u} \mathbb{L}(u-v) \dot{\mathbf{\Lambda}}(v) \cdot \dot{\mathbf{\Lambda}}(u) dv du \\
&= \phi(t) + \int_{-\infty}^{t} \int_{v}^{t} \mathbb{L}(u-v) \dot{\mathbf{\Lambda}}(v) \cdot \dot{\mathbf{\Lambda}}(u) dv du \\
&= \phi(t) + \int_{-\infty}^{t} \int_{u}^{t} \mathbb{L}(v-u) \dot{\mathbf{\Lambda}}(u) \cdot \dot{\mathbf{\Lambda}}(v) du dv.
\end{aligned} \tag{7.5.1}$$

The last two forms follow from identities similar to those in (5.2.10). The symmetry of $\mathbb{L}$ has also been used. Adding $(7.5.1)_6$ and $(7.5.1)_8$ and dividing by two gives

$$W(t) = \phi(t) + \frac{1}{2}\int_0^\infty \int_0^\infty \mathbb{L}(|s-u|)\dot{\mathbf{\Lambda}}^t(u)\cdot\dot{\mathbf{\Lambda}}^t(s)duds \tag{7.5.2}$$

after changes of variables. This is a special case of (5.2.11). By means of the relations

$$\dot{\mathbf{\Lambda}}^t(u) = \frac{\partial}{\partial t}\mathbf{\Lambda}(t-u) = -\frac{\partial}{\partial u}\mathbf{\Lambda}_r^t(u)$$

and two partial integrations, it follows from (7.5.2) that

$$W(t) = \phi(t) + \frac{1}{2}\int_0^\infty \int_0^\infty \mathbb{L}_{12}(|s-u|)\mathbf{\Lambda}_r^t(u)\cdot\mathbf{\Lambda}_r^t(s)duds, \tag{7.5.3}$$

where

$$\mathbb{L}_{12}(|s-u|) = \frac{\partial}{\partial s}\frac{\partial}{\partial u}\mathbb{L}(|s-u|) = -\frac{\partial^2}{\partial s^2}\mathbb{L}(|s-u|) = -\frac{\partial^2}{\partial u^2}\mathbb{L}(|s-u|). \tag{7.5.4}$$

It will be useful later to write (7.5.3) in the form

$$W(t) = \phi(t) + \frac{1}{2}\int_{-\infty}^t \int_{-\infty}^t \mathbb{L}_{12}(|s-u|)[\mathbf{\Lambda}(u)-\mathbf{\Lambda}(t)]\cdot[\mathbf{\Lambda}(s)-\mathbf{\Lambda}(t)]duds. \tag{7.5.5}$$

Note the relation [107]

$$\mathbb{L}_{12}(|s-u|) = -2\delta(s-u)\mathbb{L}'(|s-u|) - \mathbb{L}''(|s-u|) \tag{7.5.6}$$

in terms of the singular delta function, where the prime denotes differentiation with respect to the explicit argument. This follows by differentiating the relation

$$\frac{\partial}{\partial s}\mathbb{L}(|s-u|) = sign(s-u)\mathbb{L}'(|s-u|)$$

with respect u.

Assume that $\dot{\mathbf{\Lambda}}^t(u) = \mathbf{0}$, $\mathbf{\Lambda}_r^t(u) = \mathbf{0}$, $u < 0$. Then application of the convolution theorem and Parseval's formula (Sect. C.3) to (7.5.2) and (7.5.3) gives

$$\begin{aligned} W(t) &= \phi(t) + \frac{1}{2\pi}\int_{-\infty}^\infty \overline{\dot{\mathbf{\Lambda}}_+^t(\omega)}\cdot\frac{\mathbb{H}(\omega)}{\omega^2}\dot{\mathbf{\Lambda}}_+^t(\omega)d\omega \\ &= \phi(t) + \frac{1}{2\pi}\int_{-\infty}^\infty \overline{\mathbf{\Lambda}_{r+}^t(\omega)}\cdot\mathbb{H}(\omega)\mathbf{\Lambda}_{r+}^t(\omega)d\omega, \end{aligned} \tag{7.5.7}$$

where (7.2.22) has been used. The Fourier transforms of $\mathbb{L}(|s|)$ and $-d^2\mathbb{L}(|s|)/ds^2$ (see $(7.5.4)_2$), required for the convolution theorem, are determined as follows. Recalling (7.2.25), we observe further that by virtue of $(C.1.5)_1$,

$$\mathbb{L}_F(\omega) = 2\mathbb{L}_c(\omega) = 2\frac{\mathbb{H}(\omega)}{\omega^2}.$$

Also, the Fourier transform of $-d^2\mathbb{L}(|s|)/ds^2$ is given by

$$-(i\omega)^2\mathbb{L}_F(\omega) = 2\mathbb{H}(\omega).$$

The nonnegativity of $\mathbb{H}$ yields that the integral terms in (7.5.7) are also nonnegative. Thus, the memory integrals in (7.5.2) and (7.5.3) have this property, which gives that

$$W(t) \geq \phi(t). \tag{7.5.8}$$

The quantity $W(t)$ is in fact a free energy with zero rate of dissipation in that it meets the requirements listed in Sect. 5.1.1. Properties P1 and P2 follow from (7.5.3), while P3 is a consequence of (7.5.8). Property P4 with $D(t) = 0$ follows from $(7.5.1)_2$. The quantity $W(t)$ is greater than any other free energy, and so is often referred to as the maximum free energy, denoted by $\psi_M(t)$. However, it will be shown to be a function of the minimal state only if that state is a singleton.

8

Viscoelastic Solids and Fluids

We now consider special cases of the constitutive relations (7.1.13), namely linear viscoelastic solids and fluids with linear memory under isothermal conditions in the present chapter and an approximate version of rigid heat conductors in Chap. 9. Some of the formulas are similar to those derived in the general case, and detailed proofs are omitted or a different version is given. Other formulas are specific to completely linear materials, for example.

More use will be made, for illustrative purposes, of the abstract terminology and notation introduced in Chaps. 3 and 4 in discussing these specific materials than in the general case. This is particularly true for Chap. 9.

Combining constitutive relations for solids and fluids with the equations of motion (1.3.25) yields the dynamical equations describing the time evolution of the material under specified initial and boundary conditions. Questions of the existence, uniqueness, and stability of the solutions of these integro–partial differential equations are considered in Part IV, particularly in Chap. 24. Practical methods for obtaining explicit solutions, particularly in the quasistatic approximation, may be found in older texts such as [65] and especially [167].

8.1 Linear Viscoelastic Solids

In the general form of the theory, we are dealing with finite linear viscoelasticity [73], where the stress is given by a linear memory functional of the strain history and a nonlinear (or linear) function of the current strain. The space Γ reduces to Sym and Lin(Γ) to Lin(Sym). Thus, (7.1.21) reduces to

$$\widehat{\mathbf{S}}(t) = \widehat{\mathbf{S}}_e(\mathbf{E}(t)) + \int_0^\infty \mathbb{G}'(u)\mathbf{E}_r^t(u)du, \qquad \mathbb{G}'(u) = \frac{\rho}{\kappa}\mathbb{L}_E'(u). \tag{8.1.1}$$

G. Amendola et al., *Thermodynamics of Materials with Memory*,
https://doi.org/10.1007/978-3-030-80534-0_8

For the case of completely linear viscoelasticity, we replace $\widehat{\mathbf{S}}$ by the Cauchy stress tensor $\mathbf{T}$, and (8.1.1) becomes

$$\mathbf{T}(t) = \mathbb{G}_\infty \mathbf{E}(t) + \int_0^\infty \mathbb{G}'(u)\mathbf{E}_r^t(u)du, \tag{8.1.2}$$

where the relative strain history $\mathbf{E}_r^t$ is given by

$$\mathbf{E}_r^t(s) = \mathbf{E}^t(s) - \mathbf{E}(t). \tag{8.1.3}$$

The quantity $\mathbb{G}$ is the relaxation function of the viscoelastic material. For a viscoelastic solid, $\mathbb{G}(\infty) = \mathbb{G}_\infty$ is a positive tensor, defined by $(1.4.12)_2$. Thus, we no longer have the condition $(7.1.15)_1$. This property can be retained by using $\mathcal{G}(u) = \mathbb{G}(u) - \mathbb{G}_\infty$. We will not do so, however, for reasons of convention. This means that certain partial integrations are slightly more complicated. We are therefore adopting what was referred to in Remark 7.1.2 as the conventional choice. Note that the assumption (7.2.4) no longer applies. We shall assume that $\mathbb{G}(\cdot) - \mathbb{G}_\infty \in L^1(\mathbb{R}^+) \cap L^2(\mathbb{R}^+)$. Relations $(7.1.13)_{2,3}$ become

$$\begin{aligned}
\widehat{\mathbf{S}}(t) &= \widehat{\mathbf{S}}_e(\mathbf{E}(t)) + (\mathbb{G}_0 - \mathbb{G}_\infty)\mathbf{E}(t) + \int_0^\infty \mathbb{G}'(u)\mathbf{E}^t(u)du,\\
&= \widehat{\mathbf{S}}_e(\mathbf{E}(t)) - \mathbb{G}_\infty \mathbf{E}(t) + \int_0^\infty \mathbb{G}(u)\dot{\mathbf{E}}^t(u)du, \qquad \mathbb{G}_0 = \mathbb{G}(0),\\
\dot{\mathbf{E}}^t(u) &= \frac{\partial}{\partial t}\mathbf{E}^t(u) = -\frac{\partial}{\partial u}\mathbf{E}^t(u),
\end{aligned}$$

where we have used (7.1.14) and assumed that $\mathbf{E}^t(\infty) = \mathbf{E}(-\infty) = 0$. Applying (7.2.19) to the subspace Sym of Γ gives

$$\mathbb{G}_0 > \mathbb{G}_\infty \geq \mathbf{0}, \tag{8.1.4}$$

where $\mathbb{G}_\infty$ (or specifically its shear part may vanish for a viscoelastic fluid.

In the completely linear case, these become

$$\mathbf{T}(t) = \mathbb{G}_0 \mathbf{E}(t) + \int_0^\infty \mathbb{G}'(u)\mathbf{E}^t(u)du, \tag{8.1.5}$$

or alternatively,

$$\mathbf{T}(t) = \int_0^\infty \mathbb{G}(u)\dot{\mathbf{E}}^t(u)du. \tag{8.1.6}$$

Equation (8.1.5) is identical to (1.4.11), without the explicit $\mathbf{X}$ dependence. The forms (8.1.2), (8.1.5), and (8.1.6) correspond to (7.1.13).

We have already supposed in Sect. 1.4.3 that $\mathbb{G}'(\cdot) \in L^1(\mathbb{R}^+)$; now we further assume that $\mathbb{G}'(\cdot) \in L^1(\mathbb{R}^+) \cap L^2(\mathbb{R}^+)$, in accordance with Sect. C.1. The relaxation function is defined by

$$\mathbb{G}(s) = \mathbb{G}_0 + \int_0^s \mathbb{G}'(\xi)d\xi. \tag{8.1.7}$$

Thermodynamics implies the symmetry of $\mathbb{G}_0$ and $\mathbb{G}_\infty$, as we shall see below, but not the symmetry of $\mathbb{G}(s)$, $s \in \mathbb{R}^{++}$. However, we shall assume that $\mathbb{G}(s)$ is a fourth-order symmetric tensor, a special case of (7.1.18).

Remark 8.1.1. In a particular basis, $\mathbb{G}$ has components G_{ijkl}, all subscripts in the range 1–3. The symmetry referred to in the previous paragraph refers to the interchange of $\{ij\}$ and $\{kl\}$, as in (2.4.8). However, we see from (8.1.5), together with the symmetry of $\mathbf{T}$ and $\mathbf{E}$, respectively, that

$$G_{ijkl} = G_{jikl} = G_{ijlk}, \tag{8.1.8}$$

which generalizes (2.4.7) and (2.4.9).

From (1.3.32) in Definition 1.3.10, we see that the stress power per unit volume is $\mathbf{T} \cdot \mathbf{D}$. In the linear approximation, $\mathbf{D}$, given by (1.2.23), reduces to $\dot{\mathbf{E}}$, so that the power now becomes $\mathbf{T} \cdot \dot{\mathbf{E}}$. Thus, the inequality (4.1.7), expressing the dissipation principle, yields in this context

$$\int_0^{d_P} \dot{\mathbf{E}}(t) \cdot \mathbb{G}_0 \mathbf{E}(t) dt + \int_0^{d_P} \dot{\mathbf{E}}(t) \cdot \int_0^{\infty} \mathbb{G}'(s) \mathbf{E}^t(s) ds \geq 0, \tag{8.1.9}$$

which must hold for any cycle, where $\dot{\mathbf{E}}(t)$ is the process with duration d_P.

8.1.1 Thermodynamic Restrictions for Viscoelastic Solids

We now examine thermodynamic restrictions on the relaxation function [118, 120] by an approach equivalent to but more elementary than that developed in Sect. 7.2.1 for the general theory. Let us consider strain-tensor time dependence of the form

$$\mathbf{E}(s) = \mathbf{E}_1 \cos \omega s + \mathbf{E}_2 \sin \omega s, \quad s \leq t, \tag{8.1.10}$$

where $\omega \in \mathbb{R}^{++}$ and $\mathbf{E}_1, \mathbf{E}_2 \in \mathrm{Sym}$. The corresponding process $\tilde{P} \in \Pi$ at time t is given by

$$\tilde{P}(t) = \dot{\mathbf{E}}(t) = -\omega \mathbf{E}_1 \sin \omega t + \omega \mathbf{E}_2 \cos \omega t, \quad t \in [0, d_P), \tag{8.1.11}$$

where $d_P = 2\pi m/\omega$, m being any positive integer. Thus, we obtain a cycle, denoted by $(\tilde{\sigma}(t), \tilde{P})$.

Theorem 8.1.2. *The inequality (8.1.9) holds for any cycle $(\tilde{\sigma}(t), \tilde{P})$ only if the inequality*

$$\begin{aligned} \mathbf{E}_1 \cdot [\mathbb{G}_0^T - \mathbb{G}_0]\mathbf{E}_2 - \int_0^{\infty} [\mathbf{E}_1 \cdot \mathbb{G}'(s)\mathbf{E}_1 + \mathbf{E}_2 \cdot \mathbb{G}'(s)\mathbf{E}_2] \sin \omega s \, ds \\ - \int_0^{\infty} \mathbf{E}_1 \cdot [\mathbb{G}'(s) - \mathbb{G}'^T(s)]\mathbf{E}_2 \cos \omega s \, ds \geq 0 \end{aligned} \tag{8.1.12}$$

holds for every $\omega \in \mathbb{R}^{++}$ and every $\mathbf{E}_1, \mathbf{E}_2 \in \mathrm{Sym}$.

Proof. Substitution of (8.1.10) and (8.1.11) into (8.1.9) gives

$$\int_0^{d_P} (-\omega \mathbf{E}_1 \sin \omega t + \omega \mathbf{E}_2 \cos \omega t) \cdot \mathbb{G}_0(\mathbf{E}_1 \cos \omega t + \mathbf{E}_2 \sin \omega t) dt$$
$$+ \int_0^{d_P} \Big\{ (-\omega \mathbf{E}_1 \sin \omega t + \omega \mathbf{E}_2 \cos \omega t) \int_0^{\infty} \mathbb{G}'(s)[\mathbf{E}_1(\cos \omega t \cos \omega s + \sin \omega t \sin \omega s)$$
$$+ \mathbf{E}_2(\sin \omega t \cos \omega s - \cos \omega t \sin \omega s)] ds \Big\} dt \geq 0,$$

which, after integrating with respect to t, with $d_P = 2\pi m/\omega$ and using (A.2.3), yields (8.1.12). □

Some useful results can be derived by considering in (8.1.12) the limiting cases $\omega \to \infty$ and $\omega \to 0$.

Corollary 8.1.3. *The inequality (8.1.12) implies the symmetry of* $\mathbb{G}_0$, *i.e.,*

$$\mathbb{G}_0 = \mathbb{G}_0^T. \tag{8.1.13}$$

Proof. By virtue of the Riemann–Lebesgue lemma (C.2.13), the integrals in (8.1.12) vanish when we consider the limit $\omega \to \infty$. Hence, the arbitrariness of $\mathbf{E}_1, \mathbf{E}_2 \in \mathrm{Sym}$ gives (8.1.13). □

Corollary 8.1.4. *The inequality (8.1.12) implies the symmetry of* $\mathbb{G}_\infty$, *i.e.,*

$$\mathbb{G}_\infty = \mathbb{G}_\infty^T. \tag{8.1.14}$$

Proof. By virtue of (8.1.13), relation (8.1.12), in the limiting case $\omega \to 0$, gives

$$\mathbf{E}_1 \cdot [\mathbb{G}_\infty^T - \mathbb{G}_\infty]\mathbf{E}_2 \geq 0,$$

and the arbitrariness of $\mathbf{E}_1, \mathbf{E}_2$ leads to (8.1.14). □

By (8.1.13), we have the following result.

Corollary 8.1.5. *The inequality (8.1.12) implies that*

$$\int_0^{\infty} [\mathbf{E}_1 \cdot \mathbb{G}'(s)\mathbf{E}_1 + \mathbf{E}_2 \cdot \mathbb{G}'(s)\mathbf{E}_2] \sin \omega s \, ds$$
$$+ \int_0^{\infty} \mathbf{E}_1 \cdot [\mathbb{G}'(s) - \mathbb{G}'^{T}(s)]\mathbf{E}_2 \cos \omega s \, ds \leq 0 \tag{8.1.15}$$

for every $\omega \in \mathbb{R}^{++}$ *and every* $\mathbf{E}_1, \mathbf{E}_2 \in \mathrm{Sym}$.

Referring to (C.1.3) and (C.2.2), we put

$$\mathbb{G}'_+(\omega) = \int_0^{\infty} \mathbb{G}'(u) e^{-i\omega u} du = \mathbb{G}'_c(\omega) - i\mathbb{G}'_s(\omega), \tag{8.1.16}$$

where $\mathbb{G}'_c$ and $\mathbb{G}'_s$ denote the Fourier cosine and sine transforms of $\mathbb{G}'$. Explicitly, the sine transform is given by

$$\mathbb{G}'_s(\omega) = \int_0^{\infty} \mathbb{G}'(u) \sin \omega u \, du. \tag{8.1.17}$$

The following important result is a special case of (7.2.12).

Corollary 8.1.6. *The inequality (8.1.12) implies the negative definiteness of* $\mathbb{G}'_s \in \mathrm{Sym}$ *for every* $\omega \in \mathbb{R}^{++}$.

Proof. Putting $\mathbf{E}_1 = \mathbf{E}_2$ in (8.1.15), we obtain $\mathbb{G}'_s(\omega) \le \mathbf{0}$, $\omega \in \mathbb{R}^+$. Thus, we have

$$\mathbb{G}'_s(\omega) < \mathbf{0}, \qquad \omega \in \mathbb{R}^{++}, \tag{8.1.18}$$

while $\mathbb{G}'_s(0) = \mathbf{0}$. □

If $\mathbb{G}(\tau)$ is assumed to be symmetric for all $\tau \in \mathbb{R}^+$, then condition (8.1.18) implies that (8.1.12) and the dissipation principle (8.1.9) must hold for all histories of the form (8.1.10). More generally, one can show that (8.1.18), for $\mathbb{G}$ symmetric, implies that (8.1.9) holds for any cycle, using histories represented by Fourier series (see Proposition 7.2.2). This procedure is presented in some detail for compressible fluids in Sect. 8.9.3.

The definition of $\mathbb{G}'_s(\omega)$ can be extended to $\mathbb{R}^-$ by the relation $\mathbb{G}'_s(-\omega) = -\mathbb{G}'_s(\omega)$, $\omega \in \mathbb{R}$.

Corollary 8.1.7. *The inequality (8.1.18) implies that (cf.* (7.2.16)*)*

$$\mathbb{G}_0 - \mathbb{G}(s) > \mathbf{0}, \qquad s \in \mathbb{R}^{++}. \tag{8.1.19}$$

Proof. From the inversion formula of the Fourier sine transform $\mathbb{G}'_s(\omega)$, expressed by (see (C.1.6))

$$\mathbb{G}'(s) = \frac{2}{\pi}\int_0^\infty \sin \omega s \mathbb{G}'_s(\omega) d\omega,$$

we have, by integrating with respect to s,

$$\mathbb{G}(s) - \mathbb{G}_0 = \frac{2}{\pi}\int_0^\infty \frac{1 - \cos \omega s}{\omega} \mathbb{G}'_s(\omega) d\omega, \tag{8.1.20}$$

which, by virtue of the inequality (8.1.18), provides the desired result. □

For ease in writing let

$$\mathbb{G}'_0 := \mathbb{G}'(0).$$

Corollary 8.1.8. *We have*

$$\mathbb{G}'_0 \le \mathbf{0} \tag{8.1.21}$$

and

$$\mathbb{G}_0 - \mathbb{G}_\infty \ge \mathbf{0}. \tag{8.1.22}$$

Proof. Relation (8.1.21) can be deduced in the same way as (7.2.17), while (8.1.22) follows from (8.1.19) by taking the limit $s \to \infty$. □

Besides the assumptions that $\mathbb{G}'_0$ exists and is bounded, we now add that it is such that

$$\mathbb{G}'_0 < \mathbf{0}, \tag{8.1.23}$$

which is a special case of (7.2.18). Moreover, since $\mathbb{G}_\infty$ is positive definite, from (8.1.22) it follows that $\mathbb{G}_0$ also has this property.

These results were derived at different times through various approaches. The most pertinent references are now noted.

Coleman [68] proved the symmetry of the instantaneous elastic modulus (8.1.13) from the second law in the form of the Clausius–Duhem inequality. The symmetry of the equilibrium elastic modulus (8.1.14) was obtained by Day [86] via the Clausius inequality. Apart from the inequality being strict, (8.1.18) was first derived by Graffi [169] in the case of isotropic materials by requiring that energy be dissipated in a period of a sinusoidal strain function $\mathbf{E}(t) = \mathbf{E}\sin\omega t$. Accordingly, (8.1.18) may be rightly referred to as *Graffi's inequality*.

The connection between (8.1.18) and energy dissipation is emphasized in [233], where the energy dissipated in one period $[0, d_P]$, $d_P = 2\pi/\omega$, is shown to be

$$\int_0^{d_P} \mathbf{T}(\mathbf{E}^t) \cdot \dot{\mathbf{E}}(t)dt = -\pi\mathbf{E} \cdot \mathbb{G}'_s(\omega)\mathbf{E}. \tag{8.1.24}$$

Incidentally, that is why $-\mathbb{G}'_s(\omega)$ is often referred to as the *loss modulus*. The inequality (8.1.21) for the initial derivative of the relaxation function was proved first by Bowen and Chen [40], by having recourse to discontinuous histories, in the one-dimensional case via the Clausius–Duhem inequality. The same result was proved in [265] with C^∞ histories in the three-dimensional case. The inequality (8.1.22) traces back to Coleman [67, 68].

Apparently, the inequality (8.1.19) first appeared in [120], but it is in a sense related to a previous result by Day [86] (cf. also [321]), who showed that as a consequence of dissipativity, the relaxation function satisfies the condition

$$\mathbb{G}_0 - \mathbb{G}_\infty \geq \pm[\mathbb{G}(s) - \mathbb{G}_\infty]. \tag{8.1.25}$$

To show the connection, observe that the limit $s \to \infty$ in the expression (8.1.20) for $\mathbb{G}(s) - \mathbb{G}_0$ gives (cf. (7.2.19))

$$\mathbb{G}_\infty - \mathbb{G}_0 = \frac{2}{\pi}\int_0^\infty \frac{1}{\omega}\mathbb{G}'_s(\omega)d\omega. \tag{8.1.26}$$

Consequently,

$$\mathbb{G}(s) - \mathbb{G}_\infty = -\frac{2}{\pi}\int_0^\infty \frac{\cos\omega s}{\omega}\mathbb{G}'_s(\omega)d\omega,$$

and the obvious inequalities

$$\int_0^\infty \frac{1}{\omega}|\mathbf{E} \cdot \mathbb{G}'_s(\omega)\mathbf{E}|d\omega \geq \int_0^\infty \frac{|\cos\omega s|}{\omega}|\mathbf{E} \cdot \mathbb{G}'_s(\omega)\mathbf{E}|d\omega \geq \left|\int_0^\infty \frac{\cos\omega s}{\omega}\mathbf{E} \cdot \mathbb{G}'_s(\omega)\mathbf{E}d\omega\right|$$

for any $\mathbf{E} \in \mathrm{Sym}$ yield (8.1.25).

A relation analogous to (8.1.25) can of course be given for the general theory (Sect. 7.2.2).

Remark 8.1.9. While (8.1.13) and (8.1.14) are enforced by thermodynamics, it is not necessarily the case that $\mathbb{G}(\tau)$ is symmetric for intermediate values of τ. We will, however, assume that (cf. (7.1.18))

$$\mathbb{G}(\tau) = \mathbb{G}^{\top}(\tau), \qquad \tau \in (0, \infty). \tag{8.1.27}$$

Since $\mathbb{G}' \in L^2(\mathbb{R}^+)$, Parseval's formula (C.3.1) allows us to write the constitutive equation (8.1.5) as [104]

$$\mathbf{T}(t) = \mathbb{G}_0 \mathbf{E}(t) + \frac{2}{\pi} \int_0^\infty \mathbb{G}'_s(\omega) \mathbf{E}^t_s(\omega) d\omega \tag{8.1.28}$$

for any $\mathbf{E}^t \in L^2(\mathbb{R}^+)$. This formula is obtained by extending the integral in (8.1.5) to $\mathbb{R}$ and taking the odd extension of $\mathbf{E}^t$, using $(\text{C.1.6})_1$. We can replace $\overline{\mathbb{G}'_+}(\omega)$ by $i\mathbb{G}'_s(\omega)$ because of the oddness of $\mathbf{E}^t_s(\omega)$. Since the integrand is now even, the integration interval can be transformed to $\mathbb{R}^+$.

A more general viewpoint on this kind of manipulation was adopted earlier to yield (7.2.33) and in particular (7.2.34), which corresponds to (8.1.28).

Now it is important to generalize (8.1.28) so that it holds for all $\mathbf{E}^t \in \mathcal{E}$, where $\mathcal{E}$ is the set of histories $\mathbf{E}^t$ such that

$$\left| \int_0^\infty \mathbb{G}'(s) \mathbf{E}^t(s) ds \right| < \infty$$

for a given $\mathbb{G}' \in L^2(\mathbb{R}^+) \cap L^1(\mathbb{R}^+)$. If we denote by $\mathcal{G}$ the vector space defined by

$$\mathcal{G} = \left\{ \mathbb{F} : \mathbb{R}^+ \to \text{Lin(Sym, Sym)}; \mathbb{F} = \alpha \mathbb{G}' + \mathbf{f}, \ \forall \mathbf{f} \in \mathbf{C}_0^\infty(\mathbb{R}^+) \right\},$$

then $\mathcal{E} = \mathcal{G}'$, where $\mathcal{G}'$ is the dual of $\mathcal{G}$ or the space of all linear continuous functionals on $\mathcal{G}$. Thus, the elements of $\mathcal{E}$ have a Fourier transform in a distributional sense. Relation (8.1.28) can be carried over into the set $\mathcal{E}$.

8.2 Decomposition of Stress

Consider the constitutive equation of linear viscoelasticity, given by (8.1.2) or (8.1.5). The integrals with $\mathbf{E}^t$ and $\mathbf{E}^t_r$ suggest the introduction of certain functions that will prove useful. These are defined by

$$\begin{aligned} \check{\mathbf{I}}^t(\tau, \mathbf{E}^t) &:= \mathbb{G}(\tau)\mathbf{E}(t) + \int_0^\infty \mathbb{G}'(s+\tau)\mathbf{E}^t(s) ds \\ &= \mathbb{G}_\infty \mathbf{E}(t) + \mathbf{I}^t(\tau, \mathbf{E}^t_r), \end{aligned} \tag{8.2.1}$$

where (cf. $(7.4.2)_2$)*

* The quantity $\mathbf{I}^t$ was originally defined in the literature as the negative of the functional used here. This change in sign, which is consistent with Sect. 7.4, is introduced here and later so that its relationship with the stress functional is a little more precise.

$$\mathbf{I}^t(\tau, \mathbf{E}_r^t) = \int_0^\infty \mathbb{G}'(s+\tau)\mathbf{E}_r^t(s)ds. \tag{8.2.2}$$

It is easy to see that $\breve{\mathbf{I}}^t$ coincides with the stress resulting from the partly static history (1.4.10), namely $\hat{\mathbf{T}}(\mathbf{E}^{t(\tau)})$, where τ is the duration of the static part.

One can derive from $\breve{\mathbf{I}}^t$ both $\mathbf{E}(t)$ and $\mathbf{I}^t$ by virtue of the following relations:

$$\lim_{\tau\to\infty} \breve{\mathbf{I}}^t(\tau, \mathbf{E}^t) = \mathbb{G}_\infty \mathbf{E}(t)$$

and

$$\mathbf{I}^t(\tau, \mathbf{E}_r^t) = \breve{\mathbf{I}}^t(\tau, \mathbf{E}^t) - \lim_{\tau\to\infty} \mathbf{I}^t(\tau, \mathbf{E}^t).$$

Also, we have

$$\begin{aligned} \mathbf{I}^t(\tau, \mathbf{E}^t) &:= \int_0^\infty \mathbb{G}'(s+\tau)\mathbf{E}^t(s)ds \\ &= -\mathbb{G}(\tau)\mathbf{E}(t) + \breve{\mathbf{I}}^t(\tau, \mathbf{E}^t) = -\breve{\mathbb{G}}(\tau)\mathbf{E}(t) + \mathbf{I}^t(\tau, \mathbf{E}_r^t), \end{aligned} \tag{8.2.3}$$

where

$$\breve{\mathbb{G}}(\tau) := \mathbb{G}(\tau) - \mathbb{G}_\infty. \tag{8.2.4}$$

The time derivative of $\mathbf{I}^t(\cdot, \mathbf{E}_r^t)$ with respect to t will be of interest. This is given by

$$\dot{\mathbf{I}}^t(\tau, \mathbf{E}_r^t) = \frac{d}{dt}\mathbf{I}^t(\tau, \mathbf{E}_r^t) = \breve{\mathbb{G}}(\tau)\dot{\mathbf{E}}(t) + \mathbf{I}_{(1)}^t(\tau, \mathbf{E}_r^t), \tag{8.2.5}$$

where

$$\mathbf{I}_{(1)}^t(\tau, \mathbf{E}_r^t) = \frac{d}{d\tau}\mathbf{I}^t(\tau, \mathbf{E}_r^t) = \int_0^\infty \mathbb{G}''(s+\tau)\mathbf{E}_r^t(s)ds. \tag{8.2.6}$$

In the following we shall also use a simpler notation by writing $\breve{\mathbf{I}}^t(\tau)$, $\mathbf{I}^t(\tau)$, and $\tilde{\mathbf{I}}^t(\tau)$ instead of $\breve{\mathbf{I}}^t(\tau, \mathbf{E}^t)$, $\mathbf{I}^t(\tau, \mathbf{E}_r^t)$, and $\mathbf{I}^t(\tau, \mathbf{E}^t)$, respectively.

Let $t = 0$ be the initial instant when a process P_τ is applied to the material. The stress will be a function of the initial state σ and of this process. It can be written as follows:

$$\hat{\mathbf{T}}(\sigma, P_\tau) = \mathbb{G}_0\mathbf{E}(\tau) + \int_0^\tau \mathbb{G}'(u)\mathbf{E}^\tau(u)du + \int_\tau^\infty \mathbb{G}'(u)\mathbf{E}^\tau(u)du.$$

An integration by parts and a change of variable give

$$\hat{\mathbf{T}}(\sigma, P_\tau) = \int_0^\tau \mathbb{G}(\tau-u)\dot{\mathbf{E}}(u)du + \mathbb{G}(\tau)\mathbf{E}(0) + \int_0^\infty \mathbb{G}'(\xi+\tau)\mathbf{E}(-\xi)d\xi. \tag{8.2.7}$$

We can identify the state σ with $(\mathbf{E}(0), \mathbf{E}^0)$, where the history is $\mathbf{E}^0(\xi) = \mathbf{E}(-\xi)$, $\xi \in \mathbb{R}^{++}$, and the process P_τ with $\dot{\mathbf{E}}_\tau^P$, defined as $\dot{\mathbf{E}}(u)$, $u \in [0.\tau)$. Moreover, in (8.2.7) we can distinguish two effects by putting

$$\hat{\mathbf{T}}(\mathbf{0}, \dot{\mathbf{E}}_\tau^P) = \int_0^\tau \mathbb{G}(\tau-u)\dot{\mathbf{E}}_\tau^P(u)du$$

and noting that $\check{\mathbf{I}}^t$, given by (8.2.1), has the form at $t = 0$

$$\check{\mathbf{I}}^0(\tau, \mathbf{E}^0) = \mathbb{G}(\tau)\mathbf{E}(0) + \int_0^\infty \mathbb{G}'(\xi + \tau)\mathbf{E}^0(\xi)d\xi.$$

Thus, we can write the stress (8.1.5) in the form

$$\mathbf{T}(\tau) = \hat{\mathbf{T}}(\mathbf{0}, \dot{\mathbf{E}}^P_\tau) + \check{\mathbf{I}}^0(\tau, \mathbf{E}^0). \tag{8.2.8}$$

Here, we observe that $\hat{\mathbf{T}}(\mathbf{0}, \dot{\mathbf{E}}^P_\tau)$ denotes the stress determined by the initial zero state $\mathbf{0}$ (($\mathbf{E}(0) = 0, \mathbf{E}^0(\xi) = 0$, $\xi \in \mathbb{R}^+$) and the process $\dot{\mathbf{E}}^P_\tau$, whereas $\check{\mathbf{I}}^0(\tau, \mathbf{E}^0)$ is the stress determined by the initial state $(\mathbf{E}(0), \mathbf{E}^0)$ and the zero process $\mathbf{0}^\dagger_\tau$, with duration $\tau \in \mathbb{R}^+$; this zero process renders the first term on the right of (8.2.7) zero. Equation (8.2.8) means that $\mathbf{T}(\tau)$ can be viewed as the superposition of the two effects $\hat{\mathbf{T}}(\mathbf{0}, \dot{\mathbf{E}}^P_\tau)$, which involves only the process $\dot{\mathbf{E}}^P_\tau$, and $\mathbf{I}^0(\tau, \mathbf{E}^0)$, which involves only the state $(\mathbf{E}(0), \mathbf{E}^0)$.

8.3 Equivalence and Minimal States

We now suppose that the process is applied at time t, thus acting in the time interval $[t, t + \tau)$, where τ denotes its duration. For a linear viscoelastic solid, in the initial state $\mathbf{E}^t$ at time t, we have

$$\begin{aligned} \mathbf{T}(t + \tau) &= \mathbb{G}_0\mathbf{E}(t + \tau) + \int_0^\tau \mathbb{G}'(u)\mathbf{E}^{t+\tau}(u)du + \int_\tau^\infty \mathbb{G}'(u)\mathbf{E}^{t+\tau}(u)du \\ &= \mathbb{G}(\tau)\mathbf{E}(t) + \int_0^\tau \mathbb{G}(u)\dot{\mathbf{E}}^{t+\tau}(u)du + \int_0^\infty \mathbb{G}'(\tau + \xi)\mathbf{E}^t(\xi)d\xi. \end{aligned} \tag{8.3.1}$$

Definition 4.1.2 of equivalence yields some restrictions on the constitutive equations. The first result [31, 90] is the following theorem (cf. (7.4.3)).

Theorem 8.3.1. *Two histories $\mathbf{E}^t_1$, $\mathbf{E}^t_2$ of $\mathbf{E}$ are equivalent, relative to (8.1.5), if and only if*

$$\mathbf{E}^t_1(0) = \mathbf{E}^t_2(0), \qquad \int_0^\infty \mathbb{G}'(\xi + \tau)\mathbf{E}^t_1(\xi)d\xi = \int_0^\infty \mathbb{G}'(\xi + \tau)\mathbf{E}^t_2(\xi)d\xi \quad \forall\tau \geq 0. \tag{8.3.2}$$

Proof. We have the state $\sigma(t) = \mathbf{E}^t$. The requirement $\hat{\mathbf{T}}(\mathbf{E}^t_1, \dot{\mathbf{E}}^P_\tau) = \hat{\mathbf{T}}(\mathbf{E}^t_2, \dot{\mathbf{E}}^P_\tau)\ \forall\tau \geq 0$, taking into account (8.3.1), yields

$$\begin{aligned} &\mathbb{G}(\tau)\mathbf{E}_1(t) + \int_0^\infty \mathbb{G}'(\xi + t)\mathbf{E}^t_1(\xi)d\xi \\ &\quad = \mathbb{G}(\tau)\mathbf{E}_2(t) + \int_0^\infty \mathbb{G}'(\xi + t)\mathbf{E}^t_2(\xi)d\xi \quad \forall\tau \geq 0. \end{aligned} \tag{8.3.3}$$

Taking $\tau \to \infty$ gives $(8.3.2)_1$. Then $(8.3.2)_2$ follows immediately. □

Some important consequences of this theorem, considered in [123], will now be described.

Corollary 8.3.2. *For a viscoelastic material characterized by (8.1.5), the equivalence conditions for two histories* $\mathbf{E}_1^t$, $\mathbf{E}_2^t$ *can be expressed by*

$$\hat{\mathbf{T}}(\mathbf{E}_1^t, \mathbf{0}_\tau^\dagger) = \hat{\mathbf{T}}(\mathbf{E}_2^t, \mathbf{0}_\tau^\dagger) \qquad \forall \tau \in \mathbf{R}^+, \tag{8.3.4}$$

where $\mathbf{0}_\tau^\dagger$ *is the zero process of duration* $\tau \in \mathbb{R}^+$.

Proof. For any $\tau \in \mathbb{R}^+$ the relation (8.3.4) applied to (8.3.1) gives (8.3.3). □

By virtue of Theorem 8.3.1, we see that equivalent histories are characterized by the pair $(\mathbf{E}^t(0), \tilde{\mathbf{I}}^t)$, where $\tilde{\mathbf{I}}^t(\tau)$ is given by (8.2.3) (note the comment after (8.2.6)). Consequently, the state of a linear viscoelastic solid may be identified with the pair $(\mathbf{E}^t(0), \tilde{\mathbf{I}}^t)$ instead of the whole history $\mathbf{E}^t$. This observation was first made in [176], where the particular case with the kernel $\mathbb{G}'$ given by a sum of exponentials was studied.

Remark 8.3.3. The class σ_R of equivalent histories, by virtue of (8.2.3), can also be represented by the single function given by (8.2.1)

$$\breve{\mathbf{I}}^t(\tau) = \hat{\mathbf{T}}(\mathbf{E}^t, \mathbf{0}_\tau^\dagger) = \tilde{\mathbf{I}}^t(\tau) + \mathbb{G}(\tau)\mathbf{E}(t) \qquad \forall \tau \in \mathbb{R}^+, \tag{8.3.5}$$

where $\mathbf{E}^t$ is any history among those in σ_R, since by definition, the function $\breve{\mathbf{I}}^t(\tau)$ is the same for all $\mathbf{E}^t \in \sigma_R$. Moreover, the knowledge of $\breve{\mathbf{I}}^t$ on $\mathbb{R}^+$ provides

$$\mathbf{E}(t) = \mathbb{G}_\infty^{-1} \lim_{\tau \to \infty} \breve{\mathbf{I}}^t(\tau)$$

and hence also $\tilde{\mathbf{I}}^t$ by (8.3.5).

A minimal state is identified with an equivalence class represented by

$$\sigma_R(t) = (\mathbf{E}(t), \tilde{\mathbf{I}}^t(\cdot)) \tag{8.3.6}$$

or

$$\sigma_R(t) = \breve{\mathbf{I}}^t(\cdot).$$

The description of a state as minimal refers to the fact that it can be characterized by a minimum set of data. Examples are discussed in the next section.

8.4 State and History for Exponential-Type Relaxation Functions

It is of interest to consider materials for which the relaxation function is a linear combination of decaying exponentials, i.e.,

$$\mathbb{G}(\xi) = \mathbb{G}_\infty + \mathbf{\Lambda} \sum_{k=1}^{n} g_k \exp(-\alpha_k \xi),$$
$$\mathbb{G}'(\xi) = -\mathbf{\Lambda} \sum_{k=1}^{n} \alpha_k g_k \exp(-\alpha_k \xi), \tag{8.4.1}$$

where $\mathbf{\Lambda} \in \mathrm{Lin(Sym)}$ is positive definite and the coefficients g_k, α_k are positive, $k = 1, 2, \ldots, n$. We will sometimes refer to these as discrete-spectrum materials.

We will now show that the presence of exponentials allows us to express the state σ in terms of a finite number of quantities instead of the history $\mathbf{E}^t$, which is infinite-dimensional. The description of a state in terms of such quantities can be described as minimal, since it does not contain superfluous variables.

From $(8.4.1)_1$, we have

$$\mathbb{G}_0 = \mathbb{G}_\infty + \mathbf{\Lambda} \sum_{k=1}^{n} g_k.$$

Moreover, putting

$$\mathbf{T}_k(t) = \mathbf{\Lambda} g_k \left[\mathbf{E}(t) - \alpha_k \int_0^\infty \exp(-\alpha_k \xi) \mathbf{E}(t - \xi) d\xi \right],$$

the stress tensor, given by (8.1.5), becomes

$$\mathbf{T}(t) = \mathbb{G}_\infty \mathbf{E}(t) + \sum_{k=1}^{n} \mathbf{T}_k(t).$$

We can consider the $(n + 1)$-tuple $(\mathbf{E}, \mathbf{T}_1, \mathbf{T}_2, \ldots, \mathbf{T}_n)$, as the state at time t. Alternatively, let

$$\mathbf{E}_k(t) = \int_0^\infty \exp(-\alpha_k \xi) \mathbf{E}(t - \xi) d\xi, \tag{8.4.2}$$

giving

$$\mathbf{T}(t) = \mathbb{G}_\infty \mathbf{E}(t) + \mathbf{\Lambda} \sum_{k=1}^{n} g_k [\mathbf{E}(t) - \alpha_k \mathbf{E}_k(t)].$$

Thus, we can also consider the state as the $(n + 1)$-tuple $(\mathbf{E}, \mathbf{E}_1, \mathbf{E}_2, \ldots, \mathbf{E}_n)$.

These two modes of description of state are related to but not the same as that discussed in Sect. 8.3. Let us now consider how σ_R might be described for a viscoelastic material with a relaxation function of the form $(8.4.1)_1$.

Let $\sigma_R(t)$ be given by (8.3.6). Using $(8.2.3)_1$ and $(8.4.1)_2$, we obtain

$$\tilde{\mathbf{I}}^t(\tau) = -\mathbf{\Lambda} \sum_{i=1}^{n} \alpha_i g_i \exp(-\alpha_i \tau) \mathbf{E}_i(t),$$

where the quantities $\mathbf{E}_i$ are defined by (8.4.2). The derivatives of the function $\tilde{\mathbf{I}}^t(\tau)$ with respect to τ, at $\tau = 0$, are given by

$$\tilde{\mathbf{I}}^t_p(0) = \mathbf{\Lambda} \sum_{i=1}^{n} g_i(-\alpha_i)^{p+1}\mathbf{E}_i(t), \qquad p = 0, 1, \ldots, n-1.$$

Thus, we obtain a linear system, which can be solved for $\mathbf{E}_1(t), \ldots, \mathbf{E}_n(t)$ in terms of the quantities $\tilde{\mathbf{I}}^t(0), \tilde{\mathbf{I}}^t_1(0), \ldots, \tilde{\mathbf{I}}^t_{(n-1)}(0)$. Accordingly, putting

$$\sigma_R = (\mathbf{E}(t), \tilde{\mathbf{I}}^t(0), \tilde{\mathbf{I}}^t_1(0), \ldots, \tilde{\mathbf{I}}^t_{(n-1)}(0)),$$

the state is $(n+1)$-dimensional.

8.5 Inversion of Constitutive Relations

The inversion of the constitutive equations of linear viscoelasticity has been studied in [41] as a Wiener–Hopf problem. We now describe a more direct approach [123]. Let us consider the constitutive equation (8.1.5), namely

$$\mathbf{T}(t) = \mathbb{G}_0\mathbf{E}(t) + \int_0^\infty \mathbb{G}'(s)\mathbf{E}(t-s)ds, \tag{8.5.1}$$

where the domain of $\mathbb{G}'$ is carried over to $\mathbb{R}$ by putting $\mathbb{G}'(s) = \mathbf{0}\ \forall s < 0$. Hence, we also have

$$\mathbf{T}(t) = \mathbb{G}_0\mathbf{E}(t) + \int_{-\infty}^\infty \mathbb{G}'(s)\mathbf{E}(t-s)ds.$$

Putting, for formal convenience,

$$\mathbf{H}(t) = \mathbb{G}_0\mathbf{E}(t), \qquad \mathbb{K}(s) = \mathbb{G}'(s)\mathbb{G}_0^{-1},$$

and assuming that $\mathbf{H} \in L^1(\mathbb{R})$, we have

$$\mathbf{T}(t) = \mathbf{H}(t) + \int_{-\infty}^\infty \mathbb{K}(s)\mathbf{H}(t-s)ds \qquad \forall t \in \mathbb{R}.$$

Taking the Fourier transform and applying the convolution theorem (C.3.3) gives

$$\mathbf{T}_F(\omega) = [\mathbf{1} + \mathbb{K}_+(\omega)]\mathbf{H}_F(\omega), \tag{8.5.2}$$

where, referring to (8.1.16),

$$\mathbb{K}_+(\omega) = \mathbb{G}'_+(\omega)\mathbb{G}_0^{-1} = [\mathbb{G}'_c(\omega) - i\mathbb{G}'_s(\omega)]\mathbb{G}_0^{-1}. \tag{8.5.3}$$

It follows that

$$\mathbb{K}_+(0) = [\mathbb{G}_\infty - \mathbb{G}_0]\mathbb{G}_0^{-1} = \mathbb{G}_\infty\mathbb{G}_0^{-1} - \mathbf{1}, \tag{8.5.4}$$

since $\mathbb{G}'_+(0) = \mathbb{G}_\infty - \mathbb{G}_0$; this follows immediately from the definition of $\mathbb{G}'_+(\omega)$, given by $(8.1.16)_1$.

Remark 8.5.1. We see from Proposition C.2.1 that $\mathbb{K}_+(\omega)$ or $[\mathbf{1} + \mathbb{K}_+(\omega)]$ has all its singularities in the upper half-plane and thus is analytic in the lower half-plane. This is equivalent to the requirement that (8.5.1) be a causal relationship.

The quantity $\mathbf{1} + \mathbb{K}_+(\omega)$ is invertible for any $\omega \in \mathbb{R}$. We see this by observing that for real ω,

$$\begin{aligned} \mathrm{Im}[\mathbf{1} + \mathbb{K}_+(\omega)] &= -\mathbb{G}'_s(\omega)\mathbb{G}_0^{-1} \neq \mathbf{0} \quad \forall \omega \neq 0, \\ \mathrm{Re}[\mathbf{1} + \mathbb{K}_+(0)] &= \mathbb{G}_\infty \mathbb{G}_0^{-1} \neq \mathbf{0}. \end{aligned}$$

Accordingly, using (8.5.2), we can express $\mathbf{H}_F(\omega)$ in the form

$$\mathbf{H}_F(\omega) = \mathbf{T}_F(\omega) - [\mathbf{1} + \mathbb{K}_+(\omega)]^{-1}\mathbb{K}_+(\omega)\mathbf{T}_F(\omega).$$

Taking inverse Fourier transforms yields

$$\mathbf{H}(t) = \mathbf{T}(t) + \mathbb{G}_0 \int_{-\infty}^{\infty} \mathbb{J}'(\xi)\mathbf{T}(t-\xi)d\xi,$$

where

$$\mathbb{G}_0\mathbb{J}'(\xi) = -\frac{1}{2\pi}\int_{-\infty}^{\infty} [1 + \mathbb{K}_+(\omega)]^{-1}\mathbb{K}_+(\omega)\exp(i\omega\xi)d\xi. \tag{8.5.5}$$

Applying $\mathbb{G}_0^{-1}$, we have

$$\mathbf{E}(t) = \mathbb{G}_0^{-1}\mathbf{T}(t) + \int_{-\infty}^{\infty} \mathbb{J}'(\xi)\mathbf{T}(t-\xi)d\xi.$$

Now, this relationship must be causal; in other words, we must be able to write it in the form

$$\mathbf{E}(t) = \mathbb{J}_0\mathbf{T}(t) + \int_0^{\infty} \mathbb{J}'(\xi)\mathbf{T}(t-\xi)d\xi, \qquad \mathbb{J}_0 = \mathbb{G}_0^{-1} > \mathbf{0}, \tag{8.5.6}$$

so that

$$\mathbb{J}'(\xi) = \mathbf{0}, \qquad \xi < 0.$$

Remark 8.5.2. It follows from Proposition C.2.1 and (8.5.5) that the zeros of $\mathbf{1} + \mathbb{K}_+(\omega)$ must also be in the upper half-plane [167]. These observations apply where $\mathbf{1} + \mathbb{K}_+(\omega)$ has zeros and isolated singularities. If branch-cut singularities (see Sect. B.1.5), in particular continuous-spectrum singularities, are present, these must be in the upper half-plane for both $\mathbf{1}+\mathbb{K}_+(\omega)$ and its inverse. This is discussed further in Sect. 16.5, in particular in Remark 16.5.1.

As in (8.1.7) for the relaxation function, we define the creep function as

$$\mathbb{J}(s) = \mathbb{J}_0 + \int_0^s \mathbb{J}'(\xi)d\xi, \qquad \mathbb{J}_0 = \mathbb{J}(0). \tag{8.5.7}$$

The inverse Fourier transform of (8.5.5) gives that

$$\mathbb{G}_0^{-1}[\mathbf{1} + \mathbb{K}_+(\omega)]^{-1}\mathbb{K}_+(\omega) = -\int_0^\infty \mathbb{J}'(\xi)e^{-i\omega\xi}\,d\xi. \tag{8.5.8}$$

Let us assume that $\mathbb{J}(\infty)$ is finite. This implies that the material is a solid [167]. Combining (8.5.4) and (8.5.8) at $\omega = 0$, one deduces with the aid of $(8.5.6)_2$ that

$$\mathbb{J}(\infty) = \mathbb{J}_\infty = \mathbb{G}_\infty^{-1} > \mathbf{0}. \tag{8.5.9}$$

Recalling (8.1.22), we see that

$$\mathbb{J}_0 = \mathbb{G}_0^{-1} \leq \mathbb{G}_\infty^{-1} = \mathbb{J}_\infty, \tag{8.5.10}$$

so that the creep function $\mathbb{J}(s)$ is, at least in an overall sense, increasing, while the relaxation function $\mathbb{G}(s)$ is, at least in an overall sense, decreasing.

Indeed, we can come at this result in another way. Let us write (8.5.8) in the form

$$\mathbb{G}_0\mathbb{J}'_+(\omega) = -[\mathbf{1} + \mathbb{K}_+(\omega)]^{-1}\mathbb{K}_+(\omega) = [\mathbf{1} + \mathbb{K}_+(\omega)]^{-1} - \mathbf{1}. \tag{8.5.11}$$

Since, by virtue of (8.1.27), all quantities are symmetric in Lin(Sym), we have that $\overline{\mathbb{K}}_+(\omega) = \mathbb{K}^*(\omega)$, for example. It will be assumed that $\mathbb{K}_+(\omega)$ and $\overline{\mathbb{K}}_+(\omega)$ (or $\mathbb{K}_+^*(\omega)$) commute, so that $\mathbb{K}_+(\omega)$ is a normal transformation (see after (A.2.11)) and that they commute with $\mathbb{G}_0$. These properties will in fact hold under the assumption made in Sect. 7.1.5. Then, from (8.5.11) and (8.5.3),

$$\begin{aligned}[\mathbf{1} + \mathbb{K}_+(\omega)][\mathbf{1} + \overline{\mathbb{K}}_+(\omega)]\mathbb{G}_0\left[\overline{\mathbb{J}}'_+(\omega) - \mathbb{J}'_+(\omega)\right]& \\ = \mathbb{K}_+(\omega) - \overline{\mathbb{K}}_+(\omega) = -2i\mathbb{G}'_s(\omega)\mathbb{G}_0^{-1}&.\end{aligned} \tag{8.5.12}$$

Since $[\mathbf{1} + \mathbb{K}_+(\omega)][\mathbf{1} + \overline{\mathbb{K}}_+(\omega)]\mathbb{G}_0$ is a nonnegative invertible tensor, we deduce from (8.1.18) that

$$\mathbb{J}'_s(\omega) = \frac{1}{2i}\left[\overline{\mathbb{J}}'_+(\omega) - \mathbb{J}'_+(\omega)\right] > 0. \tag{8.5.13}$$

The analogue of Corollary 8.1.7 yields that

$$\mathbb{J}(s) - \mathbb{J}_0 > \mathbf{0}, \quad s \in \mathbb{R}^+,$$

and in particular, (8.5.10) follows.

The general linear relation (7.1.33) can be inverted in a manner similar to that outlined above.

Let $\mathbf{E}$ or $\mathbf{T}$, on $[0, t)$, be regarded as the process; then we can write

$$\mathbf{T}^P(t) = \mathbb{G}_0\mathbf{E}^P(t) + \int_0^t \mathbb{G}'(s)\mathbf{E}^P(t-s)ds + \hat{\mathbf{I}}(t) \tag{8.5.14}$$

and

$$\mathbf{E}^P(t) = \mathbb{J}_0\mathbf{T}^P(t) + \int_0^t \mathbf{J}'(\xi)\mathbf{T}^P(t-\xi)d\xi + \hat{\mathbf{Y}}(t), \tag{8.5.15}$$

where $\mathbf{E}^P$ and $\mathbf{T}^P$ denote the restrictions of $\mathbf{E}$ and $\mathbf{T}$ to $[0,t)$. Hence, we put $\mathbf{E}^P = \mathbf{0}$, $\mathbf{T}^P = \mathbf{0}$ on $(-\infty,0)$. Moreover, recalling (8.3.1), we see that the two equations (8.5.14) and (8.5.15) hold if

$$\hat{\mathbf{I}}(t) = \begin{cases} \int_0^\infty \mathbb{G}'(t+\xi)\mathbf{E}^0(\xi)d\xi, & \forall t > 0, \\ \mathbf{0}, & \forall t \leq 0, \end{cases} \tag{8.5.16}$$

and

$$\hat{\mathbf{Y}}(t) = \begin{cases} \int_0^\infty \mathbb{J}'(t+\xi)\mathbf{T}^0(\xi)d\xi, & \forall t > 0, \\ \mathbf{0}, & \forall t \leq 0. \end{cases}$$

Let us now consider (8.5.14). The domain of $\mathbb{G}'$ is carried over to $\mathbb{R}$ by putting $\mathbb{G}'(s) = \mathbf{0}\,\forall s < 0$. Again, we put $\mathbf{H}(t) = \mathbb{G}_0\mathbf{E}^P(t)$, $\mathbb{K}(s) = \mathbb{G}'(s)\mathbb{G}_0^{-1}$ and assume that

$$\mathbf{H} \in L^1(\mathbb{R}), \qquad \hat{\mathbf{I}} \in L^1(\mathbb{R}).$$

Hence,

$$\mathbf{T}^P(t) = \mathbf{H}(t) + \int_{-\infty}^{\infty} \mathbb{K}(s)\mathbf{H}(t-s)ds + \hat{\mathbf{I}}(t) \qquad \forall\, t \in \mathbb{R}.$$

After applying a Fourier transform, we can solve for $\mathbf{H}_F$, obtaining

$$\mathbf{H}_F(\omega) = \mathbf{T}^P_F(\omega) - \hat{\mathbf{I}}_F(\omega) - [\mathbf{1} + \mathbb{K}_+(\omega)]^{-1}\mathbb{K}_+(\omega)[\mathbf{T}^P_F(\omega) - \hat{\mathbf{I}}_F(\omega)].$$

Using the inverse Fourier transform, we have

$$\mathbf{E}^P(t) = \mathbb{G}_0^{-1}[\mathbf{T}^P(t) - \hat{\mathbf{I}}(t)] + \int_{-\infty}^{\infty} \mathbb{J}'(\xi)\mathbf{T}^P(t-\xi)d\xi - \int_{-\infty}^{\infty} \mathbb{J}'(\xi)\hat{\mathbf{I}}(t-\xi)d\xi \quad \forall\, t \in \mathbb{R},$$

and hence

$$\begin{aligned} \mathbf{E}^P(t) &= \mathbb{G}_0^{-1}\mathbf{T}^P(t) + \int_0^t \mathbb{J}'(\xi)\mathbf{T}^P(t-\xi)d\xi - \mathbb{G}_0^{-1}\hat{\mathbf{I}}(t) \\ &\quad - \int_0^\infty \mathbb{J}'(\xi)\hat{\mathbf{I}}(t-\xi)d\xi \quad \forall\, t \in \mathbf{R}. \end{aligned} \tag{8.5.17}$$

Equations (8.5.15) and (8.5.17) are required to provide the same values of $\mathbf{E}^P$ on $\mathbb{R}$. Thus, we find that

$$\hat{\mathbf{Y}}(t) = -\mathbb{J}_0\hat{\mathbf{I}}(t) - \int_0^\infty \mathbb{J}'(\xi)\hat{\mathbf{I}}(t-\xi)d\xi \quad \forall t \in \mathbb{R}.$$

8.6 Linear Viscoelastic Free Energies as Quadratic Functionals

We now give a representation of a free energy in the linear viscoelastic case and examine some of its properties [91, 105, 158].

Relation (7.1.9) reduces to

$$\psi(t) = \widetilde{\phi}(\mathbf{E}(t)) + \frac{1}{2}\int_0^\infty \int_0^\infty \mathbf{E}^t_r(s) \cdot \mathbb{G}_{12}(s,u)\mathbf{E}^t_r(u)dsdu, \tag{8.6.1}$$

where

$$\mathbb{G}_{12}(s,u) = \frac{\partial^2}{\partial s \partial u}\mathbb{G}(s,u). \tag{8.6.2}$$

By Definition 4.1.6, we must have $\psi \geq 0$ for all states $(\mathbf{E}(t), \mathbf{E}^t)$, in particular for $(\mathbf{0}, \mathbf{E}^t)$, where the history is arbitrary. Thus, $\mathbb{G}_{12}$ must be a nonnegative operator in the sense that the integral in (8.6.17) must be nonnegative for all histories. Therefore, we have

$$\psi(t) \geq \phi(t), \tag{8.6.3}$$

which is a special case of (5.1.25), a property that follows very generally from the second law.

Since the integral in (8.6.1) must exist for finite relative histories, we assume that (cf. (7.1.3)) as $s \to \infty$, the kernel $\mathbb{G}_{12}(s,u)$ goes to zero as

$$\mathbb{G}_{12}(s,u) \sim s^{-1-b}, \quad b > 0, \quad u \in \mathbb{R}^+, \tag{8.6.4}$$

or more strongly; similarly for the limit of large u at fixed s.

Let us define (cf. (7.1.5))

$$\mathbb{G}(s,u) = \mathbb{G}_\infty + \int_s^\infty \int_u^\infty \mathbb{G}_{12}(s',u')ds'\,du', \quad \mathbb{G}_\infty = \mathbb{G}(\infty,\infty). \tag{8.6.5}$$

Note that we are adopting the conventional choice as specified in Remark 7.1.2. Relations (7.1.7) are replaced by

$$\lim_{s\to\infty} \mathbb{G}(s,u) = \mathbb{G}_\infty, \qquad \lim_{s\to\infty} \frac{\partial}{\partial u}\mathbb{G}(s,u) = \mathbf{0}, \quad u \in \mathbb{R}^+, \tag{8.6.6}$$

with similar limits at large u holding for fixed s.

Remark 8.6.1. In fact, $(7.1.7)_1$ could be retained by using

$$\mathcal{G}(s,u) = \mathbb{G}(s,u) - \mathbb{G}_\infty$$

instead of $\mathbb{G}$.

We impose the conditions

$$\mathbb{G}(s) = \mathbb{G}(0,s) = \mathbb{G}(s,0), \qquad \mathbb{G}(0) = \mathbb{G}_0 = \mathbb{G}(0,0), \tag{8.6.7}$$

where $\mathbb{G}(s)$ is the relaxation function. This ensures the correct constitutive relations, as does $(7.1.14)_3$ in the general case. Relations

$$\mathbb{G}_1(s,0) = \mathbb{G}_2(0,s) = \mathbb{G}'(s), \tag{8.6.8}$$

where $\mathbb{G}'(s)$ is the derivative of the relaxation function $\mathbb{G}(s)$, are an immediate consequence. Note that (8.6.5) gives

$$\mathbb{G}(s,\infty) = \mathbb{G}(\infty,s) = \mathbb{G}_\infty \qquad \forall s \in \mathbb{R}^+, \tag{8.6.9}$$

from which, with (8.6.7), we deduce that

$$\mathbb{G}(\infty) = \mathbb{G}_\infty. \tag{8.6.10}$$

It follows from (8.6.5) and (8.6.4) that $\mathbb{G}_1(s,u)$ and $\mathbb{G}_2(s,u)$ vanish at large s,u, respectively, a property corresponding to (7.1.8).

Equation $(7.1.6)_3$ reduces to

$$\mathbb{G}^\top(s,u) = \mathbb{G}(u,s). \tag{8.6.11}$$

Replacing $\widetilde{\phi}(\mathbf{E}(t))$ by $\phi(t)$ and carrying out two partial integrations, we can write (8.6.1) in the form

$$\psi(t) = \phi(t) - \frac{1}{2}\mathbf{E}(t)\cdot\mathbb{G}_\infty\mathbf{E}(t) + \frac{1}{2}\int_0^\infty\int_0^\infty \dot{\mathbf{E}}^t(s)\cdot\mathbb{G}(s,u)\dot{\mathbf{E}}^t(u)dsdu. \tag{8.6.12}$$

From $(7.1.14)_3$, we have

$$\mathbb{G}(s,0) = \mathbb{G}(0,s) = \mathbb{G}(s). \tag{8.6.13}$$

In the completely linear case,

$$\phi(t) = \frac{1}{2}\mathbf{E}(t)\cdot\mathbb{G}_\infty\mathbf{E}(t). \tag{8.6.14}$$

It is nonnegative by (1.4.13). Thus, (8.6.12) becomes

$$\psi(t) = \frac{1}{2}\int_0^\infty\int_0^\infty \dot{\mathbf{E}}^t(s)\cdot\mathbb{G}(s,u)\dot{\mathbf{E}}^t(u)dsdu.$$

Also, with the aid of $(7.1.14)_5$, the form (7.1.19) becomes

$$\begin{aligned}\psi(t) &= S(t) + \frac{1}{2}\int_0^\infty\int_0^\infty \mathbf{E}^t(s)\cdot\mathbb{G}_{12}(s,u)\mathbf{E}^t(u)dsdu,\\ S(t) &= \phi(t) + \mathbf{E}(t)\cdot[\widehat{\mathbf{S}}(t) - \widehat{\mathbf{S}}_e(t) - (\mathbb{G}_0 - \mathbb{G}_\infty)\mathbf{E}(t)]\\ &\quad + \frac{1}{2}\mathbf{E}(t)\cdot(\mathbb{G}_0 - \mathbb{G}_\infty)\mathbf{E}(t).\end{aligned}$$

In the completely linear case, the form of S is given by (see (7.1.34))

$$S(t) = \mathbf{E}(t)\cdot\mathbf{T}(t) - \frac{1}{2}\mathbf{E}(t)\cdot\mathbb{G}_0\mathbf{E}(t). \tag{8.6.15}$$

These results derive from the general theory for which the equilibrium stress may be nonlinear. It is instructive, however, to work through the completely linear case in some detail.

Relation (5.1.11) reduces to the form

$$\dot{\psi}(t) + D(t) = \mathbf{T}(t)\cdot\dot{\mathbf{E}}(t), \tag{8.6.16}$$

for linear viscoelastic materials under isothermal conditions. The quantity $D(t)$ denotes the internal dissipation function, which must be nonnegative because of thermodynamic considerations.

We consider the quadratic functional form

$$\psi(t) = \frac{1}{2}\mathbf{E}(t) \cdot \mathbb{G}_{\infty}\mathbf{E}(t) + \frac{1}{2}\int_0^{\infty}\int_0^{\infty} \mathbf{E}_r^t(s) \cdot \mathbb{G}_{12}(s,u)\mathbf{E}_r^t(u)dsdu, \tag{8.6.17}$$

instead of (8.6.1).

The functional (8.6.17) must satisfy (8.6.16), a constraint that will now be explored. The time derivative of (8.6.17), using (8.1.2), (8.1.3) and noting the relations

$$\frac{d}{du}\mathbf{E}_r^t(u) = \frac{d}{du}\mathbf{E}^t(u) = -\frac{d}{dt}\mathbf{E}^t(u) = -\dot{\mathbf{E}}^t(u),$$

gives, with the aid of some integrations by parts,

$$\begin{aligned}
\dot{\psi}(t) &= \dot{\mathbf{E}}(t) \cdot \mathbb{G}_{\infty}\mathbf{E}(t) + \frac{1}{2}\int_0^{\infty}\int_0^{\infty} \Big\{[\dot{\mathbf{E}}^t(s) - \dot{\mathbf{E}}(t)] \cdot \mathbb{G}_{12}(s,u)\mathbf{E}_r^t(u) \\
&\quad + \mathbf{E}_r^t(s) \cdot \mathbb{G}_{12}(s,u)[\dot{\mathbf{E}}^t(u) - \dot{\mathbf{E}}(t)]\Big\}dsdu \\
&= \dot{\mathbf{E}}(t) \cdot \left[\mathbb{G}_{\infty}\mathbf{E}(t) + \int_0^{\infty} \mathbb{G}'(s)\mathbf{E}_r^t(s)ds\right] \\
&\quad + \frac{1}{2}\int_0^{\infty}\int_0^{\infty} \dot{\mathbf{E}}^t(s) \cdot [\mathbb{G}_1(s,u) + \mathbb{G}_2(s,u)]\dot{\mathbf{E}}^t(u)dsdu \\
&= \mathbf{T}(t) \cdot \dot{\mathbf{E}}(t) + \frac{1}{2}\int_0^{\infty}\int_0^{\infty} \dot{\mathbf{E}}^t(s) \cdot \mathbb{K}(s,u)\dot{\mathbf{E}}^t(u)dsdu \\
&= \mathbf{T}(t) \cdot \dot{\mathbf{E}}(t) + \frac{1}{2}\int_0^{\infty}\int_0^{\infty} \mathbf{E}_r^t(s) \cdot \mathbb{K}_{12}(s,u)\mathbf{E}_r^t(u)dsdu,
\end{aligned} \tag{8.6.18}$$

where $\mathbb{K}_{12}(s,u)$ denotes differentiation with respect to the arguments of[†]

$$\mathbb{K}(s,u) = \mathbb{G}_1(s,u) + \mathbb{G}_2(s,u). \tag{8.6.19}$$

Comparing (8.6.18) with (8.6.16), it follows that

$$\begin{aligned}
D(t) &= -\frac{1}{2}\int_0^{\infty}\int_0^{\infty} \dot{\mathbf{E}}^t(s) \cdot \mathbb{K}(s,u)\dot{\mathbf{E}}^t(u)dsdu \\
&= -\frac{1}{2}\int_0^{\infty}\int_0^{\infty} \mathbf{E}_r^t(s) \cdot \mathbb{K}_{12}(s,u)\mathbf{E}_r^t(u)dsdu \geq 0,
\end{aligned} \tag{8.6.20}$$

so that thermodynamics requires that $\mathbb{K}$ and $\mathbb{K}_{12}$ must yield a nonnegative integral. These expressions are special cases of (7.1.25) and (7.1.26).

Note that we can summarize the calculation in (8.6.18) by

[†] Note that the tensor $\mathbb{K}$ here is quite different from the quantity used briefly in the early part of Sect. 7.1.

$$\dot{\widetilde{\psi}}(\mathbf{E}^t, \mathbf{E}(t)) = \frac{\partial}{\partial \mathbf{E}(t)} \widetilde{\psi}(\mathbf{E}^t), \mathbf{E}(t) \cdot \dot{\mathbf{E}}(t) + \partial_t \widetilde{\psi}(\mathbf{E}^t, \mathbf{E}(t)), \tag{8.6.21}$$

where

$$\frac{\partial}{\partial \mathbf{E}(t)} \widetilde{\psi}(\mathbf{E}^t, \mathbf{E}(t)) = \tilde{\mathbf{T}}(\mathbf{E}^t, \mathbf{E}(t)) = \mathbf{T}(t) \tag{8.6.22}$$

and ∂_t indicates differentiation with respect to the t dependence in $\mathbf{E}^t$. We express this as a functional derivative in Sect. 5.1. Relation (8.6.22) follows by comparing (8.6.21) with $(8.6.18)_4$ because the rightmost term in $(8.6.18)_4$ has no dependence on $\dot{\mathbf{E}}(t)$.

Remark 8.6.2. We treat (8.6.22), (8.6.3), and the nonnegative property of D in (8.6.16) as the defining properties of a free energy, referred to as the Graffi conditions in Sect. 5.1.1.

Alternative forms of relations (8.6.17) can be given in terms of histories rather than relative histories. Partial integrations give

$$\begin{aligned} \psi(t) &= \frac{1}{2}\mathbf{E}(t) \cdot \mathbb{G}_\infty \mathbf{E}(t) - \frac{1}{2}\mathbf{E}_r^t(\infty) \cdot \mathbb{G}_\infty \mathbf{E}_r^t(\infty) \\ &+ \frac{1}{2}\int_0^\infty \int_0^\infty \dot{\mathbf{E}}^t(s) \cdot \mathbb{G}(s,u)\dot{\mathbf{E}}^t(u)dsdu \end{aligned}$$

if $\mathbf{E}(-\infty) = \mathbf{E}^t(\infty)$ is finite. The first two terms on the right cancel if $\mathbf{E}(-\infty)$ is zero.

Also,

$$\begin{aligned} \psi(t) &= \frac{1}{2}\mathbf{E}(t) \cdot \mathbb{G}_\infty \mathbf{E}(t) + \frac{1}{2}\int_0^\infty \int_0^\infty \mathbf{E}^t(s_1) \cdot \mathbb{G}_{12}(s_1, s_2)\mathbf{E}^t(s_2)ds_1 ds_2 \\ &- \mathbf{E}(t) \cdot \int_0^\infty \int_0^\infty \mathbb{G}_{12}(s_1, s_2)\mathbf{E}^t(s_2)ds_1 ds_2 \\ &+ \frac{1}{2}\mathbf{E}(t) \cdot \int_0^\infty \int_0^\infty \mathbb{G}_{12}(s_1, s_2)ds_1 ds_2 \mathbf{E}(t). \end{aligned}$$

From (8.6.5) and $(8.6.7)_2$ we have

$$\int_0^\infty \int_0^\infty \mathbb{G}_{12}(s_2, s_1)ds_1 ds_2 = \mathbb{G}_0 - \mathbb{G}_\infty.$$

Carrying out an integration in the second integral, we obtain

$$\begin{aligned} \psi(t) &= \frac{1}{2}\mathbf{E}(t) \cdot \mathbb{G}_0 \mathbf{E}(t) + \mathbf{E}(t) \cdot \int_0^\infty \mathbb{G}'(s_1)\mathbf{E}^t(s_1)ds_1 \\ &+ \frac{1}{2}\int_0^\infty \int_0^\infty \mathbf{E}^t(s_1) \cdot \mathbb{G}_{12}(s_2, s_1)\mathbf{E}^t(s_2)ds_1 ds_2, \qquad (8.6.23) \\ &= S(t) + \frac{1}{2}\int_0^\infty \int_0^\infty \mathbf{E}^t(s_1) \cdot \mathbb{G}_{12}(s_2, s_1)\mathbf{E}^t(s_2)ds_1 ds_2, \end{aligned}$$

where $S(t)$ is given by (8.6.15). Recalling Remark 7.1.2, we see that this relation is a special case of (7.1.34) if we identify $\mathbb{L}_e$ in that relation with $\mathbb{L}_\infty$. Note that

$$\frac{\partial S(t)}{\partial \mathbf{E}(t)} = \mathbf{T}(t),$$

and (8.6.22) follows immediately.

8.6.1 General Forms of a Free Energy in Terms of Stress

The representation (8.6.17) of a free energy can be given in terms of stress history, rather than strain history, using (8.5.6). Also, consider the following argument. Let us introduce a functional χ with the properties

$$\frac{\partial \chi(t)}{\partial \mathbf{T}(t)} = \mathbf{E}(t) \tag{8.6.24}$$

and

$$\dot{\chi}(t) - D_1(t) = \mathbf{E}(t) \cdot \dot{\mathbf{T}}(t). \tag{8.6.25}$$

Noting the developments from (8.6.16) onward, we see that χ can be represented as

$$\begin{gathered}\chi(t) = \frac{1}{2}\mathbf{T}(t) \cdot \mathbb{J}_\infty \mathbf{T}(t) + \frac{1}{2}\int_0^\infty \int_0^\infty \mathbf{T}_r^t(s) \cdot \mathbb{J}_{12}(s,u)\mathbf{T}_r^t(u)dsdu, \\ \mathbb{J}(s,0) = \mathbb{J}(0,s) = \mathbb{J}(s), \qquad \mathbb{J}_\infty = \mathbb{J}(\infty),\end{gathered} \tag{8.6.26}$$

in terms of the creep function defined by (8.5.5) and (8.5.7) and where $\mathbb{J}(\cdot,\cdot)$ has similar properties to those listed for $\mathbb{G}(\cdot,\cdot)$ in (8.6.6)–(8.6.11). Also, referring to (8.6.19) and (8.6.20), we see that

$$\begin{gathered}D_1(t) = \frac{1}{2}\int_0^\infty \int_0^\infty \mathbf{T}_r^t(s) \cdot \mathbb{N}_{12}(s,u)\mathbf{T}_r^t(u)dsdu, \\ \mathbb{N}(s,u) = \mathbb{J}_1(s,u) + \mathbb{J}_2(s,u).\end{gathered} \tag{8.6.27}$$

We can write (8.6.26) as

$$\begin{gathered}\chi(t) = S_1(t) + \frac{1}{2}\int_0^\infty \int_0^\infty \mathbf{T}^t(s) \cdot \mathbb{J}_{12}(s,u)\mathbf{T}^t(u)dsdu, \\ S_1(t) = \mathbf{T}(t) \cdot \mathbf{E}(t) - \frac{1}{2}\mathbf{T}(t) \cdot \mathbb{J}_0\mathbf{T}(t), \qquad \mathbb{J}_0 = \mathbb{J}(0,0) = \mathbb{J}(0),\end{gathered} \tag{8.6.28}$$

by analogy with (8.6.23) and (8.6.15). Relation (8.6.25) can be rewritten as

$$\frac{d}{dt}[\mathbf{E}(t) \cdot \mathbf{T}(t) - \chi(t)] + D_1(t) = \mathbf{T}(t) \cdot \dot{\mathbf{E}}(t). \tag{8.6.29}$$

Comparison with (8.6.16) allows us to identify the quantity ψ, defined by the Legendre transformation

$$\psi(t) = \mathbf{E}(t) \cdot \mathbf{T}(t) - \chi(t),$$

as a free energy and $D_1(t)$ as the associated rate of dissipation, denoted conventionally by $D(t)$. We have

$$\frac{\partial \psi(t)}{\partial \mathbf{E}(t)} = \mathbf{T}(t) + \mathbf{E}(t) \cdot \frac{\partial \mathbf{T}(t)}{\partial \mathbf{E}(t)} - \frac{\partial \chi(t)}{\partial \mathbf{T}(t)} \cdot \frac{\partial \mathbf{T}(t)}{\partial \mathbf{E}(t)} = \mathbf{T}(t),$$

by virtue of (8.6.24). From (8.6.26) and (8.6.28), it follows that

$$\begin{aligned}\psi(t) &= \mathbf{E}(t) \cdot \mathbf{T}(t) - \frac{1}{2}\mathbf{T}(t) \cdot \mathbb{J}_\infty \mathbf{T}(t) - \frac{1}{2}\int_0^\infty \int_0^\infty \mathbf{T}_r^t(s) \cdot \mathbb{J}_{12}(s,u)\mathbf{T}_r^t(u) ds du \\ &= \frac{1}{2}\mathbf{T}(t) \cdot \mathbb{J}_0 \mathbf{T}(t) - \frac{1}{2}\int_0^\infty \int_0^\infty \mathbf{T}^t(s) \cdot \mathbb{J}_{12}(s,u)\mathbf{T}^t(u) ds du.\end{aligned} \tag{8.6.30}$$

Proposition 8.6.3. *The kernel* $\mathbb{J}_{12}$ *must be such that the double integral on the right of* (8.6.30) *is nonpositive and that on the right of* (8.6.27) *is nonnegative.*

Proof. The nonpositivity of the integral on the right of (8.6.30) is sufficient to ensure the non-negativity of $\psi(t)$. We argue that it is also necessary. The property (8.6.3) must always hold, where $\phi(t)$ is the free energy for the special history $\mathbf{E}^t(s) = \mathbf{E}(t)$, $s \leq t$. It follows from (8.1.2) that $\mathbf{T}(t) = \mathbb{G}_\infty \mathbf{E}(t)$ for this history and indeed $\mathbf{T}(t-u) = \mathbf{T}(t)$, $u \geq 0$, so that the history of stress is also constant in this limit. Thus, we have

$$\begin{aligned}\psi(t) &\geq \frac{1}{2}\mathbf{E}(t) \cdot \mathbb{G}_\infty \mathbf{E}(t) = \frac{1}{2}\mathbf{T}(t) \cdot \mathbb{G}_\infty^{-1}\mathbf{T}(t) \\ &= \frac{1}{2}\mathbf{T}(t) \cdot \mathbb{J}_\infty \mathbf{T}(t) \geq \frac{1}{2}\mathbf{T}(t) \cdot \mathbb{J}_0 \mathbf{T}(t),\end{aligned}$$

by virtue of (8.5.9) and (8.5.10). This means that the double integral in (8.6.30) must be nonpositive.

The second law requires that $D_1(t)$ in (8.6.29) be nonnegative, which leads to the claimed property of the integral in (8.6.27). □

8.6.2 The Work Function as a Free Energy

We now consider a particular case of the expression (8.6.17), obtained on supposing that $\mathbb{G}_{12}(s_1, s_2) = \mathbb{G}_{12}(|s_1 - s_2|)$, which clearly obeys (8.6.8). Thus, we introduce the following functional (cf. (7.5.3)):

$$\psi_M(t) = \frac{1}{2}\mathbf{E}(t) \cdot \mathbb{G}_\infty \mathbf{E}(t) + \frac{1}{2}\int_0^\infty \int_0^\infty \mathbf{E}_r^t(s_1) \cdot \mathbb{G}_{12}(|s_1 - s_2|)\mathbf{E}_r^t(s_2) ds_1 ds_2. \tag{8.6.31}$$

Note that by (8.6.11), we can take $\mathbb{G}_{12}(u)$, $u \in \mathbb{R}^+$, to be symmetric, which was in any case assumed in Sect. 8.1 (see (8.1.27)).

Observe that for the choice of kernel in (8.6.31), the quantity $\mathbb{K}$, given by (8.6.19), vanishes, so that D is zero and (8.6.16) reduces to

$$\dot{\psi}_M(t) = \mathbf{T}(t) \cdot \dot{\mathbf{E}}(t). \tag{8.6.32}$$

We can therefore identify ψ_M with the work function

$$\psi_M(t) = \psi_M(t_0) + \int_{t_0}^{t} \mathbf{T}(u) \cdot \dot{\mathbf{E}}(u) du,$$

where t_0 is some fixed time, which may be $-\infty$ if the integral exists (see $(7.5.1)_1$).

It is clear that ψ_M obeys the requirements of a free energy if the memory integral in (8.6.31) is nonnegative. This indeed follows from the fact that (8.6.31) is a special case of (7.5.3) and by recalling the argument leading to (7.5.8). However, it is instructive to outline a more detailed, though equivalent, argument.

There is, however, a problem with categorizing the work function as a free energy, which arises out of Remark 18.2.

Proposition 8.6.4. *Let* $\mathbf{W} \in L^2(\mathbb{R}^+)$ *and* $\mathbb{M} \in L^1(\mathbb{R}^+)$*;* $\mathbf{W}$ *has values in* $\mathrm{Lin}(\mathbb{R}^3)$ *and* $\mathbb{M} \in \mathrm{Lin}(\mathrm{Lin}(\mathbb{R}^3))$*. Also, let* $\mathbb{M}^\top = \mathbb{M}$*. The integral*

$$J = \int_0^\infty \mathbf{W}(t) \cdot \int_0^t \mathbb{M}(t-\tau)\mathbf{W}(\tau) d\tau dt$$

is positive for every nonzero $\mathbf{W}$ *if and only if the Fourier cosine transform* $\mathbb{M}_c$ *is positive definite for* $\omega \in \mathbb{R}^{++}$.

Proof. The faltung theorem applied to causal functions (see Sect. C.3) gives that if

$$\mathbf{V}(t) = \int_0^t \mathbb{M}(t-\tau)\mathbf{W}(\tau) d\tau,$$

then $\mathbf{V}_+(\omega) = \mathbb{M}_+(\omega)\mathbf{W}_+(\omega)$. Given two functions $\mathbf{W}$ and $\mathbb{M} \in L^2(\mathbb{R}^+)$ that vanish on $\mathbb{R}^-$, then Parseval's formula (29.2.2) yields

$$\int_0^\infty \mathbf{W}(t) \cdot \mathbf{V}(t) dt = \frac{1}{2\pi} \int_{-\infty}^{\infty} \overline{\mathbf{W}_+(\omega)} \cdot \mathbf{V}_+(\omega) d\omega.$$

It follows from (C.2.2) that

$$\int_{-\infty}^{\infty} \overline{\mathbf{W}_+(\omega)} \cdot \mathbf{V}_+(\omega) d\omega = \int_{-\infty}^{\infty} [\mathbf{W}_c(\omega) \cdot \mathbb{M}_c(\omega)\mathbf{W}_c(\omega) + \mathbf{W}_s(\omega) \cdot \mathbb{M}_c(\omega)\mathbf{W}_s(\omega)] d\omega.$$

The remaining terms vanish due to either the oddness of the integrand or a cancellation that occurs by virtue of the symmetry assumption on $\mathbb{M}$. Hence, we see that $J > 0$ for every nonzero function $\mathbf{W}$ if $\mathbb{M}_c(\omega)$ is positive definite for every $\omega \in \mathbb{R}^+$. Conversely, if $J > 0$ for every nonzero choice of $\mathbf{W}$, then $\mathbf{W}_c(\omega) \cdot \mathbb{M}_c(\omega)\mathbf{W}_c(\omega) + \mathbf{W}_s(\omega) \cdot \mathbb{M}_c(\omega)\mathbf{V}_s(\omega) > 0$ for $\omega > 0$, and hence it follows that $\mathbb{M}_c(\omega)$ is positive definite for every $\omega \in \mathbb{R}^{++}$. □

Relation (7.5.6) becomes, in the current context,

$$\mathbb{G}_{12}(|s_1 - s_2|) = -2\delta(s_1 - s_2)\mathbb{G}'(|s_1 - s_2|) - \mathbb{G}''(|s_1 - s_2|). \tag{8.6.33}$$

Using this result, we have

$$\begin{aligned}
&\int_0^\infty \int_0^\infty \mathbf{W}(s_1) \cdot \mathbb{G}_{12}(|s_1 - s_2|)\mathbf{W}(s_2)ds_1ds_2 \\
&= -2\int_0^\infty \int_0^\infty \mathbf{W}(s_1) \cdot \delta(s_1 - s_2)\mathbb{G}'(|s_1 - s_2|)\mathbf{W}(s_2)ds_1ds_2 \\
&\quad - \int_0^\infty \int_0^\infty \mathbf{W}(s_1) \cdot \mathbb{G}''(|s_1 - s_2|)\mathbf{W}(s_2)ds_1ds_2 \qquad (8.6.34)\\
&= -2\int_0^\infty \mathbf{W}(s) \cdot \mathbb{G}_0'\mathbf{W}(s)ds - 2\int_0^\infty \int_0^t \mathbf{W}(t) \cdot \mathbb{G}''(t-\tau)\mathbf{W}(\tau)d\tau dt \\
&= -2\int_0^\infty \mathbf{W}(t) \cdot \int_0^t [\mathbb{G}''(t-\tau) + \mathbb{G}_0'\delta(t-\tau)]\mathbf{W}(\tau)d\tau dt.
\end{aligned}$$

Hence, by virtue of Proposition 8.6.4, it follows that $\mathbb{G}_{12}$ is a positive definite kernel if and only if the cosine transform of $\mathbb{G}''(t-\tau) + \mathbb{G}_0'\delta(t-\tau)$ is negative definite. Now, from the definition of the Fourier cosine transform (C.1.3)$_4$ we have

$$\begin{aligned}
\mathbb{G}_c''(\omega) &= \int_0^\infty \mathbb{G}''(\xi)\cos\omega\xi d\xi \\
&= \mathbb{G}'(\xi)\cos\omega\xi \,|_0^\infty + \omega\int_0^\infty \mathbb{G}'(\xi)\sin\omega\xi\, d\xi,
\end{aligned}$$

whence

$$\mathbb{G}_c''(\omega) + \mathbb{G}_0' = \omega\mathbb{G}_s'(\omega) < 0 \quad \forall\omega \in \mathbb{R}^{++}, \qquad (8.6.35)$$

by (8.1.18) and the oddness of $\mathbb{G}_s'$. Since

$$\int_0^\infty [\mathbb{G}''(\xi) + \mathbb{G}'(\xi)\delta(\xi)]\cos\omega\xi\, d\xi = \mathbb{G}_c''(\omega) + \mathbb{G}_0',$$

the quadratic form

$$\int_0^\infty \int_0^\infty \mathbf{W}(s_1) \cdot \mathbb{G}_{12}(|s_1 - s_2|)\mathbf{W}(s_2)ds_1ds_2$$

is positive definite and hence vanishes only at $\mathbf{W} = \mathbf{0}$. Putting $\mathbf{W}(s) = \mathbf{E}_r^t(s)$, we conclude that the memory integral in (8.6.31) is nonnegative. Thus, ψ_M is minimal at constant histories.

8.7 The Relaxation Property and a Work Function Norm

Let us consider the fading memory property introduced in Sect. 1.4.3 for linear viscoelastic materials. The function $\mathbb{H}$, defined in the general case by (7.2.22), is given in the present context by

$$\mathbb{H}(\omega) = -\omega\mathbb{G}_s'(\omega) = \omega^2\mathbb{G}_c(\omega) \geq \mathbf{0} \quad \forall\omega \in \mathbb{R}, \qquad (8.7.1)$$

where $\mathbb{G}'_s$ is defined by (8.1.17). It follows in general from (7.2.23) that $\mathbb{H}$ is bounded on $\mathbb{R}^+$.

From (8.6.34) and (8.6.35) it follows that (8.6.31) can be written as

$$\widetilde{\psi}_M(\mathbf{E}^t_r, \mathbf{E}(t)) = \psi_M(t)$$
$$= \frac{1}{2}\mathbf{E}(t) \cdot \mathbb{G}_\infty \mathbf{E}(t) + \frac{1}{2\pi}\int_{-\infty}^{\infty} \overline{\mathbf{E}^t_{r+}(\omega)} \cdot \mathbb{H}(\omega)\mathbf{E}^t_{r+}(\omega)d\omega, \tag{8.7.2}$$

which is of course a special case of $(7.5.7)_2$. From (7.2.29), we have

$$\mathbf{E}^t_{r+}(\omega) = \mathbf{E}^t_+(\omega) - \frac{\mathbf{E}(t)}{i\omega^-} = \mathbf{E}^t_c(\omega) - i\left(\mathbf{E}^t_s(\omega) - \frac{\mathbf{E}(t)}{\omega}\right). \tag{8.7.3}$$

The subscripts c, s indicate the cosine and sine transforms, defined by $(C.1.3)_{3,4}$. The quantity ω^- may be replaced by ω, since the terms with ω^- in the denominator multiply $\mathbb{H}(\omega)$, which vanishes quadratically at the origin. From (8.7.2) and (8.7.3), using the evenness of $\mathbb{H}$, $\mathbf{E}^t_c$ and the oddness of $\mathbf{E}^t_s$, it follows that

$$\widetilde{\psi}_M(\mathbf{E}^t_r, \mathbf{E}(t)) = \psi_M(t) = \frac{1}{2}\mathbf{E}(t) \cdot \mathbb{G}_\infty \mathbf{E}(t) + \frac{1}{\pi}\int_0^{\infty}\left[\left(\mathbf{E}^t_s(\omega) - \frac{\mathbf{E}(t)}{\omega}\right)\right.$$
$$\left. \cdot\, \mathbb{H}(\omega)\left(\mathbf{E}^t_s(\omega) - \frac{\mathbf{E}(t)}{\omega}\right) + \mathbf{E}^t_c(\omega) \cdot \mathbb{H}(\omega)\mathbf{E}^t_c(\omega)\right]d\omega. \tag{8.7.4}$$

Let us introduce the space $\mathcal{H}$ of histories $(\mathbf{E}^t_r, \mathbf{E}(t))$ on $\mathbb{R}^+$ with values in Sym, defined by

$$\mathcal{H} = \left\{(\mathbf{E}^t_r, \mathbf{E}(t)) \in L^2(\mathbb{R}^{++}) \times \text{Sym};\ \int_0^{\infty}\left[\left(\mathbf{E}^t_s(\omega) - \frac{\mathbf{E}(t)}{\omega}\right)\right.\right.$$
$$\left.\left. \cdot\, \mathbb{H}(\omega)\left(\mathbf{E}^t_s(\omega) - \frac{\mathbf{E}(t)}{\omega}\right) + \mathbf{E}^t_c(\omega) \cdot \mathbb{H}(\omega)\mathbf{E}^t_c(\omega)\right]d\omega < \infty\right\}.$$

The space $\mathcal{H}$ becomes a pre-Hilbert space with the inner product

$$(\mathbf{E}^t_1, \mathbf{E}^t_2) = \frac{1}{2}\mathbf{E}_1(t) \cdot \mathbb{G}_\infty \mathbf{E}_2(t) + \frac{1}{\pi}\int_0^{\infty}\left[\left(\mathbf{E}^t_{1s}(\omega) - \frac{\mathbf{E}_1(t)}{\omega}\right)\right.$$
$$\left. \cdot\, \mathbb{H}(\omega)\left(\mathbf{E}^t_{2s}(\omega) - \frac{\mathbf{E}_2(t)}{\omega}\right) + \mathbf{E}^t_{1c}(\omega) \cdot \mathbb{H}(\omega)\mathbf{E}^t_{2c}(\omega)\right]d\omega,$$

where $\mathbf{E}^t_1$, $\mathbf{E}^t_2$ are elements of $\mathcal{H}$ and the corresponding norm is denoted by $\|\cdot\|_{\mathcal{H}}$.

We observe that

$$\|\mathbf{E}^t\|^2_{\mathcal{H}} = \widetilde{\psi}_M(\mathbf{E}^t_r, \mathbf{E}(t)). \tag{8.7.5}$$

Using (8.1.2) rather than (8.1.5) for the stress functional in linear viscoelasticity, we can write, instead of (8.1.28),

$$\mathbf{T}(t) = \mathbb{G}_\infty \mathbf{E}(t) + \frac{2}{\pi}\int_0^{\infty} \mathbb{G}'_s(\omega)\left[\mathbf{E}^t_s(\omega) - \frac{\mathbf{E}(t)}{\omega}\right]d\omega.$$

Indeed, this follows directly from (8.1.28) on noting (8.1.26). This is a continuous functional with respect to the norm $\|\cdot\|_{\mathcal{H}}$ [104].

Let

$$T^{(a)}\mathbf{E}^t = \begin{cases} \mathbf{0} & \forall s \in [0,a], \\ \mathbf{E}^t(s-a) & \forall s \in (a,\infty) \end{cases}$$

be the translated or partly static (see (1.4.10)) history associated with any history $\mathbf{E}^t \in \mathcal{H}$. From the definitions of the Fourier sine and cosine transforms $(C.1.3)_{3,4}$, we obtain

$$(T^{(a)}\mathbf{E}^t)_s(\omega) = \mathbf{E}^t_c(\omega)\sin\omega a + \mathbf{E}^t_s(\omega)\cos\omega a,$$
$$(T^{(a)}\mathbf{E}^t)_c(\omega) = \mathbf{E}^t_c(\omega)\cos\omega a - \mathbf{E}^t_s(\omega)\sin\omega a.$$

Using these relations, we can evaluate $\widetilde{\psi}_M(T^{(a)}\mathbf{E}^t)$, given by (8.7.4), which yields the following expression for the norm, as defined by (8.7.5):

$$\|T^{(a)}\mathbf{E}^t\|^2_{\mathcal{H}} = \int_0^\infty [\mathbf{E}^t_s(\omega)\cdot\mathbb{H}(\omega)\mathbf{E}^t_s(\omega) + \mathbf{E}^t_c(\omega)\cdot\mathbb{H}(\omega)\mathbf{E}^t_c(\omega)]d\omega.$$

Thus, we see that the norm $\|T^{(a)}\mathbf{E}^t\|_{\mathcal{H}}$, defined by (8.7.5), is independent of a, and hence as $a \to \infty$, it does not approach zero, so that

$$\lim_{a\to\infty}\|T^{(a)}\mathbf{E}^t\|_{\mathcal{H}} \neq 0.$$

We observe that in the standard theory of fading memory, the norm $\|\cdot\|$ involves an influence function k [72], and the relaxation property assumes the form

$$\lim_{a\to\infty}\|T^{(a)}\mathbf{E}^t\| = 0$$

for every history $\mathbf{E}^t$ belonging to the corresponding function space. Hence, we might say that with respect to the norm $\|\cdot\|_{\mathcal{H}}$, the relaxation property does not hold for linear viscoelastic solids. However, it has the fading memory property, defined in Sect. 1.4.3, as expressed by $(1.4.12)_1$.

8.8 Viscoelastic Fluids

Fluids are with a symmetry group that is the full unimodular group. Memory effects can be included also for these materials.

A viscoelastic fluid*may remember everything that ever happened to it, yet it cannot recall any one configuration as being physically different from any other except in regard to its mass density*" [313]. Moreover, a "*fluid may have definite memory of all its past experience,* [yet] *it reacts to those experiences only by comparing them with its present configuration*" [312]. In other words, the stress in a fluid is unchanged by a change of the reference configuration. Therefore, the present configuration is used as reference.

We confine our attention to the classical theory of linear viscoelasticity. A constitutive equation of the Boltzmann type yields a hereditary law expressed by a linear relationship between the stress and the infinitesimal strain history. A fluid characterized by such a constitutive equation is a simple material in the sense of the definition given in [75, 103], and therefore and processes can be introduced as in Sect. 3.2. We shall distinguish the cases of compressible and incompressible fluids.

8.9 Compressible Viscoelastic Fluids

Consider a viscoelastic fluid, with a constitutive equation for the stress of the form

$$\mathbf{T}(\rho, \mathbf{E}_r^t) = -p(\rho)\mathbf{I} + \tilde{\mathbf{T}}(\rho, \mathbf{E}_r^t), \tag{8.9.1}$$

where $p(\mathbf{x}, t)$ is the mass density, $\mathbf{E}_r^t(\mathbf{x}, s)\ \forall s \in \mathbb{R}^{++}$ is the relative strain history, given by

$$\mathbf{E}_r^t(\mathbf{x}, s) = \mathbf{E}^t(\mathbf{x}, s) - \mathbf{E}(\mathbf{x}, t), \tag{8.9.2}$$

while p denotes the pressure, $\mathbf{I}$ is the identity second-order tensor, and $\mathbf{x}$ is the position vector, which will be omitted henceforth. The last term $\tilde{\mathbf{T}}$ is the extra stress given by

$$\tilde{\mathbf{T}}(\rho, \mathbf{E}_r^t) = \int_0^\infty \lambda_1(\rho, s)E_r^t(s)ds\,\mathbf{I} + 2\int_0^\infty \mu_1(\rho, s)\mathbf{E}_r^t(s)ds, \tag{8.9.3}$$

where $E_r^t(s) = \mathrm{tr}(\mathbf{E}_r^t)$ and the memory kernels $\lambda_1(\rho, \cdot)$, $\mu_1(\rho, \cdot)$ belong to $L^1(\mathbb{R}^+) \cap L^2(\mathbb{R}^+)$ for any fixed $\rho > 0$.

The state of such a compressible fluid can be described by means of the mass density ρ and the history of $\mathbf{E}_r^t$. Thus

$$\sigma = (\rho, \mathbf{E}_r^t).$$

The process P is expressed by means of a piecewise continuous function $\mathbf{D}^P : [0, d_P) \rightarrow \mathrm{Sym}$, defined by $\mathbf{D}^P(\tau) = \dot{\mathbf{E}}_P(\tau)$, the time derivative of the strain tensor over the time interval $[0, d_P)$, d_P being the duration of the process. We refer in this context to the paragraph before (8.1.9). The ensuing states $\sigma(t+\tau) = (\rho(t+\tau), \mathbf{E}_r^{t+\tau})\ \forall \tau \in (0, d_P]$, due to the application of any process $P = \mathbf{D}^P$, are solutions of two differential equations. That determining strain evolution has the form

$$\frac{d}{d\tau}\mathbf{E}_r^{t+\tau}(s) = \mathbf{D}^P(\tau - s) - \mathbf{D}^P(\tau), \quad 0 < s < \tau. \tag{8.9.4}$$

Moreover, referring to the conservation of mass relation $(1.3.2)_2$, we see that since

$$\nabla \cdot \mathbf{v} = \mathrm{tr}\dot{\mathbf{E}} = \mathrm{tr}\mathbf{D} = D = -\dot{\rho}/\rho, \tag{8.9.5}$$

the balance of mass in the evolving system is expressed by the equation

$$\frac{d}{d\tau}\rho(t+\tau) = -\rho(t+\tau)D^P(\tau), \tag{8.9.6}$$

with solution

$$\rho(t+\tau) = \rho(t)e^{-\int_0^\tau D^P(s)ds}, \tag{8.9.7}$$

which specifies the evolution of the density function.

We denote by Π the set of all processes $P = \mathbf{D}^P$ with finite duration. For the set Σ of states we give a definition characterized by the boundedness of the stress, putting

$$\Sigma = \left\{\sigma = (\rho, \mathbf{E}_r^t);\ |\mathbf{T}(\rho, \mathbf{E}_r^{t+\tau(c)})| < \infty \quad \forall \tau \in \mathbb{R}^+\right\}, \tag{8.9.8}$$

where t is a parameter. Here we have used the partly static history (cf. (1.4.10))

$$\mathbf{E}_r^{t+\tau(c)} = \begin{cases} \mathbf{E}_r^t(s-\tau) & \forall s \in [\tau, \infty), \\ \mathbf{0} & \forall s \in (0, \tau). \end{cases} \tag{8.9.9}$$

The extra stress, given by (8.9.3), obeys the equation

$$\lim_{\tau\to\infty} \tilde{\mathbf{T}}(\rho, \mathbf{E}_r^{t+\tau(c)}) = \tilde{\mathbf{T}}(\rho, \mathbf{0}^\dagger) = \mathbf{0}, \tag{8.9.10}$$

because of the fading memory property (Sect. 1.4.3).

The concept of equivalence of states, as defined in general by Definition 4.1.2, can be introduced for fluids as follows.

Definition 8.9.1. *Two states $\sigma_j(t) = (\rho_j(t), \mathbf{E}_{r_j}^t)\,(j = 1, 2)$ are equivalent if they give the same stress,*

$$\mathbf{T}(\rho_1(t+\tau), \mathbf{E}_{r_1}^{t+\tau}) = \mathbf{T}(\rho_2(t+\tau), \mathbf{E}_{r_2}^{t+\tau}) \quad \forall \tau \in (0, d_P], \tag{8.9.11}$$

for any process $\mathbf{D}^P : [0, d_P) \to \mathrm{Sym}$.

The equivalence class induced by such a definition comprises the minimal states σ_R of the fluid.

8.9.1 A Particular Class of Compressible Fluids

We consider the particular class of viscoelastic fluids characterized by the following kernels:

$$\lambda_1(\rho, s) = \rho\lambda'(s), \qquad \mu_1(\rho, s) = \rho\mu'(s),$$

such that $\lambda', \mu' \in L^1(\mathbb{R}^+) \cap L^2(\mathbb{R}^+)$. Also $\lambda, \mu \in L^1(\mathbb{R}^+) \cap L^2(\mathbb{R}^+)$, where

$$\lambda(t) = -\int_t^\infty \lambda'(s)ds, \qquad \mu(t) = -\int_t^\infty \mu'(s)ds. \tag{8.9.12}$$

Thus, in the constitutive equation (8.9.1), the extra stress $\tilde{\mathbf{T}}$, given by (8.9.3), assumes the form

$$\tilde{\mathbf{T}}(\rho, \mathbf{E}_r^t) = \rho \mathbf{V}(\mathbf{E}_r^t) = \rho \int_0^\infty \kappa'(s) E_r^t(s) ds\, \mathbf{I} + 2\rho \int_0^\infty \mu'(s) \breve{\mathbf{E}}_r^t(s) ds, \tag{8.9.13}$$

where $\breve{\mathbf{E}}_r^t = \mathbf{E}_r^t - \frac{1}{3} E_r^t \mathbf{I}$ is the trace-free part of $\mathbf{E}_r^t$ and

$$\kappa'(s) = \lambda'(s) + \frac{2}{3}\mu'(s).$$

We introduce a compact representation for $\tilde{\mathbf{T}}$, putting

$$\tilde{\mathbf{T}}(\rho, \mathbf{E}_r^t) = \rho \mathbf{V}(\mathbf{E}_r^t) = \rho \int_0^\infty \mathbb{G}'(s) \mathbf{E}_r^t(s), \tag{8.9.14}$$

where $\mathbb{G}'(s)$ is a fourth-order tensor-valued function $\mathbb{G}' : \mathbb{R}^+ \to \mathrm{Lin(Sym)}$, with representation in $\mathrm{Lin}(\mathbb{R}^6)$ as a diagonal matrix. The nonvanishing diagonal elements are $\kappa'(s)$ and $2\mu'(s)$. The stress tensor, given by (8.9.1), can be written as

$$\mathbf{T}(\rho, \mathbf{E}_r^t) = -p(\rho)\mathbf{I} + \rho \int_0^\infty \mathbb{G}'(s) \mathbf{E}_r^t(s) ds. \tag{8.9.15}$$

For such materials, under the hypothesis that any finite density ρ yields a finite pressure $p(\rho)$, the space (8.9.8) of possible states can be defined as $\Sigma = \mathbb{R}^+ \times \Gamma$, where

$$\Gamma = \left\{ \mathbf{E}_r^t ; \left| \int_0^\infty \mathbb{G}'(s+\tau) \mathbf{E}_r^t(s) ds \right| < \infty \quad \forall \tau \in \mathbb{R}^+ \right\}. \tag{8.9.16}$$

Moreover, the space of minimal states σ_R, denoted by Σ_R, can be characterized as follows [100] (cf. Theorem 8.3.1).

Theorem 8.9.2. *For a viscoelastic fluid of type (8.9.15), two states* $\sigma_1 = (\rho_1, \mathbf{E}_{r_1}^t)$ *and* $\sigma_2 = (\rho_2, \mathbf{E}_{r_2}^t)$ *are equivalent in the sense of Definition 8.9.1 if and only if*

$$\begin{gathered} \rho_1(t) = \rho_2(t), \quad \int_0^\infty \mu'(s+\tau) \breve{\mathbf{E}}_r^t(s) ds = \mathbf{0}, \\ \int_0^\infty \kappa'(s+\tau) E_r^t(s) ds = 0 \quad \forall \tau \in \mathbb{R}^+, \end{gathered} \tag{8.9.17}$$

where $\mathbf{E}_r^t = \mathbf{E}_{r_1}^t - \mathbf{E}_{r_2}^t$.

Proof. If (8.9.17) are satisfied, then (8.9.11) follows immediately. Conversely, if (8.9.11) holds for any process, then the expression (8.9.15) for time $t+\tau$ yields

$$\begin{aligned} \mathbf{T}(\rho, \mathbf{E}_r^{t+\tau}) &= -p(t+\tau)\mathbf{I} + \rho(t+\tau) \int_0^\infty \mathbb{G}'(s) \mathbf{E}_r^{t+\tau}(s) ds \\ &= -p(t+\tau)\mathbf{I} + \rho(t+\tau) \int_0^\infty \mathbb{G}'(u+\tau) \mathbf{E}_r^t(u) du \\ &\quad + \rho(t+\tau) \int_{-\tau}^0 \mathbb{G}'(u+\tau) \mathbf{E}_r^t(u) du. \end{aligned} \tag{8.9.18}$$

The quantity $\mathbf{E}_r^t$ in the last integral is determined by the process through (8.9.4). Then, from (8.9.11),

$$\begin{aligned} &-[p(\rho_1(t+\tau)) - p(\rho_2(t+\tau))]\mathbf{I} + \rho_1(t+\tau)\int_0^\infty \mathbb{G}'(u+\tau)\mathbf{E}_{r_1}^t(u)du \\ &-\rho_2(t+\tau)\int_0^\infty \mathbb{G}'(u+\tau)\mathbf{E}_{r_2}^t(u)du \\ &+[\rho_1(t+\tau)) - \rho_2(t+\tau)]\int_{-\tau}^0 \mathbb{G}'(u+\tau)\mathbf{E}_r^t(u)du = \mathbf{0} \end{aligned} \tag{8.9.19}$$

is satisfied for any process. The scalar part of the process, which determines $\rho_1(t+\tau)$ and $\rho_2(t+\tau)$, is specified by (8.9.7); moreover, for any fixed scalar part D^P, the trace-free part can be changed arbitrarily, affecting only the last integral in (8.9.19). Thus, from (8.9.19) we obtain

$$\rho_1(t+\tau) = \rho_2(t+\tau) \qquad \forall\tau \in \mathbb{R}^+, \tag{8.9.20}$$

which, using (8.9.7), yields

$$\rho_1(t) = \rho_2(t).$$

These results give the other two conditions in (8.9.17), on using (8.9.20) in (8.9.19).□

This theorem allows us to state that the minimal state of a linear viscoelastic fluid is an element of

$$\Sigma_R = \mathbb{R}^+ \times (\Gamma/\Gamma_0),$$

where Γ/Γ_0 denotes the usual quotient space, where Γ_0 is the set of the histories $\mathbf{E}_r^t \in \Gamma$ satisfying $(8.9.17)_{2,3}$.

A process can be considered as a function $P : \Sigma \to \Sigma$; thus, it maps the initial state $\sigma^i \in \Sigma$ into the final state $P\sigma^i = \sigma^f \in \Sigma$, and the differential equations (8.9.4)–(8.9.6) govern the evolution. Recalling (8.9.16), we can also consider $P : \Gamma \to \Gamma$, such that any initial relative strain history $\gamma^i \in \Gamma$ is associated with $P\gamma^i = \gamma^f \in \Gamma$; then only (8.9.4) governs such an evolution.

8.9.2 Representation of Free Energies for Compressible Fluids

Under the hypothesis of isothermal processes, the dissipation principle (4.1.7) and (5.1.10) (using only the mechanical components and replacing $\widetilde{\mathbf{S}}$ by the Cauchy stress tensor in this linearized theory; see also (3.4.9)) yield in the present context that starting from a state σ, the work done on any cycle is nonnegative, i.e.,

$$W(\sigma, P) = \oint_0^{d_P} \frac{1}{\rho}\mathbf{T}(\rho, \mathbf{E}_r^t) \cdot \mathbf{D}(t)dt \geq 0, \tag{8.9.21}$$

where we omit the superscript on $\mathbf{D}^P$. For a fluid characterized by the constitutive equation (8.9.15), we obtain two terms in (8.9.21), the first of which vanishes on a cycle, by virtue of (8.9.5), since

$$-\oint_0^{d_P} \frac{p(\rho)}{\rho}\mathbf{I} \cdot \mathbf{D}(t)dt = -\oint_0^{d_P} \frac{p(\rho)}{\rho} D(t)dt = \oint_0^{d_P} \frac{p(\rho)}{\rho^2}\dot{\rho}dt = 0.$$

This follows on observing that the quantity

$$\phi(\rho) = \int_{\rho_0}^{\rho} \frac{1}{\xi^2} p(\xi)d\xi, \tag{8.9.22}$$

ρ_0 being the equilibrium mass density, is a single-valued function of ρ, with the consequence that $\phi(\rho_0)$ vanishes. Thus, (8.9.21) reduces to

$$W(\sigma, P) = \oint_0^{d_P} \frac{1}{\rho}\tilde{\mathbf{T}}(\rho, \mathbf{E}_r^t) \cdot \mathbf{D}(t)dt \geq 0. \tag{8.9.23}$$

We observe that the zero state is $\sigma^\dagger = (\rho_0, \mathbf{0}^\dagger)$, where $\mathbf{0}^\dagger$ is the zero relative history.

Taking into account Definitions 4.1.6 of a free energy and 4.2.1 of a minimum free energy, we prove an important property of the free energy of a fluid characterized by (8.9.15) [100].

Theorem 8.9.3. *For materials described by (8.9.15), every free energy may be written as the sum of two terms*

$$\psi(\sigma) = \phi(\rho) + \varphi(\gamma),$$

where ϕ is given by (8.9.22) *and $\varphi : \mathcal{S}_\Gamma \to \mathbb{R}$ is defined on a set $\mathcal{S}_\Gamma$ that is Γ-invariant (namely, if $\gamma \in \mathcal{S}_\Gamma$, then $P\gamma \in \mathcal{S}_\Gamma$ for every $P \in \Pi$) and satisfies*

$$\varphi(\gamma_2) - \varphi(\gamma_1) \leq \int_0^{d_P} \mathbf{V}(\mathbf{E}_r^t) \cdot \mathbf{D}(t)dt, \tag{8.9.24}$$

where $P\gamma_1 = \gamma_2$. Moreover, if $\psi(\sigma^\dagger) = 0$, then

$$\varphi(\mathbf{0}^\dagger) = 0.$$

Proof. We recall from Definition 4.1.6 that the domain of a free energy must be invariant under the action of any process $P \in \Pi$. Let $\sigma(t) = (\rho(t), \gamma(t)) \in \mathcal{S}$ be a state and P a process with duration d_P. Then, $P\sigma(t) \in \mathcal{S}$ if $(\rho(t + d_P), \gamma(t + d_P)) \in \mathcal{S}$, where $\rho(t + d_P)$ is the solution (8.9.7) of (8.9.6) with $\tau = d_P$ and $\gamma(t + d_P) = P\gamma(t)$, the solution of (8.9.4). Therefore, $\mathcal{S}$ is invariant if and only if $\mathcal{S} = \mathbb{R}^+ \times \mathcal{S}_\Gamma$, where $\mathcal{S}_\Gamma$ is Γ-invariant.

Moreover, let P be a process of duration d_P such that $P\sigma_1 = \sigma_2$, where σ_1 is the state at time $t = 0$. The inequality defining a free energy (see (4.1.8)) yields

$$\begin{aligned}\psi(\sigma_2) - \psi(\sigma_1) \leq \int_0^{d_P} \frac{1}{\rho(t)}\mathbf{T}(t) \cdot \mathbf{D}(t)dt = -\int_0^{d_P} \frac{p(\rho(t))}{\rho(t)} D(t)dt \\ + \int_0^{d_P} \left[\int_0^{\infty} \kappa'(s)E_r^t(s)ds\, D(t) + 2\int_0^{\infty} \mu'(s)\check{\mathbf{E}}_r^t(s) \cdot \check{\mathbf{D}}(t)ds\right] dt,\end{aligned} \tag{8.9.25}$$

since

$$\breve{\mathbf{E}}_r^t(s) \cdot \mathbf{D}(t) = \breve{\mathbf{E}}_r^t(s) \cdot \breve{\mathbf{D}},$$

where $\breve{\mathbf{D}}$ is the trace-free part of $\mathbf{D}$. Using (8.9.5), the first term of the right-hand side of (8.9.25) becomes

$$-\int_0^{d_P} \frac{p(\rho(t))}{\rho(t)} D(t)dt = \int_{\rho_1}^{\rho_2} \frac{p(\rho)}{\rho^2} d\rho = \phi(\rho_2) - \phi(\rho_1) \tag{8.9.26}$$

with $\phi(\rho)$ given by (8.9.22). Substituting, we obtain

$$\begin{aligned} \psi(\sigma_2) - \psi(\sigma_1) &\le \phi(\rho_2) - \phi(\rho_1) \\ &+ \int_0^{d_P} \left[\int_0^\infty \kappa'(s) E_r^t(s) ds\, D(t) + 2 \int_0^\infty \mu'(s) \breve{\mathbf{E}}_r^t(s) \cdot \breve{\mathbf{D}}(t) ds \right] dt. \end{aligned}$$

From this relation it follows that the function $\varphi = \psi - \phi$ satisfies

$$\varphi(\gamma_2) - \varphi(\gamma_1) \le \int_0^{d_P} \left[\int_0^\infty \kappa'(s) E_r^t(s) ds\, D(t) + 2 \int_0^\infty \mu'(s) \breve{\mathbf{E}}_r^t(s) \cdot \breve{\mathbf{D}}(t) ds \right] dt,$$

which is (8.9.24).

Finally, since

$$\psi(\sigma^\dagger) = \phi(\rho_0) + \varphi(\mathbf{0}^\dagger), \quad \phi(\rho_0) = 0,$$

then $\psi(\sigma^\dagger) = 0$ if and only if $\varphi(\mathbf{0}^\dagger) = 0$. □

Therefore, the right-hand side of (8.9.24) gives the work done by starting from $\gamma \in \Gamma$,

$$W(\gamma, P) = \int_0^{d_P} \frac{1}{\rho} \tilde{\mathbf{T}}(\rho, \mathbf{E}_r^t) \cdot \mathbf{D}(t) dt = \int_0^{d_P} \mathbf{V}(\mathbf{E}_r^t) \cdot \mathbf{D}(t) dt. \tag{8.9.27}$$

We define

$$\mathcal{W}_\Gamma(\gamma) = \{W(\gamma, P);\ P \in \Pi\}.$$

It is easy to prove that the minimum free energy is given by

$$\psi_m(\sigma) = \phi(\rho) + \varphi_m(\gamma),$$

where, recalling (4.2.2),

$$\varphi_m(\gamma) = -\inf \mathcal{W}_\Gamma(\gamma). \tag{8.9.28}$$

The right-hand side is the able work.

A corresponding result can be proved for the general quadratic model discussed in Chap. 7 (and therefore for all materials considered in Part III), where instead of (8.9.15), we have $(7.1.13)_1$, and the generalization of (8.9.26) follows from $(7.1.14)_2$.

The discussion in Sect. 8.6 on free energies as quadratic functionals goes through for fluids also but with $\phi(t)$, defined by (8.6.14), replaced by $\phi(\rho)$), which is given by (8.9.22).

8.9.3 Thermodynamic Restrictions for Compressible Fluids

A procedure similar to that developed for viscoelastic solids in Theorem 8.1.2 and corollaries will now be used to derive the restrictions imposed by the dissipation principle on the constitutive equation (8.9.15).

For this purpose, taking into account (8.9.13) and (8.9.12), we can express the constitutive equation in the form

$$\begin{aligned}\mathbf{T}(t) &= -p(\rho(t))\mathbf{I} + \rho(t)\int_0^{\infty} \kappa(s)\dot{E}^t(s)ds\,\mathbf{I} + 2\rho(t)\int_0^{\infty} \mu(s)\frac{d}{dt}\breve{\mathbf{E}}^t(s)ds \\ &= -p(\rho(t))\mathbf{I} + \rho(t)\int_0^{\infty} \kappa(s)D^t(s)ds\,\mathbf{I} + 2\rho(t)\int_0^{\infty} \mu(s)\breve{\mathbf{D}}^t(s)ds,\end{aligned} \tag{8.9.29}$$

with the aid of two integrations by parts.

Theorem 8.9.4. *The constitutive equation (8.9.29) complies with the dissipation principle if and only if*

$$\kappa_c(\omega) > 0, \qquad \mu_c(\omega) > 0 \quad \forall\omega \in \mathbb{R}^{++}. \tag{8.9.30}$$

Proof. The expression for the work given by (8.9.21), on substituting (8.9.29), has two terms, the first of which, as we have already observed to derive (8.9.23), vanishes, by virtue of the balance of mass (8.9.5) and because the integral is evaluated along a cycle. To discuss the other term, we consider the periodic function

$$\mathbf{D}(t) = \mathbf{D}_1 \cos\omega t + \mathbf{D}_2 \sin\omega t \quad \forall\omega \in \mathbb{R}^{++},$$

where $\mathbf{D}_1$, $\mathbf{D}_2 \in \mathrm{Sym}$, and assume that the duration of the process P is $2\pi/\omega$ times some positive integer. As t runs over $[0, d_P)$ we obtain a cycle, since $\mathbf{D}(0) = \mathbf{D}(d_P)$ and, by virtue of (8.9.7), where the integral on $[0, d_P)$ vanishes, $\rho(0) = \rho(d_P)$.

We can put $\mathbf{D}_1 = \breve{\mathbf{D}}_1 + \frac{1}{3}D_1\mathbf{I}$, with $D_1 = \mathrm{tr}\mathbf{D}_1$; similarly for $\mathbf{D}_2$. Therefore, (8.9.21) reduces to

$$\begin{aligned}\oint_0^{d_P} \Big\{ \int_0^{\infty} &\kappa(s)\left(D_1^2 \cos\omega(t-s)\cos\omega t + D_2^2 \sin\omega(t-s)\sin\omega t\right. \\ &+ D_1D_2[\cos\omega(t-s)\sin\omega t + \sin\omega(t-s)\cos\omega t]\big)\,ds \\ &+ 2\int_0^{\infty} \mu(s)(\breve{\mathbf{D}}_1 \cdot \breve{\mathbf{D}}_1 \cos\omega(t-s)\cos\omega t + \breve{\mathbf{D}}_2 \cdot \breve{\mathbf{D}}_2 \sin\omega(t-s)\sin\omega t \\ &+ \breve{\mathbf{D}}_1 \cdot \breve{\mathbf{D}}_2[\cos\omega(t-s)\sin\omega t + \sin\omega(t-s)\cos\omega t]\big)\,ds\Big\}\,dt > 0,\end{aligned}$$

and hence, by integrating with respect to t, we have

$$\kappa_c(\omega)\left(D_1^2 + D_2^2\right) + 2\mu_c(\omega)\left(\breve{\mathbf{D}}_1 \cdot \breve{\mathbf{D}}_1 + \breve{\mathbf{D}}_2 \cdot \breve{\mathbf{D}}_2\right) > 0 \quad \forall\omega \in \mathbb{R}^{++}$$

for any nonzero $\mathbf{D}_1$ and $\mathbf{D}_2$. Thus, the results (8.9.30) follow.

To show that (8.9.30) is a sufficient condition for the validity of (8.9.21), we note that $(\sigma(0), P)$ is a cycle if and only if $\mathbf{D}$ is periodic in $[0, d_P)$ with vanishing mean value in the period. However, since any periodic history can be expressed through a Fourier series, we can write

$$\mathbf{D}^t(s) = \sum_{h=1}^{\infty} \Big\{\breve{\mathbf{A}}_h \cos h\omega(t-s) + \breve{\mathbf{B}}_h \sin h\omega(t-s) + \frac{1}{3}[A_h \cos h\omega(t-s) + B_h \sin h\omega(t-s)]\mathbf{I}\Big\},$$

where $\mathbf{A}_h$, $\mathbf{B}_h \in \mathrm{Sym}$ and $\omega = 2\pi/d_P$.

The work done on a cycle is expressed by

$$\begin{aligned} W(\gamma, P) = & \oint_0^{d_P} \int_0^{\infty} \kappa(s) \sum_{h,k=1}^{\infty} [A_h \cos h\omega(t-s) + B_h \sin h\omega(t-s)] \\ & \times (A_k \cos k\omega t + B_k \sin k\omega t) ds dt \\ & + 2 \oint_0^{d_P} \int_0^{\infty} \mu(s) \sum_{h,k=1}^{\infty} \left[\breve{\mathbf{A}}_h \cos h\omega(t-s) + \breve{\mathbf{B}}_h \sin h\omega(t-s)\right] \\ & \times (\breve{\mathbf{A}}_k \cos k\omega t + \breve{\mathbf{B}}_k \sin k\omega t) ds dt \\ = & \frac{\pi}{\omega} \sum_{k=1}^{\infty} \left[\kappa_c(k\omega)\left(A_k^2 + B_k^2\right) + 2\mu_c(k\omega)\left(\breve{\mathbf{A}}_k^2 + \breve{\mathbf{B}}_k^2\right)\right] > 0, \end{aligned}$$

because of (8.9.30); thus, the work on any nontrivial cycle satisfies the dissipation principle. □

For viscoelastic solids, the negative definiteness of the half-range sine transform of $\mathbb{G}(s)$ is required by thermodynamics, as specified by (8.1.18). More generally, we refer to (7.2.12). For viscoelastic fluids, the dissipation principle yields the positive definiteness of the half-range cosine transforms of $\kappa(s)$ and $\mu(s)$. However, from (8.9.30), taking account of the relation $f_s'(\omega) = -\omega f_c(\omega)$ (see (7.2.13)), we also have

$$\frac{1}{\omega}\kappa_s'(\omega) < 0, \qquad \frac{1}{\omega}\mu_s'(\omega) < 0 \quad \forall \omega \in \mathbb{R}^{++}.$$

Hence, it follows that

$$\mathbb{G}_c(\omega) = -\frac{1}{\omega}\mathbb{G}_s'(\omega) > \mathbf{0} \quad \forall \omega \in \mathbb{R}^{++}, \tag{8.9.31}$$

in terms of the kernel introduced in (8.9.14). Moreover, $\mathbb{G}_c$ vanishes like ω^{-2} for large ω, since by $(\mathrm{C.2.17})_1$, we have

$$\lim_{\omega \to \infty} \omega^2 \mathbb{G}_c(\omega) = -\mathbb{G}'(0),$$

where $\mathbb{G}'(0)$ is negative definite by virtue of the same argument as that leading to (8.1.21).

8.10 Incompressible Viscoelastic Fluids

For compressible fluids, the pressure p is a scalar function of the density ρ, on which also the extra stress $\tilde{\mathbf{T}}$ depends. However, for incompressible fluids the density ρ does not depend on time, and therefore, it is omitted from the constitutive equations for the stress tensor. Consequently, the pressure p is no longer a function of ρ, but now becomes an indeterminate function of time.

Incompressible viscoelastic fluids are the simplest materials with memory considered in this work, in that they are characterized by a single scalar memory kernel. Some of the results presented correspond to properties demonstrated earlier for solids and compressible fluids. Indeed, we shall abbreviate or omit certain demonstrations because closely analogous proofs have been given earlier. Others derived in this section have corresponding analogues for solids and compressible fluids that were not discussed earlier. They are conveniently included here because of the simplicity of the model. More recent work on this topic includes [11].

The constitutive equation for the stress tensor of a linear viscoelastic incompressible fluid, supposed for simplicity homogeneous, is given by

$$\mathbf{T}(\mathbf{x},t) = -p(\mathbf{x},t)\mathbf{I} + 2\int_0^\infty \mu'(s)\mathbf{E}_r^t(\mathbf{x},s)ds, \tag{8.10.1}$$

where the material function $\mu' \in L^1(\mathbb{R}^+) \cap L^2(\mathbb{R}^+)$ yields the shear relaxation function

$$\mu(s) = -\int_0^\infty \mu'(\tau)d\tau \quad \forall s \in \mathbf{R}^+. \tag{8.10.2}$$

A fluid so characterized is a simple material, with state determined only by the relative strain history, i.e.,

$$\sigma(t) = \mathbf{E}_r^t(s) \qquad \forall s \in \mathbb{R}^{++}.$$

The set of all states of the fluid is denoted by Σ. A process is a piecewise continuous map $P : [0, d_P) \to \mathrm{Sym}$ defined by

$$P(\tau) = \dot{\mathbf{E}}^P(\tau) \qquad \forall \tau \in [0, d_P). \tag{8.10.3}$$

We denote by Π the set of all processes.

It is useful, as in the compressible case, to introduce the extra stress

$$\begin{aligned}\tilde{\mathbf{T}}(\mathbf{E}_r^t) = \mathbf{T}(t) + p(t)\mathbf{I} &= 2\int_0^\infty \mu'(s)\mathbf{E}_r^t(s)ds \\ &= 2\int_0^\infty \mu(s)\dot{\mathbf{E}}^t(s)ds = 2\int_0^\infty \mu(s)\mathbf{D}(t-s)ds,\end{aligned} \tag{8.10.4}$$

which expresses the part of the stress due only to the relative strain history. We have used the relation, valid in the linear approximation, that $\mathbf{D}(t) = \dot{\mathbf{E}}(t)$, which has been invoked earlier for both solids and fluids. Note that (8.10.1) reduces to the linear version of the Navier–Stokes equation (2.2.15), if $\mu(s)$ is given by $\mu_0\delta(s)$, in terms of the delta function. This is the short-memory limit discussed in Sect. 7.1.6.

Referring to (8.9.9) and (8.9.10), we note that similar properties hold for incompressible fluids,

$$\lim_{\tau\to\infty} \tilde{\mathbf{T}}(\mathbf{E}_r^{t+\tau_{(c)}}) = \lim_{\tau\to\infty} 2\int_\tau^\infty \mu'(s)\mathbf{E}_r^{t+\tau_{(c)}}(s)ds = \tilde{\mathbf{T}}(\mathbf{0}^\dagger) = \mathbf{0}.$$

For a partly static history (see (1.4.10) and (8.9.9)), we have

$$\tilde{\mathbf{T}}(t+\tau) = \mathbf{T}(t+\tau) + p(t+\tau)\mathbf{I} = 2\int_0^\infty \mu'(\tau+\xi)\mathbf{E}_r^t(\xi)d\xi.$$

This expression suggests the following definition of the space of states:

$$\Sigma = \left\{\mathbf{E}_r^t : \mathbb{R}^{++} \to \mathrm{Sym}; \left|\int_0^\infty \mu'(\tau+\xi)\mathbf{E}_r^t(\xi)d\xi\right| < \infty \quad \forall\tau\in\mathbb{R}^+\right\},$$

where t is a parameter.

The process $P \in \Pi$ is applied to a given $\sigma(t) = \mathbf{E}_r^t(s)\ \forall s \in \mathbb{R}^{++}$ and $\tau \in [0, d_P)$ as in (8.10.3). The extra stress is given by

$$\tilde{\mathbf{T}}(t+\tau) = \mathbf{T}(t+\tau) + p(t+\tau)\mathbf{I} = 2\int_0^\tau \mu(s)\dot{\mathbf{E}}^{t+\tau}(s)ds + \mathbf{I}^t(\tau, \mathbf{E}_r^t), \qquad (8.10.5)$$

where (cf. (8.2.2); the footnote relating to that equation applies here also)

$$\mathbf{I}^t(\tau, \mathbf{E}_r^t) = 2\int_0^\infty \mu'(\xi+\tau)\mathbf{E}_r^t(\xi)d\xi, \qquad \tau \geq 0. \qquad (8.10.6)$$

Note that

$$\mathbf{I}^t(\tau, \mathbf{E}_r^t) = 2\int_{-\infty}^0 \mu'(\tau-u)\mathbf{E}_r^t(-u)du, \qquad \tau \geq 0. \qquad (8.10.7)$$

The definition of $\mathbf{I}^t$ gives the extra stress due to a partly static relative strain history. The decomposition in $(8.10.5)_2$ is similar to that in (8.9.18) for compressible fluids and (8.2.8) for solids.

Definition 8.9.1 reduces here to the following statement.

Definition 8.10.1. *Two states $\sigma_j(t) = \mathbf{E}_{r_j}^t$ $(j = 1, 2)$ are said to be equivalent if for every process P of duration d_P, the subsequent states $\sigma_j(t+\tau) = \mathbf{E}_{r_j}^{t+\tau}$ $(j = 1, 2)$ satisfy*

$$\tilde{\mathbf{T}}(\mathbf{E}_{r_1}^{t+\tau}) = \tilde{\mathbf{T}}(\mathbf{E}_{r_2}^{t+\tau}) \qquad \forall\tau \in (0, d_P]. \qquad (8.10.8)$$

Therefore, two equivalent states are indistinguishable, since they give the same subsequent extra stress. Thus, we can introduce an equivalence relation R in Σ, the quotient space Σ_R of which has as elements the equivalence classes σ_R, each of which is a set of equivalent states. These are the minimal states of the material.

Theorem 8.10.2. *For a viscoelastic fluid of type (8.10.1), two states* $\sigma_j(t) = \mathbf{E}^t_{r_j}$ $(j = 1, 2)$ *are equivalent if and only if*

$$\int_0^\infty \mu'(\xi + \tau)\left[\mathbf{E}^t_{r_1}(\xi) - \mathbf{E}^t_{r_2}(\xi)\right] d\xi = \mathbf{0} \qquad \forall \tau > 0. \tag{8.10.9}$$

Proof. The proof, simpler than that of Theorem 8.9.2 for compressible fluids, follows at once by considering the expression (8.10.5) for the extra stress and the arbitrariness of P and τ. □

The equivalent forms (8.10.8) and (8.10.9) can be expressed also in terms of the function $\mathbf{I}^t$ given by (8.10.6). Two states, i.e., two relative strain histories, are equivalent if and only if

$$\mathbf{I}^t(\tau, \mathbf{E}^t_{r_1}) = \mathbf{I}^t(\tau, \mathbf{E}^t_{r_2}) \qquad \forall \tau > 0. \tag{8.10.10}$$

Consequently, the function $\mathbf{I}^t$ represents an equivalence class or minimal state of Σ_R.

8.10.1 Thermodynamic Restrictions for Incompressible Viscoelastic Fluids

Putting $\rho = 1$ and $\mathbf{D}(t) = \dot{\mathbf{E}}(t)$ in (8.9.21), we see that the work on a path γ performed by going from an initial state σ to a final state $\hat{\rho}(\sigma, P)$ by means of a process P is given by

$$W(\sigma, P) = \int_\gamma \mathbf{T}(t) \cdot \dot{\mathbf{E}}(t) dt. \tag{8.10.11}$$

The path γ is not necessarily a cycle in general. The dissipation principle is expressed by

$$W(\sigma, P) = \oint \mathbf{T}(t) \cdot \dot{\mathbf{E}}(t) dt \geq 0,$$

where the integral is evaluated on any cycle (σ, P) and the equality sign refers only to reversible processes [297].

The analytical restrictions imposed by thermodynamics on the constitutive equation (8.10.1) have been derived in [113] and are stated by the following theorem.

Theorem 8.10.3. *The constitutive equation (8.10.1) is compatible with the dissipation principle if and only if*

$$\mu_c(\omega) > 0 \qquad \forall \omega \in \mathbb{R}, \tag{8.10.12}$$

where $\mu \in L^1(\mathbb{R}^+)$ *and* $\int_0^\infty \mu(s) ds \neq 0$.

This follows by a simplified version of the proof of Theorem 8.9.4, where κ is put equal to zero. The extension to $\mathbb{R}$ follows from the evenness of μ_c and the final assumption on the integral of μ. This extension also applies to Theorem 8.9.4 under the same assumption.

By continuity, we have

$$\int_0^\infty \mu(s)ds = \mu_c(0) > 0.$$

Moreover, since $\mu' \in L^1(\mathbb{R}^+) \cap L^2(\mathbb{R}^+)$, its Fourier transform is given by

$$\mu'_+(\omega) = \mu'_c(\omega) - i\mu'_s(\omega) \qquad \forall \omega \in \mathbb{R}$$

and belongs to $L^2(\mathbb{R})$. Since

$$\mu'_s(\omega) = -\omega\mu_c(\omega), \tag{8.10.13}$$

it follows from (8.10.1) that μ'_s vanishes linearly at the origin. Also, (8.10.12) implies that

$$\omega\mu'_s(\omega) < 0 \qquad \forall \omega \neq 0. \tag{8.10.14}$$

Moreover, from the inverse Fourier transform of $\mu_c(\omega)$ given by

$$\mu(s) = \frac{2}{\pi}\int_0^\infty \mu_c(\omega)\cos(\omega s)d\omega,$$

we obtain that

$$\mu(0) = \frac{2}{\pi}\int_0^\infty \mu_c(\omega)d\omega = -\frac{1}{\pi}\int_{-\infty}^\infty \frac{\mu'_s(\omega)}{\omega}d\omega > 0.$$

From (C.2.17), we see that

$$\mu'(0) = -\lim_{\omega\to\infty} \omega^2\mu_c(\omega) \le 0 \tag{8.10.15}$$

and that the asymptotic behavior of $\mu'_c(\omega)$ and $\mu'_s(\omega)$ is given by

$$\mu'_s(\omega) \sim \frac{\mu'(0)}{\omega}, \qquad \mu'_c(\omega) \sim -\frac{\mu''(0)}{\omega^2},$$

where it is assumed that $0 \neq |\mu'(0)| < \infty$. If $\mu'' \in L^1(\mathbb{R}^+)$, we obtain

$$\omega\mu'_s(\omega) = \mu'(0) + \mu''_c(\omega). \tag{8.10.16}$$

8.10.2 The Mechanical Work

Firstly, we consider the work done on the material by the relative strain history up to time t, when the final state is $\sigma(t) = \mathbf{E}^t_r$. Referring to (8.10.11), one has

$$\tilde{W}(\mathbf{E}^t_r) = \int_{-\infty}^t \mathbf{T}(\tau)\cdot\dot{\mathbf{E}}(\tau)d\tau = 2\int_{-\infty}^t\int_0^\infty \mu'(s)\mathbf{E}^\tau_r(s)ds\cdot\dot{\mathbf{E}}(\tau)d\tau, \tag{8.10.17}$$

by virtue of the expression (8.10.1) for the stress tensor and the constraint of incompressibility, $\mathbf{I}\cdot\dot{\mathbf{E}} = 0$. We shall be concerned with relative strain histories that yield finite work, i.e., histories such that $\tilde{W}(\mathbf{E}_r^t) < \infty$. This work function can be expressed in various forms. We refer to Sect. 8.6.2 for analogous results applicable to a viscoelastic solid. Similar relations can be obtained for compressible fluids. In the present case, where the memory kernel is one scalar functional, the manipulations are particularly simple.

Integrating by parts in (8.10.17) (see (8.10.4)) and using the definition of $\mathbf{E}_r^t$, given by (8.9.2), we have

$$\begin{aligned}\tilde{W}(\mathbf{E}_r^t) &= 2\int_{-\infty}^{t}\int_0^{\infty}\mu(s)\dot{\mathbf{E}}(\tau - s)ds \cdot \dot{\mathbf{E}}(\tau)d\tau\\ &= \int_{-\infty}^{t}\int_{-\infty}^{t}\mu(|\rho - u|)\dot{\mathbf{E}}(u)\cdot\dot{\mathbf{E}}(\rho)dud\rho,\end{aligned}$$

where a change of variables has been implemented. Changing variables again, we have

$$\tilde{W}(\mathbf{E}_r^t) = \int_0^{\infty}\int_0^{\infty}\mu(|\eta - \xi|)\dot{\mathbf{E}}^t(\xi)\cdot\dot{\mathbf{E}}^t(\eta)d\xi d\eta, \tag{8.10.18}$$

which, with two integrations by parts, yields

$$\tilde{W}(\mathbf{E}_r^t) = \int_0^{\infty}\int_0^{\infty}\mu_{12}(|\eta - \xi|)\mathbf{E}_r^t(\xi)\cdot\mathbf{E}_r^t(\eta)d\xi d\eta, \tag{8.10.19}$$

where (cf. (8.6.33))

$$\mu_{12}(|\eta - \xi|) = \frac{\partial^2}{\partial\eta\partial\xi}\mu(|\eta - \xi|) = -\mu''(|\eta - \xi|) - 2\delta(\eta - \xi)\mu'(|\eta - \xi|). \tag{8.10.20}$$

The following result corresponds to that proved in Proposition 8.6.4. The manipulations are particularly simple in the present case.

Lemma 8.10.4. *The work done on the material by the relative strain history,* $\mathbf{E}_r^t(s)$ $\forall s \in \mathbb{R}^{++}$, *is a nonnegative quantity.*

Proof. Consider the expression (8.10.19) for $\tilde{W}(\mathbf{E}_r^t)$. By substituting (8.10.20), we obtain

$$\tilde{W}(\mathbf{E}_r^t) = -\int_0^{\infty}\int_0^{\infty}\mu''(|\eta - \xi|)\mathbf{E}_r^t(\xi)\cdot\mathbf{E}_r^t(\eta)d\xi d\eta - 2\mu'(0)\int_0^{\infty}\mathbf{E}_r^t(s)\cdot\mathbf{E}_r^t(s)ds.$$

Applying the convolution theorem and Parseval's formula (Sect. C.3), together with (C.1.5), gives

$$\begin{aligned}\tilde{W}(\mathbf{E}_r^t) &= -\frac{1}{\pi}\int_{-\infty}^{\infty}[\mu_c''(\omega) + \mu'(0)]\mathbf{E}_{r+}^t(\omega)\cdot\overline{\mathbf{E}_{r+}^t}(\omega)d\omega\\ &= -\frac{1}{\pi}\int_{-\infty}^{\infty}\omega\mu_s'(\omega)\mathbf{E}_{r+}^t(\omega)\cdot\overline{\mathbf{E}_{r+}^t}(\omega)d\omega,\end{aligned} \tag{8.10.21}$$

because of (8.10.16). This expression is nonnegative by virtue of (8.10.14). □

Now we consider the work done by the process $P(\tau) = \dot{\mathbf{E}}_P(\tau)\,\forall\tau \in [0, d_P)$, applied at time t when $\sigma(t) = \mathbf{E}_r^t$ is the initial state. It is a function of the state σ and the process P, given by (8.10.11), which, taking account of (8.10.5) and (8.10.6), yields

$$
\begin{aligned}
W(\sigma, P) &= \int_t^{t+d} \mathbf{T}(\xi) \cdot \dot{\mathbf{E}}(\xi)d\xi = \int_0^d \mathbf{T}(t+\tau) \cdot \dot{\mathbf{E}}_P(\tau)d\tau \\
&= \int_0^d \left[2\int_0^\tau \mu(\tau-\xi)\dot{\mathbf{E}}_P(\xi)d\xi + \mathbf{I}^t(\tau, \mathbf{E}_r^t) \right] \cdot \dot{\mathbf{E}}_P(\tau)d\tau,
\end{aligned}
\tag{8.10.22}
$$

putting $\dot{\mathbf{E}}_P(\tau) = \dot{\mathbf{E}}(t+\tau)$.

The process P, defined for any $\tau \in [0, d_P)$, d_P being its finite duration, can be extended to $\mathbb{R}^+$ by means of the trivial extension $P(\tau) = \mathbf{0}\,\forall\tau \in [d_P, \infty)$.

Firstly, we consider the case that such a process P is applied at time $t = 0$, when the initial state is $\sigma(0) = \mathbf{E}_r^0(s)$. Putting $t = 0$ and replacing τ by t, we have

$$
\begin{aligned}
\widetilde{W}(\mathbf{E}_r^0, \dot{\mathbf{E}}) &= \int_0^\infty \left[2\int_0^t \mu(t-\tau)\dot{\mathbf{E}}(\tau)d\tau + \mathbf{I}^0(t, \mathbf{E}_r^0) \right] \cdot \dot{\mathbf{E}}(t)dt \\
&= \int_0^\infty \left[\int_0^\infty \mu(|t-\tau|)\dot{\mathbf{E}}(\tau)d\tau + \mathbf{I}^0(t, \mathbf{E}_r^0) \right] \cdot \dot{\mathbf{E}}(t)dt,
\end{aligned}
$$

where $\dot{\mathbf{E}}_P$ is now denoted by $\dot{\mathbf{E}}(t)$ $(t > 0)$.

The work due only to the process can be evaluated by supposing that the initial relative strain history is $\mathbf{E}_r^0(s) = \mathbf{0}^\dagger(s) = \mathbf{0}\,\forall s \in \mathbb{R}^{++}$; thus, the last relation yields

$$
\widetilde{W}(\mathbf{0}^\dagger, \dot{\mathbf{E}}) = \int_0^\infty \int_0^\infty \mu(|t-\tau|)\dot{\mathbf{E}}(\tau) \cdot \dot{\mathbf{E}}(t)d\tau dt. \tag{8.10.23}
$$

We now give a definition introduced by Gentili [145] for viscoelastic solids.

Definition 8.10.5. *A process P, of any duration, is said to be a finite work process if*

$$
\widetilde{W}(\mathbf{0}^\dagger, \dot{\mathbf{E}}_P) < \infty.
$$

This work is a positive quantity as the following lemma states [8].

Lemma 8.10.6. *The work in Definition 8.10.5 satisfies the inequality*

$$
\widetilde{W}(\mathbf{0}^\dagger, \dot{\mathbf{E}}_P) > 0.
$$

Proof. Using the same procedure as in (8.10.21), one can show that relation (8.10.23), the expression for the work done by P, applied at time $t = 0$ to the null relative strain history $\mathbf{0}^\dagger$, can be expressed in the form

$$
\widetilde{W}(\mathbf{0}^\dagger, \dot{\mathbf{E}}_P) = \frac{1}{\pi}\int_{-\infty}^{\infty} \mu_c(\omega)\dot{\mathbf{E}}_{P+}(\omega) \cdot \overline{\dot{\mathbf{E}}_{P+}(\omega)}d\omega > 0. \tag{8.10.24}
$$

The inequality follows by virtue of (8.10.12). □

We can characterize the set of finite work processes by means of [145],

$$\tilde{H}_\mu(\mathbb{R}^+) = \left\{ \mathbf{g} : \mathbb{R}^+ \to \mathrm{Sym};\ \int_{-\infty}^{\infty} \mu_c(\omega)\mathbf{g}_+(\omega) \cdot \overline{\mathbf{g}_+}(\omega) d\omega < \infty \right\},$$

the completion of which, with the norm induced by the inner product

$$(\mathbf{g}_1, \mathbf{g}_2)_\mu = \int_{-\infty}^{\infty} \mu_c(\omega)\mathbf{g}_{1+}(\omega) \cdot \overline{\mathbf{g}_{2+}}(\omega) d\omega,$$

yields the Hilbert space $H_\mu(\mathbb{R}^+)$ of processes.

Let us now consider the general case in which P is applied at any time $t > 0$. The expression for the work done by the process P, again extended to $\mathbb{R}^+$ by means of its trivial extension, becomes, using (8.10.22),

$$\begin{aligned} W(\mathbf{I}^t, \dot{\mathbf{E}}_P) &= \widetilde{W}(\mathbf{E}_r^t, \dot{\mathbf{E}}_P) \\ &= \int_0^{\infty} \left[\int_0^{\infty} \mu(|\tau - \xi|)\dot{\mathbf{E}}_P(\xi) d\xi + \mathbf{I}^t(\tau, \mathbf{E}_r^t) \right] \cdot \dot{\mathbf{E}}_P(\tau) d\tau, \end{aligned} \tag{8.10.25}$$

where (σ, P) has been replaced by $(\mathbf{I}^t, \dot{\mathbf{E}}_P)$ or $(\mathbf{E}_r^t, \dot{\mathbf{E}}_P)$. This becomes, in the frequency domain (see (8.10.24)),

$$W(\mathbf{I}^t, \dot{\mathbf{E}}_P) = \frac{1}{\pi} \int_{-\infty}^{\infty} \mu_c(\omega)\dot{\mathbf{E}}_{P+}(\omega) \cdot \overline{\dot{\mathbf{E}}}_{P+}(\omega) d\omega + \frac{1}{2\pi} \int_{-\infty}^{\infty} \mathbf{I}_+^t(\omega) \cdot \overline{\dot{\mathbf{E}}}_{P+}(\omega) d\omega,$$

where $\mathbf{I}_+^t(\omega)$ denotes the Fourier transform of $\mathbf{I}^t(\tau, \mathbf{E}_r^t)$, defined by (8.10.6) on $\mathbb{R}^+$. The states σ are now expressed by means of $\mathbf{I}^t$, which belong to the dual of $H_\mu(\mathbb{R}^+)$, i.e.,

$$H_\mu'(\mathbb{R}^+) = \left\{ \mathbf{I}^t : \mathbb{R}^+ \to \mathrm{Sym};\ \int_{-\infty}^{\infty} \mathbf{I}_+^t(\omega) \cdot \overline{\dot{\mathbf{E}}_{\mathbf{P}+}}(\omega) d\omega < \infty\ \forall \dot{\mathbf{E}}_P \in H_\mu(\mathbb{R}^+) \right\}.$$

The definition of equivalence for two states, that is, for two relative strain histories, can be expressed in terms of the work, as for viscoelastic solids [145]. A similar result can be given for compressible fluids.

Definition 8.10.7. *Two states* $\sigma_j(t) = \mathbf{E}_{r_j}^t$ $(j = 1, 2)$ *are said to be w-equivalent if for every* $P : [0, \tau) \to \mathrm{Sym}$ *and for every* $\tau > 0$, *they satisfy*

$$\tilde{W}(\mathbf{E}_{r_1}^t, \dot{\mathbf{E}}_P) = \tilde{W}(\mathbf{E}_{r_2}^t, \dot{\mathbf{E}}_P). \tag{8.10.26}$$

The two definitions of equivalence coincide by virtue of the following lemma.

Lemma 8.10.8. *For every fluid characterized by the constitutive equation (8.10.1), two states are w-equivalent if and only if they are equivalent in the sense of Definition 8.10.1.*

Proof. Let us consider the expression (8.10.22) for the work corresponding to two states $\sigma_j(t) = \mathbf{E}^t_{r_j}$ $(j = 1, 2)$. If these two states are equivalent, then the two expressions for the work for any P coincide, i.e., they are w-equivalent. On the other hand, if (8.10.26) is satisfied for any P of any duration, then by virtue of (8.10.25), relation (8.10.10) holds and hence also (8.10.9) and (8.10.8). Thus, two w-equivalent states are equivalent also in the sense of Definition 8.10.1. □

The general representation of a free energy, already studied for viscoelastic solids in Sect. 8.6, can be considered for fluids, as noted at the end of Sect. 8.9.2. For incompressible fluids, this representation takes a particularly simple form by putting $\mathbb{G}_\infty = 0$ in (8.6.17) and replacing $\mathbb{G}_{12}(s, u)$ by a scalar kernel.

8.10.3 Maximum Free Energy for Incompressible Fluids

We seek to show that the mechanical work as given by (8.10.18) or (8.10.19) is the maximum free energy for incompressible fluids, just as (8.6.31) has this property in the case of viscoelastic solids. Let us put

$$\psi_M(t) = \int_0^\infty \int_0^\infty \mu_{12}(|s-u|)\mathbf{E}^t_r(s) \cdot \mathbf{E}^t_r(u)dsdu.$$

We rewrite this expression as (cf. (8.6.23))

$$\psi_M(t) = \mathbf{E}(t) \cdot \left[\mu_0\mathbf{E}(t) + 2\int_0^\infty \mu'(u)\mathbf{E}^t(u)du\right] + \int_0^\infty \int_0^\infty \mu_{12}(|s-u|)\mathbf{E}^t(s) \cdot \mathbf{E}^t(u)dsdu,$$

using (8.9.2) and noting that

$$\int_0^\infty \int_0^\infty \mu_{12}(|s-u|)dsdu = \mu(0,0) = \mu(0) = \mu_0.$$

Hence, recalling Remark 8.6.2, we observe that the property of a free energy expressed by $\partial\psi_M/\partial\mathbf{E} = \tilde{\mathbf{T}}$ follows. Moreover, differentiating with respect to t, we obtain

$$\dot{\psi}_M(t) = 2\left[\mu_0\mathbf{E}(t) + \int_0^\infty \mu'(u)\mathbf{E}^t(u)du\right] \cdot \dot{\mathbf{E}}(t) + 2\mathbf{E}(t) \cdot \int_0^\infty \mu'(u)\dot{\mathbf{E}}^t(u)du + 2\int_0^\infty \int_0^\infty \mu_{12}(|s-u|)\dot{\mathbf{E}}^t(s) \cdot \mathbf{E}^t(u)dsdu.$$

The last integral, with an integration by parts, gives two terms, one of which vanishes on account of the oddness of

$$\frac{\partial\mu}{\partial s}(|s-u|) = sign(s-u)\mu'(|s-u|),$$

while the second one cancels the third term in the expression of $\dot{\psi}_M$. Thus, taking into account the constitutive equation (8.10.1), it follows that

$$\dot{\psi}_M(t) = \mathbf{T}(t) \cdot \dot{\mathbf{E}}(t), \tag{8.10.27}$$

by virtue of the constraint of incompressibility. Therefore, (8.6.16) is satisfied without dissipation, that is, with

$$D(t) = 0.$$

Integrating (8.6.16) over all past time, under the assumption that the integrals exist (as in (5.1.34)), it is easy to see that $\psi_M(t) \geq \psi(t)$, where ψ is any other free energy. It is of interest to present a demonstration that ψ_M is the maximum free energy, which avoids infinite integrals.

For this purpose, we consider an arbitrary process P, applied at time $t = 0$ to the zero state $\sigma^\dagger = \mathbf{E}_r^0(s) = \mathbf{0}\ \forall s \in \mathbb{R}^{++}$, and we denote by $\sigma(t) = \mathbf{E}_r^t = \hat{\rho}(\mathbf{0}^\dagger, P_{[0,t)})$ the final state, belonging to

$$\mathcal{D}_{\psi_M} = \left\{\mathbf{E}_r^t;\ \psi_M(\mathbf{E}_r^t) < \infty\right\},$$

the set of all relative histories that yield a finite $\psi_M(\sigma(t)) = \psi_M(\mathbf{E}_r^t)$. An integration of (8.10.27) on the time interval $(0, t)$ gives

$$\psi_M(\sigma(t)) = \int_0^t \mathbf{T}(s) \cdot \dot{\mathbf{E}}(s) ds,$$

since $\psi_M(\sigma^\dagger) = 0$. Furthermore, any other free energy $\psi(\sigma(t))$, with $\sigma(t) = \hat{\rho}(\mathbf{0}^\dagger, P_{[0,t)})$, must satisfy

$$\psi(\sigma(t)) \leq \int_0^t \mathbf{T}(s) \cdot \dot{\mathbf{E}}(s) ds,$$

because of (4.1.8) and since $\psi(\sigma^\dagger) = 0$. These last two relations yield the inequality

$$\psi_M(\sigma) \geq \psi(\sigma),$$

where $\sigma(t) = \mathbf{E}_r^t \in \mathcal{D}_{\psi_M}$ is an arbitrary final state, obtained by means of any process P. This inequality holds for any free energy functional ψ. Consequently, ψ_M is the maximum free energy for incompressible viscoelastic fluids. As noted previously, however, there is a problem with categorizing the work function as a free energy, which arises out of Remark 18.2.

9 Heat Conductors

To remove the paradox of classical Fourier theory relating to the instantaneous propagation of thermal disturbances, Cattaneo [59] suggested a generalized Fourier law, which he justified by means of statistical considerations. This constitutive equation relates the heat flux, its time derivative, and the temperature gradient. It is referred to as the *Cattaneo–Maxwell relation*, since Maxwell [254] previously obtained it but immediately eliminated the term involving the time derivative of the heat flux. It leads to a hyperbolic heat equation.

On the basis of Coleman's theory for materials with memory [67], a nonlinear model for rigid heat conductors was developed by Gurtin and Pipkin [191]. In this work, the authors derived a linearization of their theory, corresponding to infinitesimal temperature gradients, which yields a linearized constitutive equation for the heat flux in terms of the history of the temperature gradient. This linear relation is a generalization of the Cattaneo–Maxwell equation.

The Gurtin–Pipkin approach is built into the general theory developed in Chaps. 5 and 7. We refer in particular to the discussion centering on (5.1.8).

More recent work on this topic includes [4, 12, 13].

9.1 Constitutive Equations for Rigid Heat Conductors

A rigid heat conductor with memory effects within the linear theory developed in [191] and considered also in [102] is characterized by the constitutive equation

$$\mathbf{q}(\mathbf{x}, t) = -\int_0^\infty \mathbf{k}(s)\mathbf{g}^t(\mathbf{x}, s)ds, \tag{9.1.1}$$

where $\mathbf{x}$ denotes the position vector, $t \in \mathbb{R}^+$ is the time variable, and $\mathbf{g} = \nabla\theta$ is the temperature gradient, expressed in terms of θ, which denotes the absolute tem-

G. Amendola et al., *Thermodynamics of Materials with Memory*,
https://doi.org/10.1007/978-3-030-80534-0_9

perature.* Moreover, $\mathbf{g}^t(\mathbf{x}, s) = \mathbf{g}(\mathbf{x}, t - s)\ \forall s \in \mathbb{R}^{++}$ denotes the past history of the temperature gradient. We consider the heat flux relaxation function $\mathbf{k} : \mathbb{R}^+ \to \mathrm{Sym}$, such that $\mathbf{k} \in L^1(\mathbb{R}^+) \cap H^1(\mathbb{R}^+)$ [102, 115, 147] and

$$\lim_{t\to\infty} \mathbf{k}(t) = \mathbf{0}.$$

Referring to the discussion around (5.1.8) and (5.1.9), we introduce the integrated history of $\mathbf{g}$ [191], which is the function $\overline{\mathbf{g}}^t(\mathbf{x}, \cdot) : \mathbb{R}^+ \to \mathbb{R}^3$ defined by

$$\overline{\mathbf{g}}^t(\mathbf{x}, s) = \int_{t-s}^{t} \mathbf{g}(\mathbf{x}, \tau) d\tau. \tag{9.1.2}$$

Note that

$$\frac{\partial}{\partial s}\overline{\mathbf{g}}^t(\mathbf{x}, s) = \mathbf{g}^t(\mathbf{x}, s).$$

The constitutive equation (9.1.1) can be expressed in terms of $\overline{\mathbf{g}}^t$, by means of an integration by parts, yielding

$$\mathbf{q}(\mathbf{x}, t) = \int_0^\infty \mathbf{k}'(s)\overline{\mathbf{g}}^t(\mathbf{x}, s) ds. \tag{9.1.3}$$

The evolution problem for a rigid heat conductor is governed by the energy equation (see (3.3.7))

$$\dot{e}(\mathbf{x}, t) = -\nabla \cdot \mathbf{q}(\mathbf{x}, t) + r(\mathbf{x}, t), \tag{9.1.4}$$

where r denotes the external heat supply per unit volume and e is the internal energy per unit volume. This relation is also given by (5.1.2), since the Lagrangian and Eulerian descriptions coincide for rigid bodies. We take the constant mass density ρ to be unity. Equation (5.1.3) can be rewritten as

$$\theta\dot{\eta} \geq \dot{e} + \frac{1}{\theta}\mathbf{q} \cdot \mathbf{g}, \tag{9.1.5}$$

with the aid of (9.1.4), where η is the entropy per unit volume. The internal energy is assumed to be given by the constitutive equation

$$e(\mathbf{x}, t) = \alpha_0\vartheta(\mathbf{x}, t) + \int_0^\infty \alpha'(s)\vartheta^t(\mathbf{x}, s) ds, \quad \vartheta = \theta - \Theta_0, \tag{9.1.6}$$

where $\alpha' \in L^1(\mathbb{R}^+) \cap H^1(\mathbb{R}^+)$ and Θ_0 is a reference temperature, uniform in the body. The internal energy relaxation function is given by

$$\alpha(t) = \alpha_0 + \int_0^t \alpha'(\tau) d\tau \qquad \forall t \in \mathbb{R}^+, \qquad \alpha_0 \in \mathbb{R}^{++}. \tag{9.1.7}$$

* Consider (7.1.23), neglecting the first two integrals on the right and carrying out an integration by parts in the third integral. In the linear approximation, $\mathbf{d}(t)$, defined in general by (5.1.1), is given by $-\mathbf{g}/\Theta_0^2$, where Θ_0 is defined after (9.1.6). Absorbing the constant $\rho\Theta_0^2$ into the kernel, we see that this relation is the inverted form of (9.1.1).

We introduce the pseudoenergy [102]

$$\psi(\mathbf{x}, t) = \Theta_0(e - \Theta_0\eta),$$

which will play the role of the free energy in the present context. It follows from (9.1.5) that

$$\dot{\psi} - \dot{e}\vartheta\frac{\Theta_0}{\theta} + \mathbf{q}\cdot\mathbf{g}\frac{\Theta_0^2}{\theta^2} \leq 0. \tag{9.1.8}$$

The approximate theory developed in [102] requires a linearization of the Clausius–Duhem inequality (9.1.8) to the form

$$\dot{\psi}(\mathbf{x}, t) \leq \dot{e}(\mathbf{x}, t)\vartheta(\mathbf{x}, t) - \mathbf{q}(\mathbf{x}, t)\cdot\mathbf{g}(\mathbf{x}, t). \tag{9.1.9}$$

We refer to [102] for a detailed derivation of (9.1.9). This approximate form of the Clausius–Duhem relation will be used in the present chapter. By introducing the internal dissipation function $D(t) \geq 0$, we can write (9.1.9) as an equality

$$\dot{\psi}(\mathbf{x}, t) + D(t) = \dot{e}(\mathbf{x}, t)\vartheta(\mathbf{x}, t) - \mathbf{q}(\mathbf{x}, t)\cdot\mathbf{g}(\mathbf{x}, t). \tag{9.1.10}$$

A rigid heat conductor, characterized by the constitutive equations (9.1.1) and (9.1.6), is a simple material, and therefore, its behavior can be described by means of states and processes, as described in Chaps. 3 and 4. We shall introduce these concepts in a more systematic way here than was done in Chap. 8.

In the following, the dependence on $\mathbf{x}$ will be understood.

9.1.1 States in Terms of $\vartheta^t(s)$ and $\mathbf{g}^t$

We observe that in the linear theory, the internal energy depends on all the history $\vartheta^t(s) = \vartheta(t-s)\ \forall s \in \mathbb{R}^+$, that is, both on the past history $\vartheta^t(s) = \vartheta(t-s)\ \forall s \in \mathbb{R}^{++}$ and on the present value $\vartheta(t)$, while the present value of the temperature gradient does not have an equivalent role in the constitutive equation for the heat flux. We shall identify the history of any function f up to and including time t, $f^t(s) = f(t-s)\ \forall s \in \mathbb{R}^+$, with the pair $(f(t), f^t)$.

The thermodynamic state at time t and at any fixed point $\mathbf{x}$ of the body, taking into account (9.1.1) and (9.1.6), is

$$\sigma(t) = (\vartheta(t),\ \vartheta^t,\ \mathbf{g}^t), \tag{9.1.11}$$

The set of possible states is denoted by Σ.

The kinetic process of duration $d_P \in \mathbb{R}^+$ is the map, piecewise continuous on the time interval $[0, d_P)$, defined by

$$P(\tau) = (\dot{\vartheta}_P(\tau), \mathbf{g}_P(\tau)) \qquad \forall\tau \in [0, d_P), \tag{9.1.12}$$

where $\dot{\vartheta}_P(\tau)$ is the derivative of the temperature with respect to τ and the temperature gradient $\mathbf{g}_P(\tau)$ is, in particular, defined also for $\tau = 0$, corresponding to the instant when P is applied to the body. The set of all accessible processes for the body is denoted by Π. There exists in Π every type of restriction of a process P, of duration

d_P, to an interval $[\tau_1, \tau_2) \subset [0, d_P)$, denoted by $P_{[\tau_1,\tau_2)} \in \Pi$; if $[\tau_1, \tau_2) \equiv [0, t)$ we shall denote $P_{[0,t)}$ by P_t. The evolution function $\hat{\rho} : \Sigma \times \Pi \to \Sigma$ is defined by the property that $\sigma^f = \hat{\rho}(\sigma^i, P) \in \Sigma$, where $\sigma^i \in \Sigma$ is the initial state and σ^f is the final state obtained by applying the process $P \in \Pi$.

Different choices of state for a heat conductor with memory have been used in [147, 191, 279]. Following [102], we can choose as the thermodynamic state that given by (9.1.11), but where the integrated history $\overline{\mathbf{g}}^t$ takes the place of $\mathbf{g}^t$.

The set of possible states Σ is the set of states $\sigma(t) = (\vartheta(t), \vartheta^t, \mathbf{g}^t)$ such that the corresponding e and $\mathbf{q}$ are both finite, so that

$$\left| \int_0^{\infty} \alpha'(s)\vartheta^t(s)ds \right| < \infty, \qquad \left| \int_0^{\infty} \mathbf{k}(s)\mathbf{g}^t(s)ds \right| < \infty. \tag{9.1.13}$$

Let $\sigma(t) = (\vartheta(t), \vartheta^t, \mathbf{g}^t)$ be an initial state of Σ. The evolution function gives a family of states induced by a process $P(\tau) = (\dot{\vartheta}_P(\tau), \mathbf{g}_P(\tau))$ defined for every $\tau \in [0, d_P)$ and applied at the generic time t; in particular, the temperature gradient is the assigned function

$$\mathbf{g}_P : [0, d_P) \to \mathbb{R}^3, \qquad \mathbf{g}_P(\tau) = \mathbf{g}(t + \tau) \qquad \forall \tau \in [0, d_P). \tag{9.1.14}$$

The process P also determines the evolution of temperature according to

$$\vartheta_P : (0, d_P] \to \mathbb{R}, \qquad \vartheta_P(\tau) = \vartheta(t) + \int_0^{\tau} \dot{\vartheta}_P(\xi)d\xi \qquad \forall \tau \in (0, d_P]; \tag{9.1.15}$$

thus, at each instant $\tau' \equiv t + \tau \le t + d_P$, the final value of the temperature, yielded by ϑ^t and $\dot{\vartheta}_P$ and denoted by $\vartheta_f(\tau') = (\vartheta_P * \vartheta)(\tau')$, is given by

$$\vartheta_f(t + d_P - s) = (\vartheta_P * \vartheta)(t + d_P - s) = \begin{cases} \vartheta_P(d_P - s), & 0 \le s < d_P, \\ \vartheta(t + d_P - s), & s \ge d_P, \end{cases} \tag{9.1.16}$$

where the symbol $*$ denotes the continuation of histories with any process. Similarly, the final value of the temperature gradient $\mathbf{g}_f(\tau') = (\mathbf{g}_P * \mathbf{g})(\tau')\, \forall \tau' \equiv t + \tau < t + d_P$ depends on $\mathbf{g}^t$ and $\mathbf{g}_P$ and is expressed by

$$\mathbf{g}_f(t + d_P - s) = (\mathbf{g}_P * \mathbf{g})(t + d_P - s) = \begin{cases} \mathbf{g}_P(d_P - s), & 0 < s \le d_P, \\ \mathbf{g}(t + d_P - s), & s > d_P, \end{cases} \tag{9.1.17}$$

by virtue of (9.1.14).

Given two histories of the temperature and of the temperature gradient, their static continuations of duration $a \in \mathbb{R}^+$ are defined by

$$\vartheta^{t_a} = \begin{cases} \vartheta^t(s - a), & s > a, \\ \vartheta(t), & s \in [0, a], \end{cases} \qquad \mathbf{g}^{t_a} = \begin{cases} \mathbf{g}^t(s - a), & s > a, \\ \mathbf{g}(t), & s \in [0, a]. \end{cases} \tag{9.1.18}$$

The static continuations applied to (9.1.6) and (9.1.1) yield

$$
\begin{aligned}
e(t+a) &= \alpha(a)\vartheta(t) + \int_0^\infty \alpha'(a+\xi)\vartheta^t(\xi)d\xi, \\
\mathbf{q}(t+a) &= -\mathbf{K}(a)\mathbf{g}(t) - \int_0^\infty \mathbf{k}(a+\xi)\mathbf{g}^t(\xi)d\xi,
\end{aligned}
\tag{9.1.19}
$$

where we have denoted the thermal conductivity tensor by

$$
\mathbf{K}(t) = \int_0^t \mathbf{k}(\xi)d\xi. \tag{9.1.20}
$$

Consequently, by virtue of (9.1.13) and (9.1.19), the set of possible states Σ is characterized by

$$
\Gamma_\alpha = \left\{ \vartheta^t : (0,\infty) \to \mathbb{R};\ \left| \int_0^\infty \alpha'(s+\tau)\vartheta^t(s)\,ds \right| < \infty \quad \forall\tau \in \mathbb{R}^+ \right\} \tag{9.1.21}
$$

and

$$
\Gamma_k = \left\{ \mathbf{g}^t : (0,\infty) \to \mathbb{R}^3;\ \left| \int_0^\infty \mathbf{k}(s+\tau)\mathbf{g}^t(s)\,ds \right| < \infty \quad \forall\tau \in \mathbb{R}^+ \right\}, \tag{9.1.22}
$$

where t is a parameter.

In particular, if we consider the constant histories $(\vartheta, \vartheta^\dagger)$, where $\vartheta^t(s) = \vartheta^\dagger = \vartheta\,\forall s \in \mathbb{R}^{++}$, and $\mathbf{g}^t(s) = \mathbf{g}^\dagger = \mathbf{g}\,\forall s \in \mathbb{R}^{++}$, the internal energy (9.1.6) and the heat flux (9.1.1) assume the values

$$
e(t) = \alpha_\infty\vartheta, \qquad \mathbf{q}(t) = -\mathbf{K}_\infty\mathbf{g}, \tag{9.1.23}
$$

where $\alpha_\infty = \lim_{t\to\infty}\alpha(t)$ and $\mathbf{K}_\infty = \lim_{t\to\infty}\mathbf{K}(t)$ are the asymptotic values of α and $\mathbf{K}$ given by (9.1.7) and (9.1.20). These limits are assumed to be finite.

9.1.2 Constitutive Equations in Terms of States and Processes

We now present a decomposition for rigid heat conductors similar to (8.2.8) for viscoelastic solids, (8.9.18), (8.10.5) for fluids but with more detailed discussion in the present case. The constitutive equations (9.1.1) and (9.1.6) define two linear functionals $\tilde{\mathbf{q}} : \Gamma_k \to \mathbb{R}^3$ and $\tilde{e} : \mathbb{R}\times\Gamma_\alpha \to \mathbb{R}$ such that

$$
\tilde{\mathbf{q}}(\mathbf{g}^t) = -\int_0^\infty \mathbf{k}(s)\mathbf{g}^t(s)ds, \quad \tilde{e}(\vartheta(t), \vartheta^t) = \alpha_0\vartheta(t) + \int_0^\infty \alpha'(s)\vartheta^t(s)ds, \tag{9.1.24}
$$

which give the set of heat fluxes and internal energies related to any past history of the temperature gradient $\mathbf{g}^t \in \Gamma_k$ and the temperature $\vartheta^t \in \Gamma_\alpha$.

If P_τ is a process of duration τ applied at the initial time t, it is defined in the time interval $[t, t+\tau)$; if $(\vartheta(t), \vartheta^t, \mathbf{g}^t)$ is the initial state, then the final values of e and $\mathbf{q}$ are given by

$$
\begin{aligned}
e(t+\tau) &= \alpha_0\vartheta(t+\tau) + \int_0^\infty \alpha'(s)\vartheta^{t+\tau}(s)ds,\\
\mathbf{q}(t+\tau) &= -\int_0^\infty \mathbf{k}(s)\mathbf{g}^{t+\tau}(s)ds.
\end{aligned}
\tag{9.1.25}
$$

We consider each of the integrals in (9.1.25) as the sum of two integrals, the first of which is evaluated between 0 and τ and the second between τ and ∞. An integration by parts in the first one, using (9.1.15), yields

$$
\begin{aligned}
e(t+\tau) &= \alpha(\tau)\vartheta(t) + \int_0^\infty \alpha'(\tau+\xi)\vartheta^t(\xi)d\xi + \int_0^\tau \alpha(\tau-\eta)\dot{\vartheta}_{P_\tau}(\eta)d\eta\\
&= \hat{e}(\vartheta(t), \vartheta^t; 0_\tau^\dagger) + \hat{e}(0, 0^\dagger; \dot{\vartheta}_{P_\tau}),
\end{aligned}
\tag{9.1.26}
$$

where $0^\dagger$ is the zero past history for the temperature, given by $0^\dagger(s) = 0\, \forall s \in \mathbf{R}^{++}$, $0_\tau^\dagger$ denotes the zero process with duration τ, $\dot{\vartheta}_{P_\tau}(\eta) = 0_\tau^\dagger(\eta) = 0\, \forall \eta \in [0, \tau)$, and where

$$
\begin{aligned}
\hat{e}(\vartheta(t), \vartheta^t; 0_\tau^\dagger) &= \alpha(\tau)\vartheta(t) + \int_0^\infty \alpha'(\tau+\xi)\vartheta^t(\xi)d\xi,\\
\hat{e}(0, 0^\dagger; \dot{\vartheta}_{P_\tau}) &= \int_0^\tau \alpha(\tau-\eta)\dot{\vartheta}_{P_\tau}(\eta)d\eta.
\end{aligned}
\tag{9.1.27}
$$

In $(9.1.26)_2$, we have a superposition of two effects, the first of which depends on the process through $\dot{\vartheta}_{P_\tau}$, while the second is expressed in terms of the initial state through the initial data of the temperature. Explicitly, $\hat{e}(0, 0^\dagger; \dot{\vartheta}_{P_\tau})$ is due to the part of the process, characterized by $\dot{\vartheta}_{P_\tau}$, starting from the initial state with a null temperature history $(\vartheta(t), \vartheta^t) = (0, 0^\dagger)$, while $\hat{e}(\vartheta(t), \vartheta^t; 0_\tau^\dagger)$ is related to a process with $\dot{\vartheta}_{P_\tau} = 0_\tau^\dagger$, applied to the history $(\vartheta(t), \vartheta^t)$. Also, after the same manipulations,

$$
\begin{aligned}
\mathbf{q}(t+\tau) &= -\int_0^\infty \mathbf{k}(\tau+\xi)\mathbf{g}^t(\xi)d\xi - \int_0^\tau \mathbf{k}(\tau-\eta)\mathbf{g}_{P_\tau}(\eta)d\eta\\
&= \hat{\mathbf{q}}(\mathbf{g}^t; \mathbf{0}_\tau^\dagger) + \hat{\mathbf{q}}(\mathbf{0}^\dagger; \mathbf{g}_{P_\tau}),
\end{aligned}
\tag{9.1.28}
$$

where $\mathbf{0}^\dagger$ denotes the zero past history for $\mathbf{g}$, that is, $\mathbf{0}^\dagger(s) = \mathbf{0}\, \forall s \in \mathbb{R}^{++}$, and $\mathbf{0}_\tau^\dagger$ is the zero process of duration τ, i.e., $\mathbf{g}_{P_\tau}(\eta) = \mathbf{0}_\tau^\dagger(\eta) = \mathbf{0}\, \forall \eta \in [0, \tau)$, and where

$$
\begin{aligned}
\hat{\mathbf{q}}(\mathbf{g}^t; \mathbf{0}_\tau^\dagger) &= -\int_0^\infty \mathbf{k}(\tau+\xi)\mathbf{g}^t(\xi)d\xi,\\
\hat{\mathbf{q}}(\mathbf{0}^\dagger; \mathbf{g}_{P_\tau}) &= -\int_0^\tau \mathbf{k}(\tau-\eta)\mathbf{g}_{P_\tau}(\eta)d\eta.
\end{aligned}
\tag{9.1.29}
$$

Thus, also for the heat flux we have a superposition of two effects. The term $\hat{\mathbf{q}}(\mathbf{0}^\dagger; \mathbf{g}_{P_\tau})$ in $(9.1.29)_2$ expresses the heat flux due to the process P_τ characterized by $\mathbf{g}_{P_\tau}$ and applied to the initial state corresponding to a null past history of the temperature gradient $\mathbf{0}^\dagger$, whereas the quantity $\hat{\mathbf{q}}(\mathbf{g}^0; \mathbf{0}_\tau^\dagger)$ is the heat flux obtained by the process $\mathbf{g}_{P_\tau} = \mathbf{0}_\tau^\dagger$ applied to the initial state characterized by the past history $\mathbf{g}^0$.

9.1.3 Equivalent Histories and Minimal States

We now consider the concepts introduced in Definition 4.1.2 and discussed in Sects. 7.4, 8.3 (see also Theorems 8.9.2 and 8.10.2) for rigid heat conductors.

Definition 9.1.1. *Two states $\sigma_j(t) = (\vartheta_i(t), \vartheta_i^t, \mathbf{g}_j^t)$, $j = 1, 2$, of a rigid heat conductor, characterized by the constitutive equations (9.1.24), are equivalent if*

$$\begin{aligned}\tilde{e}(\vartheta_P(\tau), (\vartheta_P * \vartheta_1)^{t+\tau}) &= \tilde{e}(\vartheta_P(\tau), (\vartheta_P * \vartheta_2)^{t+\tau}),\\ \tilde{\mathbf{q}}((\mathbf{g}_P * \mathbf{g}_1)^{t+\tau}) &= \tilde{\mathbf{q}}((\mathbf{g}_P * \mathbf{g}_2)^{t+\tau})\end{aligned} \tag{9.1.30}$$

for every process $P \in \Pi$ and for every $\tau > 0$.

Thus, the definition of equivalent states ensures the same final value both for the internal energy and for the heat flux, whatever may be their continuations obtained by means of any process, of arbitrary duration, applied to both of them.

Theorem 9.1.2. *Two states $\sigma_j(t) = (\vartheta_j(t), \vartheta_j^t, \mathbf{g}_j^t)$, $j = 1, 2$, are equivalent if and only if*

$$\begin{aligned}\vartheta_1(t) = \vartheta_2(t), \quad &\int_0^\infty \alpha'(\tau+\xi)[\theta_1^t(\xi) - \theta_2^t(\xi)]d\xi = 0,\\ &\int_0^\infty \mathbf{k}(\tau+\xi)[\mathbf{g}_1^t(\xi) - \mathbf{g}_2^t(\xi)]d\xi = \mathbf{0}\end{aligned} \tag{9.1.31}$$

for every $\tau > 0$.

Proof. The equivalence conditions (9.1.30), which are required to be satisfied by the histories, must be evaluated using (9.1.15)–(9.1.17), where the duration of the process d_P is replaced by τ. Consider each integral between zero and infinity as the sum of two integrals, the first between zero and τ and the other between τ and infinity, as we have done in (9.1.26) and (9.1.28). Thus, for every P we have

$$\begin{aligned}\alpha(\tau)[\vartheta_1(t) - \vartheta_2(t)] + &\int_\tau^\infty \alpha'(s)[\theta_1^{t+\tau}(s) - \theta_2^{t+\tau}(s)]ds = 0,\\ &\int_\tau^\infty \mathbf{k}(s)[\mathbf{g}_1^{t+\tau}(s) - \mathbf{g}_2^{t+\tau}(s)]ds = \mathbf{0}.\end{aligned}$$

The arbitrariness of τ yields

$$\begin{aligned}\vartheta_1(t) = \vartheta_2(t), \quad &\int_0^\infty \alpha'(\tau+\xi)\vartheta_1^t(\xi)d\xi = \int_0^\infty \alpha'(\tau+\xi)\vartheta_2^t(\xi)d\xi,\\ &\int_\tau^\infty \mathbf{k}(\tau+\xi)\mathbf{g}_1^t(\xi)d\xi = \int_0^\infty \mathbf{k}(\tau+\xi)\mathbf{g}_2^t(\xi)d\xi,\end{aligned} \tag{9.1.32}$$

for any $\tau > 0$. Using these same relations, the converse also follows. □

We observe that the history $(0, \vartheta^t)$ characterized by a zero instantaneous value and a given past history of the temperature and past history of the temperature gradient $\mathbf{g}^t$ is equivalent to the zero history $(0, 0^\dagger)$ of ϑ and the zero past history $\mathbf{0}^\dagger$ of $\mathbf{g}$ if

$$\begin{aligned} \int_\tau^\infty \alpha'(s)\vartheta^{t+\tau}(s)ds &= \int_0^\infty \alpha'(\tau+\xi)\vartheta^t(\xi)d\xi = 0, \\ \int_\tau^\infty \mathbf{k}(s)\mathbf{g}^{t+\tau}(s)ds &= \int_0^\infty \mathbf{k}(\tau+\xi)\mathbf{g}^t(\xi)d\xi = \mathbf{0}. \end{aligned} \tag{9.1.33}$$

Thus, from (9.1.32), (9.1.33) it follows that two states $\sigma_j(t) = (\vartheta_j(t), \vartheta^t_j, \mathbf{g}^t_j)$, $j = 1, 2$, are equivalent in the sense of Definition 9.1.1 if the differences $\vartheta^t = \vartheta^t_1 - \vartheta^t_2$ and $\mathbf{g}^t = \mathbf{g}^t_1 - \mathbf{g}^t_2$ satisfy (9.1.33) with $\vartheta(t) = \vartheta_1(t) - \vartheta_2(t) = 0$; in other words, two states $\sigma_j(t)$, $j = 1, 2$, are equivalent if the state $\sigma(t) = \sigma_1(t) - \sigma_2(t) = (\vartheta(t), \vartheta^t, \mathbf{g}^t)$ is equivalent to the zero state $(0, 0^\dagger, \mathbf{0}^\dagger)$.

Furthermore [277], we see that two pairs of histories $(\vartheta_j(t), \vartheta^t_j)$, $j = 1, 2$, with $\vartheta_1(t) = \vartheta_2(t)$, and two past histories $\mathbf{g}^t_j$, $j = 1, 2$, whose differences $\vartheta^t = \vartheta^t_1 - \vartheta^t_2$ and $\mathbf{g}^t = \mathbf{g}^t_1 - \mathbf{g}^t_2$ satisfy the relations (9.1.33), represent the same state $\sigma(t)$. Consequently, this state expresses the "minimum" set of variables that give a univocal relation between the process $P(\cdot) = (\dot{\vartheta}_P(\cdot), \mathbf{g}_P(\cdot))$, defined in $[0, \tau)$, and the internal energy $e(t+\tau) = \tilde{e}(\vartheta_P(\tau), (\vartheta_P * \vartheta)^{t+\tau})$ and the heat flux $\mathbf{q}(t+\tau) = \tilde{\mathbf{q}}((\mathbf{g}_P * \mathbf{g})^{t+\tau})$ for every $\tau > 0$. Finally [90, 176], denoting by Γ_{α_0} and Γ_{k_0} the subsets of the past histories of Γ_α and Γ_k satisfying (9.1.33), respectively, and by $\Gamma_\alpha/\Gamma_{\alpha_0}$ and Γ_k/Γ_{k_0} their usual quotient spaces, the state σ of a rigid heat conductor is characterized as $(\vartheta(t), \vartheta^t, \mathbf{g}^t) \in \Sigma = \mathbb{R} \times (\Gamma_\alpha/\Gamma_{\alpha_0}) \times (\Gamma_k/\Gamma_{k_0})$.

We define

$$\tilde{I}^t_\alpha(\tau) := \int_0^\infty \alpha'(\tau+\xi)\theta^t(\xi)d\xi, \tag{9.1.34}$$

while for the heat flux we introduce

$$\tilde{\mathbf{I}}^t_k(\tau) := \int_0^\infty \mathbf{k}(\tau+\xi)\mathbf{g}^t(\xi)d\xi. \tag{9.1.35}$$

From Definition 9.1.1 and by virtue of (9.1.31), it follows that equivalent states $(\vartheta_j(t), \vartheta^t_j, \mathbf{g}^t_j)$, $j = 1, 2$, can be characterized by the triplet $(\vartheta(t), \tilde{I}^t_\alpha, \tilde{\mathbf{I}}^t_k)$, where

$$\begin{aligned} \vartheta(t) &= \vartheta_1(t) = \vartheta_2(t), \\ \tilde{I}^t_\alpha &= \tilde{I}^t_{\alpha 1} = \tilde{I}^t_{\alpha 2}, \\ \tilde{\mathbf{I}}^t_k &= \tilde{\mathbf{I}}^t_{k1} = \tilde{\mathbf{I}}^t_{k2}. \end{aligned}$$

The subscripts 1, 2 on $\tilde{I}^t_\alpha$, $\tilde{\mathbf{I}}^t_k$ refer to histories $(\vartheta^t_1, \mathbf{g}^t_1)$ and $(\vartheta^t_2, \mathbf{g}^t_2)$. Therefore, the minimal state of a rigid heat conductor can be described by $(\vartheta(t), \tilde{I}^t_\alpha, \tilde{\mathbf{I}}^t_k)$.

Let the equivalence relation between states in Σ be denoted by R. The class σ_R of equivalent states can be represented by $\sigma_R = (\vartheta(t), \tilde{I}^t_\alpha, \tilde{\mathbf{I}}^t_k)$ but also by $\sigma_R = (I^t_\alpha, \mathbf{I}^t_k)$ if, taking into account (9.1.34)–(9.1.35) with $(9.1.27)_1$, $(9.1.29)_1$, we introduce

$$I^t_\alpha(\tau) = \tilde{I}^t_\alpha(\tau) + \alpha(\tau)\vartheta(t) = \hat{e}(\vartheta(t), \vartheta^t; 0^\dagger_\tau), \; \mathbf{I}^t_k(\tau) = \tilde{\mathbf{I}}^t_k(\tau) = -\hat{\mathbf{q}}(\mathbf{g}^t; \mathbf{0}^\dagger_\tau) \quad \forall \tau > 0.$$

We observe that I_α^t and $\mathbf{I}_k^t$ are the same for all $(\theta(t), \theta^t, \mathbf{g}^t) \in \sigma_R$.

For heat conductors with discrete-spectrum relaxation functions, namely those consisting of sums of decaying exponentials, one can show as in Sect. 8.4 that the state is finite-dimensional.

9.2 Thermodynamic Constraints for Rigid Heat Conductors

We now determine the restrictions imposed on constitutive equation (9.1.1) by thermodynamics. Let us assume that $\alpha'(s)$ in (9.1.6) is zero. Then, integrating (9.1.9) over any cycle of period T, we obtain

$$\oint_0^T \mathbf{q}(t) \cdot \mathbf{g}(t) dt \leq 0. \tag{9.2.1}$$

The equality sign occurs if and only if the history of $\mathbf{g}$ in (9.1.1) is zero. Consequently, any cycle characterized by the history

$$\mathbf{g}^t(s) = \mathbf{g}_1 \cos \omega(t - s) + \mathbf{g}_2 \sin \omega(t - s),$$

where $\omega \in \mathbb{R}^{++}$ and $(\mathbf{g}_1, \mathbf{g}_2) \in \mathbb{R}^3 \times \mathbb{R}^3 \backslash \{\mathbf{0}, \mathbf{0}\}$, must satisfy (9.2.1) as an inequality, with $\mathbf{q}$ given by (9.1.1); therefore, we must have

$$\int_0^{2\pi/\omega} \int_0^\infty \mathbf{k}(s)[\mathbf{g}_1 \cos \omega(t - s) + \mathbf{g}_2 \sin \omega(t - s)] ds \cdot (\mathbf{g}_1 \cos \omega t + \mathbf{g}_2 \sin \omega t) dt > 0.$$

Integrating with respect to t, we obtain

$$\frac{\pi}{\omega} \int_0^\infty [\mathbf{k}(s)\mathbf{g}_1 \cdot \mathbf{g}_1 + \mathbf{k}(s)\mathbf{g}_2 \cdot \mathbf{g}_2] \cos \omega s \, ds > 0,$$

which, since $\mathbf{g}_1$ and $\mathbf{g}_2$ are arbitrary, yields

$$\mathbf{k}_c(\omega) = \int_0^\infty \mathbf{k}(s) \cos \omega s \, ds > \mathbf{0} \qquad \forall \omega \in \mathbb{R}^{++}, \tag{9.2.2}$$

so that

$$\mathbf{k}_s'(\omega) = -\omega \mathbf{k}_c(\omega) < \mathbf{0} \quad \forall \omega \neq 0, \qquad \mathbf{k}(0) = -\frac{2}{\pi} \int_{-\infty}^\infty \frac{1}{\omega} \mathbf{k}_s'(\omega) d\omega > \mathbf{0} \tag{9.2.3}$$

(see (7.2.19)). Also, by (C.2.17),

$$\lim_{\omega \to \infty} \omega \mathbf{k}_s'(\omega) = - \lim_{\omega \to \infty} \omega^2 \mathbf{k}_c(\omega) = \mathbf{k}'(0) \leq \mathbf{0}. \tag{9.2.4}$$

We assume the following stronger conditions:

$$\mathbf{k}_c(0) = \int_0^\infty \mathbf{k}(\xi) d\xi \equiv \mathbf{K}(\infty) > \mathbf{0}, \qquad \mathbf{k}'(0) < \mathbf{0}. \tag{9.2.5}$$

Relation (9.1.20) has been used here. The assumption $(9.2.5)_1$, in particular, yields that the heat flux $(9.1.23)_2$ resulting from a constant past history of the temperature gradient has the opposite sign to that of $\mathbf{g}$.

Analogously, one can show that the thermodynamic restriction on the memory function α' for the internal energy is expressed by [144, 147] (cf. (8.5.13))

$$\omega\alpha'_s(\omega) > 0, \qquad \omega \neq 0. \tag{9.2.6}$$

Under the hypothesis that $\alpha'' \in L^1(\mathbf{R}^+)$ and using the analogue of (8.1.20), we have

$$\alpha''_c(\omega) = \omega\alpha'_s(\omega) - \alpha'(0), \quad \alpha(t) - \alpha_0 = \frac{2}{\pi}\int_0^\infty \frac{\alpha'_s(\omega)}{\omega}[1 - \cos(\omega t)]d\omega > 0. \tag{9.2.7}$$

It follows from $(9.2.7)_2$ that

$$\alpha_\infty - \alpha_0 = \frac{2}{\pi}\int_0^\infty \frac{\alpha'_s(\omega)}{\omega} d\omega > 0. \tag{9.2.8}$$

Also, referring to (9.2.4), we have

$$\lim_{\omega\to\infty} \omega\alpha'_s(\omega) = \alpha'(0) \geq 0.$$

It will be assumed that

$$\alpha'(0) > 0. \tag{9.2.9}$$

9.3 Thermal Work

The linearized form (9.1.9) of the Clausius–Duhem inequality allows us to introduce the thermal power expressed by

$$w(t) = \dot{e}(t)\vartheta(t) - \mathbf{q}(t) \cdot \mathbf{g}(t); \tag{9.3.1}$$

hence, the thermal work done on a process $P(\tau) = (\dot{\vartheta}_P(\tau), \mathbf{g}_P(\tau))$ applied for every $\tau \in [0, d_P)$, starting from the initial state $\sigma(t)$ at time t, is expressed by

$$W(\sigma(t), P) = \int_0^{d_P} [\dot{e}(t+\tau)\vartheta_P(\tau) - \mathbf{q}(t+\tau) \cdot \mathbf{g}_P(\tau)]d\tau, \tag{9.3.2}$$

where in particular, $\vartheta_P(\tau)$ is given by (9.1.15).

Remark 9.3.1. We observe that in the first term on the right of (9.3.1), the time derivative is on the dependent field variable e rather than the independent variable ϑ, in contrast to (8.6.32), for example. This results in certain differences between the developments in this chapter and those in most of Chap. 8, dealing with solids and fluids. There are similarities, however, with Sect. 8.6.1.

9.3.1 Integrated Histories for Isotropic Heat Conductors

We consider a homogeneous and isotropic rigid heat conductor for which (9.1.6) and (9.1.3) become

$$\begin{aligned} e(t) &= \alpha_0\vartheta(t) + \int_0^\infty \alpha'(s)\vartheta^t(s)ds, \\ \mathbf{q}(t) &= \int_0^\infty k'(s)\bar{\mathbf{g}}^t(s)ds, \end{aligned} \tag{9.3.3}$$

where $\bar{\mathbf{g}}^t$ is the integrated history (9.1.2). The relaxation function for the heat flux is the function $k : \mathbb{R}^+ \to \mathbb{R}$ such that $k \in L^1(\mathbb{R}^+) \cap H^1(\mathbb{R}^+)$ satisfies the thermodynamic restrictions (9.2.2) and the consequences and assumptions (9.2.3)–(9.2.5). Similarly, $\alpha' : \mathbb{R}^+ \to \mathbb{R}$ obeys (9.2.6)–(9.2.9).

Instead of the definition (9.1.11) for the thermodynamic state of the conductor, we now choose the triplet

$$\sigma(t) = (\vartheta(t),\ \vartheta^t, \bar{\mathbf{g}}^t),$$

where the history of the temperature up to time t is again expressed by means of the pair $(\vartheta(t),\ \vartheta^t)$. The process $P : [0, d_P) \to \mathbb{R} \times \mathbb{R}^3$ is still defined by (9.1.12).

Relations (9.1.15)–(9.1.16), which express the continuation $(\vartheta_P * \vartheta)^{t+d_P}$, also remain applicable, together with the set of possible states defined by (9.1.21). However, the presence of the integrated history of the temperature gradient in the state means that we must replace (9.1.17) with the continuation $(\mathbf{g}_P * \bar{\mathbf{g}})^{t+d_P}$ defined by

$$\bar{\mathbf{g}}(t + d_P - s) = (\mathbf{g}_P * \bar{\mathbf{g}})^{t+d_P}(s) = \begin{cases} \displaystyle\int_{d_P - s}^{d_P} \mathbf{g}_P(\xi)d\xi = \bar{\mathbf{g}}_P^{d_P}(s), & \forall s \in [0, d_P), \\ \bar{\mathbf{g}}_P^{d_P}(d_P) + \bar{\mathbf{g}}^t(s - d_P), & \forall s \geq d_P. \end{cases} \tag{9.3.4}$$

The integrated history of $\mathbf{g}$ corresponding to a static continuation of a specified past history $\mathbf{g}^t$, defined in (9.1.18), is given by

$$\bar{\mathbf{g}}^{t+a}(s) = \begin{cases} \displaystyle\int_{a-s}^{a} \mathbf{g}(t)d\xi = s\mathbf{g}(t) \quad \forall s \in [0, a], \\ \displaystyle\int_0^a \mathbf{g}(t)d\xi + \int_{t-(s-a)}^{t} \mathbf{g}(\xi)d\xi = a\mathbf{g}(t) + \int_0^{s-a} \mathbf{g}^t(\rho)d\rho \quad \forall s > a; \end{cases}$$

thus, we obtain the expression (9.1.19) modified as follows:

$$\mathbf{q}(t + a) = -K(a)\mathbf{g}(t) + \int_0^\infty k'(\xi + a)\bar{\mathbf{g}}^t(\xi)d\xi,$$

where the thermal conductivity K is given by the scalar form of (9.1.20). Consequently, the function space (9.1.22) must be replaced by

$$\bar{\Gamma}_k = \left\{\bar{\mathbf{g}}^t : \mathbb{R}^+ \to \mathbf{R}^3; \left|\int_0^\infty k'(\xi + \tau)\bar{\mathbf{g}}^t(\xi)d\xi\right| < \infty\ \forall\tau \geq 0\right\}, \tag{9.3.5}$$

where t is a parameter.

Let P_τ be the restriction of a process applied at time t to the state $\sigma(t) = (\vartheta(t),\ \vartheta^t, \bar{\mathbf{g}}^t)$. By using (9.1.16) and (9.3.4), where d_P is replaced by τ, we have

$$\begin{aligned} e(t+\tau) &= \alpha_0 \vartheta_P(\tau) + \int_0^\tau \alpha'(s)\vartheta_P^\tau(s)ds + \int_\tau^\infty \alpha'(s)\vartheta(t+\tau-s)ds, \\ \mathbf{q}(t+\tau) &= \int_0^\tau k'(s)\bar{\mathbf{g}}_P^\tau(s)ds + \int_\tau^\infty k'(s)[\bar{\mathbf{g}}_P^\tau(\tau) + \bar{\mathbf{g}}^t(s-\tau)]ds. \end{aligned} \tag{9.3.6}$$

The constitutive equations (9.3.3) can be expressed in the general form

$$e(\sigma(t)) = \tilde{e}(\vartheta(t),\ \vartheta^t), \qquad \mathbf{q}(\sigma(t)) = \hat{\mathbf{q}}(\bar{\mathbf{g}}^t).$$

The equivalence relation introduced in the state space Σ by means of Definition 9.1.1 can now be given as follows.

Definition 9.3.2. *Two states $\sigma_j = (\vartheta_j,\ \vartheta_j^t, \bar{\mathbf{g}}_j^t) \in \Sigma$, $j = 1, 2$, of a rigid heat conductor characterized by the constitutive equations (9.3.3) are equivalent if for every process $P \in \Pi$ and for every $\tau > 0$,*

$$\tilde{e}(\rho(\sigma_1, P_{[0,\tau)})) = \tilde{e}(\rho(\sigma_2, P_{[0,\tau)})),\ \hat{\mathbf{q}}((\mathbf{g}_P * \bar{\mathbf{g}}_1)^{t+\tau}) = \hat{\mathbf{q}}((\mathbf{g}_P * \bar{\mathbf{g}}_2)^{t+\tau}). \tag{9.3.7}$$

The following result is the analogue of Theorem 9.1.2 and can be proved similarly.

Theorem 9.3.3. *For a heat conductor characterized by the constitutive equations (9.3.3), two states $\sigma_j = (\vartheta_j,\ \vartheta_j^t, \bar{\mathbf{g}}_j^t)$, $j = 1, 2$, are equivalent if and only if*

$$\begin{aligned} \vartheta_1(t) = \vartheta_2(t), \qquad & \int_0^\infty \alpha'(\tau+\rho)\left[\vartheta_1^t(\rho) - \vartheta_2^t(\rho)\right] d\rho = 0, \\ & \int_0^\infty k'(\tau+\rho)\left[\bar{\mathbf{g}}_1^t(\rho) - \bar{\mathbf{g}}_2^t(\rho)\right] d\rho = \mathbf{0} \end{aligned} \tag{9.3.8}$$

for every $\tau > 0$.

Consequently, a state $\sigma(t) = (\vartheta(t),\ \vartheta^t, \bar{\mathbf{g}}^t)$ is equivalent to the zero state $\sigma_0(t) = (0, 0^\dagger, \bar{\mathbf{0}}^\dagger)$, where in particular $\bar{\mathbf{0}}^\dagger(s) = \bar{\mathbf{g}}^t(s) = \mathbf{0}\ \forall s \in \mathbb{R}^+$ is the zero integrated history of $\mathbf{g}$, if

$$\begin{aligned} \vartheta(t) = 0, \qquad & \int_\tau^\infty \alpha'(s)\vartheta^{t+\tau}(s)ds = \int_0^\infty \alpha'(\tau+\xi)\vartheta^t(\xi)d\xi = 0, \\ & \int_\tau^\infty k'(s)\bar{\mathbf{g}}^t(s-\tau)ds = \int_0^\infty k'(\tau+\xi)\bar{\mathbf{g}}^t(\xi)d\xi = \mathbf{0}. \end{aligned}$$

Thus, two equivalent states σ_j, $j = 1, 2$, are such that their difference $\sigma_1(t) - \sigma_2(t) = (\vartheta_1(t) - \vartheta_2(t),\ \vartheta_1^t - \vartheta_2^t, \bar{\mathbf{g}}_1^t - \bar{\mathbf{g}}_2^t)$ is a state equivalent to the zero state, $\sigma_0(t) = (0, 0^\dagger, \bar{\mathbf{0}}^\dagger)$.

9.3.2 Finite Work Processes and *w*-Equivalence for States

The thermal work done during the application of a process $P(\tau) = (\dot{\vartheta}_P(\tau), \mathbf{g}_P(\tau))\ \forall\tau \in [0, d_P)$, starting from the initial state $\sigma(t) = (\vartheta(t),\ \vartheta^t, \overline{\mathbf{g}}^t)$ at time t, is given by (9.3.2). To evaluate it we must consider the derivative of the internal energy through $(9.3.6)_1$ and take account of the heat flux in the form $(9.3.6)_2$. From $(9.3.6)_1$, by differentiating with respect to τ and integrating by parts, we have

$$\dot{e}(t+\tau) = \alpha_0\dot{\vartheta}_P(\tau) + \alpha'(0)\vartheta_P(\tau) + \int_0^\tau \alpha''(s)\vartheta_P(\tau - s)ds + \int_\tau^\infty \alpha''(s)\,\vartheta(t+\tau-s)ds. \tag{9.3.9}$$

To derive the expression for the work due only to a process P of duration $d_P < \infty$, applied at time $t = 0$, we suppose that the initial state is $\sigma_0(0) = (0, 0^\dagger, \overline{\mathbf{0}}^\dagger)$. Denoting the ensuing fields by $(\vartheta_0,\ \vartheta_0^t, \overline{\mathbf{g}}_0^t)$, (9.1.15)–(9.1.16) with (9.3.4) yield

$$\begin{aligned}
\vartheta_0(t) &= \int_0^t \dot{\vartheta}_P(s)ds,\\
\vartheta_0^t(s) = (\vartheta_P * 0^\dagger)^t(s) &= \begin{cases} \displaystyle\int_0^{t-s} \dot{\vartheta}_P(\eta)d\eta & \forall s \in (0, t],\\ 0 & \forall s > t,\end{cases}\\
\overline{\mathbf{g}}_0^t(s) = (\mathbf{g}_P * \overline{\mathbf{0}}^\dagger)^t(s) &= \begin{cases} \overline{\mathbf{g}}_0^t(s) & \forall s \in [0, t),\\ \overline{\mathbf{g}}_0^t(t) & \forall s \geq t.\end{cases}
\end{aligned} \tag{9.3.10}$$

Let $W(\sigma_0(0), P)$ be the work obtained by applying $P(t) = (\dot{\vartheta}_P(t), \mathbf{g}_P(t))\ \forall\ t \in [0, d_P)$ to the zero state $\sigma_0(0)$, at time $t = 0$. By evaluating directly from $(9.3.3)_1$ or from (9.3.9), we have

$$\dot{e}(t) = \alpha_0\dot{\vartheta}_0(t) + \alpha'(0)\vartheta_0(t) + \int_0^t \alpha''(s)\vartheta^t(s)ds$$

and, from $(9.3.3)_2$ with $(9.3.10)_3$,

$$-\mathbf{q}(t)\cdot\mathbf{g}_P(t) = -\left[\int_0^t k'(s)\overline{\mathbf{g}}_0^t(s)ds + \int_t^\infty k'(s)\overline{\mathbf{g}}_0^t(t)ds\right]\cdot\mathbf{g}_0(t) = \int_0^t k(s)\mathbf{g}_0^t(s)ds\cdot\mathbf{g}_0(t).$$

We see that this work is given by

$$\begin{aligned}
\tilde{W}(0, 0^\dagger, \overline{\mathbf{0}}^\dagger; \dot{\vartheta}_P, \mathbf{g}_P) &= \frac{1}{2}\alpha_0\vartheta_0^2(d_P) + \alpha'(0)\int_0^{d_P}\vartheta_0^2(t)dt\\
&+ \int_0^{d_P}\int_0^t \alpha''(s)\vartheta_0^t(s)ds\vartheta_0(t)dt + \int_0^{d_P}\int_0^t k(s)\mathbf{g}_0^t(t)ds\cdot\mathbf{g}_0(t)dt.
\end{aligned} \tag{9.3.11}$$

Definition 9.3.4. *A process P of duration d_P is a finite work process if*

$$W(\sigma_0(0), P) < \infty.$$

Lemma 9.3.5. *The work done by any finite process is positive.*

Proof. In fact, by assuming that the integrands in (9.3.11) are equal to zero for any $t > d_P$, we can extend the integrals to $\mathbb{R}^+$ and apply Parseval's formula (C.3.1) to obtain

$$
\begin{aligned}
W(\sigma_0(0), P) &= \frac{1}{2}\alpha_0\vartheta_0^2(d_P) + \frac{\alpha'(0)}{2\pi}\int_{-\infty}^{\infty} |\vartheta_{0+}(\omega)|^2 d\omega \\
&+ \frac{1}{2\pi}\int_{-\infty}^{\infty} \alpha''_+(\omega)|\vartheta_{0+}(\omega)|^2 d\omega + \frac{1}{2\pi}\int_{-\infty}^{\infty} k_+(\omega)|\mathbf{g}_{0+}(\omega)|^2 d\omega \\
&= \frac{1}{2}\alpha_0\vartheta_0^2(d_P) + \frac{1}{2\pi}\int_{-\infty}^{\infty} [\alpha'(0) + \alpha''_c(\omega)]|\vartheta_+(\omega)|^2 d\omega \\
&+ \frac{1}{2\pi}\int_{-\infty}^{\infty} k_c(\omega)|\mathbf{g}_{0+}(\omega)|^2 d\omega \\
&= \frac{1}{2}\alpha_0\vartheta_0^2(d_P) + \frac{1}{2\pi}\int_{-\infty}^{\infty} \left\{\omega\alpha'_s(\omega)|\vartheta_{0+}(\omega)|^2 + k_c(\omega)|\mathbf{g}_{0+}(\omega)|^2\right\} d\omega > 0,
\end{aligned}
$$

by virtue of the oddness of the sine Fourier transform together with $(9.2.7)_1$, (9.2.6), and the scalar form of $(9.2.3)_1$. □

Hence, to characterize the set of finite work processes we consider the following sets [145]:

$$
\begin{aligned}
\tilde{H}_\alpha(\mathbb{R}^+, \mathbb{R}) &= \left\{\vartheta : \mathbb{R}^+ \to \mathbb{R};\ \int_{-\infty}^{\infty} \omega\alpha'_s(\omega)|\vartheta_{P+}(\omega)|^2 d\omega < \infty\right\}, \\
\tilde{H}_k(\mathbb{R}^+, \mathbb{R}^3) &= \left\{\mathbf{g} : \mathbb{R}^+ \to \mathbb{R}^3;\ \int_{-\infty}^{\infty} k_c(\omega)|\mathbf{g}_+(\omega)|^2 d\omega < \infty\right\}.
\end{aligned}
\tag{9.3.12}
$$

With the completion with respect to the norm corresponding to the inner product $(\vartheta_1, \vartheta_2)_\alpha = \int_{-\infty}^{\infty} \omega\alpha'_s(\omega)\vartheta_{1+}(\omega)\overline{\vartheta_{2+}(\omega)}d\omega$, we have another Hilbert space $H_\alpha(\mathbb{R}^+, \mathbb{R})$, besides $H_k(\mathbb{R}^+, \mathbb{R}^3)$.

Let $\sigma(t) = (\vartheta(t), \vartheta^t, \bar{\mathbf{g}}^t)$ be the initial state of the body at time $t > 0$, where $\vartheta^t \in \Gamma_\alpha$ and $\bar{\mathbf{g}}^t \in \bar{\Gamma}_k$, the spaces Γ_α and $\bar{\Gamma}_k$ being defined by (9.1.21) and (9.3.5), are possible histories that yield finite work during any process, as defined by (9.3.12). Any of these processes $P = (\dot{\vartheta}_P, \mathbf{g}_P)$ with a finite duration $d_P < \infty$ may be extended to $\mathbb{R}^+$ by putting $P(\tau) = (0, \mathbf{0})\ \forall\tau \geq d_P$. The expression (9.3.2) for the work, taking into account (9.3.9) for $\dot{e}(t+\tau)$ and $(9.3.6)_2$ for $\mathbf{q}(t+\tau)$, with some integrations, becomes

$$
\begin{aligned}
W(\sigma(t), P) &= \tilde{W}(\vartheta(t), \vartheta^t, \bar{\mathbf{g}}^t; \dot{\vartheta}_P, \mathbf{g}_P) \\
&= \frac{1}{2}\alpha_0\left[\vartheta_P^2(d_P) - \vartheta_P^2(0)\right] + \alpha'(0)\int_0^{\infty} \vartheta_P^2(\tau)d\tau \\
&+ \int_0^{\infty}\left[\frac{1}{2}\int_0^{\infty} \alpha''(|\tau - \eta|)\vartheta_P(\eta)d\eta + I^t_{(\alpha)}(\tau, \vartheta^t)\right]\vartheta_P(\tau)d\tau \\
&+ \int_0^{\infty}\left[\frac{1}{2}\int_0^{\infty} k(|\tau - \eta|)\mathbf{g}_P(\eta)d\eta + \mathbf{I}^t_{(k)}(\tau, \bar{\mathbf{g}}^t)\right]\cdot \mathbf{g}_P(\tau)d\tau,
\end{aligned}
\tag{9.3.13}
$$

where

$$
\begin{aligned}
I^t_{(\alpha)}(\tau,\vartheta^t) &= \int_0^\infty \alpha''(\tau+\xi)\vartheta^t(\xi)d\xi,\\
\mathbf{I}^t_{(k)}(\tau,\overline{\mathbf{g}}^t) &= -\int_0^\infty k'(\tau+\xi)\overline{\mathbf{g}}^t(\xi)d\xi,\ \tau\geq 0.
\end{aligned}
\tag{9.3.14}
$$

A definition of equivalence of states is now given in terms of the work function, which we must compare with Definition 9.3.2.

Definition 9.3.6. *Two states $\sigma_j(t) = (\vartheta_j(t), \vartheta^t_j, \overline{\mathbf{g}}^t_j)$, $j = 1, 2$, are said to be w-equivalent if they satisfy*

$$
W(\sigma_1(t), P) = W(\sigma_2(t), P) \tag{9.3.15}
$$

for every process $P : [0, \tau) \to \mathbb{R}\times\mathbb{R}^3$ and for every $\tau > 0$.

Theorem 9.3.7. *Two states are equivalent in the sense of Definition 9.3.2 if and only if they are w-equivalent.*

Proof. Two states $\sigma_j(t) = (\vartheta_j(t), \vartheta^t_j, \overline{\mathbf{g}}^t_j)$, $j = 1, 2$, equivalent in the sense of Definition 9.3.2, satisfy (9.3.7) for every process $P_{[0,\tau)}$ and for every $\tau > 0$. Hence, it follows that we have the same derivative with respect to τ of $(9.3.7)_1$, which appears in the expression (9.3.2) for the work, as well as the same heat flux. Thus, the work done by the same process applied to both $\sigma_j(t)$, $j = 1, 2$, coincide and (9.3.15) holds.

On the other hand, let two states $\sigma_j(t)$, $j = 1, 2$, be w-equivalent. Then for any P with arbitrary duration d_P, taking account of (9.3.13) and (9.1.15), we obtain

$$
\begin{aligned}
&\alpha_0\int_0^{d_P}\dot\vartheta_P(\tau)d\tau[\vartheta_1(t)-\vartheta_2(t)] + \alpha'(0)[\vartheta_1(t)-\vartheta_2(t)]\int_0^{d_P}\{[\vartheta_1(t)+\vartheta_2(t)]\\
&\quad + 2\int_0^\tau \dot\vartheta_P(\xi)d\xi\Big\}d\tau + \frac{1}{2}\int_0^\infty\int_0^\infty \alpha''(|\tau-\eta|)[\vartheta_1(t)-\vartheta_2(t)]\{[\vartheta_1(t)+\vartheta_2(t)]\\
&\quad + 2\left[\int_0^\tau \dot\vartheta_P(\rho)d\rho + \int_0^\eta \dot\vartheta_P(\xi)d\xi\right]\Big\}d\eta d\tau\\
&= -\int_0^{d_P}\Big\{\left[I^t_{(\alpha)}(\tau,\vartheta^t_1)\vartheta_1(t) - I^t_{(\alpha)}(\tau,\vartheta^t_2)\vartheta_2(t)\right]\\
&\quad + \left[I^t_{(\alpha)}(\tau,\vartheta^t_1) - I^t_{(\alpha)}(\tau,\vartheta^t_2)\right]\int_0^\tau \dot\vartheta_P(\xi)d\xi\Big\}d\tau\\
&\quad - \int_0^\infty\left[\mathbf{I}^t_{(k)}(\tau,\overline{\mathbf{g}}^t_1) - \mathbf{I}^t_{(k)}(\tau,\overline{\mathbf{g}}^t_2)\right]\cdot\mathbf{g}_P(\tau)d\tau,
\end{aligned}
$$

where the integrals with $k(|\tau-\eta|)$ cancel, since they have the same $\mathbf{g}_P$. Since in this relation $\dot\vartheta_P$ and d_P, as well as $\mathbf{g}_P$, are arbitrary, it follows that

$$
\vartheta_1(t) = \vartheta_2(t),\quad I^t_{(\alpha)}(\tau,\vartheta^t_1) = I^t_{(\alpha)}(\tau,\vartheta^t_2),\quad \mathbf{I}^t_{(k)}(\tau,\overline{\mathbf{g}}^t_1) = \mathbf{I}^t_{(k)}(\tau,\overline{\mathbf{g}}^t_2). \tag{9.3.16}
$$

The first of these conditions coincides with $(9.3.8)_1$, while the third, by virtue of $(9.3.14)_2$, yields $(9.3.8)_3$; the second equality, using $(9.3.14)_1$, yields

$$I^t_{(\alpha)}(\tau,\vartheta^t_1) - I^t_{(\alpha)}(\tau,\vartheta^t_2) = \int_0^\infty \alpha''(\tau+\xi)\left[\vartheta^t_1(\xi)-\vartheta^t_2(\xi)\right] d\xi$$
$$= \frac{d}{d\tau}\int_0^\infty \alpha'(\tau+\xi)\left[\vartheta^t_1(\xi)-\vartheta^t_2(\xi)\right] d\xi = 0.$$

Hence, the function

$$f(\tau) \equiv \int_0^\infty \alpha'(\tau+\xi)\left[\vartheta^t_1(\xi)-\vartheta^t_2(\xi)\right] d\xi$$

is equal to the constant c_1, which can be evaluated by means of

$$c_1 = \lim_{\tau\to\infty} f(\tau) = 0.$$

Thus, (9.3.16) and (9.3.8) coincide. □

9.3.3 Free Energies as Quadratic Functionals for Rigid Heat Conductors

We can express free energies obeying (9.1.10) as quadratic functionals of the independent quantities $\mathbf{g}^t$ and $(\vartheta(t),\vartheta^t)$, respectively, based on the constitutive relations (9.1.1) and (9.1.6), using a formalism analogous to that in Sect. 8.6. This yields a free energy

$$\psi = \psi_e + \psi_{\mathbf{g}},$$

where ψ_e is a quadratic functional of temperature and $\psi_{\mathbf{g}}$ a similar functional of the temperature gradient. Noting Remark 9.3.1, we see that an analogy with the formalism sketched out in Sect. 8.6.1 is the appropriate one for ψ_e.

Let us write the special case of (9.3.1) in which the contribution from the temperature gradient is neglected:

$$w(t) = \vartheta(t)\dot{e}(t). \tag{9.3.17}$$

The analogue of (8.6.30) in this context is

$$\begin{aligned}\psi^{(e)}(t) &= \frac{1}{2}\alpha_0\vartheta^2(t) - \frac{1}{2}\int_0^\infty\int_0^\infty \alpha_{12}(s,u)\vartheta^t(s)\vartheta^t(u)ds du,\\ \alpha_{12}(s,u) &= \frac{\partial^2}{\partial s\partial u}\alpha(s,u), \qquad \alpha_0 = \alpha(0,0),\end{aligned} \tag{9.3.18}$$

where we must choose $\alpha(\cdot,\cdot)$ so that the integral in $(9.3.18)_1$ exists and is nonpositive for all finite relative histories. Thus, the equivalent of condition (8.6.4) must apply. Putting

$$\alpha(s,u) = \alpha_\infty + \int_s^\infty\int_u^\infty \alpha_{12}(s',u')ds'\,du',$$

we have

$$\alpha(s,\infty) = \alpha(\infty,s) = \alpha_\infty.$$

Also, $(8.6.26)_2$ becomes

$$\alpha(s,0) = \alpha(0,s) = \alpha(s), \quad s \in \mathbb{R},$$

so that $\alpha_\infty = \alpha(\infty)$.

From (8.6.27), the rate of dissipation is given by

$$D(t) = \frac{1}{2}\int_0^\infty \int_0^\infty [\alpha_{121}(s,u) + \alpha_{122}(s,u)]\vartheta_r^t(s)\vartheta_r^t(u)dsdu. \tag{9.3.19}$$

This involves the further constraint on α that the kernel in (9.3.19) must be such that the integral is nonnegative for all relative histories of the internal energy.

By differentiation of $(9.3.18)_1$ and use of $(9.3.3)_1$, we have (cf. (8.6.29))

$$\dot{\psi}^{(e)}(t) + D(t) = \vartheta\dot{e}(t).$$

9.3.4 The Work Function

The work function or maximum free energy (or upper bound on free energies) is obtained from (9.3.18) by putting $\alpha(s,u) = \alpha(|s-u|)$, so that

$$\psi_M^{(e)}(t) = \frac{1}{2}\alpha_\infty\vartheta^2(t) - \frac{1}{2}\int_0^\infty \int_0^\infty \alpha_{12}(|s_1 - s_2|)\vartheta^t(s_1)\vartheta^t(s_2)ds_1ds_2.$$

Applying (8.10.20), we see that this agrees with the relevant terms (9.3.11) if $\vartheta^t(s)$ vanishes for $s > d_P$, using an argument similar to that leading to (7.5.2). Clearly $D(t)$, given by (9.3.19), vanishes in this case and

$$\dot{\psi}_M^{(e)}(t) = \dot{e}(t)\vartheta(t) = w(t),$$

from (9.3.17).

Recall, however, that there is a problem with categorizing the work function as a free energy, arising out of Remark 18.2.

10

Free Energies on Special Classes of Material

10.1 The General Nonisothermal Case

We present in this chapter functionals that are free energies, provided certain assumptions on the relaxation function are valid. In the first section, the general nonisothermal model introduced in Chap. 7 is considered, while in subsequent sections, these functionals are discussed for materials introduced in Chaps. 8 and 9.

In Sects. 10.1.1 and 10.1.3, we consider quadratic functionals that are free energies only for a subcategory of materials, namely those with the property

$$\mathbb{L}'(s) \leq \mathbf{0}, \qquad \mathbb{L}''(s) \geq \mathbf{0} \quad \forall\, s \in \mathbb{R}^+. \tag{10.1.1}$$

Remark 10.1.1. Note that the assumption $\mathbb{L}''(u) \geq \mathbf{0}$, $u \geq s$, implies $\mathbb{L}'(s) \leq \mathbf{0}$. It implies $\mathbb{L}'(s) < \mathbf{0}$ if $\mathbb{L}''$ is nonzero on a set of finite measure with elements $u > s$. This in particular implies that $\mathbb{L}'(0) < \mathbf{0}$ except in trivial cases.

In Sect. 10.1.2, a stronger assumption must be made.

10.1.1 The Graffi–Volterra Free Energy

Let us first present the Graffi–Volterra functional [89, 90, 174, 175]

$$\begin{aligned} \psi_G(t) &= \phi(t) - \frac{1}{2}\int_0^\infty \mathbb{L}'(s)\, \mathbf{\Lambda}_r^t(s) \cdot \mathbf{\Lambda}_r^t(s) ds \\ &= S(t) - \frac{1}{2}\int_0^\infty \mathbb{L}'(s)\, \mathbf{\Lambda}^t(s) \cdot \mathbf{\Lambda}^t(s) ds, \end{aligned} \tag{10.1.2}$$

where $\mathbf{\Lambda}_r^t$ is defined by $(5.1.21)_1$. The quantity $\phi(t)$ is the equilibrium free energy, defined in general before (5.1.24) and in the linear case by (7.1.31), while $S(t)$ is given by $(7.1.19)_2$ (in the linear case, (7.1.34) or (7.1.35)). The rate of dissipation associated with ψ_G may be derived from (5.1.11), in the form

G. Amendola et al., *Thermodynamics of Materials with Memory*,
https://doi.org/10.1007/978-3-030-80534-0_10

$$D_G(t) = \frac{1}{2}\int_0^\infty \mathbb{L}''(s)\mathbf{\Lambda}_r^t(s)\cdot\mathbf{\Lambda}_r^t(s)ds,$$

which is nonnegative under assumption $(10.1.1)_2$. From Sect. 5.1.1, we see that ψ_G is a free energy if (10.1.1) holds.

The functional ψ_G can be expressed in the form (7.1.9), where

$$\mathbb{L}(s,u) = \frac{1}{2}[\mathbb{L}(s)I_{(0,s)}(u) + \mathbb{L}(u)I_{(0,u)}(s)], \quad s>0, \quad u>0, \tag{10.1.3}$$

where $I_{\mathcal{S}}$ is the characteristic function of the set $\mathcal{S}$, so that

$$I_{(0,s)}(u) = \begin{cases} 1, & u\in(0,s), \\ 0, & u\notin(0,s). \end{cases}$$

We observe that $\mathbb{L}_{12}$ is unbounded.

Note the single-integral character of the Graffi–Volterra functional. In this context, the reference [189] is of interest. The Graffi–Volterra free energy is a special case of single-integral nonlinear one-dimensional free energies given in that work.

It is a functional of the minimal state if the material is such that the minimal states are singletons, in other words, if the minimal state is simply $(\mathbf{\Lambda}^t, \mathbf{\Lambda}(t))$. Indeed, if we have a history $\mathbf{\Lambda}_d^t \neq 0$, equivalent to the zero history, as given by (7.4.3), then (7.4.8) does not hold, because from (10.1.1), it follows that

$$\int_0^\infty \mathbb{L}'(s)\mathbf{\Lambda}_d^t(s)\cdot\mathbf{\Lambda}_d^t(s)ds > 0, \tag{10.1.4}$$

except in the trivial case that $\mathbb{L}'$ vanishes on $\mathbb{R}^+$.

Remark 10.1.2. If the minimal state is nonsingleton, the Graffi–Volterra free energy is not a functional of the minimal state.

We have the following result.

Proposition 10.1.3. *If* $\mathbb{L}_{12}(s_1,s_2)$ *is a bounded nonnegative (positive semidefinite) tensor (see discussion of* (A.2.12)*) for all* $s_1, s_2 \in \mathbb{R}^+$, *then*

$$\psi(t) \le \psi_G(t), \quad t\in\mathbb{R}, \tag{10.1.5}$$

where ψ *is any functional of the form* (7.1.19) *and* ψ_G *is the Graffi–Volterra functional* $(10.1.2)_2$.

Proof. Consider the identity

$$\begin{aligned}
&\frac{1}{2}\int_0^\infty\int_0^\infty \left[\mathbf{\Lambda}^t(s_1) - \mathbf{\Lambda}^t(s_2)\right]\cdot\mathbb{L}_{12}(s_1,s_2)\left[\mathbf{\Lambda}^t(s_1) - \mathbf{\Lambda}^t(s_2\right]ds_1ds_2 \\
&= \frac{1}{2}\int_0^\infty\int_0^\infty \mathbf{\Lambda}^t(s_1)\cdot\mathbb{L}_{12}(s_1,s_2)\mathbf{\Lambda}^t(s_1)ds_1ds_2 \\
&+ \frac{1}{2}\int_0^\infty\int_0^\infty \mathbf{\Lambda}^t(s_2)\cdot\mathbb{L}_{12}(s_1,s_2)\mathbf{\Lambda}^t(s_2)ds_1ds_2 \\
&- \int_0^\infty\int_0^\infty \mathbf{\Lambda}^t(s_1)\cdot\mathbb{L}_{12}(s_1,s_2)\mathbf{\Lambda}^t(s_2)ds_1ds_2.
\end{aligned}$$

The left-hand side is nonnegative by virtue of the assumption on $\mathbb{L}_{12}(s_1, s_2)$. The first two terms on the right yield, after integration, the integral term in $(10.1.2)_2$, and the last term is the memory term in (7.1.19), multiplied in both cases by a factor of 2. Relation (10.1.5) follows immediately. □

Remark 10.1.4. It must be emphasized that the non-negativity assumption on $\mathbb{L}_{12}(s_1, s_2)$ is not necessary in general for ψ, given by (7.1.9), to be a free energy, though $\mathbb{L}_{12}$ must be a nonnegative operator in the sense that the integral term in (7.1.9) must be nonnegative (Remark 7.1.3).

10.1.2 Dill/Staverman–Schwarzl Free Energy

Early work on the determination of the free energy of a linear viscoelastic solid by Staverman and Schwarzl [305] involved arguments based on mechanical models. These authors perceived the problem of nonuniqueness and used detailed model assumptions to deal with the issue. The resulting functional was given independently by others (for example, [36, 204]). It is often referred to as the Dill free energy [96], a terminology that will be adopted here.

The functional

$$\psi_{Dill}(t) = \phi(t) + \frac{1}{2}\int_0^\infty\int_0^\infty \mathbb{L}''(s_1+s_2)\mathbf{\Lambda}_r^t(s_2)\cdot\mathbf{\Lambda}_r^t(s_1)ds_1ds_2 \tag{10.1.6}$$

is a free energy with rate of dissipation given by

$$D_{Dill}(t) = -\int_0^\infty\int_0^\infty \mathbb{L}'''(s_1+s_2)\mathbf{\Lambda}_r^t(s_2)\cdot\mathbf{\Lambda}_r^t(s_1)ds_1ds_2 \tag{10.1.7}$$

if and only if $\mathbb{L}$ is completely monotonic [89], in other words, if it is differentiable to any order and

$$(-1)^n\mathbb{L}_{(n)}(s) \geq \mathbf{0}, \quad s\in\mathbb{R}^+, \quad n = 0, 1, 2, \ldots, \tag{10.1.8}$$

where the subscript indicates the nth derivative. Relations (10.1.6) and (10.1.7) are special cases of (7.1.9) and (7.1.26). Note that in this case, $(7.1.6)_3$ implies (7.1.18), namely the symmetry of $\mathbb{L}$.

Because of the non-negativity of $\mathbb{L}''$, we have, from Proposition 10.1.3, that

$$\psi_{Dill}(t) \leq \psi_G(t), \quad t\in\mathbb{R}.$$

The quantity ψ_{Dill} is a functional of the minimal state. Indeed, in the notation of $(7.4.2)_3$ and (7.4.5), it obeys (the superscript t can be omitted)

$$\langle\mathbf{\Lambda}_1^t, \mathbf{\Lambda}_d^t\rangle = \langle\mathbf{\Lambda}_2^t, \mathbf{\Lambda}_d^t\rangle = 0,$$

so that (7.4.7) and (7.4.9) hold.

Relation (10.1.8) will be true if and only if [88]

$$\mathbb{L}(s) = \int_0^\infty e^{-\alpha s}d\mathbb{K}(\alpha),$$

where $\mathbb{K} : [0, \infty) \mapsto \mathrm{Lin}(\Gamma^+)$ is symmetric, bounded, and nondecreasing with respect to the order relation of Sect. A.2.1. Let us assume that $\mathbb{K}$ is sectionally smooth, with a finite number of discontinuities, so that

$$\mathbb{L}(s) = \int_0^\infty \mathbb{K}'(\alpha)e^{-\alpha s}\, d\alpha + \sum_{i=0}^n \mathbb{K}_i e^{-\alpha_i s},$$

where α_i, $i = 0, 1, 2, \ldots, n$, are the points of discontinuity and

$$\mathbb{K}_i = \mathbb{K}(\alpha_i^+) - \mathbb{K}(\alpha_i^-) > 0, \quad \mathbb{K}_0 = \mathbb{K}(0), \quad \alpha_0 = 0.$$

Then,

$$\psi_{Dill}(t) = \phi(t) + \frac{1}{2}\int_0^\infty \alpha^2 \mathbf{\Lambda}^t_{rL}(\alpha) \cdot \mathbb{K}'(\alpha)\mathbf{\Lambda}^t_{rL}(\alpha)d\alpha + \frac{1}{2}\sum_{i=1}^n \alpha_i^2 \mathbf{\Lambda}^t_{rL}(\alpha_i) \cdot \mathbb{K}_i\mathbf{\Lambda}^t_{rL}(\alpha_i),$$

where $\mathbf{\Lambda}^t_{rL}$ is the Laplace transform (C.2.4)

$$\mathbf{\Lambda}^t_{rL}(\alpha) = \int_0^\infty e^{-\alpha s}\mathbf{\Lambda}^t_r(\alpha)ds,$$

while

$$D_{Dill}(t) = \int_0^\infty \alpha^3\, \mathbf{\Lambda}^t_{rL}(\alpha) \cdot \mathbb{K}'(\alpha)\mathbf{\Lambda}^t_{rL}(\alpha)d\alpha + \sum_{i=1}^n \alpha_i^3\, \mathbf{\Lambda}^t_{rL}(\alpha_i) \cdot \mathbb{K}_i\mathbf{\Lambda}^t_{rL}(\alpha_i).$$

For the restricted class of materials obeying (10.1.8), ψ_{Dill} is clearly a free energy in that it obeys all the criteria in Sect. 5.1.1. The converse, proved in [89], is more difficult. If we assume that $\mathbb{L}(\cdot)$ is infinitely differentiable and that (7.1.15) is replaced by

$$\mathbb{L}_{(n)}(\infty) = \mathbf{0}, \; n = 0, 1, \ldots,$$

then a simple demonstration can be given, based on the intuitive observations in Remark 7.1.3. It follows from (7.1.12) applied to (10.1.6) that

$$\mathbb{L}_{(n)}(2s) \geq \mathbf{0}, \; n = 0, 2, 4, \ldots, \tag{10.1.9}$$

for $s \in \mathbb{R}^+$. The factor 2 is irrelevant and can be dropped. The nonpositivity of odd derivatives can be seen by integrating (10.1.9) over the interval $[s, \infty)$. Thus, (10.1.8) holds.

Note that from $(\mathrm{C}.1.3)_1$,

$$\mathbb{L}'_+(\omega) = -\int_0^\infty \frac{\alpha\mathbb{K}'(\alpha)}{\alpha + i\omega}d\alpha - \sum_{i=1}^n \frac{\alpha_i\mathbb{K}_i}{\alpha_i + i\omega},$$

$$\mathbb{L}''_+(\omega) = \int_0^\infty \frac{\alpha^2\mathbb{K}'(\alpha)}{\alpha + i\omega}d\alpha + \sum_{i=1}^n \frac{\alpha_i^2\mathbb{K}_i}{\alpha_i + i\omega} = i\omega\mathbb{L}'_+(\omega) + \int_0^\infty \alpha\mathbb{K}'(\alpha)d\alpha + \sum_{1=1}^n \alpha_i\mathbb{K}_i.$$

Also, using the inverse Fourier transform, we have

$$\mathbb{L}''(s) = \frac{1}{2\pi}\int_{-\infty}^{\infty} \mathbb{L}''_+(\omega)e^{i\omega s}\,d\omega. \tag{10.1.10}$$

Substituting (10.1.10) for $\mathbb{L}''$ into (10.1.6), we obtain

$$\begin{aligned}\psi_{Dill}(t) &= \phi(t) + \frac{1}{4\pi}\int_{-\infty}^{\infty} \overline{\boldsymbol{\Lambda}^t_{r+}(\omega)} \cdot \mathbb{L}''_+(\omega)\overline{\boldsymbol{\Lambda}^t_{r+}}(\omega)d\omega \\ &= \phi(t) + \frac{1}{4\pi}\int_{-\infty}^{\infty} \boldsymbol{\Lambda}^t_{r+}(\omega) \cdot \overline{\mathbb{L}''_+}(\omega)\boldsymbol{\Lambda}^t_{r+}(\omega)d\omega \\ &= \phi(t) + \frac{1}{4\pi}\int_{-\infty}^{\infty} \mathrm{Re}\left\{\boldsymbol{\Lambda}^t_{r+}(\omega) \cdot \overline{\mathbb{L}''_+}(\omega)\boldsymbol{\Lambda}^t_{r+}(\omega)\right\}d\omega,\end{aligned} \tag{10.1.11}$$

where the reality of ψ_{Dill} has been invoked in writing the second and third forms.

10.1.3 Single-Integral Quadratic Functionals of $\mathbf{I}^t$

We now introduce a functional that is a free energy for materials with the property (10.1.1) and is a functional of the minimal state. We also present a family of related functionals that are free energies for more restrictive conditions on the material. These results were first reported in [91].

Consider the functional

$$\begin{aligned}\widetilde{\psi}_F(\mathbf{I}^t) &= \psi_F(t) = \phi(t) - \frac{1}{2}\int_0^{\infty} \mathbb{L}'^{-1}(\tau)\mathbf{I}^t_{(1)}(\tau) \cdot \mathbf{I}^t_{(1)}(\tau)d\tau, \\ \mathbf{I}^t_{(1)}(\tau) &= \frac{\partial}{\partial\tau}\mathbf{I}^t(\tau),\end{aligned} \tag{10.1.12}$$

where $\mathbb{L}'^{-1}(\tau)$ is the inverse of the tensor $\mathbb{L}'(\tau)$ in the algebraic sense (treating them as matrices) and $\mathbf{I}^t$ is defined by

$$\mathbf{I}^t(\tau) = \mathbf{I}^t(\tau, \boldsymbol{\Lambda}^t_r), \tag{10.1.13}$$

where the right-hand side is given by (7.4.2), with $\boldsymbol{\Lambda}^t_r$ replacing $\boldsymbol{\Lambda}^t_d$. The integral term in (10.1.12) is nonnegative by virtue of (10.1.1). The tensor $\mathbb{L}'^{-1}$ becomes singular at large τ, but it is clear from the representation (10.1.15) below that the integral exists.

The domain of definition of the functional $\widetilde{\psi}_F$ will be denoted by

$$\mathcal{H}^*_F(\mathbb{R}^+) = \left\{\mathbf{I}^t;\ \int_0^{\infty} \mathbb{L}'^{-1}(\tau)\mathbf{I}^t_{(1)}(\tau) \cdot \mathbf{I}^t_{(1)}(\tau)d\tau < \infty\right\}. \tag{10.1.14}$$

This space is very much larger than the domain of definition of the Graffi–Volterra free energy, as we see for a kernel given by an exponential or a sum of exponentials. One can choose exponentially diverging histories such that the integral in (10.1.2) will diverge, because of the quadratic dependence on $\boldsymbol{\Lambda}^t_r$, but for which $\mathbf{I}^t$ exists.

We can write ψ_F in terms of the relative history $\boldsymbol{\Lambda}^t_r$ as follows:

$$\psi_F(t) = \phi(t) + \frac{1}{2}\int_0^\infty \int_0^\infty \mathbb{L}_{12}(s_1, s_2)\mathbf{\Lambda}_r^t(s_1) \cdot \mathbf{\Lambda}_r^t(s_2)ds_1 ds_2, \tag{10.1.15}$$

where

$$\mathbb{L}(s_1, s_2) = -\int_0^\infty \mathbb{L}'(\tau + s_2)\mathbb{L}'^{-1}(\tau)\mathbb{L}'(\tau + s_1)d\tau = \mathbb{L}(s_2, s_1). \tag{10.1.16}$$

We see that $\mathbb{L}(s_1, s_2)$ obeys (7.1.7), by virtue of (7.1.15). Relation (7.1.8) holds if we assume that $\mathbb{L}''(\infty) = 0$. The important condition $(7.1.14)_3$ can be shown without difficulty. Also, the kernel in (7.1.25) is given by

$$-\frac{1}{2}\int_0^\infty \left[\mathbb{L}''(\tau + s_1)\mathbb{L}'^{-1}(\tau)\mathbb{L}'(\tau + s_2) + \mathbb{L}'(\tau + s_1)\mathbb{L}'^{-1}(\tau)\mathbb{L}''(\tau + s_2)\right] d\tau.$$

Partial integration with respect to τ gives

$$\frac{1}{2}\mathbb{L}'(s_1)\dot{\mathbb{L}}'^{-1}(0)\mathbb{L}'(s_2) + \frac{1}{2}\int_0^\infty \mathbb{L}'(\tau + s_1)\frac{d}{d\tau}\mathbb{L}'^{-1}(\tau)\mathbb{L}'(\tau + s_2)d\tau,$$

which yields a nonnegative dissipation, since under our assumptions,

$$\frac{d}{d\tau}\mathbb{L}'^{-1}(\tau) = -\mathbb{L}'^{-1}(\tau)\mathbb{L}''(\tau)\mathbb{L}'^{-1}(\tau) \tag{10.1.17}$$

is a nonpositive tensor.

In fact, a more direct demonstration of the fact that ψ_F is a free energy can be given. From (10.1.13), we have that

$$\frac{d}{dt}\mathbf{I}^t(\tau) = \dot{\mathbf{I}}^t(\tau) = \mathbb{L}(\tau)\dot{\mathbf{\Lambda}}(t) + \mathbf{I}^t_{(1)}(\tau),$$

giving

$$\frac{d}{dt}\mathbf{I}^t_{(1)}(\tau) = \ddot{\mathbf{I}}^t(\tau) = \mathbb{L}'(\tau)\dot{\mathbf{\Lambda}}(t) + \mathbf{I}^t_{(2)}(\tau), \quad \mathbf{I}^t_{(2)}(\tau) = \frac{d^2}{d\tau^2}\mathbf{I}^t(\tau), \tag{10.1.18}$$

so that

$$\begin{aligned}\frac{d}{dt}\psi_F(t) &= \mathbf{\Sigma}(t) \cdot \dot{\mathbf{\Lambda}}(t) - \int_0^\infty \mathbb{L}'^{-1}(\tau)\mathbf{I}^t_{(2)}(\tau) \cdot \mathbf{I}^t_{(1)}(\tau)d\tau \\ &= \mathbf{\Sigma}(t) \cdot \dot{\mathbf{\Lambda}}(t) + \frac{1}{2}\mathbb{L}'^{-1}(0)\mathbf{I}^t_{(1)}(0) \cdot \mathbf{I}^t_{(1)}(0) \\ &\quad + \frac{1}{2}\int_0^\infty \left[\frac{d}{d\tau}\mathbb{L}'^{-1}(\tau)\right]\mathbf{I}^t_{(1)}(\tau) \cdot \mathbf{I}^t_{(1)}(\tau)d\tau,\end{aligned} \tag{10.1.19}$$

using $(7.1.14)_2$ and the fact that $\mathbf{I}^t(0)$ is the integral term in $(7.1.13)_1$. Thus, we have, from (5.1.11),

$$D_F(t) = -\frac{1}{2}\mathbb{L}'^{-1}(0)\mathbf{I}^t_{(1)}(0) \cdot \mathbf{I}^t_{(1)}(0) - \frac{1}{2}\int_0^\infty \left[\frac{d}{d\tau}\mathbb{L}'^{-1}(\tau)\right]\mathbf{I}^t_{(1)}(\tau) \cdot \mathbf{I}^t_{(1)}(\tau)d\tau \geq 0.$$

Note that

$$D_F(t) \geq -\frac{1}{2}\int_0^\infty \left[\frac{d}{d\tau}\mathbb{L}'^{-1}(\tau)\right]\mathbf{I}^t_{(1)}(\tau)\cdot\mathbf{I}^t_{(1)}(\tau)d\tau \geq 0. \tag{10.1.20}$$

Let us assume further that there exists a nonnegative α_1 such that

$$\mathbb{L}''(\tau) + \alpha_1\mathbb{L}'(\tau) \geq \mathbf{0} \quad \forall\tau \in \mathbb{R}^+.$$

This yields

$$\frac{d}{d\tau}\mathbb{L}'^{-1}(\tau) \leq \alpha_1\mathbb{L}'^{-1}(\tau) \leq \mathbf{0},$$

using (10.1.17) and from (10.1.20) with (10.1.12),

$$D_F(t) \geq \alpha_1[\psi_F(t) - \phi(t)]. \tag{10.1.21}$$

One can write down a family of free energies [91] using a simple generalization of the above procedure [91]. These may be described as multiple-integral quadratic forms. Consider, for a given integer $n \geq 1$,

$$\begin{aligned}
\widetilde{\psi}_n(\mathbf{I}^t) &= \psi_n(t) = \phi(t) + \frac{(-1)^n}{2}\int_n \mathbb{L}^{-1}_{(n)}(\tau)\mathbf{I}^t_{(n)}(\tau_n)\cdot\mathbf{I}^t_{(n)}(\tau_n),\\
\mathbb{L}_{(n)}(\tau) &:= \frac{d^n}{d\tau^n}\mathbb{L}(\tau), \qquad \mathbf{I}^t_{(n)}(\tau) = \frac{d^n}{d\tau^n}\mathbf{I}^t(\tau),\\
\int_n &:= \int_0^\infty d\tau_1\int_{\tau_1}^\infty d\tau_2\int_{\tau_2}^\infty d\tau_3\ldots\int_{\tau_{n-1}}^\infty d\tau_n.
\end{aligned} \tag{10.1.22}$$

It is assumed that for all $\tau \in \mathbb{R}^+$,

$$(-1)^{n+1}\mathbb{L}_{(n+1)}(\tau) \geq \mathbf{0}. \tag{10.1.23}$$

It follows that

$$(-1)^m\mathbb{L}_{(m)}(\tau) \geq \mathbf{0}, \tag{10.1.24}$$

where m is any integer in the interval $0 < m \leq n$. Using a generalization of (10.1.18),

$$\frac{d}{dt}\mathbf{I}^t_{(n)}(\tau) = \mathbb{L}_{(n)}(\tau)\dot{\mathbf{\Lambda}}(t) + \mathbf{I}^t_{(n+1)}(\tau),$$

one can show that

$$\begin{aligned}
\frac{d}{dt}\psi_n(t) = \mathbf{\Sigma}(t)\cdot\dot{\mathbf{\Lambda}}(t) &+ \frac{(-1)^{n-1}}{2}\int_{n-1}\mathbb{L}^{-1}_{(n)}(\tau_{n-1})\mathbf{I}^t_{(n)}(\tau_{n-1})\cdot\mathbf{I}^t_{(n)}(\tau_{n-1})\\
&- \frac{(-1)^n}{2}\int_n\left[\frac{d}{d\tau}\mathbb{L}^{-1}_{(n)}(\tau_n)\right]\mathbf{I}^t_{(n)}(\tau_n)\cdot\mathbf{I}^t_{(n)}(\tau_n).
\end{aligned}$$

For $n = 1$, the middle term on the right is understood to yield the middle term on the right of $(10.1.19)_2$. We therefore obtain

$$D_F(t) = \frac{(-1)^n}{2}\int_{n-1} \mathbb{L}^{-1}_{(n)}(\tau_{n-1})\mathbf{I}^t_{(n)}(\tau_{n-1}) \cdot \mathbf{I}^t_{(n)}(\tau_{n-1})$$
$$+ \frac{(-1)^n}{2}\int_{n}\left[\frac{d}{d\tau_n}\mathbb{L}^{-1}_{(n)}(\tau_n)\right]\mathbf{I}^t_{(n)}(\tau_n) \cdot \mathbf{I}^t_{(n)}(\tau_n) \geq 0, \tag{10.1.25}$$

which is nonnegative since

$$(-1)^n\frac{d}{d\tau}\mathbb{L}^{-1}_{(n)}(\tau) = (-1)^{n+1}\mathbb{L}^{-1}_{(n)}(\tau)\mathbb{L}_{(n+1)}(\tau)\mathbb{L}^{-1}_{(n)}(\tau) \geq \mathbf{0},$$

by virtue of (10.1.23). Since the first term on the right of (10.1.25) is positive, we have

$$D_F(t) \geq \frac{(-1)^n}{2}\int_{n}\left[\frac{d}{d\tau_n}\mathbb{L}^{-1}_{(n)}(\tau_n)\right]\mathbf{I}^t_{(n)}(\tau_n) \cdot \mathbf{I}^t_{(n)}(\tau_n) \geq 0.$$

Again, if we assume that an $\alpha_n > 0$ exists such that

$$(-1)^{n+1}[\mathbb{L}_{(n+1)}(\tau) + \alpha_n\mathbb{L}_{(n)}(\tau)] \geq \mathbf{0} \quad \forall\tau \in \mathbb{R}^+,$$

then

$$(-1)^n\frac{d}{d\tau}\mathbb{L}^{-1}_{(n)}(\tau) \geq (-1)^n\alpha_n\mathbb{L}^{-1}_{(n)}(\tau)$$

and (10.1.21) holds, with ψ_n and α_n replacing ψ_F and α_1.

We deduce from (10.1.24) that each ψ_m, $0 < m \leq n$, is also a free energy if (10.1.23) holds.

Note that if $\mathbb{L}$ is completely monotonic, in other words, if (10.1.24) holds for all integers $m \geq 0$ (or (10.1.8)) [89], then there is an infinite sequence of free energies given by (10.1.22).

There is a free energy corresponding to $n = 0$, which does not completely fit into the above scheme. Consider

$$\widetilde{\psi}_0(\mathbf{I}^t) = \psi_0(t) = \phi(t) + \frac{1}{2}\mathbb{L}^{-1}(0)\mathbf{I}^t(0) \cdot \mathbf{I}^t(0), \tag{10.1.26}$$

where the second term is positive by virtue of the fact that $\mathbb{L}^{-1}(0) > \mathbf{0}$. By similar manipulations to those in (10.1.15) and (10.1.16), we obtain

$$\mathbb{L}(s_1, s_2) = \mathbb{L}(s_1)\mathbb{L}^{-1}(0)\mathbb{L}(s_2),$$

which can be seen to obey $(7.1.6)_2$, (7.1.7), and (7.1.8) with the aid of (7.1.15). It also obeys $(7.1.14)_3$. The kernel in (7.1.25) is

$$\frac{1}{2}[\mathbb{L}'(s_1)\mathbb{L}^{-1}(0)\mathbb{L}(s_2) + \mathbb{L}(s_1)\mathbb{L}^{-1}(0)\mathbb{L}'(s_2)].$$

Let us assume that

$$\mathbb{L}'(\tau) \leq \mathbf{0} \quad \forall\tau \in \mathbb{R}^+,$$

which implies that $\mathbb{L}(\tau) \geq \mathbf{0}$, $\tau \in \mathbb{R}^+$, and further assume that there exists a nonnegative $\alpha_0 \in \mathbb{R}^{++}$ such that

$$\mathbb{L}'(\tau) + \alpha_0 \mathbb{L}(\tau) = \mathbf{0} \quad \forall \tau \in \mathbb{R}^+,$$

so that $\mathbb{L}$ consists of a single exponential term. Then, instead of (10.1.21), we have

$$D_0(t) = 2\alpha_0 \left[\psi_0(t) - \phi(t)\right].$$

Note that we can write ψ_0 as

$$\psi_0(t) = \phi(t) + \frac{1}{2}\mathbb{L}^{-1}(0)[\boldsymbol{\Sigma}(t) - \boldsymbol{\Sigma}_e(t)] \cdot [\boldsymbol{\Sigma}(t) - \boldsymbol{\Sigma}_e(t)].$$

This functional was first considered in [43] for viscoelastic materials; it depends on the minimal state as shown in [89].

10.2 Free Energies for Restricted Classes of Solids

We now consider functionals introduced in the last two sections for the special cases of viscoelastic solids and fluids and also for rigid heat conductors. Indeed, these free energies were originally introduced for viscoelastic solids, which we now discuss.

Consider the functional

$$\widetilde{\psi}_G(\mathbf{E}^t) = \frac{1}{2}\mathbf{E}(t) \cdot \mathbb{G}_\infty \mathbf{E}(t) - \frac{1}{2}\int_0^\infty \mathbf{E}_r^t(\xi) \cdot \mathbb{G}'(\xi)\mathbf{E}_r^t(\xi)d\xi, \tag{10.2.1}$$

where instead of (10.1.1), we have the restrictions on the stress relaxation tensor

$$\mathbb{G}'(s) \leq \mathbf{0}, \qquad \mathbb{G}''(s) \geq \mathbf{0} \quad \forall s \in \mathbb{R}^+. \tag{10.2.2}$$

This is a generalization of a free energy functional determined by Graffi [175] and is known as the *Graffi–Volterra free energy* for a viscoelastic solid.

The internal dissipation function $D_G(t)$ related to this free energy can be evaluated by differentiating (10.2.1) with respect to time. Using the expression (8.1.3) for $\mathbf{E}_r^t(s)$, we obtain

$$\dot{\psi}_G(t) = \dot{\mathbf{E}}(t) \cdot \left[\mathbb{G}_0 \mathbf{E}(t) + \int_0^\infty \mathbb{G}'(s)\mathbf{E}^t(s)ds\right] - \int_0^\infty \dot{\mathbf{E}}^t(s) \cdot \mathbb{G}'(s)\mathbf{E}_r^t(s)ds. \tag{10.2.3}$$

Integration by parts gives

$$\int_0^\infty \dot{\mathbf{E}}^t(s) \cdot \mathbb{G}'(s)\mathbf{E}_r^t(s)ds = \frac{1}{2}\int_0^\infty \mathbf{E}_r^t(s) \cdot \mathbb{G}''(s)\mathbf{E}_r^t(s)ds.$$

Substituting into (10.2.3) and taking into account (8.1.5), we obtain

$$\dot{\psi}_G(t) = \mathbf{T}(t) \cdot \dot{\mathbf{E}}(t) - \frac{1}{2}\int_0^\infty \mathbf{E}_r^t(s) \cdot \mathbb{G}''(s)\mathbf{E}_r^t(s)ds,$$

which satisfies (8.6.16) with

$$D_G(t) = \frac{1}{2}\int_0^\infty \mathbf{E}_r^t(s) \cdot \mathbb{G}''(s)\mathbf{E}_r^t(s)ds \geq 0,$$

provided $\mathbb{G}''(s) \geq \mathbf{0}\,\forall s \in \mathbb{R}^+$.

Another example is the *Dill free energy* [96], also referred to as the *Staverman–Schwarzl free energy* [305] given by

$$\widetilde{\psi}_{Dill}(\mathbf{E}^t) = \frac{1}{2}\mathbf{E}(t) \cdot \mathbb{G}_\infty\mathbf{E}(t) + \frac{1}{2}\int_0^\infty\int_0^\infty \mathbf{E}_r^t(\xi_1) \cdot \mathbb{G}''(\xi_1+\xi_2)\mathbf{E}_r^t(\xi_2)d\xi_1 d\xi_2. \quad (10.2.4)$$

Its time derivative has the form

$$\begin{aligned}\dot{\psi}_{Dill}(t) &= \dot{\mathbf{E}}(t) \cdot \mathbb{G}_\infty\mathbf{E}(t)\\ &+ \frac{1}{2}\int_0^\infty\int_0^\infty \left[\dot{\mathbf{E}}_r^t(\xi_1) \cdot \mathbb{G}''(\xi_1+\xi_2)\mathbf{E}_r^t(\xi_2) + \mathbf{E}_r^t(\xi_1) \cdot \mathbb{G}''(\xi_1+\xi_2)\dot{\mathbf{E}}_r^t(\xi_2)\right] d\xi_1 d\xi_2.\end{aligned} \quad (10.2.5)$$

Hence, taking into account the expression for $\mathbf{E}_r^t$ as before and using (8.1.2), we obtain, after partial integrations,

$$\begin{aligned}\dot{\psi}_{Dill}(t) &= \left[\mathbb{G}_\infty\mathbf{E}(t) + \int_0^\infty \mathbb{G}'(s)\mathbf{E}_r^t(s)ds\right] \cdot \dot{\mathbf{E}}(t)\\ &\quad + \int_0^\infty\int_0^\infty \dot{\mathbf{E}}^t(\xi_1) \cdot \mathbb{G}'(\xi_1+\xi_2)\dot{\mathbf{E}}^t(\xi_2)d\xi_1 d\xi_2 \qquad (10.2.6)\\ &= \mathbf{T}(t) \cdot \dot{\mathbf{E}}(t) + \int_0^\infty\int_0^\infty \dot{\mathbf{E}}^t(\xi_1) \cdot \mathbb{G}'(\xi_1+\xi_2)\dot{\mathbf{E}}^t(\xi_2)d\xi_1 d\xi_2.\end{aligned}$$

Thus, (8.6.16) is satisfied with

$$D_{Dill}(t) = -\int_0^\infty\int_0^\infty \dot{\mathbf{E}}^t(\xi_1) \cdot \mathbb{G}'(\xi_1+\xi_2)\dot{\mathbf{E}}^t(\xi_2)d\xi_1 d\xi_2 \geq 0.$$

This quantity is nonnegative if and only if $\mathbb{G}$ is monotonic as defined in a more general context by (10.1.8). Note that a frequency-domain representation of ψ_{Dill} is given by (10.1.11) in the general case.

Finally, let us consider free energies that are quadratic functions of $\mathbf{I}^t$, given by (8.2.2) (see comment after (8.2.6)), at least for the categories of material obeying (10.2.2). The motivation for this is that such quantities will be functionals of the minimal state, by virtue of the discussion in Sect. 8.3. These results were first reported in [91].

The developments presented here are a special case of those in Sect. 10.1.3. Differences in the formulas occur because of the fact that $\mathbb{L}(\infty)$ vanishes by $(7.1.15)_1$, while $\mathbb{G}(\infty)$ does not. Also, the general equilibrium free energy $\phi(t)$ is replaced by its linear viscoelastic form. We define

$$\psi_F(t) = \frac{1}{2}\mathbf{E}(t) \cdot \mathbb{G}_\infty\mathbf{E}(t) - \frac{1}{2}\int_0^\infty [\mathbb{G}'(\tau)]^{-1}\mathbf{I}_{(1)}^t(\tau) \cdot \mathbf{I}_{(1)}^t(\tau)d\tau, \quad (10.2.7)$$

where $[\mathbb{G}'(\tau)]^{-1}$ is the inverse tensor of $\mathbb{G}'(\tau)$ in the algebraic sense and $\mathbf{I}^t_{(1)}(\tau)$ is the derivative with respect to τ of $\mathbf{I}^t(\tau)$. The assumption $(10.2.2)_1$ yields a nonnegative term for the integral in (10.2.7). The domain of definition is given by (10.1.14) but with $\mathbb{G}$ replacing $\mathbb{L}$.

This functional can be written in terms of the relative history $\mathbf{E}^t_r$, using (8.2.6), as follows:

$$\begin{aligned}
\psi_F(t) &= \frac{1}{2}\mathbf{E}(t)\cdot\mathbb{G}_\infty\mathbf{E}(t)\\
&\quad -\frac{1}{2}\int_0^\infty\int_0^\infty\int_0^\infty \mathbb{G}''(\tau+s_2)[\mathbb{G}'(\tau)]^{-1}\mathbb{G}''(\tau+s_1)\mathbf{E}^t_r(s_1)\cdot\mathbf{E}^t_r(s_2)ds_1ds_2d\tau\\
&= \frac{1}{2}\mathbf{E}(t)\cdot\mathbb{G}_\infty\mathbf{E}(t)+\frac{1}{2}\int_0^\infty\int_0^\infty \mathbb{G}_{12}(s_1,s_2)\mathbf{E}^t_r(s_1)\cdot\mathbf{E}^t_r(s_2)ds_1ds_2,
\end{aligned}\tag{10.2.8}$$

where we have put

$$\mathbb{G}(s_1,s_2)=-\int_0^\infty \mathbb{G}'(\tau+s_2)[\mathbb{G}'(\tau)]^{-1}\mathbb{G}'(\tau+s_1)d\tau+\mathbb{G}_\infty=\mathbb{G}(s_2,s_1)\tag{10.2.9}$$

and $\mathbb{G}_{12}$ is given by (8.6.2). Relation (8.6.9) has been used. From this expression for ψ_F, we see that the integral exists, even though $[\mathbb{G}']^{-1}$ becomes singular at large $\tau\in\mathbb{R}^+$. Note that $\mathbb{G}(s_1,s_2)$ satisfies (8.6.11), by virtue of (8.1.27).

One can show that a nonnegative dissipation rate is associated with this functional. The expression (8.6.19) with an integration by parts gives

$$\begin{aligned}
&\mathbb{K}(s_1,s_2)\\
&= -\int_0^\infty [\mathbb{G}''(\tau+s_1)[\mathbb{G}'(\tau)]^{-1}\mathbb{G}'(\tau+s_2)+\mathbb{G}'(\tau+s_1)[\mathbb{G}'(\tau)]^{-1}\mathbb{G}''(\tau+s_2)]d\tau\\
&= \mathbb{G}'(s_1)[\mathbb{G}'(0)]^{-1}\mathbb{G}'(s_2)+\int_0^\infty \mathbb{G}'(\tau+s_1)\frac{d}{d\tau}[\mathbb{G}'(\tau)]^{-1}\mathbb{G}'(\tau+s_2)d\tau,
\end{aligned}$$

so that

$$\mathbb{K}_{12}(s_1,s_2)=\mathbb{G}''(s_1)[\mathbb{G}'(0)]^{-1}\mathbb{G}''(s_2)+\int_0^\infty \mathbb{G}''(\tau+s_1)\frac{d}{d\tau}[\mathbb{G}'(\tau)]^{-1}\mathbb{G}''(\tau+s_2)d\tau,$$

which is a negative semidefinite tensor, because of the assumption on $\mathbb{G}''(\tau)$ and the relationship

$$\frac{d}{d\tau}[\mathbb{G}'(\tau)]^{-1}=-[\mathbb{G}'(\tau)]^{-1}\mathbb{G}''(\tau)[\mathbb{G}'(\tau)]^{-1}.$$

A more direct demonstration can also be given, by evaluating

$$\dot{\psi}_F(t)=\dot{\mathbf{E}}(t)\cdot\mathbb{G}_\infty\mathbf{E}(t)-\int_0^\infty [\mathbb{G}'(\tau)]^{-1}\dot{\mathbf{I}}^t_{(1)}(\tau)\cdot\mathbf{I}^t_{(1)}(\tau)d\tau,\tag{10.2.10}$$

where, by virtue of (8.2.5) and (8.2.4), we have

$$\dot{\mathbf{I}}^t_{(1)}(\tau)=\mathbb{G}'(\tau)\dot{\mathbf{E}}(t)+\mathbf{I}^t_{(2)}(\tau),$$

where

$$\mathbf{I}^t_{(2)}(\tau) = \frac{d^2}{d\tau^2}\mathbf{I}^t(\tau).$$

Substituting and integrating by parts, we have

$$\begin{aligned}\dot{\psi}_F(t) &= \left[\mathbb{G}_\infty \mathbf{E}(t) + \mathbf{I}^t(0)\right] \cdot \dot{\mathbf{E}}(t) - \int_0^\infty [\mathbb{G}'(\tau)]^{-1}\mathbf{I}^t_{(2)}(\tau) \cdot \mathbf{I}^t_{(1)}(\tau)d\tau \\ &= \mathbf{T}(t) \cdot \dot{\mathbf{E}}(t) + \frac{1}{2}[\mathbb{G}'(0)]^{-1}\mathbf{I}^t_{(1)}(0) \cdot \mathbf{I}^t_{(1)}(0) \\ &\quad + \frac{1}{2}\int_0^\infty \frac{d}{d\tau}[\mathbb{G}'(\tau)]^{-1}\mathbf{I}^t_{(1)}(\tau) \cdot \mathbf{I}^t_{(1)}(\tau)d\tau.\end{aligned}$$

Hence, we obtain

$$\begin{aligned}D_F(t) &= -\frac{1}{2}[\mathbb{G}'(0)]^{-1}\mathbf{I}^t_{(1)}(0) \cdot \mathbf{I}^t_{(1)}(0) \\ &\quad - \frac{1}{2}\int_0^\infty \frac{d}{d\tau}[\mathbb{G}'(\tau)]^{-1}\mathbf{I}^t_{(1)}(\tau) \cdot \mathbf{I}^t_{(1)}(\tau)d\tau \geq 0.\end{aligned} \tag{10.2.11}$$

Note that

$$D_F(t) \geq -\frac{1}{2}\int_0^\infty \frac{d}{d\tau}[\mathbb{G}'(\tau)]^{-1}\mathbf{I}^t_{(1)}(\tau) \cdot \mathbf{I}^t_{(1)}(\tau)d\tau. \tag{10.2.12}$$

Let us further assume that there is a constant $\alpha_1 \in \mathbb{R}^+$ such that

$$\mathbb{G}''(\tau) + \alpha_1 \mathbb{G}'(\tau) \geq \mathbf{0} \qquad \forall \tau \in \mathbb{R}^+. \tag{10.2.13}$$

This yields

$$\frac{d}{d\tau}[\mathbb{G}'(\tau)]^{-1} \leq \alpha_1 [\mathbb{G}'(\tau)]^{-1}(\tau) \leq \mathbf{0},$$

and therefore, on using (10.2.7), we see that (10.2.12) assumes the form

$$D_F(t) \geq \alpha_1 \left[\psi_F(t) - \frac{1}{2}\mathbb{G}_\infty \mathbf{E}(t) \cdot \mathbf{E}(t)\right]. \tag{10.2.14}$$

The generalization (10.1.22) applies in this case with notational changes [91].

Consider the functional (see (10.1.26))

$$\widetilde{\psi}_0(\mathbf{I}^t) = \frac{1}{2}\mathbf{E}(t) \cdot \mathbb{G}_\infty \mathbf{E}(t) + \frac{1}{2}[\breve{\mathbb{G}}(0)]^{-1}\mathbf{I}^t(0) \cdot \mathbf{I}^t(0),$$

where $\breve{\mathbb{G}}$ is defined in (8.2.4). The last term is positive, since $[\breve{\mathbb{G}}(0)]^{-1} > \mathbf{0}$, on account of (8.1.22). By similar manipulations to those in (10.2.8) and (10.2.9), we obtain

$$\mathbb{G}(s_1, s_2) = \breve{\mathbb{G}}(s_1)[\breve{\mathbb{G}}(0)]^{-1}\breve{\mathbb{G}}(s_2) + \mathbb{G}_\infty,$$

which satisfies (8.6.7). Also,

$$\mathbb{K}(s_1, s_2) = [\mathbb{G}'(s_1)]^{-1}[\breve{\mathbb{G}}(0)]^{-1}\breve{\mathbb{G}}(s_2) + [\breve{\mathbb{G}}(s_1]]^{-1}\breve{\mathbb{G}}^{-1}(0)\mathbb{G}'(s_2).$$

Let us assume

$$\mathbb{G}'(\tau) \leq \mathbf{0} \qquad \forall \tau \in \mathbb{R}^+,$$

whence it follows that $\breve{\mathbb{G}}(\tau) \geq \mathbf{0}\, \forall \tau \in \mathbb{R}^+$; moreover, if we suppose that $\alpha_0 \in \mathbb{R}^{++}$ exists and is such that

$$\mathbb{G}'(\tau) + \alpha_0 \breve{\mathbb{G}}(\tau) = \mathbf{0} \qquad \forall \tau \in \mathbb{R}^+,$$

it follows that $\breve{\mathbb{G}}$ is proportional to a single decaying exponential.

Thus, we obtain (10.2.14) as an equality, where $2\alpha_0$ replaces α_1, which yields a nonnegative dissipation.

Observe that ψ_0 can be written as

$$\psi_0(t) = \frac{1}{2}\mathbf{E}(t) \cdot \mathbb{G}_\infty \mathbf{E}(t) + \frac{1}{2}\breve{\mathbb{G}}^{-1}(0)[\mathbf{T}(t) - \mathbb{G}_\infty \mathbf{E}(t)] \cdot [\mathbf{T}(t) - \mathbb{G}_\infty \mathbf{E}(t)].$$

This functional was first considered in [43]; it depends on the minimal state as shown in [89], where it is also proved that it is a free energy if and only if $\mathbb{G}$ is proportional to an exponential and $\mathbb{G}'(0) \leq \mathbf{0}$. It can be written in the form

$$\psi_0(t)(\mathbf{E}^t) = \frac{1}{2}\mathbf{E}(t) \cdot \mathbb{G}_\infty \mathbf{E}(t) + \frac{1}{2}\left[(\mathbb{G}_0 - \mathbb{G}_\infty)^{-1/2} \int_0^\infty \mathbb{G}'(s)\mathbf{E}_r^t(s) ds\right]^2, \qquad (10.2.15)$$

where

$$\mathbb{G}(s) = \mathbb{G}_\infty + \mathbb{G}_1 \exp(-\gamma s), \qquad \gamma > 0,$$

with

$$\mathbb{G}_1 = \mathbb{G}_0 - \mathbb{G}_\infty > 0,$$

by (8.1.22) and

$$\mathbb{G}'(s) = -\gamma \mathbb{G}_1 \exp(-\gamma s) = \mathbb{G}'(0)\exp(-\gamma s).$$

This is the *Day free energy* [121, Chapter 3]. It arises most naturally as a special case of the minimum and other related free energies discussed in Sects. 11.9, 16.9.1, and 17.8. For materials characterized by a single decaying exponential, it is equal to both the minimum and maximum free energies dependent on the minimal state, the latter quantity as defined in Sect. 16.5.

One can show that the corresponding rate of dissipation has the form

$$D_0(t) = \gamma \left|(\mathbb{G}_0 - \mathbb{G}_\infty)^{-1/2} \int_0^\infty \mathbb{G}'(s)\mathbf{E}_r^t(s)\right|^2 \geq 0.$$

Remark 10.2.1. For materials with a relaxation function given by a sum of strictly decaying exponentials, as in (8.4.1), one can consider a general (algebraic) quadratic form, using the quantities $\mathbf{E}_k(t)$, $k = 1, 2, \ldots, n$, given by (8.4.2), and explore what conditions it must obey in order to be a free energy. This is done in the scalar case in Sect. 16.3.1. Such forms are functionals of the minimal state.

10.3 Free Energies for Restricted Classes of Fluids

The free energies introduced for viscoelastic solids in Sect. 10.2 and for the general case in Sect. 10.1 can be considered also for viscoelastic fluids with a few changes. We shall focus here on the simplest case, that of incompressible fluids, discussed in Sect. 8.10.

Let us firstly consider the Graffi–Volterra functional [174, 175], corresponding to (10.2.1), which, for incompressible fluids discussed in Sect. 8.10, assumes the following form:

$$\psi_G(t) = -\int_0^{\infty} \mu'(s)\mathbf{E}_r^t(s) \cdot \mathbf{E}_r^t(s)ds = -\int_0^{\infty} \mu'(s)|\mathbf{E}_r^t(s)|^2 ds. \tag{10.3.1}$$

Under the hypothesis that

$$\mu''(s) \geq 0, \quad s \in \mathbb{R}^+, \tag{10.3.2}$$

yielding that

$$\mu'(s) \leq 0, \quad s \in \mathbb{R}^+, \tag{10.3.3}$$

this functional is a free energy. These restrictions were assumed by Slemrod in [300] to study the stability question for such fluids (see [215, 216, 301]). Note that $\mu'(s)$ will vanish only if $\mu''(u)$ is zero for all $u \geq s$.

There is no equilibrium term in (10.3.1), since the first term in (8.10.1) does no work as a result of the incompressibility property.

One can show that the functional (10.3.1) satisfies (8.6.16), that is, $\dot{\psi}+D_G = \mathbf{T}\cdot\dot{\mathbf{E}}$, where D_G is the dissipation rate given by

$$D_G(t) = \int_0^{\infty} \mu''(s)\mathbf{E}_r^t(s) \cdot \mathbf{E}_r^t(s)ds \geq 0.$$

The final inequality follows from the hypothesis (10.3.2).

Consider the Dill functional in the form (cf. (10.2.4))

$$\psi_{Dill}(t) = \int_0^{\infty}\int_0^{\infty} \mu''(\xi_1+\xi_2)\mathbf{E}_r^t(\xi_1) \cdot \mathbf{E}_r^t(\xi_2)d\xi_1 d\xi_2.$$

Taking the time derivative of this quantity yields, after manipulations analogous to those in (10.2.5) and (10.2.6),

$$\dot{\psi}_{Dill}(t) = \mathbf{T}(t) \cdot \dot{\mathbf{E}}(t) + 2\int_0^{\infty}\int_0^{\infty} \mu'(\xi_1+\xi_2)\dot{\mathbf{E}}^t(\xi_1) \cdot \dot{\mathbf{E}}^t(\xi_2)d\xi_1 d\xi_2.$$

Thus, (8.6.16) is satisfied if we put

$$D_{Dill}(t) = -2\int_0^{\infty}\int_0^{\infty} \mu'(\xi_1+\xi_2)\dot{\mathbf{E}}^t(\xi_1) \cdot \dot{\mathbf{E}}^t(\xi_2)d\xi_1 d\xi_2.$$

The quantity ψ_{Dill} is a free energy (which implies that $D_{Dill}(t)$ is nonnegative) if and only if μ is monotonic (see (10.1.8)).

Another functional is that considered from (10.2.7) onward [91], denoted by ψ_F,

$$\psi_F(\mathbf{I}^t) = -\frac{1}{4}\int_0^\infty \frac{1}{\mu'(\tau)}\mathbf{I}^t_{(1)}(\tau)\cdot\mathbf{I}^t_{(1)}(\tau)d\tau, \tag{10.3.4}$$

where we have denoted by $\mathbf{I}^t_{(1)}(\tau)$ the derivative of $\mathbf{I}^t(\tau, \mathbf{E}^t_r)$, given by (8.10.6), with respect to τ, that is,

$$\mathbf{I}^t_{(1)}(\tau) = \frac{d}{d\tau}\mathbf{I}^t(\tau, \mathbf{E}^t_r) = 2\int_0^\infty \mu''(\xi+\tau)\mathbf{E}^t_r(\xi)d\xi. \tag{10.3.5}$$

The hypothesis (10.3.2) yielding (10.3.3) is assumed here also.

The domain of such a functional is (see (10.1.14))

$$H'_F(\mathbb{R}^+) = \left\{\mathbf{I}^t;\ \left|\int_0^\infty \frac{1}{\mu'(\tau)}\mathbf{I}^t_{(1)}(\tau)\cdot\mathbf{I}^t_{(1)}(\tau)d\tau\right| < \infty\right\},$$

which yields a much larger space than that considered for ψ_G.

Using the same steps as those outlined from (10.2.10) onward, we obtain

$$\dot{\psi}_F(t) + D_F(t) = \mathbf{T}(t)\cdot\dot{\mathbf{E}}(t),$$

where

$$D_F(t) = -\frac{1}{4}\frac{1}{\mu'(0)}\mathbf{I}^t_{(1)}(0)\cdot\mathbf{I}^t_{(1)}(0) - \frac{1}{4}\int_0^\infty \frac{d}{d\tau}\left[\frac{1}{\mu'(\tau)}\right]\mathbf{I}^t_{(1)}(\tau)\cdot\mathbf{I}^t_{(1)}(\tau)d\tau \geq 0,$$

because of the properties (10.3.2) and (10.3.3) of μ. We have the following inequality:

$$D_F(t) \geq -\frac{1}{4}\int_0^\infty \frac{d}{d\tau}\left[\frac{1}{\mu'(\tau)}\right]\mathbf{I}^t_{(1)}(\tau)\cdot\mathbf{I}^t_{(1)}(\tau)d\tau \geq 0. \tag{10.3.6}$$

Consider the particular case that there exists $\alpha \in \mathbb{R}^{++}$ such that

$$\mu''(\tau) + \alpha\mu'(\tau) \geq 0 \qquad \forall\tau \in \mathbb{R}^+,$$

which is the scalar version of (10.2.13). It follows that

$$-\mu''(\tau) \leq \alpha\mu'(\tau), \qquad -\frac{d}{d\tau}\left[\frac{1}{\mu'(\tau)}\right] \geq -\frac{\alpha}{\mu'(\tau)}.$$

Therefore, from (10.3.6) and (10.3.4), we have (cf. (10.2.14))

$$D_F(t) \geq \alpha\psi_F(t).$$

The other free energy functionals described for the general case in Sect. 10.1 and for solids in Sect. 10.2 are easily defined also for fluids.

10.4 Free Pseudoenergies for Restricted Classes of Rigid Heat Conductors

Several functionals are given in Sects. 10.2 and 10.3 that are free energies for restricted classes of kernels. Similar results apply to rigid heat conductors. In particular, the contribution of the temperature gradient to the free energy is given by an expression identical to each of those considered in Sect. 10.3 if we replace μ' by $k'/2$ and $\mathbf{E}_r^t$ by $\overline{\mathbf{g}}^t$. This contribution will not be discussed further in the present context. Thus, we consider only the contributions due to the internal energy e.

The kernel of $(9.3.3)_1$ is assumed to be such that

$$\alpha''(s) \leq 0 \quad \forall s \in \mathbb{R}^+, \tag{10.4.1}$$

so that

$$\alpha'(s) > 0, \quad \forall s \in \mathbb{R}^+, \tag{10.4.2}$$

if we exclude the case that $\alpha''(u)$ vanishes for all $u > s$.

The Graffi–Volterra free energy corresponding to the general quadratic form (9.3.18) (see (10.1.3)) is given by

$$\psi_G^{(e)}(t) = \frac{1}{2}\alpha_0\vartheta^2(t) + \frac{1}{2}\int_0^\infty \alpha'(s)[\vartheta^t(s)]^2 ds.$$

The time derivative of this quantity has the form

$$\begin{aligned}\dot{\psi}_G^{(e)}(t) &= \alpha_0\dot{\vartheta}(t)\vartheta(t) + \int_0^\infty \alpha'(s)\dot{\vartheta}^t(s)\vartheta^t(s)ds \\ &= \dot{e}(t)\vartheta(t) + \frac{1}{2}\int_0^\infty \alpha''(s)[\vartheta_r^t(s)]^2 ds,\end{aligned}$$

so that the rate of dissipation is given by

$$D_G^{(e)}(t) = -\frac{1}{2}\int_0^\infty \alpha''(s)[\vartheta_r^t(s)]^2 ds.$$

The Dill free energy (cf. (10.2.4)) is obtained by putting $\alpha(s,u) = \alpha(s+u)$ in (9.3.18) and (9.3.19). We must assume that $\alpha(\cdot)$ is a monotonic function (see (10.1.8)), though in this case increasing, so that the condition becomes

$$(-1)^n\alpha_{(n)}(s) \leq 0, \quad s \in \mathbb{R}^+, \quad n = 1, 2, 3, \ldots.$$

A further example is the functional [91]

$$\psi_F^{(e)}(t) = \vartheta(t)e(t) - \frac{1}{2}\alpha_\infty\vartheta^2(t) + \frac{1}{2}\int_0^\infty \frac{1}{\alpha'(\tau)}\left[I_{(\alpha)}^t(\tau, \vartheta_r^t)\right]^2 d\tau, \tag{10.4.3}$$

where $I_{(\alpha)}^t(\cdot, \vartheta_r^t)$ is given by $(9.3.14)_1$ but with ϑ_r^t replacing ϑ^t, under the hypotheses (10.4.1) and (10.4.2). The factor $1/\alpha'(\tau)$, which diverges at large τ, is multiplied by factors that have a behavior such that the integral in (10.4.3) exists; consequently, we can consider the domain

$$H'_F(\mathbb{R}^+) = \left\{ I^t;\ \left| \int_0^\infty \frac{1}{\alpha'(\tau)} \left[I^t_{(\alpha)}(\tau, \vartheta^t_r) \right]^2 d\tau \right| < \infty \right\}.$$

Equation (10.4.3) corresponds to $(8.6.30)_1$. We can also write $\psi_F^{(e)}$ in a form corresponding to (9.3.18):

$$\psi_F^{(e)}(t) = \frac{1}{2}\alpha_0 \vartheta^2(t) + \frac{1}{2}\int_0^\infty \frac{1}{\alpha'(\tau)} \left[I^t_{(\alpha)}(\tau, \vartheta^t) \right]^2 d\tau.$$

Relation (10.4.3) can be written as

$$\psi_F^{(e)}(t) = \vartheta(t)e(t) - \frac{1}{2}\alpha_\infty \vartheta^2(t) - \frac{1}{2}\int_0^\infty \int_0^\infty \alpha_{12}(s_1, s_2)\vartheta^t_r(s_1)\vartheta^t_r(s_2) ds_1 ds_2,$$

where

$$\alpha(s_1, s_2) = -\int_0^\infty \frac{1}{\alpha'(\tau)} \alpha'(\tau + s_1)\alpha'(\tau + s_2) d\tau,$$
$$\alpha_{12}(s_1, s_2) = \frac{\partial^2}{\partial s_2 \partial s_1} \alpha(s_1, s_2),$$

satisfying

$$\alpha(s_1, 0) = \alpha(s_1),$$
$$\alpha(0, s_2) = \alpha(s_2),$$
$$\alpha(0, 0) = \alpha(0).$$

Following steps similar to those from (10.2.10) to (10.2.11), we can show that $\psi_F^{(e)}$ obeys the relation

$$\dot{\psi}_F^{(e)}(t) + D^{(e)}(t) = \dot{e}(t)\vartheta(t),$$

where

$$D_F^{(e)}(t) = \frac{1}{2}\int_0^\infty \frac{d}{d\tau}\left[\frac{1}{\alpha'(\tau)}\right] \left[I^t_{(\alpha)}(\tau, \vartheta^t_r) \right]^2 d\tau + \frac{1}{2}\frac{1}{\alpha'(0)} \left[I^t_{(\alpha)}(0, \vartheta^t_r) \right]^2 \geq 0,$$

since

$$\frac{d}{d\tau}\left[\frac{1}{\alpha'(\tau)}\right] = -\frac{\alpha''(\tau)}{(\alpha'(\tau))^2} \geq 0,$$

so that $\psi_F^{(e)}$ is a free energy.

Analogues of the other functionals described in Sects. 10.1 and 10.2 are easily determined.

Remark 10.4.1. For the remainder of Part III, we will consider free energies that emerge from extremum conditions, in particular, the minimum and maximum free energies. No special restriction is imposed on the material, of the kind introduced in the earlier chapters of Part III. However, in order to derive expressions for these free energies, an assumption is made of analyticity of $\mathbb{L}'_+(\omega)$ in an open set including the real axis, which is equivalent to restricting $\mathbb{L}'(s)$ so that it decays exponentially at large s.

11

The Minimum Free Energy

Breuer and Onat [42] considered the following question: what is the maximum amount of work recoverable from a body that has undergone a specified strain history? They found that the answer for linear viscoelastic memory materials is provided by the solution of an integral equation of Wiener–Hopf type, which is in fact a special case of the result given in Sect. 5.2. They gave a detailed solution by elementary means for a material with relaxation function in the form of a finite sum of decaying exponentials. The nonuniqueness problem was also explicitly exposed by these authors [43].

Day [84] presented an alternative formulation of the thermodynamics of materials with memory. In [85], he revisited the problem considered in [42] within a more rigorous framework, introducing the concept of a (time) reversible extension and discussing the maximum recoverable work in terms of this concept. An expression for the minimum free energy of a standard linear solid (linear viscoelastic solid with a relaxation function that has only one decaying exponential) was given in [87].

A general expression for the minimum free energy of a linear viscoelastic solid under isothermal conditions was given in [158]. This was for a scalar constitutive relation. A generalization to the full tensor case was presented in [92]. These results were used in the context of the viscoelastic Saint-Venant problem in [93]. Detailed explicit expressions for the minimum free energy and related quantities were given in [92, 158] for discrete-spectrum materials, namely those for which the relaxation function is a sum of exponentials. The minimum free energies of compressible and incompressible viscoelastic fluids were determined in [5, 8, 100], while materials with finite memory were considered in [111]. The maximum recoverable work or equivalently the minimum free energy for rigid heat conductors was considered in [6, 21, 22].

We now derive a general expression for the minimum free energy and the associated rate of dissipation for a material described by the linear memory model. The results discussed above are special cases of this, with the exception of the approximate treatment of rigid heat conductors based on (9.1.9), which corresponds to the results of Sect. 8.6.1, as noted in Sect. 9.3.3.

G. Amendola et al., *Thermodynamics of Materials with Memory*,
https://doi.org/10.1007/978-3-030-80534-0_11

11.1 Factorization of Positive Definite Tensors

It will be required to factorize the quantity $\mathbb{H}$, given by (7.2.22), in order to determine an expression for the minimum and other free energies. This was first discussed in [158] for the scalar case, where an appropriate factorization of the one-dimensional counterpart of $\mathbb{H}(\omega)$ was introduced. Such a particular factorization does not apply to fourth-order tensors, so that the extension of the result of [158] to the general case, which was given in [92], is not trivial.

In this section, we show that $\mathbb{H}$ can always be factorized. Use will be made of a result by Gohberg and Kreĭn [156] for tensor-valued functions. Given a nonsingular continuous tensor-valued function $\mathbb{K}(\omega) \in \mathrm{Lin}(\Gamma(\Omega))$, $\omega \in \mathbb{R}$ ($\mathbb{K}$ not connected with the quantity in (7.1.1)), we say that $\mathbb{K}$ has a left [right] factorization if it can be represented in the form

$$\mathbb{K}(\omega) = \mathbb{K}_{(+)}(\omega)\mathbb{K}_{(-)}(\omega), \qquad [\mathbb{K}(\omega) = \mathbb{K}_{(-)}(\omega)\mathbb{K}_{(+)}(\omega)], \tag{11.1.1}$$

where the tensor functions $\mathbb{K}_{(\pm)}$ admit analytic continuations, analytic in the interior and continuous up to the boundary of the corresponding complex half-planes $\Omega^{\pm}$, and are such that

$$\det \mathbb{K}_{(\pm)}(\zeta) \neq \mathbf{0}, \qquad \zeta \in \Omega^{\pm}.$$

We say that $\mathbb{K}$ belongs to $\mathcal{F}_{m\times m}$, $\mathcal{F}^{+}_{m\times m}$, and $\mathcal{F}^{-}_{m\times m}$, respectively, if there exists a constant tensor $\mathbb{C}_0$ and a tensor function $\mathbb{F}(t)$ such that

$$\mathbb{K}(\omega) = \mathbb{C}_0 + \int_{-\infty}^{\infty} \mathbb{F}(t)e^{-i\omega t}\,dt,$$
$$\mathbb{K}(\omega) = \mathbb{C}_0 + \int_{0}^{\infty} \mathbb{F}(t)e^{-i\omega t}\,dt, \quad \mathbb{K}(\omega) = \mathbb{C}_0 + \int_{-\infty}^{0} \mathbb{F}(t)e^{-i\omega t}\,dt. \tag{11.1.2}$$

Note that if $\mathbb{K} \in \mathcal{F}^{\pm}_{m\times m}$, it has the analytic properties ascribed to $\mathbb{K}_{(\pm)}$ above. The main result we use is Theorem 8.2 of [156], which can be stated in our context as follows:

Theorem 11.1.1. *(Gohberg–Kreĭn) In order that the nonsingular Hermitian tensor function* $\mathbb{K} \in \mathcal{F}_{m\times m}$ *possesses a representation of the form*

$$\mathbb{K}(\omega) = \mathbb{K}_{(+)}(\omega)\mathbb{K}^{*}_{(+)}(\omega), \tag{11.1.3}$$

in which the tensor function $\mathbb{K}_{(+)}$ *is in* $\mathcal{F}^{+}_{m\times m}$ *and satisfies* $\det \mathbb{K}_{(+)}(\zeta) \neq 0$ *for* $\zeta \in \Omega^{+}$, *it is necessary and sufficient that* $\mathbb{K}(\omega)$ *be positive definite for every* $\omega \in \mathbb{R}$.

Observe that comparison of (11.1.3) with $(11.1.1)_1$ yields

$$\mathbb{K}_{(-)}(\omega) = \mathbb{K}^{*}_{(+)}(\omega).$$

It follows from the assumption (7.2.20) that

$$\lim_{\omega\to 0} \frac{\mathbb{H}(\omega)}{\omega^2} = \mathbb{H}_0, \tag{11.1.4}$$

where $\mathbb{H}_0$ is symmetric and positive definite. Consider now the tensor

$$\mathbb{K}(\omega) := \frac{\omega_0^2 + \omega^2}{\omega^2}\mathbb{H}(\omega), \tag{11.1.5}$$

where $\omega_0 \neq 0$ is some given frequency. The tensor $\mathbb{K}(\omega)$ is symmetric, real (therefore Hermitian), and positive definite $\forall\omega \in \mathbb{R}$; moreover, it is such that

$$\lim_{\omega\to 0} \mathbb{K}(\omega) = \omega_0^2\mathbb{H}_0, \qquad \lim_{\omega\to\infty} \mathbb{K}(\omega) = \mathbb{H}_\infty.$$

In order to apply Theorem 11.1.1, we have to show that $\mathbb{K} \in \mathcal{F}_{m\times m}$, i.e., that the representation (11.1.2) applies.

Proposition 11.1.2. *If* $\mathbb{L}$ *and* $\mathbb{L}''$ *are tensor functions, integrable on* $[0,\infty)$*, and* $\mathbb{L}'_0$ *is finite, the tensor-valued function* $\mathbb{K}$*, related to* $\mathbb{L}$ *through (7.2.22) and (11.1.5), belongs to* $\mathcal{F}_{m\times m}$*.*

Proof. Observe that from (7.2.22),

$$\mathbb{K}(\omega) = -\left[\omega\mathbb{L}'_s(\omega) + \frac{\omega_0^2}{\omega}\mathbb{L}'_s(\omega)\right]. \tag{11.1.6}$$

Integration by parts of the integral in (7.2.1) and of a corresponding integral for $\mathbb{L}''_+(\omega)$ in terms of $\mathbb{L}''(s)$ yields

$$-\frac{1}{\omega}\mathbb{L}'_s(\omega) = \mathbb{L}_c(\omega), \qquad \omega\mathbb{L}'_s(\omega) = \mathbb{L}'(0) + \mathbb{L}''_c(\omega),$$

so that (11.1.6) becomes

$$\mathbb{K}(\omega) = \left[-\mathbb{L}'(0) - \mathbb{L}''_c(\omega) + \omega_0^2\mathbb{L}_c(\omega)\right]. \tag{11.1.7}$$

Consider now the tensors

$$\mathbb{C}_0 = -\mathbb{L}'(0), \quad \mathbb{F}(t) = \frac{1}{2}\left[-\mathbb{L}''(t) + \omega_0^2\mathbb{L}(t)\right], \quad t \in \mathbb{R}, \tag{11.1.8}$$

where $\mathbb{L}$ and $\mathbb{L}''$ are extended on the real line as even functions, so that from (C.1.5), $\mathbb{L}_F = 2\mathbb{L}_c$ and $\mathbb{L}''_F = 2\mathbb{L}''_c$. Then, in view of (11.1.8), (11.1.7) is equivalent to $(11.1.2)_1$ and the assertion is proved. □

Since $\mathbb{K}(\omega)$ is Hermitian and positive definite for every $\omega \in \mathbb{R}$, it satisfies Theorem 11.1.1. In particular, it has a representation of the form (left factorization)

$$\mathbb{K}(\omega) = \mathbb{K}_{(+)}(\omega)\mathbb{K}^*_{(+)}(\omega), \tag{11.1.9}$$

with $\mathbb{K}_{(+)}(\omega) \in \mathcal{F}^+_{m\times m}$ and

$$\det \mathbb{K}_{(+)}(\zeta) \neq 0 \qquad \text{for} \quad \zeta \in \Omega^+.$$

Moreover, such a factorization is unique up to a multiplication on the right of $\mathbb{K}_{(+)}$ by a constant unitary tensor.

Similarly, $\mathbb{K}$ has a right factorization of the type

$$\mathbb{K}(\omega) = \mathbb{K}_{(-)}(\omega)\mathbb{K}^*_{(-)}(\omega) \tag{11.1.10}$$

with corresponding properties. In fact, since $\mathbb{K}(\omega)$ is an even function of ω, we can replace ω by $-\omega$ on the right of (11.1.9). Now, $\mathbb{K}_{(+)}(-\omega) \in \mathcal{F}^-_{m\times m}$ with nonzero determinant in $\Omega^{(-)}$, so that $\mathbb{K}_{(-)}(\omega) = \mathbb{K}_{(+)}(-\omega)$.

By virtue of (11.1.5) and (11.1.10), $\mathbb{H}(\omega)$ can be factorized as follows:

$$\mathbb{H}(\omega) = \mathbb{H}_+(\omega)\mathbb{H}_-(\omega), \tag{11.1.11}$$

where

$$\mathbb{H}_+(\omega) = \frac{\omega}{\omega - i\omega_0}\mathbb{K}_{(-)}(\omega), \qquad \mathbb{H}_-(\omega) = \frac{\omega}{\omega + i\omega_0}\mathbb{K}^*_{(-)}(\omega). \tag{11.1.12}$$

Alternatively, the left factorization (11.1.9) may be used, though the right factorization is more convenient in the present context. Representation (11.1.12) gives that

$$\mathbb{H}_\pm(\omega) = \mathbb{H}^*_\mp(\omega). \tag{11.1.13}$$

We have introduced in the present work an assumption that is stronger than those required in Theorem 11.1.1, namely that $\mathbb{H}$ is analytic in an open set including the real axis $\mathbb{R}$. Since $\mathbb{H}_+$ has the singularities of $\mathbb{H}$ in $\Omega^{(+)}$, then $\mathbb{H}_\pm$ will share this property.

In general, $\mathbb{H}_+$ and $\mathbb{H}_-$ do not commute, and various general results can be proved without assuming that they do. However, in order to derive explicit forms for various free energies, we must make an assumption that implies commutativity or, put another way, that $\mathbb{H}_\pm$ are normal tensors. This arises out of an assumption made in Sect. 11.6.

The notation for $\mathbb{H}_+(\omega)$ and $\mathbb{H}_-(\omega)$ follows the convention used in [158], i.e., the sign indicates the half-plane in which the singularities of the tensor lie. These factors also have the property that any zeros in their determinant occur also in the indicated half-plane. This latter property will not apply when factorizations leading to free energies other than the minimum are discussed. We adopt, however, the convention that $f_\pm$ has all its singularities in $\Omega^\pm$, respectively. This is in particular the convention adopted in Appendix C for Fourier-transformed quantities.

Recalling (11.1.4), we see that $\mathbb{H}_\pm$ each vanish linearly at the origin.

Note that we require (7.1.18) to ensure that $\mathbb{H}$, and therefore $\mathbb{K}$, is symmetric (i.e., Hermitian for real tensors).

The quantity $\mathbb{H}_\infty$, defined by (7.2.24), is given by

$$\mathbb{H}_\infty = \mathbb{H}_+(\infty)\mathbb{H}_-(\infty) = \mathbb{H}_{+\infty}\mathbb{H}_{-\infty}. \tag{11.1.14}$$

If $\mathbb{H}_{\pm\infty}$ can be chosen to be Hermitian, which is possible at least in the commutative case considered in Sect. 11.6, then they are both equal to the square root of the nonnegative tensor $\mathbb{H}_\infty$.

11.1.1 The Scalar Case

We can derive explicit forms for the factors if H is a scalar function. It is real and nonnegative on $\mathbb{R}$, vanishing quadratically at the origin. It is an even function of ω and therefore a function of ω^2, in view of its analyticity about the origin. Its singularities are as ascribed to $\mathbb{H}$ (and $\mathbb{L}'_s$ before (7.2.22)).

We define the function

$$K(\omega) = \log[H(\omega)T(\omega)], \qquad T(\omega) = \frac{\omega^2 + \omega_0^2}{H_\infty \omega^2},$$

where ω_0 may be chosen arbitrarily on $\mathbb{R}$. Then, K is a well-defined analytic function on $\mathbb{R}$, vanishing like ω^{-2} for large values of ω. Consider the quantity

$$M(z) = \frac{1}{2\pi i}\int_{-\infty}^{\infty} \frac{K(\omega)}{\omega - z} d\omega, \quad z \in \Omega \backslash R,$$

which goes to zero like z^{-1} at large z (see (B.2.13)). For $z \in \Omega^{(+)}$, $M(z) = M_-(z)$ is a function analytic in $\Omega^{(+)}$, while for $z \in \Omega^{(-)}$, $M(z) = M_+(z)$, which is analytic in $\Omega^{(-)}$ (Sect. B.2.1). The Plemelj formulas (B.2.14) take the form

$$M_-(\omega) = \frac{1}{2}K(\omega) + \frac{1}{2\pi i} P \int_{-\infty}^{\infty} \frac{K(\omega')}{\omega' - \omega} d\omega',$$
$$M_+(\omega) = -\frac{1}{2}K(\omega) + \frac{1}{2\pi i} P \int_{-\infty}^{\infty} \frac{K(\omega')}{\omega' - \omega} d\omega',$$

giving that

$$M_-(\omega) - M_+(\omega) = K(\omega), \quad \omega \in R,$$

where $M_+(\omega)\,(M_-(\omega))$ are the limiting values of $M_+(z)\,(M_-(z))$ as z approaches the real axis from below (above). Then, if we put

$$\begin{aligned} H_+(\omega) &= \frac{\omega h_\infty}{\omega - i\omega_0} e^{-M_+(\omega)}, \\ H_-(\omega) &= \frac{\omega h_\infty}{\omega + i\omega_0} e^{+M_-(\omega)}, \qquad h_\infty = H_\infty^{1/2}, \end{aligned} \tag{11.1.15}$$

it follows that $H_+(z)$ is analytic and free of zeros in $\Omega^{(-)}$; similarly for $H_-(z)$ in $\Omega^{(+)}$. Also,

$$H_+(\omega)H_-(\omega) = H(\omega). \tag{11.1.16}$$

Noting that $\overline{M_\pm}(\omega) = M_\pm(-\omega) = -M_\mp(\omega)$, we see that

$$\begin{aligned} H_\pm(\omega) &= H_\mp(-\omega) = \overline{H_\mp}(\omega), \\ H(\omega) &= |H_\pm(\omega)|^2, \quad \omega \in \mathbb{R}. \end{aligned} \tag{11.1.17}$$

According to the general result on the uniqueness of the factorization, noted above, a scalar factorization should be unique up to multiplication by a phase factor $e^{i\alpha}$, where α is a constant. Relation $(11.1.17)_2$ reduces this arbitrariness to multiplication by a factor ± 1.

It follows from (11.1.15) that

$$H_\pm(\infty) = h_\infty = H_\infty^{1/2}, \tag{11.1.18}$$

so that $H_\pm(\infty)$ are real and equal.

11.2 Derivation of the Form of the Minimum Free Energy

We shall be seeking to find the continuation $\mathbf{\Lambda}^t(u)$, $u \in \mathbb{R}^{--}$, that maximizes the recoverable work (Theorem 4.2.3). For this purpose, our attention will be confined to the family of continuations that vanish at large u, since it can be shown [92, 145] that the maximum recoverable work obtained by searching in this set is equal to that obtained by a wider search in the general set of bounded recoverable works.

The set of continuations for which

$$W(\infty) = \int_{-\infty}^{\infty} \mathbf{\Sigma}(u) \cdot \dot{\mathbf{\Lambda}}(u) du \tag{11.2.1}$$

exists and $\mathbf{\Lambda}(\infty)$ vanishes, we label $\mathcal{C}_0$. Using (7.5.5), we can write (11.2.1) as

$$\begin{aligned} W(\infty) &= \frac{1}{2} \int_{-\infty}^{\infty} \int_{-\infty}^{\infty} \mathbb{L}_{12}(|s-u|) \mathbf{\Lambda}(u) \cdot \mathbf{\Lambda}(s) du ds \\ &= \frac{1}{2} \int_{-\infty}^{\infty} \int_{-\infty}^{\infty} \mathbb{L}_{12}(|s-u|) \mathbf{\Lambda}^t(u) \cdot \mathbf{\Lambda}^t(s) du ds, \end{aligned} \tag{11.2.2}$$

where the latter form is obtained by changes of integration variables. The superscript t is now an arbitrary parameter, which we interpret again as the current time. Applying the convolution theorem and Parseval's formula to (7.5.5) for $t = \infty$, we obtain, as in Sect. 7.5,

$$\begin{aligned} W(\infty) &= \frac{1}{2\pi} \int_{-\infty}^{\infty} \overline{\mathbf{\Lambda}_F^t}(\omega) \cdot \mathbb{H}(\omega) \mathbf{\Lambda}_F^t(\omega) d\omega \\ &= \frac{1}{2\pi} \int_{-\infty}^{\infty} \overline{\mathbf{\Lambda}_{rF}^t}(\omega) \cdot \mathbb{H}(\omega) \mathbf{\Lambda}_{rF}^t(\omega) d\omega \\ &= \frac{1}{2\pi} \int_{-\infty}^{\infty} \left[\overline{\mathbf{\Lambda}_{r+}^t}(\omega) + \overline{\mathbf{\Lambda}_{r-}^t}(\omega)\right] \cdot \mathbb{H}(\omega) \left[\mathbf{\Lambda}_{r+}^t(\omega) + \mathbf{\Lambda}_{r-}^t(\omega)\right] d\omega \end{aligned} \tag{11.2.3}$$

by virtue of (C.1.4). We have used the notation of Sect. 7.2.3. Relations $(11.2.3)_1$ and $(11.2.3)_2$ are equal by virtue of (7.2.31), (C.2.19), and the fact that $\mathbb{H}(\omega)$ vanishes for $\omega = 0$. For continuations in $\mathcal{C}_0$, the recoverable work from the state at time t (see (5.2.3)) is given by

$$W_R(t) = -\int_t^{\infty} \mathbf{\Sigma}(u) \cdot \dot{\mathbf{\Lambda}}(u) du = W(t) - W(\infty). \tag{11.2.4}$$

To obtain the minimum free energy, we seek to maximize $W_R(t)$. Since $W(t)$ is a given quantity, this is equivalent to minimizing $W(\infty)$.

We now give three derivations of the form of the minimum free energy. The first uses a variational technique developed in [92, 158]. Also, a quite different and simplified version of this approach is presented. The third approach is based on the solution of (5.2.8) in the linear memory case, where it reduces to a linear Wiener–Hopf equation.

11.2.1 A Variational Approach

Let $\mathbf{\Lambda}_o^t$ be the optimal future continuation (so that $\dot{\mathbf{\Lambda}}_o$ is the optimal process) and

$$\mathbf{\Lambda}_{ro}^t(s) = \mathbf{\Lambda}_o^t(s) - \mathbf{\Lambda}(t), \quad s \in \mathbb{R}^-. \tag{11.2.5}$$

Let $\mathbf{\Lambda}_m^t$ denote the Fourier transform of $\mathbf{\Lambda}_{ro}^t$, so that $\mathbf{\Lambda}_m^t(\omega) = \mathbf{\Lambda}_{ro-}^t(\omega)$. Put

$$\mathbf{\Lambda}_{r-}^t(\omega) = \mathbf{\Lambda}_m^t(\omega) + \mathbf{k}_-(\omega), \tag{11.2.6}$$

where $\mathbf{k}_-(\omega)$ is arbitrary apart from the fact that it must have the same analytic properties as $\mathbf{\Lambda}_{r-}^t(\omega)$, i.e., $\mathbf{k}_-(z)$ must be analytic in Ω^+, and vanish like z^{-1} at large z. Then, varying $\mathbf{k}_-$, we find that $W(\infty)$ will be minimized by $\mathbf{\Lambda}_m^t$ if

$$\int_{-\infty}^{\infty} Re\left\{\mathbb{H}(\omega)\left[\mathbf{\Lambda}_{r+}^t(\omega) + \mathbf{\Lambda}_m^t(\omega)\right] \cdot \overline{\mathbf{k}_-(\omega)}\right\} d\omega = 0.$$

The restriction to the real part of the integral may be removed, since the imaginary part vanishes by virtue of the symmetric range of integration and (C.1.7). Using the factorization (11.1.11), we can rewrite this condition in the form

$$\begin{aligned}
&\int_{-\infty}^{\infty} \mathbb{H}(\omega)\left[\mathbf{\Lambda}_{r+}^t(\omega) + \mathbf{\Lambda}_m^t(\omega)\right] \cdot \overline{\mathbf{k}_-(\omega)} d\omega \\
&= \int_{-\infty}^{\infty} \mathbb{H}_+(\omega)\left[\mathbb{H}_-(\omega)\mathbf{\Lambda}_{r+}^t(\omega) + \mathbb{H}_-(\omega)\mathbf{\Lambda}_m^t(\omega)\right] \cdot \overline{\mathbf{k}_-(\omega)} d\omega = 0.
\end{aligned} \tag{11.2.7}$$

Consider now the quantity $\mathbb{H}_-(\omega)\mathbf{\Lambda}_{r+}^t(\omega)$, the components of which are continuous, indeed analytic on $\mathbb{R}$, by virtue of the analyticity properties of $\mathbb{H}_-(\omega)$ and $\mathbf{\Lambda}_+^t(\omega)$. The Plemelj formula $(B.2.15)_2$ gives that

$$\mathbf{P}^t(\omega) = \mathbb{H}_-(\omega)\mathbf{\Lambda}_{r+}^t(\omega) = \mathbf{p}_-^t(\omega) - \mathbf{p}_+^t(\omega), \tag{11.2.8}$$

where

$$\mathbf{p}^t(z) = \frac{1}{2\pi i}\int_{-\infty}^{\infty} \frac{\mathbf{P}^t(\omega')}{\omega' - z} d\omega', \quad \mathbf{p}_\pm^t(\omega) := \lim_{\alpha \to 0^\mp} \mathbf{p}^t(\omega + i\alpha). \tag{11.2.9}$$

Moreover, $\mathbf{p}^t(z) = \mathbf{p}_+^t(z)$ is analytic for $z \in \Omega^{(-)}$, and $\mathbf{p}^t(z) = \mathbf{p}_-^t(z)$ is analytic for $z \in \Omega^{(+)}$. Both are analytic on the real axis by virtue of the argument leading up to Remark B.2.2. We write them in the form

$$\mathbf{p}^t_{\pm}(\omega) = \frac{1}{2\pi i}\int_{-\infty}^{\infty} \frac{\mathbf{P}^t(\omega')}{\omega' - \omega^{\mp}} d\omega', \tag{11.2.10}$$

where the notation $\omega^{\pm}$ of (C.2.10) and (C.2.11) has been used. If we can determine explicit formulas for $\mathbf{p}^t_{\pm}(\omega)$, $\omega \in \mathbb{R}$ (or $\omega \in \Omega^{(\mp)}$), then they can be analytically continued into $\Omega^{(\pm)}$, respectively, defined everywhere except at singularities, unless a blocking branch cut prevents this (Sect. B.1). Examples will be given later. Using (11.2.8) in (11.2.7), we obtain

$$\int_{-\infty}^{\infty} \mathbb{H}_+(\omega)\left[\mathbf{p}^t_-(\omega) - \mathbf{p}^t_+(\omega) + \mathbb{H}_-(\omega)\boldsymbol{\Lambda}^t_m(\omega)\right] \cdot \overline{\mathbf{k}_-}(\omega) d\omega = 0. \tag{11.2.11}$$

Note that the integral

$$\int_{-\infty}^{\infty} \mathbb{H}_+(\omega)\mathbf{p}^t_+(\omega) \cdot \overline{\mathbf{k}_-}(\omega) d\omega$$

vanishes identically by virtue of (B.1.14), because the integrand is analytic on $\Omega^{(-)}$, by Remark B.1.2, and vanishes like z^{-2} at large z. Therefore, (11.2.11) becomes

$$\int_{-\infty}^{\infty} \mathbb{H}_+(\omega)\left[\mathbf{p}^t_-(\omega) + \mathbb{H}_-(\omega)\boldsymbol{\Lambda}^t_m(\omega)\right] \cdot \overline{\mathbf{k}_-}(\omega) d\omega = 0. \tag{11.2.12}$$

This will be true for arbitrary $\overline{\mathbf{k}_-}(\omega)$ only if the expression in brackets is a function that is analytic in $\Omega^{(-)}$. However, $\boldsymbol{\Lambda}^t_m(\omega)$ must be analytic in Ω^+. Remembering that $\mathbf{p}^t_-(\omega)$ and $\mathbb{H}_-(\omega)$ are also analytic in Ω^+, we see that the expression in brackets must be analytic in both the upper and the lower half-planes and on the real axis. Thus, it is analytic over the entire complex plane. Now, $\mathbf{p}^t_-(\omega)$ clearly vanishes like ω^{-1} at infinity, as also must $\boldsymbol{\Lambda}^t_m(\omega)$ by (C.2.16) if $\boldsymbol{\Lambda}^t_o$ is to be nonzero and finite at $s = 0$. Therefore, the function in brackets is analytic everywhere, zero at infinity, and consequently must vanish everywhere by Liouville's theorem (Sect. B.1.3). Thus,

$$\mathbf{p}^t_-(\omega) + \mathbb{H}_-(\omega)\boldsymbol{\Lambda}^t_m(\omega) = \mathbf{0} \qquad \forall\omega \in \mathbb{R}, \tag{11.2.13}$$

whence

$$\boldsymbol{\Lambda}^t_m(\omega) = -[\mathbb{H}_-(\omega)]^{-1}\,\mathbf{p}^t_-(\omega). \tag{11.2.14}$$

Using this relation and (11.1.11) in $(11.2.3)_3$, we find that the optimal value of $W(\infty)$ is

$$W_{opt}(\infty) = \frac{1}{2\pi}\int_{-\infty}^{\infty} |\mathbf{p}^t_+(\omega)|^2\, d\omega. \tag{11.2.15}$$

Note that from (11.1.11), $(7.5.7)_2$, and (11.2.8),

$$\begin{aligned} W(t) &= \phi(t) + \frac{1}{2\pi}\int_{-\infty}^{\infty} |\mathbf{p}^t_-(\omega) - \mathbf{p}^t_+(\omega)|^2\, d\omega \\ &= \phi(t) + \frac{1}{2\pi}\int_{-\infty}^{\infty} \left[|\mathbf{p}^t_-(\omega)|^2 + |\mathbf{p}^t_+(\omega)|^2\right] d\omega, \end{aligned} \tag{11.2.16}$$

since the cross terms vanish by Proposition B.1.3. Thus, from (11.2.4), (11.2.15), and (11.2.16), we have

$$\psi_m(t) = \phi(t) + \frac{1}{2\pi}\int_{-\infty}^{\infty} |\mathbf{p}_-^t(\omega)|^2 \, d\omega. \tag{11.2.17}$$

Using (11.2.14), we can write this as

$$\begin{aligned}
\psi_m(t) &= \phi(t) + \frac{1}{2\pi}\int_{-\infty}^{\infty} \overline{\mathbf{\Lambda}_m^t(\omega)} \cdot \mathbb{H}(\omega)\mathbf{\Lambda}_m^t(\omega)d\omega \\
&= \phi(t) + \frac{1}{2}\int_{-\infty}^{0}\int_{-\infty}^{0} \mathbb{L}_{12}(|s-u|)\mathbf{\Lambda}_{ro}^t(u)\cdot\mathbf{\Lambda}_{ro}^t(s)duds \\
&= \phi(t) + \frac{1}{2}\int_{-\infty}^{0}\int_{-\infty}^{0} \mathbb{L}(|s-u|)\dot{\mathbf{\Lambda}}_o^t(u)\cdot\dot{\mathbf{\Lambda}}_o^t(s)duds \\
&= \phi(t) + \frac{1}{2\pi}\int_{-\infty}^{\infty} \overline{\dot{\mathbf{\Lambda}}_m^t(\omega)} \cdot \frac{\mathbb{H}(\omega)}{\omega^2}\dot{\mathbf{\Lambda}}_m^t(\omega)d\omega,
\end{aligned} \tag{11.2.18}$$

where $\mathbf{\Lambda}_{ro}^t$ is defined by (11.2.5). This last form can be seen to be a special case of (5.2.11).

From (5.1.13), we have

$$\dot{\psi}_m(t) + D_m(t) = \mathbf{\Sigma}(t)\cdot\dot{\mathbf{\Lambda}}(t), \tag{11.2.19}$$

where D_m is the rate of dissipation corresponding to the minimum free energy and must be nonnegative by the second law. Let us assume that the material is undisturbed in the distant past. Integrating (11.2.19) up to time t gives a special case of (5.1.34):

$$\psi_m(t) + \mathfrak{D}_m(t) = W(t), \tag{11.2.20}$$

where

$$\mathfrak{D}_m(t) = \int_{-\infty}^{t} D_m(s)ds$$

is the total dissipation up to time t, corresponding to the minimum free energy. Since ψ_m is less than or equal to any other free energy, it follows from (11.2.20) that $\mathfrak{D}_m(t)$ is the largest estimate of dissipation in the material element. We have, from (11.2.15)–(11.2.17),

$$\mathfrak{D}_m(t) = W(t) - \psi_m(t) = \frac{1}{2\pi}\int_{-\infty}^{\infty} |\mathbf{p}_+^t(\omega)|^2 \, d\omega = W_{opt}(\infty) \geq 0. \tag{11.2.21}$$

Also,

$$D_m(t) = \dot{\mathfrak{D}}_m(t).$$

In order to give an explicit expression for D_m, we note certain properties of $\mathbf{p}_\pm^t$. From (11.2.10) and $(7.2.30)_2$, it follows that

$$\begin{aligned}
\frac{d}{dt}\mathbf{p}_+^t(\omega) &= -i\omega\mathbf{p}_+^t(\omega) - \mathbf{K}(t), \\
\frac{d}{dt}\mathbf{p}_-^t(\omega) &= -i\omega\mathbf{p}_-^t(\omega) - \mathbf{K}(t) - \frac{\mathbb{H}_-(\omega)\dot{\mathbf{\Lambda}}(t)}{i\omega}, \\
\mathbf{K}(t) &= \frac{1}{2\pi}\int_{-\infty}^{\infty} \mathbb{H}_-(\omega)\mathbf{\Lambda}_{r+}^t(\omega)d\omega.
\end{aligned} \tag{11.2.22}$$

The quantity $\mathbf{K}$ is in fact real if (11.6.3) below holds, which is true for commuting factors. The relation

$$\frac{1}{2\pi i}\int_{-\infty}^{\infty}\frac{\mathbb{H}_{-}(\omega')}{i\omega'^{-}(\omega'-\omega^{+})}d\omega' = \frac{\mathbb{H}_{-}(\omega)}{i\omega} \tag{11.2.23}$$

has been used. This follows by remembering that $\mathbb{H}_{-}(\omega)$ vanishes linearly at the origin and by closing the contour on $\Omega^{(+)}$, on which half-plane $\mathbb{H}_{-}$ is analytic. If $(\omega' - \omega^{-})$ occurs in the denominator, the integral vanishes. Furthermore,

$$\begin{aligned}&\lim_{|\omega|\to\infty}\omega\mathbf{p}^t_{\pm}(\omega) = i\mathbf{K}(t),\\ &\frac{1}{2\pi}\int_{-\infty}^{\infty}\mathbf{p}^t_{\pm}(\omega)d\omega = \mp\frac{1}{2}\mathbf{K}(t).\end{aligned} \tag{11.2.24}$$

The first relation follows from (11.2.10) and the second from (B.1.13) and the first relation, remembering the analyticity properties of $\mathbf{p}^t_{\pm}$. Differentiating (11.2.21) with respect to t, we find the explicitly nonnegative form for the rate of dissipation:

$$D_m(t) = |\mathbf{K}(t)|^2, \tag{11.2.25}$$

where $\mathbf{K}$ is given by $(11.2.22)_3$.

Remark 11.2.1. The following, simpler, derivation of (11.2.14) and (11.2.15) (yielding (11.2.17)) can be given. Let us write $(11.2.3)_3$ as

$$W(\infty) = \frac{1}{2\pi}\int_{-\infty}^{\infty}|\mathbf{p}^t_{-}(\omega) - \mathbf{p}^t_{+}(\omega) + \mathbb{H}_{-}(\omega)\mathbf{\Lambda}^t_{r-}(\omega)|^2\, d\omega.$$

Putting

$$\mathbf{p}^t_{1-}(\omega) = \mathbf{p}^t_{-}(\omega) + \mathbb{H}_{-}(\omega)\mathbf{\Lambda}^t_{r-}(\omega),$$

where $\mathbf{p}^t_{1-}$ is analytic on $\Omega^{(+)}$, we have

$$W(\infty) = \frac{1}{2\pi}\int_{-\infty}^{\infty}|\mathbf{p}^t_{1-}(\omega) - \mathbf{p}^t_{+}(\omega)|^2\, d\omega = \frac{1}{2\pi}\int_{-\infty}^{\infty}[|\mathbf{p}^t_{1-}(\omega)|^2 + |\mathbf{p}^t_{+}(\omega)|^2]\, d\omega,$$

by Proposition B.1.3. Only $\mathbf{p}^t_{1-}$ depends on $\mathbf{\Lambda}^t_{r-}$, so that the minimum must be given by the condition

$$\mathbf{p}^t_{1-}(\omega) = \mathbf{0},$$

which is (11.2.14). Relation (11.2.15) follows immediately.

11.2.2 The Wiener–Hopf Method

The first-order variation of $(11.2.2)_2$ due to $\mathbf{\Lambda}^t(u) \to \mathbf{\Lambda}^t(u) + \delta\mathbf{\Lambda}^t(u)$ has the form

$$\delta W(\infty) = \int_{-\infty}^{\infty}\int_{-\infty}^{\infty}\mathbb{L}_{12}(|s-u|)\mathbf{\Lambda}^t(u)\cdot\delta\mathbf{\Lambda}^t(s)duds,$$

where the symmetry of $\mathbb{L}_{12}(|s-u|)$ has been used. We vary only the future continuation, so that

$$\delta\boldsymbol{\Lambda}^t(s) = \mathbf{0}, \qquad s \in \mathbb{R}^+.$$

The resulting $\delta W(\infty)$ is put equal to zero, yielding the optimization condition

$$\int_{-\infty}^{\infty} \frac{\partial^2}{\partial s \partial u} \mathbb{L}(|s-u|)\boldsymbol{\Lambda}^t(u)du = \mathbf{0}, \qquad s \in \mathbb{R}^-.$$

Removing the derivative with respect to s gives a constant on the right-hand side, which can be shown to be zero by observing that from $(7.1.15)_2$,

$$\lim_{s\to-\infty} \int_{-\infty}^{\infty} \frac{\partial}{\partial u} \mathbb{L}(|s-u|)\boldsymbol{\Lambda}^t(u)du = \mathbf{0}.$$

Thus, we obtain the Wiener–Hopf equation

$$\begin{aligned} \int_{-\infty}^{\infty} \frac{\partial}{\partial u} \mathbb{L}(|s-u|)\boldsymbol{\Lambda}^t(u)du &= \int_{-\infty}^{\infty} \frac{\partial}{\partial u} \mathbb{L}(|s-u|)\boldsymbol{\Lambda}_r^t(u)du \\ &= -\int_{-\infty}^{\infty} \frac{\partial}{\partial s} \mathbb{L}(|s-u|)\boldsymbol{\Lambda}_r^t(u)du = -\mathbf{R}(s), \quad \mathbf{R}(s) = \mathbf{0} \quad \forall s \in \mathbb{R}^-. \end{aligned} \tag{11.2.26}$$

Relation $(7.1.15)_1$ allows $\boldsymbol{\Lambda}^t(u)$ to be replaced by $\boldsymbol{\Lambda}_r^t(u)$. This is an integral equation for the optimal continuation $\boldsymbol{\Lambda}_{ro}^t$, defined by (11.2.5).* The quantity $\mathbf{R}$ on $\mathbb{R}^+$ is for the moment undetermined. Taking Fourier transforms of (11.2.26) and multiplying across by ω, we obtain, with the aid of the convolution theorem (C.3.3) together with (7.2.22) and (7.2.25),

$$2i\mathbb{H}(\omega)\left[\boldsymbol{\Lambda}_{r+}^t(\omega) + \boldsymbol{\Lambda}_m^t(\omega)\right] = \omega\mathbf{R}_+(\omega), \tag{11.2.27}$$

where $\boldsymbol{\Lambda}_m^t(\omega)$ is the Fourier transform of $\boldsymbol{\Lambda}_{ro}^t$ on $\mathbb{R}^-$ and is the quantity we wish to determine. It is analytic on $\Omega^{(+)}$ and by assumption also on $\mathbb{R}$ (Sect. C.2). Similarly, $\mathbf{R}_+(\omega)$ is analytic on $\Omega^{(-)}$ and by assumption also on $\mathbb{R}$.

The factorization (11.1.11) is now used. We multiply (11.2.27) by $[2i\mathbb{H}_+(\omega)]^{-1}$ to obtain

$$\mathbb{H}_-(\omega)\left[\boldsymbol{\Lambda}_{r+}^t(\omega) + \boldsymbol{\Lambda}_m^t(\omega)\right] = \frac{\omega}{2i}[\mathbb{H}_+(\omega)]^{-1}\mathbf{R}_+(\omega). \tag{11.2.28}$$

Substituting (11.2.8) into (11.2.28), we obtain

$$\mathbf{A}(\omega) = \mathbf{p}_-^t(\omega) + \mathbb{H}_-(\omega)\boldsymbol{\Lambda}_m^t(\omega) = \mathbf{p}_+^t(\omega) + \frac{\omega}{2i}[\mathbb{H}_+(\omega)]^{-1}\,\mathbf{R}_+(\omega). \tag{11.2.29}$$

---

* Carrying out a partial integration in (11.2.26), we have the form

$$\int_{-\infty}^{\infty} \mathbb{L}(|s-u|)\dot{\boldsymbol{\Lambda}}^t(u)du = \mathbf{R}(s).$$

This is a special case of (5.2.8), as can be seen by splitting the integral at $u = s$, changing the integration variable, and recalling $(7.1.17)_4$.

The function $\mathbf{A}(\omega)$ is analytic on Ω^- by virtue of the first relation and analytic on Ω^+ by virtue of the second. It is therefore analytic over the entire complex plane. By Liouville's theorem, it must be a polynomial. However, for $|\omega| \to \infty$, $\mathbf{A}(\omega) \to 0$ like $1/\omega$, on applying the argument presented after (11.2.12). Hence, it must vanish everywhere, so that

$$\mathbb{H}_-(\omega)\mathbf{\Lambda}^t_m(\omega) + \mathbf{p}^t_-(\omega) = \mathbf{0},$$

which is (11.2.13). The right-hand side of (11.2.29) also vanishes, which yields a relationship for $\mathbf{R}_+$.

The above solution can be extended to a more general set of histories. We write (11.2.26) in the form

$$\begin{aligned}\int_{-\infty}^{0} \frac{\partial}{\partial s}\mathbb{L}(|s-u|)\mathbf{\Lambda}^t_r(u)du - \mathbf{J}^t(s) = \mathbf{R}(s), \quad s \in \mathbb{R},\\ \mathbf{J}^t(s) = -\int_0^{\infty} \frac{\partial}{\partial s}\mathbb{L}(|s-u|)\mathbf{\Lambda}^t_r(u)du.\end{aligned} \tag{11.2.30}$$

Observe that

$$\mathbf{J}^t(s) = \mathbf{I}^t(-s, \mathbf{\Lambda}^t_r), \quad s \in \mathbb{R}^-, \tag{11.2.31}$$

where $\mathbf{I}^t(\cdot, \mathbf{\Lambda}^t_r)$ is defined by (7.4.2). Let us assume that $\mathbf{J}^t \in L^1(\mathbb{R}) \cap L^2(\mathbb{R})$. Denoting the Fourier transform of $\mathbf{J}^t$ by $\mathbf{J}^t_F \in L^2(\mathbb{R})$, we obtain, instead of (11.2.27),

$$2i\mathbb{H}(\omega)\mathbf{\Lambda}^t_m(\omega) - \omega\mathbf{J}^t_F(\omega) = \omega\mathbf{R}_+(\omega).$$

The argument proceeds as outlined above but where $\mathbf{P}^t(\omega)$ in (11.2.8) is now defined by

$$\mathbf{P}^t(\omega) = -\frac{\omega}{2i}[\mathbb{H}_+(\omega)]^{-1}\,\mathbf{J}^t_F(\omega) = \mathbf{p}^t_-(\omega) - \mathbf{p}^t_+(\omega). \tag{11.2.32}$$

It is assumed that $\mathbf{J}^t_F$ is analytic on $\mathbb{R}$. The Fourier transform on $\mathbb{R}$ of a function that is continuous at $t = 0$ behaves like ω^{-2} at large frequencies, by virtue of (C.2.16) (putting $m = 0$ in (C.2.18)), so that $\mathbf{J}^t_F$ has this property. Thus, $\mathbf{P}^t(\omega) \sim \omega^{-1}$ at large ω, as required for the convergence of the integral in (11.2.9).

This formulation is valid for histories $\mathbf{\Lambda}^t$ that do not have a Fourier transform but where $\mathbf{J}^t$ exists and has a Fourier transform [110, 145].

The quantity $\mathbf{\Lambda}^t_m$ is the Fourier transform of the optimal future continuation $\mathbf{\Lambda}^t_{ro}$ introduced in (11.2.6). Consider (11.2.30) for $s < 0$. We differentiate this relation, multiplying by $\mathbf{\Lambda}^t_{ro}$, and integrate over $\mathbb{R}^-$ to obtain

$$\begin{aligned}\int_{-\infty}^{0}\int_{-\infty}^{0} \mathbf{\Lambda}^t_{ro}(s)\cdot\mathbb{L}_{12}(|s-u|)\mathbf{\Lambda}^t_{ro}(u)duds &= \int_{-\infty}^{0} \mathbf{J}^t(s)\cdot\mathbf{\Lambda}^t_{ro}(s)ds\\ &= \int_{-\infty}^{0} \mathbf{I}(-s, \mathbf{\Lambda}^t_r)\cdot\mathbf{\Lambda}^t_{ro}(s)ds.\end{aligned}$$

Relation $(11.2.18)_2$ gives that

$$\psi(t) = \phi(t) + \frac{1}{2}\int_{-\infty}^{0} \mathbf{I}(-s, \mathbf{\Lambda}^t_r)\cdot\mathbf{\Lambda}^t_{ro}(s)ds$$

and, from (11.2.17),

$$\frac{1}{2\pi}\int_{-\infty}^{\infty}|\mathbf{p}_-^t(\omega)|^2 d\omega = \frac{1}{2}\int_{-\infty}^{0}\mathbf{I}(-s,\mathbf{\Lambda}_r^t)\cdot\mathbf{\Lambda}_{ro}^t(s)ds. \tag{11.2.33}$$

11.2.3 Histories Rather Than Relative Histories

In early work on the minimum and other free energies [92, 110, 158], histories, rather than relative histories, were used. By essentially identical arguments to those above, one can show the following result. Let

$$\mathbf{Q}^t(\omega) := \mathbb{H}_-(\omega)\mathbf{\Lambda}_+^t(\omega) = \mathbf{q}_-^t(\omega) - \mathbf{q}_+^t(\omega),$$

where

$$\mathbf{q}_\pm^t(\omega) = \frac{1}{2\pi i}\int_{-\infty}^{\infty}\frac{\mathbf{Q}^t(\omega')}{\omega'-\omega^\mp}d\omega'.$$

Then,

$$\psi(t) = S(t) + \frac{1}{2\pi}\int_{-\infty}^{\infty}|\mathbf{q}_-^t(\omega)|^2 d\omega, \tag{11.2.34}$$

where $S(t)$ is given by $(7.1.19)_2$, or, in the completely linear case, (7.1.34). The free energies in Sect. 16.4 can also be expressed as functionals of histories rather than relative histories. The situation for free energies in dielectrics is in a sense reversed, as we see from Sect. 22.3. There are two disadvantages to this approach, one noted earlier, namely that $S(t)$ is not a nonnegative quantity. The other is that $\mathbf{\Lambda}_+^t(\omega)$ behaves like ω^{-1} at large ω, while $\mathbf{\Lambda}_{r+}^t(\omega)$ behaves like ω^{-2}.

11.2.4 Confirmation That ψ_m Is a Free Energy

We now ascertain that ψ_m has the characteristic properties of a free energy.

Proposition 11.2.2. *The functional $\psi_m(t)$, given by* (11.2.17), *obeys the Graffi conditions, given by P1–P4 in Sect. 5.1.1 or* (5.1.30)–(5.1.33).

Proof. Property P2 follows from the fact that $\mathbf{\Lambda}_{r+}^t$, and therefore $\mathbf{P}^t$, defined by (11.2.8), vanishes for a static history. Property P3 is immediately apparent, while P4 follows from the fact that D_m, given by (11.2.25), is nonnegative. Property P1 can be proved as follows. Using (7.2.29), we can write

$$\mathbf{p}_-^t(\omega) = \frac{1}{2\pi i}\int_{-\infty}^{\infty}\frac{\mathbb{H}_-(\omega_1)\left[\mathbf{\Lambda}_+^t(\omega_1) - \frac{\mathbf{\Lambda}(t)}{i\omega_1^-}\right]}{\omega_1-\omega^+}d\omega_1, \tag{11.2.35}$$

giving

$$\frac{\partial\mathbf{p}_-^t(\omega)}{\partial\mathbf{\Lambda}(t)} = -\frac{1}{2\pi i}\int_{-\infty}^{\infty}\frac{\mathbb{H}_-(\omega_1)}{i\omega_1^-(\omega_1-\omega^+)}d\omega_1 = -\frac{\mathbb{H}_-(\omega)}{i\omega}, \tag{11.2.36}$$

where we have used (11.2.23). Also,

$$\frac{\partial \overline{\mathbf{p}_-^t(\omega)}}{\partial \boldsymbol{\Lambda}(t)} = \frac{\mathbb{H}_+(\omega)^\top}{i\omega}, \tag{11.2.37}$$

by virtue of (11.1.13). Thus,

$$\begin{aligned}
\frac{\partial}{\partial \boldsymbol{\Lambda}(t)} \frac{1}{2\pi} \int_{-\infty}^{\infty} \overline{\mathbf{p}_-^t(\omega)} \cdot \mathbf{p}_-^t(\omega) d\omega &= \frac{\partial}{\partial \boldsymbol{\Lambda}(t)} \frac{1}{2\pi} \int_{-\infty}^{\infty} \overline{\mathbf{p}_-^t(\omega)}^\top \mathbf{p}_-^t(\omega) d\omega \\
&= \frac{1}{2\pi i} \int_{-\infty}^{\infty} \frac{\mathbb{H}_+(\omega)}{\omega} \mathbf{p}_-^t(\omega) d\omega - \overline{\frac{1}{2\pi i} \int_{-\infty}^{\infty} \frac{\mathbb{H}_+(\omega)}{\omega} \mathbf{p}_-^t(\omega) d\omega} \\
&= \mathrm{Re}\left\{ \frac{1}{\pi i} \int_{-\infty}^{\infty} \frac{\mathbb{H}_+(\omega)}{\omega} \mathbf{p}_-^t(\omega) d\omega \right\}.
\end{aligned}$$

Recall that $\mathbb{H}_-$ vanishes linearly at the origin. Also, using (11.2.8), the frequency integral in (7.2.34), which must be real, can be written as

$$\begin{aligned}
-\frac{i}{\pi} \int_{-\infty}^{\infty} \frac{\mathbb{H}(\omega)}{\omega} \boldsymbol{\Lambda}_{r+}^t(\omega) d\omega &= -\frac{i}{\pi} \int_{-\infty}^{\infty} \frac{\mathbb{H}_+(\omega)}{\omega} [\mathbf{p}_-^t(\omega) - \mathbf{p}_+^t(\omega)] d\omega \\
&= -\frac{i}{\pi} \int_{-\infty}^{\infty} \frac{\mathbb{H}_+(\omega)}{\omega} \mathbf{p}_-^t(\omega) d\omega,
\end{aligned} \tag{11.2.38}$$

because the term involving $\mathbf{p}_+^t$ vanishes by Cauchy's theorem. Since the last quantity is real, P1 follows. □

It is shown in [92] that ψ_m is a free energy also under the definition of Coleman and Owen [75, 76]. This was for linear isothermal systems, but the proof applies to the present, more general, case.

11.2.5 Double Frequency Integral Form

We can write (11.2.17) in a more explicit form by carrying out the integration over ω. The following relationships are required. Firstly, we have

$$\begin{aligned}
\overline{\mathbb{H}_-(\omega_1) \boldsymbol{\Lambda}_r^t(\omega_1)} \cdot \mathbb{H}_-(\omega_2) \boldsymbol{\Lambda}_r^t(\omega_2) &= \overline{\boldsymbol{\Lambda}_r^t(\omega_1)} \cdot \mathbb{H}_-^*(\omega_1) \mathbb{H}_-(\omega_2) \boldsymbol{\Lambda}_r^t(\omega_2) \\
&= \overline{\boldsymbol{\Lambda}_r^t(\omega_1)} \cdot \mathbb{H}_+(\omega_1) \mathbb{H}_-(\omega_2) \boldsymbol{\Lambda}_r^t(\omega_2),
\end{aligned}$$

where (11.1.13), (A.2.7), and (A.2.8) have been used. Also, recalling (C.2.10) and (C.2.11), we can write for real ω_1, ω_2, and ω,

$$\begin{aligned}
\overline{\omega^-} &= \overline{\lim_{\alpha \to 0^+} (\omega - i\alpha)} = \omega^+, \\
\omega_1 - \omega^- &= \lim_{\alpha \to 0^+} (\omega_1 - \omega + i\alpha) = \omega_1^+ - \omega, \\
\omega_2 - \omega^+ &= \lim_{\alpha \to 0^+} (\omega_2 - \omega - i\alpha) = \omega_2^- - \omega,
\end{aligned}$$

where the limits are taken after any integrations are carried out. Finally,

$$\frac{1}{2\pi i}\int_{-\infty}^{\infty}\frac{d\omega}{(\omega_1^+ - \omega)(\omega_2^- - \omega)} = \frac{1}{\omega_1^+ - \omega_2^-},$$

by closing the contour on either $\Omega^{(+)}$ or $\Omega^{(-)}$.

Using these results, we can write the expression (11.2.17) for $\psi_m(t)$, where $\mathbf{p}_-^t(\omega)$ is given by (11.2.10) or (11.2.35), in the form

$$\begin{aligned}\psi_m(t) &= \phi(t) + \frac{i}{4\pi^2}\int_{-\infty}^{\infty}\int_{-\infty}^{\infty}\frac{\overline{\mathbf{\Lambda}_{r+}^t(\omega_1)}\cdot\mathbb{M}_m(\omega_1,\omega_2)\mathbf{\Lambda}_{r+}^t(\omega_2)}{\omega_1^+ - \omega_2^-}d\omega_1 d\omega_2,\\ \mathbb{M}_m(\omega_1,\omega_2) &= \mathbb{H}_+(\omega_1)\mathbb{H}_-(\omega_2).\end{aligned} \tag{11.2.39}$$

The notation in the denominator of the integral in $(11.2.39)_1$ means that if we integrate first over ω_1, it becomes $(\omega_1 - \omega_2^-)$, or if over ω_2 first, then it is $(\omega_1^+ - \omega_2)$. Also, using $(11.2.22)_3$, $D_m(t)$, given by (11.2.25), can be expressed as

$$D_m(t) = \frac{1}{4\pi^2}\int_{-\infty}^{\infty}\int_{-\infty}^{\infty}\overline{\mathbf{\Lambda}_{r+}^t(\omega_1)}\cdot\mathbb{M}_m(\omega_1,\omega_2)\mathbf{\Lambda}_{r+}^t(\omega_2)d\omega_1 d\omega_2.$$

Let us write the double integral in (11.2.39) as

$$\begin{aligned}P_-(t) &= \frac{1}{2\pi}\int_{-\infty}^{\infty}|\mathbf{p}_-^t(\omega)|^2 d\omega = \frac{i}{4\pi^2}\int_{-\infty}^{\infty}\int_{-\infty}^{\infty}\frac{A^t(\omega_1,\omega_2)}{\omega_1^+ - \omega_2^-}d\omega_1 d\omega_2,\\ A^t(\omega_1,\omega_2) &= \overline{\mathbf{\Lambda}_{r+}^t(\omega_1)}\cdot\mathbb{M}_m(\omega_1,\omega_2)\mathbf{\Lambda}_{r+}^t(\omega_2).\end{aligned} \tag{11.2.40}$$

In the same way, we obtain

$$P_+(t) = \frac{1}{2\pi}\int_{-\infty}^{\infty}|\mathbf{p}_+^t(\omega)|^2 d\omega = -\frac{i}{4\pi^2}\int_{-\infty}^{\infty}\int_{-\infty}^{\infty}\frac{A^t(\omega_1,\omega_2)}{\omega_1^- - \omega_2^+}d\omega_1 d\omega_2. \tag{11.2.41}$$

From (11.2.21), we see that this is the total dissipation up to time t. One can show that

$$\begin{aligned}R_-(t) &= \frac{i}{4\pi^2}\int_{-\infty}^{\infty}\int_{-\infty}^{\infty}\frac{B^t(\omega_1,\omega_2)}{\omega_1^+ - \omega_2^-}d\omega_1 d\omega_2 = 0,\\ B^t(\omega_1,\omega_2) &= \overline{\mathbf{\Lambda}_{r+}^t(\omega_1)}\cdot\mathbb{M}_m(\omega_2,\omega_1)\mathbf{\Lambda}_{r+}^t(\omega_2),\end{aligned} \tag{11.2.42}$$

by integrating over ω_2 for example and closing the contour on $\Omega^{(-)}$, since $\mathbb{H}_+$ and $\mathbf{\Lambda}_{r+}^t$ have no singularity in the lower half-plane. Furthermore, using the same procedure, one obtains

$$\begin{aligned}R_+(t) &= -\frac{i}{4\pi^2}\int_{-\infty}^{\infty}\int_{-\infty}^{\infty}\frac{B^t(\omega_1,\omega_2)}{\omega_1^- - \omega_2^+}d\omega_1 d\omega_2\\ &= \frac{1}{2\pi}\int_{-\infty}^{\infty}\overline{\mathbf{\Lambda}_{r+}^t(\omega)}\cdot\mathbb{H}(\omega)\mathbf{\Lambda}_{r+}^t(\omega)d\omega\\ &= P_-(t) + P_+(t),\end{aligned} \tag{11.2.43}$$

by virtue of $(7.5.7)_2$ and (11.2.16). Relation $(11.2.42)_1$ allows us to write (11.2.40) in the explicitly convergent form

$$P_-(t) = \frac{i}{4\pi^2}\int_{-\infty}^{\infty}\int_{-\infty}^{\infty}\frac{A^t(\omega_1,\omega_2)-B^t(\omega_1,\omega_2)}{\omega_1-\omega_2}d\omega_1 d\omega_2, \tag{11.2.44}$$

which is convenient for numerical evaluation. We can replace the $(\omega_1-\omega_2)$ in the denominator by $(\omega_1^+-\omega_2^-)$, which gives (11.2.40), or by $(\omega_1^--\omega_2^+)$, which gives the same result by way of (11.2.43) and (11.2.41). Relation (11.2.44) implies that the kernel

$$\mathbb{D}(\omega_1,\omega_2) = i\frac{[\mathbb{H}_+(\omega_1)\mathbb{H}_-(\omega_2)-\mathbb{H}_+(\omega_2)\mathbb{H}_-(\omega_1)]}{\omega_1-\omega_2}$$

is nonnegative in the operator sense, i.e., it must yield a nonnegative value for the integral, for all histories. Using very localized choices of $\mathbf{\Lambda}_+^t(\omega)$, we deduce that the "diagonal elements" of $\mathbb{D}(\omega_1,\omega_2)$ are nonnegative, as in Remark 7.1.3. Using a prime to denote differentiation, we can write these as

$$\mathbb{D}(\omega) = i[\mathbb{H}_+'(\omega)\mathbb{H}_-(\omega)-\mathbb{H}_+(\omega)\mathbb{H}_-'(\omega)] \geq \mathbf{0}, \qquad \omega\in\mathbb{R}. \tag{11.2.45}$$

Proposition 11.2.3. *We have*

$$\begin{aligned}\int_{-\infty}^{t}P_+(u)du &= -\frac{1}{4\pi^2}\int_{-\infty}^{\infty}\int_{-\infty}^{\infty}\frac{A^t(\omega_1,\omega_2)}{(\omega_1^--\omega_2^+)^2}d\omega_1 d\omega_2\\ &= -\frac{1}{4\pi^2}\int_{-\infty}^{\infty}\int_{-\infty}^{\infty}\left[\frac{A^t(\omega_1,\omega_2)}{(\omega_1^+-\omega_2^-)^2}+\frac{B^t(\omega_1,\omega_2)}{(\omega_1^--\omega_2^+)^2}\right]d\omega_1 d\omega_2\\ &\quad -\frac{1}{2\pi}\int_{-\infty}^{\infty}\overline{\mathbf{\Lambda}_{r+}^t(\omega)}\cdot\mathbb{D}(\omega)\mathbf{\Lambda}_{r+}^t(\omega)d\omega,\end{aligned} \tag{11.2.46}$$

where $\mathbb{D}(\cdot)$ *is defined by* (11.2.45) *and the integral on the left is assumed to exist for all finite values of* t.

Proof. Relation $(11.2.46)_1$ follows immediately, by time differentiation, using (7.2.30) and the relations

$$\int_{-\infty}^{\infty}\frac{\mathbb{H}_-(\omega_2)}{\omega_2^-(\omega_1^--\omega_2^+)^2}d\omega_2 = \int_{-\infty}^{\infty}\frac{\mathbb{H}_+(\omega_1)}{\omega_1^+(\omega_1^--\omega_2^+)^2}d\omega_1 = \mathbf{0},$$

which follow from the fact that $\mathbb{H}_\pm$ vanish linearly at the origin and Cauchy's theorem, closing the first integral on $\Omega^{(+)}$ and the second on $\Omega^{(-)}$. There can be no term independent of t, since the integral on the left of (11.2.46) vanishes as $t\to-\infty$. Equation $(11.2.46)_2$ can be verified similarly, on noting a cancellation between the derivatives of the single and double integral terms. Relations such as

$$\begin{aligned}\frac{1}{2\pi i}\int_{-\infty}^{\infty}\frac{\mathbb{H}_-(\omega_2)}{i\omega_2^-(\omega_1^+-\omega_2^-)^2}d\omega_2 &= \frac{d}{d\omega_1}\left[\frac{\mathbb{H}_-(\omega_1)}{i\omega_1}\right],\\ \frac{1}{2\pi i}\int_{-\infty}^{\infty}\frac{\mathbb{H}_+(\omega_1)}{i\omega_1^+(\omega_1^+-\omega_2^-)^2}d\omega_1 &= -\frac{d}{d\omega_2}\left[\frac{\mathbb{H}_+(\omega_2)}{i\omega_2}\right],\end{aligned}$$

are required. These follow from (B.1.3). The minus sign in the second relation is a consequence of the fact that the contour must be completed on $\Omega^{(-)}$ (Sect. B.1.1). □

The quantity $\mathbb{D}(\cdot)$ occurs in (11.7.10) in the context of the minimum free energy for sinusoidal histories.

11.3 Characterization of the Minimal State in the Frequency Domain

In this section, we show that the quantity $\mathbf{p}_-^t$, defined by (11.2.10), and occurring in $\psi_m(t)$, given by (11.2.17), is a function of the minimal state. Two histories $\mathbf{\Lambda}_1^t$ and $\mathbf{\Lambda}_2^t$ are equivalent if their difference $\mathbf{\Lambda}^t = \mathbf{\Lambda}_1^t - \mathbf{\Lambda}_2^t$ satisfies (7.4.3). Relation $(7.4.3)_2$ can be written in the form

$$\mathbf{F}^t(\tau) := \int_\tau^\infty \mathbb{L}'(s)\mathbf{\Lambda}^{t+\tau}(s)ds = \mathbf{0}, \qquad \forall\tau \geq 0. \tag{11.3.1}$$

We adopt the assumptions made before (7.4.5), so that $(7.4.3)_1$ and (7.4.4) are replaced by

$$\mathbf{\Lambda}_1(t+\tau) = \mathbf{\Lambda}_2(t+\tau), \quad \tau \geq 0. \tag{11.3.2}$$

Condition (11.3.2) gives that $\mathbf{\Lambda}^{t+\tau}(s)$ is equal to zero for $\tau \geq s$. Let us identify $\mathbb{L}'$ with its odd extension on $\mathbb{R}$, so that (7.2.25) applies. Then, $\mathbf{F}^t(\tau)$ can be rewritten in terms of Fourier transforms:

$$\begin{aligned}\mathbf{F}^t(\tau) = \int_{-\infty}^\infty \mathbb{L}'(s)\mathbf{\Lambda}^{t+\tau}(s)ds &= \frac{1}{2\pi}\int_{-\infty}^\infty \overline{\mathbb{L}'_F}(\omega)\mathbf{\Lambda}_+^{t+\tau}(\omega)d\omega \\ &= -\frac{i}{\pi}\int_{-\infty}^\infty \mathbb{L}'_s(\omega)\mathbf{\Lambda}_+^{t+\tau}(\omega)d\omega.\end{aligned}$$

Moreover, note that

$$\mathbf{\Lambda}_+^{t+\tau}(\omega) = \int_0^\infty \mathbf{\Lambda}^{t+\tau}(s)e^{-i\omega s}ds = \int_\tau^\infty \mathbf{\Lambda}^t(s-\tau)e^{-i\omega(s-\tau)}ds\, e^{-i\omega\tau} = \mathbf{\Lambda}_+^t(\omega)e^{-i\omega\tau},$$

which yields

$$\mathbf{F}^t(\tau) = \frac{i}{\pi}\int_{-\infty}^\infty \frac{\mathbb{H}(\omega)}{\omega}\mathbf{\Lambda}_+^t(\omega)e^{-i\omega\tau}\,d\omega. \tag{11.3.3}$$

This will be taken as the definition of $\mathbf{F}^t(\tau)$ for $\tau \in \mathbb{R}$.

Remembering the factorization of $\mathbb{H}(\omega)$ given by (11.1.11), (11.3.3) can be rewritten as

$$\mathbf{F}^t(\tau) = \frac{i}{\pi}\int_{-\infty}^\infty \frac{\mathbb{H}_+(\omega)}{\omega}\mathbb{H}_-(\omega)\mathbf{\Lambda}_+^t(\omega)e^{-i\omega\tau}\,d\omega, \tag{11.3.4}$$

and the substitution of (11.2.8) into (11.3.4) yields

$$\mathbf{F}^t(\tau) = \frac{i}{\pi}\int_{-\infty}^\infty \frac{\mathbb{H}_+(\omega)}{\omega}\left[\mathbf{p}_-^t(\omega) - \mathbf{p}_+^t(\omega)\right]e^{-i\omega\tau}\,d\omega. \tag{11.3.5}$$

Observe that $\mathbb{H}_+(z)$, $\mathbf{p}_+^t(z)$, and $e^{-iz\tau}$ ($\tau > 0$) are analytic functions in the lower half-plane $z \in \Omega^{(-)}$, their product converging strongly to zero at infinity, so that by Cauchy's theorem, (11.3.5) reduces to (cf. (11.2.38))

$$\mathbf{F}^t(\tau) = \frac{i}{\pi}\int_{-\infty}^{\infty} \frac{\mathbb{H}_+(\omega)}{\omega}\mathbf{p}_-^t(\omega)e^{-i\omega\tau}\,d\omega. \tag{11.3.6}$$

This form of $\mathbf{F}^t$ allows us to prove the following theorem characterizing a minimal state in the frequency domain.

Theorem 11.3.1. *For every material with linear memory and a symmetric (required for factorization) relaxation function* $\mathbb{L}$, *a given history* $\mathbf{\Lambda}^t$ *is equivalent to the zero history* $\mathbf{0}^\dagger$ *if and only if the* $\mathbf{p}_-^t$ *related to* $\mathbf{\Lambda}^t$ *by* (11.2.10) *with* (11.2.8) *is such that*

$$\mathbf{p}_-^t(\omega) = \mathbf{0}, \qquad \forall\omega \in \mathbb{R}.$$

Observe that the theorem in effect states that

$$\mathbf{F}^t(\tau) = \mathbf{0}\ \forall\tau \geq 0 \qquad \Longleftrightarrow \qquad \mathbf{p}_-^t(\omega) = \mathbf{0}\ \forall\omega \in \mathbb{R}. \tag{11.3.7}$$

Proof. The statement relating to the left-pointing arrow of (11.3.7) follows trivially from (11.3.6). In order to prove the statement relating to the right arrow, let us invert the Fourier transform in (11.3.6) to obtain

$$\begin{aligned}\mathbf{f}^t(\omega) &= \frac{i}{\pi}\frac{\mathbb{H}_+(\omega)}{\omega}\mathbf{p}_-^t(\omega) = \frac{1}{2\pi}\int_{-\infty}^{\infty}\mathbf{F}^t(\tau)e^{iw\tau}\,d\tau \\ &= \frac{1}{2\pi}\int_{-\infty}^{0}\mathbf{F}^t(\tau)e^{iw\tau}d\tau = \frac{1}{2\pi}\int_{0}^{\infty}\mathbf{F}^t(-u)e^{-iwu}\,du.\end{aligned} \tag{11.3.8}$$

It follows from Proposition C.2.1 that $\mathbf{f}^t$ (the Fourier transform of $\mathbf{F}^t(-u)$, a function that is zero on $\mathbb{R}^{--}$) is analytic in $\Omega^{(-)}$. The zeros in rows or columns of $\mathbb{H}_+$ (zeros of $\det\mathbb{H}_+$) cannot cancel singularities of $\mathbf{p}_-^t$, since all such zeros are in $\Omega^{(+)}$. Also, any branch-cut singularity of $\mathbb{H}_+$ is in $\Omega^{(+)}$, and those of $\mathbf{p}_-^t$ are in $\Omega^{(-)}$, so there can be no neutralization of such singularities. Thus $\mathbf{p}_-^t$ must be analytic in $\Omega^{(-)}$ and therefore in Ω. It goes to zero at infinity and must therefore be zero everywhere by Liouville's theorem (Sect. B.1.3).

□

Note that Theorem 11.3.1 is in effect saying that if a history is equivalent to the zero history, then the minimum free energy vanishes, or (7.4.8) holds. We now show that this implies the following result.

Proposition 11.3.2. *If two states are equivalent, then* (7.4.6) *holds, so that the minimum free energy, given by* (11.2.17), *is a functional of the minimal state. Furthermore,* (7.4.8) *is true.*

Proof. Let $(\mathbf{\Lambda}_1^t, \mathbf{\Lambda}(t))$ and $(\mathbf{\Lambda}_2^t, \mathbf{\Lambda}(t))$ be equivalent states. Then the difference of their histories is equivalent to the zero history, as argued before (7.4.5). Also, let $\mathbf{p}_-^t(\omega, \mathbf{\Lambda}_1^t)$ indicate this quantity defined by (11.2.10) for a history $\mathbf{\Lambda}_1^t$ and similarly for $\mathbf{p}_-^t(\omega, \mathbf{\Lambda}_2^t)$. Then,

$$\mathbf{p}_-^t(\cdot, \mathbf{\Lambda}_1^t) - \mathbf{p}_-^t(\cdot, \mathbf{\Lambda}_2^t) = \mathbf{p}_-^t(\cdot, \mathbf{\Lambda}_1^t - \mathbf{\Lambda}_2^t) = \mathbf{0}$$

by the linearity of $\mathbf{p}_-^t$ and Theorem 11.3.1. Thus, (7.4.6) and (7.4.8) follow immediately. Also, (7.4.12) must be true, since (7.4.6) and (7.4.8) hold. □

The observations around (7.4.10) and (7.4.11) are relevant in the present context.

11.4 The Space of States and Processes

We recall from Definition 4.2.1 of the minimum free energy that it must be defined on the entire space of states. From (11.2.17), this is the space of relative histories and current values $(\mathbf{\Lambda}_r^t, \mathbf{\Lambda}(t))$ such that $\mathbf{p}_-^t$, defined by (11.2.10) with (11.2.8), belongs to $L^2(\mathbb{R})$. As we have seen in Sect. 11.2.2, the Fourier transform $\mathbf{\Lambda}_{r+}^t$ need not exist. However, $\mathbf{P}^t$, given by (11.2.32) in terms of $\mathbf{J}_F^t$, must be finite for $\omega \in \mathbb{R}$.

Recalling $(11.2.18)_{3,4}$, we define the space of processes to be [145]

$$\begin{aligned} \mathcal{H}_\Gamma(\mathbb{R}^-) = \Big\{ \boldsymbol{\phi} : \mathbb{R}^- \mapsto \Gamma; \frac{1}{2} \int_{-\infty}^0 \int_{-\infty}^0 \boldsymbol{\phi}(s) \cdot \mathbb{L}(|s-u|) \boldsymbol{\phi}(u) ds du \\ = \frac{1}{2\pi} \int_{-\infty}^{\infty} \overline{\boldsymbol{\phi}_-(\omega)} \cdot \frac{\mathbb{H}(\omega)}{\omega^2} \boldsymbol{\phi}_-(\omega) d\omega < \infty \Big\}. \end{aligned} \tag{11.4.1}$$

The dual of this space is defined as

$$\mathcal{H}'_\Gamma(\mathbb{R}^-) = \left\{ \mathbf{f} : \mathbb{R}^- \mapsto \Gamma; \left| \int_{-\infty}^0 \mathbf{f}(u) \boldsymbol{\phi}(u) du \right| < \infty \forall \boldsymbol{\phi} \in \mathcal{H}_\Gamma(\mathbb{R}^-) \right\}.$$

The space of histories is defined as those for which

$$\mathbf{J}^t \in \mathcal{H}'_\Gamma(\mathbb{R}^-), \tag{11.4.2}$$

where $\mathbf{J}^t(\cdot)$ is related to $\mathbf{I}(\cdot, \mathbf{\Lambda}_r^t)$ by (11.2.31).

We now prove the following result of Gentili [145].

Proposition 11.4.1. *Given the relative history* $\mathbf{\Lambda}_r^t : \mathbb{R}^+ \mapsto \Gamma$, *then* $\mathbf{J}^t \in \mathcal{H}'_\Gamma(\mathbb{R}^-)$ *if and only if* $\mathbf{p}_-^t \in L^2(\mathbb{R})$.

Proof. For any $\boldsymbol{\phi} \in \mathcal{H}_\Gamma(\mathbb{R}^-)$,

$$\begin{aligned} \int_{-\infty}^0 \mathbf{J}^t(s) \cdot \boldsymbol{\phi}(s) ds &= \frac{1}{2\pi} \int_{-\infty}^{\infty} \mathbf{J}_F^t(\omega) \cdot \overline{\boldsymbol{\phi}_-(\omega)} d\omega \\ &= \frac{1}{2\pi} \int_{-\infty}^{\infty} \frac{\omega}{2i} [\mathbb{H}_+(\omega)]^{-1} \mathbf{J}_F^t(\omega) \cdot \frac{2i}{\omega} \mathbb{H}_+^\top(\omega) \overline{\boldsymbol{\phi}_-(\omega)} d\omega \\ &= -\frac{i}{\pi} \int_{-\infty}^{\infty} \mathbf{P}^t(\omega) \cdot \frac{1}{\omega} \overline{\mathbb{H}_-(\omega) \boldsymbol{\phi}_-(\omega)} d\omega, \end{aligned}$$

where (11.2.32) has been used. The $\mathbf{p}_+^t$ term in $\mathbf{P}^t$ yields zero by Cauchy's theorem and Remark B.1.2, so we have

$$\begin{aligned} \int_{-\infty}^0 \mathbf{J}^t(s) \cdot \boldsymbol{\phi}(s) ds &= -\frac{i}{\pi} \int_{-\infty}^{\infty} \mathbf{p}_-^t(\omega) \cdot \frac{1}{\omega} \overline{\mathbb{H}_-(\omega) \boldsymbol{\phi}_-(\omega)} d\omega \\ &\le \frac{1}{2\pi} \int_{-\infty}^{\infty} |\mathbf{p}_-^t(\omega)|^2 d\omega + \frac{1}{2\pi} \int_{-\infty}^{\infty} \frac{1}{\omega^2} |\mathbb{H}_-(\omega) \boldsymbol{\phi}_-(\omega)|^2 d\omega. \end{aligned}$$

If $\mathbf{p}_-^t \in L^2(\mathbb{R})$, the first term is bounded. Also, the second term is bounded by (11.4.1). It follows that (11.4.2) holds. If $\mathbf{J}^t \in \mathcal{H}'_\Gamma(\mathbb{R}^-)$, then from (11.2.33), we see that $\mathbf{p}_-^t \in L^2(\mathbb{R})$. □

11.5 Limiting Properties of the Optimal Future Continuation

We draw attention to certain properties of the optimal future continuation. From (11.2.14) and $(11.2.24)_1$, it follows that

$$\mathbf{\Lambda}^t_m(\omega) \underset{\omega\to\infty}{\longrightarrow} \frac{\mathbb{H}^{-1}_{\frac{1}{2}}\mathbf{K}(t)}{i\omega}, \tag{11.5.1}$$

where

$$\mathbb{H}_{\frac{1}{2}} = \mathbb{H}_-(\infty)\ (= \mathbb{H}_+(\infty)).$$

The last relation holds for commutative factors, by virtue of (11.6.2) below. Thus, by (C.2.16), the relative optimal continuation, given by (11.2.5), has the form at $s = 0$

$$\mathbf{\Lambda}^t_{ro}(0) = -\mathbb{H}^{-1}_{\frac{1}{2}}\mathbf{K}(t), \tag{11.5.2}$$

and the optimal continuation $\mathbf{\Lambda}^t_o(0)$ is given by

$$\mathbf{\Lambda}^t_o(0) = \mathbf{\Lambda}(t) - \mathbb{H}^{-1}_{\frac{1}{2}}\mathbf{K}(t). \tag{11.5.3}$$

Therefore, the optimal continuation involves a sudden discontinuity at time t, the magnitude of which is related to the rate of dissipation, as we see from (11.2.25).

Also, putting

$$\mathbb{H}_1(\omega) = \frac{1}{\omega}\mathbb{H}_-(\omega), \tag{11.5.4}$$

we have, with the aid of (11.2.10),

$$\mathbf{\Lambda}^t_m(\omega) \approx -\frac{[\mathbb{H}_1(0)]^{-1}}{2\pi i\omega^+}\int_{-\infty}^{\infty}\mathbb{H}_1(\omega')\mathbf{\Lambda}^t_{r+}(\omega')d\omega'$$

as $\omega \to 0$. The quantity ω in the denominator is replaced by ω^+, since the singularities of $\mathbf{\Lambda}^t_m(\omega)$ must be in $\Omega^{(-)}$. Then, with the aid of (C.2.11), we have

$$\mathbf{\Lambda}^t_{ro}(-\infty) = \frac{[\mathbb{H}_1(0)]^{-1}}{2\pi}\int_{-\infty}^{\infty}\mathbb{H}_1(\omega')\mathbf{\Lambda}^t_{r+}(\omega')d\omega' \tag{11.5.5}$$

and

$$\mathbf{\Lambda}^t_o(-\infty) = \mathbf{\Lambda}^t_{ro}(-\infty) + \mathbf{\Lambda}(t).$$

This quantity is in general nonzero.

11.6 Time-Independent Eigenspaces

We now make the assumption (7.1.36), which immediately yields

$$\mathbb{L}'(t) = \sum_{k=1}^{m} L'_k(t)\mathbb{B}_k, \quad t \in \mathbb{R}.$$

Taking Fourier transforms gives

$$\mathbb{L}'_+(\omega) = \sum_{k=1}^{m} L'_{k+}(\omega)\mathbb{B}_k, \qquad \omega \in \mathbb{R}.$$

This relation can also be written at $-\omega$. Adding and subtracting give that $\mathbb{L}'_c$ and $\mathbb{L}'_s$ can also be represented in this manner, and furthermore,

$$\mathbb{H}(\omega) = -\omega\mathbb{L}'_s(\omega) = \sum_{k=1}^{m} H_k(\omega)\mathbb{B}^k,$$
$$H_k(\omega) > 0, \qquad \omega \in \mathbb{R}\setminus\{0\}, \quad k = 1, \ldots, m.$$

Using the technique of Sect. 11.1.1, the quantities H_k can be factorized into H_{k+} and H_{k-}, where from (11.1.17),

$$H_{k\pm}(\omega) = H_{k\mp}(-\omega) = \overline{H_{k\mp}(\omega)}; \quad H_k(\omega) = |H_{k\pm}(\omega)|^2. \tag{11.6.1}$$

We put

$$H_k(\infty) = H_{k\infty}, \quad k = 1, 2, \ldots, m.$$

Since the $\{\mathbb{B}^k\}$ are orthonormal projectors, it follows that

$$\mathbb{H}(\omega) = \sum_{k=1}^{m} H_{k+}(\omega)H_{k-}(\omega)\mathbb{B}^k = \mathbb{H}_+(\omega)\mathbb{H}_-(\omega),$$

where

$$\mathbb{H}_\pm(\omega) = \sum_{k=1}^{m} H_{k\pm}(\omega)\mathbb{B}^k = \mathbb{H}_\pm^\top(\omega).$$

The last relation follows from the symmetry of the $\mathbb{B}^k$. Thus, $\mathbb{H}_\pm$ are symmetric for all frequencies. We see that the factors $\mathbb{H}_+(\omega)$ and $\mathbb{H}_-(\omega)$ commute, so that they are normal transformations; see the comment relating to (A.2.11). Recalling (11.1.18) and the comment after (11.1.14), we see that

$$\mathbb{H}_\pm(\infty) = \mathbb{H}_\infty^{1/2}. \tag{11.6.2}$$

The quantities $\mathbb{H}_\pm$ also commute when evaluated at different frequencies, by virtue of (A.2.11). It follows that products of these factors at the same or different frequencies are symmetric. From (11.6.1), we have†

$$\mathbb{H}_\pm(\omega) = \mathbb{H}_\mp(-\omega) = \overline{\mathbb{H}_\mp(\omega)}, \tag{11.6.3}$$

† These relations allow us to show that

$$\overline{\mathbf{p}^t_\pm(\omega)} = \mathbf{p}^t_\pm(-\omega), \qquad \omega \in \mathbb{R},$$

with the aid of (11.2.10) and (C.1.7).

which also hold for complex ω, where the rightmost term involves the complex conjugate of the functional form, leaving ω unchanged, or

$$\overline{\mathbb{H}_{\mp}}(\omega) = \overline{\mathbb{H}_{\mp}(\overline{\omega})}.$$

These relations are consistent with but more detailed than (11.1.11). They reduce the nonuniqueness of the factorization to an arbitrariness of sign on each eigenspace. Also, $\mathbb{M}_m$, given by $(11.2.39)_2$, can be expanded on this basis:

$$\mathbb{M}_m(\omega_1, \omega_2) = \sum_{k=1}^{m} M_k^{(m)}(\omega_1, \omega_2)\mathbb{B}^k,$$

$$M_k^{(m)}(\omega_1, \omega_2) = H_{k+}(\omega_1)H_{k-}(\omega_2).$$

In the basis $\{\mathbf{B}_k\}$ (defined after (7.1.36)) and $\{\mathbb{B}_k\}$, the individual components of each of the relevant quantities obey the relationships that hold in the scalar case. We can expand any member of Γ in this basis; in particular,

$$\boldsymbol{\Lambda}(t) = \sum_{k=1}^{m} \Lambda_k(t)\mathbf{B}^k, \qquad \boldsymbol{\Lambda}^t(s) = \sum_{k=1}^{m} \Lambda_k^t(s)\mathbf{B}^k,$$

$$\boldsymbol{\Lambda}_r^t(s) = \sum_{k=1}^{m} \Lambda_{kr}^t(s)\mathbf{B}^k, \qquad \boldsymbol{\Lambda}_{r\pm}^t(\omega) = \sum_{k=1}^{m} \Lambda_{kr\pm}^t(\omega)\mathbf{B}^k.$$

Scalar quadratic forms, such as the memory-dependent part of a free energy or a rate of dissipation, are given by the sum of contributions to this quantity from each eigenspace. In particular,

$$\psi_m(t) - \phi(t) = \sum_{k=1}^{m} \psi_{mk}^{(k)}(t),$$

where $\psi_{mk}^{(k)}$ is the minimum free energy relating to the scalar problem for H_k. Such relations follow readily from the orthonormality and time (frequency) independence of the basis.

In the particular example discussed in Sect. 7.1.5, we can write any free energy (not just the minimum free energy) in the form

$$\psi(t) - \phi(t) = 5\psi_S(t) + \psi_{B1}(t) + \psi_{B2}(t) + 3\psi_q(t),$$

where $\psi_S(t)$, $\psi_{B1}(t)$, $\psi_{B2}(t)$, and $\psi_q(t)$ are the memory-dependent parts of the free energies corresponding to the scalar problems with relaxation function derivatives G_S', G_1', G_2', and V_m' and constants $G_{S\infty}, G_1$, and G_2. The coefficient 5 reflects the degeneracy or symmetry of the five-dimensional representation of the rotation group corresponding to shear deformation in a mechanically isotropic material, while the coefficient 3 results from thermal isotropy. For completely linear materials, we can include the equilibrium terms explicitly in these formulas.

11.7 The Minimum Free Energy for Sinusoidal Histories

Consider a history and current value $(\mathbf{\Lambda}^t, \mathbf{\Lambda}(t))$ defined by

$$\mathbf{\Lambda}(t) = \mathbf{C}e^{i\omega_- t} + \overline{\mathbf{C}}e^{-i\omega_+ t}, \quad \mathbf{\Lambda}^t(s) = \mathbf{\Lambda}(t-s), \tag{11.7.1}$$

where $\mathbf{C}$ is an amplitude in Γ and $\overline{\mathbf{C}}$ is its complex conjugate. Furthermore,

$$\omega_- = \omega_0 - i\eta, \quad \omega_+ = \overline{\omega_-}, \quad \omega_0 \in \mathbb{R}, \quad \eta \in \mathbb{R}^{++}.$$

The quantity η is introduced to ensure finite results in certain quantities. The quantity $\mathbf{\Lambda}^t_+$ has the form

$$\mathbf{\Lambda}^t_+(\omega) = \mathbf{C}\frac{e^{i\omega_- t}}{i(\omega + \omega_-)} + \overline{\mathbf{C}}\frac{e^{-i\omega_+ t}}{i(\omega - \omega_+)},$$

and the Fourier transform of the relative history $\mathbf{\Lambda}^t_r(s) = \mathbf{\Lambda}^t(s) - \mathbf{\Lambda}(t)$, namely $\mathbf{\Lambda}^t_{r+}(\omega)$, is given by

$$\mathbf{\Lambda}^t_{r+}(\omega) = \mathbf{\Lambda}^t_+(\omega) - \frac{\mathbf{\Lambda}(t)}{i\omega^-} = -\mathbf{C}\frac{\omega_-}{\omega^-}\frac{e^{i\omega_- t}}{i(\omega + \omega_-)} + \overline{\mathbf{C}}\frac{\omega_+}{\omega^-}\frac{e^{-i\omega_+ t}}{i(\omega - \omega_+)}. \tag{11.7.2}$$

From (7.1.17) and $(7.1.14)_5$, the generalized stress has the form

$$\begin{aligned}
\mathbf{\Sigma}(t) &= \mathbf{\Sigma}_0(t) + \mathbf{\Sigma}_h(t), \\
\mathbf{\Sigma}_h(t) &= \int_0^\infty \mathbb{L}'(s)\mathbf{\Lambda}^t(s)\,ds \\
&= \mathbb{L}'_+(\omega_-)\mathbf{C}e^{i\omega_- t} + \mathbb{L}'_+(-\omega_+)\overline{\mathbf{C}}e^{-i\omega_+ t}, \\
\mathbf{\Sigma}_0(t) &= \mathbf{\Sigma}_e(t) + \mathbb{L}_0\mathbf{\Lambda}(t).
\end{aligned} \tag{11.7.3}$$

The expression for $\mathbf{\Sigma}_h(t)$ reduces to that given in $(7.2.6)_3$ as $\eta \to 0$.

The work $W(t)$ done on the material to achieve the state $(\mathbf{\Lambda}^t, \mathbf{\Lambda}(t))$ is given by $(7.5.1)_1$, which in this context becomes

$$\begin{aligned}
W(t) &= \phi(t) + \frac{1}{2}\mathbf{\Lambda}(t)\cdot\mathbb{L}_0\mathbf{\Lambda}(t) \\
&+ \frac{1}{2}\left[\mathbf{C}\cdot\mathbb{L}'_+(\omega_-)\mathbf{C}e^{2i\omega_- t} + \overline{\mathbf{C}}\cdot\mathbb{L}'_+(-\omega_+)\overline{\mathbf{C}}e^{-2i\omega_+ t}\right] \\
&+ \overline{\mathbf{C}}\cdot[\omega_-\mathbb{L}'_+(-\omega_+) - \omega_+\mathbb{L}'_+(\omega_-)]\mathbf{C}\frac{e^{i(\omega_- - \omega_+)t}}{(\omega_- - \omega_+)},
\end{aligned} \tag{11.7.4}$$

where the symmetry of $\mathbb{L}_+$ has been used. This quantity diverges as $\eta \to 0$, as would be expected on physical grounds. Taking the limit $\eta \to 0$ in the terms that are convergent, we can write this in the form

$$\begin{aligned}
W(t) &= \phi(t) + \frac{1}{2}\mathbf{\Lambda}(t)\cdot\mathbb{L}_0\mathbf{\Lambda}(t) \\
&+ \frac{1}{2}\left[\mathbf{C}\cdot\mathbb{L}'_+(\omega_0)\mathbf{C}e^{2i\omega_0 t} + \overline{\mathbf{C}}\cdot\mathbb{L}'_+(-\omega_0)\overline{\mathbf{C}}e^{-2i\omega_0 t}\right] \\
&+ \overline{\mathbf{C}}\cdot\left[\mathbb{L}'_c(\omega_0) - \omega_0\frac{d}{d\omega_0}\mathbb{L}'_c(\omega_0) - 2\omega_0 t\mathbb{L}'_s(\omega_0)\right]\mathbf{C} \\
&- \mathbf{C}\cdot\mathbb{L}'_s(\omega_0)\mathbf{C}\frac{\omega_0}{\eta},
\end{aligned} \tag{11.7.5}$$

on using (7.2.1). The divergence is associated with $\mathbb{L}'_s$, which is physically reasonable.

We shall require the relation

$$\overline{\mathbb{H}_{\pm}(\omega)} = \mathbb{H}_{\pm}(-\overline{\omega}) \tag{11.7.6}$$

for complex ω, which follows from (11.6.3). The minimum free energy $\psi_m(t)$ is given by (11.2.17). Using (11.7.2), we evaluate the integral in (11.2.9) by closing the contour on $\Omega^{(+)}$ to obtain

$$\mathbf{p}^t_+(\omega) = -\left[\frac{e^{i\omega_- t}}{i(\omega+\omega_-)}\mathbb{H}_-(-\omega_-)\mathbf{C} + \frac{e^{-i\omega_+ t}}{i(\omega-\omega_+)}\mathbb{H}_-(\omega_+)\overline{\mathbf{C}}\right] \tag{11.7.7}$$

and

$$\mathbf{p}^t_-(\omega) = \mathbb{H}_-(\omega)\mathbf{\Lambda}^t_{r+}(\omega) + \mathbf{p}^t_+(\omega).$$

The expression for $\psi_m(t)$ can be obtained from $(11.2.21)_2$ combined with (11.7.4). From (11.7.7), we obtain

$$\begin{aligned}\frac{1}{2\pi}\int_{-\infty}^{\infty}|\mathbf{p}^t_+(\omega)|^2 d\omega = &-\frac{ie^{2i\omega_- t}}{2\omega_-}\mathbf{C}\cdot\mathbb{H}^2_-(-\omega_-)\mathbf{C} + \frac{ie^{-2i\omega_+ t}}{2\omega_+}\overline{\mathbf{C}}\cdot\mathbb{H}^2_-(\omega_+)\overline{\mathbf{C}} \\ &- \frac{2ie^{i(\omega_- -\omega_+)t}}{(\omega_- - \omega_+)}\overline{\mathbf{C}}\cdot\mathbb{H}_-(\omega_+)\mathbb{H}_-(-\omega_-)\mathbf{C},\end{aligned} \tag{11.7.8}$$

where (11.7.6) has been used. It will be observed that the last term diverges in the limit $\eta \to 0$. The quantity given by (11.7.8) in the limit $\eta \to 0$ is in fact the total dissipation over the history, given by (11.2.21), so this divergence is an expression of a physically obvious fact. Its derivative is the rate of dissipation.

Taking the limit $\eta \to 0$ in the convergent terms, we obtain

$$\begin{aligned}\frac{1}{2\pi}\int_{-\infty}^{\infty}|\mathbf{p}^t_+(\omega)|^2 d\omega = &-\frac{ie^{2i\omega_0 t}}{2\omega_0}\mathbf{C}\cdot\mathbb{H}^2_-(-\omega_0)\mathbf{C} + \frac{ie^{-2i\omega_0 t}}{2\omega_0}\overline{\mathbf{C}}\cdot\mathbb{H}^2_-(\omega_0)\overline{\mathbf{C}} \\ &- \overline{\mathbf{C}}\cdot[2t\mathbb{H}(\omega_0) + \mathbb{D}(\omega_0)]\mathbf{C} + \frac{1}{\eta}\overline{\mathbf{C}}\cdot\mathbb{H}(\omega_0)\mathbf{C},\end{aligned}$$

where $\mathbb{D}$ is defined by (11.2.45).

From (11.7.4), (11.7.8), and $(11.2.21)_2$, taking the limit $\eta \to 0$, we obtain

$$\begin{aligned}\psi_m(t) = &\ \phi(t) + \frac{1}{2}\mathbf{\Lambda}(t)\cdot\mathbb{L}_0\mathbf{\Lambda}(t) \\ &+ \mathbf{C}\cdot\mathbb{B}_1(\omega_0)\mathbf{C}e^{2i\omega_0 t} + \overline{\mathbf{C}}\cdot\overline{\mathbb{B}}_1(\omega_0)\overline{\mathbf{C}}e^{-2i\omega_0 t} + \overline{\mathbf{C}}\cdot\mathbb{B}_2(\omega_0)\mathbf{C},\end{aligned} \tag{11.7.9}$$

where

$$\begin{aligned}\mathbb{B}_1(\omega_0) &= \frac{1}{2}\left[\mathbb{L}'_+(\omega_0) + \frac{i}{\omega_0}\mathbb{H}^2_-(-\omega_0)\right], \\ \mathbb{B}_2(\omega_0) &= \mathbb{L}'_c(\omega_0) - \omega_0\frac{d}{d\omega_0}\mathbb{L}'_c(\omega_0) + \mathbb{D}(\omega_0).\end{aligned} \tag{11.7.10}$$

The divergent terms and those proportional to t cancel. The most interesting contribution to ψ_m is the rightmost term of (11.7.9), which gives the average over a time cycle

$$(\psi_m)_{av} = \overline{\mathbf{C}} \cdot \mathbb{B}_2(\omega_0)\mathbf{C}.$$

We can express this quantity in terms of $\mathbb{H}$ and its factors, using (7.2.26). Note that $\mathbb{B}_2$ must be a nonnegative quantity in general for all $\omega \in \mathbb{R}$. We recall from (11.2.45) that $\mathbb{D}$ is nonnegative for all $\omega \in \mathbb{R}$.

The rate of dissipation is given by (11.2.25) and $(11.2.22)_3$. Using (11.7.2) and closing on Ω^+, we find that

$$\mathbf{K}(t) = \mathbb{H}_-(-\omega_0)\mathbf{C}e^{i\omega_0 t} + \mathbb{H}_-(\omega_0)\overline{\mathbf{C}}e^{-i\omega_0 t}$$

on taking $\eta \to 0$. Therefore,

$$D(t) = \mathbf{C} \cdot \mathbb{H}_-^2(-\omega_0)\mathbf{C}e^{2i\omega_0 t} + \overline{\mathbf{C}} \cdot \mathbb{H}_-^2(\omega_0)\overline{\mathbf{C}}e^{-2i\omega_0 t} + 2\overline{\mathbf{C}} \cdot \mathbb{H}(\omega)\mathbf{C}. \qquad (11.7.11)$$

One may check that (11.2.19) holds, using (11.7.1), (11.7.3) in the limit $\eta \to 0$, (11.7.9), and (11.7.11).

Similar results apply to the free energies and dissipations in Sect. 16.4.

11.8 Example: Viscoelastic Materials

We write out the main results of the present chapter for perhaps the most important special case, namely the isothermal theory, where temperature variation over time and space is neglected. Only the mechanical equations are relevant.‡

Recall (8.1.16), (8.1.18), and the fact that $\mathbb{G}'_s$ is an odd function of ω. Following (7.2.22) and (11.1.11), let us define $\mathbb{H} \in \mathrm{Lin(Sym)}$ by

$$\mathbb{H}(\omega) = -\omega\mathbb{G}'_s(\omega) = \mathbb{H}_+(\omega)\mathbb{H}_-(\omega) \geq \mathbf{0}, \quad \omega \in \mathbb{R}, \qquad (11.8.1)$$

where

$$\mathbb{H}(\infty) = -\mathbb{G}'(0). \qquad (11.8.2)$$

We have kept the same notation for simplicity. The work function, giving the amount of mechanical work required to achieve the state $(\mathbf{E}^t, \mathbf{E}(t))$, has the form (cf. (7.5.1), (7.5.3), and (7.5.7))

‡ The results of Sect. 13.1.1 overlap to some degree with those in this section. The former results are, however, derived specifically for completely linear viscoelastic solids (as discussed in Sects. 8.1–8.7), using a somewhat different but equivalent methodology to that developed in the present chapter.

$$
\begin{aligned}
\widetilde{W}(\mathbf{E}^t, \mathbf{E}(t)) = W(t) &= \phi(-\infty) + \int_{-\infty}^{t} \widehat{\mathbf{S}}(u) \cdot \dot{\mathbf{E}}(u) du \\
&= \phi(t) + \frac{1}{2} \int_0^\infty \int_0^\infty \mathbb{G}_{12}(|s-u|) \mathbf{E}_r^t(u) \cdot \mathbf{E}_r^t(s) du ds \\
&= \phi(t) + \frac{1}{2\pi} \int_{-\infty}^{\infty} \overline{\mathbf{E}_{r+}^t(\omega)} \cdot \mathbb{H}(\omega) \mathbf{E}_{r+}^t(\omega) d\omega,
\end{aligned}
\tag{11.8.3}
$$

$$
\mathbf{E}_{r+}^t(\omega) = \mathbf{E}_+^t(\omega) - \frac{\mathbf{E}(t)}{i\omega^-},
$$

where the constitutive equation (8.1.1) has been used. Again, keeping some notation from (11.2.8),

$$
\mathbb{H}_-(\omega) \mathbf{E}_{r+}^t(\omega) = \mathbf{p}_-^t(\omega) - \mathbf{p}_+^t(\omega),
$$

where

$$
\mathbf{p}_\pm^t(\omega) = \frac{1}{2\pi i} \int_{-\infty}^{\infty} \frac{\mathbb{H}_-(\omega') \mathbf{E}_{r+}^t(\omega')}{\omega' - \omega^\mp} d\omega', \tag{11.8.4}
$$

which is the analogue of (11.2.10). We note that

$$
\mathbf{E}_{r+}^t(\omega) \sim \omega^{-2} \tag{11.8.5}
$$

at large ω and write the analogue of (7.2.30) as

$$
\frac{d}{dt} \mathbf{E}_+^t(\omega) = -i\omega \mathbf{E}_{r+}^t(\omega), \qquad \frac{d}{dt} \mathbf{E}_{r+}^t(\omega) = -i\omega \mathbf{E}_{r+}^t(\omega) - \frac{\dot{\mathbf{E}}(t)}{i\omega^-}.
$$

The Fourier transform of the relative optimal future continuation has the form

$$
\mathbf{E}_m^t(\omega) = -[\mathbb{H}_-(\omega)]^{-1} \mathbf{p}_-^t(\omega), \tag{11.8.6}
$$

which is a special case of (11.2.14). Its time-domain version $\mathbf{E}_{ro}^t$ is given by

$$
\mathbf{E}_{ro}^t(u) = \mathbf{E}_o^t(u) - \mathbf{E}(t) = \frac{1}{2\pi} \int_{-\infty}^{\infty} \mathbf{E}_m^t(\omega) e^{i\omega u} d\omega. \tag{11.8.7}
$$

The work function, given by (11.8.3), retains the form of (11.2.16),

$$
W(t) = \phi(t) + \frac{1}{2\pi} \int_{-\infty}^{\infty} \left[|\mathbf{p}_-^t(\omega)|^2 + |\mathbf{p}_+^t(\omega)|^2 \right] d\omega, \tag{11.8.8}
$$

but with $\mathbf{p}_\pm^t$ defined by (11.8.4), while the minimum free energy (11.2.17) retains the form

$$
\psi_m(t) = \phi(t) + \frac{1}{2\pi} \int_{-\infty}^{\infty} |\mathbf{p}_-^t(\omega)|^2 d\omega. \tag{11.8.9}
$$

The relations in (11.2.18) become

$$\begin{aligned}\psi_m(t) &= \phi(t) + \frac{1}{2\pi}\int_{-\infty}^{\infty} \overline{\mathbf{E}^t_m(\omega)} \cdot \mathbb{H}(\omega)\mathbf{E}^t_m(\omega)d\omega \\ &= \phi(t) + \frac{1}{2}\int_{-\infty}^{0}\int_{-\infty}^{0} \mathbb{G}_{12}(|s-u|)\mathbf{E}^t_{ro}(u)\cdot\mathbf{E}^t_{ro}(s)duds \\ &= \phi(t) + \frac{1}{2}\int_{-\infty}^{0}\int_{-\infty}^{0} \mathbb{G}(|s-u|)\dot{\mathbf{E}}^t_o(u)\cdot\dot{\mathbf{E}}^t_o(s)duds \\ &= \phi(t) + \frac{1}{2\pi}\int_{-\infty}^{\infty} \overline{\dot{\mathbf{E}}^t_m(\omega)} \cdot \frac{\mathbb{H}(\omega)}{\omega^2}\dot{\mathbf{E}}^t_m(\omega)d\omega,\end{aligned} \tag{11.8.10}$$

where $\mathbf{E}^t_{ro}$ is defined by (11.8.7). Relation (11.2.19) reduces to

$$\dot{\psi}_m(t) + D_m(t) = \widehat{\mathbf{S}}(t)\cdot\dot{\mathbf{E}}(t), \tag{11.8.11}$$

where D_m is the rate of mechanical dissipation corresponding to the minimum free energy and must be nonnegative by virtue of the second law. Let us assume that the material is undisturbed in the distant past. Integrating (11.8.11) up to time t gives a special case of (11.2.20):

$$\psi_m(t) + \mathcal{D}_m(t) = W(t),$$

where

$$\mathcal{D}_m(t) = \int_{-\infty}^{t} D_m(s)ds$$

is the total mechanical dissipation up to time t, corresponding to the minimum free energy. We have, from (11.8.8) and (11.8.9) (see (11.2.21)),

$$\mathcal{D}_m(t) = W(t) - \psi_m(t) = \frac{1}{2\pi}\int_{-\infty}^{\infty} |\mathbf{p}^t_+(\omega)|^2\, d\omega \geq 0. \tag{11.8.12}$$

The rate of dissipation has the form (11.2.25) or

$$D_m(t) = |\mathbf{K}(t)|^2, \tag{11.8.13}$$

where $\mathbf{K}$ is given by a special case of $(11.2.22)_3$:

$$\mathbf{K}(t) = \frac{1}{2\pi}\int_{-\infty}^{\infty} \mathbb{H}_-(\omega)\mathbf{E}^t_{r+}(\omega)d\omega. \tag{11.8.14}$$

Certain properties of the optimal future continuation are derived in Sect. 11.5. We summarize them here for the isothermal case. Relation (11.5.1) becomes

$$\mathbf{E}^t_m(\omega) \underset{\omega\to\infty}{\longrightarrow} \frac{\mathbb{H}^{-1}_{\frac{1}{2}}\mathbf{K}(t)}{i\omega},$$

while (11.5.2) and (11.5.3) take the form

$$\mathbf{E}^t_{ro}(0) = -\mathbb{H}^{-1}_{\frac{1}{2}}\mathbf{K}(t), \qquad \mathbf{E}^t_o(0) = \mathbf{E}(t) - \mathbb{H}^{-1}_{\frac{1}{2}}\mathbf{K}(t). \tag{11.8.15}$$

At low frequencies and large times, we have, from (11.5.4) and (11.5.5),

$$\mathbf{E}^t_{ro}(-\infty) = \frac{\mathbb{H}_1(0)^{-1}}{2\pi} \int_{-\infty}^{\infty} \mathbb{H}_1(\omega')\mathbf{E}^t_{r+}(\omega')d\omega'$$

and

$$\mathbf{E}^t_o(-\infty) = \mathbf{E}^t_{ro}(-\infty) + \mathbf{E}(t). \tag{11.8.16}$$

If we apply the assumptions of Sect. 11.6, that the eigenspaces of $\mathbb{G}$ are time-independent, any problem on each of the individual eigenspaces is in effect a scalar problem, like the one dealt with in the next section, where, however, the subscript indicating the eigenspace is omitted.

11.9 Explicit Forms of the Minimum Free Energy for Discrete-Spectrum Materials

We now consider the general results of earlier sections for a particular class of response functions, namely discrete-spectrum scalar models. Here, as above and in later sections dealing with the scalar case, we shall continue to use notation often associated with viscoelastic materials. We replace $\mathbf{E}(t)$ by $E(t)$ and $\mathbb{G}(s)$ by $G(s)$. However, it must be emphasized that the results apply to any time-independent eigenspace of $\mathbb{G}$ (see the comment at the end of Sect. 11.8) or indeed of $\mathbb{L}$, as introduced in Sect. 11.6.

Let the relaxation function $G(t)$ have the form

$$G(t) = G_\infty + \sum_{i=1}^{n} G_i e^{-\alpha_i t}, \quad G_\infty \geq 0, \tag{11.9.1}$$

where n is a positive integer, the inverse decay times α_i, $i = 1, 2, \ldots, n$, are positive, and the coefficients G_i are also generally assumed to be positive. We arrange that $\alpha_1 < \alpha_2 < \alpha_3 \ldots$ It follows that

$$G'(t) = \sum_{i=1}^{n} g_i e^{-\alpha_i t}, \qquad g_i = -\alpha_i G_i < 0,$$

and

$$G'_+(\omega) = \sum_{i=1}^{n} \frac{g_i}{\alpha_i + i\omega}, \qquad G'_c(\omega) = \sum_{i=1}^{n} \frac{\alpha_i g_i}{\alpha_i^2 + \omega^2}, \qquad G'_s(\omega) = \omega \sum_{i=1}^{n} \frac{g_i}{\alpha_i^2 + \omega^2},$$

recalling (8.1.16). Thus, from $(11.8.1)_1$,

$$H(\omega) = -\omega^2 \sum_{i=1}^{n} \frac{g_i}{\alpha_i^2 + \omega^2} \geq 0, \tag{11.9.2}$$

and (11.8.2) can easily be checked. Observe that $f(z) = H(\omega)$, $z = -\omega^2$, has simple poles at α_i^2, $i = 1, 2, \ldots, n$. It will therefore have zeros at γ_i^2, $i = 2, 3, \ldots, n$, where

$$\alpha_1^2 < \gamma_2^2 < \alpha_2^2 < \gamma_3^2 \cdots, \tag{11.9.3}$$

by virtue of Remark B.1.1. It will have no more than one zero between each pole because $H(\infty)$ is a finite constant. The function $f(z)$ also vanishes at $\gamma_1 = 0$. Therefore, H is a rational function of the form

$$H(\omega) = H_\infty \prod_{i=1}^{n} \left\{ \frac{\gamma_i^2 + \omega^2}{\alpha_i^2 + \omega^2} \right\}, \tag{11.9.4}$$

and either by inspection or by applying the general formula (11.1.15), one can show that

$$\begin{aligned} H_+(\omega) &= h_\infty \prod_{i=1}^{n} \left\{ \frac{\omega - i\gamma_i}{\omega - i\alpha_i} \right\}, \\ H_-(\omega) &= h_\infty \prod_{i=1}^{n} \left\{ \frac{\omega + i\gamma_i}{\omega + i\alpha_i} \right\}, \qquad h_\infty = [H_\infty]^{1/2}. \end{aligned} \tag{11.9.5}$$

By considering the residue at each pole, we find that

$$\begin{aligned} H_-(\omega) &= h_\infty \left[1 + i \sum_{i=1}^{n} \frac{R_i}{\omega + i\alpha_i} \right], \qquad H_+(\omega) = \overline{H_-(\omega)}, \\ R_i &= (\gamma_i - \alpha_i) \prod_{\substack{j=1 \\ j \neq i}}^{n} \left\{ \frac{\gamma_j - \alpha_i}{\alpha_j - \alpha_i} \right\}. \end{aligned} \tag{11.9.6}$$

It follows from (11.9.6) and the fact that H_- vanishes at $\omega = 0$ that

$$\sum_{i=1}^{n} \frac{R_i}{\alpha_i} = -1. \tag{11.9.7}$$

Therefore, we can also write $H_-(\omega)$ in the form

$$H_-(\omega) = -h_\infty \omega \sum_{i=1}^{n} \frac{R_i}{\alpha_i(\omega + i\alpha_i)}. \tag{11.9.8}$$

The quantity $p_-^t(\omega)$, defined by the scalar version of (11.8.4), may be evaluated by closing on $\Omega^{(-)}$, giving, with the aid of $(\text{B.1.15})_3$,

$$p_-^t(\omega) = ih_\infty \sum_{i=1}^{n} \frac{R_i E_{r+}^t(-i\alpha_i)}{\omega + i\alpha_i}. \tag{11.9.9}$$

The quantities $E_{r+}^t(-i\alpha_i)$ are real. Also,

$$\begin{aligned} p_+^t(\omega) &= p_-^t(\omega) - H_-(\omega) E_{r+}^t(\omega) \\ &= ih_\infty \sum_{i=1}^{n} R_i \frac{[E_{r+}^t(-i\alpha_i) - E_{r+}^t(\omega)]}{\omega + i\alpha_i} - h_\infty E_{r+}^t(\omega), \end{aligned}$$

which has singularities at those of $E^t_{r+}(\omega)$ in $\Omega^{(+)}$ but none in $\Omega^{(-)}$. These explicit relations for $p^t_\pm$ allow their analytic continuation to the whole complex plane, excluding singular points, as discussed in Sect. B.1.2.

From (11.8.6), (11.9.5), and (11.9.9),

$$E^t_m(\omega) = -i\sum_{i=1}^n J_i(\omega)R_i E^t_{r+}(-i\alpha_i),$$
$$J_i(\omega) = \frac{\prod_{j=1, j\neq i}^n(\omega+i\alpha_j)}{\prod_{j=1}^n(\omega+i\gamma_j)} = \sum_{l=1}^n \frac{Q_{il}}{\omega+i\gamma_l}, \tag{11.9.10}$$

where

$$Q_{il} = \frac{\prod_{j=1, j\neq i}^n(\gamma_l-\alpha_j)}{\prod_{j=1, j\neq l}^n(\gamma_l-\gamma_j)},$$

so that

$$E^t_m(\omega) = -i\sum_{l=1}^n \frac{B^t_l}{\omega+i\gamma_l}, \qquad B^t_l = \sum_{i=1}^n R_i Q_{il} E^t_{r+}(-i\alpha_i). \tag{11.9.11}$$

We conclude that the relative optimal deformation as defined in (11.8.7) has the form

$$E^t_{ro}(s) = -\sum_{l=1}^n B^t_l e^{\gamma_l s} = -B^t_1 - \sum_{l=2}^n B^t_l e^{\gamma_l s}, \quad s<0,$$

since $\gamma_1 = 0$. It follows that

$$E^t_{ro}(-\infty) = -B^t_1,$$

which is a special case of (11.5.5), as may be seen by using (11.9.8) to determine the form of $H_1(\omega)$ (defined by the scalar version of (11.5.4)) and $(11.9.5)_2$ to write $H_1(0)$. By considering $\omega J_i(\omega)$ for large ω, it can be deduced that

$$\sum_{l=1}^n Q_{il} = 1,$$

so that

$$E^t_{ro}(0) = -\sum_{l=1}^n B^t_l = -\sum_{i=1}^n R_i E^t_{r+}(-i\alpha_i). \tag{11.9.12}$$

Relation (11.9.12) follows from $(11.9.11)_2$ and (C.2.16). From (11.8.14) and (11.9.6), we have

$$K(t) = h_\infty\left[\sum_{i=1}^n R_i E^t_{r+}(-i\alpha_i)\right], \tag{11.9.13}$$

since the constant term in H_- yields zero by (B.1.15) and (11.8.5). Observe that (11.9.12) and (11.9.13) agree with (11.8.15).

We deduce from (11.9.9) and (11.8.9) that

$$\begin{aligned}\psi_m(t) &= \phi(t) + H_\infty \sum_{i,j=1}^{n} \frac{R_i R_j}{\alpha_i + \alpha_j} E^t_{r+}(-i\alpha_i) E^t_{r+}(-i\alpha_j) \\ &= \phi(t) + \frac{1}{2}\int_0^\infty ds_1 \int_0^\infty ds_2 E^t_r(s_1) G_{12}(s_1, s_2) E^t_r(s_2),\end{aligned} \tag{11.9.14}$$

where the reality of $E^t_{r+}(-i\alpha_i)$ has been used. The kernel G_{12} is given by

$$G_{12}(s_1, s_2) = 2H_\infty \sum_{i,j=1}^{n} \frac{R_i R_j}{\alpha_i + \alpha_j} e^{-\alpha_i s_1 - \alpha_j s_2}.$$

The solution of this partial differential equation under conditions (8.6.6) is

$$G(s_1, s_2) = G_\infty + 2H_\infty \sum_{i,j=1}^{n} \frac{R_i R_j}{(\alpha_i + \alpha_j)\alpha_i \alpha_j} e^{-\alpha_i s_1 - \alpha_j s_2}.$$

The relation (8.6.13) reducing in the scalar case to $G(0, s) = G(s)$, where the latter quantity is given by (11.9.1), can be confirmed with the aid of the identity

$$\sum_{j=1}^{n} \frac{R_j}{(\alpha_i + \alpha_j)\alpha_j} = -\frac{g_i}{2R_i H_\infty}, \tag{11.9.15}$$

which follows from (11.9.7) and the identity

$$\sum_{j=1}^{n} \frac{R_j}{\alpha_i + \alpha_j} = -1 + \frac{\alpha_i g_i}{2R_i H_\infty},$$

which in turn can be deduced by comparing the product $H_+(\omega)H_-(\omega)$ given by (11.9.6) near poles of $H_+(\omega)$ or $H_-(\omega)$ with $H(\omega)$ given by (11.9.2).

It is shown in [158] that (11.9.14) agrees with the expression by Breuer and Onat [42] for the maximum recoverable work. Noting (11.8.13) and (11.9.13), we see that

$$D_m(t) = H_\infty \left[\sum_{i=1}^{n} R_i E^t_{r+}(-i\alpha_i)\right]^2 = H_\infty \left[\int_0^\infty \sum_{i=1}^{n} R_i e^{-\alpha_i s} E^t_r(s) ds\right]^2. \tag{11.9.16}$$

For $n = 1$,

$$\psi_m(t) = \phi(t) + \frac{1}{2} H_\infty \alpha |E^t_{r+}(-i\alpha)|^2 = \phi(t) + \frac{1}{2} H_\infty \alpha \left[\int_0^\infty E^t_r(s) e^{-\alpha s} ds\right]^2, \tag{11.9.17}$$

which can be shown to agree with the result of Day [87]; see also (10.2.15). Finally, (11.9.16) becomes

$$D_m(t) = H_\infty |\alpha E^t_{r+}(-i\alpha)|^2 = H_\infty \left[\alpha \int_0^\infty e^{-\alpha s} E^t_r(s) ds\right]^2. \tag{11.9.18}$$

12

Representation of the Minimum Free Energy in the Time Domain

12.1 The Minimum Free Energy in Terms of Time-Domain Relative Histories

Consider formula (11.2.17) for the minimum free energy. We wish to derive an expression for the integral term involving time-domain quantities. The method applies also to the family of free energies derived in Chap. 16. Let us first define the quantities $\mathbf{Y}_{\pm}^t, \mathbf{Y}^t \in \Gamma$:

$$\mathbf{Y}_{-}^t(s) = \begin{cases} \dfrac{1}{2\pi}\int_{-\infty}^{\infty} \mathbf{p}_{-}^t(\omega)e^{i\omega s}\,d\omega, & s \in \mathbb{R}^{-}, \\ \mathbf{0}, & s \in \mathbb{R}^{++}, \end{cases}$$
$$\mathbf{Y}_{+}^t(s) = \begin{cases} -\dfrac{1}{2\pi}\int_{-\infty}^{\infty} \mathbf{p}_{+}^t(\omega)e^{i\omega s}\,d\omega, & s \in \mathbb{R}^{+}, \\ \mathbf{0}, & s \in \mathbb{R}^{--}, \end{cases} \tag{12.1.1}$$
$$\mathbf{Y}^t(s) = \frac{1}{2\pi}\int_{-\infty}^{\infty} \mathbb{H}_{-}(\omega)\boldsymbol{\Lambda}_{r+}^t(\omega)e^{i\omega s}\,d\omega = \mathbf{Y}_{+}^t(s) + \mathbf{Y}_{-}^t(s), \quad s \in \mathbb{R}.$$

The second and fourth relations follow from the analyticity properties of $\mathbf{p}_{\pm}^t$ and Proposition C.2.1. Since $\mathbf{p}_{\pm}^t \in L^2(\mathbb{R})$, it follows that $\mathbf{Y}_{\pm}^t \in L^2(\mathbb{R}^{\pm})$. Thus, we have

$$\mathbf{Y}_{-}^t(s) = \mathbf{Y}^t(s), \quad s \in \mathbb{R}^{-}, \qquad \mathbf{Y}_{+}^t(s) = \mathbf{Y}^t(s), \quad s \in \mathbb{R}^{+}. \tag{12.1.2}$$

Recalling Parseval's formula (C.3.1), we have, from $(12.1.1)_5$ and $(7.5.7)_2$, that

$$W(t) = \phi(t) + \int_{-\infty}^{\infty} |\mathbf{Y}^t(s)|^2 ds = \phi(t) + \int_{-\infty}^{0} |\mathbf{Y}^t(s)|^2 ds + \int_{0}^{\infty} |\mathbf{Y}^t(s)|^2 ds,$$

which is the time-domain version of $(11.2.16)_2$. It follows from (11.2.17) that

$$\psi_m(t) = \phi(t) + \int_{-\infty}^{0} |\mathbf{Y}^t(s)|^2 ds \tag{12.1.3}$$

G. Amendola et al., *Thermodynamics of Materials with Memory*,
https://doi.org/10.1007/978-3-030-80534-0_12

and, from (11.2.21),

$$\mathcal{D}_m(t) = \int_0^\infty |\mathbf{Y}^t(s)|^2 ds.$$

We define $\mathbb{U}_\pm \in \mathrm{Lin}(\Gamma)$ as follows:

$$\mathbb{U}_-(s) = \begin{cases} \dfrac{1}{2\pi}\int_{-\infty}^{\infty} \frac{\mathbb{H}_-(\omega)}{\omega} e^{i\omega s} d\omega, & s \in \mathbb{R}^-, \\ \mathbf{0}, & s \in \mathbb{R}^{++}, \end{cases}$$
$$\mathbb{U}_+(s) = \begin{cases} \dfrac{1}{2\pi}\int_{-\infty}^{\infty} \frac{\mathbb{H}_+(\omega)}{\omega} e^{i\omega s} d\omega, & s \in \mathbb{R}^+, \\ \mathbf{0}, & s \in \mathbb{R}^{--}, \end{cases} \tag{12.1.4}$$

where the second and fourth relations again follow from the analyticity properties of $\mathbb{H}_\pm$.

We shall not assume that $\mathbb{H}_\pm$ commute in this chapter. The property $\mathbb{H}_+(\omega) = \mathbb{H}_-^*(\omega)$, as given by (11.1.13), yields

$$\mathbb{U}_-^*(-s) = \mathbb{U}_+(s). \tag{12.1.5}$$

Let $\mathbb{U} \in \mathrm{Lin}(\Gamma)$ be given by

$$\mathbb{U}(s) = \frac{1}{2\pi}\int_{-\infty}^{\infty} \frac{\mathbb{H}(\omega)}{\omega^2} e^{i\omega s} d\omega = \frac{1}{\pi}\int_0^\infty \frac{\mathbb{H}(\omega)}{\omega^2} \cos\omega s d\omega, \qquad s \in \mathbb{R},$$

since $\mathbb{H}$ is an even function. Note that from (7.2.22) and (7.2.14),

$$\mathbb{U}'(s) = \frac{1}{\pi}\int_0^\infty \mathbb{L}'_s(\omega) \sin\omega s d\omega = \frac{1}{2}\mathbb{L}'(s),$$

yielding

$$\mathbb{U}(s) - \mathbb{U}(0) = \frac{1}{2}[\mathbb{L}(s) - \mathbb{L}_0].$$

We see, using (7.2.19) and (7.2.22), that

$$\mathbb{U}(0) = \frac{1}{2}\mathbb{L}_0,$$

so

$$\mathbb{U}(s) = \frac{1}{2}\mathbb{L}(s). \tag{12.1.6}$$

Since, recalling (11.1.14), we see that $\mathbb{H}_\pm(\omega)/\omega \in L^2(\mathbb{R})$, it follows that $\mathbb{U}_\pm \in L^2(\mathbb{R}^\pm)$. Similarly, $\mathbb{U} \in L^2(\mathbb{R})$. They go to zero at large times. Applying the Faltung theorem (Sect. C.3) to (11.1.11), we deduce that

$$\mathbb{U}(s) = \begin{cases} \displaystyle\int_{-\infty}^{0} \mathbb{U}_+(s-u)\mathbb{U}_-(u)du, & s \in \mathbb{R}^+, \\ \displaystyle\int_{-\infty}^{s} \mathbb{U}_+(s-u)\mathbb{U}_-(u)du, & s \in \mathbb{R}^-, \end{cases} \tag{12.1.7}$$

where $(12.1.4)_{2,4}$ have been used. Applying the Faltung theorem to the fifth relation of (12.1.1) yields

$$\mathbf{Y}^t(s) = \begin{cases} -i \int_s^\infty \mathbb{U}'_-(s-u)\boldsymbol{\Lambda}_r^t(u)du, & s \in \mathbb{R}^+, \\ -i \int_0^\infty \mathbb{U}'_-(s-u)\boldsymbol{\Lambda}_r^t(u)du, & s \in \mathbb{R}^-. \end{cases} \tag{12.1.8}$$

Recalling (C.2.12), we see from (12.1.4) that $\mathbb{U}'_\pm(s)$ will contain a singular term $\delta(s)$, because of the fact that $\mathbb{H}_\pm(\infty)$ are in general nonzero. The nonsingular part belongs to $L^2(\mathbb{R}^\pm)$ and in particular vanishes at large times. Now, referring to (12.1.3) and $(12.1.8)_2$, we see that

$$\begin{aligned} \int_{-\infty}^0 |\mathbf{Y}^t(s)|^2 ds &= \int_{-\infty}^0 \int_0^\infty \overline{\mathbb{U}'_-}(s-u)\boldsymbol{\Lambda}_r^t(u)du \cdot \int_0^\infty \mathbb{U}'_-(s-v)\boldsymbol{\Lambda}_r^t(v)dvds \\ &= \int_{-\infty}^0 \int_0^\infty \int_0^\infty \boldsymbol{\Lambda}_r^t(u) \cdot \overline{\mathbb{U}'_-}(s-u)^\top \mathbb{U}'_-(s-v)\boldsymbol{\Lambda}_r^t(v)dudvds \\ &= \frac{1}{2} \int_0^\infty \int_0^\infty \boldsymbol{\Lambda}_r^t(u) \cdot \mathbb{K}(u,v)\boldsymbol{\Lambda}_r^t(v)dudv, \end{aligned} \tag{12.1.9}$$

where, on invoking (A.2.8), $\mathbb{K} \in \mathrm{Lin}(\Gamma)$ can be written as

$$\mathbb{K}(u,v) = 2\int_{-\infty}^0 \mathbb{U}'^*_-(s-u)\mathbb{U}'_-(s-v)\,ds = \frac{\partial}{\partial u}\frac{\partial}{\partial v}\mathbb{L}(u,v),$$

in which

$$\begin{aligned} \mathbb{L}(u,v) &= 2\int_{-\infty}^0 \mathbb{U}^*_-(s-u)\mathbb{U}_-(s-v)\,ds \\ &= 2\int_{-\infty}^0 \mathbb{U}_+(u-s)\mathbb{U}_-(s-v)\,ds, \end{aligned} \tag{12.1.10}$$

by virtue of (12.1.5). It follows from (12.1.9) that the components of $\mathbb{K}$ and therefore $\mathbb{L}$ are real, so that

$$\mathbb{L}^*(u,v) = \mathbb{L}^\top(v,u).$$

Taking the Hermitian conjugate of (12.1.10) then yields

$$\mathbb{L}^\top(u,v) = \mathbb{L}(v,u),$$

which is a special case of $(7.1.6)_3$. Relations (12.1.6) and (12.1.7) and the symmetry of $\mathbb{L}(u)$ give

$$\mathbb{L}(u,0) = \mathbb{L}(0,u) = \mathbb{L}(u),$$

which is $(7.1.14)_3$. The other relations of (7.1.6)–(7.1.8) follow from the fact that $\mathbb{U}_\pm$ and $\mathbb{U}'_\pm$ vanish at large times. Finally, from $(11.2.22)_3$, (11.2.25), $(12.1.1)_5$, and (12.1.8), we have that the rate of dissipation is given by

$$D_m = |\mathbf{Y}^t(0)|^2 = \left| \int_0^\infty \mathbb{U}'(-u)\boldsymbol{\Lambda}_r^t(u)du \right|^2. \tag{12.1.11}$$

The delta function in $\mathbb{U}'$ does not contribute to D_m because $\boldsymbol{\Lambda}_r^t(0) = \mathbf{0}$.

12.2 The Minimum Free Energy Expressed in Terms of $\mathbf{I}^t$

The functional $\mathbf{I}^t$, given by (see (7.4.2)$_2$)

$$\mathbf{I}^t(\tau, \mathbf{\Lambda}^t) = \mathbf{I}^t(\tau) = \int_0^\infty \mathbb{L}'(s+\tau)\mathbf{\Lambda}_r^t(s)ds, \tag{12.2.1}$$

is a functional (in fact, the defining functional) of the minimal state. We can in fact characterize the minimal state by the pair $(\mathbf{I}^t, \mathbf{\Lambda}(t))$. Since ψ_m is also a functional of the minimal state (Remark 4.2.11), it is reasonable to suppose that we can express it in terms of $\mathbf{I}^t$. Where the second argument of $\mathbf{I}^t$ in (12.2.1) is clear, we omit it, writing $\mathbf{I}^t(\tau)$. The developments in this section are based on [110, 145] but most substantially on [91].

For the following derivation, we put

$$\mathbb{L}'(u) = 0, \qquad \mathbf{\Lambda}_r^t(u) = 0, \quad u \in \mathbb{R}^{--}. \tag{12.2.2}$$

Also, observe that

$$\begin{aligned}\int_{-\infty}^\infty \mathbb{L}'(u+\tau)e^{-i\omega u}du &= \int_{-\infty}^\infty \mathbb{L}'(v)e^{-i\omega v}dv\, e^{i\omega\tau}\\ &= \int_0^\infty \mathbb{L}'(v)e^{-i\omega v}dv\, e^{i\omega\tau} = \mathbb{L}'_+(\omega)\, e^{i\omega\tau},\end{aligned}$$

which is an example of (C.1.8). Parseval's formula (C.3.1) gives that

$$\mathbf{I}^t(\tau) = \frac{1}{2\pi}\int_{-\infty}^\infty \overline{\mathbb{L}'_+}(\omega)\mathbf{\Lambda}_{r+}^t(\omega)e^{-i\omega\tau}d\omega, \qquad \tau \ge 0.$$

We have

$$\mathbf{I}^t(\tau) = \frac{1}{2\pi}\int_{-\infty}^\infty [\overline{\mathbb{L}'}_+(\omega) + \lambda\mathbb{L}'_+(\omega)]\mathbf{\Lambda}_{r+}^t(\omega)e^{-i\omega\tau}\, d\omega, \qquad \tau \ge 0, \tag{12.2.3}$$

since the added term gives zero, which can be seen by integrating over a contour around Ω^-, noting that the exponential goes to zero as $\mathrm{Im}\omega \to -\infty$ (see (7.2.33)). If $\lambda = -1$, this formula agrees with (11.3.3). We put, as in (11.2.31),

$$\mathbf{J}^t(\tau) = \mathbf{I}^t(-\tau), \qquad \tau \le 0. \tag{12.2.4}$$

Then, using the notation of (C.1.3)$_2$ and relation (12.2.3),

$$\begin{aligned}\mathbf{J}_-^t(\omega) &= \int_{-\infty}^0 \mathbf{J}^t(\tau)e^{-i\omega\tau}\, d\tau\\ &= \frac{1}{2\pi i}\int_{-\infty}^\infty \frac{[\overline{\mathbb{L}'_+}(\omega') + \lambda\mathbb{L}'_+(\omega')]\mathbf{\Lambda}_{r+}^t(\omega')}{\omega' - \omega^+}d\omega'.\end{aligned} \tag{12.2.5}$$

We must choose ω^+ so that the integration over the exponential converges. Similarly, let $\mathbf{J}^t$ be defined by (12.2.3) and (12.2.4) for $\tau > 0$. In this case, it depends on λ. We have

$$\begin{aligned}\mathbf{J}_+^t(\omega,\lambda) &= \int_0^\infty \mathbf{J}^t(\tau,\lambda)e^{-i\omega\tau}\,d\tau,\\ &= -\frac{1}{2\pi i}\int_{-\infty}^\infty \frac{[\overline{\mathbb{L}'_+}(\omega')+\lambda\mathbb{L}'_+(\omega')]\mathbf{\Lambda}_{r+}^t(\omega')}{\omega'-\omega^-}d\omega'\end{aligned} \tag{12.2.6}$$

and

$$\mathbf{J}_F^t(\omega,\lambda) = \mathbf{J}_+^t(\omega,\lambda)+\mathbf{J}_-^t(\omega) = [\overline{\mathbb{L}'_+}(\omega)+\lambda\mathbb{L}'_+(\omega)]\mathbf{\Lambda}_{r+}^t(\omega), \tag{12.2.7}$$

by (C.1.4) and the Plemelj formula $(\text{B.2.15})_2$. We see from (7.2.23) that

$$\mathbf{J}_F^t(\omega,\lambda)\sim\omega^{-3},\quad \lambda\neq 1,\qquad \mathbf{J}_F^t(\omega,1)\sim\omega^{-4},\quad \lambda=1$$

at large ω, since, as observed after (7.2.29), $\mathbf{\Lambda}_{r+}^t(\omega)\sim\omega^{-2}$. Note that the derivations of (12.2.5) and (12.2.6) are essentially the same as the manipulations leading to (C.2.20).

For $\lambda=-1$, $(12.2.7)_2$ becomes

$$\mathbf{J}_F^t(\omega,-1) = -2i\frac{\mathbb{H}(\omega)}{\omega}\mathbf{\Lambda}_{r+}^t(\omega). \tag{12.2.8}$$

Note that $\mathbf{J}_F^t(\omega,-1)$ (or $\mathbf{J}^t(s,-1)$, $s\in\mathbb{R}$ in the time domain) corresponds to $\mathbf{I}^t(-\tau)$, defined by the odd extension of $\mathbb{L}'$ to $\mathbb{R}$, or

$$\mathbf{I}^t(-\tau) = -\int_0^\infty \frac{\partial}{\partial\tau}\mathbb{L}(|\tau-u|)\mathbf{\Lambda}_r^t(u)du,\qquad \tau\in\mathbb{R}.$$

The integral can be extended over $\mathbb{R}$ by virtue of $(12.2.2)_2$. Taking the Fourier transform immediately yields (12.2.8) with the aid of the convolution theorem and (7.2.22).

Also, for $\lambda=-1$, we obtain from (12.2.5) and (11.2.8) that

$$\mathbf{J}_-^t(\omega) = -\frac{1}{\pi}\int_{-\infty}^\infty \frac{\mathbb{H}(\omega')\mathbf{\Lambda}_{r+}^t(\omega')}{\omega'(\omega'-\omega^+)}d\omega' = -\frac{1}{\pi}\int_{-\infty}^\infty \frac{\mathbb{H}_+(\omega')[\mathbf{p}_-^t(\omega')-\mathbf{p}_+^t(\omega')]}{\omega'(\omega'-\omega^+)}d\omega'.$$

Now, the $\mathbf{p}_+^t(\omega')$ term vanishes on integrating over Ω^-, and we have (recall (11.3.6))

$$\mathbf{J}_-^t(\omega) = -\frac{1}{\pi}\int_{-\infty}^\infty \frac{\mathbb{H}_+(\omega')\mathbf{p}_-^t(\omega')}{\omega'(\omega'-\omega^+)}d\omega'.$$

Since $\mathbf{p}_-^t$ is a functional of the minimal state by Theorem 11.3.1, it follows that $\mathbf{J}_-^t$ also has this property, which of course it must have, as a Fourier transform of $\mathbf{I}^t(s)$, $s>0$. There is no corresponding result for $\mathbf{J}_+^t$.

We put

$$\begin{aligned}\mathbf{P}^t(\omega,\lambda) &= -\frac{\omega}{2i}[\mathbb{H}_+(\omega)]^{-1}\,\mathbf{J}_F^t(\omega,\lambda)\\ &= -\frac{\omega}{2i}[\mathbb{H}_+(\omega)]^{-1}\,[\mathbf{J}_-^t(\omega)+\mathbf{J}_+^t(\omega,\lambda)]\\ &= \mathbf{p}_-^t(\omega)-\mathbf{p}_+^t(\omega,\lambda).\end{aligned}$$

This is the same quantity as in (11.2.8) (or more generally (11.2.32) when $\mathbf{\Lambda}^t_{r+}$ does not exist) if $\lambda = -1$. We see that $\mathbf{p}^t_-$ is part of the term proportional to $[\mathbb{H}_+(\omega)]^{-1}\mathbf{J}^t_-(\omega)$ and is thus independent of λ. The quantities $\mathbf{p}^t_\pm(\omega)$ are analytic on $\Omega^\mp$, respectively. They can be written, with the aid of the Plemelj formulas, in the form

$$\mathbf{p}^t_\pm(\omega,\lambda) = \frac{1}{2\pi i}\int_{-\infty}^{\infty}\frac{\mathbf{P}^t(\omega',\lambda)}{\omega'-\omega^\mp}d\omega',$$

the integral being convergent, since $\mathbf{J}^t_F(\omega)$ decays like ω^{-3} for large ω or more strongly if $\lambda = 1$. The λ dependence applies only to $\mathbf{p}^t_+$.

The minimum free energy is given by (11.2.17). Our objective is to write this quantity in the time domain, as a quadratic form in $\mathbf{I}^t$.

Let $\mathbf{Y}^t(\cdot,\lambda)$ be the inverse Fourier transform of $\mathbf{P}^t$. Then,

$$\mathbf{P}^t(\omega,\lambda) = \int_{-\infty}^{\infty}\mathbf{Y}^t(s,\lambda)e^{-i\omega s}\,ds$$

and

$$\begin{aligned}\mathbf{p}^t_+(\omega,\lambda) &= -\int_0^{\infty}\mathbf{Y}^t(s,\lambda)e^{-i\omega s}\,ds,\\ \mathbf{p}^t_-(\omega) &= \int_{-\infty}^{0}\mathbf{Y}^t(s,\lambda)e^{-i\omega s}\,ds.\end{aligned}\tag{12.2.9}$$

These correspond to (12.1.1) for $\lambda = -1$. The quantity $\mathbf{Y}^t$ can be written more explicitly, by means of the Faltung theorem. However, we first need to write $\mathbf{P}^t$ as the product of two functions both in $L^2(\mathbb{R})$. Let us divide and multiply by $(\omega^-)^2$, omitting the superscript for factors in the numerator where it is irrelevant. Thus,

$$\mathbf{P}^t(\omega,\lambda) = -\frac{[\mathbb{H}_+(\omega)]^{-1}}{2i\omega^-}\left[\omega^2\mathbf{J}^t_F(\omega,\lambda)\right],$$

where both factors are in $L^2(\mathbb{R})$ and where the first factor has all its singularities in $\Omega^{(+)}$, by virtue of the property that the zeros of $\det\mathbb{H}_+$ are in $\Omega^{(+)}$. We put

$$\mathbf{J}^t_{(2)}(s,\lambda) = \frac{d^2}{ds^2}\mathbf{J}^t(s,\lambda) = -\frac{1}{2\pi}\int_{-\infty}^{\infty}\omega^2\mathbf{J}^t_F(\omega,\lambda)e^{i\omega s}\,d\omega.$$

Also, let us define the complex quantity $\mathbb{M}\in\mathrm{Lin}(\Gamma)$ by

$$\mathbb{M}(s) = \frac{1}{2\pi}\int_{-\infty}^{\infty}\frac{[\mathbb{H}_+(\omega)]^{-1}}{2i\omega^-}e^{i\omega s}\,d\omega,\qquad s\in\mathbb{R}.\tag{12.2.10}$$

The integrand has a quadratic singularity near the origin, due to the explicit pole term and the factor ω in $\mathbb{H}_+(\omega)$, which is taken, for consistency, to be ω^-. This gives a finite contribution, which is easily calculated. The quantity $\mathbb{M}$ vanishes* for $s\in\mathbb{R}^{--}$, by Proposition C.2.1. The Hermitian conjugate of $\mathbb{M}$ has the form

* This is not the case for the more general class of factorizations introduced in Chap. 16, for which the zeros of H_+ (the scalar theory is considered in that chapter) may be in either half-plane. The method described here and the final result do not generalize readily to such factorizations.

$$\mathbb{M}^*(s) = -\frac{1}{2\pi}\int_{-\infty}^{\infty} \frac{[\mathbb{H}_-(\omega)]^{-1}}{2i\omega^+} e^{-i\omega s}\, d\omega, \quad s \in \mathbb{R},$$

by virtue of (11.1.13). We see that $\mathbb{M}^*(-s)$ vanishes on $\mathbb{R}^{++}$, since $\mathbb{H}_-(\omega)$ has all its zeros in $\Omega^{(-)}$. It follows that $\mathbb{M}^*(s)$ vanishes on $\mathbb{R}^{--}$, as is the case for $\mathbb{M}(s)$. By the convolution theorem (C.3.3), we have

$$\mathbf{Y}^t(s,\lambda) = \int_{-\infty}^{s} \mathbb{M}(s-u)\mathbf{J}^t_{(2)}(u,\lambda)du. \tag{12.2.11}$$

From (12.2.9), it is clear that only nonpositive arguments of $\mathbf{J}^t_{(2)}$ contribute to $\mathbf{p}^t_-$. For this domain, we can drop the λ dependence, as for (12.2.5). Also, from (12.2.9) and Parseval's formula (C.3.1) (cf. (12.1.3)),

$$\frac{1}{2\pi}\int_{-\infty}^{\infty} |\mathbf{p}^t_-(\omega)|^2 d\omega = \int_{-\infty}^{0} |\mathbf{Y}^t(s)|^2 ds.$$

Extending the upper limit in (12.2.11) to zero and interchanging integrations, we have, using (A.2.8),

$$\psi_m(t) = \phi(t) + \int_{-\infty}^{0}\int_{-\infty}^{0} \mathbf{J}^t_{(2)}(u)\cdot \mathbb{N}(u,v)\mathbf{J}^t_{(2)}(v)dudv, \tag{12.2.12}$$

where $\mathbb{N} \in \mathrm{Lin}(\Gamma)$ is given by

$$\mathbb{N}(u,v) = \int_{-\infty}^{0} \mathbb{M}^*(s-u)\mathbb{M}(s-v)ds = \int_{\max(u,v)}^{0} \mathbb{M}^*(s-u)\mathbb{M}(s-v)ds.$$

Since ψ_m is real, we have

$$\mathbb{N}^*(u,v) = \mathbb{N}(v,u).$$

Note that

$$\mathbb{N}(u,0) = \mathbb{N}(0,v) = \mathbf{0} \ \ \forall\, u,v \in \mathbb{R}^-$$

and also

$$\mathbb{N}(u,-\infty) = \mathbb{N}(-\infty,v) = \mathbf{0} \ \ \forall\, u,v \in \mathbb{R}^{--}.$$

Therefore, we can write (12.2.12) as

$$\psi_m(t) = \phi(t) + \int_{-\infty}^{0}\int_{-\infty}^{0} \mathbf{J}^t_{(1)}(u)\cdot \mathbb{N}_{12}(u,v)\mathbf{J}^t_{(1)}(v)dudv,$$

$$\mathbb{N}_{12}(u,v) = \frac{\partial^2}{\partial u \partial v}\mathbb{N}(u,v).$$

It follows from (12.2.10) that the quantity $\mathbb{N}_{12}$ will have singular delta distribution terms. This can be avoided by using (12.2.12). Referring to (12.2.4), it can be written in terms of $\mathbf{I}^t$ as

$$\psi_m(t) = \phi(t) + \int_{0}^{\infty}\int_{0}^{\infty} \mathbf{I}^t_{(2)}(u)\cdot \mathbb{K}(u,v)\mathbf{I}^t_{(2)}(v)dudv, \tag{12.2.13}$$

where

$$\mathbf{I}^t_{(2)}(s) = \frac{d^2}{ds^2}\mathbf{I}^t(s)$$

and

$$\mathbb{K}(u, v) = \mathbb{N}(-u, -v) = \int_0^{\min(u,v)} \mathbb{M}^*(u - s)\mathbb{M}(v - s)ds.$$

Note that the same observation applies to (12.2.13) as made after (10.1.14). This large domain of definition of the minimum free energy is reflected in the abstract Definition 4.2.1.

We now derive an expression for the rate of dissipation. Relations (11.2.25) and $(11.2.24)_1$ give

$$D_m(t) = |\mathbf{K}(t)|^2, \qquad \mathbf{K}(t) = -i \lim_{\omega\to\infty} \omega\mathbf{p}^t_-(\omega).$$

From $(12.2.9)_2$, we deduce with the aid of (C.2.16) that (cf. (12.1.11))

$$\mathbf{K}(t) = \mathbf{Y}^t(0) = \int_{-\infty}^0 \mathbb{M}(-u)\mathbf{J}^t_{(2)}(u)du = \int_0^\infty \mathbb{M}(u)\mathbf{I}^t_{(2)}(u)du$$

on using (12.2.11) and (12.2.4). Therefore,

$$D_m(t) = \left|\int_0^\infty \mathbb{M}(u)\mathbf{I}^t_{(2)}(u)du\right|^2 = \int_0^\infty \int_0^\infty \mathbf{I}^t_{(2)}(u) \cdot \mathbb{M}^*(u)\mathbb{M}(v)\mathbf{I}^t_{(2)}(v)dudv.$$

It is easy to compare (12.1.4) and (12.2.10), which determine the kernels of the quadratic forms for the minimum free energy in terms of $\mathbf{\Lambda}^t_r$ and $\mathbf{I}^t_{(2)}$, respectively.

13

Minimum Free Energy for Viscoelastic Solids, Fluids, and Heat Conductors

We now develop formulas for the minimum free energy and related quantities applicable to three categories of linear materials (completely linear for solids and heat conductors). The methods differ in detail from those in Chap. 11, though they are equivalent. As in Chaps. 8 and 9, we make more use of the abstract terminology and notation introduced in Chaps. 3 and 4.

13.1 Maximum Recoverable Work for Solids

Let $\mathbb{G}(|s|)$ denote the extension of $\mathbb{G}(s)$ on $\mathbb{R}$ by means of an even function. Furthermore, we suppose that any process $P \in \Pi$, defined over $[0, d_P)$, is identified with its trivial extension on $\mathbb{R}^+$, by putting

$$P(t) = \begin{cases} P(t), & t \in [0, d_P), \\ \mathbf{0}, & t \in [d_P, +\infty). \end{cases} \tag{13.1.1}$$

Let $\sigma_0 = \sigma(0) = \mathbf{E}^0$ be the history at the initial time $t = 0$. We wish to evaluate the work $W(\sigma_0, P)$ as given by (4.1.4) but with $\boldsymbol{\Sigma}$ and $\boldsymbol{\Lambda}$ replaced by the Cauchy stress $\mathbf{T}$ and the strain $\mathbf{E}$ [110]. The process $P \in \Pi$ is such that $P(t) = \dot{\mathbf{E}}(t)\, \forall t \in [0, d_p)$. Also, $\mathbf{E}^t = \hat{\rho}(\mathbf{E}^0, P_t)$, and recalling (1.4.11) or (8.1.5), together with (8.3.1), we write the stress as

$$\begin{aligned} \mathbf{T}(\mathbf{E}^t) &= \mathbb{G}_0\mathbf{E}(t) + \int_0^\infty \mathbb{G}'(s)\mathbf{E}^t(s)ds \\ &= \mathbb{G}_0\mathbf{E}(t) + \int_0^t \mathbb{G}'(s)\mathbf{E}^t(s)ds + \mathbf{I}^0(t, \mathbf{E}^0), \end{aligned} \tag{13.1.2}$$

where $\mathbf{I}^0(t, \mathbf{E}^0)$ is given by $(8.2.3)_1$ with $t = 0$ and τ substituted by t, that is,

$$\mathbf{I}^0(t, \mathbf{E}^0) = \int_0^\infty \mathbb{G}'(s + t)\mathbf{E}^0(s)ds \qquad \forall t \in \mathbb{R}^+. \tag{13.1.3}$$

G. Amendola et al., *Thermodynamics of Materials with Memory*,
https://doi.org/10.1007/978-3-030-80534-0_13

A slightly different notation was used in (8.5.16) for the extension of $\mathbf{I}^0$ to $\mathbb{R}$. Note that from (13.1.1), it follows that the limit $\mathbf{E}(\infty) = \lim_{t\to\infty} \mathbf{E}(t)$ exists. We have

$$\begin{aligned} W(\sigma, P) &= \int_0^\infty \left[\mathbb{G}_0 \mathbf{E}(t) + \int_0^t \mathbb{G}'(s)\mathbf{E}^t(s)ds \right] \cdot \dot{\mathbf{E}}(t)dt \\ &+ \int_0^\infty \mathbf{I}_0(t, \mathbf{E}^0) \cdot \dot{\mathbf{E}}(t)dt = \int_0^\infty \mathbb{G}_0 \mathbf{E}(t) \cdot \dot{\mathbf{E}}(t)dt \\ &+ \int_0^\infty \left[\mathbb{G}(s)\mathbf{E}(t-s)|_0^t + \int_0^t \mathbb{G}(s) \cdot \dot{\mathbf{E}}(t-s)ds \right] \cdot \dot{\mathbf{E}}(t)dt \\ &+ \int_0^\infty \mathbf{I}^0(t, \mathbf{E}^0) \cdot \dot{\mathbf{E}}(t)dt \\ &= \frac{1}{2} \int_0^\infty \int_0^\infty \mathbb{G}(|t-\tau|)\dot{\mathbf{E}}(\tau) \cdot \dot{\mathbf{E}}(t)d\tau dt + \int_0^\infty \check{\mathbf{I}}^0(t, \mathbf{E}^0) \cdot \dot{\mathbf{E}}(t)dt, \end{aligned} \tag{13.1.4}$$

where, referring to $(8.2.3)_2$, we have put

$$\check{\mathbf{I}}^0(t, \mathbf{E}^0) = \mathbb{G}(t)\mathbf{E}(0) + \mathbf{I}^0(t, \mathbf{E}^0). \tag{13.1.5}$$

The maximum recoverable work from the state $\sigma_0 = \mathbf{E}^0$ can be evaluated by considering the maximum of $-W(\sigma_0, P)$ with respect to the set of functions

$$\mathbf{E}(t) = \mathbf{E}^{(m)}(t) + \varepsilon \mathbf{e}(t) \qquad \forall t \in \mathbb{R}^+, \tag{13.1.6}$$

where ε is a real parameter and $\mathbf{e}$ is an arbitrary smooth function such that $\mathbf{e}(0) = \mathbf{0}$. If $\dot{\mathbf{E}}^{(m)}$ denotes the process corresponding to the maximum recoverable work, then the derivative of $-W$ for $\varepsilon = 0$ must be equal to zero, so that

$$\begin{aligned} &\frac{d}{d\varepsilon}[-W(\sigma, P)]\,|_{\varepsilon=0} \\ &= -\int_0^\infty \int_0^\infty \mathbb{G}(|t-\tau|)\dot{\mathbf{E}}^{(m)}(t) \cdot \dot{\mathbf{e}}(\tau)d\tau dt - \int_0^\infty \check{\mathbf{I}}^0(t, \mathbf{E}^0) \cdot \dot{\mathbf{e}}(t)dt = 0. \end{aligned}$$

Hence, because of the arbitrariness of $\dot{\mathbf{e}}(t)$, we obtain

$$\int_0^\infty \mathbb{G}(|t-\tau|)\dot{\mathbf{E}}^{(m)}(\tau)d\tau = -\check{\mathbf{I}}^0(t, \mathbf{E}^0) \qquad \forall t \in \mathbb{R}^+. \tag{13.1.7}$$

This relation (13.1.7) is an integral equation of Wiener–Hopf type, the solution of which gives the process $\dot{\mathbf{E}}^{(m)}$, related to the maximum recoverable work. Thus, from Theorem 4.2.3 together with relations $(13.1.4)_3$ and (13.1.7), we obtain an expression for the minimum free energy,

$$\psi_m(\mathbf{E}^0) = \frac{1}{2} \int_0^\infty \int_0^\infty \mathbb{G}(|t-\tau|)\dot{\mathbf{E}}^{(m)}(\tau) \cdot \dot{\mathbf{E}}^{(m)}(t)d\tau dt, \tag{13.1.8}$$

where $\dot{\mathbf{E}}^{(m)}$ must be evaluated as the solution of the Wiener–Hopf equation (13.1.7).

It is important to prove that the solution of (13.1.7) exists and is unique.

We can write (13.1.8) in terms of the frequency-domain quantities, by analogy with the steps from (7.5.2) to $(7.5.7)_1$. This yields

$$\begin{aligned}\psi_m(\mathbf{E}^0) &= -\frac{1}{2\pi}\int_{-\infty}^{\infty}\frac{\mathbb{G}'_s(\omega)}{\omega}\overline{\dot{\mathbf{E}}_+^{(m)}}(\omega)\cdot\dot{\mathbf{E}}_+^{(m)}(\omega)d\omega \\ &= \frac{1}{2\pi}\int_{-\infty}^{\infty}\mathbb{G}_c(\omega)\overline{\mathbf{E}_+^{(m)}}(\omega)\cdot\mathbf{E}_+^{(m)}(\omega)d\omega \geq 0.\end{aligned} \tag{13.1.9}$$

The kernel $\mathbb{G}(|t|)$ is a positive operator, as can be deduced from (8.1.18) by writing (13.1.8) in the frequency domain. We introduce the set

$$\mathcal{G} = \left\{\dot{\mathbf{E}} : [0,+\infty) \to \mathrm{Sym};\ \int_0^{\infty}\int_0^{\infty}\mathbb{G}(|t-\tau|)\dot{\mathbf{E}}(\tau)\cdot\dot{\mathbf{E}}(t)d\tau dt < \infty\right\}. \tag{13.1.10}$$

An inner product on $\mathcal{G}$ can be defined by means of

$$\left(\dot{\mathbf{E}}_1, \dot{\mathbf{E}}_2\right) = \int_0^{\infty}\int_0^{\infty}\mathbb{G}(|t-\tau|)\dot{\mathbf{E}}_1(\tau)\cdot\dot{\mathbf{E}}_2(t)d\tau dt.$$

The norm on $\mathcal{G}$ is given by

$$\begin{aligned}||\dot{\mathbf{E}}||^2 &= \int_0^{\infty}\int_0^{\infty}\mathbb{G}(|t-\tau|)\dot{\mathbf{E}}(\tau)\cdot\dot{\mathbf{E}}(t)d\tau dt \\ &= \mathbb{G}_\infty\left|\int_0^{\infty}\dot{\mathbf{E}}(\tau)d\tau\right|^2 + \int_0^{\infty}\int_0^{\infty}\check{\mathbb{G}}(|t-\tau|)\dot{\mathbf{E}}(\tau)\cdot\dot{\mathbf{E}}(t)d\tau dt,\end{aligned} \tag{13.1.11}$$

where $\check{\mathbb{G}}(s) = \mathbb{G}(s) - \mathbb{G}_\infty$, which is assumed to belong to $L^1(\mathbf{R}^+)$. Therefore, the processes $\dot{\mathbf{E}}$ are such that

$$\int_0^{\infty}\dot{\mathbf{E}}(t)dt = \mathbf{E}(\infty) - \mathbf{E}(0) < \infty,$$
$$\int_0^{\infty}\int_0^{\infty}\check{\mathbb{G}}(|t-\tau|)\dot{\mathbf{E}}(\tau)\cdot\dot{\mathbf{E}}(t)d\tau dt < \infty.$$

If $\check{\mathbb{G}}$ has the exponential form

$$\check{\mathbb{G}}(s) = Ae^{-\alpha s}\mathbf{I},$$

where $\mathbf{I}$ is the unit operator in Lin(Sym) and A is a scalar constant, then the space $\mathcal{G}$ is well defined. In fact, replacing $\mathbb{G}$ in (13.1.8) by $\check{\mathbb{G}}$, we see from $(13.1.9)_2$ (using the evenness of the integrand as a function of ω) that

$$\int_0^{\infty}\int_0^{\infty}\check{\mathbb{G}}(|t-\tau|)\dot{\mathbf{E}}(t)\cdot\dot{\mathbf{E}}(\tau)d\tau dt = \frac{A}{\pi}\int_0^{\infty}\frac{\alpha}{\alpha^2+\omega^2}\left|\dot{\mathbf{E}}_+(\omega)\right|^2 d\omega,$$

so that $\dot{\mathbf{E}} \in H^{-1}(\mathbb{R}^+)$, the dual of the Sobolev space $H^1(\mathbb{R}^+)$. In general, we can affirm that $\mathcal{G} \supset L^1(\mathbb{R}^+) \cap L^2(\mathbb{R}^+)$.

Remark 13.1.1. By means of the norm of $\mathcal{G}$, it is possible to provide the set of processes Π with a topology. In particular, the closure of Π, using the norm (13.1.11), is the Hilbert space $\mathcal{G}$.

We can write Eq. (13.1.7) as

$$\mathcal{A}\dot{\mathbf{E}} = \mathbf{I}^0,$$

where $\mathcal{A}$ is an operator from $\mathcal{G}$ to its dual $\mathcal{G}'$. It is bounded and coercive. Thus, from the Lax–Milgram theorem, we have the result expressed by the following theorem [110].

Theorem 13.1.2. *For any* $\mathbf{I}^0 \in \mathcal{G}'$, *Eq. (13.1.7) has a unique solution* $\dot{\mathbf{E}} \in \mathcal{G}$ *such that*

$$\|\dot{\mathbf{E}}\|_{\mathcal{G}} \leq K\|\mathbf{I}^0\|_{\mathcal{G}'}.$$

In other words, there exists an isomorphism between $\mathcal{G}$ and $\mathcal{G}'$.

We now recall Theorem 4.2.8 and Corollary 4.2.10. Since the elastic free energy at a given current strain is less than or equal to any viscoelastic free energy with the same current strain (relation (5.1.32)), we take ϕ_M to be $\phi(\infty)$, where

$$\phi(t) = \frac{1}{2}\mathbb{G}_\infty \mathbf{E}(t) \cdot \mathbf{E}(t)$$

is the elastic or equilibrium free energy corresponding to the strain $\mathbf{E}(t)$.

We seek the history that maximizes the functional

$$\begin{aligned} W(\sigma, P) - \phi(\infty) &= \frac{1}{2}\int_0^\infty \int_0^\infty \mathbb{G}(|t-\tau|)\dot{\mathbf{E}}(\tau) \cdot \dot{\mathbf{E}}(t) d\tau dt \\ &+ \int_0^\infty \breve{\mathbf{I}}^0(t, \mathbf{E}^0) \cdot \dot{\mathbf{E}}(t) dt - \phi(\infty), \end{aligned} \tag{13.1.12}$$

where $\breve{\mathbf{I}}^0(t, \mathbf{E}^0)$ is given by (13.1.5). Repeating the steps leading to (13.1.7), noting that in (13.1.6) $\mathbf{e}(0) = \mathbf{0}$ and $\mathbf{e}(\infty) = \int_0^\infty \dot{\mathbf{e}}(t)dt$, we obtain

$$\int_0^\infty \mathbb{G}(|t-\tau|)\dot{\mathbf{E}}^{(m)}(\tau)d\tau = -\breve{\mathbf{I}}^0(t, \mathbf{E}^0) + \mathbb{G}_\infty \mathbf{E}^{(m)}(\infty) \qquad \forall t \in \mathbb{R}^+. \tag{13.1.13}$$

The solution $\dot{\mathbf{E}}^{(m)}$ of (13.1.7) can be expressed as the sum

$$\dot{\mathbf{E}}^{(m)} = \dot{\mathbf{E}}^{(m_1)} + \dot{\mathbf{E}}^{(m_2)}, \tag{13.1.14}$$

where $\dot{\mathbf{E}}^{(m_1)}$ denotes the solution of (13.1.13) and $\dot{\mathbf{E}}^{(m_2)}$ satisfies the equation

$$\begin{aligned} \int_0^\infty \mathbb{G}(|t-\tau|)\dot{\mathbf{E}}^{(m_2)}(\tau)d\tau &= -\mathbb{G}_\infty \mathbf{E}^{(m_1)}(\infty) \\ &= -\mathbb{G}_\infty \left[\mathbf{E}^{(m_1)}(0) + \int_0^\infty \dot{\mathbf{E}}^{(m_1)}(\tau)d\tau\right], \end{aligned}$$

derived by subtracting (13.1.13) from (13.1.7). This equation, by virtue of Theorem 13.1.2, has a unique solution $\dot{\mathbf{E}}^{(m_2)} \in \mathcal{G}$, which we denote formally by

$$\dot{\mathbf{E}}^{(m_2)}(t) = -\mathbf{E}^{(m_1)}(\infty)\delta(t-\infty).$$

Such a function can be deduced as the limit of the sequence

$$\dot{\mathbf{E}}^{(m_2)}(t) = \begin{cases} \mathbf{0}, & t < n, \\ -\frac{\mathbf{E}^{(m_1)}(\infty)}{n}, & n < t < 2n, \\ \mathbf{0}, & t > 2n. \end{cases}$$

We note that (13.1.14) corresponds to continuations given formally by

$$\mathbf{E}^{(m)}(t) = \mathbf{E}^{(m_1)}(t) - \mathbf{E}^{(m_1)}(\infty)H(t-\infty),$$

where $H(s)$ denotes the Heaviside step function.

Remark 13.1.3. We obtain the same maximum recoverable work using both functionals $W(\sigma, P)$ and $W(\sigma, P) - \phi_M(\hat{\rho}(\sigma, P))$; however, the processes that yield this maximum recoverable work are different. In fact, using $W(\sigma, P)$, the optimal process is $\dot{\mathbf{E}}^{(m)}$, which satisfies (13.1.7); however, using $W(\sigma, P) - \phi_M(\hat{\rho}(\sigma, P))$, the optimal process is $\dot{\mathbf{E}}^{(m_1)}$, which satisfies (13.1.13).

We observe that (13.1.13), integrating by parts and using (13.1.5) with (13.1.3), can be written as

$$\int_{-\infty}^{\infty} \frac{\partial}{\partial t}\mathbb{G}(|t+\tau|)\mathbf{E}^0(\tau)d\tau = \mathbf{0} \qquad \forall t \in \mathbb{R}^+, \tag{13.1.15}$$

where the quantity $\mathbf{E}^0(\tau)$ is the given history on $\mathbb{R}^+$, while on $\mathbb{R}^{--}$ it is defined as

$$\mathbf{E}^0(\tau) = \mathbf{E}^{(m_1)}(-\tau) \qquad \forall \tau \in \mathbb{R}^{--}. \tag{13.1.16}$$

Thus, from Theorem 4.2.8 and Corollary 4.2.10 together with (13.1.12) and (13.1.13), we have

$$\psi_m(\mathbf{E}^0) = \frac{1}{2}\int_0^{\infty}\int_0^{\infty} \mathbb{G}(|t-\tau|)\dot{\mathbf{E}}^{(m_1)}(\tau)\cdot\dot{\mathbf{E}}^{(m_1)}(t)d\tau dt - \phi(\infty) + \mathbb{G}_\infty\mathbf{E}(\infty)\cdot\mathbf{E}(0),$$

where $\dot{\mathbf{E}}^{(m_1)}$ is the solution of Eq. (13.1.13). Integrating by parts and taking account of (13.1.15)–(13.1.16) with the expression $(13.1.2)_1$ for the stress tensor, this relation becomes

$$\psi_m(\mathbf{E}^0) = S(0) + \frac{1}{2}\int_0^{\infty}\int_0^{\infty} \frac{\partial^2}{\partial t \partial \tau}\mathbb{G}(|t-\tau|)\mathbf{E}^{(m_1)}(\tau)\cdot\mathbf{E}^{(m_1)}(t)d\tau dt, \tag{13.1.17}$$

where $S(t)$ is given by (8.6.15).

Remark 13.1.4. Thus, we see that the second formulation above, based on (13.1.12) and leading to (13.1.15), corresponds to the developments based on (11.2.26) and indeed all the formulations in Chap. 11. Note that the form (13.1.17) corresponds to $(11.8.10)_2$, the latter applying to relative histories; indeed, the two are related in the same way as (11.2.34) and (11.2.17). On the other hand, the developments arising out of the form (13.1.7) are somewhat different. The relationship between the two is addressed in Remark 13.1.3.

13.1.1 Minimum Free Energy for Solids

We refer to the footnote in Sect. 11.8 in the context of this topic.

The Fourier transform of $\mathbb{G}'$ is the function $\mathbb{G}'_+(\omega)$, which is analytic on $\Omega^{(-)}$, by Proposition C.2.1. It is further assumed (Hypothesis C.2.3) that $\mathbb{G}'_+$ is analytic on $\mathbb{R}$ and therefore on Ω^-. It is defined by analytic continuation from the region of analyticity in regions of Ω^+ where the Fourier integral does not converge.

The quantity $\mathbb{G}'_+$ is related to $\mathbb{G}'_c$ and $\mathbb{G}'_s$ by (C.2.2). The function $\mathbb{G}'_s$ has singularities in both $\Omega^{(+)}$ and $\Omega^{(-)}$ that are mirror images of one another; thus, it must be analytic at the origin, where it goes to zero like ω^n with the integer $n \geq 1$; we assume that $n = 1$ (see (7.2.20) and (7.2.21)).

It is important for the following considerations to consider the function $\mathbb{H}(\omega)$ introduced in (8.7.1). We note that this quantity is an even function of frequency and positive, except at the origin, where it vanishes linearly.

From (C.2.16) and (C.2.17),

$$i \lim_{\omega\to\infty} \omega\mathbb{G}'_F(\omega) = \lim_{\omega\to\infty} \omega\mathbb{G}'_s(\omega) = \mathbb{G}'(0), \tag{13.1.18}$$

whence we obtain

$$\mathbb{H}(\infty) = -\mathbb{G}'(0) > 0,$$

by virtue of (8.1.23).

We extend $\mathbb{G}(s)$, defined for any $s \in \mathbb{R}^+$, by means of the even function $\mathbb{G}(|s|)\ \forall s \in \mathbb{R}^+$; then, it follows that $\mathbb{G}'(|s|)$ is an odd function with Fourier transform (see (C.1.6))

$$\mathbb{G}'_F(\omega) = -2i\mathbb{G}'_s(\omega). \tag{13.1.19}$$

We shall consider the Fourier transforms $\mathbf{E}^t_+(\omega)$ and $\mathbf{E}^t_-(\omega)$ of the strain tensor history and continuation $\mathbf{E}^t(s)\ \forall s \in \mathbb{R}$, defined by (see (C.1.3))

$$\mathbf{E}^t_+(\omega) = \int_0^\infty \exp(-i\omega s)\mathbf{E}^t(s)ds \quad \text{and} \quad \mathbf{E}^t_-(\omega) = \int_{-\infty}^0 \exp(-i\omega s)\mathbf{E}^t(s)ds, \tag{13.1.20}$$

respectively. The history and continuation in some cases do not belong to L^2. Thus, if $\mathbf{E}(-\infty) \neq \mathbf{0}$, then defining

$$\mathbf{E}^t_d(s) = \mathbf{E}^t(s) - \mathbf{E}^t(\infty) = \mathbf{E}^t(s) - \mathbf{E}(-\infty),$$

we assume that $\mathbf{E}^t_d \in L^2(\mathbb{R}^+)$. In this case, we have, recalling (C.2.10),

$$\begin{aligned}\mathbf{E}^t_+(\omega) &= \int_0^\infty \exp(-i\omega s)\mathbf{E}^t_d(s)ds + \mathbf{E}(-\infty)\int_0^\infty \exp(-i\omega s)ds \\ &= \mathbf{E}^t_{d+}(\omega) + \frac{\mathbf{E}(-\infty)}{i\omega^-}, \qquad \omega^- = \lim_{\alpha\to 0^+}(\omega - i\alpha),\end{aligned} \tag{13.1.21}$$

where the limit in the last relation is taken after any integrations over ω have been carried out. Analogously, we also have

$$\mathbf{E}_-^t(\omega) = \mathbf{E}_{d-}^t(\omega) - \frac{\mathbf{E}(\infty)}{i\omega^+}, \quad \omega^+ = \lim_{\alpha\to 0^+}(\omega + i\alpha), \tag{13.1.22}$$

if $\mathbf{E}(\infty) \neq \mathbf{0}$.

The quantity $\mathbf{E}_+^t$ is analytic on $\Omega^{(-)}$ and $\mathbf{E}_-^t$ is analytic on $\Omega^{(+)}$ by Proposition C.2.1, while their analyticity on $\mathbb{R}$ is assumed by Hypothesis C.2.3. Poles that may exist near the origin are moved slightly off the real axis as prescribed in (13.1.21) and (13.1.22). We also assume that they are analytic at infinity, so that if $\mathbf{E}^t(0)$ is finite, which is of course assumed, $\mathbf{E}_\pm^t$ go to zero at large ω like ω^{-1} in all directions, by virtue of (C.2.16).

Under the hypothesis that the strain-tensor history has a continuous derivative (see (7.2.27)), we have

$$\frac{d}{dt}\mathbf{E}_+^t(\omega) = -i\omega\mathbf{E}_+^t(\omega) + \mathbf{E}(t).$$

Moreover, from (13.1.21), it follows that

$$i\lim_{\omega\to 0}\omega\mathbf{E}_+^t(\omega) = \lim_{\omega\to 0}\omega\mathbf{E}_s^t(\omega) = \mathbf{E}(-\infty),$$

and, as in (13.1.18),

$$i\lim_{\omega\to\infty}\omega\mathbf{E}_+^t(\omega) = \lim_{\omega\to\infty}\omega\mathbf{E}_s^t(\omega) = \mathbf{E}(t).$$

Observe that from the definition $(13.1.20)_1$, we have, for $\omega \in \mathbb{R}$,

$$\overline{\mathbf{E}_+^t}(\omega) = \mathbf{E}_+^t(-\omega),$$

and hence

$$\mathbf{E}_+^t(\omega) - \overline{\mathbf{E}_+^t}(\omega) = -2i\int_0^\infty \sin\omega s\mathbf{E}^t(s)ds = -2i\mathbf{E}_s^t(\omega).$$

Taking account of these last two relations and of the definition of $\mathbb{H}$, given by (8.7.1), the expression (8.1.28) for the stress tensor (with integrand extended to $\mathbb{R}$) becomes

$$\begin{aligned}\mathbf{T}(t) &= \mathbb{G}_0\mathbf{E}(t) + \frac{1}{2\pi i}\int_{-\infty}^{\infty}\frac{\mathbb{H}(\omega)}{\omega}\left[\mathbf{E}_+^t(\omega) - \mathbf{E}_+^t(-\omega)\right]d\omega \\ &= \mathbb{G}_0\mathbf{E}(t) + \frac{1}{\pi i}\int_{-\infty}^{\infty}\frac{\mathbb{H}(\omega)}{\omega}\mathbf{E}_+^t(\omega)d\omega,\end{aligned}$$

where we have used the property that $\mathbb{H}(\omega)/\omega$ is an odd function of ω.

We now seek to solve the Wiener–Hopf equation (13.1.15). Replacing the parameter t by $-t$, the latter form can be written as

$$\int_{-\infty}^{\infty}\frac{\partial}{\partial t}\mathbb{G}(|t-\tau|)\mathbf{E}^0(\tau)d\tau = \mathbf{R}(t),$$

$$\mathbf{R}(t) = \mathbf{0} \quad \forall t \in \mathbb{R}^-, \qquad \mathbf{E}^0(\tau) = \mathbf{E}^{(m_1)}(-\tau), \quad \forall\tau \in \mathbb{R}^{++},$$

where $\mathbf{E}^0$ coincides with the given history on $\mathbb{R}^+$ and $\mathbf{R}(t)$ is defined by this relation on $\mathbb{R}^{++}$. Taking the Fourier transform, we obtain, with the aid of (8.7.1) and (13.1.19),

$$2i\mathbb{H}(\omega)\left[\mathbf{E}^0_+(\omega) + \mathbf{E}^{(m)}(\omega)\right] = \omega\mathbf{R}_+(\omega), \tag{13.1.23}$$

where $\mathbf{E}^{(m)}(\omega)$, which denotes the Fourier transform of $\mathbf{E}^0(\tau)$ defined on $\mathbb{R}^{--}$ by (13.1.16), is the quantity we seek to determine. The quantity $\mathbf{R}_+(\omega)$ is analytic on $\Omega^{(-)}$ and by assumption also on the real axis.

The tensor $\mathbb{H}$ is isomorphic to a matrix in $\mathbb{R}^6 \times \mathbb{R}^6$ and can be factorized (this is discussed in Sect. 11.1) in the form

$$\mathbb{H}(\omega) = \mathbb{H}_+(\omega)\mathbb{H}_-(\omega), \tag{13.1.24}$$

where $\mathbb{H}_\pm$ is analytic and has no zeros in its determinant on $\Omega^\mp$. We multiply (13.1.23) by $[\mathbb{H}_+(\omega)]^{-1}$ to obtain

$$\mathbb{H}_-(\omega)\left[\mathbf{E}^0_+(\omega) + \mathbf{E}^{(m)}(\omega)\right] = \frac{\omega}{2i}[\mathbb{H}_+(\omega)]^{-1}\,\mathbb{R}_+(\omega). \tag{13.1.25}$$

Now, using the Plemelj formulas (B.2.15), we have

$$\begin{aligned}
\mathbf{Q}(\omega) &= \mathbb{H}_-(\omega)\mathbf{E}^0_+(\omega) = \mathbf{q}_-(\omega) - \mathbf{q}_+(\omega),\\
\mathbf{q}_\pm(\omega) &= \lim_{z\to\omega^\mp} \mathbf{q}(z),\\
\mathbf{q}(z) &= \frac{1}{2\pi i}\int_{-\infty}^{\infty} \frac{\mathbf{Q}(\omega')}{\omega' - z}d\omega',
\end{aligned} \tag{13.1.26}$$

where $\mathbf{q}_-$ is analytic on $W^{(+)}$ and $\mathbf{q}_+$ is analytic on $\Omega^{(-)}$. Moreover, $\mathbb{H}_-(\omega)$ is analytic in Ω^+ and $\mathbf{E}^0_+$ in Ω^-; hence, it follows that $\mathbf{Q}(\omega)$ is analytic on $\mathbb{R}$. Closing the contour in $(13.1.26)_3$ on the half-plane that does not contain z, we pick up the singularities of $\mathbf{Q}$ in that half-plane; therefore, both $\mathbf{q}_\pm$ are analytic on $\mathbb{R}$. Finally, $\mathbf{q}_-$ is defined in $\Omega^{(-)}$ by analytic continuation from Ω^+, while $\mathbf{q}_+$ is correspondingly defined in $\Omega^{(+)}$.

By substituting $(13.1.26)_1$ into relation (13.1.25), we obtain the equality

$$\mathbf{K}(\omega) = \mathbf{q}_-(\omega) + \mathbb{H}_-(\omega)\mathbf{E}^{(m)}(\omega) = \mathbf{q}_+(\omega) + \frac{\omega}{2i}[\mathbb{H}_+(\omega)]^{-1}\,\mathbf{R}_+(\omega).$$

The function $\mathbf{K}$ must be analytic both on Ω^- by virtue of its first definition and on Ω^+ by virtue of the second one; therefore, it is analytic over all the complex plane. Thus, using Liouville's theorem (Sect. B.1.3), $\mathbf{K}(\omega)$ must vanish everywhere, so that

$$\mathbf{q}_-(\omega) + \mathbb{H}_-(\omega)\mathbf{E}^{(m)}(\omega) = \mathbf{0}, \tag{13.1.27}$$

since $\mathbf{K}(\omega) \to \mathbf{0}$ like $1/\omega$, as $|\omega| \to \infty$, because $\mathbf{q}_-$ and $\mathbf{E}^{(m)}$ have this property.

We can now consider the expression (13.1.17) for the minimum free energy, which, using Parseval's formula and the convolution theorem (Sect. C.3), becomes

$$\psi_m(\mathbf{E}^0) = S(0) + \frac{1}{2\pi}\int_{-\infty}^{\infty} \mathbb{H}(\omega)\mathbf{E}^{(m)}(\omega)\cdot\overline{\mathbf{E}^{(m)}}(\omega)d\omega.$$

This is analogous to the derivation of $(7.5.7)_2$ from (7.5.3). Relations (13.1.24) and (13.1.27) yield the required expression for the free energy:

$$\widetilde{\psi}_m(\mathbf{E}^0) = S(0) + \frac{1}{2\pi}\int_{-\infty}^{\infty} |\mathbf{q}_-(\omega)|^2 d\omega. \tag{13.1.28}$$

Note that this can be put in the form [92]

$$\begin{aligned} \widetilde{\psi}_m(\mathbf{E}^0) &= \frac{1}{2}\mathbf{E}(0)\cdot \mathbb{G}_\infty \mathbf{E}(0) + \frac{1}{2\pi}\int_{-\infty}^{\infty} |\mathbf{p}-(\omega)|^2 d\omega, \\ \mathbf{p}_-(\omega) &= \mathbf{q}_-(\omega) - \frac{\mathbb{H}_-(\omega)\mathbf{E}(0)}{i\omega^+}. \end{aligned} \tag{13.1.29}$$

If the expressions (13.1.28) and (13.1.29) are to be useful, it must be possible to carry out the factorization (13.1.24) explicitly. This issue is addressed in a general manner in Sect. 11.6, based on the results in Sect. 11.1. Here, we content ourselves with the observation that for an isotropic solid, $\mathbb{G}'$ is diagonal and given by

$$\begin{aligned} [\mathbb{G}'(s)]_{ijkl} &= \mu'(s)(\delta_{ik}\delta_{jl} + \delta_{il}\delta_{jk} - \frac{2}{3}\delta_{ij}\delta_{kl}) + k'(s)\delta_{ij}\delta_{kl}, \\ k'(s) &= \frac{2}{3}\mu'(s) + \lambda'(s), \end{aligned} \tag{13.1.30}$$

where μ' and λ' are scalar relaxation function derivatives corresponding to the Lamé moduli introduced in Theorem 2.4.3, while k' corresponds to the modulus of compression (see (2.4.16)). Thus, recalling (8.7.1), we have

$$\begin{aligned} [\mathbb{H}(\omega)]_{ijkl} &= H^{(\mu)}(\omega)(\delta_{ik}\delta_{jl} + \delta_{il}\delta_{jk} - \frac{2}{3}\delta_{ij}\delta_{kl}) + H^{(k)}(\omega)\delta_{ij}\delta_{kl}, \\ H^{(\mu)}(\omega) &= -\omega\mu'_s(\omega), \qquad H^{(k)}(\omega) = -\omega k'_s(\omega). \end{aligned} \tag{13.1.31}$$

If we can factorize the scalar quantities $H^{(\mu)}$, $H^{(k)}$ by

$$H^{(\mu)}(\omega) = H_+^{(\mu)}(\omega)H_-^{(\mu)}(\omega), \qquad H^{(k)}(\omega) = H_+^{(k)}(\omega)H_-^{(k)}(\omega),$$

where the singularities and zeros of $H_+^{(\mu)}$, $H_+^{(k)}$ are in $\Omega^{(+)}$ and those of $H_-^{(\mu)}$, $H_-^{(k)}$ are in $\Omega^{(-)}$, then it is easy to show that by putting

$$[\mathbb{H}_\pm(\omega)]_{ijkl} = \frac{1}{\sqrt{2}}H_\pm^{(\mu)}(\omega)(\delta_{ik}\delta_{jl} + \delta_{il}\delta_{jk} - \frac{2}{3}\delta_{ij}\delta_{kl}) + \frac{1}{\sqrt{3}}H_\pm^{(k)}(\omega)\delta_{ij}\delta_{kl},$$

relation (13.1.24) holds. This procedure is a special case of that outlined in Sect. 11.6. In particular, the factors $\mathbb{H}_\pm(\omega)$ commute. The factorization of a scalar positive function is discussed in Sect. 11.1.1.

Finally, we note that the rate of dissipation for a viscoelastic solid is given by (11.8.13). Also, a discontinuity occurs between the optimal continuation and the history, as shown by (11.8.15). Furthermore, the optimal continuation does not vanish at large times, as indicated by (11.8.16).

13.1.2 Minimum Free Energies in Terms of Stress History

The minimum free energy, where stress rather than strain is the independent field variable, can be given without difficulty. The analogous developments to those given above can be carried through to give a formula corresponding to (13.1.28). This will have the form

$$\widetilde{\psi}_m(\mathbf{T}^0) = S_{\mathcal{T}}(0) + \frac{1}{2\pi}\int_{-\infty}^{\infty} |\mathbf{q}_-^T(\omega)|^2 d\omega,$$

where, recalling (8.6.15),

$$S_{\mathcal{T}}(0) = \mathbf{T}(t)\cdot\mathbf{E}(t) - \frac{1}{2}\mathbb{J}_0\mathbf{T}(t)\cdot\mathbf{T}(t).$$

The quantity $\mathbb{J}_0$ is given by (8.5.6). Also, to determine $\mathbf{q}_-^T$, we observe that the core issue is to determine what replaces $\mathbb{H}$ given by (8.7.1) and factorized as in (13.1.24). Let us denote it by $\mathbb{H}^T$. Recalling (8.5.13), we see that

$$\mathbb{H}^T(\omega) = \omega\mathbb{J}_s'(\omega).$$

Using $(8.5.10)_1$, (8.5.12), and (8.7.1), we can write

$$\mathbb{H}^T(\omega) = [\mathbf{1} + \mathbb{K}_+(\omega)]^{-1}\mathbb{J}_0\mathbb{H}(\omega)\mathbb{J}_0[\mathbf{1} + \mathbb{K}_+(\omega)]^{-1},$$

giving

$$\mathbb{H}_+^T(\omega) = [\mathbf{1} + \mathbb{K}_+(\omega)]^{-1}\mathbb{J}_0\mathbb{H}_+(\omega),$$
$$\mathbb{H}_-^T(\omega) = \mathbb{J}_0\mathbb{H}_-(\omega)[\mathbf{1} + \overline{\mathbb{K}}_+(\omega)]^{-1},$$

with $\mathbb{H}_\pm(\omega)$ as introduced in (13.1.24). Note that by virtue of Remarks 8.5.1 and 8.5.2, the factors $\mathbb{H}_+^T(\omega)$ and $\mathbb{H}_-^T(\omega)$ have their singularities and zeros in the upper and lower half-planes, respectively. The zeros of $\mathbf{1} + \mathbb{K}_+(\omega)$ are the isolated singularities of $\mathbb{H}_+^T(\omega)$, while the isolated singularities of $\mathbf{1} + \mathbb{K}_+(\omega)$ cancel those of $\mathbb{H}_+(\omega)$. This may be confirmed with the aid of (8.5.3), on noting that the isolated singularities of $\mathbb{H}_+(\omega)$ are the same as those of $\mathbb{G}_+'(\omega)$. The zeros of $\mathbb{H}_+^T(\omega)$ are those of $\mathbb{H}_+(\omega)$. Corresponding observations apply to $\mathbb{H}_-^T(\omega)$, $\mathbf{1} + \overline{\mathbb{K}}_+(\omega)$, and $\mathbb{H}_-(\omega)$.

Thus, recalling (13.1.26), we have

$$\mathbf{q}_\pm^T(\omega) = \lim_{z\to\omega^\mp}\mathbf{q}^T(z), \qquad \mathbf{q}(z) = \frac{1}{2\pi i}\int_{-\infty}^{\infty}\frac{\mathbb{H}_-^T(\omega')\mathbf{T}_+^0(\omega')}{\omega' - z}d\omega'.$$

13.2 Maximum Recoverable Work for Fluids

We now present developments for fluids analogous to those in Sect. 13.1 for solids. To determine the optimal process that maximizes the recoverable work, we consider the trivial extension on $\mathbb{R}$ of a process $P(t) = \mathbf{D}(t)$, of duration $d_P < \infty$, as given by (13.1.1). Observe that with such a process, there exists $\mathbf{E}(\infty) = \lim_{t\to\infty}\mathbf{E}(t)$.

Consider the work $W(\gamma_0, P)$ done by a process P, applied at time $t = 0$, to the state $\gamma_0 = \mathbf{E}_r^0$. Firstly, we observe that the extra stress, given by (8.9.14), becomes, with an integration by parts, remembering that $\mathbb{G}(\infty) = \mathbf{0}$ for fluids,

$$\begin{aligned}\frac{1}{\rho}\tilde{\mathbf{T}}(\rho, \mathbf{E}_r^t) = \mathbf{V}(\mathbf{E}_r^t) &= \int_0^t \mathbb{G}(s)\dot{\mathbf{E}}^t(s)ds + \mathbb{G}(t)\mathbf{E}(0) + \int_t^\infty \mathbb{G}'(s)\mathbf{E}^t(s)ds \\ &= \int_0^t \mathbb{G}(s)\dot{\mathbf{E}}^t(s) + \mathbf{I}^0(t, \mathbf{E}_r^0), \\ \dot{\mathbf{E}}^t(s) &= \frac{\partial}{\partial t}\mathbf{E}^t(s) = -\frac{\partial}{\partial s}\mathbf{E}^t(s),\end{aligned} \tag{13.2.1}$$

where we have put (cf. (8.2.2))

$$\begin{aligned}\mathbf{I}^0(t, \mathbf{E}_r^0) &= \int_0^\infty \mathbb{G}'(t+\tau)\mathbf{E}_r^0(\tau)d\tau \\ &= \int_0^\infty \mathbb{G}(t+\tau)\mathbf{D}(-\tau)d\tau = \int_{-\infty}^0 \mathbb{G}(t-u)\mathbf{D}(u)du, \\ \mathbf{D}(u) &= \dot{\mathbf{E}}(u).\end{aligned} \tag{13.2.2}$$

Thus, substituting $(13.2.1)_2$ into (8.9.27), we obtain

$$W(\gamma, P) = \frac{1}{2}\int_0^\infty \int_0^\infty \mathbb{G}(|t-\tau|)\mathbf{D}(t)\cdot\mathbf{D}(\tau)d\tau dt + \int_0^\infty \mathbf{I}^0(t, \mathbf{E}_r^0)\cdot\mathbf{D}(t)dt, \tag{13.2.3}$$

after a change of variable in the first term and the use of the manipulation leading to (7.5.2).

The optimal process $P^{(m)}$, denoted by $\mathbf{D}^{(m)}$, is the time derivative of the optimal future strain $\mathbf{E}^{(m)}$. We seek the supremum of $-W(\gamma_0, P)$ with respect to the functions given by (13.1.6). Substituting into (13.2.3) the time derivative of these functions, i.e., $\mathbf{D}(t) = \mathbf{D}^{(m)}(t) + \varepsilon\dot{\mathbf{e}}(t)$, and evaluating the derivative with respect to ε, we have

$$\begin{aligned}&\left.\frac{d[-W(\gamma_0, P)]}{d\varepsilon}\right|_{\varepsilon=0} \\ &= -\int_0^\infty \left[\int_0^\infty \mathbb{G}(|t-\tau|)\mathbf{D}^{(m)}(\tau)d\tau + \mathbf{I}^0(t, \mathbf{E}_r^0)\right]\cdot\dot{\mathbf{e}}(t)dt = 0,\end{aligned}$$

whence, because of the arbitrariness of $\dot{\mathbf{e}}(t)$, we obtain

$$\int_0^\infty \mathbb{G}(|t-\tau|)\mathbf{D}^{(m)}(\tau)d\tau = -\mathbf{I}^0(t, \mathbf{E}_r^0) \qquad \forall t \in \mathbb{R}^{++}, \tag{13.2.4}$$

which is in agreement with (13.1.7) in the light of (8.2.1), given that $\mathbb{G}_\infty$ vanishes for fluids.* This is a Wiener–Hopf equation for the quantity $\mathbf{D}^{(m)}$ that maximizes the recoverable work. By substituting (13.2.4) into (13.2.3), we obtain

* Note that (13.1.7) and (13.1.13) are in agreement for $\mathbb{G}_\infty = 0$, so that the problems discussed in Remarks 13.1.3 and 13.1.3 do not exist for fluids.

$$\varphi_m(\mathbf{E}_r^0) = \frac{1}{2}\int_0^\infty \int_0^\infty \mathbb{G}(|t-\tau|)\mathbf{D}^{(m)}(t)\cdot\mathbf{D}^{(m)}(\tau)d\tau dt, \tag{13.2.5}$$

by virtue of (8.9.28). Consider the completion $\mathcal{G}$ of the set (cf. (13.1.10))

$$\tilde{\mathcal{G}} = \left\{\mathbf{D} : [0,\infty) \to \mathrm{Sym}; \int_0^\infty \int_0^\infty \mathbb{G}(|t-\tau|)\mathbf{D}(t)\cdot\mathbf{D}(\tau)d\tau dt < \infty\right\},$$

by means of the norm

$$\begin{aligned} \|\cdot\|_{\mathcal{G}}^2 &= \int_0^\infty \int_0^\infty \mathbb{G}(|t-\tau|)\mathbf{D}(t)\cdot\mathbf{D}(\tau)d\tau dt \\ &= \frac{1}{\pi}\int_{-\infty}^\infty \mathbb{G}_c(\omega)\mathbf{D}_+^{(m)}(\omega)\cdot\overline{\mathbf{D}_+^{(m)}}(\omega)d\omega. \end{aligned} \tag{13.2.6}$$

The second form, on the frequency domain, is derived in the same way as (13.1.9). The thermodynamic restrictions (8.9.31) imply that both of these integrals are non-negative.

The inner product is

$$(\mathbf{D}_1,\mathbf{D}_2) = \int_0^\infty \int_0^\infty \mathbb{G}(|t-\tau|)\mathbf{D}_1(t)\cdot\mathbf{D}_2(\tau)d\tau dt,$$

with which $\mathcal{G}$ becomes a Hilbert space. The set of processes Π is a subset of $\mathcal{G}$.

Thus, for fluids, we have the following results, also given for viscoelastic solids (Remark 13.1.1 and subsequent observations).

Remark 13.2.1. Using the norm of $\mathcal{G}$, we can provide the set Π of processes with a topology; moreover, by means of the norm (13.2.6), the closure of Π is the Hilbert space $\mathcal{G}$.

Equation (13.2.4) can be put in the form

$$\mathcal{A}\mathbf{D} = \mathbf{I}^0,$$

where $\mathcal{A}$ is a bounded and coercive operator from $\mathcal{G}$ to its dual $\mathcal{G}'$; therefore, also for fluids, using the Lax–Milgram theorem, we have the following result.

Theorem 13.2.2. *For any $\mathbf{I}^0 \in \mathcal{G}'$, Eq. (13.2.4) has a unique solution $\mathbf{D} \in \mathcal{G}$ such that*

$$\|\mathbf{D}\|_{\mathcal{G}} \le K\|\mathbf{I}^0\|_{\mathcal{G}'}.$$

In other words, there exists an isomorphism between $\mathcal{G}$ and $\mathcal{G}'$.

Proposition 13.2.3. *Two states, characterized by $\mathbf{E}_{r_j}^0$ $(j = 1, 2)$, are equivalent in the sense of (8.9.17)$_{2,3}$ if and only if*

$$\mathbf{I}^0(t,\mathbf{E}_{r_1}^0) = \mathbf{I}^0(t,\mathbf{E}_{r_2}^0) \qquad \forall t \in \mathbb{R}^+. \tag{13.2.7}$$

Proof. If (13.2.7) holds for any $t \in \mathbb{R}^+$ and $\mathbf{E}_r^0 = \mathbf{E}_{r_1}^0 - \mathbf{E}_{r_2}^0$, then $\mathbf{I}^0(t, \mathbf{E}_r^0) = \mathbf{0}$ for any $t \in \mathbb{R}^+$ or, from (13.2.2),

$$\int_0^\infty \mathbb{G}'(t+\tau)\mathbf{E}_r^0(\tau)d\tau = \mathbf{0} \qquad \forall t \in \mathbb{R}^+,$$

which is equivalent to $(8.9.17)_{2,3}$. The converse is trivial. □

Remark 13.2.4. This proposition allows us to consider a bijective map between $\mathcal{G}'$ and the quotient space $\Gamma_{(m)} = \Gamma/\Gamma_0$. Thus, it is possible to identify any class of equivalent relative strain histories with $\mathbf{I}^0$.

Therefore, the minimum free energy can be represented as a function defined on $\Gamma_{(m)}$.

13.2.1 The Minimum Free Energy for Fluids

We consider the Wiener–Hopf equation (13.2.4). The solution $\mathbf{D}^{(m)}$ of this equation can be determined by virtue of the thermodynamic properties of the kernel $\mathbb{G}$. The maximum recoverable work coincides with the minimum free energy

$$\widetilde{\psi}_m(\rho, \mathbf{E}_r^t) = \phi(\rho) + \varphi_m(\mathbf{E}_r^t), \tag{13.2.8}$$

where ϕ is defined by (8.9.22) and (see (13.2.5))

$$\varphi_m(\mathbf{E}_r^t) == \frac{1}{2}\int_0^\infty \int_0^\infty \mathbb{G}(|\tau - s|)\mathbf{D}^{(m)}(\tau)\cdot \mathbf{D}^{(m)}(s)d\tau ds. \tag{13.2.9}$$

Let us introduce the function $\mathbf{r} : \mathbb{R} \to \mathrm{Sym}$, with the property that

$$\mathbf{r}(\tau) = 0 \qquad \forall \tau \in \mathbb{R}^{++}.$$

Its value on $\mathbb{R}^-$ is not yet determined. Using (13.2.2), we can write (13.2.4) in the form

$$\int_0^\infty \mathbb{G}(|\tau - s|)\mathbf{D}^{(m)}(s)ds = -\int_{-\infty}^0 \mathbb{G}(\tau - u)\mathbf{D}(u)du + \mathbf{r}(\tau) \qquad \forall \tau \in \mathbb{R}, \tag{13.2.10}$$

where $\mathrm{supp}(\mathbf{D}^{(m)}) \subseteq \mathbb{R}^+$, $\mathrm{supp}(\mathbf{D})) \subseteq \mathbb{R}^-$, and $\mathrm{supp}(\mathbf{r}) \subseteq \mathbb{R}^-$. Taking the Fourier transform of this equation yields, with the aid of $(C.1.5)_1$,

$$2\mathbb{G}_c(\omega)[\mathbf{D}_+^{(m)}(\omega) + \mathbf{D}_-(\omega)] = \mathbf{r}_-(\omega), \tag{13.2.11}$$

in the notation of (C.1.3), where

$$\begin{aligned} \mathbf{D}_-(\omega) &= \int_{-\infty}^0 \mathbf{D}(u)e^{-i\omega u}du = \int_{-\infty}^0 \dot{\mathbf{E}}(u)e^{-i\omega u}du \\ &= \mathbf{E}(0) + i\omega \int_0^\infty \mathbf{E}(-u)e^{i\omega u}du \\ &= i\omega\left[\overline{\mathbf{E}_+^0}(\omega) + \frac{\mathbf{E}(0)}{i\omega^+}\right] = i\omega\overline{\mathbf{E}_{r+}^0}(\omega). \end{aligned} \tag{13.2.12}$$

We have from (8.7.1) that

$$\mathbb{G}_c(\omega) = \frac{1}{\omega^2}\mathbb{H}(\omega) \geq 0. \tag{13.2.13}$$

This is true in general. However, we are dealing here with the special case of an isotropic material for which $\mathbb{H}$ has the form (13.1.31) and the factors commute.

Referring to Sect. 11.1, we see that the non-negativity property of $\mathbb{G}_c$ implies that it can be factorized as follows:

$$\mathbb{G}_c(\omega) = \mathbb{G}_{(+)}(\omega)\mathbb{G}_{(-)}(\omega) = \mathbb{G}_{(-)}(\omega)\mathbb{G}_{(+)}(\omega), \tag{13.2.14}$$

where the singularities of $\mathbb{G}_{(\pm)}(\omega)$ and the zeros of its determinant are all in $\Omega^{(\pm)}$, respectively. The factors obey the relations

$$\overline{\mathbb{G}_{(\pm)}}(\omega) = \mathbb{G}_{(\mp)}(\omega),$$

by virtue of (11.1.13) for symmetric tensors. We have

$$\mathbb{G}_{(+)}(\omega) = \frac{1}{i\omega}\mathbb{H}_+(\omega), \quad \mathbb{G}_{(-)}(\omega) = -\frac{1}{i\omega}\mathbb{H}_-(\omega), \tag{13.2.15}$$

where the quantities $\mathbb{H}_\pm$ are the factors of $\mathbb{H}$. Using the factorization (13.2.14), (13.2.11) yields

$$\mathbb{G}_{(+)}(\omega)\mathbf{D}_+^{(m)}(\omega) = -\mathbb{G}_{(+)}(\omega)\mathbf{D}_-(\omega) + \frac{1}{2}\mathbb{G}_{(-)}^{-1}(\omega)\mathbf{r}_-(\omega),$$

since $\det \mathbb{G}_{(-)}(\omega) \neq 0\, \forall \omega \in \mathbb{R}$. The Plemelj formulas (B.2.15) give

$$\mathbb{G}_{(+)}(\omega)\mathbf{D}_-(\omega) = \mathbf{p}_{(-)}(\omega) - \mathbf{p}_{(+)}(\omega),$$

where the functions $\mathbf{p}_{(\pm)}(z)$, analytic for $z \in \Omega^{\mp}$, respectively, are defined by means of

$$\begin{aligned} \mathbf{p}(z) &= \frac{1}{2\pi i}\int_{-\infty}^{\infty} \frac{\mathbb{G}_{(+)}(\omega)\mathbf{D}_-(\omega)}{\omega - z} d\omega \quad z \in \Omega\backslash\mathbb{R}, \\ \mathbf{p}_{(\pm)}(\omega) &= \lim_{\alpha\to 0^{\mp}} \mathbf{p}(\omega + i\alpha). \end{aligned} \tag{13.2.16}$$

Remark 13.2.5. Note that by virtue of (13.2.12) and (13.2.15),

$$\mathbb{G}_{(+)}(\omega)\mathbf{D}_-(\omega) = \mathbb{H}_+(\omega)\overline{\mathbf{E}_{r+}^0}(\omega) = \overline{\mathbb{H}_-(\omega)\mathbf{E}_{r+}^0(\omega)},$$

which is interesting to compare with (11.2.8) and indeed (13.1.26), though the latter is expressed in terms of nonrelative histories. We see that the quantity $\mathbf{p}(z)$, defined by (13.2.16), is $-\overline{\mathbf{p}}(\overline{z})$, as given by (11.2.9).

Thus, we obtain

$$\mathbb{G}_{(+)}(\omega)\mathbf{D}_+^{(m)}(\omega) = -\mathbf{p}_{(-)}(\omega) + \mathbf{p}_{(+)}(\omega) + \frac{1}{2}\mathbb{G}_{(-)}^{-1}(\omega)\mathbf{r}_-(\omega),$$

where $\mathbb{G}_{(+)}(z)\mathbf{D}_+^{(m)}(z)$ and $\mathbf{p}_{(+)}(z)$ are analytic for $z \in \Omega^-$, while $\mathbb{G}_{(-)}^{-1}(z)\mathbf{r}_-(z)$ and $\mathbf{p}_{(-)}(z)$ are analytic for $z \in \Omega^+$. Therefore, the function

$$\mathbf{L}(\omega) = \mathbb{G}_{(+)}(\omega)\mathbf{D}_+^{(m)}(\omega) - \mathbf{p}_{(+)}(\omega) = -\mathbf{p}_{(-)}(\omega) + \frac{1}{2}\mathbb{G}_{(-)}^{-1}(\omega)\mathbf{r}_-(\omega)$$

must be equal to zero everywhere, by Liouville's theorem (Sect. B.1.3), because it is analytic on the whole complex plane and vanishes at infinity. Thus, in particular, we have

$$\mathbf{D}_+^{(m)}(\omega) = \mathbb{G}_{(+)}^{-1}(\omega)\mathbf{p}_{(+)}(\omega). \tag{13.2.17}$$

Remark 13.2.6. Since $\mathbb{H}_+(\infty)$ is not zero in general, it follows from $(13.2.15)_1$ that

$$\lim_{\omega\to\infty} \mathbf{D}_+^{(m)}(\omega) \neq \mathbf{0},$$

so that $\mathbf{D}^{(m)}(\tau)$ has an initial delta function-type singularity as $\tau \to 0^+$, since the Fourier transform of a constant yields a delta function, as given by (C.2.12). Thus, the optimal continuation $\mathbf{E}^{(m)}$, where $\mathbf{D}^{(m)} = \dot{\mathbf{E}}^{(m)}$, has an initial discontinuity as $\tau \to 0^+$, and hence $\mathbf{E}^m(0^+) \neq \mathbf{E}(0^-) = \mathbf{E}(0)$ (see (11.8.15)).

Remark 13.2.7. Since $\det \mathbb{G}_{(+)}(0) \neq \mathbf{0}$, we obtain

$$\mathbf{E}^{(m)}(\infty) - \mathbf{E}(0^-) = \int_{0^-}^{\infty} \mathbf{D}^{(m)}(\tau)d\tau = \mathbf{D}_+^{(m)}(0) = \mathbb{G}_{(+)}^{-1}(0)\mathbf{p}_{(+)}(0),$$

where we have emphasized that the integral includes the discontinuity. Thus, the optimal continuation tends to the finite limit

$$\lim_{\tau\to\infty} \mathbf{E}^{(m)}(\tau) = \mathbf{E}^{(m)}(\infty) = \mathbf{E}(0) + \mathbb{G}_{(+)}^{-1}(0)\mathbf{p}_{(+)}(0),$$

which corresponds to (11.8.16) for viscoelastic solids.

From (13.2.9), we have, recalling $(13.2.6)_2$,

$$\begin{aligned} \varphi_m(\mathbf{E}_r^t) &= \frac{1}{2}\int_{-\infty}^{\infty} \mathbb{G}(|\tau - s|)\mathbf{D}^{(m)}(\tau) \cdot \mathbf{D}^{(m)}(s)d\tau ds \\ &= \frac{1}{2}\int_{-\infty}^{\infty} \mathbb{G}_c(\omega)\mathbf{D}_+^{(m)}(\omega) \cdot \overline{\mathbf{D}_+^{(m)}}(\omega)d\omega \\ &= \frac{1}{2\pi}\int_{-\infty}^{\infty} |\mathbf{p}_{(+)}(\omega)|^2 d\omega, \end{aligned} \tag{13.2.18}$$

where the last form follows from (13.2.17).

Finally, the minimum free energy (13.2.8) is given by

$$\psi_m(\rho, \mathbf{E}_r^t) = \phi(\rho) + \frac{1}{2\pi}\int_{-\infty}^{\infty} |\mathbf{p}_{(+)}(\omega)|^2 d\omega. \tag{13.2.19}$$

The integral term in relation (13.2.19) is the same as that in (13.1.29), as outlined in Remark 13.2.5.

Observe that

$$\varphi_m(\mathbf{E}_r^t) = \hat{\varphi}_m(\gamma_m), \tag{13.2.20}$$

where γ_m is an element of $\Gamma_{(m)}$, defined in Remark 13.2.4, and $\hat{\varphi}_m$ is a functional of this quantity, by virtue of (13.2.4) and (13.2.9) and of the results expressed by Proposition 13.2.3. Consequently, such an equivalence class γ_m can be represented by the quantity $\mathbf{p}_{(+)}$.

We recall the following theorem [100], which makes this idea precise.

Theorem 13.2.8. *For every viscoelastic material with a symmetric relaxation function, a given relative strain history $\mathbf{E}_r^t$ is equivalent to $\mathbf{0}^\dagger$ in the sense of (8.9.17)$_{2,3}$ if and only if the quantity $\mathbf{p}_{(+)}(\omega)$, defined by* (13.2.16), *is such that*

$$\mathbf{p}_{(+)}(\omega) = \mathbf{0} \qquad \forall \omega \in \mathbf{R}.$$

A general proof of this result is given in Sect. 11.3. Theorem 13.2.8, with the expression (13.2.18) for $\varphi_m(\mathbf{E}_r^t)$, allows us to introduce a norm in $\Gamma_{(m)}$ by means of

$$\|\gamma_m\|^2 = \hat{\varphi}_m(\gamma_m).$$

Therefore, it follows that the minimum free energy (13.2.19) induces a norm in the space of the minimal states $\mathbf{\Sigma}_{(m)}$, defined as

$$\|\sigma_m\|^2 = \hat{\psi}(\sigma_m) = \phi(\rho) + \hat{\varphi}_m(\gamma_m),$$

where $\sigma_m = (\sigma, \gamma_m)$ and $\hat{\psi}(\sigma_m) = \psi(\rho, \mathbf{E}_r^t)$.

13.3 The Minimum Free Energy for Incompressible Fluids

We now consider the case of incompressible fluids that are discussed in Sect. 8.10. Considering processes acting on states at time t, the Wiener–Hopf equation corresponding to (13.1.7) and (13.2.4) for the optimal process $\dot{\mathbf{E}}_t^{(m)}$ is given by (see (13.2.5) and (13.2.6))

$$2\int_0^\infty \mu(|\tau - \xi|)\dot{\mathbf{E}}_t^{(m)}(\xi)d\xi = -\mathbf{I}^t(\tau, \mathbf{E}_r^t) \qquad \forall \tau \in \mathbb{R}^+, \tag{13.3.1}$$

where μ is defined by (8.10.2). Substituting (13.3.1) into (8.10.25), we obtain that the maximum recoverable work, which is equal to the minimum free energy, is given by (cf. (13.2.5))

$$\begin{aligned}\psi_m(t) &= \int_0^\infty \int_0^\infty \mu(|\tau - \xi|)\dot{\mathbf{E}}_t^{(m)}(\xi) \cdot \dot{\mathbf{E}}_t^{(m)}(\tau)d\xi d\tau \\ &= \frac{1}{\pi}\int_{-\infty}^\infty \mu_c(\omega)\dot{\mathbf{E}}_{t+}^{(m)}(\omega) \cdot \overline{\dot{\mathbf{E}}_{t+}^{(m)}}(\omega)d\omega.\end{aligned}$$

The second relation follows as in (13.2.6) and earlier analogous cases. Consider the function

$$K(\omega) = 2(\omega_0^2 + \omega^2)\mu_c(\omega),$$

where ω_0 is a fixed positive frequency. Since, by virtue of (C.2.17)$_1$, $K(\omega)$ is not zero for any real ω, including infinity, it can be factorized as follows (see Theorem 10.1.2):

$$K(\omega) = K_{(+)}(\omega)K_{(-)}(\omega),$$

where the singularities and zeros of $K_{(\pm)}$ are in $\Omega^{(\pm)}$, respectively. Thus, we have

$$2\mu_c(\omega) = \mu_{(+)}(\omega)\mu_{(-)}(\omega), \tag{13.3.2}$$

where

$$\mu_{(+)}(\omega) = \frac{1}{\omega_0 + i\omega}K_{(+)}(\omega), \qquad \mu_{(-)}(\omega) = \frac{1}{\omega_0 - i\omega}K_{(-)}(\omega).$$

Proceeding as in Sect. 13.2.1, using (8.10.7) in the equivalent of (13.2.10), we obtain, instead of (13.2.17),

$$\dot{\mathbf{E}}^{(m)}_{t+}(\omega) = \frac{1}{\mu_{(+)}(\omega)}\mathbf{p}^t_{(+)}(\omega).$$

The minimum free energy is given by

$$\psi_m(t) = \frac{1}{2\pi}\int_{-\infty}^{\infty} |\mathbf{p}^t_{(+)}(\omega)|^2 d\omega,$$

where

$$\begin{aligned} \mathbf{p}^t(z) &= \frac{1}{2\pi i}\int_{-\infty}^{\infty} \frac{\mu_{(+)}(\omega)\dot{\mathbf{E}}^t_-(\omega)}{\omega - z}d\omega, \qquad z \in \Omega\backslash\mathbb{R}, \\ \mathbf{p}_{(\pm)}(\omega) &= \lim_{\alpha\to 0^\mp} \mathbf{p}(\omega + i\alpha), \end{aligned} \tag{13.3.3}$$

and

$$2\mu_{(+)}(\omega)\dot{\mathbf{E}}^t_-(\omega) = \mathbf{p}^t_{(-)}(\omega) - \mathbf{p}^t_{(+)}(\omega).$$

An alternative notation is now introduced. Referring to (13.2.12), we see also in the present context that

$$\dot{\mathbf{E}}^t_-(\omega) = \int_{-\infty}^{0} \dot{\mathbf{E}}^t(u)e^{-i\omega u}du = i\omega\overline{\mathbf{E}^t_{r+}}(\omega). \tag{13.3.4}$$

Relation (13.2.13) reduces in this case to

$$H(\omega) = 2\omega^2\mu_c(\omega) = -2\omega\mu'_s(\omega) < 0 \qquad \forall\omega \in \mathbb{R}, \tag{13.3.5}$$

where we have used (8.10.13). This function has an even signature, is analytic on the real axis, and vanishes at the origin quadratically. It follows from (13.3.2) that (see (13.2.15))

$$H(\omega) = H_+(\omega)H_-(\omega), \quad H_+(\omega) = i\omega\mu_{(+)}(\omega), \quad H_-(\omega) = -i\omega\mu_{(-)}(\omega).$$

The quantity $H_+(\omega)$ has no singularities or zeros in $\Omega^{(-)}$ and is analytic in Ω^-; similarly, $H_-(\omega)$ is analytic in Ω^+ with no singularities or zeros in $\Omega^{(+)}$. Observe that (8.10.15) yields

$$H(\infty) = H_+(\infty)H_-(\infty) = -2\mu'(0) > 0. \tag{13.3.6}$$

Thus, from (13.3.3) and (13.3.4), we have

$$\begin{aligned}
-\overline{\mathbf{p}^t}(z) &= \frac{1}{2\pi i}\int_{-\infty}^{\infty}\frac{H_-(\omega)\mathbf{E}^t_{r+}(\omega)}{\omega - z}d\omega, \qquad z \in \Omega\backslash\mathbb{R},\\
\mathbf{p}^t_{\pm}(\omega) &= -\lim_{\alpha\to 0^{\mp}}\overline{\mathbf{p}^t}(\omega + i\alpha),\\
H_-(\omega)\mathbf{E}^t_{r+}(\omega) &= \mathbf{p}^t_-(\omega) - \mathbf{p}^t_+(\omega),
\end{aligned} \tag{13.3.7}$$

and

$$\psi_m(t) = \frac{1}{2\pi}\int_{-\infty}^{\infty}|\mathbf{p}^t_-(\omega)|^2 d\omega. \tag{13.3.8}$$

Comparing with (11.8.4), we see that the function $\mathbf{p}^t$ in (13.3.3) is, as indicated in Remark 13.2.5, the negative complex conjugate of that introduced in Chap. 11. The latter corresponds to the quantity defined by $(13.3.7)_1$. The expression for the rate of dissipation is given by (11.8.13) and (11.8.14) for scalar $H(\omega)$.

13.3.1 The Minimum Free Energy in Terms of $\mathbf{I}^t$

This topic, which is the analogue of that discussed in Sect. 12.2, is included here (but not earlier in the chapter) because incompressible fluids provide a particularly simple illustration of the technique. From Remark 4.2.11, we see that ψ_m can be considered as a function of the minimal state. Therefore, it is natural to expect that it should be expressible in terms of $\mathbf{I}^t$, as in Sect. 12.2.

Let

$$\mathbf{J}^t(s) = \mathbf{I}^t(-s, \mathbf{E}^t_r) \qquad \forall s \le 0, \tag{13.3.9}$$

where $\mathbf{I}^t$ is defined by (8.10.5). Thus,

$$\mathbf{J}^t(s) = 2\int_0^{\infty}\mu'(u-s)\mathbf{E}^t_r(u)du = -2\int_0^{\infty}\frac{\partial}{\partial s}\mu(|s-u|)\mathbf{E}^t_r(u)du \quad \forall s \le 0.$$

This function can be extended to $\mathbb{R}$ by putting

$$\mathbf{J}^t(\tau) = \mathbf{I}^t(-\tau, \mathbf{E}^t_r) = -2\int_0^{\infty}\frac{\partial}{\partial\tau}\mu(|\tau-u|)\mathbf{E}^t_r(u)du \quad \forall\tau \in \mathbb{R}.$$

Applying the convolution theorem, we obtain

$$\mathbf{J}^t_F(\omega) = -\frac{4i}{\omega}H(\omega)\mathbf{E}^t_{r+}(\omega),$$

by virtue of (13.3.5) and (C.1.6). We see that $\mathbf{J}^t_F(\omega)$ decays like ω^{-3} at large ω, because of the asymptotic behavior of $H(\omega)$, expressed by (13.3.6) and the fact that $\mathbf{E}^t_{r+}(\omega)$ obeys (11.8.5). Let

$$\mathbf{P}^t(\omega) = -\frac{\omega}{4i}[H_+(\omega)]^{-1}\,\mathbf{J}^t_F(\omega) = H_-(\omega)\mathbf{E}^t_{r+}(\omega). \tag{13.3.10}$$

We consider the function $\mathbf{\Pi}^t(s)$, the Fourier transform of which is $\mathbf{P}^t(\omega)$, i.e.,

$$\mathbf{P}^t(\omega) = \int_{-\infty}^{\infty} \mathbf{\Pi}^t(s) e^{-i\omega s} ds.$$

From $(13.3.7)_3$, it follows that

$$\mathbf{p}^t_-(\omega) = \int_{-\infty}^{0} \mathbf{\Pi}^t(s) e^{-i\omega s} ds, \quad \mathbf{p}^t_+(\omega) = -\int_{0}^{\infty} \mathbf{\Pi}^t(s) e^{-i\omega s} ds, \tag{13.3.11}$$

which is the analogue, in this context, of (12.2.9). We can write $\mathbf{P}^t(\omega)$ as

$$\mathbf{P}^t(\omega) = -\left\{ \frac{[H_+(\omega)]^{-1}}{4i\omega^-} \right\} \left[\omega^2 \mathbf{J}^t_F(\omega) \right], \tag{13.3.12}$$

which is (13.3.10) multiplied and divided by $(\omega^-)^2$. The latter quantity can be replaced by ω^2 in the numerator. We observe that the two expressions in parentheses belong to $L^2(\mathbb{R})$ and that the singularities of $[H_+(\omega)]^{-1}$ are in $\Omega^{(+)}$. The inverse Fourier transform of the first factor will be denoted by

$$M(s) = \frac{1}{2\pi} \int_{-\infty}^{\infty} \frac{[H_+(\omega)]^{-1}}{4i\omega^-} e^{i\omega s} d\omega \qquad \forall s \in \mathbb{R}. \tag{13.3.13}$$

Also,

$$\mathbf{J}^t_{(2)}(s) = -\frac{1}{2\pi} \int_{-\infty}^{\infty} \omega^2 \mathbf{J}^t_F(\omega) e^{i\omega s} d\omega = \frac{d^2}{ds^2} \mathbf{J}^t(s).$$

We note that the integrand in (13.3.13) has a quadratic singularity at the origin because of the factor ω contained in $H_+(\omega)$. This becomes ω^- to give the correct analytic behavior of $H_+(\omega)$. The singularity can be integrated to give a finite value. Moreover, $M(s) = 0 \, \forall s \in \mathbb{R}^{--}$. Thus, in (13.3.12), we have the product of two Fourier transforms, which, by virtue of the convolution theorem (C.3.3), gives

$$\mathbf{\Pi}^t(s) = \int_{-\infty}^{s} M(s-u) \mathbf{J}^t_{(2)}(u) du. \tag{13.3.14}$$

Parseval's formula (C.3.1), applied to the right of (13.3.8) and $(13.3.11)_1$, yields

$$\frac{1}{2\pi} \int_{-\infty}^{\infty} |\mathbf{p}^t_-(\omega)|^2 d\omega \equiv \frac{1}{2\pi} \int_{-\infty}^{\infty} \mathbf{p}^t_-(\omega) \cdot \overline{\mathbf{p}}^t_-(\omega) d\omega = \int_{-\infty}^{0} |\mathbf{\Pi}^t(s)|^2 ds.$$

Now, $\mathbf{J}^t_{(2)}(s)$ contributes to $\mathbf{p}^t_-(\omega)$, by means of (13.3.14), using $(13.3.11)_1$, only for nonpositive values of its argument. By the causal property of $M(s)$, the upper limit in (13.3.14) can be extended to zero. Thus, (13.3.8) becomes

$$\begin{aligned} \psi_m(t) &= \int_{-\infty}^{0} \left[\int_{-\infty}^{0} M(s-u) \mathbf{J}^t_{(2)}(u) du \right] \cdot \left[\int_{-\infty}^{0} M(s-v) \mathbf{J}^t_{(2)}(v) dv \right] ds \\ &= \int_{-\infty}^{0} \int_{-\infty}^{0} L(u,v) \mathbf{J}^t_{(2)}(u) \cdot \mathbf{J}^t_{(2)}(v) du dv, \end{aligned}$$

where

$$\begin{aligned} L(u,v) &= \int_{-\infty}^{0} M(s-u)M(s-v)ds \\ &= \int_{\max(u,v)}^{0} M(s-u)M(s-v)ds, \quad \forall u,v \in \mathbb{R}^{--}, \end{aligned} \tag{13.3.15}$$

again by virtue of the causal property of $M(s)$. We note that

$$L(u,v) = L(v,u) \qquad \forall u,v \in \mathbb{R}^{--}$$

and, by virtue of $(13.3.15)_2$,

$$L(u,0) = L(0,v) = 0 \qquad \forall u,v \in \mathbb{R}^{--}.$$

Moreover, since $M \in L^2(\mathbb{R}^+)$,

$$L(u,-\infty) = L(-\infty,v) = 0 \qquad \forall u,v \in \mathbb{R}^{--}.$$

Finally, using (13.3.9), we have

$$\psi_m(t) = \int_0^{\infty}\int_0^{\infty} K(\xi,\eta)\mathbf{I}^t_{(2)}(\xi,\mathbf{E}^t_r)\cdot\mathbf{I}^t_{(2)}(\eta,\mathbf{E}^t_r)d\xi d\eta,$$

where

$$K(\xi,\eta) = L(-\xi,-\eta) = \int_0^{\min(\xi,\eta)} M(\xi-\rho)M(\eta-\rho)d\rho.$$

Remark 13.3.1. An explicit expression for the minimum free energy of incompressible fluids associated with a discrete-spectrum model [8], that is, where the kernel μ is expressed by a linear combination of strictly decaying exponentials,

$$2\mu(t) = \sum_{i=1}^{n} \mu_i e^{-\alpha_i t} \qquad \forall t \geq 0,$$

where n is a positive integer and $\mu_i, \alpha_i \in \mathbb{R}^{++}$, $i = 1,2,\ldots n$, can be obtained by a close analogy with the calculation in Sect. 11.9, replacing $G(t)$ by $2\mu(t)$ and putting $G_\infty = 0$. The fact that we are now dealing with tensor strain histories is a very minor complication.

13.4 The Maximum Recoverable Work for Heat Conductors

To derive the expression for the maximum recoverable work, we consider the expression (9.3.13), taking $\vartheta_P(d_P) = 0$, so that

$$
\begin{aligned}
W(\sigma(t), P) = & -\frac{1}{2}\alpha_0\vartheta^2(t) + \alpha'(0)\int_0^\infty \vartheta_P^2(\tau)d\tau \\
& + \frac{1}{2}\int_0^\infty\int_0^\infty \alpha''(|\tau-\eta|)\vartheta_P(\eta)\vartheta_P(\tau)d\eta d\tau \\
& + \frac{1}{2}\int_0^\infty\int_0^\infty k(|\tau-\eta|)\mathbf{g}_P(\eta)\cdot\mathbf{g}_P(\tau)d\eta d\tau \\
& + \int_0^\infty I_{(\alpha)}^t(\tau, \vartheta^t)\vartheta_P(\tau)d\tau + \int_0^\infty \mathbf{I}_{(k)}^t(\tau, \bar{\mathbf{g}}^t)\cdot\mathbf{g}_P(\tau)d\tau.
\end{aligned}
\tag{13.4.1}
$$

Let $P^{(m)}(\tau) = (\dot{\vartheta}^{(m)}(\tau), \mathbf{g}^{(m)}(\tau))$ be the optimal process, to which corresponds the maximum recoverable work. We denote by $\vartheta^{(m)}$ and $\mathbf{g}^{(m)}$ the optimal future temperature and temperature gradient. To evaluate the maximum of $-W(\sigma, P)$, let us consider the ensuing fields expressed in terms of $\vartheta^{(m)}$ and $\mathbf{g}^{(m)}$ as follows:

$$
\vartheta_P(\tau) = \vartheta^{(m)}(\tau) + \gamma\varphi(\tau), \qquad \mathbf{g}_P(\tau) = \mathbf{g}^{(m)}(\tau) + \delta\mathbf{e}(\tau) \quad \forall\tau\in\mathbb{R}^+,
$$

where γ and δ are real parameters and φ and $\mathbf{e}$ are arbitrary smooth functions with $\varphi(0) = 0$, $\mathbf{e}(0) = \mathbf{0}$. Thus, we obtain

$$
\begin{aligned}
& -W(\sigma, P) \\
& = -\tilde{W}(\vartheta(t), \vartheta^t, \bar{\mathbf{g}}^t; \dot{\vartheta}^{(m)} + \gamma\dot{\varphi}, \mathbf{g}^{(m)} + \delta\mathbf{e}) \\
& = \frac{1}{2}\alpha_0\vartheta^2(t) - \alpha'(0)\int_0^\infty \left\{[\vartheta^{(m)}(\tau)]^2 + 2\vartheta^{(m)}(\tau)\varphi(\tau)\gamma + \varphi^2(\tau)\gamma^2\right\}d\tau \\
& - \frac{1}{2}\int_0^\infty\int_0^\infty \alpha''(|\tau-\eta|)\left\{\vartheta^{(m)}(\eta)\vartheta^{(m)}(\tau) + [\vartheta^{(m)}(\eta)\varphi(\tau) + \varphi(\eta)\vartheta^{(m)}(\tau)]\gamma\right. \\
& \left. + \varphi(\eta)\varphi(\tau)\gamma^2\right\}d\eta d\tau - \frac{1}{2}\int_0^\infty\int_0^\infty k(|\tau-\eta|)\left\{\mathbf{g}^{(m)}(\eta)\cdot\mathbf{g}^{(m)}(\tau)\right. \\
& \left. + [\mathbf{g}^{(m)}(\eta)\cdot\mathbf{e}(\tau) + \mathbf{e}(\eta)\cdot\mathbf{g}^{(m)}(\tau)]\delta + \mathbf{e}(\eta)\cdot\mathbf{e}(\tau)\delta^2\right\}d\eta d\tau \\
& - \int_0^\infty I_{(\alpha)}^t(\tau, \vartheta^t)[\vartheta^{(m)}(\tau) + \varphi(\tau)\gamma]d\tau - \int_0^\infty \mathbf{I}_{(k)}^t(\tau, \bar{\mathbf{g}}^t)\cdot[\mathbf{g}^{(m)}(\tau) + \mathbf{e}(\tau)\delta]d\tau.
\end{aligned}
$$

Taking derivatives with respect to γ and δ gives

$$
\begin{aligned}
\frac{\partial}{\partial\gamma}[-W(\sigma, P)]\bigg|_{\gamma=0} = & -2\alpha'(0)\int_0^\infty \vartheta^{(m)}(\tau)\varphi(\tau)d\tau \\
& - \frac{1}{2}\int_0^\infty\int_0^\infty \alpha''(|\tau-\eta|)[\vartheta^{(m)}(\eta)\varphi(\tau) + \varphi(\eta)\vartheta^{(m)}(\tau)]d\eta d\tau \\
& - \int_0^\infty I_{(\alpha)}^t(\tau, \vartheta^t)\varphi(\tau)d\tau = 0, \\
\frac{\partial}{\partial\delta}[-W(\sigma, P)]\bigg|_{\delta=0} = & -\frac{1}{2}\int_0^\infty\int_0^\infty k(|\tau-\eta|)[\mathbf{g}^{(m)}(\eta)\cdot\mathbf{e}(\tau) \\
& + \mathbf{e}(\eta)\cdot\mathbf{g}^{(m)}(\tau)]d\eta d\tau - \int_0^\infty \mathbf{I}_{(k)}^t(\tau, \bar{\mathbf{g}}^t)\cdot\mathbf{e}(\tau)d\tau = 0.
\end{aligned}
$$

These relations can be written as

$$\int_0^\infty \varphi(\tau)\left[-2\alpha'(0)\vartheta^{(m)}(\tau) - \int_0^\infty \alpha''(|\tau-\eta|)\vartheta^{(m)}(\eta)d\eta - I^t_{(\alpha)}(\tau,\vartheta^t)\right]d\tau = 0,$$
$$\int_0^\infty \mathbf{e}(\tau)\cdot\left[\int_0^\infty k(|\tau-\eta|)\mathbf{g}^{(m)}(\eta)d\eta + \mathbf{I}^t_{(k)}(\tau,\bar{\mathbf{g}}^t)\right]d\tau = 0.$$

Since φ and $\mathbf{e}$ are arbitrary, we have

$$\int_0^\infty \alpha''(|\tau-\eta|)\vartheta^{(m)}(\eta)d\eta + 2\alpha'(0)\vartheta^{(m)}(\tau) = I^t_{(\alpha)}(\tau,\vartheta^t),$$
$$\int_0^\infty k(|\tau-\eta|)\mathbf{g}^{(m)}(\eta)d\eta = -\mathbf{I}^t_{(k)}(\tau,\bar{\mathbf{g}}^t) \quad \forall\tau\in\mathbb{R}^+. \tag{13.4.2}$$

This system is composed of two Wiener–Hopf integral equations, of the second and of the first kind, respectively. They are solvable, by virtue of the thermodynamic properties of the integral kernels and of certain factorization properties, as used in earlier chapters. The solutions $\vartheta^{(m)}$ and $\mathbf{g}^{(m)}$ yield the maximum recoverable work. It is of interest to recall Remark 9.3.1 in this context.

By substituting the expressions (13.4.2) of $I^t_{(\alpha)}$ and $\mathbf{I}^t_{(k)}$ into (13.4.1), we see that the maximum recoverable work is given by

$$\begin{aligned} W_R(\sigma) &= \frac{1}{2}\alpha_0\vartheta^2(t) + \alpha'(0)\int_0^\infty [\vartheta^{(m)}(\tau)]^2 d\tau \\ &+ \frac{1}{2}\int_0^\infty\int_0^\infty \alpha''(|\tau-\eta|)\vartheta^{(m)}(\eta)\vartheta^{(m)}(\tau)d\eta d\tau \\ &+ \frac{1}{2}\int_0^\infty\int_0^\infty k(|\tau-\eta|)\mathbf{g}^{(m)}(\eta)\cdot\mathbf{g}^{(m)}(\tau)d\eta d\tau. \end{aligned}$$

Its form in the frequency domain can be derived (as noted in the context of (13.1.9)), with the aid of $(9.2.7)_1$, giving

$$W_R(\sigma) = \frac{1}{2}\alpha_0\vartheta^2(t) + \frac{1}{2\pi}\int_{-\infty}^\infty \omega\alpha'_s(\omega)|\vartheta_+^{(m)}(\omega)|^2 d\omega + \frac{1}{2\pi}\int_{-\infty}^\infty k_c(\omega)|\mathbf{g}_+^{(m)}(\omega)|^2 d\omega. \tag{13.4.3}$$

13.4.1 The Minimum Free Energy for Heat Conductors

Let us identify ϑ^t and $\bar{\mathbf{g}}^t$ with their causal extensions to $\mathbb{R}$, putting $\vartheta^t(s) = 0$ and $\bar{\mathbf{g}}^t(s) = \mathbf{0}\,\forall s\in\mathbb{R}^{--}$; moreover, we extend the kernels $\alpha''(s)$ and $k'(s)$ to $\mathbb{R}^{--}$ as an even function denoted by $\alpha''^{(e)}(s)$ and an odd function denoted by $k'^{(o)}(s)$, respectively, i.e.,

$$\alpha''^{(e)}(s) = \begin{cases}\alpha''(s) & \forall s\geq 0,\\ \alpha''(-s) & \forall s<0,\end{cases} \qquad k'^{(o)}(s) = \begin{cases}k'(s) & \forall s\geq 0,\\ -k'(-s) & \forall s<0.\end{cases}$$

These extensions allow us to write (9.3.14) as follows:

$$
\begin{aligned}
I^t_{(\alpha)}(\tau, \vartheta^t) &= \int_{-\infty}^{\infty} \alpha''^{(e)}(\tau+\xi)\vartheta^t(\xi)d\xi, \\
\mathbf{I}^t_{(k)}(\tau, \bar{\mathbf{g}}^t) &= -\int_{-\infty}^{\infty} k'^{(o)}(\tau+\xi)\bar{\mathbf{g}}^t(\xi)d\xi \quad \forall\tau \geq 0.
\end{aligned}
\tag{13.4.4}
$$

Putting

$$
\begin{aligned}
I^{t(n)}_{(\alpha)}(\tau, \vartheta^t) &= \int_{-\infty}^{\infty} \alpha''^{(e)}(\tau+\xi)\vartheta^t(\xi)d\xi, \\
\mathbf{I}^{t(n)}_{(k)}(\tau, \bar{\mathbf{g}}^t) &= -\int_{-\infty}^{\infty} k'^{(o)}(\tau+\xi)\bar{\mathbf{g}}^t(\xi)d\xi \quad \forall\tau < 0,
\end{aligned}
$$

we can extend the functions in (13.4.4) to $\mathbb{R}$:

$$
\begin{aligned}
I^{t(\mathbb{R})}_{(\alpha)}(\tau, \vartheta^t) &= \int_{-\infty}^{\infty} \alpha''^{(e)}(\tau+\xi)\vartheta^t(\xi)d\xi = \begin{cases} I^t_{(\alpha)}(\tau, \vartheta^t) & \forall\tau \geq 0, \\ I^{t(n)}_{(a)}(\tau, \vartheta^t) & \forall\tau < 0, \end{cases} \\
\mathbf{I}^{t(\mathbb{R})}_{(k)}(\tau, \bar{\mathbf{g}}^t) &= -\int_{-\infty}^{\infty} k'^{(o)}(\tau+\xi)\bar{\mathbf{g}}^t(\xi)d\xi = \begin{cases} \mathbf{I}^t_{(k)}(\tau, \bar{\mathbf{g}}^t) & \forall\tau \geq 0, \\ \mathbf{I}^{t(n)}_{(k)}(\tau, \bar{\mathbf{g}}^t) & \forall\tau < 0. \end{cases}
\end{aligned}
\tag{13.4.5}
$$

Let $\vartheta^t_n(s) = \vartheta^t(-s)$ and $\bar{\mathbf{g}}^t_n(s) = \bar{\mathbf{g}}^t(-s)$, defined for any $s \leq 0$, be extended to $\mathbb{R}$ by putting $\vartheta^t_n(s) = 0$ and $\bar{\mathbf{g}}^t_n(s) = \mathbf{0}\ \forall s > 0$. Their Fourier transforms are given by

$$
\vartheta^t_{nF}(\omega) = \vartheta^t_{n-}(\omega) = \overline{\vartheta^t_+}(\omega), \qquad \bar{\mathbf{g}}^t_{nF}(\omega) = \bar{\mathbf{g}}^t_{n-}(\omega) = \overline{\bar{\mathbf{g}}^t_+}(\omega).
$$

We can now rewrite (13.4.5) in terms of these functions:

$$
\begin{aligned}
I^{t(\mathbb{R})}_{(\alpha)}(\tau, \vartheta^t) &= \int_{-\infty}^{\infty} \alpha''^{(e)}(\tau-s)\vartheta^t_n(s)ds, \\
\mathbf{I}^{t(\mathbb{R})}_{(k)}(\tau, \bar{\mathbf{g}}^t) &= -\int_{-\infty}^{\infty} k'^{(o)}(\tau-s)\bar{\mathbf{g}}^t_n(s)ds.
\end{aligned}
$$

The Fourier transforms of these relations are

$$
\begin{aligned}
I^{t(\mathbb{R})}_{(\alpha)F}(\omega, \vartheta^t) &= -2\alpha''_c(\omega)\vartheta^t_{nF}(\omega) \\
&= -2[h(\omega) - \alpha'(0)]\overline{\vartheta^t_+}(\omega), \qquad h(\omega) = \omega\alpha'_s(\omega), \\
\mathbf{I}^{t(\mathbb{R})}_{(k)F}(\omega, \bar{\mathbf{g}}^t) &= -2ik'_s(\omega)\bar{\mathbf{g}}^t_{n_F}(\omega) = 2i\omega k_c(\omega)\overline{\bar{\mathbf{g}}^t_+}(\omega),
\end{aligned}
\tag{13.4.6}
$$

by virtue of $(9.2.7)_1$. The quantities $k_c(\omega)$ and $h(\omega)$ are nonnegative by virtue of $(9.2.3)_1$ and (9.2.6). Therefore, they can be factorized in the form

$$
h(\omega) = h_{(+)}(\omega)h_{(-)}(\omega), \qquad k_c(\omega) = k_{(+)}(\omega)k_{(-)}(\omega), \tag{13.4.7}
$$

where $h_{(+)}(\omega)$ has no singularities in Ω^- or zeros in $\Omega^{(-)}$, while $h_{(-)}(\omega)$ is analytic in Ω^+ and has no zeros in $\Omega^{(+)}$.

Use of the Plemelj formulas (Sect. B.2.1) yields

$$\frac{1}{2h_{(-)}(\omega)} I^{t(\mathbb{R})}_{(\alpha)_F}(\omega, \vartheta^t) = p^t_{\alpha(-)}(\omega) - p^t_{\alpha(+)}(\omega),$$
$$\frac{1}{2k_{(-)}(\omega)} \mathbf{I}^{t(\mathbb{R})}_{(k)_F}(\omega, \overline{\mathbf{g}}^t) = \mathbf{p}^t_{k(-)}(\omega) - \mathbf{p}^t_{k(+)}(\omega),$$

with $p^t_{\alpha(\pm)}(\omega)$ and $\mathbf{p}^t_{k(\pm)}(\omega)$ defined by

$$p^t_\alpha(z) = \frac{1}{2\pi i}\int_{-\infty}^{\infty} \frac{I^{t(\mathbb{R})}_{(\alpha)_F}(\omega, \vartheta^t)/[2h_{(-)}(\omega)]}{\omega - z} d\omega,$$
$$p^t_{\alpha(\pm)}(\omega) = \lim_{\alpha\to 0^{\mp}} p^t_\alpha(\omega + i\alpha),$$
$$\mathbf{p}^t_{(k)}(z) = \frac{1}{2\pi i}\int_{-\infty}^{\infty} \frac{\mathbf{I}^{t(\mathbb{R})}_{(k)_F}(\omega, \overline{\mathbf{g}}^t)/[2k_{(-)}(\omega)]}{\omega - z} d\omega,$$
$$p^t_{k(\pm)}(\omega) = \lim_{\alpha\to 0^{\mp}} p^t_{(k)}(\omega + i\alpha).$$

Substituting $(13.4.6)_{1,4}$ into these last two relations, respectively, and using (13.4.7), we obtain

$$p^t_{\alpha(\pm)}(\omega) = -\lim_{z\to\omega^{\mp}} \frac{1}{2\pi i}\int_{-\infty}^{\infty} \frac{h_{(+)}(\omega')\overline{\vartheta^t_+}(\omega')}{\omega' - z} d\omega'$$
$$+ \lim_{z\to\omega^{\mp}} \frac{\alpha'(0)}{2\pi i}\int_{-\infty}^{\infty} \frac{\overline{\vartheta^t_+}(\omega')/h_{(-)}(\omega')}{\omega' - z} d\omega',$$
$$\mathbf{p}^t_{k(\pm)}(\omega) = \lim_{z\to\omega^{\mp}} \frac{1}{2\pi i}\int_{-\infty}^{\infty} \frac{i\omega' k_{(+)}(\omega')\overline{\mathbf{g}^t_+}(\omega')}{\omega' - z} d\omega',$$

from which

$$\overline{p^t_{\alpha(\pm)}}(\omega) = \lim_{z\to\omega^{\pm}} \frac{1}{2\pi i}\int_{-\infty}^{\infty} \frac{h_{(-)}(\omega')\vartheta^t_+(\omega')}{\omega' - z} d\omega'$$
$$- \lim_{z\to\omega^{\pm}} \frac{\alpha'(0)}{2\pi i}\int_{-\infty}^{\infty} \frac{\vartheta^t_+(\omega')/h_{(+)}(\omega')}{\omega' - z} d\omega', \qquad (13.4.8)$$
$$\overline{\mathbf{p}^t_{(k)(\pm)}}(\omega) = i\lim_{z\to\omega^{\pm}} \frac{1}{2\pi i}\int_{-\infty}^{\infty} \frac{\omega' k_{(-)}(\omega')\overline{\mathbf{g}}^t_+(\omega')}{\omega' - z} d\omega'.$$

Since in the second integral of $(13.4.8)_1$, $\vartheta^t_+(\omega)$ and $h_{(+)}(\omega)$ are analytic in Ω^-, by closing the contour on $\Omega^{(-)}$, we see that this integral vanishes. Hence, $(13.4.8)_1$ reduces to

$$\overline{p^t_{\alpha(\pm)}}(\omega) = \lim_{z\to\omega^{\pm}} \frac{1}{2\pi i}\int_{-\infty}^{\infty} \frac{h_{(-)}(\omega')\vartheta^t_+(\omega')}{\omega' - z} d\omega'.$$

We put

$$\overline{p^t_{(\alpha)(\pm)}}(\omega) = p^t_{\alpha\mp}(\omega), \qquad \overline{\mathbf{p}^t_{(k)(\pm)}}(\omega) = i\mathbf{p}^t_{k\mp}(\omega).$$

Proceeding as in Sect. 13.2.1, we can solve the Wiener–Hopf equations (13.4.2) to obtain

$$\vartheta_-^{(m)}(\omega) = \overline{\vartheta_+^{(m)}}(\omega) = -\frac{p_{\alpha-}^t(\omega)}{h_{(-)}(\omega)}, \quad \mathbf{g}_-^{(m)}(\omega) = \overline{\mathbf{g}_+^{(m)}}(\omega) = -\frac{\mathbf{p}_{k-}^t(\omega)}{k_{(-)}(\omega)},$$

and (13.4.3) yields the required expression

$$\psi_m(t) = \frac{1}{2}\alpha_0\vartheta^2(t) + \frac{1}{2\pi}\int_{-\infty}^{\infty} |p_{\alpha-}^t(\omega)|^2 d\omega + \frac{1}{2\pi}\int_{-\infty}^{\infty} |\mathbf{p}_{k-}^t(\omega)|^2 d\omega. \qquad (13.4.9)$$

The expression for the rate of dissipation corresponding to ψ_m is determined from (9.3.1) and the first law (5.1.11), where the right-hand side of the latter is replaced by $w(t)$, so that we have

$$\dot{\psi}_m(t) + D_m(t) = \dot{e}(t)\vartheta(t) - \mathbf{q}(t)\cdot\mathbf{g}(t). \qquad (13.4.10)$$

Proceeding by analogy with the steps leading to (11.2.25), we obtain

$$\begin{aligned} D_m(t) &= |K_\alpha(t)|^2 + |K_k(t)|^2, \\ K_\alpha(t) &= \frac{1}{2\pi}\int_{-\infty}^{\infty} h_{(-)}(\omega)\vartheta_+^t(\omega)d\omega, \\ K_k(t) &= \frac{1}{2\pi}\int_{-\infty}^{\infty} \omega k_{(-)}(\omega)\bar{\mathbf{g}}_+^t(\omega)d\omega. \end{aligned} \qquad (13.4.11)$$

13.4.2 The Discrete-Spectrum Model for Heat Conductors

Let us assume for the kernel functions α and k the following form:

$$\alpha(t) = \begin{cases} \alpha_\infty - \sum_{i=1}^n h_i e^{-\alpha_i t} & \forall t \geq 0, \\ 0 & \forall t < 0, \end{cases} \qquad k(t) = \begin{cases} \sum_{i=1}^n g_i e^{-k_i t} & \forall t \geq 0, \\ 0 & \forall t < 0, \end{cases} \qquad (13.4.12)$$

where n is a positive integer, and the coefficients h_i and g_i, $i = 1, 2, \ldots, n$, are assumed to be positive as are the inverse decay times α_i and k_i, $i = 1, 2, \ldots, n$, which we label so that $\alpha_j < \alpha_{j+1}$ and $k_j < k_{j+1}$, $j = 1, 2, \ldots, n-1$. Note that the conditions (9.2.8) and (9.2.9) and the scalar form of (9.2.5) are satisfied, since

$$\alpha_\infty - \alpha(0) = \sum_{i=1}^n h_i > 0, \quad \alpha'(0) = \sum_{i=1}^n \alpha_i h_i > 0, \quad k'(0) = -\sum_{i=1}^n k_i g_i < 0.$$

The kernel $\alpha(t)$ behaves like the creep function introduced in Sect. 8.5 for viscoelastic solids. This is a consequence of the observation in Remark 9.3.1. From (13.4.12), we obtain

$$\alpha'_+(\omega) = \sum_{i=1}^n \alpha_i h_i \frac{\alpha_i - i\omega}{\alpha_i^2 + \omega^2}, \quad k_+(\omega) = \sum_{i=1}^n g_i \frac{k_i - i\omega}{k_i^2 + \omega^2},$$

and

$$\alpha_s'(\omega) = \omega \sum_{i=1}^{n} \frac{\alpha_i h_i}{\alpha_i^2 + \omega^2}, \quad k_c(\omega) = \sum_{i=1}^{n} \frac{k_i g_i}{k_i^2 + \omega^2}. \tag{13.4.13}$$

Using $(13.4.13)_1$, the quantity h defined by $(13.4.6)_3$ becomes

$$h(\omega) = \omega^2 \sum_{i=1}^{n} \frac{\alpha_i h_i}{\omega^2 + \alpha_i^2} \geq 0 \quad \forall \omega \in \mathbb{R}, \quad h(\infty) = \sum_{i=1}^{n} \alpha_i h_i = \alpha'(0) > 0.$$

The function $f(z) = h(\omega)$ with $z = -\omega^2$ has n simple poles at $\alpha_i^2, i = 1, 2, \ldots, n$, and n simple zeros at $\gamma_1 = 0$ and γ_j^2, $j = 2, 3, \ldots, n$, such that

$$\alpha_1^2 < \gamma_2^2 < \alpha_2^2 < \cdots < \alpha_{n-1}^2 < \gamma_n^2 < \alpha_n^2.$$

Thus, we also have

$$h(\omega) = h(\infty) \prod_{i=1}^{n} \left\{ \frac{\gamma_i^2 + \omega^2}{\alpha_i^2 + \omega^2} \right\},$$

with factorization $(13.4.7)_1$ given by

$$h_{(-)}(\omega) = [h(\infty)]^{1/2} \prod_{i=1}^{n} \left\{ \frac{\omega + i\gamma_i}{\omega + i\alpha_i} \right\}, \quad h_{(+)}(\omega) = [h(\infty)]^{1/2} \prod_{i=1}^{n} \left\{ \frac{\omega - i\gamma_i}{\omega - i\alpha_i} \right\},$$

whence we also have

$$h_{(-)}(\omega) = [h(\infty)]^{1/2} \left(1 + i \sum_{i=1}^{n} \frac{R_i}{\omega + i\alpha_i} \right), \quad R_i = (\gamma_i - \alpha_i) \prod_{j=1, j\neq i}^{n} \left\{ \frac{\gamma_j - \alpha_i}{\alpha_j - \alpha_i} \right\}.$$

Let us now consider the kernel k. Using $(13.4.13)_2$, we introduce the function

$$K_k(\omega) = \omega^2 k_c(\omega) = \sum_{i=1}^{n} k_i g_i \frac{\omega^2}{k_i^2 + \omega^2} \geq 0 \quad \forall \omega \in \mathbb{R},$$

$$K_k(\infty) = K_\infty = \sum_{i=1}^{n} k_i g_i > 0.$$

We can factorize $K_k(\omega)$ in the same way as $h(\omega)$, obtaining

$$K_k(\omega) = K_{(+)}(\omega) K_{(-)}(\omega),$$

$$K_{(-)}(\omega) = K_\infty^{1/2} \prod_{i=1}^{n} \left\{ \frac{\omega + i\nu_i}{\omega + ik_i} \right\}, \qquad K_{(+)}(\omega) = \overline{K_{(-)}}(\omega),$$

where $\nu_1^2 = 0$ and ν_j^2, $j = 2, 3, \ldots, n$, denote the zeros of $f(z) = K(\omega)$ with $z = -\omega^2$. The quantity $\omega k_{(-)}(\omega)$, which appears in $(13.4.8)_2$, can be taken to be equal to $K_{(-)}(\omega)$. We put

$$\omega k_{(-)}(\omega) = K_\infty^{1/2} \prod_{i=1}^{n} \left\{ \frac{\omega + i\nu_i}{\omega + ik_i} \right\} = K_\infty^{1/2} \left(1 + i \sum_{i=1}^{n} \frac{S_i}{\omega + ik_i} \right),$$

where

$$S_r = (\nu_r - k_r) \prod_{i=1, i \neq r}^{n} \left\{ \frac{\nu_i - k_r}{k_i - k_r} \right\}, \qquad r = 1, 2, \ldots, n.$$

We obtain from (13.4.9) (cf. (11.9.14)) that

$$\psi_m(t) = \frac{1}{2}\alpha_0 \vartheta^2(t) + \frac{1}{2}\int_0^\infty \int_0^\infty 2\left[h(\infty) \sum_{r,l=1}^{n} \frac{R_r R_l}{\alpha_r + \alpha_l} e^{-(\alpha_r s_1 + \alpha_l s_2)} \vartheta^t(s_1)\, \vartheta^t(s_2) \right.$$
$$\left. + K_\infty \sum_{r,l=1}^{n} \frac{S_r S_l}{k_r + k_l} e^{-(k_r s_1 + k_l s_2)} \bar{\mathbf{g}}^t(s_1) \cdot \bar{\mathbf{g}}^t(s_2) \right] ds_1 ds_2. \tag{13.4.14}$$

Using (13.4.14) in (13.4.10) (or (13.4.11) directly), we find that (cf. (11.9.16))

$$D_m(t) = h(\infty) \left[\int_0^\infty \sum_{i=1}^{n} R_i e^{-\alpha_i s} \vartheta_r^t(s) ds \right]^2 + K_\infty \left[\int_0^\infty \sum_{i=1}^{n} S_i e^{-k_i s} \bar{\mathbf{g}}^t(s) ds \right]^2.$$

14

The Minimum Free Energy for a Continuous-Spectrum Material

14.1 Introduction

We now examine how the formulas emerging from the methodology developed in Chap. 11 apply to materials other than those exhibiting a discrete-spectrum response, in particular for materials with a branch-cut-type singularity. We confine our considerations to the case that the cut is on the imaginary axis. Such a material is said to have a continuous-spectrum response, i.e., those materials for which the relaxation function is given by an integral of a density function multiplying a strictly decaying exponential. The results reported in this chapter were first presented in [94].

The methodology described in Chap. 11 involves factorizing the quantity $\mathbb{H}$, defined in general by (7.2.22), in order to solve the relevant variational problem or the equivalent Wiener–Hopf equation for the optimal future continuation required to determine the minimum free energy.

We know from Proposition 16.5.2 below that for materials with branch cuts, minimal states are singletons and the maximum free energy is the work function, though this latter assertion is problematic because of Remark 18.2. In fact, we give a proof of the singleton property in Sect. 14.6 for the special kinds of material considered in this chapter, namely those with a single branch cut on the imaginary axis.

Another method was used in [42] for the discrete-spectrum case. This involved making a natural assumption on the form of the optimal future continuation and solving algebraic equations for the various parameters. The need for factorization did not arise. This method is also developed in the present chapter for continuous-spectrum materials. The assumption involved in this case is also a natural one, namely that the optimal future continuation has a singularity structure determined only by that of the Fourier transform of the relaxation function derivative.

We consider only the scalar case and use the notation of mechanics. However, it is emphasized again that the treatment applies equally to any scalar subspace of Γ, with or without a nonzero equilibrium free energy.

G. Amendola et al., *Thermodynamics of Materials with Memory*,
https://doi.org/10.1007/978-3-030-80534-0_14

The layout of the chapter is as follows. In Sect. 14.2, fundamental relations are summarized and continuous-spectrum materials introduced. The factorization procedure is discussed in depth for the continuous-spectrum case in Sect. 14.3. The form of the minimum free energy is derived in Sect. 14.4. The alternative approach referred to in the previous paragraph is discussed in detail in Sect. 14.5. The concept of a minimal state for continuous-spectrum materials is explored in Sect. 14.6.

14.2 Continuous-Spectrum Materials

We consider a constitutive equation of the form

$$S(t) = S_e(t) + \int_0^\infty G'(s)E_r^t(s)ds, \tag{14.2.1}$$

which is a scalar version of (8.1.1). Let

$$F(\omega) = G'_+(\omega) = \int_0^\infty G'(s)e^{-i\omega s}ds = G'_c(\omega) - iG'_s(\omega).$$

The notation F is introduced to simplify later formulas. We shall require the property of F that

$$\lim_{\omega\to\infty} i\omega F(\omega) = G'(0), \tag{14.2.2}$$

which is a special case of (7.2.23). Properties of $G'_s(\omega)$ include the scalar version of (8.1.18) and the oddness of sine transforms:

$$G'_s(\omega) \le 0 \quad \forall\, \omega \in \mathbb{R}^{++}, \qquad G'_s(-\omega) = -G'_s(\omega) \quad \forall\, \omega \in \mathbb{R}. \tag{14.2.3}$$

The scalar version of $(11.8.1)_1$ is given by

$$\begin{aligned} H(\omega) &= \frac{\omega}{2i}\left[F(\omega) - \overline{F}(\omega)\right] = -\omega G'_s(\omega) = H(-\omega) \ge 0 \quad \forall \omega \in \mathbb{R}, \\ H(\omega) &= H_1(\omega^2), \end{aligned} \tag{14.2.4}$$

where H_1 is the function H expressed in terms of ω^2, by virtue of the evenness property and the analyticity of $H(\omega)$ on the real axis. It is assumed (see (7.2.20)) that the behavior is in fact quadratic, i.e., $H(\omega)/\omega^2$ tends to a finite nonzero quantity as ω tends to zero. The singularities of H are the same as those of F in $\Omega^{(+)}$ and of $\overline{F}$ in $\Omega^{(-)}$. Relation (11.8.2) reduces to

$$H_\infty = \lim_{|\omega|\to\infty} H(\omega) = -G'(0) \ge 0. \tag{14.2.5}$$

We assume for present purposes that $G'(0)$ is nonzero, so that H_∞ is a finite positive number. Then $H(\omega) \in \mathbb{R}^{++}\ \forall\, \omega \in \mathbb{R}$, $\omega \ne 0$.

We adopt the following continuous-spectrum form for the relaxation function derivative:

$$G'(t) = \int_a^b k(\alpha)e^{-\alpha t}d\alpha, \quad t \in \mathbb{R}^+, \quad b > a > 0. \tag{14.2.6}$$

It is assumed that $k \in L^1([a, b])$. The upper limit b may be infinite. We take $a > 0$ because of the need to avoid singularities on the real axis. The Fourier transform of (14.2.6) has the form

$$F(\omega) = \int_a^b \frac{k(\alpha)}{\alpha + i\omega} d\alpha = \int_{ia}^{ib} \frac{k(-iu)}{u - \omega} du, \qquad \omega \in \mathbb{R}. \tag{14.2.7}$$

This formula can be extended by analytic continuation to Ω, excluding singular points (Sect. B.1.2). We restrict the density function k to be Hölder continuous on (a, b), as defined by (B.2.3). It may be singular at the endpoints with a power less than unity, as specified by (B.2.2). We assume that

$$k(\alpha) \leq 0, \quad \alpha \in [a, b].$$

This assumption is not essential but is the simplest that ensures compatibility with thermodynamic constraint $(14.2.3)_1$. Note that it renders G completely monotonic in the sense discussed in [88], for example. Also, it is easily shown that F has no zeros on the finite part of $\Omega \backslash [ia, ib]$. Taking the complex conjugate of (14.2.7), we have

$$\overline{F}(\omega) = \int_a^b \frac{k(\alpha)}{\alpha - i\omega} d\alpha = \int_{-ia}^{-ib} \frac{k(iu)}{u - \omega} du, \qquad \omega \in \mathbb{R}, \tag{14.2.8}$$

which can similarly be continued into Ω.

The quantity F has a branch cut on $[ia, ib]$ and $\overline{F}$ on $[-ib, -ia]$. As ω tends to $i\alpha$, where $\alpha \in \mathbb{R} \backslash [-b, -a]$, we have

$$\overline{F}(i\alpha) = F(-i\alpha) = \int_a^b \frac{k(\beta)}{\beta + \alpha} d\beta = K(\alpha), \tag{14.2.9}$$

while if $\alpha \in (a, b)$, we have, by virtue of the Plemelj formulas (B.2.6),

$$\begin{aligned} F_R(i\alpha) &= R(\alpha) + iI(\alpha), \\ F_L(i\alpha) &= R(\alpha) - iI(\alpha), \end{aligned} \tag{14.2.10}$$

with

$$R(\alpha) = P \int_a^b \frac{k(\beta)}{\beta - \alpha} d\beta, \qquad I(\alpha) = -\pi k(\alpha) \geq 0, \tag{14.2.11}$$

where $F_R(i\alpha)$ and $F_L(i\alpha)$ are the limiting values of $F(\omega)$, approaching from the right and the left, respectively, as one moves from ia to ib (F_R corresponds to F^- in (B.2.6)). Similarly,

$$\begin{aligned} \overline{F}_R(-i\alpha) &= R(\alpha) + iI(\alpha) = \overline{F_L(i\alpha)}, \\ \overline{F}_L(-i\alpha) &= R(\alpha) - iI(\alpha) = \overline{F_R(i\alpha)}, \end{aligned} \tag{14.2.12}$$

for $\alpha \in (a, b)$, where $\overline{F}_R(-i\alpha)$ and $\overline{F}_L(-i\alpha)$ are the limiting values of $\overline{F}(\omega)$ from the right and left, respectively, as one moves from $-ia$ to $-ib$. The symbol P in (14.2.11) indicates a principal value.

From (14.2.4), (14.2.7), and (14.2.8), we have

$$H(\omega) = -\omega^2 \int_a^b \frac{k(\alpha)}{\alpha^2 + \omega^2} d\alpha. \tag{14.2.13}$$

Let us consider the behavior of $F(\omega)$ at the endpoints ia and ib for various limiting behaviors of $k(\alpha)$ as α approaches a or b. If $k(a) = 0$, then $F(\omega)$ has a definite finite nonzero limit as $\omega \to ia$. A similar statement applies to the limit $\omega \to ib$ if $k(b) = 0$. If

$$k(a) = k_a < 0$$

and k is Hölder continuous near and at a, then $F(\omega)$ has a logarithmic singularity at $\omega = ia$. As ω approaches this endpoint along any path off $[ia, ib]$, then by (B.2.7),

$$F(\omega) = k_a \log \frac{1}{a + i\omega} + F_1(\omega), \tag{14.2.14}$$

where $F_1(a)$ is well defined. Similarly, by (B.2.8), if

$$k(b) = k_b < 0$$

and k is Hölder continuous near and at b, then as ω approaches ib not along $[ia, ib]$, we have

$$F(\omega) = -k_b \log \frac{1}{b + i\omega} + F_2(\omega), \tag{14.2.15}$$

where $F_2(b)$ is well defined. For points on (ia, ib), referring to (B.2.9) and (B.2.11), we see that relations (14.2.14) and (14.2.15) are replaced by

$$R(\alpha) \to \begin{cases} k_a \log \frac{1}{\alpha - a}, & \alpha \to ia^+, \\ -k_b \log \frac{1}{b - \alpha}, & \alpha \to ib^-, \end{cases} \tag{14.2.16}$$

where $R(\alpha)$ is given by (14.2.11). If $k(\alpha)$ has dominant behavior as $\alpha \to a^+$ along (a, b) of the form

$$k(\alpha) \underset{\alpha \to a^+}{\longrightarrow} \frac{k_1}{(\alpha - a)^\gamma}, \quad 0 < \gamma < 1, \quad k_1 < 0,$$

then from (B.2.10), for $\omega \notin (ia, ib)$,

$$F(\omega) \underset{\omega \to ia}{\longrightarrow} \frac{Ak_1}{(a + i\omega)^\gamma}. \tag{14.2.17}$$

The detailed form of A is given in [274]. A similar observation applies to the case that k has such behavior at b. For points on (ia, ib), (B.2.11) gives that relation (14.2.17) is replaced by

$$R(\alpha) \underset{\alpha \to a^+}{\longrightarrow} \frac{A_1 k_1}{(\alpha - a)^\gamma}, \tag{14.2.18}$$

where again the form of A_1 may be found in [274]. A similar observation applies at b.

14.3 Factorization of H for a Continuous-Spectrum Material

Relations (11.1.15) and (11.1.17) give, for the factors in (11.1.16),

$$H_+(\omega) = \frac{\omega h_\infty}{\omega - i\omega_0} e^{-M_+(\omega)}, \quad h_\infty = H_\infty^{1/2}, \quad H_-(\omega) = H_+(-\omega) = \overline{H_+}(\omega), \tag{14.3.1}$$

where M_+ is given by

$$\begin{aligned} M_+(\omega) &= \lim_{\beta \to 0^-} M(\omega + i\beta), \\ M(z) &= \frac{1}{2\pi i} \int_{-\infty}^{\infty} \frac{\log[T(\omega')H(\omega')]}{\omega' - z} d\omega', \quad z \in \Omega \backslash \mathbb{R}, \end{aligned} \tag{14.3.2}$$

with

$$T(\omega) = \frac{\omega^2 + \omega_0^2}{H_\infty \omega^2}, \tag{14.3.3}$$

and we have

$$H(\omega) = H_+(\omega)H_-(\omega), \tag{14.3.4}$$

where H_+ has no singularities or zeros in $\Omega^{(-)}$ and is thus analytic in Ω^-. Similarly, H_- is analytic in Ω^+ with no zeros in $\Omega^{(+)}$. Therefore, the singularities of F must all occur in H_+ and those of $\overline{F}$ in H_-. There may be other singularities in $H_\pm$ that cancel on multiplication.

Using (14.3.3) and $(14.2.4)_1$, one can write

$$\begin{aligned} \log[T(\omega)H(\omega)] &= \log\left[-i\frac{\omega - i\omega_0}{H_\infty} F(\omega)\right] + \log U(\omega), \\ U(\omega) &= \frac{1}{2}\left[1 - \frac{\overline{F}(\omega)}{F(\omega)}\right]\left[\frac{\omega + i\omega_0}{\omega}\right]. \end{aligned} \tag{14.3.5}$$

Referring to the discussion of (B.1.9), we choose the standard branch of the logarithm function, namely that which vanishes for argument unity. The function U is complex but nonzero on the real line and approaches unity for large ω, by virtue of (14.2.2); similarly for the argument of the first term on the right of $(14.3.5)_1$. This term has all its singularities in $\Omega^{(+)}$, so that if we close the integral in $(14.3.2)_2$ on $\Omega^{(-)}$, then by Cauchy's integral formula (B.1.2) for a clockwise contour, its contribution to $M(z)$ is simply the negative of itself. Thus, we have, from (14.3.1),

$$\begin{aligned} H_+(\omega) &= \frac{-i\omega}{h_\infty} F(\omega) e^{-N_+(\omega)}, \quad \omega \in \mathbb{R}, \\ N_+(\omega) &= \lim_{\beta \to 0^-} N(\omega + i\beta), \\ N(z) &= \frac{1}{2\pi i} \int_{-\infty}^{\infty} \frac{\log U(\omega')}{\omega' - z} d\omega', \quad z \in \Omega \backslash \mathbb{R}. \end{aligned} \tag{14.3.6}$$

We deduce that

$$
\begin{aligned}
H_-(\omega) &= \frac{i\omega}{h_\infty}\overline{F}(\omega)e^{-N_-(\omega)}, \quad \omega \in \mathbb{R}, \\
N_-(\omega) &= \lim_{\beta\to 0^-} \overline{N}(\omega - i\beta), \\
\overline{N}(z) &= -\frac{1}{2\pi i}\int_{-\infty}^{\infty} \frac{\log \overline{U}(\omega')}{\omega' - z} d\omega', \quad z \in \Omega\backslash\mathbb{R}.
\end{aligned}
\tag{14.3.7}
$$

These exhibit the correct behavior of $H_\pm$ at small ω, by inspection, and at large ω, by virtue of (14.2.2) and (14.2.5). This behavior is anyway clear from (14.3.1).

There are two motivations for the extraction of the factor F in (14.3.6) and $\overline{F}$ in (14.3.7), rather than using the form (14.3.1), which is the natural first approach. Firstly, the elimination of the apparent singularity at $\omega = \pm i\omega_0$ is straightforward, as we will see after (14.3.8) below. If these factors are not extracted, the mechanism for this elimination is not so clear.

Singularities in F (in this case, a branch cut) must also occur in H_+, while those of $\overline{F}$ must be in H_- (though in fact, $H_\pm$ may have other singularities). In particular, the singular behavior of F in $\Omega^{(+)}$, given by (14.2.14), (14.2.15), and (14.2.17), must occur also in H_+, while a similar statement applies to H_- and $\overline{F}$. Such behavior is difficult to show with the form (14.3.1) but is explicitly present in (14.3.6) and (14.3.7).

The function $U(i\alpha)$, $\alpha > 0$, in (14.3.5) is real for $\alpha \notin [a, b]$. It is discontinuous across $[a, b]$. We define, for $\alpha \in [a, b]$,

$$
\begin{aligned}
U_R(i\alpha) &= \lim_{\omega\to\omega_R} U(\omega), \quad \omega_R = \alpha e^{i\pi/2}, \\
U_L(i\alpha) &= \lim_{\omega\to\omega_L} U(\omega), \quad \omega_L = \alpha e^{-3i\pi/2}.
\end{aligned}
$$

Since F is never infinite on $\Omega^{(+)}\backslash[ia, ib]$ and its imaginary part does not vanish on this region, as noted earlier, the function $U(\omega)$ is nonzero on $\Omega^{(+)}\backslash[ia, ib]$ and approaches unity as $\omega \to \infty$. Thus, $\log U(\omega)$ has a branch cut on $[ia, ib]$ through its dependence on $F(\omega)$ and $\overline{F}(\omega)$ and no other singularity in $\Omega^{(+)}$. Referring to (B.1.10), we see that the factor $\log[(\omega + i\omega_0)/\omega]$ is assigned a branch cut on $[0, -i\omega_0]$. Moving the line of integration in $(14.3.6)_3$ to the infinite half-circle in $\Omega^{(+)}$ while going around the branch cut, we obtain, with a change of integration variable,

$$
\begin{aligned}
N(z) &= \frac{1}{2\pi i}\int_a^b \frac{\Delta(\alpha)}{\alpha + iz} d\alpha, \\
\Delta(\alpha) &= \log U_R(i\alpha) - \log U_L(i\alpha),
\end{aligned}
\tag{14.3.8}
$$

where the branch of the logarithm function is as specified earlier. Its imaginary part lies in $[-\pi, \pi]$. Note that the factor $[(\omega + i\omega_0)/\omega]$ in $U(\omega)$ cancels out of $\Delta(\alpha)$. It can henceforth be omitted. Thus, we put

$$
\begin{aligned}
Y(\omega) &= \frac{1}{2}\left[1 - \frac{\overline{F}(\omega)}{F(\omega)}\right], \\
\Delta(\alpha) &= \log Y_R(i\alpha) - \log Y_L(i\alpha),
\end{aligned}
\tag{14.3.9}
$$

where from (14.2.9) and (14.2.10), for $\alpha \in (a, b)$,

$$\begin{aligned} Y_R(i\alpha) &= \frac{1}{2}\left[1 - \frac{K(\alpha)}{R(\alpha) + iI(\alpha)}\right] = \overline{Y_L(i\alpha)}, \\ Y_L(i\alpha) &= \frac{1}{2}\left[1 - \frac{K(\alpha)}{R(\alpha) - iI(\alpha)}\right] = \overline{Y_R(i\alpha)}. \end{aligned} \tag{14.3.10}$$

For future use, we note that

$$\overline{Y}_R(-i\alpha) = Y_R(i\alpha), \qquad \overline{Y}_L(-i\alpha) = Y_L(i\alpha), \tag{14.3.11}$$

by virtue of (14.2.9), (14.2.10), (14.2.12), and (14.3.9). Also,

$$|Y_R(i\alpha)|^2 = |Y_L(i\alpha)|^2,$$

so that we can write

$$\Delta(\alpha) = 2iA(\alpha), \quad A(\alpha) = \arg Y_R(i\alpha), \quad -\pi \leq A(\alpha) \leq \pi, \tag{14.3.12}$$

and

$$\begin{aligned} H_+(\omega) &= -\frac{i\omega}{h_\infty} F(\omega) e^{-N^+(\omega)}, \\ N^+(\omega) &= \frac{1}{\pi}\int_a^b \frac{A(\alpha)}{\alpha + i\omega} d\alpha, \end{aligned} \tag{14.3.13}$$

while

$$\begin{aligned} H_-(\omega) &= \frac{i\omega}{h_\infty} \overline{F}(\omega) e^{-N^-(\omega)}, \\ N^-(\omega) &= \frac{1}{\pi}\int_a^b \frac{A(\alpha)}{\alpha - i\omega} d\alpha. \end{aligned} \tag{14.3.14}$$

In the notation of (14.2.11), we have

$$V(\alpha) = 2\left\{[R(\alpha)]^2 + [I(\alpha)]^2\right\} \mathrm{Re} Y_R(i\alpha) = [R(\alpha)]^2 + [I(\alpha)]^2 - K(\alpha)R(\alpha),$$

where $K(\alpha)$, given by (14.2.9), is real and negative for $\alpha > -a$. Also,

$$W(\alpha) = 2\left\{[R(\alpha)]^2 + [I(\alpha)]^2\right\} \mathrm{Im} Y_R(i\alpha) = K(\alpha)I(\alpha) \leq 0,$$

for $\alpha > -a$, from which it follows that $-\pi \leq A(\alpha) \leq 0$. Then,

$$\begin{aligned} A(\alpha) &= \begin{cases} -B(\alpha), & V(\alpha) \geq 0, \\ -\pi + B(\alpha), & V(\alpha) < 0, \end{cases} \\ B(\alpha) &= \arctan\left|\frac{W(\alpha)}{V(\alpha)}\right|, \quad 0 \leq B(\alpha) \leq \frac{\pi}{2}. \end{aligned} \tag{14.3.15}$$

Some exploration of the sign of V is presented in [94].

14.3.1 Properties of the Factorization Formulas

It is of interest to consider the limits of $H_\pm$, given by (14.3.13) and (14.3.14), as ω approaches the branch cuts on $[ia, ib]$ and $[-ib, -ia]$. Consider (14.3.14) as $\omega \to -i\alpha$, $\alpha \in (a, b)$, from the left, i.e., from the fourth quadrant and from the right. Noting $(14.2.12)_2$ and (14.3.14), we obtain

$$\begin{aligned} H_{-L}(-i\alpha) &= \frac{\alpha}{h_\infty}[R(\alpha) - iI(\alpha)]P(\alpha)e^{-iA(\alpha)} = \overline{H_{-R}(-i\alpha)}, \\ P(\alpha) &= \exp\left\{-\frac{1}{\pi}P\int_a^b \frac{A(\beta)}{\beta - \alpha}d\beta\right\}, \end{aligned} \tag{14.3.16}$$

where the Plemelj formulas (B.2.6) have been used. Also, from (14.2.9) and (14.3.13),

$$\begin{aligned} H_{+L}(-i\alpha) &= -\frac{\alpha}{h_\infty}K(\alpha)Q(\alpha) = H_{+R}(-i\alpha), \\ Q(\alpha) &= \exp\left\{-\frac{1}{\pi}\int_a^b \frac{A(\beta)}{\beta + \alpha}d\beta\right\}. \end{aligned} \tag{14.3.17}$$

Multiplying $H_{\pm L}$ together, we obtain the limit of $H(\omega)$ as $\omega \to -i\alpha$, $\alpha \in (a, b)$, namely

$$H_L(-i\alpha) = -\frac{\alpha^2}{H_\infty}[R(\alpha) - iI(\alpha)]K(\alpha)P(\alpha)Q(\alpha)e^{-iA(\alpha)}. \tag{14.3.18}$$

Also, from $(14.2.4)_1$, (14.2.9), and (14.2.12), we have

$$H_L(-i\alpha) = \frac{\alpha}{2}[R(\alpha) - K(\alpha) - iI(\alpha)]. \tag{14.3.19}$$

Equating the arguments of these two expressions for $H_L(-i\alpha)$ gives

$$\arg[R(\alpha) - K(\alpha) - iI(\alpha)] = -A(\alpha) + \arg[R(\alpha) - iI(\alpha)], \tag{14.3.20}$$

or, taking complex conjugates,

$$A(\alpha) = \arg\left[1 - \frac{K(\alpha)}{R(\alpha) + iI(\alpha)}\right],$$

which is of course simply $(14.3.12)_2$. Equating the magnitudes of the two expressions given by (14.3.18) and (14.3.19), we obtain

$$-2\alpha K(\alpha)P(\alpha)Q(\alpha) = H_\infty\sqrt{\frac{[R(\alpha) - K(\alpha)]^2 + I^2(\alpha)}{R^2(\alpha) + I^2(\alpha)}}. \tag{14.3.21}$$

With the aid of (14.3.20), we can write (14.3.16) in the form

$$\begin{aligned} H_{-L}(-i\alpha) &= \frac{\alpha}{h_\infty}[R(\alpha) - K(\alpha) - iI(\alpha)]\sqrt{\frac{R^2(\alpha) + I^2(\alpha)}{(R(\alpha) - K(\alpha))^2 + I^2(\alpha)}}P(\alpha) \\ &= -\frac{h_\infty}{2}\frac{R(\alpha) - K(\alpha) - iI(\alpha)}{K(\alpha)Q(\alpha)}. \end{aligned} \tag{14.3.22}$$

The second form is a consequence of (14.3.21). By virtue of $(14.3.16)_2$, we see that

$$H_{-R}(-i\alpha) = -\frac{h_\infty}{2}\frac{R(\alpha) - K(\alpha) + iI(\alpha)}{K(\alpha)Q(\alpha)}. \tag{14.3.23}$$

Using $(14.2.4)_1$ and (14.2.12), we see that

$$H_R(-i\alpha) = \frac{\alpha}{2}[R(\alpha) - K(\alpha) + iI(\alpha)] = \overline{H_L(-i\alpha)}, \tag{14.3.24}$$

the last relation following from (14.3.19).

Finally, we observe that $(14.2.4)_1$, (14.3.4), (14.3.13), and (14.3.14) give

$$Z(\omega) = \frac{H_\infty}{2i\omega}\left[\frac{1}{\overline{F}(\omega)} - \frac{1}{F(\omega)}\right] = \exp\left[-\frac{1}{\pi}\int_a^b \frac{A(\alpha)d\alpha}{\alpha + i\omega} - \frac{1}{\pi}\int_a^b \frac{A(\alpha)d\alpha}{\alpha - i\omega}\right]. \tag{14.3.25}$$

Let us show this directly, noting that the left-hand side does not vanish at the origin and is unity at infinity, by virtue of (14.2.2) and (14.2.5). Consider the contour C, taken clockwise at infinity except that it excludes the positive imaginary axis above ia and the negative imaginary axis below $-ia$. The quantity Z is finite and nonzero within C. Then, we see that by Cauchy's integral formula (B.1.2),

$$Z(\omega) = \exp\left\{-\frac{1}{2\pi i}\int_C \frac{\log[Z(u)]\,du}{u - \omega}\right\},$$

where ω is in the interior of C and

$$\begin{aligned}
-\frac{1}{2\pi i}\int_C \frac{\log[Z(u)]\,du}{u - \omega} &= \frac{1}{2\pi i}\int_{ia}^{ib} \frac{\log Z_R(u) - \log Z_L(u)}{u - \omega}du \\
&\quad + \frac{1}{2\pi i}\int_{-ia}^{-ib} \frac{\log Z_R(u) - \log Z_L(u)}{u - \omega}du \\
&= \frac{1}{2\pi i}\int_a^b \frac{\log Z_R(i\alpha) - \log Z_L(i\alpha)}{\alpha + i\omega}d\alpha \\
&\quad + \frac{1}{2\pi i}\int_a^b \frac{\log Z_R(-i\alpha) - \log Z_L(-i\alpha)}{\alpha - i\omega}d\alpha.
\end{aligned}$$

Invoking an argument similar to that leading to (14.3.8) and (14.3.9), we have, from $(14.3.9)_1$ and $(14.3.16)_1$,

$$Z(\omega) = \frac{H_\infty}{i\omega}\frac{Y(\omega)}{\overline{F}(\omega)} = -\frac{H_\infty}{i\omega}\frac{\overline{Y}(\omega)}{F(\omega)}, \qquad \omega \in \mathbb{R},$$

by virtue of (14.2.9). On the cut in the upper half-plane, $\overline{F}$ is analytic and $-\overline{F}(i\alpha)\alpha = -K(\alpha)\alpha > 0$, so that we have

$$\log Z_R(i\alpha) - \log Z_L(i\alpha) = \log Y_R(i\alpha) - \log Y_L(i\alpha) = 2iA(\alpha).$$

On the cut in the lower half-plane where $-F(-i\alpha)\alpha = -K(\alpha)\alpha > 0$, one obtains

$$\begin{aligned}\log Z_R(-i\alpha) - \log Z_L(-i\alpha) &= \log \overline{Y}_R(-i\alpha) - \log \overline{Y}_L(-i\alpha) \\ &= \log Y_R(i\alpha) - \log Y_L(i\alpha) = 2iA(\alpha),\end{aligned}$$

where (14.3.11) has been used. Thus, (14.3.25) is true. This amounts to a verification of (14.3.4).

14.4 The Minimum Free Energy

In the scalar case, (11.8.6) has the form

$$E^t_m(\omega) = -\frac{p^t_-(\omega)}{H_-(\omega)}, \tag{14.4.1}$$

where E^t_m is the Fourier transform of the optimal future continuation and $p^t_-(\omega)$ is the scalar version of one of the quantities defined by (11.8.4), given here by

$$p^t_-(\omega) = \frac{1}{2\pi i}\int_{-\infty}^{\infty} \frac{H_-(\omega')E^t_{r+}(\omega')}{\omega' - \omega^+}d\omega'.$$

Referring to (11.8.9) and (11.8.10), we see that the minimum free energy has the form

$$\psi_m(t) = \phi(t) + \frac{1}{2\pi}\int_{-\infty}^{\infty} H(\omega)|E^t_m(\omega)|^2\, d\omega = \phi(t) + \frac{1}{2\pi}\int_{-\infty}^{\infty} |p^t_-(\omega)|^2\, d\omega. \tag{14.4.2}$$

Note that p^t_- can be written as

$$\begin{aligned}p^t_-(\omega) &= \frac{1}{2\pi}\int_a^b \frac{\Delta_h(\alpha)E^t_{r+}(-i\alpha)}{\alpha - i\omega}d\alpha, \\ \Delta_h(\alpha) &= -i[H_{-L}(-i\alpha) - H_{-R}(-i\alpha)],\end{aligned} \tag{14.4.3}$$

by closing the contour on $\Omega^{(-)}$ around the branch cut and changing variables. The quantity H_{-L} is given by (14.3.22) and H_{-R} by (14.3.23). Thus, using (14.3.21), we have

$$\begin{aligned}\Delta_h(\alpha) &= -\frac{2\alpha}{h_\infty}I(\alpha)P(\alpha)\sqrt{\frac{R^2(\alpha) + I^2(\alpha)}{[R(\alpha) - K(\alpha)]^2 + I^2(\alpha)}} \\ &= h_\infty \frac{I(\alpha)}{K(\alpha)Q(\alpha)} \le 0, \quad \alpha \in [a, b].\end{aligned} \tag{14.4.4}$$

The second form has the advantage that the need to evaluate a principal value integral is avoided. The definitions of the various quantities are summarized for convenience in Table 14.1. Explicit formulas for these quantities are given for various choices of k in [94].

Using (14.4.2) and (14.4.3), we can write the minimum free energy in the form

$$\psi_m(t) = \phi(t) + \frac{1}{2}\int_0^{\infty}\int_0^{\infty} E^t_r(s)G_{12}(s,u)E^t_r(u)ds du,$$

Formula	Equation reference
$F(\omega) = \int_a^b \frac{k(\alpha)}{\alpha+i\omega} d\alpha, \quad \omega \in \mathbb{R}$	(14.2.7)
$K(\alpha) = \int_a^b \frac{k(\beta)}{\beta+\alpha} d\beta, \quad \alpha \in \mathbb{R}\backslash[-b,-a]$	(14.2.9)
$R(\alpha) = P\int_a^b \frac{k(\beta)}{\beta-\alpha} d\beta, \quad I(\alpha) = -\pi k(\alpha), \quad \alpha \in (a,b)$	(14.2.11)
$A(\alpha) = \arg\left[1 - \frac{K(\alpha)}{R(\alpha)+iI(\alpha)}\right], \quad -\pi \le A(\alpha) \le 0$	(14.3.10), (14.3.12), (14.3.15)
$P(\alpha) = \exp\left[-\frac{1}{\pi} P\int_a^b \frac{A(\beta)}{\beta-\alpha} d\beta\right]$	(14.3.16)
$Q(\alpha) = \exp\left[-\frac{1}{\pi} \int_a^b \frac{A(\beta)}{\beta+\alpha} d\beta\right]$	(14.3.17)

Table 14.1. Definitions of the various quantities in the formula (14.4.4)

where

$$G_{12}(s,u) = \frac{1}{2\pi^2}\int_a^b \int_a^b \frac{\Delta_h(\alpha)e^{-\alpha s}\Delta_h(\beta)e^{-\beta u}}{\alpha+\beta} d\alpha d\beta,$$

and we understand the subscripts to mean differentiation with respect to the first and second variables. It follows that

$$G(s,u) = G(\infty,\infty) + \frac{1}{2\pi^2}\int_a^b \int_a^b \frac{\Delta_h(\alpha)e^{-\alpha s}\Delta_h(\beta)e^{-\beta u}}{(\alpha+\beta)\alpha\beta} d\alpha d\beta,$$

where

$$G(\infty,\infty) = G(s,\infty) = G(\infty,s), \quad s \in \mathbb{R}^+, \tag{14.4.5}$$

yielding $G_1(s,\infty) = G_2(\infty,s) = 0$. It is also required that (cf. (8.6.13))

$$G(s,0) = G(0,s) = G(s), \quad s \in \mathbb{R}^+. \tag{14.4.6}$$

We deduce from (14.4.5) and (14.4.6) that

$$G(\infty,\infty) = G(\infty) = G_\infty.$$

Observe that both G and G_{12} are positive quantities. To show that (14.4.6) holds, observe that, for $z \in \Omega^{(+)}$,

$$\frac{1}{2\pi i}\int_{-\infty}^{\infty} \frac{H_-(\omega')}{(\omega'-z)\omega'} d\omega' = \frac{H_-(z)}{z} = \frac{i}{2\pi}\int_a^b \frac{\Delta_h(\beta)}{(\beta-iz)\beta} d\beta,$$

where the first relation is essentially a scalar version of (11.2.23) and the second results in the same manner as (14.4.3), from the leftmost integral. It follows from (14.3.17) and (11.1.17) extended to the complex plane (yielding $H_-(i\alpha) = H_+(-i\alpha)$) that

$$\frac{1}{2\pi}\int_a^b \frac{\Delta_h(\beta)}{(\beta+\alpha)\beta} d\beta = \frac{1}{h_\infty} K(\alpha)Q(\alpha).$$

From $(14.2.11)_2$, $(14.4.4)_2$, and

$$G(s) = G_\infty - \int_a^b \frac{k(\alpha)}{\alpha} e^{-\alpha s} d\alpha,$$

which follows from (14.2.6), we deduce that (14.4.6) holds.

The rate of dissipation associated with the minimum free energy is given by (11.8.13), which, in this context, yields

$$\begin{aligned} D_m(t) &= \left[\frac{1}{2\pi} \int_{-\infty}^{\infty} H_-(\omega) E^t_{r+}(\omega) d\omega \right]^2 \\ &= \left[\frac{1}{2\pi} \int_a^b \Delta_h(\alpha) E^t_{r+}(-i\alpha) d\alpha \right]^2 \\ &= \left[\frac{1}{2\pi} \int_0^{\infty} \int_a^b \Delta_h(\alpha) E^t_r(u) e^{-\alpha u} d\alpha du \right]^2 \geq 0. \end{aligned}$$

14.5 An Alternative Approach

Another approach to finding the minimum free energy of a continuous-spectrum material is outlined in this section. Its most remarkable feature is that it does not require explicit factorization of the function H. It was motivated initially by the method by Breuer and Onat [42], who proposed an ansatz for the optimal continuation in the discrete-spectrum case and solved the problem by this means. A similar ansatz can be written down without difficulty for the continuous-spectrum case. However, it turns out that no such explicit assumption is required.

We start from the scalar form of (11.2.26), the Wiener–Hopf equation for the optimal future continuation, which we write in our present notation as

$$\int_{-\infty}^{\infty} \frac{\partial}{\partial s} G(|s-u|) E^t_r(u) du = R(u), \quad s \in \mathbb{R}, \text{ where } R(s) = 0,\ s \in \mathbb{R}^-. \tag{14.5.1}$$

The Fourier transform of (14.5.1) is the scalar version of (11.2.27):

$$2iH(\omega)[E^t_{r+}(\omega) + E^t_m(\omega)] = \omega R_+(\omega),$$

where E^t_m is the Fourier transform of the optimal future continuation introduced in (14.4.1) (which is not used here). It is analytic in Ω^+. We absorb the factor $2i$ in R_+ and seek not $E^t_m(\omega)$ but

$$\Xi^t_m(\omega) = i\omega \overline{F}(\omega) E^t_m(\omega), \tag{14.5.2}$$

which is also analytic in Ω^+. The reason for this change of unknown is that we end up with formulas that are directly comparable with earlier results, in particular (14.4.1), based on the factorization of H with factors F and $\overline{F}$ extracted, as in (14.3.6), (14.3.7), and later formulas.

Thus, recalling (14.2.4), we consider the relation

$$H(\omega)\left[E^t_{r+}(\omega) + \frac{\Xi^t_m(\omega)}{i\omega\overline{F}(\omega)}\right] = H(\omega)E^t_{r+}(\omega) + \overline{Y}(\omega)\Xi^t_m(\omega) = R_+(\omega),$$

where Y is defined by (14.3.9). We consider the discontinuity of both sides across the cut $(-ib, -ia)$. The quantities E^t_{r+} and R_+ are analytic in Ω^- and therefore have no discontinuity across the cut. Using (14.3.19) and (14.3.24), we obtain

$$\overline{Y}_L(-i\alpha)\Xi^t_L(-i\alpha) - \overline{Y}_R(-i\alpha)\Xi^t_R(-i\alpha) = \begin{cases} i\alpha I(\alpha)E^t_{r+}(-i\alpha), & \alpha \in (a,b), \\ 0, & \alpha \notin (a,b), \end{cases} \tag{14.5.3}$$

where Ξ^t_L and Ξ^t_R are the limits of Ξ^t_m on $[-ib, -ia]$ from the left and right, respectively. If it were assumed that Ξ^t_m could be written as a Cauchy integral of the form (B.2.1) over $[-ib, -ia]$, which amounts to the continuous version of the Breuer–Onat ansatz, then (14.5.3) could be put in the form of a singular integral equation. As remarked earlier, this is unnecessary. The only and very natural assumption needed is that the only singularity of E^t_m and therefore Ξ^t_m is a branch cut on $[-ib, -ia]$. Note that

$$\Xi^t_m(\omega) \approx \frac{1}{\omega} \tag{14.5.4}$$

for large frequencies, which follows from (14.2.2) and the behavior of E^t_m in (14.5.2), which can be deduced from (C.2.16). Relation (14.5.3) is a Hilbert problem, which we can write in the form

$$\begin{aligned} &\Xi^{t+}(\alpha) = C_1(\alpha)\Xi^{t-}(\alpha) + C_2(\alpha), \\ &\Xi^{t+}(\alpha) = \Xi^t_L(-i\alpha), \qquad \Xi^{t-}(\alpha) = \Xi^t_R(-i\alpha), \\ &C_1(\alpha) = \frac{\overline{Y}_R(-i\alpha)}{\overline{Y}_L(-i\alpha)}, \qquad C_2(\alpha) = \frac{i\alpha I(\alpha)E_{r+}(-i\alpha)}{\overline{Y}_L(-i\alpha)}. \end{aligned} \tag{14.5.5}$$

Note that from (14.3.11),

$$C_1(a) = C_1(b) = 1. \tag{14.5.6}$$

This is clear for singular endpoints as given by (14.2.16) and (14.2.18). For the non-singular case, $I(a)$ and $I(b)$ vanish.

Equation (14.5.5) will now be solved for $\Xi^t(z) = \Xi^t_m(-iz)$, which has a branch cut on $[a,b]$ and where $\Xi^{t+}(\alpha)$ and $\Xi^{t-}(\alpha)$ are the limits of this function from the left and the right of the cut. The solution is subject to (14.5.4) and to the condition that it be bounded except possibly at a or b, where it may diverge logarithmically or as a power less than unity. This latter property reflects the assumptions made relating to the density function k. The general solution is [274, page 237]

$$\begin{aligned} &\Xi^t(z) = \frac{X(z)}{2\pi i}\int_a^b \frac{C_2(\beta)}{X^+(\beta)(\beta - z)}d\beta + X(z)P(z), \\ &X(z) = \Pi(z)e^{N(iz)}, \\ &N(iz) = \frac{1}{2\pi i}\int_a^b \frac{\log C_1(\lambda)}{\lambda - z}d\lambda, \\ &\Pi(z) = (z-a)^{\lambda_1}(z-b)^{\lambda_2}, \end{aligned}$$

where λ_1 and λ_2 are integers and $P(z)$ is an arbitrary polynomial of degree not less than $\kappa - 1$ with

$$\kappa = -\lambda_1 - \lambda_2.$$

Observe that by virtue of (14.3.11), $N(iz)$ is the quantity defined by (14.3.8) and (14.3.9). The quantity $X^+(\beta)$ is the limit of $X(z)$ as $z \to \beta \in (a, b)$ from the positive half-plane. Near $z = a, b$, the quantity N is finite because of (14.5.6) and the observation after (B.2.9), so that

$$X(z) \approx \begin{cases} K_1(z-a)^{\lambda_1}, & z \to a, \\ K_2(z-b)^{\lambda_2}, & z \to b, \end{cases}$$

where K_1 and K_2 are constants. To ensure no divergence in Ξ^t of order unity or stronger, we must have the integers $\lambda_1, \lambda_2 \geq 0$ and $\kappa \leq 0$. For $\kappa < 0$, solutions vanishing at infinity are possible only if restrictions are placed on C_2 [274], which depends only on given physical parameters. Thus, we must have $\kappa = 0$ and $\lambda_1 = \lambda_2 = 0$. The polynomial P is zero because of (14.5.4). Therefore,

$$X(z) = e^{N(iz)}$$

and

$$\Xi^t(i\omega) = \Xi^t_m(\omega) = \frac{X(i\omega)}{2\pi i}\int_a^b \frac{C_2(\beta)}{X^+(\beta)(\beta - i\omega)}d\beta. \tag{14.5.7}$$

Observe that from (14.3.14),

$$X(i\omega) = \frac{i\omega\overline{F}(\omega)}{h_\infty H_-(\omega)} \tag{14.5.8}$$

and

$$X^+(\beta) = \frac{\beta\overline{F}_L(-i\beta)}{h_\infty H_{-L}(-i\beta)} = \frac{1}{P(\beta)}e^{iA(\beta)}, \qquad \beta \in (a, b), \tag{14.5.9}$$

where (14.2.12) and (14.3.16) have been used. Now, from (14.3.10) and (14.3.11),

$$\overline{Y}_L(-i\beta) = \frac{1}{2}\left[1 - \frac{K(\beta)}{R(\beta) - iI(\beta)}\right] = \frac{1}{2}\sqrt{\frac{[R(\beta) - K(\beta)]^2 + I^2(\beta)}{R^2(\beta) + I^2(\beta)}}\,e^{-iA(\beta)},$$

by virtue of (14.3.20). Thus, from $(14.5.5)_5$ and (14.4.4),

$$\begin{aligned}\frac{C_2(\beta)}{X^+(\beta)} &= 2i\beta P(\beta)I(\beta)\sqrt{\frac{R^2(\beta) + I^2(\beta)}{[R(\beta) - K(\beta)]^2 + I^2(\beta)}}E^t_{r+}(-i\beta) \\ &= -ih_\infty \Delta_h(\beta)E^t_{r+}(-i\beta).\end{aligned} \tag{14.5.10}$$

Then, we finally obtain, from (14.5.2), (14.5.7), and (14.5.10),

$$E^t_m(\omega) = -\frac{1}{2\pi H_-(\omega)}\int_a^b \frac{\Delta_h(\beta)E^t_{r+}(-i\beta)}{\beta - i\omega}d\beta,$$

which agrees with (14.4.1) and (14.4.3).

Observe that the quantity X, given by (14.5.8), is closely related to the factor H_-. This is how the factors of H enter the formulas. The quantity X is the solution of the homogeneous part of the Hilbert problem (14.5.5):

$$X^+(\alpha) = C_1(\alpha)X^-(\alpha). \tag{14.5.11}$$

We note that the factorization problem of H can be expressed as a homogeneous Hilbert problem on the real axis:

$$H_-(w) = H(\omega)[H_+(w)]^{-1}.$$

It is straightforward to show that this is equivalent to (14.5.11), by taking the limit of this relation on both sides of the branch cut on $[-ib, -ia]$ in H_- and H and using (14.5.9).

14.6 Minimal States

Finally, let us explore the concept of minimal states in the context of continuous-spectrum materials. Two viscoelastic states $(E_1^t, E_1(t))$ and $(E_2^t, E_2(t))$ are equivalent or in the same minimal state if (see (7.4.3))

$$E_1(t) = E_2(t), \quad \int_0^\infty G'(s+\tau)\left[E_1^t(s) - E_2^t(s)\right] ds = 0 \;\forall\, \tau \geq 0. \tag{14.6.1}$$

Proposition 14.6.1. *For the relaxation function derivative given by* (14.2.6), *where k is negative on (a, b), except possibly at a finite number of isolated points, and for histories with E_+^t analytic on $\mathbb{R}$, the minimal states are singletons. In other words, $(E^t, E(t))$ is the minimal state.*

Proof. We define $(E_d^t, E_d(t))$ as

$$E_d(t) = E_1(t) - E_2(t),$$
$$E_d^t(s) = E_1^t(s) - E_2^t(s), \quad s \in \mathbb{R}^+.$$

Then (14.6.1) becomes

$$E_d(t) = 0,$$
$$\int_0^\infty G'(s+\tau)E_d^t(s)ds = \int_a^b k(\alpha)e^{-\alpha\tau}E_{d+}^t(-i\alpha)d\alpha = 0, \quad \forall\, \tau \geq 0.$$

The function

$$A(\tau) = \int_a^b k(\alpha)e^{-\alpha\tau}E_{d+}^t(-i\alpha)d\alpha$$

can be analytically continued (Sect. B.1.2) to the complex τ plane. It is analytic (and therefore zero) for $\mathrm{Re}\tau > 0$. Taking the inverse Laplace transform (see (C.2.5)), we deduce that $k(\alpha)E^t_+(-i\alpha)$ vanishes for $\alpha \in \mathbb{R}^+$. Thus, since $k(\alpha)$ does not vanish for $\alpha \in (a, b)$, except at most at a finite number of isolated points, we have

$$E^t_{d+}(-i\alpha) = 0,$$

over (a, b) or some open subinterval of this region, which in turn implies that $E^t_{d+}(\omega)$ vanishes in the region of analyticity connected to $(-ib, -ia)$. This certainly includes Ω^- and in particular the real axis. We conclude that

$$E_d(t) = 0, \qquad E^t_d(s) = 0, \quad s \in \mathbb{R}^{++}.$$

□

A generalization of Proposition 14.6.1 is given by Proposition 16.5.2.

15

The Minimum Free Energy for a Finite-Memory Material

15.1 Introduction

In this chapter, based on work reported in [111], we derive an expression for the minimum free energy corresponding to a relaxation function with the special property that its derivative is nonzero over only a finite interval of time. It will be seen that there are special features associated with the analytic behavior of the frequency-space representation of such relaxation functions that render this a nontrivial extension, with unique features, of the general treatments presented in Chaps. 7 and 14. This property of finite memory is of interest in the first instance because finite and infinite memories are not necessarily experimentally distinguishable; also, the assumption of infinite memory can lead to paradoxical results for certain problems.

Specifically, the problem of a viscoelastic membrane in a frictional medium [111] illustrates that results running counter to physical intuition emerge from the assumption of infinite memory. In fact, a result is quoted that shows that while (time) exponential decay in the displacement occurs in the elastic problem, this is not so in the viscoelastic problem if the viscoelastic function does not decay exponentially. One would expect that any viscoelastic function would simply enhance the elastic exponential decay because of the dissipative effects associated with viscoelasticity.

In Sect. 15.2, it is shown that the singularity structure of the Fourier transform of the relaxation function derivative is quite different from the infinite-memory case in that it is an entire function with essential singularities of exponential type at infinity, rather than poles and branch points generally in the finite complex plane, as is the case for infinite-memory materials. The latter may of course include essential singularities at infinity, though these have been excluded for simplicity from earlier (and indeed later) chapters.

In Sect. 15.4, the crucial factorization required to determine the minimum free energy is discussed for specific (step function) examples of finite-memory materials, while explicit forms of the minimum free energy and the corresponding work function are given in Sect. 15.5 for two choices of finite-memory relaxation functions.

The scalar case is again dealt with in this chapter.

G. Amendola et al., *Thermodynamics of Materials with Memory*,
https://doi.org/10.1007/978-3-030-80534-0_15

In the examples considered, we find that the function H, while nonnegative, may be zero for nonzero frequencies. This contradicts one of the assumptions of Theorem 11.1.1, and in fact, we shall see that in one example, there is a lack of uniqueness in the factorization.

15.2 Finite Memory

We adopt the constitutive equation (14.2.1). Consider the case that

$$G'(t) = 0, \quad t > d > 0.$$

It will be assumed that

$$G'(d) \neq 0,$$

for reasons of simplicity. If $G'(d)$ vanishes and $G''(d) \neq 0$, the formulas given below are altered slightly (Remark C.2.5). We have

$$G'_+(\omega) = \int_0^d G'(s)e^{-i\omega s}\, ds,$$

and the inverse relationship is

$$G'(s) = \frac{1}{2\pi}\int_{-\infty}^{\infty} G'_+(\omega)e^{i\omega s}\, d\omega.$$

The function $e^{-i\omega b}$ on Ω, where $b > 0$, diverges exponentially in $\Omega^{(+)}$ as $\mathrm{Im}(\omega) \to +\infty$. Similarly, $e^{i\omega b}$, where $b > 0$, diverges exponentially in $\Omega^{(-)}$ as $\mathrm{Im}(\omega) \to -\infty$. The functions $e^{\pm i\omega b}$ decay exponentially as $\mathrm{Im}\omega \to \pm\infty$, respectively. As noted in Sect. B.1.4, we shall refer to them as analytic in $\Omega^{(\pm)}$.

From Proposition C.2.4, we conclude that G'_+ is an entire function with an essential singularity at infinity. Its dominant behavior is an exponential divergence given by

$$G'_+(\omega) \xrightarrow[\mathrm{Im}\omega\to+\infty]{} -\frac{G'(d)}{i\omega}e^{-i\omega d}. \tag{15.2.1}$$

Also, for large $|\mathrm{Re}\omega|$,

$$G'_+(\omega) \sim \frac{G'(0)}{i\omega}\left(1 - e^{-i\omega d}\right), \tag{15.2.2}$$

so that

$$G'_+(\omega) \xrightarrow[\mathrm{Im}\omega\to-\infty]{} \frac{G'(0)}{i\omega}.$$

The quantities of central interest are H, given by (14.2.4), and its factors $H_\pm$ defined by (14.3.4). The function H has exponential divergences as $\mathrm{Im}(\omega)$ approaches infinity in both $\Omega^{(+)}$ and $\Omega^{(-)}$; the factor H_+ has exponential divergences only in $\Omega^{(+)}$ and H_- only in $\Omega^{(-)}$. Indeed, using (15.2.1),

$$H(\omega) \approx \begin{cases} \frac{1}{2}G'(d)e^{-i\omega d}, & \mathrm{Im}\omega \to +\infty, \\ \frac{1}{2}G'(d)e^{i\omega d}, & \mathrm{Im}\omega \to -\infty, \end{cases} \tag{15.2.3}$$

so that

$$H_+(\omega) \xrightarrow[\mathrm{Im}\omega\to+\infty]{} Ae^{-i\omega d}, \qquad H_-(\omega) \xrightarrow[\mathrm{Im}\omega\to-\infty]{} \overline{A}e^{i\omega d}, \tag{15.2.4}$$

where A is a constant. From (15.2.2), it follows that along the real axis or any axis parallel to it,

$$\lim_{\omega\to\infty} H(\omega) = -G'(0)\left(1 - \lim_{\omega\to\infty} \cos\omega d\right).$$

This limit does not exist. There is, however, no resultant singularity.

The factorization given by (11.1.15) is not useful in this case. Special techniques must be employed, as outlined in Sect. 15.4.

15.3 The History Dependence of the Minimum Free Energy

The minimum free energy $\psi_m(t)$ is given by (14.4.2) or the scalar version of (11.8.9) with p_-^t (and p_+^t) defined by (11.8.4). If (recall $(12.1.1)_5$)

$$Y^t(s) = \frac{1}{2\pi}\int_{-\infty}^{\infty} H_-(\omega)E_{r+}^t(\omega)e^{i\omega s}\,d\omega, \tag{15.3.1}$$

then

$$H_-(\omega)E_+^t(\omega) = \int_{-\infty}^{\infty} Y^t(s)e^{-i\omega s}\,d\omega = p_-^t(\omega) - p_+^t(\omega)$$

and, using (12.1.1) with (12.1.2),

$$p_+^t(\omega) = -\int_0^{\infty} Y^t(s)e^{-i\omega s}\,ds, \qquad p_-^t(\omega) = \int_{-\infty}^{0} Y^t(s)e^{-i\omega s}\,ds. \tag{15.3.2}$$

One can deduce from $(15.2.4)_2$ that $Y^t(s)$, given by (15.3.1), vanishes for $s < -d$ by closing the contour on $\Omega^{(-)}$, recalling that H_- has no singularities at finite points. Therefore, $(15.3.2)_2$ becomes

$$p_-^t(\omega) = \int_{-d}^{0} Y^t(s)e^{-i\omega s}\,ds. \tag{15.3.3}$$

Proposition 15.3.1. *For a material with finite memory of duration d, the minimum free energy $\psi_m(t)$ depends only on that part of the history for which G' is nonzero, i.e., $E_r^t(s)$, $0 \le s \le d$, while the work function $W(t)$ may depend on the entire history of strain.*

Proof. *We have from* (15.3.1) *that*

$$Y^t(s) = \frac{1}{2\pi} \int_{-\infty}^{\infty} H_-(\omega) \int_0^{\infty} E_r^t(u) e^{i\omega(s-u)} du d\omega.$$

It follows from $(15.2.4)_2$ *that*

$$\int_{-\infty}^{\infty} H_-(\omega) e^{i\omega(s-u)} d\omega = 0 \;\; \forall\, s + d < u,$$

so that

$$Y^t(s) = \frac{1}{2\pi} \int_{-\infty}^{\infty} H_-(\omega) \int_0^{s+d} E_r^t(u) e^{i\omega(s-u)} du d\omega.$$

It is clear now that $p_-^t(\omega)$, *given by* (15.3.3), *depends only on* $E_r^t(s)$, $0 \le s \le d$. *However,* $p_+^t(\omega)$, *given by* $(15.3.2)_1$, *may depend on the entire history of strain. The result follows from* (14.4.2) *and the scalar version of* (11.8.8). □

A consequence of Proposition 15.3.1 is that a time-domain representation of the minimum free energy, with history-dependent part given by (12.1.9), reduces to the form

$$\psi_m(t) = \phi(t) + \frac{1}{2} \int_0^d \int_0^d E_r^t(s_1) \frac{\partial^2}{\partial s_1 \partial s_2} G(s_1, s_2) E_r^t(s_2) ds_1 ds_2$$

rather than this expression with infinite integrations as in the general case.

The condition that a state $(E_1^t, E_1(t))$ is in the same minimal state as $(E^t, E(t))$ in a finite-memory material can be adapted from (14.6.1) to give

$$E_1(t) = E(t), \qquad \int_0^{d-\tau} G'(s+\tau) \left[E_1^t(s) - E^t(s) \right] ds = 0 \;\; \forall\, \tau \in [0, d]. \qquad (15.3.4)$$

Thus, $E_1^t(s)$, $s > d$, can be chosen arbitrarily, so that the state is certainly nonsingleton.

Remark 15.3.2. It is interesting to consider Proposition 15.3.1 against the background of this observation on minimal states. We see from (15.3.4) that a functional of the minimal state can depend only on the history in the interval $[0, d]$, since values outside of this interval can be varied arbitrarily without altering the minimal state. Thus, p_-^t and ψ_m must have this property, which is in effect Proposition 15.3.1.

15.4 Factorization of $H(\omega)$

Let us now address the problem of factorization of $H(\omega)$ as given by (14.3.4) in the finite-memory case, for specified forms of the relaxation function. It turns out that progress on this issue is possible for step function choices of G'. Explicit factorizations seem not to be possible for more general and realistic relaxation functions.

Consider the ansatz

$$H_+(\omega) = e^{-i\omega d/2}\,[H(\omega)]^{1/2}, \qquad H_-(\omega) = e^{i\omega d/2}\,[H(\omega)]^{1/2}, \qquad \omega \in \mathbb{R}, \quad (15.4.1)$$

where $[H(\omega)]^{1/2}$ is assumed to be analytic at all finite points in Ω. We note that H vanishes at $\omega = 0$, where it has a quadratic zero that does not produce a branch point. It is assumed that any other zero of H in Ω is of even power type. The relations in (15.4.1) are consistent with (15.2.3) and (15.2.4).

Let us now look at specific cases. Consider first the choice

$$G'(t) = \begin{cases} -K_0, & 0 \le t < d, \\ 0, & t \ge d, \end{cases} \qquad (15.4.2)$$

where, referring to (8.1.4),

$$K_0 = \frac{G_0 - G_\infty}{d} > 0.$$

Then,

$$G'_+(\omega) = \frac{iK_0}{\omega}\left(1 - e^{-i\omega d}\right),$$

$$H(\omega) = K_0\,(1 - \cos\omega d) = 2K_0 \sin^2\frac{\omega d}{2},$$

which has zeros at $\omega d = 2n\pi$ for all integer values of n and is thus not positive definite for nonzero ω. Also,

$$[H(\omega)]^{1/2} = \sqrt{2K_0}\sin\frac{\omega d}{2},$$

so that from (15.4.1),

$$\begin{aligned} H_+(\omega) &= \sqrt{\frac{K_0}{2}}\left(1 - e^{-i\omega d}\right), \\ H_-(\omega) &= \sqrt{\frac{K_0}{2}}\left(1 - e^{i\omega d}\right), \end{aligned} \qquad (15.4.3)$$

where a factor i has been omitted.

Relation (15.4.1) is not useful in the next example. Consider the case where

$$G'(t) = \begin{cases} -K_0, & 0 \le t < d/2, \\ -K_1, & d/2 \le t < d, \\ 0, & t \ge d, \end{cases} \qquad 0 < K_1 < K_0. \qquad (15.4.4)$$

We have that

$$\frac{K_0 + K_1}{2} = \frac{G_0 - G_\infty}{d}$$

and

$$G'_+(\omega) = \frac{i}{\omega}\left[K_0 - (K_0 - K_1)e^{-i\omega d/2} - K_1 e^{-i\omega d}\right].$$

It follows that

$$H(\omega) = K_0 - (K_0 - K_1)\cos\frac{\omega d}{2} - K_1 \cos \omega d \geq 0 \;\; \forall\, \omega \in \mathbb{R},$$

which again vanishes for discrete nonzero values of ω obeying $\cos(\omega d/2) = \cos(\omega d) = 1$. We look for factors of the form

$$\begin{aligned} H_+(\omega) &= b_0 + b_1 e^{-i\omega d/2} + b_2 e^{-i\omega d}, \\ H_-(\omega) &= b_0 + b_1 e^{i\omega d/2} + b_2 e^{i\omega d}, \end{aligned}$$

where b_0, b_1, and b_2 are real. Using (14.3.4) and comparing coefficients give the conditions

$$\begin{aligned} b_0^2 + b_1^2 + b_2^2 &= K_0, \\ b_1(b_0 + b_2) &= \frac{K_1 - K_0}{2}, \\ b_0 b_2 &= -\frac{K_1}{2}, \end{aligned} \tag{15.4.5}$$

with solution

$$\begin{aligned} b_1 &= \epsilon_1 \sqrt{\frac{K_0 - K_1}{2}}, \qquad \epsilon_1 = \pm 1, \\ b_0 &= \frac{1}{2}\left\{-b_1 + \epsilon_2 \sqrt{b_1^2 + 2K_1}\right\}, \\ b_2 &= \frac{1}{2}\left\{-b_1 - \epsilon_2 \sqrt{b_1^2 + 2K_1}\right\}, \\ \epsilon_2 &= \pm 1. \end{aligned} \tag{15.4.6}$$

Observe that

$$b_0 + b_1 + b_2 = 0, \tag{15.4.7}$$

as required for $H_\pm$ to have zeros at $\omega = 0$. If we choose

$$K_0 = 2K_1,$$

then these reduce to

$$\begin{aligned} b_1 &= \epsilon_1 \frac{\sqrt{K_0}}{2}, \\ b_0 &= \frac{\sqrt{K_0}}{4}\left(-\epsilon_1 + \epsilon_2 \sqrt{5}\right), \\ b_2 &= \frac{\sqrt{K_0}}{4}\left(-\epsilon_1 - \epsilon_2 \sqrt{5}\right). \end{aligned} \tag{15.4.8}$$

Since $H_\pm$ are arbitrary up to a constant real phase factor, we choose $\epsilon_1 = 1$. However, there remain two possible solutions, corresponding to an interchange of b_0 and b_2. A choice will be made between these two possibilities below. This is the lack of uniqueness of the factorization referred to in the introduction to this chapter.

Some discussion of more general step behavior in relaxation functions may be found in [111].

15.5 Explicit Forms of the Minimum Free Energy

For the relaxation function derivative given by (15.4.2), we have that the stress, given by (14.2.1), has the form

$$S(t) = S_e(t) - K_0 \int_0^d E_r^t(s)ds = S_e(t) - K_0 \int_{t-d}^t [E(v) - E(t)]\, dv,$$

and the work function is given by the scalar version of $(11.8.3)_1$. Assuming that $E(v)$ vanishes at large negative v, we have the identities

$$\begin{aligned}
&\int_{-\infty}^t \int_{u-d}^{u-c} (E(v) - E(u))\dot{E}(u)dvdu \\
&\quad = \int_{-\infty}^t \int_{u-d}^{u-c} [E(v) - E(u)]\frac{\partial}{\partial u}[E(u) - E(v)]dvdu \\
&\quad = -\frac{1}{2}\int_{-\infty}^t \int_{u-d}^{u-c} \frac{\partial}{\partial u}[E(v) - E(u)]^2 dvdu \\
&\quad = -\frac{1}{2}\int_{-\infty}^t \frac{d}{du}\int_{u-d}^{u-c} [E(v) - E(u)]^2 dvdu \\
&\qquad + \frac{1}{2}\int_{-\infty}^t \left\{[E(u-c) - E(u)]^2 - [E(u-d) - E(u)]^2\right\} du \\
&\quad = -\frac{1}{2}\int_{t-c}^{t-d} [E(v) - E(t)]^2 dv \\
&\qquad + \frac{1}{2}\int_{\infty}^t \left\{[E(u-c) - E(u)]^2 - [E(u-d) - E(u)]^2\right\} du,
\end{aligned} \tag{15.5.1}$$

where c is a constant. We can show, putting c equal to zero, that

$$W(t) = \phi(t) + \frac{K_0}{2}\int_0^d [E_r^t(s)]^2 ds + \frac{K_0}{2}\int_0^\infty \left[E_r^t(s+d) - E_r^t(s)\right]^2 ds. \tag{15.5.2}$$

Similarly, for G' given by (15.4.4),

$$\begin{aligned}
S(t) &= S_e(t) - K_0 \int_0^{d/2} E_r^t(s)ds - K_1 \int_{d/2}^d E_r^t(s)ds \\
&= S_e(t) - K_0 \int_{t-d/2}^t [E(u) - E(t)]du - K_1 \int_{t-d}^{t-d/2} [E(u) - E(t)]\, du.
\end{aligned}$$

Using (15.5.1) for $c = 0$ and $c = d/2$, we can show that the work function has the form

$$
\begin{aligned}
W(t) &= \phi(t) + \frac{K_0}{2}\int_0^{d/2}\left[E_r^t(s)\right]^2 ds + \frac{K_1}{2}\int_{d/2}^{d}\left[E_r^t(s)\right]^2 ds \\
&\quad + \frac{K_0 - K_1}{2}\int_0^{\infty}\left[E_r^t(s) - E_r^t\left(s + \frac{d}{2}\right)\right]^2 ds \\
&\quad + \frac{K_1}{2}\int_0^{\infty}\left[E_r^t(s) - E_r^t(s+d)\right]^2 ds \qquad (15.5.3)\\
&= \phi(t) + K_0\int_0^{\infty}\left[E_r^t(s)\right]^2 ds + (K_1 - K_0)\int_0^{\infty} E_r^t(s)E_r^t\left(s + \frac{d}{2}\right) ds \\
&\quad - K_1\int_0^{\infty} E_r^t(s)E_r^t(s+d)ds.
\end{aligned}
$$

We now write down the form of the minimum free energy and associated quantities for the explicit factorizations considered in Sect. 15.4. Consider first (15.4.2), where the factors are given by (15.4.3). The quantity Y^t, defined by (15.3.1), has the form

$$
Y^t(s) = \frac{1}{2\pi}\sqrt{\frac{K_0}{2}}\int_{-\infty}^{\infty}\left(1 - e^{i\omega d}\right)E_{r+}^t(\omega)e^{i\omega s}\, d\omega = \sqrt{\frac{K_0}{2}}\left[E_c^t(s) - E_c^t(s+d)\right],
$$

where E_c^t is the function E_r^t with $E_c^t(s) = 0,\ s < 0$. This property can be demonstrated by closing the contour of integration on $\Omega^{(-)}$ for $s < 0$. From (15.3.3), we see that

$$
p_-^t(\omega) = -\sqrt{\frac{K_0}{2}}\int_{-d}^{0} E_c^t(s+d)e^{-i\omega s}\, ds = -\sqrt{\frac{K_0}{2}}e^{i\omega d}\int_0^{d} E_c^t(u)e^{-i\omega u}\, du,
$$

and, referring to $(15.3.2)_1$,

$$
p_+^t(\omega) = -\sqrt{\frac{K_0}{2}}\int_0^{\infty}\left[E_c^t(s) - E_c^t(s+d)\right]e^{-i\omega s}\, ds.
$$

From Parseval's formula (C.3.1) and $(14.4.2)_2$, we see that

$$
\psi_m(t) = \phi(t) + \frac{K_0}{2}\int_0^{d}\left[E_c^t(s)\right]^2 ds
$$

and

$$
\frac{1}{2\pi}\int_{-\infty}^{\infty}|p_+^t(\omega)|^2 d\omega = \frac{K_0}{2}\int_0^{\infty}\left[E_c^t(s) - E_c^t(s+d)\right]^2 ds.
$$

This is the total dissipation corresponding to the minimum free energy, as given by the scalar version of (11.8.12). Differentiating it with respect to t gives the rate of dissipation associated with the minimum free energy. We can switch to a derivative with respect to s, using (5.1.14). Then the integration can be carried out to give

$$
D_m(t) = \frac{K_0}{2}\left[E_c^t(d)\right]^2.
$$

This formula can also be derived using the scalar version of (11.8.13) and the observation that $K(t) = Y^t(0)$. The scalar version of (11.8.8) gives

$$W(t) = \phi(t) + \frac{1}{2\pi}\int_{-\infty}^{\infty} [|p_-^t(\omega)|^2 + |p_+^t(\omega)|^2]d\omega, \tag{15.5.4}$$

which yields an expression in agreement with (15.5.2), on noting that we can replace E_c^t by E_r^t.

For the second case, where G' is defined by (15.4.4), we have

$$\begin{aligned} Y^t(s) &= \frac{1}{2\pi}\int_{-\infty}^{\infty} \left(b_0 + b_1 e^{i\omega d/2} + b_2 e^{i\omega d}\right) E_{r+}^t(\omega) e^{i\omega s}\, d\omega \\ &= b_0 E_c^t(s) + b_1 E_c^t\left(s + \frac{d}{2}\right) + b_2 E_c^t(s+d), \end{aligned}$$

so that

$$p_-^t(\omega) = b_1 \int_{-d/2}^{0} E_c^t\left(s + \frac{d}{2}\right) e^{-i\omega s}\, ds + b_2 \int_{-d}^{0} E_c^t(s+d) e^{-i\omega s}\, ds,$$

and

$$p_+^t(\omega) = -\int_0^{\infty} \left[b_0 E_c^t(s) + b_1 E_c^t\left(s + \frac{d}{2}\right) + b_2 E_c^t(s+d)\right] e^{-i\omega s}\, ds. \tag{15.5.5}$$

Thus,

$$\begin{aligned} \psi_m(t) &= \phi(t) + b_1^2 \int_0^{d/2} \left[E_c^t(s)\right]^2 ds + b_2^2 \int_0^{d} \left[E_c^t(s)\right]^2 ds \\ &\quad + 2b_1 b_2 \int_0^{d/2} E_c^t(s) E_c^t\left(s + \frac{d}{2}\right) ds \end{aligned} \tag{15.5.6}$$

after changing variables in the integrations. Formula (15.5.3) can also be reproduced from (15.5.4) and (15.5.5) with the aid of the conditions in (15.4.5). The second form of (15.5.3) is best used for this purpose.

One can show as above that the rate of dissipation is

$$D_m(t) = \left| b_1 E_c^t\left(\frac{d}{2}\right) + b_2 E_c^t(d)\right|^2 .$$

Relation (15.5.6) can be written in the form

$$\begin{aligned} \psi_m(t) &= \phi(t) + a_1 \int_0^{d/2} \left[E_c^t(s)\right]^2 ds + a_2 \int_{d/2}^{d} \left[E_c^t(s)\right]^2 ds, \\ &\quad - b_1 b_2 \int_0^{d/2} \left[E_c^t(s) - E_c^t\left(s + \frac{d}{2}\right)\right]^2 ds, \\ a_1 &= b_0^2 - b_1 b_2 = \frac{K_0}{2}, \qquad a_2 = b_2^2 + b_1 b_2 = \frac{K_1}{2}. \end{aligned}$$

The last two relations may be verified by manipulating (15.4.5). The first relation can be transformed into (15.5.6) by using in particular the fact that

$$b_0^2 - b_2^2 - 2b_1b_2 = b_1^2,$$

which is a consequence of (15.4.7). We see therefore that the largest choice of b_2 minimizes ψ_m and thus choose $\epsilon_2 = -1$ in (15.4.6). This is the choice referred to after (15.4.8).

The results of this section are consistent with Proposition 15.3.1.

16

Free Energies for the Case of Isolated Singularities

We now focus on the case of materials characterized by memory kernels with only isolated singularities in the frequency domain and derive explicit expressions for a family of free energies, including the minimum free energy discussed in Chap. 11 and the maximum free energy. All of these will be shown to be functionals of the minimal state.

We will briefly deal with the general case of isolated singularities here, where poles can be of any finite order, as discussed in Sect. 7.3.1. The poles may be on the positive imaginary axis or, if not on this axis, must occur in pairs that are mirror images in that axis. These restrictions are consequences of the reality condition (7.3.4). The number of poles is taken to be finite. For such materials, any quantity of interest in the frequency domain is a rational function of frequency and can be expressed as a ratio of two polynomials. In principle, the general result can be obtained by first writing down the formula for simple poles (poles of order 1) and then allowing zeros and poles to coagulate, yielding the final form, though substantial algebra may be involved. We therefore focus mainly on the simple pole case.

The treatment of materials with only simple poles is usually algebraically easier than for those with higher order poles. Such simple poles correspond to decaying exponentials in the time domain (see (7.3.2)). Materials with this type of singularity, restricted to poles on the positive imaginary axis, corresponding to strictly decaying exponentials (Sect. 7.3.1), are discrete-spectrum materials.

The developments in this chapter are based on [101, 110, 160, 161]. In the interests of simplicity, certain results presented in this context, in the first edition, are omitted.

Remark 16.0.1. In the light of the fact that on each eigenspace, the formulas are given by the scalar version, we shall treat only the scalar case in this and some later chapters. In order to use them in the tensor case, one simply attaches a subscript k with values in the range $1, 2, \ldots, m$ to each field quantity and to all parameters, both integer and continuous, where k refers to the particular eigenspace in question.

G. Amendola et al., *Thermodynamics of Materials with Memory*,
https://doi.org/10.1007/978-3-030-80534-0_16

The memory term of any free energy is given by the sum over all eigenspaces of the formulas in each eigenspace.

We shall use the mechanics symbols E and T and the terms strain and stress to represent the independent and dependent scalar fields, though emphasizing that these can be any of the eigenvalues of $\boldsymbol{\Lambda}$ or $\boldsymbol{\Sigma}$. Indeed, similar results can be given with little extra difficulty, for viscoelastic fluids, nonisothermal problems, electromagnetism, nonsimple materials, etc. There is no connection between T and the quantity used in Sect. 5.1.2 and later to indicate a time period. A typical eigenvalue of $\mathbb{L}'$ will be denoted by G'.

16.1 Constitutive Relations, Histories, and Free Energy Properties for the Scalar Case

For convenience, the scalar version of certain tensor relations, derived earlier, will now be recorded. Firstly, we consider the scalar version of certain formulae in Sects. 5.1 and 7.1.1. To begin with, no restriction will be placed on the type of singularities in the frequency-domain memory function.

The current value of the strain function is $E(t)$, while the strain history and relative history are given by

$$E^t(s) = E(t-s), \qquad E_r^t(s) = E^t(s) - E(t), \qquad s \in \mathbb{R}^+. \tag{16.1.1}$$

It is assumed here that

$$\lim_{s\to\infty} E^t(s) = \lim_{u\to-\infty} E(u) = 0, \tag{16.1.2}$$

which simplifies certain formulae. The state of the material, in the most basic sense, is specified by $(E^t, E(t))$ or $(E_r^t, E(t))$. Another definition of state, corresponding to that in Sect. 7.4, will be introduced in Sect. 16.5.1.

Let $T(t)$ be the stress* at time t. Then the constitutive relations with linear memory terms have the equivalent forms

$$\begin{aligned}
T(t) &= T_e(t) + \int_0^\infty G'(u)E_r^t(u)du \\
&= T_0(t) + \int_0^\infty G'(u)E^t(u)du \\
&= T_e(t) - G_\infty E(t) + \int_0^\infty G(u)\dot{E}^t(u)du \\
&= T_e(t) + \int_0^\infty \widetilde{G}(u)\dot{E}^t(u)du, \\
T_0(t) &= T_e(t) + (G_0 - G_\infty)E(t), \\
G'(u) &= \frac{d}{du}G(u), \qquad \widetilde{G}(u) = G(u) - G_\infty, \\
\dot{E}^t(u) &= \frac{\partial}{\partial t}E^t(u) = -\frac{\partial}{\partial u}E^t(u) = -\frac{\partial}{\partial u}E_r^t(u), \qquad \ddot{E}^t(u) = -\frac{\partial}{\partial u}\dot{E}^t(u),
\end{aligned} \tag{16.1.3}$$

* This choice is more common than (14.2.1) in a linear context.

where $T_e(t)$ is the stress function for the equilibrium limit ($E_r^t(u) = 0, \quad u \in \mathbb{R}^+$), and the quantity $G(\cdot) : \mathbb{R}^+ \mapsto \mathbb{R}^+$ is the relaxation function of the material. We define

$$G_\infty = G(\infty), \qquad G_0 = G(0), \qquad \widetilde{G}(0) = G_0 - G_\infty = \widetilde{G}_0. \tag{16.1.4}$$

The assumption is made that

$$\widetilde{G},\ G' \in L^1(\mathbb{R}^+) \cap L^2(\mathbb{R}^+), \tag{16.1.5}$$

which will be relevant in the context of taking the Fourier transform of these quantities. For a completely linear material, $T_e(t)$ is given by the form

$$T_e(t) = G_\infty E(t). \tag{16.1.6}$$

16.1.1 Frequency-Domain Quantities for the Scalar Case

The scalar versions of certain formulae in Sect. 7.2, together with some others, are now listed. We have

$$\begin{aligned} \widetilde{G}_+(\omega) &= \int_0^\infty \widetilde{G}(s) e^{-i\omega s} ds = \widetilde{G}_c(\omega) - i\widetilde{G}_s(\omega), \\ G'_+(\omega) &= \int_0^\infty G'(s) e^{-i\omega s} ds = G'_c(\omega) - iG'_s(\omega). \end{aligned} \tag{16.1.7}$$

Applying a partial integration in (16.1.7)$_3$ yields that

$$G'_+(\omega) = -\widetilde{G}_0 + i\omega\widetilde{G}_+(\omega), \tag{16.1.8}$$

so that

$$G'_c(\omega) = -\widetilde{G}_0 + \omega\widetilde{G}_s(\omega), \qquad G'_s(\omega) = -\omega\widetilde{G}_c(\omega),$$

and

$$G_0 + G'_+(\omega) = G_\infty + i\omega\widetilde{G}_+(\omega) = \mathcal{M}_+(\omega), \quad \omega \in \Omega^-, \tag{16.1.9}$$

where $\mathcal{M}_+(\omega)$ is the complex modulus of the material [167]. Its real and imaginary parts for $\omega \in \mathbb{R}$ are given by

$$\begin{aligned} \mathcal{M}_+(\omega) &= \mathcal{R}(\omega) + i\mathcal{I}(\omega), \\ \mathcal{R}(\omega) &= G_0 + G'_c(\omega) = G_\infty + \omega\widetilde{G}_s(\omega), \\ \mathcal{I}(\omega) &= -G'_s(\omega) = \omega\widetilde{G}_c(\omega). \end{aligned} \tag{16.1.10}$$

Note the properties of $\widetilde{G}_+$ and G'_+ that (see (C.2.16) and (C.2.17))

$$\begin{aligned} \lim_{\omega\to\infty} i\omega\widetilde{G}_+(\omega) &= \lim_{\omega\to\infty} \omega\widetilde{G}_s(\omega) = \widetilde{G}_0, \\ \lim_{\omega\to\infty} i\omega G'_+(\omega) &= \lim_{\omega\to\infty} \omega G'_s(\omega) = -\lim_{\omega\to\infty} \omega^2\widetilde{G}_c(\omega) = G'(0). \end{aligned} \tag{16.1.11}$$

Further properties of $\widetilde{G}_c(\omega)$ and $G'_s(\omega)$ include (see (7.2.12))

$$\widetilde{G}_c(\omega) \geq 0, \quad G_s'(\omega) \leq 0 \quad \forall\, \omega \in \mathbb{R}^{++}, \tag{16.1.12}$$

which are consequences of the second law. We also have

$$\widetilde{G}_0 > 0, \qquad G_\infty > 0,$$

the latter relation being true for a viscoelastic solid. The quantity $H(\omega)$ is defined by

$$H(\omega) = -\omega G_s'(\omega) = \omega^2 \widetilde{G}_c(\omega) = \omega \mathfrak{I}(\omega) \geq 0, \quad \omega \in \mathbb{R}, \tag{16.1.13}$$

where the inequality is a consequence of (16.1.12). From $(16.1.13)_2$, it follows that the quantity $H(\omega)$ goes to zero quadratically at the origin. Using $(16.1.11)_{4,5}$ and (16.1.13), one can show that

$$H_\infty = \lim_{\omega \to \infty} H(\omega) = -G'(0) \geq 0.$$

We assume for present purposes that $G'(0)$ is nonzero so that H_∞ is a finite positive number. Then $H(\omega) \in \mathbb{R}^{++}\ \forall\, \omega \in \mathbb{R},\ \omega \neq 0$. By virtue of the general theorem for tensor quantities in Sect. 11.1 (and Sect. 11.1.1 for the scalar case), the quantity $H(\omega)$ can always be expressed as the product of two factors

$$H(\omega) = H_+(\omega) H_-(\omega), \tag{16.1.14}$$

where $H_\pm(\omega)$ are analytic on $\Omega_\mp$, respectively. If $G(s)$, $s \in \mathbb{R}^+$, is extended to the even function $G(|s|)$ on $\mathbb{R}$, then $dG(|s|)/ds$ is an odd function with the Fourier transform (see (13.1.19))

$$G_F'(\omega) = -2iG_s'(\omega) = \frac{2i}{\omega} H(\omega). \tag{16.1.15}$$

We list here some scalar versions of relations involving histories in the frequency domain, given in Sect. 7.2.3.

The Fourier transform of $E^t(s)$ and $E_r^t(s)$, given by (16.1.1) for $s \in \mathbb{R}^+$, is denoted by $E_+^t(\omega)$ and $E_{r+}^t(\omega)$. These have the same analyticity properties as $\widetilde{G}_+(\omega)$. However, $E_r^t(s)$ does not have the property (16.1.5), so that $E_{r+}^t(\omega)$ must be defined with care (Sect. C.2.3). For a constant history, $E^t(s) = E(t)$, $s \in \mathbb{R}^+$, we have

$$E_+^t(\omega) = \frac{E(t)}{i\omega^-},$$

where the notation ω^- (and ω^+) is defined in Sect. C.2.3. Thus, we have

$$E_{r+}^t(\omega) = E_+^t(\omega) - \frac{E(t)}{i\omega^-}. \tag{16.1.16}$$

Also,

$$\frac{d}{dt} E_+^t(\omega) = \dot{E}_+^t(\omega) = -i\omega E_+^t(\omega) + E(t) = -i\omega E_{r+}^t(\omega), \tag{16.1.17}$$

and

$$\begin{aligned}
\frac{d}{dt}\dot{E}^t_+(\omega) &= -i\omega\dot{E}^t_+(\omega) + \dot{E}(t),\\
\frac{d}{dt}E^t_{r+}(\omega) &= \dot{E}^t_{r+}(\omega) = -i\omega E^t_{r+}(\omega) - \frac{\dot{E}(t)}{i\omega^-}.
\end{aligned} \tag{16.1.18}$$

Similarly to (16.1.11), we have, for large ω,

$$E^t_+(\omega) \sim \frac{E(t)}{i\omega}, \qquad E^t_{r+}(\omega) \sim \frac{A(t)}{\omega^2}, \tag{16.1.19}$$

where $A(t)$ is independent of ω. Also, from (16.1.17),

$$\dot{E}^t_+(\omega) \sim \frac{A(t)}{i\omega}, \tag{16.1.20}$$

for large ω. Relation (16.1.17) is convenient for converting formulae from those in terms of $E^t_{r+}(\omega)$ to equivalent expressions in terms of $\dot{E}^t_+(\omega)$ or *vice versa*.

Applying Parseval's formula (C.3.1) to $(16.1.3)_{1,4}$, we obtain

$$\begin{aligned}
T(t) &= T_e(t) + \frac{1}{2\pi}\int_{-\infty}^{\infty} \overline{G'_+(\omega)}E^t_{r+}(\omega)\, d\omega\\
&= T_e(t) + \frac{1}{2\pi}\int_{-\infty}^{\infty} \overline{\widetilde{G}_+(\omega)}\dot{E}^t_+(\omega)\, d\omega.
\end{aligned} \tag{16.1.21}$$

The integral in these expressions must be real. This can be demonstrated by changing the integration variable from ω to $-\omega$. Using (16.1.17), relation $(16.1.21)_1$ can be shown to agree with $(16.1.21)_2$ with the aid of (16.1.8) and the property $\int_{-\infty}^{\infty} E^t_{r+}(\omega)d\omega = 0$, which follows from Cauchy's theorem (B.1.1).

The line of argument now followed is that described in Sect. 7.2.4, but with both forms given in (16.1.21). If we replace $\overline{G'_+(\omega)}$ in $(16.1.21)_1$ by $[\overline{G'_+(\omega)} + F(\omega)]$ where $F(\omega)$ is analytic on Ω^- and goes to zero at large frequencies at least as ω^{-1}, the relationship still holds. A similar statement applies to $(16.1.21)_2$. These follow by a simple application of Cauchy's theorem. In particular, we have

$$\begin{aligned}
T(t) &= T_e(t) + \frac{1}{2\pi}\int_{-\infty}^{\infty} [\overline{G'_+(\omega)} + \lambda_1 G'_+(\omega)]E^t_{r+}(\omega)d\omega\\
&= T_e(t) + \frac{1}{2\pi}\int_{-\infty}^{\infty} [\overline{\widetilde{G}_+(\omega)} + \lambda_2\widetilde{G}_+(\omega)]\dot{E}^t_+(\omega)d\omega,
\end{aligned} \tag{16.1.22}$$

where λ_1 and λ_2 are arbitrary complex constants. For $\lambda_1 = -1$ and $\lambda_2 = 1$, we have

$$\begin{aligned}
T(t) &= T_e(t) + \frac{1}{\pi i}\int_{-\infty}^{\infty} \frac{H(\omega)}{\omega}E^t_{r+}(\omega)d\omega\\
&= T_e(t) + \frac{1}{\pi}\int_{-\infty}^{\infty} \frac{H(\omega)}{\omega^2}\dot{E}^t_+(\omega)d\omega.
\end{aligned} \tag{16.1.23}$$

Using the same type of argument as for (16.1.22), one can also replace $(16.1.21)_2$ by

$$T(t) = T_e(t) + \frac{1}{2\pi} \int_{-\infty}^{\infty} \overline{\overline{G}_+(\omega)} \dot{E}^t_F(\omega)\, d\omega,$$
$$E^t_F(\omega) = E^t_+(\omega) + E^t_-(\omega), \tag{16.1.24}$$

where $E^t_-(\omega)$ is the Fourier transform of a strain continuation into the future, $E^t(s), \quad s \in \mathbb{R}^{--}$. This quantity is arbitrary, subject to the requirement that the integral in (16.1.24) exists.

16.1.2 Defining Properties of Free Energies

The scalar version of the defining properties of a free energy, given in the general tensor case in Sect. 5.1.1, is now stated, in a somewhat altered form.

We denote a particular free energy at time t by $\psi(t) = \tilde{\psi}(E^t, E(t))$, where $\tilde{\psi}$ is understood to be a functional of E^t and a function of $E(t)$. It is assumed that ψ is differentiable with respect to $E(t)$ and Fréchet differentiable with respect to E^t. Then,

P1

$$\frac{\partial}{\partial E(t)} \tilde{\psi}(E^t, E(t)) = \frac{\partial}{\partial E(t)} \psi(t) = T(t). \tag{16.1.25}$$

P2 For any history E^t,

$$\tilde{\psi}(E^t, E(t)) \geq \tilde{\phi}(E(t)) \text{ or } \psi(t) \geq \phi(t), \tag{16.1.26}$$

where $\phi(t)$ is the equilibrium value of the free energy $\psi(t)$, defined as

$$\tilde{\phi}(E(t)) = \phi(t) = \tilde{\psi}(E^t, E(t)), \quad E^t(s) = E(t) \;\forall\; s \in \mathbb{R}^+. \tag{16.1.27}$$

Thus, equality in (16.1.26) is achieved for equilibrium conditions. This property combines P2 and P3 of Sect. 5.1.1.

P3 For any $(E^t, E(t))$, we have the first law (balance of energy)

$$\dot{\psi}(t) + D(t) = T(t)\dot{E}(t), \tag{16.1.28}$$

where $D(t) \geq 0$ is the rate of dissipation of energy associated with $\psi(t)$. This non-negativity requirement on $D(t)$ is an expression of the second law.

The basic condition is P3. Relations P1 and P2 follow from P3 [67].

Integrating (16.1.28) over $(-\infty, t]$ yields that

$$\psi(t) + \mathcal{D}(t) = W(t), \qquad \mathcal{D}(t) \geq 0, \tag{16.1.29}$$

where

$$W(t) = \int_{-\infty}^{t} T(u)\dot{E}(u)du, \qquad \mathcal{D}(t) = \int_{-\infty}^{t} D(u)du \geq 0. \tag{16.1.30}$$

We assume that these integrals are finite. The quantity $W(t)$ is the work function, while $\mathcal{D}(t)$ is the total dissipation resulting from the entire history of deformation of the body.

The quantity $T_e(t)$ in (16.1.3) is given by

$$T_e(t) = \frac{\partial \phi(t)}{\partial E(t)}. \tag{16.1.31}$$

For a completely linear material,

$$\tilde{\phi}(E(t)) = \phi_l(t) = \frac{1}{2} G_\infty E^2(t). \tag{16.1.32}$$

Observe that, in this case, (16.1.31) gives (16.1.6).

16.2 Materials with Only Isolated Singularities

Consider the kth time-independent eigenspace of $\mathbb{L}'$ (Sect. 7.1.5), where the eigenvalue in the frequency domain is given by

$$L'_{k+}(\omega) = \sum_{l=1}^{S_k} \sum_{j=1}^{m_{lk}} \frac{r_{ljk}}{(\zeta_{lk} - \omega)^j}, \quad \zeta_{lk} \in \Omega^{(+)}. \tag{16.2.1}$$

To simplify notation (see Remark 16.0.1), the subscript k will be omitted, henceforth, so that we are dealing essentially with a scalar problem. Relation (16.2.1) corresponds to the time-domain form

$$L'(s) = G'(s) = \sum_{l=1}^{S} g_l(s) e^{i\zeta_l s}, \qquad g_l(s) = \sum_{j=0}^{m_l - 1} c_j^l s^j, \qquad c_{j-1}^l = \frac{r_{lj}(i)^j}{(j-1)!}, \tag{16.2.2}$$

where S is the number of points at which singularities occur and m_l is the highest order singularity at the point ζ_l. We have used (7.3.3). Individual values of r_{lj} may of course be zero, though not for $j = m_l$, which defines the highest power at a given singular point.

In order to satisfy (7.3.4), we write (16.2.1) in the form

$$G'_+(\omega) = \sum_{l=1}^{I} \sum_{j=1}^{m_l} \frac{g_{lj}}{(\zeta_l - \omega)^j} + \sum_{l=I+1}^{T} \sum_{j=1}^{m_l} \left\{ \frac{r_{lj}}{(\zeta_l - \omega)^j} + \frac{\overline{r}_{lj}}{(\overline{\zeta}_l + \omega)^j} \right\}, \qquad \omega \in \mathbb{R},$$

$$\zeta_l = i\alpha_l,\ \alpha_l \in \mathbb{R}^{++},\ l = 1, 2, \ldots, I,\ g_{lj} = (-1)^j \overline{g}_{lj},$$

$$\zeta_l \in \Omega^{(+)}, \quad l = I+1, \ldots, T, \quad r_{lj} = (-1)^j \overline{r}_{lj}. \tag{16.2.3}$$

Thus, there are I singularities on the imaginary axis and $M = T - I$ pairs of singularities at $(\zeta_l, -\overline{\zeta}_l)$, $l = I+1, \ldots, T$, so that $S = I + 2M$.

We consider $H(\omega)$ (see (16.1.13)) for $\omega \in \mathbb{R}$, given as follows:

$$H(\omega) = H_\infty \frac{\prod_{l=1}^{Z} [(\omega - \eta_l)(\omega - \overline{\eta}_l)]^{n_l}}{\prod_{l=1}^{S} \left[(\omega - \zeta_l)(\omega - \overline{\zeta}_l)\right]^{m_l}} > 0 \tag{16.2.4}$$

with factors

$$H_+(\omega) = h_\infty \frac{\prod_{l=1}^{Z}(\omega - \eta_l)^{n_l}}{\prod_{l=1}^{S}(\omega - \zeta_l)^{m_l}},$$

$$\zeta_l \in \Omega^{(+)},\ n_1 = 1,\ \eta_1 = 0, \quad \eta_l \in \Omega^{(+)},\ l = 2, 3, \ldots, Z, \tag{16.2.5}$$

$$H_-(\omega) = \overline{H}_+(\omega),$$

where h_∞ is real and $h_\infty^2 = H_\infty$. The quantity Z is the number of zeros of the function H in $\Omega^{(+)}$. Also,

$$\sum_{l=1}^{Z} n_l = \sum_{l=1}^{S} m_l = q.$$

The reality condition (7.3.4) implies that $G'_s(-\omega) = -G'_s(\omega)$ or $H(-\omega) = H(\omega)$ for real ω, which in turn is guaranteed if for each ζ_l and η_i off the imaginary axis, there exist ζ_n and η_j such that

$$\overline{\zeta}_l = -\zeta_n, \quad m_l = m_n, \qquad \overline{\eta}_i = -\eta_j, \quad n_i = n_j. \tag{16.2.6}$$

This is a statement that each singularity and zero in a given half-plane must have a corresponding singularity and zero each of which is a reflection of the first in the imaginary axis. Note that

$$(\omega - \zeta_l)(\omega - \overline{\zeta}_l)(\omega - \zeta_n)(\omega - \overline{\zeta}_n) = (\omega^2 + |\zeta_l|^2)^2 - 4\omega^2(\mathrm{Re}\zeta_l)^2,$$
$$(\omega - \eta_i)(\omega - \overline{\eta}_i)(\omega - \eta_j)(\omega - \overline{\eta}_j) = (\omega^2 + |\eta_i|^2)^2 - 4\omega^2(\mathrm{Re}\eta_i)^2,$$

if (16.2.6) holds. If a singularity or zero is on the imaginary axis, then property (16.2.6) is satisfied for $i = j$ and $(\omega - \zeta_l)(\omega - \overline{\zeta}_l) = \omega^2 + |\zeta_l|^2$, with a similar result for zeros. These relations ensure that $H(\omega)$ is an even function. Also, observe that

$$(\omega - \overline{\zeta}_l)(\omega - \overline{\zeta}_n) = (\omega - \overline{\zeta}_l)(\omega + \zeta_l) = (-\omega - \zeta_l)(-\omega - \zeta_n),$$

with a similar property applying to zeros. These yield the property

$$\overline{H}_\pm(\omega) = H_\pm(-\omega).$$

The constitutive equation of a material is determined by H and the parameters of the term without memory. In particular, for completely linear materials, these parameters reduce to one, namely G_∞.

The special case of a material where $n_l = m_l = 1$ for $l = 1, 2, \ldots, Z$ and $S = Z$ is of particular interest. It will be described as a simple pole material. In this context, we put

$$S = Z = n$$

and

$$G'(s) = \sum_{l=1}^{n} g_l e^{i\zeta_l s}, \qquad G'_+(\omega) = \sum_{l=1}^{n} \frac{ig_l}{\zeta_l - w}. \tag{16.2.7}$$

The quantity $G'_+(\omega)$ can be expressed in the form (16.2.2), with $m_l = 1$ and $g_l = c_0^l$.

Strictly, the fact that all poles are simple does not necessarily imply that the zeros will also be simple. However, there are always materials with only simple zeros as close as desired to a material with higher powers, by virtue of continuity arguments. For such simple pole materials,

$$H(\omega) = H_\infty \frac{\prod_{l=1}^{n}(\omega - \eta_l)(\omega - \overline{\eta}_l)}{\prod_{l=1}^{n}(\omega - \zeta_l)(\omega - \overline{\zeta}_l)} > 0, \tag{16.2.8}$$

with factors

$$\begin{aligned} H_+(\omega) &= h_\infty \frac{\prod_{l=1}^{n}(\omega - \eta_l)}{\prod_{l=1}^{n}(\omega - \zeta_l)}, \\ \zeta_l \in \Omega^{(+)}, \ \eta_1 &= 0, \quad \eta_l \in \Omega^{(+)}, \ l = 2, 3, \ldots, n, \\ H_-(\omega) &= \overline{H}_+(\omega). \end{aligned} \tag{16.2.9}$$

Note that, starting with a simple pole material characterized by H in the form (16.2.8), we can modify this into the form describing a general material with isolated singularities, namely (16.2.4), by merging singularities and zeros. Thus, the simple pole case is completely general, at least in principle. A special case of a simple pole materials is a discrete-spectrum material, where the poles and zeros are on the positive imaginary axis. This category was used in various contexts in earlier chapters, in particular in Sect. 11.9.

16.3 Free Energies as Discrete Quadratic Forms

In this section, we divert briefly to discuss the most general form of free energy that is a functional of the minimal state, for materials with linear memory constitutive relations characterized in the frequency domain by simple poles. The free energy is expressed in terms of discrete quadratic forms of E^t_{r+}, evaluated at the various singularities.

From (16.2.7), replacing g_l with c^l, and the reality of G', we have

$$\begin{aligned} G'(s) &= \sum_{l=1}^{n} c^l e^{i\zeta_l s} = \sum_{l=1}^{n} \overline{c^l} e^{-i\overline{\zeta_l} s} \\ &= \frac{1}{2}\sum_{l=1}^{n}\left[c^l e^{i\zeta_l s} + \overline{c^l} e^{-i\overline{\zeta_l} s}\right]. \end{aligned}$$

Thus, the requirement that G' be real means that if $\text{Re}\zeta_l \neq 0$, there must also be a term with $e^{-i\overline{\zeta_l} s}$. If $\text{Re}\zeta_l = 0$, the two terms merge into one strictly decaying exponential. We obtain from (16.1.3), (16.1.16), and (16.1.17) that

$$
\begin{aligned}
T(t) &= T_e(t) + \sum_{l=1}^{n} c^l E^t_{r+}(-\zeta_l) \\
&= T_e(t) + \sum_{l=1}^{n} \overline{c^l} E^t_{r+}(\overline{\zeta_l}) \\
&= T_e(t) + \frac{1}{2}\sum_{l=1}^{n}\left[c^l E^t_{r+}(-\zeta_l) + \overline{c^l} E^t_{r+}(\overline{\zeta_l})\right],
\end{aligned}
\tag{16.3.1}
$$

$$
\overline{E^t_+(-\zeta_l)} = E^t_+(\overline{\zeta_l}), \qquad \frac{d}{dt}E^t_+(\zeta_l) = -i\zeta_l E^t_{r+}(\zeta_l).
$$

It follows from (7.4.3) that two states are equivalent if and only if the differences between the current strains and histories, denoted by $(E^t_d, E_d(t))$, have the following properties, for a scalar theory:

$$
E_d(t) = 0, \qquad \int_0^\infty G'(s+\tau)E^t_d(s)ds = 0, \quad \tau \in \mathbb{R}^+.
$$

Now,

$$
G'(s+\tau) = \sum_{l=1}^{n} c^l e^{i\zeta_l(s+\tau)}.
$$

The arbitrariness of the factor $e^{i\zeta_l\tau}$ allows us to put each individual term in the summation over l to zero. Therefore, the final conditions of equivalence are

$$
E_d(t) = 0, \qquad \int_0^\infty e^{i\zeta_l s}E^t_d(s)ds = E^t_{d+}(-\zeta_l) = 0, \qquad l = 1,2,\ldots,n. \tag{16.3.2}
$$

Remark 16.3.1. These are necessary and sufficient conditions for equivalence. Thus, the quantities

$$
E(t), \quad E^t_+(-\zeta_l), \quad l = 1,2,\ldots,n,
$$

uniquely define the minimal state for materials with only simple poles.

Define a vector $\mathbf{e}$ in $\mathbb{R}^n$, with components

$$
e_l(t) = i\zeta_l E^t_{r+}(-\zeta_l), \qquad \overline{e_l}(t) = -i\overline{\zeta_l}E^t_{r+}(\overline{\zeta_l}), \qquad l = 1,2,\ldots,n. \tag{16.3.3}
$$

Using (16.1.16), we see that

$$
\frac{\partial}{\partial E(t)}e_l = \frac{\partial}{\partial E(t)}\overline{e_l} = 1, \quad l = 1,2,\ldots,n. \tag{16.3.4}
$$

Consider the quantity

$$
\psi(t) = \phi(t) + \frac{1}{2}\mathbf{e}^*\mathbf{C}\mathbf{e} = \phi(t) + \frac{1}{2}\overline{\mathbf{e}}\cdot\mathbf{C}\mathbf{e}, \tag{16.3.5}
$$

where $\phi(t)$ is the equilibrium free energy and $\mathbf{C}$ is a Hermitian (see (A.2.9)) positive definite tensor with components C_{kl}, $k,l = 1,2,\ldots,n$. The quantity $\mathbf{e}^*$ is the Hermitian conjugate of the vector $\mathbf{e}$; in other words, if $\mathbf{e}$ is a column vector, $\mathbf{e}^* = \overline{\mathbf{e}}^\top$, which is a row vector. We write (16.3.5) in full as

$$\psi(t) = \phi(t) + \frac{1}{2}\sum_{k=1}^{n}\sum_{l=1}^{n}\overline{e_k}C_{kl}e_l.$$

The Hermitian property of C_{kl} gives the relation

$$\overline{C_{lk}} = C_{kl}. \tag{16.3.6}$$

Clearly, $\psi(t)$ has property P2, given by (16.1.26), of a free energy. For a stationary history, $E^t(s) = E(t)$, $s \in \mathbb{R}^+$, yielding for $\omega \in \Omega$ that $E^t_+(\omega) = E(t)/(i\omega)$ so that $e_l(t) = 0$, $l = 1, 2, \ldots, n$. Therefore, $\psi(t)$ obeys (16.1.27). To demonstrate property P1, given by (16.1.25), we first note that

$$\frac{\partial}{\partial E(t)}\psi(t) = T_e(t) + \frac{1}{2}\sum_{k=1}^{n}\sum_{l=1}^{n}C_{kl}e_l + \frac{1}{2}\sum_{k=1}^{n}\sum_{l=1}^{n}\overline{e_k}C_{kl}, \tag{16.3.7}$$

which follows from (16.3.4). The relationship (16.1.31) has been used. Comparing (16.3.7) and $(16.3.1)_3$, we see that

$$\begin{aligned}\sum_{k=1}^{n}C_{kl} &= -i\frac{c^l}{\zeta_l}, \quad l = 1, 2, \ldots, n,\\ \sum_{l=1}^{n}C_{kl} &= \sum_{l=1}^{n}\overline{C_{lk}} = i\frac{\overline{c^k}}{\overline{\zeta_k}}, \quad k = 1, 2, \ldots, n.\end{aligned} \tag{16.3.8}$$

Using $(16.1.18)_3$, we have

$$\dot{e}_l(t) = i\zeta_l e_l(t) + \dot{E}(t), \qquad \dot{\overline{e_l}}(t) = -i\overline{\zeta_l}\overline{e_l}(t) + \dot{E}(t). \tag{16.3.9}$$

Referring to (16.3.1), one can show that

$$\begin{aligned}\dot{\psi}(t) + D(t) &= T(t)\dot{E}(t), \qquad D(t) = \frac{1}{2}\overline{\mathbf{e}}\cdot\mathbf{\Gamma}\mathbf{e},\\ \Gamma_{kl} &= i(\overline{\zeta_k} - \zeta_l)C_{nl}, \qquad k, l = 1, 2, \ldots, n,\end{aligned} \tag{16.3.10}$$

where Γ_{nl} are the elements of the matrix $\mathbf{\Gamma}$. The final requirement, given after (16.1.28), is that $D(t) \geq 0$, so that $\mathbf{\Gamma}$ will be assumed to be at least semidefinite. Thus, under these constraints, the quantity $\psi(t)$, given by (16.3.5), is a free energy.

Remark 16.3.2. It follows from (16.3.2), or more particularly Remark 16.3.1, that free energies of the form (16.3.5), for a given history, are functions of the minimal state. They manifestly obey (7.4.7) and therefore (7.4.6).

Remark 16.3.3. The set of quadratic free energies expressible in the form (16.3.5), where $\mathbf{C}$ has the required positivity properties, is identical to the set of free energies associated with the equivalence class containing $(\mathbf{e}, E(t))$, which are functions of the minimal state. This is clear from Remark 16.3.1. It is a convex set, by virtue of Proposition 4.1.7.

16.3.1 Discrete-Spectrum Materials

For the discrete-spectrum case discussed in, for example, Sect. 11.9, these formulas simplify significantly. The results are obtained by replacing the singularity positions ζ_l by $i\alpha_l$, $l = 1, 2, \ldots, n$, where $\alpha_l \in \mathbb{R}^{++}$. Such materials have been studied by Graffi and Fabrizio [175–177] and Fabrizio et al. [105].

Consider a material with relaxation function of the form (11.9.1). From this relation and $(16.1.3)_1$, we have that

$$T(t) = T_e(t) + \sum_{i=1}^{n} G_i \dot{E}_+^t(-i\alpha_i). \tag{16.3.11}$$

As previously, two states are equivalent if and only if the difference between the states has the properties

$$E_d(t) = 0, \qquad E_d^{(i)}(t) = \int_0^\infty e^{-\alpha_i s} E_d^t(s) ds = E_{d+}^t(-i\alpha_i) = 0,$$
$$i = 1, 2, \ldots, n, \qquad \alpha_i \in \mathbb{R}^+.$$

The vector $\mathbf{e}$ in $\mathbb{R}^n$, defined by (16.3.3), becomes

$$e_i(t) = -\alpha_i E_{r+}^t(-i\alpha_i) = E(t) - \alpha_i E_+^t(-i\alpha_i) = \frac{d}{dt} E_+^t(-i\alpha_i), \quad i = 1, 2, \ldots, n. \tag{16.3.12}$$

The quantities $E_+^t(-i\alpha_i)$ are real. Consider the function

$$\psi(t) = \phi(t) + \frac{1}{2} \mathbf{e} \cdot \mathbf{C}\mathbf{e}, \tag{16.3.13}$$

where $\phi(t)$ is the equilibrium free energy and $\mathbf{C}$ is a symmetric positive definite matrix with components C_{ij}, $i, j = 1, 2, \ldots, n$. The arguments after (16.3.6) apply, with minor changes, to this special case. Condition (16.3.8) reduces to

$$\sum_{j=1}^{n} C_{ij} = G_i, \quad i = 1, 2, \ldots, n, \tag{16.3.14}$$

where the quantities G_i are the coefficients occurring in (11.9.1). With the aid of (16.3.12) or by invoking (16.3.9), we obtain

$$\dot{e}_i(t) = \dot{E}(t) - \alpha_i e_i(t), \quad i = 1, 2, \ldots, n.$$

It is easy to show that

$$\begin{aligned} &\dot{\psi}(t) + D(t) = T(t)\dot{E}(t), \\ &D(t) = \frac{1}{2} \mathbf{e} \cdot \mathbf{\Gamma} \mathbf{e}, \qquad \Gamma_{ij} = (\alpha_i + \alpha_j) C_{ij}, \end{aligned} \tag{16.3.15}$$

where Γ_{ij} are the elements of the matrix $\mathbf{\Gamma}$. The requirement in P3 that $D(t) \geq 0$ implies that $\mathbf{\Gamma}$ must be at least positive semidefinite.

Relation $(11.9.14)_1$ yields an explicit form for $\mathbf{C}$ corresponding to the minimum free energy. Comparing this relation with $(11.9.16)_1$, we see that $(16.3.15)_3$ holds.

One can show that for the Dill free energy, $\mathbf{C}$ is diagonal [175].

16.4 The Minimum and Related Free Energies

We return to the developments described in Sect. 16.2.

A much larger class of factorizations of H is now considered. These are obtained by interchanging the corresponding zeros of H_+ and H_- (i.e., the complex conjugates of each other), excluding of course the simple zero at the origin. Such interchanges leave the singularity structure unchanged. There are

$$N = 2^{n-1}$$

distinct factorizations of this kind, which we distinguish by the label $f = 1, 2, \ldots, N$. In the case $f = 1$, no zeros are interchanged, and for $f = N$, all zeros are interchanged. For real ω, we have

$$\begin{aligned} &H(\omega) = H_+^f(\omega)H_-^f(\omega), \\ &H_+^f(\omega) = h_\infty \frac{\prod_{l=1}^n (\omega - \eta_l)^{a_l^f} (\omega - \overline{\eta}_l)^{b_l^f}}{\prod_{l=1}^n (\omega - \zeta_l)}, \\ &H_-^f(\omega) = \overline{H}_+^f(\omega), \\ &a_l^f + b_l^f = 1,\ l = 2, 3, \ldots, n,\ a_l^f,\ b_l^f \geq 0. \end{aligned} \tag{16.4.1}$$

The integers a_l^f and b_l^f can take value 0 or 1. Note that the permutation f is completely specified by the $(n-1)$-dimensional vector a_l^f (or b_l^f), $l = 2, 3, \ldots, n$.

We can consider the different factorizations in the following way. The quantity

$$X_l(\omega) = \frac{\omega - \overline{\eta}_l}{\omega - \eta_l} \tag{16.4.2}$$

for real ω is a phase transformation. Then,

$$H_+^f(\omega) = \left\{ \prod_{l=2}^n X_l^{b_l^f}(\omega) \right\} H_+(\omega).$$

The corresponding H_-^f is obtained by taking the complex conjugate.

Note that if one of a pair of zeros related by (16.2.6) is interchanged but not the other, then we no longer have the property

$$\overline{H}_\pm^f(\omega) = H_\pm^f(-\omega).$$

Remark 16.4.1. It will now be shown explicitly that to each factorization of H, there corresponds a free energy, and moreover, this free energy is a functional of the minimal state.

Noting (11.2.8) and (11.2.9), we write

$$P^{(ft)}(\omega) = H_-^f(\omega)E_{r+}^t(\omega) = p_-^{(ft)}(\omega) - p_+^{(ft)}(\omega), \tag{16.4.3}$$

where $p_-^{(ft)}(\omega)$ is analytic in Ω^+, going to zero at large ω as ω^{-1}, while $p_+^{(ft)}(\omega)$ is analytic in Ω^- with similar large-ω behavior. They are given by (see (11.2.10))

$$\begin{aligned} p_\pm^{(ft)}(\omega) &= \frac{1}{2\pi i}\int_{-\infty}^{\infty}\frac{P^{(ft)}(\omega')}{\omega'-\omega^\mp}d\omega' \\ &= \frac{1}{2\pi i}\int_{-\infty}^{\infty}\frac{H_-^f(\omega')E_{r+}^t(\omega')}{\omega'-\omega^\mp}d\omega'. \end{aligned} \tag{16.4.4}$$

The analyticity of these quantities on $\mathbb{R}$ follows by the argument leading up to Remark B.2.2. They are defined over the entire complex plane by analytic continuation, except at singularities (and in regions blocked by branch cuts, which are excluded in the present context, unless otherwise stated).

We can show, using (16.1.17), that (*cf.* (11.2.22))

$$\begin{aligned} \frac{d}{dt}p_+^{(ft)}(\omega) &= -i\omega p_+^{(ft)}(\omega) - K_f(t), \\ \frac{d}{dt}p_-^{(ft)}(\omega) &= -i\omega p_-^{(ft)}(\omega) - K_f(t) - \frac{H_-^f(\omega)\dot{E}(t)}{i\omega}, \end{aligned} \tag{16.4.5}$$

where

$$K_f(t) = \frac{1}{2\pi}\int_{-\infty}^{\infty} H_-^f(\omega)E_{r+}^t(\omega)d\omega = \lim_{\omega\to\infty}[-i\omega p_-^{ft}(\omega)]. \tag{16.4.6}$$

Also, we have

$$\begin{aligned} \lim_{|\omega|\to\infty}\omega p_\pm^{(ft)}(\omega) &= iK_f(t), \\ \frac{1}{2\pi}\int_{-\infty}^{\infty}p_\pm^{(ft)}(\omega)d\omega &= \mp\frac{1}{2}K_f(t), \end{aligned} \tag{16.4.7}$$

by similar arguments to those yielding (11.2.24). Finally, we have (*cf.* (11.2.36) and (11.2.37))

$$\frac{\partial p_-^{(ft)}(\omega)}{\partial E(t)} = -\frac{1}{2\pi i}\int_{-\infty}^{\infty}\frac{H_-^f(\omega_1)}{i\omega_1^-(\omega_1-\omega^+)}d\omega_1 = -\frac{H_-^f(\omega)}{i\omega} \tag{16.4.8}$$

and

$$\frac{\partial \overline{p_-^{(ft)}}(\omega)}{\partial E(t)} = \frac{H_+^f(\omega)}{i\omega}. \tag{16.4.9}$$

Let us define the quantity

$$\psi_f(t) = \phi(t) + \int_{-\infty}^{\infty}|p_-^{(ft)}(\omega)|^2 d\omega, \tag{16.4.10}$$

which will be shown to be the free energy corresponding to the factorization (16.4.1). The factorizations will be assigned a code f in such a way that $f = 1$ corresponds to the minimum free energy (no exchange of zeros) and $f = N$ to the maximum free energy, which is discussed in detail in Sect. 16.5. It is shown in Sect. 16.6 that the total dissipation is given by

$$\mathfrak{D}_f(t) = \frac{1}{2\pi}\int_{-\infty}^{\infty} |p_+^{ft}(\omega)|^2\, d\omega, \tag{16.4.11}$$

and the associated rate of dissipation has the form

$$D_f(t) = |K_f(t)|^2, \tag{16.4.12}$$

which can be proved by differentiating (16.4.11), using (16.4.5) and (16.4.7).

Having introduced in (16.4.1) a factorization such that the zeros of $H_\pm^f$ are not necessarily in $\Omega^\pm$, we now generalize to allow the possibility of non-isolated (branch-cut) singularities in H along with the isolated singularities and zeros. After factorization, of course, the branch-cut singularities in $H_\pm^f$ must be in $\Omega^{(\pm)}$, respectively.

We now show that $p_-^{(\mathbf{f}t)}$ is a functional of the minimal state.

Lemma 16.4.2. *For every material with linear memory, a given history E^t is equivalent to the zero history if and only if the quantity $p_-^{(ft)}$ related to E^t, by* (16.4.4), *is such that*

$$p_-^{(ft)}(\omega) = 0 \qquad \forall \omega \in \mathbb{R}.$$

Proof. We wish to show that the quantity $F^t(\tau)$, $\tau \geq 0$, given by (11.3.1) and (11.3.3) for the general tensor case, vanishes if and only if $p_-^{(ft)}(\omega) = 0$ for all $\omega \in \mathbb{R}$. The steps leading from (11.3.4) to (11.3.6) go through for any factorization of H, so we obtain

$$F^t(s) = \frac{i}{\pi}\int_{-\infty}^{\infty} \frac{H_+^f(\omega)}{\omega} p_-^{(ft)}(\omega) e^{-i\omega s}\, d\omega. \tag{16.4.13}$$

This gives immediately that F^t vanishes if $p_-^{(ft)}$ is zero for all $\omega \in \mathbb{R}$, which is one of the conclusions sought. We now wish to show that if $F^t(s) = 0$, $s \in \mathbb{R}^+$, then $p_-^{(ft)}$ vanishes. Noting (11.3.8), we see that F^t vanishing on $\mathbb{R}^+$ implies that

$$\frac{H_+^f(\omega)}{\omega} p_-^{(ft)}(\omega)$$

is analytic on $\Omega^{(-)}$. This can be true only if $H_+^f(\omega)/\omega$ vanishes at each of the isolated singularities of $p_-^{(ft)}$ in $\Omega^{(-)}$. Thus, $H_+^f(\omega)$ must vanish at these points. For each cut-type singularity in $p_-^{(ft)}$, we must have a compensating singularity of the same kind in $H_+^f(\omega)/\omega$. Since $H_+^f(\omega)/\omega$ is analytic in $\Omega^{(-)}$, $p_-^{(ft)}$ cannot have cuts in this half-plane. As for isolated singularities, the case that no permutation of zeros has taken place ($f = 1$) is covered by Theorem 11.3.1.

If permutations of zeros are allowed, then H_+^f has zeros in $\Omega^{(-)}$. Consider the definition of $p_-^{(ft)}$, given by (16.4.4). Evaluating the integral by closing the contour on $\Omega^{(-)}$, referring to $(\text{B.1.15})_3$, shows that $p_-^{(ft)}$ may have singularities in $\Omega^{(-)}$, where H_-^f has singularities, and these must be isolated, as concluded above. If a zero in H_+^f cancels a singularity in H_-^f, then the zero and the singularity would cancel in H. In other words, there would be no singularity or zero to begin with. Therefore, the only option is that $p_-^{(ft)}$ is free of singularities in $\Omega^{(-)}$ and, by Liouville's theorem, must be zero. □

If $p_-^{(ft)}$ is a function of the minimal state, then in general $p_+^{(ft)} = p_-^{(ft)} - H_-^f E_{r+}^t$ will not be, because of the occurrence of the transformed history, which will not necessarily vanish when the history is equivalent to the zero history.

Observe that by applying the argument that yields (16.4.13) (see also (11.2.38)) to $(16.1.23)_1$, using (16.4.3), the quantity T, at time t, can be written as

$$T(t) = T_e(t) + \frac{1}{\pi i}\int_{-\infty}^{\infty} \frac{H_+^f(\omega)}{\omega} p_-^{(ft)}(\omega)d\omega \tag{16.4.14}$$

for each permutation f. All of these are of course equivalent.

These various formulae apply in particular to the case where no exchange of zeros takes place, which is denoted by $f = 1$. In this case, the formulae in fact apply to all materials, not just those with isolated singularities. This may be seen from Chaps. 11–14.

We can write $\psi_f(t)$ in the form [160] (*cf.* results in Sect. 11.2.5)

$$\psi_f(t) = \phi(t) + \frac{i}{4\pi^2}\int_{-\infty}^{\infty}\int_{-\infty}^{\infty} \frac{\overline{E_{r+}^t}(\omega_1)H_+^f(\omega_1)H_-^f(\omega_2)E_{r+}^t(\omega_2)}{\omega_1^+ - \omega_2^-} d\omega_1 d\omega_2. \tag{16.4.15}$$

The total dissipation, given by (16.4.11), can be shown by similar manipulations to have the form

$$\mathcal{D}_f(t) = -\frac{i}{4\pi^2}\int_{-\infty}^{\infty}\int_{-\infty}^{\infty} \frac{\overline{E}_{r+}^t(\omega_1)H_+^f(\omega_1)H_-^f(\omega_2)E_{r+}^t(\omega_2)}{\omega_1^- - \omega_2^+} d\omega_1 d\omega_2, \tag{16.4.16}$$

while $D_f(t)$, given by (16.4.12), can be expressed as

$$D_f(t) = \frac{1}{4\pi^2}\int_{-\infty}^{\infty} \overline{E_{r+}^t}(\omega_1)H_+^f(\omega_1)H_-^f(\omega_2)E_{r+}^t(\omega_2)d\omega_1 d\omega_2. \tag{16.4.17}$$

All the free energies (16.4.15) are on the boundary of the convex set of free energies associated with a given state of the material, since the rate of dissipation given by (16.4.17) is a nonnegative rather than a positive definite functional. Also, they are all functionals of the minimal state by virtue of Lemma 16.4.2. The factorization (16.2.9) yields the minimum free energy $\psi_m(t)$. Each exchange of zeros, starting from these factors, will be shown to yield a free energy which is greater than or equal to the previous quantity.

Note that there are several (indeed many, for a large number of isolated singularities) different zero exchange pathways leading from the minimum to the maximum free energy.

We can simplify (16.4.1) for discrete-spectrum materials. Let us introduce an n-dimensional vector with components ϵ_i^f, $i = 1, 2, \ldots, n$, where each ϵ_i^f can take values ± 1. We define $\rho_i^f = \epsilon_i^f \gamma_i$ and write

$$H_+^f(\omega) = h_\infty \prod_{i=1}^{n}\left\{\frac{\omega - i\rho_i^f}{\omega - i\alpha_i}\right\}, \qquad H_-^f(\omega) = h_\infty \prod_{i=1}^{n}\left\{\frac{\omega + i\rho_i^f}{\omega + i\alpha_i}\right\}. \tag{16.4.18}$$

The interchange of zeros means switching a given γ_i to $-\gamma_i$ in both $H_+(\omega)$ and $H_-(\omega)$. If no zeros are interchanged, we have the minimum free energy, for which the factors are given by (11.9.5). The case where all the zeros are interchanged [110] is labeled $f = N$. The resulting factors are given by

$$H_+^N(\omega) = h_\infty \prod_{i=1}^{n} \left\{ \frac{\omega + i\gamma_i}{\omega - i\alpha_i} \right\}, \qquad H_-^N(\omega) = h_\infty \prod_{i=1}^{n} \left\{ \frac{\omega - i\gamma_i}{\omega + i\alpha_i} \right\}. \tag{16.4.19}$$

16.5 Equivalent States and the Maximum Free Energy

We consider the maximum free energy of a given state, defined as the minimum energy required to achieve this state (for example, [75, 76, 101, 104, 110, 266]) from an initial state (see Theorem 4.2.12), where this initial state is taken to be the zero state.

If the minimal state σ_R is a singleton, then in the present context, there is no nontrivial set over which the minimization can take place. There is only one state, which defined by the given history, and the required work is the work function.

If, on the other hand, σ_R is not a singleton, then there will be a state $\sigma_{\min} \in \sigma_R$ that minimizes the required work, and this minimal work will be less than the work function.

Thus, there are two distinct cases here: (1) the maximum free energy is equal to the work function; this occurs when the set of minimal states, defined by the equivalence conditions (7.4.3), is a singleton, and (2) the maximum free energy is less than the work function, which is true in the case of materials for which the space of minimal states contains more than one member.

There are problems with identifying the work function as the maximum free energy, which are discussed in Sect. 18.2. These can be avoided by identifying the maximum free energy as the limit of a sequence.

A difference between isolated and branch-cut singularities is that the former always have infinite behavior associated with them, while the latter are characterized by generally finite discontinuities, though in fact, infinities may occur at branch points and indeed on the cut. However, there is the following clear-cut distinction, which is important in the present context.

Remark 16.5.1. If a function $F : \Omega \mapsto \mathbb{R}$ has isolated singularities at a set of points, then $1/F$ will have zeros at these points, while if F has a branch cut between two branch points, then $1/F$ will also have a branch cut between these two branch points. The converses of these two statements also hold.

We consider a material characterized by isolated singularities or branch cuts or both and a factorization in which zeros, if present, may be interchanged. The notation of $(16.4.1)_1$ will be used.

16.5.1 Minimal States

The concept of a minimal state, as discussed in Sect. 7.4, can be expressed for the scalar model as follows: two viscoelastic states $(E_1^t, E_1(t))$, $(E_2^t, E_2(t))$ are equivalent or in the same equivalence class or minimal state if

$$
\begin{gathered}
E_1(t) = E_2(t), \\
\int_0^\infty G'(s+\tau)\left[E_1^t(s) - E_2^t(s)\right] ds = I^t(\tau, E_1^t) - I^t(\tau, E_2^t) = 0 \;\; \forall\, \tau \geq 0, \\
I^t(\tau, E^t) = \int_0^\infty G'(s+\tau) E_r^t(s) ds = \int_0^\infty \widetilde{G}(s+\tau)\dot{E}^t(s) ds = I^t(\tau).
\end{gathered}
\tag{16.5.1}
$$

The abbreviated notation $I^t(\tau)$ will be used henceforth. Note the property

$$
\lim_{\tau \to \infty} I^t(\tau) = 0. \tag{16.5.2}
$$

It follows from $(16.1.3)_1$ and (16.5.1) that

$$
I^t(0) = T(t) - T_e(t). \tag{16.5.3}
$$

A functional of $(E^t, E(t))$ that yields the same value for all members of the same minimal state is referred to as a functional of the minimal state, or a minimal state variable, as outlined in Sect. 7.4. We will adopt the abbreviation FMS to indicate such a functional. The quantity $I^t(\tau)$ is an FMS, in fact, the defining example of an FMS.

For discrete-spectrum materials,

$$
I^t(\tau) = \sum_{i=1}^{n} G_i \dot{E}_+^t(-i\alpha_i) e^{-\alpha_i \tau},
$$

which is consistent with (16.3.11) and (16.5.3). Various formulae relating to the functional I^t are given in [163].

The condition that minimal states are nonsingleton is that the integral equation

$$
\int_0^\infty G'(s+\tau) E_d^t(s) ds = 0, \qquad \tau \in \mathbb{R}^+, \tag{16.5.4}
$$

for $E_d^t(s) = E_1^t(s) - E_2^t(s)$, has nonzero solutions. The other requirement $(16.5.1)_1$ will be enforced below (see (16.5.14)). Putting $E_d^t(s) = 0$, $s \in \mathbb{R}^-$, and $\tau = -u$, we can write (16.5.4) as

$$
\int_{-\infty}^{\infty} \frac{\partial}{\partial u} G(|u-s|) E_d^t(s) ds = 0, \qquad u \in \mathbb{R}^-. \tag{16.5.5}
$$

This is a Wiener–Hopf equation, which can be solved by a standard technique. We put

$$\int_{-\infty}^{\infty} \frac{\partial}{\partial u} G(|u-s|) E_d^t(s) ds = \begin{cases} J(u), & u \in \mathbb{R}^{++} \\ 0, & u \in \mathbb{R}^{-}, \end{cases} \tag{16.5.6}$$

where $J(u)$ is a quantity to be determined. Taking the Fourier transform of both sides, we obtain, with the aid of the convolution theorem and (16.1.15),

$$\frac{2i}{\omega} H(\omega) E_{d+}^t(\omega) = J_+(\omega). \tag{16.5.7}$$

Using $(16.4.1)_1$ and (16.4.3), we can write (16.5.7) in the form

$$\frac{2i}{\omega} \left\{ H_+^f(\omega) \left[p_{d-}^{ft}(\omega) - p_{d+}^{ft}(\omega) \right] \right\} = J_+(\omega), \tag{16.5.8}$$

where the subscript d implies that E_{d+}^t is used in (16.4.3) and (16.4.4). The value of the superscript f will be assigned below. Because $p_-^{ft}(\omega)$ is an FMS (Lemma 16.4.2), we have

$$p_{d-}^{ft}(\omega) = 0. \tag{16.5.9}$$

It then follows from (16.5.8) that

$$p_{d+}^{ft}(\omega) = -\frac{\omega}{2i} \frac{J_+(\omega)}{H_+^f(\omega)}. \tag{16.5.10}$$

Using (16.5.10) in (16.5.7), we obtain

$$H(\omega) E_{d+}^t(\omega) = -H_+^f(\omega) p_{d+}^{ft}(\omega), \tag{16.5.11}$$

or

$$E_{d+}^t(\omega) = -\frac{p_{d+}^{ft}(\omega)}{H_-^f(\omega)}. \tag{16.5.12}$$

Now, E_{d+}^t, if it is nonzero, must be analytic in Ω^-. Therefore, from Remark 16.5.1, H_-^f can have no branch-cut singularities, and we adopt the form $(16.4.1)_2$ for $H_\pm^f$. It follows from $(16.4.1)_1$ and $(16.4.1)_3$ that H_+^f and H have the same property. Thus, we have the following result.

Proposition 16.5.2. *For a material such that G'_+ has no essential singularities, the set of minimal states has more than one member only if G'_+ possesses no branch-cut singularities.*

In other words, G'_+ can have only isolated singularities.

Also from (16.5.12), the zeros of $H_-^f(\omega)$ must be in $\Omega^{(+)}$, which means that the factorization must be associated with $f = N$, namely where all the zeros are interchanged.

Thus, if we can find a quantity $E_{d+}^t(\omega)$ which satisfies (16.5.9), it satisfies (16.5.11) and (16.5.12) by virtue of (16.4.4), applied to this history difference. Relation (16.5.11) is equivalent to (16.5.7), with $J_+(\omega)$ determined by (16.5.10). Therefore, a solution to (16.5.6) or (16.5.5) is provided by any choice of $E_d^t(s)$ where the corresponding $E_{d+}^t(\omega)$ satisfies (16.5.9). Now, from (16.4.4),

$$p_{d-}^{Nt}(\omega) = \frac{1}{2\pi i} \int_{-\infty}^{\infty} \frac{H_{-}^{N}(\omega')E_{d+}^{t}(\omega')}{\omega' - \omega^{+}} d\omega' = 0. \tag{16.5.13}$$

If there are non-isolated singularities in the material, the only solution is the trivial one, $E_{d+}^{t}(\omega) = 0$. Thus, we can focus on the case of a material with only isolated singularities.

16.5.1.1 Explicit Examples of Minimal States

The simplifying assumption will now be made that $E_{d+}^{t}(\omega)$ is a rational function. More generally, it could also have branch cuts in $\Omega^{(+)}$. We confine the discussion to discrete-spectrum materials.

At large ω, we must have

$$E_{d+}^{t}(\omega) \sim \frac{1}{\omega^2}, \tag{16.5.14}$$

by virtue of (16.1.19) and $(16.5.1)_1$. If the zeros of $E_{d+}^{t}(\omega)$ cancel the poles in $H_{-}^{N}(\omega)$, given by $(16.4.19)_2$, then, by taking the contour around $\Omega^{(-)}$, we see that (16.5.13) is obeyed. Thus, nontrivial solutions to (16.5.7) are given by

$$E_{d+}^{t}(\omega) = \frac{E_0(t)}{\omega - i\chi_0} \prod_{j=1}^{n} \left\{ \frac{\omega + i\alpha_j}{\omega - i\chi_j} \right\} \frac{1}{\omega - i\chi_{n+1}}, \tag{16.5.15}$$

where the constants χ_i, $i = 0, 1, \ldots, n+1$, indicate the positions of singularities on the imaginary axis in $\Omega^{(+)}$. These are arbitrary positive quantities, though they may include some or all of the zeros of $H_{-}^{N}(\omega)$, given by $(16.4.19)_2$, if these are not cancelled by some other means in (16.5.12). The factor $E_0(t)$, which determines the time dependence of $E_{d+}^{t}(\omega)$, is also arbitrary. We can write (16.5.15) in the form

$$\begin{aligned}
E_{d+}^{t}(\omega) &= -iE_0(t) \sum_{i=0}^{n+1} \frac{A_i}{\omega - i\chi_i}, \\
A_i &= \frac{\chi_i + \alpha_i}{\chi_i - \chi_0} \prod_{\substack{j=1 \\ j \neq i}}^{n} \left\{ \frac{\chi_i + \alpha_j}{\chi_i - \chi_j} \right\} \frac{1}{\chi_i - \chi_{n+1}}, \qquad i = 1, 2, \ldots, n, \\
A_0 &= \prod_{j=1}^{n} \left\{ \frac{\chi_0 + \alpha_j}{\chi_0 - \chi_j} \right\} \frac{1}{\chi_0 - \chi_{n+1}}, \\
A_{n+1} &= \frac{1}{\chi_{n+1} - \chi_0} \prod_{j=1}^{n} \left\{ \frac{\chi_{n+1} + \alpha_j}{\chi_{n+1} - \chi_j} \right\},
\end{aligned} \tag{16.5.16}$$

where, to satisfy (16.5.14), we must have

$$\sum_{i=0}^{n+1} A_i = 0.$$

Taking the inverse transform of $(16.5.16)_1$, we obtain that

$$E_d^t(s) = E_0(t) \sum_{i=0}^{n+1} A_i e^{-\chi_i s} = E_d^t(\chi_j, j = 0, 1, \ldots, n+1; s).$$

A given history $E_1^t(s)$ belongs to the minimal state with members

$$E^t(\chi_j, j = 0, 1, \ldots, n+1; s) = E_1^t(s) + E_d^t(\chi_j, j = 0, 1, \ldots, n+1; s),$$

where the parameters χ_j may take any positive value.

If (16.5.4) is true for $\widetilde{G}$ given by (11.9.1), we must have

$$\sum_{j=0}^{n+1} \frac{A_j}{\chi_j + \alpha_i} = 0, \quad i = 1, 2 \ldots, n,$$

which is simply a statement that $E_{d+}^t(\omega)$, given by $(16.5.16)_1$, vanishes at ω equal to each $-i\alpha_i$.

16.5.1.2 The Maximum Free Energy

We now seek the choice of state $(E_1(t), E_{r1}^t)$ such that the work done to achieve this state is least among members of the minimal state that has $(E_r^t, E(t))$ as a member. We have from $(7.5.7)_2$ and (16.4.3), recalling also Proposition B.1.3 and (11.2.16),

$$\begin{aligned} W_1(t) &= \phi(t) + \frac{1}{2\pi} \int_{-\infty}^{\infty} \overline{E_{r1+}^t}(\omega) H(\omega) E_{r1+}^t(\omega) d\omega \\ &= \phi(t) + \frac{1}{2\pi} \int_{-\infty}^{\infty} \left[|p_{1-}^{(Nt)}(\omega)|^2 + |p_{1+}^{(Nt)}(\omega)|^2 \right] d\omega, \end{aligned} \tag{16.5.17}$$

where

$$p_{1\pm}^{(Nt)}(\omega) = \frac{1}{2\pi i} \int_{-\infty}^{\infty} \frac{H_-^N(\omega') E_{r1+}^t(\omega') d\omega'}{\omega' - \omega^{\mp}}. \tag{16.5.18}$$

Now, recalling (16.5.9), we see that $p_{1-}^{(Nt)}$ is fixed by virtue of the fact that it is equal to $p_-^{(Nt)}$, defined by (16.5.18), but with E_{r1+}^t replaced by E_{r+}^t. However, $p_{1+}^{(Nt)}$ can be varied, and the choice that minimizes W_1 is clearly

$$p_{m+}^{Nt)}(\omega) = 0, \quad \omega \in \mathbb{R},$$

where $p_{m+}^{(Nt)}$ is the optimal choice, corresponding to an optimal relative history E_m^t. Noting that

$$p_{d+}^{(Nt)}(\omega) = p_{m+}^{(Nt)}(\omega) - p_+^{(Nt)}(\omega) = -p_+^{(Nt)}(\omega),$$

where $p_+^{(Nt)}$ is given by (16.5.18) with E_{r1+}^t replaced by E_{r+}^t, we see that (16.5.12) gives

$$
\begin{aligned}
E^t_{m+}(\omega) &= E^t_{r+}(\omega) + \left[H^N_-(\omega)\right]^{-1} p^{(Nt)}_+(\omega) \\
&= [H^N_-(\omega)]^{-1}[p^{(Nt)}_-(\omega) - p^{(Nt)}_+(\omega) + p^{(Nt)}_+(\omega)] \\
&= \left[H^N_-(\omega)\right]^{-1} p^{(Nt)}_-(\omega),
\end{aligned}
\tag{16.5.19}
$$

with the help of the Plemelj formulas (B.2.14) applied to $H^N_-(\omega)E^t_{r+}(\omega)$.

Proposition 16.5.4. *For materials such that G'_F has only isolated singularities, the maximum free energy is given by*

$$
\psi_M(t) = \phi(t) + \frac{1}{2\pi}\int_{-\infty}^{\infty} |p^{(Nt)}_-(\omega)|^2 d\omega < W(t), \tag{16.5.20}
$$

and the Fourier-transformed optimal history associated with this quantity has the form

$$
E^t_{m+}(\omega) = \left[H^N_-(\omega)\right]^{-1} p^{(Nt)}_-(\omega), \tag{16.5.21}
$$

where

$$
p^{(Nt)}_-(\omega) = \frac{1}{2\pi i}\int_{-\infty}^{\infty} \frac{H^N_-(\omega')E^t_{r+}(\omega')d\omega'}{\omega' - \omega^+}.
$$

The associated rate of dissipation is given by

$$
D_M(t) = |K_M(t)|^2, \quad K_M(t) = \frac{1}{2\pi}\int_{-\infty}^{\infty} H^N_-(\omega)E^t_{r+}(\omega)d\omega. \tag{16.5.22}
$$

If G'_+ has branch-cut singularities, then the maximum free energy is equal to the work function (but see Sect. 18.2).

Relation (16.5.21), which is the final form of (16.5.19), substituted into $(16.5.17)_1$ with E^t_{r1+} replaced by E^t_{m+}, yields (16.5.20). Just as for (11.2.16) and indeed $(16.5.17)_2$, we can write (*cf.* (11.2.20))

$$
W(t) = \phi(t) + \frac{1}{2\pi}\int_{-\infty}^{\infty} \left[|p^{(Nt)}_-(\omega)|^2 + |p^{(Nt)}_+(\omega)|^2\right] d\omega = \psi_M(t) + \mathfrak{D}_M(t),
$$

where

$$
\mathfrak{D}_M(t) = \frac{1}{2\pi}\int_{-\infty}^{\infty} \left|p^{(Nt)}_+(\omega)\right|^2 d\omega,
$$

which is the total dissipation up to time t associated with the maximum free energy. The rate of dissipation (16.5.22) is found by differentiating $\mathfrak{D}_M(t)$ with respect to time, using (16.4.5) and (16.4.7) for $f = N$.

Relation $(16.5.22)_1$ confirms (16.1.28) or property P3 of the Graffi conditions in Sect. 16.1.2. Property P2 follows immediately from (16.5.20), while P1 may be demonstrated in a manner similar to that applied to (11.2.17) in Sect. 11.2.4, noting (16.4.8) and (16.4.9).

The quantity ψ_M, given by (16.5.20), is the maximum in a family of free energies introduced below in Sect. 16.4 that are functionals of the minimal state. As indicated by (16.5.20), it is less than $W(t)$.

16.6 Scalar Product Notation for ψ_f and Related Quantities as Quadratic Functionals

A convenient and compact notation is now introduced and used to confirm that ψ_f is a free energy.

Based on (7.5.3) and (7.5.7), we define the scalar product for E_1, E_2:

$$\begin{aligned}(E_1, E_2) &= \frac{1}{2}\int_{-\infty}^{\infty}\int_{-\infty}^{\infty} G_{12}(|s-u|)E_1^t(u)E_2^t(s)duds \\ &= \frac{1}{2\pi}\int_{-\infty}^{\infty} H(\omega)E_{1F}^t(\omega)\overline{E_{2F}^t}(\omega)d\omega \\ &= (E_2, E_1),\end{aligned} \tag{16.6.1}$$

where

$$E_F^t(\omega) = \int_{-\infty}^{\infty} e^{-i\omega s}E^t(s)ds = E_+^t(\omega) + E_-^t(\omega),$$

and (16.1.15) has been used. If quantities defined in frequency space are in parentheses, it is understood that the second form of (16.6.1) is to be used. The norm of E is defined by

$$||E||^2 = (E, E) = ||E_F||^2 = (E_F, E_F) \geq 0.$$

Some formulae are expressed in terms of relative histories. The quantity $||E_r||^2$ corresponds to the integral terms in (7.5.3) or $(7.5.7)_2$ if E_r^t vanishes on $\mathbb{R}^-$ or, equivalently, if $E_{r-}^t(\omega)$ is zero. We have in fact

$$W(t) = \phi(t) + ||E_r^t||^2. \tag{16.6.2}$$

It is also convenient to introduce a conventional scalar product

$$\langle g, h\rangle = \frac{1}{2\pi}\int_{-\infty}^{\infty} \overline{g}(\omega)\cdot h(\omega)d\omega$$

as well as (16.6.1). Observe that g and h are orthogonal in this scalar product under conditions specified by Proposition B.1.3. We write

$$\langle g, g\rangle = ||g||_L^2,$$

indicating the Lebesgue L^2 norm. Note that, from (16.4.10),

$$\psi_f(t) = \phi(t) + ||p_-^{(ft)}||_L^2. \tag{16.6.3}$$

The Fourier transform of the relative optimal history/continuum is given by

$$E_f^t(\omega) = -\frac{p_-^{(ft)}(\omega)}{H_-^f(\omega)} = E_{f+}^t(\omega) + E_{f-}^t(\omega). \tag{16.6.4}$$

This quantity is a generalization to all f of the quantity given by (11.2.14). The inverse Fourier transform of $E_f^t(\omega)$, which we denote by $E_m^t(s)$, is nonzero on $\mathbb{R}^{\pm}$ for $f = 2, 3, \ldots, N-1$. It is nonzero only on $\mathbb{R}^-$ for $f = 1$ and on $\mathbb{R}^+$ for $f = N$.

Relation (16.6.3) can be put in the form

$$\psi_f(t) = \phi(t) + ||E_f^t||^2.$$

16.6.1 Confirmation That ψ_f Is a Free Energy

Let us now show that ψ_f has the required properties of a free energy as listed in Sect. 16.1.2. Properties P1 and P2 follow from (16.6.3); see (16.4.8) and (16.4.14). Property P3 or (16.1.28) becomes in this context

$$T(t)\dot{E}(t) = \dot{W}(t) = \dot{\psi}_f(t) + D_f(t), \quad D_f(t) \geq 0. \tag{16.6.5}$$

Using the frequency-domain version of (16.6.2) and (16.4.3), we can write

$$\begin{aligned} W(t) &= \phi(t) + \|p_-^{(ft)} - p_+^{(ft)}\|_L^2 \\ &= \phi(t) + \|p_-^{(ft)}\|_L^2 + \|p_+^{(ft)}\|_L^2 \\ &= \psi_f(t) + \|p_+^{(ft)}\|_L^2, \end{aligned}$$

where the orthogonality of $p_-^{ft)}$, $p_+^{(ft)}$ follows from Proposition B.1.3. We therefore identify the total dissipation as (*cf.* (16.4.11))

$$\mathcal{D}_f(t) = \|p_+^{(ft)}\|_L^2$$

and the rate if dissipation as (16.4.12) or

$$D_f(t) = \frac{d}{dt}\|p_+^{(ft)}\|_L^2) = |K_f|^2 \geq 0. \tag{16.6.6}$$

16.7 Asymptotic Behavior and Discontinuities

Let us denote the optimal history/continuation E_m^t by $E_o^{(ft)}$, and let us separate it into history $E_{oh}^{(ft)}$ and continuation $E_{oc}^{(ft)}$, defined by

$$\begin{aligned} E_{oh}^{(ft)} &= E_o^{(ft)}, \quad s \geq 0, \\ E_{oc}^{(ft)} &= E_o^{(ft)}, \quad s < 0. \end{aligned}$$

These quantities are the inverse Fourier transforms of E_{f+}^t and E_{f-}^t, respectively, which are defined by (16.6.4). Recalling (C.2.10) and (C.2.11), we write

$$\begin{aligned} E_{oh}^{(ft)}(\infty) &= i \lim_{\omega\to 0} \omega E_{f+}^t(\omega), \\ E_{oc}^{(ft)}(-\infty) &= -i \lim_{\omega\to 0} \omega E_{f-}^t(\omega). \end{aligned} \tag{16.7.1}$$

Also, from (C.2.16),

$$\begin{aligned} E_{oh}^{(ft)}(0^+) &= i \lim_{\omega\to \infty} \omega E_{f+}^t(\omega), \\ E_{oc}^{(ft)}(0^-) &= -i \lim_{\omega\to \infty} \omega E_{f-}^t(\omega). \end{aligned} \tag{16.7.2}$$

The quantities $E_{oh}^{(ft)}(\infty)$ and $E_{oc}^{(ft)}(-\infty)$ are generally nonzero. In fact, from (16.6.4) and (16.7.1), it is clear that their difference is given by

$$E_{oh}^{(ft)}(\infty) - E_{oc}^{(ft)}(-\infty) = \left[H_1^{(f)}(0)\right]^{-1} \frac{1}{2\pi} \int_{-\infty}^{\infty} H_1^{(f)}(\omega) E_{r+}^{t}(\omega) d\omega,$$
$$H_1^{(f)}(\omega) = \frac{H_-^f(\omega)}{\omega}. \tag{16.7.3}$$

We deduce that

$$E_f^t(\omega) \underset{\omega\to\infty}{\longrightarrow} -\frac{K_f(t)}{i\omega h_\infty}.$$

Relation (16.7.2) then gives

$$E_{oh}^{(ft)}(0^+) - E_{oc}^{(ft)}(0^-) = -\frac{K_f(t)}{h_\infty}. \tag{16.7.4}$$

Remark 16.7.2. We see from (16.6.6) that the discontinuity at the origin is closely related to the rate of dissipation $D_f(t)$.

In the case of the minimum free energy, $E_{oh}^{(ft)}(0^+)$ is zero and we obtain (11.5.2), identifying $(-E_{oc}^{(ft)}(0^-))$ as the optimal continuation at the origin. In the case of the maximum free energy, $E_{oc}^{(ft)}(0^-)$ is zero and there is a jump of magnitude $K_f(t)/h_\infty$ in the optimal history at the origin, recalling that we are dealing with relative histories and $E_r^t(s)$ vanishes at the origin. The result also follows from (16.5.21).

Equations (16.7.3) and (16.7.4) leave $E_{oh}^{(ft)}$ and $E_{oc}^{(ft)}$ arbitrary to within an additive constant, the same constant in each quantity, for the intermediate cases, though not for the optimal continuation/history leading to the minimum and maximum free energies.

16.8 Partial Orderings of the ψ_f

We return to the discussion of the free energies ψ_f. A partial ordering will be established among these quantities. A special case of this result was first given in [110], and the general derivation presented here is taken from [160].

Starting from ψ_m, one can show that interchanging one zero at a time, the resulting ψ_f is not less than the previous one. This leads to a chain of inequalities up to the maximum free energy ψ_M. We recall the quantity $X_l(\omega)$ given by (16.4.2). The function

$$I_l(\omega_1, \omega_2) = \frac{X_l(\omega_1)}{X_l(\omega_2)} = \left(\frac{\omega_1 - \overline{\eta_l}}{\omega_1 - \eta_l}\right)\left(\frac{\omega_2 - \eta_l}{\omega_2 - \overline{\eta_l}}\right) \tag{16.8.1}$$

is the interchange operator, transferring $\overline{\eta_l}$ from $H_-^f(\omega_2)$ to $H_+^f(\omega_1)$ and η_l from $H_+^f(\omega_1)$ to $H_-^f(\omega_2)$. It is required to show that

$$\frac{i}{4\pi^2} \int_{-\infty}^{\infty} \frac{\overline{E_{r+}^t(\omega_1) H_+^f(\omega_1)} [I_l(\omega_1, \omega_2) - 1] H_-^f(\omega_2) E_{r+}^t(\omega_2)}{\omega_1^+ - \omega_2^-} d\omega_1 d\omega_2 \geq 0. \tag{16.8.2}$$

We have

$$I_l(\omega_1, \omega_2) - 1 = -2i \frac{(\mathrm{Im}\eta_l)(\omega_1 - \omega_2)}{(\omega_1 - \eta_l)(\omega_2 - \overline{\eta_l})}.$$

From (16.2.5), $\mathrm{Im}\eta_l > 0$ for all l, so that the integral in (16.8.2) becomes

$$\frac{\mathrm{Im}\eta_l}{2\pi^2} \left| \int_{-\infty}^{\infty} \frac{H_-^f(\omega) E_{r+}^t(\omega)}{\omega - \overline{\eta_l}} d\omega \right|^2 > 0.$$

The first and subsequent choices of zeros exchanged determine a pathway of nondecreasing free energies, starting from the minimum free energy ψ_m and ending, when all zeros have been exchanged, with the maximum free energy ψ_M (Sect. 16.5). If there are p factors, excluding zero, then, we have $p!$ such pathways.

16.9 Explicit Forms for ψ_f

Expressions for $p_\pm^{(ft)}$ and K_f are now derived, in terms of the parameters of the isolated singularities, that allow straightforward evaluation of ψ_f and D_f through (16.6.3) and (16.6.6).

The quantities $p_\pm^{(ft)}(\omega)$, given by (16.4.4), have the form

$$\begin{aligned} p_-^{(ft)}(\omega) &= \sum_{l=1}^{n} Z_l(\omega, \omega', \overline{\zeta_l}) \Big|_{\omega' = \overline{\zeta_l}}, \\ Z_l(\omega, \omega', \overline{\zeta_l}) &= \left[\frac{H_-^f(\omega')(\omega' - \overline{\zeta_l}) E_{r+}^t(\omega')}{\omega - \omega'} \right], \\ p_+^{(ft)}(\omega) &= p_-^{(ft)}(\omega) - H_-^f(\omega) E_+^t(\omega), \end{aligned} \tag{16.9.1}$$

as may be seen from (B.1.3) and by closing the contour on $\Omega^{(-)}$, so that only the singularities $\overline{\xi}_l$ of H_-^f, all of which lie in this half-plane, are picked up. Observe that $p_-^{(ft)}(\omega)$ depends on E_r^t only through the quantities $E_{r+}^t(\overline{\zeta_l})$, so that it is a minimal state variable by Remark 16.3.1 and indeed by Lemma 16.4.2.

The free energy $\psi_f(t)$, given by (16.6.3), has the form

$$\psi_f(t) = \phi(t) + i \sum_{k,l=1}^{n} \left\{ \frac{(\omega_1 - \zeta_k)(\omega_2 - \overline{\zeta_l}) A_f^t(\omega_1, \omega_2)}{\omega_1 - \omega_2} \right\} \Bigg|_{\substack{\omega_1 = \zeta_n \\ \omega_2 = \overline{\zeta_l}}}, \tag{16.9.2}$$

where

$$A_f^t(\omega_1, \omega_2) = \overline{E_{r+}^t}(\omega_1) H_+^f(\omega_1) H_-^f(\omega_2) E_{r+}^t(\omega_2).$$

The quantity $K_f(t)$, given by $(16.4.5)_3$, is best evaluated using $(16.4.7)_1$ and (16.9.1) to give

$$K_f(t) = -i \sum_{l=1}^{n} \left[H_-^f(\omega')(\omega' - \overline{\zeta_l}) E_{r+}^t(\omega') \right] \Big|_{\omega' = \overline{\zeta_l}}.$$

Using $(16.6.6)_2$, we obtain

$$D_f(t) = |K_f(t)|^2 = \sum_{k,l=1}^{n} \left\{ (\omega_1 - \zeta_k)(\omega_2 - \overline{\zeta_l}) A_f^t(\omega_1, \omega_2) \right\} \Bigg|_{\substack{\omega_1 = \zeta_n \\ \omega_2 = \overline{\zeta_l}}} . \tag{16.9.3}$$

The quantity K_f is a linear function of $E_{r+}^t(\overline{\zeta_l})$ and therefore of the vector elements $\overline{e_{lj}}$, defined by $(16.3.3)_3$. This means that D_f will be a quadratic form as in (16.3.10), where $\mathbf{\Gamma}$ is positive semidefinite rather than strictly positive definite (Remark A.2.1).

Recalling Proposition 4.1.7, we see that the N free energies defined by different factorizations are all in a convex set $\mathcal{F}$.

The boundary of the convex set $\mathcal{F}$ will be indicated by the breakdown in the positive definiteness of either $\mathbf{C}$ or $\mathbf{\Gamma}$ or both. Thus, we are on the boundary of $\mathcal{F}$ in the sense that the positive definiteness of $\mathbf{\Gamma}^f$ is breaking down to positive semidefiniteness. In this sense, the free energies formed from the different factorizations are extrema, which has of course been well recognized with regard to the minimum and maximum free energies.

16.9.1 Explicit Forms of the Minimum and Related Free Energies for Discrete-Spectrum Materials

To obtain the minimum free energy, one chooses the factorization of (11.9.4) given by (11.9.5), while the factorizations (16.4.18) yield $\psi^f(t)$, $f = 1, 2, \ldots, N$. We have

$$\begin{gathered} H_-^f(\omega) = ih_\infty \omega \sum_{i=1}^{n} \frac{R_i^f}{\alpha_i(\alpha_i - i\omega)}, \quad H_+^f(\omega) = \overline{H}_-^f(\omega), \\ R_i^f = (\rho_i^f - \alpha_i) \prod_{\substack{j=1 \\ j \neq i}}^{n} \left\{ \frac{\rho_j^f - \alpha_i}{\alpha_j - \alpha_i} \right\}. \end{gathered} \tag{16.9.4}$$

We have the relation [16]

$$G_i = 2H_\infty \sum_{j=1}^{n} \frac{R_i^f R_j^f}{(\alpha_i + \alpha_j)\alpha_i \alpha_j},$$

which reduces to (11.9.15) for $f = 1$. The quantity $p_-^{ft}(\omega)$ is given by (16.9.1). Using (16.3.3), we obtain in this special case

$$p_-^{ft}(\omega) = -ih_\infty \sum_{i=1}^{n} \frac{R_i^f e_i(t)}{\alpha_i(\omega + i\alpha_i)}. \tag{16.9.5}$$

Using (16.9.5) in $(16.4.6)_2$, we obtain

$$K_f(t) = -h_\infty \left[\sum_{i=1}^{n} \frac{R_i^f e_i(t)}{\alpha_i} \right]. \tag{16.9.6}$$

It follows from (16.9.5) and (16.6.3) that

$$\psi_f(t) = \phi(t) + H_\infty \sum_{i,j=1}^{n} \frac{R_i^f R_j^f}{(\alpha_i + \alpha_j)\alpha_i\alpha_j} e_i(t)e_j(t), \tag{16.9.7}$$

where the reality of the quantities $e_i(t)$ has been used. This can be recast in the form (16.3.13) with

$$C_{ij}^f = 2H_\infty \frac{R_i^f R_j^f}{\alpha_i\alpha_j(\alpha_i + \alpha_j)}. \tag{16.9.8}$$

From (16.4.12) and (16.9.6), we see that

$$D_f(t) = H_\infty \left[\sum_{i=1}^{n} \frac{R_i^f}{\alpha_i} e_i(t) \right]^2 = \frac{1}{2} \sum_{i,j=1}^{n} \Gamma_{ij} e_i(t)e_j(t), \tag{16.9.9}$$

where

$$\Gamma_{ij} == 2H_\infty \frac{R_i^f R_j^f}{\alpha_i\alpha_j},$$

which obeys $(16.3.15)_3$. Again, referring to Remark A.2.1, it may be seen that $\mathbf{\Gamma}$ is positive semidefinite.

Consider materials characterized by a relaxation function with only one decaying exponential, so that we have

$$\widetilde{G}(s) = G_1 e^{-\alpha s}, \quad G'(s) = -\alpha G_1 e^{-\alpha s}, \quad H(\omega) = \frac{\omega^2 \alpha G_1}{\alpha^2 + \omega^2}. \tag{16.9.10}$$

All free energies for such materials are equal to the Day free energy functional [87]. From (16.9.7) and (16.9.9), this free energy and the corresponding rate of dissipation are equal to

$$\psi_{Day}(t) = \phi(t) + \frac{G_1}{2} e^2(t), \quad D_{Day}(t) = \alpha G_1 e^2(t), \quad e(t) = \dot{E}_+^t(-i\alpha). \tag{16.9.11}$$

Returning to the general case and following the developments of Sect. 11.9, we find that the quantity $p_+^{(ft)}(\omega)$ has the form

$$\begin{aligned} p_+^{(ft)}(\omega) &= p_-^{(ft)}(\omega) - H_+^f(\omega)E_{r+}^t(\omega) \\ &= ih_\infty \sum_{i=1}^{n} \frac{R_i^f \left[E_{r+}^t(-i\alpha_i) - E_{r+}^t(\omega)\right]}{\omega + i\alpha_i} - h_\infty E_+^t(\omega). \end{aligned}$$

Also, from (16.6.4), using similar manipulations to those in (11.9.10) and subsequent relations,

$$E_f^t(\omega) = i \sum_{l=1}^{n} \frac{B_l^{(ft)}}{\omega + i\rho_l},$$

$$B_l^{(ft)} = \sum_{i=1}^{n} R_i^f Q_{il}^f E_+^t(-i\alpha_i), \qquad Q_{il}^f = \frac{\prod_{\substack{j=1 \\ j\neq i}}^{n} (\rho_l - \alpha_j)}{\prod_{\substack{j=1 \\ j\neq l}}^{n} (\rho_l - \rho_j)}.$$

In the case $f = 1$, where no exchange of zeros has taken place, $E_f^t(\omega) = -E_m^t(\omega)$, given by (11.9.11). Recalling (16.6.4), we see that

$$
\begin{aligned}
E^t_{f-}(\omega) &= -\frac{B_{1-}^{(ft)}}{i\omega^+} + i\sum_{\rho_l>0}\frac{B_l^{(ft)}}{\omega + i\rho_l},\\
E^t_{f+}(\omega) &= -\frac{B_{1+}^{(ft)}}{i\omega^-} + i\sum_{\rho_l<0}\frac{B_l^{(ft)}}{\omega + i\rho_l},\\
B_{1+}^{(ft)} &+ B_{1-}^{(ft)} = B_1^{(ft)},
\end{aligned}
\tag{16.9.12}
$$

where $B_1^{(ft)}$ corresponds to the first zero, which is $\rho_1^f = \gamma_1 = 0$. In the time domain, we have

$$
\begin{aligned}
E^{ft}_{oc}(s) &= B_{1-}^{(ft)} + \sum_{\rho_l>0} B_l^{(ft)}\, e^{\rho_l s}, \quad s<0,\\
E^{ft}_{oh}(s) &= -B_{1+}^{(ft)} - \sum_{\rho_l<0} B_l^{(ft)}\, e^{\rho_l s}, \quad s>0.
\end{aligned}
$$

We know from (16.7.4) that $E^{ft}_{oh}(0^+) \neq E^{ft}_{oc}(0^-)$ and from (16.7.3) that $E^{ft}_{oh}(\infty) \neq E^{ft}_{oc}(-\infty)$. Equations (16.7.3) and (16.7.4) are relations that must be obeyed by the coefficients $B_{1\pm}^{(ft)}$ and $B_l^{(ft)}$. It was noted at the end of Sect. 16.7 that E^{ft}_{oh} and E^{ft}_{oc} are arbitrary to within a constant. Thus, we cannot determine uniquely the quantities $B_{1\pm}^{(ft)}$ that occur in (16.9.12).

16.10 The Central Free Energy and Related Dissipation

In this section, we derive the form of a free energy and its associated rate of dissipation, using the free energies $\psi_f(t)$, $f = 1, 2, \ldots, N$. It can be viewed as an average of these quantities, so that we will refer to it as the central free energy. The results presented were first given in [160], where it was proposed as a candidate for the physical free energy. Such a viewpoint may retain a certain validity, but the whole question of what is or is not a physical free energy is best approached as outlined in Sect. 17.1.

The convexity of $\mathfrak{F}$ means that we can form (see (4.1.10)) a family of free energies given by

$$
\psi(t) = \sum_f \lambda_f \psi_f(t), \qquad \sum_f \lambda_f = 1, \qquad \lambda_f \geq 0, \quad f = 1, 2, \ldots, N, \tag{16.10.1}
$$

where the sum is in general over all N factorizations. Clearly, this free energy must be bounded by the minimum and maximum free energies, so that $\psi_m(t) \leq \psi(t) \leq \psi_M(t)$, the latter quantity being discussed in Sect. 16.5.

Recalling (16.4.15), we can write

$$\psi(t) = \phi(t) + \frac{j}{4\pi^2}\int_{-\infty}^{\infty}\int_{-\infty}^{\infty}\frac{\overline{E^t_{r+}(\omega_1)}L(\omega_1,\omega_2)E^t_{r+}(\omega_2)}{\omega_1^+ - \omega_2^-}d\omega_1 d\omega_2,$$
$$L(\omega_1,\omega_2) = \sum_f \lambda_f H^f_+(\omega_1)H^f_-(\omega_2). \tag{16.10.2}$$

The central free energy is given by the special case of (16.10.2):

$$\psi_c(t) = \phi(t) + \frac{i}{4\pi^2}\int_{-\infty}^{\infty}\int_{-\infty}^{\infty}\frac{\overline{E^t_{r+}(\omega_1)}L_c(\omega_1,\omega_2)E^t_{r+}(\omega_2)}{\omega_1^+ - \omega_2^-}d\omega_1 d\omega_2,$$
$$L_c(\omega_1,\omega_2) = \frac{1}{N}\sum_f H^f_+(\omega_1)H^f_-(\omega_2). \tag{16.10.3}$$

Recalling the interchange operator given by (16.8.1), we define

$$I(\omega_1,\omega_2) = \frac{1}{N}\prod_{l=2}^{n}\{1 + I_l(\omega_1,\omega_2)\}.$$

The kernel L_c can be expressed in terms of this quantity in the form

$$\begin{aligned} L_c(\omega_1,\omega_2) &= \frac{1}{N}H_+(\omega_1)\prod_{l=2}^{n}\{1 + I_l(\omega_1,\omega_2)\}H_-(\omega_2) \\ &= \frac{1}{N}H_\infty\omega_1\omega_2\frac{\prod_{l=2}^{n}\{(\omega_1-\eta_l)(\omega_2-\overline{\eta_l}) + (\omega_2-\eta_l)(\omega_1-\overline{\eta_l})\}}{\prod_{l=1}^{n}[(\omega_1-\zeta_l)(\omega_2-\overline{\zeta_l})]}, \end{aligned}$$

where, from (16.2.9),

$$H_+(\omega) = h_\infty\frac{\prod_{l=1}^{n}(\omega-\eta_l)}{\prod_{l=1}^{n}(\omega-\zeta_l)}, \qquad H_-(\omega) = \overline{H_+(\omega)}.$$

We can also write L_c in the form

$$L_c(\omega_1,\omega_2) = H_\infty\omega_1\omega_2\frac{\prod_{l=2}^{n}\left\{\omega_1\omega_2 + |\eta_l|^2 - (\omega_1+\omega_2)\mathrm{Re}\eta_l\right\}}{\prod_{l=1}^{n}[(\omega_1-\zeta_l)(\omega_2-\overline{\zeta_l})]},$$

where N cancels out.

The corresponding rate of dissipation can be determined to be

$$D_c(t) = \frac{1}{4\pi^2}\int_{-\infty}^{\infty}\int_{-\infty}^{\infty}\overline{E^t_{r+}(\omega_1)}L_c(\omega_1,\omega_2)E^t_{r+}(\omega_2)d\omega_1 d\omega_2.$$

The quantities ψ_c and D_c obey (*cf.* (16.6.5))

$$\dot{\psi}_c(t) + D_c(t) = T(t)\dot{E}(t),$$

which can be demonstrated with the aid of (16.1.18), (16.1.23), and contour integration. Also, the relation

$$L_c(\omega, \omega) = H(\omega)$$

is required, which follows either directly or from $(16.10.3)_2$. One can show that (*cf.* (11.2.42))

$$\int_{-\infty}^{\infty}\int_{-\infty}^{\infty} \frac{\overline{E^t_{r+}(\omega_1)} L_c(\omega_2, \omega_1) E^t_{r+}(\omega_2)}{\omega_1^+ - \omega_2^-} d\omega_1 d\omega_2 = 0$$

by completing the ω_1 integration into a contour over $\Omega^{(+)}$. Thus, we can express $\psi_c(t)$ in an explicitly finite form:

$$\psi_c(t) = \phi(t) + \frac{i}{4\pi^2}\int_{-\infty}^{\infty}\int_{-\infty}^{\infty} \frac{\overline{E^t_{r+}(\omega_1)} \left[L_c(\omega_1, \omega_2) - L_c(\omega_2, \omega_1)\right] E^t_{r+}(\omega_2)}{\omega_1 - \omega_2} d\omega_1 d\omega_2.$$

It is also possible to give more explicit expressions for $\psi_c(t)$, $D_c(t)$ using the same approach as in Sect. 16.9, but applied directly to the double integration in (16.10.3). We obtain formulas similar to (16.9.2) and (16.9.3). Let

$$A^t_c(\omega_1, \omega_2) = \overline{E^t_{r+}(\omega_1)} L_c(\omega_1, \omega_2) E^t_{r+}(\omega_2).$$

Then, we have

$$\begin{aligned}
\psi_c(t) &= \phi(t) + i \sum_{k,l=1}^{n} \left. \frac{(\omega_1 - \zeta_k)(\omega_2 - \overline{\zeta_l}) A^t_c(\omega_1, \omega_2)}{\omega_1 - \omega_2} \right|_{\substack{\omega_1 = \zeta_k \\ \omega_2 = \overline{\zeta_l}}} \\
&= \phi(t) + iH_\infty \sum_{k,l=1}^{n} \frac{\overline{E^t_{r+}(\zeta_k)} N_c(\zeta_k, \overline{\zeta_l}) E^t_{r+}(\overline{\zeta_l})}{\zeta_k - \overline{\zeta_l}}, \\
N_c(\zeta_k, \overline{\zeta_l}) &= \frac{\zeta_k \overline{\zeta_l} \prod_{j=2}^{n} \{\zeta_k \overline{\zeta_l} + |\eta_j|^2 - (\zeta_k + \overline{\zeta_l}) \mathrm{Re}\eta_j\}}{\prod_{\substack{j=1 \\ j \neq k}}^{n} (\zeta_k - \zeta_j) \prod_{\substack{j=1 \\ j \neq l}}^{n} (\overline{\zeta_l} - \overline{\zeta_j})}.
\end{aligned}$$

Also,

$$D_c(t) = H_\infty \sum_{k,l=1}^{n} \overline{E^t_{r+}(\zeta_k)} N_c(\zeta_k, \overline{\zeta_l}) E^t_{r+}(\overline{\zeta_l}).$$

We can write these formulas in the time domain as follows:

$$\begin{aligned}
\psi_c(t) &= \phi(t) + \frac{1}{2}\int_0^\infty ds_1 \int_0^\infty ds_2 E^t_r(s_1) F_c(s_1, s_2) E^t_r(s_2), \\
F_c(s_1, s_2) &= 2iH_\infty \sum_{k,l=1}^{n} \frac{N_c(\zeta_k, \overline{\zeta_l})}{\zeta_k - \overline{\zeta_l}} e^{i(\zeta_k s_1 - \overline{\zeta_l} s_2)},
\end{aligned}$$

and

$$\begin{aligned}
D_c(t) &= \frac{1}{2}\int_0^\infty ds_1 \int_0^\infty ds_2 E^t_r(s_1) M_c(s_1, s_2) E^t_r(s_2), \\
M_c(s_1, s_2) &= 2H_\infty \sum_{k,l=1}^{n} N_c(\zeta_k, \overline{\zeta_l}) e^{i(\zeta_k s_1 - \overline{\zeta_l} s_2)}.
\end{aligned}$$

Let all the singularities be on the imaginary axis. The zeros will also have this property and obey (11.9.3). We put

$$\zeta_k = i\alpha_k, \quad k = 1, 2, \ldots, n, \qquad \eta_l = i\gamma_l, \quad l = 1, 2, \ldots, n,$$
$$\alpha_k > 0, \quad k = 1, 2, \ldots, n, \quad \gamma_1 = 0, \quad \gamma_l > 0, \quad l = 2, 3, \ldots, n.$$

Then,

$$F_c(s_1, s_2) = 2H_\infty \sum_{k,l=1}^{n} \frac{N_c(i\alpha_k, -i\alpha_l)}{\alpha_k + \alpha_l} e^{-\alpha_k s_1 - \alpha_l s_2},$$
$$N_c(i\alpha_k, -i\alpha_l) = \frac{\alpha_k \alpha_l \prod_{j=2}^{n} \{\alpha_k \alpha_l + |\gamma_j|^2\}}{\prod_{\substack{j=1 \\ j \neq k}}^{n} (\alpha_k - \alpha_j) \prod_{\substack{j=1 \\ j \neq l}}^{n} (\alpha_l - \alpha_j)}.$$

The quantity $\mathbf{C}$ in (16.3.13) is given for this case by a matrix with components

$$(\mathbf{C}_c)_{kl} = 2H_\infty \frac{N_c(i\alpha_k, -i\alpha_l)}{\alpha_k + \alpha_l},$$

while from (16.3.15), we deduce that

$$(\boldsymbol{\Gamma}_c)_{kl} = 2H_\infty N_k(i\alpha_k, -i\alpha_l).$$

For one singularity at $i\alpha$, we obtain Day's [87] formulas (11.9.17) and (11.9.18) for the minimum free energy, which is also the maximum free energy and all the intermediate free energies ψ_f.

In [160], detailed formulas for the case of two poles are presented, both the case in which the two poles and the single zero (other than the one at the origin) are on the positive imaginary axis and also that in which both poles are off the imaginary axis and are conjugate pairs in the sense noted after (16.2.3).

16.11 Plots of Free Energies

In this section, certain free energy functionals for materials with memory are plotted for sinusoidal strain histories. We deal with discrete-spectrum materials in the completely linear case. These plots show an interesting proximity between apparently unconnected functionals. A more comprehensive range of plots and detailed expressions for the free energies may be found in [16].

Consider a history and current value $(E^t, E(t))$ defined by

$$E(t) = E_0 e^{i\omega_- t} + \overline{E_0} e^{-i\omega_+ t}, \quad E^t(s) = E(t-s), \tag{16.11.1}$$

where E_0 is an amplitude and $\overline{E_0}$ its complex conjugate. Furthermore,

$$\omega_- = \omega_0 - i\eta, \quad \omega_+ = \overline{\omega}_-, \quad \omega_0, \eta \in \mathbb{R}^{++}. \tag{16.11.2}$$

The parameter η is introduced to ensure finite results in certain quantities. In the limit $\eta \to 0$, any real algebraic quadratic form in $E(t)$ or real functional quadratic form in $E^t(s)$ can be written as

$$V = ME_0^2 e^{2i\omega_0 t} + \overline{M}\,\overline{E_0}^2 e^{-2i\omega_0 t} + N|E_0|^2. \tag{16.11.3}$$

The quantity N must be real. If V is restricted to be nonnegative, as in the present context, then the conditions

$$N \geq 0, \qquad 2|M| \leq N$$

must apply. The first relation follows by taking a time average over a cycle and the second by expressing the first two terms in (16.11.3) in polar form and combining them into a cosine function.

Our choice of inverse decay times α_i, $i = 1, 2, \ldots, n$, is based on classical theoretical arguments. We adopt the following formula, which emerges from simple Brownian motion molecular theories (see references in [133, 157], for example):

$$\alpha_r = \alpha_n \kappa_r, \qquad \kappa_r = \frac{\sin^2\left[\dfrac{r\pi}{2(n+1)}\right]}{\sin^2\left[\dfrac{n\pi}{2(n+1)}\right]}, \quad r = 1, 2, \ldots, n. \tag{16.11.4}$$

We take $n = 4$ [16]. The coefficients G_i in (11.9.1) are all taken to be the same [133]:

$$G_i = G_1, \quad i = 1, 2, \ldots, n. \tag{16.11.5}$$

We take E_0 in (16.11.1) to be real, so that

$$E(t) = 2E_0 \cos \omega_0 t.$$

It is assumed that

$$G_0 = E_0 = E_1 = \omega_0 = 1,$$

for a fully dimensionless version. Also, [16],

$$G_1 = \frac{1}{n}[1 - G_\infty] \in \left[0, \frac{1}{n}\right]. \tag{16.11.6}$$

In Figs. 16.1, 16.2, and 16.3, free energies and rates of dissipation are plotted against time. In Fig. 16.1, the minimum, maximum, and two of the intermediate free energies are shown for sinusoidal histories, while the corresponding rates of dissipation are presented in Fig. 16.2. Several free energies are plotted in Fig. 16.3, including the Graffi–Volterra functional, which is not an FMS, and $\psi_c(t)$, denoted on the plot as $\psi_p(t)$, which is the original notation.

The main information content is the vertical ordering of the free energies. In particular, the close proximity (Fig. 16.3) of the minimum free energy $\psi_m(t)$ and $\psi_F(t)$ is noteworthy. They appear to be almost equal. There is no apparent algebraic evidence for equality. The almost equally close correspondence between $\psi_{Dill}(t)$ and

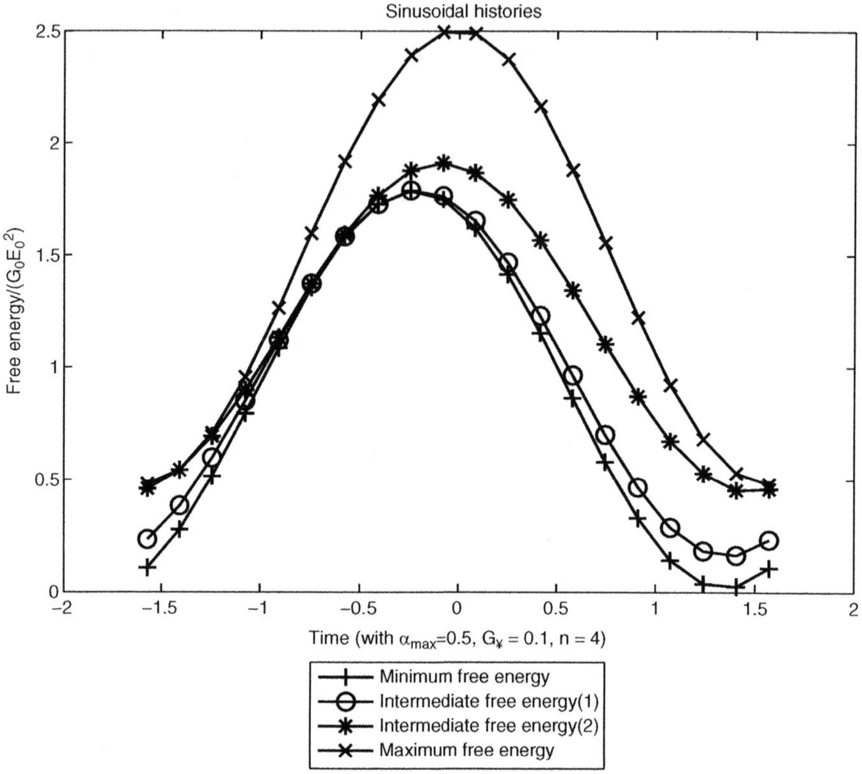

Fig. 16.1. Minimum, maximum, and intermediate free energies

$\psi_m(t)$, $\psi_F(t)$ is also interesting. This feature does not appear to be a consequence of the special assumptions made in this section, in particular (16.11.5). Even when this is altered, the close proximity of these quantities is retained.

In Fig. 16.3, $\psi_G(t)$ crosses over the curve for $\psi_M(t)$. There is no contradiction here since $\psi_M(t)$ is the maximum of free energies which are functionals of the minimal state, while $\psi_G(t)$ does not have this property.

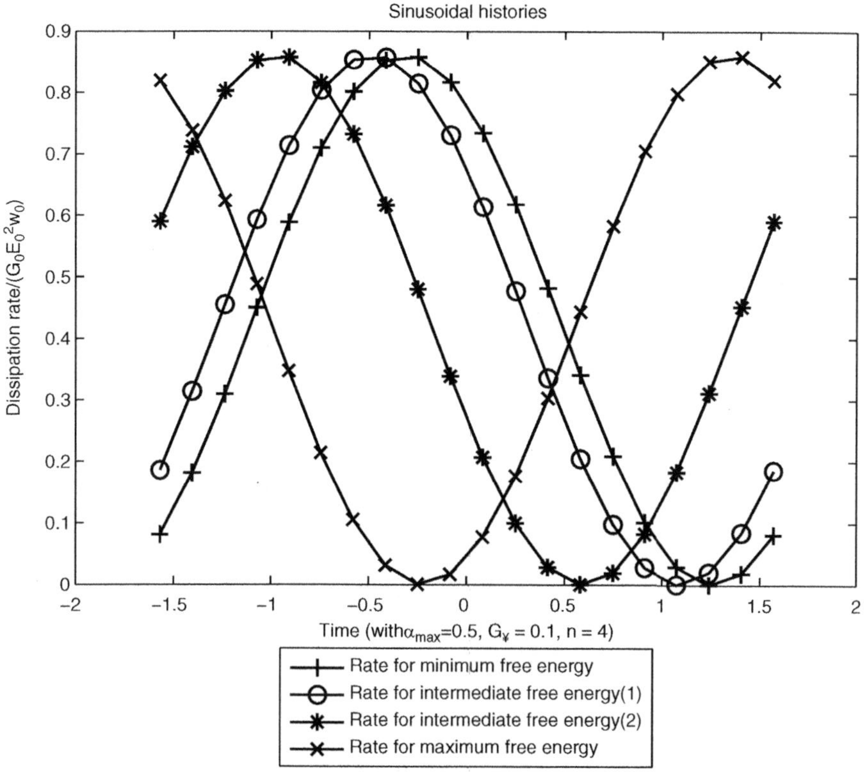

Fig. 16.2. Rates of dissipation relating to the free energies on Fig. 16.1

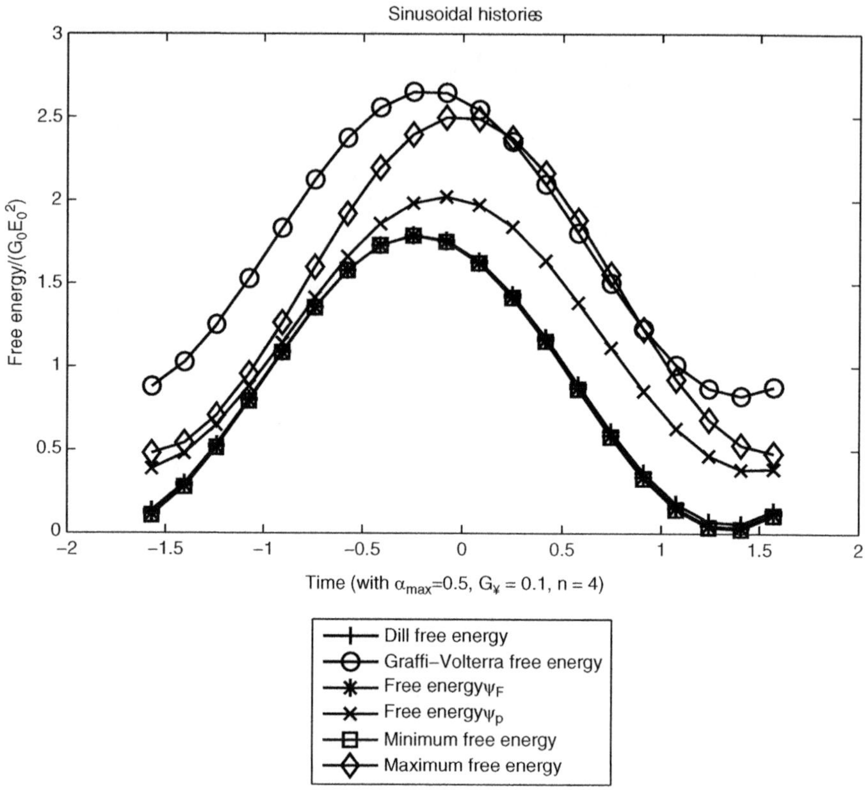

Fig. 16.3. Free energies as listed

17

Constructing Free Energies for Materials with Memory

We now discuss methods for deriving new functionals which have the properties of a free energy. Perhaps the central difficulty in constructing such quantities arises in making choices that ensure both a nonnegative quadratic form for the free energy and the rate of dissipation. Typically, if one chooses a suitable functional for the free energy, the associated rate of dissipation does not have the non-negativity property. A technique is presented in this chapter (see also [162]) which in effect reverses this procedure. We choose a nonnegative functional for the rate of dissipation and derive formulae which give the associated free energy functional in terms of the dissipation rate kernel. It emerges that the resulting free energy has the required non-negativity property.

The main topics dealt with in this chapter are based on [162] and are a development of the discussion in Sect. 7.1.3. Also, Sects. 17.1 and 17.6 are based on results in [18].

Also discussed in this chapter is the issue of approximating a general discrete-spectrum relaxation function by a relaxation function with one decaying exponential, corresponding to the Day free energy.

Finally, we consider single-integral free energies in terms of the functional I^t, which is the functional of the minimal state defined by (16.5.1).

17.1 Two Equivalent Interpretations of the Set of Free Energies

Before deriving the main results of this chapter, we discuss two possible interpretations of the set of free energy functionals associated with a given constitutive equation relating stress and strain.

Let us identify a particular material with memory, which will be referred to as material I. It is assumed to exhibit linear behavior. The stress-strain or constitutive relation of this material is known, in other words, its relaxation function is given. There are generally many free energies and corresponding dissipation functionals

G. Amendola et al., *Thermodynamics of Materials with Memory*,
https://doi.org/10.1007/978-3-030-80534-0_17

associated with material I. All of these generate the same stress and therefore have the same relaxation function. As in earlier chapters, we denote this convex set by $\mathcal{F}$, which is of course dependent on the choice of strain history. The physical free energy for material I, yielding the observed rate of dissipation, is a member of $\mathcal{F}$, as well as all free energies generating the given stress.

It is shown in [162, 164] (see also Sect. 17.2) that any material with memory can be uniquely characterized by specifying the kernel of its physical rate of dissipation functional. This quantity determines the associated free energy kernel and the relaxation function, which in turn yields the stress-strain or constitutive relation, for a given strain history. The work function can be deduced from these quantities. Also, the dissipation kernel determines the amount of dissipation under deformation. We will consider the set of all such kernels associated with materials with a specified constitutive relation; this set will be denoted by $\mathcal{K}$. For a given choice of strain history, $\mathcal{K}$ generates a set of free energies $\mathcal{F}$, corresponding to our chosen constitutive relation. It will emerge that the boundaries of $\mathcal{K}$ and $\mathcal{F}$ are at least roughly determined by the relaxation function of the constitutive relation.

The following alternative viewpoint is now described. We interpret the set of kernels $\mathcal{K}$ as specifying all the distinct linear materials with the same constitutive relation but different dissipation rates as a result of deformation. These can be labeled by individual members of $\mathcal{K}$. One of them yields the physical free energy in $\mathcal{F}$ for material I. Other members of $\mathcal{F}$ would traditionally be regarded as approximations to or bounds on (notably the minimum and maximum free energy) this physical free energy. Instead, we now regard these, or more specifically the corresponding kernels in $\mathcal{K}$, as describing different actual materials with the same constitutive relation, but different dissipation properties. For the material labeled by a particular kernel, the relevant member of $\mathcal{F}$ for a given strain history is the physical free energy for that material. Particular examples may not currently exist as real materials but it seems reasonable to assume that they could be manufactured, to a close approximation, now or in the future.

Both of the above viewpoints are valid and can be adopted as context demands. We will refer to the more traditional viewpoint, where $\mathcal{K}$ is the set of kernels producing the physical free energy of material I as well as approximations to and bounds on this quantity, as Interpretation 1 or I1. The viewpoint that each $K(s, u) \in \mathcal{K}$ fully describes a separate material, each equally of interest and with the same constitutive equation, will be referred to as I2.

Under I2, the set $\mathcal{K}$ is defined not by the choice of material I but by the constitutive equation of the materials. If we replace material I by another material with the same stress-strain relation, the set $\mathcal{K}$ remains unchanged. We do not focus on one specific example but rather treat all materials labeled by members of $\mathcal{K}$ on an equal footing.

The Day free energy, a discrete-spectrum material with one decay time, is the only free energy that is a functional of the minimal state. It is therefore the unique physical free energy for that material. This quantity and the associated dissipation are explored for a choice of relaxation function approximately equal to that for the more general set of materials under consideration.

17.2 Unique Characterization of Materials with Memory

Materials with linear memory constitutive relations, i.e., a linear functional of the strain history, are characterized by a relaxation function, if we leave aside the matter of stored and dissipated energy. Following the discussion in Sects. 7.1.3 and 17.1, it is proposed that a material is characterized by the kernel of the rate of dissipation functional. This will be shown to yield a unique free energy and relaxation function. It also answers the question posed in Sect. 7.1.3 about two alternatives, by choosing the second option, though the first option remains of interest since all free energies obtained to date are examples of this.

There are generally many rates of dissipation and free energy kernels which yield a given relaxation function. Such non-uniqueness means that there is no simple method of identifying which is the physical free energy and rate of dissipation of a particular material. Ideally, the behavior of this kernel should be determined by measurements carried out on the material of interest, in which case it would be the physical kernel. This would lead to a unique, valid, free energy and provides a formula for the relaxation function. Thus, it is a complete characterization of the material. Of course, we are not referring to a real material, but rather a mathematical model, approximately describing some aspects of the behavior of a real material.

The kernel of the rate of dissipation functional is, however, difficult to measure, particularly in a non-isothermal context. For an isothermal problem, a quadratic functional using this kernel is the amount of heat produced per unit time by work on the material. The issue of measurement is briefly discussed in Sect. 7.1.3.

The standard approach to determining free energy functionals is to specify the relaxation function and seek free energy functionals either explicitly dependent on this quantity, or through a factorization process on a function derivable from it (see option 1 in Sect. 7.1.3). The first method is applicable only if this relaxation function is a monotonically decaying quantity, while the second approach yields the minimum and related free energies, which lie on the boundaries of $\mathcal{F}$. However, since we generally cannot determine the physically correct choice, this approach provides a complete description of constitutive behavior in the sense of stress-strain relations, but gives at best a partial characterization of energy storage and dissipation.

Our assertion is that a material with memory should be characterized by the kernel of the rate of dissipation, $\underset{\sim}{K}(\cdot,\cdot)$, defined in (17.3.12). Using a simple formula, the kernel of a unique free energy, $\underset{\sim}{G}(\cdot,\cdot)$, with the correct non-negativity property, can then be deduced, from which in turn the relaxation function can be obtained.

It is assumed that, as a separate exercise, the equilibrium free energy has been fully determined. For completely linear materials, as given by (16.1.32) and (16.1.6), this amounts to measuring G_∞, which is part of the task of determining the relaxation function.

Thus, all properties of the material are uniquely defined.

These developments are discussed in Sects. 17.4 and 17.5; see also [164].

17.3 Quadratic Models for Free Energies

Materials with linear constitutive relations will have free energies that are quadratic functionals. Let us first consider the simplest models for such materials.

17.3.1 A Single-Integral Model

Consider the form

$$\psi(t) = \phi(t) + \frac{1}{2}\int_0^\infty C(s)[E_r^t(s)]^2 ds. \tag{17.3.1}$$

The integral term in (17.3.1) must be nonnegative for all histories by virtue of condition P2 or (16.1.26) of the Graffi conditions, which requires that $C(s)$ be a nonnegative quantity for all $s \in \mathbb{R}^+$. If (16.1.25) is to yield the constitutive relations $(16.1.3)_1$ for any arbitrary choice of history of strain, we must put $C(s) = -G'(s)$ where $G'(s)$ is assumed to be nonnegative, giving

$$\begin{aligned}\psi(t) &= \phi(t) - \frac{1}{2}\int_0^\infty G'(s)[E_r^t(s)]^2 ds \\ &= S(t) - \frac{1}{2}\int_0^\infty G'(s)[E^t(s)]^2 ds, \\ S(t) &= \phi(t) + (T(t) - T_0(t))E(t) + \frac{1}{2}(G_0 - G_\infty)E^2(t).\end{aligned} \tag{17.3.2}$$

The quantity $S(t)$ is the quantity defined by $(7.1.19)_2$, modified in the manner specified before (7.1.35), for scalar theories. The rate of dissipation associated with ψ can be determined from (16.1.28) to have the form

$$D(t) = \frac{1}{2}\int_0^\infty G''(s)[E_r^t(s)]^2 ds \geq 0, \tag{17.3.3}$$

provided it is assumed that $G''(s) \geq 0, \ \ \forall s \in \mathbb{R}^+$. We conclude that there is only one example of a single-integral quadratic free energy in terms of strain history and this quantity is a free energy only if the conditions

$$G'(s) \leq 0, \qquad G''(s) \geq 0, \ \ \forall s \in \mathbb{R}^+ \tag{17.3.4}$$

hold. This is of course the Graffi–Volterra free energy functional discussed in Sect. 10.1.1.

Remark 17.3.1. The first condition in (17.3.4) yields the non-negativity of the integral term in the free energy and ensures that Graffi condition P2 is satisfied, while the second condition relates to P3, ensuring that the rate of dissipation is nonnegative or the second law holds. The second condition implies the first condition, as can be seen from the relation

$$G'(s) = -\int_s^\infty G''(u)du.$$

However, the first condition does not in general imply the second. Thus, for functional forms which are free energies only for materials obeying (17.3.4), the requirement that the rate of dissipation be nonnegative is sufficient to ensure the nonnegativity of the free energy, but not *vice-versa*. This is not surprising since P2 actually follows from P3 [67].

17.3.2 A Double Integral Model

For a scalar theory with a linear memory constitutive relation for the stress, the most general form of a free energy is

$$\begin{aligned}\psi(t) &= \phi(t) + \frac{1}{2}\int_0^\infty\int_0^\infty E_r^t(s)\mathcal{G}(s,u)E_r^t(u)dsdu \\ &= S(t) + \frac{1}{2}\int_0^\infty\int_0^\infty E^t(s)\mathcal{G}(s,u)E^t(u)dsdu \\ &= \phi(t) - \phi_l(t) + \frac{1}{2}\int_0^\infty\int_0^\infty \dot{E}^t(s)G(s,u)\dot{E}^t(u)dsdu \qquad (17.3.5)\\ &= \phi(t) + \frac{1}{2}\int_0^\infty\int_0^\infty \dot{E}^t(s)\widetilde{G}(s,u)\dot{E}^t(u)dsdu,\end{aligned}$$

$$\mathcal{G}(s,u) = \frac{\partial^2}{\partial s\partial u}G(s,u) = G_{12}(s,u), \qquad \widetilde{G}(s,u) = G(s,u) - G_\infty,$$

where $S(t)$ is defined by $(17.3.2)_3$, $\phi_l(t)$ by (16.1.32) and $T_0(t)$ by $(16.1.3)_5$. This is the scalar version of the full tensor developments of Sect. 7.1. There is no loss of generality in taking

$$\widetilde{G}(s,u) = \widetilde{G}(u,s), \qquad \mathcal{G}(s,u) = \mathcal{G}(u,s). \qquad (17.3.6)$$

The following properties of G will be assumed to hold for $s, u \in \mathbb{R}^+$:

$$\begin{aligned}G_\infty &= G(\infty,u) = G(s,\infty),\\ G_1(s,\infty) &= G_2(\infty,u) = 0, \qquad (17.3.7)\\ G_1(\infty,s) &= G_2(u,\infty) = 0 \quad \forall\ s,u \in \mathbb{R}^+.\end{aligned}$$

Furthermore, we have

$$G_0 = G(0,0). \qquad (17.3.8)$$

The relaxation function $G(u)$ is given by

$$G(u) = G(0,u) = G(u,0) \quad \forall\ u \in \mathbb{R}^+, \qquad (17.3.9)$$

yielding

$$G'(u) = G_2(0,u) = G_1(u,0). \qquad (17.3.10)$$

Relation (17.3.9) is the basic constraint ensuring that (16.1.25) holds. Conversely, if (16.1.25) is valid for all histories, then (17.3.9) must be true.

Relation $(17.3.7)_{1,2}$ for $s = u = 0$ and (17.3.8) agree with (16.1.4), by virtue of (17.3.9). The Graffi condition P2, given by (16.1.26), requires that the kernels $\mathcal{G}$ and $\widetilde{G}$ must be such that the integral terms in $(17.3.5)_{1,4}$ are nonnegative.

Remark 17.3.2. We will consider all free energies associated with a given constitutive equation. Thus, the quantity $G(u)$ is the same for all choices of free energy, *i.e.*, for all choices of $G(s,u)$.

Referring to the quantity $S(t)$ in (17.3.2) and $(17.3.5)_2$, we see that

$$\frac{\partial}{\partial E(t)} S(t) = T(t),$$

which is condition P1 or (16.1.25).

The rate of dissipation can be deduced from (16.1.28) and $(16.1.3)_{10,11}$ to be

$$\begin{aligned} D(t) &= -\frac{1}{2}\int_0^\infty \int_0^\infty \dot{E}^t(s)K(s,u)\dot{E}^t(u)dsdu \\ &= -\frac{1}{2}\int_0^\infty \int_0^\infty E_r^t(s)\mathcal{K}(s,u)E_r^t(u)dsdu, \end{aligned} \tag{17.3.11}$$

where

$$K(s,u) = G_1(s,u) + G_2(s,u), \qquad \mathcal{K}(s,u) = \mathcal{G}_1(s,u) + \mathcal{G}_2(s,u). \tag{17.3.12}$$

The quantity G must be such that the integral in (17.3.11) is non-positive, as required by P3 of the Graffi conditions. We have

$$\mathcal{K}(s,u) = \frac{\partial^2}{\partial s \partial u} K(s,u) = K_{12}(s,u). \tag{17.3.13}$$

The quantities K and $\mathcal{K}$ can also be taken to be symmetric in their arguments, *i.e.*,

$$K(s,u) = K(u,s), \qquad \mathcal{K}(s,u) = \mathcal{K}(u,s). \tag{17.3.14}$$

The non-negativity requirements on $\widetilde{G}$, $\mathcal{G}$, $-K$, and $-\mathcal{K}$ imply in particular that

$$\begin{aligned} &\widetilde{G}(s,s) \geq 0, \qquad \mathcal{G}(s,s) \geq 0, \\ &K(s,s) \leq 0, \qquad \mathcal{K}(s,s) \leq 0, \quad s \in \mathbb{R}^+. \end{aligned} \tag{17.3.15}$$

In the approach developed in this chapter, the quantity $K(s,u)$ will play a more fundamental role than $G(s,u)$.

Seeking to express $\mathcal{D}(t)$ given by $(16.1.30)_2$ as a general quadratic functional form similar to those in (17.3.5) or (17.3.11), we put

$$\begin{aligned} \mathcal{D}(t) &= \frac{1}{2}\int_0^\infty \int_0^\infty \dot{E}^t(s)Q(s,u)\dot{E}^t(u)dsdu \\ &= \frac{1}{2}\int_0^\infty \int_0^\infty E_r^t(s)\mathcal{Q}(s,u)E_r^t(u)dsdu. \end{aligned} \tag{17.3.16}$$

There are two equivalent alternatives for the developments outlined below, the first being to use $\widetilde{G}(s,u)$, $K(s,u)$, $\dot{E}^t(s)$ and the second to use $\mathcal{G}(s,u)$, $\mathcal{K}(s,u)$, $E_r^t(s)$. Both have been widely adopted in discussing the minimum and related free energies. The first approach will be favored in the present context, though in earlier chapters, the other formulation is widely used, so there is a need to move between the two notations.

Remark 17.3.3. The Principle of Causality must apply to all physical systems and in the present context means that quantities such as $T(t)$, $\psi(t)$, $D(t)$, *etc.* can only depend on $E(s)$, $s \leq t$. Therefore, we can write $(16.1.3)_4$ and $(17.3.5)_4$, for example, in the form

$$T(t) = T_e(t) + \int_{-\infty}^{\infty} \widetilde{G}(s)\dot{E}^t(s)ds,$$

$$\psi(t) = \phi(t) + \frac{1}{2}\int_{-\infty}^{\infty}\int_{-\infty}^{\infty} \dot{E}^t(s)\widetilde{G}(s,u)\dot{E}^t(u)dsdu,$$

provided that

$$\dot{E}^t(s)\widetilde{G}(s,u)\dot{E}^t(u) = 0, \quad s,u < 0, \qquad \widetilde{G}(s)\dot{E}^t(s) = 0, \quad s < 0. \tag{17.3.17}$$

The simplest way to enforce (17.3.17) is to take either option (a) or (b), given by

$$\begin{aligned} &(a)\ \ \dot{E}^t(s) = 0, \quad \widetilde{G}(s,u), \widetilde{G}(s) \ \ arbitrary, \ \ for \ \ s,u < 0; \\ &(b)\ \ \widetilde{G}(s,u), \widetilde{G}(s) = 0, \quad \dot{E}^t(s) \ \ arbitrary, \ \ for \ \ s,u < 0. \end{aligned} \tag{17.3.18}$$

The arbitrariness of $\widetilde{G}(\cdot,\cdot)$, $\widetilde{G}(\cdot)$, and $E^t(\cdot)$ for negative arguments is subject to the requirement that certain integrals of these quantities (for example, Fourier transforms) converge.

17.3.3 The Work Function

This quantity, given by $(16.1.30)_1$, can be put in the following forms (see Sect. 7.5):

$$\begin{aligned} W(t) &= \phi(t) + \frac{1}{2}\int_0^{\infty}\int_0^{\infty} \dot{E}^t(s)\widetilde{G}(|s-u|)\dot{E}^t(u)duds \\ &= \phi(t) + \frac{1}{2}\int_0^{\infty}\int_0^{\infty} E_r^t(s)\frac{\partial^2}{\partial s \partial u}G(|s-u|)E_r^t(u)duds \\ &= \phi(t) + \frac{1}{2\pi}\int_{-\infty}^{\infty} \frac{H(\omega)}{\omega^2}|\dot{E}_+^t(\omega)|^2 d\omega \\ &= \phi(t) + \frac{1}{2\pi}\int_{-\infty}^{\infty} H(\omega)|E_{r+}^t(\omega)|^2 d\omega. \end{aligned} \tag{17.3.19}$$

Both versions of the frequency domain formulation are manifestly nonnegative. One follows from the other by invoking (16.1.17). We see that it can be cast in the forms $(17.3.5)_{1,4}$ by putting

$$\widetilde{G}(s,u) = \widetilde{G}(|s-u|), \qquad G_{12}(s,u) = \frac{\partial^2}{\partial s \partial u}G(|s-u|). \tag{17.3.20}$$

Remark 17.3.4. The quantity $W(t)$ can be formally regarded as a free energy, but with zero dissipation, which is clear from (16.1.29). Because of the vanishing dissipation, it must be the maximum free energy associated with the material or greater than this quantity, an observation which follows from (16.1.29) for any free energy $\psi(t)$. Indeed, we have in general the requirement that

$$\psi(t) \leq W(t). \tag{17.3.21}$$

A problem with the treatment of $W(t)$ as a free energy is raised in Chap. 18.

From (16.1.29), (17.3.16), (17.3.19) and (17.3.20), we deduce that

$$\begin{aligned} Q(s,u) &= \widetilde{G}(|s-u|) - \widetilde{G}(s,u), \\ \mathcal{Q}(s,u) &= G_{12}(|s-u|) - \mathcal{G}(s,u), \end{aligned} \tag{17.3.22}$$

from which it follows that

$$\begin{aligned} Q_1(s,u) + Q_2(s,u) &= -G_1(s,u) - G_2(s,u) = -K(s,u), \\ \mathcal{Q}_1(s,u) + \mathcal{Q}_2(s,u) &= -\mathcal{G}_1(s,u) - \mathcal{G}_2(s,u) = -\mathcal{K}(s,u). \end{aligned} \tag{17.3.23}$$

Relations (17.3.22) also yield

$$Q(s,0) = Q(0,u) = 0 \quad \forall\ s,u \in \mathbb{R}^+, \tag{17.3.24}$$

and

$$\int_0^\infty \mathcal{Q}(s,v)dv = \int_0^\infty \mathcal{Q}(v,u)dv = 0 \quad \forall\ s,u \in \mathbb{R}^+, \tag{17.3.25}$$

by virtue of (17.3.7), (17.3.9) and (17.3.10). A consequence of (17.3.25) is that $(17.3.16)_2$ can be replaced by

$$\mathcal{D}(t) = \frac{1}{2}\int_0^\infty\int_0^\infty E^t(s)\mathcal{Q}(s,u)E^t(u)dsdu.$$

A requirement similar to (17.3.15) must be imposed on Q. In the light of $(17.3.22)_1$, we have

$$Q(s,s) = \widetilde{G}_0 - \widetilde{G}(s,s) = G_0 - G(s,s) \geq 0, \quad \forall\ s \in \mathbb{R}^+.$$

The corresponding relation for $\mathcal{Q}$ presents difficulties in that the quantities involved have a delta function singularity, which arises from the differentiations in $(17.3.22)_2$.

17.4 Time Domain Representation of Free Energies in Terms of the Kernel $K(\cdot,\cdot)$

We now present the results on which the assertions of Sect. 17.2 are based.

Two equivalent versions of the argument will now be presented, one in the time domain, the other in the frequency domain. A formalism is developed in this section for the time domain and in Sect. 17.5 for the frequency domain, which extend the developments of Sect. 7.1.3 and allow us to apply the new strategy.

We treat $(17.3.12)_1$ as a first order partial differential equation for $G(s,u),\ s,u \in \mathbb{R}^+$, where $K(s,u),\ s,u \in \mathbb{R}^+$ is presumed to be known and has the property that

$$\int_0^\infty \int_0^\infty f(s)K(s,u)f(u)dsdu \le 0 \tag{17.4.1}$$

for all choices of f such that the integral exists. We use the variables (7.1.29):

$$x = s + u, \quad y = s - u, \tag{17.4.2}$$

in terms of which $(17.3.12)_1$ becomes

$$\frac{\partial}{\partial x} G_n(x,y) = \frac{1}{2} K_n(x,y), \quad G_n(x,y) = G(s,u), \quad K_n(x,y) = K(s,u),$$

with general solution

$$G_n(x,y) = G_n(x_0,y) + \frac{1}{2}\int_{x_0}^{x} K_n(x',y)dx', \tag{17.4.3}$$

where x_0 is an arbitrary nonnegative real quantity. This is the scalar version of (7.1.30). It follows from $(17.3.6)_1$ and $(17.3.14)_1$ that

$$G_n(x,y) = G_n(x,-y) = G_n(x,|y|), \qquad K_n(x,y) = K_n(x,-y) = K_n(x,|y|). \tag{17.4.4}$$

Observe that (17.3.9) yields

$$G(u) = G_n(u,u) = G_n(u,-u) = G_n(u,|u|), \quad u \in \mathbb{R}^+. \tag{17.4.5}$$

Putting

$$x' = s' + u', \quad y = s' - u' = s - u,$$

we have

$$s' = \frac{1}{2}(x' + y), \quad u' = \frac{1}{2}(x' - y).$$

In particular, these yield (s,u) corresponding to $x' = x$. Letting $x_0 \to \infty$ gives a solution of the form

$$G_n(x,y) = G_\infty - \frac{1}{2}\int_x^\infty K_n(x',y)dx'. \tag{17.4.6}$$

We have

$$\begin{aligned} G(s,u) = G_n(x,y) &= G_\infty - \frac{1}{2}\int_{s+u}^\infty K_n(x', s-u)dx' \\ &= G_\infty - \frac{1}{2}\int_{s+u}^\infty K(\frac{1}{2}(x' + s - u), \frac{1}{2}(x' - s + u))dx'. \end{aligned}$$

Let us change the x' variable of integration to $z \ge 0$, defined by

$$x' = 2z + s + u, \tag{17.4.7}$$

so that

$$\widetilde{G}(s,u) = -\int_0^\infty K(z+s, z+u)dz. \tag{17.4.8}$$

A similar relation for $\mathcal{G}(s,u)$ can be deduced from $(17.3.5)_5$ and (17.3.13).

We assume that $K(s,u)$ in $(17.3.12)_1$ is given, so that the rate of dissipation for a particular strain history of the material, specified by (17.3.11), is uniquely known. The following proposition will now be proved.

Proposition 17.4.1. *For free energy functionals and rates of dissipation of the forms* (17.3.5) *and* (17.3.11), *respectively, if we assume that the rate of dissipation is nonnegative (which is the second law), then it follows that the integral term in* $(17.3.5)_4$ *is nonnegative.*

Thus, the property noted at the end of Remark 17.3.1 is confirmed.

Proof. Multiplying (17.4.8) by $\dot{E}^t(s)$ and $\dot{E}^t(u)$ and integrating yields

$$\begin{aligned}\psi(t) &= \phi(t) - \frac{1}{2}\int_0^\infty\int_0^\infty \dot{E}^t(s)\int_0^\infty K(z+s,z+u)dz\dot{E}^t(u)dsdu\\ &= \phi(t) - \frac{1}{2}\int_0^\infty\int_0^\infty\int_0^\infty \dot{E}^t(s)K(z+s,z+u)\dot{E}^t(u)dsdudz,\end{aligned} \tag{17.4.9}$$

on interchanging integrations. Now

$$\int_0^\infty\int_0^\infty \dot{E}^t(s)K(z+s,z+u)\dot{E}^t(u)dsdu$$
$$= \int_z^\infty\int_z^\infty \dot{E}^t(v-z)K(v,w)\dot{E}^t(w-z)dvdw.$$

Let us put

$$f(v) = \begin{cases}\dot{E}^t(v-z), & v \geq z\\ 0, & 0 \leq v < z\end{cases}$$

for arbitrary choices of $\dot{E}^t$. By virtue of assumption (17.4.1), it follows that the integral term in (17.4.9) is nonnegative. □

We identify the relaxation function, in accordance with (17.3.9), as

$$G(s) = G(s,0) = G_\infty - \int_0^\infty K(z+s,z)dz = G_\infty - \int_0^\infty K(z,z+s)dz, \tag{17.4.10}$$

which ensures that the condition P1 or (16.1.25) is satisfied. It follows from (17.4.10) that

$$\widetilde{G}_0 = -\int_0^\infty K(z,z)dz, \tag{17.4.11}$$

using the notation of (16.1.4).

We take $s \geq u$ and $y = s - u$. Choosing $x_0 = y$ in (17.4.3), we have

$$\begin{aligned} G(s,u) = G_n(s+u, s-u) &= G(s-u) + \frac{1}{2}\int_{s-u}^{s+u} K_n(x', s-u)dx' \\ &= G(s-u) + \int_{-u}^{0} K(z+s, z+u)dz, \end{aligned} \tag{17.4.12}$$

where (17.4.5) and (17.4.7) have been used. Comparing the two solutions (17.4.8) and (17.4.12), we obtain

$$\begin{aligned} G(s-u) &= G_\infty - \int_{-u}^{\infty} K(z+s, z+u)dz \\ &= G_\infty - \int_{0}^{\infty} K(w, w+s-u)dw, \end{aligned}$$

which agrees with (17.4.10). A similar result follows if we take $u \geq s$ and $y = u - s$. We can write (17.4.12) in a manner covering both cases as follows:

$$\begin{aligned} G(s,u) &= G(|s-u|) + \int_{-\min(s,u)}^{0} K(z+s, z+u)dz \\ &= G(|s-u|) + \int_{0}^{\min(s,u)} K(s-v, u-v)dv. \end{aligned}$$

This is the solution obeying the boundary conditions (17.3.9). Substituting this relation into the quadratic forms in $(17.3.5)_4$ and $(17.3.19)_1$, we obtain

$$\psi(t) = W(t) + \frac{1}{2}\int_0^\infty ds \int_0^\infty du \dot{E}^t(s) \int_0^{\min(s,u)} dv K(s-v, u-v)\dot{E}^t(u). \tag{17.4.13}$$

The integral term in (17.4.13) is non-positive by virtue of (17.3.21). It is of course the negative of the total dissipation $(-\mathfrak{D}(t))$, by virtue of (16.1.29).

17.4.1 Some Examples

1. *Discrete-spectrum materials:* The kernels $\widetilde{G}$ and K have the form

$$\begin{aligned} \widetilde{G}(s,u) &= \sum_{i,j=1}^{n} C_{ij} e^{-\alpha_i s - \alpha_j u}, \\ K(s,u) &= -\sum_{i,j=1}^{n} (\alpha_i + \alpha_j) C_{ij} e^{-\alpha_i s - \alpha_j u}, \end{aligned} \tag{17.4.14}$$

since if these are substituted into $(17.3.5)_4$ and $(17.3.11)_1$, they yield (16.3.13) and $(16.3.15)_{2,3}$. It is clear that (17.4.8) applied to $(17.4.14)_2$ yields $(17.4.14)_1$. Let the kernel $K(\cdot,\cdot)$ have the form

$$K(s,u) = -\sum_{i,j=1}^{n} \Gamma_{ij} e^{-\alpha_i s - \alpha_j u}, \tag{17.4.15}$$

where the symmetric matrix $\boldsymbol{\Gamma}$ with components $\Gamma_{ij},\ i,j=1,2,\ldots,n$ is nonnegative. The material is characterized by $\boldsymbol{\Gamma}$ and the vector $\boldsymbol{\alpha}$, with components $\alpha_i,\ i=1,2,\ldots,n$. These parameters are assumed to be known. Then, from $(17.3.11)_1$,

$$D(t)=\frac{1}{2}\sum_{i,j=1}^{n}\Gamma_{ij}e_i(t)e_j(t)=\frac{1}{2}\mathbf{e}\cdot\boldsymbol{\Gamma}\mathbf{e},$$
$$e_i(t)=\dot{E}_+^t(-i\alpha_i),\quad i=1,2,\ldots,n,\qquad \mathbf{e}=(e_1(t),e_2(t),\ldots,e_n(t)). \tag{17.4.16}$$

This expression for $D(t)$ agrees with (16.3.15). The final relation is a definition of the vector quantity $\mathbf{e}(t)$, which is that introduced earlier by (16.3.12). Applying (17.4.8), we obtain

$$\widetilde{G}(s,u)=\sum_{i,j=1}^{n}\frac{\Gamma_{ij}}{\alpha_i+\alpha_j}e^{-\alpha_i s-\alpha_j u},$$

so that from $(17.3.5)_2$

$$\psi(t)=\phi(t)+\frac{1}{2}\sum_{i,j=1}^{n}\frac{\Gamma_{ij}}{\alpha_i+\alpha_j}e_i(t)e_j(t)=\phi(t)+\frac{1}{2}\mathbf{e}\cdot\mathbf{C}\mathbf{e}, \tag{17.4.17}$$

where the matrix $\mathbf{C}$ has components of the form (see $(16.3.15)_3$)

$$C_{ij}=\frac{\Gamma_{ij}}{\alpha_i+\alpha_j},\quad i,j=1,2,\ldots,n.$$

The last form of (17.4.17) is of course (16.3.13). Then the relaxation function is given by

$$\widetilde{G}(0,u)=\widetilde{G}(u)=\sum_{j=1}^{n}G_je^{-\alpha_j u},\qquad G_j=\sum_{i=1}^{n}\frac{\Gamma_{ij}}{\alpha_i+\alpha_j},\quad j=1,\ldots,n.$$

2. *The Dill free energy:* Using $(17.3.5)_4$, we put

$$K(s,u)=2G'(s+u),$$

so that, from (17.4.8),

$$\widetilde{G}(s,u)=\widetilde{G}(s+u).$$

This yields the functional

$$\psi_{Dill}(t)=\phi(t)+\frac{1}{2}\int_0^\infty\int_0^\infty\widetilde{G}(s_1+s_2)\dot{E}^t(s_2)\dot{E}^t(s_1)ds_1ds_2, \tag{17.4.18}$$

which is a free energy with rate of dissipation (*cf.* $(17.3.11)_1$) given by

$$D_{Dill}(t)=-\int_0^\infty\int_0^\infty G'(s_1+s_2)\dot{E}^t(s_2)\dot{E}^t(s_1)ds_1ds_2 \tag{17.4.19}$$

if and only if $G(\cdot)$ is completely monotonic, as defined in [89].
For discrete-spectrum materials, these are given by

$$\psi_{Dill}(t) = \phi(t) + \frac{1}{2}\sum_{i=1}^{n} G_i e_i^2(t), \qquad D_{Dill}(t) = \sum_{i=1}^{n} \alpha_i G_i e_i^2(t). \tag{17.4.20}$$

3. *Short-term memory free energy:* We now construct a new free energy. Let us consider $K(s, u)$ given by the product form $k(s)k(u)$ which guarantees non-negativity of the rate of dissipation. Consider

$$k(s) = ae^{-\lambda s^2}, \qquad a, \lambda > 0, \tag{17.4.21}$$

which could be taken to model sharply declining or short-term memory behavior. Relation (17.4.21) yields

$$K(s,u) = -a^2 \exp\left[-\lambda(s^2+u^2)\right] = -a^2 \exp\left[-\frac{1}{2}\lambda(x^2+y^2)\right],$$

where the variables x and y are defined by (17.4.2). We obtain, from (17.4.6),

$$\begin{aligned}\widetilde{G}(s,u) &= \frac{1}{2}\int_x^\infty a^2 \exp\left[-\frac{1}{2}\lambda((x')^2+y^2)\right] dx' \\ &= \frac{1}{2}\sqrt{\frac{\pi}{2\lambda}}a^2\left[1-\Phi\left(\sqrt{\frac{\lambda}{2}}(s+u)\right)\right]\exp\left[-\frac{1}{2}\lambda(s-u)^2\right],\end{aligned} \tag{17.4.22}$$

where $\Phi(\cdot)$ is the probability integral [168]:

$$\Phi(z) = \frac{2}{\sqrt{\pi}}\int_0^z e^{-u^2}du.$$

From (17.4.10), (17.4.11), and (17.4.22), we have

$$\begin{aligned}\widetilde{G}(s) &= \frac{1}{2}\sqrt{\frac{\pi}{2\lambda}}a^2\left[1-\Phi\left(\sqrt{\frac{\lambda}{2}}s\right)\right]\exp\left[-\frac{1}{2}\lambda s^2\right], \\ \widetilde{G}_0 &= \frac{1}{2}\sqrt{\frac{\pi}{2\lambda}}a^2.\end{aligned}$$

17.5 Frequency Domain Representations of Free Energies in Terms of the Kernel $K_{+-}(\cdot,\cdot)$

The first discussion of the topic developed in this section was given in [158]. Also, special cases of the formulae given below, namely those relating to the minimum and associated free energies, were presented in [161] (see also (16.4.15)–(16.4.17)). One of our aims here is to provide generalizations of these formulae. However, our main goal is to derive certain results which will prove more convenient for determining new free energy functionals.

Let us now consider free energies with general kernels. We define

$$Z_{+-}(\omega_1, \omega_2) = \int_0^\infty \int_0^\infty Z(s, u) e^{-i\omega_1 s + i\omega_2 u} ds du, \tag{17.5.1}$$

where $Z(s, u)$ represents any one of the kernels $\widetilde{G}(s, u)$, $K(s, u)$, or $Q(s, u)$. Note that

$$\overline{Z_{+-}}(\omega_1, \omega_2) = Z_{+-}(-\omega_1, -\omega_2) = Z_{+-}(\omega_2, \omega_1), \tag{17.5.2}$$

so that $Z_{+-}(\omega, \omega)$ is real. The quantity $Z_{+-}(\omega_1, \omega_2)$ is analytic in the lower half of the ω_1 complex plane and in the upper half of the ω_2 plane.

Note that

$$\begin{aligned} &Z_{+-}(\omega_1, \omega_2) = Z_{+-}(-\omega_2, -\omega_1), \\ &\overline{Z_{+-}(\omega_1, \omega_2)} = Z_{+-}(-\overline{\omega_1}, -\overline{\omega_2}) = Z_{+-}(\overline{\omega_2}, \overline{\omega_1}), \quad \omega_1, \omega_2 \in \Omega, \end{aligned} \tag{17.5.3}$$

where the property $Z(s, u) = Z(u, s)$ has been used. These relations hold if ω_1, ω_2 are points of analyticity of $Z_{+-}(\omega_1, \omega_2)$. It follows from (17.5.3) that $Z_{+-}(\omega_1, \omega_2)$ is real if $\overline{\omega_2} = \omega_1$. In particular, $Z_{+-}(\omega_0, \omega_0)$ is real if ω_0 is real. Thus, $Z_{+-}(\omega_1, \omega_2)$ is given by analytic continuation from the real axis for $\omega_1 \in \Omega^{(-)}$ and $\omega_2 \in \Omega^{(+)}$.

Inverting Fourier transforms in (17.5.1) yields that

$$Z(s, u) = \frac{1}{4\pi^2} \int_{-\infty}^\infty \int_{-\infty}^\infty Z_{+-}(\omega_1, \omega_2) e^{i\omega_1 s - i\omega_2 u} d\omega_1 d\omega. \tag{17.5.4}$$

We shall not explicitly discuss the frequency domain version of the kernels $\mathcal{G}(s, u)$, $\mathcal{K}(s, u)$, and $\mathcal{Q}(s, u)$ in this section. These occur in quadratic forms expressed in terms of $E^t_{r+}(\omega)$ rather than $\dot{E}^t_+(\omega)$. We can, however, easily switch to this type of functional by using (16.1.17). Relations (17.3.12)$_1$, (17.3.23)$_1$, (17.3.24), and (17.5.1), yield that

$$\begin{aligned} i(\omega_1 - \omega_2)\widetilde{G}_{+-}(\omega_1, \omega_2) &= K_{+-}(\omega_1, \omega_2) + \widetilde{G}_+(\omega_1) + \overline{\widetilde{G}_+(\omega_2)} \\ &= K^{(1)}(\omega_1, \omega_2), \\ i(\omega_1 - \omega_2)Q_{+-}(\omega_1, \omega_2) &= -K_{+-}(\omega_1, \omega_2), \end{aligned} \tag{17.5.5}$$

where $\widetilde{G}_+(\omega)$ is defined by (16.1.7)$_1$. From (17.3.9), (16.1.7)$_1$, and (17.5.1), it follows that

$$\widetilde{G}_{+-}(\omega_1, \omega_2) \sim \begin{cases} \dfrac{\widetilde{G}_+(\omega_1)}{-i\omega_2} & \text{as } \omega_2 \to \infty, \\ \dfrac{\widetilde{G}_+(\omega_2)}{i\omega_1} & \text{as } \omega_1 \to \infty. \end{cases} \tag{17.5.6}$$

By an analogous argument to that yielding (17.5.6), we can deduce similar properties for $K_{+-}(\omega_1, \omega_2)$.

We can write (17.3.5)$_4$, (17.3.11)$_1$, and (17.3.16)$_1$ in the forms

$$
\begin{aligned}
\psi(t) &= \phi(t) + \frac{1}{8\pi^2}\int_{-\infty}^{\infty}\int_{-\infty}^{\infty}\overline{\dot{E}^t_+(\omega_1)}\widetilde{G}_{+-}(\omega_1,\omega_2)\dot{E}^t_+(\omega_2)d\omega_1 d\omega_2,\\
D(t) &= -\frac{1}{8\pi^2}\int_{-\infty}^{\infty}\int_{-\infty}^{\infty}\overline{\dot{E}^t_+(\omega_1)}K_{+-}(\omega_1,\omega_2)\dot{E}^t_+(\omega_2)d\omega_1 d\omega_2,\\
\mathcal{D}(t) &= \frac{1}{8\pi^2}\int_{-\infty}^{\infty}\int_{-\infty}^{\infty}\overline{\dot{E}^t_+(\omega_1)}Q_{+-}(\omega_1,\omega_2)\dot{E}^t_+(\omega_2)d\omega_1 d\omega_2,
\end{aligned}
\tag{17.5.7}
$$

where $\dot{E}^t_+(\omega)$ is defined by (16.1.17). Note that $(17.5.7)_2$ can be written in the form

$$
D(t) = -\frac{1}{8\pi^2}\int_{-\infty}^{\infty}\int_{-\infty}^{\infty}\overline{E^t_{r+}(\omega_1)}\omega_1\omega_2 K_{+-}(\omega_1,\omega_2)E^t_{r+}(\omega_2)d\omega_1 d\omega_2,
$$

by virtue of (16.1.17). A similar observation applies to $(17.5.7)_{1,3}$. Using the analyticity properties of the kernels (see (16.1.24)), we can write (17.5.7) in the form

$$
\begin{aligned}
\psi(t) &= \phi(t) + \frac{1}{8\pi^2}\int_{-\infty}^{\infty}\int_{-\infty}^{\infty}\overline{\dot{E}^t_F(\omega_1)}\widetilde{G}_{+-}(\omega_1,\omega_2)\dot{E}^t_F(\omega_2)d\omega_1 d\omega_2,\\
D(t) &= -\frac{1}{8\pi^2}\int_{-\infty}^{\infty}\int_{-\infty}^{\infty}\overline{\dot{E}^t_F(\omega_1)}K_{+-}(\omega_1,\omega_2)\dot{E}^t_F(\omega_2)d\omega_1 d\omega_2,\\
\mathcal{D}(t) &= \frac{1}{8\pi^2}\int_{-\infty}^{\infty}\int_{-\infty}^{\infty}\overline{\dot{E}^t_F(\omega_1)}Q_{+-}(\omega_1,\omega_2)\dot{E}^t_F(\omega_2)d\omega_1 d\omega_2,
\end{aligned}
\tag{17.5.8}
$$

where $E^t_F(\omega)$ is defined by $(16.1.24)_2$. Referring to Remark 17.3.3, we see that (17.5.8) corresponds to case (b) of (17.3.18), as is also true for (16.1.24).

As for the time domain expressions, we conclude from P2 and P3 (see (16.1.26), (16.1.28)) that the quadratic functionals in $(17.5.7)_1$ and $(17.5.7)_3$ must be nonnegative, while those in $(17.5.7)_2$ must be nonpositive. A similar statement applies to (17.5.8). We seek therefore to write down the analogue of (17.3.15). It follows from (17.5.8) that

$$
\widetilde{G}_{+-}(\omega,\omega) \geq 0, \qquad K_{+-}(\omega,\omega) \leq 0 \;\; \forall\, \omega \in \mathbb{R}. \tag{17.5.9}
$$

Relations (17.5.9) do not follow from (17.5.7) because of the analyticity properties of $\dot{E}^t_+(\omega)$ which, as we will see in Sect. 17.5.2, allow considerable non-uniqueness in the kernels of (17.5.7), so that in particular the diagonal quantities ($\omega_1 = \omega_2$) are not unique. Such non-uniqueness is not present in (17.5.8). The quantity $Q_{+-}(\omega_1,\omega_2)$ is excluded because it contains singularities at $\omega_1 = \omega_2$, which arises from the fact that $(17.3.19)_{3,4}$ can be written in the form (17.5.8) but with a delta function.

It follows from $(17.5.5)_2$ that

$$
\psi(t) = \phi(t) - \frac{i}{8\pi^2}\int_{-\infty}^{\infty}\int_{-\infty}^{\infty}\frac{\overline{\dot{E}^t_+(\omega_1)}K^{(1)}(\omega_1,\omega_2)\dot{E}^t_+(\omega_2)}{\omega_1^+ - \omega_2^-}d\omega_1 d\omega_2. \tag{17.5.10}
$$

The notation in the denominator of this last form means that if we integrate first over ω_1, it becomes $(\omega_1 - \omega_2^-)$ or if ω_2 first then it is $(\omega_1^+ - \omega_2)$. This choice of denominator, rather than $(\omega_1^- - \omega_2^+)$, is initially assigned by using the example given by (16.4.15). It will be justified below by means of a general argument. The terms in $K^{(1)}$ depending only on one variable (ω_1 or ω_2) do not contribute to the integral. To see this, consider, for example, the term

$$-\frac{i}{8\pi^2}\int_{-\infty}^{\infty}\int_{-\infty}^{\infty}\frac{\overline{\dot{E}^t_+(\omega_1)}\widetilde{G}_+(\omega_1)\dot{E}^t_+(\omega_2)}{\omega_1^+-\omega_2^-}d\omega_1 d\omega_2 \tag{17.5.11}$$

and carry out the integral over ω_2, completing the contour over $\Omega^{(-)}$, on which $\dot{E}^t_+(\omega_2)$ is analytic. The denominator becomes $\omega_1^+ - \omega_2$. The infinite part of the contour yields a vanishing contribution, by virtue of (16.1.20), so the result is zero. We note, however, for later use that if ω_1^+ were replaced by ω_1^-, the result would be

$$-\frac{1}{4\pi}\int_{-\infty}^{\infty}\overline{\dot{E}^t_+(\omega_1)}\widetilde{G}_+(\omega_1)\dot{E}^t_+(\omega_1)d\omega_1. \tag{17.5.12}$$

Thus, (17.5.10) can be replaced by

$$\begin{aligned}\psi(t) &= \phi(t) - \frac{i}{8\pi^2}\int_{-\infty}^{\infty}\int_{-\infty}^{\infty}\frac{\overline{\dot{E}^t_+(\omega_1)}K_{+-}(\omega_1,\omega_2)\dot{E}^t_+(\omega_2)}{\omega_1^+-\omega_2^-}d\omega_1 d\omega_2\\ &= \phi(t) - \frac{i}{8\pi^2}\int_{-\infty}^{\infty}\int_{-\infty}^{\infty}\frac{\overline{E^t_{r+}(\omega_1)}\omega_1\omega_2 K_{+-}(\omega_1,\omega_2)E^t_{r+}(\omega_2)}{\omega_1^+-\omega_2^-}d\omega_1 d\omega_2,\end{aligned} \tag{17.5.13}$$

on using (16.1.17). Let us apply the Plemelj formulae to the integral in $(17.5.13)_1$ over ω_1 to obtain

$$\begin{aligned}\psi(t) = \phi(t) &- \frac{i}{8\pi^2}P\int_{-\infty}^{\infty}\int_{-\infty}^{\infty}\frac{\overline{\dot{E}^t_+(\omega_1)}K_{+-}(\omega_1,\omega_2)\dot{E}^t_+(\omega_2)}{\omega_1-\omega_2}d\omega_1 d\omega_2\\ &- \frac{1}{8\pi}\int_{-\infty}^{\infty}\overline{\dot{E}^t_+(\omega)}K_{+-}(\omega,\omega)\dot{E}^t_+(\omega)d\omega,\end{aligned} \tag{17.5.14}$$

where the symbol "P" indicates a principal value integral over ω_1. Also, consider the integral (see $(17.5.5)_3$)

$$\mathcal{D}_1(t) = \frac{i}{8\pi^2}\int_{-\infty}^{\infty}\int_{-\infty}^{\infty}\frac{\overline{\dot{E}^t_+(\omega_1)}K_{+-}(\omega_1,\omega_2)\dot{E}^t_+(\omega_2)}{\omega_1^- - \omega_2^+}d\omega_1 d\omega_2.$$

Differentiating with respect to t, we find, with the aid of $(16.1.18)_1$, that

$$\dot{\mathcal{D}}_1(t) = D(t), \tag{17.5.15}$$

where $D(t)$ is given by $(17.5.7)_2$. The relation

$$\int_{-\infty}^{\infty}\frac{\overline{\dot{E}^t_+(\omega_1)}K_{+-}(\omega_1,\omega_2)}{\omega_1^- - \omega_2}d\omega_2 = 0, \tag{17.5.16}$$

and a similar one involving integration over ω_1, have been used. Equation (17.5.16) follows by closing the contour on $\Omega^{(+)}$ and recalling that a property similar to (17.5.6) applies to $K_{+-}(\omega_1, \omega_2)$. Relation (17.5.15) implies that $\mathcal{D}_1(t) = \mathcal{D}(t) + k$, where k is a constant which can be shown to be zero because $\mathcal{D}_1(-\infty) = 0 = \mathcal{D}(-\infty)$. This follows by observing that $\dot{E}^t_+$, given by (16.1.17), tends to zero as $t \to -\infty$, because of assumption (16.1.2). We conclude that

$$\begin{aligned}\mathcal{D}(t) = \mathcal{D}_1(t) &= \frac{i}{8\pi^2} \int_{-\infty}^{\infty} \int_{-\infty}^{\infty} \frac{\overline{\dot{E}^t_+(\omega_1)} K_{+-}(\omega_1, \omega_2) \dot{E}^t_+(\omega_2)}{\omega_1^- - \omega_2^+} d\omega_1 d\omega_2 \\ &= \frac{i}{8\pi^2} \int_{-\infty}^{\infty} \int_{-\infty}^{\infty} \frac{\overline{E^t_{r+}(\omega_1)} \omega_1 \omega_2 K_{+-}(\omega_1, \omega_2) E^t_{r+}(\omega_2)}{\omega_1^- - \omega_2^+} d\omega_1 d\omega_2.\end{aligned} \tag{17.5.17}$$

In a similar manner to the derivation of (17.5.14), we have

$$\begin{aligned}\mathcal{D}(t) = &\frac{i}{8\pi^2} P \int_{-\infty}^{\infty} \int_{-\infty}^{\infty} \frac{\overline{\dot{E}^t_+(\omega_1)} K_{+-}(\omega_1, \omega_2) \dot{E}^t_+(\omega_2)}{\omega_1 - \omega_2} d\omega_1 d\omega_2 \\ &- \frac{1}{8\pi} \int_{-\infty}^{\infty} \overline{\dot{E}^t_+(\omega)} K_{+-}(\omega, \omega) \dot{E}^t_+(\omega) d\omega.\end{aligned}$$

Therefore,

$$\begin{aligned}\psi(t) + \mathcal{D}(t) &= \phi(t) - \frac{1}{4\pi} \int_{-\infty}^{\infty} \overline{\dot{E}^t_+(\omega)} K_{+-}(\omega, \omega) \dot{E}^t_+(\omega) d\omega \\ &= \phi(t) + \frac{1}{2\pi} \int_{-\infty}^{\infty} \overline{\dot{E}^t_+(\omega)} \frac{H(\omega)}{\omega^2} \dot{E}^t_+(\omega) d\omega,\end{aligned}$$

by virtue of (16.1.29) and (17.3.19) for arbitrary histories. It follows from this result together with $(16.1.7)_1$ and (16.1.13) that

$$\begin{aligned}K_{+-}(\omega, \omega) &= -2\frac{H(\omega)}{\omega^2} = -2\widetilde{G}_c(\omega) \\ &= -\frac{G'_+(\omega)}{i\omega} + \frac{\overline{G'_+(\omega)}}{i\omega} = -\widetilde{G}_+(\omega) - \overline{\widetilde{G}_+(\omega)},\end{aligned} \tag{17.5.18}$$

where (16.1.8) has also been invoked. This relation can in fact be shown directly. From (17.5.1), we can write

$$K_{+-}(\omega, \omega) = \int_0^{\infty} \int_0^{\infty} K(s, u) e^{-i\omega(s-u)} ds du.$$

Transforming to the variables x, y defined by (17.4.2), we obtain

$$\begin{aligned}K_{+-}(\omega, \omega) &= -\frac{1}{2} \int_0^{\infty} dx \int_{-x}^{x} dy K_n(x, y) e^{-i\omega y} \\ &= - \int_0^{\infty} dx \int_0^{x} dy K_n(x, y) \cos \omega y,\end{aligned}$$

where (17.4.4) has been used. We can write this in the form

$$K_{+-}(\omega,\omega) = -\int_0^\infty dy \int_y^\infty dx K_n(x,y)\cos\omega y$$
$$= -2\int_0^\infty \widetilde{G}(y)\cos\omega y dy = -2\widetilde{G}_c(w),$$

by virtue of (17.4.6) and (17.4.5). This proves the result.

Remark 17.5.1. We see from this derivation that the constraint in the time domain equivalent to (17.5.18) is (17.4.5) combined with (17.4.6) (which yield (17.4.10)). Relation (17.4.5) (or (17.3.9)) is of course the property P1 or (16.1.25). The frequency domain form (17.5.18) is more useful for deriving explicit forms of free energies, as we shall see.

Remark 17.5.2. Relation (17.5.18) can be used to prove P1 or (16.1.25), from $(17.5.13)_2$, by differentiating the latter equation with respect to $E(t)$.

Note that (17.5.18) implies that $K^{(1)}(\omega,\omega)$, defined by $(17.5.5)_2$, vanishes. The quantity $K_{+-}(\omega,\omega)$ is independent of the choice of free energy. Observe that (17.5.18) is consistent with $(17.5.9)_2$. Both the non-positivity of $K_{+-}(\omega,\omega)$ and the non-negativity of $H(\omega)$ are direct consequences of the second law (see P3 after (16.1.28), (16.1.12) and (16.1.13)).

If we were to take the other choice of denominator, namely $\omega_1^- - \omega_2^+$, in (17.5.13) and $\omega_1^+ - \omega_2^-$ in (17.5.17), there would be a positive sign on $2H(\omega)/\omega^2$ in $(17.5.18)_1$, which contradicts the second law, as expressed by $(17.5.9)_2$.

We now show that (17.4.9) is the time domain version of $(17.5.13)_1$. Substituting (17.5.4) for $Z(s,u) = K(s,u)$ into $(17.4.8)_1$ yields

$$\widetilde{G}(s,u) = -\frac{1}{4\pi^2}\int_{-\infty}^{\infty}\int_{-\infty}^{\infty}\int_0^\infty K_{+-}(\omega_1,\omega_2)e^{i\omega_1(s+z) - i\omega_2(u+z)}dz d\omega_1 d\omega_2$$
$$= -\frac{i}{4\pi^2}\int_{-\infty}^{\infty}\int_{-\infty}^{\infty}\frac{K_{+-}(\omega_1,\omega_2)}{\omega_1^+ - \omega_2^-}e^{i\omega_1 s - i\omega_2 u}d\omega_1 d\omega_2. \tag{17.5.19}$$

The denominator results from putting

$$e^{i(\omega_1 - \omega_2)z} = e^{i(\omega_1 - \omega_2^-)z} = e^{i(\omega_1^+ - \omega_2)z}, \tag{17.5.20}$$

which ensures that the integral over z exists. If $\widetilde{G}(s,u)$, given by (17.5.19), is inserted into $(17.3.5)_4$, we obtain $(17.5.13)_1$. The argument based on (17.5.20) is perhaps the simplest way of showing that the denominator in (17.5.13) is the correct choice, rather than the alternative in (17.5.17).

Using (16.1.25) to obtain the form of the stress function from $(17.5.13)_1$, we find, by virtue of (16.1.17), that

$$T(t) = T_e(t) - \frac{i}{4\pi^2}\int_{-\infty}^{\infty}\int_{-\infty}^{\infty}\frac{K_{+-}(\omega_1,\omega_2)\dot{E}_+^t(\omega_2)}{\omega_1^+ - \omega_2^-}d\omega_1 d\omega_2.$$

This involves combining two terms which can be shown to be equal with the help of (17.5.2). Carrying out the ω_1 integration by closing the contour on $\Omega^{(-)}$ and using (17.5.18), we obtain $(16.1.23)_2$.

Thus, if $K_{+-}(\omega_1, \omega_2)$ is given for all ω_1 and $\omega_2 \in \mathbb{R}$, the material is completely characterized by (17.5.13) and (17.5.18).

Relations (16.4.15) and (16.4.17) are special cases of $(17.5.13)_2$ and $(17.5.7)_2$ (the latter expressed in terms of E^t_{r+}), where

$$\omega_1\omega_2 K_{+-}(\omega_1, \omega_2) = -2H^f_+(\omega_1)H^f_-(\omega_2),$$

while (16.4.16) is a special case of $(17.5.17)_2$. Also, replacements corresponding to (17.5.22) below can be implemented here. The particular case referred to in Remark 17.5.3 is obtained by subtracting a term $H^f_+(\omega_2)H^f_-(\omega_1)$ from the kernels in (16.4.15), which yields an alternative expression involving non-singular integrals:

$$\psi_f(t) = \phi(t) + \frac{i}{4\pi^2}\int_{-\infty}^{\infty}\int_{-\infty}^{\infty} \frac{\overline{E^t_{r+}(\omega_1)}N(\omega_1, \omega_2)E^t_{r+}(\omega_2)}{\omega_1 - \omega_2} d\omega_1 d\omega_2,$$
$$N(\omega_1, \omega_2) = H^f_+(\omega_1)H^f_-(\omega_2) - H^f_+(\omega_2)H^f_-(\omega_1).$$

This is a special case of (17.5.26) below, with $\dot{E}^t_+(\omega)$ replaced by $-i\omega E^t_{r+}(\omega)$, by virtue of (16.1.17). Using the same relation, we can interchangeably express the above formulae in terms of $\dot{E}^t_+(\omega)$ or $E^t_{r+}(\omega)$.

17.5.1 Example: Discrete-Spectrum Materials

This was already discussed in the context of the time domain theory, as expressed by (17.4.14). From (17.5.1) and (17.4.15), it follows that

$$\begin{aligned} K_{+-}(\omega_1, \omega_2) &= -\sum_{i,j=1}^{n} \frac{\Gamma_{ij}}{(\alpha_i + i\omega_1)(\alpha_j - i\omega_2)}, \\ \widetilde{G}_{+-}(\omega_1, \omega_2) &= \sum_{i,j=1}^{n} \frac{\Gamma_{ij}}{(\alpha_i + \alpha_j)(\alpha_i + i\omega_1)(\alpha_j - i\omega_2)}, \\ \widetilde{G}_{+}(\omega) &= \sum_{i=1}^{n} \frac{G_i}{\alpha_i + i\omega} = \sum_{i,j=1}^{n} \frac{\Gamma_{ij}}{(\alpha_i + \alpha_j)(\alpha_i + i\omega_1)}. \end{aligned} \tag{17.5.21}$$

The formula for $\widetilde{G}_+(\omega)$ also follows from (17.5.6). Observe that

$$\begin{aligned} K_{+-}(\omega, \omega) &= -\sum_{i,j=1}^{n} \frac{\Gamma_{ij}}{(\alpha_i + \alpha_j)} \left\{ \frac{1}{\alpha_i + i\omega} + \frac{1}{\alpha_j - i\omega} \right\} \\ &= -\sum_{i,j=1}^{n} \frac{\Gamma_{ij}}{(\alpha_i + \alpha_j)} \left\{ \frac{1}{\alpha_i + i\omega} + \frac{1}{\alpha_i - i\omega} \right\} = -\frac{2H(\omega)}{\omega^2}, \end{aligned}$$

by virtue of $(17.5.21)_{3,4}$. This agrees with (17.5.18).

The positivity of the individual G_j is not in itself a requirement. What is important is that $\widetilde{G}_c(\omega)$ be nonnegative, as required by (16.1.12). Now, from $(17.5.21)_4$,

$$\widetilde{G}_c(\omega) = \sum_{i,j}^{n} \frac{\alpha_i \Gamma_{ij}}{\alpha_i^2 + \omega^2} = \sum_{i,j}^{n} \frac{\alpha_j \Gamma_{ij}}{\alpha_j^2 + \omega^2}$$

$$= \frac{1}{2} \sum_{i,j}^{n} \frac{(\omega^2 + \alpha_i \alpha_j) \Gamma_{ij}}{(\alpha_i^2 + \omega^2)(\alpha_j^2 + \omega^2)}.$$

This is the sum of two terms, one proportional to ω^2 in the numerator, and the other proportional to $\alpha_i \alpha_j$. Both of these can be seen to be separately nonnegative, on recalling that Γ is a nonnegative matrix, so we have the desired property (16.1.12).

17.5.2 Non-uniqueness of the Kernels

We now consider how the kernels in (17.5.7) are not unique. This phenomenon is the frequency domain version of Causality constraints outlined in Remark 17.3.3. We deal here with case (a) of (17.3.18). Using the same argument as that leading to (16.1.22), we can express (17.5.7) in different forms. Consider the replacement

$$Z_{+-}(\omega_1, \omega_2) \to Y(\omega_1, \omega_2) = Z_{+-}(\omega_1, \omega_2) + z_{2+}(\omega_1, \omega_2) + z_{1-}(\omega_1, \omega_2), \qquad (17.5.22)$$

where $Z_{+-}(\omega_1, \omega_2)$ becoming $Y(\omega_1, \omega_2)$ represents either $K_{+-}(\omega_1, \omega_2)$ becoming $L(\omega_1, \omega_2)$ or $\widetilde{G}_{+-}(\omega_1, \omega_2)$ becoming $R(\omega_1, \omega_2)$, both of which will be used later. The functions $z_{2+}(\omega_1, \omega_2)$ and $z_{1-}(\omega_1, \omega_2)$ correspondingly represent $k_{2+}(\omega_1, \omega_2)$ and $k_{1-}(\omega_1, \omega_2)$ for $K_{+-}(\omega_1, \omega_2)$ and $g_{2+}(\omega_1, \omega_2)$ and $g_{1-}(\omega_1, \omega_2)$ for $\widetilde{G}_{+-}(\omega_1, \omega_2)$, respectively.

The function $z_{2+}(\omega_1, \omega_2)$ has singularities on the ω_2 complex plane only in $\Omega^{(+)}$ and $z_{1-}(\omega_1, \omega_2)$ has singularities on the ω_1 complex plane only in $\Omega^{(-)}$.

Such substitutions leave the relations in (17.5.7) unchanged, as may be seen by closing the relevant integral on $\Omega^{(\pm)}$ as appropriate, and invoking Cauchy's theorem. It is required, however, that the contributions from the infinite parts of the contours vanish. Noting (16.1.20) in this context, we see that the quantities $k_{2+}(\omega_1, \omega_2)$, $k_{1-}(\omega_1, \omega_2)$ and $g_{2+}(\omega_1, \omega_2)$, $g_{1-}(\omega_1, \omega_2)$ must decay to zero at large ω_1 or ω_2.

Similar substitutions can be made for Q_{+-} in $(17.5.7)_3$.

Interesting special cases are as follows:

$$Y(\omega_1, \omega_2) = Z_{+-}(\omega_1, \omega_2) + a_1 Z_{+-}(-\omega_1, \omega_2) + a_2 Z_{+-}(\omega_1, -\omega_2)$$
$$+ a_3 Z_{+-}(-\omega_1, -\omega_2),$$

where the a_i, $i = 1, 2, 3$, are arbitrary complex constants which may be different for each quantity represented by Z. Similar remarks apply to $Q_{+-}(\omega_1, \omega_2)$.

Remark 17.5.3. If we choose $a_1 = a_2 = 0$ and $a_3 = -1$, then from (17.5.2), it follows that $Y(\omega, \omega) = 0$.

Thus, we can write a general form of $(17.5.7)_{1,2}$, incorporating substitutions of the kind outlined above, as follows:

$$\begin{aligned}\psi(t) &= \phi(t) + \frac{1}{8\pi^2}\int_{-\infty}^{\infty}\int_{-\infty}^{\infty}\overline{\dot{E}_+^t(\omega_1)}R(\omega_1,\omega_2)\dot{E}_+^t(\omega_2)d\omega_1 d\omega_2 \\ D(t) &= -\frac{1}{8\pi^2}\int_{-\infty}^{\infty}\int_{-\infty}^{\infty}\overline{\dot{E}_+^t(\omega_1)}L(\omega_1,\omega_2)\dot{E}_+^t(\omega_2)d\omega_1 d\omega_2,\end{aligned} \tag{17.5.23}$$

in the notation specified after (17.5.22), so that, for example,

$$L(\omega_1,\omega_2) = K_{+-}(\omega_1,\omega_2) + k_{2+}(\omega_1,\omega_2) + k_{1-}(\omega_1,\omega_2).$$

If (17.5.7) is replaced by (17.5.8), the non-uniqueness of the kernels no longer holds, as noted after (17.5.9). Equation $(17.5.23)_1$ can be replaced by

$$\psi(t) = \phi(t) - \frac{i}{8\pi^2}\int_{-\infty}^{\infty}\int_{-\infty}^{\infty}\frac{\overline{\dot{E}_+^t(\omega_1)}L(\omega_1,\omega_2)\dot{E}_+^t(\omega_2)}{\omega_1^+ - \omega_2^-}d\omega_1 d\omega_2, \tag{17.5.24}$$

by virtue of the relation

$$-\frac{i}{8\pi^2}\int_{-\infty}^{\infty}\int_{-\infty}^{\infty}\frac{\overline{\dot{E}_+^t(\omega_1)}[k_{2+}(\omega_1,\omega_2) + k_{1-}(\omega_1,\omega_2)]\dot{E}_+^t(\omega_2)}{\omega_1^+ - \omega_2^-}d\omega_1 d\omega_2 = 0,$$

so that (17.5.24) reduces to $(17.5.13)_1$. The term $k_{2+}(\omega_1,\omega_2)$ yields zero by integrating the variable ω_2 over a contour enclosing $\Omega^{(-)}$, which is a generalization of the argument relating to (17.5.11). Similarly for the term $k_{1-}(\omega_1,\omega_2)$, which is zero by virtue of the integration over ω_1.

However, for $\mathfrak{D}(t)$, we have

$$\begin{aligned}\mathfrak{D}(t) &= \frac{i}{8\pi^2}\int_{-\infty}^{\infty}\int_{-\infty}^{\infty}\frac{\overline{\dot{E}_+^t(\omega_1)}L(\omega_1,\omega_2)\dot{E}_+^t(\omega_2)}{\omega_1^- - \omega_2^+}d\omega_1 d\omega_2 \\ &\quad + \frac{1}{4\pi}\int_{-\infty}^{\infty}\overline{\dot{E}_+^t(\omega)}\,[k_{2+}(\omega,\omega) + k_{1-}(\omega,\omega)]\,\dot{E}_+^t(\omega)d\omega.\end{aligned} \tag{17.5.25}$$

This follows by a generalization of the argument leading to (17.5.12). Applying the Plemelj formulae to (17.5.24) and (17.5.25), we find that the condition (17.5.18) re-emerges.

Note that if

$$L(\omega_1,\omega_2) = K_{+-}(\omega_1,\omega_2) - K_{+-}(-\omega_1,-\omega_2),$$

as specified by Remark 17.5.3, then the integrals in (17.5.24) and (17.5.25) are non-singular because $L(\omega,\omega)$ vanishes, by virtue of (17.5.2). Thus,

$$\psi(t) = \phi(t) - \frac{i}{8\pi^2}\int_{-\infty}^{\infty}\int_{-\infty}^{\infty}\frac{\overline{\dot{E}_+^t(\omega_1)}L(\omega_1,\omega_2)\dot{E}_+^t(\omega_2)}{\omega_1 - \omega_2}d\omega_1 d\omega_2. \tag{17.5.26}$$

The general forms of free energies or dissipation functionals can be specialized in two ways: specifying histories or choosing particular functional forms for the rate of dissipation kernels $K(\cdot,\cdot)$. We now explore both of these approaches.

17.6 General Dissipative Materials for Specified Histories

In this section, we choose general kernels and particular histories.

As noted in Sect. 17.2, a given material with memory typically has a set of many free energy functionals associated with it, all members of which yield the same constitutive relations. Explicit formulae are derived in this chapter for the free energy and total dissipation of such a material in the cases of step function and sinusoidal/exponential histories. Expressions for the fraction of stored and dissipated energy are deduced.

17.6.1 Free Energy and Dissipation Functionals for Particular Histories

Two important dimensionless parameters are

$$\beta = \frac{G_\infty}{G_0}, \quad \chi = \frac{G_0 - G_\infty}{G_0} = 1 - \beta = -\frac{1}{G_0}\int_0^\infty K(z,z)dz, \quad \beta, \chi \in [0,1]. \quad (17.6.1)$$

The relation (17.3.15) has been used. These parameters provide simple measures of the memory contribution and therefore, the amount of energy loss due to material deformation. The smaller the quantity β or the larger the parameter χ, the greater the energy loss.

The fraction of energy stored and dissipated, respectively, for any given history, can be determined according to the formulae

$$F_s(t) = \frac{\psi(t)}{W(t)}, \qquad F_d(t) = \frac{\mathcal{D}(t)}{W(t)}, \qquad F_s(t) + F_d(t) = 1. \qquad (17.6.2)$$

The interpretation of these quantities is discussed further in [18].

Expressions for the free energy, total dissipation, and the ratios $F_s(t)$, $F_d(t)$ will be given in the case of a general rate of dissipation kernel $K(s,u)$, and three different types of strain history. These quantities are important characteristic properties of the material described by this kernel.

Any choice of $K(s,u) \in \mathcal{K}$ will describe the stress-strain and energy behavior of one specific material. Matters are more difficult when we seek to reverse this process and determine the kernel $K(s,u)$ which describe a pre-chosen material, referred to as material I in Sect. 17.1.

Let the quantity $K_{+-}(\omega_1,\omega_2)$, determined from each $K(s,u) \in \mathcal{K}$ by (17.5.1), form a set $\mathcal{K}_F$.

As already noted, the functionals $\psi_f(t)$, $f = 1,2,\ldots,N$ lie on the boundary of $\mathcal{F}$, in particular the minimum and maximum free energies which provide lower and upper bounds; similarly for $K^f_{+-}(\omega_1,\omega_2)$ with respect to $\mathcal{K}_F$. The factors $H^f_\pm(\omega)$ and therefore all these quantities are deduced from the parameters of the relaxation function. The size of the set $\mathcal{F}$ (and $\mathcal{K}_F$ or $\mathcal{K}$) is, in this sense, determined by the relaxation function.

We seek to give detailed expressions for free energies and related quantities for general choices of the kernel $K(s,u) \in \mathcal{K}$ and histories with step function and SE behavior.

17.6.1.1 Step Function Histories

This is the simplest non-constant behavior of the strain history, given as follows. Consider $E(u)$, $u \leq t$ where t is the current time, assumed to be positive. Let

$$E(u) = \begin{cases} 0, & u \leq 0, \\ E_0, & 0 < u \leq t, \end{cases} \tag{17.6.3}$$

giving

$$\dot{E}(u) = \frac{dE(u)}{du} = E_0\delta(u),$$

in terms of the singular delta function. Thus,

$$\dot{E}^t(s) = E_0\delta(t-s).$$

It follows from this relation and (17.3.11) that

$$D(t) = -\frac{E_0^2}{2}K(t,t). \tag{17.6.4}$$

Also, from (17.4.8), $(17.3.5)_4$ and (16.1.32), we have

$$\begin{aligned} \psi(t) &= \phi(t) - \frac{E_0^2}{2}\int_0^\infty K(t+z, t+z)dz \\ &= \frac{1}{2}G_\infty E_0^2 - \frac{E_0^2}{2}\int_t^\infty K(y,y)dy, \quad t \geq 0. \end{aligned} \tag{17.6.5}$$

The integral term is nonnegative, by virtue of (17.3.15). Relation (17.4.11) yields that

$$\frac{1}{2}G_0E_0^2 = \frac{1}{2}G_\infty E_0^2 - \frac{E_0^2}{2}\int_0^\infty K(y,y)dy. \tag{17.6.6}$$

Thus,

$$\psi(t) \leq \frac{1}{2}G_0E_0^2.$$

It follows from (17.3.19) that

$$W(t) = \frac{1}{2}G_0E_0^2,$$

so that (17.3.21) is satisfied. Relation (16.1.29) gives that

$$\mathcal{D}(t) = -\frac{E_0^2}{2}\int_0^t K(y,y)dy.$$

The finite range of the integral is easily understood, from a physical point of view. For the infinite period specified by $(17.6.3)_1$, there is no dissipation. At time $t = 0$, dissipation begins. Referring to (17.6.2), we see that

$$F_d(t) = -\frac{1}{G_0}\int_0^t K(y,y)dy, \qquad F_s(t) = 1 - F_d(t).$$

Differentiating with respect to time, we obtain

$$\frac{d}{dt}F_d(t) = -\frac{1}{G_0}K(t,t), \qquad \frac{d}{dt}F_s(t) = -\frac{d}{dt}F_d(t),$$

so that $F_d(t)$ is monotonically increasing and $F_s(t)$ is monotonically decreasing. Note that

$$F_d(\infty) = \frac{G_0 - G_\infty}{G_0} = \chi = 1 - \beta, \qquad F_s(\infty) = \frac{G_\infty}{G_0} = \beta,$$

in terms of the quantities introduced in (17.6.1). Thus, χ measures the energy dissipation and β the energy storage for any material in $\mathcal{K}$, due to a sudden step change in strain.

The quantity $F_d(t)$ is zero at $t = 0$ and increases monotonically to χ as $t \to \infty$, while $F_s(t) = 1$ at $t = 0$ and decreases monotonically to β at large t. Note that $K(y, y)$ must tend to zero as $y \to \infty$ to yield convergent integrals in (17.6.5) or (17.6.6).

Observe from (17.6.4) that, if $D(t)$ can be determined, this yields a measurement of $K(t, t)$. To obtain measurements of $K(s, u)$, $s, u \in R^+$, one needs to consider histories with two steps (Sect. 7.1.3). Of course, this is not a very practical technique since step function histories are difficult to approximate closely.

17.6.1.2 SSE Histories

Consider the history and current value $(E^t, E(t))$ defined by (16.11.1). Furthermore, $\omega_\pm$ are given by (16.11.2). For $\eta = 0$, we have purely sinusoidal behavior, while for $\omega_0 = 0$, the history is exponentially growing. The derivative $\dot{E}^t(s)$ has the form

$$\dot{E}^t(s) = i\omega_- E_0 e^{i\omega_-(t-s)} - i\omega_+ \overline{E_0} e^{-i\omega_+(t-s)}. \tag{17.6.7}$$

Also, the quantity $E^t_+(\omega)$ is given by

$$E^t_+(\omega) = E_0 \frac{e^{i\omega_- t}}{i(\omega + \omega_-)} + \overline{E_0}\frac{e^{-i\omega_+ t}}{i(\omega - \omega_+)}, \tag{17.6.8}$$

while

$$\dot{E}^t_+(\omega) = E_0\omega_- \frac{e^{i\omega_- t}}{\omega + \omega_-} - \overline{E_0}\omega_+ \frac{e^{-i\omega_+ t}}{\omega - \omega_+} = -i\omega E^t_{r+}(\omega), \tag{17.6.9}$$

by virtue of (16.1.17).

Using (16.1.3) and (17.6.7), we find that the stress is given by

$$T(t) = \mathcal{M}_+(\omega_-)E_0 e^{i\omega_- t} + \mathcal{M}_+(-\omega_+)\overline{E_0} e^{-i\omega_+ t}, \tag{17.6.10}$$

where $\mathcal{M}_+(\omega)$ is defined by (16.1.9). Referring to (17.6.8) and (17.6.9), we see that

$$E_+^t(-i\alpha) = E_0 \frac{e^{i\omega_- t}}{\alpha + i\omega_-} + \overline{E_0} \frac{e^{-i\omega_+ t}}{\alpha - i\omega_+},$$
$$\dot{E}_+^t(-i\alpha) = iE_0 \frac{\omega_- e^{i\omega_- t}}{\alpha + i\omega_-} - i\overline{E_0} \frac{\omega_+ e^{-i\omega_+ t}}{\alpha - i\omega_+}.$$

It follows from (17.6.9) that

$$\begin{aligned} E_{r+}^t(-i\alpha_i) &= -iE_0 \frac{\omega_-}{\alpha_i} \frac{e^{i\omega_- t}}{\alpha_i + i\omega_-} + i\overline{E_0} \frac{\omega_+}{\alpha_i} \frac{e^{-i\omega_+ t}}{\alpha_i - i\omega_+} \\ &= -\frac{e_i(t)}{\alpha_i}, \quad i = 1, 2, \ldots, n, \end{aligned}$$

where the e_i are given by (17.4.16) with the aid of (16.1.17).

The real quadratic form

$$\begin{aligned} V(t) &= ME_0^2 e^{2i\omega_- t} + \overline{M}\, \overline{E_0}^2 e^{-2i\omega_+ t} + N|E_0|^2 e^{i(\omega_- - \omega_+)t} \\ &= 2Re[ME_0^2 e^{2i\omega_- t}] + N|E_0|^2 e^{i(\omega_- - \omega_+)t} \\ &= \left[2Re\left(ME_0^2 e^{2i\omega_0 t}\right) + N|E_0|^2\right] e^{2\eta t} \end{aligned} \tag{17.6.11}$$

will be denoted by

$$V(t) = \{M, N\}. \tag{17.6.12}$$

which is a generalization of (16.11.3). The quantity N is real. All free energies, total dissipations, rates of dissipation and work functions can be represented in the form $V(t)$, for histories given by (16.11.1). Note that

$$\begin{aligned} \int_{-\infty}^t V(s)ds &= \left\{\frac{M}{2i\omega_-}, \frac{N}{i(\omega_- - \omega_+)}\right\}, \\ \dot{V}(t) &= \{2i\omega_- M, i(\omega_- - \omega_+)N\}. \end{aligned} \tag{17.6.13}$$

Remark 17.6.1. For $\eta = 0$, the quantity $V = \{M, N\}$ will be completely periodic if M and N are finite quantities, independent of t.

We have, from (17.6.7) for $s = 0$ and (17.6.10),

$$\begin{aligned} T(t)\dot{E}(t) &= i\omega_- \mathcal{M}_+(\omega_-)E_0^2 e^{2i\omega_- t} - i\omega_+ \mathcal{M}_+(-\omega_+)\overline{E_0}^2 e^{-2i\omega_+ t} \\ &\quad + i\left[\omega_- \mathcal{M}_+(-\omega_+) - \omega_+ \mathcal{M}_+(\omega_-)\right]|E_0|^2 e^{i(\omega_- - \omega_+)t} \\ &= \{i\omega_- \mathcal{M}_+(\omega_-), i\left[\omega_- \mathcal{M}_+(-\omega_+) - \omega_+ \mathcal{M}_+(\omega_-)\right]\}. \end{aligned} \tag{17.6.14}$$

Using $(16.1.30)_1$, $(17.6.13)_1$, and (17.6.14), we see that

$$W(t) = \{M_W, N_W\},$$

where

$$M_W = \frac{1}{2}\mathcal{M}_+(\omega_-), \qquad N_W = \frac{\omega_-\mathcal{M}_+(-\omega_+) - \omega_+\mathcal{M}_+(\omega_-)}{\omega_- - \omega_+}. \tag{17.6.15}$$

The term N_W diverges in the purely sinusoidal limit as $\eta \to 0$.

A general free energy $(17.3.5)_1$ for histories of the form (16.11.1) is given by

$$\psi(t) = \left\{M_\psi, N_\psi\right\}. \tag{17.6.16}$$

where

$$\begin{aligned} M_\psi &= \frac{1}{2}\left[G_\infty - \omega_-^2\widetilde{G}_{+-}(\omega_-, -\omega_-)\right] \\ &= \frac{1}{2}\left[\mathcal{M}_+(\omega_-) + \frac{i\omega_-}{2}K_{+-}(\omega_-, -\omega_-)\right], \end{aligned} \tag{17.6.17}$$

by virtue of (16.1.9), (17.5.1), and (17.5.5). Also, from (16.1.9), $(17.5.3)_1$ and subsequent observations,

$$\begin{aligned} N_\psi &= G_\infty + \frac{1}{2}\left\{|\omega_-|^2\widetilde{G}_{+-}(\omega_-, \omega_+) + |\omega_+|^2\widetilde{G}_{+-}(-\omega_+, -\omega_-)\right\} \\ &= G_\infty + |\omega_-|^2\widetilde{G}_{+-}(\omega_-, \omega_+) \\ &= G_\infty + \frac{|\omega_-|^2}{i(\omega_- - \omega_+)}\left[K_{+-}(\omega_-, \omega_+) + \widetilde{G}_+(\omega_-) + \overline{\widetilde{G}_+(\omega_+)}\right] \\ &= \frac{-i|\omega_-|^2K_{+-}(\omega_-, \omega_+) + \omega_-\mathcal{M}_+(-\omega_+) - \omega_+\mathcal{M}_+(\omega_-)}{\omega_- - \omega_+}, \end{aligned} \tag{17.6.18}$$

again using (17.5.5). From (17.3.11), (17.5.3), and (17.6.7), we find that

$$D(t) = \{M_D, N_D\}, \tag{17.6.19}$$

where

$$\begin{aligned} M_D &= \frac{\omega_-^2}{2}K_{+-}(\omega_-, -\omega_-), \\ N_D &= -\frac{|\omega_-|^2}{2}\{K_{+-}(\omega_-, \omega_+) + K_{+-}(-\omega_+, -\omega_-)\} \\ &= -|\omega_-|^2K_{+-}(\omega_-, \omega_+). \end{aligned} \tag{17.6.20}$$

Let

$$\mathcal{D}(t) = \{M_{\mathcal{D}}, N_{\mathcal{D}}\}.$$

Then, from $(17.6.13)_1$,

$$M_{\mathcal{D}} = -\frac{i\omega_-}{4}K_{+-}(\omega_-, -\omega_-), \qquad N_{\mathcal{D}} = i|\omega_-|^2\frac{K_{+-}(\omega_-, \omega_+)}{\omega_- - \omega_+}. \tag{17.6.21}$$

We see from (17.6.15), (17.6.17), (17.6.18), and (17.6.21) that (16.1.29) is obeyed. Note that $N_{\mathcal{D}}$ diverges in the sinusoidal limit.

The ratios (17.6.2) are given by

$$F_d(t) = \frac{V_{\mathcal{D}}(t)}{V_W(t)}, \qquad F_s(t) = 1 - F_d(t),$$

where $V_{\mathcal{D}}(t)$ has the form (17.6.11) with $M = M_{\mathcal{D}}$ and $N = N_{\mathcal{D}}$ given by (17.6.21). Also, $V_W(t)$ is similarly defined, with M_W and N_W given by (17.6.15).

The factor $e^{2\eta t}$, giving the exponential part of the history, cancels out of the ratios, yielding for $F_d(t)$,

$$F_d(t) = \frac{M_{\mathcal{D}} E_0^2 e^{2i\omega_0 t} + \overline{M_{\mathcal{D}} E_0^2 e}^{-2i\omega_0 t} + N_{\mathcal{D}} |E_0|^2}{M_W E_0^2 e^{2i\omega_0 t} + \overline{M_W E_0^2 e}^{-2i\omega_0 t} + N_W |E_0|^2}. \tag{17.6.22}$$

Now, we have

$$M_W E_0^2 e^{2i\omega_0 t} + \overline{M_W E_0^2 e}^{-2i\omega_0 t} + N_W |E_0|^2 > 0, \tag{17.6.23}$$

which is a consequence of the fact that $W(t)$, given by (17.3.19), is positive, as is the cancelled factor $e^{2\eta t}$. Also, by averaging over any interval of duration π/ω_0, the oscillatory terms vanish and we deduce that $N_W > 0$. The expression in (17.6.23) may be written as

$$N_W |E_0|^2 \left[1 + \frac{2|M_W| \cos(2\omega_0 t + \lambda)}{N_W} \right], \qquad \lambda = \arg[M_W E_0^2],$$

where the term in brackets must be positive. It follows that, for all t,

$$N_W > 2|M_W| \cos(2\omega_0 t + \lambda). \tag{17.6.24}$$

We can therefore write (17.6.22) as the numerator multiplying the factor

$$\frac{1}{N_W |E_0|^2}[1 + A],$$

where A is an infinite expansion of powers of terms involving $e^{\pm 2i\omega_0 t}$. This expansion is convergent by virtue of the inequality (17.6.24). If we take the average of $F_d(t)$ over any time interval of duration $\frac{\pi}{\omega_0}$, it reduces to

$$\begin{aligned} F_{dc} &= \frac{N_{\mathcal{D}}}{N_W} = \frac{i|\omega_-|^2 K_{+-}(\omega_-, \omega_+)}{\omega_- \mathcal{M}_+(-\omega_+) - \omega_+ \mathcal{M}_+(\omega_-)}, \\ F_{sc} &= 1 - F_{dc} = \frac{N_\psi}{N_W} \\ &= \frac{-i|\omega_-|^2 K_{+-}(\omega_-, \omega_+) + \omega_- \mathcal{M}_+(-\omega_+) - \omega_+ \mathcal{M}_+(\omega_-)}{\omega_- \mathcal{M}_+(-\omega_+) - \omega_+ \mathcal{M}_+(\omega_-)}. \end{aligned} \tag{17.6.25}$$

17.6.1.3 Purely Sinusoidal Histories

For this case, the quantities $W(t)$ and $\mathcal{D}(t)$ diverge, as pointed out after (17.6.15) and (17.6.21). We can write an approximate version of (11.7.5) as

$$W(t) = \{M_W, N_W\},$$

$$M_W = \frac{1}{2}[G_0 + G'_+(\omega_0)]$$

$$N_W = G_0 + G'_c(\omega_0) - \omega_0 \frac{\partial}{\partial \omega_0} G'_c(\omega_0) - G'_s(\omega_0)\omega_0 \left(2t + \frac{1}{\eta}\right).$$

In the limit $\eta \to 0$, the quantity N_W does not meet the criterion specified in Remark 17.6.1, so that $W(t)$ does not obey condition P4.

In this limit, where $\eta \to 0$, (17.6.11) and (17.6.12) become

$$\{M, N\} = ME_0^2 e^{2i\omega_0 t} + \overline{M}\,\overline{E_0}^2 e^{-2i\omega_0 t} + N|E_0|^2. \tag{17.6.26}$$

Relation (17.6.14) converges to a finite result of the form

$$T(t)\dot{E}(t) = \{i\omega_0 \mathcal{M}_+(\omega_0), 2H(\omega_0)\},$$

where (16.1.10) and (16.1.13) have been invoked. Also, (17.6.17) and (17.6.18) yield

$$\begin{aligned}
M_\psi &= \frac{1}{2}\left\{\mathcal{M}_+(\omega_0) + \frac{i\omega_0}{2} K_{+-}(\omega_0, -\omega_0)\right\}, \\
N_\psi &= \mathcal{R}(\omega_0) - \omega_0 \frac{d}{d\omega_0} \mathcal{R}(\omega_0) + U(\omega_0), \\
U(\omega_0) &= \frac{i\omega_0^2}{2}\left[-\frac{\partial}{\partial \omega_1} K_{+-}(\omega_1, \omega_2) + \frac{\partial}{\partial \omega_2} K_{+-}(\omega_1, \omega_2)\right]\Bigg|_{\omega_1=\omega_2=\omega_0},
\end{aligned} \tag{17.6.27}$$

where $\mathcal{R}(\omega)$ is defined by (16.1.10). Finally, from (17.6.19) and (17.6.20), we deduce that

$$D(t) = \left\{\frac{\omega_0^2}{2} K_{+-}(\omega_0, -\omega_0), 2H(\omega_0)\right\}, \tag{17.6.28}$$

where (17.5.18) has been used. Applying $(17.6.13)_2$ in the sinusoidal limit, one can show that (16.1.28) is obeyed.

Comparison with Special Cases

We now compare these results with previously given particular examples of sinusoidal histories. The minimum free energy is discussed in detail for such histories in Sect. 11.7. Precisely analogous formulae apply to all the $\psi_f(t)$. Thus, we have, in the notation (17.6.26), using the complex modulus given by (16.1.9), rather than $G'(\omega_0)$ or $\widetilde{G}(\omega_0)$,

$$\psi_f(t) = \left\{\frac{1}{2}\left[\mathcal{M}_+(\omega_0) + \frac{i}{\omega_0}\left[H^f_-(-\omega_0)\right]^2\right], R(\omega_0) - \omega_0 \frac{d}{d\omega_0} R(\omega_0) + Q_f(\omega_0)\right\}, \tag{17.6.29}$$

where $Q_f(\omega)$ is given by

$$Q_f(\omega) = i\left[\frac{d}{d\omega} H^f_+(\omega) H^f_-(\omega) - H^f_+(\omega) \frac{d}{d\omega} H^f_-(\omega)\right] \geq 0, \quad \omega \in \mathbb{R}. \tag{17.6.30}$$

The rate of dissipation is given by (16.4.12) and (16.4.6). It can be shown that

$$K_f(t) = H^f_-(-\omega_0)E_0 e^{i\omega_0 t} + H^f_-(\omega_0)\overline{E_0}e^{-i\omega_0 t},$$

yielding

$$D_f(t) = \{[H^f_-(-\omega_0)]^2, 2H(\omega_0)\} = \{[H^f_+(\omega_0)]^2, 2H(\omega_0)\}, \tag{17.6.31}$$

where (16.4.1) has been used.

We see that relations (17.6.29) and (17.6.31) are the special cases of (17.6.27) and (17.6.28) for $\omega_1\omega_2 K_{+-}(\omega_1, \omega_2) = -2H^f_+(\omega_1)H^f_+(\omega_2)$. In particular, the quantity $Q_f(\omega)$ in (17.6.30) is equal to $U(\omega)$ in (17.6.27) for this choice of kernel.

17.6.1.4 Exponential Histories

This can be treated either by direct calculation, or as a special case of the general formulae of Sect. 17.6.1.2, where $\omega_0 \to 0$. We consider a history and current value $(E^t, E(t))$ given by (16.11.1) with $\omega_0 = 0$, so that

$$E(t) = E_e e^{\eta t}, \qquad E^t(s) = E(t-s), \qquad E_e = E_0 + \overline{E_0}.$$

The stress function, given by (17.6.10), has the form

$$T(t) = \mathcal{M}_+(-i\eta)E(t), \qquad \mathcal{M}_+(-i\eta) = G_\infty + \eta\widetilde{G}_+(-i\eta), \tag{17.6.32}$$

where the forms of $\mathcal{M}_+(-i\eta)$ can be deduced from (16.1.9). This quantity is real. From (17.6.32) or as special cases of (17.6.14) and (17.6.15), we have

$$\begin{aligned} T(t)\dot{E}(t) &= \eta\mathcal{M}_+(-i\eta)E^2(t) = \frac{1}{2}\mathcal{M}_+(-i\eta)\frac{d}{dt}E^2(t), \\ W(t) &= \frac{1}{2}\mathcal{M}_+(-i\eta)E^2(t). \end{aligned} \tag{17.6.33}$$

Also, (17.6.16), (17.6.17), and (17.6.18) reduce to

$$\psi(t) = \frac{1}{2}[\mathcal{M}_+(-i\eta) + \frac{\eta}{2}K_{+-}(-i\eta, i\eta)]E^2(t). \tag{17.6.34}$$

The rate of dissipation and total dissipation are special cases of (17.6.19) and (17.6.21), given by

$$D(t) = -\frac{\eta^2}{2}K_{+-}(-i\eta, i\eta)E^2(t), \qquad \mathcal{D}(t) = -\frac{\eta}{4}K_{+-}(-i\eta, i\eta)E^2(t). \tag{17.6.35}$$

The results for the various quadratic quantities above can be summarized in a simple formula. Putting $\omega_0 = 0$ in (17.6.11), we have

$$V(t) = V_0 e^{2\eta t}, \qquad V_0 = ME_0^2 + \overline{M}\,\overline{E_0}^2 + N|E_0|^2.$$

It can be shown using (17.3.5), (17.3.11), and (17.3.19), relating, respectively, to $\psi(t)$, $D(t)$, and $W(t)$, that

$$M = \overline{M} = \frac{N}{2}, \tag{17.6.36}$$

for exponential histories. This relation must therefore hold true for $\mathcal{D}(t)$, by virtue of (16.1.29). Equation (17.6.36) can also be shown using various explicit formulae from (17.6.15) to (17.6.21). It gives that $V_0 = ME_e^2$ or

$$V(t) = ME^2(t). \tag{17.6.37}$$

Each result in (17.6.33)–(17.6.35) has the form (17.6.37) where, for example, if $M = M(\omega_-, \omega_+)$ in the general sinusoidal/exponential case, this is replaced by $M = M(-i\eta, i\eta)$. The property (17.6.37) was first noted in [16].

The quantities F_s and F_d, defined by (17.6.2), are time-independent and given by

$$F_s = \frac{\mathcal{M}_+(-i\eta) + \frac{\eta}{2}K_{+-}(-i\eta, i\eta)}{\mathcal{M}_+(-i\eta)}, \qquad F_d = -\frac{\eta}{2}\frac{K_{+-}(-i\eta, i\eta)}{\mathcal{M}_+(-i\eta)}.$$

Note that

$$F_s = F_{sc}, \qquad F_d = F_{dc},$$

where F_{sc} and F_{dc} are the quantities introduced in (17.6.25), with $\omega_\pm$ replaced by $\pm i\eta$, as in the comment after (17.6.37).

Various detailed expressions for step function and SE histories are presented in [18]. Plots of some of these functions are also presented.

17.7 Product Formulae in the Time and Frequency Domains

We now choose general histories and special forms of kernels, using results obtained in Sect. 17.5 to help determine new nonnegative rates of dissipation and from these to deduce new free energies.

In the time and frequency domains, we have the corresponding conditions (17.4.10) and (17.5.18), which constrain the choice of kernels for the rate of dissipation. This category is in fact very general and will be explored in this section.

The simplest method of creating nonnegative quadratic functionals is to assume that the relevant kernels have the form of sums of products.

Also, a family of free energy functionals is introduced, which is a generalization of the category discussed in Sect. 16.4 consisting of the minimum and related free energies.

It should be emphasized that all developments in the time and frequency domains are equivalent.

17.7.1 The Time Domain

Let us take

$$K(s,u) = -\sum_{i,j=1}^{m} A_{ij}k_i(s)k_j(u), \tag{17.7.1}$$

for some positive integer m, where it is assumed that all quantities $\int_0^\infty k_i(s)\dot{E}^t(s)ds$, $i = 1, 2, \ldots, m$ exist, for the class of histories of interest. The matrix $\mathbf{A}$ is assumed to be nonnegative. The rate of dissipation is given by

$$D(t) = \frac{1}{2}\int_0^\infty \int_0^\infty \sum_{i,j=1}^{m} \dot{E}^t(s)A_{ij}k_i(s)k_j(u)\dot{E}^t(u)dsdu \geq 0. \tag{17.7.2}$$

We obtain from (17.4.9) that

$$\psi(t) = \phi(t) + \frac{1}{2}\int_0^\infty \int_0^\infty \int_0^\infty \dot{E}^t(s)\sum_{i,j=1}^{m} A_{ij}k_i(z+s)k_j(z+u)dz\dot{E}^t(u)dsdu. \tag{17.7.3}$$

From (17.4.10), it follows that

$$\widetilde{G}(s) = \int_0^\infty \sum_{i,j=1}^{m} A_{ij}k_i(z)k_j(z+s)dz. \tag{17.7.4}$$

The simplest case of (17.7.1) is where $m = 1$. Absorbing $\sqrt{A_{11}}$ into $k(s)$, we obtain

$$K(s,u) = -k(s)k(u), \qquad D(t) = \frac{1}{2}\left[\int_0^\infty k(s)\dot{E}^t(s)ds\right]^2, \tag{17.7.5}$$

and

$$\widetilde{G}(s,u) = \int_0^\infty k(z+s)k(z+u)dz,$$

giving

$$\psi(t) = \phi(t) + \frac{1}{2}\int_0^\infty \int_0^\infty \int_0^\infty \dot{E}^t(s)k(z+s)k(z+u)\dot{E}^t(u)dsdudz. \tag{17.7.6}$$

Relation (17.7.4) becomes

$$\widetilde{G}(s) = \int_0^\infty k(z)k(z+s)dz. \tag{17.7.7}$$

The quantity $\widetilde{G}(s)$ is the given relaxation function characterizing the material, while the free energy determined by (17.7.6) is one of (usually) many functionals that generate this quantity. Relation (17.7.7) can be regarded as a nonlinear integral equation for $k(\cdot)$ in terms of $\widetilde{G}(s)$.

The quantity $k(s)$ determines $K(s,u)$ in accordance with $(17.7.5)_1$ and therefore the corresponding $D(t)$ and $\psi(t)$. We return to this case in Sect. 17.7.2,

Two further special cases will be considered, that where

$$A_{ij} = 1 \quad \text{or} \quad A_{ij} = \delta_{ij}, \quad i, j = 1, 2, \ldots, m, \tag{17.7.8}$$

where δ_{ij} is the standard Kronecker delta notation. For $(17.7.8)_1$, relation (17.7.2) becomes

$$D(t) = \frac{1}{2}\left[\sum_{i=1}^{m} \int_0^\infty k_i(s)\dot{E}^t(s)ds\right]^2, \tag{17.7.9}$$

which is clearly nonnegative. It follows from (17.7.3) that the corresponding free energy has the form

$$\psi(t) = \phi(t) + \frac{1}{2}\int_0^\infty \int_0^\infty \int_0^\infty \dot{E}^t(s) \sum_{i,j=1}^{m} k_i(z+s)k_j(z+u)dz\dot{E}^t(u)dsdu. \tag{17.7.10}$$

For example, consider the results of Sect. 16.9.1. Referring to (16.9.9), we see that, for $D_f(t)$, the quantity $k_i(s)$ in (17.7.9) is given by

$$k_i(s) = \frac{\sqrt{2}h_\infty R_i^f e^{-\alpha_i s}}{\alpha_i}, \quad i = 1, 2, \ldots, n,$$

where $m = n$, so that (17.7.10) gives (16.9.7).

For $(17.7.8)_2$, we obtain

$$K(s, u) = -\sum_{i=1}^{m} k_i(s)k_i(u). \tag{17.7.11}$$

The rate of dissipation has the form

$$D(t) = \frac{1}{2}\sum_{i=1}^{m}\left[\int_0^\infty k_i(s)\dot{E}^t(s)ds\right]^2 \geq 0. \tag{17.7.12}$$

It follows from (17.7.3) that

$$\psi(t) = \phi(t) + \frac{1}{2}\int_0^\infty \int_0^\infty \int_0^\infty \dot{E}^t(s) \sum_{i=1}^{m} k_i(z+s)) k_i(z+u)) dz\dot{E}^t(u)dsdu. \tag{17.7.13}$$

Also, from (17.7.4),

$$\widetilde{G}(s) = \int_0^\infty \sum_{i=1}^{m} k_i(z)k_i(z+s)dz. \tag{17.7.14}$$

Putting

$$k_i(s) = \sqrt{2\alpha_i G_i} e^{-\alpha_i s}$$

in (17.7.12) gives $(17.4.20)_2$ while (17.7.13) yields $(17.4.20)_1$, relating to the Dill free energy.

17.7.1.1 New Category of Free Energies: Time Domain

We now sketch a systematic approach to the construction of a new family of free energies. This can be developed only in general terms for the time domain representation. The frequency representation is considered in Sect. 17.7.2.1 below, and it emerges that one can give explicit formulae for the new free energies in that formalism.

Remark 17.7.1. Let us assume that $\widetilde{G}(s)$ can be decomposed into m components $\widetilde{G}_i(s), \quad i = 1, 2, \ldots, m$, so that

$$\widetilde{G}(s) = \sum_{i=1}^{m} \widetilde{G}_i(s), \tag{17.7.15}$$

where each $\widetilde{G}_i(s)$ is the relaxation function of a well-defined material, obeying the laws of thermodynamics, in particular (16.1.12). The term sub-material will be used in reference to each of these. The strain function is assumed to be the same in each sub-material.

For example, in the case of discrete-spectrum materials, $\widetilde{G}(s)$ is given by (11.9.1). We could take $\widetilde{G}_i(\omega)$ to be any partial sum of the terms in that expression, *e.g.*,

$$\widetilde{G}_i(s) = \sum_{k=n_i}^{m_i} G_k e^{-\alpha_k s}, \qquad n \geq m_i > n_i \geq 1. \tag{17.7.16}$$

The choice of terms in this relation need not reflect our original order.

We put

$$\widetilde{G}_i(s) = \int_0^{\infty} k_i(z) k_i(z+s) dz, \quad i = 1, 2, \ldots m, \tag{17.7.17}$$

yielding a relation of the form (17.7.14). For each i, (17.7.17) can be regarded as a nonlinear integral equation for $k_i(s)$ in terms of the known quantity $\widetilde{G}_i(s)$. These equations can be solved in the same way as (17.7.7) (see Sect. 17.7.2.1 below). The solutions yield the form of $K(s,u)$ as given by (17.7.11), and therefore allow us to determine $D(t)$ and $\psi(t)$.

17.7.2 The Frequency Domain

Referring to (17.5.1) and (17.7.1), we have

$$K_{+-}(\omega_1, \omega_2) = -\sum_{i,j=1}^{m} A_{ij} k_{i+}(\omega_1) k_{j-}(\omega_2) = -\sum_{i,j=1}^{m} A_{ij} \overline{k_{i-}}(\omega_1) k_{j-}(\omega_2), \tag{17.7.18}$$

where

$$k_{i-}(\omega) = \int_0^{\infty} k_i(s) e^{i\omega s} ds, \qquad k_{i+}(\omega) = \overline{k_{i-}}(\omega). \tag{17.7.19}$$

Condition $(17.5.18)_1$ gives that

$$\sum_{i,j=1}^{m} A_{ij}\overline{k_{i-}}(\omega)k_{j-}(\omega) = 2\frac{H(\omega)}{\omega^2}. \tag{17.7.20}$$

In the case where $m = 1$, we have

$$K_{+-}(\omega_1, \omega_2) = -k_+(\omega_1)k_-(\omega_2) \tag{17.7.21}$$

and relation (17.7.20) becomes

$$k_+(\omega)k_-(\omega) = |k_-(\omega)|^2 = 2\frac{H(\omega)}{\omega^2}. \tag{17.7.22}$$

We solve (17.7.7) at this point and show that it is equivalent to (17.7.22). Substituting the inverse transforms of (17.7.19) into (17.7.7) and carrying out two time domain integrations yields

$$\begin{aligned}\widetilde{G}_+(\omega) &= -\frac{1}{4\pi^2}\int_{-\infty}^{\infty}\int_{-\infty}^{\infty}\frac{k_-(\omega_1)k_+(\omega_2)}{(\omega_1^- - \omega_2^+)(\omega^- - \omega_2^+)}d\omega_1 d\omega_2 \\ &= -\frac{1}{2\pi i}\int_{-\infty}^{\infty}\frac{k_+(\omega')k_-(\omega')}{\omega' - \omega^-}d\omega'.\end{aligned} \tag{17.7.23}$$

The choice ω_2^+ in the first form is dictated by the need for the time domain integrals to converge. The final form is obtained by integrating ω_1 over $\Omega^{(+)}$. Taking the complex conjugate of $(17.7.23)_2$ and applying the Plemelj formulae give

$$\overline{\widetilde{G}_+(\omega)} + \widetilde{G}_+(\omega) = 2\widetilde{G}_c(\omega) = 2\frac{H(\omega)}{\omega^2} = k_+(\omega)k_-(\omega) = |k_-(\omega)|^2, \tag{17.7.24}$$

which agrees with (17.7.22).

Relation (17.7.22) or (17.7.24) is of course the factorization problem for $H(\omega)$. The solution is given either by the factors (16.1.14), which yield the minimum free energy for all materials, or those given by (16.4.1) for materials with only isolated singularities. Thus, for some materials, the solution of (17.7.22) is non-unique. We put

$$k_-(\omega) = \sqrt{2}\frac{H_-^f(\omega)}{\omega}. \tag{17.7.25}$$

It follows from (17.7.21) that

$$K_{+-}(\omega_1, \omega_2) = -2\frac{H_+^f(\omega_1)H_-^f(\omega_2)}{\omega_1\omega_2}. \tag{17.7.26}$$

Equation (17.7.26) yields the rate of dissipation $D_f(t)$ given by (16.4.17) and $\psi_f(t)$ of the form (16.4.15), both expressed in terms of $\dot{E}_+^t$ with the aid of (16.1.17).

We now assume that $\mathbf{A}$ is the unit matrix so that

$$K_{+-}(\omega_1, \omega_2) = -\sum_{i=1}^{m}\overline{k_{i-}}(\omega_1)k_{i-}(\omega_2). \tag{17.7.27}$$

Relations $(17.5.7)_2$ and (17.5.13) give

$$D(t) = \frac{1}{8\pi^2} \sum_{i=1}^{m} \left| \int_{-\infty}^{\infty} k_{i-}(\omega) \dot{E}_+^t(\omega) \right|^2 d\omega,$$

$$\psi(t) = \phi(t) + \frac{i}{8\pi^2} \int_{-\infty}^{\infty} \int_{-\infty}^{\infty} \frac{\overline{\dot{E}_+^t}(\omega_1) \sum_{i=1}^m \overline{k_{i-}}(\omega_1) k_{i-}(\omega_2) \dot{E}_+^t(\omega_2)}{\omega_1^+ - \omega_2^-} d\omega_1 d\omega_2.$$

Condition $(17.5.18)_1$ gives that

$$\sum_{i=1}^{m} \overline{k_{i-}}(\omega) k_{i-}(\omega) = \sum_{i=1}^{m} |k_{i-}(\omega)|^2 = 2\frac{H(\omega)}{\omega^2}. \tag{17.7.28}$$

An immediate example of (17.7.27) satisfying (17.7.28) is obtained by taking $m = N$ and

$$k_{i-}(\omega) = k_{f-}(\omega) = \frac{\sqrt{2\lambda_f} H_-^f(\omega)}{\omega}, \tag{17.7.29}$$

giving

$$K_{+-}(\omega_1, \omega_2) = -2 \sum_{f=1}^{N} \lambda_f \frac{H_+^f(\omega_1) H_-^f(\omega_2)}{\omega_1 \omega_2}.$$

Other solutions of (17.7.28) may of course exist.

17.7.2.1 New Category of Free Energies: Frequency Domain

The developments in this section are the frequency domain version of those in Sect. 17.7.1.1. Here, we use (17.7.27) and (17.7.28) to determine a family of free energies, corresponding to (17.7.11) and (17.7.14) in Sect. 17.7.1.1.

It follows from Remark 17.7.1 that the quantity $H(\omega)$ can be decomposed into m components $H_i(\omega)$, $i = 1, 2, \ldots, m$, corresponding to (17.7.15) so that

$$H(\omega) = \sum_{i=1}^{m} H_i(\omega), \tag{17.7.30}$$

where each $H_i(\omega)$ is related to $\widetilde{G}_i(s)$ in accordance with (16.1.13) and in particular is nonnegative for $\omega \in \mathbb{R}$. It is the quantity $H(\omega)$ for the sub-material with relaxation function $\widetilde{G}_i(s)$.

For example, in the case of discrete-spectrum materials, $H(\omega)$ is given by (11.9.2). We could take $H_i(\omega)$ to be any partial sum of the terms in that expression, *e.g.*,

$$H_i(\omega) = \omega^2 \sum_{k=n_i}^{m_i} \frac{\alpha_k G_k}{\alpha_k^2 + \omega^2}, \qquad n \geq m_i > n_i \geq 1,$$

corresponding to (17.7.16).

In general, each $H_i(\omega)$ can be factorized as in (16.1.14) or, for a material with only isolated singularities, (16.2.9). Thus, we write

$$H_i(\omega) = H^f_{i+}(\omega)H^f_{i-}(\omega), \qquad H^f_{i\pm}(\omega) = H^f_{i\mp}(-\omega) = \overline{H^f_{i\mp}(\omega)}.$$

For materials with only isolated singularities, the factorization indicated by the superscript f will vary with the subscript i.

Starting from (17.7.17) and applying the procedure leading to (17.7.24) and (17.7.25) for each i, we obtain (17.7.28) by virtue of (17.7.30). Relation (17.7.25) for each i and (17.7.30) can be written as

$$k_{i-}(\omega) = \sqrt{2}\frac{H^f_{i-}(\omega)}{\omega},$$
$$\sum_{i=1}^{m}\overline{k_{i-}}(\omega)k_{i-}(\omega) = 2\sum_{i=1}^{m}\frac{H^f_{i+}(\omega)H^f_{i-}(\omega)}{\omega^2} = 2\frac{H(\omega)}{\omega^2},$$

for the range of possible choices of f corresponding to each i, where $H^f_{i\pm}(\omega)$ are a particular pair of factors of $H_i(\omega)$. Thus, we have

$$K_{+-}(\omega_1,\omega_2) = -2\sum_{i=1}^{m}\frac{H^f_{i+}(\omega_1)H^f_{i-}(\omega_2)}{\omega_1\omega_2}. \tag{17.7.31}$$

The superscript f will now be dropped to simplify notation.

If $(17.7.31)_1$ is inserted into (17.5.7) expressed in terms of $E^t_{r+}(\omega)$ by means of (16.1.17), we obtain

$$\begin{aligned} D(t) &= \sum_{i=1}^{m}|K_i(t)|^2, \\ K_i(t) &= \frac{1}{2\pi}\int_{-\infty}^{\infty} H_{i-}(\omega)E^t_{r+}(\omega)d\omega. \end{aligned} \tag{17.7.32}$$

These relations are analogous to (16.4.12) and (16.4.6) with $H_{i-}(\omega)$ playing the role of $H^f_-(\omega)$. Substituting $(17.7.31)_1$ into $(17.5.13)_2$ gives a sum of terms identical to (16.4.15) but with $H_{i\pm}$ replacing $H^f_\pm$. Substituting $(17.7.31)_1$ into $(17.5.17)_2$ yields a sum of terms of the form (16.4.16) with $H_{i\pm}$ rather than $H^f_\pm$. Again, each term can be put in the form (16.4.11) with the above replacements. The free energy and total dissipation can therefore be written as

$$\psi(t) = \phi(t) + \frac{1}{2\pi}\sum_{i=1}^{m}\int_{-\infty}^{\infty}|p^t_{i-}(\omega)|^2 d\omega, \qquad \mathcal{D}(t) = \sum_{i=1}^{m}\int_{-\infty}^{\infty}|p^t_{i+}(\omega)|^2 d\omega,$$

where

$$p^t_{i-}(\omega) = \frac{1}{2\pi i}\int_{-\infty}^{\infty}\frac{H_{i-}(\omega')E^t_{r+}(\omega')}{\omega'-\omega^+}d\omega',$$
$$p^t_{i+}(\omega) = \frac{1}{2\pi i}\int_{-\infty}^{\infty}\frac{H_{i-}(\omega')E^t_{r+}(\omega')}{\omega'-\omega^-}d\omega'.$$

These new free energies clearly obey (16.1.26). The rate of dissipation is nonnegative, by virtue of (17.7.32). Also, we can demonstrate that (16.1.25) holds by an argument analogous to that in Proposition 11.2.2.

The ordering of these free energy functionals is discussed in some detail in [162].

17.8 Approximation of a Discrete-Spectrum Material by a Day Functional

This and the following section of the present chapter deal with topics that are not closely related to those in earlier sections.

For materials with relaxation function containing one decaying exponential, the associated Day functional is the physical free energy. For general discrete-spectrum materials, we seek a best fit of the relaxation function with one decaying exponential to that for the general case.

For the case $n = 1$, the relations (16.3.13) and (16.3.15) reduce to the formulae for the Day free energy and rate of dissipation (see (10.2.15), (16.9.10) and [87]). A relaxation function with only one decaying exponential, has the form (see (10.2.15), (16.9.10), and (16.9.11))

$$G_D(t) = G_\infty + G_d e^{-\alpha t}, \qquad G_d = G_0 - G_\infty. \tag{17.8.1}$$

This behavior has, in some contexts, been referred to as the standard linear model [167]. The complex modulus, defined by $(16.1.9)_2$, is given by

$$\mathcal{M}_+(\omega) = G_\infty + i\omega \sum_{i=1}^{n} \frac{G_i}{\alpha_i + i\omega},$$

so that for $n = 1$

$$\mathcal{M}_+(\omega) = \frac{(\alpha + i\omega)G_\infty + i\omega G_d}{\alpha + i\omega}. \tag{17.8.2}$$

By virtue of (16.3.14) and $(16.3.15)_4$, we must have

$$C_{11} = G_d, \qquad \Gamma_{11} = 2\alpha G_d.$$

The Day free energy functional is given by

$$\psi_D(t) = \phi(t) + \frac{G_d}{2} e_1^2(t), \tag{17.8.3}$$

in terms of $e_1(t)$ defined by (16.3.12). The corresponding rate of dissipation is

$$D_D(t) = \alpha G_d e_1^2(t).$$

From (17.6.15) and (17.8.2), we can determine M_W and N_W for this case. In particular,

$$N_W = G_\infty + G_d \frac{(\omega_0^2 + \eta^2)(\alpha + \eta)}{\eta[(\alpha + \eta)^2 + \omega_0^2]}.$$

The kernels (17.4.14) reduce to

$$\widetilde{G}(s,u) = G_d e^{-\alpha(s+u)}, \qquad K(s,u) = -2\alpha G_d e^{-\alpha(s+u)}, \tag{17.8.4}$$

and (17.5.21) becomes

$$\widetilde{G}_{+-}(\omega_1,\omega_2) = \frac{G_d}{(\alpha + i\omega_1)(\alpha - i\omega_2)},$$
$$K_{+-}(\omega_1,\omega_2) = -\frac{2\alpha G_d}{(\alpha + i\omega_1)(\alpha - i\omega_2)},$$

which yield explicit functions for $M_{\mathfrak{D}}$ and $N_{\mathfrak{D}}$, given by (17.6.21). Thus, we obtain that

$$M_{\mathfrak{D}} = \frac{iG_d}{2} \frac{(w_0 - i\eta)\alpha}{(\alpha + \eta + i\omega_0)^2}, \qquad N_{\mathfrak{D}} = G_d \frac{\alpha(\omega_0^2 + \eta^2)}{\eta[(\alpha + \eta)^2 + \omega_0^2]}.$$

Relations (16.9.9) and (16.9.7) are particular cases of the general formulae (16.3.13) and (16.3.15), and must have the same limit for $n = 1$. This can be seen by noting that

$$H_\infty = -G'(0) = \alpha G_d,$$

and

$$R_1^1 = -\alpha,$$

which follows from the generalization of (11.9.7) to R_i^f, for $n = 1$.

The fundamental point made in this chapter is that a material with memory is completely characterized by choosing $K(s,u)$ rather than the relaxation function. However, for the $n = 1$ case, specifying the relaxation function parameters fixes uniquely the kernel $K(s,u)$ and therefore determines the material completely. Thus, the set $\mathcal{K}$ for $n = 1$ is a singleton given by $(17.8.4)_2$, and there is only one material with the constitutive relation generated by the relaxation function (17.8.1), with parameters as specified.

Remark 17.8.1. If a material behavior can be adequately simulated by one decay constant α, together with coefficients G_∞ and G_d, then $\psi_D(t)$, given by (17.8.3), is the physical free energy for that material.

We now seek the Day relaxation function which is closest to the actual relaxation function for the materials in $\mathcal{K}$.

The α_i, $i = 1,2,\ldots,n$, in (11.9.1) are given by (16.11.4), while the G_i, $i = 1,2,\ldots,n$, are determined by the assumption that the G_i are all equal [18]. We choose G_0, G_∞ and α such that the resulting $n = 1$ relaxation function approximates (11.9.1) as closely as possible.

We choose G_0 and G_∞ to be the same for the materials with relaxation function given by (11.9.1) and (17.8.1), respectively.

Choosing the optimal value of α is somewhat more difficult. Consider

$$\begin{aligned} f(\alpha) &= \frac{1}{G_d^2}\int_0^\infty [G_D(s) - G(s)]^2\, ds \\ &= \frac{1}{G_d^2}\int_0^\infty \left[G_d e^{-\alpha s} - \sum_{i=1}^n G_i e^{-\alpha_i s}\right]^2 ds, \end{aligned} \tag{17.8.5}$$

which is a $L^2(\mathbb{R}^+)$ norm of the difference between the relaxation functions for the Day case and that in (11.9.1) . Then, we put

$$f_m = \min_{\alpha > 0} f(\alpha), \tag{17.8.6}$$

and this minimum is achieved at α_d, so that

$$f(\alpha_d) = f_m \leq f(\alpha) \quad \forall \alpha \in \mathbb{R}^+. \tag{17.8.7}$$

The quantity α_d is the inverse time decay constant that will be used in the Day free energy and dissipation. The function $f(\alpha)$, given by (17.8.5), can be explicitly calculated. Let us replace α by α_0. Then

$$\begin{aligned} f(\alpha_0) &= \sum_{i,j=0}^n \frac{c_i c_j}{\alpha_i + \alpha_j}, \quad \alpha_0 \in (0, 1], \\ c_i &= \begin{cases} 1, & i = 0, \\ -\dfrac{G_i}{G_d}, & i = 1, 2, \ldots, n. \end{cases} \end{aligned} \tag{17.8.8}$$

This can be written in the form

$$f(\alpha_0) = \frac{1}{2\alpha_0} + 2\sum_{i=1}^n \frac{c_i}{\alpha_0 + \alpha_j} + \sum_{i,j=1}^n \frac{c_i c_j}{\alpha_i + \alpha_j}, \tag{17.8.9}$$

so that

$$f'(\alpha_0) = -\frac{1}{2\alpha_0^2} - 2\sum_{i=1}^n \frac{c_i}{(\alpha_0 + \alpha_j)^2}.$$

Thus, the α_0 satisfying (17.8.7) is the solution of the equation

$$1 + 4\sum_{i=1}^n \frac{c_i \alpha_0^2}{(\alpha_0 + \alpha_j)^2} = 0. \tag{17.8.10}$$

The solution gives a minimum value of $f(\alpha)$ if $f''(\alpha_0) > 0$. This quantity α_0 is equal to α_d.

It makes little difference what value of n is adopted. We choose $n = 5$ as an example. The numerical values of α_r, $r = 1, 2, 3, 4, 5$, given by (16.11.4), are 0.0718, 0.2679, 0.5359, 0.8038 and 1.0.

The appropriate Day relaxation function is determined by (17.8.5) - (17.8.7) where $f(\alpha)$ has the form (17.8.9). Since the G_i are all equal, $(17.8.8)_2$ becomes

$$c_i = \begin{cases} 1, & i = 0, \\ -\frac{1}{n}, & i = 1, 2, \ldots, n, \end{cases}$$

with the aid of (16.11.6) and $(17.8.1)_2$. We look for a solution to (17.8.10) where $\alpha_0 \in (0, 1]$. This is equal to 0.3225 [18], with optimum choice $f(\alpha_d) = f_m = 0.0968$.

Thus, an approximate form of the physical free energy for all the materials associated with the relaxation function $\widetilde{G}(s)$ of the form (11.9.1), with parameter values as stated, is given by (17.8.3) with $\alpha_1 = \alpha_d$. There are in general many such materials so the approximation is probably not very accurate in most cases.

17.9 Single-Integral Free Energies in Terms of I^t Derivatives

Single-integral free energy functionals that are expressible as quadratic forms of the state functional I^t (see (16.5.1)) are considered in this section. The discussion is based on [163].

This form is shown to include the functional ψ_F. There is also, however, a further category of functionals of this kind for materials with non-singleton minimal states. These latter functionals are difficult to construct, since basic inequalities relating to thermodynamics must be explicitly imposed; they are therefore not so useful for practical applications, in contrast to ψ_F.

The following notation will be useful:

$$I_k^t(\tau) = \frac{d^k}{d\tau^k} I^t(\tau), \qquad k = 1, 2, \ldots. \tag{17.9.1}$$

Then,

$$I_1^t(\tau) = \int_0^\infty G'(\tau + u)\dot{E}^t(u)du, \qquad I_2^t(\tau) = \int_0^\infty G''(\tau + u)\dot{E}^t(u)du. \tag{17.9.2}$$

Also,

$$\frac{\partial}{\partial t} I_1^t(s) = G'(s)\dot{E}(t) + I_2^t(s), \qquad \frac{\partial}{\partial t} I_2^t(s) = G''(s)\dot{E}(t) + I_3^t(s). \tag{17.9.3}$$

Just as in (16.5.2), we have

$$\lim_{\tau \to \infty} I_k^t(\tau) = 0, \quad k = 1, 2, 3, \ldots. \tag{17.9.4}$$

Consider the functional

$$\psi(t) = \phi(t) + \frac{1}{2}\int_0^\infty L(\tau)[I_1^t(\tau)]^2 d\tau. \tag{17.9.5}$$

This quantity is assumed to be a free energy. We now explore the constraints on $L(\tau)$ imposed by this requirement.

The relation (16.1.28) must hold. Using (16.1.31), $(17.9.3)_1$, and (17.9.4), we deduce that

$$\dot{\psi}(t) = \dot{E}(t)\left[T_e(t) + \int_0^\infty G'(\tau)L(\tau)I_1^t(\tau)d\tau\right] + \int_0^\infty I_2^t(\tau)L(\tau)I_1^t(\tau)d\tau$$
$$= T(t)\dot{E}(t) - \frac{1}{2}L(0)[I_1^t(0)]^2 - \frac{1}{2}\int_0^\infty L'(\tau)[I_1^t(\tau)]^2d\tau,$$

provided that the condition

$$\int_0^\infty G'(\tau)L(\tau)I_1^t(\tau)d\tau = T(t) - T_e(t)$$

holds. With the help of (16.1.3), (16.5.3), and $(17.9.2)_1$, this can be written as

$$\int_0^\infty [G'(\tau)L(\tau) + 1]I_1^t(\tau)d\tau$$
$$= \int_0^\infty \int_0^\infty [G'(\tau)L(\tau) + 1]G'(\tau + u)\dot{E}^t(u)d\tau du = 0,$$

which must be true for arbitrary histories. Let us write the resulting condition as an integral equation of the form

$$\int_0^\infty G'(\tau + u)f(\tau)d\tau = 0 \quad \forall u \in \mathbb{R}^+, \qquad f(\tau) = G'(\tau)L(\tau) + 1. \tag{17.9.6}$$

An alternative pathway to (17.9.6) is to express (17.9.5) in the form (17.3.5) with

$$\widetilde{G}(s,u) = \int_0^\infty G'(\tau + s)L(\tau)G'(\tau + u)d\tau,$$

and to impose the constraint (17.3.9), written in terms of $\widetilde{G}(u)$. Condition (17.9.6) has the same form as (16.5.4), leading to

$$\frac{2i}{\omega}H(\omega)f_+(\omega) = J_+(\omega),$$

where $J_+(\omega)$ is an unknown function, analytic in $\Omega^{(-)}$. This corresponds to (16.5.7).

If the material has only isolated singularities, there are many nontrivial solutions of (17.9.6) given by a form similar to (16.5.15), as shown in [163].

If the material has branch cut singularities, then $f(\tau) = 0$, $\tau \in \mathbb{R}^+$ is the only solution of (17.9.6), so that

$$L(\tau) = -\frac{1}{G'(\tau)}, \quad \tau \in \mathbb{R}^+,$$

and the only possibility for a free energy given by a single-integral quadratic form is the quantity ψ_F (see Sect. 10.1.3). In the scalar theory, this functional and the associated rate of dissipation have the forms

$$\psi_F(t) = \phi(t) - \frac{1}{2}\int_0^\infty \frac{[I_1^t(\tau)]^2}{G'(\tau)} d\tau, \tag{17.9.7}$$

and

$$\begin{aligned} D_F(t) &= -\frac{1}{2}\frac{[I_1^t(0)]^2}{G'(0)} - \frac{1}{2}\int_0^\infty \left[\frac{d}{d\tau}\frac{1}{G'(\tau)}\right][I_1^t(\tau)]^2 d\tau \\ &= -\frac{1}{2}\frac{[I_1^t(0)]^2}{G'(0)} + \frac{1}{2}\int_0^\infty G''(\tau)\left[\frac{I_1^t(\tau)}{G'(\tau)}\right]^2 d\tau. \end{aligned}$$

These quantities are nonnegative and $\psi_F(t)$ is a valid free energy if conditions (17.3.4) hold, not only for materials with branch point singularities, but for all materials. It is a relatively simple functional, convenient for applications.

The case of double integral quadratic forms is studied in depth in [163]. It is shown that the only such form that is a free energy is that for the minimum free energy discussed in Sect. 12.2.

18

Minimal States and Periodic Histories

This chapter deals with the following two topics [17, 165].

Using a standard representation of a free energy associated with a linear memory constitutive relation, a new condition, involving linear functionals, is derived which, if satisfied, ensures that the free energy is a functional of the minimal state. Using this condition and results on constructing free energy functionals in Chap. 17, it is shown that if the kernel of the rate of dissipation functional is given by sums of products, the associated free energy functional is a FMS. Because this condition is linear rather than a quadratic, it is easier to explore and to apply in new contexts.

Also, it is argued that for a free energy functionals to be physically realistic, it should exhibit strict periodic behavior for histories that have been periodic for all past times. The work function does not have this property, which causes difficulties in ascribing free energies to materials with singleton minimal states. A method of avoiding this difficulty is outlined.

18.1 A New Linear Condition for Determining If a Free Energy Is a FMS

A viscoelastic state is defined in general by the history and current value of strain $(E^t, E(t))$. The concept of a minimal state is introduced in Sect. 7.4. It can be expressed as follows (in slightly simplified form, ignoring the constant term in $(7.4.1)_2$: two viscoelastic states $(E_1^t, E_1(t))$, $(E_2^t, E_2(t))$ are equivalent or in the same equivalence class or minimal state if for

$$E_1(t+u) = E_2(t+u), \quad \forall u \geq 0, \tag{18.1.1}$$

we have

$$T_1(t+u) = T_2(t+u), \quad \forall u \geq 0. \tag{18.1.2}$$

It is of interest to express these conditions for linear materials. In this case, the defining state functional of minimal states I^t, used earlier in various contexts, is given by

G. Amendola et al., *Thermodynamics of Materials with Memory*,
https://doi.org/10.1007/978-3-030-80534-0_18

$$I^t(\tau) = \int_0^\infty G'(s+\tau)E_r^t(s)ds = \int_0^\infty \widetilde{G}(s+\tau)\dot{E}^t(s)ds, \qquad (18.1.3)$$

for any history E^t and relative history E_r^t. Let $(E_1^t, E_1(t))$ and $(E_2^t, E_2(t))$ be two histories and current values, which we wish to check for equivalence. Then

$$\begin{aligned} I_i^t(\tau) &= \int_0^\infty G'(s+\tau)E_{ir}^t(s)ds = \int_0^\infty \widetilde{G}(s+\tau)\dot{E}_i^t(s)ds, \quad i = 1,2, \\ I_d^t(\tau) &= \int_0^\infty G'(s+\tau)E_{dr}^t(s)ds = \int_0^\infty \widetilde{G}(s+\tau)\dot{E}_d^t(s)ds, \end{aligned} \qquad (18.1.4)$$

where

$$E_d^t(s) = E_2^t(s) - E_1^t(s), \quad \forall\, s \geq 0, \qquad E_d(t) = E_2(t) - E_1(t). \qquad (18.1.5)$$

The condition that E_1^t and E_2^t are equivalent histories then takes the form

$$E_d(t) = 0, \qquad I_d^t(\tau) = I_2^t(\tau) - I_1^t(\tau) = 0, \quad \forall \tau \geq 0. \qquad (18.1.6)$$

Note that (18.1.6) is also the condition that E_d^t is equivalent to the zero history. We point out that the notation $I_i^t(\cdot)$, $i = 1, 2$, in (18.1.4) has a different meaning to that in Sect. 17.9.1.

The interesting situation is where (18.1.6) applies for $E_d^t(s) \neq 0$, which corresponds to non-singleton minimal states. It will therefore be assumed that the relaxation function of the material under consideration has only isolated singularities.

For such materials, there is a maximum free energy that is less than the work function $W(t)$ and also a range of related intermediate free energies. Furthermore, the free energy functional is positive semidefinite (see (7.4.9)).

A free energy $\psi(t) = \widetilde{\psi}(E^t, E(t))$ is a FMS if it has the property that any two members $(E_1^t, E(t))$, $(E_2^t, E(t))$ of the same minimal state yield equal values of $\widetilde{\psi}$, or

$$\widetilde{\psi}(E_1^t, E(t)) = \widetilde{\psi}(E_2^t, E(t)), \qquad (18.1.7)$$

which is the scalar form of (7.4.6).

Remark 18.1.1. In the discussion (18.1.1)–(18.1.6), the history E_1^t can be chosen arbitrarily, while E_2^t is restricted to a degree by the conditions (18.1.6). This observation is important for Proposition 18.1.1 below.

Relations (18.1.1) and (18.1.2) for $u = 0$ become

$$E_1(t) = E_2(t), \qquad T_1(t) = T_2(t),$$

or

$$E_d(t) = 0, \qquad I_d^t(0) = \int_0^\infty G'(s)E_d^t(s)ds = 0. \qquad (18.1.8)$$

Observe that they yield the equality

$$T_1(t) = T_2(t), \qquad (18.1.9)$$

where T_1 and T_2 are the stress function T, for histories and current values $(E_1^t, E_1(t))$ and $(E_2^t, E_2(t))$, respectively.

Relation (18.1.7) is a constraint on quadratic functionals of the history. We now derive a linear condition that is equivalent to (18.1.7).

Proposition 18.1.1. *If* (18.1.8) *holds for* E_d^t*, defined by* $(18.1.5)_1$*, then the condition* (18.1.7) *is equivalent to either*

$$\begin{aligned}&\int_0^\infty \mathcal{G}(s,u)E_d^t(s)ds = \int_0^\infty \mathcal{G}(u,s)E_d^t(s)ds = 0, \quad or\\ &\int_0^\infty \widetilde{G}(s,u)\dot{E}_d^t(s)ds = \int_0^\infty \widetilde{G}(u,s)\dot{E}_d^t(s)ds = 0, \quad \forall u \in \mathbb{R}^+.\end{aligned} \tag{18.1.10}$$

Proof. If (18.1.7) is applied to $(17.3.5)_2$, using (18.1.9), it reduces to

$$\int_0^\infty\int_0^\infty E_1^t(s)\mathcal{G}(s,u)E_1^t(u)dsdu = \int_0^\infty\int_0^\infty E_2^t(s)\mathcal{G}(s,u)E_2^t(u)dsdu,$$

or

$$\int_0^\infty\int_0^\infty \mathcal{G}(s,u)[E_1^t(s)E_1^t(u) - E_2^t(s)E_2^t(u)]dsdu = 0. \tag{18.1.11}$$

Now

$$\begin{aligned}E_1^t(s)E_1^t(u) - E_2^t(s)E_2^t(u) = (E_1^t(s) - E_2^t(s))(E_1^t(u) + E_2^t(u))\\ + E_2^t(s)E_1^t(u) - E_2^t(u)E_1^t(s).\end{aligned}$$

Noting that

$$\int_0^\infty\int_0^\infty \mathcal{G}(s,u)[E_2^t(s)E_1^t(u) - E_2^t(u)E_1^t(s)]dsdu = 0,$$

by virtue of $(17.3.6)_2$ and an interchange of integration variables, we see that (18.1.11) can be replaced by

$$\int_0^\infty\int_0^\infty \mathcal{G}(s,u)(E_1^t(s) - E_2^t(s))(E_1^t(u) + E_2^t(u))dsdu = 0.$$

This relationship can be written as

$$\int_0^\infty\int_0^\infty \mathcal{G}(s,u)E_d^t(s)(2E_1^t(u) + E_d^t(u))dsdu = 0,$$

where the history E_1^t can be arbitrarily chosen (Remark 18.1.1). Therefore, (18.1.7) implies $(18.1.10)_1$, while $(18.1.10)_2$ follows from (17.3.6). Also, since the steps of the proof are reversible, $(18.1.10)_{1,2}$ imply (18.1.7). □

If we integrate $(18.1.10)_2$ over $[u_1, \infty)$, the result is

$$\int_0^\infty G_2(u_1,s)E_d^t(s)ds = 0, \tag{18.1.12}$$

for all $u_1 \geq 0$. In particular, for $u_1 = 0$, we have

$$\int_0^\infty G_2(0, s)E_d^t(s)ds = 0. \tag{18.1.13}$$

Relation $(18.1.10)_{3,4}$ can be shown to be equivalent to $(18.1.10)_{1,2}$, using integration by parts in (18.1.12).

These results are derived under the constraints (18.1.8). For the stronger assumption (18.1.6) for $\tau \geq 0$, it remains of course true that (18.1.7) and (18.1.10) are equivalent. The condition (18.1.10) states that the linear functionals

$$\begin{aligned} &\int_0^\infty \mathcal{G}(s,u)E^t(s)ds = \int_0^\infty \mathcal{G}(u,s)E^t(s)ds, \quad \text{and} \\ &\int_0^\infty \widetilde{G}(s,u)\dot{E}^t(s)ds = \int_0^\infty \widetilde{G}(u,s)\dot{E}^t(s)ds, \quad \forall\, u \in \mathbb{R}^+ \end{aligned} \tag{18.1.14}$$

are FMSs. One can replace $E^t(s)$ by $E_r^t(s)$ in $(18.1.14)_1$.

Proposition 18.1.1 can be expressed as a statement that the free energy corresponding to $G(s, u)$ is a FMS if and only if, given that (18.1.13) holds, then (18.1.12) is true for all $u_1 > 0$.

If E_d^t is equivalent to the zero history, then (18.1.7) yields that (see (7.4.9))

$$\widetilde{\psi}(E_d^t, 0) = \int_0^\infty \int_0^\infty E_d^t(s)\mathcal{G}(s,u)E_d^t(u)dsdu = 0. \tag{18.1.15}$$

Condition (18.1.10) yields, more generally, that if either (but not both) of the histories E_d^t is replaced by an arbitrary history, $(18.1.15)_2$ still holds.

18.1.1 Some Examples of Application of the New Condition

The Graffi–Volterra free energy: This functional is given by (17.3.2). The rate of dissipation associated with ψ_{GV} has the form (17.3.3). The functional $\psi_{GV}(t)$ is a free energy only if the conditions (17.3.4) hold.
We can formally express $\psi_{GV}(t)$ in the form (17.3.5) by putting

$$\mathcal{G}(s,u) = -\frac{1}{2}[G'(s) + G'(u)]\delta(s-u), \quad s, u \in \mathbb{R}^+,$$

where $\delta(s-u)$ is the singular delta function. Thus,

$$\int_0^\infty \mathcal{G}(s,u)E_d^t(u)du = -G'(s)E_d^t(s) \neq 0,$$

so that (18.1.10) is not obeyed and ψ_{GV} is not a FMS. This is shown by another method in Sect. 10.1.1 and earlier in [89].

The Dill free energy: This functional is given by (17.4.18) is a free energy with rate of dissipation given by (17.4.19) if and only if $G(\cdot)$ is completely monotonic, as defined for the general tensor case by (10.1.8). Thus, $\widetilde{G}(s, u) = \widetilde{G}(s + u)$ and it follows immediately from Proposition 18.1.1 that the Dill free energy is a FMS, because (18.1.10) has the same form as $(18.1.6)_{2,3}$.

18.1.2 Corresponding Frequency Domain Results

Relation $(18.1.10)_3$, for $\widetilde{G}(s,u)$ given by (17.4.8), becomes

$$\int_0^\infty \int_0^\infty K(z+s, z+u)\dot{E}_d^t(s)ds dz = 0, \quad u \geq 0. \tag{18.1.16}$$

Taking the Fourier transform of the free variable u in $(18.1.10)_4$, and applying Parseval's formula to the integrated product, we obtain

$$\int_{-\infty}^\infty \widetilde{G}_{+-}(\omega,\omega_1)\dot{E}_{d+}^t(\omega_1)d\omega_1 = 0, \quad \omega \in \mathbb{R}, \tag{18.1.17}$$

where E_{d+}^t is the Fourier transform of E_d^t, defined by (18.1.5). Then, (17.5.5) gives

$$\int_{-\infty}^\infty \frac{K_{+-}(\omega,\omega_1)\dot{E}_{d+}^t(\omega_1)}{\omega_1-\omega^\pm}d\omega_1 = -\widetilde{G}_+(\omega)\int_{-\infty}^\infty \frac{\dot{E}_{d+}^t(\omega_1)}{\omega_1-\omega^\pm}d\omega_1$$
$$-\int_{-\infty}^\infty \frac{\overline{\widetilde{G}_+(\omega_1)}\dot{E}_{d+}^t(\omega_1)}{\omega_1-\omega^\pm}d\omega_1, \quad \forall\ \omega \in \mathbb{R}. \tag{18.1.18}$$

The versions with ω^+ and ω^- will be shown to be equivalent. Consider the first integral on the right of (18.1.18), for ω^+. Closing the contour on $\Omega^{(-)}$, it is clear that

$$\int_{-\infty}^\infty \frac{\dot{E}_{d+}^t(\omega_1)}{\omega_1-\omega^+}d\omega_1 = 0. \tag{18.1.19}$$

Condition $(18.1.6)_3$ has the form [163]

$$I_d^t(\tau) = \frac{1}{2\pi}\int_{-\infty}^\infty [\overline{\widetilde{G}_+(\omega_1)} + \lambda\widetilde{G}_+(\omega_1)]\dot{E}_{d+}^t(\omega_1)e^{-i\omega_1\tau}d\omega_1 = 0, \quad \tau \geq 0, \tag{18.1.20}$$

where λ is any complex constant. It follows that

$$\int_0^\infty I_d^t(\tau)e^{-i\omega\tau}d\tau = I_{d+}^t(\omega) = 0, \quad \omega \in \mathbb{R}.$$

From the complex conjugate of this relationship, $\overline{I_{d+}^t}(\omega) = 0$, we deduce that

$$\int_{-\infty}^\infty \frac{[\overline{\widetilde{G}_+(\omega_1)} + \lambda\widetilde{G}_+(\omega_1)]\dot{E}_{d+}^t(\omega_1)}{\omega_1-\omega^+}d\omega_1 = 0, \quad \forall\ \omega \in \mathbb{R}, \tag{18.1.21}$$

where the choice ω^+ is dictated by convergence requirements. The enforcement of (18.1.21) ensures that $\dot{E}_{d+}^t(\omega)$ is the Fourier transform of a history equivalent to zero. Combining (18.1.19), multiplied by $\widetilde{G}_+(\omega)$, with (18.1.21) for $\lambda = 0$, we obtain from (18.1.18) that

$$\int_{-\infty}^\infty \frac{K_{+-}(\omega,\omega_1)\dot{E}_{d+}^t(\omega_1)}{\omega_1-\omega^+}d\omega_1 = 0. \tag{18.1.22}$$

From (18.1.22), (17.5.18) and the Plemelj formulae, we obtain

$$\frac{1}{2\pi i}\int_{-\infty}^{\infty}\frac{K_{+-}(\omega,\omega_1)\dot{E}^t_{d+}(\omega_1)}{\omega_1-\omega^-}d\omega_1 = 2\frac{H(\omega)}{\omega^2}\dot{E}^t_{d+}(\omega), \tag{18.1.23}$$

which is equal to the right-hand side of (18.1.18), for ω^- in the denominator. Either (18.1.22) or (18.1.23) express condition (18.1.17) or (18.1.10).

18.1.3 Application of Product Formulae in the Time and Frequency Domains

For the time domain, the rate of dissipation (17.3.11) becomes (17.7.2) and the free energy has the form (17.7.3). Also, relation (17.7.4) holds.

The condition that $\dot{E}^t_d$ is equivalent to the zero history takes the form

$$\sum_{i,j=1}^{m}\int_0^{\infty}A_{ij}k_i(z)\left[\int_0^{\infty}k_j(z+s+u)\dot{E}^t_d(s)ds\right]dz = 0, \quad \forall\, u \geq 0 \tag{18.1.24}$$

by virtue of (17.7.4) and (18.1.6). We see that, since (18.1.24) holds for arbitrary u, the individual constraints

$$\int_0^{\infty}k_i(y+s)\dot{E}^t_d(s)ds = 0, \quad \forall\, y \geq 0, \quad i = 1,2,\ldots,m, \tag{18.1.25}$$

must be obeyed. Condition (18.1.16) becomes, for $K(s,u)$ given by (17.7.1),

$$\begin{aligned}&\sum_{i,j=1}^{m}\int_0^{\infty}\int_0^{\infty}A_{ij}k_i(z+s)k_j(z+u)\dot{E}^t_d(s)dsdz\\ &= \sum_{i,j=1}^{m}\int_0^{\infty}A_{ij}k_j(z+u)\left[\int_0^{\infty}k_i(z+s)\dot{E}^t_d(s)ds\right]dz = 0, \quad u \geq 0,\end{aligned} \tag{18.1.26}$$

which is true by virtue of (18.1.25). Conversely, since u is arbitrary, we see that (18.1.26) implies the conditions (18.1.25).

Relation (18.1.26) is equivalent to the statement that any free energy functional with rate of dissipation kernel given by (17.7.1) is a FMS, by virtue of Proposition 18.1.1. The conditions (18.1.25) are a statement that the quantities

$$\int_0^{\infty}k_i(z+s)\dot{E}^t(s)ds, \quad \forall\, z \geq 0, \quad i = 1,2,\ldots,m \tag{18.1.27}$$

are FMSs.

For the frequency domain, we consider the forms (17.7.18) and (17.7.19). The quantities $k_{i-}(\omega)$ have all their singularities in $\Omega^{(-)}$ and are analytic on an open set including $\Omega^{(+)}$, just as for the functions $H^f_-(\omega)$. We put

$$\begin{aligned}&i\frac{k_{i-}(\omega)}{\omega}\dot{E}^t_+(\omega) = q^t_{i-}(\omega) - q^t_{i+}(\omega),\\ &q^t_{i\pm}(\omega) = \frac{1}{2\pi i}\int_{-\infty}^{\infty}\frac{k_{i-}(\omega')\dot{E}^t_+(\omega')}{\omega'-\omega^{\mp}}d\omega', \quad i = 1,2\ldots,m.\end{aligned}$$

Condition $(17.5.18)_1$ gives that

$$\sum_{i,j=1}^{m} A_{ij}\overline{k_{i-}}(\omega)k_{j-}(\omega) = 2\frac{H(\omega)}{\omega^2}.$$

This equation implies that the singularities of $k_{j\pm}(\omega)$ are a subset (or the full set) of those in $H(\omega)$. The possibility of extra singularities occurring in individual terms, which drop out of the summation due to cancellation between terms, is excluded by assumption.

Relations $(17.5.7)_2$ and (17.5.13) yield

$$\begin{aligned} D(t) &= \frac{1}{8\pi^2}\int_{-\infty}^{\infty}\int_{-\infty}^{\infty}\overline{\dot{E}^t_+}(\omega_1)\sum_{i,j=1}^{m} A_{ij}\overline{k_{i-}}(\omega_1)k_{j-}(\omega_2)\dot{E}^t_+(\omega_2)d\omega_1 d\omega_2, \\ \psi(t) &= \phi(t) \\ &+ \frac{i}{8\pi^2}\int_{-\infty}^{\infty}\int_{-\infty}^{\infty}\frac{\overline{\dot{E}^t_+}(\omega_1)\sum_{i,j=1}^{m} A_{ij}\overline{k_{i-}}(\omega_1)k_{j-}(\omega_2)\dot{E}^t_+(\omega_2)}{\omega_1^+ - \omega_2^-}d\omega_1 d\omega_2. \end{aligned} \tag{18.1.28}$$

The case $m = 1$, as in Sect. 17.7.2, is precisely that discussed in Sect. 16.4. For $m > 1$, we are dealing with more general classes of free energies.

The following result is proved in [17], using arguments similar to those in Lemma 16.4.2 and Theorem 11.3.1.

Proposition 18.1.2. *The quantities*

$$q^t_{j-}(\omega) = \frac{1}{2\pi i}\int_{-\infty}^{\infty}\frac{k_{j-}(\omega_1)\dot{E}^t_+(\omega_1)}{\omega_1 - \omega^+}d\omega_1, \qquad j = 1, 2, \ldots, m,$$

are FMSs.

Also, from [17],

Proposition 18.1.3. *Any free energy of the form* $(18.1.28)_2$ *is a FMS.*

Proof. Relation (18.1.22) takes the form

$$\sum_{i,j=1}^{m} A_{ij}\overline{k_{i-}}(\omega)\int_{-\infty}^{\infty}\frac{k_{j-}(\omega_1)\dot{E}^t_{d+}(\omega_1)}{\omega_1 - \omega^+}d\omega_1 = 0.$$

This is always satisfied for $\dot{E}^t_{d+}$ equivalent to the zero history, by virtue of Proposition 18.1.2. Thus, $(18.1.28)_2$ is a FMS. □

These results show that, for all materials described by (17.7.1) or (17.7.18), the resulting free energy is a FMS. Proposition 18.1.3 suggests the conjecture in [17] that, other than certain degenerate cases, all finite quadratic functionals are FMSs. An example of a degenerate form of (17.3.5) is given by the Graffi–Volterra free energy.

We note that Parseval's formula, applied to (18.1.25), gives (*cf.* (18.1.20))

$$\int_{-\infty}^{\infty} k_{i-}(\omega)\dot{E}^t_{d+}(\omega)e^{-i\omega y}d\omega = 0, \quad \forall\, y \geq 0, \quad i = 1, 2 \ldots, m.$$

Taking the Fourier transform of the complex conjugate of this relation gives

$$q^t_{di-}(\omega) = \frac{1}{2\pi i}\int_{-\infty}^{\infty} \frac{k_{i-}(\omega_1)\dot{E}^t_{d+}(\omega_1)}{\omega_1 - \omega^+}d\omega_1 = 0, \qquad i = 1, 2, \ldots, m,$$

where ω^+ is the choice required for convergence. Thus, the results of the time and frequency domains are in agreement.

The cases (17.7.29) are discussed in [17], as are the new categories based on (17.7.30), or equivalently, (17.7.15). All the free energies generated by these methods are FMSs, by virtue of Proposition 18.1.3.

18.2 Free Energies for Singleton Minimal States

In the light of the analysis of Sect. 5.1.2, we assume that any free energy function exhibits strict periodic behavior for histories that have been periodic for all past times. This is not the case for the work function, though it has the usual defining properties of a free energy. Forms given in fairly recent years for the minimum (Chap. 11) and related free energies of linear materials with memory (see Sect. 16.3) have this property, as have the various functionals in Chap. 10.

Materials for which the minimal states are all singletons are those for which at least some of the singularities of the Fourier transform of the relaxation function are not isolated. For such materials, the maximum free energy is the work function, and free energies intermediate between the minimum free energy and the work function should be given by a linear relation involving these two quantities. Discussion of these topics may be found in Sect. 16.5. Thus,

$$\psi(t) = c_1\psi_m(t) + c_2 W(t), \qquad c_1, c_2 \in \mathbb{R}^+, \qquad c_1 + c_2 = 1,$$

where $\psi_m(t)$ is the minimum free energy for such materials. All such functionals, except the minimum free energy, therefore do not have strict periodic behavior for periodic histories, which contradicts our assumption.

A way out of the difficulty is explored which involves approximating the relaxation function by a form for which the minimal states are no longer singletons. A representation can then be given of an arbitrary free energy as a linear combination of the minimum, maximum, and intermediate free energies derived in Chap. 16, where the maximum free energy is less than the work function. This representation obeys our periodicity assumption.

Remark 18.2.1. The relation (5.1.35) will be regarded as a basic property of any free energy functional, sufficiently important to label it as P4, following P3, defined by (16.1.28).

Therefore, we have

P4 For a long established periodic history, where transient effects have died away, a free energy functional must obey the condition

$$\psi(t+T)=\psi(t), \quad t\in\mathbb{R}. \tag{18.2.1}$$

where T is the core period. From (5.1.37) it follows that (18.2.1) does not hold for $W(t)$ if the material is dissipative.

The special property of $\psi(t)$, not shared by $\mathcal{D}(t)$ and $W(t)$, that it is finite for infinite periodic histories is a reflection of the fact that it is related to energy stored in the body, which must be finite. In particular, the minimum free energy, for a given state, is equal to the maximum recoverable work from that state.

Remark 17.3.2 is assumed to apply here also.

18.2.1 Approximating Continuous-Spectrum Behavior by Discrete-Spectrum Formulae

The form of the minimum free energy for continuous-spectrum materials is discussed in detail in Chap. 14, with explicit formulae given for all quantities of interest. We wish to explore another approach here, which is to approximate the continuous-spectrum case by a discrete-spectrum material.

Let the quantity $H(\omega)$ be given by (14.2.13), but with $k(s)$ replaced by $g(s)$. This formula can be approximated by a quadrature formula

$$H(\omega)=\omega^2\sum_{i=1}^{n}\frac{\alpha_i\kappa_i g(\alpha_i)}{\alpha_i^2+\omega^2}, \tag{18.2.2}$$

where the constants α_i, $i=1,2,\ldots,n$ are values of $\alpha\in[a,b]$ and κ_i are the quadrature coefficients associated with the particular method chosen (trapezoidal, Simpson, Gauss, *etc.*). Comparing with (11.9.2), we put

$$G_i=\kappa_i g(\alpha_i), \qquad i=1,2,\ldots,n \tag{18.2.3}$$

to obtain a discrete-spectrum material approximating the continuous-spectrum material under discussion.

In the numerical work described in Sect. 18.2.5, we use the simplest option, namely the trapezoidal rule, so that

$$\begin{aligned}\alpha_i&=a+\frac{(i-1)(b-a)}{n-1}, \qquad i=1,2,\ldots,n,\\ \kappa_i&=\begin{cases}\dfrac{b-a}{n-1}, & i=2,3,\ldots,n-1,\\[2mm] \dfrac{b-a}{2(n-1)}, & i=1,n.\end{cases}\end{aligned} \tag{18.2.4}$$

18.2.2 The Minimum Free Energy for Continuous-Spectrum Materials

The form $(17.3.5)_1$ is used in Chap. 14. We will use $(17.3.5)_3$ here, giving

$$\begin{aligned}\psi_m(t) &= \phi(t) + \frac{1}{2}\int_0^\infty \int_0^\infty \dot{E}^t(s)\widetilde{G}_m(s,u)\dot{E}^t(u)dsdu,\\ \widetilde{G}_m(s,u) &= \frac{1}{2\pi^2}\int_a^b \int_a^b \frac{\Delta(\alpha)e^{-\alpha s}\Delta(\beta)e^{-\beta s}}{(\alpha+\beta)\alpha\beta}d\alpha d\beta,\end{aligned} \tag{18.2.5}$$

where

$$\begin{aligned}\Delta(\alpha) &= \pi h_\infty \frac{\alpha g(\alpha)}{K(\alpha)Q(\alpha)},\\ K(\alpha) &= -\int_a^b \frac{\beta g(\beta)}{\beta+\alpha}d\beta, \quad \alpha \notin [-b,-a],\\ L(\beta) &= -P\int_a^b \frac{\lambda g(\lambda)}{\lambda-\beta}d\lambda, \quad \beta \in (a,b),\\ Q(\alpha) &= \exp\left\{-\frac{1}{\pi}\int_a^b \frac{A(\beta)}{\beta+\alpha}d\beta\right\},\\ A(\beta) &= \arg\left[1-\frac{K(\beta)}{L(\beta)+i\pi\alpha g(\beta)}\right], \quad -\pi \le A(\beta) \le 0.\end{aligned} \tag{18.2.6}$$

The symbol P indicates a principal value integral. If we evaluate $(18.2.5)_2$ by quadrature, the result is

$$\widetilde{G}_m(s,u) = \frac{1}{2\pi^2}\sum_{i,j=1}^{n} \frac{\kappa_i\Delta(\alpha_i)\kappa_j\Delta(\alpha_j)}{(\alpha_i+\alpha_j)\alpha_i\alpha_j}e^{-\alpha_i s - \alpha_j u}, \tag{18.2.7}$$

where the κ_i were introduced in (18.2.2), and specified for the trapezoidal rule by (18.2.4). Since the range of α and β in (18.2.5) are both $[a,b]$, which is divided into $n-1$ intervals, we have replaced β_j by α_j in (18.2.7). Comparing (18.2.7) with (16.9.8) for $f=1$ and invoking (17.4.14), we see that

$$\frac{\kappa_i\Delta(\alpha_i)}{\sqrt{2}\pi} = \sqrt{2}h_\infty R_i, \tag{18.2.8}$$

where the quantities R_i are the R_i^f for $f=1$, as given by (16.9.4), or more specifically by $(11.9.6)_3$. The validity of this relation will be demonstrated for particular material parameters and strain histories in Sect. 18.2.5.

18.2.3 Proposed Method for Approximating Free Energies for Materials with Singleton Minimal States

It is assumed that we can approximate continuous-spectrum materials as discrete-spectrum materials, by means of (18.2.2). Arbitrary free energies greater than or equal to the minimum free energy are approximated by (16.10.1).

We need to confirm that this approximation improves as n gets larger, so that there is a convergent process for increasing n. The value of n finally chosen to approximate the continuous-spectrum material should be such that convergence is essentially complete, yielding a virtually unique choice of $\psi(t)$. Based on numerical results described in Sect. 18.2.5 (see also [17, 165]), it will be taken to be 8. For this value of n, there are 126 free energies $\psi_f(t)$ lying between $\psi_m(t)$ and $\psi_M(t)$, and indeed those given by (16.10.1).

Remark 18.2.2. It is asserted that all these free energies can be treated as free energy functionals of the continuous-spectrum material. The difficulty with this assertion is that if some other value than $n = 8$ is chosen, for example, $n = 9$, the intermediate free energies $\psi_f(t)$, $f = 2, 3, \ldots, M - 1$ are given by quite different formulae. However, the minimum and maximum free energies converge to fixed functionals. Numerical confirmation that this is true to a good approximation is presented in Sect. 18.2.5.

18.2.4 Free Energy Functionals for Sinusoidal/Exponential Histories Which Vanish for $t < 0$

A change of variable in $(16.1.3)_4$ gives

$$T(t) = T_e(t) + \int_{-\infty}^{t} \widetilde{G}(t-s)\dot{E}(s)ds.$$

If the strain history is zero for $t < 0$, then

$$T(t) = T_e(t) + \int_{0}^{t} \widetilde{G}(t-s)\dot{E}(s)ds, \qquad t \geq 0,$$

and vanishes for $t < 0$. By a double change of variables, we can express $(17.3.5)_3$ (and analogously $(17.3.5)_{1,2}$) in the form

$$\psi(t) = \phi(t) + \frac{1}{2}\int_{-\infty}^{t}\int_{-\infty}^{t} \dot{E}(s)\widetilde{G}(t-s, t-u)\dot{E}(u)dsdu,$$

so that in relation to histories which vanish for negative times, we have

$$\psi(t) = \phi(t) + \frac{1}{2}\int_{0}^{t}\int_{0}^{t} \dot{E}(s)\widetilde{G}(t-s, t-u)\dot{E}(u)dsdu, \qquad t \geq 0. \tag{18.2.9}$$

Again, it is equal to zero for $t < 0$.

For discrete-spectrum materials, in particular those defined by (18.2.3), we give expressions for the work function and the free energy functionals $\psi_f(t)$, $\psi_{Dill}(t)$, and $\psi_F(t)$, in the case of a history of the strain that vanishes at negative times. Firstly, we note that the stress is given by

$$\begin{aligned} T(t) &= T_e(t) + \sum_{i=1}^{n} G_i S_i(t), \\ S_i(t) &= e^{-\alpha_i t}\int_{0}^{t} e^{\alpha_i s}\dot{E}(s)ds, \quad t \geq 0. \end{aligned} \tag{18.2.10}$$

Similarly, (18.2.9) becomes

$$\psi(t) = \phi(t) + \frac{1}{2}\sum_{i,j=1}^{n} C_{ij} S_i(t) S_j(t), \tag{18.2.11}$$

in the light of (17.4.14).

Let us now specialize to the case where the history, at positive times, is sinusoidal with frequency ω_0, but combined with exponential behavior. Consider a history and current value $(E^t, E(t))$ defined by

$$\begin{aligned} E(t) &= -iE_1[e^{i\omega_- t} - e^{-i\omega_+ t}] = 2E_1 e^{\eta t}\sin(\omega_0 t), \quad t > 0, \\ &= 0, \quad t \leq 0, \\ E^t(s) &= E(t-s), \end{aligned} \tag{18.2.12}$$

where $2E_1 e^{\eta t}$ is the amplitude. The parameter η determines the exponential factor. In the present context, it will be allowed to have both positive, negative and zero values. Equation (18.2.12) yields that

$$\dot{E}(t) = E_1[\omega_- e^{i\omega_- t} + \omega_+ e^{-i\omega_+ t}], \quad t > 0. \tag{18.2.13}$$

Thus,

$$S_i(t) = 2E_1 Re\left\{\frac{\omega_-}{\alpha_i + i\omega_-}\left[e^{i\omega_- t} - e^{-\alpha_i t}\right]\right\}. \tag{18.2.14}$$

Also, from (18.2.10) and (18.2.13),

$$\begin{aligned} W(t) &= \phi(t) + \sum_{i=1}^{n}\int_0^t G_i S_i(u)\dot{E}(u)du \\ &= \phi(t) + 2E_1^2 \sum_{i=1}^{n} G_i Re\left\{\frac{\omega_-}{2i(\alpha_i + i\omega_-)}(e^{2i\omega_- t} - 1)\right. \\ &\quad + \frac{|\omega_-|^2}{2\eta(\alpha_i + i\omega_-)}(e^{2\eta t} - 1) \\ &\quad \left. + \frac{2\omega_-}{(\alpha_i + i\omega_-)} Re\left[\frac{\omega_-}{(\alpha_i - i\omega_-)}(e^{(i\omega_- - \alpha_i)t} - 1)\right]\right\}. \end{aligned} \tag{18.2.15}$$

The quantity $\psi_f(t)$ becomes

$$\psi_f(t) = \phi(t) + H_\infty \sum_{i,j=1}^{n} \frac{R_i^f R_j^f}{(\alpha_i + \alpha_j)\alpha_i \alpha_j} S_i(t) S_j(t), \tag{18.2.16}$$

by virtue of (16.9.8) and (18.2.11), where $S_i(t)$ has the form (18.2.14).

If $\eta = -|\eta| < 0$, the exponential behavior of the history is decaying for large times rather than increasing.

The functional $\psi_M(t)$ rapidly approaches $W(t)$ for increasing exponential histories, as n gets larger. Matters are more complex for decaying exponential histories. As t increases, we have from (18.2.15) for $\eta < 0$ that

$$\begin{aligned} W(t) \to -2E_1^2 \sum_{i=1}^{n} \int_0^t G_i Re \Bigg\{ & \frac{\omega_-}{2i(\alpha_i + i\omega_-)} + \frac{|\omega_-|^2}{2\eta(\alpha_i + i\omega_-)} \\ & + \frac{2\omega_-}{(\alpha_i + i\omega_-)} Re \left[\frac{\omega_-}{(\alpha_i - i\omega_-)} \right] \Bigg\} = \Lambda \neq 0, \end{aligned} \tag{18.2.17}$$

where Λ is independent of t. In contrast,

$$S_i(t) \longrightarrow 0 \quad \text{as} \quad t \to \infty, \tag{18.2.18}$$

so that the approximate equality $\psi_M(t) \approx W(t)$ does not hold at large values of t.

If $\dot{E}(u) = 0$ for $u < 0$, relation (18.2.9) gives that

$$\psi_{Dill}(t) = \phi(t) + \frac{1}{2} \int_0^t \int_0^t \dot{E}(s) \widetilde{G}(2t - s - u) \dot{E}(u) ds du, \quad t \geq 0,$$

and vanishes for $t < 0$. From this, we have

$$\psi_{Dill}(t) = \phi(t) + \frac{1}{2} \sum_{i=1}^{n} G_i S_i^2(t), \quad t \geq 0. \tag{18.2.19}$$

The functional $\psi_F(t)$ in the scalar case is defined by (17.9.7) where $I_1^t(\tau)$ is defined by (17.9.1). If $\dot{E}(u)$ is equal to zero for $u < 0$, then

$$I^t(\tau) = \int_0^t \widetilde{G}(t - u + \tau) \dot{E}(u) du, \quad t \geq 0,$$

and zero for $t < 0$, giving

$$I^t(\tau) = \sum_{i=1}^{n} G_i S_i(t) e^{-\alpha_i \tau}, \quad t \geq 0,$$

so that

$$\begin{aligned} \psi_F(t) &= \phi(t) + \frac{1}{2} \sum_{i,j=1}^{n} C_{ij}^F S_i(t) S_j(t), \quad t \geq 0, \\ C_{ij}^F &= \alpha_i \alpha_j G_i G_j \int_0^\infty \frac{e^{-(\alpha_i + \alpha_j)\tau}}{\sum_{l=1}^n \alpha_l G_l e^{-\alpha_l \tau}}. \end{aligned} \tag{18.2.20}$$

The functionals $\psi_{Dill}(t)$ and $\psi_F(t)$, given by (18.2.19) and (18.2.20), were plotted for various parameters and found to be remarkably close to $\psi_m(t)$. They could therefore be replaced by $\psi_m(t)$ and omitted from further explicit consideration.

18.2.5 Numerical Results Relevant to the Method for Approximating Continuous-Spectrum Materials

The proposals put forward in Sect. 18.2.3 indicate how the problem noted at the beginning of Sect. 18.2 can be avoided. In this section, we will present numerical results relating to these proposals, based on particular materials defined below and the histories introduced in Sect. 18.2.4. More detailed numerical results are presented in [17] or [165].

Consider the discrete-spectrum approximation to the continuous-spectrum theory as given by (18.2.2) and (18.2.3), yielding (18.2.16). This produces, for $f = 1$, a minimum free energy which should approximate the continuous-spectrum formulae in [94] more and more closely, as we increase the number of sub-divisions of the interval $[a, b]$ in (18.2.2). This can be checked using (18.2.8), and turns out to be remarkably accurate, even for low values of n.

The discrete-spectrum approximation also gives a maximum free energy, which converges rapidly to $W(t)$, as n increases, at least for diverging exponential histories [16]. Indeed, for $n = 4$, the two quantities are virtually equal, at least for the parametric values in that reference. This property does not always hold.

Regarding the behavior of the maximum free energy for increasingly large values of n in (18.2.2), it is established that convergence to a unique functional, not necessarily $W(t)$, occurs. Also, $n = 8$ is the value at which, for practical purposes, this is achieved. This approach also yields 126 intermediate free energies $\psi_f(t)$, $f = 2, 3, \dots, 127$. All the free energies ψ_f, for any n, have the property P4.

The material parameters and those of the histories are now specified. Dimensions will be removed from all expressions for free energies by dividing by $G_0 E_1^2$ [16]. This amounts to taking

$$G_0 = E_1 = 1.$$

Also, G_∞, in the dimensionless version, is equal to G_∞ / G_0 in terms of dimensional quantities, and varies in the range $[0, 1]$. It is chosen to be 0.1. We take $g(\alpha)$ in (18.2.6) to be a constant quantity g_0, so that (18.2.3), for the trapezoidal rule, becomes

$$G_i = \begin{cases} \dfrac{b-a}{n-1} g_0, & i = 2, 3, \dots, n-1 \\ \dfrac{b-a}{2(n-1)} g_0, & i = 1, n, \end{cases}$$

where a and b are assigned values 0.5 and 1.5, respectively. Noting that $(b - a)g_0 = 1 - G_0$, we have

$$G_i = \begin{cases} \dfrac{1 - G_\infty}{n-1}, & i = 2, 3, \dots, n-1, \\ \dfrac{1 - G_\infty}{2(n-1)}, & i = 1, n. \end{cases}$$

Also, the quantities α_i are given by $(18.2.4)_1$.

We take $\omega_0 = 1$ in (16.11.2) and consider values of η/ω_0 in the range $[0.1, 1.0]$ and $[-1.0, -0.1]$.

Let us first check the validity of (18.2.8) for these materials. Since $g(\alpha)$ is constant, the quantity $K(\alpha)$, defined by $(18.2.6)_2$, has the form

$$K(\alpha) = -g_0\left[b - a - \alpha\log\left(\frac{b+\alpha}{a+\alpha}\right)\right], \qquad \alpha \notin [-b, -a],$$

while $L(\alpha)$, given by $(18.2.6)_3$, becomes

$$L(\beta) = -g_0\left[b - a + \beta\log|\frac{b-\beta}{a-\beta}|\right], \qquad \beta \in (a, b).$$

We define

$$C_i = \frac{\kappa_i \Delta(\alpha_i)}{2\pi h_\infty}, \quad i = 1, 2, \ldots, n,$$

where h_∞ is given by (11.1.18). Observe that (18.2.8) can be written as $C_i = R_i$, $i = 1, 2, \ldots, n$. Values of C_i and R_i, given by (16.9.4) for $f = 1$, corresponding to $\epsilon_i^f = 1$, $i = 2, 3, \ldots, n$, were compared for all values of n considered, namely $n = 3, 4, \ldots, 8$. In all cases, the quantities C_i and R_i were virtually identical.

Recall that (18.2.8) follows from a comparison of (18.2.7) (in the light of (17.4.14)) and (16.9.8). The numerical results are a good check on these results or on the formulae (18.2.5), (18.2.6) on the one hand, and (16.9.7) for $f = 1$, on the other. It should be emphasized that the derivations of these two sets of results are entirely different. The level of agreement between them also provides support for the discretization procedure outlined in Sect. 18.2.1 and in (18.2.7).

The notation $\psi_f^{(n)}(t)$ will be used to denote (18.2.16) and $W^{(n)}(t)$ for (18.2.15). The superscript (n) on these quantities was omitted up to now to minimize cumbersome notation.

Next, we consider whether $\psi_M^{(n)}(t)$ approaches $W^{(n)}(t)$, as n increases. This issue has been addressed earlier by (18.2.17) and (18.2.18). The data plotted on Fig. 18.1 indicates that this is true to a high degree of accuracy for $\eta > 0$. Indeed, the convergence is rapid. For $n = 8$, they are virtually identical, and are very close from $n = 4$ onwards.

However, for $\eta < 0$, there is rapid convergence on the interval $t \in (0.0, 1.0)$ approximately, while for larger times, we have a perceptible difference for $n = 8$. This is consistent with the observation made after (18.2.18).

In general, $\psi_M^{(n)}(t)$ converges to $W^{(n)}(t)$ on a subset $\mathcal{C}$ of $\mathbb{R}^+$, where, in some but not all cases, $\mathcal{C} = \mathbb{R}^+$.

The next question to explore is whether $\psi_M^{(n)}(t)$ converges to a definite value, which may or may not be very close to $W^{(n)}(t)$ for $t \in \mathbb{R}^+$. Convergence occurs for $t \in \mathcal{C}$ and for $\eta > 0$, we have $\mathcal{C} = \mathbb{R}^+$. This is confirmed on Fig. 18.2, which also illustrates that convergence is virtually complete for $n = 4$, as noted in the context of Fig. 18.1.

For $\eta < 0$, we see from Fig. 18.3 that convergence to a definite function $\psi_M^{(8)}(t)$ occurs over the range of times shown. However, the convergence is not uniform. It is very rapid up to about $t = 1$, where convergence to $W^{(8)}(t)$ is taking place. For larger times, the convergence is more gradual, but is essentially complete for $n = 8$. We conclude that there is evidence that $\psi_M^{(n)}(t)$ converges to a definite functional, not always equal to $W^{(8)}(t)$.

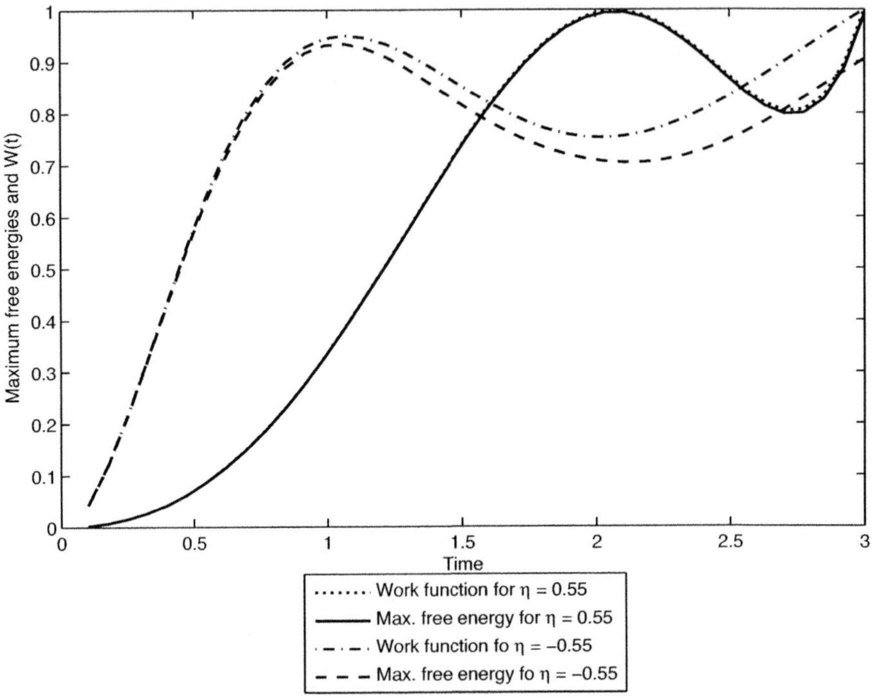

Fig. 18.1. Maximum free energies for $n = 8$ and the work function, for $\eta = 0.55$ and -0.55. These quantities are divided by $W^{(n)}(3)$

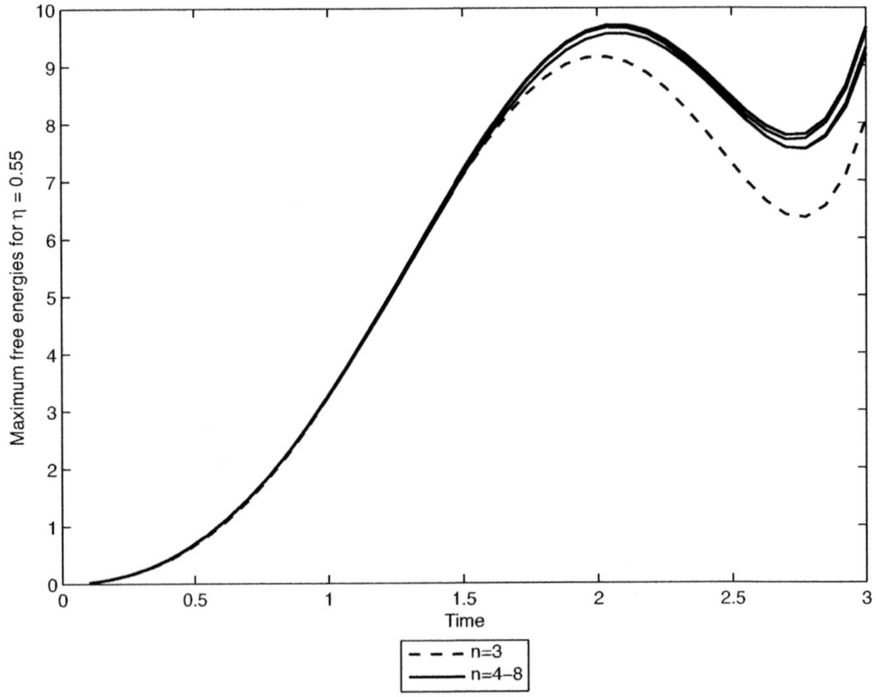

Fig. 18.2. Maximum free energies for n = 3–8 and $\eta = 0.55$

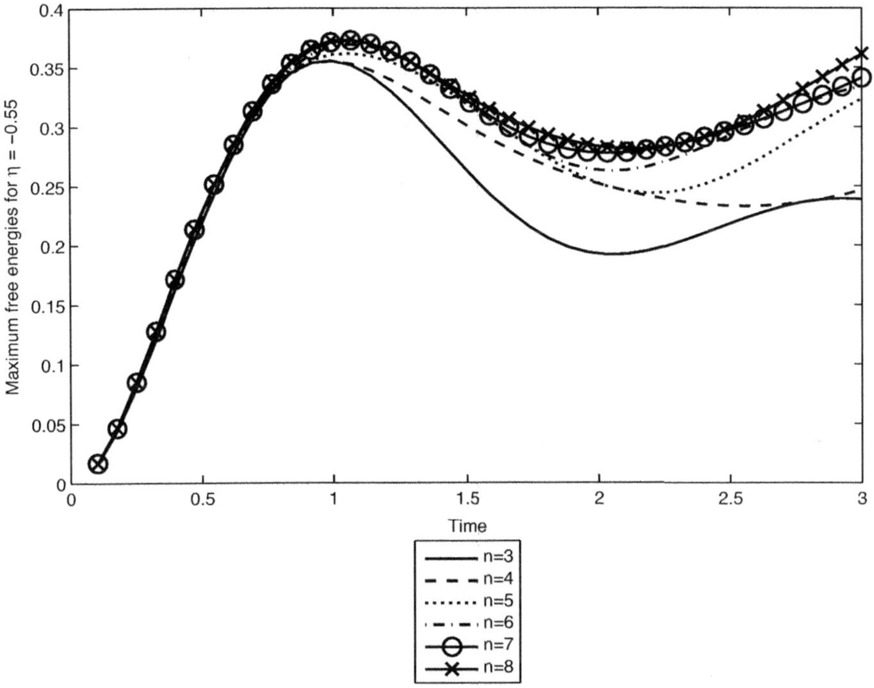

Fig. 18.3. Maximum free energies for n = 3–8 and $\eta = -0.55$

19

Second-Order Approximation for Heat Conduction: Dissipation Principle and Free Energies

19.1 Introduction

This chapter is based on [20].

In the context of new models of heat conduction, the second-order approximation of Tzou's theory, derived by Quintanilla and Racke [289, 290] and generalized in [126], is considered as a fading memory material, for which it is of interest to determine free energy functionals. The kernel* does not satisfy the convexity restrictions (17.3.4), which allow us to obtain various traditional free energies for materials with fading memory. It is therefore necessary to restrict the study to the minimum and related free energies, which do not require these restrictions. We first derive an explicit expression for the minimum free energy. Also, simple modifications of this expression give an intermediate free energy and the maximum free energy for the material. These derivations differ in certain important respects from those in mechanics.

After Cattaneo's pioneering work [59] on heat conduction with a finite propagation speed, many theories with this property, known also as theories of second sound, have been considered in order to remove the difficulty of infinite propagation speed associated with Fourier's law.

Thus, Tzou, to replace Fourier's law, has suggested in [315] the following constitutive equation for any $\mathbf{x}$ of a body $\mathcal{B} \subset \mathcal{R}^3$

$$\mathbf{q}(\mathbf{x}, t + \tau_{\mathbf{q}}) = -k(\mathbf{x})\nabla\theta(\mathbf{x}, t + \tau_\theta), \qquad \tau_{\mathbf{q}} > 0, \qquad \tau_\theta > 0, \qquad \mathbf{q} \in \mathcal{R}^3. \tag{19.1.1}$$

The coefficient $k(\mathbf{x})$ is positive. Also, $\tau_{\mathbf{q}}$ and τ_θ are delay times related to the heat flux and the gradient of temperature, respectively. Obviously, by assuming $\tau_{\mathbf{q}} = 0$ and $\tau_\theta = 0$, Fourier's law follows.

The general form (19.1.1) cannot be used since, as has been recently observed in [125], it is not in agreement with the second law of thermodynamics. Consequently, several constitutive equations for the heat flux have been obtained by means of Tay-

* The description "kernel" has been used in earlier chapters in a somewhat different sense to its present usage, which is analogous to a relaxation function in linear viscoelasticity.

G. Amendola et al., *Thermodynamics of Materials with Memory*,
https://doi.org/10.1007/978-3-030-80534-0_19

lor's expansions of (19.1.1) up to different orders with respect to the parameters $\tau_{\mathbf{q}}$ and τ_θ. In this manner, some models have been proposed for which the required compatibility with thermodynamics is satisfied.

It is interesting to observe that the expansion of the left-hand side of (19.1.1) to first order with respect to the parameter $\tau_{\mathbf{q}}$, while putting $\tau_\theta = 0$, yields Cattaneo's theory [59]

$$\tau_{\mathbf{q}}\dot{\mathbf{q}}(\mathbf{x}, t) + \mathbf{q}(\mathbf{x}, t) = -k\nabla\theta(\mathbf{x}, t). \tag{19.1.2}$$

If the second-order approximation is considered both for the heat flux and for the temperature gradient, then we obtain the following equation:

$$\mathbf{q}(\mathbf{x}, t) + \tau_{\mathbf{q}}\dot{\mathbf{q}}(\mathbf{x}, t) + \frac{\tau_{\mathbf{q}}^2}{2}\ddot{\mathbf{q}}(\mathbf{x}, t) = -k\left[\nabla\theta(\mathbf{x}, t) + \tau_\theta\nabla\dot{\theta}(\mathbf{x}, t) + \frac{\tau_\theta^2}{2}\nabla\ddot{\theta}(\mathbf{x}, t)\right], \tag{19.1.3}$$

which was derived in [289, 290]. Several authors have considered this equation and studied mathematical problems of existence, uniqueness, and stability in relation to it. Some restrictions on the constants $\tau_{\mathbf{q}}$ and τ_θ (see, for example, [288, 319]) have been established by such studies.

In [126], Eq. (19.1.3) has been expressed as a fading memory model, so the heat flux depends on the history of the temperature gradient, as for Gurtin and Pipkin's theory [191]. Moreover, a restriction on $\tau_{\mathbf{q}}$ and τ_θ given by

$$\left(2 - \sqrt{3}\right)\tau_\theta < \tau_{\mathbf{q}} < \left(2 + \sqrt{3}\right)\tau_\theta \tag{19.1.4}$$

has been obtained from the requirement of compatibility with thermodynamics.

19.2 A Fading Memory Constitutive Equation and the Second Law

We now express (19.1.3) as a relation between the heat flux and the history of the temperature gradient. The Cattaneo–Maxwell equation (19.1.2) can be written in the form (recall also (9.1.1))

$$\mathbf{q}(\mathbf{x}, t) = -\int_{-\infty}^{t} \kappa(\mathbf{x}, t - s)\nabla\theta(\mathbf{x}, s)ds, \tag{19.2.1}$$

where the kernel has the form

$$\kappa(\mathbf{x}, s) = \frac{1}{\tau_{\mathbf{q}}}k(\mathbf{x})e^{-\frac{1}{\tau_{\mathbf{q}}}s}. \tag{19.2.2}$$

For simplicity, we take the kernels to be scalar quantities, though indeed, the results for tensor kernels can also be given, using the methods outlined below. The dependence on $\mathbf{x}$ will be omitted henceforth. Relation (19.1.3) can also be rewritten in the form [126]

$$\mathbf{q}(t) = -\frac{k}{\tau_{\mathbf{q}}^2}\left[\tau_\theta^2\nabla\theta(t) + 2\left(\tau_{\mathbf{q}} - \tau_\theta\right)\int_{-\infty}^{t} \kappa(s)\nabla\theta^t(s)ds\right], \tag{19.2.3}$$

where the kernel κ is given by

$$\kappa(s) = e^{-\frac{1}{\tau_{\mathbf{q}}}s}\left[\frac{\tau_\theta}{\tau_{\mathbf{q}}}\cos\frac{s}{\tau_{\mathbf{q}}} + \sin\frac{s}{\tau_{\mathbf{q}}}\right] \tag{19.2.4}$$

and $\theta^t(s) = \theta(t-s)$ is the history of the temperature. This generalizes (19.2.1).

As noted above, the kernel $\kappa(s)$, given by (19.2.4), does not in general obey the sign restrictions $\kappa(s) \geq 0$, $\kappa'(s) \leq 0$, $s \in \mathbb{R}^+$, which are true for (19.2.2).

For materials with memory such as (19.2.3), the state $\sigma(t)$ is defined by the past history of the temperature gradient. Thus,

$$\sigma(t) = \nabla\theta(t-s) := \nabla\theta^t(s),$$

while a process P of duration $d_P \in R^+$ is defined by

$$P(t) = \nabla\theta(t)\,, \quad t \in [0, d_P)\,.$$

We denote by Σ the state space and by Π the set of processes. The mapping $\hat{\rho} : \Sigma \times \Pi \to \Sigma$ is the evolution function such that $\sigma^f = \hat{\rho}(\sigma^i, P)$, where σ^i and σ^f are, respectively, the initial and final states, while P is the process connecting these.

Definition 19.2.1. *A closed cycle is defined by a pair* (σ, P) *such that*

$$\hat{\rho}(\sigma, P) = \sigma.$$

Second Law of Thermodynamics On any closed cycle, we have

$$\oint \mathbf{q}(t)\cdot\nabla\theta(t)dt \leq 0. \tag{19.2.5}$$

19.3 Fundamental Relations

Let us record in this section the basic equations required in order to derive the form of the minimum and other related free energies. We write (19.2.3) in the form

$$\begin{aligned}
\mathbf{q}(t) &= K_0\mathbf{h}(t) + \int_{-\infty}^{t} K'(t-u)\mathbf{h}(u)du \\
&= K_0\mathbf{h}(t) + \int_0^{\infty} K'(s)\mathbf{h}^t(s)ds, \\
K'(s) &= k\frac{2(\tau_{\mathbf{q}} - \tau_\theta)}{\tau_{\mathbf{q}}^2}\kappa(s), \quad \mathbf{h}^t(s) = \mathbf{h}(t-s), \quad \mathbf{h}(u) = -\nabla\theta(u), \\
K(s) &= K_0 + \int_0^s K'(u)du, \qquad K_0 = K(0) = k\frac{\tau_\theta^2}{\tau_{\mathbf{q}}^2},
\end{aligned} \tag{19.3.1}$$

where $\kappa(s)$ is given by (19.2.4). Let us write the kernel $K'(s)$ as follows:

$$K'(s) = Ae^{i\zeta s} + \overline{A}e^{-i\overline{\zeta}s}, \tag{19.3.2}$$

where

$$\begin{aligned} \zeta &= \alpha(1+i), \qquad A = k\alpha(1-\rho)(\rho - i), \\ \alpha &= \frac{1}{\tau_{\mathbf{q}}}, \qquad \rho = \frac{\tau_\theta}{\tau_{\mathbf{q}}}, \qquad K_0 = k\rho^2. \end{aligned} \tag{19.3.3}$$

In the frequency domain, we have

$$\begin{aligned} K'_+(\omega) &= \int_0^\infty K'(s)e^{-i\omega s}ds \\ &= \frac{iA}{\zeta - \omega} - \frac{i\overline{A}}{\overline{\zeta} + \omega}, \end{aligned}$$

which indicates a structure of two isolated simple poles placed symmetrically about the upper imaginary axis in the complex plane of frequencies. Unless indicated otherwise, the frequency ω will be assumed to be real.

In Chap. 16, a general theory of minimum and related free energies is developed for materials where the memory functions in the frequency domain have arbitrary, within allowed constraints, isolated singularities of any finite order. The present theory is developed in notation closely matching viscoelastic materials with linear memory. Note, for example, the relations in (19.3.1) above, where there is a close analogy based on assigning to $\mathbf{h}^t(s)$ a role corresponding to the strain history.

There is however a significant difference between the theory for viscoelastic materials and the present case, illustrated in particular by (19.2.5). The negative of the gradient of temperature, $\mathbf{h}(t)$ in the notation introduced in $(19.3.1)_4$, is not differentiated with respect to time, as is the case for the strain function in the rate of work done on viscoelastic materials. Therefore, the analogy with viscoelastic materials is not precise and is a source only of general guidance.

The form of the condition imposed by the second law may be determined from (19.2.5), as shown in [126]. This can be expressed as

$$K_0 + K'_c(\omega) > 0, \qquad K'_c(\omega) = \int_0^\infty K'(s)\cos\omega s ds, \tag{19.3.4}$$

in the notation of (19.3.1). Also, using $(19.3.1)_1$, we see that the total thermal work done on the rigid conductor has the form

$$
\begin{aligned}
\int_{-\infty}^{t} \mathbf{q}(u)\cdot\mathbf{h}(u)du &= K_0\int_{-\infty}^{t}|\mathbf{h}(u)|^2\,du + \int_{-\infty}^{t}\int_{-\infty}^{u}\mathbf{h}(u)\cdot K'(u-s)\mathbf{h}(s)dsdu\\
&= K_0\int_{-\infty}^{t}|\mathbf{h}(u)|^2\,du + \frac{1}{2}\int_{-\infty}^{t}\int_{-\infty}^{t}\mathbf{h}(u)\cdot K'(|u-s|)\mathbf{h}(s)dsdu\\
&= K_0\int_{0}^{\infty}\left|\mathbf{h}^t(u)\right|^2 du + \frac{1}{2}\int_{0}^{\infty}\int_{0}^{\infty}\mathbf{h}^t(u)\cdot K'(|u-s|)\mathbf{h}^t(s)dsdu\\
&= \frac{K_0}{2\pi}\int_{-\infty}^{\infty}\left|\mathbf{h}_+^t(\omega)\right|^2 d\omega + \frac{1}{2\pi}\int_{-\infty}^{\infty}\overline{\mathbf{h}_+^t}(\omega)\cdot K_c'(\omega)\mathbf{h}_+^t(\omega)d\omega\\
&= \frac{1}{2\pi}\int_{-\infty}^{\infty}\overline{\mathbf{h}_+^t}(\omega)\cdot H(\omega)\mathbf{h}_+^t(\omega)d\omega \geq 0,\\
\mathbf{h}_+^t(\omega) &= \int_0^{\infty}\mathbf{h}^t(u)e^{-i\omega u}du,\\
H(\omega) &= K_0 + K_c'(\omega) = K_0 + \frac{1}{2}\left[K_+'(\omega) + \overline{K}_+'(\omega)\right],\\
|\mathbf{h}(u)|^2 &= \mathbf{h}(u)\cdot\mathbf{h}(u).
\end{aligned}
\tag{19.3.5}
$$

The various forms given in (19.3.5) follow by virtue of a standard development as outlined in, for example, Sect. 7.5. Note that the quantity $H(\omega)$ is nonnegative by virtue of $(19.3.4)_1$, which is an expression of the second law. It can therefore be concluded that

$$
\int_{-\infty}^{t}\mathbf{q}(u)\cdot\mathbf{h}(u)du \geq 0.
$$

We can write $(19.3.1)_1$ in the frequency domain as

$$
\begin{aligned}
\mathbf{q}(t) &= K_0\mathbf{h}(t) + \frac{1}{2\pi}\int_{-\infty}^{\infty}\overline{K_+'}(\omega)\mathbf{h}_+^t(\omega)d\omega\\
&= \frac{K_0}{\pi}\int_{-\infty}^{\infty}\mathbf{h}_+^t(\omega)d\omega + \frac{1}{2\pi}\int_{-\infty}^{\infty}\overline{K_+'}(\omega)\mathbf{h}_+^t(\omega)d\omega.
\end{aligned}
\tag{19.3.6}
$$

Observe that the first term on the right of $(19.3.6)_2$ has a factor of 2 over what one might expect. This term emerges from the time domain version by integrating over the lower frequency complex plane in a contour yielding zero, noting however that the infinite part of the contour makes a nonzero contribution, which can be evaluated using

$$
\mathbf{h}_+^t(\omega) \sim \frac{\mathbf{h}(t)}{i\omega},
$$

at large ω. This contribution must be subtracted from the contour integral and the result shown in $(19.3.6)_2$ emerges. Using a manipulation discussed in Sect. 7.2.4, we can replace $(19.3.6)_2$ by

$$
\begin{aligned}
\mathbf{q}(t) &= \frac{K_0}{\pi}\int_{-\infty}^{\infty}\mathbf{h}_+^t(\omega)d\omega + \frac{1}{2\pi}\int_{-\infty}^{\infty}[\overline{K_+'}(\omega) + K_+'(\omega)]\mathbf{h}_+^t(\omega)d\omega\\
&= \frac{1}{\pi}\int_{-\infty}^{\infty}H(\omega)\mathbf{h}_+^t(\omega)d\omega.
\end{aligned}
\tag{19.3.7}
$$

On substituting (19.3.2) into $(19.3.1)_1$, we obtain

$$\mathbf{q}(t) = K_0\mathbf{h}(t) + A\mathbf{h}^t_+(-\zeta) + \overline{A}\mathbf{h}^t_+(\overline{\zeta}). \tag{19.3.8}$$

Observe that $\mathbf{h}^t_+(\overline{\zeta}) = \overline{\mathbf{h}^t_+(-\zeta)}$. The relation

$$\frac{d}{dt}\mathbf{h}^t_+(-\zeta) = \mathbf{h}(t) + i\zeta\mathbf{h}^t_+(-\zeta) \tag{19.3.9}$$

will be used later.

The property that the histories $\mathbf{h}^t_1(s)$ and $\mathbf{h}^t_2(s)$ are in the same minimal state has the standard form which follows from $(19.3.1)_1$, namely

$$\mathbf{h}_1(t) = \mathbf{h}_2(t), \quad \int_0^\infty K'(s+u)\mathbf{h}^t_1(u)du = \int_0^\infty K'(s+u)\mathbf{h}^t_2(u)du, \quad s \geq 0.$$

It can be shown that the second relation is equivalent to the requirements (see Remark 16.3.1)

$$\mathbf{h}^t_{1+}(-\zeta) = \mathbf{h}^t_{2+}(-\zeta), \qquad \mathbf{h}^t_{1+}(\overline{\zeta}) = \mathbf{h}^t_{2+}(\overline{\zeta}),$$

so that an equivalence class of histories is characterized by the triplet $V = (\mathbf{h}^t_+(-\zeta), \mathbf{h}^t_+(\overline{\zeta}), \mathbf{h}(t))$. A functional of a minimal state containing $\mathbf{h}^t(s)$ must have the property that its dependence on $\mathbf{h}^t(s)$ is expressible as a dependence only on V. This is in particular true of the minimum free energy discussed below.

We can show that

$$\begin{aligned}
H(\omega) &= \frac{N(\omega)}{D(\omega)}, \\
N(\omega) &= a\omega^4 + b\omega^2 + c, \quad a = K_0, \quad b = i(\overline{A}\zeta - A\zeta) - K_0(\zeta^2 + \overline{\zeta}^2), \\
c &= [K_0\,|\zeta|^2 - i(\overline{A}\zeta - A\overline{\zeta})]\,|\zeta|^2\,, \\
D(\omega) &= (\zeta - \omega)(\zeta + \omega)(\overline{\zeta} - \omega)(\overline{\zeta} + \omega) \\
&= (\zeta^2 - \omega^2)(\overline{\zeta}^2 - \omega^2) = |\zeta - \omega|^2\,|\zeta + \omega|^2 \geq 0.
\end{aligned} \tag{19.3.10}$$

The polynomial $N(\omega)$ is clearly real for real ω. It must also be nonnegative, which is the constraint imposed by the second law, as discussed in the context of (19.3.4) and (19.3.5). Note that it is positive for large values of ω. To ensure non-negativity for all real ω, the form of $N(\omega)$ given by $(19.3.10)_2$, expressed as a function of ω^2, cannot have real roots. These roots are given by

$$\eta^2 = \frac{1}{2a}[-b \pm \sqrt{b^2 - 4ac}], \tag{19.3.11}$$

so that we must have

$$b^2 - 4ac \leq 0.$$

For ζ and A given by (19.3.3), we obtain

$$a = k\rho^2 > 0, \quad b = -2k\alpha^2(\rho - 1)^2 < 0, \quad c = 4\alpha^4 k > 0, \tag{19.3.12}$$

and

$$\begin{aligned} b^2 - 4ac &= 4k^2\alpha^4[(\rho-1)^4 - 4\rho^2] \\ &= 4k^2\alpha^4(\rho^2+1)(\rho-2+\sqrt{3})(\rho-2-\sqrt{3}) \le 0. \end{aligned} \tag{19.3.13}$$

This inequality yields

$$2-\sqrt{3} < \rho < 2+\sqrt{3},$$

which is (19.1.4), as expected.

We can write $N(\omega)$ in a form similar to $D(\omega)$:

$$\begin{aligned} N(\omega) &= K_0(\eta-\omega)(\eta+\omega)(\overline{\eta}-\omega)(\overline{\eta}+\omega) \\ &= K_0(\eta^2-\omega^2)(\overline{\eta}^2-\omega^2) = K_0\,|\eta-\omega|^2\,|\eta+\omega|^2 \ge 0. \end{aligned}$$

The quantity $-b/(2a)$ is always positive for values specified by (19.3.12). The roots given by (19.3.11) are therefore in the first and fourth quadrants of the complex plane. These have the form

$$\begin{aligned} \eta^2 &= x+iy, \qquad x = \frac{\alpha^2}{\rho^2}(\rho-1)^2, \\ y &= \frac{\alpha^2}{\rho^2}\sqrt{\rho^2+1}\sqrt{4\rho-\rho^2-1}. \end{aligned} \tag{19.3.14}$$

It follows that

$$x^2+y^2 = 4\frac{\alpha^4}{\rho^2}.$$

If $\eta = u+iv$, then

$$\begin{aligned} u^2 &= \frac{1}{2}(\sqrt{x^2+y^2}+x) = \frac{1}{2}\frac{\alpha^2}{\rho^2}(\rho^2+1), \\ v^2 &= \frac{1}{2}(\sqrt{x^2+y^2}-x) = \frac{1}{2}\frac{\alpha^2}{\rho^2}(4\rho-\rho^2-1). \end{aligned}$$

We choose $u > 0$ and $v > 0$ so that η is in the first quadrant and

$$u = \frac{1}{\sqrt{2}}\frac{\alpha}{\rho}\sqrt{\rho^2+1}, \qquad v = \frac{1}{\sqrt{2}}\frac{\alpha}{\rho}\sqrt{4\rho-\rho^2-1}.$$

The thermodynamic constraint expressed by the inequality in (19.3.13) can be written in the form

$$4\rho-\rho^2-1 \ge 0,$$

which will be useful later.

The quantity $H(\omega)$ can be factorized as follows:

$$\begin{aligned} H(\omega) &= H_+(\omega)H_-(\omega), \\ H_+(\omega) &= \sqrt{K_0}\frac{(\omega-\eta)(\omega+\overline{\eta})}{(\omega-\zeta)(\omega+\overline{\zeta})}, \\ H_-(\omega) &= \sqrt{K_0}\frac{(\omega+\eta)(\omega-\overline{\eta})}{(\omega+\zeta)(\omega-\overline{\zeta})} = \overline{H_+(\omega)}, \end{aligned} \tag{19.3.15}$$

where the singularities and zeros of $H_+(\omega)$ are in the upper half of the complex plane, while $H_-(\omega) = \overline{H_+(\omega)}$ has all singularities and zeros in the lower half.

We note in passing that the number of nontrivial zeros of $H(\omega)$ in the present case is twice that in the corresponding viscoelastic problem. This is related to the fact that $H(0)$ is not zero.

19.4 The Minimum Free Energy for Second-Order Heat Conduction

We now use the results of Sect. 19.3 to derive an explicit expression for the minimum free energy in the case of second-order heat conduction.

With the aid of the Plemelj formulae, one can show that the thermal work given by the negative of (19.3.5) can be expressed in the form

$$\begin{aligned} \int_{-\infty}^{t} \mathbf{q}(u)\cdot\mathbf{h}(u)du &= \frac{1}{2\pi}\int_{-\infty}^{\infty}[\left|\mathbf{p}_+^t(\omega)\right|^2 + \left|\mathbf{p}_-^t(\omega)\right|^2]d\omega, \\ \mathbf{p}_\pm^t(\omega) &= \frac{1}{2\pi i}\int_{-\infty}^{\infty}\frac{H_-(\omega_1)\mathbf{h}_+^t(\omega_1)}{\omega_1-\omega^\mp}d\omega_1, \\ \left|\mathbf{p}_\pm^t(\omega)\right|^2 &= \overline{\mathbf{p}_\pm^t}(\omega)\cdot\mathbf{p}_\pm^t(\omega). \end{aligned} \tag{19.4.1}$$

It is convenient to use histories of the temperature gradient in the present work, rather than relative histories. In particular, histories are used in the definition of $\mathbf{p}_\pm^t(\omega)$, so that these quantities are most closely related to those in [158]†. We note however that the quantities $\mathbf{p}_\pm^t(\omega)$ used here differ somewhat from those in [158] because of the fact that $H(0)$ is not zero in the present model.

Using any of the arguments presented in Sect. 11.2, we can deduce that the minimum free energy is given by

$$\psi_m(t) = \frac{1}{2\pi}\int_{-\infty}^{\infty}\left|\mathbf{p}_-^t(\omega)\right|^2 d\omega \geq 0. \tag{19.4.2}$$

The time derivative of relation $(19.4.1)_1$ is in fact the energy balance equation or First Law of Thermodynamics in the context of the minimum free energy. We can write it as

$$\begin{aligned} \dot{\psi}_m(t) + D_m(t) &= \mathbf{q}(t)\cdot\mathbf{h}(t), \\ D_m(t) &= \frac{d}{dt}\frac{1}{2\pi}\int_{-\infty}^{\infty}\left|\mathbf{p}_+^t(\omega)\right|^2 d\omega \geq 0, \end{aligned} \tag{19.4.3}$$

where $D_m(t)$ is the rate of dissipation related to the minimum free energy. By following steps closely related to those in [158], we can show that

---

† Note that the quantities $\mathbf{p}_\pm^t(\omega)$ here are designated as $\mathbf{q}_\pm^t(\omega)$ in [158]; see also Sect. 11.2.3. We avoid this notation because of possible confusion with the magnitude of the heat flow vector $\mathbf{q}(t)$.

$$
\begin{aligned}
D_m(t) &= |\mathbf{J}(t)|^2, \\
\mathbf{J}(t) &= \frac{1}{2\pi}\int_{-\infty}^{\infty} H_-(\omega)\mathbf{h}_+^t(\omega)d\omega + \frac{1}{2}\sqrt{K_0}\mathbf{h}(t) \\
&= \frac{1}{2\pi}\int_{-\infty}^{\infty} H_-(\omega)\left[\mathbf{h}_+^t(\omega) - \frac{\mathbf{h}(t)}{i\omega^-}\right]d\omega + H_-(0)\mathbf{h}(t).
\end{aligned}
\tag{19.4.4}
$$

The term with $H_-(0)$ is not present in the developments of [158]. The non-negativity of $D_m(t)$, which is clear from (19.4.4), is an expression of the second law.

We can write $H_\pm(\omega)$, given by (19.3.15), in the form

$$
\begin{aligned}
H_+(\omega) &= \sqrt{K_0}\left[1 - \frac{M}{\omega-\zeta} + \frac{\overline{M}}{\omega+\overline{\zeta}}\right], \\
H_-(\omega) &= \sqrt{K_0}\left[1 + \frac{M}{\omega+\zeta} - \frac{\overline{M}}{\omega-\overline{\zeta}}\right], \\
M &= \frac{(\eta-\zeta)(\overline{\eta}+\zeta)}{(\zeta+\overline{\zeta})}].
\end{aligned}
\tag{19.4.5}
$$

One can show that

$$
\begin{aligned}
M &= \frac{1}{2\alpha}[|\eta|^2 + \zeta(\eta-\overline{\eta}) - \zeta^2] = \frac{\alpha}{\rho}[1-e+i(e-\rho)], \\
e &= \sqrt{2\rho - \frac{\rho^2+1}{2}}.
\end{aligned}
\tag{19.4.6}
$$

Closing the contour of $(19.3.7)_2$ on $\Omega^{(-)}$ and using (19.4.5), we obtain

$$
\begin{aligned}
\mathbf{q}(t) = K_0\mathbf{h}(t) &- 2iK_0\left[1 + \frac{M}{2\zeta} - \frac{\overline{M}}{\zeta-\overline{\zeta}}\right]M\mathbf{h}_+^t(-\zeta) \\
&+ 2iK_0\left[1 - \frac{M}{\overline{\zeta}-\zeta} + \frac{\overline{M}}{2\overline{\zeta}}\right]\overline{M}\mathbf{h}_+^t(\overline{\zeta}).
\end{aligned}
\tag{19.4.7}
$$

Comparing with (19.3.8), it will be observed that

$$
A = -2iK_0\left[M + \frac{M^2}{2\zeta} - \frac{|M|^2}{\zeta-\overline{\zeta}}\right],
$$

where A is defined in (19.3.3). This relation can be confirmed with the help of (19.4.6).

We deduce from $(19.4.1)_2$ and (19.4.5) that

$$
\mathbf{p}_-^t(\omega) = \sqrt{K_0}\left[M\frac{\mathbf{h}_+^t(-\zeta)}{\omega+\zeta} - \overline{M}\frac{\mathbf{h}_+^t(\overline{\zeta})}{\omega-\overline{\zeta}}\right],
\tag{19.4.8}
$$

by closing the contour on $\Omega^{(-)}$. Let us define

$$\begin{aligned}\mathbf{h}_+^t(-\zeta) &= \mathbf{h}_r^t(\alpha) + i\mathbf{h}_i^t(\alpha),\\ \mathbf{h}_r^t(\alpha) &= \int_0^\infty \mathbf{h}^t(s)e^{-\alpha s}\cos(\alpha s)ds,\\ \mathbf{h}_i^t(\alpha) &= \int_0^\infty \mathbf{h}^t(s)e^{-\alpha s}\sin(\alpha s)ds,\end{aligned} \tag{19.4.9}$$

with the aid of $(19.3.1)_1$

Using (19.4.2) and (19.4.8), the minimum free energy can be written in the form

$$\psi_m(t) = iK_0\left\{\frac{[|M|^2\left|\mathbf{h}_+^t(-\zeta)\right|]^2}{i\alpha} - \frac{[M\mathbf{h}_+^t(-\zeta)]^2}{2\zeta} + \frac{[\overline{M}\mathbf{h}_+^t(\overline{\zeta})]^2}{2\overline{\zeta}}\right\}, \tag{19.4.10}$$

where

$$[M\mathbf{h}_+^t(-\zeta)]^2 = M^2\mathbf{h}_+^t(-\zeta)\cdot\mathbf{h}_+^t(-\zeta),$$

and similarly for the complex conjugate term and others introduced below. Let us define

$$M = \frac{\alpha}{\rho}(\lambda + i\mu), \qquad \lambda = 1 - e, \qquad \mu = e - \rho.$$

Then $\psi_m(t)$ can be expressed as

$$\begin{aligned}\psi_m(t) &= \frac{1}{2}\int_0^\infty\int_0^\infty \mathbf{h}^t(s)\cdot G(s,s')\mathbf{h}^t(s')dsds',\\ G(s,s') &= 2K_0 Re\left\{e^{i(\zeta s - \overline{\zeta}s')}\frac{|M|^2}{\alpha} - ie^{i\zeta(s+s')}\frac{M^2}{\zeta}\right\}\\ &= 2K_0 e^{-\alpha(s+s')}\frac{\alpha}{\rho^2}\Big\{(\lambda^2+\mu^2)\cos\alpha(s-s')\\ &\quad -\frac{1}{2}(\lambda^2-\mu^2-2\lambda\mu)\cos\alpha(s+s')\\ &\quad +\frac{1}{2}(\lambda^2-\mu^2+2\lambda\mu)\sin\alpha(s+s')\Big\}\\ &= 2K_0 e^{-\alpha(s+s')}\frac{\alpha}{\rho^2}\{2[2\rho - e(1+\rho)]\cos\alpha(s-s')\\ &\quad -[3\rho-\rho^2-2e]\cos\alpha(s+s')\\ &\quad +[1+2e\rho-3\rho]\sin\alpha(s+s')\},\end{aligned} \tag{19.4.11}$$

in terms of the quantities α, ρ, and K_0 defined in (19.3.3) and e which is introduced in (19.4.6). Either by expanding the trigonometric functions in (19.4.11) or by using (19.4.9) in (19.4.10), we can also express the minimum free energy as the quadratic form

$$\begin{aligned}\psi_m(t) &= \frac{K_0\alpha}{\rho^2}\mathbf{h}^\top\mathbf{A}\mathbf{h}, \qquad \mathbf{h} = (\mathbf{h}_r^t(\alpha), \mathbf{h}_i^t(\alpha)),\\ \mathbf{A} &= \frac{1}{2}\begin{pmatrix}\lambda^2+3\mu^2+2\lambda\mu & \lambda^2-\mu^2+2\lambda\mu\\ \lambda^2-\mu^2+2\lambda\mu & 3\lambda^2+\mu^2-2\lambda\mu\end{pmatrix} \geq 0.\end{aligned} \tag{19.4.12}$$

The non-negativity property of **A** can be shown for general values of λ and μ. In the present context,

$$\mathbf{A} = \begin{pmatrix} \rho(1+\rho-2e) & 1-3\rho+2e\rho \\ \\ 1-3\rho+2e\rho & 7\rho-\rho^2-4e-2e\rho \end{pmatrix}. \tag{19.4.13}$$

The rate of dissipation associated with $\psi_m(t)$ is given by (19.4.4), which, in the light of (19.4.5), becomes

$$\begin{aligned} D_m(t) &= K_0\left[\mathbf{h}(t) - iM\mathbf{h}^t_+(-\zeta) + i\overline{M}\mathbf{h}^t_+(\overline{\zeta})\right]^2 \\ &= K_0\Big\{\mathbf{h}^2(t) - 2iM\mathbf{h}^t_+(-\zeta)\cdot\mathbf{h}(t) + 2i\overline{M}\mathbf{h}^t_+(\overline{\zeta})\cdot\mathbf{h}(t) \\ &\qquad +2\,|M|^2\left|\mathbf{h}^t_+(-\zeta)\right|^2 - [M\mathbf{h}^t_+(-\zeta)]^2 - [\overline{M}\mathbf{h}^t_+(\overline{\zeta})]^2\Big\}. \end{aligned} \tag{19.4.14}$$

It is of interest to confirm that $(19.4.3)_1$ holds for the above formulae. Using (19.3.9) and (19.4.7), we see that

$$\begin{aligned} \dot{\psi}_m(t) &= iK_0\left\{\frac{|M|^2}{i\alpha}[\mathbf{h}^t_+(\overline{\zeta}) + \mathbf{h}^t_+(-\zeta)] - \frac{M^2\mathbf{h}^t_+(-\zeta)}{\zeta} + \frac{\overline{M}^2\mathbf{h}^t_+(\overline{\zeta})}{\overline{\zeta}}\right\}\cdot\mathbf{h}(t) \\ &\quad - K_0\left\{2\,|M|^2\left|\mathbf{h}^t_+(-\zeta)\right|^2 - [M\mathbf{h}^t_+(-\zeta)]^2 - [\overline{M}\mathbf{h}^t_+(\overline{\zeta})]^2\right\} \\ &= \mathbf{q}(t)\cdot\mathbf{h}(t) - D_m(t). \end{aligned} \tag{19.4.15}$$

We observe that this relation is satisfied in a somewhat different way from that for viscoelastic problems. For these other materials, the quantity $\dot{\psi}_m(t)$ generates the rate of dissipation $D_m(t)$ and the rate of work. In the present case, $\dot{\psi}_m(t)$ generates those parts of the rate of dissipation and the rate of work $\mathbf{q}(t)\cdot\mathbf{h}(t)$, which are quadratic in M and $\overline{M}$, while the terms in the rate of dissipation which are linear in M and $\overline{M}$ or independent of these quantities provide the remainder of the rate of work.

In relation to this difference, consider the quantity $G(s, s')$ in (19.4.11). The quantity $G(s, 0) = G(0, s)$ in viscoelasticity is the relaxation function of the material. In the present case, using (19.4.7) and $(19.4.11)_2$, we see that

$$\begin{aligned} \mathbf{q}(t) = K_0\mathbf{h}(t) &+ \int_0^\infty G(0,s)\mathbf{h}^t(s)ds - 2iK_0M\int_0^\infty e^{i\zeta s}\mathbf{h}^t(s)ds \\ &+ 2iK_0\overline{M}\int_0^\infty e^{-i\overline{\zeta}s}\mathbf{h}^t(s)ds, \end{aligned}$$

so that $G(0, s)$ is that part of the relaxation function that is quadratic in M and $\overline{M}$.

The quantity $D_m(t)$ can be expressed as

$$
\begin{aligned}
D_m(t) &= K_0|\mathbf{h}(t)|^2 + 4K_0\frac{\alpha}{\rho}\mathbf{h}(t)\cdot[\mu\mathbf{h}_r^t(\alpha)+\lambda\mathbf{h}_i^t(\alpha)] \\
&\quad + \frac{1}{2}\int_0^\infty\int_0^\infty \mathbf{h}^t(s)\cdot L(s,s')\mathbf{h}^t(s')ds ds', \\
L(s,s') &= 4K_0 Re\left\{e^{i(\zeta s-\bar{\zeta}s')}|M|^2 - e^{i\zeta(s+s')}M^2\right\} \\
&= 4K_0 e^{-\alpha(s+s')}\frac{\alpha^2}{\rho^2}\left\{(\lambda^2+\mu^2)\cos\alpha(s-s')\right. \\
&\quad \left.-(\lambda^2-\mu^2)\cos\alpha(s+s')+2\lambda\mu\sin\alpha(s+s')\right\} \\
&= 4K_0 e^{-\alpha(s+s')}\frac{\alpha^2}{\rho^2}\left\{2[2\rho-e(1+\rho)]\cos\alpha(s-s')\right. \\
&\quad -[1-\rho^2+2e(\rho-1)]\cos\alpha(s+s') \\
&\quad \left.+[2e(1+\rho)+(\rho-1)^2-4\rho]\sin\alpha(s+s')\right\}.
\end{aligned}
\tag{19.4.16}
$$

Note that

$$
L(s,s') = -\frac{\partial}{\partial s}G(s,s') - \frac{\partial}{\partial s'}G(s,s').
$$

As for the minimum free energy given by (19.4.12), we can also express the rate of dissipation as the quadratic form

$$
\begin{aligned}
D_m(t) &= K_0|\mathbf{h}(t)|^2 + 4K_0\frac{\alpha}{\rho}(\mathbf{m}^\top\mathbf{h})\cdot\mathbf{h}(t) + 2\frac{K_0\alpha^2}{\rho^2}\mathbf{h}^\top\mathbf{B}\mathbf{h}, \\
\mathbf{m} &= (\mu,\lambda) = (e-\rho, 1-e), \\
\mathbf{B} &= 2\begin{pmatrix}\mu^2 & \lambda\mu \\ \lambda\mu & \lambda^2\end{pmatrix} \\
&= \begin{pmatrix} 4\rho+\rho^2-1-4e\rho & 2e(1+\rho)+(\rho-1)^2-4\rho \\ 2e(1+\rho)+(\rho-1)^2-4\rho & 1-4e+4\rho-\rho^2\end{pmatrix}.
\end{aligned}
\tag{19.4.17}
$$

This matrix is positive semidefinite. The full expression for $D_m(t)$, given by (19.4.16), is not manifestly nonnegative. However, the earlier form (19.4.14) is clearly so.

19.5 Free Energies Related to the Minimum Free Energy

We now proceed to derive certain other free energies from the expression for the minimum free energy, using a procedure analogous to that in Sect. 16.8.

Replacing the zero at $\omega = \overline{\eta}$ of $H_-(\omega)$ by the zero η of $H_+(\omega)$ gives another free energy, which we now examine. This is the quantity, which we denote by $\psi_1(t)$, related to the factorization of $H(\omega)$ given by

$$
\begin{aligned}
H(\omega) &= H_+^{(1)}(\omega)H_-^{(1)}(\omega),\\
H_-^{(1)}(\omega) &= H_-(\omega)\frac{\omega-\eta}{\omega-\overline{\eta}} = \sqrt{K_0}\frac{(\omega-\eta)(\omega+\eta)}{(\omega+\zeta)(\omega-\overline{\zeta})} = \sqrt{K_0}\frac{\omega^2-\eta^2}{(\omega+\zeta)(\omega-\overline{\zeta})},\\
H_+^{(1)}(\omega) &= H_+(\omega)\frac{\omega-\overline{\eta}}{\omega-\eta} = \sqrt{K_0}\frac{(\omega-\overline{\eta})(\omega+\overline{\eta})}{(\omega-\zeta)(\omega+\overline{\zeta})} = \sqrt{K_0}\frac{\omega^2-\overline{\eta}^2}{(\omega-\zeta)(\omega+\overline{\zeta})}.
\end{aligned}
\tag{19.5.1}
$$

The zeros of $H_+^{(1)}(\omega)$ are no longer only in the upper half of the complex plane, while those of $H_-^{(1)}(\omega) = \overline{H_+^{(1)}(\omega)}$ are no longer only in the lower half.

We must have $\psi_1(t) \geq \psi_m(t)$, since the latter is the minimum of the set of free energies for the material. It follows also from the general property that exchanging zeros in this manner results in a nondecreasing sequence (Sect. 16.8).

Instead of (19.4.5), we have

$$
\begin{aligned}
H_-^{(1)}(\omega) &= \sqrt{K_0}\left[1+\frac{M_1}{\omega+\zeta}-\frac{M_2}{\omega-\overline{\zeta}}\right],\\
H_+^{(1)}(\omega) &= \sqrt{K_0}\left[1-\frac{\overline{M_2}}{\omega-\zeta}+\frac{\overline{M_1}}{\omega+\overline{\zeta}}\right],\\
M_1 &= \frac{(\eta^2-\zeta^2)}{\zeta+\overline{\zeta}}, \qquad M_2 = \frac{(\eta^2-\overline{\zeta}^2)}{\zeta+\overline{\zeta}}.
\end{aligned}
\tag{19.5.2}
$$

Note that $M_2 \neq \overline{M}_1$, in contrast to the minimum free energy case (19.4.5). Replacing (19.4.6), we have

$$
\begin{aligned}
M_1 &= \frac{(\eta^2-\zeta^2)}{2\alpha} = \frac{\alpha}{\rho}\left[\frac{\beta}{\rho}-1+i\left(\frac{e_1}{\rho}-\rho\right)\right],\\
M_2 &= \frac{(\eta^2-\overline{\zeta}^2)}{2\alpha} = \frac{\alpha}{\rho}\left[\frac{\beta}{\rho}-1+i\left(\frac{e_1}{\rho}+\rho\right)\right],\\
\beta &= \frac{\rho^2+1}{2}, \qquad e_1 = \sqrt{\beta(2\rho-\beta)},
\end{aligned}
\tag{19.5.3}
$$

where (19.3.14) has been used. Closing the contour of $(19.3.7)_2$ on $\Omega^{(-)}$ and using (19.4.5), we obtain

$$
\begin{aligned}
\mathbf{q}(t) = K_0\mathbf{h}(t) &- 2iK_0\left[1+\frac{\overline{M_2}}{2\zeta}-\frac{\overline{M_1}}{\zeta-\overline{\zeta}}\right]M_1\mathbf{h}_+^t(-\zeta)\\
&+ 2iK_0\left[1-\frac{\overline{M_2}}{\overline{\zeta}-\zeta}+\frac{\overline{M_1}}{2\overline{\zeta}}\right]M_2\mathbf{h}_+^t(\overline{\zeta}).
\end{aligned}
$$

Comparing with (19.3.8), it will be observed that

$$A = -2iK_0\left[M_1 + \frac{M_1\overline{M_2}}{2\zeta} - \frac{|M_1|^2}{\zeta - \overline{\zeta}}\right]$$
$$= -2iK_0\left[\overline{M_2} - \frac{|M_2|^2}{\zeta - \overline{\zeta}} + \frac{M_1\overline{M_2}}{2\zeta}\right], \tag{19.5.4}$$

where A is defined in (19.3.3). Note that (19.5.4) implies the condition

$$M_1 - \frac{|M_1|^2}{\zeta - \overline{\zeta}} = \overline{M_2} - \frac{|M_2|^2}{\zeta - \overline{\zeta}}, \tag{19.5.5}$$

which in fact follows from $(19.5.2)_{4,5}$. Equation $(19.5.4)_1$ can be confirmed with the help of (19.5.3).

We deduce from $(19.4.1)_2$ and (19.5.2) that

$$\mathbf{p}^t_-(\omega) = \sqrt{K_0}\left[M_1\frac{\mathbf{h}^t_+(-\zeta)}{\omega + \zeta} - M_2\frac{\mathbf{h}^t_+(\overline{\zeta})}{\omega - \overline{\zeta}}\right], \tag{19.5.6}$$

by closing the contour on $\Omega^{(-)}$. Using (19.4.2) and (19.5.6), $\psi_1(t)$ can be written in the form

$$\psi_1(t) = iK_0\left\{\frac{(|M_1|^2 + |M_2|^2)\left|\mathbf{h}^t_+(-\zeta)\right|^2}{2i\alpha} - \frac{M_1\overline{M_2}[\mathbf{h}^t_+(-\zeta)]^2}{2\zeta} + \frac{\overline{M_1}M_2[\mathbf{h}^t_+(\overline{\zeta})]^2}{2\overline{\zeta}}\right\}.$$

Let us define

$$M_1 = \frac{\alpha}{\rho}(\lambda_1 + i\mu_1), \qquad M_2 = \frac{\alpha}{\rho}(\lambda_1 + i\mu_2),$$
$$\lambda_1 = \frac{\beta}{\rho} - 1 = \frac{(\rho-1)^2}{2\rho}, \qquad \mu_1 = \frac{e_1}{\rho} - \rho, \qquad \mu_2 = \frac{e_1}{\rho} + \rho.$$

Then, $\psi_1(t)$ can be expressed as

$$\psi_1(t) = \frac{1}{2}\int_0^\infty\int_0^\infty \mathbf{h}^t(s)\cdot G_1(s,s')\mathbf{h}^t(s')ds\,ds',$$
$$G_1(s,s') = 2K_0 Re\left\{e^{i(\zeta s - \overline{\zeta}s')}\frac{|M_1|^2 + |M_2|^2}{2\alpha} - ie^{i\zeta(s+s')}\frac{M_1\overline{M_2}}{\zeta}\right\}$$
$$= 2K_0 e^{-\alpha(s+s')}\frac{\alpha}{\rho^2}\Big\{[\lambda_1^2 + \frac{1}{2}(\mu_1^2 + \mu_2^2)]\cos\alpha(s-s')$$
$$-\frac{1}{2}[\lambda_1^2 + \mu_1\mu_2 - \lambda_1(\mu_1 - \mu_2)]\cos\alpha(s+s')$$
$$+\frac{1}{2}[\lambda_1^2 + \mu_1\mu_2 + \lambda_1(\mu_1 - \mu_2)]\sin\alpha(s+s')\Big\}$$
$$= 2K_0 e^{-\alpha(s+s')}\frac{\alpha}{\rho^2}\big\{(1+\rho^2)\cos\alpha(s-s')$$
$$-(1-\rho)\cos\alpha(s+s') + \rho(1-\rho)\sin\alpha(s+s')\big\}, \tag{19.5.7}$$

in terms of the quantities α, ρ, and K_0 defined in (19.3.3). Following the steps leading to (19.4.12), we can also express $\psi_1(t)$ as the quadratic form

$$\psi_1(t) = \frac{K_0\alpha}{\rho^2}\mathbf{h}^\top \mathbf{A}_1 \mathbf{h},$$

using the notation of $(19.4.12)_2$, where

$$\mathbf{A}_1 = \frac{1}{2}\begin{pmatrix} \Lambda + \Delta(\lambda_1 + \Delta) & \Lambda + \lambda_1\Delta \\ \Lambda + \lambda_1\Delta & 3\Lambda + \Delta(\Delta - \lambda_1) \end{pmatrix} \geq 0,$$

$$\Lambda = \lambda_1^2 + \mu_1\mu_2, \qquad \Delta = \mu_1 - \mu_2.$$

The non-negativity property of $\mathbf{A}_1$ can be shown for general values of λ_1, μ_1, and μ_2. In the present context,

$$\mathbf{A}_1 = \begin{pmatrix} \rho(1+\rho) & \rho(1-\rho) \\ \rho(1-\rho) & \rho^2 - \rho + 2 \end{pmatrix}. \tag{19.5.8}$$

The rate of dissipation associated with $\psi_1(t)$ is given by the analogue of (19.4.4) for the factorization (19.5.1), which, in the light of (19.5.2), becomes

$$\begin{aligned} D_1(t) &= K_0|\mathbf{h}(t) - iM_1\mathbf{h}_+^t(-\zeta) + iM_2\mathbf{h}_+^t(\overline{\zeta})|^2 \\ &= K_0\Big\{|\mathbf{h}(t)|^2 - 2Re[iM_1\mathbf{h}_+^t(-\zeta) - iM_2\mathbf{h}_+^t(\overline{\zeta})]\cdot\mathbf{h}(t) \\ &+(|M_1|^2 + |M_2|^2)\left|\mathbf{h}_+^t(-\zeta)\right|^2 - M_1\overline{M_2}[\mathbf{h}_+^t(-\zeta)]^2 - \overline{M_1}M_2[\mathbf{h}_+^t(\overline{\zeta})]^2\Big\}. \end{aligned} \tag{19.5.9}$$

It is of interest to confirm that the analogue of $(19.4.3)_1$ holds for the above formulae. Using (19.3.9) and (19.4.7), we see that

$$\begin{aligned} \dot{\psi}_1(t) &= iK_0\left\{\frac{|M_1|^2 + |M_2|^2}{2i\alpha}[\mathbf{h}_+^t(\overline{\zeta}) + \mathbf{h}_+^t(-\zeta)]\right. \\ &\qquad \left. - \frac{M_1\overline{M_2}\mathbf{h}_+^t(-\zeta)}{\zeta} + \frac{\overline{M_1}M_2\mathbf{h}_+^t(\overline{\zeta})}{\overline{\zeta}}\right\}\cdot\mathbf{h}(t) \\ &\qquad - K_0\{(|M_1|^2 + |M_2|^2)\left|\mathbf{h}_+^t(-\zeta)\right|^2 \\ &\qquad - M_1\overline{M_2}[\mathbf{h}_+^t(-\zeta)]^2 - \overline{M_1}M_2[\mathbf{h}_+^t(\overline{\zeta})]^2\} \\ &= \mathbf{q}(t)\cdot\mathbf{h}(t) - D_1(t). \end{aligned} \tag{19.5.10}$$

The demonstration of $(19.5.10)_2$ requires the use of condition (19.5.5). The observation after (19.4.15) also applies here. The quantity $D_1(t)$ can be expressed as

$$
\begin{aligned}
D_1(t) &= K_0|\mathbf{h}(t)|^2 + 2K_0\frac{\alpha}{\rho}\mathbf{h}(t)\cdot[(\mu_1-\mu_2)\mathbf{h}_r^t(\alpha)+2\lambda_1\mathbf{h}_i^t(\alpha)] \\
&\quad + \frac{1}{2}\int_0^\infty\int_0^\infty \mathbf{h}^t(s)\cdot L_1(s,s')\mathbf{h}^t(s')dsds', \\
L_1(s,s') &= 2K_0 Re\left\{e^{i(\zeta s-\overline{\zeta}s')}(|M_1|^2+|M_2|^2) - 2e^{i\zeta(s+s')}M_1\overline{M_2}\right\} \\
&= 2K_0 e^{-\alpha(s+s')}\frac{\alpha^2}{\rho^2}\left\{(2\lambda_1^2+\mu_1^2+\mu_2^2)\cos\alpha(s-s')\right. \\
&\quad \left. -2(\lambda_1^2+\mu_1\mu_2)\cos\alpha(s+s')+2\lambda_1(\mu_1-\mu_2)\sin\alpha(s+s')\right\} \\
&= 4K_0 e^{-\alpha(s+s')}\frac{\alpha^2}{\rho^2}\left\{(\rho^2+1)\cos\alpha(s-s')\right. \\
&\quad \left. +(\rho^2-1)\cos\alpha(s+s')-(\rho-1)^2\sin\alpha(s+s')\right\}.
\end{aligned}
\tag{19.5.11}
$$

Note that

$$
L_1(s,s') = -\frac{\partial}{\partial s}G_1(s,s') - \frac{\partial}{\partial s'}G_1(s,s').
$$

We can also express the rate of dissipation in a form similar to (19.4.17):

$$
\begin{aligned}
D_1(t) &= K_0|\mathbf{h}(t)|^2 + 2K_0\frac{\alpha}{\rho}(\mathbf{m}^\top\mathbf{h})\cdot\mathbf{h}(t) + 2\frac{K_0\alpha^2}{\rho^2}\mathbf{h}^\top\mathbf{B}_1\mathbf{h}, \\
\mathbf{m} &= (\mu_1-\mu_2, 2\lambda_1) = \left(-2\rho, 2\left(\frac{\beta}{\rho}-1\right)\right), \\
\mathbf{B}_1 &= \frac{1}{2}\begin{pmatrix} (\mu_1-\mu_2)^2 & 2\lambda_1(\mu_1-\mu_2) \\ 2\lambda_1(\mu_1-\mu_2) & 4\lambda_1^2+(\mu_1+\mu_2)^2 \end{pmatrix} \\
&= \frac{1}{2}\begin{pmatrix} 4\rho^2 & -2(\rho-1)^2 \\ -2(\rho-1)^2 & 4 \end{pmatrix}.
\end{aligned}
$$

This matrix is positive definite. The full expression for $D_1(t)$, given by (19.5.11), is not manifestly nonnegative. However, the earlier form (19.5.9) is clearly so. From (19.4.13) and (19.5.8), we find that

$$
\mathbf{A}_1 - \mathbf{A} = \begin{pmatrix} 2e\rho & 2\rho(1-e)-(\rho-1)^2 \\ 2\rho(1-e)-(\rho-1)^2 & 2e(2+\rho)-4e^2 \end{pmatrix},
\tag{19.5.12}
$$

which can be shown to be nonnegative.

The other intermediate free energy is obtained by exchanging the other zeros in $H_\pm(\omega)$, so that the factor $(\omega-\eta)/(\omega-\overline{\eta})$ in the second line of (19.5.1) is replaced by $(\omega + \overline{\eta})/(\omega + \eta)$. The resulting formulae are similar, but with η replaced by $\overline{\eta}$ or e_1 replaced by $(-e_1)$, which follows by virtue of (19.3.14) and the last relation of (19.5.3). This actually produces the same free energy, because (19.5.7) and (19.5.8) do no depend on the square root quantity e_1, given in (19.5.3), but only on its square.

The maximum free energy is obtained by exchanging both zeros in $H_{\pm}(\omega)$. This amounts to changing the sign of e in the formulae (17.3.2) and (19.4.13). The resulting matrix corresponding to $\mathbf{A}$ in (19.4.13) is easily shown to be greater than or equal to $\mathbf{A}$. It can also be shown to be greater than or equal to $\mathbf{A}_1$, given by (19.5.8), using a procedure similar to the demonstration that $\mathbf{A}_1 - \mathbf{A}$ in (19.5.12) is nonnegative.

20

Free Energies for Nonlinear Materials with Memory

20.1 Introduction

Expressions are obtained in this chapter for free energies of materials with a certain type of nonlinear constitutive relation. These developments are based on certain aspects of results presented in [166].

We recall the integrated form of the energy balance relation, given by (16.1.29), where $\mathcal{D}(t)$ and $W(t)$ are the total dissipation and the work function , respectively, of the material, defined by

$$
\begin{aligned}
&\mathcal{D}(t) = \widetilde{\mathcal{D}}(E^t, E(t)) = \int_{-\infty}^{t} D(u)du \geq 0, \qquad \dot{\mathcal{D}}(t) = D(t), \\
&W(t) = \int_{-\infty}^{t} T(u)\dot{E}(u)du = \widetilde{W}(E^t, E(t)) = W(t; E(t)),
\end{aligned}
\tag{20.1.1}
$$

while $T(t) = \widetilde{T}(E^t, E(t))$ is the stress. It is assumed that these integrals exist. We note that $W(t)$ behaves similarly to a free energy functional with zero dissipation associated with it (except that it does not have property P4, stated in Sect. 18.2). From (20.1.1), it follows that

$$
\dot{W}(t) = T(t)\dot{E}(t), \tag{20.1.2}
$$

which is P3 for $D(t) = 0$. It is also clear from (16.1.26) and $(20.1.1)_3$ that

$$
W(t) \geq \psi(t) \geq \phi(t), \tag{20.1.3}
$$

where $\psi(t)$ is any free energy functional, so that $W(t)$ is either the maximum free energy or greater than this quantity. Also, $\phi(t)$ is $\psi(t)$ for a static history.

Let us now briefly demonstrate that it also obeys P1, given by (16.1.25) (using an intuitive and modified version of the argument in [67]) and P2. Relation (20.1.3) is equivalent to (16.1.26), though it must be shown that equality is achieved only for

G. Amendola et al., *Thermodynamics of Materials with Memory*,
https://doi.org/10.1007/978-3-030-80534-0_20

static histories. We can write

$$\dot{W}(t) = \frac{\partial}{\partial E(t)}\widetilde{W}(E^t, E(t))\dot{E}(t) + \delta\widetilde{W},$$

where the rightmost term is a Fréchet differential of $\widetilde{W}$, defined within a suitable Hilbert space (for example, (5.1.17)). Thus, (20.1.2) can be written in the form

$$\left[\frac{\partial}{\partial E(t)}\widetilde{W}(E^t, E(t)) - \widetilde{T}(E^t, E(t))\right]\dot{E}(t) = -\delta\widetilde{W}.$$

The quantity $\dot{E}(t)$ can take arbitrary values, so that (16.1.25) or P1 must hold for $\widetilde{\psi}$ replaced by $\widetilde{W}$, giving

$$\frac{\partial}{\partial E(t)}W(t) = T(t). \tag{20.1.4}$$

Also, the quantity $\delta\widetilde{W}$ must vanish. Using P1 for $\psi(t)$, we obtain

$$\frac{\partial}{\partial E(t)}\psi(t) = \frac{\partial}{\partial E(t)}W(t) = T(t).$$

It follows that

$$\frac{\partial \mathfrak{D}(t)}{\partial E(t)} = 0,$$

giving

$$\widetilde{\mathfrak{D}}(E^t, E(t)) = \widetilde{\mathfrak{D}}(E^t).$$

For the static history $E^\dagger$, the quantity $\widetilde{T}(E^t, E(t))$ will be denoted by $\widetilde{T}_e(E(t)) = T_e(t)$, a function only of the current strain. We see that

$$\frac{d\widetilde{\phi}(E(t))}{dE(t)} = \frac{d\phi(t)}{dE(t)} = \widetilde{T}_e(E(t)) = T_e(t), \tag{20.1.5}$$

giving

$$\widetilde{\phi}(E(t)) = \int_{E_0}^{E(t)} \widetilde{T}_e(\epsilon)d\epsilon,$$

where E_0 is chosen to be some convenient value of strain. Equation $(20.1.1)_3$ for the static history yields that

$$\widetilde{W}(E^\dagger, E(t)) = \int_{-\infty}^{t} \frac{d\phi(u)}{dE(u)}\dot{E}(u)du = \phi(t),$$

so that equality in $(20.1.3)_2$ is achieved for the static history.

20.2 A Generalized Quadratic Model

We seek explicit forms for free energies of materials which yield constitutive relations with memory functionals that are nonlinear in a particular sense. The equilibrium term of these relations, which has no memory effects, may be unrestrictedly nonlinear.

The free energy $\widetilde{\psi}(E^t, E(t))$ can be approximated by the scalar version of (7.1.1):

$$\begin{aligned}
\psi(t) &= \phi(t) + \frac{1}{2}\int_0^\infty\int_0^\infty E_r^t(u)G_{12}(u,v;E(t))E_r^t(v)dudv \\
&= \phi(t) + \frac{1}{2}\int_0^\infty\int_0^\infty \dot{E}^t(u)\widetilde{G}(u,v;E(t))\dot{E}^t(v)dudv, \\
&= \phi(t) - \phi_l(t) + \frac{1}{2}\int_0^\infty\int_0^\infty \dot{E}^t(u)G(u,v;E(t))\dot{E}^t(v)dudv, \qquad (20.2.1)\\
\dot{E}^t(u) &= \frac{\partial}{\partial t}E^t(u) = -\frac{\partial}{\partial u}E^t(u) = -\frac{\partial}{\partial u}E_r^t(u) \\
\phi_l(t) &= \frac{1}{2}G_\infty(E(t))E^2(t),
\end{aligned}$$

where $G_{12}(u, v; E(t))$ and $\widetilde{G}(u, v; E(t))$ may be assumed to be symmetric in the interchange of u and v. Also,

$$\begin{aligned}
G_{12}(u,v;E(t)) &= \frac{\partial^2}{\partial u\partial v}G(u,v;E(t)) \\
\widetilde{G}(u,v;E(t)) &= G(u,v;E(t)) - G_\infty(E(t)),
\end{aligned} \qquad (20.2.2)$$

and

$$\lim_{u\to\infty} G(u,v;E(t)) = G_\infty(E(t)), \qquad \lim_{u\to\infty}\frac{\partial}{\partial v}G(u,v;E(t)) = 0, \quad u \in \mathbb{R}^+. \qquad (20.2.3)$$

Relation $(20.2.3)_1$ is a definition of the quantity $G_\infty(E(t))$ used in $(20.2.1)_7$ and $(20.2.2)_2$. Note that it is independent of v, which is a restriction on $G(u, v; E(t))$. The quantity $G_\infty(E(t))$ is unique to the material and therefore not dependent on the choice of kernel $G(u, v; E(t))$. Also,

$$\lim_{u\to\infty}\frac{\partial}{\partial u}G(u,v;E(t)) = 0, \quad v \in \mathbb{R}^+, \qquad \lim_{v\to\infty}\frac{\partial}{\partial v}G(u,v;E(t)) = 0, \quad u \in \mathbb{R}^+.$$

Note that $(20.2.1)_1$ follows from $(20.2.1)_2$, by virtue of $(20.2.1)_6$. We have $G_\infty(E) > 0$ for all values of E.

If we choose $\phi(t)$ in (20.2.1) to be equal to $\phi_l(t)$, then

$$\psi(t) = \frac{1}{2}\int_0^\infty\int_0^\infty \dot{E}^t(u)G(u,v;E(t))\dot{E}^t(v)dudv.$$

Remark 20.2.1. The quantity $\phi(t)$ is always taken to be a nonnegative function of $E(t)$, zero only if $E(t) = 0$, while $(20.1.3)_2$ gives that the integral terms in (20.2.1) are nonnegative. From $(20.2.1)_1$, it is clear that the integral terms vanish for the static history. They are positive for all other histories. The kernels $\widetilde{G}$ and G_{12} must be such that this property holds. The linear functional terms are omitted from the expression because they may take any sign.

If the effect on the kernels of the variation of $E(t)$ can be neglected, we obtain the standard quadratic form for free energies, yielding linear memory terms in the constitutive relations. For example, if $E(t)$ varies only to a small extent from a constant strain E_c, the kernel can be well approximated by $G(u, v, E_c)$, which yields a linear memory constitutive relation.

We refer to a theory based on such a kernel as a linear memory theory. The associated free energy will be referred to as a linear memory free energy, though it is in fact a quadratic functional; similarly for the related rate of dissipation. These are of course the functionals that have been discussed in the previous chapters of Part III.

Let us consider the kernel $G(u, v, E_1)$ for any arbitrary time-independent choice of the strain E_1 (which may be different from E_c) as that for a particular, known, linear memory free energy. We can in general write

$$G(u, v; E(t)) = G(u, v, E_1) + G^{(E)}(u, v; E(t)), \qquad G^{(E)}(u, v; E_1) = 0. \tag{20.2.4}$$

The quantity $G(u, v, E_1)$ is assumed to be nonzero, since it is important that a term independent of $E(t)$ is present, and the limiting linear memory constitutive equation exists.

The simplest possibility for $G(u, v, E(t))$ is where it is a polynomial in $E(t) - E_c$, including a term independent of $E(t)$. The difference $E(t) - E_c$ is not now regarded as negligible. If the highest power of the polynomial is N, we refer to the model as of order N.

An example is the model of order 2, given by

$$G(u, v; E(t)) = G^{(0)}(u, v) + (E(t) - E_c)^2 G^{(2)}(u, v). \tag{20.2.5}$$

The term proportional to $E(t) - E_c$ is omitted because there are positivity requirements on the kernel, and this linear term can take any sign. The quantities $G^{(i)}(u, v)$, $i = 0, 2$, must be such that $G(u, v; E(t))$ has the property imposed by $(20.1.3)_2$ and referred to in Remark 20.2.1, for any choice of $E(t)$. It follows that both must be suitable kernels for linear memory free energies, in the sense of the non-negativity requirements deriving from $(20.1.3)_2$.

Equation (20.2.1) can be written in the form

$$\begin{aligned}
\psi(t) &= \sigma(t) + \frac{1}{2}\int_0^\infty\int_0^\infty E^t(u) G_{12}(u, v; E(t)) E^t(v)\,du\,dv, \\
\sigma(t) &= \phi(t) - \frac{1}{2}\left[G_\infty(E(t)) - G_0(E(t))\right] E^2(t) \\
&\quad + E(t)\int_0^\infty G'(u; E(t)) E^t(u)\,du,
\end{aligned} \tag{20.2.6}$$

where

$$G(u; E(t)) = G(0, u; E(t)) = G(u, 0; E(t)), \qquad u \in \mathbb{R}^+. \tag{20.2.7}$$

The left-hand side of this relation will be referred to as the relaxation function, which is a unique characteristic property of the material. There are typically many choices of the free energy kernel $G(u, v; E(t))$ with $G(0, u; E(t))$ equal to $G(u; E(t))$. The relaxation function is generally assumed to be a nonnegative quantity.

The prime on $G(u; E(t))$ in $(20.2.6)_2$ indicates differentiation with respect to its first argument. Furthermore,

$$G_0(E(t)) = G(0; E(t)), \qquad G_\infty(E(t)) = \lim_{u\to\infty} G(u, E(t)).$$

Note that $G_\infty(E(t))$ is the same as the quantity introduced in $(20.2.3)_1$. For $\phi(t) = \phi_l(t)$,

$$\sigma(t) = \frac{1}{2} G_0(E(t))E^2(t) + E(t)\int_0^\infty G'(u; E(t))E^t(u)du.$$

Corresponding to (20.2.4) and taking account of (20.2.7), we have

$$G(u; E(t)) = G(u; E_1) + G^{(E)}(u, E(t)), \qquad G^{(E)}(u; E_1) = 0. \tag{20.2.8}$$

The model of order 2, given by (20.2.5), corresponds to the form of the relaxation function

$$\begin{aligned} G(u; E(t)) &= G^{(0)}(u) + (E(t) - E_c)^2 G^{(2)}(u), \\ G^{(i)}(u) &= G^{(i)}(u, 0), \quad i = 0, 2. \end{aligned} \tag{20.2.9}$$

The quantities $G^{(i)}(u)$ must be such that $G(u; E(t))$ has the required properties for a relaxation function, for any value of $E(t)$. This implies that both these quantities are valid relaxation functions for a linear memory material.

We will use the model of order 2, given by (20.2.5) and (20.2.9) as our material model for discussion of our free energy functional in the nonlinear case. The extension of this discussion to models of order $2N$, where $N = 1, 2, 3, \ldots$, is straightforward.

Remark 20.2.2. For even powered polynomial models, and indeed any model where $G(u, v; E(t))$ diverges positively for large values of $|E(t)|$, a limit most be imposed on the size of $|E(t)|$, which we denote by $E_l > 0$. For example, this could be slightly below a critical strain at which failure or a phase transition occurs.

Equivalent expressions for the total work done on the material are now presented. We seek quadratic forms for $W(t)$ similar to (20.2.1) or (20.2.6) and for which the kernel K of the rate of dissipation, given by (20.3.7) below, vanishes. Also, the linear memory result must readily emerge. The only choices obeying these requirements are given by

$$\begin{aligned} W(t) &= \phi(t) + \frac{1}{2}\int_0^\infty\int_0^\infty G_{12}(|u-v|; E(t))E^t_r(u)E^t_r(v)dudv \\ &= \sigma(t) + \frac{1}{2}\int_0^\infty\int_0^\infty G_{12}(|u-v|; E(t))E^t(u)E^t(v)dudv \\ &= \phi(t) + \frac{1}{2}\int_0^\infty\int_0^\infty \widetilde{G}(|u-v|; E(t))\dot{E}^t(u)\dot{E}^t(v)dudv, \end{aligned} \tag{20.2.10}$$

$$\widetilde{G}(u, E(t)) = G(u, E(t)) - G_\infty(E(t)),$$

where the kernel is the relaxation function introduced in (20.2.7). We also define the functional

$$W(t;E_1)=\phi_1(t)+\frac{1}{2}\int_0^\infty\int_0^\infty G_{12}(|u-v|;E_1)E_r^t(u)E_r^t(v)dudv, \tag{20.2.11}$$

where E_1 is a given strain, independent of t, and $\phi_1(t)$ is $\phi(t)$ with any occurrence of $E(t)$ in $G_\infty(\cdot)$ or $G_0(\cdot)$ replaced by E_1.

20.3 Dissipation

Using $(20.2.1)_2$ and $(20.2.10)_3$, we see that

$$G(u,v;E(t))+\varDelta(u,v)=G(|u-v|;E(t)), \tag{20.3.1}$$

where $\varDelta(u,v)$ is the kernel of the total dissipation. Relations (20.2.7), (20.3.1), and $(20.2.3)_1$ give that

$$\begin{aligned}\varDelta(u,0)=\varDelta(0,u)=0,\quad u\in\mathbb{R}^+,\\ \lim_{u\to\infty}\varDelta(u,v)=\lim_{u\to\infty}\varDelta(v,u)=0,\quad v\in\mathbb{R}^+.\end{aligned} \tag{20.3.2}$$

The relations in $(20.3.2)_{1,2}$ have been given previously in (17.3.24). The total dissipation has the form

$$\begin{aligned}\mathfrak{D}(t)&=\frac{1}{2}\int_0^\infty\int_0^\infty \varDelta(u,v)\dot{E}^t(u)\dot{E}^t(v)dudv\\ &=\frac{1}{2}\int_0^\infty\int_0^\infty \varDelta_{12}(u,v)E_r^t(u)E_r^t(v)dudv,\end{aligned} \tag{20.3.3}$$

where the second form requires $(20.2.1)_6$. Using (20.3.2), we find that $(20.3.3)_2$ becomes

$$\mathfrak{D}(t)=\frac{1}{2}\int_0^\infty\int_0^\infty \varDelta_{12}(u,v)E^t(u)E^t(v)dudv.$$

Since $\varDelta(u,v)$ in (20.3.1) is independent of $E(t)$, we can put

$$\varDelta(u,v)=G(|u-v|;E_1)-G(u,v;E_1), \tag{20.3.4}$$

for any choice of E_1 and in particular for the parameter introduced in (20.2.4). Using this in (20.3.1), we obtain

$$G(u,v;E(t))=G(u,v;E_1)+G(|u-v|;E(t))-G(|u-v|;E_1), \tag{20.3.5}$$

or, recalling (20.2.4) and (20.2.8),

$$G^{(E)}(u,v;E(t))=G^{(E)}(|u-v|;E(t))=G(|u-v|;E(t))-G(|u-v|;E_1).$$

These are significant restrictions on $G(u, v; E(t))$. They ensure the uniqueness of the constitutive relations. Note that $\Delta(u, v)$ depends in general on the parameter E_1. We rewrite (20.2.5) as

$$G(u, v; E(t)) = G^{(0)}(u, v) + (E(t) - E_c)^2 G^{(2)}(|u - v|), \tag{20.3.6}$$

where $G^{(2)}(u)$ is the same quantity as in (20.2.9).

Using $(20.3.3)_1$, (20.3.2), and (20.3.4), we find that the rate of dissipation, given by the derivative of $\mathcal{D}(t)$, has the form

$$\begin{aligned} D(t) = D(t; E_1) &= -\frac{1}{2}\int_0^\infty\int_0^\infty \dot{E}^t(u)K(u, v; E_1)\dot{E}^t(v)dudv \\ &= -\frac{1}{2}\int_0^\infty\int_0^\infty E_r^t(u)K_{12}(u, v; E_1)E_r^t(v)dudv, \\ K(u, v; E_1) &= G_1(u, v; E_1) + G_2(u, v; E_1). \end{aligned} \tag{20.3.7}$$

The kernel $K(u, v, E_1)$ is independent of $E(t)$. However, we see from $(20.3.7)_2$ that $D(t; E_1)$ may depend on $E(t)$.

20.4 General Form of a Free Energy for Nonlinear Materials

By virtue of (20.3.5), we find that the general form of the free energy functional is

$$\begin{aligned} \psi(t) &= \psi_l(t; E_1) + W(t; E(t)) - W(t; E_1), \\ \psi_l(t; E_1) &= \phi_1(t) + \frac{1}{2}\int_0^\infty\int_0^\infty G_{12}(u, v; E_1)E_r^t(u)E_r^t(v)dudv. \end{aligned} \tag{20.4.1}$$

The quantity $\psi_l(t; E_1)$ is a quadratic functional of the form (20.2.1), yielding a linear memory free energy, while $\phi_1(t)$ is defined after (20.2.11). For the model of order 2, we see from (20.3.6) that it has the form

$$G(u, v; E_1) = G^{(0)}(u, v) + (E_1 - E_c)^2 G^{(2)}(|u - v|).$$

The quantity $W(t; E_1)$ is defined by $(20.2.11)_3$. The free energy $\psi(t)$ will in general depend on the choice of E_1, so that we denote it by $\psi(t; E_1)$.

Remark 20.4.1. It is assumed that the quantity $\psi_l(t; E_1)$ is a valid linear memory free energy with kernel $G_{12}(u, v; E_1)$, which is independent of $E(t)$. The associated total dissipation is

$$\mathcal{D}_l(t; E_1) = W(t; E_1) - \psi_l(t; E_1) \geq 0, \tag{20.4.2}$$

while the rate of dissipation $D_l(t; E_1)$ is given by (20.3.7) and must be nonnegative, by virtue of (16.1.28), for $\psi_l(t; E_1)$.

The total dissipation relating to $\psi(t; E_1)$ can be seen from (20.4.1) to be equal to $\mathcal{D}_l(t; E_1)$, so that

$$\mathcal{D}(t) = \mathcal{D}(t; E_1) = W(t; E_1) - \psi_l(t; E_1) \geq 0,$$

which is independent of $E(t)$. The time derivative yields $D(t; E_1)$, as given by (20.3.7).

We must show that $\psi(t; E_1)$, given by (20.4.1), obeys the properties P1–P3 listed in Sect. 16.1.2. Property P1 follows from (20.1.4).The time derivative of (20.4.1) gives

$$\dot{\psi}(t; E_1) + D(t; E_1) = T(t)\dot{E}(t),$$

$$D(t; E_1) = D_l(t; E_1) = \dot{\mathcal{D}}_l(t; E_1) \geq 0,$$

as noted after (20.4.2). It follows that P3 is true also for $\psi(t; E_1)$.

Property P2 can be shown as follows. It is essentially a statement that the integral terms in $(20.2.1)_{1,2}$ are nonnegative. Substituting (20.3.6) into the integral terms of (20.2.1) and recalling that $G^{(0)}(u, v)$ and $G^{(2)}(|u - v|)$ are valid linear memory kernels yielding a free energy and a work function, respectively, we conclude that these are both nonnegative integrals, so that P2 holds.

20.4.1 Generalizing Specified Linear Memory Free Energies

It is of interest to discuss natural generalizations of standard free energy functional forms in the linear memory case to the nonlinear materials of interest here. These are given by (20.4.1), where $\psi_l(t; E_1)$ is taken to be the linear memory free energy of interest. In particular, examples where $\psi_l(t; E_1)$ is put equal to the Graffi–Volterra and Dill functionals, together with ψ_F, are presented in [166]. These are free energies in the linear memory case, provided the relaxation function obeys certain decay conditions (see Sect. 10.1.2 and (17.3.4)). The only modification necessary to the familiar expressions for these quantities is to insert dependence on the parameter E_1 into the relaxation function $G(u)$. The resulting generalized free energy functionals $\psi_{GV}(t; E_1)$ $\psi_D(t; E_1)$ and $\psi_F(t; E_1)$ will depend on this parameter.

A similar procedure can be applied to the minimum free energy ψ_m and, for materials with only isolated singularities, the related free energies ψ_f introduced in Sect. 16.4. In the case of $\psi_m(t; E_1)$, there is the extra stage of seeking the value of E_1, which minimizes this functional or maximizes the associated total dissipation [166].

20.5 Constitutive Relations

From P1, we obtain the following forms of the stress function:

$$
\begin{aligned}
T(t) &= \widetilde{T}_e(E(t)) + \int_0^\infty G'(u; E(t))E_r^t(u)du \\
&\quad + \frac{1}{2}\int_0^\infty \int_0^\infty E_r^t(u)G_{12E}(|u-v|; E(t))E_r^t(v)dudv \\
&= \widetilde{T}_e(E(t)) + \int_0^\infty \widetilde{G}(u; E(t))\dot{E}^t(u)du \\
&\quad + \frac{1}{2}\int_0^\infty \int_0^\infty \dot{E}^t(u)\widetilde{G}_E(|u-v|; E(t))\dot{E}^t(v)dudv \\
&= \widetilde{T}_0(E^t, E(t)) + \frac{1}{2}\int_0^\infty \int_0^\infty E^t(u)G_{12E}(|u-v|; E(t))E^t(v)dudv,
\end{aligned} \tag{20.5.1}
$$

where $\widetilde{T}_e(E(t))$ is defined by (20.1.5). Also,

$$
\begin{aligned}
\widetilde{G}(u; E(t)) &= G(u; E(t)) - G_\infty(E(t)), \\
\widetilde{T}_0(E^t, E(t)) &= \frac{\partial \sigma(t)}{\partial E(t)} = \widetilde{T}_e(E(t)) - [G_\infty(E(t)) - G_0(E(t))]\, E(t) \\
&\quad + \int_0^\infty G'(u; E(t))E^t(u)du + E(t)\int_0^\infty G'_E(u; E(t))E^t(u)du,
\end{aligned}
$$

where the second relation follows from (20.2.6). Furthermore,

$$
\begin{aligned}
G_E(|u-v|; E(t)) &= \frac{\partial}{\partial E(t)} G(|u-v|; E(t)), \\
G_{12E}(|u-v|; E(t)) &= \frac{\partial}{\partial E(t)} G_{12}(|u-v|; E(t)).
\end{aligned}
$$

The forms given by (20.5.1) differ from what is known as *finite linear viscoelasticity* [73, 121] by the extra terms involving the derivative with respect to $E(t)$ of the kernels in the quadratic memory terms. They are also clearly derivable from a free energy functional.

21
Free Energies for Nonlocal Materials

When nonlocal materials are considered, the classical laws of thermodynamics must be modified by expressing these laws in terms of internal powers or by introducing directly into them suitable extra fluxes, characteristic of the material under consideration [117], as discussed in some detail in Sect. 3.7. The first formulation, in terms of internal powers, is more general than the second one, since it is defined a priori by means of the constitutive equations, by taking into account the power balance laws; in the second method, there is the problem of introducing a posteriori the vector fluxes in order that compatibility with the laws of thermodynamics be satisfied.

In Sects. 21.1 and 21.2 of this chapter, we shall be concerned with two further types of nonlocal materials other than those considered in Sect. 3.7. These are second-gradient thermoviscoelastic fluids and heat flux in rigid conductors with nonlocal behavior, in both cases with memory. The latter case is related to an example discussed in Sect. 3.7, namely the material described by (3.7.34). Our main emphasis here will be on proposing and discussing free energy functionals that generalize the forms given in Chaps. 10 and 11 for simple materials.

Finally, in Sect. 21.3, a general theory of nonlocal materials, based on [15], is described. As in earlier chapters, the terms nonlocal and nonsimple are taken to be interchangeable; similarly for local and simple.

21.1 Second-Gradient Thermoviscoelastic Fluids

Incompressible second-order fluids with memory and thermal effects, characterized by a constitutive equation for the stress tensor $\mathbf{T}$ expressed by

$$\begin{aligned} \mathbf{T}(t) &= [-p(t) + \alpha\vartheta(t)]\mathbf{I} + 2\int_0^{+\infty} \mu'(s)\mathbf{E}_r^t(s)ds - \int_0^{+\infty} \varkappa'(s)\nabla\cdot\left[\nabla\mathbf{E}_r^t(s)\right]ds, \\ \vartheta &= \theta - \theta_0, \end{aligned} \tag{21.1.1}$$

G. Amendola et al., *Thermodynamics of Materials with Memory*,
https://doi.org/10.1007/978-3-030-80534-0_21

are considered in this section, where p and θ are the pressure and the absolute temperature, θ_0 is the constant ambient absolute temperature, while μ and $\varkappa$ are two smooth positive functions, which belong to $L^1(\mathbb{R}^+) \cap H^1(\mathbb{R}^+)$. The quantity $\mathbf{E}_r^t$ is the relative strain history defined by (8.1.3) in terms of the linear strain tensor. Equation (21.1.1) generalizes the theory introduced in [139] in which memory effects are absent.

Since the constitutive equation assumed for $\mathbf{T}$ is not local, these fluids are nonsimple materials.

For convenience of reference, some basic equations from Sect. 3.7 are repeated here. Referring to Sect. 3.7.1, we note that the first law assumes the form

$$\rho\dot{e}(t) = \rho h(t) + \mathcal{P}_m^i(t), \tag{21.1.2}$$

where e is the internal energy, h is the specific internal heat (or thermal) power, defined as the rate at which heat is absorbed per unit mass, and $\mathcal{P}_m^i$ denotes the internal mechanical power. The heat balance law (see (3.7.13)) is given by

$$\rho h = -\nabla \cdot \mathbf{q} + \rho r, \tag{21.1.3}$$

where r is the heat supply. The heat flux $\mathbf{q}$ is determined by the Fourier law $\mathbf{q} = -k_0\nabla\theta$. Moreover, the second law yields the existence of the entropy η such that (see (3.7.24))

$$\rho\dot{\eta} \geq -\nabla \cdot \left(\frac{\mathbf{q}}{\theta}\right) + \rho\frac{r}{\theta},$$

whence it follows that

$$\rho\dot{\eta} \geq \rho\frac{h}{\theta} - \frac{k_0}{\theta^2}(\nabla\theta)^2.$$

Introducing the free energy $\psi = e - \theta\eta$, this inequality becomes

$$\dot{\psi} \leq -\eta\dot{\theta} + \frac{1}{\rho}\mathcal{P}_m^i + \frac{k_0}{\rho\theta}(\nabla\theta)^2. \tag{21.1.4}$$

The mechanical balance of power is given by

$$\rho\frac{d}{dt}\left(\frac{1}{2}\mathbf{v}^2\right) = (\nabla \cdot \mathbf{T}) \cdot \mathbf{v} + \rho\mathbf{b} \cdot \mathbf{v} = [\nabla \cdot (\mathbf{T}\mathbf{v})] - \mathbf{T} \cdot \nabla\mathbf{v} + \rho\mathbf{b} \cdot \mathbf{v}, \tag{21.1.5}$$

where $\mathbf{b}$ denotes the body forces. This is obtained by taking the scalar product of $\mathbf{v}$ with $(2.2.14)_1$. The integral of (21.1.5) over part or all of the fluid generalizes (2.2.16). To derive the expression for $\mathcal{P}_m^i$ related to the material under consideration, we substitute the constitutive equation (21.1.1) into (21.1.5), to obtain

$$\begin{aligned}
&\rho\frac{d}{dt}\left(\frac{1}{2}\mathbf{v}^2\right) + 2\int_0^{+\infty} \mu'(s)\mathbf{E}_r^t(s) \cdot \nabla\mathbf{v}(t)ds + \int_0^{+\infty} \varkappa'(s)\nabla\mathbf{E}_r^t(s) \cdot \nabla\nabla\mathbf{v}(t)ds \\
&= \nabla \cdot \left(\left\{(-p + \alpha\theta)\mathbf{I} + 2\int_0^{+\infty} \mu'(s)\mathbf{E}_r^t(s)ds - \int_0^{+\infty} \varkappa'(s)\nabla \cdot \left[\nabla\mathbf{E}_r^t(s)\right]ds\right\}\mathbf{v}(t)\right. \\
&\left.+ \int_0^{+\infty} \varkappa'(s)\nabla\mathbf{E}_r^t(s)ds\nabla\mathbf{v}(t)\right) + \rho\mathbf{b} \cdot \mathbf{v},
\end{aligned}$$

which must be compared with (3.7.2). Since the divergence term becomes a surface integral when integrated over the body, or any part of it, we associate this term with the external mechanical power and deduce that the internal mechanical power is given by

$$\mathcal{P}^i_m(t) = 2\int_0^{+\infty} \mu'(s)\mathbf{E}^t_r(s)\cdot\nabla\mathbf{v}(t)ds + \int_0^{+\infty} \varkappa'(s)\nabla\mathbf{E}^t_r(s)\cdot\nabla\nabla\mathbf{v}(t)ds. \tag{21.1.6}$$

To characterize the behavior of our fluid, let us introduce the state

$$\sigma^T(t) = (\theta, \sigma(t)) = \left(\theta, \mathbf{E}^t_r(s), \nabla\mathbf{E}^t_r(s)\right) \tag{21.1.7}$$

and the process P^T, of duration d, given by a piecewise continuous map defined as

$$P^T(\tau) = (\dot\theta_P, P) = (\dot\theta_P, \nabla\mathbf{v}_P(\tau), \nabla\nabla\mathbf{v}_P(\tau)) \qquad \forall\tau\in[0,d). \tag{21.1.8}$$

Now consider a free energy ψ of the form

$$\psi(t) = \psi_1(\theta) + \psi_2\left(\mathbf{E}^t_r(s), \nabla\mathbf{E}^t_r(s)\right).$$

Substituting into (21.1.4), we obtain

$$\left[\frac{\partial\psi_1(\theta)}{\partial\theta} + \eta\right]\dot\theta + \dot\psi_2\left(\mathbf{E}^t_r(s), \nabla\mathbf{E}^t_r(s)\right) \le \frac{1}{\rho}\mathcal{P}^i_m + \frac{k_0}{\rho\theta}(\nabla\theta)^2.$$

This inequality, taking account of (21.1.6), is satisfied if

$$\begin{aligned}
\eta &= -\frac{\partial\psi_1(\theta)}{\partial\theta},\\
\dot\psi_2\left(\mathbf{E}^t_r(s), \nabla\mathbf{E}^t_r(s)\right) &\le \frac{1}{\rho}\left[2\int_0^{+\infty} \mu'(s)\mathbf{E}^t_r(s)\cdot\nabla\mathbf{v}(t)ds\right.\\
&\quad\left. + \int_0^{+\infty} \varkappa'(s)\nabla\mathbf{E}^t_r(s)\cdot\nabla\nabla\mathbf{v}(t)ds\right],\\
\frac{k_0}{\rho\theta}(\nabla\theta)^2 &\ge 0.
\end{aligned} \tag{21.1.9}$$

Taking account of the incompressibility of the fluid, we absorb the density into the kernels and write $(21.1.9)_2$ as follows:

$$\dot\psi_2\left(\mathbf{E}^t_r(s), \nabla\mathbf{E}^t_r(s)\right) \le \mathcal{A}(\sigma, P), \tag{21.1.10}$$

where $\mathcal{A}$ is the *mechanical action*, defined by

$$\begin{aligned}
\mathcal{A}(t) = \mathcal{A}(\sigma, P) &= 2\int_0^{+\infty} \mu'(s)\mathbf{E}^t_r(s)ds\cdot\nabla\mathbf{v}(t) + \int_0^{+\infty} \varkappa'(s)\nabla\mathbf{E}^t_r(s)ds\cdot\nabla\nabla\mathbf{v}(t)\\
&= 2\int_0^{+\infty} \mu'(s)\mathbf{E}^t_r(s)ds\cdot\dot{\mathbf{E}}(t) + \int_0^{+\infty} \varkappa'(s)\nabla\mathbf{E}^t_r(s)ds\cdot\nabla\dot{\mathbf{E}}(t).
\end{aligned} \tag{21.1.11}$$

The state, as indicated in (21.1.7), is given by $\sigma(t) = (\mathbf{E}_r^t(s), \nabla\mathbf{E}_r^t(s))$, while P is as specified by (21.1.8).

By introducing a nonnegative function $D_2(\mathbf{x}, t)$, the *internal rate of dissipation*, we can transform the inequality (21.1.10) into an equality of the form

$$\dot{\psi}_2(t) + D_2(t) = \mathcal{A}(t). \tag{21.1.12}$$

The *total mechanical action* $\mathcal{B}(\sigma, P)$ of the material during the application of a process P of duration d is defined as

$$\mathcal{B}(\sigma, P) = \int_t^{t+d} \mathcal{A}(\xi)d\xi.$$

For any cycle (σ, P), we have

$$\mathcal{B}(\sigma, P) \geq 0, \tag{21.1.13}$$

by virtue of the dissipation principle (see (4.1.7)), which follows from (21.1.12) and the nonnegativity of D_2, this latter requirement being in effect the second law. The equality sign in (21.1.13) occurs if and only if the cycle is reversible.

A generalization of the method noted in Sect. 8.10.1 (described in more detail in the proof of Theorem 8.9.4) allows us to show that

$$\mu_p(\omega) > 0, \qquad \varkappa_p(\omega) > 0 \quad \forall\omega \in \mathbb{R}, \tag{21.1.14}$$

under the hypotheses that $\mu_p(0) \neq 0$ and $\varkappa_p(0) \neq 0$. Moreover, it follows that

$$\lim_{\omega\to+\infty} \omega\mu_s'(\omega) = -\lim_{\omega\to+\infty} \omega^2\mu_p(\omega) = \mu'(0) \leq 0,$$
$$\lim_{\omega\to+\infty} \omega\varkappa_s'(\omega) = -\lim_{\omega\to+\infty} \omega^2\varkappa_p(\omega) = \varkappa'(0) \leq 0;$$

in particular, we shall assume

$$\mu'(0) < 0, \qquad \varkappa'(0) < 0. \tag{21.1.15}$$

We now focus on $\psi_2(t)$, omitting the subscript 2.

21.1.1 The Nonlocal Graffi–Volterra Free Energy for Thermoviscoelastic Fluids

We firstly consider a generalization of the Graffi–Volterra functional (10.3.1) in the form

$$\psi_{\mathcal{G}}(t) = -\int_0^{+\infty} \mu'(s)\mathbf{E}_r^t(s)\cdot\mathbf{E}_r^t(s)ds - \frac{1}{2}\int_0^{+\infty} \varkappa'(s)\nabla\mathbf{E}_r^t(s)\cdot\nabla\mathbf{E}_r^t(s)ds, \tag{21.1.16}$$

which is nonnegative if the conditions

$$\mu'(s) < 0, \qquad \varkappa'(s) < 0 \quad \forall s \in \mathbb{R}^+ \tag{21.1.17}$$

are satisfied. Other constraints must also be imposed on each kernel in order to obtain a nonnegative rate of dissipation, that is,

$$\mu''(s) \geq 0, \qquad \varkappa''(s) \geq 0 \quad \forall s \in \mathbb{R}^+, \tag{21.1.18}$$

which generalize (10.3.2). As pointed out in the context of Sect. 10.3, (21.1.18) implies a weaker form of (21.1.17). The derivative with respect to time of (21.1.16), by integrating by parts and using

$$\frac{d}{dt}\mathbf{E}_r^t(s) = -\dot{\mathbf{E}}(t) - \frac{d}{ds}\mathbf{E}_r^t(s)$$

with the analogous relation for $\frac{d}{dt}\nabla\mathbf{E}_r^t(s)$, gives

$$\begin{aligned}
\dot{\psi}_{\mathcal{G}}(t) &= -2\int_0^{+\infty} \mu'(s)\mathbf{E}_r^t(s)\cdot\frac{d}{dt}\mathbf{E}_r^t(s)ds - \int_0^{+\infty} \varkappa'(s)\nabla\mathbf{E}_r^t(s)\cdot\frac{d}{dt}\nabla\mathbf{E}_r^t(s)ds \\
&= 2\int_0^{+\infty} \mu'(s)\mathbf{E}_r^t(s)ds\cdot\nabla\mathbf{v}(t) + \int_0^{+\infty} \varkappa'(s)\nabla\mathbf{E}_r^t(s)ds\cdot\nabla\nabla\mathbf{v}(t) \\
&\quad - \int_0^{+\infty} \mu''(s)\left[\mathbf{E}_r^t(s)\right]^2 ds - \frac{1}{2}\int_0^{+\infty} \varkappa''(s)\left[\nabla\mathbf{E}_r^t(s)\right]^2 ds \\
&= \mathcal{A}(t) - D_{\mathcal{G}}(t),
\end{aligned}$$

by virtue of (21.1.11). Therefore, (21.1.10) and (21.1.12) are satisfied by

$$D_{\mathcal{G}}(t) = \int_0^{+\infty} \mu''(s)\left[\mathbf{E}_r^t(s)\right]^2 ds + \frac{1}{2}\int_0^{+\infty} \varkappa''(s)\left[\nabla\mathbf{E}_r^t(s)\right]^2 ds \geq 0,$$

which expresses the nonnegative internal dissipation related to the Graffi–Volterra functional.

Note that

$$\frac{\partial}{\partial\mathbf{E}(t)}\psi_{\mathcal{G}}(t) = 2\int_0^{\infty} \mu'(s)\mathbf{E}_r^t(s)ds,$$

so that the standard (for simple materials) relation (5.1.30) does not apply. However, we have

$$\frac{\partial}{\partial\nabla\mathbf{E}(t)}\psi_{\mathcal{G}}(t) = \int_0^{\infty} \kappa'(s)\nabla\mathbf{E}_r^t(s)ds,$$

giving, with the aid of (21.1.1),

$$T(t) = \frac{\partial}{\partial\mathbf{E}(t)}\psi_{\mathcal{G}}(t) - \nabla\cdot\frac{\partial}{\partial\nabla\mathbf{E}(t)}\psi_{\mathcal{G}}(t). \tag{21.1.19}$$

This is the generalization of (5.1.30) to the case of the nonsimple materials considered in this section and the next. It applies to all the free energy functionals introduced in the present chapter. Note that we can write (21.1.19) in the form

$$T(t) = \frac{\delta}{\delta\mathbf{E}(t)}\psi_{\mathcal{G}}(t), \tag{21.1.20}$$

where the quantity on the right is an example of a variational derivative in the sense of the calculus of variations. A generalization of this approach, applicable to cases with higher than second gradients, is presented in Sect. 21.3. We will see that (21.1.20) applies to the broader theory, where the variational derivative is the form appropriate to the calculus of variations, where higher gradients are present.

21.1.2 A Single-Integral Free Energy in Terms of the Minimal State

The free energy ψ_F (Sects. 10.1.3 and 10.3) can be adapted to our fluid. For this purpose, let us introduce the functions

$$\mathbf{I}^t(\tau, \mathbf{E}_r^t) = 2\int_0^{+\infty} \mu'(\tau+\eta)\mathbf{E}_r^t(\eta)d\eta,$$

$$\mathfrak{I}^t(\tau, \nabla\mathbf{E}_r^t) = \int_0^{+\infty} \varkappa'(\tau+\eta)\nabla\mathbf{E}_r^t(\eta)d\eta,$$

which express a minimal state of the fluid just as (8.10.6) does for the simple case. Consider the functional

$$\psi_{\mathcal{F}}(t) = -\frac{1}{4}\int_0^{+\infty} \frac{1}{\mu'(\tau)}|\mathbf{I}^t_{(1)}(\tau, \mathbf{E}_r^t)|^2 d\tau - \frac{1}{2}\int_0^{+\infty} \frac{1}{\varkappa'(\tau)}\left|\mathfrak{I}^t_{(1)}(\tau, \nabla\mathbf{E}_r^t)\right|^2 d\tau,$$

where $\mathbf{I}^t_{(1)}$ and $\mathfrak{I}^t_{(1)}$ are the derivatives with respect to τ of $\mathbf{I}^t$ and $\mathfrak{I}^t$, as in (10.3.5). The absolute value squared notation indicates scalar products of $\mathbf{I}^t_{(1)}$ and $\mathfrak{I}^t_{(1)}$ with themselves in the appropriate vector space.

Using the same algebra as in Chap. 10, it follows that

$$\dot{\psi}_{\mathcal{F}}(t) = 2\int_0^{+\infty} \mu'(\eta)\mathbf{E}_r^t(\eta)d\eta \cdot \nabla\mathbf{v}(t) + \int_0^{+\infty} \varkappa'(\eta)\nabla\mathbf{E}_r^t(\eta)d\eta \cdot \nabla\nabla\mathbf{v}(t) - D_{\mathcal{F}}(t),$$

where the internal dissipation has the form

$$\begin{aligned} D_{\mathcal{F}}(t) = &\frac{1}{2}\int_0^{+\infty} \frac{\mu''(\tau)}{2[\mu'(\tau)]^2}\left|\mathbf{I}^t_{(1)}(\tau, \mathbf{E}_r^t)\right|^2 d\tau - \frac{1}{4\mu'(0)}\left|\mathbf{I}^t_{(1)}(0, \mathbf{E}_r^t)\right|^2 \\ &+ \frac{1}{2}\int_0^{+\infty} \frac{\varkappa''(\tau)}{[\varkappa'(\tau)]^2}\left|\mathfrak{I}^t_{(1)}(\tau, \nabla\mathbf{E}_r^t)\right|^2 d\tau - \frac{1}{2\varkappa'(0)}\left|\mathfrak{I}^t_{(1)}(0, \nabla\mathbf{E}_r^t)\right|^2 \geq 0, \end{aligned}$$

because of the hypotheses (21.1.15) and (21.1.18).

One can show without difficulty that (21.1.19) holds in this case also.

21.1.3 The Nonlocal Minimum Free Energy for Thermoviscoelastic Fluids

Let us write (21.1.12) for the minimum free energy in the form

$$\dot{\psi}_m(t) + D_m(t) = \mathcal{A}(t), \tag{21.1.21}$$

where $D_m(t)$ is the rate of internal dissipation associated with ψ_m. It will be assumed that the material under consideration is undisturbed in the distant past in such a way that (21.1.21) can be integrated from $t = -\infty$, giving (cf. (11.2.20))

$$\psi_m(t) + \mathcal{D}_m(t) = \mathcal{B}(t), \qquad \mathcal{B}(t) = \int_{-\infty}^{t} \mathcal{A}(\xi)d\xi.$$

We identify $\mathcal{A}(t)$ with the right-hand side of (4.1.9) in the abstract theory and conclude from Theorem 4.2.3 that the minimum free energy is that associated with the maximum of the quantity

$$\mathcal{B}_R(t) = -\int_{t}^{\infty} \mathcal{A}(\xi)d\xi,$$

where future continuations are constrained so that the integral on the right exists. Proceeding as in Sect. 11.2, noting in particular the observation after (11.2.4), we seek ψ_m on the basis of the constraint that $\mathcal{B}(\infty)$ is minimum.

Generalizing the procedure leading to (8.10.21), we conclude that

$$\mathcal{B}(t) = \frac{1}{2\pi}\int_{-\infty}^{\infty} H_\mu(\omega)\mathbf{E}^t_{r+}(\omega)\cdot\overline{\mathbf{E}^t_{r+}(\omega)}d\omega + \frac{1}{2\pi}\int_{-\infty}^{\infty} H_\varkappa(\omega)\nabla\mathbf{E}^t_{r+}(\omega)\cdot\overline{\nabla\mathbf{E}^t_{r+}(\omega)}d\omega,$$
$$H_\mu(\omega) = -2\omega\mu_s'(\omega) = 2\omega^2\mu_p(\omega) \geq 0,$$
$$H_\varkappa(\omega) = -\omega\varkappa_s'(\omega) = \omega^2\varkappa_p(\omega) \geq 0,$$

where (21.1.14) has been invoked. The quantities H_μ and $H_\varkappa$ can be factorized according to (11.1.16). Note that H_μ is precisely that quantity given by (13.3.5). We define $\mathbf{p}^t_{\mu\pm}(\omega)$ and $\mathbf{p}^t_{\varkappa\pm}(\omega)$ by substituting $H_{\mu-}$ and $H_{\varkappa-}$, respectively, for $\mathbb{H}_-$, in (11.2.10) with (11.2.8), where $\mathbf{\Lambda}^t_{r+}$ is replaced by $\mathbf{E}^t_{r+}$ in the H_μ term and by $\nabla\mathbf{E}^t_{r+}$ in the $H_\varkappa$ term. Following any* one of the lines of argument described in Sect. 11.2 (or indeed by a somewhat different but equivalent route as in Sect. 13.3), we conclude that

$$\psi_m(t) = \frac{1}{2\pi}\int_{-\infty}^{\infty} |\mathbf{p}^t_{\mu-}(\omega)|^2 d\omega + \frac{1}{2\pi}\int_{-\infty}^{\infty} |\mathbf{p}^t_{\varkappa-}(\omega)|^2 d\omega,$$
$$D_m(t) = |\mathbf{K}_\mu(t)|^2 + |\mathbf{K}_\varkappa(t)|^2,$$

where $\mathbf{K}_\mu(t)$ and $\mathbf{K}_\varkappa(t)$ are defined by $(11.2.22)_3$ with the same substitutions as specified above for $\mathbf{p}^t_{\mu\pm}(\omega)$ and $\mathbf{p}^t_{\varkappa\pm}(\omega)$. The result proved in Sect. 11.2.4 can be used to demonstrate that (21.1.19) holds.

21.2 Heat Flux in a Rigid Conductor with Nonlocal Behavior

We now consider quite a different material but one for which the equations have significant similarities to those of Sect. 21.1.

---

* The simplest being that in Remark 11.2.1.

Guyer and Krumhansl [192] (see also [66]) have obtained from kinetic theory, with a suitable approximation, the generalization (3.7.34) of the Cattaneo–Maxwell equation for the heat flux, given here with a slight change of notation:

$$\tau\dot{\mathbf{q}}(\mathbf{x},t)+\mathbf{q}(\mathbf{x},t)=-k\nabla\theta(\mathbf{x},t)+\alpha[\nabla\cdot\nabla\mathbf{q}(\mathbf{x},t)+2\nabla\nabla\cdot\mathbf{q}(\mathbf{x},t)],$$

where τ is the relaxation time, $k>0$ is the heat conduction coefficient, and α is a constant. An argument similar to that used by Guyer and Krumhansl (see also [66]) yields the new equation

$$\tau\dot{\mathbf{q}}(\mathbf{x},t)+\mathbf{q}(\mathbf{x},t)=-h_1\mathbf{g}(\mathbf{x},t)+h_2\nabla\cdot\nabla\mathbf{g}(\mathbf{x},t), \tag{21.2.1}$$

where, besides the effect of $\mathbf{g}$, we also have the contribution of the divergence of its gradient $\nabla\mathbf{g}$. This new constitutive equation is equivalent, in the Gurtin–Pipkin [191] sense, to

$$\mathbf{q}(\mathbf{x},t)=-\int_0^{+\infty}\frac{h_1}{\tau}e^{-\gamma s}\mathbf{g}^t(\mathbf{x},s)ds+\nabla\cdot\int_0^{+\infty}\frac{h_2}{\tau}e^{-\gamma s}\nabla\mathbf{g}^t(\mathbf{x},s)ds,\quad \gamma=\frac{1}{\tau}. \tag{21.2.2}$$

In fact, its time derivative, taking account of $\dot{\mathbf{g}}^t(s)=-\frac{d}{ds}\mathbf{g}^t(s)$ and integrating by parts, can be expressed in the form

$$\begin{aligned}\dot{\mathbf{q}}(\mathbf{x},t)&=\frac{h_1}{\tau}\int_0^{+\infty}e^{-\gamma s}\frac{d}{ds}\mathbf{g}^t(\mathbf{x},s)ds-\nabla\cdot\frac{h_2}{\tau}\int_0^{+\infty}e^{-\gamma s}\frac{d}{ds}\nabla\mathbf{g}^t(\mathbf{x},s)ds\\&=\gamma\left[\int_0^{+\infty}\frac{h_1}{\tau}e^{-\gamma s}\mathbf{g}^t(\mathbf{x},s)ds-\nabla\cdot\int_0^{+\infty}\frac{h_2}{\tau}e^{-\gamma s}\nabla\mathbf{g}^t(\mathbf{x},s)ds\right]\\&\quad-\frac{1}{\tau}[h_1\mathbf{g}(\mathbf{x},t)-h_2\nabla\cdot\nabla\mathbf{g}(\mathbf{x},t)]\\&=\frac{1}{\tau}[-\mathbf{q}(\mathbf{x},t)-h_1\mathbf{g}(\mathbf{x},t)+h_2\nabla\cdot\nabla\mathbf{g}(\mathbf{x},t)],\end{aligned}$$

so that (21.2.1) is satisfied by assuming $\gamma=\frac{1}{\tau}$.

In this section we study a constitutive equation more general [9] than (21.2.2), of the form

$$\mathbf{q}(\mathbf{x},t)=\int_0^{+\infty}K_1'(s)\bar{\mathbf{g}}^t(\mathbf{x},s)ds-\nabla\cdot\int_0^{+\infty}K_2'(s)\nabla\bar{\mathbf{g}}^t(\mathbf{x},s)ds, \tag{21.2.3}$$

where K_1 and $K_2\in L^1(\mathbb{R}^+)\cap Ł^2(\mathbb{R}^+)$ are two smooth positive functions and

$$\bar{\mathbf{g}}^t(\mathbf{x},s)=\int_0^s\mathbf{g}^t(\mathbf{x},\xi)d\xi=\int_{t-s}^t\mathbf{g}(\mathbf{x},\lambda)d\lambda.$$

As proved in Sect. 3.7, using the heat balance law (21.1.3), the second law for nonlocal materials can now be written as (see (3.7.33))

$$\dot{\eta}\geq\mathcal{P}_\eta^i=\frac{h}{\theta}+\frac{1}{\theta^2}\mathbf{q}\cdot\nabla\theta-\nabla\cdot\mathbf{\Phi}, \tag{21.2.4}$$

where $\mathbf{\Phi}$ is a suitable vector depending on the material. Also, since we are dealing with a rigid conductor, the first law (21.1.2) reduces to

$$\dot{e} = h.$$

The inequality (21.2.4) becomes

$$\dot{\psi} \leq -\eta\dot{\theta} - \frac{1}{\theta}\mathbf{q}\cdot\mathbf{g} + \theta\nabla\cdot\mathbf{\Phi}, \tag{21.2.5}$$

on introducing the free energy $\psi = e - \theta\eta$. Following the viewpoint of [102], we consider a linearization in a neighborhood of the ambient temperature θ_0 and the null values of $\bar{\mathbf{g}}^t$ and $\nabla\bar{\mathbf{g}}^t$; therefore, in particular, we suppose that $\theta(x,t) = \theta_0 + \epsilon\theta_1(x,t)$, where $\epsilon \ll 1$ and $\theta_1(x,t)$ is the relative temperature. Then the linearization of (21.2.5) yields the inequality

$$\dot{\psi} \leq -\eta\dot{\theta} - \frac{1}{\theta_0}\mathbf{q}\cdot\mathbf{g} + \theta_0\nabla\cdot\mathbf{\Phi}. \tag{21.2.6}$$

The behavior of a homogeneous and isotropic rigid heat conductor with the constitutive equation (21.2.3) is characterized by states

$$\sigma^T(t) = (\theta, \sigma(t)) = \left(\theta, \bar{\mathbf{g}}^t, \nabla\bar{\mathbf{g}}^t\right)$$

and processes that are piecewise continuous maps $P^T : [0,d) \to \mathbb{R}\times\mathbb{R}^3$, given by

$$P^T(\tau) = (\dot{\theta}_P, P) = (\dot{\theta}_P, \mathbf{g}_P(\tau), \nabla\mathbf{g}_P(\tau)) \qquad \forall\tau\in[0,d),$$

where d is the duration of the process.

Consider a free energy ψ of the form

$$\psi\left(\theta, \bar{\mathbf{g}}^t, \nabla\bar{\mathbf{g}}^t\right) = \psi_1(\theta) + \psi_2\left(\bar{\mathbf{g}}^t, \nabla\bar{\mathbf{g}}^t\right).$$

Then the inequality (21.2.6) can be satisfied if we put

$$\begin{aligned} &\frac{\partial}{\partial\theta}\psi_1(\theta) = -\eta(\theta), \\ &\dot{\psi}_2\left(\bar{\mathbf{g}}^t, \nabla\bar{\mathbf{g}}^t\right) \leq -\frac{1}{\theta_0}\mathbf{q}\cdot\mathbf{g} + \theta_0\nabla\cdot\mathbf{\Phi}. \end{aligned} \tag{21.2.7}$$

In the following, it will be supposed that

$$\theta_0 = 1. \tag{21.2.8}$$

From the linearized constitutive equation (21.2.3), it follows that

$$\begin{aligned} \mathbf{q}(t)\cdot\mathbf{g}(t) = &\int_0^{+\infty} K_1'(s)\bar{\mathbf{g}}^t(s)ds\cdot\mathbf{g}(t) \\ &+ \int_0^{+\infty} K_2'(s)\nabla\bar{\mathbf{g}}^t(s)ds\cdot\nabla\mathbf{g}(t) - \nabla\cdot\left\{\left[\int_0^{+\infty} K_2'(s)\nabla\bar{\mathbf{g}}^t(s)ds\right]\mathbf{g}(t)\right\}. \end{aligned} \tag{21.2.9}$$

Taking

$$\mathbf{\Phi}(t) = -\left[\int_0^{+\infty} K_2'(s)\nabla\bar{\mathbf{g}}^t(s)ds\right]\mathbf{g}(t) \tag{21.2.10}$$

and using (21.2.8), the inequality $(21.2.7)_2$ reduces to

$$\begin{aligned}\dot{\psi}_2(t) &\leq -\mathbf{q}\cdot\mathbf{g} + \nabla\cdot\mathbf{\Phi} \\ &\leq -\int_0^{+\infty} K_1'(s)\bar{\mathbf{g}}^t(s)ds\cdot\mathbf{g}(t) - \int_0^{+\infty} K_2'(s)\nabla\bar{\mathbf{g}}^t(s)ds\cdot\nabla\mathbf{g}(t).\end{aligned} \tag{21.2.11}$$

Introducing *the entropy action*

$$\mathcal{A}(t) = \mathcal{A}(\sigma, P) = -\int_0^{+\infty} K_1'(s)\bar{\mathbf{g}}^t(s)ds\cdot\mathbf{g}(t) - \int_0^{+\infty} K_2'(s)\nabla\bar{\mathbf{g}}^t(s)ds\cdot\nabla\mathbf{g}(t), \tag{21.2.12}$$

the inequality $(21.2.11)_2$ can be rewritten as

$$\dot{\psi}_2\left(\bar{\mathbf{g}}^t, \nabla\bar{\mathbf{g}}^t\right) \leq \mathcal{A}(t).$$

Let us introduce the *internal rate of dissipation* $D_2(\mathbf{x}, t)$, a nonnegative function obeying

$$\dot{\psi}_2(t) + D_2(t) = \mathcal{A}(t) \equiv -\mathbf{q}(t)\cdot\mathbf{g}(t) + \nabla\cdot\mathbf{\Phi}(t), \tag{21.2.13}$$

by virtue of (21.2.9) and (21.2.12).

The *total entropy action* $\mathcal{B}(\sigma, P)$ of the material during the application of a process P of duration d is defined as

$$\mathcal{B}(\sigma, P) = \int_t^{t+d} A(\xi)d\xi.$$

For any cycle (σ, P), we have

$$\mathcal{B}(\sigma, P) \geq 0,$$

where the equality sign occurs if and only if the cycle is reversible. This follows from $(21.2.13)_1$ and is another example of the dissipation principle (4.1.7). It is of course analogous to (21.1.13).

By a simple generalization of the method used to show (9.2.2), we can prove that if $K_{1_p}(\omega)$ and $K_{2_p}(\omega)$ are the half-range Fourier cosine transforms of the kernels $K_1(s)$ and $K_2(s)$, then [9]

$$K_{1_p}(\omega) > 0, \qquad K_{2_p}(\omega) > 0 \quad \forall\omega \in \mathbb{R},$$

under the hypotheses that $K_{1_p}(0) \neq 0$ and $K_{2_p}(0) \neq 0$. Note that K_1 and K_2 correspond to $\mathbf{k}$ in Chap. 9 for purposes of this generalization. Furthermore, we obtain

$$\lim_{\omega\to+\infty} \omega K_{i_s}'(\omega) = -\lim_{\omega\to+\infty} \omega^2 K_{i_p}(\omega) = K_i'(0) \leq 0 \quad (i = 1, 2),$$

but it will be assumed that

$$K_i'(0) < 0 \quad (i = 1, 2).$$

21.2.1 The Graffi–Volterra Free Energy for Nonlocal Rigid Conductors

The Graffi–Volterra free energy can be generalized to our nonsimple heat conductors with the form

$$\psi_{\mathcal{G}}(t) = -\frac{1}{2}\int_0^{+\infty} K_1'(s)\bar{\mathbf{g}}^t(s)\cdot\bar{\mathbf{g}}^t(s)ds - \frac{1}{2}\int_0^{+\infty} K_2'(s)\nabla\bar{\mathbf{g}}^t(s)\cdot\nabla\bar{\mathbf{g}}^t(s)ds, \quad (21.2.14)$$

which is nonnegative if we assume that

$$K_i'(s) < 0 \quad \forall s \in \mathbb{R}^+ \ (i = 1, 2). \quad (21.2.15)$$

Other constraints will be imposed on these kernel functions, namely

$$K_i''(s) \geq 0 \quad \forall s \in \mathbb{R}^+ \ (i = 1, 2), \quad (21.2.16)$$

to ensure a nonnegative rate of dissipation. We note the application of the observation after (21.1.18) in this context. Taking account of

$$\frac{d}{dt}\bar{\mathbf{g}}^t(s) = \mathbf{g}(t) - \frac{d}{ds}\bar{\mathbf{g}}^t(s)$$

and the analogous relation for $\frac{d}{dt}\nabla\bar{\mathbf{g}}^t(s)$, the derivative with respect to time of (21.2.14) can be expressed as

$$\begin{aligned}
\dot{\psi}_{\mathcal{G}}(t) &= -\int_0^{+\infty} K_1'(s)\bar{\mathbf{g}}^t(s)\cdot\frac{d}{dt}\bar{\mathbf{g}}^t(s)ds - \int_0^{+\infty} K_2'(s)\nabla\bar{\mathbf{g}}^t(s)\cdot\frac{d}{dt}\nabla\bar{\mathbf{g}}^t(s)ds \\
&= -\int_0^{+\infty} K_1'(s)\bar{\mathbf{g}}^t(s)ds\cdot\mathbf{g}(t) - \int_0^{+\infty} K_2'(s)\nabla\bar{\mathbf{g}}^t(s)ds\cdot\nabla\mathbf{g}(t) \\
&\quad -\frac{1}{2}\int_0^{+\infty} K_1''(s)\left[\bar{\mathbf{g}}^t(s)\right]^2 ds - \frac{1}{2}\int_0^{+\infty} K_2''(s)\left[\nabla\bar{\mathbf{g}}^t(s)\right]^2 ds \\
&= \mathcal{A}(t) - D_{\mathcal{G}}(t),
\end{aligned} \quad (21.2.17)$$

with the aid of an integration by parts and (21.2.12). The last form confirms $(21.2.13)_1$, provided we take

$$D_{\mathcal{G}}(t) = \frac{1}{2}\int_0^{+\infty} K_1''(s)\left[\bar{\mathbf{g}}^t(s)\right]^2 ds + \frac{1}{2}\int_0^{+\infty} K_2''(s)\left[\nabla\bar{\mathbf{g}}^t(s)\right]^2 ds \geq 0.$$

This is the nonnegative internal rate of dissipation corresponding to the Graffi–Volterra functional.

We can derive from $(21.2.17)_3$ the relation $(21.2.11)_1$, or $(21.2.13)_2$, in terms of the extra flux, since

$$\begin{aligned}
\dot{\psi}_{\mathcal{G}}(t) &= -\left[\int_0^{+\infty} K_1'(s)\bar{\mathbf{g}}^t(s)ds - \nabla\cdot\int_0^{+\infty} K_2'(s)\nabla\bar{\mathbf{g}}^t(s)ds\right]\cdot\mathbf{g}(t) \\
&\quad -\frac{1}{2}\int_0^{+\infty} K_1''(s)\left[\bar{\mathbf{g}}^t(s)\right]^2 ds - \frac{1}{2}\int_0^{+\infty} K_2''(s)\left[\nabla\bar{\mathbf{g}}^t(s)\right]^2 ds \\
&\quad +\nabla\cdot\left\{\left[-\int_0^{+\infty} K_2'(s)\nabla\bar{\mathbf{g}}^t(s)ds\right]\mathbf{g}(t)\right\} \\
&= -\mathbf{q}(t)\cdot\mathbf{g}(t) - D_{\mathcal{G}}(t) + \nabla\cdot\mathbf{\Phi}(t) \leq -\mathbf{q}(t)\cdot\mathbf{g}(t) + \nabla\cdot\mathbf{\Phi}(t),
\end{aligned}$$

where (21.2.10) has been invoked.

21.2.2 A Nonlocal Free Energy in Terms of the Minimal State

We now consider the analogy, for a rigid conductor, of the free energy $\psi_{\mathcal{F}}$, discussed in Sect. 21.1.2. Let us introduce the quantities

$$\mathbf{I}^t(\tau, \bar{\mathbf{g}}^t) = \int_0^{+\infty} K_1'(\tau + \eta)\bar{\mathbf{g}}^t(\eta)d\eta,$$

$$\mathfrak{I}^t(\tau, \nabla\bar{\mathbf{g}}^t) = \int_0^{+\infty} K_2'(\tau + \eta)\nabla\bar{\mathbf{g}}^t(\eta)d\eta,$$

which define a minimal state. In particular, we have

$$\mathbf{I}^t(0, \bar{\mathbf{g}}^t) = \int_0^{+\infty} K_1'(\eta)\bar{\mathbf{g}}^t(\eta)d\eta, \qquad \mathfrak{I}^t(0, \nabla\bar{\mathbf{g}}^t) = \int_0^{+\infty} K_2'(\eta)\nabla\bar{\mathbf{g}}^t(\eta)d\eta.$$

Now, denoting by $\mathbf{I}^t_{(1)}$ and $\mathfrak{I}^t_{(1)}$ the derivatives with respect to τ of $\mathbf{I}^t$ and $\mathfrak{I}^t$, let us consider the functional

$$\psi_{\mathcal{F}}(t) = -\frac{1}{2}\int_0^{+\infty} \frac{1}{K_1'(\tau)}\left|\mathbf{I}^t_{(1)}(\tau, \bar{\mathbf{g}}^t)\right|^2 d\tau - \frac{1}{2}\int_0^{+\infty} \frac{1}{K_2'(\tau)}\left|\mathfrak{I}^t_{(1)}(\tau, \nabla\bar{\mathbf{g}}^t)\right|^2 d\tau,$$

under the hypotheses that $K_i'(s)$ and $K_i''(s)(i = 1, 2)$ satisfy (21.2.15) and (21.2.16). The absolute value squared notation is analogous to that in Sect. 21.1.2. In order to verify that this expression gives a possible free energy $\psi_2(\bar{\mathbf{g}}^t, \nabla\bar{\mathbf{g}}^t)$ for our body, we evaluate its derivative with respect to t, which can be put in the following form:

$$\begin{aligned}
\dot{\psi}_{\mathcal{F}}(t) = &-\left[\int_0^{+\infty} K_1'(\eta)\bar{\mathbf{g}}^t(\eta)d\eta \cdot \mathbf{g}(t) + \int_0^{+\infty} K_2'(\eta)\nabla\bar{\mathbf{g}}^t(\eta)d\eta \cdot \nabla\mathbf{g}(t)\right] \\
&- \frac{1}{2}\int_0^{+\infty} \frac{K_1''(\tau)}{[K_1'(\tau)]^2}\left|\mathbf{I}^t_{(1)}(\tau, \bar{\mathbf{g}}^t)\right|^2 d\tau - \frac{1}{2}\int_0^{+\infty} \frac{K_2''(\tau)}{\left[K_2'(\tau)\right]^2}\left|\mathfrak{I}^t_{(1)}(\tau, \nabla\bar{\mathbf{g}}^t)\right|^2 d\tau \\
&+ \frac{1}{2}\frac{1}{K_1'(0)}\left|\mathbf{I}^t_{(1)}(0, \bar{\mathbf{g}}^t)\right|^2 + \frac{1}{2}\frac{1}{K_2'(0)}\left|\mathfrak{I}^t_{(1)}(0, \nabla\bar{\mathbf{g}}^t)\right|^2.
\end{aligned}$$

Referring to (21.2.12), we conclude that

$$\dot{\psi}_{\mathcal{F}}(t) \le \mathcal{A}(t),$$

so that $\dot{\psi}_{\mathcal{F}}(t)$ is indeed a valid free energy.

Remark 21.2.1. The other free energies considered in Chaps. 10 and 11 and indeed those introduced in the later chapters can also be generalized in a similar manner to these nonsimple materials.

For the heat flux problem, an explicit formula for the minimum free energy is given in [9]. Alternatively, the argument summarized in Sect. 21.1.3 can be applied to obtain a similar formula.

21.3 Free Energies in a General Nonlocal Theory of a Material with Memory

A general theory of nonlocal materials, with linear constitutive equations and memory effects, is now developed within a thermodynamic framework. These developments are based on certain results in [15]. In contrast to various earlier chapters, the full tensor version is presented. The formulations in Sects. 21.1 and 21.2 are generalized to apply to a range of field theories with linear memory constitutive equations.

Referring to the two approaches described at the beginning of this chapter, we attempt to consider both in parallel. It is shown that these approaches are consistent with an hypothesis on the form of a generalized work function, which will be introduced below. This hypothesis is the basis of the general theory presented here.

21.3.1 Derivation of the Field Equations

The approach adopted here is a generalization to nonlocal materials of the developments of Sect. 5.1, though where the distinction between material and spatial coordinates (see (1.2.3)) is neglected. This distinction is important in nonlinear mechanics, which is considered in [15].

Let us begin by stating the First and Second Laws of Thermodynamics. The internal energy per unit mass and the entropy per unit mass at $(\mathbf{x}, t)$, both scalar quantities, are denoted, respectively, by $e(\mathbf{x}, t)$ and $\eta(\mathbf{x}, t)$, as in Sect. 5.1. Similarly for the local absolute temperature, $\theta(\mathbf{x}, t) \in \mathbb{R}^+$, the heat flux vector $\mathbf{q}(\mathbf{x}, t) \in \mathbb{R}^3$ and the coldness $\kappa(\mathbf{x}, t)$, given by $1/\theta > 0$. We define $\mathbf{g}, \mathbf{d} \in \mathbb{R}^3$ as in Sect. 5.1.1, by

$$\mathbf{g} = \nabla\theta, \qquad \mathbf{d} = \nabla\kappa = -\frac{1}{\theta^2}\mathbf{g}.$$

The energy balance equation or First Law of Thermodynamics has the form

$$p - \rho\dot{e} - \nabla \cdot \mathbf{q} - \nabla \cdot \boldsymbol{\Phi}_e + \rho r = 0, \tag{21.3.1}$$

where p is a power term related to the case of simple materials and $\boldsymbol{\Phi}_e \in \mathbb{R}^3$ is an extra flux, the *interstitial work flux*, associated with the internal energy and the first law to take account of nonlocal behavior [97, 252]. This generalizes (5.1.2). The quantity ρ is the mass density, while r is the external radiation absorbed per unit time, per unit mass at $(\mathbf{x}, t)$.

In the alternative formulation [127], the first law for thermomechanical systems can be written as

$$\rho\dot{e} = \mathcal{P}^i_m + \mathcal{P}^i_h, \qquad \mathcal{P}^i_h = -\nabla \cdot \mathbf{q} + \rho r, \tag{21.3.2}$$

where $\mathcal{P}^i_m$ is the internal mechanical power which, for simple materials, reduces to p. The correspondence between the two formulations is expressed by $\mathcal{P}^i_m = p - \nabla \cdot \boldsymbol{\Phi}_e$.

The second law can be written as

$$D = \dot{\eta} + \frac{1}{\rho}\nabla \cdot \left(\frac{\mathbf{q}}{\theta}\right) + \frac{1}{\rho}\nabla \cdot \boldsymbol{\Phi}_\eta - \kappa r \geq 0, \tag{21.3.3}$$

where the quantity $D(\mathbf{x}, t)$ is the total rate of entropy production per unit mass, while Φ_η is an extra flux of entropy [268] associated with nonlocal behavior.

The Helmholtz free energy per unit mass is defined by

$$\psi = e - \theta\eta.$$

In terms of this quantity, we can write (21.3.1) as

$$\dot{\psi} + \theta D = \frac{p}{\rho} - \eta\dot{\theta} - \frac{1}{\rho}\nabla \cdot \Phi_e + \frac{\theta}{\rho}\nabla \cdot \Phi_\eta - \frac{1}{\rho\theta}\mathbf{q} \cdot \mathbf{g}. \tag{21.3.4}$$

In the alternative formulation, we write (21.3.3) in the form [127]

$$\dot{\eta} - \frac{1}{\rho}\mathcal{A}^i_{en} = D \geq 0,$$

where $\mathcal{A}^i_{en}$ is the *internal entropy action* given by

$$\mathcal{A}^i_{en} = -\nabla \cdot \frac{\mathbf{q}}{\theta} + \rho\frac{r}{\theta} - \nabla \cdot \Phi_\eta = \mathcal{A}^e_{en}.$$

The latter quantity is the *external entropy action* [127]. Instead of (21.3.4), we have

$$\dot{\psi} + \theta D = \frac{1}{\rho}\left(\mathcal{P}^i - \theta\mathcal{A}^i_{en}\right) - \eta\dot{\theta}, \qquad \mathcal{P}^i = \mathcal{P}^i_m + \mathcal{P}^i_h,$$

where $\mathcal{P}^i_m$ and $\mathcal{P}^i_h$ are introduced in (21.3.2). For a general framework explaining actions in systems, see [75].

Further discussion of thermodynamics is given in [15], essentially a generalization of developments in Sect. 5.1, involving an alternative choice of free energy. We now, for simplicity, consider an isothermal theory, neglecting the terms $\eta\dot{\theta}$ and that proportional to $\mathbf{q} \cdot \mathbf{g}$ in (21.3.4), so that the alternative choice of free energy is not relevant. The average temperature and density are taken to be unity.

A convenient compact notation is now introduced. Let the field variables be $\boldsymbol{\Lambda} : \mathbb{R} \mapsto \Gamma$, $\boldsymbol{\Sigma} : \mathbb{R} \mapsto \Gamma$, where Γ is a given finite-dimensional vector space. We will treat $\boldsymbol{\Lambda}$ as the independent and $\boldsymbol{\Sigma}$ as the dependent field quantity. One can write (21.3.4)as[†]

$$\dot{\psi} + D = \dot{W}(t) = \boldsymbol{\Sigma} \cdot \dot{\boldsymbol{\Lambda}} - \nabla \cdot \Phi, \tag{21.3.5}$$

where p is equal to $\boldsymbol{\Sigma} \cdot \dot{\boldsymbol{\Lambda}}$, which has a form typical of mechanical systems and, with some modifications, electromagnetic systems. If the thermal variables are reintroduced in (21.3.5), heat flow systems can also be changed into this form.

We now seek to develop a nonlocal theory with linear constitutive equations which generalizes the results of Sects. 21.1 and 21.2 (also [9, 14]), so that they apply to a wide range of linear nonsimple theories.

[†] Note the two quite different uses of the dot product in (21.3.5). The first is in Γ, while the second is in $\mathbb{R}^3$, indicating a divergence.

The introduction of nonsimple behavior can be achieved (see [238, 313], for example) by adding n more independent field quantities that consist of gradient operators acting on $\mathbf{\Lambda}$.

The new independent field variables are $\nabla\mathbf{\Lambda}$, $\nabla^2\mathbf{\Lambda}$, $\nabla^3\mathbf{\Lambda}$,..., $\nabla^n\mathbf{\Lambda}$. The operators ∇^r, $r = 2, 3, \ldots, n$, are completely symmetric. The number of independent quantities among their 3^r components on a given basis is $(r + 1)(r + 2)/2$ (see [131], page 39). The new histories are denoted by

$$\nabla\mathbf{\Lambda}^t(\mathbf{x}, s), \quad \nabla^2\mathbf{\Lambda}^t(\mathbf{x}, s), \ldots, \quad \nabla^n\mathbf{\Lambda}^t(\mathbf{x}, s), \quad s \in \mathbb{R}^+,$$

where the standard notation for the history

$$\mathbf{\Lambda}^t(\mathbf{x}, s) = \mathbf{\Lambda}(\mathbf{x}, t - s) \tag{21.3.6}$$

has been used.

The Fundamental Hypotheses

To completely determine the first and second laws as given by (21.3.5) and (21.3.3), one must specify $\dot{W}$, ψ, and D. Indeed, it emerges that if we specify $\dot{W}$ and ψ, then the requirement of consistency with the second law yields the form of D and indeed the detailed form of the constitutive equations for the material. We first give the form of $\dot{W}$ and show how it is consistent in general with (21.3.5).

For a wide class of nonsimple materials, the generalized rate of work can be put in the form

$$\dot{W} = \mathbf{\Sigma}_0 \cdot \dot{\mathbf{\Lambda}} + \mathbf{\Sigma}_1 \cdot \nabla\dot{\mathbf{\Lambda}} + \mathbf{\Sigma}_2 \cdot \nabla^2\dot{\mathbf{\Lambda}} + \cdots + \mathbf{\Sigma}_n \cdot \nabla^n\dot{\mathbf{\Lambda}}, \tag{21.3.7}$$

where the tensor functions $\mathbf{\Sigma}_r$, $r = 0, 1, 2, \ldots, n$, are fundamental to the theory. In particular, for simple mechanical materials, $\mathbf{\Sigma}_r = \mathbf{0}$, $r = 1, 2, \ldots, n$, and $\mathbf{\Sigma}_0$ is proportional to the stress tensor‡. The quantity $\mathbf{\Sigma}_r$ in a given basis will have the same number of subscripts with the same symmetry properties as $\nabla^r\dot{\mathbf{\Lambda}}$ and each element of this tensor quantity belongs to Γ. The inner product notation $\mathbf{\Sigma}_r \cdot \nabla^r\dot{\mathbf{\Lambda}}$ means that each ∇ is connected to a specific dimension of $\mathbf{\Sigma}_r$ in the sense that, on a particular basis, the components of a given ∇ are summed over a particular subscript of $\mathbf{\Sigma}_r$ and each term in the summation is a scalar product $\Gamma \times \Gamma \mapsto \mathbb{R}$. Thus, on a particular basis, $\mathbf{\Sigma}_2$, for example, has components which we denote by $\Sigma_{(2)ij} = \Sigma_{(2)ji}$, $i, j = 1, 2, 3$, and ∇^2 will be represented as $\dfrac{\partial^2}{\partial x_i \partial x_j} = \partial_i\partial_j$, giving that $\mathbf{\Sigma}_2 \cdot \nabla^2\dot{\mathbf{\Lambda}} = \Sigma_{(2)ij} \cdot \partial_i\partial_j\dot{\mathbf{\Lambda}} \in \mathbb{R}$. The dot product in this last expression is the scalar product $\Gamma \times \Gamma \mapsto \mathbb{R}$ between $\Sigma_{(2)ij}$ and $\dot{\mathbf{\Lambda}}$.

Proposition 21.3.1. *For $\dot{W}$ given by the right-hand side of* (21.3.5)$_2$, *where $\nabla \cdot \Phi$ is explicitly a divergence of a vector quantity Φ, then $\mathbf{\Sigma}$ in* (21.3.5) *is uniquely determined to be*

$$\mathbf{\Sigma} = \mathbf{\Sigma}_0 - \nabla \cdot \mathbf{\Sigma}_1 + \nabla^2 \cdot \mathbf{\Sigma}_2 + \cdots + (-1)^n\nabla^n \cdot \mathbf{\Sigma}_n, \tag{21.3.8}$$

where the notation $\nabla^r \cdot \mathbf{\Sigma}_r \in \Gamma$ means that each ∇ forms a divergence with respect to its corresponding tensor index of $\mathbf{\Sigma}_r$ for $r = 1, 2, \ldots, n$.

‡ The quantity $\mathbf{0}$ will be used for the zero in any vector space with dimensions greater than unity.

Proof. We seek to extract a scalar term $\mathbf{G}\cdot\dot{\mathbf{\Lambda}}$, where $\mathbf{G} \in \Gamma$ from the quantity $\mathbf{\Sigma}_r\cdot\nabla^r\dot{\mathbf{\Lambda}}$, for example. Given the available differential operators and field quantities, $\mathbf{G}$ must have the form

$$\mathbf{G} = \alpha\nabla^r \cdot \mathbf{\Sigma}_r, \tag{21.3.9}$$

where $\alpha \in \mathbb{R}$ is a constant. Indeed, there is no finite manipulation that could yield an irrational number in this role, so that it can be assumed to be a rational number.

The basic tool of transformation of expressions such as (21.3.7) is the product rule of calculus[§]

$$(\nabla^p \cdot \mathbf{\Sigma}_r)\cdot(\nabla^q\dot{\mathbf{\Lambda}}) = \nabla\cdot\left[(\nabla^p \cdot \mathbf{\Sigma}_r)\cdot(\nabla^{q-1}\dot{\mathbf{\Lambda}})\right] - (\nabla^{p+1}\cdot\mathbf{\Sigma}_r)\cdot(\nabla^{q-1}\dot{\mathbf{\Lambda}}), \tag{21.3.10}$$

or the inverse of this, transferring a ∇ operator from $\mathbf{\Sigma}_r$ to $\dot{\mathbf{\Lambda}}$. The latter transformation is not of interest here. Consider the right-hand side of (21.3.10). The divergence term is included in the overall divergence term, and we can reapply the operation to the second term, which approaches the desired form (21.3.9). By systematically proceeding in this manner, we uniquely obtain the relation

$$\begin{aligned}\mathbf{\Sigma}_r\cdot\nabla^r\dot{\mathbf{\Lambda}} &= (-1)^r(\nabla^r\cdot\mathbf{\Sigma}_r)\cdot\dot{\mathbf{\Lambda}} + \nabla\cdot\Phi_r,\\ \Phi_r &= \mathbf{\Sigma}_r\cdot(\nabla^{r-1}\dot{\mathbf{\Lambda}}) - (\nabla\cdot\mathbf{\Sigma}_r)\cdot(\nabla^{r-2}\dot{\mathbf{\Lambda}}) + \dots\\ &\quad + (-1)^s(\nabla^s\cdot\mathbf{\Sigma}_r)\cdot(\nabla^{r-s-1}\dot{\mathbf{\Lambda}}) + \dots + (-1)^{r-1}(\nabla^{r-1}\cdot\mathbf{\Sigma}_r)\cdot\dot{\mathbf{\Lambda}}.\end{aligned}$$

Thus, we have

$$\Phi' = \Phi_1 + \Phi_2 + \dots + \Phi_n, \tag{21.3.11}$$

and (21.3.5) applies, where $\mathbf{\Sigma}$ is given by (21.3.8) and $\nabla\cdot\Phi' = \nabla\cdot\Phi$. □

The simplest assumption regarding Φ is to take it equal to Φ' as given by (21.3.11).

Remark 21.3.1. The quantity $\mathbf{\Sigma}(t)$ is the dependent variable with an immediate physical interpretation. For example, in mechanics, it is proportional to the physical stress tensor. The quantities $\mathbf{\Sigma}_i(t)$, $i = 0, 1, 2, \dots$, are also physically measurable quantities and expressions for them will be given below.

The fundamental constitutive assumption, which characterizes the material, is that the free energy ψ depends in a specified way on the independent field variables at the current time t and on their histories. Thus, we write

$$\begin{aligned}\psi(t) =&\widetilde{\psi}(\mathbf{x}, \mathbf{\Lambda}^t(\mathbf{x},\cdot), \mathbf{\Lambda}(\mathbf{x},t), \nabla\mathbf{\Lambda}^t(\mathbf{x},\cdot), \nabla\mathbf{\Lambda}(\mathbf{x},t),\\ &\nabla^2\mathbf{\Lambda}^t(\mathbf{x},\cdot), \nabla^2\mathbf{\Lambda}(\mathbf{x},t), \dots, \nabla^n\mathbf{\Lambda}^t(\mathbf{x},\cdot), \nabla^n\mathbf{\Lambda}(\mathbf{x},t)).\end{aligned} \tag{21.3.12}$$

[§] For the expression in the first bracket, the dot product implies that each ∇ in ∇^p, on a given basis, is summed over its corresponding index in $\mathbf{\Sigma}_r$. The dot product between the two terms in brackets means that q of the remaining dimensions in $\mathbf{\Sigma}_r$ are summed over the indices in ∇^q and a scalar product $\Gamma\times\Gamma \mapsto \mathbb{R}$ occurs for each multiplication of objects in Γ. Thus, for example, using the notation introduced earlier, we have that $(\nabla^2\cdot\mathbf{\Sigma}_5)\cdot(\nabla^2\dot{\mathbf{\Lambda}}) \in \mathbb{R}^3$ has components $\partial_i\partial_j\Sigma_{(5)ijklm}\cdot\partial_k\partial_l\dot{\mathbf{\Lambda}}$ in a given basis, where the remaining dot product refers to the scalar product $\Gamma\times\Gamma\mapsto\mathbb{R}$ between $\Sigma_{(5)ijklm}$ and $\dot{\mathbf{\Lambda}}$.

The dependence on $\mathbf{x}$, over and above that in the independent field variables, allows the possibility of spatial inhomogeneity. We will henceforth mainly omit all space variables.

It is assumed that $\widetilde{\psi}$ is differentiable with respect to the independent variables at the current time and Fréchet differentiable with respect to the histories within a suitable Hilbert space $\mathcal{H}$ (fading memory principle; see Sect. 5.1 and also [67, 71]). The time derivative $\dot{\psi}$ is given by

$$\dot{\psi} = \partial_t \psi + \partial_h \psi, \tag{21.3.13}$$

where ∂_t indicates differentiation with respect to field variables at the current time t, while $\partial_h \psi$ indicates differentiation with respect to the remainder of the time dependence, which will consist of Fréchet differentials of ψ at the histories of the independent field quantities, in the direction of the time derivatives of these histories.

The quantity $\partial_t \psi$ is given by

$$\begin{aligned}\partial_t \psi = {} & \frac{\partial \psi}{\partial \mathbf{\Lambda}(t)} \cdot \dot{\mathbf{\Lambda}}(t) + \frac{\partial \psi}{\partial \nabla \mathbf{\Lambda}(t)} \cdot \nabla \dot{\mathbf{\Lambda}}(t) \\ & + \frac{\partial \psi}{\partial \nabla^2 \mathbf{\Lambda}(t)} \cdot \nabla^2 \dot{\mathbf{\Lambda}}(t) + \cdots + \frac{\partial \psi}{\partial \nabla^n \mathbf{\Lambda}(t)} \cdot \nabla^n \dot{\mathbf{\Lambda}}(t).\end{aligned} \tag{21.3.14}$$

The derivatives with respect to field quantities are assumed to be continuous in their arguments. Consider the relation $(21.3.5)_1$. Using (21.3.13) with (21.3.14) to substitute for $\dot{\psi}$ and (21.3.7) for $\dot{W}(t)$, we generalize a standard argument [71]; see also (5.1.18), based on the indeterminacy of the signs of $\dot{\mathbf{\Lambda}}(t)$, $\nabla \dot{\mathbf{\Lambda}}(t)$, $\nabla^2 \dot{\mathbf{\Lambda}}(t), \ldots, \nabla^n \dot{\mathbf{\Lambda}}(t)$. This yields the detailed constitutive relations

$$\begin{aligned}\mathbf{\Sigma}_0(t) &= \frac{\partial \psi}{\partial \mathbf{\Lambda}(t)}, & \mathbf{\Sigma}_1(t) &= \frac{\partial \psi}{\partial \nabla \mathbf{\Lambda}(t)}, \\ \mathbf{\Sigma}_2(t) &= \frac{\partial \psi}{\partial \nabla^2 \mathbf{\Lambda}(t)}, \ldots, & \mathbf{\Sigma}_n(t) &= \frac{\partial \psi}{\partial \nabla^n \mathbf{\Lambda}(t)},\end{aligned} \tag{21.3.15}$$

together with an expression for the rate of dissipation

$$D = -\partial_h \psi \geq 0.$$

Thus, the Fréchet differentials, which crucially involve the history dependence of ψ, yield the rate of dissipation of the material. Using (21.3.15) in (21.3.14), we see from (21.3.7) that

$$\partial_t \psi(t) = \dot{W}(t). \tag{21.3.16}$$

An immediate consequence of (21.3.15) and (21.3.8) is that

$$\begin{aligned}\mathbf{\Sigma}(t) = {} & \frac{\partial \psi}{\partial \mathbf{\Lambda}(t)} - \nabla \cdot \frac{\partial \psi}{\partial \nabla \mathbf{\Lambda}(t)} + \nabla^2 \cdot \frac{\partial \psi}{\partial \nabla^2 \mathbf{\Lambda}(t)} + \ldots \\ & + (-1)^n \nabla^n \cdot \frac{\partial \psi}{\partial \nabla^n \mathbf{\Lambda}(t)} = \frac{\delta \psi}{\delta \mathbf{\Lambda}},\end{aligned} \tag{21.3.17}$$

where $\dfrac{\delta\psi}{\delta\mathbf{\Lambda}}$ is a variational derivative, as in the calculus of variations, with respect to the independent field variables at the current time, $\nabla^r\mathbf{\Lambda}(t)$, $r = 0, 1, 2, \ldots, n$.

Let us introduce another compact notation on an encompassing vector space $\mathcal{B}$ containing all the field variables. We define the independent variable

$$\mathbf{C}(t) = (\mathbf{\Lambda}(t), \nabla\mathbf{\Lambda}(t), \nabla^2\mathbf{\Lambda}(t), \ldots, \nabla^n\mathbf{\Lambda}(t)) \in \mathcal{B} \tag{21.3.18}$$

and the dependent variable

$$\mathbf{D}(t) = (\mathbf{\Sigma}_0(t), \mathbf{\Sigma}_1(t), \mathbf{\Sigma}_2(t), \ldots, \mathbf{\Sigma}_n(t)) \in \mathcal{B}. \tag{21.3.19}$$

Relations (21.3.7) and (21.3.16) yield that

$$\partial_t\psi = \dot{W}(t) = \mathbf{D}(t) \odot \dot{\mathbf{C}}(t), \tag{21.3.20}$$

where the symbol $\odot$ is the dot product in $\mathcal{B}$, implying a scalar product of the individual components in their respective vector spaces. Let us denote the dimension of $\mathcal{B}$ by m. The sequence of operators ∇ and ∇_a are defined as

$$\begin{aligned} \nabla &= (1, \nabla, \nabla^2, \ldots, \nabla^n), \\ \nabla_a &= (1, -\nabla, +\nabla^2, \ldots, (-1)^n\nabla^n). \end{aligned}$$

Then (21.3.18) can be written symbolically as

$$\mathbf{C}(t) = \nabla\mathbf{\Lambda}(t).$$

Also, from (21.3.8) and (21.3.17),

$$\mathbf{\Sigma}(t) = \frac{\delta\psi}{\delta\mathbf{\Lambda}(t)} = \nabla_a \odot \mathbf{D}(t). \tag{21.3.21}$$

We can write the dependence of ψ on the independent field variables given by (21.3.12) as

$$\psi = \widetilde{\psi}(\mathbf{C}^t, \mathbf{C}(t)),$$

where $\mathbf{C}^t$ is the past history, given by (see (21.3.6))

$$\mathbf{C}^t(s) = \mathbf{C}(t-s), \quad s \in \mathbb{R}^{++}.$$

The relative history is defined as

$$\mathbf{C}^t_r(s) = \mathbf{C}^t(s) - \mathbf{C}(t).$$

It will be assumed that

$$\lim_{s\to\infty} \mathbf{C}^t(s) = \mathbf{0} \tag{21.3.22}$$

Note that

$$\begin{aligned} \frac{\partial}{\partial t}\mathbf{C}^t(s) &= -\frac{\partial}{\partial s}\mathbf{C}^t(s) = -\frac{\partial}{\partial s}\mathbf{C}^t_r(s), \\ \frac{\partial}{\partial t}\mathbf{C}^t_r(s) &= \frac{\partial}{\partial t}\mathbf{C}^t(s) - \dot{\mathbf{C}}(t). \end{aligned}$$

From $(21.3.5)_1$ and (21.3.20), we can write the first law as

$$\dot{\psi} + D = \mathbf{D} \odot \dot{\mathbf{C}}. \tag{21.3.23}$$

By virtue of the requirement that D be nonnegative (see (21.3.3)), it will always be true that

$$\dot{\psi} \leq \dot{W} = \mathbf{D} \odot \dot{\mathbf{C}}. \tag{21.3.24}$$

From (21.3.23), we see that if $\dot{\mathbf{C}}(s) = \mathbf{0}$, $s \in [t, \infty)$ (which is true if $\dot{\mathbf{\Lambda}}(s) = \mathbf{0}$, $s \in [t, \infty)$), then $\dot{\psi}(s) \leq 0 \ \forall \ s \geq t$. Thus, over this interval, $\psi(s)$ is nonincreasing if the independent field variables are constant in time. It follows that [67]

$$\widetilde{\psi}(\mathbf{C}^t, \mathbf{C}(t)) = \psi(t) \geq \phi(t) = \widetilde{\phi}(\mathbf{C}(t)), \tag{21.3.25}$$

where $\widetilde{\phi}(\mathbf{C}(t))$ is the equilibrium form of ψ, or this functional for constant histories, specifically those given by $\mathbf{C}^t(s) = \mathbf{C}(t)$, $s \in \mathbb{R}^{++}$.

Using (21.3.19), we can write (21.3.15) in compact form as

$$\frac{\partial}{\partial \mathbf{C}(t)} \widetilde{\psi}(\mathbf{C}^t, \mathbf{C}(t)) = \tilde{\mathbf{D}}(\mathbf{C}^t, \mathbf{C}(t)) = \mathbf{D}(t). \tag{21.3.26}$$

Requirements for a Free Energy

Let us gather together the properties which must be associated with a functional $\widetilde{\psi}$ if it is to be a free energy for a material with nonlocal behavior.

P1 We have

$$\mathbf{D}(t) = \frac{\partial \psi}{\partial \mathbf{C}(t)},$$

which is (21.3.26). Then, from (21.3.17),

$$\frac{\delta \psi}{\delta \mathbf{\Lambda}(t)} = \mathbf{\Sigma}(t).$$

P2 Let $\mathbf{C}^\dagger$ be a static history, equal to $\mathbf{C}(t)$ at the current and all past times. Then,

$$\psi(\mathbf{C}^\dagger, \mathbf{C}(t)) = \widetilde{\phi}(\mathbf{C}(t)),$$

which is the definition of $\widetilde{\phi}(\mathbf{C}(t)) = \phi(t)$. Also,

$$\widetilde{\psi}(\mathbf{C}^t, \mathbf{C}(t)) \geq \widetilde{\phi}(\mathbf{C}(t)),$$

which is (21.3.25).

P3

$$\frac{d}{dt} \widetilde{\psi}(\mathbf{C}^t, \mathbf{C}(t)) \leq \mathbf{D}(t) \odot \dot{\mathbf{C}}(t),$$

which is the second law. This is (21.3.24).

These may be referred to as the nonlocal Graffi conditions. The extra condition P4, introduced by (18.2.1), also applies to nonlocal materials.

A substantial advantage of the compact notation, which emerges below, is that it exposes the fact that the nonlocal theory is closely analogous to the local tensor case. This applies to all aspects of the theory, including procedures for constructing free energy functionals. Note in particular that the nonlocal Graffi conditions are identical to the local conditions given in Sect. 5.1.1, except for P1 which connects the theory to the quantity $\mathbf{\Sigma}(t)$.

21.3.2 A Nonlocal Quadratic Model for Free Energies

The simplest choice of ψ is obtained by means of a functional Taylor expansion of $\widetilde{\psi}(\mathbf{C}^t, \mathbf{C}(t))$, stopping at the quadratic term. This procedure is formally identical to that for simple materials in Sect. 7.1. We put

$$\psi(t) = \widetilde{\phi}(\mathbf{C}(t)) + \frac{1}{2}\int_0^\infty \int_0^\infty \mathbf{C}_r^t(s) \odot \mathbb{M}(s,u)\mathbf{C}_r^t(u)dsdu. \tag{21.3.27}$$

The operator $\mathbb{M} \in Lin(\mathcal{B})$ is at least positive semidefinite in the sense that the function $\mathbb{M}(\cdot,\cdot)$ is such that the integral is nonnegative. It is assumed that $\mathbb{M}(\cdot,\cdot) \in L^1(\mathbb{R}^+ \times \mathbb{R}^+)$ and, in particular,

$$\lim_{s\to\infty} \mathbb{M}(s,u) = \mathbf{0}, \quad u \in \mathbb{R}^+.$$

A similar relation holds for the limit of large u at fixed s. There is no loss of generality in taking

$$\mathbb{M}^\top(s,u) = \mathbb{M}(u,s), \tag{21.3.28}$$

where the transpose refers to $Lin(\mathcal{B})$. Any space dependence in $\mathbb{M}$ is neglected, so that we are dealing with a homogeneous material.

Defining $\mathbb{L}(s,u)$ as in (7.1.5), we write (21.3.27) as

$$\begin{aligned}
\psi(t) &= \widetilde{\phi}(\mathbf{C}(t)) + \frac{1}{2}\int_0^\infty \int_0^\infty \mathbf{C}_r^t(s) \odot \mathbb{L}_{12}(s,u)\mathbf{C}_r^t(u)dsdu \\
&= \frac{1}{2}\int_0^\infty \int_0^\infty \dot{\mathbf{C}}^t(s) \odot \mathbb{L}(s,u)\dot{\mathbf{C}}^t(u)dsdu, \\
\dot{\mathbf{C}}^t(u) &= \frac{\partial}{\partial t}\mathbf{C}^t(u) = -\frac{\partial}{\partial u}\mathbf{C}^t(u) = -\frac{\partial}{\partial u}\mathbf{C}_r^t(u), \\
\lim_{s\to\infty} \mathbb{L}(s,u) &= \lim_{u\to\infty} \mathbb{L}(s,u) = \mathbf{0}.
\end{aligned}$$

It is possible to develop the theory without restriction, for a general equilibrium term ϕ. However, for simplicity, we specialize here to the case of a fully linear theory by taking

$$\widetilde{\phi}(\mathbf{C}(t)) = \phi(t) = \frac{1}{2}\mathbf{C}(t) \odot \mathbb{L}_\infty \mathbf{C}(t) \geq 0. \tag{21.3.29}$$

It follows that $\mathbb{L}$ must be a nonnegative operator in the same sense as $\mathbb{M}$. Also, from (21.3.29), we have $\mathbb{L}_\infty \geq \mathbf{0}$.

Constitutive relations are given as follows. Applying (21.3.26) to (17.3.5) yields, with the aid of (21.3.28),

$$\begin{aligned}\mathbf{D}(t) &= \mathbb{L}_\infty \mathbf{C}(t) + \int_0^\infty \mathbb{L}'(u)\mathbf{C}_r^t(u)du \\ &= \mathbb{L}_0 \mathbf{C}(t) + \int_0^\infty \mathbb{L}'(u)\mathbf{C}^t(u)du \\ &= \int_0^\infty \mathbb{L}(u)\dot{\mathbf{C}}^t(u)du,\end{aligned}$$

where $\mathbf{C}^t(0) = \mathbf{C}(t)$.

Consider the case where (21.3.18) and (21.3.19) reduce to $\mathbf{C}(t) = (\boldsymbol{\Lambda}(t), \nabla\boldsymbol{\Lambda}(t))$ and $\mathbf{D}(t) = (\boldsymbol{\Sigma}_0(t), \boldsymbol{\Sigma}_1(t))$. More explicitly, we write

$$\mathbf{C}(t) = \nabla\boldsymbol{\Lambda}(t) = \begin{pmatrix} \boldsymbol{\Lambda}(t) \\ \partial_1\boldsymbol{\Lambda}(t) \\ \partial_2\boldsymbol{\Lambda}(t) \\ \partial_3\boldsymbol{\Lambda}(t) \end{pmatrix}, \qquad \mathbf{D}(t) = \begin{pmatrix} \boldsymbol{\Sigma}_0(t) \\ \boldsymbol{\Sigma}_{(1)1}(t) \\ \boldsymbol{\Sigma}_{(1)2}(t) \\ \boldsymbol{\Sigma}_{(1)3}(t) \end{pmatrix},$$

where

$$\partial_i = \frac{\partial}{\partial x_i}, \quad i = 1, 2, 3.$$

Also,

$$\nabla_a \odot = (1 \quad -\partial_1 \quad -\partial_2 \quad -\partial_3),$$

yielding in particular (see (21.3.21))

$$\boldsymbol{\Sigma}(t) = \nabla_a \odot \mathbf{D}(t) = \boldsymbol{\Sigma}_0(t) - \partial_1\boldsymbol{\Sigma}_{(1)1}(t) - \partial_2\boldsymbol{\Sigma}_{(1)2}(t) - \partial_3\boldsymbol{\Sigma}_{(1)3}(t).$$

The tensor $\mathbb{L}$ is representable as a 4×4 matrix of transformations in $Lin(\Gamma)$.

The rate of dissipation D can be determined by (21.3.23), as outlined in Sect. 7.1.2, to obtain

$$D(t) = -\frac{1}{2}\int_0^\infty \int_0^\infty \dot{\mathbf{C}}^t(s) \odot [\mathbb{L}_1(s,u) + \mathbb{L}_2(s,u)]\dot{\mathbf{C}}^t(u)dsdu,$$

when the terms proportional to $\dot{\mathbf{C}}(t)$ are omitted. With further partial integrations, we can also write D in the form

$$D(t) = -\frac{1}{2}\int_0^\infty \int_0^\infty \mathbf{C}_r^t(s) \odot [\mathbb{L}_{121}(s,u) + \mathbb{L}_{122}(s,u)]\mathbf{C}_r^t(u)dsdu.$$

It is clear that the correspondence between this nonlocal theory and that presented in Chap. 7 is exact. This applies to all developments in the frequency domain as well as the time domain. Every general property, including the forms of all the free energy and dissipation functionals, can be taken over unchanged. The same applies to the definition of minimal states. More details and examples are given in [15].

22

The Minimum and Related Free Energies for Dielectric Materials

22.1 Introduction and General Relations

Free energies in electromagnetism, as in mechanics, are not in general uniquely determined by the constitutive equations but form a convex set with a minimum and a maximum element. A formula for the minimum free energy of a dielectric material with a linear memory-dependent constitutive equation was first given in [35], based on results in mechanics [158]. This was generalized in [151, 152], which form the basis of the present chapter.

We note also [10], dealing with the case of a conductor. This topic was discussed from a general point of view in Sect. 6.3.2.2. There is a close analogy with mechanics (particularly if the integrated history is used), which allows the minimum and other free energies to be determined.

The developments in this work are similar to the corresponding theory for mechanics and heat flow, described in earlier chapters. However, there are important differences, which we will discuss. The application of thermodynamics to electromagnetism is described in Chap. 6.

Some results in this context were given independently [153–155] for dielectric materials with memory, but with quite different methods, notation, and terminology. These papers emerged out of ongoing work over a decade or so, exploring various physical issues in optics, quite distant from the continuum thermodynamics environment described here. References [151, 152] seek to bring these two streams together. In particular, the correspondence between the two terminologies is noted.

Functionals that are free energies only for materials with kernels obeying monotonicity conditions are discussed in [91, 121, 122] for example and in Chap. 10. With the exception of ψ_F, these emerge from the older literature. They are less relevant for dielectrics than for mechanics, in that the required monotonicity restrictions may not so frequently apply. This issue also arises in Chap. 19.

Consider a rigid dielectric material, subject to a varying electromagnetic field. Let the body under consideration occupy a volume $\mathcal{B} \subset \mathbb{R}^3$. The electric field on this region is $\mathbf{E}(\mathbf{x}, t)$, with electric displacement denoted by $\mathbf{D}(\mathbf{x}, t)$. The magnetic field

G. Amendola et al., *Thermodynamics of Materials with Memory*,
https://doi.org/10.1007/978-3-030-80534-0_22

is $\mathbf{H}(\mathbf{x}, t)$ and the magnetic induction $\mathbf{B}(\mathbf{x}, t)$. The quantities $\mathbf{\Lambda} : \mathbb{R}^3 \times \mathbb{R} \mapsto \mathbb{R}^6$ and $\mathbf{\Sigma} : \mathbb{R}^3 \times \mathbb{R} \mapsto \mathbb{R}^6$, respectively, the electromagnetic vector and the electromagnetic induction, are defined by

$$\mathbf{\Lambda} = (\mathbf{E}, \mathbf{H}), \qquad \mathbf{\Sigma} = (\mathbf{D}, \mathbf{B}).$$

We shall denote $\mathbb{R}^6$ by $\mathcal{V}$. The quantity $\mathbf{\Lambda}$ will be treated as the independent variable. The rate of work done by the electromagnetic field on the body, per unit volume, is ([122] and references therein)

$$p = \dot{\mathbf{\Sigma}} \cdot \mathbf{\Lambda}, \tag{22.1.1}$$

where the dot product here represents a scalar product in $\mathcal{V}$.

Thermodynamic relations can be developed analogously to those in Sect. 5.1. Let us consider the isothermal case where θ is independent of space and time variables. It is put equal to unity. Then, $\mathbf{q}$ is zero. The quantity r is assumed to be negligible. Also, we take ρ to be constant and put it equal to unity. Then, recalling (5.1.11), we have the energy balance equation or First Law of Thermodynamics

$$\dot{\psi} + D = \dot{\mathbf{\Sigma}} \cdot \mathbf{\Lambda}, \tag{22.1.2}$$

where D is the entropy. The second law is imposed through the requirement that

$$D = \dot{\mathbf{\Sigma}} \cdot \mathbf{\Lambda} - \dot{\psi} \geq 0. \tag{22.1.3}$$

Remark 22.1.1. In a mechanics context, the time differentiation in (22.1.2) would be on the $\mathbf{\Lambda}$, corresponding to the strain tensor, rather than on the $\mathbf{\Sigma}$, corresponding to the stress tensor. Thus, the theory developed here is analogous to the case in mechanics where the stress is treated as the independent variable and, as we shall see, the memory functions involved will behave similarly to creep rather than relaxation functions, in the sense that they tend to increase rather than decrease with time. The phenomenon of creep is discussed in Sect. 8.5.

Let us define the free enthalpy as

$$\mathcal{F} = \psi - \mathbf{\Sigma} \cdot \mathbf{\Lambda}, \tag{22.1.4}$$

which is analogous to the Gibbs free energy in mechanics. The quantity $\mathbf{\Sigma} \cdot \mathbf{\Lambda}$ is of course unique, so that to each free energy, there is a corresponding free enthalpy. In terms of this quantity, (22.1.2) becomes

$$\dot{\mathcal{F}} + D = -\mathbf{\Sigma} \cdot \dot{\mathbf{\Lambda}}, \qquad D \geq 0, \tag{22.1.5}$$

the latter being the second law. A constitutive assumption is now made by requiring that the free enthalpy $\mathcal{F}$ depends in a specified way on the history and current value of $\mathbf{\Lambda}$. We put

$$\mathcal{F}(t) = \widetilde{\mathcal{F}}_d_{s>0}\left(\mathbf{\Lambda}_r^t(s), \mathbf{\Lambda}(t)\right), \tag{22.1.6}$$

where the relative history $\mathbf{\Lambda}_r^t$ is defined by

$$\mathbf{\Lambda}_r^t(s) = \mathbf{\Lambda}^t(s) - \mathbf{\Lambda}(t). \tag{22.1.7}$$

A relative future continuation is also defined by (22.1.7) for $s \in \mathbb{R}^{--}$.

We define the equilibrium free enthalpy $\mathcal{F}_e(t)$ to be given by (22.1.6) for the static history $\boldsymbol{\Lambda}^t(s) = \boldsymbol{\Lambda}(t),\ s \in \mathbb{R}^+$, or equivalently with $\boldsymbol{\Lambda}_r^t(s) = \mathbf{0},\ s \in \mathbb{R}^+$. This quantity depends only on $\boldsymbol{\Lambda}(t)$, so that

$$\mathcal{F}_e(t) = \widetilde{\mathcal{F}}_e(\boldsymbol{\Lambda}(t)).$$

It can be deduced from (22.1.5), by means of a fading memory argument [69], that

$$\mathcal{F}_e(t) \le \mathcal{F}(t), \quad \forall\, t \in \mathbb{R}, \tag{22.1.8}$$

giving that the equilibrium free enthalpy is less than or equal to the free enthalpy for an arbitrary history. The notation $\mathcal{F}_e(t)$ will be used in most cases rather than $\widetilde{\mathcal{F}}_e(\boldsymbol{\Lambda}(t))$.

We can write (22.1.6) in the form

$$\mathcal{F}(t) = \mathcal{F}_e(t) + \widetilde{\mathcal{F}}_M\left(\boldsymbol{\Lambda}_r^t(s), \boldsymbol{\Lambda}(t)\right),$$

where the second term on the right must be nonnegative by virtue of (22.1.8). It contains the memory contributions.

Let us state the characteristic properties of a free enthalpy, provable within a general framework [35, 69, 70]:

P1

$$\frac{\partial \mathcal{F}(t)}{\partial \boldsymbol{\Lambda}(t)} = -\boldsymbol{\Sigma}(t). \tag{22.1.9}$$

P2 For any relative history $\boldsymbol{\Lambda}_a^t$ and current value $\boldsymbol{\Lambda}_a(t)$,

$$\widetilde{\mathcal{F}}(\boldsymbol{\Lambda}_a^t, \boldsymbol{\Lambda}_a(t)) \ge \widetilde{\mathcal{F}}_e(\boldsymbol{\Lambda}_a(t)),$$

which is (22.1.8).

P3 Condition (22.1.5) holds.

These will be referred to as the Graffi conditions by analogy with those for a free energy in mechanics given, for example, in Sect. 5.1.1. The additional condition P4, given by (18.2.1), must also apply.

22.2 A Linear Memory Model for Dielectric Materials

By analogy with the developments of Sect. 7.1 for the free enthalpy rather than the free energy, we obtain, instead of (7.1.9) and (7.1.10),

$$\begin{aligned}\mathcal{F}(t) &= \mathcal{F}_e(t) - \frac{1}{2}\int_0^\infty\int_0^\infty \boldsymbol{\Lambda}_r^t(s)\cdot \mathbb{L}_{12}(s,u)\boldsymbol{\Lambda}_r^t(u)\,ds\,du \\ &= \mathcal{F}_e(t) - \frac{1}{2}\int_0^\infty\int_0^\infty \dot{\boldsymbol{\Lambda}}^t(s)\cdot \widetilde{\mathbb{L}}(s,u)\dot{\boldsymbol{\Lambda}}^t(u)\,ds\,du,\end{aligned} \tag{22.2.1}$$

where

$$\dot{\mathbf{\Lambda}}^t(u) = \frac{\partial}{\partial t}\mathbf{\Lambda}^t(u) = -\frac{\partial}{\partial u}\mathbf{\Lambda}^t(u) = -\frac{\partial}{\partial u}\mathbf{\Lambda}_r^t(u),$$
$$\widetilde{\mathbb{L}}(s,u) = \mathbb{L}(s,u) - \mathbb{L}_\infty. \tag{22.2.2}$$

The negative sign on the integral is required to maintain (22.1.8), given that the double integral is nonpositive (see Proposition 8.6.3; also, note the sign on the right-hand side of (22.1.5), which contrasts with that in (5.1.11); this sign difference also appears in (22.1.9) and (5.1.30). Relation (22.1.9) gives

$$\begin{aligned}\mathbf{\Sigma}(t) &= \widetilde{\mathbf{\Sigma}}_e(\mathbf{\Lambda}(t)) + \int_0^\infty \mathbb{L}'(u)\mathbf{\Lambda}_r^t(u)du \\ &= \mathbf{\Sigma}_0(t) + \int_0^\infty \mathbb{L}'(u)\mathbf{\Lambda}^t(u)du \\ &= \widetilde{\mathbf{\Sigma}}_e(\mathbf{\Lambda}(t)) + \int_0^\infty \widetilde{\mathbb{L}}(u)\dot{\mathbf{\Lambda}}^t(u)du,\end{aligned} \tag{22.2.3}$$

where

$$\mathbb{L}(u) = \mathbb{L}(0,u), \qquad \mathbb{L}_0 = \mathbb{L}(0,0) = \mathbb{L}(0),$$
$$\widetilde{\mathbb{L}}(u) = \mathbb{L}(u) - \mathbb{L}_\infty, \qquad \mathbb{L}'(u) = \frac{d}{du}\mathbb{L}(u),$$

and

$$\widetilde{\mathbf{\Sigma}}_e(\mathbf{\Lambda}(t)) = \mathbf{\Sigma}_e(t) = -\frac{d\mathfrak{F}_e(\mathbf{\Lambda}(t))}{d\mathbf{\Lambda}(t)},$$
$$\mathbf{\Sigma}_0(t) = \widetilde{\mathbf{\Sigma}}_e(\mathbf{\Lambda}(t)) + (\mathbb{L}_0 - \mathbb{L}_\infty)\mathbf{\Lambda}(t). \tag{22.2.4}$$

It is shown below (see (22.2.24)) that $\mathbb{L}_\infty$, the equilibrium modulus, is a positive definite matrix. It is assumed that $\mathbf{\Lambda}(-\infty) = \mathbf{\Lambda}^t(\infty)$ vanishes. Also,

$$\mathbb{L}_\infty = \mathbb{L}(s,\infty) = \mathbb{L}(0,\infty) = \mathbb{L}(\infty).$$

The quantity $\mathbb{L}_0$ is the instantaneous modulus and, on the basis of physical evidence [35],

$$\mathbb{L}_0 > 0. \tag{22.2.5}$$

It will be assumed here, as in (7.1.18), that

$$\mathbb{L}(u) = \mathbb{L}^\top(u), \quad u \in \mathbb{R}^+. \tag{22.2.6}$$

We deduce from (22.1.5), (22.2.2), and the time derivative of $(22.2.1)_2$ that

$$\begin{aligned}D(t) &= \frac{1}{2}\int_0^\infty\int_0^\infty \dot{\mathbf{\Lambda}}^t(s)\cdot[\mathbb{L}_1(s,u) + \mathbb{L}_2(s,u)]\,\dot{\mathbf{\Lambda}}^t(u)dsdu \\ &= \frac{1}{2}\int_0^\infty\int_0^\infty \mathbf{\Lambda}_r^t(s)\cdot[\mathbb{L}_{112}(s,u) + \mathbb{L}_{212}(s,u)]\,\mathbf{\Lambda}_r^t(u)dsdu.\end{aligned} \tag{22.2.7}$$

The first expression results from two partial integrations. Further partial integrations, using (22.2.2), gives the second form.

Relation (22.2.3) allows for general nonlinear behavior in the equilibrium term $\boldsymbol{\Sigma}_e$. We now however specialize to the case of linear behavior. Following [35, 122], we write $(22.2.3)_2$ as

$$\boldsymbol{\Sigma}(t) = \mathbb{L}_0\boldsymbol{\Lambda}(t) + \int_0^\infty \mathbb{L}'(u)\boldsymbol{\Lambda}^t(u)du, \tag{22.2.8}$$

so that, from $(22.2.4)_3$,

$$\boldsymbol{\Sigma}_e(t) = \mathbb{L}_\infty\boldsymbol{\Lambda}(t), \tag{22.2.9}$$

and equations $(22.2.3)_{1,3}$ become

$$\begin{aligned}\boldsymbol{\Sigma}(t) &= \mathbb{L}_\infty\boldsymbol{\Lambda}(t) + \int_0^\infty \mathbb{L}'(u)\boldsymbol{\Lambda}_r^t(u)du \\ &= \int_0^\infty \mathbb{L}(u)\dot{\boldsymbol{\Lambda}}^t(u)du.\end{aligned}$$

It follows from $(22.2.4)_2$ and (22.2.9) that

$$\mathcal{F}_e(t) = -\frac{1}{2}\boldsymbol{\Lambda}(t)\cdot\mathbb{L}_\infty\boldsymbol{\Lambda}(t), \tag{22.2.10}$$

provided we add the condition that $\mathcal{F}_e(t)$ must vanish when $\boldsymbol{\Lambda}(t) = \mathbf{0}$. Relation $(22.2.1)_1$ becomes

$$\begin{aligned}\mathcal{F}(t) &= -\frac{1}{2}\boldsymbol{\Lambda}(t)\cdot\mathbb{L}_\infty\boldsymbol{\Lambda}(t) - \frac{1}{2}\int_0^\infty\int_0^\infty \boldsymbol{\Lambda}_r^t(s)\cdot\mathbb{L}_{12}(s,u)\boldsymbol{\Lambda}_r^t(u)dsdu \\ &= \frac{1}{2}\boldsymbol{\Lambda}(t)\cdot\mathbb{L}_0\boldsymbol{\Lambda}(t) - \boldsymbol{\Sigma}(t)\cdot\boldsymbol{\Lambda}(t) \\ &\quad - \frac{1}{2}\int_0^\infty\int_0^\infty \boldsymbol{\Lambda}^t(s)\cdot\mathbb{L}_{12}(s,u)\boldsymbol{\Lambda}^t(u)dsdu.\end{aligned} \tag{22.2.11}$$

Since the integral term in $(22.2.11)_2$ is independent of $\boldsymbol{\Lambda}(t)$, we must have

$$\frac{\partial}{\partial\boldsymbol{\Lambda}(t)}\left[\frac{1}{2}\boldsymbol{\Lambda}(t)\cdot\mathbb{L}_0\boldsymbol{\Lambda}(t) - \boldsymbol{\Sigma}(t)\cdot\boldsymbol{\Lambda}(t)\right] = -\boldsymbol{\Sigma}(t),$$

which is easily checked. From (22.1.4) and (22.2.11), we deduce that

$$\begin{aligned}\psi(t) &= \phi_0(t) - \frac{1}{2}\int_0^\infty\int_0^\infty \boldsymbol{\Lambda}^t(s)\cdot\mathbb{L}_{12}(s,u)\boldsymbol{\Lambda}^t(u)dsdu \\ &= U(t) - \frac{1}{2}\int_0^\infty\int_0^\infty \boldsymbol{\Lambda}_r^t(s)\cdot\mathbb{L}_{12}(s,u)\boldsymbol{\Lambda}_r^t(u)dsdu,\end{aligned} \tag{22.2.12}$$

where

$$\phi_0(t) = \frac{1}{2}\boldsymbol{\Lambda}(t)\cdot\mathbb{L}_0\boldsymbol{\Lambda}(t), \qquad U(t) = \boldsymbol{\Sigma}(t)\cdot\boldsymbol{\Lambda}(t) - \frac{1}{2}\boldsymbol{\Lambda}(t)\cdot\mathbb{L}_\infty\boldsymbol{\Lambda}(t). \tag{22.2.13}$$

Using (22.1.1), we see that the total work done by the electromagnetic field up to time t is

$$W(t) = \int_{-\infty}^{t} \dot{\Sigma}(u) \cdot \Lambda(u) du = \Sigma(t) \cdot \Lambda(t) - \int_{-\infty}^{t} \Sigma(u) \cdot \dot{\Lambda}(u) du. \tag{22.2.14}$$

It is assumed here and below that field quantities vanish at large negative times sufficiently strongly so that various required integrals exist. Integrating $(22.1.2)_1$ on $(-\infty, t]$, we have

$$\psi(t) + \mathcal{D}(t) = W(t), \tag{22.2.15}$$

where

$$\mathcal{D}(t) = \int_{-\infty}^{t} D(u) du \geq 0 \tag{22.2.16}$$

is the total dissipation up to time t.

The physical content of (22.2.8) is hidden to some degree by the generality of the notation. It is worthwhile writing out in detail the relations implied by this expression. These are the most general within the category of an isothermal linear constitutive relation. We have

$$\begin{aligned}
\mathbf{D}(t) &= \mathbb{G}_{E0}\mathbf{E}(t) + \mathbb{G}_{H0}\mathbf{H}(t) \\
&\quad + \int_0^{\infty} \mathbb{G}'_E(u)\mathbf{E}^t(u)du + \int_0^{\infty} \mathbb{G}'_H(u)\mathbf{H}^t(u)du, \\
\mathbf{B}(t) &= \mathbb{M}_{E0}\mathbf{E}(t) + \mathbb{M}_{H0}\mathbf{H}(t) \\
&\quad + \int_0^{\infty} \mathbb{M}'_E(u)\mathbf{E}^t(u)du + \int_0^{\infty} \mathbb{M}'_H(u)\mathbf{H}^t(u)du, \\
&\mathbb{G}_{E0}, \mathbb{G}_{H0}, \mathbb{M}_{E0}, \mathbb{M}_{H0} \in Lin(R^3), \\
&\mathbb{G}'_E, \mathbb{G}'_H, \mathbb{M}'_E, \mathbb{M}'_H : \mathbb{R}^+ \mapsto Lin(R^3), \\
&\mathbb{G}_{E0} = \mathbb{G}_E(0), \quad \mathbb{G}_{H0} = \mathbb{G}_H(0), \\
&\mathbb{M}_{E0} = \mathbb{M}_E(0), \quad \mathbb{M}_{H0} = \mathbb{M}_H(0).
\end{aligned} \tag{22.2.17}$$

It follows from (22.2.6) that

$$\begin{aligned}
\mathbb{G}_{E0} &= \mathbb{G}_{E0}^{\top}, \quad \mathbb{G}'_E(u) = [\mathbb{G}'_E(u)]^{\top}, \\
\mathbb{M}_{H0} &= \mathbb{M}_{H0}^{\top}, \quad \mathbb{M}'_H(u) = [\mathbb{M}'_H(u)]^{\top}, \\
\mathbb{G}_{H0} &= \mathbb{M}_{E0}^{\top}, \quad \mathbb{G}'_H(u) = [\mathbb{M}'_E(u)]^{\top}, \quad u \in \mathbb{R}^+.
\end{aligned}$$

Remark 22.2.1. The concept of minimal states may be developed in electromagnetism (for example, [151, 152]) as in mechanics (Sect. 7.4).

22.2.1 The Kernel $\mathbb{L}(u)$ for Dielectric Materials

We can write the Fourier transforms of $\mathbb{L}'$ and $\widetilde{\mathbb{L}}$ in (22.2.3) as

$$\begin{aligned}
\mathbb{L}'_+(\omega) &= \mathbb{L}'_c(\omega) - i\mathbb{L}'_s(\omega), \\
\widetilde{\mathbb{L}}_+(\omega) &= \widetilde{\mathbb{L}}_c(\omega) - i\widetilde{\mathbb{L}}_s(\omega).
\end{aligned}$$

By partial integration, one can show that

$$\mathbb{L}'_+(\omega) = -(\mathbb{L}_0 - \mathbb{L}_\infty) + i\omega\widetilde{\mathbb{L}}_+(\omega), \tag{22.2.18}$$

giving, in particular, that

$$\mathbb{L}'_s(\omega) = -\omega\widetilde{\mathbb{L}}_c(\omega). \tag{22.2.19}$$

The notation $\mathbb{L}'_F$ will be reserved for a somewhat different use in (22.2.30) below.

If the system is in a given state at time t_0 and returns to this state at time $t_0 + T$, then we have a cycle, as discussed for the electromagnetic context in Sect. 6.3.3. In fact, for materials with memory, this situation can only exist if the independent variables have exhibited periodic behavior over a sufficiently long period of time to allow transient effects to die away so that the system is in a fully periodic state. In particular, $\Sigma(t)$, $\psi(t)$, and $\mathcal{F}(t)$ will be periodic functions. Integrating (22.1.3) or (22.1.5) over a cycle gives

$$\int_{t_0}^{t_0+T} \dot{\Sigma}(u) \cdot \Lambda(u)du \geq 0 \quad \text{or} \quad \int_{t_0}^{t_0+T} \Sigma(u) \cdot \dot{\Lambda}(u)du \leq 0, \tag{22.2.20}$$

which is a statement of the second law.

Consequences of these inequalities can be derived ([121, 122] and Sect. 7.2.1), by considering the case where $\Lambda(t)$ has sinusoidal behavior. In particular, it follows that

$$\mathbb{L}'_s(\omega_0) > 0 \;\text{ or }\; \widetilde{\mathbb{L}}_c(\omega_0) < 0, \qquad 0 < \omega_0 < \infty, \tag{22.2.21}$$

which are equivalent by virtue of (22.2.19). These inequalities have the opposite sign to those for the relaxation function in mechanics.

Following the steps outlined in Sect. 7.2.2, we obtain the analogue of (7.2.15), or

$$\mathbb{L}(u) - \mathbb{L}(0) = \frac{2}{\pi}\int_0^\infty \frac{1-\cos(\omega u)}{\omega}\mathbb{L}'_s(\omega)d\omega > 0.$$

It follows that

$$\mathbb{L}(u) \geq \mathbb{L}(0) = \mathbb{L}_0. \tag{22.2.22}$$

In particular,

$$\mathbb{L}(\infty) - \mathbb{L}(0) = \mathbb{L}_\infty - \mathbb{L}_0 = \frac{2}{\pi}\int_0^\infty \frac{\mathbb{L}'_s(\omega)}{\omega}d\omega > 0. \tag{22.2.23}$$

Then, from (22.2.5) and (22.2.23), we also have that

$$\mathbb{L}_\infty > 0. \tag{22.2.24}$$

Relations (22.2.22) and (22.2.23) indicate that $\mathbb{L}(u)$ behaves similarly to a creep function in mechanics rather than a relaxation function (see Remark 22.1.1). It is interesting to note that this follows from (22.2.21), which itself is a consequence of (22.1.1), leading to the negative sign on the left-hand side of (22.1.5) and more specifically the nonpositivity of the integral in $(22.2.20)_2$.

A quantity central to our considerations is defined by

$$\mathbb{H}(\omega) = \omega\mathbb{L}'_s(\omega) = -\omega^2\widetilde{\mathbb{L}}_c(\omega), \tag{22.2.25}$$

where (22.2.19) has been invoked. It is a nonnegative, even tensor function of the frequency, which vanishes quadratically at $\omega = 0$. The relation

$$i \lim_{\omega\to\infty} \omega \mathbb{L}'_+(\omega) = \lim_{\omega\to\infty} \omega \mathbb{L}'_s(\omega) = \mathbb{L}'(0)$$

yields

$$\mathbb{L}'(0) = \mathbb{H}(\infty) = \mathbb{H}_\infty. \tag{22.2.26}$$

22.2.2 The Work Function for Dielectric Materials

The integral term in $(22.2.14)_2$ has exactly the mechanics form, so that, by virtue of the developments of Sect. 7.5 for example, we obtain

$$\begin{aligned} W(t) &= U(t) - \frac{1}{2}\int_{-\infty}^{t}\int_{-\infty}^{t} \mathbb{L}_{12}(|u-s|)(\mathbf{\Lambda}(u)-\mathbf{\Lambda}(t))\cdot(\mathbf{\Lambda}(s)-\mathbf{\Lambda}(t))dsdu \\ &= U(t) - \frac{1}{2}\int_{0}^{\infty}\int_{0}^{\infty} \mathbb{L}_{12}(|u-s|)\mathbf{\Lambda}_r^t(u)\cdot\mathbf{\Lambda}_r^t(s)dsdu \\ &= U(t) - \frac{1}{2}\int_{-\infty}^{\infty}\int_{-\infty}^{\infty} \mathbb{L}_{12}(|u-s|)\mathbf{\Lambda}_r^t(u)\cdot\mathbf{\Lambda}_r^t(s)dsdu, \end{aligned} \tag{22.2.27}$$

in terms of relative histories, where $U(t)$ is defined by (22.2.13) and where the last relation assumes that $\mathbf{\Lambda}_r^t(u)$ vanishes for $u < 0$. We can write $W(t)$ in terms of histories as follows:

$$\begin{aligned} W(t) &= \phi_0(t) - \frac{1}{2}\int_{0}^{\infty}\int_{0}^{\infty} \mathbb{L}_{12}(|s-u|)\mathbf{\Lambda}^t(u)\cdot\mathbf{\Lambda}^t(s)duds \\ &= \phi_0(t) - \frac{1}{2}\int_{-\infty}^{\infty}\int_{-\infty}^{\infty} \mathbb{L}_{12}(|s-u|)\mathbf{\Lambda}^t(u)\cdot\mathbf{\Lambda}^t(s)duds, \end{aligned} \tag{22.2.28}$$

where $(22.2.28)_2$ requires that $\mathbf{\Lambda}^t(u)$ vanishes for $u < 0$. Relations $(22.2.28)_1$ and $(22.2.27)_2$ are special cases of (22.2.12).

In terms of frequency domain quantities, we find that

$$\begin{aligned} W(t) &= U(t) + \frac{1}{2\pi}\int_{-\infty}^{\infty} \overline{\mathbf{\Lambda}}_{r+}^t(\omega)\cdot\mathbb{H}(\omega)\mathbf{\Lambda}_{r+}^t(\omega)d\omega, \\ &= \phi_0(t) + \frac{1}{2\pi}\int_{-\infty}^{\infty} \overline{\mathbf{\Lambda}}_{+}^t(\omega)\cdot\mathbb{H}(\omega)\mathbf{\Lambda}_{+}^t(\omega)d\omega. \end{aligned} \tag{22.2.29}$$

These relations follow from application of the convolution theorem and Parseval's formula, together with the fact that the Fourier transform of the even function $\frac{d^2}{ds^2}\mathbb{L}(|s|)$ for $s \in \mathbb{R}$, is given by $(-i\omega)^2\widetilde{\mathbb{L}}_F(\omega)$ (see near the end of Sect. 7.5 and $(17.3.20)_{3,4}$) , where

$$\widetilde{\mathbb{L}}_F(\omega) = 2\widetilde{\mathbb{L}}_c(\omega) = -2\frac{\mathbb{H}(\omega)}{\omega^2}. \tag{22.2.30}$$

22.3 The Minimum Free Energy for Dielectric Materials

The tensor $\mathbb{H}$ can be regarded as a matrix in $\mathcal{V}\times\mathcal{V}$. The quantity $\mathbb{H}(\omega)$ can always be factorized as follows (see Sect. 11.1, in particular (11.1.11)):

$$\mathbb{H}(\omega) = \mathbb{H}_+(\omega)\mathbb{H}_-(\omega), \qquad \mathbb{H}_\pm(\omega) = \mathbb{H}^*_\mp(\omega), \tag{22.3.1}$$

where all the zeros of $\det(\mathbb{H}_\pm(\omega))$ and the singularities of $\mathbb{H}_\pm(\omega)$ are in $\Omega^{(\pm)}$, respectively. The factorization is unique up to multiplication by a constant unitary matrix on the right of $\mathbb{H}_+(\omega)$. The quantity $\mathbb{H}(\omega)$ is even in ω so that it is a function of ω^2. It has an isolated singularity at a point $\omega^2 = -\alpha^2$ if any one of its elements has a pole at this point. Then $\mathbb{H}_\pm(\omega)$ has a pole at $\omega = \pm i\alpha$, respectively. Also, $\mathbb{H}(\omega)$ may have non-isolated singularities, i.e., branch cuts. The quantity $\det(\mathbb{H}(\omega))$ will be zero at the point ω if at least one element in each row (column) of $\mathbb{H}(\omega)$ is zero at this point. In (22.3.1), $\mathbb{H}^*_\mp(\omega)$ is the Hermitian conjugate of $\mathbb{H}_\mp(\omega)$. It is assumed that $\mathbb{H}_\pm(\omega)$ commute with each other.

The quantity $\mathbb{H}_\infty$, defined by (22.2.26), is given by

$$\mathbb{H}_\infty = \mathbb{H}_+(\infty)\mathbb{H}_-(\infty) = \mathbb{H}_{+\infty}\mathbb{H}_{-\infty}.$$

If $\mathbb{H}_{\pm\infty}$ can be chosen to be Hermitian, which is possible at least in the commutative case, then they are both equal to the square root of the nonnegative tensor $\mathbb{H}_\infty$. We therefore put

$$\mathbb{H}_{+\infty} = \mathbb{H}_{-\infty} = \mathbb{H}_{sr}. \tag{22.3.2}$$

The quantity $\mathbb{H}(\infty)$ vanishes for the dielectric discussed in Sect. 22.4.

It will be assumed that $\boldsymbol{\Lambda}(\infty) = \mathbf{0}$. From $(22.2.27)_1$, one obtains

$$\begin{aligned} W(\infty) &= \int_{-\infty}^{\infty} \dot{\Sigma}(u)\cdot\boldsymbol{\Lambda}(u)du = -\int_{-\infty}^{\infty} \Sigma(u)\cdot\dot{\boldsymbol{\Lambda}}(u)du \\ &= -\frac{1}{2}\int_{-\infty}^{\infty}\int_{-\infty}^{\infty} \mathbb{L}_{12}(|s-u|)\boldsymbol{\Lambda}(u)\cdot\boldsymbol{\Lambda}(s)duds \\ &= -\frac{1}{2}\int_{-\infty}^{\infty}\int_{-\infty}^{\infty} \mathbb{L}_{12}(|s-u|)\boldsymbol{\Lambda}^t(u)\cdot\boldsymbol{\Lambda}^t(s)duds \\ &= \frac{1}{2\pi}\int_{-\infty}^{\infty} \overline{\boldsymbol{\Lambda}^t_F}(\omega)\cdot\mathbb{H}(\omega)\boldsymbol{\Lambda}^t_F(\omega)d\omega \\ &= \frac{1}{2\pi}\int_{-\infty}^{\infty} \left(\overline{\boldsymbol{\Lambda}^t_+}(\omega)+\overline{\boldsymbol{\Lambda}^t_-}(\omega)\right)\cdot\mathbb{H}(\omega)\left(\boldsymbol{\Lambda}^t_+(\omega)+\boldsymbol{\Lambda}^t_-(\omega)\right)d\omega, \end{aligned} \tag{22.3.3}$$

since $U(\infty)$ vanishes. In $(22.3.3)_4$, the quantity t is now an arbitrary parameter, which can be reidentified as the current time. Equation $(22.3.3)_5$ follows from $(22.3.3)_4$. The recoverable work from the state at time t is given by

$$W_R(t) = -\int_t^{\infty} \dot{\Sigma}(u)\cdot\boldsymbol{\Lambda}(u)du = W(t) - W(\infty). \tag{22.3.4}$$

To obtain the minimum free energy, we seek to maximize this quantity (for example, [69, 70, 104] and Theorem 4.2.3). The optimization is carried out by varying the future continuation. Equivalently, one can minimize $W(\infty)$, given by (22.3.3), since $W(t)$ is not affected by the optimization process.

With the aid of the Plemelj formulae [274] (see also (B.2.15)), we write

$$\begin{aligned} \mathbf{Q}^t(\omega) = \mathbb{H}_-(\omega)\mathbf{\Lambda}_+^t(\omega) &= \mathbf{q}_-^t(\omega) - \mathbf{q}_+^t(\omega) \\ \mathbf{q}_\pm^t(\omega) &= \lim_{z\to\omega^\mp} \mathbf{q}^t(z) \\ \mathbf{q}^t(z) &= \frac{1}{2\pi i}\int_{-\infty}^{\infty} \frac{\mathbf{Q}^t(\omega')}{\omega' - z} d\omega', \end{aligned} \tag{22.3.5}$$

where $\mathbf{q}_-^t(\omega)$ is analytic on $\Omega^{(+)}$, going to zero at large ω as ω^{-1} and $\mathbf{q}_+^t(\omega)$ is analytic on $\Omega^{(-)}$ with similar behavior at large ω. The quantities $\mathbf{q}_\pm^t(\omega)$ are analytic on an open region including $\mathbb{R}$. In $\Omega^{(-)}$, away from singularities, $\mathbf{q}_-^t(\omega)$ is defined by analytic continuation from Ω^+, while $\mathbf{q}_+^t(\omega)$ is correspondingly defined in $\Omega^{(+)}$. We will write them as

$$\mathbf{q}_\pm^t(\omega) = \frac{1}{2\pi i}\int_{-\infty}^{\infty} \frac{\mathbf{Q}^t(\omega')}{\omega' - \omega^\mp} d\omega', \qquad \omega \in \mathbb{R}. \tag{22.3.6}$$

From $(22.2.29)_2$ and $(22.3.5)_{1,2}$, we have

$$\begin{aligned} W(t) &= \phi_0(t) + \frac{1}{2\pi}\int_{-\infty}^{\infty} |\mathbf{q}_-^t(\omega) - \mathbf{q}_+^t(\omega)|^2 d\omega \\ &= \phi_0(t) + \frac{1}{2\pi}\int_{-\infty}^{\infty} \left[|\mathbf{q}_-^t(\omega)|^2 + |\mathbf{q}_+^t(\omega)|^2\right] d\omega, \end{aligned} \tag{22.3.7}$$

since the cross terms vanish by virtue of Proposition B.1.3.

Different methods of deriving the formula for the minimum free energy are outlined in Sect. 11.2 and Sect. 16.5 for the maximum free energy. A version of the latter method is used here. With the aid of (22.3.1), let us write $(22.3.3)_6$ as

$$W(\infty) = \frac{1}{2\pi}\int_{-\infty}^{\infty} |\mathbf{q}_-^t(\omega) - \mathbf{q}_+^t(\omega) + \mathbb{H}_-(\omega)\mathbf{\Lambda}_-^t(\omega)|^2\, d\omega.$$

Putting

$$\mathbf{q}_{1-}^t(\omega) = \mathbf{q}_-^t(\omega) + \mathbb{H}_-(\omega)\mathbf{\Lambda}_-^t(\omega),$$

where $\mathbf{q}_{1-}^t(\omega)$ is analytic on $\Omega^{(+)}$, we have

$$\begin{aligned} W(\infty) &= \frac{1}{2\pi}\int_{-\infty}^{\infty} |\mathbf{q}_{1-}^t(\omega) - \mathbf{q}_+^t(\omega)|^2\, d\omega \\ &= \frac{1}{2\pi}\int_{-\infty}^{\infty} [|(\mathbf{q}_{1-}^t(\omega)|^2 + |\mathbf{q}_+^t(\omega)|^2]\, d\omega. \end{aligned}$$

Only $\mathbf{q}_{1-}^t(\omega)$ depends on $\mathbf{\Lambda}_-^t(\omega)$. Therefore, the minimum must be given by choosing a value of $\mathbf{\Lambda}_-^t(\omega)$ such that

$$\mathbf{q}_{1-}^t(\omega) = \mathbf{0},$$

as the optimal continuation $\mathbf{\Lambda}_{m-}^t(\omega)$. It follows that

$$\mathbf{\Lambda}^t_{m-}(\omega) = -\,[\mathbb{H}_-(\omega)]^{-1}\,\mathbf{q}^t_-(\omega) = -\frac{[\mathbb{H}_-(\omega)]^{-1}}{2\pi i}\int_{-\infty}^{\infty}\frac{\mathbb{H}_-(\omega')\mathbf{\Lambda}^t_+(\omega')}{\omega'-\omega^+}d\omega'. \tag{22.3.8}$$

The resulting minimum value of $W(\infty)$ is

$$W_{opt}(\infty) = \frac{1}{2\pi}\int_{-\infty}^{\infty}|\mathbf{q}^t_+(\omega)|^2 d\omega. \tag{22.3.9}$$

The maximum value of $W_R(t)$, given by (22.3.4), is the minimum free energy and has the form

$$\begin{aligned}\psi_m(t) &= \phi_0(t) + \frac{1}{2\pi}\int_{-\infty}^{\infty}|\mathbf{q}^t_-(\omega)|^2 d\omega \\ &= \phi_0(t) + \frac{1}{2\pi}\int_{-\infty}^{\infty}\overline{\mathbf{\Lambda}^t_{m-}}(\omega)\cdot\mathbb{H}(\omega)\mathbf{\Lambda}^t_{m-}(\omega)d\omega,\end{aligned} \tag{22.3.10}$$

which follows from (22.3.4), (22.3.7), and (22.3.9). It can be confirmed that this quantity has the required properties of a free energy, by the method outlined in Sect. 11.2.4.

With the aid of (22.3.6) and (7.2.27), we obtain [92, 158]

$$\begin{aligned}\frac{d}{dt}\mathbf{q}^t_+(\omega) &= -i\omega\mathbf{q}^t_+(\omega) - \mathbf{K}(t), \\ \frac{d}{dt}\mathbf{q}^t_-(\omega) &= -i\omega\mathbf{q}^t_-(\omega) - \mathbf{K}(t) + \mathbb{H}_-(\omega)\mathbf{\Lambda}(t), \\ \mathbf{K}(t) &= \frac{1}{2\pi}\int_{-\infty}^{\infty}\mathbb{H}_-(\omega)\mathbf{\Lambda}^t_{r+}(\omega)d\omega,\end{aligned} \tag{22.3.11}$$

and

$$\begin{aligned}&\lim_{|\omega|\to\infty}\omega\mathbf{q}^t_+(\omega) = i\mathbf{K}(t), \\ &\lim_{|\omega|\to\infty}\omega\mathbf{q}^t_-(\omega) = i(\mathbf{K}(t) - \mathbb{H}_{sr}\mathbf{\Lambda}(t)), \\ &\frac{1}{2\pi}\int_{-\infty}^{\infty}\mathbf{q}^t_+(\pm\omega)d\omega = -\frac{1}{2}\mathbf{K}(t), \\ &\frac{1}{2\pi}\int_{-\infty}^{\infty}\mathbf{q}^t_-(\pm\omega)d\omega = \frac{1}{2}(\mathbf{K}(t) - \mathbb{H}_{sr}\mathbf{\Lambda}(t)).\end{aligned} \tag{22.3.12}$$

From (22.3.7), (22.3.9), (22.3.10), and (22.2.15), we deduce that the total dissipation corresponding to the minimum free energy is given by

$$\mathcal{D}_m(t) = \int_{-\infty}^{t} D_m(u)du = \frac{1}{2\pi}\int_{-\infty}^{\infty}|\mathbf{q}^t_+(\omega)|^2 d\omega = W_{opt}(\infty) \geq 0. \tag{22.3.13}$$

If we differentiate this relation with respect to t, using (22.3.11) and (22.3.12), the result is

$$D_m(t) = |\mathbf{K}(t)|^2. \tag{22.3.14}$$

From (22.3.8) and $(22.3.12)_2$, it follows that

$$\mathbf{\Lambda}^t_{m-}(\omega) \underset{\omega\to\infty}{\longrightarrow} -\frac{\mathbf{\Lambda}(t) - [\mathbb{H}_{sr}]^{-1}\,\mathbf{K}(t)}{i\omega},$$

which indicates a discontinuity between the history $\mathbf{\Lambda}^t(s),\ s \in \mathbb{R}^+$, leading to $\mathbf{\Lambda}(t)$ and the optimal continuation $\mathbf{\Lambda}^t_{opt}(s)$, at $s = 0$, given by

$$\mathbf{\Lambda}^t_{opt}(0) = \mathbf{\Lambda}(t) - [\mathbb{H}_{sr}]^{-1}\,\mathbf{K}(t).$$

This discontinuity has the form

$$\mathbf{\Lambda}^t(0) - \mathbf{\Lambda}^t_{opt}(0) = [\mathbb{H}_{sr}]^{-1}\,\mathbf{K}(t) \tag{22.3.15}$$

and is related to the rate of dissipation (22.3.14).

If $\mathbb{H}_{sr}$ vanishes (which is true for the case dealt with in Sect. 22.4), then the discontinuity, as given by (22.3.15), becomes infinite. The associated free energy, given by (22.3.10), and the rate of dissipation, given by (22.3.14), are finite quantities, however.

We can reexpress these results in terms of relative histories. Instead of (22.3.5), we write

$$\begin{aligned}\mathbf{P}^t(\omega) &= \mathbb{H}_-(\omega)\mathbf{\Lambda}^t_{r+}(\omega) = \mathbf{p}^t_-(\omega) - \mathbf{p}^t_+(\omega) = \mathbf{Q}^t(\omega) - \frac{\mathbb{H}_-(\omega)}{i\omega}\mathbf{\Lambda}(t),\\ \mathbf{p}^t_\pm(\omega) &= \frac{1}{2\pi i}\int_{-\infty}^{\infty}\frac{\mathbf{P}^t(\omega')}{\omega' - \omega^\mp}d\omega' = \mathbf{q}^t_\pm(\omega) + \frac{1}{2\pi}\int_{-\infty}^{\infty}\frac{\mathbb{H}_-(\omega')}{\omega'(\omega' - \omega^\mp)}d\omega'\mathbf{\Lambda}(t).\end{aligned} \tag{22.3.16}$$

By closing the contour on Ω^+, we find that

$$\mathbf{p}^t_-(\omega) = \mathbf{q}^t_-(\omega) - \frac{\mathbb{H}_-(\omega)}{i\omega}\mathbf{\Lambda}(t), \qquad \mathbf{p}^t_+(\omega) = \mathbf{q}^t_+(\omega). \tag{22.3.17}$$

With the aid of (22.2.29), relation (22.3.7) is replaced by

$$W(t) = U(t) + \frac{1}{2\pi}\int_{-\infty}^{\infty}\left[|\mathbf{p}^t_-(\omega)|^2 + |\mathbf{p}^t_+(\omega)|^2\right]d\omega. \tag{22.3.18}$$

The minimum free energy has the form

$$\psi_m(t) = U(t) + \frac{1}{2\pi}\int_{-\infty}^{\infty}|\mathbf{p}^t_-(\omega)|^2 d\omega, \tag{22.3.19}$$

which is an alternative form of $(22.3.10)_1$. Equations (22.3.13) and (22.3.14) are unchanged.

Using $(22.3.10)_1$ and $(22.3.19)_1$, we can write $\psi_m(t)$ in the form (*cf.* (11.2.39) and also Sect. 17.5)

$$\begin{aligned}\psi_m(t) &= \phi_0(t) + \frac{i}{4\pi^2}\int_{-\infty}^{\infty}\int_{-\infty}^{\infty}\frac{\overline{\mathbf{\Lambda}^t_+(\omega_1)}\cdot\mathbb{M}_m(\omega_1,\omega_2)\mathbf{\Lambda}^t_+(\omega_2)}{\omega_1^+ - \omega_2^-}d\omega_1 d\omega_2\\ &= U(t) + \frac{i}{4\pi^2}\int_{-\infty}^{\infty}\int_{-\infty}^{\infty}\frac{\overline{\mathbf{\Lambda}^t_{r+}(\omega_1)}\cdot\mathbb{M}_m(\omega_1,\omega_2)\mathbf{\Lambda}^t_{r+}(\omega_2)}{\omega_1^+ - \omega_2^-}d\omega_1 d\omega_2,\\ &\mathbb{M}_m(\omega_1,\omega_2) = \mathbb{H}_+(\omega_1)\mathbb{H}_-(\omega_2),\end{aligned} \tag{22.3.20}$$

by carrying out the integration with respect to ω over Ω^+ or over Ω^- (see Sect. 11.2.5). Also, $D_m(t)$, given by (22.3.14), can be expressed as

$$D_m(t) = \frac{1}{4\pi^2}\int_{-\infty}^{\infty}\int_{-\infty}^{\infty} \overline{\mathbf{\Lambda}^t_{r+}(\omega_1)} \cdot \mathbb{M}_m(\omega_1,\omega_2)\mathbf{\Lambda}^t_{r+}(\omega_2)d\omega_1 d\omega_2. \tag{22.3.21}$$

From (22.3.13), we deduce that

$$\begin{aligned}\mathcal{D}_m(t) &= -\frac{i}{4\pi^2}\int_{-\infty}^{\infty}\int_{-\infty}^{\infty} \frac{\overline{\mathbf{\Lambda}^t_{+}(\omega_1)} \cdot \mathbb{M}_m(\omega_1,\omega_2)\mathbf{\Lambda}^t_{+}(\omega_2)}{\omega_1^- - \omega_2^+}d\omega_1 d\omega_2 \\ &= -\frac{i}{4\pi^2}\int_{-\infty}^{\infty}\int_{-\infty}^{\infty} \frac{\overline{\mathbf{\Lambda}^t_{r+}(\omega_1)} \cdot \mathbb{M}_m(\omega_1,\omega_2)\mathbf{\Lambda}^t_{r+}(\omega_2)}{\omega_1^- - \omega_2^+}d\omega_1 d\omega_2.\end{aligned} \tag{22.3.22}$$

Alternative forms of (22.3.20) and (22.3.22) are derived in [151], including an explicitly convergent form of the free energy, similar to (11.2.44). It is also shown that ψ_m is a functional of the minimal state.

The free enthalpy corresponding to the minimum free energy may be deduced from (22.1.4), (22.2.13), $(22.3.10)_1$, and (22.3.19) to be

$$\begin{aligned}\mathcal{F}_m(t) = \psi_m(t) - \mathbf{\Sigma}(t)\cdot\mathbf{\Lambda}(t) &= -S(t) + \frac{1}{2\pi}\int_{-\infty}^{\infty}|\mathbf{q}^t_-(\omega)|^2 d\omega \\ &= -\phi_\infty(t) + \frac{1}{2\pi}\int_{-\infty}^{\infty}|\mathbf{p}^t_-(\omega)|^2 d\omega,\end{aligned} \tag{22.3.23}$$

where (see (7.1.35) and (22.2.10))

$$S(t) = \mathbf{\Sigma}(t)\cdot\mathbf{\Lambda}(t) - \phi_0(t), \qquad \phi_\infty(t) = \frac{1}{2}\mathbf{\Lambda}(t)\cdot\mathbb{L}_\infty\mathbf{\Lambda}(t) = -\mathcal{F}_e(t). \tag{22.3.24}$$

It is easy to show that $\mathcal{F}_m(t)$ obeys the Graffi conditions listed at the end of Sect. 22.1. Property P2 is immediately apparent, while P3 is equivalent to (22.1.2). The relation (22.2.15) holds for the minimum free energy. The time derivative of (22.2.15) gives (22.1.2), on recalling the derivation of (22.3.14). Property P1 can be proved with the aid of $(22.3.23)_2$, by showing that

$$\frac{\partial S(t)}{\partial\mathbf{\Lambda}(t)} = \mathbf{\Sigma}(t), \tag{22.3.25}$$

which is a special case of (7.1.20).

A representation of the minimum free energy, given by (22.3.10), in terms of time domain quantities, analogous to results in Chap. 12, is deduced in [151]. Also, the form of ψ_m and related quantities for sinusoidal histories are deduced in that paper.

22.4 Free Energies for Non-magnetic Materials

Consider a rigid non-magnetic isotropic dielectric material subject to a varying electric field. Let the body under consideration occupy a volume $\mathcal{B} \subset \mathbb{R}^3$. A typical point in $\mathcal{B}$ is $\mathbf{x}$, while t is a given time. The electric field on this region is $\mathbf{E}(\mathbf{x},t)$, with electric displacement denoted by $\mathbf{D}(\mathbf{x},t)$. The magnetic field and induction contributions are neglected.

Let $\psi(t)$ be any free energy of the material, for the isothermal case, and $D(t)$ the rate of dissipation. The laws of thermodynamics are as outlined in (22.1.2)–(22.1.5), where $\dot{\boldsymbol{\Sigma}}\cdot\boldsymbol{\Lambda}$ is replaced by $\dot{\mathbf{D}}\cdot\mathbf{E}$, so that we define the free enthalpy in this context as

$$\mathcal{F}=\psi-\mathbf{D}\cdot\mathbf{E}. \tag{22.4.1}$$

We define the equilibrium free enthalpy $\mathcal{F}_e(t)$ to be that given for the static history $\mathbf{E}^t(s)=\mathbf{E}(t),\ s\in\mathbb{R}^+$. Therefore,

$$\mathcal{F}_e(t)=\widetilde{\mathcal{F}}_e(\mathbf{E}(t)).$$

A special case of the linear model introduced in Sect. 22.2 is now considered. All kernels are scalar quantities, reflecting an electromagnetic isotropy. Relation (22.2.11) becomes

$$\begin{aligned}
\mathcal{F}(t)&=\mathcal{F}_e(t)-\frac{1}{2}\int_0^\infty\int_0^\infty \dot{\mathbf{E}}^t(s)\cdot\widetilde{G}(s,u)\dot{\mathbf{E}}^t(u)dsdu,\\
&=\frac{1}{2}G_0|\mathbf{E}(t)|^2-\mathbf{D}(t)\cdot\mathbf{E}(t)-\frac{1}{2}\int_0^\infty\int_0^\infty \mathbf{E}^t(s)\cdot G_{12}(s,u)\mathbf{E}^t(u)dsdu\\
&=\mathcal{F}_e(t)-\frac{1}{2}\int_0^\infty\int_0^\infty \mathbf{E}_r^t(s)\cdot G_{12}(s,u)\mathbf{E}_r^t(u)dsdu\\
\mathcal{F}_e(t)&=-\frac{1}{2}G_\infty|\mathbf{E}(t)|^2,\qquad |\mathbf{E}(t)|^2=\mathbf{E}(t)\cdot\mathbf{E}(t),\qquad \widetilde{G}(s,u)=G(s,u)-G_\infty,
\end{aligned} \tag{22.4.2}$$

where $(22.4.2)_3$ follows from $(22.4.2)_1$ by partial integrations. The linear constitutive relation in this case is given by

$$\begin{aligned}
\mathbf{D}(t)&=G_\infty\mathbf{E}(t)+\int_0^\infty G'(u)\mathbf{E}_r^t(u)du=G_0\mathbf{E}(t)+\int_0^\infty G'(u)\mathbf{E}^t(u)du\\
&=G_\infty\mathbf{E}(t)+\int_0^\infty \widetilde{G}(u)\dot{\mathbf{E}}^t(u)du=\int_0^\infty G(u)\dot{\mathbf{E}}^t(u)du,
\end{aligned} \tag{22.4.3}$$

where

$$\begin{aligned}
G(u)&=G(0,u),\qquad G_0=G(0,0)=G(0),\\
\widetilde{G}(u)&=G(u)-G_\infty,\qquad G'(u)=\frac{d}{du}G(u).
\end{aligned} \tag{22.4.4}$$

Relation $(22.4.3)_2$ is a special case of $(22.2.17)_1$. In writing the final form of (22.4.3), we are assuming that $\mathbf{E}(-\infty)=\mathbf{E}^t(\infty)$ vanishes; we furthermore assume that it goes to zero sufficiently strongly so that various integrals, introduced below, exist. The quantity G_∞ is related to $G(s,u)$ through

$$G_\infty=G(s,\infty)=G(0,\infty)=G(\infty).$$

The relations in (22.2.7) reduce to

$$
\begin{aligned}
\mathcal{D}(t) &= \frac{1}{2}\int_0^\infty\int_0^\infty \dot{\mathbf{E}}^t(s)\cdot[G_1(s,u)+G_2(s,u)]\,\dot{\mathbf{E}}^t(u)dsdu \\
&= \frac{1}{2}\int_0^\infty\int_0^\infty \mathbf{E}_r^t(s)\cdot[G_{112}(s,u)+G_{212}(s,u)]\,\mathbf{E}_r^t(u)dsdu.
\end{aligned}
$$

From (22.4.1) and (22.4.2), we deduce that

$$
\begin{aligned}
\psi(t) &= \phi_0(t) - \frac{1}{2}\int_0^\infty\int_0^\infty \mathbf{E}^t(s)\cdot G_{12}(s,u)\mathbf{E}^t(u)dsdu \\
&= U(t) - \frac{1}{2}\int_0^\infty\int_0^\infty \mathbf{E}_r^t(s)\cdot G_{12}(s,u)\mathbf{E}_r^t(u)dsdu, \\
&= U(t) - \frac{1}{2}\int_0^\infty\int_0^\infty \dot{\mathbf{E}}^t(s)\cdot \widetilde{G}(s,u)\dot{\mathbf{E}}^t(u)dsdu,
\end{aligned}
\tag{22.4.5}
$$

where

$$
\phi_0(t) = \frac{1}{2}G_0|\mathbf{E}(t)|^2, \qquad U(t) = \mathbf{D}(t)\cdot\mathbf{E}(t) - \frac{1}{2}G_\infty|\mathbf{E}(t)|^2. \tag{22.4.6}
$$

The total work done by the electric field up to time t is

$$
W(t) = \int_{-\infty}^t \dot{\mathbf{D}}(u)\cdot\mathbf{E}(u)du = \mathbf{D}(t)\cdot\mathbf{E}(t) - \int_{-\infty}^t \mathbf{D}(u)\cdot\dot{\mathbf{E}}(u)du. \tag{22.4.7}
$$

Relations (22.2.15) and (22.2.16) hold in the present context.

We take G_0 to be equal to the vacuum permittivity ϵ_0 [153–155], giving

$$
\mathbf{D}(t) = \epsilon_0\mathbf{E}(t) + \mathbf{P}(t),
$$

where, using $(22.4.3)_2$, one sees that the polarization $\mathbf{P}(t)$ is given by the causal relationship

$$
\mathbf{P}(t) = \int_0^\infty G'(u)\mathbf{E}^t(u)du = \int_{-\infty}^t G'(t-s)\mathbf{E}(s)ds.
$$

Then, $(22.4.7)_1$ becomes

$$
W(t) = \phi_0(t) + W_{int}(t), \qquad \phi_0(t) = \frac{1}{2}\epsilon_0|\mathbf{E}(t)|^2,
$$

where $W_{int}(t)$ is given by

$$
W_{int}(t) = \int_{-\infty}^t \dot{\mathbf{P}}(s)\cdot\mathbf{E}(s)ds,
$$

is the accumulation of energy (density) transferred to the medium at the point under consideration, from the beginning of the pulse–medium interaction until time t. We can write (22.2.15) as

$$
\psi_{int}(t) + \mathcal{D}(t) = W_{int}(t),
$$

where $\psi_{int}(t)$ is given by the integral term in $(22.4.5)_1$.

22.4.1 The Kernel $G(u)$ for Non-magnetic Materials

We can write the Fourier transforms of $G'(u)$ and $\widetilde{G}(u)$ in (22.4.3) as

$$G'_+(\omega) = G'_c(\omega) - iG'_s(\omega) = \chi_+(\omega) = \chi_c(\omega) - i\chi_s(\omega),$$
$$\widetilde{G}_+(\omega) = \widetilde{G}_c(\omega) - i\widetilde{G}_s(\omega).$$

The quantity $\chi_+(\omega)$ is the susceptibility, denoted by $\chi(\omega)$ in [153–155]. Equation (22.2.18) reduces to

$$G'_+(\omega) = -(G_0 - G_\infty) + i\omega\widetilde{G}_+(\omega),$$

giving, in particular, that

$$G'_s(\omega) = \chi_s(\omega) = -\omega\widetilde{G}_c(\omega) > 0, \qquad 0 < \omega < \infty.$$

We see that

$$\overline{G_+'}(\omega) = \overline{\chi_+}(\omega) = G'_+(-\omega) = \chi_+(-\omega). \tag{22.4.8}$$

As a special case of (22.2.23), we have $G_\infty > G_0$. Also, $G_0 = \epsilon_0 > 0$, so that $G_\infty > 0$.

The quantity $\mathbb{H}$, defined by (22.2.25), reduces to a scalar quantity H defined by

$$H(\omega) = \omega G'_s(\omega) = \omega\chi_s(\omega) = -\omega^2\widetilde{G}_c(\omega). \tag{22.4.9}$$

It is a nonnegative, even function of the frequency and goes to zero quadratically at the origin. The relation

$$i\lim_{\omega\to\infty}\omega G'_+(\omega) = \lim_{\omega\to\infty}\omega G'_s(\omega) = G'(0)$$

yields

$$G'(0) = H(\infty) = H_\infty.$$

For the model considered later, $H(\omega)$ goes to zero at large ω so that $H_\infty = 0$.

Relations $(22.2.27)_2$ and $(22.2.28)_1$ reduce to

$$\begin{aligned} W(t) &= U(t) - \frac{1}{2}\int_0^\infty\int_0^\infty G_{12}(|u-s|)\mathbf{E}^t_r(u)\cdot\mathbf{E}^t_r(s)\,ds\,du \\ &= \phi_0(t) - \frac{1}{2}\int_0^\infty\int_0^\infty G_{12}(|u-s|)\mathbf{E}^t(u)\cdot\mathbf{E}^t(s)\,ds\,du, \end{aligned}$$

where $U(t)$ and $\phi_0(t)$ are defined by (22.4.6). In terms of frequency domain quantities, we find that (22.2.29) becomes

$$\begin{aligned} W(t) &= U(t) + \frac{1}{2\pi}\int_{-\infty}^{\infty}\overline{\mathbf{E}}^t_{r+}(\omega)\cdot H(\omega)\mathbf{E}^t_{r+}(\omega)\,d\omega, \\ &= \phi_0(t) + \frac{1}{2\pi}\int_{-\infty}^{\infty}\overline{\mathbf{E}}^t_{+}(\omega)\cdot H(\omega)\mathbf{E}^t_{+}(\omega)\,d\omega. \end{aligned} \tag{22.4.10}$$

22.4.2 Factorization of $H(w)$ for Non-magnetic Dielectrics

We consider materials such that $G'_+(\omega)$ (or $\widetilde{G}_+(\omega)$) has no branch cuts and a finite number of isolated singularities in $\Omega^{(+)}$. Thus, $H(\omega)$ has only isolated singularities in $\Omega^{(\pm)}$, which are mirror images of each other in the real axis. This means that it can be put in the form of a ratio of polynomials. We will take the singularities to be a finite number of simple poles. The quantity $H(\omega)$ has a finite number of zeros in $\Omega^{\pm}(\omega)$, also mirror images of each other. It is real and nonnegative on $\mathbb{R}$, an even function of ω and therefore a function of ω^2, in view of its analyticity about the origin. It vanishes quadratically at the origin. This nonnegative quantity can be factorized in general as outlined in [158] and Sect. 11.1.1, according to (11.1.16) and (11.1.17).

The quantity $H_+(\omega)$ ($H_-(\omega)$) has all its singularities in $\Omega^{(+)}$ ($\Omega^{(-)}$) and all its zeros in Ω^+ (Ω^-). There are many other factorizations, obtained by interchanging some or all of the zeros of $H_+(\omega)$ and $H_-(\omega)$, while retaining the same singularity structure. The different factorizations will be labeled by the subscript or superscript f and are given by (16.4.1).

Each factorization generally yields a different free energy $\psi_f(t)$, $f = 1, 2, \ldots, N$, where N is the total number of factorizations of $H(\omega)$, though there may be exceptions. The factorization with no exchange of zeros, which is that given by (11.1.16), yields the minimum free energy $\psi_m(t)$.

Remark 22.4.1. Each exchange of zeros, starting from these factors, can be shown to yield a free energy, which is greater than or equal to the previous quantity (Sect. 16.8).

Note that the zeros of $H^f_{\pm}(\omega)$ at the origin play no part in these exchanges.

We denote by $\psi_M(t)$ the free energy obtained by interchanging all the zeros. This can be identified as the maximum free energy among all those that are functionals of the minimal state. It is discussed for a mechanics context in Sect. 16.5.

The subscript $f = N$ is chosen to denote the maximum free energy.

The most general free energy arising from these factorizations is given by (16.10.1).

22.4.3 The Free Energy for the Non-magnetic Case Associated with a Particular Factorization

Consider a particular factorization $H^f_+(\omega)$, $H^f_-(\omega)$ of $H(\omega)$. We define

$$
\begin{aligned}
H^f_-(\omega)\mathbf{E}^t_+(\omega) &= \mathbf{q}^{ft}_-(\omega) - \mathbf{q}^{ft}_+(\omega), \\
\mathbf{q}^{ft}_{\pm}(\omega) &= \frac{1}{2\pi i}\int_{-\infty}^{\infty} \frac{H^f_-(\omega')\mathbf{E}^t_+(\omega')}{\omega' - \omega^{\mp}} d\omega',
\end{aligned}
\tag{22.4.11}
$$

where the $\mathbf{q}^{ft}_{\pm}(\omega)$ are analytic in $\Omega^{\mp}$, respectively. The singularities of $\mathbf{q}^{ft}_-(\omega)$ are the same as those of $H^f_-(\omega)$, as may be perceived by closing the contour in $(22.4.11)_2$ on $\Omega^{(-)}$. Singularities on the real axis are excluded by assumption.

The free energy associated with this particular factorization is a generalization of the form (22.3.10), given by

$$\psi_f(t) = \phi_0(t) + \frac{1}{2\pi}\int_{-\infty}^{\infty} |\mathbf{q}_-^{ft}(\omega)|^2 d\omega. \tag{22.4.12}$$

Relations (22.3.11) and (22.3.12) for the scalar quantities $H_\pm(\omega)$ apply here also.

Remark 22.4.2. For the material considered in Sect. 22.4.4 below, the quantity $H(\omega)$ vanishes for large values of ω. Thus, the scalar version of $\mathbb{H}_{sr}$ in (22.3.2) and (22.3.12) is zero. Also,

$$\int_{-\infty}^{\infty} \frac{H_-^f(\omega)}{\omega} = 0,$$

as can be seen by closing the contour on $\Omega^{(+)}$, so that $(22.3.11)_4$ can be replaced by

$$\mathbf{K}_f(t) = \frac{1}{2\pi}\int_{-\infty}^{\infty} H_-^f(\omega)\mathbf{E}_+^t(\omega)d\omega. \tag{22.4.13}$$

The work function, given by (22.4.10), takes the form

$$W(t) = \phi_0(t) + \frac{1}{2\pi}\int_{-\infty}^{\infty} \left[|\mathbf{q}_-^{ft}(\omega)|^2 + |\mathbf{q}_+^{ft}(\omega)|^2\right] d\omega. \tag{22.4.14}$$

From (22.2.15), (22.4.12), and (22.4.14), we deduce that the total dissipation corresponding to the free energy $\psi_f(t)$ is given by

$$\mathfrak{D}_f(t) = \int_{-\infty}^{t} D_f(u)du = \frac{1}{2\pi}\int_{-\infty}^{\infty} |\mathbf{q}_+^{ft}(\omega)|^2 d\omega.$$

Differentiating this relation with respect to t and using $(22.3.11)_1$ and $(22.3.12)_3$ for scalar $\mathbb{H}$ give

$$D_f(t) = |\mathbf{K}_f(t)|^2. \tag{22.4.15}$$

Since $\mathbf{K}_f(t)$, given by (22.4.13), can vanish for nonzero histories, $D_f(t)$ is a positive semidefinite rather than a positive definite quadratic form.

The results for relative histories as defined for the general case by (22.3.16)–(22.3.19) can be adopted without difficulty to the present case, similarly for (22.3.20)–(22.3.25).

22.4.4 A Detailed Dielectric Model

We have, from (22.4.9),

$$H(\omega) = \omega\chi_s(\omega) = -\frac{\omega}{2i}(\chi_+(\omega) - \overline{\chi}_+(\omega)) \geq 0, \quad \omega \in \mathbb{R}. \tag{22.4.16}$$

As in [153–155], we take the susceptibility to be modeled by a sum of Lorentz oscillators:

$$\chi_+(\omega) = \sum_{n=1}^{N} \frac{f_n\omega_{p_n}^2}{\omega_n^2 - i\gamma_n\omega - \omega^2}, \tag{22.4.17}$$

in terms of the oscillator strength f_n, the plasma frequency ω_{p_n}, the resonant frequency ω_n , and the damping rate γ_n of each oscillator. All of these are positive parameters.

This behavior, while often used to describe memory behavior of dielectrics, is not usually applied to viscoelastic materials. The singularity structure given by (22.4.17) corresponds to exponential decay with sinusoidal behavior in the time domain (see also Chap. 19), while viscoelastic materials are generally modeled by simple exponential decay, which in the frequency domain yields simple poles on the positive imaginary axis.

It is assumed that

$$\omega_n^2 > \frac{\gamma_n^2}{4}.$$

The singularities of $\chi_+(\omega)$ are of course isolated. They are simple poles at

$$\begin{aligned} \zeta_n &= \frac{1}{2}\left[i\gamma_n + \sigma_n\right], \qquad \zeta_n' = \frac{1}{2}\left[i\gamma_n - \sigma_n\right] = -\overline{\zeta}_n, \\ \sigma_n &= \sqrt{4\omega_n^2 - \gamma_n^2}, \qquad n = 1, 2, \ldots, N. \end{aligned} \tag{22.4.18}$$

Thus, they occur in pairs, equidistant from the positive imaginary axis. Separating the poles in (22.4.17), we can write

$$\chi_+(\omega) = -\sum_{n=1}^{N} \frac{f_n\omega_{p_n}^2}{\sigma_n}\left\{\frac{1}{\omega - \zeta_n} - \frac{1}{\omega - \zeta_n'}\right\}. \tag{22.4.19}$$

In the time domain, this translates into

$$G'(s) = i\sum_{n=1}^{N} \frac{f_n\omega_{p_n}^2}{\sigma_n}(e^{i\zeta_n s} - e^{i\zeta_n' s}), \quad s \in \mathbb{R}^+, \tag{22.4.20}$$

which are decaying exponentials multiplying sine functions [122], so that there will be oscillatory behavior superimposed on the exponential decay. From (22.4.17) and (22.4.16), we have

$$H(\omega) = \sum_{n=1}^{N} \frac{f_n\omega_{p_n}^2\gamma_n\omega^2}{(\omega_n^2 - \omega^2)^2 + \gamma_n^2\omega^2},$$

which can be written as a ratio of two (factored) polynomials:

$$\begin{aligned} H(\omega) &= H_1\omega^2 \frac{\prod_{l=1}^{N-1}\left[(\omega - \eta_l)(\omega - \eta_l')(\omega - \overline{\eta}_l)(\omega - \overline{\eta}_l')\right]}{\prod_{l=1}^{N}\left[(\omega - \zeta_l)(\omega - \zeta_l')(\omega - \overline{\zeta}_l)(\omega - \overline{\zeta}_l')\right]}, \\ H_1 &= \sum_{n=1}^{N} f_n\omega_{p_n}^2\gamma_n, \quad \omega \in \mathbb{R}, \end{aligned} \tag{22.4.21}$$

where the denominator of $(22.4.21)_1$ uses the notation of (22.4.18), while the numerator is factored to yield the zeros of $H(\omega)$; these must occur also in pairs, as in (22.4.18), so that $\eta_l' = -\overline{\eta}_l$ for each l. We have explicitly included the fact that $H(\omega)$ vanishes quadratically at the origin. Note that the smaller number of zeros reflects the fact that $H(\omega)$ behaves as ω^{-2} for large ω.

The factorization can be carried out by inspection. We obtain

$$
\begin{aligned}
&H_-(\omega) = h_1\omega\frac{\prod_{l=1}^{N-1}(\omega-\overline{\eta_l})(\omega-\overline{\eta_l'})}{\prod_{l=1}^{N}(\omega-\overline{\zeta_l})(\omega-\overline{\zeta_l'})},\\
&\overline{\zeta_l}\in\Omega^{(-)},\ l=1,2,\ldots,N,\qquad \overline{\eta_l}\in\Omega^{(-)},\ l=1,2,\ldots,N-1,\\
&H_+(\omega)=\overline{H}_-(\omega)=H_-(-\omega),
\end{aligned}
\tag{22.4.22}
$$

where $h_1^2 = H_1$.

The most general case of a rational function, which is considered in [160] and Chap. 16, can be obtained from (22.4.21) by allowing singularities to coalesce. As a result of this, some zeros may also coalesce. Indeed, we cannot exclude the possibility that some of the zeros in (22.4.21) have a power higher than unity, even if all the singularities are simple poles. For simplicity, it is assumed that this does not happen for our choice of parameters.

Let us define for $l = 1, 2, \ldots, 2N-2$,

$$
\alpha_l = \begin{cases} \overline{\zeta_{\frac{l+1}{2}}}, & l \text{ odd},\\[4pt] \overline{\zeta'_{\frac{l}{2}}}, & l \text{ even}, \end{cases}
\qquad
\beta_l = \begin{cases} \overline{\eta_{\frac{l+1}{2}}}, & l \text{ odd},\\[4pt] \overline{\eta'_{\frac{l}{2}}}, & l \text{ even}, \end{cases}
$$

and

$$
\alpha_{2N-1} = \overline{\zeta_N}, \qquad \alpha_{2N} = \overline{\zeta'_N}.
$$

We can write (22.4.19) as

$$
\chi_+(\omega) = \sum_{l=1}^{2N}\frac{g_l}{\omega-\overline{\alpha_l}} = -\sum_{l=1}^{2N}\frac{g_l}{\omega+\alpha_l}, \qquad \overline{\alpha_l}\in\Omega^{(+)},
\tag{22.4.23}
$$

where (22.4.8) has been used and

$$
g_l = \begin{cases} -\dfrac{f_i\omega_{pi}^2}{\sigma_i}, & i=\dfrac{l+1}{2},\ l \text{ odd},\\[8pt] \dfrac{f_i\omega_{pi}^2}{\sigma_i}, & i=\dfrac{l}{2},\quad l \text{ even}. \end{cases}
$$

In this notation, (22.4.20) becomes

$$
G'(s) = i\sum_{l=1}^{2N} g_l e^{i\overline{\alpha_l}s} = -i\sum_{l=1}^{2N} g_l e^{-i\alpha_l s}.
\tag{22.4.24}
$$

We define the quantities

$$\mathbf{e}_l(t) = \mathbf{E}^t_+(\alpha_l), \qquad l = 1, 2, \ldots, 2N,$$

for which

$$\overline{\mathbf{e}_l}(t) = \overline{\mathbf{E}^t_+(\alpha_l)} = \mathbf{E}^t_+(-\overline{\alpha_l}), \qquad l = 1, 2, \ldots, 2N.$$

For $\omega = \alpha_l$, we have

$$\dot{\mathbf{e}}_l(t) = -i\alpha_l \mathbf{e}_l(t) + \mathbf{E}(t), \tag{22.4.25}$$

which is a special case of (7.2.27).

Let us now consider the free energies $\psi_f(t)$. We can put (22.4.22) in the form

$$\begin{gathered} H_-(\omega) = h_1\omega \frac{\prod_{l=1}^{2N-2}(\omega - \beta_l)}{\prod_{l=1}^{2N}(\omega - \alpha_l)} = h_1\omega \sum_{i=1}^{2N} \frac{R_i}{\omega - \alpha_i}, \\ \alpha_l \in \Omega^{(-)},\ l = 1, 2, \ldots, 2N, \quad \beta_l \in \Omega^{(-)},\ l = 1, 2, \ldots, 2N-2, \end{gathered} \tag{22.4.26}$$

where

$$R_i = (\omega - \alpha_i)H_-(\omega)\Big|_{\omega=\alpha_i} = h_1\alpha_i \frac{\prod_{l=1}^{2N-2}(\alpha_i - \beta_l)}{\prod_{l=1, l\neq i}^{2N}(\alpha_i - \alpha_l)}.$$

Also,

$$\begin{aligned} H_+(\omega) &= h_1\omega \frac{\prod_{l=1}^{2N-2}(\omega - \overline{\beta_l})}{\prod_{l=1}^{2N}(\omega - \overline{\alpha_l})} = -h_1\omega \frac{\prod_{l=1}^{2N-2}(\omega + \gamma_l)}{\prod_{l=1}^{2N}(\omega + \alpha_l)} \\ &= h_1\omega \sum_{i=1}^{2N} \frac{\overline{R_i}}{\omega - \overline{\alpha_i}} = h_1\omega \sum_{i=1}^{2N} \frac{R_i}{\omega + \alpha_i}. \end{aligned} \tag{22.4.27}$$

The relationship

$$\sum_{i=1}^{2N} R_i = 0 \tag{22.4.28}$$

must hold since $H_-(\omega)$ tends to $h_1\omega^{-1}$ at large ω. Therefore, we can also write $(22.4.26)_2$ and $(22.4.27)_{3,4}$ as

$$\begin{gathered} H_-(\omega) = h_1 \sum_{i=1}^{2N} \frac{\alpha_i R_i}{\omega - \alpha_i}, \\ H_+(\omega) = h_1 \sum_{i=1}^{2N} \frac{\overline{\alpha_i R_i}}{\omega - \overline{\alpha_i}} = h_1 \sum_{i=1}^{2N} \frac{\alpha_i R_i}{\omega + \alpha_i}. \end{gathered}$$

We identify also a much larger class of factorizations of $H(\omega)$, determined by interchanging particular β_l in $(22.4.26)_1$ with $\overline{\beta_l}$ in $(22.4.27)_1$. These different factorizations are labeled by the subscript or superscript f. Thus,

$$H_-^f(\omega) = h_1\omega\frac{\prod_{l=1}^{2N-2}(\omega-\zeta_l^f)}{\prod_{l=1}^{2N}(\omega-\alpha_l)} = h_1\omega\sum_{i=1}^{2N}\frac{R_i^f}{\omega-\alpha_i} = h_1\sum_{i=1}^{2N}\frac{\alpha_i R_i^f}{\omega-\alpha_i},$$
$$\alpha_l\in\Omega^{(-)},\ l=1,2,\ldots,2N,\qquad \beta_l\in\Omega^{(-)},\ l=1,2,\ldots,2N-2,$$
$$\zeta_l^f = \lambda_l^f\beta_l + (1-\lambda_l^f)\overline{\beta_l},\qquad \lambda_l^f = 0 \text{ or } 1, \tag{22.4.29}$$
$$R_i^f = (\omega-\alpha_i)H_-^f(\omega)\Big|_{\omega=\alpha_i} = h_1\alpha_i\frac{\prod_{l=1}^{2N-2}(\alpha_i-\zeta_l^f)}{\prod_{l=1,\,l\neq i}^{2N}(\alpha_i-\alpha_l)}.$$

Observe that (22.4.28) also holds for the R^f, a property that has been used in writing $(22.4.29)_3$. There are 2^{2N-2} different factorizations. Referring to the discussion in Sects. 22.4.2 and 22.4.3, we note that if all the λ_l^f are equal to one, then $\psi_f(t)$ is the minimum free energy $\psi_m(t)$, while if all are zero, we obtain the maximum free energy $\psi_M(t)$. All other possibilities yield functionals that are intermediate between these two extremes. These observations follow from Remark 22.4.1. Observe that

$$H(\omega) = H_+^f(\omega)H_-^f(\omega) = H_1\omega^2\sum_{i,j=1}^{2N}\frac{\overline{R_i^f}R_j^f}{(\omega-\overline{\alpha_i})(\omega-\alpha_j)}$$
$$= H_1\omega^2\sum_{i,j=1}^{2N}\frac{\overline{R_i^f}R_j^f}{\overline{\alpha_i}-\alpha_j}\left\{\frac{1}{\omega-\overline{\alpha_i}}-\frac{1}{\omega-\alpha_j}\right\}.$$

Now, we have the relation

$$\chi_+(\omega) = \frac{1}{\pi}\int_{-\infty}^{\infty}\frac{H(\omega')}{\omega'(\omega'-\omega^-)}d\omega',$$

which follows by applying an integration over $\mathbb{R}$ to (22.4.16) divided by $\omega'(\omega'-\omega^-)$ and completing the contour over $\Omega^{(-)}$. The integral over $\overline{\chi_+}(\omega)$ vanishes. Thus, we have, on integrating over Ω^+,

$$\chi_+(\omega) = \frac{H_1}{\pi}\int_{-\infty}^{\infty}\frac{\omega'}{\omega'-\omega^-}\sum_{i,j=1}^{2N}\frac{\overline{R_i^f}R_j^f}{\overline{\alpha_i}-\alpha_j}\left\{\frac{1}{\omega'-\overline{\alpha_i}}-\frac{1}{\omega'-\alpha_j}\right\}d\omega'$$
$$= -2iH_1\sum_{i,j=1}^{2N}\frac{\overline{\alpha_i}\,\overline{R_i^f}R_j^f}{(\overline{\alpha_i}-\alpha_j)(\omega-\overline{\alpha_i})}. \tag{22.4.30}$$

The quantity $\mathbf{q}_-^{ft}(\omega)$, defined by (22.4.11), may be evaluated by closing the contour on $\Omega^{(-)}$, giving

$$\mathbf{q}_-^{ft}(\omega) = \frac{1}{2\pi i}\int_{-\infty}^{\infty}\frac{H_-^f(\omega')\mathbf{E}_+^t(\omega')}{\omega'-\omega^+}d\omega' = h_1\sum_{i=1}^{n}\frac{\alpha_i R_i^f\mathbf{e}_i(t)}{\omega-\alpha_i}. \tag{22.4.31}$$

From $(22.4.11)_1$, we have

$$\mathbf{q}_+^{ft}(\omega) = \mathbf{q}_-^{ft}(\omega) - H_-^f(\omega)\mathbf{E}_+^t(\omega)$$
$$= h_1 \sum_{i=1}^{n} R_i^f \frac{[\alpha_i \mathbf{E}_+^t(\alpha_i) - \omega \mathbf{E}_+^t(\omega)]}{\omega - \alpha_i},$$

which has singularities at those of $\mathbf{E}_+^t(\omega)$ in $\Omega^{(+)}$ but none in $\Omega^{(-)}$. These explicit relations allow analytic continuation of $\mathbf{q}_\pm^t(\omega)$ to the whole complex plane, excluding singular points. From (22.4.25), (22.4.29), and (22.4.31), we see that the quantity $\mathbf{p}_-^{ft}(\omega)$ is given by

$$\mathbf{p}_-^{ft}(\omega) = \frac{1}{2\pi i} \int_{-\infty}^{\infty} \frac{H_-^f(\omega')\mathbf{E}_{r+}^t(\omega')}{\omega' - \omega^+} d\omega' = ih_1 \sum_{i=1}^{n} \frac{R_i^f \dot{\mathbf{e}}_i(t)}{\omega - \alpha_i}. \tag{22.4.32}$$

The optimal history/continuation in the frequency domain, which is a special case of (22.3.8), is given by formulae that are generalizations of those in Sect. 16.9.1.

We deduce from $(22.4.12)_2$ and (22.4.31) that

$$\begin{aligned} \psi_f(t) &= \phi_0(t) + iH_1 \sum_{i,j=1}^{2N} \frac{\overline{\alpha_i}\alpha_j \overline{R_i^f} R_j^f}{\overline{\alpha_i} - \alpha_j} \overline{\mathbf{e}_i}(t) \cdot \mathbf{e}_j(t) \\ &= \phi_0(t) - \frac{1}{2} \int_0^{\infty} \int_0^{\infty} \mathbf{E}^t(s_1) \cdot G_{12}^f(s_1, s_2) \mathbf{E}^t(s_2) ds_1 ds_2, \end{aligned} \tag{22.4.33}$$

where

$$G_{12}^f(s_1, s_2) = -2iH_1 \sum_{i,j=1}^{2N} \frac{\overline{\alpha_i}\alpha_j \overline{R_i^f} R_j^f}{\overline{\alpha_i} - \alpha_j} e^{i\overline{\alpha_i} s_1 - i\alpha_j s_2}.$$

Also, from (22.4.32),

$$\begin{aligned} \psi_f(t) &= U(t) + iH_1 \sum_{i,j=1}^{2N} \frac{\overline{R_i^f} R_j^f}{\overline{\alpha_i} - \alpha_j} \overline{\dot{\mathbf{e}}_i}(t) \cdot \dot{\mathbf{e}}_j(t) \\ &= U(t) - \frac{1}{2} \int_0^{\infty} \int_0^{\infty} \dot{\mathbf{E}}^t(s_1) \cdot \widetilde{G}^f(s_1, s_2) \dot{\mathbf{E}}^t(s_2) ds_1 ds_2, \end{aligned} \tag{22.4.34}$$

where

$$\widetilde{G}^f(s_1, s_2) = -2iH_1 \sum_{i,j=1}^{2N} \frac{\overline{R_i^f} R_j^f}{\overline{\alpha_i} - \alpha_j} e^{i\overline{\alpha_i} s_1 - i\alpha_j s_2}. \tag{22.4.35}$$

We write $(22.4.33)_1$ and $(22.4.34)_1$ in the form

$$\begin{aligned} \psi_f(t) &= \phi_0(t) + \frac{1}{2} \sum_{i,j=1}^{2N} C'_{fij} \overline{\mathbf{e}_i}(t) \cdot \mathbf{e}_j(t), \\ &= U(t) + \frac{1}{2} \sum_{i,j=1}^{2N} C_{fij} \overline{\dot{\mathbf{e}}_i}(t) \cdot \dot{\mathbf{e}}_j(t), \qquad C_{fij} = 2iH_1 \frac{\overline{R_i^f} R_j^f}{\overline{\alpha_i} - \alpha_j}. \end{aligned}$$

Comparing $(22.4.30)_2$ with (22.4.23), we see that

$$g_i = -2iH_1 \sum_{j=1}^{2N} \frac{\overline{\alpha_i} \overline{R_i^f} R_j^f}{\overline{\alpha_i} - \alpha_j} = -2iH_1 \sum_{j=1}^{2N} \frac{\alpha_j \overline{R_i^f} R_j^f}{\overline{\alpha_i} - \alpha_j}, \tag{22.4.36}$$

where (22.4.28), for the R_i^f, has been used. The relation $(22.4.4)_1$, which yields that $G_2^f(0, s) = G'(s)$, where the latter quantity is given by (22.4.24), can be confirmed with the aid of (22.4.35) and (22.4.36).

From (22.4.13) and $(22.4.29)_{2,3}$, we have

$$\mathbf{K}_f(t) = -ih_1 \left[\sum_{i=1}^{n} \alpha_i R_i^f \mathbf{e}_i(t) \right] = h_1 \left[\sum_{i=1}^{n} R_i^f \dot{\mathbf{e}}_i(t) \right],$$

so that

$$\begin{aligned} D_f(t) &= H_1 \left| \sum_{i=1}^{n} \alpha_i R_i^f \mathbf{e}_i(t) \right|^2 = H_1 \left| \sum_{i=1}^{n} R_i^f \dot{\mathbf{e}}_i(t) \right|^2 \\ &= H_1 \sum_{i,j=1}^{2N} \overline{R_i^f} R_j^f \overline{\dot{\mathbf{e}}_i}(t) \cdot \dot{\mathbf{e}}_j(t) \\ &= H_1 \sum_{i,j=1}^{2N} \overline{\alpha_i} \alpha_j \overline{R_i^f} R_j^f \overline{\mathbf{e}_i}(t) \cdot \mathbf{e}_j(t). \end{aligned} \tag{22.4.37}$$

Note that $D_f(t)$ vanishes if

$$\sum_{i=1}^{2N} \alpha_i R_i^f \mathbf{e}_i(t) = \mathbf{0}.$$

Solutions to this equation will exist for nonzero values of $\mathbf{e}_i(t)$. Therefore, $(22.4.37)_3$ is a positive semidefinite rather than a positive definite quadratic form, so that the associated matrix will have some zero eigenvalues. If one of these zero eigenvalues were to become slightly negative, then the second law would no longer hold. Thus, the free energy $\psi_f(t)$ is on the boundary of the set of free energies. A more general related observation was made after (22.4.15).

23 Fractional Derivative Models of Materials with Memory

23.1 Introduction

Materials with constitutive equations expressed in terms of fractional derivatives [47] are of increasing interest in recent years (see [214, 287]). It is well known that such materials can be considered in the class of materials with memory and may describe elastic, fluid, viscoelastic, and electromagnetic materials, but also other kinds of phenomena, such as heat flux models.

The fractional derivative central to the present study is that of Caputo [48, 51]. We consider thermomechanical models with memory within this fractional derivative framework, and compare them with the classical Volterra theory, which is that described and used throughout most of the present work. It emerges that the two viewpoints are formally similar [250]. Indeed, fractional models are those for which the viscoelastic memory kernel (or relaxation function) $G(s)$ is given by

$$G(s) = \frac{k_0}{s^\alpha}\,, \quad \alpha \in (0,1)\ ,\ k_0 > 0. \tag{23.1.1}$$

However, in contrast to the Volterra theory for fluids with memory, this kernel is not $L^1(0,\infty)$, which implies significant dissimilarity in the behavior of solutions of the dynamical equations, compared with the traditional theory. The differences are more evident for solid materials. Therefore, the fractional and Volterra models provide viewpoints which are not reconcilable. Various of these differences between the two models are noted in [128].

An analysis of the thermodynamic restrictions provides compatibility conditions on the kernels. These conditions, combined with analogies with the Volterra theory, yield certain free energies, which enable the definition of a topology on the history space. A similar analysis can be carried out for the phenomenon of heat propagation with memory.

The derivation of the minimum free energy in this context is presented in particular detail because it requires careful treatment of the factorization problem.

G. Amendola et al., *Thermodynamics of Materials with Memory*,
https://doi.org/10.1007/978-3-030-80534-0_23

For solid viscoelastic materials, some experimental observations are particularly in agreement with models using fractional derivatives, because of the power law behavior of the relaxation function given by (23.1.1) (see [50, 167, 202, 280, 293, 296, 303]). The creep function for such models also has a power law form. Such experimental backing has motivated many studies of materials with fading memory given by a fractional derivative, including [26, 48, 140, 200, 214, 249, 261] and in the frequency domain [202, 303] .

Many experimental observations on a variety of materials subject to a constant load show plastic behavior, which can be described by the fractional derivative approach. However, this is not predicted by Volterra models, which under constant load describe elastic materials. Moreover, when the load is removed, fractional models predict recovery of a portion of the deformation, unlike for the case of classical viscous fluids. Thus, the fractional derivative approach allows us to describe materials displaying both elastic and viscous/plastic behavior.

The fractional theory can also be applied to heat diffusion. It seems natural to generalize the Fourier law and the Cattaneo–Maxwell equation, using a fractional derivative instead of the time derivative. This approach allows us to describe a wider range of phenomena and gives a good description of oscillating behavior [1, 223, 285, 299].

In recent decades, the fractional calculus has been widely used, as indicated by the many mathematical volumes dealing with this topic (e.g., Baleanu et al. [28], Caponetto [53], Caputo [46], Diethelm [95], Hilfer [198], Jiao et al. [214], Kilbas et al. [219], Kyriakova [222], Mainardi [248], McBride [259], Miller and Ross [264], Petras [286], Samko et al. [295], Podlubny [287], Sabatier et al. [294], Torres and Malinowska [309], Ying and Chen [327]), by the many meetings dedicated to it and the plethora of articles appearing in mathematical (e.g., Kilbas and Marzan [218], Heinsalu et al. [196], Luchko and Gorenflo [246]) and non-mathematical journals.

The use of derivative of fractional order has also spread into many other fields of science besides mathematics and physics (e.g., Laskin [229], Naber [275], Baleanu et al. [29], Zavada [328], Baleanu et al. [30], Caputo and Fabrizio [55]) such as biology (e.g., Cesarone et al. [62], Caputo and Cametti [52]), economics (e.g., Caputo [54]), demography (e.g., Jumarie [217]), geophysics (e.g., Iaffaldano [207]) and medicine (e.g., El Sahed [99], Magin [247]). However, its somewhat cumbersome mathematical definition and the consequent complications in the solution of fractional order differential equations have led to some difficulties.

23.2 Fractional Derivatives

In this section, the original Caputo fractional derivative is introduced, together with certain new fractional derivative formulae without singularities in the kernel.

23.2.1 The Caputo Fractional Derivative

We begin with an argument which provides motivation for the formula, introduced below [47].

Let $f(z)$ be analytic on an open subset O of the complex plane, where O includes the real axis. Then, an immediate consequence of Cauchy's Integral Formula, (B.1.2) is

$$f^{(n)}(z) = \frac{n!}{2\pi i}\oint_C \frac{f(z')}{(z'-z)^{n+1}}dz' = \frac{(n-1)!}{2\pi i}\oint_C \frac{f'(z')}{(z'-z)^{n}}dz',$$

where $f^{(n)}(z)$ is the nth derivative of $f(z)$, while $f'(z) = f^{(1)}(z)$. The contour $C \subset O$ includes the point z. Let us replace n by $\alpha \in \mathbb{R}^+$. We take z to be on the real axis, denoted by t, and the contour to be tightly around the branch cut joining the branch point t to infinity. This step forces the exclusion of integrals with $\alpha \in (1, \infty)$. The cut is assumed to lie along the semi-infinite interval $(-\infty, t)$. The integration variable is changed to τ. We have therefore

$$f^{(\alpha)}(t) = \frac{\Gamma(\alpha)}{2\pi i}\oint_C \frac{f'(\tau)}{|\tau - t|^{\alpha}e^{i\theta}}d\tau, \tag{23.2.1}$$

where θ is the argument of the denominator, which varies as we move around the contour, anti-clockwise. Below the cut, we have $\theta = -\pi\alpha$, while above the cut, it is $\theta = \pi\alpha$. The quantity $\Gamma(\cdot)$ is the Gamma function, given for any $\beta > 0$, by

$$\Gamma(\beta) = \int_0^{\infty} r^{\beta-1}e^{-r}\,dr. \tag{23.2.2}$$

Thus,

$$f^{(\alpha)}(t) = \frac{\Gamma(\alpha)\sin(\pi\alpha)}{\pi}\int_{-\infty}^{t}\frac{f'(\tau)}{|\tau - t|^{\alpha}}d\tau.$$

Using the standard formula

$$\Gamma(1-z)\Gamma(z) = \frac{\pi}{\sin(\pi z)}, \tag{23.2.3}$$

relation (23.2.1) becomes

$$f^{(\alpha)}(t) = \frac{1}{\Gamma(1-\alpha)}\int_{-\infty}^{t}\frac{f'(\tau)}{|\tau - t|^{\alpha}}d\tau. \tag{23.2.4}$$

The analyticity property assumed for f can be weakened but must ensure that the derivative and integral in (23.2.4) exist.

The general form of the Caputo α fractional derivative, defined for any $\alpha \in (0, 1)$, is given by

$${}_a^C D_t^{\alpha} f(t) = \frac{1}{\Gamma(1-\alpha)}\int_a^t \frac{f'(\tau)}{(t-\tau)^{\alpha}}d\tau, \tag{23.2.5}$$

where $a \in (-\infty, t)$, $f \in H^1(a, b)$, where $b > t$. We can take $a = -\infty$, since if necessary it is always possible to extend f to the interval $(-\infty, a)$ by the zero function. Thus, we have

$$ {}_{-\infty}^{\;\;C}D_t^{\alpha} f(t) = \frac{1}{\Gamma(1-\alpha)} \int_{-\infty}^{t} \frac{f'(\tau)}{(t-\tau)^{\alpha}} d\tau = F(t), \tag{23.2.6} $$

which agrees with the form (23.2.4). We assume that $f(t) \to 0$, as t tends to $-\infty$. Now, (23.2.6) can be written as

$$ F(t) = \frac{\alpha}{\Gamma(1-\alpha)} \int_{-\infty}^{t} \frac{f(t) - f(\tau)}{(t-\tau)^{1+\alpha}} d\tau, \tag{23.2.7} $$

or, by a change of variable,

$$ F(t) = -\frac{\alpha}{\Gamma(1-\alpha)} \int_{0}^{\infty} \frac{f(t-s) - f(t)}{s^{1+\alpha}} ds, \tag{23.2.8} $$

which are equivalent representations of the Caputo derivative. Thus, the definition (23.2.5) may be rewritten as

$$ \begin{aligned} {}_{-\infty}^{\;\;C}D_t^{\alpha} f(t) &= \frac{\alpha}{\Gamma(1-\alpha)} \int_{-\infty}^{t} \frac{f(t) - f(\tau)}{(t-\tau)^{1+\alpha}} d\tau \\ &= -\frac{\alpha}{\Gamma(1-\alpha)} \int_{0}^{\infty} \frac{f(t-s) - f(t)}{s^{1+\alpha}} ds. \end{aligned} \tag{23.2.9} $$

Remark 23.2.1. As α tends to zero, we see from (23.2.6) that

$$ {}_{-\infty}^{\;\;C}D_t^{\alpha} f(t) \to f(t). \tag{23.2.10} $$

If α tends to 1, the integral in (23.2.5) diverges. However, there is a limit which gives a finite value (as must be true, given the derivation of (23.2.4)). Before the limit $\alpha \to 1$ is taken, let us reverse the shrinking of the contour described above, so that the integral exists for all nonnegative powers of the denominator. In particular, as $\alpha \to 1$, we have

$$ {}_{-\infty}^{\;\;C}D_t^{\alpha} f(t) = f'(t). \tag{23.2.11} $$

23.2.2 Fractional Derivatives Without Singular Kernels

We present a new definition of fractional derivative with a smooth kernel which takes on two different representations for the temporal and spatial variables, respectively. For the first, operating on time variables, it is natural to use the Laplace transform. For the other representation, related to the spatial variables by a nonlocal fractional derivative, it is more convenient to use the Fourier transform. The interest in this new approach with a regular kernel in terms of spatial variables arose from the perception that there is a class of nonlocal models, which have the ability to describe material heterogeneities and fluctuations on different scales, which cannot be described adequately by classical local theories or by fractional models with singular kernels.

The original definition of fractional derivative appears to be particularly convenient for mechanical and electromagnetic phenomena involving plasticity, fatigue, damage, and electromagnetic hysteresis. When these effects are not present it seems more appropriate to use the new temporal fractional derivative to describe, for example, the behavior of classical viscoelastic materials, thermal media, and electromagnetic systems.

The new nonlocal fractional derivative for spatial variables can describe material heterogeneities and structures with different scales, which cannot be described adequately by classical local theories.

23.2.2.1 A New Fractional Time Derivative

We refer to the usual Caputo fractional time derivative as UFD$_t$. For order α, it is given by (23.2.5), or in somewhat simplified notation, by

$$D_t^{(\alpha)} f(t) = \frac{1}{\Gamma(1-\alpha)} \int_a^t \frac{f'(\tau)}{(t-\tau)^\alpha} d\tau \tag{23.2.12}$$

with $\alpha \in [0,1]$ and $a \in (-\infty, t)$, $f \in H^1(a,b)$, $b > a$. By replacing the kernel $(t-\tau)^{-\alpha}$ with the function $exp(-\frac{\alpha}{1-\alpha}(t-\tau))$ and $\frac{1}{\Gamma(1-\alpha)}$ with $\frac{M(\alpha)}{1-\alpha}$, we obtain the following definition of a new fractional time derivative, denoted by NFD$_t$, and given by the formula

$$\mathfrak{D}_t^{(\alpha)} f(t) = \frac{M(\alpha)}{1-\alpha} \int_a^t f'(\tau) \exp\left[-\frac{\alpha(t-\tau)}{1-\alpha}\right] d\tau = N(t), \tag{23.2.13}$$

where $M(\alpha)$ is a normalization coefficient with the property that $M(0) = M(1) = 1$. According to the definition (23.2.13), the NFD$_t$ is zero when $f(t)$ is constant, as is the UFD$_t$. However, in contrast to the UFD$_t$, the kernel in (23.2.13) does not have a singularity for $t = \tau$.

The NFD$_t$ can also be applied to functions that do not belong to $H^1(a,b)$. Indeed, the definition (23.2.13) can be formulated also for $f \in L^1(-\infty, b)$ and for any $\alpha \in (0,1)$, which can be seen by expressing (23.2.13) in the form

$$\mathfrak{D}_t^{(\alpha)} f(t) = \frac{\alpha M(\alpha)}{(1-\alpha)^2} \int_{-\infty}^t (f(t) - f(\tau)) \exp\left[-\frac{\alpha(t-\tau)}{1-\alpha}\right] d\tau.$$

If we put

$$\sigma = \frac{1-\alpha}{\alpha} \in [0,\infty] \ , \ \alpha = \frac{1}{1+\sigma} \in (0,1),$$

definition (23.2.13) of NFD$_t$ assumes the form

$$\tilde{\mathfrak{D}}_t^{(\sigma)} f(t) = \frac{(1+\sigma)M^*(\sigma)}{\sigma} \int_a^t f'(\tau) \exp\left[-\frac{(t-\tau)}{\sigma}\right] d\tau, \tag{23.2.14}$$

where $\sigma \in (0,\infty)$ and $M^*(\sigma)$ is the corresponding normalization term to $M(\alpha)$, and is such that $M^*(0) = M^*(\infty) = 1$.

In the context of mechanical models discussed later, for $a \to -\infty$, the relation (23.2.14) corresponds to a relaxation function given by

$$G(s) = \frac{(1+\sigma)M^*(\sigma)}{\sigma} \exp\left[-\frac{s}{\sigma}\right],$$

instead of that given by (23.1.1) for the Caputo fractional derivative. This describes the well-known Maxwell material (for example [167]), which is a very simple model of a viscoelastic fluid.

For $\alpha \to 1$, we have $\sigma \to 0$. Also,

$$\lim_{\sigma \to 0} \frac{1+\sigma}{\sigma} \exp\left[-\frac{(t-\tau)}{\sigma}\right] = \delta(t-\tau).$$

Therefore (see [150] and [197]),

$$\begin{aligned}\lim_{\alpha \to 1} \mathcal{D}_t^{(\alpha)} f(t) &= \lim_{\alpha \to 1} \frac{M(\alpha)}{1-\alpha} \int_a^t f'(\tau) \exp\left[-\frac{\alpha(t-\tau)}{1-\alpha}\right] d\tau \\ &= \lim_{\sigma \to 0} \frac{(1+\sigma)M^*(\sigma)}{\sigma} \int_a^t f'(\tau) \exp\left[-\frac{(t-\tau)}{\sigma}\right] d\tau = f'(t).\end{aligned} \tag{23.2.15}$$

Also, for $\alpha \to 0$ we have $\sigma \to +\infty$. Hence,

$$\begin{aligned}\lim_{\alpha \to 0} \mathcal{D}_t^{(\alpha)} f(t) &= \lim_{\alpha \to 0} \frac{M(\alpha)}{1-\alpha} \int_a^t f'(\tau) \exp\left[-\frac{\alpha(t-\tau)}{1-\alpha}\right] d\tau \\ &= \lim_{\sigma \to +\infty} \frac{(1+\sigma)M^*(\sigma)}{\sigma} \int_a^t f'(\tau) \exp\left[-\frac{(t-\tau)}{\sigma}\right] d\tau = f(t) - f(a).\end{aligned} \tag{23.2.16}$$

If $f(a) = 0$, then

$$\lim_{\alpha \to 0} \mathcal{D}_t^{(\alpha)} f(t) = f(t). \tag{23.2.17}$$

Thus, (23.2.15) and (23.2.17) indicate that for $\alpha = 0$ and 1, the NFD_t behaves as expected in the integer limit.

Simulations comparing the UFD_t and NFD_t for particular choices of $f(\cdot)$ were described in [56]. These suggested that for $\alpha = 0.66$ the classical NFD_t is very similar to the UFD_t. However, for models with α close to 0, we see different behavior. For $\alpha = 0.1$ differences between NFD_t and UFD_t become apparent. In particular the classical UFD_t is more affected by past history than NFD_t, which exhibits rapid stabilization.

For integer n where $n \geq 1$, and $\alpha \in [0, 1]$ the fractional time derivative $\mathcal{D}_t^{(\alpha+n)} f(t)$ of order $(n + \alpha)$ is defined by

$$\mathcal{D}_t^{(\alpha+n)} f(t) := \mathcal{D}_t^{(\alpha)}(\mathcal{D}_t^{(n)} f(t)). \tag{23.2.18}$$

Theorem 23.2.1. *If the function $f(t)$ is such that*

$$f^{(s)}(a) = 0, \quad s = 1, 2, \ldots, n,$$

then we have

$$\mathcal{D}_t^{(n)}(\mathcal{D}_t^{(\alpha)} f(t)) = \mathcal{D}_t^{(\alpha)}(\mathcal{D}_t^{(n)} f(t)).$$

Proof. Consider the case $n = 1$. From definition (23.2.18) of $\mathcal{D}_t^{(\alpha+1)} f(t)$, we obtain

$$\mathcal{D}_t^{(\alpha)}\left(\mathcal{D}_t^{(1)} f(t)\right) = \frac{M(\alpha)}{1-\alpha}\int_a^t f''(\tau)\exp\left[-\frac{\alpha(t-\tau)}{1-\alpha}\right]d\tau.$$

By means of an integration by parts, using the property $f'(a) = 0$, we obtain

$$\begin{aligned}\mathcal{D}_t^{(\alpha)}\left(\mathcal{D}_t^{(1)} f(t)\right) &= \frac{M(\alpha)}{(1-\alpha)}\int_a^t \left(\frac{d}{d\tau} f'(\tau)\right)\exp-\left[\frac{\alpha(t-\tau)}{1-\alpha}\right]d\tau \\ &= \frac{M(\alpha)}{(1-\alpha)}\left[\int_a^t \frac{d}{d\tau}\left\{f'(\tau)\exp\left[-\frac{\alpha(t-\tau)}{1-\alpha}\right]\right\}d\tau \right. \\ &\quad \left. -\frac{\alpha}{1-\alpha}\int_a^t f'(\tau)\exp\left[-\frac{\alpha(t-\tau)}{1-\alpha}\right]d\tau\right] \\ &= \frac{M(\alpha)}{(1-\alpha)}\left[f'(t) - \frac{\alpha}{1-\alpha}\int_a^t f'(\tau)\exp\left[-\frac{\alpha(t-\tau)}{1-\alpha}\right]d\tau\right].\end{aligned}$$

Also,

$$\begin{aligned}\mathcal{D}_t^{(1)}(\mathcal{D}_t^{(\alpha)} f(t)) &= \frac{d}{dt}\left\{\frac{M(\alpha)}{1-\alpha}\int_a^t f'(\tau)\exp\left[-\frac{\alpha(t-\tau)}{1-\alpha}\right]d\tau\right\} \\ &= \frac{M(\alpha)}{1-\alpha}\left[f'(t) - \frac{\alpha}{1-\alpha}\int_a^t f'(\tau)\exp\left[-\frac{\alpha(t-\tau)}{1-\alpha}\right]d\tau\right].\end{aligned}$$

It is easy to generalize the proof for any $n > 1$. □

The property asserted in Theorem 23.2.1 is implied by the notation on the left of (23.2.18). Also, note that (23.2.18), together with (23.2.15) and (23.2.16) yield that

$$\lim_{\alpha\to 0}\mathcal{D}_t^{(\alpha+n)} f(t) = f^{(n)}(t), \qquad \lim_{\alpha\to 1}\mathcal{D}_t^{(\alpha+n)} f(t) = f^{(n+1)}(t),$$

which, again, is expected behavior for a non-integer derivative.

In the following, we suppose the function $M(\alpha) = 1$. We can rewrite the definition (23.2.14) in the form

$$\tilde{\mathcal{D}}_t^{(\nu)} f(t) = V(\nu)\int_a^t f'(\tau)\exp[-\nu(t-\tau)]d\tau \tag{23.2.19}$$

obtained from (23.2.13) or (23.2.14) with $\nu = 1/\sigma > 0$, where $V(\nu) = (\nu+1)M^*(1/\nu)$. Then, we have the following theorem.

Theorem 23.2.2. *If the function $f \in W^{1,1}(a,b)$, then the integral in (23.2.19) exists for $t \in [a,b]$ and $\tilde{\mathcal{D}}_t^{(\nu)} f(t) \in L^1\,[a,b]$.*

Proof. Let us write

$$\tilde{\mathfrak{D}}_t^{(\nu)} f(t) = V(\nu)\int_a^t f'(\tau)\exp[-\nu(t-\tau)]d\tau =$$
$$= V(\nu)\int_{-\infty}^{\infty} p_\nu(t-s)q(s)ds, \tag{23.2.20}$$

where $p_\nu(y) = \exp(-\nu y)$, *when* $0 < y < b - a$, *with* $p_\nu(y) = 0$ *when* $y < 0$ *or* $y > b - a$. *Also,* $q(y) = f'(y)$ *when* $a \le y \le b$. *Finally,* $q(y) = 0$ *when* $y < a$ *or* $y > b$. *Hence, under the hypotheses of the theorem, the functions* p_ν, $q \in L^1(a,b)$. *Then, by the classical results on Lebesgue integrals (see [326]), the integral (23.2.19) exists almost everywhere in* $t \in [a,b]$ *and* $\tilde{\mathfrak{D}}_t^{(\nu)} f(t) \in L^1\,[a,b]$.

The definition (23.2.19) can be generalized by choosing p_ν in (23.2.20) to be any function with the properties assigned to the exponential kernel in Theorem 23.2.2, where $p_\nu(0)$ is a finite number, chosen for convenience to be unity. This latter property ensures that it constitutes a non-singular kernel. It can be shown without difficulty that Theorem 23.2.1 also applies to the generalized definition.

23.2.2.2 Some Results for Given Histories

It is of interest to see the fractional derivatives of elementary functions according to the new definition (23.2.13). We begin with $\sin\omega t$ and $\cos\omega t$. It is convenient to first consider $f(t) = \exp(i\omega t)$, which combined both of these. In fact, we have

$$\mathfrak{D}_t^{(\alpha)}\exp(i\omega t) = \mathfrak{D}_t^{(\alpha)}(\cos\omega t) + i\mathfrak{D}_t^{(\alpha)}(\sin\omega t),$$

and

$$\begin{aligned}\mathfrak{D}_t^{(\alpha)}\exp(i\omega t) &= i\omega E(\alpha)\int_0^t \exp(-\nu(t-s)+i\omega s)ds\\ &= \frac{i\omega E(\alpha)}{i\omega+\nu}[\exp(i\omega t)-\exp(-\nu t)]\\ &= i\omega E(\alpha)\frac{(\cos\omega t + i\sin\omega t - \exp(-\nu t))(\nu - i\omega)}{\nu^2+\omega^2},\\ \nu &= \frac{\alpha}{1-\alpha},\end{aligned}$$

where $E(\alpha) = \dfrac{M(\alpha)}{1-\alpha}$. We have, from these relations that

$$\begin{aligned}\mathfrak{D}_t^{(\alpha)}(\sin\omega t) &= \frac{E(\alpha)\omega}{\nu^2+\omega^2}[\nu\cos\omega t + \omega\sin\omega t - \nu\exp(-\nu t)]\\ &= \frac{E(\alpha)\omega}{\nu^2+\omega^2}[\sqrt{\nu^2+\omega^2}\sin(\omega t+\lambda) - \nu\exp(-\nu t)]\\ &= E(\alpha)\cos\lambda[\sin(\omega t+\lambda) - \sin\lambda\exp(-\nu t)],\end{aligned}$$

where λ is such that

$$\tan\lambda = \frac{\nu}{\omega}, \quad \sin\lambda = \frac{\nu}{\sqrt{\nu^2+\omega^2}}, \quad \cos\lambda = \frac{\omega}{\sqrt{\nu^2+\omega^2}}.$$

Thus, the new derivative of $\sin\omega t$ yields a change of phase by amount λ, while the amplitude becomes

$$E(\alpha)\cos\lambda = \frac{\omega E(\alpha)}{\sqrt{\nu^2+\omega^2}}.$$

Also,

$$\mathcal{D}_t^{(\alpha)}(\cos\omega t) = E(\alpha)\cos\lambda[\cos(\omega t+\lambda) - \cos\lambda\exp(-\nu t)],$$

which also exhibits a phase change and the same amplitude variation noted for the case of $\sin\omega t$.

The new derivative, for an exponential history, has the form

$$\begin{aligned}\mathcal{D}_t^{(\alpha)}(\exp\omega t) &= \frac{E(\alpha)\omega}{\nu+\omega}\{\exp(\omega t) - \exp(-\nu t)\}\\ &= \frac{E(\alpha)\omega}{\nu+\omega}\exp(\omega t)\{1-\exp[-(\omega+\nu t)]\}.\end{aligned}$$

Finally, for a linear history, defined by

$$f(t) = \begin{cases} t, & t \geq 0,\\ 0, & t < 0,\end{cases}$$

we obtain

$$\begin{aligned}\mathcal{D}_t^{(\alpha)}t &= \frac{M(\alpha)}{1-\alpha}\int_0^t \exp(-\nu(t-s))ds\\ &= \frac{M(\alpha)}{\alpha}[1-\exp(-\nu t)], \ 0 < \alpha \leq 1.\end{aligned}$$

23.2.2.3 The Laplace Transform of the NFD$_t$

The Laplace transform of the NFD$_t$, defined by Eq. (23.2.13), will be of interest. We have

$$\begin{aligned}N_L(p) &= \int_0^\infty \exp(-pt)\mathcal{D}_t^\alpha f(t)dt = \int_0^\infty \exp(-pt)N(t)dt,\\ N_L^{(n)}(p) &= \int_0^\infty \exp(-pt)\mathcal{D}_t^{\alpha+n} f(t)(t)dt = \int_0^\infty \exp(-pt)N^{(n)}(t)dt.\end{aligned}$$

One can show that

$$\begin{aligned}&\int_0^\infty \exp(-pt)f'(t)dt = [f_L'](p) = pf_L(p) - f(0),\\ &\int_0^\infty \exp(-pt)\exp\left[-\frac{\alpha t}{1-\alpha}\right]dt = \frac{1-\alpha}{p+\alpha(1-p)}.\end{aligned}$$

Because of the convolution form of $N(t)$ in (23.2.13), we have

$$N_L(p) = M(\alpha)\frac{pf_L(p) - f(0)}{p + \alpha(1-p)}. \tag{23.2.21}$$

Similarly,

$$N_L^{(1)}(p) = M(\alpha)\frac{p^2 f_L(p) - pf(0) - f'(0)}{p + \alpha(1-p)},$$

and, more generally,

$$N_L^{(n)}(p) = M(\alpha)\frac{p^{n+1} f_L(p) - p^n f(0) - p^{n-1} f'(0) \ldots - f^{(n)}(0)}{p + \alpha(1-p)}.$$

23.2.2.4 Fractional Gradient Operator

We introduce a new concept of fractional gradient, which can describe nonlocal dependence in constitutive equations [324, 325].

Let us consider a set $\Omega \in \mathbb{R}^3$ and a scalar function $u(\cdot) : \Omega \to \mathbb{R}$. We define the fractional gradient of order $\alpha \in [0, 1]$ as follows

$$\nabla^{(\alpha)} u(\mathbf{x}) = \frac{\alpha}{(1-\alpha)\sqrt{\pi^\alpha}} \int_\Omega \nabla u(\mathbf{y}) \exp\left[-\frac{\alpha^2(\mathbf{x}-\mathbf{y})^2}{(1-\alpha)^2}\right] d\mathbf{y} \tag{23.2.22}$$

with $\mathbf{x}, \mathbf{y} \in \boldsymbol{\Omega}$. The nonlocal property of this fractional derivative relates to the integration of $\mathbf{y}$ over $\boldsymbol{\Omega}$. A rotationally invariant three-dimensional Normal (Gaussian) distribution has been chosen to describe this nonlocality.

It is easy to prove from definition (23.2.22) that

$$\nabla^{(1)} u(\mathbf{x}) = \nabla u(\mathbf{x}),$$

using the relation

$$\lim_{\alpha \to 1} \frac{\alpha}{(1-\alpha)\sqrt{\pi^\alpha}} \exp\left[-\frac{\alpha^2(\mathbf{x}-\mathbf{y})^2}{(1-\alpha)^2}\right] = \delta(\mathbf{x}-\mathbf{y}).$$

Thus, when $\alpha = 1$, $\nabla^{(\alpha)} u(\mathbf{x})$ loses the nonlocality property. Also, we clearly have

$$\nabla^{(0)} u(\mathbf{x}) = 0.$$

This fractional gradient is easily generalized to the case of a vector $\mathbf{u}(\mathbf{x})$, where the gradient is assumed to exist on Ω. We define the fractional gradient of this vector by

$$\nabla^{(\alpha)} \mathbf{u}(\mathbf{x}) = \frac{\alpha}{(1-\alpha)\sqrt{\pi^\alpha}} \int_\Omega \nabla \mathbf{u}(\mathbf{y}) \exp\left[-\frac{\alpha^2(\mathbf{x}-\mathbf{y})^2}{(1-\alpha)^2}\right] d\mathbf{y}.$$

Thus, a material with a nonlocal property may be described by fractional constitutive equations. As an example we consider an elastic nonlocal material, defined by the following constitutive equation between the stress tensor $\mathbf{T}$ and $\nabla^{(\alpha)}\mathbf{u}(\mathbf{x})$

$$\mathbf{T}(\mathbf{x}, t) = \mathbf{A}\nabla^{(\alpha)}\mathbf{u}(\mathbf{x}, t)\,, \quad \alpha \in (0, 1]\,,$$

where $\mathbf{A}$ is a fourth order symmetric tensor. The nonlocal property is clear from the detailed form

$$\mathbf{T}(\mathbf{x},t) = \frac{\alpha \mathbf{A}}{(1-\alpha)\sqrt{\pi^\alpha}} \int_\Omega \nabla \mathbf{u}(\mathbf{y}) \exp\left[-\frac{\alpha^2(\mathbf{x}-\mathbf{y})^2}{(1-\alpha)^2}\right] d\mathbf{y}.$$

Likewise, we can introduce the fractional divergence, defined for a smooth $\mathbf{u}(\cdot) : \Omega \to \mathbb{R}^3$ by

$$\nabla^{(\alpha)} \cdot \mathbf{u}(\mathbf{x}) = \frac{\alpha}{(1-\alpha)\sqrt{\pi^\alpha}} \int_\Omega \nabla \cdot \mathbf{u}(\mathbf{y}) \exp\left[-\frac{\alpha^2(\mathbf{x}-\mathbf{y})^2}{(1-\alpha)^2}\right] d\mathbf{y}. \tag{23.2.23}$$

Theorem 23.2.3. *From definitions (23.2.22) and (23.2.23), we have for any $u(\mathbf{x}) : \Omega \to \mathbb{R}$, such that*

$$\nabla u(\mathbf{x}) \cdot \mathbf{n}|_{\partial\Omega} = 0, \tag{23.2.24}$$

the following identity

$$\nabla \cdot \nabla^{(\alpha)} u(\mathbf{x}) = \nabla^{(\alpha)} \cdot \nabla u(\mathbf{x}). \tag{23.2.25}$$

Proof. Using (23.2.22), we obtain

$$\begin{aligned}
\nabla \cdot \nabla^{(\alpha)} u(\mathbf{x}) &= \frac{\alpha}{(1-\alpha)\sqrt{\pi^\alpha}} \int_\Omega \nabla u(\mathbf{y}) \cdot \nabla_{\mathbf{x}} \exp\left[-\frac{\alpha^2(\mathbf{x}-\mathbf{y})^2}{(1-\alpha)^2}\right] d\mathbf{y} \\
&= -\frac{\alpha}{(1-\alpha)\sqrt{\pi^\alpha}} \int_\Omega \nabla u(\mathbf{y}) \cdot \nabla \exp\left[-\frac{\alpha^2(\mathbf{x}-\mathbf{y})^2}{(1-\alpha)^2}\right] d\mathbf{y} \\
&= \frac{\alpha}{(1-\alpha)\sqrt{\pi^\alpha}} \int_\Omega \nabla \cdot \nabla u(\mathbf{y}) \exp\left[-\frac{\alpha^2(\mathbf{x}-\mathbf{y})^2}{(1-\alpha)^2}\right] d\mathbf{y} \\
&\quad - \frac{\alpha}{(1-\alpha)\sqrt{\pi^\alpha}} \int_{\partial\Omega} \nabla u(\mathbf{y}) \cdot \mathbf{n} \exp\left[-\frac{\alpha^2(\mathbf{x}-\mathbf{y})^2}{(1-\alpha)^2}\right] d\mathbf{y}.
\end{aligned}$$

Hence, for the boundary condition (23.2.24), the identity (23.2.25) is proved, because

$$\nabla^{(\alpha)} \cdot \nabla u(\mathbf{x}) = \frac{\alpha}{(1-\alpha)\sqrt{\pi^\alpha}} \int_\Omega \nabla \cdot \nabla u(\mathbf{y}) \exp\left[-\frac{\alpha^2(\mathbf{x}-\mathbf{y})^2}{(1-\alpha)^2}\right] d\mathbf{y}.$$

23.2.2.5 Fourier Transform of the Fractional Gradient and Divergence

For a smooth function $u(\mathbf{x}) : \mathbb{R}^3 \to \mathbb{R}$, the Fourier transform of the fractional gradient is defined by

$$(\nabla^{(\alpha)} u)_F(\boldsymbol{\xi}) = \int_{\mathbb{R}^3} \nabla^{(\alpha)} u(\mathbf{x}) \exp\left[-i\boldsymbol{\xi} \cdot \mathbf{x}\right]\, d\mathbf{x}.$$

This quantity is given by

$$\begin{aligned}
(\nabla^\alpha u)_F(\boldsymbol{\xi}) &= \frac{\alpha}{(1-\alpha)\sqrt{\pi^\alpha}} \left(\int_{\mathbb{R}^3} \nabla u(\mathbf{y}) \exp\left[-\frac{\alpha^2(\mathbf{x}-\mathbf{y})^2}{(1-\alpha)^2}\right] d\mathbf{y} \right)_F (\boldsymbol{\xi}) \\
&= \frac{\alpha}{(1-\alpha)\sqrt{\pi^\alpha}} (\nabla u)_F(\boldsymbol{\xi}) \left(\exp\left[-\frac{\alpha^2 \mathbf{x}^2}{(1-\alpha)^2}\right] \right)_F (\boldsymbol{\xi}),
\end{aligned}$$

where the well-known formula for the Fourier transform of a convolution product has been used. From

$$\left(\exp\left[-\frac{\alpha^2\mathbf{x}^2}{(1-\alpha)^2}\right]\right)_F(\xi) = \frac{(1-\alpha)\sqrt{\pi}}{\alpha}\exp\left[-\frac{(1-\alpha)^2\xi^2}{4\alpha^2}\right],$$

we obtain

$$(\nabla^\alpha u)_F(\xi) = \sqrt{\pi}^{1-\alpha}(\nabla u)_F(\xi)\exp\left[-\frac{(1-\alpha)^2\xi^2}{4\alpha^2}\right].$$

The Fourier transform of fractional divergence is defined by

$$(\nabla^\alpha\cdot\mathbf{u})_F(\xi) = \frac{\alpha}{(1-\alpha)\sqrt{\pi^\alpha}}\left(\int_\Omega \nabla\cdot\mathbf{u}(\mathbf{y})\exp\left[-\frac{\alpha^2(\mathbf{x}-\mathbf{y})^2}{(1-\alpha)^2}\right]d\mathbf{y}\right)_F(\xi),$$

from which we have

$$(\nabla^\alpha\cdot\mathbf{u})_F(\xi) = \sqrt{\pi}^{1-\alpha}\,(\nabla\cdot\mathbf{u})_F\,(\xi)\exp\left[-\frac{(1-\alpha)^2\xi^2}{4\alpha^2}\right].$$

23.2.2.6 Fractional Laplacian

In the study of partial differential equations, the Laplacian is of considerable interest. It is therefore of interest to determine the factional Laplacian. Based on the definitions of fractional gradient and divergence, we suggest a representation of the fractional Laplacian for a smooth function $f(\mathbf{x})$: $\Omega \to \mathbb{R}^3$, such that $\nabla f(\mathbf{x})\cdot\mathbf{n}|_{\partial\Omega} = 0$, of the form

$$(\nabla^2)^\alpha f(\mathbf{x}) = \frac{\alpha}{(1-\alpha)\sqrt{\pi^\alpha}}\int_\Omega \nabla\cdot\nabla f(\mathbf{y})\exp\left[-\frac{\alpha^2(\mathbf{x}-\mathbf{y})^2}{(1-\alpha)^2}\right]d\mathbf{y}.$$

By the use of Theorem 23.2.3, we have

$$(\nabla^2)^\alpha f(\mathbf{x}) = \nabla\cdot\nabla^\alpha f(\mathbf{x}) = \nabla^\alpha\cdot\nabla f(\mathbf{x}).$$

Suppose that

$$f(\mathbf{x}) = 0 \;\; on \;\; \partial\Omega.$$

Then, we can extend the function $f(\mathbf{x})$ to $\mathbb{R}^3$ by taking it to be zero on $\mathbb{R}^3\backslash\Omega$. This allows us to consider the Fourier transform

$$\begin{aligned}\left((\nabla^2)^\alpha f\right)_F(\xi) &= \frac{\alpha}{(1-\alpha)\sqrt{\pi^\alpha}}\left(\int_{\mathbb{R}^3}\nabla^2 f(\mathbf{y})\exp\left[-\frac{\alpha^2(\mathbf{x}-\mathbf{y})^2}{(1-\alpha)^2}\right]d\mathbf{y}\right)_F(\xi)\\ &= \frac{\alpha}{(1-\alpha)\sqrt{\pi^\alpha}}(\nabla\cdot\nabla f)_F(\xi)\left(\exp\left[-\frac{\alpha^2\mathbf{x}^2}{(1-\alpha)^2}\right]\right)_F(\xi) \qquad (23.2.26)\\ &= 4\pi\,|\xi|^2\, f_F(\xi)\,\sqrt{\pi}^{1-\alpha}\exp\left[-\frac{(1-\alpha)^2\xi^2}{4\alpha^2}\right].\end{aligned}$$

Finally, if $\alpha = 1$ we obtain from (23.2.26)

$$\begin{aligned}\left(\nabla^2 f\right)_F(\xi) &= \lim_{\alpha\to 1} 4\pi\,|\xi|^2\, f_F(\xi)\,\sqrt{\pi^{1-\alpha}}\exp\left[-\frac{(1-\alpha)^2\xi^2}{\alpha^2}\right] \\ &= 4\pi\,|\xi|^2\, f_F(\xi).\end{aligned}$$

23.2.2.7 Memory Operators

Fractional derivatives are memory operators which usually represent dissipation of energy [128, 248, 287] or damage [55] in the medium, as in the case of anelastic media or diffusion in porous media. In general, they are in agreement with the Second Law of Thermodynamics [109, 128].

Their validity rests not only on the fact that they represent appropriately a variety of phenomena, but also, in the case of the Caputo derivative, because they have the "elegant and rigorous property" that when the order of differentiation is integer, they coincide with the classic derivative of that order. However, this property is not relevant to the effects they represent in physical phenomena. It may be that using other differential operators, possibly simpler but without this property, one may obtain the same results as for Caputo fractional derivatives.

The effects of the fractional memory formalism for the new fractional derivative (NFD_t), compared with the Caputo derivative (UFD_t) on a linear trend are presented in [56].

A distributed order fractional memory operator may be introduced, which is simpler and easier to handle than the Caputo derivative [326]. It is defined by

$$\begin{aligned}{}_aP_b f(t) &= \int_a^b g(\alpha)\mathfrak{D}_t^{(\alpha)} f(t)d\alpha \\ &= \int_a^b g(\alpha)\int_0^t \exp(-\frac{\alpha}{1-\alpha}(t-\tau))f'(\tau)d\tau d\alpha,\end{aligned} \tag{23.2.27}$$

where $g(\alpha)$ is a weight function and $0 < a < b < 1$. We now take the Laplace transform of (23.2.27). Following the method of Caputo [45, 46, 56], one may interchange the order of integration of α and t. Thus, we obtain

$$\begin{aligned}({}_aP_b f(t))_L(p) &= \int_0^\infty\int_a^b \left[g(\alpha)\mathfrak{D}_t^{(\alpha)} f(t)\right]\exp(-pt)d\alpha dt \\ &= \int_0^\infty\int_a^b \left[g(\alpha)\int_0^t \exp(-\frac{\alpha}{1-\alpha}(t-\tau))f'(\tau)d\tau\right]\exp(-pt)d\alpha dt \\ &= \int_a^b \left\{\int_0^\infty\left[\int_0^t \exp(-\frac{\alpha}{1-\alpha}(t-\tau))f'(\tau)d\tau\right]\exp(-pt)dt\right\} g(\alpha)d\alpha.\end{aligned}$$

By virtue of the convolution form of the integral over τ, we have, as in (23.2.21),

$$\begin{aligned}({}_aP_b f)_L(p) &= \frac{r(p)F(p)}{p}\int_a^b \frac{g(\alpha)(1-\alpha)}{r(p)+\alpha}d\alpha, \\ F(p) &= pf_L(p) - f(0), \quad r(p) = \frac{p}{1-p},\end{aligned}$$

which represents the filtering properties of the operator and is simpler than that obtained using the Caputo derivative.

As an example we may consider the simple case $g(\alpha) = 1$, which gives

$$
\begin{aligned}
({}_aP_bf)_L(p) &= \frac{r(p)}{p}F(p)\int_a^b \frac{1-\alpha}{\alpha + r(p)}d\alpha \\
&= \frac{r(p)}{p}F(p)\left[\int_a^b \frac{1}{\alpha + r(p)}d\alpha - \int_a^b \frac{\alpha}{\alpha + r(p)}d\alpha\right].
\end{aligned}
\tag{23.2.28}
$$

This can be written as follow

$$
\begin{aligned}
({}_aP_bf)_L(p) &= \frac{1}{1-p}F(p)\left\{\frac{1}{1-p}\log\frac{b+r(p)}{a+r(p)} - (b-a)\right\} \\
&= F(p)\left\{\frac{1}{(1-p)^2}\log\frac{b+r(p)}{a+r(p)} - \frac{1}{1-p}(b-a)\right\}.
\end{aligned}
$$

23.3 The Fractional Derivative Memory Model

Fractional derivatives and their connection with power law relaxation functions are now discussed. Constraints imposed by thermodynamics are derived, and the Graffi–Volterra free energy for fractional derivative models is introduced.

23.3.1 Power Laws and Fractional Derivatives

Let us assume that the viscoelastic memory kernel or relaxation function $\mathbb{G}(s)$ is given by the power law form (23.1.1)

$$
\mathbb{G}(s) = \frac{\mathbb{C}}{\Gamma(1-\alpha)s^{\alpha}}, \qquad \alpha \in (0,1), \tag{23.3.1}
$$

where $\Gamma(\cdot)$ is the gamma function and $\mathbb{C}$ is a fourth order tensor. Relaxation functions of this type are discussed briefly in [167, page 32], where references to older works are given. We have

$$
\mathbb{G}(\infty) = \mathbb{G}_\infty = \mathbf{0}, \qquad \lim_{s\to 0}\mathbb{G}(s) = \mathbb{G}_0 = \infty. \tag{23.3.2}
$$

The property $\mathbb{G}_\infty = \mathbf{0}$ in the classical Volterra theory corresponds to that for viscoelastic fluids. In the fractional model though the kernel (23.3.1) is not $L^1(0,\infty)$, we will see (Remark 23.3.1 below) that this is true for values of α close to 1. However, for solid viscoelastic materials, some experimental observations are in approximate agreement with predictions based on (23.3.1) [167, 280, 293, 296], notably the property that the loss angle [167] is independent of frequency, as indicated by (23.5.2) below.

It follows from (23.3.1) that

$$\mathbb{G}'(s) = -\frac{\alpha\,\mathbb{C}}{\Gamma(1-\alpha)s^{1+\alpha}}. \tag{23.3.3}$$

The creep function for materials characterized by (23.3.1) or (23.3.3) also has a power law form [167].

Observe from (23.3.1) that there are no dimensional parameters in the theory, other than the overall coefficient $\mathbb{C}$ and the time dimensional quantity s. This implies that various quantities can be simply determined to within a multiplying constant, by means of dimensional analysis.

It is in fact advisable to introduce an extra parameter ξ with the dimension of time. This can usually be absorbed into $\mathbb{C}$ or η, if the parameter α is no being explicitly varied. For solids, we put

$$\mathbb{C} = \mathbb{C}_1\xi^{\alpha}, \tag{23.3.4}$$

so that $\mathbb{C}_1$ has dimensions of stress. For fluids, we have

$$\eta = \eta_1\xi^{\alpha},$$

where η_1 has dimensions of stress.

The use of the Caputo fractional derivative, defined for any $\alpha \in (0, 1)$ by any of the forms (23.2.6)–(23.2.9), is equivalent to adopting (23.3.1) as the relaxation function of the material. This model may be applied to both fluids and solids.

We follow here the classical papers [249] on fractional derivatives in defining the constitutive equation of viscoelasticity by

$$\mathbf{T}(\mathbf{x}, t) = \frac{\mathbb{C}(\mathbf{x})}{\Gamma(1-\alpha)} \int_a^t \frac{\dot{\mathbf{E}}(\mathbf{x}, \tau)}{(t-\tau)^{\alpha}} d\tau. \tag{23.3.5}$$

Let us take $a = -\infty$, since if necessary it is always possible to extend $\mathbf{E}$ to the interval $(-\infty, a)$ by the null tensor. Thus, (23.3.5) can be written as

$$\mathbf{T}(\mathbf{x}, t) = \frac{\alpha\mathbb{C}(\mathbf{x})}{\Gamma(1-\alpha)} \int_{-\infty}^t \frac{\mathbf{E}(\mathbf{x}, t) - \mathbf{E}(\mathbf{x}, \tau)}{(t-\tau)^{1+\alpha}} d\tau, \tag{23.3.6}$$

or, by a change of variable,

$$\mathbf{T}(\mathbf{x}, t) = -\frac{\alpha\mathbb{C}(\mathbf{x})}{\Gamma(1-\alpha)} \int_0^{\infty} \frac{\mathbf{E}_r^t(\mathbf{x}, s)}{s^{1+\alpha}} ds, \tag{23.3.7}$$

which are equivalent representations of the Caputo derivative. Using the notation of Sect. 23.2.1, relations (23.3.5)–(23.3.7) may be put in the form

$$\mathbf{T}(\mathbf{x}, t) = {}^{C}_{-\infty}D_t^{\alpha}[\mathbb{C}(\mathbf{x})\mathbf{E}(\mathbf{x}, t)]. \tag{23.3.8}$$

The constitutive equations (23.3.6) or (23.3.7) allow us to define the domain of definition of these functionals by a fractional Sobolev space, now called a Gagliardo space [142], defined for any $x \in \Omega$,

$$W^{\alpha,1}(-\infty,\infty) = \left\{ \mathbf{E}(t) \in L^1(-\infty,\infty), \ \frac{\mathbf{E}(t)-\mathbf{E}(\tau)}{(t-\tau)^{1+\alpha}} \in L^1((-\infty,t)\times(-\infty,\infty)) \right\},$$

with norm given by

$$||\mathbf{E}||_{W^{\alpha,1}_{(-\infty,\infty)}} = \left(\int_0^{\infty} |\mathbf{E}(t)|\, dt + \int_{-\infty}^{\infty} \frac{\alpha}{\Gamma(1-\alpha)} \int_{-\infty}^{t} \frac{|\mathbf{E}(\mathbf{x},t)-\mathbf{E}(\mathbf{x},\tau)|}{(t-\tau)^{1+\alpha}}\, d\tau\, dt \right).$$

In this framework, the constitutive equation of an incompressible viscoelastic fluid is entirely analogous to (23.3.5), (23.3.6) or (23.3.7), with the formal difference that instead of the tensor $\mathbf{C}(\mathbf{x})$, we now have a scalar constant η, which is related to the viscosity of the fluid. So, the constitutive functional is given by the stress

$$\mathbf{T}(\mathbf{x},t) = -p(\mathbf{x},t)\mathbf{I} + \mathbf{T}_E(\mathbf{x},t),$$

where p denotes the pressure and $\mathbf{T}_E$ the extra-stress defined by

$$\begin{aligned} \mathbf{T}_E(\mathbf{x},t) &= \frac{2\eta}{\Gamma(1-\alpha)} \int_{-\infty}^{t} \frac{\dot{\mathbf{E}}(\mathbf{x},\tau)}{(t-\tau)^{\alpha}} d\tau \\ &= \frac{\eta}{\Gamma(1-\alpha)} \int_{-\infty}^{t} \frac{\nabla\mathbf{v}(\mathbf{x},\tau) + (\nabla\mathbf{v}(\mathbf{x},\tau))^T}{(t-\tau)^{\alpha}} d\tau, \end{aligned} \tag{23.3.9}$$

where $\mathbf{v}$ is the fluid velocity. Alternative forms are given by

$$\mathbf{T}_E(\mathbf{x},t) = \frac{2\alpha\eta}{\Gamma(1-\alpha)} \int_{-\infty}^{t} \frac{\mathbf{E}(\mathbf{x},t)-\mathbf{E}(\mathbf{x},\tau)}{(t-\tau)^{1+\alpha}} d\tau = -\frac{2\alpha\eta}{\Gamma(1-\alpha)} \int_0^{\infty} \frac{\mathbf{E}^t_r(\mathbf{x},s)}{s^{1+\alpha}} ds. \tag{23.3.10}$$

Remark 23.3.1. Note that when α is close to 0, the model well represents a viscoelastic solid. When α is close to 1, we have a viscoelastic fluid. These features of the model are implied by the (23.3.8), in view of the properties of the Caputo fractional derivative given by (23.2.10) and (23.2.11), leading to

$$\mathbf{T}(\mathbf{x},t) = \begin{cases} \mathbf{C}_1(\mathbf{x})\mathbf{E}(\mathbf{x},t) & \text{solids,} \\ \xi\eta_1(\mathbf{x})\dot{\mathbf{E}}(\mathbf{x},t) & \text{fluids,} \end{cases}$$

where we have used (23.3.4).

Remark 23.3.2. Another important feature of a solid is the existence of only one null strain $\mathbf{E}_0$ (or reference configuration) such that the space of histories is a subset of

$$\mathcal{G}^t_S = \left\{ \mathbf{E}^t(\cdot) : [0,\infty) \to Sym(\mathbb{R}^3);\ \mathbf{E}^t \in L^1(0,\infty);\ \lim_{s\to\infty} \mathbf{E}^t(s) = \mathbf{E}_0 \right\},$$

where, from (23.3.5) or (23.3.6), if $\mathbf{E}^t(s) = \mathbf{E}_0$ we have $\mathbf{T}(\mathbf{E}^t(s)) = \mathbf{0}$.

For a fluid, the set of histories belongs to

$$\mathcal{G}^t_F = \left\{ \mathbf{E}^t(\cdot) : [0,\infty) \to Sym(\mathbb{R}^3);\ \mathbf{E}^t \in L^1(0,\infty) \right\}.$$

The main difference between the Volterra and fractional derivative models, is evident in the study of solid materials, if we examine stress behavior for $t \to \infty$. Indeed, in the context of the Volterra theory, if the system is subject to a constant strain $\mathbf{E}_0$, then stress will tend to $\mathbf{G}_\infty \mathbf{E}_0$. This can be proved by assuming a constant strain $\mathbf{E}_0(\mathbf{x})$ for $t > t_0$, so that the following limit is obtained

$$\begin{aligned}\lim_{t\to\infty} \mathbf{T}(\mathbf{x},t) &= \mathbb{G}_\infty(\mathbf{x})\mathbf{E}_0(\mathbf{x}) \\ &\quad + \lim_{t\to\infty} \int_{t-t_0}^{\infty} \mathbb{G}'(\mathbf{x},s)(\mathbf{E}(\mathbf{x},t-s) - \mathbf{E}_0(\mathbf{x}))ds \\ &= \mathbb{G}_\infty(\mathbf{x})\mathbf{E}_0(\mathbf{x}), \quad t > t_0.\end{aligned} \tag{23.3.11}$$

On the other hand, in the fractional theory we find that the stress will go to zero. Indeed,

$$\lim_{t\to\infty} \frac{\alpha \mathbb{C}(\mathbf{x})}{\Gamma(1-\alpha)} \int_{-\infty}^{t_0} \frac{\mathbf{E}_0(\mathbf{x}) - \mathbf{E}(\mathbf{x},\tau)}{(t-\tau)^{1+\alpha}} d\tau = \mathbf{0}, \quad t > t_0.$$

In this case, the material undergoes a kind of plastic deformation [49].

Remark 23.3.3. It is interesting to observe that Volterra sought to describe properties related to dislocation phenomena in terms of memory effects. However, it is now known that the model with the standard constitutive relation ((8.1.5), for example) is not capable of describing plastic effects produced by dislocations.

23.4 Thermodynamical Constraints and Free Energies

The issue of compatibility of fractional derivative models with thermodynamics is explored in this section. Only isothermal processes will be considered, so that the Second Law of Thermodynamics reduces to the Dissipation Principle

$$\rho(\mathbf{x})\dot{\psi}(\mathbf{x},t) \le \mathbf{T}(\mathbf{x},t) \cdot \dot{\mathbf{E}}(\mathbf{x},t), \tag{23.4.1}$$

where ψ denotes a free energy and ρ is the mass density. This is equivalent to property P3 given by (16.1.28). We have from (23.4.1) that on any cyclic process of period $T = 2\pi/\omega$ with $\omega \in \mathbb{R}^{++}$,

$$\int_0^T \mathbf{T}(\mathbf{x},t) \cdot \dot{\mathbf{E}}(\mathbf{x},t)\, dt \ge 0. \tag{23.4.2}$$

In particular, for periodic strain processes of the form

$$\mathbf{E}(\mathbf{x},t) = \mathbf{E}_1(\mathbf{x}) \cos \omega t + \mathbf{E}_2(\mathbf{x}) \sin \omega t,$$

it follows from (23.4.2) (see [108, 124]) that for all $\mathbf{E}_1, \mathbf{E}_2 \in Sym(V)$,

$$\begin{aligned}&\int_0^\infty \left(\mathbf{E}_1 \cdot \frac{\mathbb{C}}{s^{1+\alpha}}\mathbf{E}_1 + \mathbf{E}_2 \cdot \frac{\mathbb{C}}{s^{1+\alpha}}\mathbf{E}_2\right) \sin \omega s\, ds \\ &\quad + \int_0^\infty \mathbf{E}_1 \cdot \frac{\mathbb{C} - \mathbb{C}^T}{s^{1+\alpha}}\mathbf{E}_2 \cos \omega s\, ds \le 0\,, \quad \text{for all } \omega \in \mathbb{R}^+.\end{aligned} \tag{23.4.3}$$

The second term can vary arbitrarily in sign and magnitude for different choices of $\mathbf{E}_1$ and $\mathbf{E}_2$, so that it can be concluded that the fourth order tensor $\mathbb{C}$ is symmetric. Also, in the first term, the quantity $\mathbf{E} \cdot \mathbb{C}\mathbf{E}$ must have a definite signature for the inequality to be obeyed in a simple manner. We choose $\mathbf{E} \cdot \mathbb{C}\mathbf{E} \geq 0$ so that $\mathbb{C}$ is at least positive semidefinite. Thus, the condition (23.4.3) becomes

$$\int_0^\infty \frac{1}{s^{1+\alpha}} \sin \omega s \, ds \leq 0 \text{ for all } \omega \in \mathbb{R}^+. \tag{23.4.4}$$

Remark 23.4.1. For a fluid defined by (23.3.10), we obtain from the second law the same inequality (23.4.4). Alternatively, if (23.3.9) is used, the equivalent condition emerges

$$\int_0^\infty \frac{1}{s^\alpha} \cos \omega s \, ds \geq 0 \;\; for\ all \;\; \omega \in \mathbb{R}. \tag{23.4.5}$$

We observe that (23.4.5) has the form (16.1.12) for the case of a fractional derivative relaxation function. It will be confirmed that this property is actually true in Remark 23.5.1.

23.4.1 The Graffi–Volterra Free Energy

We now consider a particular free energy within fractional derivative theory. Let us first take the case of a solid, described by Eq. (23.3.7). This functional is denoted by ψ^S. Any free energy functional must satisfy the inequality (23.4.1). Thus, we must have

$$\rho(\mathbf{x})\dot{\psi}^S(\mathbf{x}, t) \leq \mathbf{T}(\mathbf{x}, t) \cdot \dot{\mathbf{E}}(\mathbf{x}, t). \tag{23.4.6}$$

For simplicity, let us take the $\mathbb{C}(\mathbf{x})$ to be a scalar quantity $C(\mathbf{x})$. Using the identity

$$\frac{d}{dt}\mathbf{E}(\mathbf{x}, t) = -\frac{d}{ds}\mathbf{E}_r^t(\mathbf{x}, s) - \frac{d}{dt}\mathbf{E}_r^t(\mathbf{x}, s),$$

we have

$$\begin{aligned}
\mathbf{T}(\mathbf{x}, t) \cdot \dot{\mathbf{E}}(\mathbf{x}, t) &= -\frac{\alpha C(\mathbf{x})}{\Gamma(1-\alpha)} \int_0^\infty \frac{\mathbf{E}_r^t(\mathbf{x}, s)}{s^{1+\alpha}} ds \cdot \dot{\mathbf{E}}(\mathbf{x}, t) \\
&= \frac{\alpha C(\mathbf{x})}{\Gamma(1-\alpha)} \int_0^\infty \frac{\mathbf{E}_r^t(\mathbf{x}, s)}{s^{1+\alpha}} \cdot \frac{d}{dt}\mathbf{E}_r^t(\mathbf{x}, s))ds \\
&\quad + \frac{\alpha C(\mathbf{x})}{\Gamma(1-\alpha)} \int_0^\infty \frac{\mathbf{E}_r^t(\mathbf{x}, s)}{s^{1+\alpha}} \cdot \frac{d}{ds}\mathbf{E}_r^t(\mathbf{x}, s)ds.
\end{aligned} \tag{23.4.7}$$

Let us assume that $\psi^S(\mathbf{x}, t)$ is given by

$$\psi^S(\mathbf{x}, t) = \frac{\alpha C(\mathbf{x})}{2\rho(\mathbf{x})\Gamma(1-\alpha)} \int_0^\infty \frac{|\mathbf{E}_r^t(\mathbf{x}, s)|^2}{s^{1+\alpha}} ds. \tag{23.4.8}$$

This is the Graffi–Volterra free energy for fractional derivative models. It is discussed in a more general context in Sects. 17.3.1 and 10.1.1, where it is shown to be the only free energy that is a single-integral quadratic form. The inequality (23.4.6) is satisfied because by (23.4.7) we obtain

$$\rho(\mathbf{x})\dot{\psi}^S(\mathbf{x},t) = \mathbf{T}(\mathbf{x},t)\cdot\dot{\mathbf{E}}(\mathbf{x},t) - \frac{\alpha C(\mathbf{x})}{2\Gamma(1-\alpha)}\int_0^\infty \frac{\frac{d}{ds}|\mathbf{E}_r^t(\mathbf{x},s))|^2}{s^{1+\alpha}}ds$$
$$= \mathbf{T}(\mathbf{x},t)\cdot\dot{\mathbf{E}}(\mathbf{x},t) - \frac{\alpha C(\mathbf{x})(1+\alpha)}{2\Gamma(1-\alpha)}\int_0^\infty \frac{|\mathbf{E}_r^t(\mathbf{x},s)|^2}{s^{2+\alpha}}ds.$$

Hence

$$\rho(\mathbf{x})\dot{\psi}^S(\mathbf{x},t) = \mathbf{T}(\mathbf{x},t)\cdot\dot{\mathbf{E}}(\mathbf{x},t) - D(\mathbf{x},t), \tag{23.4.9}$$

where $D(\mathbf{x},t)\geq 0$ denotes the rate of dissipation, given by

$$D(\mathbf{x},t) = \frac{\alpha C(\mathbf{x})(1+\alpha)}{2\Gamma(1-\alpha)}\int_0^\infty \frac{|\mathbf{E}_r^t(\mathbf{x},s)|^2}{s^{2+\alpha}}ds. \tag{23.4.10}$$

The set of histories $\mathcal{H}_S^t$ available with this model is defined by

$$\mathcal{H}_S^t = \left\{\mathbf{E}^t : [0,\infty)\to Sym(V);\ \psi^S(\mathbf{E}(t),\mathbf{E}^t(\cdot))<\infty\right\}.$$

For a viscoelastic fluid, we use the constitutive Eq. (23.3.10), which is more convenient than (23.3.9). This gives

$$\mathbf{T}_E(\mathbf{x},t)\cdot\dot{\mathbf{E}}(\mathbf{x},t) = \frac{2\alpha\eta}{\Gamma(1-\alpha)}\int_{-\infty}^t \frac{\mathbf{E}(\mathbf{x},t)-\mathbf{E}(\mathbf{x},\tau)}{(t-\tau)^{1+\alpha}}d\tau\cdot\dot{\mathbf{E}}(\mathbf{x},t)$$
$$= \frac{2\alpha\eta}{\Gamma(1-\alpha)}\int_{-\infty}^t \frac{\mathbf{E}(\mathbf{x},t)-\mathbf{E}(\mathbf{x},\tau)}{(t-\tau)^{1+\alpha}}\cdot\frac{d}{dt}(\mathbf{E}(\mathbf{x},t)-\mathbf{E}(\mathbf{x},\tau))d\tau$$
$$= \frac{\alpha\eta}{\Gamma(1-\alpha)}\left[\frac{d}{dt}\int_{-\infty}^t \frac{|\mathbf{E}(\mathbf{x},t)-\mathbf{E}(\mathbf{x},\tau)|^2}{(t-\tau)^{1+\alpha}}d\tau\right.$$
$$\left. -(1+\alpha)\int_{-\infty}^t \frac{|\mathbf{E}(\mathbf{x},t)-\mathbf{E}(\mathbf{x},\tau)|^2}{(t-\tau)^{2+\alpha}}d\tau\right].$$

The Graffi–Volterra for fluids is given in this case by

$$\psi^F(\mathbf{x},t) = \frac{\alpha\eta}{\rho(\mathbf{x})\Gamma(1-\alpha)}\int_{-\infty}^t \frac{|\mathbf{E}(\mathbf{x},t)-\mathbf{E}(\mathbf{x},\tau)|^2}{(t-\tau)^{1+\alpha}}d\tau$$

or, using the variable $s=t-\tau$, the equivalent form

$$\psi^F(\mathbf{x},t)) = \frac{\alpha\eta}{\rho(\mathbf{x})\Gamma(1-\alpha)}\int_0^\infty \frac{|(\mathbf{E}_r^t(\mathbf{x},s)|^2}{s^{1+\alpha}}ds.$$

Of course, it has a similar form to (23.4.8) for solids. We may define the set of histories $\mathcal{H}_F^t$ available with this model by

$$\mathcal{H}_F^t = \left\{\mathbf{E}^t : [0,\infty)\to Sym(V);\ \psi^F(\mathbf{E}(t),\mathbf{E}^t(\cdot))<\infty\right\}.$$

The rate of dissipation $D(\mathbf{x}, t)$ is given by

$$D(\mathbf{x}, t) = \frac{\alpha\eta(1+\alpha)}{\Gamma(1-\alpha)} \int_{-\infty}^{t} \frac{|\mathbf{E}(\mathbf{x}, t) - \mathbf{E}(\mathbf{x}, \tau)|^2}{(t-\tau)^{2+\alpha}} d\tau \geq 0,$$

or

$$D(\mathbf{x}, t) = \frac{\alpha\eta(1+\alpha)}{\Gamma(1-\alpha)} \int_{0}^{\infty} \frac{|\mathbf{E}_r^t(\mathbf{x}, s)|^2}{s^{2+\alpha}} ds \geq 0.$$

23.5 Frequency-Domain Quantities for Scalar Fractional Derivative Materials

In the next two sections, we deal, for simplicity, with the scalar theory, for which the relaxation function and strain history are scalar quantities. Also, we generally omit the space variable $\mathbf{x}$.

23.5.1 Complex Modulus for the Fractional Derivative Model

It follows from (23.3.3) and $(16.1.7)_2$ that the quantity $G'_+(\omega)$ does not exist for fractional derivative forms, while $\widetilde{G}_+(\omega)$, given by $(16.1.7)_1$, is finite. Therefore, the complex modulus, defined by $(16.1.9)_2$, is finite. Recalling (23.3.1) and (23.3.2), we see that it is given by

$$M(\omega) = i\omega \int_0^\infty G(s) e^{-i\omega s} ds = i\omega \frac{k}{\Gamma(1-\alpha)} \int_0^\infty s^{-\alpha} e^{-i\omega s} ds,$$

where the coefficient k corresponds to $\mathbb{C}$ (or C) and η in Sect. 23.3. From dimensional analysis, we can determine that

$$M(\omega) = k\, c\, \omega^\alpha,$$

where c is a dimensionless constant to be determined. Putting $z = i\omega$ and rotating it to a point on the positive real axis, the integral can be evaluated in terms of the Gamma function. Rotating back, we find that ([167, page 33])

$$M(\omega) = i\omega G_+(\omega) = k\, c\, \omega^\alpha \quad \forall \omega \geq 0, \qquad c = \exp\left(i\frac{\alpha\pi}{2}\right).$$

The real and imaginary parts of $M(\omega)$ have the form

$$\begin{aligned} M_1(\omega) &= \omega G_s(\omega) = k\,\omega^\alpha \cos\left(\frac{\alpha\pi}{2}\right), \\ M_2(\omega) &= \omega G_c(\omega) = k\,\omega^\alpha \sin\left(\frac{\alpha\pi}{2}\right) \quad \forall \omega \geq 0. \end{aligned} \qquad (23.5.1)$$

Remark 23.5.1. The positivity of $G_c(\omega)$ is clear from (23.5.1), for $\omega \in \mathbb{R}^+$ and therefore for all real ω. This confirms the thermodynamic constraint $(16.1.12)_1$.

The loss angle γ, defined by the relationship

$$\frac{M_2(\omega)}{M_1(\omega)} = \tan\gamma,$$

is given by

$$\gamma = \frac{\alpha\pi}{2}, \tag{23.5.2}$$

which is independent of ω, as noted earlier.

It is apparent from (23.5.1) that the only singularity in $G_+(\omega)$ is a branch cut from the origin to infinity. Thus, minimal states for power law materials are singletons, in accordance with Proposition 16.5.2. The branch cut must lie in Ω^+. Apart from this constraint, we can choose it to be in any direction. Each choice yields a different function. We choose it to be along the positive imaginary axis. An expression for the minimum free energy of this material is derived in Chap. 14. Thus, we can reach the negative real axis, without crossing singularities, by a rotation $e^{-i\pi}$ of ω. This gives

$$M(-\omega) = -i\omega G_+(-\omega) = k|\omega|^{\alpha}\exp\left(-i\frac{\alpha\pi}{2}\right) = \overline{M}(\omega), \quad \forall\omega \geq 0, \tag{23.5.3}$$

which has a branch cut along the negative imaginary axis.

From $(16.1.21)_4$, we see that the frequency-domain version of the constitutive relation has the form

$$\begin{aligned} T(t) &= \frac{1}{2\pi}\int_{-\infty}^{\infty}\overline{M}(\omega)E^t_{r+}(\omega)d\omega \\ &= \frac{k}{2\pi}\int_0^{\infty} e^{-i\frac{\alpha\pi}{2}}\omega^{\alpha}E^t_{r+}(\omega)d\omega + \frac{k}{2\pi}\int_{-\infty}^{0} e^{i\frac{\alpha\pi}{2}}(-\omega)^{\alpha}E^t_{r+}(\omega)d\omega \\ &= -\frac{k\sin(\alpha\pi)}{\pi}\int_0^{\infty} r^{\alpha}E^t_{r+}(-ir)dr. \end{aligned} \tag{23.5.4}$$

The last forms are obtained by moving the contour to closely surround the cut on the negative imaginary axis. The first term of $(23.5.4)_2$ becomes the integral over $[0, -i\infty)$ on the right side of the negative imaginary axis, while in the second term becomes the integral over $(-i\infty, 0]$ on the left side. Relation (16.1.17) has been invoked in writing the last equation.

Using (16.1.13), $(23.5.1)_2$ and (23.5.3), we deduce that the function H is defined over $\mathbb{R}$ by

$$H(\omega) = a|\omega|^{\alpha+1}, \quad \forall\omega \in \mathbb{R}, \qquad \text{where } a = k\sin\left(\frac{\alpha\pi}{2}\right). \tag{23.5.5}$$

Using $(16.1.23)_1$, we can write $T(t)$ in the form

$$
\begin{aligned}
T(t) &= \frac{1}{\pi i}\int_{-\infty}^{\infty}\frac{H(\omega)}{\omega}E^t_{r+}(\omega)d\omega \\
&= \frac{a}{\pi i}\int_0^{\infty}\omega^{\alpha}E^t_{r+}(\omega)d\omega - \frac{a}{\pi i}\int_{-\infty}^{0}(-\omega)^{\alpha}E^t_{r+}(\omega)d\omega \qquad (23.5.6)\\
&= -k\frac{\sin(\alpha\pi)}{\pi}\int_0^{\infty}r^{\alpha}E^t_{r+}(-ir)dr = k\frac{\sin(\alpha\pi)}{\pi}\int_0^{\infty}r^{\alpha-1}\dot{E}^t_{+}(-ir)dr.
\end{aligned}
$$

The penultimate form is obtained by transforming the integrals according to the changes described in relation to (23.5.4), while the final form uses (16.1.17).

23.5.2 The Work Function for Fractional Derivative Materials

Relation (17.3.19) for the work function becomes

$$
W(t) = \frac{k}{2\Gamma(1-\alpha)}\int_0^{\infty}\int_0^{\infty}\frac{\dot{E}^t(s)\dot{E}^t(u)}{|s-u|^{\alpha}}duds,
$$

which involves an integrable singularity. Using (23.5.5), relations (17.3.19) become

$$
\begin{aligned}
W(t) &= \frac{1}{2\pi}\int_{-\infty}^{\infty}\frac{H(\omega)}{\omega^2}|\dot{E}^t_{+}(\omega)|^2\,d\omega = \frac{1}{2\pi}\int_{-\infty}^{\infty}H(\omega)|E^t_{r+}(\omega)|^2d\omega \\
&= \frac{a}{2\pi}\int_{-\infty}^{\infty}|\omega|^{\alpha-1}|\dot{E}^t_{+}(\omega)|^2\,d\omega = \frac{a}{2\pi}\int_{-\infty}^{\infty}|\omega|^{\alpha+1}|E^t_{r+}(\omega)|^2d\omega.
\end{aligned} \qquad (23.5.7)
$$

The basic property $\dot{W}(t) = T(t)\dot{E}(t)$ can be shown using $(23.5.6)_1$, (23.5.7), $(16.1.18)_2$ and the evenness of $H(\omega)$.

23.6 The Minimum Free Energy for Fractional Derivative Models

We now derive the form for the minimum free energy and the corresponding rate of dissipation for fractional derivative materials. These are for general histories. Simple explicit formulae are also given for sinusoidal and exponential histories. The derivations are for the scalar case.

23.6.1 General Form of the Minimum Free Energy

The factors $H_{\pm}(\omega)$ of $H(\omega)$ have the form

$$
\begin{aligned}
H_{+}(\omega) &= \sqrt{a}\,\omega^{\eta}e^{i\lambda(\omega)}, \\
H_{-}(\omega) &= \sqrt{a}\,\omega^{\eta}e^{-i\lambda(\omega)}, \qquad \eta = \frac{\alpha+1}{2}\in(\frac{1}{2},1),
\end{aligned} \qquad (23.6.1)
$$

for $\omega > 0$. The phase $\lambda(\omega)$ remains to be determined. On the complex plane, $H_{\pm}(\omega)$ are analytic continuations of these quantities, except at their singularities, which are now described.

The singularities of $H_+(\omega)$ are chosen to be along the positive imaginary axis, similarly to $G_+(\omega)$ as described before (23.5.3). Then $H_-(\omega)$ is the complex conjugate of this function with singularities consisting of a branch cut along the negative imaginary axis.

If we had chosen different straight line branch cuts, the resulting factors would differ from those we determine here by a constant phase factor. This can be seen from (23.6.1) by considering a transformed complex plane with typical point $\omega_1 = \omega \exp(i\delta)$, where δ is a constant. It is relevant that $\lambda(\omega)$ is later shown to be independent of ω. The phase factor δ would not affect the resulting formula for the minimum free energy.

The branch cuts described are simple consequences of the factors ω^η in (23.6.1). The product of $H_\pm(\omega)$ will lead to branch cuts in $H(\omega)$. The factors of $H(\omega)$ obey the following relationships, for general complex ω:

$$\overline{H_\pm(\omega)} = \overline{H_\pm}(\overline{\omega}) = H_\mp(\overline{\omega}) = H_\pm(-\overline{\omega}),$$

from which it follows that $H_+(\omega)$ ($H_-(\omega)$) must be real on the negative (positive) imaginary axis; indeed, this property applies to any point on the imaginary axis where these factors exist. Thus,

$$\lambda(re^{i\frac{\pi}{2}}) = \lambda(re^{-i\frac{\pi}{2}}) = \frac{\pi\eta}{2}. \tag{23.6.2}$$

The discontinuity in $H_-(\omega)$ across the branch cut along the imaginary axis is given by the discontinuity in

$$H_-(\omega) = \frac{H(\omega)}{H_+(\omega)}, \qquad H_+(re^{-i\frac{\pi}{2}}) = \sqrt{a}r^\eta.$$

Thus, the discontinuity is determined by $H(\omega)$, divided by a real quantity of the form $H_+(r\exp(-i\frac{\pi}{2}))$. Therefore, the singularities in $H_-(\omega)$ are determined by those in $H(\omega)$. Similar observations apply to the singularities of $H_+(\omega)$.

Proposition 23.6.1. *The phase $\lambda(\omega)$ is independent of ω.*

Proof. The phase factors $e^{\pm i\lambda(\omega)}$ are continued analytically to the whole complex plane. By the argument just outlined, they cannot contribute new singularities over and above those determined by $H(\omega)$, which is independent of $\lambda(\omega)$. Thus, we conclude that the phase factors will yield no singularities and must therefore be entire functions. However, this means either that they contribute essential singularities at infinity, which must be excluded in the same way as singularities on the finite plane, or they are constant. Therefore, the quantity λ is independent of ω. □

The factorization (16.1.14) clearly allows us to replace $H_\pm(\omega)$ by $-H_\pm(\omega)$.

For $\omega \in \mathbb{R}^{++}$, we have

$$\begin{aligned} H_+(\omega) &= \sqrt{a}\,\omega^\eta\, e^{i\lambda}, \\ H_-(\omega) &= \sqrt{a}\,\omega^\eta\, e^{-i\lambda}, \end{aligned} \tag{23.6.3}$$

where, from (23.6.2),

$$\lambda = \frac{\pi\eta}{2}. \tag{23.6.4}$$

We put

$$\omega = re^{i\theta}$$

so that $\theta = \arg(\omega)$. The behavior of $H_+(\omega)$ as ω approaches the positive imaginary axis from the first and second quadrants, respectively, are given by

$$H_+(\omega) = \begin{cases} \sqrt{a}\, r^\eta\, e^{i\frac{\pi(\alpha+1)}{4} + i\lambda}, & \text{1st quadrant, } \theta = \frac{\pi}{2}, \\ \sqrt{a}\, r^\eta\, e^{-3i\frac{\pi(\alpha+1)}{4} + i\lambda}, & \text{2nd quadrant, } \theta = -\frac{3\pi}{2}. \end{cases} \tag{23.6.5}$$

Similarly, the behavior of $H_-(\omega)$ as it approaches the negative imaginary axis from the fourth and third quadrants, respectively, are given by

$$H_-(\omega) = \begin{cases} \sqrt{a}\, r^\eta\, e^{-i\frac{\pi(\alpha+1)}{4} - i\lambda}, & \text{4th quadrant, } \theta = -\frac{\pi}{2}, \\ \sqrt{a}\, r^\eta\, e^{3i\frac{\pi(\alpha+1)}{4} - i\lambda}, & \text{3rd quadrant, } \theta = \frac{3\pi}{2}. \end{cases} \tag{23.6.6}$$

In the light of (23.6.4), the limiting values (23.6.5) and (23.6.6) reduce to

$$H_+(\omega) = \begin{cases} \sqrt{a}\, r^\eta\, e^{2i\lambda}, & \text{1st quadrant, } \theta = \frac{\pi}{2}, \\ \sqrt{a}\, r^\eta\, e^{-2i\lambda}, & \text{2nd quadrant, } \theta = -\frac{3\pi}{2}, \end{cases}$$

and

$$H_-(\omega) = \begin{cases} \sqrt{a}\, r^\eta\, e^{-2i\lambda}, & \text{4th quadrant, } \theta = -\frac{\pi}{2}, \\ \sqrt{a}\, r^\eta\, e^{2i\lambda}, & \text{3rd quadrant, } \theta = \frac{3\pi}{2}. \end{cases}$$

Let us use (23.6.3) to extend $H_\pm(\omega)$ to $\omega \in \mathbb{R}$. We take $\theta = -\pi$ to obtain $H_+(\omega)$, and $\theta = \pi$ for $H_-(\omega)$. This gives, for $\omega < 0$,

$$\begin{aligned} H_+(\omega) &= \sqrt{a}\, r^\eta\, e^{-i\eta\pi + i\lambda} = \sqrt{a}|\omega|^\eta e^{-i\lambda}, \\ H_-(\omega) &= \sqrt{a}\, r^\eta\, e^{i\eta\pi - i\lambda} = \sqrt{a}|\omega|^\eta e^{i\lambda}. \end{aligned} \tag{23.6.7}$$

These formulae are consistent with (22.1.8). Note that on multiplying the factors in (23.6.3) and (23.6.7) together, we obtain (23.5.5) for $\omega \in \mathbb{R}$.

The quantity $p_-^t(\omega)$ has the form

$$\begin{aligned} p_-^t(\omega) = {} & \frac{\sqrt{a}}{2\pi i} e^{-i\lambda} \int_0^\infty \frac{\omega_1^\eta E_{r+}^t(\omega_1)}{\omega_1 - \omega^+} d\omega_1 \\ & + \frac{\sqrt{a}}{2\pi i} e^{i\lambda} \int_{-\infty}^0 \frac{(-\omega_1)^\eta E_{r+}^t(\omega_1)}{\omega_1 - \omega^+} d\omega_1, \end{aligned}$$

on using (23.6.3) and (23.6.7). The singularities of $E^t_{r+}(\omega)$ are in $\mathbb{C}^{(+)}$. Recalling the position of the cut in $H_-(\omega)$, we see that the integrations over the real axis can be moved into the lower half-plane to closely surround the branch cut, yielding

$$\begin{aligned} p^t_-(\omega) &= \frac{\sqrt{a}}{2\pi i}\int_0^\infty \frac{r_1^\eta E^t_{r+}(-ir_1)}{r_1 - i\omega}\left(e^{-2i\lambda} - e^{2i\lambda}\right)dr_1 \\ &= -\frac{\sqrt{a}\sin 2\lambda}{\pi}\int_0^\infty \frac{r_1^\eta E^t_{r+}(-ir_1)}{r_1 - i\omega}dr_1, \end{aligned}$$

where $E^t_{r+}(-ir_1)$ is a real quantity given by

$$E^t_{r+}(-ir_1) = \int_0^\infty E^t_r(u)e^{-r_1 u}du. \tag{23.6.8}$$

Using

$$\psi_m(t) = \frac{1}{2\pi}\int_{-\infty}^\infty |p^t_-(\omega)|^2 d\omega$$

and Cauchy's integral formula, we obtain

$$\begin{aligned} \psi_m(t) &= \kappa \int_0^\infty\int_0^\infty \frac{(r_1 r_2)^\eta E^t_{r+}(-ir_1)E^t_{r+}(-ir_2)}{r_1 + r_2}dr_1 dr_2 \\ &= \kappa \int_0^\infty\int_0^\infty\int_0^\infty\int_0^\infty \frac{(r_1 r_2)^\eta E^t_r(u)E^t_r(v)e^{-r_1 u - r_2 v}}{r_1 + r_2}dr_1 dr_2 du dv, \\ \kappa &= \frac{a}{\pi^2}\sin^2\left[\frac{\pi}{2}(\alpha + 1)\right] = \frac{k}{\pi^2}\sin\frac{\alpha\pi}{2}\cos^2\frac{\alpha\pi}{2}. \end{aligned} \tag{23.6.9}$$

Using (16.1.17), we can also write (23.6.9) in the form

$$\begin{aligned} \psi_m(t) &= \kappa \int_0^\infty\int_0^\infty \frac{(r_1 r_2)^{\eta - 1}\dot{E}^t_+(-ir_1)\dot{E}^t_+(-ir_2)}{r_1 + r_2}dr_1 dr_2 \\ &= \kappa \int_0^\infty\int_0^\infty\int_0^\infty\int_0^\infty \frac{(r_1 r_2)^{\eta - 1}\dot{E}^t(u)\dot{E}^t(v)e^{-r_1 u - r_2 v}}{r_1 + r_2}dr_1 dr_2 du dv. \end{aligned} \tag{23.6.10}$$

The quantity η is defined by $(23.6.1)_3$.

The minimum free energy may be written in the form (see (11.9.14))

$$\psi_m(t) = \frac{1}{2}\int_0^\infty\int_0^\infty \dot{E}^t(s)G_m(s,u)\dot{E}^t(u)ds du,$$

where the equilibrium term vanishes, since $G_\infty = 0$. We must have

$$G(u) = G(0,u) = G(u,0) = \frac{k}{\Gamma(1-\alpha)u^\alpha}, \tag{23.6.11}$$

where $G(u)$ is the relaxation function. Thus, we have

$$G_m(u,v) = 2\kappa \int_0^\infty \int_0^\infty \frac{(r_1 r_2)^{\eta - 1} e^{-r_1 u - r_2 v}}{r_1 + r_2} dr_1 dr_2.$$

Recalling (23.3.2), we see that the properties (17.3.7) are valid in this case. Relation (23.6.11) is confirmed since

$$\begin{aligned} 2\kappa \int_0^\infty \int_0^\infty & \frac{(r_1 r_2)^{\eta - 1} e^{-r_2 v}}{r_1 + r_2} dr_1 dr_2 \\ &= 2\kappa \int_0^\infty \int_0^\infty \frac{(r_1 r_2)^{\eta - 1} e^{-r_1 v}}{r_1 + r_2} dr_1 dr_2 \\ &= \frac{k \sin \alpha\pi}{\pi} \int_0^\infty r_1^{\alpha - 1} e^{-r_1 v} dr_1 \\ &= \frac{k \sin \alpha\pi \Gamma(\alpha)}{\pi v^\alpha} = \frac{k}{\Gamma(1-\alpha) v^\alpha}, \end{aligned}$$

by virtue of the formula, for complex z ([168, page 285]),

$$\int_0^\infty \frac{r^{\mu - 1}}{r + z} dr = \frac{\pi z^{\mu - 1}}{\sin \pi\mu}, \qquad |\arg z| < \pi, \qquad 0 < \mu < 1, \tag{23.6.12}$$

together with the relations (23.2.2) and (23.2.3). We now seek to determine the rate of dissipation corresponding to the minimum free energy. Relation $(16.1.18)_2$ yields that

$$\frac{d}{dt} E^t_{r+}(-ir) = -r E^t_{r+}(-ir) - \frac{\dot{E}(t)}{r},$$

so that, from (23.6.9), we have

$$\begin{aligned} \dot{\psi}_m(t) = -\kappa & \left| \int_0^\infty r_1^\eta E^t_{r+}(-ir_1) dr_1 \right|^2 \\ & - 2\kappa \dot{E}(t) \int_0^\infty \int_0^\infty \frac{r_1^\eta E^t_{r+}(-ir_1) r_2^{\eta - 1}}{r_1 + r_2} dr_1 dr_2. \end{aligned} \tag{23.6.13}$$

Using (23.6.12) and the expression $(23.6.9)_3$ for κ, the last term of (23.6.13) becomes

$$-\frac{k \sin \alpha\pi}{\pi} \int_0^\infty r_1^\alpha E^t_{r+}(-ir_1) dr_1 \dot{E}(t) = T(t)\dot{E}(t),$$

by virtue of (23.5.6). Consequently, recalling (16.1.28), it follows that the rate of dissipation is given by the negative of the first term on the right-hand side of (23.6.13), where we have used (16.1.17). These relations can be deduced also from (16.4.12). The quantity $D_m(t)$ can be written in the form (17.3.11) with

$$K_m(s,u) = -\frac{2\kappa \Gamma^2(\eta)}{(su)^\eta},$$

$$\begin{aligned} D_m(t) &= \kappa\left|\int_0^\infty r^{\eta} E^t_{r+}(-ir)dr\right|^2 = \kappa\left|\int_0^\infty r^{\eta-1}\dot{E}^t_{+}(-ir)dr\right|^2 \\ &= \kappa\left|\int_0^\infty\int_0^\infty r^{\eta-1}e^{-ru}\dot{E}^t(u)drdu\right|^2 \\ &= \kappa\Gamma^2(\eta)\left|\int_0^\infty \frac{\dot{E}^t(u)}{u^{\eta}}du\right|^2 \geq 0. \end{aligned} \tag{23.6.14}$$

23.6.2 The Minimum Free Energy for Simple Histories

Free energies and rates of dissipation for sinusoidal and increasing exponential histories are discussed in [15, 16]. Sinusoidal histories are useful in many practical contexts, though the total dissipation and the work function, defined by (16.1.30), are infinite. Increasing exponential histories provide a simple example where all quantities are finite. Also, the algebra involved is similar to, though simpler than, the sinusoidal case.

23.6.2.1 Sinusoidal Histories

Formulae relating to general materials for sinusoidal histories are presented in [15, 16] and earlier papers. Also, for exponential histories, similar general results are introduced in [16]. Here, we consider the specific cases (23.6.10) and (23.6.14) directly for the relevant forms of the strain history.

Consider a history and current value $(E^t, E(t))$ defined by

$$E(t) = E_0 e^{i\omega_0 t} + \overline{E}_0 e^{-i\omega_0 t}, \quad E^t(s) = E(t-s), \tag{23.6.15}$$

where E_0 is an amplitude and $\overline{E}_0$ its complex conjugate. The quantities E^t_+ and $\dot{E}^t_+$ have the form

$$\begin{aligned} E^t_+(\omega) &= E_0\frac{e^{i\omega_0 t}}{i(\omega+\omega_0)} + \overline{E}_0\frac{e^{-i\omega_0 t}}{i(\omega-\omega_0)}, \\ \dot{E}^t_+(\omega) &= \omega_0 E_0\frac{e^{i\omega_0 t}}{\omega+\omega_0} - \omega_0\overline{E}_0\frac{e^{-i\omega_0 t}}{\omega-\omega_0}. \end{aligned} \tag{23.6.16}$$

From $(17.6.8)_2$, we find that

$$\dot{E}^t_+(-ir) = \omega_0 E_0\frac{e^{i\omega_0 t}}{\omega_0 - ir} + \omega_0\overline{E}_0\frac{e^{-i\omega_0 t}}{\omega_0 + ir}, \tag{23.6.17}$$

where r is real. The final form of (23.5.6), together with (23.6.12) and (23.6.17) give

$$T(t) = k\omega_0^{\alpha}\left[e^{i\pi\alpha/2}E_0 e^{i\omega_0 t} + e^{-i\pi\alpha/2}\overline{E}_0 e^{-i\omega_0 t}\right]. \tag{23.6.18}$$

Any real algebraic quadratic form in $E(t)$ or real functional quadratic form in $E^t(s)$ can be written in the form (16.11.3), denoted by V. Recalling (17.6.12) we introduce the abbreviated notation

$$V = \{A, B\}. \tag{23.6.19}$$

Thus, we can write $\psi_m(t)$, given by (23.6.10), as

$$\psi_m(t) = \{A, B\},$$

where, by dimensional arguments, it follows that

$$A = k\rho\,\omega_0^{\alpha}, \qquad B = k\chi\,\omega_0^{\alpha}. \tag{23.6.20}$$

The dimensionless quantities ρ and χ are to be determined. We find, using (23.6.17), that

$$A = -\kappa\omega_0^2 \int_0^{\infty}\int_0^{\infty} \frac{(r_1 r_2)^{\eta-1}}{(r_1 + i\omega_0)(r_2 + i\omega_0)(r_1 + r_2)} dr_1 dr_2,$$
$$B = 2\kappa\omega_0^2 Re \int_0^{\infty}\int_0^{\infty} \frac{(r_1 r_2)^{\eta-1}}{(r_1 + i\omega_0)(r_2 - i\omega_0)(r_1 + r_2)} dr_1 dr_2.$$

The following integrals, together with (23.6.12), will be required in the calculations below. For complex y, z, we have

$$\int_0^{\infty} \frac{r^{\mu-1}}{(r+y)(r+z)} dr = \frac{\pi}{\sin\mu\pi}\left[\frac{y^{\mu-1} - z^{\mu-1}}{z-y}\right], \quad |\arg y|, |\arg z| < \pi,$$
$$\int_0^{\infty} \frac{r^{\mu-1}}{(r+z)^2} dr = -\frac{\pi(\mu-1)}{\sin\mu\pi} z^{\mu-2}, \quad 0<\mu<2, \qquad |\arg z| < \pi. \tag{23.6.21}$$

Relation $(23.6.21)_1$ is given in [168, page 289], while $(23.6.21)_2$ is a special case of a result in [168, page 285]. The latter can also be obtained by differentiating (23.6.12) with respect to z.

With the aid of $(23.6.21)_1$, we find that

$$A = -\frac{\kappa\omega_0^2\pi}{\sin\eta\pi}\int_0^{\infty} \frac{r_2^{\eta-1}\left[(i\omega_0)^{\eta-1} - r_2^{\eta-1}\right]}{r_2^2 + \omega_0^2} dr_2,$$
$$B = 2\frac{\kappa\omega_0^2\pi}{\sin\eta\pi} Re \int_0^{\infty} \frac{r_2^{\eta-1}\left[(i\omega_0)^{\eta-1} - r_2^{\eta-1}\right]}{(r_2 - i\omega_0)^2} dr_2.$$

These integrals can also be evaluated using (23.6.21). With the use of results derivable from $(23.6.21)_1$, we find that ρ in $(23.6.20)_1$ has the form

$$\rho = -\frac{1}{4}\sin\alpha\pi\left[1 - \frac{1}{\sin\frac{\pi\alpha}{2}} - i\frac{1-\sin\frac{\alpha\pi}{2}}{\cos\frac{\alpha\pi}{2}}\right]$$
$$= \frac{1}{2}\left(1 - \sin\frac{\pi\alpha}{2}\right)e^{i\frac{\pi\alpha}{2}}.$$

Also, using $(23.6.21)_2$, one can show that the quantity χ in $(23.6.20)_2$ is given by

$$\chi = k(1-\alpha)\cos\frac{\alpha\pi}{2} \geq 0.$$

Observe that

$$\dot{\psi}_m(t) = \{2\,i\,\omega_0\,A, 0\} = \left\{2\,i\,k\rho\,\omega_0^{\alpha+1}, 0\right\}.$$

The rate of dissipation given by $(23.6.14)_2$ is now considered. Let us put

$$\int_0^\infty r_1^{\eta-1}\dot{E}_+^t(-ir_1)dr_1 = K_0 e^{i\omega_0 t} + \overline{K}_0 e^{-i\omega_0 t}.$$

Then

$$D_m(t) = \kappa\left\{K_0^2, 2|K_0|^2\right\} = k\omega_0^{\alpha+1}\{D_1, D_2\}, \tag{23.6.22}$$

in the notation (23.6.19), where from (23.6.17) and (23.6.12),

$$K_0 = i\omega_0\int_0^\infty \frac{r^{\eta-1}}{r+i\omega_0}dr = (i\omega_0)^\eta\frac{\pi}{\sin\eta\pi}.$$

In the rightmost form of (23.6.22), D_1 and D_2 are dimensionless constants, which are now determined. The coefficient in this term emerges from dimensional analysis. Relation $(23.6.22)_1$ becomes

$$D_m(t) = \frac{\kappa\pi^2\omega_0^{2\eta}}{\sin^2\eta\pi}\left\{e^{i\pi\eta}, 2\right\} = k\sin\frac{\alpha\pi}{2}\omega_0^{\alpha+1}\left\{\exp\left(i\pi\frac{\alpha+1}{2}\right), 2\right\},$$

which gives

$$D_1 = \sin\frac{\alpha\pi}{2}\exp\left[i\frac{\pi(\alpha+1)}{2}\right], \qquad D_2 = 2\sin\frac{\alpha\pi}{2}.$$

From (23.6.15) and (23.6.18), we see that the rate of input of mechanical energy is given by

$$T(t)\dot{E}(t) = k\omega_0^{\alpha+1}\{W_1, W_2\},$$
$$W_1 = \exp\left[i\frac{\pi(\alpha+1)}{2}\right], \qquad W_2 = 2\sin\frac{\alpha\pi}{2}.$$

By virtue of (16.1.28), we must have

$$2i\rho + D_1 = W_1, \qquad D_2 = W_2,$$

which are easily confirmed. Relation (16.1.29) involves divergent quantities, namely $\mathcal{D}(t)$ and $W(t)$, for sinusoidal histories [15, 16].

23.6.2.2 Exponential History

Consider a history and current value $(E^t, E(t))$ given by

$$E(t) = E_1 e^{\gamma t}, \qquad E^t(s) = E(t-s), \quad s \in \mathbb{R}^+, \tag{23.6.23}$$

where E_1 is a constant amplitude. Some of the formulae for this history may be obtained from the sinusoidal case by simple substitutions [16]. However, we present direct derivations here. Instead of (23.6.16), we have

$$E_+^t(\omega) = \frac{E_1 e^{\gamma t}}{(\gamma + i\omega)} = \frac{E(t)}{(\gamma + i\omega)},$$
$$\dot{E}_+^t(\omega) = \frac{\gamma E(t)}{\gamma + i\omega}, \qquad \dot{E}_+^t(-ir) = \frac{\gamma E(t)}{\gamma + r}.$$

The stress function becomes

$$T(t) = kE(t)\gamma^{\alpha}, \tag{23.6.24}$$

on using (23.6.12). The minimum free energy $\psi_m(t)$, given by $(23.6.10)_1$, has the form

$$\psi_m(t) = E_1^2 e^{2\gamma t} A_\gamma = E^2(t) A_\gamma, \tag{23.6.25}$$

where A_γ is given by

$$\begin{aligned} A_\gamma &= \kappa\gamma^2 \int_0^\infty \int_0^\infty \frac{(r_1 r_2)^{\eta - 1}}{(r_1+\gamma)(r_2+\gamma)(r_1+r_2)} dr_1 dr_2, \\ &= \frac{\kappa\gamma^2\pi}{\sin\eta\pi} \int_0^\infty \frac{r_2^{\eta - 1}\left[\gamma^{\eta - 1} - r_2^{\eta - 1}\right]}{(r_2+\gamma)(r_2-\gamma)} dr_2, \\ &= \frac{1}{2} k\gamma^{\alpha}(1 - \sin\frac{\alpha\pi}{2}) \geq 0, \end{aligned}$$

with the aid of (23.6.21) and the integral ([168, page 289])

$$\int_0^\infty \frac{r^{\mu - 1}}{(r+\gamma)(r-\gamma)} dr = -\frac{\pi\gamma^{\mu - 2}}{2\sin\mu\pi}(1 + \cos\mu\pi), \qquad \gamma > 0, \;\; 0 < Re\mu < 2.$$

This formula is in fact in a similar category to (23.6.21) but where the parameters are in a different range. We conclude that

$$\dot{\psi}_m(t) = k E_1^2 \gamma^{\alpha + 1} e^{2\gamma t}\left(1 - \sin\frac{\alpha\pi}{2}\right).$$

It follows from $(23.6.9)_3$, (23.6.12), and (23.6.14) that

$$D_m(t) = k E_1^2 \gamma^{\alpha + 1} e^{2\gamma t} \sin\frac{\alpha\pi}{2}. \tag{23.6.26}$$

Finally, from (23.6.23) and (23.6.24) we find that

$$T(t)\dot{E}(t) = k E_1^2 \gamma^{\alpha + 1} e^{2\gamma t}, \tag{23.6.27}$$

and (16.1.28) is obeyed. Indeed, we can also consider (16.1.29) in the case of exponential histories, since there are no convergence difficulties. We write $\psi_m(t)$, given by (23.6.25), and the integrated forms of (23.6.26) and (23.6.27), as

$$\psi_m(t) = F_m W(t), \qquad F_m = 1 - \sin\frac{\alpha\pi}{2}, \qquad W(t) = \frac{1}{2} k E_1^2 \gamma^\alpha e^{2\gamma t},$$
$$\mathcal{D}_m(t) = S_m W(t), \qquad S_m = \sin\frac{\alpha\pi}{2}.$$

Thus, (16.1.29) is also clearly obeyed.

23.6.2.3 The Physical Free Energy

The physical free energy, discussed in several different contexts in the present work is defined by the property that its associated rate of dissipation is the true rate for the material. We tentatively identify the physical free energy of a fractional derivative material, as the minimum free energy by virtue of the following argument.

Two functionals have been identified as being associated with such materials, the Graffi–Volterra and the minimum free energies. The Graffi–Volterra functional is a degenerate version of the full two variable quadratic form discussed in earlier chapters. As such, it cannot be included as a free energy with deeper physical meaning, though it is a simple functional with the correct positivity properties and therefore very useful as a mathematical tool.

Invoking the property P4 introduced in (18.2.1), we see that the work function cannot be a valid free energy. Also, it has degenerate features somewhat similar to the Graffi–Volterra functional. Therefore, the minimum free energy, given by the elementary explicit expression (23.6.10), is the only non-degenerate free energy functional associated with the material. It must therefore be identified as the physical free energy of this material. Furthermore, the physical rate of dissipation has the form (23.6.14).

23.7 Application to Viscoelastic Systems

We now consider the dynamical equations for viscoelastic solids and fluids within the framework of fractional derivative models. An energy theorem is proved in both of these cases.

23.7.1 Viscoelastic Solids

Let $\Omega \subset \mathbb{R}^3$ be a smooth bounded domain of a linear viscoelastic solid, whose constitutive equation is given by the fractional model with constitutive relation given by (23.3.7). The initial boundary value problem is defined by the differential system in the domain $Q = \Omega \times (0, T)$ by

$$\begin{aligned} \rho_0(\mathbf{x})\frac{\partial^2 \mathbf{u}(\mathbf{x},t)}{\partial t^2} &= \nabla\cdot\mathbf{T}(\mathbf{x},t) + \rho_0(\mathbf{x})\mathbf{f}(\mathbf{x},t) \\ &= -\frac{\alpha}{\Gamma(1-\alpha)}\nabla\cdot\left[\mathbb{C}(\mathbf{x})\int_0^\infty \frac{\mathbf{E}_r^t(\mathbf{x},s)}{s^{1+\alpha}}ds\right] \\ &\quad + \rho_0(\mathbf{x})\mathbf{f}(\mathbf{x},t), \end{aligned} \tag{23.7.1}$$

where $\rho_0(\mathbf{x})$ denotes the mass density, $\mathbf{u}(\mathbf{x}, t)$ the displacement such that $\mathbf{E} = \frac{1}{2}(\nabla\mathbf{u} + \nabla\mathbf{u}^T)$ and $\mathbf{f}(\mathbf{x}, t)$ the body forces. The initial conditions are

$$\mathbf{u}(\mathbf{x}, 0) = \mathbf{u}_0(\mathbf{x}) \, , \quad \left.\frac{\partial \mathbf{u}(\mathbf{x}, t)}{\partial t}\right|_{t=0} = \mathbf{v}_0(\mathbf{x})$$

along with the boundary conditions

$$\mathbf{u}(\mathbf{x}, t)|_{\partial\Omega} = \mathbf{u}^0(\mathbf{x}), \tag{23.7.2}$$

where $\mathbf{u}_0$ and $\mathbf{v}_0$ are given functions.

Using the definition of fractional derivative given in (23.3.8), Eq. (23.7.1) can be rewritten in the form

$$\rho_0(\mathbf{x})\frac{\partial^2 \mathbf{u}(\mathbf{x}, t)}{\partial t^2} = \nabla \cdot \left[{}^{C}_{-\infty}D^{\alpha}_{t}\mathbb{C}(\mathbf{x})\mathbf{E}(\mathbf{x}, t)\right] + \rho_0(\mathbf{x})\mathbf{f}(\mathbf{x}, t). \tag{23.7.3}$$

Now, our purpose is to obtain an energy theorem for the problem (23.7.1) and (23.7.2). To this end, we multiply (23.7.3) by the first time derivative of $\mathbf{u}(\mathbf{x}, t)$. Then, after an integration on $Q = \Omega \times (0, T)$, we obtain

$$\begin{aligned} &\int_0^T \int_\Omega \rho_0(\mathbf{x})\frac{\partial^2 \mathbf{u}(\mathbf{x}, t)}{\partial t^2} \cdot \frac{\partial \mathbf{u}(\mathbf{x}, t)}{\partial t} d\mathbf{x}\, dt \\ &= \int_0^T \int_\Omega \left\{ (\nabla \cdot \left[{}^{C}_{-\infty}D^{\alpha}_{t}\mathbb{C}(\mathbf{x})\mathbf{E}(\mathbf{x}, t)\right] \cdot \frac{\partial \mathbf{u}(\mathbf{x}, t)}{\partial t} + \rho_0(\mathbf{x})\mathbf{f}(\mathbf{x}, t) \cdot \frac{\partial \mathbf{u}(\mathbf{x}, t)}{\partial t} \right\} d\mathbf{x}\, dt, \end{aligned} \tag{23.7.4}$$

where $d\mathbf{x}$ is the three-dimensional space volume element. Hence, using the divergence theorem and the boundary condition (23.7.2), it follows from (23.7.4) that

$$\begin{aligned} &\int_0^T \frac{\partial}{\partial t}\frac{1}{2}\int_\Omega \rho_0(\mathbf{x})\left(\frac{\partial \mathbf{u}(\mathbf{x}, t)}{\partial t}\right)^2 d\mathbf{x}\, dt \\ &= \int_0^T \int_\Omega \left[-\mathbb{C}(\mathbf{x})\left({}^{C}_{-\infty}D^{\alpha}_{t}\mathbf{E}(\mathbf{x}, t)\right) \cdot \frac{\partial \mathbf{E}(\mathbf{x}, t)}{\partial t} + \rho_0(\mathbf{x})\mathbf{f}(\mathbf{x}, t) \cdot \frac{\partial \mathbf{u}(\mathbf{x}, t)}{\partial t} \right] d\mathbf{x}\, dt \\ &= \int_0^T \int_\Omega \left[-\mathbf{T}(\mathbf{x}, t) \cdot \dot{\mathbf{E}}(\mathbf{x}, t) + \rho_0(\mathbf{x})\mathbf{f}(\mathbf{x}, t) \cdot \frac{\partial \mathbf{u}(\mathbf{x}, t)}{\partial t} \right] d\mathbf{x}\, dt. \end{aligned}$$

Then, from (23.4.6) or (23.4.9), we obtain

$$\begin{aligned} &\int_0^T \frac{\partial}{\partial t}\frac{1}{2}\int_\Omega \rho_0(\mathbf{x})\left[\left(\frac{\partial \mathbf{u}(\mathbf{x}, t)}{\partial t}\right)^2 + \Psi^S(\mathbf{x}, t)\right] d\mathbf{x}\, dt \\ &\leq \int_0^T \int_\Omega \rho_0(\mathbf{x})\mathbf{f}(\mathbf{x}, t) \cdot \frac{\partial \mathbf{u}(\mathbf{x}, t)}{\partial t} d\mathbf{x}\, dt, \end{aligned}$$

where $\Psi^S(\mathbf{x}, t)$ is the Graffi–Volterra free energy functional for solids, given by (23.4.8). Finally, carrying out the time integration, we find that

$$\frac{1}{2}\int_{\Omega}\rho_0(\mathbf{x})\left[\left(\frac{\partial\mathbf{u}(\mathbf{x},t)}{\partial t}\right)^2+\Psi^S(\mathbf{x},t)\right]d\mathbf{x}$$
$$\leq\frac{1}{2}\int_{\Omega}\rho_0(\mathbf{x})\left(\mathbf{v}_0(\mathbf{x})^2+\Psi^S(\mathbf{x},0)\right)dx+\int_0^T\int_{\Omega}\rho_0(\mathbf{x})\mathbf{f}(\mathbf{x},t)\cdot\frac{\partial\mathbf{u}(\mathbf{x},t)}{\partial t}d\mathbf{x}\,dt.$$

It is easily checked that the same inequalities hold for any free energy for the system. This in effect means any positive functional obeying (23.4.6) or (23.4.9). A similar observation applies to the case of fluids, which is now discussed.

23.7.2 Viscoelastic Fluids

The initial boundary value problem for a viscoelastic incompressible fluid described by the velocity $\mathbf{v}(\mathbf{x},t)$, the pressure $p(\mathbf{x},t)$ and the constant density ρ_0, is defined by the differential system

$$\begin{aligned}\rho_0\frac{\partial\mathbf{v}(\mathbf{x},t)}{\partial t}&=-\nabla p+\nabla\cdot\mathbf{T}_E(\mathbf{x},t)+\rho_0\mathbf{f}(\mathbf{x},t)\\&=-\nabla p+\frac{2\eta}{\Gamma(1-\alpha)}\nabla\cdot\int_{-\infty}^t\frac{\mathbf{E}(\mathbf{x},t)-\mathbf{E}(\mathbf{x},\tau)}{(t-\tau)^{1+\alpha}}d\tau+\rho_0\mathbf{f}(\mathbf{x},t)\\&=-\nabla p+\frac{\eta}{\Gamma(1-\alpha)}\nabla\cdot\int_{-\infty}^t\frac{\nabla\mathbf{v}(\mathbf{x},\tau)}{(t-\tau)^{\alpha}}d\tau+\rho_0\mathbf{f}(\mathbf{x},t),\\&\nabla\cdot\mathbf{v}(\mathbf{x},t)=0,\end{aligned}\tag{23.7.5}$$

with initial and boundary conditions

$$\mathbf{v}(\mathbf{x},0)=\mathbf{v}_0(\mathbf{x})\ ,\qquad \mathbf{v}(\mathbf{x},t)|_{\partial\Omega}=\mathbf{0}.\tag{23.7.6}$$

We again seek an energy theorem. It follows from $(23.7.5)_3$ and (23.7.6), together with a standard step involving integration by parts, that

$$\begin{aligned}&\int_0^T\frac{d}{dt}\int_{\Omega}\rho_0\,\mathbf{v}^2(\mathbf{x},t)d\mathbf{x}dt\\&=\int_0^T\int_{\Omega}\left[\nabla\mathbf{T}_E(\mathbf{x},t)\cdot\mathbf{v}(\mathbf{x},t)+\rho_0\mathbf{f}(\mathbf{x},t)\cdot\mathbf{v}(\mathbf{x},t)\right]\,d\mathbf{x}\,dt\\&=\int_0^T\int_{\Omega}\left[\mathbf{T}_E(\mathbf{x},t)\cdot\dot{\mathbf{E}}(\mathbf{x},t)+\rho_0\mathbf{f}(\mathbf{x},t)\cdot\mathbf{v}(\mathbf{x},t)\right]\,d\mathbf{x}\,dt.\end{aligned}$$

Using the inequality (23.4.6), we obtain

$$\int_0^T\frac{d}{dt}\int_{\Omega}\left[\rho_0\,\mathbf{v}^2(\mathbf{x},t)+\Psi^F(\mathbf{x},t)\right]d\mathbf{x}dt\leq\int_0^T\int_{\Omega}\rho_0\mathbf{f}(\mathbf{x},t)\cdot\mathbf{v}(\mathbf{x},t)d\mathbf{x}dt.$$

Hence, we have

$$\begin{aligned}&\int_{\Omega}\left[\rho_0\,\mathbf{v}^2(\mathbf{x},t)+\Psi^F(\mathbf{x},t))\right]d\mathbf{x}\\&\leq\int_{\Omega}\left[\rho_0\,\mathbf{v}^2(\mathbf{x},0)+\Psi^F(\mathbf{x},0))\right]d\mathbf{x}+\int_0^T\int_{\Omega}\rho_0\mathbf{f}(\mathbf{x},t)\cdot\mathbf{v}(\mathbf{x},t)d\mathbf{x}dt.\end{aligned}$$

23.8 Application to Rigid Heat Conductors

In this section, we use the fractional method to describe the behavior of heat conductors. The literature on this topic, including [1, 141, 299], is motivated by experimental findings such as those reported in [223, 285].

The classical Fourier constitutive equation is given by

$$\mathbf{q}(t) = -k_c\mathbf{g}(t),$$

where $\mathbf{q}$ is the heat flux , $\mathbf{g}$ the temperature gradient and $k_c > 0$ is the thermal conductivity, which may be space dependent. In order to obtain a finite wave propagation speed, Cattaneo [59] proposed the modified constitutive relation

$$-\gamma\dot{\mathbf{q}}(t) = \mathbf{q}(t) + k_c\mathbf{g}(t), \quad \gamma > 0.$$

This is the Cattaneo–Maxwell or, for brevity the Cattaneo equation. It can be put in the form of an integral equation, describing a material with thermal memory

$$\begin{aligned} &\mathbf{q}(t) = -\int_{-\infty}^{t} k(t-s)\mathbf{g}(s)ds = -\int_0^{\infty} k(s)\mathbf{g}^t(s)ds, \\ &\mathbf{g}^t(s) = \mathbf{g}(t-s), \qquad k(s) = k_0 e^{-\lambda s}, \ \lambda = \frac{1}{\gamma}, \ k_c = \gamma k_0. \end{aligned} \tag{23.8.1}$$

Relation (23.8.1) can be expressed in terms of the NFD$_t$, given by (23.2.13) or (23.2.14), where

$$\sigma = \gamma, \qquad k_0 = \frac{1+\sigma}{\sigma}M^*(\sigma), \qquad \alpha = \frac{1}{1+\sigma}.$$

Thus, $(23.8.1)_1$ can be written as

$$\mathbf{q}(t) = -\mathcal{D}_t^{(\alpha)}\overline{\mathbf{g}}(t), \tag{23.8.2}$$

where

$$\overline{\mathbf{g}}(t) = \int_0^t \mathbf{g}(u)du, \quad \dot{\overline{\mathbf{g}}}(t) = \mathbf{g}(t).$$

23.8.1 UFD$_t$ Fractional Cattaneo Equation

Let us replace the NFD$_t$ in (23.8.2) by the UFD$_t$ as defined by (23.2.12) or (23.2.6). Thus, we put

$$k(s) = \frac{k_0}{\Gamma(1-\alpha)}\frac{1}{s^\alpha},$$

giving

$$\begin{aligned} \mathbf{q}(t) &= -k_0 D_t^{(\alpha)}\overline{\mathbf{g}}(t) = -\frac{k_0}{\Gamma(1-\alpha)}\int_{-\infty}^{t}\frac{\dot{\overline{\mathbf{g}}}(\tau)}{(t-\tau)^\alpha}d\tau \\ &= \frac{\alpha k_0}{\Gamma(1-\alpha)}\int_{-\infty}^{t}\frac{\overline{\mathbf{g}}(\tau)-\overline{\mathbf{g}}(t)}{(t-\tau)^{1+\alpha}}d\tau = \frac{\alpha k_0}{\Gamma(1-\alpha)}\int_0^{\infty}\frac{\overline{\mathbf{g}}_r^t(s)}{s^{1+\alpha}}ds, \end{aligned}$$

where, recalling (9.1.2), we introduce $\overline{\mathbf{g}}_r^t(s)$, defined as

$$\overline{\mathbf{g}}_r^t(s) = \overline{\mathbf{g}}(t-s) - \overline{\mathbf{g}}(t) = -\int_{t-s}^{t} \mathbf{g}(u)du,$$

to emphasize an analogy between heat flow and viscoelasticity whereby $-\overline{\mathbf{g}}(t)$, $\mathbf{q}(t)$ correspond to $\mathbf{E}(t)$, $\mathbf{T}(t)$. It can be concluded from this and (23.4.8) that the functional

$$\psi(t) = \frac{\alpha k_0}{2\Gamma(1-\alpha)} \int_0^{\infty} \frac{\overline{\mathbf{g}}_r^t(s) \cdot \overline{\mathbf{g}}_r^t(s)}{s^{1+\alpha}} ds$$

is the Graffi-Volterra free energy for fractional Cattaneo models, with corresponding rate of dissipation deduced from (23.4.10) to be

$$D(t) = \frac{\alpha(1+\alpha)k_1}{2\Gamma(1-\alpha)} \int_0^{\infty} \frac{\overline{\mathbf{g}}_r^t(s) \cdot \overline{\mathbf{g}}_r^t(s)}{s^{2+\alpha}} ds.$$

Indeed, the minimum free energy for isotropic materials of this kind can be written down immediately from (23.6.9) and (23.6.10), replacing $\dot{E}^t(u)\dot{E}^t(v)$ by $\mathbf{g}^t(u) \cdot \mathbf{g}^t(v)$. Thus, we obtain

$$\psi_m(t) = \kappa \int_0^{\infty} \int_0^{\infty} \int_0^{\infty} \int_0^{\infty} \frac{(r_1 r_2)^{\eta - 1} \mathbf{g}^t(u) \cdot \mathbf{g}^t(v) e^{-r_1 u - r_2 v}}{r_1 + r_2} dr_1 dr_2 du dv,$$
$$\kappa = \frac{k_0}{\pi^2} \sin\frac{\alpha\pi}{2} \cos^2\frac{\alpha\pi}{2}.$$

The quantity η is defined by (23.6.1). The coefficient k_1 replaces $\mathbb{C}$ in Sect. 23.3 or k in Sects. 23.5 and 23.6. Similarly, the corresponding rate of dissipation is deduced from $(23.6.14)_3$ by replacing $\dot{E}^t(u)$ with $-\mathbf{g}^t(u)$, yielding

$$D_m(t) = \kappa\Gamma^2(\eta) \left| \int_0^{\infty} \frac{\mathbf{g}^t(u)}{u^{\eta}} du \right|^2 \geq 0.$$

23.8.2 The NFD$_t$ Model

Analogous results can be derived for the NFD$_t$ model given by (23.8.2), which is of course the Cattaneo equation. Free energies for this relationship are those for a simple memory function $k(s)$ described in (23.8.1). This has one decaying exponential and goes to zero at large times. The Graffi-Volterra free energy corresponding to this form is given by

$$\psi(t) = \frac{1}{2}\lambda k_0 \int_0^{\infty} e^{-\lambda s} \overline{\mathbf{g}}_r^t(s) \cdot \overline{\mathbf{g}}_r^t(s) ds,$$

while the corresponding rate of dissipation is

$$D(t) = \lambda^2 k_0 \int_0^{\infty} e^{-\lambda s} \overline{\mathbf{g}}_r^t(s) \cdot \overline{\mathbf{g}}_r^t(s) ds.$$

Also, for a relaxation function consisting of one decaying exponential, the minimum free energy has the form

$$\psi_D(t) = \frac{1}{2}\lambda^2 k_0 \left| \int_0^\infty e^{-\lambda s} \overline{\mathbf{g}}_r^t(s) ds \right|^2,$$

$$D_D(t) = \lambda^3 k_0 \left| \int_0^\infty e^{-\lambda s} \overline{\mathbf{g}}_r^t(s) ds \right|^2.$$

These are a special case of the Day free energy and rate of dissipation where the relaxation function goes to zero at large times.

23.9 Application to Electromagnetic Systems

We now explore some properties of two electromagnetic bodies, which are characterized by constitutive equations expressed as fractional models. The form of a particular free energy will be derived in both cases. These are examples or generalizations of the Graffi–Volterra free energy discussed in Sect. 23.4.1. This section is based on [19].

23.9.1 Visco-Ferromagnetic Materials

Let us consider a visco-ferromagnetic material characterized by the following constitutive equation

$$\mathbf{B}(\mathbf{x}, t) = \frac{\mathbf{C}(\mathbf{x})}{\Gamma(1-\alpha)} \int_a^t \frac{\mathbf{H}_\tau(\mathbf{x}, \tau)}{(t-\tau)^\alpha} d\tau, \quad \mathbf{H}_\tau(\mathbf{x}, \tau) = \frac{\partial}{\partial \tau}\mathbf{H}(\mathbf{x}, \tau), \tag{23.9.1}$$

where the magnetic induction $\mathbf{B}(\mathbf{x}, t)$ and the magnetic field are $\mathbf{H}(\mathbf{x}, t)$ are defined for any point $\mathbf{x} \in \Omega$, the smooth bounded domain occupied by the material; moreover, the quantity $\mathbf{C}(\mathbf{x})$ is a second-order positive tensor, defined for any point $\mathbf{x} \in \Omega$.

This equation is expressed in terms of the α–Caputo fractional derivative (23.2.5); see also [47]

$${}_CD_t^\alpha \mathbf{H}(\mathbf{x}, t) = \frac{1}{\Gamma(1-\alpha)} \int_a^t \frac{\mathbf{H}_\tau(\mathbf{x}, \tau)}{(t-\tau)^\alpha} d\tau.$$

Using this definition, Eq. (23.9.1) assumes the following form

$$\mathbf{B}(\mathbf{x}, t) = \mathbf{C}(\mathbf{x})\, {}_CD_t^\alpha \mathbf{H}(\mathbf{x}, t).$$

Taking $a = -\infty$ and carrying a time integration by parts, (23.9.1) assumes the more useful form

$$\mathbf{B}(\mathbf{x}, t) = -\frac{\alpha}{\Gamma(1-\alpha)} \int_{-\infty}^t \mathbf{C}(\mathbf{x}) \frac{\mathbf{H}^{(r)}(\mathbf{x}, \tau)}{(t-\tau)^{\alpha+1}} d\tau, \mathbf{H}^{(r)}(\mathbf{x}, \tau) = \mathbf{H}(\mathbf{x}, \tau) - \mathbf{H}(\mathbf{x}, t),$$

where $\mathbf{H}^{(r)}$ denotes the relative history of the magnetic field.

The dissipation law states that, for any simple electromagnetic material, there exists at least one state functional, denoted by $\psi(\mathbf{x}, t$, and referred to as a free energy, which satisfies the following fundamental requirement:

$$\frac{\partial}{\partial t}\psi(\mathbf{x}, t) \leq \mathbf{B}(\mathbf{x}, t) \cdot \mathbf{H}_t(\mathbf{x}, t). \tag{23.9.2}$$

By introducing the rate of dissipation $\mathcal{D}(\mathbf{x}, t) \geq 0$, the dissipation law can be written as

$$\frac{\partial}{\partial t}\psi(\mathbf{x}, t) + \mathcal{D}(\mathbf{x}, t) = \mathbf{B}(\mathbf{x}, t) \cdot \mathbf{H}_t(\mathbf{x}, t). \tag{23.9.3}$$

Taking account of (23.9.1), we can apply an integration by parts to the scalar product on the right-hand side of both Eqs. (23.9.2) and (23.9.3); the expression so derived allows us to identify the following functional as a free energy:

$$\psi(t) = \frac{\alpha}{2\Gamma(1-\alpha)}\int_0^{\infty} \mathbf{C}(\mathbf{x})\frac{\mathbf{H}^{(r)}(\mathbf{x}, t-s) \cdot \mathbf{H}^{(r)}(\mathbf{x}, t-s)}{s^{\alpha+1}} ds.$$

This is a particular example of the *Graffi-Volterra* free energy. The related rate of dissipation is given by

$$\mathcal{D}(\mathbf{x}, t) = \frac{\alpha(\alpha+1)}{2\Gamma(1-\alpha)}\int_0^{\infty} \mathbf{C}(\mathbf{x})\frac{\mathbf{H}^{(r)}(\mathbf{x}, t-s) \cdot \mathbf{H}^{(r)}(\mathbf{x}, t-s)}{s^{\alpha+2}} ds.$$

23.9.2 Nonlocal Visco-Ferromagnetic Materials

The behavior of nonlocal visco-ferromagnetic materials can be described by means of a constitutive equation expressed in terms of the fractional operator $\mathcal{M}_{\mathbf{x}}^{\beta}$ of order $\beta \in (\frac{1}{2}, 1)$, acting on a function $\mathbf{f}(\mathbf{x})$, introduced in [56] and expressed by

$$\mathcal{M}_{\mathbf{x}}^{\beta}\mathbf{F}(\mathbf{x}) = \frac{\beta\pi^{-\frac{\beta}{2}}}{1-\beta}\int_{\Omega} \mathbf{f}(\mathbf{y})e^{-\frac{\beta^2}{1-\beta^2}(\mathbf{x}-\mathbf{y})^2} d\mathbf{y},$$

where $\mathbf{x}, \mathbf{y} \in \Omega$.

By replacing $\mathbf{f}(\mathbf{x})$ with $\mathbf{C}(\mathbf{x}, \mathbf{y})\mathbf{H}(\mathbf{x}, \tau)$, we can introduce the following new constitutive equation for the ferromagnetic induction

$$\begin{aligned}\mathbf{B}(\mathbf{x}, t) &= \mathcal{M}_{\mathbf{x}}^{\beta} D_t^{\alpha}\left[\mathbf{C}(\mathbf{x}, \mathbf{y})\mathbf{H}(\mathbf{x}, t)\right] \\ &= \frac{\beta\pi^{-\frac{\beta}{2}}}{(1-\beta)\Gamma(1-\alpha)}\int_{\Omega}\int_a^t \frac{1}{(t-\tau)^{\alpha}}\mathbf{C}(\mathbf{x}, \mathbf{y})\mathbf{H}_{\tau}(\mathbf{y}, \tau)e^{-\frac{\beta^2}{1-\beta^2}(\mathbf{x}-\mathbf{y})^2} d\tau d\mathbf{y},\end{aligned} \tag{23.9.4}$$

where the scalar α, assumed to be in the interval $\left(0, \frac{1}{2}\right)$, is the degree of the Caputo fractional derivative, denoted by D_t^{α}, while the second-order tensor $\mathbf{C}(\mathbf{x}, \mathbf{y})$ is taken to be symmetric in $\mathbf{x}$ and $\mathbf{y}$ and positively defined.

We note that, for nonsimple materials characterized by (23.9.4), the magnetic induction $\mathbf{B}(\mathbf{x}, t)$ at any fixed point $\mathbf{x} \in \Omega$ depends on the values of the magnetic field $\mathbf{H}(\mathbf{y}, t)$ $\forall \mathbf{y} \in \Omega$.

The total internal power can be obtained by integrating over Ω the scalar product in the right-hand side of both Eqs. (23.9.2) and (23.9.3); thus, by virtue of (23.9.4), it is given by

$$\int_\Omega \mathbf{B}(\mathbf{x},t)\cdot\mathbf{H}_t(\mathbf{x},t)d\mathbf{x} = \int_\Omega \mathcal{M}_{\mathbf{x}}^{\beta} D_t^{\alpha}\left[\mathbf{C}(\mathbf{x},\mathbf{y})\mathbf{H}(\mathbf{x},t)\right]\cdot\mathbf{H}_t(\mathbf{x},t)d\mathbf{x} \tag{23.9.5}$$

$$= \frac{\beta\pi^{-\frac{\beta}{2}}}{(1-\beta)\Gamma(1-\alpha)}\int_\Omega\left[\int_\Omega\int_a^t \mathbf{C}(\mathbf{x},\mathbf{y})\frac{\mathbf{H}_\tau(\mathbf{y},\tau)}{(t-\tau)^\alpha}e^{-\frac{\beta^2}{1-\beta^2}(\mathbf{x}-\mathbf{y})^2}d\tau d\mathbf{y}\right]\cdot\mathbf{H}_t(\mathbf{x},t)d\mathbf{x}$$

and the global free energy has the form:

$$\psi(\Omega,t) \equiv \int_\Omega \psi(\mathbf{x},t)d\mathbf{x}. \tag{23.9.6}$$

If we assume $\mathbf{H}(\cdot,\tau)=\mathbf{0}$ for $\tau\le a$, the inequality (23.9.2), by using (23.9.5) and (23.9.6), becomes

$$\dot\psi(\Omega,t) \equiv \frac{\partial}{\partial t}\int_\Omega \psi(\mathbf{x},t)d\mathbf{x} \le \int_\Omega \mathbf{B}(\mathbf{x},t)\cdot\mathbf{H}_t(\mathbf{x},t)d\mathbf{x}$$

$$= \frac{\beta\pi^{-\frac{\beta}{2}}}{(1-\beta)\Gamma(1-\alpha)}\int_\Omega\int_\Omega\int_{-\infty}^t \mathbf{C}(\mathbf{x},\mathbf{y})\frac{\left[\mathbf{H}(\mathbf{y},\tau)-\mathbf{H}(\mathbf{y},t)\right]_\tau}{(t-\tau)^\alpha}e^{-\frac{\beta^2}{1-\beta^2}(\mathbf{x}-\mathbf{y})^2}d\tau d\mathbf{y}$$
$$\cdot\frac{\partial}{\partial t}\left[\mathbf{H}(\mathbf{x},t)-\mathbf{H}(\mathbf{x},\tau)\right]d\mathbf{x},$$

whence, with a time integration by parts, it follows that

$$\frac{\partial}{\partial t}\int_\Omega \psi(\mathbf{x},t)d\mathbf{x} \le \int_\Omega \mathbf{B}(\mathbf{x},t)\cdot\mathbf{H}_t(\mathbf{x},t)d\mathbf{x}$$

$$= \frac{\alpha\beta\pi^{-\frac{\beta}{2}}}{(1-\beta)\Gamma(1-\alpha)}\int_\Omega\int_\Omega\int_{-\infty}^t \mathbf{C}(\mathbf{x},\mathbf{y})\frac{\mathbf{H}(\mathbf{y},t)-\mathbf{H}(\mathbf{y},\tau)}{(t-\tau)^{\alpha+1}}$$
$$\cdot\frac{\partial}{\partial t}\left[\mathbf{H}(\mathbf{x},t)-\mathbf{H}(\mathbf{x},\tau)\right]e^{-\frac{\beta^2}{1-\beta^2}(\mathbf{x}-\mathbf{y})^2}d\tau d\mathbf{y}d\mathbf{x}.$$

This inequality is satisfied by the following functionals:

$$\psi(\Omega,t) = \frac{\alpha\beta\pi^{-\frac{\beta}{2}}}{2(1-\beta)\Gamma(1-\alpha)}\int_\Omega\int_\Omega\int_{-\infty}^t \frac{1}{(t-\tau)^{\alpha+1}}\left[\mathbf{H}(\mathbf{y},t)-\mathbf{H}(\mathbf{y},\tau)\right]$$
$$\cdot\mathbf{C}(\mathbf{x},\mathbf{y})\left[\mathbf{H}(\mathbf{x},t)-\mathbf{H}(\mathbf{x},\tau)\right]e^{-\frac{\beta^2}{1-\beta^2}(\mathbf{x}-\mathbf{y})^2}d\tau d\mathbf{y}d\mathbf{x},$$

which gives the global free energy $\psi(\Omega,t)$, and

$$\mathcal{D}(\Omega,t) = \frac{\alpha(\alpha+1)\beta\pi^{-\frac{\beta}{2}}}{2(1-\beta)\Gamma(1-\alpha)}\int_\Omega\int_\Omega\int_{-\infty}^t \frac{1}{(t-\tau)^{\alpha+2}}\left[\mathbf{H}(\mathbf{y},t)-\mathbf{H}(\mathbf{y},\tau)\right]$$
$$\cdot\mathbf{C}(\mathbf{x},\mathbf{y})\left[\mathbf{H}(\mathbf{x},t)-\mathbf{H}(\mathbf{x},\tau)\right]e^{-\frac{\beta^2}{1-\beta^2}(\mathbf{x}-\mathbf{y})^2}d\tau d\mathbf{y}d\mathbf{x},$$

which is the expression for the related global rate of dissipation $\mathcal{D}(\Omega,t)$.

Part IV

The Dynamical Equations for Materials with Memory

24

Existence and Uniqueness

24.1 Introduction to Existence and Uniqueness

The study of differential problems related to materials with fading memory began with the work of Graffi [170, 172]. Later on, these studies were considered by many authors, and in particular, a new important description of such phenomena was given by Dafermos in [79, 80], using semigroup theory, where besides existence and uniqueness of the solution, the interesting problem of asymptotic stability was also examined.

In 1979, Fichera [135] considered the question of well-posedness for the so-called quasistatic problem, by exhibiting various counterexamples, proposed also in subsequent work [137, 138]. This problem was solved by assigning initial conditions for the quasistatic problem as Fichera suggested.

A recent new formulation of these problems for materials with memory effects has been introduced in [91] and [106], where instead of using the past history to characterize the initial state, the notion of a minimal state $\mathbf{I}^0$, defined in $(7.4.2)_2$, is preferred.

This new point of view starts from the observation that if we consider a viscoelastic medium, the differential equation in terms of the displacement vector $\mathbf{u}$ is given by

$$\rho\ddot{\mathbf{u}}(\mathbf{x},t) = \nabla \cdot \left[\mathbb{G}_0(\mathbf{x})\mathbf{E}(\mathbf{x},t) + \int_0^\infty \mathbb{G}'(\mathbf{x},s)\mathbf{E}^t(\mathbf{x},s)ds\right] + \rho\mathbf{f}(\mathbf{x},t) \qquad (24.1.1)$$
$$\forall\,(\mathbf{x},t) \in \Omega \times (0,\infty),$$

where $\mathbf{E}^t(\mathbf{x},s) = \mathbf{E}(\mathbf{x},t-s)$ is the history of the strain tensor $\mathbf{E} = \frac{\nabla\mathbf{u}+(\nabla\mathbf{u})^T}{2}$. We take $\rho = 1$ in the present chapter. The initial conditions*

* In this chapter, we adopt the widely used (in this context) notation that $\Omega \subset \mathbb{R}^3$ is the region occupied by the body under consideration. In Part III, it is used to denote the complex frequency plane. It is also used to denote the complex plane in Chap. 27.

G. Amendola et al., *Thermodynamics of Materials with Memory*,
https://doi.org/10.1007/978-3-030-80534-0_24

$$\begin{aligned}
&\mathbf{u}(\mathbf{x},0)=\mathbf{u}_0(\mathbf{x}), \qquad \dot{\mathbf{u}}(\mathbf{x},\mathbf{0})=\dot{\mathbf{u}}_0(\mathbf{x}) \qquad \forall \mathbf{x}\in\Omega,\\
&\mathbf{u}^{t=0}(\mathbf{x},s)=\mathbf{u}^0(\mathbf{x},s) \qquad \forall(\mathbf{x},s)\in\Omega\times[\mathbf{0},\infty)
\end{aligned} \tag{24.1.2}$$

must be associated with (24.1.1), together with boundary conditions, which, for example, can be expressed by

$$\mathbf{u}(\mathbf{x},t)=\mathbf{0} \qquad \forall(\mathbf{x},t)\in\partial\Omega\times[\mathbf{0},\infty). \tag{24.1.3}$$

We now observe that two initial histories $\mathbf{u}_1^0(\mathbf{x},s)$, $\mathbf{u}_2^0(\mathbf{x},s)$ yield the same solution if the initial data satisfy the relation

$$\int_0^\infty \mathbb{G}'(\tau+s)\mathbf{E}_1^0(s)ds=\int_0^\infty \mathbb{G}'(\tau+s)\mathbf{E}_2^0(s)ds \quad \forall\tau\geq 0,$$

as well as conditions (24.1.2). In fact, it is easy to show that (24.1.1) can be written in the form

$$\ddot{\mathbf{u}}(\mathbf{x},t)=\nabla\cdot\left[\mathbb{G}_0(\mathbf{x})\mathbf{E}(\mathbf{x},t)+\int_0^t \mathbb{G}'(\mathbf{x},s)\mathbf{E}^t(\mathbf{x},s)ds\right]+\nabla\cdot\mathbf{F}(\mathbf{x},t)+\mathbf{f}(\mathbf{x},t),$$

where

$$\mathbf{F}(\mathbf{x},t)=\int_t^\infty \mathbb{G}'(\mathbf{x},s)\mathbf{E}^t(\mathbf{x},s)ds=\int_0^\infty \mathbb{G}'(\mathbf{x},t+\mathbf{s})\mathbf{E}^0(\mathbf{x},s)ds.$$

Thus, two different initial histories $\mathbf{u}_1^0(\mathbf{x},s)$ and $\mathbf{u}_2^0(\mathbf{x},s)$ to which corresponds the same quantity $\mathbf{F}(\mathbf{x},t)$ yield the same solution to the problem (24.1.1)–(24.1.3). Consequently, these different histories can be considered as the same state of the material.

Therefore, from considerations related to the differential problem, we arrive at the notion of equivalence between initial histories, characterized by

$$\mathbf{E}_1(0)=\mathbf{E}_2(0), \quad \int_0^\infty \mathbb{G}'(\tau+s)\mathbf{E}_1^0(s)ds=\int_0^\infty \mathbb{G}'(\tau+s)\mathbf{E}_2^0(s)ds \quad \forall\tau\geq 0, \tag{24.1.4}$$

as discussed in Sect. 7.4 and used in later chapters. Note that $\mathbf{E}_1(0)$ and $\mathbf{E}_2(0)$ in $(24.1.4)_1$ correspond to $\mathbf{u}(\mathbf{x},0)$ in (24.1.2).

For certain classes of material, (24.1.4) can be satisfied by different histories, while for others, it is true only in the trivial case of equal histories. This distinction is discussed in detail in Sect. 16.5.

The origin of the idea that different histories represent the same state of the material can be found in the concept of equivalence between states introduced by Noll in his axiomatic formulation of continuum mechanics [277].

The introduction of this notion of equivalence provides meaningful insights into the study of stability problems and, even in some measure, into the choice of spaces used to prove existence and uniqueness theorems. We remark here that in the study of conditions for stability, the topology chosen for the spaces of states affects the results to a remarkable degree.

In the literature, stability is almost always studied in relation to the topology associated with the Graffi–Volterra free energy (Sects. 10.1.1, 10.2, etc.) defined by

$$\psi_G(\mathbf{E}^t) = \frac{1}{2}\mathbb{G}_\infty \mathbf{E}(t) \cdot \mathbf{E}(t) + \frac{1}{2}\int_0^\infty \mathbb{G}'(s)\big[\mathbf{E}^t(s) - \mathbf{E}(t)\big] \cdot \big[\mathbf{E}^t(s) - \mathbf{E}(t)\big]\, ds, \quad (24.1.5)$$

where the dependence on $\mathbf{x}$ is understood.

It is easy to prove that the norm related to the free energy (24.1.5) introduces an unnatural separation between histories. In fact, two equivalent histories have a nonzero distance between them in this norm, which contrasts with the impossibility of distinguishing the future effects of the two histories by calculating the stress or determining the solution. We refer in this context to (10.1.4) and Remark 10.1.2.

Furthermore, it is possible to prove that the topology related to the free energy ψ_G appears coarser with respect to other topologies defined on minimal states, which, in the study of stability, can yield curious results.

It therefore appears natural to consider a formulation of these problems based on the notion of a minimal state for the material, defined by $\sigma = (\mathbf{E}(t), \mathbf{I}^t)$, where $\mathbf{I}^t$ represents the equivalence class of the relative strain histories $\mathbf{E}_r^t(s) = \mathbf{E}^t(s) - \mathbf{E}(t)$ given by (see (8.2.2) and the comment after (8.2.6))

$$\mathbf{I}^t(\tau) = \int_0^\infty \mathbb{G}'(\tau + s)\mathbf{E}_r^t(s) ds \qquad \forall \tau \geq 0.$$

We shall therefore develop this formulation and seek to express both function spaces and free energies in terms of minimal states.

For this purpose, free energies that are functionals of the minimal state, as defined by (7.4.6), are particularly of interest, notably ψ_F defined in Sect. 10.1.3 (later for various specific materials) and the minimum free energy introduced in Chap. 11.

A further relevant point here is that the space of definition of the free energy ψ_G is much smaller than that of ψ_F and the minimum free energy, so that stability results will have more restricted application (see observations after (10.1.14) and (12.2.13)).

24.2 Dynamics of Viscoelastic Solids

24.2.1 Existence and Uniqueness of Solutions

It is worthwhile recalling the traditional initial boundary value problem for a viscoelastic material.

In the linear case, the constitutive equation for the stress tensor $\mathbf{T}$ in a viscoelastic material is given by (8.1.5), and if we include space dependence, it becomes

$$\mathbf{T}(\mathbf{x}, t) = \mathbb{G}_0(\mathbf{x})\mathbf{E}(\mathbf{x}, t) + \int_0^\infty \mathbb{G}'(\mathbf{x}, s)\mathbf{E}^t(\mathbf{x}, s) ds, \quad (24.2.1)$$

where $\mathbf{E}$ is the infinitesimal strain tensor given by $\mathbf{E} = \frac{1}{2}[\nabla\mathbf{u} + (\nabla\mathbf{u})^T]$, where $\mathbf{u}$ denotes the displacement field. The quantity $\mathbb{G}_\infty(\mathbf{x})$ is defined by

$$\mathbb{G}_\infty(\mathbf{x}) = \mathbb{G}_0(\mathbf{x}) + \int_0^\infty \mathbb{G}'(\mathbf{x}, s)ds.$$

The symmetry of the tensors $\mathbb{G}_0(\mathbf{x})$ and $\mathbb{G}_\infty(\mathbf{x})$ is required by thermodynamics and expressed by (8.1.13) and (8.1.14), respectively, while that of $\mathbb{G}'(\mathbf{x}, s)$ is ensured by assumption (8.1.27). For solids, $\mathbb{G}_\infty(\mathbf{x})$ is positive definite, so that from (8.1.22), $\mathbb{G}_0(\mathbf{x})$ has the same property. We shall also assume that $-\mathbb{G}'(\mathbf{x}, s)$ is a positive tensor.

The constitutive equation (24.2.1) can be written as

$$\mathbf{T}(\mathbf{x}, t) = \mathbb{G}_0(\mathbf{x})\nabla\mathbf{u}(\mathbf{x}, t) + \int_0^\infty \mathbb{G}'(\mathbf{x}, s)\nabla\mathbf{u}(\mathbf{x}, t - s)ds, \tag{24.2.2}$$

by virtue of (8.1.8).

Therefore, on a smooth bounded domain $\Omega \subset \mathbb{R}^3$, the differential problem is expressed by the equation

$$\ddot{\mathbf{u}}(\mathbf{x}, t) = \nabla \cdot \left[\mathbb{G}_0(\mathbf{x})\nabla\mathbf{u}(\mathbf{x}, t) + \int_0^\infty \mathbb{G}'(\mathbf{x}, s)\nabla\mathbf{u}(\mathbf{x}, t - \mathrm{s})ds\right] + \mathbf{f}(\mathbf{x}, t) \tag{24.2.3}$$

together with the initial conditions

$$\mathbf{u}(\mathbf{x}, 0) = \mathbf{u}_0(\mathbf{x}), \quad \dot{\mathbf{u}}(\mathbf{x}, \mathbf{0}) = \dot{\mathbf{u}}_0(\mathbf{x}), \quad \mathbf{u}^{t=0}(\mathbf{x}, s) = \mathbf{u}^0(\mathbf{x}, s)\ \forall s \in \mathbb{R}^+$$

and, for example, Dirichlet boundary conditions

$$\mathbf{u}(\mathbf{x}, t)|_{\partial\Omega} = 0,$$

the quantity $\mathbf{f}$ in (24.2.3) being the body forces.

In this framework the kernel $\mathbb{G}$, defined by (8.1.7), is assumed continuous in $\mathbf{x} \in \Omega$ and such that for any $\mathbf{x} \in \Omega$,

(a) $\mathbb{G}(\mathbf{x}, \cdot) \in H^1(0, \infty)$,
(b) $\mathbb{G}_0(\mathbf{x}) > \mathbf{0}$ as stated above.

Classical theorems on existence, uniqueness, and continuous dependence can be found in the work of Dafermos [79] (see [114] also).

24.2.2 Quasistatic Problem in Linear Viscoelasticity: Fichera's Problem

The Fichera problem consists in the study of the quasistatic problem for a viscoelastic material. Fichera observed that such a problem cannot be resolved, as in linear elasticity, without providing the datum on the initial history and so working on the time interval $[0, \infty)$. Other authors [119] studied the same problem on the interval $(-\infty, \infty)$, but they had to assign suitable decay conditions for $t \to -\infty$ on the solutions, which is the same as giving initial conditions.

We recall that Fichera exhibited some counterexamples related to the well-posedness of Cauchy's problem, in consequence of his related correspondence with Capriz and Gurtin and also with Morro. The first counterexample, proposed by

Fichera in [135] for the one-dimensional problem, is characterized by the relaxation function $G(s)$ given by

$$G(s) = 1 - \lambda + \lambda \exp(-s),$$

where λ is a real-valued parameter. With such a relaxation function, Fichera showed that the spectrum of (24.2.2) is $\lambda \geq 1$; consequently, it follows that for λ satisfying this condition, we cannot have uniqueness for the solution to the quasistatic problem in $L^1(\mathbb{R}; H_0^1(\Omega))$.

However, since the condition $\lambda > 1$ implies $G(s) < 0$ for large enough values of s, while the value $\lambda = 1$, which yields $G_\infty = 0$, corresponds to fluids, the counterexample is not significant [136].

Thus, with a second counterexample, characterized by

$$G(s) = \frac{1}{2} - s \exp(-s). \tag{24.2.4}$$

Fichera proved that the requirement $G(s) > 0$ does not guarantee existence and uniqueness; in fact, with such a relaxation function, nonuniqueness can be easily seen.

However, this counterexample is not very interesting, since the relaxation function (24.2.4) does not satisfy the constraint (8.1.18) imposed by the Second Law of Thermodynamics.

A third counterexample, satisfying this constraint, was then given by Fichera, on assuming a relaxation function

$$G(s) = G_\infty + (G_0 - G_\infty) \exp(-\lambda s),$$

with $G_\infty, G_0 - G_\infty, \lambda \in \mathbb{R}^{++}$. However, the strain $E(t) = \exp[-(\lambda G_\infty / G_0)t]$ is an eigensolution of

$$G_0 E(t) + \int_0^\infty G'(s) E(t - s) ds = 0,$$

thus showing that the solution to the one-dimensional version of the quasistatic problem is nonunique.

The Fichera elaboration of these counterexamples effected an interesting improvement in such studies. Later on, the problem of existence and uniqueness of the solution to several problems related to viscoelastic materials was thus considered, especially by Italian researchers on continuum mechanics (see, for example, [184]). We only observe that we agree with what Fichera wrote to Morro, i.e., that the difficulties related to the problem of existence and uniqueness arise because the infinite extent of the memory causes the problem to be affected by the topology of the space of solutions.

Relation (24.2.2) can be written in the form

$$\mathbf{T}(\mathbf{x}, t) = \mathbb{G}(\mathbf{x}, t) \nabla \mathbf{u}(\mathbf{x}, 0) + \int_0^t \mathbb{G}(\mathbf{x}, t - \tau) \nabla \dot{\mathbf{u}}(\mathbf{x}, \tau) d\tau + \tilde{\mathbf{I}}^0(\mathbf{x}, t),$$

with the aid of an integration by parts, where $\tilde{\mathbf{I}}$ is defined in $(8.2.3)_1$ (see comment after (8.2.6)). Thus, the equation of motion (24.2.3) becomes

$$\ddot{\mathbf{u}}(\mathbf{x},t) = \nabla \cdot \int_0^t \mathbb{G}(\mathbf{x},t-\tau)\nabla\dot{\mathbf{u}}(\mathbf{x},\tau)d\tau + \mathbf{b}(\mathbf{x},t), \tag{24.2.5}$$

where

$$\mathbf{b}(\mathbf{x},t) = \mathbf{f}(\mathbf{x},t) + \nabla \cdot [\tilde{\mathbf{I}}^0(\mathbf{x},t) + \mathbb{G}(\mathbf{x},t)\nabla\mathbf{u}(\mathbf{x},0)]$$

is a given function in $\Omega \times \mathbb{R}^+$. Putting $\mathbf{v} = \dot{\mathbf{u}}$, the differential equation (24.2.5) can be written as

$$\dot{\mathbf{v}}(\mathbf{x},t) = \nabla \cdot \int_0^t \mathbb{G}(\mathbf{x},t-\tau)\nabla\mathbf{v}(\mathbf{x},\tau)d\tau + \mathbf{b}(\mathbf{x},t) \quad \forall(\mathbf{x},t) \in \Omega \times \mathbb{R}^+. \tag{24.2.6}$$

This is a differential equation for the unknown function $\mathbf{v}$. We must add the initial and boundary conditions. In the general case we have initial conditions expressed by a given function $\mathbf{v}(\mathbf{x},0) = \mathbf{v}_0(\mathbf{x})$; however, we assume the data

$$\mathbf{v}(\mathbf{x},0) = \mathbf{0} \quad \forall\mathbf{x} \in \bar{\Omega}, \qquad \mathbf{v}(\mathbf{x},t) = \mathbf{0} \quad \forall(\mathbf{x},t) \in \partial\Omega \times \mathbb{R}^+, \tag{24.2.7}$$

since it is well known that $\mathbf{v}_0(\mathbf{x})$ can always be supposed equal to zero with a suitable change of the sources, as elaborated after (24.2.29) below.

Let $\mathbf{w}$ be a smooth vector function on $\bar{\Omega} \times \mathbb{R}^+$, vanishing on the boundary $\partial\Omega$ and such that $\mathbf{w}(\mathbf{x},T) = \mathbf{0}\ \forall\mathbf{x} \in \Omega$. Using the expression $(8.2.1)_1$ for $\check{\mathbf{I}}$, inner multiplying by $\mathbf{w}$, and integrating on $\Omega \times [0,T]$, we obtain

$$\begin{aligned}\int_0^T \int_\Omega \dot{\mathbf{v}}(\mathbf{x},t) \cdot \mathbf{w}(\mathbf{x},t)d\mathbf{x}dt + \int_0^T \int_\Omega \int_0^t \mathbb{G}(\mathbf{x},t-\tau)\nabla\mathbf{v}(\mathbf{x},\tau) \cdot \nabla\mathbf{w}(\mathbf{x},t)d\tau d\mathbf{x}dt \\ + \int_0^T \int_\Omega \check{\mathbf{I}}^0(\mathbf{x},t) \cdot \nabla\mathbf{w}(\mathbf{x},t)d\mathbf{x}dt - \int_0^T \int_\Omega \mathbf{f}(\mathbf{x},t) \cdot \mathbf{w}(\mathbf{x},t)d\mathbf{x}dt = 0,\end{aligned} \tag{24.2.8}$$

which corresponds to the virtual work principle and allows us to provide a new definition of weak solution.

Now we consider the quasistatic problem, which is connected with Eqs. (24.2.8) and (24.2.7), on the time domain $(0,\infty)$, and is given in a weak sense by the system

$$\begin{aligned}&\int_0^\infty \int_\Omega \int_0^t \mathbb{G}(\mathbf{x},t-\tau)\nabla\mathbf{v}(\mathbf{x},\tau) \cdot \nabla\mathbf{w}(\mathbf{x},t)d\tau d\mathbf{x}dt \\ &\quad - \int_0^\infty \int_\Omega [\mathbf{f}(\mathbf{x},t) \cdot \mathbf{w}(\mathbf{x},t) - \check{\mathbf{I}}^0(\mathbf{x},t) \cdot \nabla\mathbf{w}(\mathbf{x},t)]d\mathbf{x}dt = 0, \\ &\mathbf{v}(\mathbf{x},0) = \mathbf{0}\ \forall\mathbf{x} \in \bar{\Omega}, \quad \mathbf{v}(\mathbf{x},t) = \mathbf{0}\ \forall(\mathbf{x},t) \in \partial\Omega \times \mathbb{R}^+.\end{aligned} \tag{24.2.9}$$

In order to obtain a rigorous definition of solution according to the equality (24.2.9), one must fix the domain of the functions $\mathbf{v}(\mathbf{x},t)$ and $\mathbf{w}(\mathbf{x},t)$. We introduce the function space

$$\mathcal{H}_{\mathbb{G}}(\mathbb{R}^+; H_0^1(\Omega)) =$$
$$\left\{\mathbf{v} \in L^2_{loc}(\mathbb{R}^+; H_0^1(\Omega));\ \int_0^\infty \int_0^\infty \int_\Omega \mathbb{G}(\mathbf{x},|\tau-\tau'|)\nabla\mathbf{v}(\mathbf{x},\tau') \cdot \nabla\mathbf{v}(\mathbf{x},\tau)d\mathbf{x}d\tau' d\tau < \infty\right\},$$

while the states $\breve{\mathbf{I}}^0(\mathbf{x},t)$ are elements of the space $\mathcal{H}'_{\mathbb{G}}(\mathbb{R}^+;H^1_0(\Omega))$.

The space $\mathcal{H}_{\mathbb{G}}(\mathbb{R}^+;H^1_0(\Omega))$ is a Hilbert space with respect to the inner product

$$(\mathbf{v}_1,\mathbf{v}_2)_{\mathcal{H}_{\mathbb{G}}} = \int_0^\infty \int_0^\tau \int_\Omega [\mathbb{G}_\infty(\mathbf{x}) + \breve{\mathbb{G}}(\mathbf{x},\tau-\tau')][\nabla\mathbf{v}_1(\mathbf{x},\tau')\cdot\nabla\mathbf{v}_2(\mathbf{x},\tau) + \nabla\mathbf{v}_1(\mathbf{x},\tau)\cdot\nabla\mathbf{v}_2(\mathbf{x},\tau')]d\mathbf{x}d\tau'\,d\tau,$$

where $\breve{\mathbb{G}}(\mathbf{x},s) = \mathbb{G}(\mathbf{x},s) - \mathbb{G}_\infty(\mathbf{x})$.

Definition 24.2.1. *A function* $\mathbf{v} \in \mathcal{H}_{\mathbb{G}}(\mathbb{R}^+;H^1_0(\Omega))$ *is called a weak solution of the quasistatic problem related to Eqs. (24.2.6)–(24.2.7) with data* $\mathbf{f} \in \mathcal{H}'_{\mathbb{G}}(\mathbb{R}^+;H^{-1}(\Omega))$ *and* $\breve{\mathbf{I}}^0 \in \mathcal{H}'_{\mathbb{G}}(\mathbb{R}^+;H^1_0(\Omega))$ *if it satisfies the identity (24.2.9) for any* $\mathbf{w} \in \mathcal{H}_{\mathbb{G}}(\mathbb{R}^+;H^1_0(\Omega))$.

We can now prove the following theorem [7].

Theorem 24.2.2. *Assume that the kernel* $\breve{\mathbb{G}}(\mathbf{x},\cdot) \in L^1(\mathbb{R}^+;L^\infty(\Omega))$ *satisfies the thermodynamic condition* $\breve{\mathbb{G}}_c(\mathbf{x},\omega) > 0$*, for any* $(\mathbf{x},\omega) \in \bar{\Omega}\times\mathbb{R}$*. Then there exists a unique weak solution* $\mathbf{v} \in \mathcal{H}_{\mathbb{G}}(\mathbb{R}^+;H^1_0(\Omega))$ *of the problem (24.2.6)–(24.2.7) for any* $\breve{\mathbf{I}}^0 \in \mathcal{H}'_{\mathbb{G}}(\mathbb{R}^+;H^1_0(\Omega))$ *and* $f \in L^2(\mathbf{R}^+;H^{-1}(\Omega))$.

Proof. The Fourier transform applied to the system (24.2.6)–(24.2.7) under the quasistatic hypothesis yields

$$\begin{aligned} \nabla\cdot[\mathbb{G}_+(\mathbf{x},\omega)\nabla\mathbf{v}_+(\mathbf{x},\omega)] &= -\mathbf{f}_+(\mathbf{x},\omega) - \nabla\cdot\breve{\mathbf{I}}^0_+(\mathbf{x},\omega), \\ \mathbf{v}_+(\mathbf{x},\omega)|_{\partial\Omega} &= \mathbf{0}. \end{aligned} \tag{24.2.10}$$

For any fixed $\omega \in \mathbb{R}$, we consider the sesquilinear form

$$a(\mathbf{v}_+(\mathbf{x},\omega),\mathbf{w}_+(\mathbf{x},\omega)) = \int_\Omega \mathbb{G}_+(\mathbf{x},\omega)\nabla\mathbf{v}_+(\mathbf{x},\omega)\cdot\overline{\nabla\mathbf{w}_+(\mathbf{x},\omega)}d\mathbf{x}, \tag{24.2.11}$$

which is a bounded and coercive form in $H^1_0(\Omega)$. It is easy to see that it is bounded. To show coercivity, we must prove that for any fixed $\omega \in \mathbb{R}$, the inequality

$$|a(\mathbf{v}_+(\mathbf{x},\omega),\mathbf{v}_+(\mathbf{x},\omega))| \geq k(\omega)\|\mathbf{v}_+(\mathbf{x},\omega)\|_{H^1_0}$$

holds for all $\mathbf{v}_+ \in H^1_0(\Omega)$, where $k(\omega)$ is a positive constant. From the definition (24.2.11) of a, and the relation $\mathbb{G}_+(\mathbf{x},\omega) = \mathbb{G}_c(\mathbf{x},\omega) - i\mathbb{G}_s(\mathbf{x},\omega)$, we obtain, for any $\omega \in \mathbb{R}$,

$$\begin{aligned} |a(\mathbf{v}_+(\mathbf{x},\omega),\mathbf{v}_+(\mathbf{x},\omega))| &\geq \int_\Omega \mathbb{G}_c(\mathbf{x},\omega)\nabla\mathbf{v}_+(\mathbf{x},\omega)\cdot\overline{\nabla\mathbf{v}_+(\mathbf{x},\omega)}d\mathbf{x} \\ &\geq k(\omega)\|\mathbf{v}_+(\mathbf{x},\omega)\|_{H^1_0}, \end{aligned}$$

where $k(\omega) = \inf\{|\mathbb{G}_c(\mathbf{x},\omega)|, \mathbf{x} \in \Omega\}$.

Hence, it follows that for any fixed $\omega \in \mathbb{R}$, the problem (24.2.10) admits a solution $\mathbf{v}_+ \in H^1_0(\Omega)$ if the supply

$$\mathbf{F}_+(\mathbf{x},\omega) = \mathbf{f}_+(\mathbf{x},\omega) + \nabla \cdot \breve{\mathbf{I}}^0_+(\mathbf{x},\omega)$$

belongs to $H^{-1}(\Omega)$.

Now we study the behavior of $\mathbf{v}_+$ when $\omega \to \infty$. Parseval's formula applied to (24.2.8), after an integration by parts, yields

$$\int_{-\infty}^{\infty}\int_\Omega \left\{\mathbf{v}_+(\mathbf{x},\omega)\cdot\overline{[i\omega\mathbf{w}_+(\mathbf{x},\omega)-\mathbf{w}_0(\mathbf{x})]} - \mathbb{G}_+(\mathbf{x},\omega)\nabla v_+(\mathbf{x},\omega)\cdot\overline{\nabla\mathbf{w}_+(\mathbf{x},\omega)}\right\} d\mathbf{x}d\omega$$
$$= -\int_{-\infty}^{\infty}\int_\Omega \left\{\mathbf{f}_+(\mathbf{x},\omega)\cdot\overline{\mathbf{w}_+(\mathbf{x},\omega)} - \breve{\mathbf{I}}^0_+(\mathbf{x},\omega)\cdot\overline{\nabla\mathbf{w}_+(\mathbf{x},\omega)}\right\} d\mathbf{x}d\omega. \tag{24.2.12}$$

We recall that for any vector function $\mathbf{f} \in H^{-1}(\Omega)$, there exists at least a tensor function $\mathbf{A} \in L^2(\Omega)$ such that for all $\mathbf{v} \in H^1(\Omega)$,

$$\int_0^{\infty}\int_\Omega \mathbf{f}(\mathbf{x},t)\cdot\mathbf{v}(\mathbf{x},t)d\mathbf{x}dt = \int_0^{\infty}\int_\Omega \mathbf{A}(\mathbf{x},t)\cdot\nabla\mathbf{v}(\mathbf{x},t)d\mathbf{x}dt,$$

and moreover, that in our case,

$$\mathbf{v}(\mathbf{x},0) = \frac{1}{\pi}\int_{-\infty}^{\infty} \mathbf{v}_+(\mathbf{x},\omega)d\omega = \mathbf{0}.$$

Thus, (24.2.12), written for $\mathbf{w}(\mathbf{x},t) = \mathbf{v}(\mathbf{x},t)$, gives

$$\int_{-\infty}^{\infty}\int_\Omega \left\{-\mathbf{v}_+(\mathbf{x},\omega)\cdot\overline{i\omega\mathbf{v}_+(\mathbf{x},\omega)} + \mathbb{G}_+(\mathbf{x},\omega)\nabla\mathbf{v}_+(\mathbf{x},\omega)\cdot\overline{\nabla\mathbf{v}_+(\mathbf{x},\omega)}\right\} d\mathbf{x}d\omega$$
$$= \int_{-\infty}^{\infty}\int_\Omega [\mathbf{A}_+(\mathbf{x},\omega) - \breve{\mathbf{I}}^0_+(\mathbf{x},\omega)]\cdot\overline{\nabla\mathbf{v}_+(\mathbf{x},\omega)}d\mathbf{x}d\omega. \tag{24.2.13}$$

The first term of this equation is an odd function of ω, and therefore, its integral over $\mathbb{R}$ vanishes. The integral of the second term reduces to

$$\int_{-\infty}^{\infty}\int_\Omega \mathbb{G}_+(\mathbf{x},\omega)\nabla\mathbf{v}_+(\mathbf{x},\omega)\cdot\overline{\nabla\mathbf{v}_+(\mathbf{x},\omega)}d\mathbf{x}d\omega$$
$$= \int_{-\infty}^{\infty}\int_\Omega \mathbb{G}_c(\mathbf{x},\omega)\nabla\mathbf{v}_+(\mathbf{x},\omega)\cdot\overline{\nabla\mathbf{v}_+(\mathbf{x},\omega)}d\mathbf{x}d\omega,$$

since $\mathbb{G}$ is symmetric, $\mathbb{G}_+ = \mathbb{G}_c - i\mathbb{G}_s$, and $\mathbb{G}_s$ is an odd function. Using the Schwarz inequality in the integral on the right-hand side of (24.2.13), it follows that

$$\int_{-\infty}^{\infty}\int_\Omega \mathbb{G}_+(\mathbf{x},\omega)\nabla\mathbf{v}_+(\mathbf{x},\omega)\cdot\overline{\nabla\mathbf{v}_+(\mathbf{x},\omega)}d\mathbf{x}d\omega$$
$$\leq \left\{\int_{-\infty}^{\infty}\int_\Omega \mathbb{G}_+^{-1}(\mathbf{x},\omega)[\mathbf{A}_+(\mathbf{x},\omega) - \breve{\mathbf{I}}^0_+(\mathbf{x},\omega)]\cdot\overline{[\mathbf{A}_+(\mathbf{x},\omega) - \breve{\mathbf{I}}^0_+(\mathbf{x},\omega)]}d\mathbf{x}d\omega\right\}^{\frac{1}{2}}$$
$$\times\left\{\int_{-\infty}^{\infty}\int_\Omega \mathbb{G}_+(\mathbf{x},\omega)\nabla\mathbf{v}_+(\mathbf{x},\omega)\cdot\overline{\nabla\mathbf{v}_+(\mathbf{x},\omega)}d\mathbf{x}d\omega\right\}^{\frac{1}{2}},$$

whence

$$\int_{-\infty}^{\infty}\int_{\Omega}\mathbb{G}_c(\mathbf{x},\omega)\nabla\mathbf{v}_+(\mathbf{x},\omega)\cdot\overline{\nabla\mathbf{v}_+(\mathbf{x},\omega)}d\mathbf{x}d\omega$$
$$\leq\int_{-\infty}^{\infty}\int_{\Omega}\mathbb{G}_+^{-1}(\mathbf{x},\omega)[\mathbf{A}_+(\mathbf{x},\omega)-\check{\mathbf{I}}_+^0(\mathbf{x},\omega)]\cdot\overline{[\mathbf{A}_+(\mathbf{x},\omega)-\check{\mathbf{I}}_+^0(\mathbf{x},\omega)]}d\mathbf{x}d\omega. \tag{24.2.14}$$

Therefore, if $\mathbf{A}-\check{\mathbf{I}}^0\in\mathcal{H}'_{\mathbb{G}}(\mathbb{R}^+;L^2(\Omega))$, then $\mathbf{v}\in\mathcal{H}_{\mathbb{G}}(\mathbb{R}^+;H_0^1(\Omega))$. From (24.2.14), applying the Poincaré theorem, it follows that there exists a constant $C(\Omega)$ such that

$$C(\Omega)\int_{-\infty}^{\infty}\int_{\Omega}\mathbb{G}_c(\mathbf{x},\omega)\mathbf{v}_+(\mathbf{x},\omega)\cdot\overline{\mathbf{v}_+(\mathbf{x},\omega)}d\mathbf{x}d\omega$$
$$\leq\int_{-\infty}^{\infty}\int_{\Omega}\mathbb{G}_+^{-1}(\mathbf{x},\omega)[\mathbf{A}_+(\mathbf{x},\omega)-\check{\mathbf{I}}_+^0(\mathbf{x},\omega)]\cdot\overline{[\mathbf{A}_+(\mathbf{x},\omega)-\check{\mathbf{I}}_+^0(\mathbf{x},\omega)]}d\mathbf{x}d\omega.$$

Hence, if the supplies $\mathbf{f}$ and $\mathbf{I}^0$ satisfy $\mathbf{f}\in\mathcal{H}'_{\mathbb{G}}(\mathbb{R}^+;H^{-1}(\Omega))$ and $\mathbf{I}^0\in\mathcal{H}'_{\mathbb{G}}(\mathbb{R}^+;L^2(\Omega))$, the function $\mathbf{v}$ belongs to $\mathcal{H}_{\mathbb{G}}(\mathbb{R}^+;H^1(\Omega))$ and it is a virtual work solution of the problem (24.2.6)–(24.2.7) in the sense of Definition 24.2.1. □

24.2.3 Dynamical Problem in Linear Viscoelasticity

Let $\Omega\subset\mathbb{R}^3$ be a smooth bounded domain occupied by a continuous linear viscoelastic solid whose constitutive equation is given by (24.2.2). The dynamical problem for such a material is expressed by the equation

$$\ddot{\mathbf{u}}(\mathbf{x},t)=\nabla\cdot\mathbf{T}(\mathbf{x},t)+\mathbf{f}(\mathbf{x},t)\qquad\forall\mathbf{x}\in\Omega,\ \forall t>0. \tag{24.2.15}$$

Boundary and initial conditions must be added. For this purpose, on assuming elastic boundary conditions on $\partial\Omega$, we put

$$\begin{aligned}&\mathbf{T}(\mathbf{x},t)\mathbf{n}(\mathbf{x})+\alpha(\mathbf{x})\mathbf{u}(\mathbf{x},t)=\mathbf{0}\qquad\forall\mathbf{x}\in\partial\Omega,\ \forall t>0,\\&\mathbf{u}(\mathbf{x},0)=\mathbf{u}_0(\mathbf{x}),\ \dot{\mathbf{u}}(\mathbf{x},0)=\mathbf{v}_0(\mathbf{x}),\ \mathbf{u}(\mathbf{x},\tau)=\mathbf{u}^0(\mathbf{x},-\tau)\ \forall\mathbf{x}\in\Omega,\ \forall\tau\leq 0,\end{aligned} \tag{24.2.16}$$

where $\mathbf{n}$ is the outward normal on $\partial\Omega$, and the scalar function α belongs to $L^2(\partial\Omega)\cap L^\infty(\partial\Omega)$ and a.e. in $\partial\Omega$ satisfies

$$\alpha(\mathbf{x})\geq\alpha_m>0. \tag{24.2.17}$$

Also $\mathbf{u}^0$ is the initial history of the displacement vector [114].

The instantaneous elastic modulus $\mathbb{G}_0$ and the Boltzmann relaxation function derivative $\mathbb{G}'$ in (24.2.2) are symmetric fourth-order tensors such that

$$\mathbb{G}_0\in C(\bar{\Omega}),\qquad\mathbb{G}'\in L^1(\mathbb{R}^+;\bar{\Omega})\cap L^2(\mathbb{R}^+;\Omega); \tag{24.2.18}$$

consequently, the relaxation function $\mathbb{G}$, given by (8.1.7), is well defined; moreover, it is continuous in $\bar{\Omega}\times\mathbb{R}^+$ and differentiable in $\Omega\times\mathbb{R}^{++}$ with

$$\mathbb{G}_\infty(\mathbf{x}) = \lim_{t\to\infty} \mathbb{G}(\mathbf{x}, t) = \mathbb{G}_0(\mathbf{x}) + \int_0^\infty \mathbb{G}'(\mathbf{x}, s)ds.$$

Since the body is a solid, we require that $\mathbb{G}_\infty$ be uniformly positive definite in Ω, i.e., that there exists a scalar, denoted $g_{\infty_m} > 0$, such that

$$0 < g_{\infty_m}|\mathbf{A}|^2 \le \inf_{\mathbf{x}\in\Omega} \mathbf{A}\cdot\mathbb{G}_\infty(\mathbf{x})\mathbf{A} \qquad \forall\mathbf{A}\in \text{Sym}\backslash\{\mathbf{0}\}. \tag{24.2.19}$$

Also, by virtue of the thermodynamic constraint (8.1.18), $-\mathbb{G}_s(\mathbf{x}, \omega)$ is uniformly positive definite in Ω for any $\omega > 0$, i.e.,

$$g_m(\omega)|\mathbf{A}|^2 \le -\inf_{\mathbf{x}\in\Omega} \mathbf{A}\cdot\mathbb{G}'_s(\mathbf{x}, \omega)\mathbf{A} \qquad \forall\mathbf{A}\in \text{Sym}, \tag{24.2.20}$$

where $g_m(\omega)$ denotes a continuous function $g_m : \mathbb{R}^{++} \to \mathbb{R}^{++}$ From the inequality (24.2.20) it follows that (see (8.1.21) and $(C.2.17)_2$)

$$\mathbb{G}'(\mathbf{x}, 0) = \lim_{\omega\to\infty} \omega\mathbb{G}'_s(\mathbf{x}, \omega)$$

is uniformly negative semidefinite in Ω. A more restrictive property of definiteness is here assumed by supposing that there exists $g'_0 > 0$ such that

$$g'_0|\mathbf{A}|^2 \le -\inf_{\mathbf{x}\in\Omega} \mathbf{A}\cdot\mathbb{G}'_s(\mathbf{x}, 0)\mathbf{A} \qquad \forall\mathbf{A}\in \text{Sym}.$$

Using the Fourier inversion formula to obtain (8.1.20) and then applying the Riemann–Lebesgue lemma (C.2.13) (see also (7.2.19)) yields the asymptotic relation

$$\mathbb{G}_\infty(\mathbf{x}) - \mathbb{G}_0(\mathbf{x}) = \frac{2}{\pi}\int_0^\infty \frac{\mathbb{G}'_s(\mathbf{x}, \omega)}{\omega}d\omega. \tag{24.2.21}$$

Thus

$$\frac{\mathbb{G}'_s(\mathbf{x}, \omega)}{\omega} \in L^1(\mathbb{R}) \qquad \forall\mathbf{x}\in\Omega,$$

which follows from (24.2.20) and the boundedness of the left-hand side of (24.2.21). These relationships also give that $\mathbb{G}_0 - \mathbb{G}_\infty$ is uniformly positive in Ω. Therefore, $\mathbb{G}_0$ has this property so there exists a scalar, denoted by $g_{0_m} > 0$, such that

$$0 < g_{0_m}|\mathbf{A}|^2 \le \inf_{\mathbf{x}\in\Omega} \mathbf{A}\cdot\mathbb{G}_0(\mathbf{x})\mathbf{A} \qquad \forall\mathbf{A}\in \text{Sym}\backslash\{\mathbf{0}\}. \tag{24.2.22}$$

The requirements (24.2.18), (24.2.20), and (24.2.22) on the relaxation function ensure that $\omega\mathbb{G}'_s(\mathbf{x}, \cdot) \in L^\infty(\mathbb{R})$ and $\mathbb{G}'_s(\mathbf{x}, \cdot)$ belong to $L^1(\mathbb{R})\cap L^2(\mathbb{R})$ for every $\mathbf{x}\in\Omega$ (see [25, Theorem 6.5 d]). The quantities $\mathbb{G}_0$, $\mathbb{G}_\infty$, and $\mathbb{G}'_s(\cdot, \omega)$ are uniformly bounded in Ω, i.e., for every $\mathbf{A}\in$ Sym,

$$\begin{gathered}\sup_{\mathbf{x}\in\Omega} \mathbb{G}_0(\mathbf{x})\mathbf{A}\cdot\mathbf{A} \le g_{0_M}|\mathbf{A}|^2, \quad \sup_{\mathbf{x}\in\Omega} \mathbb{G}_\infty(\mathbf{x})\mathbf{A}\cdot\mathbf{A} \le g_{\infty_M}|\mathbf{A}|^2, \\ \sup_{\mathbf{x}\in\Omega} -\mathbb{G}'_s(\mathbf{x}, \omega)\mathbf{A}\cdot\mathbf{A} \le g_M(\omega)|\mathbf{A}|^2,\end{gathered} \tag{24.2.23}$$

where the constants g_{0_M}, g_{∞_M} and the continuous function $g_M : \mathbb{R}^{++} \to \mathbb{R}^{++}$ are related to $\mathbb{G}_0$, $\mathbb{G}_\infty$, and $\mathbb{G}'_s(\cdot, \omega)$, respectively.

24.2.3.1 Weak Solution in a Variational Sense

In order to formulate the concept of a weak solution, we consider Hamilton's variational principle, related to the following differential problem.

The expression for the stress tensor (24.2.2) can be written in the form

$$\mathbf{T}(\tau) = \int_0^\tau \mathbb{G}'(s)\nabla\mathbf{u}(\tau - s)ds + \mathbb{G}_0\nabla\mathbf{u}(\tau) + \tilde{\mathbf{I}}^0(\tau), \tag{24.2.24}$$

where (see (8.2.3) and after (8.2.6)),

$$\tilde{\mathbf{I}}^0(\tau) = \int_0^\infty \mathbb{G}'(\tau + s)\nabla\mathbf{u}^{t=0}(s)ds.$$

The equation of motion can be expressed in terms of the unknown function $\mathbf{u}(\mathbf{x}, t) \in C^{2,2}(\Omega \times \mathbb{R}^+; \mathbb{R}^3)$ as follows:

$$L\mathbf{u} := \ddot{\mathbf{u}} - \nabla \cdot (\mathbb{G}_0\nabla\mathbf{u} + \mathbb{G}' * \nabla\mathbf{u}) - \mathbf{f} = \mathbf{0} \text{ in } \Omega \times \mathbb{R}^{++}.$$

Here, the symbol * denotes the convolution product (C.3.4), so that

$$(\mathbb{G}' * \nabla\mathbf{u})(\tau) = \int_0^\tau \mathbb{G}'(\tau - s)\nabla\mathbf{u}(s)ds$$

and

$$\mathbf{f}(\tau) = \mathbf{b}(\tau) + \nabla \cdot \tilde{\mathbf{I}}^0(\tau),$$

where the body force $\mathbf{b}$ and $\tilde{\mathbf{I}}^0$ are known functions in $\bar{\Omega} \times \mathbb{R}^+$. Moreover, suppose that the boundary $\partial\Omega$ of Ω consists of two subsets $\partial\Omega_u$ and $\partial\Omega_T$ such that

$$\partial\Omega_u \cup \partial\Omega_T = \partial\Omega, \qquad \partial\Omega_u \cap \partial\Omega_T = \emptyset.$$

Then we consider a mixed value problem P defined as follows:

$$\begin{aligned} &L\mathbf{u} = \mathbf{0} \quad \text{in } \Omega \times \mathbb{R}^{++}, \\ &\mathbf{u} = \mathbf{0} \quad \text{on } \partial\Omega_u \times \mathbb{R}^+, \qquad \mathbf{u}(0) = \mathbf{u}_0, \quad \dot{\mathbf{u}}(0) = \dot{\mathbf{u}}_0 \quad \text{in } \bar{\Omega}, \\ &\mathbf{T}\mathbf{n} = \hat{\mathbf{t}} - \tilde{\mathbf{I}}^0\mathbf{n} \quad \text{on } \partial\Omega_T \times \mathbb{R}^+, \end{aligned} \tag{24.2.25}$$

where $\hat{\mathbf{t}}$ is a known function.

Let us introduce the functional

$$\begin{aligned} F(\mathbf{u}) = \frac{1}{2}\int_\Omega [&\dot{\mathbf{u}} * \dot{\mathbf{u}} + \nabla\mathbf{u} * \mathbb{G}_0\nabla\mathbf{u} + \nabla\mathbf{u} * \mathbb{G}' * \nabla\mathbf{u} \\ &- 2\mathbf{u} * \mathbf{f} - 2\mathbf{u} \cdot \dot{\mathbf{u}}_0]d\mathbf{x} - \int_{\partial\Omega_T} \mathbf{u} * \left(\hat{\mathbf{t}} - \tilde{\mathbf{I}}^0\mathbf{n}\right)d\mathbf{x}, \end{aligned}$$

which provides a variational formulation of this problem, as the following theorem states.

Theorem 24.2.3. *Let K be the set of functions in $C^{2,2}(\Omega\times\mathbb{R}^+;\mathbb{R}^3)$ that satisfy conditions (24.2.25)$_2$. Then the functional $F(\mathbf{u})$ has a stationary point $\mathbf{u}\in K$ if and only if $\mathbf{u}$ is a solution of the mixed problem P.*

Proof. Let $\mathbf{u}\in K$ and let $\boldsymbol{\varphi}\in K$ be a function that vanishes on $\partial\Omega_u\times\mathbb{R}^+$ and at $\tau=0$ with its time derivative on $\bar{\Omega}$. Then we have $\mathbf{u}+\nu\boldsymbol{\varphi}\in K$ for any $\nu\in\mathbb{R}$ and

$$\begin{aligned}
dF(\mathbf{u}|\boldsymbol{\varphi}) &= \left.\frac{dF(\mathbf{u}+\nu\boldsymbol{\varphi})}{d\nu}\right|_{\nu=0}\\
&= \int_\Omega\Bigg[\int_0^\tau \nabla\boldsymbol{\varphi}(\tau-s)\cdot\mathbb{G}_0\nabla\mathbf{u}(s)ds\\
&\quad+\int_{D_\tau}\nabla\boldsymbol{\varphi}(\xi)\cdot\mathbb{G}'(\tau-\xi-\eta)\nabla\mathbf{u}(\eta)d\xi d\eta\\
&\quad+\int_0^\tau\dot{\boldsymbol{\varphi}}(\tau-s)\cdot\dot{\mathbf{u}}(s)ds-\int_0^\tau\boldsymbol{\varphi}(\tau-s)\cdot\mathbf{f}(s)ds\\
&\quad-\boldsymbol{\varphi}(\tau)\cdot\dot{\mathbf{u}}_0\Bigg]d\mathbf{x}-\int_{\partial\Omega_T}\int_0^\tau\boldsymbol{\varphi}(\tau-s)\cdot\left(\hat{\mathbf{t}}-\tilde{\mathbf{I}}^0\mathbf{n}\right)(s)\,dsd\mathbf{x},
\end{aligned}\tag{24.2.26}$$

where

$$D_\tau=\left\{(\xi,\eta)\in\mathbb{R}^2;\ 0\le\xi+\eta\le\tau\right\}$$

and $\mathbf{n}$ is the outward normal to $\partial\Omega_T$. Integrating by parts, we have

$$\begin{aligned}
\int_\Omega\int_0^\tau\nabla\boldsymbol{\varphi}(\tau-s)\cdot\mathbb{G}_0\nabla\mathbf{u}(s)dsd\mathbf{x} &= -\int_\Omega\int_0^\tau\boldsymbol{\varphi}(\tau-s)\cdot\nabla\cdot[\mathbb{G}_0\nabla\mathbf{u}(s)]dsd\mathbf{x}\\
&\quad+\int_{\partial\Omega_T}\int_0^\tau\boldsymbol{\varphi}(\tau-s)\cdot\mathbb{G}_0\nabla\mathbf{u}(s)\mathbf{n}\,dsd\mathbf{x};
\end{aligned}$$

analogously, we obtain

$$\begin{aligned}
&\int_\Omega\int_{D_\tau}\nabla\boldsymbol{\varphi}(\xi)\cdot\mathbb{G}'(\tau-\xi-\eta)\nabla\mathbf{u}(\eta)d\xi d\eta d\mathbf{x}\\
&\quad=-\int_\Omega\int_0^\tau\boldsymbol{\varphi}(\tau-s)\cdot\nabla\cdot\int_0^s\mathbb{G}'(s-\eta)\nabla\mathbf{u}(\eta)d\eta dsd\mathbf{x}\\
&\quad\quad+\int_{\partial\Omega_T}\int_0^\tau\boldsymbol{\varphi}(\tau-s)\cdot\int_0^s\mathbb{G}'(s-\eta)\nabla\mathbf{u}(\eta)\mathbf{n}d\eta dsd\mathbf{x}
\end{aligned}$$

and finally

$$\int_\Omega\int_0^\tau\dot{\boldsymbol{\varphi}}(\tau-s)\cdot\dot{\mathbf{u}}(s)dsd\mathbf{x}=\int_\Omega\left[\int_0^\tau\boldsymbol{\varphi}(\tau-s)\cdot\ddot{\mathbf{u}}(s)ds+\boldsymbol{\varphi}(\tau)\cdot\dot{\mathbf{u}}_0\right]d\mathbf{x}.$$

Substituting these results, we obtain

$$\begin{aligned}
&dF(\mathbf{u}|\boldsymbol{\varphi})=\\
&\int_\Omega\int_0^\tau\boldsymbol{\varphi}(\tau-s)\cdot\left[\ddot{\mathbf{u}}(s)-\nabla\cdot\mathbb{G}_0\nabla\mathbf{u}(s)-\nabla\cdot\int_0^s\mathbb{G}'(s-\eta)\nabla\mathbf{u}(\eta)d\eta-\mathbf{f}(s)\right]dsd\mathbf{x}\\
&-\int_{\partial\Omega_T}\left\{\int_0^\tau\boldsymbol{\varphi}(\tau-s)\cdot\left[\hat{\mathbf{t}}(s)-\left(\tilde{\mathbf{I}}^0(s)+\mathbb{G}_0\nabla\mathbf{u}(s)+\int_0^s\mathbb{G}'(s-\eta)\nabla\mathbf{u}(\eta)d\eta\right)\mathbf{n}\right]ds\right\}d\mathbf{x}.
\end{aligned}$$

From the arbitrariness of φ on $\Omega \times \mathbb{R}^{++}$ and $\partial\Omega_T \times \mathbb{R}^{++}$ it follows that

$$dF(\mathbf{u}|\varphi) = 0 \qquad \forall \varphi \in K \tag{24.2.27}$$

only if the conditions in (24.2.25) are satisfied.

Conversely, the validity of (24.2.25) yields that F is stationary at the solution $\mathbf{u}$. □

In terms of (24.2.27), we can give the following definition, useful for proving existence, uniqueness, and stability results.

Definition 24.2.4. *A function* $\mathbf{u} \in \mathcal{H}(\Omega, \mathbb{R}^+) = H^1(\mathbb{R}^+; L^2(\Omega)) \cap L^2(\mathbb{R}^+; H_0^1(\Omega))$ *is a weak solution of the initial boundary value problem (24.2.25), with data* $\mathbf{u}_0 \in H^1(\Omega)$ *and* $\dot{\mathbf{u}}_0 \in L^2(\Omega)$*, if it satisfies the stationarity condition (24.2.27) for all* $\varphi \in \mathcal{H}(\Omega, \mathbb{R}^+)$.

Remark 24.2.5. Using (24.2.26) and (24.2.27) we obtain an equivalent definition of weak solution by means of

$$\begin{aligned} dF(\mathbf{u}|\varphi) = \int_\Omega \Big[& \int_0^T \nabla\varphi(\tau - s) \cdot \mathbb{G}_0 \nabla\mathbf{u}(s) ds \\ & + \int_{D_\tau} \nabla\varphi(\xi) \cdot \mathbb{G}'(\tau - \xi - \eta) \nabla\mathbf{u}(\eta) d\xi d\eta \\ & + \int_0^T \dot{\varphi}(\tau - s) \cdot \dot{\mathbf{u}}(s) ds - \int_0^T \varphi(\tau - s) \cdot \mathbf{f}(s) ds \\ & - \varphi(\tau) \cdot \dot{\mathbf{u}}_0 \Big] d\mathbf{x} - \int_{\partial\Omega_T} \int_0^T \varphi(\tau - s) \cdot \left(\hat{\mathbf{t}} - \tilde{\mathbf{I}}^0 \mathbf{n}\right)(s)\, ds d\mathbf{x} = 0. \end{aligned}$$

24.2.3.2 Virtual Power Solution

In the dynamics of viscoelastic solids it is possible to give a different definition of solution by using, instead of a variational principle as in Definition 24.2.4, the classical principle of virtual power. This alternative definition seems more consistent with the use of the natural space of solutions related to the domain of definition of the free energy given by Breuer and Onat in [42].

To this purpose we observe that the equation of motion

$$\ddot{\mathbf{u}} = \nabla \cdot \mathbf{T} + \mathbf{f},$$

combined with (24.2.24), written in the form

$$\mathbf{T}(t) = \mathbb{G}(t) \nabla\mathbf{u}(0) + \int_0^t \mathbb{G}(t - \tau) \nabla\dot{\mathbf{u}}(\tau)) d\tau + \tilde{\mathbf{I}}^0(\tau),$$

yields (see (24.2.6))

$$\dot{\mathbf{v}}(t) = \nabla \cdot \int_0^t \mathbb{G}(t - \tau) \nabla\mathbf{v}(\tau) d\tau + \mathbf{b}(t) \qquad \text{in } \Omega \times \mathbb{R}^{++}, \tag{24.2.28}$$

where we put $\mathbf{v} = \dot{\mathbf{u}}$ and

$$\mathbf{b}(t) = \mathbf{f}(t) + \nabla \cdot \left[\tilde{\mathbf{T}}^0(t) + \mathbb{G}(t)\nabla\mathbf{u}(0)\right].$$

In this equation, the unknown function is now $\mathbf{v}$; we associate with it the following initial and boundary conditions:

$$\mathbf{v}(\mathbf{x}, 0) = \mathbf{v}_0(\mathbf{x}) = \mathbf{0} \quad \forall \mathbf{x} \in \Omega, \qquad \mathbf{v}(\mathbf{x}, t)|_{\partial\Omega} = \mathbf{0} \quad \forall (\mathbf{x}, t) \in \partial\Omega \times \mathbb{R}^+. \tag{24.2.29}$$

There is no loss of generality in assuming the homogeneous condition. Indeed, let us drop the assumption that $\mathbf{v}_0(\mathbf{x})$ is zero (see [232]). Now define the function $\mathbf{z}$ such that $\mathbf{z} \in C^\infty(\mathbb{R}^+; H_0^1(\Omega))$ and

$$\mathbf{z}(\mathbf{x}, 0) = -\mathbf{v}_0(\mathbf{x}), \qquad \dot{\mathbf{z}}(\mathbf{x}, 0) = -\mathbf{b}(\mathbf{x}, 0).$$

Then $\tilde{\mathbf{v}} = \mathbf{v} + \mathbf{z}$ satisfies (24.2.28) with homogeneous initial boundary conditions and with $\mathbf{b}(\mathbf{x}, t)$ replaced by

$$\mathbf{b}(\mathbf{x}, t) + \dot{\mathbf{z}}(\mathbf{x}, t) - \nabla \cdot \int_0^t \mathbb{G}(\mathbf{x}, t - s)\nabla\mathbf{z}(\mathbf{x}, s)ds.$$

Without changing notation strictly required by this observation, we inner multiply (24.2.28) by a smooth function $\varphi(\mathbf{x}, t)$, defined on $\Omega \times \mathbb{R}^+$ with properties

$$\varphi(\mathbf{x}, t)|_{\partial\Omega} = \mathbf{0}, \qquad \varphi(\mathbf{x}, T) = \mathbf{0},$$

and integrate on $\Omega \times [0, T]$ to obtain

$$\begin{aligned} \int_0^T \int_\Omega \mathbf{v}(t) \cdot \dot{\varphi}(t) d\mathbf{x} dt = & \int_0^T \int_\Omega \int_0^t \nabla\varphi(t) \cdot \mathbb{G}(t - \tau)\nabla\mathbf{v}(\tau) d\tau d\mathbf{x} dt \\ & + \int_0^T \int_\Omega \nabla\varphi(t) \cdot \left[\tilde{\mathbf{T}}^0(t) + \mathbb{G}(t)\nabla\mathbf{u}(0)\right] d\mathbf{x} dt \\ & - \int_0^T \int_\Omega \varphi(t) \cdot \mathbf{f}(t) d\mathbf{x} dt. \end{aligned} \tag{24.2.30}$$

The weak formulation of the dynamics of a viscoelastic body expressed by (24.2.30) calls to mind the principle of virtual power. Now, the unknown function is the velocity $\mathbf{v}$, of which only the first-order derivatives $\nabla\mathbf{v}$ are involved; moreover, the test function φ is any vector function on $\Omega \times \mathbb{R}^+$ that vanishes on $\partial\Omega$ and also at time T.

In order to give a definition of the weak solution $\mathbf{v}$ associated with (24.2.30), we must introduce function spaces for $\mathbf{v}$ and φ. For this purpose, it is convenient to refer to the minimum free energy due to Breuer and Onat [42].

We consider the following function spaces:

$$\begin{aligned} & \mathcal{H}_G(\mathbb{R}^+; H_0^1(\Omega)) \\ & = \left\{ \mathbf{v} \in L^2_{loc}(\mathbb{R}^+; H_0^1(\Omega)); \int_0^\infty \int_0^\infty \int_\Omega \nabla\mathbf{v}(\tau) \cdot \left[\mathbb{G}_\infty + \breve{\mathbb{G}}(\tau - \eta)\right] \nabla\mathbf{v}(\eta) d\mathbf{x} d\eta d\tau < \infty \right\} \end{aligned}$$

and

$$\mathcal{F}(Q) = H^{1/2}(\mathbb{R}^+; L^2(\Omega)) \cap \mathcal{H}_G(\mathbb{R}^+; H_0^1(\Omega)),$$

where $Q = \Omega \times \mathbb{R}^+$ and $\breve{\mathbb{G}}(\mathbf{x}, \eta) = \mathbb{G}(\mathbf{x}, \eta) - \mathbb{G}_\infty(\mathbf{x})$ as defined in (8.2.4).

These spaces are Hilbert spaces with the inner products

$$(\mathbf{v}_1, \mathbf{v}_2)_{\mathcal{H}_G} = \int_0^\infty \int_0^\tau \int_\Omega \Big\{\nabla\mathbf{v}_2(\tau) \cdot \Big[\mathbb{G}_\infty + \breve{\mathbb{G}}(\tau - \eta)\Big]\nabla\mathbf{v}_1(\eta) + \nabla\mathbf{v}_1(\tau) \cdot \Big[\mathbb{G}_\infty + \breve{\mathbb{G}}(\tau - \eta)\Big]\nabla\mathbf{v}_2(\eta)\Big\} d\mathbf{x} d\eta d\tau$$

and

$$\begin{aligned}(\mathbf{v}_1, \mathbf{v}_2)_{\mathcal{F}} &= -\frac{1}{2}\int_0^\infty \int_\Omega [\mathbf{v}_1(\mathbf{x}, t) \cdot \dot{\mathbf{v}}_2(\mathbf{x}, t) + \mathbf{v}_2(\mathbf{x}, t) \cdot \dot{\mathbf{v}}_1(\mathbf{x}, t)] d\mathbf{x} dt + (\mathbf{v}_1, \mathbf{v}_2)_{\mathcal{H}_G} \\ &= \frac{1}{2}\int_\Omega \mathbf{v}_1(\mathbf{x}, 0) \cdot \mathbf{v}_2(\mathbf{x}, 0) d\mathbf{x} + (\mathbf{v}_1, \mathbf{v}_2)_{\mathcal{H}_G}.\end{aligned}$$

Moreover, the states $\tilde{\mathbf{I}}^0$ are elements of the dual space

$$\mathcal{H}_G^*(\mathbb{R}^+; L^2(\Omega)) = \Big\{\tilde{\mathbf{I}}^0 \in L^2_{loc}(\mathbb{R}^+; L^2_{loc}(\Omega)); \int_0^\infty \int_\Omega \tilde{\mathbf{I}}^0(\tau) \cdot \nabla\mathbf{v}(\tau) d\mathbf{x} d\tau < k\|\mathbf{v}\|_{\mathcal{H}_G} \quad \forall \mathbf{v} \in \mathcal{H}_G(\mathbb{R}^+; H_0^1(\Omega))\Big\}.$$

We can now give the following definition of a virtual power solution.

Definition 24.2.6. *A function* $\mathbf{v} \in \mathcal{F}(Q)$ *is said to be a virtual power solution of the problem (24.2.28), (24.2.29) with data* $\mathbf{v}_0 = \mathbf{0}$ *and* $\mathbf{f} \in H^{-1/2}(\mathbb{R}^+; L^2(\Omega)) \cap \mathcal{H}_G^*(\mathbb{R}^+; H^{-1}(\Omega))$, $\tilde{\mathbf{I}}^0 \in \mathcal{H}_G^*(\mathbb{R}^+; L^2(\Omega))$ *if it satisfies the identity (24.2.30) for any* $\varphi \in \mathcal{F}(Q)$.

24.2.3.3 Existence and Uniqueness

We now give a proof of existence and uniqueness of the solution for the dynamical problem.

Using the constitutive equation (24.2.2), the following system can be derived:

$$\begin{aligned}&\ddot{\mathbf{u}}(\mathbf{x}, t) = \nabla \cdot \{\mathbb{G}_0(\mathbf{x})\nabla\mathbf{u}(\mathbf{x}, t) + [\mathbb{G}' * \nabla\mathbf{u}](\mathbf{x}, t)\} \\ &\qquad + \nabla \cdot \mathbf{T}_0(\mathbf{x}, t) + \mathbf{f}(\mathbf{x}, t) \qquad \forall \mathbf{x} \in \Omega,\ \forall t > 0, \\ &\{\mathbb{G}_0(\mathbf{x})\nabla\mathbf{u}(\mathbf{x}, t) + [\mathbb{G}' * \nabla\mathbf{u}](\mathbf{x}, t) + \mathbf{T}_0(\mathbf{x}, t)\}\mathbf{n}(\mathbf{x}) \\ &\qquad + \alpha(\mathbf{x})\mathbf{u}(\mathbf{x}, t) = \mathbf{0}\ \forall \mathbf{x} \in \partial\Omega,\ \forall t > 0, \\ &\mathbf{n}(\mathbf{x}, 0) = \mathbf{u}_0(\mathbf{x}), \qquad \mathbf{v}(\mathbf{x}, 0) = \mathbf{v}_0(\mathbf{x}) \qquad \forall \mathbf{x} \in \Omega,\end{aligned} \tag{24.2.31}$$

where

$$\mathbf{T}_0(\mathbf{x}, t) = \int_t^\infty \mathbb{G}'(\mathbf{x}, s)\nabla\mathbf{u}^0(\mathbf{x}, s - t) ds$$

and

$$[\mathbb{G}' * \nabla \mathbf{u}](\mathbf{x}, t) = \int_0^t \mathbb{G}'(\mathbf{x}, s) \nabla \mathbf{u}^t(\mathbf{x}, s) d\, s.$$

This system is an integrodifferential mixed problem with a radiation boundary condition.

A variational formulation of this problem can be given by introducing the function space

$$\mathcal{H}(\Omega, \mathbb{R}^+) = H^1(\mathbb{R}^+; L^2(\Omega)) \cap L^2(\mathbb{R}^+; H^1(\Omega)).$$

This is a Hilbert space with the inner product

$$\begin{aligned}(\varphi_1, \varphi_2)_{\mathcal{H}} = &\int_0^\infty \int_\Omega [\nabla \varphi_1(\mathbf{x}, t) \cdot \nabla \varphi_2(\mathbf{x}, t) + \dot{\varphi}_1(\mathbf{x}, t) \cdot \dot{\varphi}_2(\mathbf{x}, t)]\, d\mathbf{x} dt \\ &+ \int_0^\infty \int_{\partial\Omega} \varphi_1(\mathbf{x}, t) \cdot \varphi_2(\mathbf{x}, t) d\mathbf{x} dt,\end{aligned} \tag{24.2.32}$$

comparable to the usual inner product

$$\int_0^\infty \int_\Omega [\nabla \varphi_1(\mathbf{x}, t) \cdot \nabla \varphi_2(\mathbf{x}, t) + \varphi_1(\mathbf{x}, t) \cdot \varphi_2(\mathbf{x}, t) + \dot{\varphi}_1(\mathbf{x}, t) \cdot \dot{\varphi}_2(\mathbf{x}, t)] d\mathbf{x} dt,$$

by virtue of the estimate

$$||\mathbf{f}||^2_\Omega \le K_1 ||\nabla \mathbf{f}||^2_\Omega + K_2 ||\mathbf{f}||^2_{\partial\Omega}, \tag{24.2.33}$$

which holds for functions $\mathbf{f} \in H^1(\Omega)$, where Ω is a bounded regular open subset of $\mathbb{R}^3$, K_1 and K_2 being constants depending on Ω [298].

Definition 24.2.7. *A function* $\mathbf{u} \in \mathcal{H}(\Omega, \mathbf{R}^+)$ *is a weak solution of the initial boundary value problem (24.2.31) with data* $\mathbf{f}$, $\nabla \cdot \mathbf{T}_0 \in L^2(\mathbb{R}^+; L^2(\Omega))$, $\mathbf{u}^0 \in H^1(\Omega)$ *and* $\mathbf{v}^0 \in L^2(\Omega)$ *if* $\mathbf{u}(\mathbf{x}, 0) = \mathbf{u}_0(\mathbf{x})$ *almost everywhere in* Ω *and*

$$\begin{aligned}&\int_0^\infty \int_\Omega \left\{ \left[\mathbb{G}_0(\mathbf{x}) \nabla \mathbf{u}(\mathbf{x}, t) + \int_0^t \mathbb{G}'(\mathbf{x}, s) \nabla \mathbf{u}^t(\mathbf{x}, s) d\, s \right] \cdot \nabla \varphi(\mathbf{x}, t) \right. \\ &\qquad \left. - \dot{\mathbf{u}}(\mathbf{x}, t) \cdot \dot{\varphi}(\mathbf{x}, t) \right\} d\mathbf{x} dt + \int_0^\infty \int_{\partial\Omega} \alpha(\mathbf{x}) \mathbf{u}(\mathbf{x}, t) \cdot \varphi(\mathbf{x}, t) d\mathbf{x} dt \\ &= \int_\Omega \mathbf{v}_0(\mathbf{x}) \cdot \varphi(\mathbf{x}, 0) d\mathbf{x} + \int_0^\infty \int_\Omega [\mathbf{f}(\mathbf{x}, t) \cdot \varphi(\mathbf{x}, t) - \mathbf{T}_0(\mathbf{x}, t) \cdot \nabla \varphi(\mathbf{x}, t)] d\mathbf{x} dt\end{aligned} \tag{24.2.34}$$

for all $\varphi \in \mathcal{H}(\Omega, \mathbb{R}^+)$.

Before proving the theorem of existence and uniqueness of a weak solution of the evolution problem (24.2.31), we consider the following.

24.2.3.4 Transformed Problem

The dynamical problem under consideration can be transformed by means of the Fourier transform by introducing $H_F(\Omega, \mathbb{R}^+)$, the space of Fourier transforms of the functions of $H(\Omega, \mathbb{R}^+)$. For causal time functions we have

$$\varphi_F \in L^2(\mathbb{R}; H^1(\Omega)), \quad i\omega\varphi_F - \varphi_0 \in L^2(\mathbb{R}; H^1(\Omega)),$$

where

$$\varphi_0(\mathbf{x}) = \lim_{t\to 0^+} \varphi(\mathbf{x}, t).$$

By applying Parseval's formula to (24.2.32), an inner product in $H_F(\Omega, \mathbb{R}^+)$ can be naturally defined by means of

$$\begin{aligned}(\varphi_{1_F}, \varphi_{2_F})_{\mathcal{H}_F} = \frac{1}{2\pi}\int_{-\infty}^{\infty}\int_{\Omega}&\{\nabla\varphi_{1_F}(\mathbf{x},\omega)\cdot\overline{\nabla\varphi_{2_F}(\mathbf{x},\omega)}\\ &+ [i\omega\varphi_{1_F}(\mathbf{x},\omega) - \varphi_{1_0}(\mathbf{x})]\cdot\overline{[i\omega\varphi_{2_F}(\mathbf{x},\omega) - \varphi_{2_0}(\mathbf{x})]}\}d\mathbf{x}d\omega\\ &+ \frac{1}{2\pi}\int_{-\infty}^{\infty}\int_{\partial\Omega}\varphi_{1_F}(\mathbf{x},\omega)\cdot\overline{\varphi_{2_F}(\mathbf{x},\omega)}d\mathbf{x}d\omega.\end{aligned}$$

Thus, there exists a natural isomorphism between $\mathcal{H}(\Omega, \mathbb{R}^+)$ and $\mathcal{H}_F(\Omega, \mathbb{R}^+)$ [310]. Let us introduce the sesquilinear form

$$\begin{aligned}a(\mathbf{u}_F, \varphi_F) = &\frac{1}{2\pi}\int_{-\infty}^{\infty}\int_{\Omega} -[i\omega\mathbf{u}_F(\mathbf{x},\omega) - \mathbf{u}(\mathbf{x},0)]\cdot\overline{[i\omega\varphi_F(\mathbf{x},\omega) - \varphi(\mathbf{x},0)]}d\mathbf{x}d\omega\\ &+ \frac{1}{2\pi}\int_{-\infty}^{\infty}\int_{\Omega}[\mathbb{G}_0 + \mathbb{G}'_F(\mathbf{x},\omega)]\nabla\mathbf{u}_F(\mathbf{x},\omega)\cdot\overline{\nabla\varphi_F(\mathbf{x},\omega)}d\mathbf{x}d\omega\\ &+ \frac{1}{2\pi}\int_{-\infty}^{\infty}\int_{\partial\Omega}\alpha\mathbf{u}_F(\mathbf{x},\omega)\cdot\overline{\varphi_F(\mathbf{x},\omega)}d\mathbf{x}d\omega - \int_{\Omega}\dot{\mathbf{u}}(\mathbf{x},0)\cdot\varphi(\mathbf{x},0)d\mathbf{x},\end{aligned}$$

which allows us to write (24.2.34) as

$$a(\mathbf{u}_F, \varphi_F) = \frac{1}{2\pi}\int_{-\infty}^{\infty}\int_{\partial\Omega}\left[\mathbf{f}_F(\mathbf{x},\omega)\cdot\overline{\varphi_F(\mathbf{x},\omega)} - \mathbf{T}_{0F}(\mathbf{x},\omega)\cdot\overline{\nabla\varphi_F(\mathbf{x},\omega)}\right]d\mathbf{x}d\omega, \quad (24.2.35)$$

by virtue of Parseval's formula. Therefore, $\mathcal{H}_F(\Omega, \mathbf{R})$ is the natural space for the Fourier transform of the weak solution of (24.2.31).

Lemma 24.2.8. *A function $\mathbf{u}_F \in \mathcal{H}_F(\Omega, \mathbb{R})$ is the Fourier transform of a weak solution of the initial boundary value problem (24.2.31) in the sense of Definition 24.2.7 if and only if the equality (24.2.35) holds for all $\varphi_F \in \mathcal{H}_F(\Omega, \mathbb{R})$.*

Proof. The sesquilinear form a, taking account of

$$\varphi(\mathbf{x}, 0) = \frac{1}{\pi}\int_{-\infty}^{\infty}\varphi_F(\mathbf{x},\omega)d\omega, \quad \dot{\varphi}(\mathbf{x}, 0) = \frac{1}{\pi}\int_{-\infty}^{\infty}[i\omega\varphi_F(\mathbf{x},\omega) - \varphi(\mathbf{x},0)]d\omega,$$

can be written as

$$
\begin{aligned}
a(\mathbf{u}_F, \boldsymbol{\varphi}_F) = & -\frac{1}{2\pi} \int_{-\infty}^{\infty} \int_{\Omega} [\omega^2 \mathbf{u}_F(\mathbf{x}, \omega) + \dot{\mathbf{u}}(\mathbf{x}, 0) + i\omega \mathbf{u}(\mathbf{x}, 0)] \cdot \overline{\boldsymbol{\varphi}_F(\mathbf{x}, \omega)} d\mathbf{x} d\omega \\
& + \frac{1}{2\pi} \int_{-\infty}^{\infty} \int_{\Omega} [\mathbb{G}_0 + \mathbb{G}'_F(\mathbf{x}, \omega)] \nabla \mathbf{u}_F(\mathbf{x}, \omega) \cdot \overline{\nabla \boldsymbol{\varphi}_F(\mathbf{x}, \omega)} d\mathbf{x} d\omega \\
& + \frac{1}{2\pi} \int_{-\infty}^{\infty} \int_{\partial\Omega} \alpha \mathbf{u}_F(\mathbf{x}, \omega) \cdot \overline{\boldsymbol{\varphi}_F(\mathbf{x}, \omega)} d\mathbf{x} d\omega.
\end{aligned}
\tag{24.2.36}
$$

From (24.2.35), using (24.2.36) and putting $\boldsymbol{\varphi}_F(\mathbf{x}, \omega) = \boldsymbol{\varphi}_1(\mathbf{x}) \varphi_2(\omega)$, where $\boldsymbol{\varphi}_1 \in H^1(\Omega)$ and $\varphi_2 \in L^2(\mathbb{R})$ is an arbitrary function, it follows that for almost all $\omega \in \mathbb{R}$,

$$
\begin{aligned}
& - \int_{\Omega} [\omega^2 \mathbf{u}_F(\mathbf{x}, \omega) + \dot{\mathbf{u}}_0(\mathbf{x}) + i\omega \mathbf{u}_0(\mathbf{x})] \cdot \overline{\boldsymbol{\varphi}_1(\mathbf{x})} d\mathbf{x} \\
& + \int_{\Omega} [\mathbb{G}_0 + \mathbb{G}'_F(\mathbf{x}, \omega)] \nabla \mathbf{u}_F(\mathbf{x}, \omega) \cdot \overline{\nabla \boldsymbol{\varphi}_1(\mathbf{x})} d\mathbf{x} + \int_{\partial\Omega} \alpha \mathbf{u}_F(\mathbf{x}, \omega) \cdot \overline{\boldsymbol{\varphi}_1(\mathbf{x})} d\mathbf{x} \\
& = \int_{\partial\Omega} \Big[\mathbf{f}_F(\mathbf{x}, \omega) \cdot \overline{\boldsymbol{\varphi}_1(\mathbf{x})} d\mathbf{x} - \mathbf{T}_{0F}(\mathbf{x}, \omega) \cdot \overline{\nabla \boldsymbol{\varphi}_1(\mathbf{x})} \Big] d\mathbf{x}
\end{aligned}
\tag{24.2.37}
$$

holds for every $\boldsymbol{\varphi}_1 \in H^1(\Omega)$. This identity yields that $\mathbf{u}_F(\cdot, \omega)$ is a generalized solution in $H^1(\Omega)$ for the elliptic problem

$$
\begin{aligned}
& - \omega^2 \mathbf{u}_F(\mathbf{x}, \omega) - \nabla \cdot \{[\mathbb{G}_0(\mathbf{x}) + \mathbb{G}'_F(\mathbf{x}, \omega)] \nabla \mathbf{u}_F(\mathbf{x}, \omega)\} \\
& \quad = \dot{\mathbf{u}}_0(\mathbf{x}) + i\omega \mathbf{u}_0(\mathbf{x}) + \mathbf{f}_F(\mathbf{x}, \omega) + \nabla \cdot \mathbf{T}_{0F}(\mathbf{x}, \omega) \qquad \forall \mathbf{x} \in \Omega, \\
& [\mathbb{G}_0(\mathbf{x}) + \mathbb{G}'_F(\mathbf{x}, \omega)] \nabla \mathbf{u}_F(\mathbf{x}, \omega) \mathbf{n}(\mathbf{x}) + \alpha(\mathbf{x}) \mathbf{u}_F(\mathbf{x}, \omega) \\
& \quad = -\mathbf{T}_{0F}(\mathbf{x}, \omega) \mathbf{n}(\mathbf{x}) \qquad \forall \mathbf{x} \in \partial\Omega.
\end{aligned}
$$

This problem, with zero initial data $(24.2.31)_3$, becomes

$$
\begin{aligned}
& - \omega^2 \mathbf{u}_F(\mathbf{x}, \omega) - \nabla \cdot \{[\mathbb{G}_0(\mathbf{x}) + \mathbb{G}'_F(\mathbf{x}, \omega)] \nabla \mathbf{u}_F(\mathbf{x}, \omega)\} \\
& \quad = \mathbf{f}_F(\mathbf{x}, \omega) + \nabla \cdot \mathbf{T}_{0F}(\mathbf{x}, \omega) \qquad \forall \mathbf{x} \in \Omega, \\
& [\mathbb{G}_0(\mathbf{x}) + \mathbb{G}'_F(\mathbf{x}, \omega)] \nabla \mathbf{u}_F(\mathbf{x}, \omega) \mathbf{n}(\mathbf{x}) + \alpha(\mathbf{x}) \mathbf{u}_F(\mathbf{x}, \omega) = \mathbf{0} \quad \forall \mathbf{x} \in \partial\Omega.
\end{aligned}
\tag{24.2.38}
$$

Remark 24.2.9. Since $\mathbb{G}_\infty$ and $\mathbb{G}'_s(\cdot, \omega)$ are bounded and positive definite, we have, using (24.2.20) and $(24.2.23)_3$ with (24.2.19) and $(24.2.23)_2$,

$$
\begin{aligned}
g_m(\omega) \|\nabla \mathbf{u}_F(\omega)\|^2 & \leq - \int_{\Omega} \mathbb{G}'_s(\mathbf{x}, \omega) \nabla \mathbf{u}_F(\mathbf{x}, \omega) \cdot \overline{\nabla \mathbf{u}_F(\mathbf{x}, \omega)} d\mathbf{x} \\
& \leq g_M(\omega) \|\nabla \mathbf{u}_F(\omega)\|^2 \quad \forall \omega > 0, \\
g_{\infty_m} \|\nabla \mathbf{u}_F(0)\|^2 & \leq \int_{\Omega} \mathbb{G}_\infty(\mathbf{x}) \nabla \mathbf{u}_F(\mathbf{x}, 0) \cdot \overline{\nabla \mathbf{u}_F(\mathbf{x}, 0)} d\mathbf{x} \leq g_{\infty_M} \|\nabla \mathbf{u}_F(0)\|^2 \omega = 0.
\end{aligned}
$$

So the problem (24.2.38) is Fredholm solvable in $H^1(\Omega)$ for every source in $L^2(\Omega)$ (see Theorem 4.1, page 186 of [224]); moreover, as a consequence of Fredholm's theorems, the existence theorem follows from the uniqueness theorem. □

Theorem 24.2.10. *(Uniqueness) For every $\omega \in \mathbb{R}$ the problem (24.2.38) has almost one solution $\mathbf{u}_F(\cdot, \omega) \in H^1(\Omega)$.*

Proof. To prove uniqueness, one must show that for every $\omega \in \mathbb{R}$, the problem

$$
\begin{aligned}
& -\omega^2\mathbf{u}_F(\mathbf{x},\omega) - \nabla\cdot\{[\mathbb{G}_0(\mathbf{x}) + \mathbb{G}'_F(\mathbf{x},\omega)]\nabla\mathbf{u}_F(\mathbf{x},\omega)\} = \mathbf{0} && \forall\mathbf{x}\in\Omega,\\
& [\mathbb{G}_0(\mathbf{x}) + \mathbb{G}'_F(\mathbf{x},\omega)]\nabla\mathbf{u}_F(\mathbf{x},\omega)\mathbf{n}(\mathbf{x}) + \alpha(\mathbf{x})\mathbf{u}_F(\mathbf{x},\omega) = \mathbf{0} && \forall\mathbf{x}\in\partial\Omega,
\end{aligned}
$$

has only the trivial solution. Thus, (24.2.37) can be written as

$$
\begin{aligned}
&\int_\Omega \{\omega^2\mathbf{u}_F(\mathbf{x},\omega)\cdot\overline{\varphi(\mathbf{x})} - [\mathbb{G}_0(\mathbf{x}) + \mathbb{G}'_F(\mathbf{x},\omega)]\nabla\mathbf{u}_F(\mathbf{x},\omega)\cdot\overline{\nabla\varphi(\mathbf{x})}\}d\mathbf{x}\\
&\quad - \int_{\partial\Omega}\alpha\mathbf{u}_F(\mathbf{x},\omega)\cdot\overline{\varphi(\mathbf{x})}d\mathbf{x} = 0.
\end{aligned} \tag{24.2.39}
$$

Hence, if $\omega = 0$, putting $\varphi = \mathbf{u}_F(\cdot, 0)$ and noting that $\mathbb{G}_0(\mathbf{x}) + \mathbb{G}'_s(\mathbf{x}, \cdot)$ is a continuous function of ω such that

$$
\lim_{\omega\to 0}[\mathbb{G}_0(\mathbf{x}) + \mathbb{G}'_F(\mathbf{x},\omega)] = \lim_{\omega\to 0}[\mathbb{G}_0(\mathbf{x}) + \mathbb{G}'_c(\mathbf{x},\omega)] = \mathbb{G}_\infty(\mathbf{x}), \tag{24.2.40}
$$

it follows that

$$
\int_\Omega \mathbb{G}_\infty(\mathbf{x})\nabla\mathbf{u}_F(\mathbf{x},0)\cdot\overline{\nabla\mathbf{u}_F(\mathbf{x},0)}d\mathbf{x} + \int_{\partial\Omega}\alpha(\mathbf{x})|\mathbf{u}_F(\mathbf{x},0)|^2 d\mathbf{x} = 0.
$$

The symmetry and positive definiteness of $\mathbb{G}_\infty$ together with the positivity of α give

$$
\|\mathrm{sym}\nabla\mathbf{u}_F(\cdot,0)\| = 0, \qquad \|\mathbf{u}_F(\cdot,0)\|_{\partial\Omega} = 0.
$$

Then $\mathbf{u}_F(\cdot, 0) \in H^1_0(\Omega)$ and Korn's inequality [98] yields $\|\mathbf{u}_F(\cdot, 0)\| = 0$.

Let $\omega \neq 0$ and $\varphi_F = \mathbf{u}_F(\cdot, \omega)$; then the imaginary part of (24.2.39) gives

$$
\int_\Omega \mathbb{G}'_s(\mathbf{x},\omega)\nabla\mathbf{u}_F(\mathbf{x},\omega)\cdot\overline{\nabla\mathbf{u}_F(\mathbf{x},\omega)}d\mathbf{x} = 0.
$$

Thus, the assumption (24.2.20) ensures that $\|\mathrm{sym}\nabla\mathbf{u}_F(\cdot, \omega)\| = 0$. Therefore, since $\mathbf{f}_F = \mathbf{0}$ and $\mathbf{T}_{0F} = \mathbf{0}$, (24.2.39) yields

$$
\int_\Omega \omega^2\mathbf{u}_F(\mathbf{x},\omega)\cdot\overline{\varphi(\mathbf{x})}d\mathbf{x} = 0 \qquad \forall\varphi\in C_0^\infty(\Omega),
$$

that is, $\|\mathbf{u}_F(\cdot, \omega)\| = 0$. □

Remark 24.2.11. From Theorem 24.2.10 and Remark 24.2.9 it follows that the differential operator $\mathcal{T}(\omega)$ defined by the system (24.2.38) is an isomorphism of $H^1(\Omega)$ into $L^2(\Omega)$; moreover, it is a continuous function of ω, and the inverse operator $\mathcal{T}^{-1}(\omega)$ is a continuous function of ω (see Lemma 44.1 of [310]).

This remark allows us to prove the following theorem.

Theorem 24.2.12. *For every $\omega \in \mathbb{R}$ the problem (24.2.38) has one and only one solution $\mathbf{u}_F(\cdot,\omega) \in H^1(\Omega)$. Also, the following inequality holds:*

$$||\nabla \mathbf{u}_F(\omega)|| + ||\omega \mathbf{u}_F(\omega)|| + ||\mathbf{u}_F(\omega)||_{\partial\Omega} \leq \mathcal{A}(\omega)[||\mathbf{f}_F(\omega)|| + ||\omega \mathbf{T}_{0F}(\omega)||], \tag{24.2.41}$$

with $\mathcal{A} \in L^\infty(\mathbb{R})$.

Proof. Let $\mathbf{u}_F$ be a solution of the problem (24.2.38). Then we have the equality

$$\begin{aligned}
&\int_\Omega \{-|\omega \mathbf{u}_F(\mathbf{x},\omega)|^2 + [\mathbb{G}_0(\mathbf{x}) + \mathbb{G}'_F(\mathbf{x},\omega)]\nabla \mathbf{u}_F(\mathbf{x},\omega)\cdot \overline{\nabla \mathbf{u}_F(\mathbf{x},\omega)}\}d\mathbf{x}\\
&\quad + \int_{\partial\Omega} \alpha(\mathbf{x})|\mathbf{u}_F(\mathbf{x},\omega)|^2 d\mathbf{x} \\
&= \int_{\partial\Omega}\Big[\mathbf{f}_F(\mathbf{x},\omega)\cdot\overline{\mathbf{u}_F(\mathbf{x},\omega)}d\mathbf{x} - \mathbf{T}_{0F}(\mathbf{x},\omega)\cdot\overline{\nabla \mathbf{u}_F(\mathbf{x},\omega)}\Big]d\mathbf{x}.
\end{aligned} \tag{24.2.42}$$

Firstly, consider ω close to 0. By virtue of (24.2.40), there exists ω_1 such that if $|\omega| < \omega_1$,

$$\inf_{\mathbf{x}\in\Omega} ||\mathbb{G}_0(\mathbf{x}) + \mathbb{G}'_F(\mathbf{x},\omega)|| \geq \inf_{\mathbf{x}\in\Omega} ||\mathbb{G}_0(\mathbf{x}) + \mathbb{G}'_c(\mathbf{x},\omega)|| \geq \frac{1}{2} g_{\infty_m} > 0; \tag{24.2.43}$$

moreover, using (24.2.33), we can find ω_2 such that if $|\omega| < \omega_2 \leq \omega_1$,

$$\begin{aligned}
||\omega \mathbf{u}_F(\omega)||^2 &\leq \omega_2^2[K_1||\nabla \mathbf{u}_F(\omega)||^2 + K_2||\mathbf{u}_F(\omega)||^2_{\partial\Omega}]\\
&\leq \frac{1}{4} g_{\infty_m}||\nabla \mathbf{u}_F(\omega)||^2 + \frac{1}{2}\alpha_m||\mathbf{u}_F(\omega)||^2_{\partial\Omega}.
\end{aligned} \tag{24.2.44}$$

Then if $|\omega| < \omega_2$, the real part of (24.2.42), (24.2.43), and (24.2.44) give

$$\begin{aligned}
&\frac{1}{4} g_{\infty_m}||\nabla \mathbf{u}_F(\omega)||^2 + \frac{1}{2}\alpha_m||\mathbf{u}_F(\omega)||^2_{\partial\Omega}\\
&\leq \int_\Omega [\mathbb{G}_0(\mathbf{x}) + \mathbb{G}'_c(\mathbf{x},\omega)]\nabla \mathbf{u}_F(\mathbf{x},\omega)\cdot\overline{\nabla \mathbf{u}_F(\mathbf{x},\omega)}d\mathbf{x}\\
&\quad + \int_{\partial\Omega} \alpha(\mathbf{x})|\mathbf{u}_F(\mathbf{x},\omega)|^2 d\mathbf{x} - \omega^2||\mathbf{u}_F(\omega)||^2\\
&\leq ||\mathbf{f}_F(\omega)||\,||\mathbf{u}_F(\omega)|| + ||\mathbf{T}_{0F}(\omega)||\,||\nabla \mathbf{u}_F(\omega)||\\
&\leq [K_1||\mathbf{f}_F(\omega)|| + ||\mathbf{T}_{0F}(\omega)||]||\nabla \mathbf{u}_F(\omega)|| + K_2||\mathbf{f}_F(\omega)||\,||\mathbf{u}_F(\omega)||_{\partial\Omega};
\end{aligned}$$

hence, with straightforward calculations, it follows that

$$||\nabla \mathbf{u}_F(\omega)|| + ||\mathbf{u}_F(\omega)||_{\partial\Omega} \leq \lambda[||\mathbf{f}_F(\omega)|| + ||\mathbf{T}_{0F}(\omega)||] \quad \forall |\omega| < \omega_2, \tag{24.2.45}$$

where λ depends on g_{∞_m}, α_m, K_1, and K_2.

Thus, we get the inequality (24.2.41) for $|\omega| < \omega_2$ from (24.2.45) and from the classical inequality [224]

$$||\mathbf{u}_F||^2_{\partial\Omega} \leq ||\nabla \mathbf{u}_F||^2 + K_3||\mathbf{u}_F||^2, \tag{24.2.46}$$

which holds for functions of $\mathbf{u}_F \in H^1(\Omega)$, where Ω is a bounded and regular domain and K_3 a constant depending on Ω.

Consider ω close to ∞. The imaginary and real parts of (24.2.42) yield

$$\begin{aligned} g_m(\omega)\|\nabla\mathbf{u}_F\|^2 &\le \|\mathbf{f}_F(\omega)\|\|\mathbf{u}_F(\omega)\| + \|\mathbf{T}_{0F}(\omega)\|\|\nabla\mathbf{u}_F(\omega)\|,\\ \|\omega\mathbf{u}_F(\omega)\|^2 &\le \beta\|\nabla\mathbf{u}_F(\omega)\|^2 + \alpha_M\|\mathbf{u}_F(\omega)\|^2_{\partial\Omega}\\ &\quad + \|\mathbf{f}_F(\omega)\|\|\mathbf{u}_F(\omega)\| + \|\mathbf{T}_{0F}(\omega)\|\|\nabla\mathbf{u}_F(\omega)\|, \end{aligned} \tag{24.2.47}$$

where

$$\alpha_M = \operatorname*{ess\,sup}_{\mathbf{x}\in\partial\Omega}\alpha(\mathbf{x}), \qquad \beta = \sup_{(\mathbf{x},\omega)\in\Omega\times\mathbb{R}} |\mathbb{G}_0(\mathbf{x}) + \mathbb{G}'_c(\mathbf{x},\omega)| < \infty.$$

The notation $|\cdot|$ indicates here a norm in the finite-dimensional space Lin {Sym}, such as that given by (A.2.5). From the inequalities (24.2.46) and (24.2.47) we have

$$\begin{aligned} \|\nabla\mathbf{u}_F\|^2 &\le \frac{1}{\omega g_m(\omega)}\|\mathbf{f}_F(\omega)\|\|\omega\mathbf{u}_F(\omega)\| + \frac{1}{g_m(\omega)}\|\mathbf{T}_{0F}(\omega)\|\|\nabla\mathbf{u}_F(\omega)\|,\\ \|\omega\mathbf{u}_F(\omega)\|^2 &\le \frac{\alpha_M+\beta+g_m(\omega)}{g_m(\omega)}\left[\frac{1}{\omega}\|\mathbf{f}_F(\omega)\|\|\omega\mathbf{u}_F(\omega)\| + \|\mathbf{T}_{0F}(\omega)\|\|\nabla\mathbf{u}_F(\omega)\|\right]\\ &\quad + \alpha_M K_3\|\mathbf{u}_F(\omega)\|^2, \end{aligned} \tag{24.2.48}$$

and for $\omega^2 > 2\alpha_M K_3$, $(24.2.48)_2$ gives

$$\frac{1}{2}\|\omega\mathbf{u}_F(\omega)\|^2 \le \frac{\alpha_M+\beta+g_m(\omega)}{g_m(\omega)}\left[\frac{1}{\omega}\|\mathbf{f}_F(\omega)\|\|\omega\mathbf{u}_F(\omega)\| + \|\mathbf{T}_{0F}(\omega)\|\|\nabla\mathbf{u}_F(\omega)\|\right]. \tag{24.2.49}$$

Moreover, after some calculations, we obtain from $(24.2.48)_1$ and (24.2.49)

$$\|\nabla\mathbf{u}_F(\omega)\| + \|\omega\mathbf{u}_F(\omega)\| \le 2\frac{2[\alpha_M+\beta+g_m(\omega)]+1}{\omega g_m(\omega)}[\|\mathbf{f}_F(\omega)\| + \|\omega\mathbf{T}_{0F}(\omega)\|]. \tag{24.2.50}$$

The positive definiteness of $\mathbb{G}'_0$ ensures that for $|\omega| \ge \omega_3 > \sqrt{2\alpha_M K_3}$,

$$2\frac{2[\alpha_M+\beta+g_m(\omega)]+1}{\omega g_m(\omega)} < 4\frac{2\alpha_M+2\beta+3}{g'_0}, \tag{24.2.51}$$

and for $|\omega| \ge \omega_3$, (24.2.41) follows from (24.2.50), (24.2.51), and (24.2.46).

Finally, the continuity of the inverse operator $\mathcal{T}^{-1}(\omega)$ ensures that (24.2.41) holds in the compact set $\omega_2 \le |\omega| \le \omega_3$. □

We can now prove the existence and uniqueness theorem of the evolution problem.

Theorem 24.2.13. *For any linear viscoelastic solid obeying (24.2.2) with relaxation function $\mathbb{G}$ satisfying the constitutive assumptions (24.2.18)–(24.2.23), the evolution problem (24.2.31), with α satisfying (24.2.17), $\mathbf{f} \in L^2(\mathbb{R}^+; L^2(\Omega))$, $\mathbf{T}_0 \in H^1_0(\mathbb{R}^+; H^1_0(\Omega))$, and initial data equal to zero, has one and only one weak solution $\mathbf{u} \in H(\Omega, \mathbb{R}^+)$.*

Proof. By virtue of Theorem 24.2.12 and the assumptions on the data, we have

$$\int_{-\infty}^{\infty}\left[\|\nabla \mathbf{u}_F(\omega)\|^2 + \|\omega \mathbf{u}_F(\omega)\|^2 + \|\mathbf{u}_F(\omega)\|^2_{\partial\Omega}\right]d\omega$$
$$\leq \int_{-\infty}^{\infty} A^2(\omega)[\|\mathbf{f}_F(\omega)\| + \|\omega \mathbf{T}_{0F}(\omega)\|]^2 d\omega < \infty$$

and hence $\mathbf{u}_F \in \mathcal{H}_F(\Omega, \mathbb{R}^+)$. Moreover, $\mathbf{u}_F$ is the Fourier transform of the solution $\mathbf{u} \in \mathcal{H}(\Omega, \mathbb{R}^+)$ of the problem (24.2.31), because of the isomorphism between $\mathcal{H}(\Omega, \mathbb{R}^+)$ and $\mathcal{H}_F(\Omega, \mathbb{R}^+)$. □

24.2.3.5 Domain-of-Dependence Inequality

It is possible to define the maximum free energy potential as a consequence of the hypotheses assumed for the constitutive equation (24.2.2) using the results related to thermodynamic potentials [114]. Moreover, in [114] it is proved that the energy propagates through the space with a finite speed, since a priori there exists a domain-of-dependence inequality for the evolution problem (24.2.15) and (24.2.16), with an initial history $\mathbf{u}^0$ that has finite maximum free energy. Some results about the existence of a domain-of-dependence inequality were also derived in [57, 245]; however, in these works the maximum propagation speed of disturbances depends on time.

The thermodynamic potentials can be considered functionals of $\mathbf{E}$. Any history $\mathbf{E}^t(\mathbf{x},\cdot)\colon \mathbb{R}^+ \to \mathrm{Sym}$, for which the stress $\mathbf{T}$ is bounded is called an admissible history characterized by

$$\left|\int_0^{\infty} \mathbb{G}'(\mathbf{x}, s)\mathbf{E}^t(\mathbf{x}, s)ds\right| < \infty. \tag{24.2.52}$$

Moreover, we note that any such histories may be considered linear continuous functions of the space $\mathcal{F}$ and that we can take the set of admissible histories equal to the larger $\mathcal{F}'$, the space of all continuous functionals on $\mathcal{F}$. With a straightforward calculation $\mathcal{F}'$ turns out to be the set of histories $\mathbf{E}^t \in \mathcal{D}'$, the dual of $C_0^\infty(\mathbb{R}^+)$, such that (24.2.52) holds.

We consider the set of all admissible histories $\mathcal{G} \subset \mathcal{F}'$, where any free energy is well defined and its subset $\mathcal{G}_0$ is equal to $\left\{\mathbf{E}^t : \mathbb{R}^+ \to L^2(\Omega); \|\mathbf{E}^t(\cdot)\| \in L^2(\mathbb{R}^+)\right\}$.

Recall that [114] for any $\mathbf{E}^t \in \mathcal{G}_0$, since $\mathbb{G}'(\mathbf{x},\cdot) \in L^1(\mathbb{R}^+) \cap L^2(\mathbb{R}^+)$ and $\mathbb{G}'_s(\mathbf{x},\cdot) \in L^1(\mathbb{R}) \cap L^2(\mathbb{R})$, Parseval's formula allows us to give the constitutive equation for the stress tensor (24.2.1) the form

$$\mathbf{T}(\mathbf{x}, t) = \mathbb{G}_\infty(\mathbf{x})\mathbf{E}(\mathbf{x}, t) + \frac{2}{\pi}\int_0^{\infty} \mathbb{G}'_s(\mathbf{x}, \omega)\left[\mathbf{E}^t_s(\mathbf{x}, \omega) - \frac{\mathbf{E}(\mathbf{x}, t)}{\omega}\right]d\omega, \tag{24.2.53}$$

which is a special case of (7.2.34), written on $\mathbb{R}^+$ rather than $\mathbb{R}$. Moreover, under the hypotheses assumed for the relaxation function $\mathbb{G}$, we can consider the free energy density ψ_M in the form

$$\psi_M(\mathbf{E}^t(\mathbf{x},\cdot)) = \frac{1}{2}\mathbb{G}_\infty(\mathbf{x})\mathbf{E}(\mathbf{x},t)\cdot\mathbf{E}(\mathbf{x},t)$$
$$-\frac{1}{2\pi}\int_0^\infty \omega\mathbb{G}'_s(\mathbf{x},\omega)\left[\mathbf{E}^t_s(\mathbf{x},\omega)-\frac{\mathbf{E}(\mathbf{x},t)}{\omega}\right]\cdot\overline{\left[\mathbf{E}^t_s(\mathbf{x},\omega)-\frac{\mathbf{E}(\mathbf{x},t)}{\omega}\right]}d\omega$$

for every history $\mathbf{E}^t \in \mathcal{G}_0$.

Proposition 24.2.14. *The functional ψ_M defines a norm, namely*

$$|\mathbf{E}^t(\mathbf{x},\cdot)|^2_M := 2\psi_M(\mathbf{E}^t(\mathbf{x},\cdot)), \tag{24.2.54}$$

and the space $\mathcal{H}_M$, obtained as a completion of $\mathcal{G}_0$ relative to this norm, is a Banach space. Moreover, the stress tensor $\mathbf{T}$ *is well defined and is continuous on $\mathcal{H}_M$, in the sense that*

$$|\mathbf{T}(\mathbf{x},t)|^2 \le [|\mathbb{G}_0(\mathbf{x})-\mathbb{G}_\infty(\mathbf{x})| + |\mathbb{G}_\infty(\mathbf{x})|]|\mathbf{E}^t(\mathbf{x},\cdot)|^2_M. \tag{24.2.55}$$

Proof. For any symmetric and positive definite tensor $\mathbf{A}$, we can define the symmetric and positive tensor $\sqrt{\mathbf{A}}$ (Sect. A.2.1) such that $\sqrt{\mathbf{A}}\sqrt{\mathbf{A}} = \mathbf{A}$ and $|\sqrt{\mathbf{A}}| = \sqrt{|\mathbf{A}|}$. Thus, from (24.2.53) we have

$$|\mathbf{T}(\mathbf{x},t)|^2 \le (1+\alpha)|\mathbb{G}_\infty(\mathbf{x})|\mathbf{E}(\mathbf{x},t)\cdot\mathbb{G}_\infty(\mathbf{x})\mathbf{E}(\mathbf{x},t)$$
$$-\left(1+\frac{1}{\alpha}\right)\frac{2}{\pi}\left|\int_0^\infty -\frac{\mathbb{G}'_s(\mathbf{x},\omega)}{\omega}d\omega\right|\frac{1}{\pi}\int_0^\infty \omega\mathbb{G}'_s(\mathbf{x},\omega)\left[\mathbf{E}^t_s(\mathbf{x},\omega)-\frac{\mathbf{E}(\mathbf{x},t)}{\omega}\right]$$
$$\times\overline{\left[\mathbf{E}^t_s(\mathbf{x},\omega)-\frac{\mathbf{E}(\mathbf{x},t)}{\omega}\right]}d\omega,$$

since $\mathbb{G}_\infty$ and $\mathbb{G}'_s$ are positive definite. Hence, by putting $\alpha = \frac{|\mathbb{G}_0(\mathbf{x})-\mathbb{G}_\infty(\mathbf{x})|}{|\mathbb{G}_\infty(\mathbf{x})|}$ and using (24.2.21) with (24.2.54), the relation (24.2.55) follows. □

Also, for every $\mathbf{E}^t \in \mathcal{H}_M$, we have (see (8.6.32)) [114]

$$\dot{\psi}_M(\mathbf{E}^t(\mathbf{x},\cdot)) = \mathbf{T}(\mathbf{x},t)\cdot\dot{\mathbf{E}}(\mathbf{x},t). \tag{24.2.56}$$

Let us denote by $e_M(t)$ the total maximum mechanical energy at time t, related to the maximum free energy ψ_M, expressed in terms of $\mathbf{u}$ by means of

$$e_M(t) = \int_\Omega\left[\frac{1}{2}|\dot{\mathbf{u}}(\mathbf{x},t)|^2 + \psi_M(\nabla\mathbf{u}^t(\mathbf{x},\cdot))\right]d\mathbf{x} + \frac{1}{2}\int_{\partial\Omega}\alpha(\mathbf{x})|\mathbf{u}(\mathbf{x},t)|^2 d\mathbf{x}.$$

Proposition 24.2.15. *For every history $\mathbf{u}^t$, where $\mathbf{E}^t \in \mathcal{H}_M$, there exists a positive constant k such that*

$$\|\dot{\mathbf{u}}(t)\|^2 + \|\nabla\mathbf{u}(t)\|^2 + \|\mathbf{u}(t)\|^2 \le k e_M(t). \tag{24.2.57}$$

Proof. Using the positive definiteness of $\mathbb{G}_\infty$ and $-\mathbb{G}'_s$, we see that

$$\int_\Omega \psi_M(\mathbf{E}^t(\mathbf{x},\cdot))d\mathbf{x} \geq \frac{1}{2} g_{\infty_m} \|sym\nabla\mathbf{u}(t)\|^2;$$

therefore,

$$\|\dot{\mathbf{u}}(t)\|^2 + g_{\infty_m}\|sym\nabla\mathbf{u}(t)\|^2 + \alpha_m\|\mathbf{u}(t)\|^2_{\partial\Omega} \leq 2e_M(t),$$

from which, using Korn's inequality, (24.2.57) follows. □

To derive a domain-of-dependence inequality we introduce the following definition.

Definition 24.2.16. *A function* $\mathbf{u} \in \mathcal{H}(\Omega,(0,\tau))$ *is a weak solution of the initial boundary value problem (24.2.15) and (24.2.16) in the space-time domain* $\Omega \times (0,\tau)$ *with initial conditions* $(24.2.31)_3$ *and* $\mathbf{f} \in L^2(\Omega,(0,\tau))$ *if* $\mathbf{u}(\mathbf{x},0) = \mathbf{u}^0(\mathbf{x},0)$ *almost everywhere in* Ω *and*

$$\int_0^\tau \int_\Omega \Big\{\Big[\mathbb{G}_0(\mathbf{x})\nabla\mathbf{u}(t) + \int_0^\infty \mathbb{G}'(\mathbf{x},s)\mathbf{u}^t(\mathbf{x},s)ds\Big]\cdot\nabla\boldsymbol{\phi}(\mathbf{x},t)$$
$$-\dot{\mathbf{u}}(\mathbf{x},t)\cdot\dot{\boldsymbol{\phi}}(\mathbf{x},t)\Big\}d\mathbf{x}dt + \int_0^\tau\int_{\partial\Omega}\alpha(\mathbf{x})\mathbf{u}(\mathbf{x},t)\cdot\boldsymbol{\phi}(\mathbf{x},t)d\mathbf{x}dt$$
$$= \int_\Omega \mathbf{v}_0(\mathbf{x})\cdot\boldsymbol{\phi}(\mathbf{x},0)d\mathbf{x} + \int_0^\tau\int_\Omega \mathbf{f}(\mathbf{x},t)\cdot\boldsymbol{\phi}(\mathbf{x},t)d\mathbf{x}dt,$$

for every $\boldsymbol{\phi}(\mathbf{x},t) \in \mathcal{H}(\Omega,(0,\tau))$.

Proposition 24.2.17. *Let* $\mathbf{u} \in \mathcal{H}(\Omega,(0,\tau))$ *be a weak solution in the sense of Definition 24.2.16. Then*

$$\operatorname*{ess\,sup}_{(\mathbf{x},t)\in\Omega\times(0,\tau)} \frac{2|\mathbf{T}(\mathbf{x},t)\dot{\mathbf{u}}(\mathbf{x},t)|}{|\dot{\mathbf{u}}(\mathbf{x},t)|^2 + |\mathbf{E}^t(\mathbf{x},\cdot)|^2_M} = \gamma(\tau) \leq \sqrt{|\mathbb{G}_0 - \mathbb{G}_\infty| + |\mathbb{G}_\infty|}. \qquad (24.2.58)$$

Proof. A classical algebraic inequality and (24.2.55) give

$$2|\mathbf{T}(\mathbf{x},t)\cdot\dot{\mathbf{u}}(\mathbf{x},t)| \leq \frac{1}{\beta}[|\mathbb{G}_0(\mathbf{x}) - \mathbb{G}_\infty(\mathbf{x})| + |\mathbb{G}_\infty(\mathbf{x})|]|\mathbf{E}^t(\mathbf{x},\cdot)|^2_M + \beta|\dot{\mathbf{u}}(\mathbf{x},t)|^2, \quad (24.2.59)$$

where $\beta > 0$. Letting $\beta^2 = |\mathbb{G}_0(\mathbf{x}) - \mathbb{G}_\infty(\mathbf{x})| + |\mathbb{G}_\infty(\mathbf{x})|$, (24.2.59) gives (24.2.58). □

Theorem 24.2.18. *(Domain-of-Dependence Inequality) Any solution of the problem (24.2.15)–(24.2.16) satisfies*

$$\int_{\Omega\cap S_r(\mathbf{x}_0)}\Big[|\dot{\mathbf{u}}(\mathbf{x},\tau)|^2 + |\mathbf{E}^\tau(\mathbf{x},\cdot)|^2_M\Big]d\mathbf{x} + \int_{\partial\Omega\cap S_r(\mathbf{x}_0)}\alpha(\mathbf{x})|\mathbf{u}(\mathbf{x},\tau)|^2d\mathbf{x}$$
$$\leq \int_{\Omega\cap S_{r+\gamma\tau}(\mathbf{x}_0)}\Big[|\dot{\mathbf{u}}(\mathbf{x},0)|^2 + |\mathbf{E}^0(\mathbf{x},\cdot)|^2_M\Big]d\mathbf{x} + \int_{\partial\Omega\cap S_{r+\gamma\tau}(\mathbf{x}_0)}\alpha(\mathbf{x})|\mathbf{u}(\mathbf{x},0)|^2d\mathbf{x}$$
$$+2\int_{\Omega\cap S_{r+\gamma(\tau-t)}(\mathbf{x}_0)}\mathbf{f}(\mathbf{x},t)\cdot\dot{\mathbf{u}}(\mathbf{x},t)d\mathbf{x}, \qquad (24.2.60)$$

where $S_r(\mathbf{x}_0) = \{\mathbf{x};\ |\mathbf{x}-\mathbf{x}_0| < r\}$ *and* γ *is defined in (24.2.58).*

Proof. Let $\mathbf{u} : [0, \tau] \to H^1(\Omega)$ be a weak solution of the problem (24.2.15) and (24.2.16) and $\phi \in C_0^\infty(\mathbb{R}^3 \times \mathbb{R})$. We consider the function

$$e_\phi(t) = \int_\Omega \left[\frac{1}{2} |\dot{\mathbf{u}}(\mathbf{x}, t)|^2 + \psi_M(\mathbf{E}^t(\mathbf{x}, \cdot)) \right] \phi(\mathbf{x}, t) d\mathbf{x} + \frac{1}{2} \int_{\partial\Omega} \alpha(\mathbf{x}) |\dot{\mathbf{u}}(\mathbf{x}, t)|^2 d\mathbf{x},$$

whose first derivative can be seen to be given by

$$\begin{aligned}
\frac{d}{dt} e_\phi(t) &= \int_\Omega \left[\ddot{\mathbf{u}}(\mathbf{x}, t) - \nabla \cdot \mathbf{T}(\mathbf{E}^t(\mathbf{x}, \cdot)) \right] \cdot \dot{\mathbf{u}}(\mathbf{x}, t) \phi(\mathbf{x}, t) d\mathbf{x} \\
&+ \int_\Omega \left\{ \left[\frac{1}{2} |\dot{\mathbf{u}}(\mathbf{x}, t)|^2 + \psi_M(\mathbf{E}^t(\mathbf{x}, \cdot)) \right] \dot{\phi}(\mathbf{x}, t) - \mathbf{T}(\mathbf{E}^t(\mathbf{x}, \cdot)) \dot{\mathbf{u}}(\mathbf{x}, t) \cdot \nabla \phi(\mathbf{x}, t) \right\} d\mathbf{x} \\
&+ \frac{1}{2} \int_{\partial\Omega} \alpha(\mathbf{x}) |\dot{\mathbf{u}}(\mathbf{x}, t)|^2 \dot{\phi}(\mathbf{x}, t) d\mathbf{x},
\end{aligned} \tag{24.2.61}$$

using (24.2.56) with an integration by parts. Let ϕ have the form

$$\phi(\mathbf{x}, t) = \phi_\delta(|\mathbf{x} - \mathbf{x}_0| - r - \gamma(\tau - t)),$$

where γ is given by (24.2.58), $\phi_\delta \in C^\infty(\mathbb{R})$, and

$$\phi_\delta(s) = \begin{cases} 1 & \forall s \leq -\delta, \\ & \\ 0 & \forall s > \delta, \end{cases} \qquad 0 \leq \phi_\delta(s) < 1, \quad \phi_\delta'(s) \leq 0 \quad \forall s \in \mathbb{R},$$

whence

$$\nabla \phi(\mathbf{x}, t) = \nabla |\mathbf{x} - \mathbf{x}_0| \phi_\delta', \qquad \dot{\phi}(\mathbf{x}, t) = \gamma \phi_\delta'.$$

Thus, using (24.2.15), it follows from (24.2.61) that

$$\begin{aligned}
\frac{d}{dt} e_\phi(t) &= \int_\Omega \mathbf{f}(\mathbf{x}, t) \cdot \dot{\mathbf{u}}(\mathbf{x}, t) \phi_\delta(|\mathbf{x} - \mathbf{x}_0| - r - \gamma(\tau - t)) d\mathbf{x} \\
&+ \int_\Omega \left\{ \gamma \left[\frac{1}{2} |\dot{\mathbf{u}}(\mathbf{x}, t)|^2 + \psi_M(\mathbf{E}^t(\mathbf{x}, \cdot)) \right] - \mathbf{T}(\mathbf{E}^t(\mathbf{x}, \cdot)) \dot{\mathbf{u}}(\mathbf{x}, t) \cdot \nabla |\mathbf{x} - \mathbf{x}_0| \right\} \\
&\qquad \phi_\delta'(|\mathbf{x} - \mathbf{x}_0| - r - \gamma(\tau - t)) d\mathbf{x} \\
&+ \frac{1}{2} \int_{\partial\Omega} \alpha(\mathbf{x}) |\dot{\mathbf{u}}(\mathbf{x}, t)|^2 \gamma \phi_\delta'(|\mathbf{x} - \mathbf{x}_0| - r - \gamma(\tau - t)) d\mathbf{x} \\
&\leq \int_\Omega \mathbf{f}(\mathbf{x}, t) \cdot \dot{\mathbf{u}}(\mathbf{x}, t) \phi_\delta(|\mathbf{x} - \mathbf{x}_0| - r - \gamma(\tau - t)) d\mathbf{x},
\end{aligned}$$

whence a time integration yields

$$e_\phi(\tau) - e_\phi(0) \leq \int_0^\tau \int_\Omega \mathbf{f}(\mathbf{x}, t) \cdot \dot{\mathbf{u}}(\mathbf{x}, t) \phi_\delta(|\mathbf{x} - \mathbf{x}_0| - r - \gamma(\tau - t)) d\mathbf{x} dt. \tag{24.2.62}$$

Finally, since, when $\delta \to 0$, ϕ_δ tends boundedly to the characteristic function for $S_{r+\gamma(\tau-t)}(\mathbf{x}_0)$, the limit of (24.2.62) gives (24.2.60). □

The free energy ψ_M exists for any relaxation function satisfying the minimal set of properties required by thermodynamics. When further properties are assumed for the relaxation function $\mathbb{G}$, other free energy functionals can be considered for linear viscoelastic materials, namely those discussed in Chap. 10.

In particular, if the relaxation function $\mathbb{G}$ is compatible with thermodynamics and obeys (10.2.2), the Graffi–Volterra functional [173]

$$\psi_G(\mathbf{E}^t(\mathbf{x},\cdot)) = \frac{1}{2}\mathbb{G}_\infty(\mathbf{x})\mathbf{E}(\mathbf{x},t)\cdot\mathbf{E}(\mathbf{x},t) - \frac{1}{2}\int_0^\infty \mathbb{G}'(\mathbf{x},s)[\mathbf{E}^t(\mathbf{x},s)-\mathbf{E}(\mathbf{x},t)]\cdot[\mathbf{E}^t(\mathbf{x},s)-\mathbf{E}(\mathbf{x},t)]ds$$

can be used. It defines a norm, as the following proposition states.

Proposition 24.2.19. *The functional ψ_G defines a norm $|\mathbf{E}^t(\mathbf{x},\cdot)|_G^2 := 2\psi_G(\mathbf{E}^t(\mathbf{x},\cdot))$, and the space $\mathcal{H}_G$, obtained as a completion of $\mathcal{G}_0$ relative to this norm is a Banach space. Moreover, the stress tensor $\mathbf{T}$ is well defined and is continuous on $\mathcal{H}_G$, in the sense that*

$$|\mathbf{T}(\mathbf{x},t)|^2 \le [|\mathbb{G}_0(\mathbf{x})-\mathbb{G}_\infty(\mathbf{x})| + |\mathbb{G}_\infty(\mathbf{x})|]|\mathbf{E}^t(\mathbf{x},\cdot)|_G^2. \tag{24.2.63}$$

Proof. From (24.2.2), since $\mathbb{G}_\infty$ and $-\mathbb{G}'$ are positive definite, we have

$$\begin{aligned}|\mathbf{T}(\mathbf{x},t)|^2 \le{}& (1+\beta)|\mathbb{G}_\infty(\mathbf{x})|\mathbf{E}(\mathbf{x},t)\cdot\mathbb{G}_\infty(\mathbf{x})\mathbf{E}(\mathbf{x},t)\\ &-\left(1+\frac{1}{\beta}\right)\left|\int_0^\infty -\mathbb{G}'(\mathbf{x},s)ds\right|\\ &\times\int_0^\infty \mathbb{G}'(\mathbf{x},s)[\mathbf{E}^t(\mathbf{x},s)-\mathbf{E}(\mathbf{x},t)]\cdot[\mathbf{E}^t(\mathbf{x},s)-\mathbf{E}(\mathbf{x},t)]ds,\end{aligned} \tag{24.2.64}$$

with $\beta > 0$. Putting $\beta = \frac{|\mathbb{G}_0(\mathbf{x})-\mathbb{G}_\infty(\mathbf{x})|}{|\mathbb{G}_\infty(\mathbf{x})|}$, the inequality (24.2.63) follows from (24.2.64). □

We note that the inequality (24.2.63) leads to a domain-of-dependence inequality for weak solutions of (24.2.15) and (24.2.16) with initial history $\mathbf{u}^0 \in \mathcal{H}_G$.

It is interesting to consider the scalar relaxation function

$$G(\mathbf{x},s) = G_\infty(\mathbf{x}) + \sum_{k=1}^{n} A_k(\mathbf{x})e^{-\alpha_k s}, \tag{24.2.65}$$

where we have a sum of exponential functions with α_k $(k = 1,2,\ldots,N)$ positive constants, G_∞ and A_k $(k = 1,2,\ldots,N)$ positive functions belonging to $C(\bar{\Omega})\cap C^1(\Omega)$. For such viscoelastic materials (indeed for any material with completely monotonic relaxation functions, as defined by (10.1.8)), the Dill free energy functional (see (10.2.4))

$$\psi_{Dill}(\mathbf{E}^t(\mathbf{x},\cdot)) = \frac{1}{2}G_\infty(\mathbf{x})\mathbf{E}(\mathbf{x},t)\cdot\mathbf{E}(\mathbf{x},t) + \frac{1}{2}\int_0^\infty\int_0^\infty G''(\mathbf{x},s_1+s_2)[\mathbf{E}^t(\mathbf{x},s_1)-\mathbf{E}(\mathbf{x},t)]\cdot[\mathbf{E}^t(\mathbf{x},s_2)-\mathbf{E}(\mathbf{x},t)]ds_1ds_2$$

can be introduced.

Proposition 24.2.20. *The functional ψ_{Dill} defines a norm, namely $\|\mathbf{E}^t(\mathbf{x},\cdot)\|^2_{Dill} := 2\psi_{Dill}(\mathbf{E}^t(\mathbf{x},\cdot))$, and the space $\mathcal{H}_{Dill}$ obtained as a completion of $\mathcal{G}_0$ relative to this norm is a Banach space. Moreover, the stress tensor $\mathbf{T}$ is well defined and is continuous on $\mathcal{H}_{Dill}$, in the sense that*

$$|\mathbf{T}(\mathbf{x},t)|^2 \leq G_0(\mathbf{x})|\mathbf{E}^t(\mathbf{x},\cdot)|^2_{Dill}. \tag{24.2.66}$$

Proof. The constitutive equation (24.2.2) becomes

$$\mathbf{T}(\mathbf{x},t) = G_\infty(\mathbf{x})\mathbf{E}(\mathbf{x},t) - \sum_{k=1}^{n} \alpha_k A_k(\mathbf{x})\mathbf{E}_k(\mathbf{x},t),$$

where $\mathbf{E}_k(\mathbf{x},t) = \int_0^\infty e^{-\alpha_k s}[\mathbf{E}^t(\mathbf{x},s) - \mathbf{E}(\mathbf{x},t)]ds$. Hence, we have

$$|\mathbf{T}(\mathbf{x},t)|^2 \leq \left(G_\infty(\mathbf{x}) + \sum_{k=1}^{n} A_k(\mathbf{x})\right)\left(G_\infty(\mathbf{x})\mathbf{E}^2(\mathbf{x},t) + \sum_{k=1}^{n} \alpha_k^2 A_k(\mathbf{x})\mathbf{E}_k^2(\mathbf{x},t)\right). \tag{24.2.67}$$

The inequality (24.2.66) follows from (24.2.67) and the relations

$$G_\infty(\mathbf{x}) + \sum_{k=1}^{n} A_k(\mathbf{x}) = G_0(\mathbf{x}), \qquad \sum_{k=1}^{n} \alpha_k^2 A_k(\mathbf{x}) e^{-\alpha_k(s_1+s_2)} = G''(\mathbf{x}, s_1 + s_2).$$

□

Remark 24.2.21. For such models characterized by a scalar relaxation function, we have

$$|G_0(\mathbf{x}) - G_\infty(\mathbf{x})| + |G_\infty(\mathbf{x})| = G_0(\mathbf{x});$$

moreover, (24.2.55) and (24.2.63) become

$$|\mathbf{T}(\mathbf{x},t)|^2 \leq G_0(\mathbf{x})|\mathbf{E}^t(\mathbf{x},\cdot)|^2_M, \quad |\mathbf{T}(\mathbf{x},t)|^2 \leq G_0(\mathbf{x})|\mathbf{E}^t(\mathbf{x},\cdot)|^2_G.$$

The spaces $\mathcal{H}_G$ and $\mathcal{H}_{Dill}$ are larger than $\mathcal{H}_M$, because they, for example, include bounded periodic histories, which are not in $\mathcal{H}_M$. Thus, the domain-of-dependence inequality for the evolution problem (24.2.15)–(24.2.16) can be proved in relation to a larger class of initial data for relaxation functions appropriate to ψ_G (i.e., obeying (10.2.2)) and for (24.2.65), which indeed also obeys (10.2.2). However, if initial histories belong to $\mathcal{G}_0$, the best estimate of speed of propagation is defined in (24.2.58), since ψ_M is the maximum free energy.

24.2.3.6 Hyperbolicity

We recall the definition of hyperbolicity for differential operators of the following type:

$$\begin{aligned} L\mathbf{u}(\mathbf{x},t) &= \mathbf{0} \qquad \forall(\mathbf{x},t) \in \Omega \times (0,\tau_0), \\ \mathbf{u}(\mathbf{x},0) &= \mathbf{u}_0(\mathbf{x}) \ \forall \mathbf{x} \in \Omega. \end{aligned} \tag{24.2.68}$$

Definition 24.2.22. *The operator L is hyperbolic if for every smooth initial datum* $\mathbf{u}_0$ *satisfying* $\mathbf{u}_0(\mathbf{x}) = \mathbf{0}$ *for* $\mathbf{x} \in \Omega \backslash S_r(\mathbf{x}_0)$*, the problem (24.2.68) has a unique smooth solution* $\mathbf{u}(\mathbf{x}, t)$ *having a finite signal speed, i.e., there exists a positive scalar constant* c *such that* $\mathbf{u}$ *at time* t *vanishes outside the set* $S_{r+ct} \cap \Omega$ *[83, 199].*

By substituting the initial data $\mathbf{u}_0$ with the history of $\mathbf{u}$ at time $t = 0$, $\mathbf{u}^{t=0}(\mathbf{x}, s) = \mathbf{u}^0(\mathbf{x}, s)$, this definition can be extended to an integrodifferential system.

Theorem 24.2.23. *Under the hypotheses of Theorem 24.2.18, the problem (24.2.15) and (24.2.16), with sources* $\mathbf{f} = \mathbf{0}$ *in* $\Omega \times (0, \tau_0)$*, has a unique solution* $\mathbf{u}$*, which at time* t *vanishes outside the set* $\Omega \cap S_{r+ct}(\mathbf{x}_0)$*, where* $c = \sqrt{|\mathbb{G}_0 - \mathbb{G}_\infty| + |\mathbb{G}_\infty|}$.

Proof. Let $\mathbf{u}$ be a weak solution of (24.2.15) and (24.2.16) and $\bar{\mathbf{x}} \in \Omega \backslash (S_{2r+ct}(\mathbf{x}_0) \cap \Omega)$. The hypotheses on the source and the initial history together with (24.2.60) yield

$$\int_{\Omega \cap S_r(\bar{\mathbf{x}})} \left[|\dot{\mathbf{u}}(\mathbf{x}, t)|^2 + |\mathbf{E}^t(\mathbf{x}, \cdot)|_M^2 \right] d\mathbf{x} + \int_{\partial\Omega \cap S_r(\bar{\mathbf{x}})} \alpha(\mathbf{x}) |\mathbf{u}(\mathbf{x}, t)|^2 d\mathbf{x} = 0,$$

which, by virtue of (24.2.57), ensures that $\mathbf{u}(\mathbf{x}, t) = \mathbf{0}$ in $\Omega \cap S_r(\bar{\mathbf{x}})$.

Since $\bar{\mathbf{x}}$ is an arbitrary point of $\Omega \backslash (S_{2r+ct}(\mathbf{x}_0) \cap \Omega)$, it follows that $\mathbf{u}(\mathbf{x}, t) = \mathbf{0}$ outside $S_{r+ct}(\mathbf{x}_0) \cap \Omega$. □

We observe that the existence of an upper bound for the propagation speed of disturbances is very important, because in the absence of this upper bound, the hyperbolicity of the integrodifferential problem cannot be obtained as a consequence of the domain-of-dependence inequality.

25

Controllability of Thermoelastic Systems with Memory

25.1 The Controllability Problem: Generalities and Types

The evolution of any material system is described by means of partial differential equations. With a suitable choice of controls, which may be source terms or boundary conditions, we can act on a given state of the material.

If we fix an initial state and a final one for the system, we can look for a control, to be applied at any time $t \in [0, T]$, such that the solution of the differential equations matches both the initial state at time $t = 0$ and the final one at time $t = T$.

Such a study is said to be a *controllability problem* for the system.

Here, we present a class of problems related to thermoelastic systems of memory type dealing with "hyperbolic-like" dynamics. For this purpose some results obtained in the controllability context are reviewed (see, e.g., [271, 272]).

We start by recalling some useful definitions (for more details see, e.g., [323]).

Let H and U be two Hilbert spaces and let $A : H \to H$ and $B : U \to H$ be two linear operators. Consider the linear differential system

$$z_t(t) = Az(t) + Bf(t), \quad z(0) = z^0 \in H, \quad t \geq 0, \tag{25.1.1}$$

where the subscript $_t$ expresses the time derivative $\frac{d}{dt}$. We denote by

$$R(T; z^0) = \{z(T); f \in L^2(0, T; U)\}, \quad T > 0,$$

the set of reachable final states.

Let us list the following types of controllability:

1. *Approximate controllability:* System (25.1.1) is said to be approximately controllable at time T if the set of reachable states $R(T; z^0)$ is dense in H for every $z^0 \in H$.
2. *Exact controllability:* System (25.1.1) is said to be exactly controllable at time T if $R(T; z^0) = H$ for all $z^0 \in H$. That is, system (25.1.1) can be driven from any state to any state belonging to the same space of states as the system evolves.

G. Amendola et al., *Thermodynamics of Materials with Memory*,
https://doi.org/10.1007/978-3-030-80534-0_25

3. *Null controllability:* System (25.1.1) is said to be null controllable at time T if $0 \in R(T; z^0)$ for all $z^0 \in H$. This means that an arbitrary state can be transferred to the null state at time T. Null controllability is a physically interesting notion, since the null state is an equilibrium point for system (25.1.1). In the case of linear systems, null controllability is equivalent to the controllability to trajectories. System (25.1.1) is said to be *controllable to trajectories* at time $T > 0$ if for any initial datum z^0 in a suitable space H, there exists a control function $f \in L^2(0, T; U)$ such that the corresponding solution z of (25.1.1) is defined on $[0, T]$ and satisfies
$$z(T) = \hat{z}(T),$$
where $\hat{z}$ is a solution of (25.1.1) defined on $[0, T]$ and associated with given initial datum $\hat{z}_0$ in the same space H and a given function $\hat{f} \in L^2(0, T; U)$.

Obviously, every exactly controllable system is null and approximately controllable too. When the linear infinite-dimensional system (25.1.1) is time-reversible, null and exact controllability are equivalent notions, as in the finite-dimensional case. In general, the converse is not true.

Let us consider, for instance, the heat equation with distributed control in the domain Ω,

$$\begin{aligned} z_t(x,t) - \Delta z(x,t) &= f(x,t) \quad && \text{in } \Omega \times (0,T), \\ z(x,t) &= 0 && \text{on } \partial\Omega \times (0,T), \\ z(x,0) &= 0 && \text{in } \Omega, \end{aligned}$$

with the control f belonging to $L^2(\Omega \times (0,T))$. As proved, e.g., in [323, Theorem 2.7 (i), p. 214], the set of reachable states satisfies $R(T; 0) = H_0^1(\Omega) \subsetneq L^2(\Omega) = H$.

We are interested in the study of the exact controllability of a thermoelastic system with thermal memory.

As is well known, in the classical linear theory of thermoelasticity, the Fourier law is used to describe heat conduction in a body (see, e.g., [58]). This theory is unable to account for the thermal memory effect that may prevail in some materials, particularly at low temperatures. The heat equation resulting from the Fourier law is of parabolic type and predicts that a thermal disturbance at one point of the body is instantly felt everywhere in the body, as noted at the beginning of Chap. 9. Following that discussion, the Fourier law for the heat flux is replaced by the Gurtin–Pipkin law (9.1.1) [191], which contains the memory effect referred to above. In this case the thermoelastic system is fully hyperbolic (see, e.g., [82, 129, 191]), which, in particular, implies a finite speed of propagation of thermal disturbances, a property that is physically more realistic.

Let Ω be a bounded, open, and connected subset of $\mathbb{R}^n, n \geq 1$, with smooth boundary $\Gamma = \partial\Omega$. Assume $T > 0$ and put $Q = (0,T) \times \Omega, \Sigma = (0,T) \times \Gamma$. The differential system is as follows:*

---

* It is interesting to see how the dynamical equations $(25.1.2)_{1,2}$ emerge from (2.4.12) and a linear approximation to constitutive equations (7.1.21), the latter with heat flux and cold-

$$
\begin{aligned}
\boldsymbol{u}_{tt}(t) - \mu\Delta\boldsymbol{u}(t) - (\mu+\lambda)\nabla\mathrm{div}\,\boldsymbol{u}(t) + \alpha\nabla\theta(t) &= \mathbf{0} && \text{in } Q,\\
\theta_t(t) - (k*\Delta\theta)(t) + \beta\mathrm{div}\,\boldsymbol{u}_t(t) &= 0 && \text{in } Q,\\
\boldsymbol{u}(t) = \boldsymbol{f}(t), \quad (k*\theta)(t) &= g(t) && \text{on } \Sigma,\\
\boldsymbol{u}(0) = \boldsymbol{u}_0, \quad \boldsymbol{u}_t(0) = \boldsymbol{v}_0, \quad \theta(0) &= \theta_0 && \text{in } \Omega,
\end{aligned} \tag{25.1.2}
$$

where $\boldsymbol{u}$ is the *displacement vector*, θ is the *relative temperature*, μ and λ are the *Lamé coefficients* satisfying $\mu > 0$ and $\lambda + \mu > 0, k \in L^1(\mathbb{R}^+)$. The constants $\alpha, \beta > 0$ are coupling parameters depending on the properties of the material. By $k * \theta$ we are denoting the convolution product, that is, (C.3.4) or

$$
(k*\theta)(t) = \int_0^t k(t-\tau)\theta(\tau)d\tau.
$$

Controllability of linear differential systems with memory is a subject that has attracted considerable attention in the literature.

Some results are concerned with models that exhibit hyperbolic characteristics, where the notion of exact controllability is a more natural property for study. In this case, the control time T has to be sufficiently large due to the finite speed of related propagation. For a more careful review of some known results on analogous problems, we refer to [271] and the references therein.

With respect to control systems with memory we may recall the following results. In [234] Leugering proved reachability for a plate equation with memory. Lasiecka established exact controllability with boundary control for a Kirchhoff plate and a viscoelastic Kirchhoff plate with a general memory kernel depending on time and space variables (see [230]). In [220, 221] Kim studied controllability problems for systems with large memory by a unique continuation property, which is proved by adapting an idea of Bardos, Lebeau, and Rauch (see [34]). Barbu and Iannelli studied control for the heat equation with memory (see [33]). In particular, they showed the exact controllability of the one-dimensional linear equation for a sufficiently large interval of time. In [282] Pandolfi considered the Gurtin–Pipkin equation with control under the Dirichlet boundary condition and he proved exact controllability as a consequence of the known exact controllability of the wave equation, making use of cosine operator theory. Recently, in [61], Cavaterra, Lorenzi, and Yamamoto found a Carleman estimate for a hyperbolic integrodifferential equation appearing in the viscoelastic case.

25.2 Exact Controllability Under an Assumption on the Smallness of *k*

Let us denote by $\nu = (\nu_1, \ldots, \nu_n)$ the unit normal on Γ directed toward the exterior of Ω. Let $x^0 \in \mathbb{R}^n$ and

ness gradient interchanged, as discussed after (5.1.8). Equation $(25.1.2)_1$ is a generalization of (2.4.15). The second relation of (25.1.2) also relies on the linear version of (3.3.7) (or (5.1.2)) but neglecting the mechanical work term.

$$\boldsymbol{m}(x) = x - x^0 = (x_1 - x_1^0, \ldots, x_n - x_n^0),$$
$$\Gamma_0 = \{x \in \Gamma; \boldsymbol{m}(x) \cdot \boldsymbol{\nu}(x) > 0\}, \quad \Sigma_0 = (0, T) \times \Gamma_0.$$

In [271] the authors studied the control problem for the system (25.1.2) by means of boundary mechanical and thermal controls $(\boldsymbol{f}, g) \in [L^2(\Sigma_0)]^n \times L^2(\Sigma_0)$, which can be applied also only in a portion Γ_0 of the boundary Γ. In such a case, a "strong" assumption on the memory kernel is required in order to prove the exact controllability of the system. In particular, the size of the memory kernel k, in a suitable norm, has to be sufficiently small.

Hence, we have the following theorem.

Theorem 25.2.1. *Let Ω be an open, bounded, and connected subset in $\mathbb{R}^n$ with boundary Γ of class C^2. Suppose that $k(t) \in H^2(\mathbb{R}^+)$ is a positive function and β is a positive constant. Then there exist $\epsilon > 0$ and $T_0 > 0$ such that for any initial and final states $(\boldsymbol{u}_0, \boldsymbol{v}_0, \theta_0), (\boldsymbol{u}_T, \boldsymbol{v}_T, \theta_T) \in [L^2(\Omega)]^n \times [H^{-1}(\Omega)]^n \times H^{-1}(\Omega)$, there exists a boundary control $(\boldsymbol{f}, g) \in [L^2(\Sigma_0)]^n \times L^2(\Sigma_0)$ such that the solution of the system* (25.1.2) *satisfies*

$$\boldsymbol{u}(0) = \boldsymbol{u}_0, \quad \boldsymbol{u}_t(0) = \boldsymbol{v}_0, \quad \theta(0) = \theta_0,$$
$$\boldsymbol{u}(T) = \boldsymbol{u}_T, \quad \boldsymbol{u}_t(T) = \boldsymbol{v}_T, \quad \theta(T) = \theta_T,$$

for any $T > T_0$, provided that $\|k'\|_{H^1(\mathbb{R}^+)} + \beta < \epsilon$.

The procedure to prove this result follows a direct approach; it can be reduced, in particular, to obtain an inverse inequality for the adjoint system of (25.1.2) (see, e.g., [235]).

Thus, firstly, it is necessary to evaluate the adjoint system of (25.1.2). By applying the transposition method (see, e.g., [236]), we see that it is given by

$$\begin{aligned}
\boldsymbol{\varphi}_{tt}(t) - \mu\Delta\boldsymbol{\varphi}(t) - (\mu + \lambda)\nabla \operatorname{div} \boldsymbol{\varphi}(t) + \beta\nabla\psi_t(t) &= \boldsymbol{0} \quad \text{in } Q,\\
-\psi_t(t) - \int_t^T k(s - t)\Delta\psi(s)\, ds - \alpha \operatorname{div} \boldsymbol{\varphi}(t) &= 0 \quad \text{in } Q,\\
\boldsymbol{\varphi}(t) = \boldsymbol{0}, \quad \psi(t) &= 0 \quad \text{on } \Sigma,\\
\boldsymbol{\varphi}(T) = \boldsymbol{\varphi}_0, \quad \boldsymbol{\varphi}_t(T) = \boldsymbol{\varphi}_1, \quad \psi(T) &= \psi_0 \quad \text{in } \Omega,
\end{aligned} \tag{25.2.1}$$

where we assume

$$(\boldsymbol{\varphi}_0, \boldsymbol{\varphi}_1, \psi_0) \in [H_0^1(\Omega)]^n \times [L^2(\Omega)]^n \times H_0^1(\Omega).$$

Introducing, for later convenience,

$$\boldsymbol{w}(t) = \boldsymbol{\varphi}(T - t), \qquad \eta(t) = \psi(T - t),$$

the system (25.2.1) becomes

$$\begin{aligned}
\boldsymbol{w}_{tt}(t) - \mu\Delta\boldsymbol{w}(t) - (\mu+\lambda)\nabla\operatorname{div}\boldsymbol{w}(t) - \beta\nabla\eta_t(t) = \mathbf{0} \quad & \text{in } Q,\\
\eta_t(t) - \int_0^t k(t-s)\Delta\eta(s)ds - \alpha\operatorname{div}\boldsymbol{w}(t) = 0 \quad & \text{in } Q,\\
\boldsymbol{w}(t) = \mathbf{0}, \quad \eta(t) = 0 \quad & \text{on } \Sigma,\\
\boldsymbol{w}(0) = \boldsymbol{w}_0, \quad \boldsymbol{w}_t(0) = \boldsymbol{w}_1, \quad \eta(0) = \eta_0 \quad & \text{in } \Omega.
\end{aligned} \tag{25.2.2}$$

The energy function is given by

$$\mathcal{E}(t) = \frac{1}{2}\int_\Omega \left[|\boldsymbol{w}_t(t)|^2 + \mu|\nabla\boldsymbol{w}(t)|^2 + (\mu+\lambda)|\operatorname{div}\boldsymbol{w}(t)|^2 + \frac{\beta}{\alpha}|\eta_t(t)|^2 + k_0\frac{\beta}{\alpha}|\nabla\eta(t)|^2 \right] dx,$$

where $k_0 = k(0)$. The following lemmas, relating to direct and inverse inequalities, can be proved using energy estimates and multiplier techniques (for more details, see [271]).

Lemma 25.2.2 (Direct Inequality). *Let Ω be a bounded open set of $\mathbb{R}^n$ with boundary Γ of class C^2. Assume that $k \in H^2(\mathbb{R})$ with $k_0 > 0$ and*

$$(\boldsymbol{w}_0, \boldsymbol{w}_1, \eta_0) \in [H_0^1(\Omega)]^n \times [L^2(\Omega)]^n \times H_0^1(\Omega).$$

Then there exists a positive constant C such that

$$\int_\Sigma \left[\mu\left|\frac{\partial\boldsymbol{w}(t)}{\partial\nu}\right|^2 + (\mu+\lambda)|\operatorname{div}\boldsymbol{w}(t)|^2 + k_0\frac{\beta}{\alpha}\left|\frac{\partial\eta(t)}{\partial\nu}\right|^2 \right] d\Sigma \le C\mathcal{E}(0),$$

for all solutions of the system (25.2.2).

Lemma 25.2.3 (Inverse Inequality). *Let Ω be a bounded open set of $\mathbb{R}^n$ with boundary Γ of class C^2. Assume that $k \in H^2(\mathbb{R})$ and $k_0 > 0$. Then there exist $T_0, \gamma > 0$ such that for any $T > T_0$, the solution of the system* (25.2.2) *satisfies*

$$\int_{\Sigma_0} \left[\mu\left|\frac{\partial\boldsymbol{w}(t)}{\partial\nu}\right|^2 + (\mu+\lambda)|\operatorname{div}\boldsymbol{w}(t)|^2 + k_0\frac{\beta}{\alpha}\left|\frac{\partial\eta(t)}{\partial\nu}\right|^2 \right] d\Sigma \ge \gamma\mathcal{E}(0),$$

provided that $\|k'\|_{H^1(\mathbb{R})} + \beta < \epsilon$, for some ϵ small enough.

Using classical procedures (see, e.g., [235]), one can show that for any $T > T_0$, the system (25.1.2) is exactly boundary controllable (for more details, see [271]).

Remark 25.2.4. In Lemma 25.2.3 the integral term is evaluated on the boundary $\Sigma_0 \subseteq \Sigma$. Then the boundary control functions can act also only in a portion Γ_0 of the boundary Γ.

Remark 25.2.5. The hypothesis on the size of the relaxation function k, required to obtain the inverse inequality in Lemma 25.2.3, is not optimal, and it depends directly on the estimates of the multiplier terms.

25.3 Exact Controllability with No Restriction on the Size of k

We now summarize certain results presented in [272], where exact controllability for the system (25.1.2) is proved, with mechanical and thermal controls on the whole boundary Γ, but where no restriction on the size of the memory kernel k is required. These conditions depend, in particular, on the different techniques applied to find the inverse inequality for the problem. In fact, such an inequality is not obtained by multiplier techniques, as in the previous Sect. 25.2 (see also [271]), but is shown by contradiction, by means of the introduction of a resolvent kernel and the application of a unique continuation property, which is an extension of that already considered by Kim in [220].

Consequently, the following statement holds.

Theorem 25.3.1. *Let Ω be an open, bounded, and connected subset in $\mathbb{R}^n$ with boundary Γ of class C^2. Assume that the memory kernel k satisfies the following conditions:*

$$k \in H^2(\mathbb{R}^+) \cap C^2(\mathbb{R}^+), \quad k_0 = k(0) > 0, \quad \frac{|k'(0)|}{k_0} < \frac{1}{1+\sqrt{\alpha\beta}}. \tag{25.3.1}$$

The coupling parameters α and β are chosen such that

$$\alpha\beta < \frac{\mu+\lambda}{(n-1)^2} \tag{25.3.2}$$

holds when $n \geq 2$. Then there exists $T_0 > 0$ such that for any $T > T_0$ and for any initial and final states $(\boldsymbol{u}_0, \boldsymbol{v}_0, \theta_0)$, $(\boldsymbol{u}_T, \boldsymbol{v}_T, \theta_T) \in [L^2(\Omega)]^n \times [H^{-1}(\Omega)]^n \times H^{-1}(\Omega)$, there exists a boundary control $(\boldsymbol{f}, g) \in [L^2(\Sigma)]^n \times L^2(\Sigma)$ such that the solution of the system (25.1.2) *satisfies*

$$\begin{aligned} \boldsymbol{u}(0) &= \boldsymbol{u}_0,\ \boldsymbol{u}_t(0) = \boldsymbol{v}_0, \qquad \theta(0) = \theta_0, \\ \boldsymbol{u}(T) &= \boldsymbol{u}_T,\ \boldsymbol{u}_t(T) = \boldsymbol{v}_T, \qquad \theta(T) = \theta_T. \end{aligned}$$

The proof of this theorem will be given later on, after a few considerations and some lemmas.

Before finding the direct and the inverse inequalities, we rewrite the adjoint system (25.2.2) by introducing the resolvent kernel of k, denoted by r, and such that

$$k_0 r(t) + (k' * r)(t) = -\frac{k'(t)}{k_0}. \tag{25.3.3}$$

Observe that

$$r_0 = r(0) = -\frac{k'(0)}{k_0^2}, \qquad r'(0) = \frac{[k'(0)]^2}{k_0^3} - \frac{k''(0)}{k_0^2}. \tag{25.3.4}$$

Introducing

$$v(t) = k_0\eta(t) + (k' * \eta)(t) \tag{25.3.5}$$

and taking the convolution product with r (see Remark C.3.1) yields

$$[\eta * (k_0 r + k' * r)](t) = (v * r)(t).$$

Substituting (25.3.3) in this equation and using (25.3.5) gives

$$\eta(t) = \frac{v(t)}{k_0} + (r * v)(t); \tag{25.3.6}$$

hence, by differentiating with respect to t, we have

$$\eta_{tt}(t) = \frac{v_{tt}(t)}{k_0} + r_0 v_t(t) + r'(0)v(t) + (r'' * v)(t). \tag{25.3.7}$$

Differentiating $(25.2.2)_2$ with respect to t, we obtain

$$\eta_{tt}(t) - k_0 \Delta\eta(t) - (k' * \Delta\eta)(t) - \alpha \operatorname{div} \boldsymbol{w}_t(t) = 0.$$

Using (25.3.6), this can be rewritten as

$$\eta_{tt}(t) - \Delta v(t) - \alpha \operatorname{div} \boldsymbol{w}_t(t) = 0. \tag{25.3.8}$$

By virtue of (25.3.6) and (25.3.7), relations $(25.2.2)_1$ and (25.3.8) become

$$\begin{aligned}
\boldsymbol{w}_{tt}(t) - \mu\Delta\boldsymbol{w}(t) - (\mu+\lambda)\nabla \operatorname{div} \boldsymbol{w}(t) - \frac{\beta}{k_0}\nabla v_t(t) &= \boldsymbol{R}(t) && \text{in } Q,\\
v_{tt}(t) - k_0\Delta v(t) - \alpha k_0 \operatorname{div} \boldsymbol{w}_t(t) &= S(t) && \text{in } Q,\\
\boldsymbol{w}(t) = \boldsymbol{0}, \quad v(t) &= 0 && \text{on } \Sigma,\\
\boldsymbol{w}(0) = \boldsymbol{w}_0, \quad \boldsymbol{w}_t(0) = \boldsymbol{w}_1, \quad v(0) &= k_0\eta_0 && \text{in } \Omega,
\end{aligned} \tag{25.3.9}$$

where

$$\begin{aligned}
\boldsymbol{R}(t) &= \beta r_0 \nabla v(t) + \beta(r' * \nabla v)(t),\\
S(t) &= -r_0 k_0 v_t(t) - r'(0)k_0 v(t) - k_0(r'' * v)(t).
\end{aligned}$$

The energy function, related to problem (25.3.9), is

$$\begin{aligned}
\mathcal{E}(t;\boldsymbol{w},v) = \frac{1}{2}\int_\Omega \Big[&|\boldsymbol{w}_t(t)|^2 + \mu|\nabla\boldsymbol{w}(t)|^2 + (\mu+\lambda)|\operatorname{div}\boldsymbol{w}(t)|^2\\
&+ \frac{\beta}{\alpha k_0^2}|v_t(t)|^2 + \frac{\beta}{\alpha k_0}|\nabla v(t)|^2 \Big]\, dx.
\end{aligned} \tag{25.3.10}$$

Firstly, the direct inequality related to problem (25.3.9) can be proved, by means of multiplier techniques (see, e.g., [271]).

Lemma 25.3.2 (Direct Inequality). *Let Ω be a bounded open set of $\mathbb{R}^n$ with boundary Γ of class C^2. Assume that the kernel k satisfies conditions* $(25.3.1)_{1,2}$ *and*

$$(\boldsymbol{w}_0, \boldsymbol{w}_1, v_0) \in [H_0^1(\Omega)]^n \times [L^2(\Omega)]^n \times H_0^1(\Omega).$$

Then there exists a positive constant C such that

$$\int_\Sigma \left[\mu \left| \frac{\partial \boldsymbol{w}(t)}{\partial \nu} \right|^2 + (\mu + \lambda) |\operatorname{div} \boldsymbol{w}(t)|^2 + \frac{\beta}{\alpha k_0} \left| \frac{\partial v(t)}{\partial \nu} \right|^2 \right] d\Sigma \le C\mathcal{E}(0) \tag{25.3.11}$$

for all solutions of the system (25.3.9).

Remark 25.3.3. From the inequality (25.3.11) and (25.3.5), we obtain the following boundary regularity of the solution:

$$\frac{\partial \boldsymbol{w}(t)}{\partial \nu} \in [L^2(\Sigma)]^n, \qquad \frac{\partial \eta(t)}{\partial \nu} \in L^2(\Sigma).$$

Let us now study the system (25.3.9) with $\boldsymbol{R} = \boldsymbol{0}$ and $S = 0$; it can be written as

$$\begin{aligned} \tilde{\boldsymbol{w}}_{tt}(t) - \mu \Delta \tilde{\boldsymbol{w}}(t) - (\mu + \lambda) \nabla \operatorname{div} \tilde{\boldsymbol{w}}(t) - \tilde{\beta} \nabla \tilde{v}_t(t) = \boldsymbol{0} \quad \text{in } Q, \\ \tilde{v}_{tt}(t) - k_0 \Delta \tilde{v}(t) - \tilde{\alpha} \operatorname{div} \tilde{\boldsymbol{w}}_t(t) = 0 \quad \text{in } Q, \end{aligned} \tag{25.3.12}$$

where $\tilde{\alpha} = \alpha k_0, \tilde{\beta} = \frac{\beta}{k_0}$. Obviously, $\alpha\beta = \tilde{\alpha}\tilde{\beta}$. Using the multipliers method (see, e.g., [272]), we can obtain the inverse inequality for such a system, as shown by the following lemma.

Lemma 25.3.4. *Let Ω be a bounded open set of $\mathbb{R}^n$ with boundary Γ of class C^2. Assume that the memory kernel k satisfies conditions* $(25.3.1)_{1,2}$ *and the coupling parameters α and β satisfy condition* (25.3.2)*; if $n = 1$, no assumption is needed on α and β. Then there exists $T_0 > 0$ such that for any $T > T_0$, the solution of the system* (25.3.12) *satisfies*

$$\int_\Sigma \left[\mu \left| \frac{\partial \tilde{\boldsymbol{w}}(t)}{\partial \nu} \right|^2 + (\mu + \lambda) |\operatorname{div} \tilde{\boldsymbol{w}}(t)|^2 + k_0 \frac{\tilde{\beta}}{\tilde{\alpha}} \left| \frac{\partial \tilde{v}(t)}{\partial \nu} \right|^2 \right] d\Sigma \ge \gamma \mathcal{E}(0),$$

where γ is a positive constant depending on T_0.

Remark 25.3.5. The choice $\boldsymbol{R} = \boldsymbol{0}$ and $S = 0$ removes temporarily the presence of the relaxation function r and allows us to prove the inverse inequality for the system (25.3.12) without assumptions on the size of the memory kernel.

Now we return to the system (25.3.9), and by setting

$$\mathcal{L}_1(\boldsymbol{w}(t), v(t)) = \boldsymbol{w}_{tt}(t) - \mu \Delta \boldsymbol{w}(t) - (\mu + \lambda) \nabla \operatorname{div} \boldsymbol{w}(t) - \frac{\beta}{k_0} \nabla v_t(t),$$

$$\mathcal{L}_2(v(t), \boldsymbol{w}(t)) = v_{tt}(t) - k_0 \Delta v(t) - \alpha k_0 \operatorname{div} \boldsymbol{w}_t(t),$$

relations $(25.3.9)_{1,2}$ can be rewritten as

$$\begin{aligned} \mathcal{L}_1(\boldsymbol{w}(t), v(t)) &= \beta r_0 \nabla v(t) + \beta (r' * \nabla v)(t), \\ \mathcal{L}_2(v(t), \boldsymbol{w}(t)) &= -r_0 k_0 v_t(t) - r'(0) k_0 v(t) - k_0 (r'' * v)(t). \end{aligned} \tag{25.3.13}$$

Following a standard procedure, we write the solution $z = (\boldsymbol{w}, v)$ of (25.3.9) as $z = z^L + z^N$, with $z^L = (\boldsymbol{w}^L, v^L)$ and $z^N = (\boldsymbol{w}^N, v^N)$, where z^L solves the following system:

$$\begin{aligned} \mathcal{L}_1(\boldsymbol{w}^L(t), v^L(t)) &= \mathbf{0} && \text{in } Q,\\ \mathcal{L}_2(v^L(t), \boldsymbol{w}^L(t)) &= 0 && \text{in } Q,\\ \boldsymbol{w}^L(t) = \mathbf{0}, \quad v^L(t) &= 0 && \text{on } \Sigma,\\ \boldsymbol{w}^L(0) = \boldsymbol{w}_0, \quad \boldsymbol{w}_t^L(0) &= \boldsymbol{w}_1 && \text{in } \Omega,\\ v^L(0) = k_0\eta_0, \quad v_t^L(0) &= v_1 && \text{in } \Omega, \end{aligned} \tag{25.3.14}$$

and z^N is the solution of

$$\begin{aligned} \mathcal{L}_1(\boldsymbol{w}^N(t), v^N(t)) &= \beta r_0 \nabla v(t) + \beta(r' * \nabla v)(t) && \text{in } Q,\\ \mathcal{L}_2(v^N(t), \boldsymbol{w}^N(t)) &= -r_0 k_0 v_t(t) - r'(0)k_0 v(t) - k_0(r'' * v)(t) && \text{in } Q,\\ \boldsymbol{w}^N(t) = \mathbf{0}, \quad v^N(t) &= 0 && \text{on } \Sigma,\\ \boldsymbol{w}^N(0) = \mathbf{0}, \quad \boldsymbol{w}_t^N(0) &= \mathbf{0} && \text{in } \Omega,\\ v^N(0) = 0, \quad v_t^N(0) &= 0 && \text{in } \Omega. \end{aligned} \tag{25.3.15}$$

Lemma 25.3.6. *Let Ω_0 be an open subset of Ω such that $\overline{\Omega_0} \subset \Omega$. Assume that the memory kernel satisfies conditions* (25.3.1) *and the coupling parameters α and β satisfy the condition* (25.3.2)*; if $n = 1$, no assumption is needed on α and β. Suppose that $\boldsymbol{w} \in C(0,T;[L^2(\Omega)]^n) \cap C^1(0,T;[H^{-1}(\Omega)]^n)$ and $v \in C(0,T;L^2(\Omega)) \cap C^1(0,T;H^{-1}(\Omega))$ satisfy the system* (25.3.13) *in the sense of distributions in $\Omega \times (0,T)$. Suppose that if $\boldsymbol{w}(x,t) = \mathbf{0}$ and $v(x,t) = 0$ in $(\Omega\backslash\Omega_0) \times (0,T)$, then $\boldsymbol{w} \in C(0,T;[H_0^1(\Omega)]^n) \cap C^1(0,T;[L^2(\Omega)]^n)$ and $v \in C(0,T;H_0^1(\Omega)) \cap C^1(0,T;L^2(\Omega))$, provided that $T > T_0$, where T_0 is given in Lemma 25.3.4.*

Proof. Let us denote by $\star$ the convolution taken only in the space variables. We introduce $\boldsymbol{w}_\varepsilon^L = \boldsymbol{\rho}_\varepsilon \star \boldsymbol{w}^L, \boldsymbol{w}_\varepsilon^N = \boldsymbol{\rho}_\varepsilon \star \boldsymbol{w}^N, v_\varepsilon^L = \rho_\varepsilon \star v^L, v_\varepsilon^N = \rho_\varepsilon \star v^N$, where $\boldsymbol{\rho}_\varepsilon, \rho_\varepsilon$ are the Friedrichs mollifiers in $\mathbb{R}^n$. We choose ε small enough that supp $\boldsymbol{w}_\varepsilon$, supp $v_\varepsilon \subset \Omega \times [0,T]$, where $\boldsymbol{w}_\varepsilon = \boldsymbol{w}_\varepsilon^L + \boldsymbol{w}_\varepsilon^N$ and $v_\varepsilon = v_\varepsilon^L + v_\varepsilon^N$. Then $(\boldsymbol{w}_\varepsilon, v_\varepsilon)$ satisfies (25.3.13). Multiplying $(25.3.13)_1$ by $\boldsymbol{w}_{\varepsilon t}$ and $(25.3.13)_2$ by $\frac{\beta}{\alpha k_0^2} v_{\varepsilon t}$, we obtain

$$\begin{aligned} \frac{d}{dt}\mathcal{E}(t; \boldsymbol{w}_\varepsilon, v_\varepsilon) = {} & \frac{\beta k'(0)}{\alpha k_0^3} \int_\Omega |v_{\varepsilon t}|^2 dx - \frac{\beta r'(0)}{\alpha k_0} \int_\Omega v_\varepsilon v_{\varepsilon t}\, dx \\ & - \frac{\beta}{\alpha k_0} \int_\Omega (r'' * v_\varepsilon) v_{\varepsilon t}\, dx + \beta r_0 \int_\Omega \nabla v_\varepsilon \cdot \boldsymbol{w}_{\varepsilon t}\, dx \\ & + \beta \int_\Omega (r' * \nabla v_\varepsilon) \cdot \boldsymbol{w}_{\varepsilon t}\, dx, \end{aligned} \tag{25.3.16}$$

where the energy function $\mathcal{E}$ is defined as in (25.3.10). In particular, the last term in (25.3.16) can be rewritten as

$$\begin{aligned}\int_\Omega (r' * \nabla v_\varepsilon)(t) \cdot \boldsymbol{w}_{\varepsilon t}(t)\,dx &= -\int_\Omega (r' * v_\varepsilon)(t)\,\mathrm{div}\,\boldsymbol{w}_{\varepsilon t}(t)\,dx \\ &= -\frac{d}{dt}\int_\Omega (r' * v_\varepsilon)(t)\,\mathrm{div}\,\boldsymbol{w}_\varepsilon(t)\,dx \\ &\quad + \int_\Omega [r'(0)v_\varepsilon(t) + (r'' * v_\varepsilon)(t)]\,\mathrm{div}\,\boldsymbol{w}_\varepsilon(t)\,dx.\end{aligned}$$

Recalling the systems (25.3.14) and (25.3.15), we have

$$\begin{aligned} \mathcal{L}_1(\boldsymbol{w}_\varepsilon^L(t), v_\varepsilon^L(t)) &= \boldsymbol{0} && \text{in } Q,\\ \mathcal{L}_2(v_\varepsilon^L(t), \boldsymbol{w}_\varepsilon^L(t)) &= 0 && \text{in } Q,\\ \boldsymbol{w}_\varepsilon^L(t) = \boldsymbol{0}, \quad v_\varepsilon^L(t) &= 0 && \text{on } \Sigma,\\ \boldsymbol{w}_\varepsilon^L(0) = \boldsymbol{\rho}_\varepsilon \star \boldsymbol{w}_0, \quad \boldsymbol{w}_{\varepsilon_t}^L(0) &= \boldsymbol{\rho}_\varepsilon \star \boldsymbol{w}_1 && \text{in } \Omega,\\ v_\varepsilon^L(0) = k_0\rho_\varepsilon \star \eta_0, \quad v_{\varepsilon_t}^L(0) &= \rho_\varepsilon \star v_1 && \text{in } \Omega \end{aligned} \tag{25.3.17}$$

and

$$\begin{aligned} \mathcal{L}_1(\boldsymbol{w}_\varepsilon^N(t), v_\varepsilon^N(t)) &= \beta r_0 \nabla v_\varepsilon(t) + \beta(r' * \nabla v_\varepsilon)(t) && \text{in } Q,\\ \mathcal{L}_2(v_\varepsilon^N(t), \boldsymbol{w}_\varepsilon^N(t)) &= -r_0k_0v_{\varepsilon t}(t) - r'(0)k_0v_\varepsilon(t) - k_0(r'' * v_\varepsilon)(t) && \text{in } Q,\\ \boldsymbol{w}_\varepsilon^N(t) = \boldsymbol{0}, \quad v_\varepsilon^N(t) &= 0 && \text{on } \Sigma,\\ \boldsymbol{w}_\varepsilon^N(0) = \boldsymbol{0}, \quad \boldsymbol{w}_{\varepsilon_t}^N(0) &= \boldsymbol{0} && \text{in } \Omega,\\ v_\varepsilon^N(0) = 0, \quad v_{\varepsilon_t}^N(0) &= 0 && \text{in } \Omega. \end{aligned}$$

We consider (25.3.16) for $(\boldsymbol{w}_\varepsilon^N, v_\varepsilon^N)$ and integrate on $[0, t]$ to obtain

$$\begin{aligned} \mathcal{E}(t; \boldsymbol{w}_\varepsilon^N, v_\varepsilon^N) = &\frac{\beta k'(0)}{\alpha k_0^3}\int_0^t\int_\Omega v_{\varepsilon t}(t)v_{\varepsilon_t}^N(t)\,dx\,dt \\ &- \frac{\beta r'(0)}{\alpha k_0}\int_0^t\int_\Omega v_\varepsilon(t)v_{\varepsilon_t}^N(t)\,dx\,dt \\ &- \frac{\beta}{\alpha k_0}\int_0^t\int_\Omega (r'' * v_\varepsilon)(t)v_{\varepsilon_t}^N(t)\,dx\,dt \\ &+ \beta r_0\int_0^t\int_\Omega \nabla v_\varepsilon(t)\cdot \boldsymbol{w}_{\varepsilon_t}^N(t)\,dx\,dt \\ &+ \underbrace{\beta\int_0^t\int_\Omega (r' * \nabla v_\varepsilon)(t)\cdot \boldsymbol{w}_{\varepsilon_t}^N(t)\,dx\,dt}_{=:I_1(t)}. \end{aligned} \tag{25.3.18}$$

The integral I_1 can be rewritten as

$$
\begin{aligned}
I_1(t) &= \int_0^t \frac{d}{dt}\left[\int_\Omega (r' * \nabla v_\varepsilon)(t) \cdot \boldsymbol{w}_\varepsilon{}^N(t)\, dx\right] dt + \\
&\quad - \int_0^t \int_\Omega [r'(0)\nabla v_\varepsilon(t) + (r'' * \nabla v_\varepsilon)(t)] \cdot \boldsymbol{w}_\varepsilon{}^N(t)\, dx\, dt \\
&= \int_\Omega (r' * \nabla v_\varepsilon)(t) \cdot \boldsymbol{w}_\varepsilon{}^N(t)\, dx \\
&\quad - \int_0^t \int_\Omega [r'(0)\nabla v_\varepsilon(t) + (r'' * \nabla v_\varepsilon)(t)] \cdot \boldsymbol{w}_\varepsilon{}^N(t)\, dx\, dt.
\end{aligned}
$$

Defining $\mathcal{M}^N(T) = \sup\left\{\mathcal{E}(t; \boldsymbol{w}_\varepsilon^N, v_\varepsilon^N);\ 0 \le t \le T\right\}$, we find that

$$
\begin{aligned}
|I_1(t)| \le \frac{1}{\sqrt{\mu+\lambda}} &\left[\|r'(t)\|_{L^2} \left(\int_0^T \|v_\varepsilon(t)\|^2\, dt \right)^{1/2} + |r'(0)| \left(\int_0^T \|v_\varepsilon(t)\|^2\, dt \right)^{1/2} \right. \\
&\left. + \left(\int_0^T \int_\Omega |(r'' * v_\varepsilon)(t)|^2 dx\, dt \right)^{1/2} \right] \left[\mathcal{M}^N(T)\right]^{1/2}.
\end{aligned}
$$

From this inequality and considering (25.3.18), it follows that

$$
\begin{aligned}
\mathcal{M}^N(T) \le\ & \frac{|k'(0)|}{k_0} \left[\int_0^T \mathcal{E}(t; \boldsymbol{w}_\varepsilon, v_\varepsilon) dt \right]^{1/2} \left[\mathcal{M}^N(T)\right]^{1/2} \\
& + |r'(0)| \sqrt{\frac{\beta}{\alpha}} \left[\int_0^T \int_\Omega |v_\varepsilon(t)|^2\, dx\, dt \right]^{1/2} \left[\mathcal{M}^N(T)\right]^{1/2} \\
& + \sqrt{\frac{\beta}{\alpha}} \left[\int_0^T \int_\Omega |(r'' * v_\varepsilon)(t)|^2\, dx\, dt \right]^{1/2} \left[\mathcal{M}^N(T)\right]^{1/2} \\
& + |r_0| \sqrt{k_0 \alpha \beta} \left[\int_0^T \mathcal{E}(t; \boldsymbol{w}_\varepsilon, v_\varepsilon)\, dt \right]^{1/2} \left[\mathcal{M}^N(T)\right]^{1/2} \\
& + \frac{\beta}{\sqrt{\mu+\lambda}} \left[\|r'(t)\|_{L^2} \left(\int_0^T \|v_\varepsilon(t)\|^2\, dt \right)^{1/2} + |r'(0)| \left(\int_0^T \|v_\varepsilon(t)\|^2\, dt \right)^{1/2} \right. \\
& \left. + \left(\int_0^T \int_\Omega |(r'' * v_\varepsilon)(t)|^2\, dx\, dt \right)^{1/2} \right] \left[\mathcal{M}^N(T)\right]^{1/2}.
\end{aligned}
$$

From the Young inequality and (25.3.4), we obtain

$$
\begin{aligned}
&\left[1 - \frac{|k'(0)|}{2k_0} \left(1 + \sqrt{\alpha\beta}\right) - \epsilon \right] \mathcal{M}^N(T) \\
&\quad \le \left[\frac{|k'(0)|}{2k_0} + \frac{|k'(0)|}{2k_0^2} \sqrt{\alpha\beta} \right] \int_0^T \mathcal{E}(t; \boldsymbol{w}_\varepsilon, v_\varepsilon)\, dt \\
&\quad + C_\epsilon \left[\int_0^T \int_\Omega |(r'' * v_\varepsilon)(t)|^2\, dx\, dt + \int_0^T \int_\Omega |v_\varepsilon(t)|^2\, dx\, dt \right].
\end{aligned}
$$

By choosing $\frac{|k'(0)|}{2k_0}\left(1+\sqrt{\alpha\beta}\right)$ small enough, namely, for instance,

$$\frac{|k'(0)|}{k_0} < \frac{1}{\left(1+\sqrt{\alpha\beta}\right)},$$

one finds that

$$\begin{aligned}\int_\Sigma \left(\left|\frac{\partial \boldsymbol{w}_\varepsilon^N(t)}{\partial \nu}\right|^2 + \left|\frac{\partial v_\varepsilon^N(t)}{\partial \nu}\right|^2\right) d\Gamma\, dt &\le C\,T\,\mathcal{M}^N(T)\\ &\le C\,T\left[\frac{|k'(0)|}{2k_0} + \frac{|k'(0)|}{2k_0^2}\sqrt{\alpha\beta}\right]\int_0^T \mathcal{E}(t;\boldsymbol{w}_\varepsilon, v_\varepsilon)\,dt \\ &\quad + C\,T\,C_\epsilon\left[\int_0^T\int_\Omega |(r''*v_\varepsilon)(t)|^2\,dx\,dt + \int_0^T\int_\Omega |v_\varepsilon(t)|^2\,dx\,dt\right]\end{aligned} \tag{25.3.19}$$

for any $\varepsilon > 0$. Since $\operatorname{supp}\boldsymbol{w}_\varepsilon, \operatorname{supp} v_\varepsilon \subset \Omega\times[0,T]$, we obtain

$$\frac{\partial \boldsymbol{w}_\varepsilon^L(t)}{\partial \nu} = -\frac{\partial \boldsymbol{w}_\varepsilon^N(t)}{\partial \nu}, \quad \frac{\partial v_\varepsilon^L(t)}{\partial \nu} = -\frac{\partial v_\varepsilon^N(t)}{\partial \nu} \qquad \text{on } \Sigma. \tag{25.3.20}$$

Applying Lemma 25.3.4 to the system (25.3.17) and using (25.3.19) and (25.3.20), we obtain

$$\begin{aligned}\mathcal{E}(0;\boldsymbol{w}_\varepsilon^L, v_\varepsilon^L) &\le \gamma_1 \int_\Sigma \left(\left|\frac{\partial \boldsymbol{w}_\varepsilon^N(t)}{\partial \nu}\right|^2 + \left|\frac{\partial v_\varepsilon^N(t)}{\partial \nu}\right|^2\right) d\Gamma\, dt\\ &\le C_0 T\left[\frac{|k'(0)|}{2k_0} + \frac{|k'(0)|}{2k_0^2}\sqrt{\alpha\beta}\right]\int_0^T \mathcal{E}(t;\boldsymbol{w}_\varepsilon, v_\varepsilon)\,dt \\ &\quad + C_0 T\,C_\epsilon\left[\int_0^T\int_\Omega |(r''*v_\varepsilon)(t)|^2\,dx\,dt + \int_0^T\int_\Omega |v_\varepsilon(t)|^2\,dx\,dt\right],\end{aligned} \tag{25.3.21}$$

where γ_1, c_0 are positive constants depending on T_0. Since

$$\mathcal{E}(0;\boldsymbol{w}_\varepsilon^L, v_\varepsilon^L) = \mathcal{E}(0;\boldsymbol{w}_\varepsilon, v_\varepsilon), \tag{25.3.22}$$

we conclude by virtue of (25.3.16) that

$$\begin{aligned}\mathcal{E}(t;\boldsymbol{w}_\varepsilon, v_\varepsilon) = \mathcal{E}(0;\boldsymbol{w}_\varepsilon, v_\varepsilon) &+ \frac{\beta k'(0)}{\alpha\, k_0^3}\int_0^t\int_\Omega |v_{\varepsilon t}(t)|^2\,dx\,dt\\ &- \frac{\beta\, r'(0)}{\alpha\, k_0}\int_0^t\int_\Omega v_\varepsilon(t) v_{\varepsilon t}(t)\,dx\,dt\\ &- \frac{\beta}{\alpha\, k_0}\int_0^t\int_\Omega (r''*v_\varepsilon)(t) v_{\varepsilon t}(t)\,dx\,dt\\ &+ \beta r_0 \int_0^t\int_\Omega \nabla v_\varepsilon(t)\cdot \boldsymbol{w}_{\varepsilon t}(t)\,dx\,dt\\ &+ \beta\int_0^t\int_\Omega (r'*\nabla v_\varepsilon)(t)\cdot \boldsymbol{w}_{\varepsilon t}(t)\,dx\,dt.\end{aligned}$$

Application of the Gronwall inequality yields

$$\mathcal{E}(t; \boldsymbol{w}_\varepsilon, v_\varepsilon) \leq \mathcal{E}(0; \boldsymbol{w}_\varepsilon, v_\varepsilon)\, c\, e^{ct}.$$

From (25.3.21) and (25.3.22), one obtains

$$\mathcal{E}(t; \boldsymbol{w}_\varepsilon, v_\varepsilon) \leq C_0 T \left[\frac{|k'(0)|}{2\,k_0} + \frac{|k'(0)|}{2\,k_0^2} \sqrt{\alpha\beta} \right] \int_0^T \mathcal{E}(t; \boldsymbol{w}_\varepsilon, v_\varepsilon)\, dt$$
$$+ C_0 T\, C_\epsilon \left[\int_0^T \int_\Omega |(r'' * v_\varepsilon)(t)|^2\, dx\, dt + \int_0^T \int_\Omega |v_\varepsilon(t)|^2\, dx\, dt \right].$$

Defining $\mathcal{M}(T) = \sup\{\mathcal{E}(t; \boldsymbol{w}_\varepsilon, v_\varepsilon);\ 0 \leq t \leq T\}$ and taking $\frac{k'(0)}{k_0}$ small enough, we have

$$\mathcal{M}(T) \leq C_0 T\, C_\epsilon \left[\int_0^T \int_\Omega |(r'' * v_\varepsilon)(t)|^2\, dx\, dt + \int_0^T \int_\Omega |v_\varepsilon(t)|^2\, dx\, dt \right]$$

for any $\varepsilon > 0$. This yields that for $\varepsilon \to 0, \boldsymbol{w} \in C(0, T; [H_0^1(\Omega)]^n) \cap C^1(0, T; [L^2(\Omega)]^n)$ and $v \in C(0, T; H_0^1(\Omega)) \cap C^1(0, T; L^2(\Omega))$. □

Lemma 25.3.7. *Assume that the memory kernel satisfies conditions* (25.3.1) *and the coupling parameters* α, β *satisfy* (25.3.2)*; if* $n = 1$*, no assumption is needed on* α *and* β*. Let* T_0 *be a positive constant, as given in Lemma 25.3.6. For any* $T \geq T_0$*, let*

$$\boldsymbol{w} \in C(0, T; [H_0^1(\Omega)]^n) \cap C^1(0, T; [L^2(\Omega)]^n),$$
$$v \in C(0, T; H_0^1(\Omega)) \cap C^1(0, T; L^2(\Omega))$$

be the solution of the system (25.3.9) *in* Q *such that*

$$\frac{\partial \boldsymbol{w}(t)}{\partial \nu} = \boldsymbol{0}, \quad \frac{\partial v(t)}{\partial \nu} = 0 \qquad \text{on } \Sigma.$$

Then

$$\boldsymbol{w}(t) = \boldsymbol{0}, \quad v(t) = 0 \qquad \text{in } Q.$$

Proof. Since

$$\boldsymbol{w}(t) = \boldsymbol{0}, \quad \frac{\partial \boldsymbol{w}(t)}{\partial \nu} = \boldsymbol{0}, \quad v(t) = 0, \quad \frac{\partial v(t)}{\partial \nu} = 0 \qquad \text{on } \Sigma,$$

we have

$$\nabla \boldsymbol{w}(t) = \boldsymbol{0}, \quad \nabla v(t) = \boldsymbol{0} \qquad \text{on } \Sigma.$$

Let us consider a bounded open subset Ω_1 of $\mathbb{R}^n$ with smooth boundary such that $\overline{\Omega} \subset \Omega_1$ and $T > \operatorname{diam} \Omega_1 > \operatorname{diam} \Omega$. The solution $(\boldsymbol{w}, v)$ can be extended to $\Omega_1 \times [0, T]$ such that $(\boldsymbol{w}, v) \equiv (\boldsymbol{0}, 0)$ in $(\Omega_1 \backslash \Omega) \times [0, T]$ and $(\boldsymbol{w}, v)$ satisfies the system (25.3.13) in the sense of distributions in $\Omega_1 \times (0, T)$.

We introduce the following sets:

$$\begin{aligned}
\mathcal{S}_0 &= C(0,T;[H_0^1(\Omega_1)]^n \times H_0^1(\Omega_1)) \cap C^1(0,T;[L^2(\Omega_1)]^n \times L^2(\Omega_1)),\\
\mathcal{S}_1 &= \{(\boldsymbol{w},v) \in \mathcal{S}_0; (\boldsymbol{w},v) \text{ is a solution of (25.3.13) in } \Omega_1 \times (0,T),\\
&\qquad (\boldsymbol{w},v) \equiv (\boldsymbol{0},0) \text{ in } (\Omega_1 \backslash \Omega) \times [0,T]\},\\
\mathcal{S}_2 &= C(0,T;\ [H_0^1(\Omega_1) \cap H^2(\Omega_1)]^n \times [H_0^1(\Omega_1) \cap H^2(\Omega_1)])\\
&\qquad \cap C^1(0,T;\ [H_0^1(\Omega_1)]^n \times H_0^1(\Omega_1)) \cap C^2(0,T;[L^2(\Omega_1)]^n \times L^2(\Omega_1)).
\end{aligned}$$

In particular, $\mathcal{S}_1$ is a Banach space equipped with the norm of $\mathcal{S}_0$, denoted by $\|\cdot\|_{\mathcal{S}_1}$. In order to prove that $\mathcal{S}_1$ is of finite dimension, we show that

$$\mathcal{H} = \{(\boldsymbol{w},v) \in \mathcal{S}_1;\ \|(\boldsymbol{w},v)\|_{\mathcal{S}_1} \leq 1\}$$

is compact. In fact, by the procedure used in [231], for any $(\boldsymbol{w},v) \in \mathcal{H}$, we apply the regularity result of Lemma 25.3.6 to each $(\partial_h \boldsymbol{w}, \partial_h v)$, $h = 1,\ldots,n$, where

$$\partial_h \boldsymbol{w} = \frac{\partial \boldsymbol{w}}{\partial x_h}, \qquad \partial_h v = \frac{\partial v}{\partial x_h}, \qquad h = 1,\ldots,n.$$

Then $(\partial_h \boldsymbol{w}, \partial_h v) \in \mathcal{S}_1, h = 1,\ldots,n$, and there exists a positive constant C such that

$$\|(\partial_h \boldsymbol{w}, \partial_h v)\|_{\mathcal{S}_1} \leq C, \quad h = 1,\ldots,n.$$

We find that $\mathcal{H}$ is bounded in $C(0,T;[H_0^1(\Omega_1) \cap H^2(\Omega_1)]^n \times [H_0^1(\Omega_1) \cap H^2(\Omega_1)]) \cap C^1(0,T;[H_0^1(\Omega_1)]^n \times H_0^1(\Omega_1))$. It is also bounded in $C^2(0,T;\ [L^2(\Omega_1)]^n \times L^2(\Omega_1))$, by virtue of (25.3.13). Moreover, we have

$$\|(\boldsymbol{w}^\nu, v^\nu)\|_{\mathcal{S}_2} \leq C$$

for any sequence $(\boldsymbol{w}^\nu, v^\nu) \in \mathcal{H}$. But $\mathcal{S}_2 \hookrightarrow \mathcal{S}_1$, and so there exists a strongly convergent subsequence of $(\boldsymbol{w}^\nu, v^\nu)$ in $\mathcal{S}_1$. Hence $\mathcal{H}$ is compact. This implies that $\mathcal{S}_1$ is finite-dimensional. From Lemma 25.3.6, it follows that the operator $\partial_1 = \frac{\partial}{\partial x_1}$ is a linear operator from $\mathcal{S}_1$ into $\mathcal{S}_1$. Since $\mathcal{S}_1$ is finite-dimensional, there exists a basis $\{z_1,\ldots,z_m\}$ of $\mathcal{S}_1$ such that

$$\partial_1 z_h = \sum_{i=1}^{m} \alpha_{hi} z_i, \qquad h = 1,\ldots,m, \qquad \text{in } \Omega_1 \times (0,T)$$

for some constants $\{\alpha_{hi}\}, h, i = 1,\ldots,m$, or equivalently, putting

$$Z = [z_1,\ldots,z_m]^{\mathrm{T}}, \qquad A = \begin{bmatrix} \alpha_{11} & \cdots & \alpha_{1m} \\ \vdots & & \vdots \\ \alpha_{m1} & \cdots & \alpha_{mm} \end{bmatrix},$$

it can be written as

$$\partial_1 Z = AZ \quad \text{in } \Omega_1 \times (0,T),$$

and its solution is $Z = e^{A x_1} Z(0)$. As in [221], we consider any $(x^0, t^0) \in \overline{\Omega} \times (0,T)$, with $x^0 = (x_1^0, \ldots, x_n^0) \in \mathbb{R}^n$. Let $\xi^0 = (x_2^0, \ldots, x_n^0, t^0) \in \mathbb{R}^n$. There exists an $\hat{x}_1^0$ such that $(\hat{x}_1^0, \xi^0) \in (\Omega_1 \backslash \Omega) \times (0,T)$ and the line segment connecting $(\hat{x}_1^0, \xi^0)$ and (x_1^0, ξ^0) is contained in $\Omega_1 \times (0,T)$. Let us introduce two positive constants δ_1 and δ_2 such that

$$I_{\delta_1} \times B_{\delta_2} \subset \Omega_1 \times (0,T),$$

where $I_{\delta_1} = (\hat{x}_1^0 - \delta_1, x_1^0 + \delta_1)$ and $B_{\delta_2} = \{\xi \in \mathbb{R}^n; |\xi - \xi^0| < \delta_2\}$. Furthermore, we can have that $[\hat{x}_1^0 - \delta_1, \hat{x}_1^0] \times B_{\delta_2} \subset (\Omega_1 \backslash \overline{\Omega}) \times (0,T)$. Since $Z = 0$ in $(\Omega_1 \backslash \Omega) \times (0,T)$, one finds that $Z = 0$ in $(\hat{x}_1^0 - \delta_1, \hat{x}_1^0) \times [0,T]$. From the regularity properties of Z in I_{δ_1}, we obtain that $Z = 0$ in I_{δ_1}. This implies that $Z = 0$ in a neighborhood of (x_1^0, ξ^0). Finally, $Z = 0$ in $\Omega_1 \times (0,T)$ and consequently

$$\boldsymbol{w}(t) = \boldsymbol{0} \quad \text{and} \quad v(t) = 0 \quad \text{in } \Omega \times (0,T).$$

□

In the next lemma we show the inverse inequality related to (25.3.13).

Lemma 25.3.8. *Assume that the memory kernel satisfies conditions* (25.3.1) *and the coupling parameters α and β satisfy* (25.3.2)*; if $n = 1$, no assumption is needed on α and β. Let T_0 be a positive constant, as given in Lemma 25.3.6. Then, for any $T > T_0$, there exists a positive constant γ_0 such that the solution $(\boldsymbol{w}, v)$ of the system* (25.3.13) *satisfies*

$$\int_\Sigma \left[\left| \frac{\partial \boldsymbol{w}(t)}{\partial \nu} \right|^2 + \left| \frac{\partial v(t)}{\partial \nu} \right|^2 \right] d\Sigma \geq \gamma_0 \mathcal{E}(0). \tag{25.3.23}$$

Proof. We suppose that (25.3.23) is false. Hence, there exists a sequence

$$(\boldsymbol{w}_0^m, \boldsymbol{w}_1^m, v_0^m, v_1^m) \in [H_0^1(\Omega)]^n \times [L^2(\Omega)]^n \times H_0^1(\Omega) \times L^2(\Omega)$$

such that

$$\begin{aligned} &\|\nabla \boldsymbol{w}_0^m\|^2 + \|\boldsymbol{w}_1^m\|^2 + \|\nabla v_0^m\|^2 + \|v_1^m\|^2 = 1, \quad \text{for all } m \in \mathbb{R}^+, \\ &\int_\Sigma \left[\left| \frac{\partial \boldsymbol{w}^m(t)}{\partial \nu} \right|^2 + \left| \frac{\partial v^m(t)}{\partial \nu} \right|^2 \right] d\Sigma < c\, \mathcal{E}(0), \quad \text{for all } c \in \mathbb{R}^+, \end{aligned} \tag{25.3.24}$$

and as $m \to \infty$,

$$\begin{aligned} &(\boldsymbol{w}_0^m, \boldsymbol{w}_1^m) \rightharpoonup (\boldsymbol{w}_0^\infty, \boldsymbol{w}_1^\infty) \text{ in } [H_0^1(\Omega)]^n \times [L^2(\Omega)]^n, \\ &(v_0^m, v_1^m) \rightharpoonup (v_0^\infty, v_1^\infty) \text{ in } H_0^1(\Omega) \times L^2(\Omega), \\ &(\boldsymbol{w}^m, \boldsymbol{w}_t^m) \overset{*}{\rightharpoonup} (\boldsymbol{w}^\infty, \boldsymbol{w}_t^\infty) \text{ in } L^\infty(0,T; [H_0^1(\Omega)]^n) \times L^\infty(0,T; [L^2(\Omega)]^n), \\ &(v^m, v_t^m) \overset{*}{\rightharpoonup} (v^\infty, v_t^\infty) \text{ in } L^\infty(0,T; H_0^1(\Omega)) \times L^\infty(0,T; L^2(\Omega)), \end{aligned} \tag{25.3.25}$$

where each $(\boldsymbol{w}^m, v^m)$ is a solution of (25.3.13). The inequality $(25.3.24)_2$ implies that

$$\int_{\Sigma}\left[\left|\frac{\partial \boldsymbol{w}^m(t)}{\partial \nu}\right|^2+\left|\frac{\partial v^m(t)}{\partial \nu}\right|^2\right] d\Sigma \to 0, \quad \text{as } m \to \infty.$$

Then, as $m \to \infty$,

$$\begin{aligned} \frac{\partial \boldsymbol{w}^m}{\partial \nu} &\to \frac{\partial \boldsymbol{w}^\infty}{\partial \nu} = \boldsymbol{0} \quad \text{in } [L^2(\Sigma)]^n, \\ \frac{\partial v^m}{\partial \nu} &\to \frac{\partial v^\infty}{\partial \nu} = 0 \quad \text{in } L^2(\Sigma). \end{aligned} \tag{25.3.26}$$

By application of Lemma 25.3.7, we obtain that $\boldsymbol{w}^\infty = \boldsymbol{0}, v^\infty = 0$, and from $(25.3.25)_2$, it follows that

$$v^m \to v^\infty = 0 \quad \text{in } L^1(0,T; L^2(\Omega)). \tag{25.3.27}$$

Now let us write the solution $z^m = (\boldsymbol{w}^m, v^m)$ to (25.3.9) as

$$z^m = z^{mL} + z^{mN}, \tag{25.3.28}$$

with $z^{mL} = (\boldsymbol{w}^{mL}, v^{mL})$ and $z^{mN} = (\boldsymbol{w}^{mN}, v^{mN})$, where z^{mL} solves the system

$$\begin{aligned} \mathcal{L}_1(\boldsymbol{w}^{mL}(t), v^{mL}(t)) &= \boldsymbol{0} && \text{in } Q, \\ \mathcal{L}_2(v^{mL}(t), \boldsymbol{w}^{mL}(t)) &= 0 && \text{in } Q, \\ \boldsymbol{w}^{mL}(t) = \boldsymbol{0}, \quad v^{mL}(t) &= 0 && \text{on } \Sigma, \\ \boldsymbol{w}^{mL}(0) = \boldsymbol{w}_0^{mL}, \quad \boldsymbol{w}_t^{mL}(0) &= \boldsymbol{w}_1^{mL} && \text{in } \Omega, \\ v^{mL}(0) = v_0^{mL}, \quad v_t^{mL}(0) &= v_1^{mL} && \text{in } \Omega, \end{aligned}$$

and z^{mN} is the solution of

$$\begin{aligned} \mathcal{L}_1(\boldsymbol{w}^{mN}(t), v^{mN}(t)) &= \beta r_0 \nabla v^m(t) + \beta(r' * \nabla v^m)(t) && \text{in } Q, \\ \mathcal{L}_2(v^{mN}(t), \boldsymbol{w}^{mN}(t)) &= -r_0 k_0 v_t^m(t) - r'(0) k_0 v^m(t) - k_0(r'' * v^m)(t) && \text{in } Q, \\ \boldsymbol{w}^{mN}(t) = \boldsymbol{0}, \quad v^{mN}(t) &= 0 && \text{on } \Sigma, \\ \boldsymbol{w}^{mN}(0) = \boldsymbol{0}, \quad \boldsymbol{w}_t^{mN}(0) = \boldsymbol{0}, \quad v^{mN}(0) = 0, \quad v_t^{mN}(0) &= 0 && \text{in } \Omega. \end{aligned} \tag{25.3.29}$$

Using (25.3.27) and (25.3.29), we have

$$\boldsymbol{w}^{mN} \to \boldsymbol{w}^{\infty N} = 0 \quad \text{in } L^\infty(0,T; H_0^1(\Omega)).$$

Hence, as $m \to \infty$,

$$\begin{aligned} \frac{\partial \boldsymbol{w}^{mN}}{\partial \nu} &\to \frac{\partial \boldsymbol{w}^{\infty N}}{\partial \nu} = \boldsymbol{0} \quad \text{in } [L^2(\Sigma)]^n, \\ \frac{\partial v^{mN}}{\partial \nu} &\to \frac{\partial v^{\infty N}}{\partial \nu} = 0 \quad \text{in } L^2(\Sigma). \end{aligned} \tag{25.3.30}$$

From (25.3.26), (25.3.28), and (25.3.30), we obtain

$$\frac{\partial \boldsymbol{w}^{mL}}{\partial \nu} \to \frac{\partial \boldsymbol{w}^{\infty L}}{\partial \nu} = \boldsymbol{0} \qquad \text{in } [L^2(\Sigma)]^n,$$
$$\frac{\partial v^{mL}}{\partial \nu} \to \frac{\partial v^{\infty L}}{\partial \nu} = 0 \qquad \text{in } L^2(\Sigma).$$

By Lemma 25.3.6, as $m \to \infty$,

$$\|\nabla \boldsymbol{w}_0^m\|^2 + \|\boldsymbol{w}_1^m\|^2 + \|\nabla v_0^m\|^2 + \|v_1^m\|^2 \to 0,$$

and this contradicts $(25.3.24)_1$. □

Remark 25.3.9. From (25.3.5) it follows that

$$\int_\Sigma \left|\frac{\partial v(t)}{\partial \nu}\right|^2 d\Sigma \le 2\int_\Sigma \left[k_0^2 \left|\frac{\partial \eta(t)}{\partial \nu}\right|^2 + \left|\left(k' * \frac{\partial \eta}{\partial \nu}\right)(t)\right|^2\right] d\Sigma$$
$$\le 2\left(k_0^2 + \|k'\|^2_{L^1(0,T)}\right)\int_\Sigma \left|\frac{\partial \eta(t)}{\partial \nu}\right|^2 d\Sigma.$$

Then, there exists a positive constant γ_1 such that the observability inequality

$$\int_\Sigma \left[\left|\frac{\partial \boldsymbol{w}(t)}{\partial \nu}\right|^2 + \left|\frac{\partial \eta(t)}{\partial \nu}\right|^2\right] d\Sigma \ge \gamma_1 \mathcal{E}(0)$$

holds.

Consider now the system (25.1.2). Taking initial data $(\boldsymbol{u}_0, \boldsymbol{u}_1, \theta_0) \in [L^2(\Omega)]^n \times [H^{-1}(\Omega)]^n \times H^{-1}(\Omega)$ and assuming that $(\boldsymbol{f}, g) \in [L^2(\Sigma)]^n \times L^2(\Sigma)$, we observe that $\boldsymbol{u} \in L^\infty(0,T;[L^2(\Omega)]^n)$, $\theta \in H^{-1}(0,T;L^2(\Omega))$, with $k * \theta \in L^\infty(0,T;L^2(\Omega))$ (see, e.g., [272]).

Then we are able to prove that the system (25.1.2) is exactly controllable.

Proof of Theorem 25.3.1. In order to simplify notation, let us consider the reverse thermoelastic system

$$\begin{aligned} &\boldsymbol{u}_{tt}(t) - \mu\Delta \boldsymbol{u}(t) - (\mu+\lambda)\nabla \operatorname{div} \boldsymbol{u}(t) + \alpha\nabla\theta(t) = \boldsymbol{0} \quad \text{in } Q,\\ &\theta_t(t) - k * \Delta\theta(t) + \beta \operatorname{div} \boldsymbol{u}_t(t) = 0 \qquad \text{in } Q, \end{aligned} \tag{25.3.31}$$

with final conditions

$$\boldsymbol{u}(T) = z_0, \quad \boldsymbol{u}_t(T) = z_1, \quad \theta(T) = \theta_0 \quad \text{in } \Omega \tag{25.3.32}$$

and boundary conditions

$$\boldsymbol{u}(t) = \frac{\partial \varphi(t)}{\partial \nu} \quad \text{on } \Sigma_0, \qquad (k * \theta)(t) = k_0 \frac{\beta}{\alpha}\frac{\partial \psi(t)}{\partial \nu} \quad \text{on } \Sigma. \tag{25.3.33}$$

The system (25.3.9) is reversible. That is, considering

$$\boldsymbol{w}(T) = \boldsymbol{w}_0, \quad \boldsymbol{w}_t(T) = \boldsymbol{w}_1, \quad \eta(T) = \eta_0 \quad \text{in } \Omega,$$

instead of the last line of (25.3.9), we obtain again standard estimates for the solution. Then there exists only one solution to the system (25.2.1) when

$$\varphi(0) = \varphi_0, \quad \varphi_t(0) = \varphi_1, \quad \psi(0) = \psi_0 \quad \text{in} \quad \Omega$$

are assumed instead of the last line of (25.2.1), with

$$(\varphi_0, \varphi_1, \psi_0) \in [H_0^1(\Omega)]^n \times [L^2(\Omega)]^n \times L^2(\Omega).$$

Applying the transposition method to the system (25.3.31) and using (25.2.1), we obtain

$$\begin{aligned} 0 = &\int_\Omega \boldsymbol{u}_t(0) \cdot \varphi_0 dx - \int_\Omega u(0) \cdot [\varphi_1 + \beta \nabla \psi_0]\, dx + \int_\Omega \theta(0)\psi_0\, dx \\ &+ (2\mu + \lambda) \int_\Sigma \boldsymbol{f}(t) \cdot \frac{\partial \varphi(t)}{\partial \nu} d\Sigma + \int_\Sigma g(t) \frac{\partial \psi(t)}{\partial \nu} d\Sigma. \end{aligned}$$

Let us denote by Λ the operator

$$\Lambda(\varphi_0, \varphi_1, \psi_0) = (-\boldsymbol{u}_t(0), \boldsymbol{u}(0), -\theta(0) - \beta \operatorname{div} \boldsymbol{u}(0)).$$

From the above identity we have

$$\begin{aligned} \langle \Lambda(\varphi_0, \varphi_1, \psi_0), (\varphi_0, \varphi_1, \psi_0)\rangle &= (2\mu + \lambda) \int_\Sigma \boldsymbol{f}(t) \cdot \frac{\partial \varphi(t)}{\partial \nu} d\Sigma + \int_\Sigma g(t) \frac{\partial \psi(t)}{\partial \nu}\, d\Sigma \\ &= (2\mu + \lambda) \int_\Sigma \left| \frac{\partial \varphi(t)}{\partial \nu} \right|^2 d\Sigma + k_0 \frac{\beta}{\alpha} \int_\Sigma \left| \frac{\partial \psi(t)}{\partial \nu} \right|^2 d\Sigma. \end{aligned}$$

By virtue of Lemma 25.3.8 and Remark 25.3.9, this implies that Λ is an isomorphism from $[H_0^1(\Omega)]^n \times [L^2(\Omega)]^n \times H_0^1(\Omega)$ onto $[H^{-1}(\Omega)]^n \times [L^2(\Omega)]^n \times H^{-1}(\Omega)$. Therefore, applying the Lax–Milgram lemma, it follows that there exists

$$(\varphi_1, \varphi_0, \psi_0) \in [H_0^1(\Omega)]^n \times [L^2(\Omega)]^n \times H_0^1(\Omega)$$

such that

$$\langle \Lambda(\varphi_0, \varphi_1, \psi_0), (\tilde{\varphi}_0, \tilde{\varphi}_1, \tilde{\psi}_0)\rangle = \int_\Omega \boldsymbol{u}_1 \cdot \tilde{\varphi}_0\, dx - \int_\Omega u_0 \cdot [\tilde{\varphi}_1 + \beta \nabla \tilde{\psi}_0]\, dx + \int_\Omega \theta_0 \tilde{\psi}_0\, dx.$$

This means that for $\boldsymbol{f}$ and g given as in (25.3.33), the solution of (25.3.31) and (25.3.32) also satisfies

$$\boldsymbol{u}(0) = \boldsymbol{u}_0, \quad \boldsymbol{u}_t(0) = \boldsymbol{u}_1, \quad \theta(0) = \theta_0.$$

Hence, the theorem follows. □

26

The Saint-Venant Problem for Viscoelastic Materials

The Saint-Venant problem was analyzed in Sect. 2.4.2 for linear elastic materials. The same problem has been studied also for viscoelastic materials (see, for example, [64, 93]). The theory is developed here in a similar general and systematic manner.

26.1 Problem Description

Let us consider a prismatic cylinder B with plane ends and select a rectangular system of coordinates such that one end lies in the (x_1, x_2)-plane and contains the origin O. Let ∂B be the boundary of B and denote by L the length of the cylinder, so that the lateral boundary surface of the cylinder is $\pi = \partial D_{x_3} \times [0, L]$, where D_{x_3} is the bounded uniform cross-section at distance x_3 from the plane end containing the origin and ∂D_{x_3} is the boundary of this cross-section. We assume that the boundaries of such cross-sections are sufficiently smooth to admit application of the divergence theorem in the plane of the cross-section.

Let the cylinder B consist of a linearly viscoelastic material that is at rest at all times $t < 0$. Thus, the stress–strain relation becomes

$$\mathbf{S}(\mathbf{u}) = \mathbb{G}(0)\mathbf{E} + \int_0^t \mathbb{G}'(t-s)\mathbf{E}(s)\,ds, \tag{26.1.1}$$

where

$$\mathbf{E}(\mathbf{u}) = \frac{1}{2}\left(\nabla \mathbf{u} + \nabla \mathbf{u}^T\right), \qquad \mathbb{G}'(u) = \frac{\partial}{\partial u}\mathbb{G}(u). \tag{26.1.2}$$

In the above relations, $\mathbf{S}(\mathbf{u})$ is the stress tensor* associated with the displacement vector $\mathbf{u}$, $\mathbf{E}(\mathbf{u})$ is the strain tensor, and $\mathbb{G}(\cdot)$ denotes the relaxation tensor. We assume that $\mathbb{G}(\cdot)$ is a symmetric tensor, as given by (8.1.27), and moreover,

$$\mathbb{G} = \mathbb{G}(x_1, x_2, t) \tag{26.1.3}$$

* This is the same as the tensor $\mathbf{T}$, given by (24.2.1), for example; we refer to the observation after (2.4.5) and to the usage in Sect. 2.4.

G. Amendola et al., *Thermodynamics of Materials with Memory*,
https://doi.org/10.1007/978-3-030-80534-0_26

is a smooth function on $D \times [0, \infty)$. We also assume that $\mathbb{G}(0)$ is positive definite in $D \times [0, \infty)$ (Corollory 8.1.8).

Let $\mathbf{s}(\mathbf{u})$ be the surface traction at regular points of ∂B corresponding to the stress field $\mathbf{S}(\mathbf{u})$, defined by (see (1.3.24))

$$\mathbf{s}(\mathbf{u}) = \mathbf{S}(\mathbf{u})\mathbf{n}, \tag{26.1.4}$$

where $\mathbf{n}$ is the outward unit normal vector to ∂B.

We say that a vector field $\mathbf{u}$ is a quasistatic equilibrium displacement field for B if the mapping $\mathbf{u} : [0, \infty) \to C^1\left(\bar{B}\right) \cap C^2(B)$ is continuous and, moreover, we have

$$\operatorname{div} \mathbf{S}(\mathbf{u}) = \mathbf{0} \tag{26.1.5}$$

on B.

The Saint-Venant problem for B amounts to the determination of a quasistatic equilibrium displacement field $\mathbf{u}$ on B, subject to the boundary conditions

$$\begin{gathered} \mathbf{s}(\mathbf{u}) = \mathbf{0} \quad \text{on } \pi, \\ \mathbf{s}(\mathbf{u}) = \mathbf{s}^{(1)} \quad \text{on } D_0, \quad \mathbf{s}(\mathbf{u}) = \mathbf{s}^{(2)} \quad \text{on } D_L, \end{gathered} \tag{26.1.6}$$

for each time $t \in [0, \infty)$. Here $\mathbf{s}^{(1)}$ and $\mathbf{s}^{(2)}$ are functions preassigned on D_0 and D_L, respectively, for each time $t \in [0, \infty)$.

The following are necessary and sufficient conditions for the existence of a solution to this problem.

$$\int_{D_0} \mathbf{s}^{(1)} da + \int_{D_L} \mathbf{s}^{(2)} da = \mathbf{0}, \quad \int_{D_0} \mathbf{r} \times \mathbf{s}^{(1)} da + \int_{D_L} \mathbf{r} \times \mathbf{s}^{(2)} da = \mathbf{0},$$

where $\mathbf{r}$ is the position vector of a point with respect to the origin O. Under suitable smoothness hypotheses on π and on the prescribed tractions, a solution of the Saint-Venant problem exists and is continuous with respect to t on $[0, \infty)$ [134].

By introducing relations (26.1.1) and (26.1.2) into (26.1.4) and (26.1.5), we deduce that the solution $\mathbf{u}$ of the Saint-Venant problem satisfies the boundary value problem $(\mathcal{S})$ defined by the following equations written in terms of components:†

$$\mathcal{S}_i(\mathbf{u}) \equiv \left[G_{ijmn}(0) u_{m,n} + \int_0^t G'_{ijmn}(t-s) u_{m,n}(s) ds \right]_{,j} = 0, \tag{26.1.7}$$

in $B \equiv D \times (0, L)$, with the lateral boundary conditions

$$\mathcal{B}_i(\mathbf{u}) \equiv \left[G_{i\alpha mn}(0) u_{m,n} + \int_0^t G'_{i\alpha mn}(t-s) u_{m,n}(s) ds \right] n_\alpha = 0, \tag{26.1.8}$$

on $\pi \equiv \partial D \times (0, L)$ and the end boundary conditions

$$S_{3i}(\mathbf{u}) = -s_i^{(1)} \quad \text{on } D_0, \qquad S_{3i}(\mathbf{u}) = s_i^{(2)} \quad \text{on } D_L \tag{26.1.9}$$

for each $t \in [0, \infty)$.

† Italic indices take values 1, 2, 3 while Greek indices range over 1, 2.

In what follows we shall denote by $\mathcal{S}$ the set of all solutions of the Saint-Venant problem. Our main aim is to obtain such solutions. This is a three-dimensional problem, so that it is very difficult to obtain explicit expressions for $\mathbf{u}$. But using the Saint-Venant principle (Remark 2.4.11), we can replace the local end conditions (26.1.9) by equivalent equilibrated global conditions. This leads to the relaxed Saint-Venant problem (Definition 2.4.10), the solutions of which can be expressed in terms of plane strain states in the plane of the cross-section. We can approximate the solution of the local Saint-Venant problem by a solution of the relaxed problem. So our first task is to present a method for generating solutions of the relaxed Saint-Venant problem [63].

26.2 A Generalized Plane Strain State

By a generalized plane strain state in the planar region $D \subset \mathbb{R}^2$ we mean the following form of displacements:

$$\mathbf{v} = \mathbf{v}(x_1, x_2, t) \qquad \forall (x_1, x_2) \in D, \ \forall t \in [0, \infty),$$

which are denoted in this plane by $\mathbf{v}$ instead of $\mathbf{u}$. Such a state, in conjunction with the stress–strain-displacement relations, implies that the stress tensor is a function of x_1, x_2, and t, i. e., $\mathbf{T} = \mathbf{T}(x_1, x_2, t)$. (We use $\mathbf{T}$ in the plane strain context rather than $\mathbf{S}$ for later convenience.) In terms of components,

$$T_{ij}(\mathbf{v}) = G_{ijk\beta}(0) v_{k,\beta} + \int_0^t G'_{ijk\beta}(t-s) v_{k,\beta}(s)\, ds. \tag{26.2.1}$$

A vector field $\mathbf{v}$ is an admissible displacement field if $\mathbf{v}$ is continuous with respect to the time variable on $[0, \infty)$ and moreover,

1. $\mathbf{v}$ is independent of x_3,
2. $\mathbf{v} \in C^1\left(\bar{D}\right) \cap C^2(D)$.

The generalized plane strain problem for $D \cup \partial D$ consists in finding an admissible displacement $\mathbf{v}$ satisfying the equations of equilibrium

$$T_{\alpha i,\alpha}(\mathbf{v}) + f_i = 0 \qquad \text{in } D \times [0, \infty) \tag{26.2.2}$$

and the boundary conditions

$$T_{\alpha i}(\mathbf{v}) n_\alpha = p_i \qquad \text{on } \partial D \times [0, \infty), \tag{26.2.3}$$

when the body force $\mathbf{f}(x_1, x_2, t)$ and the boundary traction $\mathbf{p}(x_1, x_2, t)$ are prescribed. Body forces were neglected in the last section but will be required in the plane problem (see (26.3.6) below). By substituting (26.2.1) into (26.2.2) and (26.2.3), we obtain the displacement plane boundary value problem $(\mathcal{P})$ for $D \cup \partial D$, defined by

$$\mathcal{P}_i(\mathbf{v}) \equiv \left[G_{i\alpha k\beta}(0)v_{k,\beta}\right]_{,\alpha} + \int_0^t \left[G'_{i\alpha k\beta}(t-s)v_{k,\beta}(s)\right]_{,\alpha} ds = -f_i \quad \text{in } D,$$

$$\mathcal{T}_i(\mathbf{v}) \equiv \left[G_{i\alpha k\beta}(0)v_{k,\beta} + \int_0^t G'_{i\alpha k\beta}(t-s)v_{k,\beta}(s)\,ds\right]n_\alpha = p_i \qquad \text{on } \partial D,$$

for each $t \in [0, \infty)$.

Necessary and sufficient conditions for the existence of a solution $\mathbf{v}$ of the boundary value problem ($\mathcal{P}$) for $D \cup \partial D$ are

$$\int_D f_i da + \int_{\partial D} p_i ds = 0,$$
$$\int_D \epsilon_{3\alpha\beta} x_\alpha f_\beta da + \int_{\partial D} \epsilon_{3\alpha\beta} x_\alpha p_\beta ds = 0. \tag{26.2.4}$$

Under suitable regularity hypotheses on ∂D and the given data, a solution of the generalized plane strain problem ($\mathcal{P}$) exists for each $t \in [0, \infty)$ [134].

In what follows we shall denote by $\mathcal{P}$ the set of all plane displacement solutions on the cross-section of the cylinder.

26.3 Analysis of the Saint-Venant Problem by Plane Cross-Section Solutions

The problem ($\mathcal{S}$) defined by relations (26.1.7)–(26.1.9) is less tractable. It is therefore important to study the possibility of reducing the system (26.1.7) and the lateral boundary conditions (26.1.8) to a generalized plane strain problem, which is easier to manage.

Thus, we consider this system with lateral boundary conditions on the cross-section $D \cup \partial D$, which is in effect the plane boundary value problem

$$\mathcal{S}_i(\mathbf{u}) = 0 \qquad \text{in } D, \tag{26.3.1}$$

and

$$\mathcal{B}_i(\mathbf{u}) = 0 \qquad \text{on } \partial D, \tag{26.3.2}$$

treating $x_3 \in (0, L)$ and $t \in [0, \infty)$ as parameters. Our task is to understand when the solution $\mathbf{u} \in \mathcal{S}$ of the Saint-Venant problem resides in the set $\mathcal{P}$ of plane displacements associated with the cross-section D of the cylinder. To this end, let us introduce the vector-valued linear functionals $\mathcal{R}$ and $\mathcal{M}$, the components of which are given by

$$\mathcal{R}_i(\mathbf{u}) = \int_D S_{3i}(\mathbf{u})\,da, \quad \mathcal{M}_i(\mathbf{u}) = \int_D \epsilon_{ijk} x_j S_{3k}(\mathbf{u})\,da, \tag{26.3.3}$$

which represent the resultant force and resultant moment about O of the tractions acting on the cross-section D of the cylinder. We can further write

$$\mathcal{M}_\alpha(\mathbf{u}) = \int_D \epsilon_{3\alpha\beta} x_\beta S_{33}(\mathbf{u}) da - x_3 \epsilon_{3\alpha\beta} \mathcal{R}_\beta(\mathbf{u}),$$
$$\mathcal{M}_3(\mathbf{u}) = \int_D \epsilon_{3\alpha\beta} x_\alpha S_{3\beta}(\mathbf{u})\, da. \tag{26.3.4}$$

Let us rewrite the plane boundary value problem (26.3.1) and (26.3.2) in the following form:

$$\mathcal{P}_i(\mathbf{u}) + \left[G_{i\alpha k3}(0) u_{k,3} + \int_0^t G'_{i\alpha k3}(t-s) u_{k,3}(s) ds\right]_{,\alpha} + S_{3i,3}(\mathbf{u}) = 0 \quad \text{in } D,$$
$$\mathcal{T}_i(\mathbf{u}) = -\left[G_{i\alpha k3}(0) u_{k,3} + \int_0^t G'_{i\alpha k3}(t-s) u_{k,3}(s) ds\right] n_\alpha \quad \text{on } \partial D. \tag{26.3.5}$$

Therefore, the boundary value problem (26.3.1) and (26.3.2) can be viewed as a generalized plane strain boundary value problem with the following given data:

$$f_i = \left[G_{i\alpha k3}(0) u_{k,3} + \int_0^t G'_{i\alpha k3}(t-s) u_{k,3}(s)\, ds\right]_{,\alpha} + S_{3i,3}(\mathbf{u}),$$
$$p_i = -\left[G_{i\alpha k3}(0) u_{k,3} + \int_0^t G'_{i\alpha k3}(t-s) u_{k,3}(s)\, ds\right] n_\alpha. \tag{26.3.6}$$

To satisfy the necessary and sufficient conditions (26.2.4) for a solution $\mathbf{u} \in \mathcal{P}$, we must have

$$\int_D S_{3i,3}(\mathbf{u})\, da = 0, \qquad \int_D \epsilon_{3\alpha\beta} x_\alpha S_{3\beta,3}(\mathbf{u})\, da = 0. \tag{26.3.7}$$

Using the hypothesis (26.1.3) it is easy to see that

$$S_{3i,3}(\mathbf{u}) = S_{3i}(\mathbf{u}_{,3}),$$

so that relations (26.3.7) take the form

$$\int_D S_{3i}(\mathbf{u}_{,3}) da = 0, \qquad \int_D \epsilon_{3\alpha\beta} x_\alpha S_{3\beta}(\mathbf{u}_{,3})\, da = 0. \tag{26.3.8}$$

On the other hand, relations (26.3.7) prove that $\mathcal{R}_i(\mathbf{u})$ and $\mathcal{M}_3(\mathbf{u})$, given by (26.3.3), are independent of x_3. Moreover, using the equilibrium equations (26.1.5) and the boundary condition $(26.1.6)_1$, we obtain

$$\begin{aligned}\int_D x_\alpha S_{33}(\mathbf{u}_{,3}) da &= \int_D x_\alpha S_{33,3}(\mathbf{u}) da = -\int_D x_\alpha S_{\rho3,\rho}(\mathbf{u})\, da \\ &= -\int_D \left[x_\alpha S_{\rho3}(\mathbf{u})\right]_{,\rho} da + \int_D S_{3\alpha}(\mathbf{u})\, da \\ &= -\int_{\partial D} x_\alpha s_3(\mathbf{u})\, ds + \mathcal{R}_\alpha(\mathbf{u}) = \mathcal{R}_\alpha(\mathbf{u}),\end{aligned} \tag{26.3.9}$$

and hence it follows from (26.3.4) that the $\mathcal{M}_\alpha(\mathbf{u})$ are independent of x_3. By direct differentiation with respect to x_3, we deduce from (26.3.9) that

$$\int_D x_\alpha S_{33}\mathbf{u}_{,33})\,da = 0. \tag{26.3.10}$$

Relations (26.3.8) and (26.3.10) allow us to distinguish two classes of semi-inverse solutions to the Saint-Venant problem, which can be expressed in terms of a plane displacement.

26.4 Primary Solution Class

Let us denote by C_I the class of solutions of the Saint-Venant problem for which $\mathbf{v} = \mathbf{u}_{,3}$ is a rigid displacement, in other words, the strain tensor generated by $\mathbf{v}$ is zero. From the above analysis it follows that for any $\mathbf{u}^0 \in C_I$ we have $\mathcal{R}_\alpha(\mathbf{u}^0) = 0$, and moreover, $\mathcal{R}_3(\mathbf{u}^0)$ and $\mathcal{M}_i(\mathbf{u}^0)$ are independent of x_3. On the other hand, since $\mathbf{u}^0 \in C_I$, it follows that its components can be expressed as

$$\begin{aligned} u_\alpha^0 &= -\frac{1}{2}a_\alpha(t)x_3^2 - \epsilon_{3\alpha\beta}a_4(t)x_\beta x_3 + w_\alpha(x_1, x_2, t), \\ u_3^0 &= [a_1(t)x_1 + a_2(t)x_2 + a_3(t)]x_3 + w_3(x_1, x_2, t), \end{aligned} \tag{26.4.1}$$

combined with an arbitrary rigid displacement field. Here, $a_s(t)(s = 1, 2, 3, 4)$ are arbitrary functions of t, continuous on $[0, \infty)$, and w_i are the components of an arbitrary plane displacement vector $\mathbf{w}$. The corresponding stress field is

$$\begin{aligned} S_{ij}(\mathbf{u}^0) &= G_{ij33}(0)\big[a_\rho(t)x_\rho + a_3(t)\big] - a_4(t)G_{ij\alpha 3}(0)\epsilon_{3\alpha\beta}x_\beta \\ &+ \int_0^t \big\{G'_{ij33}(t-s)\big[a_\rho(s)x_\rho + a_3(s)\big] - a_4(s)G'_{ij\alpha 3}(t-s)\epsilon_{3\alpha\beta}x_\beta\big\}\,ds + T_{ij}(\mathbf{w}), \end{aligned} \tag{26.4.2}$$

where

$$T_{ij}(\mathbf{w}) = G_{ijk\beta}(0)w_{k,\beta} + \int_0^t G'_{ijk\beta}(t-s)w_{k,\beta}(s)\,ds.$$

The boundary value problem described by relations (26.3.5) becomes

$$\begin{aligned} \mathcal{S}_i(\mathbf{u}^0) &= \mathcal{P}_i(\mathbf{w}) + \left[G_{i\alpha 33}(0)a_\rho(t)x_\rho + \int_0^t G'_{i\alpha 33}(t-s)a_\rho(s)x_\rho ds\right]_{,\alpha} \\ &+ \left[G_{i\alpha 33}(0)a_3(t) + \int_0^t G'_{i\alpha 33}(t-s)a_3(s)\,ds\right]_{,\alpha} \\ &- \left[\epsilon_{3\alpha\beta}G_{i\alpha\rho 3}(0)a_4(t)x_\beta + \int_0^t \epsilon_{3\alpha\beta}G'_{i\alpha\rho 3}(t-s)a_4(s)x_\beta ds\right]_{,\alpha} \\ &= 0 \qquad \text{in } D, \end{aligned}$$

$$\mathcal{B}_i\left(\mathbf{u}^0\right) = \mathcal{T}_i(\mathbf{w}) + \left[G_{i\alpha 33}(0)a_\rho(t)x_\rho + \int_0^t G'_{i\alpha 33}(t-s)a_\rho(s)x_\rho ds\right]n_\alpha$$
$$+\left[G_{i\alpha 33}(0)a_3(t) + \int_0^t G'_{i\alpha 33}(t-s)a_3(s)\,ds\right]n_\alpha \tag{26.4.3}$$
$$-\left[\epsilon_{3\alpha\beta}G_{i\alpha\rho 3}(0)a_4(t)x_\beta + \int_0^t \epsilon_{3\alpha\beta}G'_{i\alpha\rho 3}(t-s)a_4(s)x_\beta ds\right]n_\alpha$$
$$= 0 \qquad \text{on } \partial D,$$

and hence $\mathbf{w}$ is the solution of a plane boundary value problem. The necessary and sufficient conditions for the existence of the solution $\mathbf{w}$ are satisfied for any functions $a_s(t)(s = 1, 2, 3, 4)$.

Let us denote by $\mathbf{w}^{(j)}$ the solution of the boundary value problem (26.4.3) when $a_i = \delta_{ij}, a_4 = 0$ and by $\mathbf{w}^{(4)}$ the solution of the same problem when $a_i = 0, a_4 = 1$. Clearly,

$$\mathbf{w} = \sum_{s=1}^{4} a_s \otimes \mathbf{w}^{(s)},$$

on introducing the notation

$$(f \otimes g)(t) = f(0)g(t) + \int_0^t \dot{f}(t-s)g(s)\,ds.$$

Thus, it follows that $\mathbf{w}^{(s)}$ is the solution of the boundary value problem $\mathcal{P}^{(s)}(s = 1, 2, 3, 4)$ characterized by the equations

$$\mathcal{P}_i\left(\mathbf{w}^{(\beta)}\right) + \left(G_{i\alpha 33}(t)x_\beta\right)_{,\alpha} = 0 \qquad (\beta = 1, 2),$$
$$\mathcal{P}_i\left(\mathbf{w}^{(3)}\right) + G_{i\alpha 33,\alpha}(t) = 0, \tag{26.4.4}$$
$$\mathcal{P}_i\left(\mathbf{w}^{(4)}\right) - \epsilon_{3\rho\beta}\left(G_{i\alpha\rho 3}(t)x_\beta\right)_{,\alpha} = 0 \qquad \text{in } D,$$

and the boundary conditions

$$\mathcal{T}_i\left(\mathbf{w}^{(\beta)}\right) + G_{i\alpha 33}(t)x_\beta n_\alpha = 0 \qquad (\beta = 1, 2),$$
$$\mathcal{T}_i\left(\mathbf{w}^{(3)}\right) + G_{i\alpha 33}(t)n_\alpha = 0, \tag{26.4.5}$$
$$\mathcal{T}_i\left(\mathbf{w}^{(4)}\right) - \epsilon_{3\rho\beta}G_{i\alpha\rho 3}(t)x_\beta n_\alpha = 0 \qquad \text{on } \partial D.$$

In what follows we shall assume that the displacement fields $\mathbf{w}^{(s)}$ are determined by one of the characteristic methods of the plane deformation problem. Then we can write (26.4.1) in the form

$$\mathbf{u}^0 = \sum_{s=1}^{4} a_s \otimes \mathbf{u}^{(s)}, \tag{26.4.6}$$

where

$$\begin{aligned} u_\alpha^{(\beta)} &= -\frac{1}{2}x_3^2\delta_{\alpha\beta} + w_\alpha^{(\beta)}, \quad u_3^{(\beta)} = x_\beta x_3 + w_3^{(\beta)} \quad (\beta = 1, 2), \\ u_\alpha^{(3)} &= w_\alpha^{(3)}, \qquad u_3^{(3)} = x_3 + w_3^{(3)}, \\ u_\alpha^{(4)} &= \epsilon_{3\beta\alpha} x_\beta x_3 + w_\alpha^{(4)}, \qquad u_3^{(4)} = w_3^{(4)}. \end{aligned} \tag{26.4.7}$$

Moreover, by virtue of (26.4.2) and (26.4.6), we can write

$$\mathbf{S}\big(\mathbf{u}^0\big) = \sum_{s=1}^{4} a_s \otimes \mathbf{S}\big(\mathbf{u}^{(s)}\big),$$

where

$$\begin{aligned} S_{ij}\big(\mathbf{u}^{(\alpha)}\big) &= T_{ij}\big(\mathbf{w}^{(\alpha)}\big) + G_{ij33}(t)x_\alpha, \\ S_{ij}\big(\mathbf{u}^{(3)}\big) &= T_{ij}\big(\mathbf{w}^{(3)}\big) + G_{ij33}(t), \\ S_{ij}\big(\mathbf{u}^{(4)}\big) &= T_{ij}\big(\mathbf{w}^{(4)}\big) - \epsilon_{3\rho\beta} G_{ij\rho 3}(t) x_\beta. \end{aligned} \tag{26.4.8}$$

Relations (26.4.4), (26.4.5), and (26.4.8) give

$$\begin{aligned} S_{\alpha i,\alpha}\big(\mathbf{u}^{(s)}\big) &= 0 \qquad \text{in } D, \\ S_{\alpha i}\big(\mathbf{u}^{(s)}\big)n_\alpha &= 0 \quad \text{on } \partial D, \end{aligned}$$

and moreover, we have

$$\begin{aligned} \int_D S_{3\alpha}\big(\mathbf{u}^{(s)}\big)da &= \int_D \Big[S_{3\alpha}\big(\mathbf{u}^{(s)}\big) + x_\alpha S_{\rho 3,\rho}\big(\mathbf{u}^{(s)}\big)\Big]da \\ &= \int_{\partial D} x_\alpha S_{\rho 3}\big(\mathbf{u}^{(s)}\big)n_\rho ds = 0 \qquad (s = 1, 2, 3, 4). \end{aligned} \tag{26.4.9}$$

Thus we conclude that for any $\mathbf{u}^0 \in C_I$,

$$\begin{aligned} &\mathcal{R}_\alpha\big(\mathbf{u}^0\big) = 0, \quad \mathcal{R}_3\big(\mathbf{u}^0\big) = \sum_{s=1}^{4} D_{3s} \otimes a_s, \\ &\mathcal{M}_\alpha\big(\mathbf{u}^0\big) = \sum_{s=1}^{4} \epsilon_{3\alpha\beta} D_{\beta s} \otimes a_s, \quad \mathcal{M}_3\big(\mathbf{u}^0\big) = \sum_{s=1}^{4} D_{4s} \otimes a_s, \end{aligned} \tag{26.4.10}$$

where

$$D_{\beta s} = \int_D x_\beta S_{33}\left(\mathbf{u}^{(s)}\right)da, \quad D_{3s} = \int_D S_{33}\left(\mathbf{u}^{(s)}\right)da,$$
$$D_{4s} = \int_D \epsilon_{3\alpha\beta}x_\alpha S_{3\beta}\left(\mathbf{u}^{(s)}\right)da \qquad (s = 1, 2, 3, 4). \tag{26.4.11}$$

Note that the solutions of the Saint-Venant problem in the class C_I correspond to the end loads

$$s_i^{(1)} = -\sum_{s=1}^{4} S_{3i}\left(\mathbf{u}^{(s)}\right) \otimes a_s, \qquad s_i^{(2)} = \sum_{s=1}^{4} S_{3i}\left(\mathbf{u}^{(s)}\right) \otimes a_s.$$

26.5 Secondary Solution Class

In order to define the secondary solution class, we observe that the relation (26.3.10) is identically satisfied if $\mathbf{u}_{,33}$ is a rigid displacement. Let us denote by $\hat{\mathbf{a}}$ the four-dimensional vector field $(a_1(t), a_2(t), a_3(t), a_4(t))$ and write $\mathbf{u}^0\{\hat{\mathbf{a}}\}$ for the displacement vector $\mathbf{u}^0$ defined by (26.4.6), indicating its dependence on the functions $a_s(t)(s = 1, 2, 3, 4)$.

We introduce the set C_{II} as the class of all solutions $\mathbf{u}^*$ of the Saint-Venant problem for which $\mathcal{R}_i(\mathbf{u}^*)$ and $\mathcal{M}_3(\mathbf{u}^*)$ are independent of x_3 and where $\mathbf{u}^*_{,33}$ is a rigid displacement. Then it follows that $\mathbf{u}^*_{,3} \in C_I$, and hence we can write

$$\mathbf{u}^*_{,3} = \mathbf{u}^0\left\{\hat{\mathbf{b}}\right\}.$$

This yields the following expression for the displacement $\mathbf{u}^*$:

$$\mathbf{u}^* = \int_0^{x_3} \mathbf{u}^0\left\{\hat{\mathbf{b}}\right\}dx_3 + \mathbf{u}^0\{\hat{\mathbf{c}}\} + \mathbf{w}^*(x_1, x_2, t), \tag{26.5.1}$$

where $\mathbf{w}^*$ is a vector field in $\mathcal{P}$ while $\hat{\mathbf{b}}$ and $\hat{\mathbf{c}}$ are arbitrary four-dimensional vector fields, depending only on the time variable $t \in [0, \infty)$.

The stress state corresponding to the displacement described by (26.5.1) is

$$\mathbf{S}(\mathbf{u}^*) = \sum_{s=1}^{4} (c_s + x_3 b_s) \otimes \mathbf{S}\left(\mathbf{u}^{(s)}\right) + \mathbf{k} + \mathbf{T}(\mathbf{w}^*), \tag{26.5.2}$$

where the components of the tensor $\mathbf{k}$ are given by

$$k_{ij} = \sum_{s=1}^{4} G_{ijk3} \otimes b_s \otimes w_k^{(s)}. \tag{26.5.3}$$

Since $\mathcal{R}_i(\mathbf{u}^*)$ and $\mathcal{M}_3(\mathbf{u}^*)$ are independent of x_3, relations (26.4.9), (26.4.11), (26.5.2), and (26.5.3) furnish

$$\sum_{s=1}^{4} D_{3s} \otimes b_s = 0, \qquad \sum_{s=1}^{4} D_{4s} \otimes b_s = 0. \tag{26.5.4}$$

The plane boundary value problem defined by (26.3.5) reduces to the following plane boundary value problem for $\mathbf{w}^*$:

$$\begin{aligned} &\mathcal{P}_i(\mathbf{w}^*) + k_{\alpha i,\alpha} + \sum_{s=1}^{4} b_s \otimes S_{3i}\left(\mathbf{u}^{(s)}\right) = 0 \qquad \text{in } D, \\ &\mathcal{T}_i(\mathbf{w}^*) + k_{\alpha i} n_\alpha = 0 \qquad \text{on } \partial D. \end{aligned} \tag{26.5.5}$$

The necessary and sufficient conditions for the existence of the solution $\mathbf{w}^*$ are satisfied by virtue of (26.4.11) and (26.5.4).

Thus, we can conclude that any $\mathbf{u}^* \in C_{II}$ has the form (26.5.1), where $\hat{\mathbf{b}}$ satisfies the conditions (26.5.4) and $\mathbf{w}^*$ can be obtained from the generalized plane strain problem described by (26.5.5). Moreover, we have

$$\begin{aligned} &\mathcal{R}_\alpha(\mathbf{u}^*) = \sum_{s=1}^{4} D_{\alpha s} \otimes b_s, \; \mathcal{R}_3(\mathbf{u}^*) = \sum_{s=1}^{4} D_{3s} \otimes c_s + \int_D [k_{33} + T_{33}(\mathbf{w}^*)]\, da, \\ &\mathcal{M}_\alpha(\mathbf{u}^*) = \epsilon_{3\alpha\beta} \left\{ \sum_{s=1}^{4} D_{\beta s} \otimes c_s + \int_D x_\beta [k_{33} + T_{33}(\mathbf{w}^*)] da \right\}, \\ &\mathcal{M}_3(\mathbf{u}^*) = \sum_{s=1}^{4} D_{4s} \otimes c_s + \int_D \epsilon_{3\alpha\beta} x_\alpha \left[k_{3\beta} + T_{3\beta}(\mathbf{w}^*)\right] da. \end{aligned} \tag{26.5.6}$$

For such a solution in C_{II} the corresponding end loads are

$$\begin{aligned} s_i^{(1)} &= -\sum_{s=1}^{4} S_{3i}\left(\mathbf{u}^{(s)}\right) \otimes c_s - k_{3i} - T_{3i}(\mathbf{w}^*), \\ s_i^{(2)} &= \sum_{s=1}^{4} S_{3i}\left(\mathbf{u}^{(s)}\right) \otimes (c_s + L b_s) + k_{3i} + T_{3i}(\mathbf{w}^*). \end{aligned}$$

26.6 Solution of the Relaxed Saint-Venant Problem

The relaxed Saint-Venant problem for the viscoelastic cylinder B consists in the determination of a quasistatic equilibrium displacement field $\mathbf{u}$ that satisfies the lateral boundary condition

$$\mathbf{s}(\mathbf{u}) = \mathbf{0} \qquad \text{on } \pi$$

and the global end conditions

$$\mathcal{R}_i(\mathbf{u}) = -R_i(t), \quad \mathcal{M}_i(\mathbf{u}) = -M_i(t) \qquad \text{on } x_3 = 0, \tag{26.6.1}$$

where R_i and M_i are continuous functions preassigned on $[0, \infty)$. Similar conditions have to be assigned on the end located at $x_3 = L$ in such a way that the global equilibrium conditions for the cylinder are satisfied.

Now we proceed to determine a solution of the relaxed Saint-Venant problem. The above analysis is used in order to decompose the relaxed problem into two problems as follows:

(P_1) the *extension-bending-torsion problem* when $R_\alpha = 0$,
(P_2) the *flexure*, when $F_3 = 0$ and $M_i = 0$.

Due to the linearity of the theory, it follows that the sum of the solutions of the above problems (P_1) and (P_2) gives a solution of the general Saint-Venant problem, so that

$$\mathbf{u} = \mathbf{u}^I + \mathbf{u}^{II}.$$

Let us first determine a solution $\mathbf{u}^I$ of the extension-bending-torsion problem (P_1). In view of the previous analyses we see that a solution has the form

$$\mathbf{u}^I = \mathbf{u}^0 = \sum_{s=1}^{4} a_s \otimes \mathbf{u}^{(s)}. \tag{26.6.2}$$

Then (26.4.10) and (26.6.1) give the integral system

$$\sum_{s=1}^{4} D_{\alpha s} \otimes a_s = \epsilon_{3\alpha\beta} M_\beta, \quad \sum_{s=1}^{4} D_{3s} \otimes a_s = -R_3, \quad \sum_{s=1}^{4} D_{4s} \otimes a_s = -M_3 \tag{26.6.3}$$

for the determination of the unknown functions $a_s(t)(s = 1, 2, 3, 4)$.

Let us denote by $\mathcal{D}(t)$ the 4×4 matrix whose components are $D_{rs}(t)(r, s = 1, 2, 3, 4)$. We set $K_I(R_3, M_1, M_2, M_3) = (-M_2, M_1, -R_3, -M_3)^T$ and $\hat{a} = \hat{\mathbf{a}}^T$, so that the integral system (26.6.3) can be written in the matrix form

$$\mathcal{D}(0)\hat{a}(t) + \int_0^t \dot{\mathcal{D}}(t-s)\hat{a}(s)\,ds = K_I(t). \tag{26.6.4}$$

Observe that $\mathbf{S}\left(\mathbf{u}^{(s)}\right)$ at $t = 0$ coincides with the stress in the auxiliary problems for an elastic material, characterized by the positive definite elasticity tensor $\mathbb{G}(0)$. Therefore, we see that $\mathcal{D}(0)$ coincides with the corresponding matrix for an elastic material with the elastic tensor $\mathbb{G}(0)$ and hence is invertible [206]. Thus, we can write (26.6.4) in the form

$$\hat{a}(t) + \int_0^t [\mathcal{D}(0)]^{-1}\dot{\mathcal{D}}(t-s)\hat{a}(s)\,ds = [\mathcal{D}(0)]^{-1}K_I(t). \tag{26.6.5}$$

Since $R_3(t)$ and $M_i(t)$ are continuous functions on $[0, \infty)$, the Volterra integral equation (26.6.5) has one and only one solution $\hat{a}(t)$, continuous on $[0, \infty)$, which can be obtained by the method of successive approximations [77].

Therefore, the solution of the problem (P_1) is given by (26.6.2), where $\mathbf{u}^{(s)}$ is defined by (26.4.7) and the unknown functions $a_s(t)$ are determined by solving (26.6.5).

Let us now proceed to determine a solution of the problem (P_2). In view of the above discussion, we seek a solution in the class C_{II}, that is, we assume that

$$\mathbf{u}^{II} = \mathbf{u}^*,$$

where

$$u_\alpha^* = -\frac{1}{6}b_\alpha(t)x_3^3 - \frac{1}{2}c_\alpha(t)x_3^2 - \frac{1}{2}b_4(t)\epsilon_{3\alpha\beta}x_\beta x_3^2$$
$$- c_4(t)\epsilon_{3\alpha\beta}x_\beta x_3 + \sum_{s=1}^{4}(c_s + x_3 b_s) \otimes w_\alpha^{(s)} + w_\alpha^*,$$
$$u_3^* = \frac{1}{2}\Big[b_\rho(t)x_\rho + b_3(t)\Big]x_3^2 + \Big[c_\rho(t)x_\rho + c_3(t)\Big]x_3 + \sum_{s=1}^{4}(c_s + x_3 b_s) \otimes w_3^{(s)} + w_3^*,$$
(26.6.6)

and the unknown functions $b_s(t)$ satisfy the conditions (26.5.4).

From (26.5.6) and (26.6.1), we obtain

$$\sum_{s=1}^{4} D_{\alpha s} \otimes b_s = -R_\alpha, \tag{26.6.7}$$

so that the integral systems (26.5.4) and (26.6.7) furnish the functions $b_s(t)(s = 1, 2, 3, 4)$. Therefore, we can assume that the functions $b_s(t)$ are known. Then the vector $\mathbf{w}^*$ can be determined from the generalized plane strain problem described by (26.5.5). Further, from (26.5.6) and (26.6.1), we obtain the following integral system for the determination of the unknown functions $c_s(t)(s = 1, 2, 3, 4)$:

$$\sum_{s=1}^{4} D_{\alpha s} \otimes c_s = -\int_D x_\alpha[k_{33} + T_{33}(\mathbf{w}^*)]da,$$
$$\sum_{s=1}^{4} D_{3s} \otimes c_s = -\int_D [k_{33} + T_{33}(\mathbf{w}^*)]da, \tag{26.6.8}$$
$$\sum_{s=1}^{4} D_{4s} \otimes c_s = -\int_D \epsilon_{3\alpha\beta}x_\alpha\Big[k_{3\beta} + T_{3\beta}(\mathbf{w}^*)\Big]da.$$

Thus, a solution of the flexure problem is given by (26.6.6), where the unknown functions $b_s(t)$ and $c_s(t)$ are determined by means of the Volterra integral equations defined by (26.5.4), (26.6.7), and (26.6.8), respectively, while the vector field $\mathbf{w}^*$ is determined as a solution of the generalized plane strain problem described by the relations (26.5.5).

26.7 The Saint-Venant Problem for an Isotropic and Homogeneous Cylinder

We now consider an isotropic and homogeneous cylinder characterized by the relaxation tensor (see (2.4.13) and (13.1.30))

$$G_{ijkl}(t) = \lambda(t)\delta_{ij}\delta_{kl} + \mu(t)\left(\delta_{ik}\delta_{jl} + \delta_{il}\delta_{jk}\right),$$

where $\lambda(t)$ and $\mu(t)$ are the relaxation functions, assumed to be continuously differentiable on $[0, \infty)$.

It is a simple matter to verify that the solutions $\mathbf{w}^{(s)}(s = 1, 2, 3, 4)$ of the boundary value problems $\mathcal{P}^{(s)}(s = 1, 2, 3, 4)$ are

$$\begin{aligned}
w_1^{(1)} &= -\frac{1}{4}\lambda \otimes (\lambda + \mu)^{-1}\left(x_1^2 - x_2^2\right),\\
w_2^{(1)} &= -\frac{1}{2}\lambda \otimes (\lambda + \mu)^{-1} x_1 x_2, \quad w_3^{(1)} = 0,\\
w_1^{(2)} &= -\frac{1}{2}\lambda \otimes (\lambda + \mu)^{-1} x_1 x_2,\\
w_2^{(2)} &= \frac{1}{4}\lambda \otimes (\lambda + \mu)^{-1}\left(x_1^2 - x_2^2\right), \quad w_3^{(2)} = 0,\\
w_1^{(3)} &= -\frac{1}{2}\lambda \otimes (\lambda + \mu)^{-1} x_1, \quad w_2^{(3)} = -\frac{1}{2}\lambda \otimes (\lambda + \mu)^{-1} x_2,\\
w_3^{(3)} &= 0, \quad w_1^{(4)} = 0, \quad w_2^{(4)} = 0, \quad w_3^{(4)} = \varphi(x_1, x_2, t),
\end{aligned} \tag{26.7.1}$$

where $\varphi(x_1, x_2, x_3)$ is the solution of the following boundary value problem:

$$\begin{aligned}
(\mu \otimes \varphi_{,\alpha})_{,\alpha} &= \epsilon_{3\alpha\beta}\left(\mu x_\beta\right)_{,\alpha} && \text{in } D,\\
\mu \otimes \varphi_{,\alpha} n_\alpha &= \mu \epsilon_{3\alpha\beta} x_\beta n_\alpha && \text{on } \partial D.
\end{aligned} \tag{26.7.2}$$

If we choose the origin of the coordinate system to be at the mass center of the cross-section D_0, then the components D_{rs} of the matrix $\mathcal{D}(t)$ are

$$\begin{aligned}
D_{\alpha\beta} &= \left[(\lambda + 2\mu) - \lambda \otimes (\lambda + \mu)^{-1} \otimes \lambda\right] I_{\alpha\beta}, \quad D_{\alpha 3} = D_{3\alpha} = 0,\\
D_{33} &= \left[(\lambda + 2\mu) - \lambda \otimes (\lambda + \mu)^{-1} \otimes \lambda\right] A, \quad D_{4\alpha} = D_{\alpha 4} = 0,\\
D_{34} &= D_{43} = 0, \quad D_{44} = \int_D \mu \otimes \left(\epsilon_{3\alpha\beta} x_\alpha \varphi_{,\beta} + x_1^2 + x_2^2\right) da,
\end{aligned} \tag{26.7.3}$$

where

$$I_{\alpha\beta} = \int_D x_\alpha x_\beta da, \quad A = \int_D da.$$

Moreover, if we choose the axes Ox_1 and Ox_2 of the coordinate system such that they coincide with the principal axes of inertia of the cross-section D_0, then we see that the matrix $\mathcal{D}(t)$ is diagonal, and hence the extension-bending-torsion problem decomposes into four problems, which can be treated separately. In fact, the solution of the extension-bending-torsion problem is

$$\begin{aligned}
u_1 &= -\frac{1}{2}a_1 x_3^2 - a_4 x_2 x_3 + \sum_{i=1}^{3} a_i \otimes w_1^{(i)},\\
u_2 &= -\frac{1}{2}a_2 x_3^2 + a_4 x_1 x_3 + \sum_{i=1}^{3} a_i \otimes w_2^{(i)},\\
u_3 &= (a_1 x_1 + a_2 x_2 + a_3) x_3 + a_4 \varphi(x_1, x_2, t),
\end{aligned} \tag{26.7.4}$$

where the $w_\alpha^{(i)}$ are given by (26.7.1), φ is the solution of the boundary value problem (26.7.2), and $a_s(t)(s = 1, 2, 3, 4)$ are the solutions of the following Volterra integral equations:

$$D_{11} \otimes a_1 = -M_2(t), \qquad D_{22} \otimes a_2 = M_1(t),$$
$$D_{33} \otimes a_3 = -R_3(t), \qquad D_{44} \otimes a_4 = -M_3(t).$$

The stress components corresponding to the displacements (26.7.4) are

$$S_{\alpha\beta}(\mathbf{u}) = \lambda \otimes (a_1 x_1 + a_2 x_2 + a_3)\delta_{\alpha\beta} + \sum_{i=1}^{3} a_i \otimes T_{\alpha\beta}\left(\mathbf{w}^{(i)}\right),$$
$$S_{\alpha 3}(\mathbf{u}) = S_{3\alpha}(\mathbf{u}) = \mu \otimes \left(a_4 \otimes \varphi_{,\alpha} - \epsilon_{3\alpha\beta} x_\beta a_4\right),$$
$$S_{33}(\mathbf{u}) = (\lambda + 2\mu) \otimes (a_1 x_1 + a_2 x_2 + a_3) + \sum_{i=1}^{3} a_i \otimes T_{33}\left(\mathbf{w}^{(i)}\right).$$

Let us now consider the flexure problem. In view of (26.5.4) and (26.7.3) we deduce that $b_3 = b_4 = 0$, while (26.6.7) gives

$$D_{11} \otimes b_1 = -R_1, \qquad D_{22} \otimes b_2 = -R_2.$$

Furthermore, we obtain that $k_{\alpha\beta} = 0$, $S_{3\alpha}\left(\mathbf{u}^{(\beta)}\right) = 0$, and hence the plane boundary value problem described by (26.5.5) implies that $w_1^* = w_2^* = 0$. This further implies that $k_{33} = 0$ and $T_{33}(\mathbf{w}^*) = 0$; hence, (26.6.8) yields that $c_1(t) = c_2(t) = c_3(t) = 0$. Therefore, the solution of the flexure problem takes the form

$$u_\alpha = -\frac{1}{6} b_\alpha(t) x_3^3 - c_4(t)\epsilon_{3\alpha\beta} x_\beta x_3 + x_3 \left[b_1 \otimes w_\alpha^{(1)} + b_2 \otimes w_\alpha^{(2)}\right],$$
$$u_3 = \frac{1}{2}[b_1(t) x_1 + b_2(t) x_2] x_3^2 + c_4 \otimes \varphi + \psi(x_1, x_2, t),$$

where $\psi(x_1, x_2, t)$ is the solution of the boundary value problem

$$(\mu \otimes \psi_{,\alpha})_{,\alpha} + \sum_{\alpha=1}^{2} b_\alpha \otimes \left[\left(\mu \otimes w_\rho^{(\alpha)}\right)_{,\rho} + T_{33}\left(\mathbf{w}^{(\alpha)}\right)\right] + (\lambda + 2\mu) \otimes b_\alpha x_\alpha = 0 \quad \text{in } D,$$
$$\mu \otimes \psi_{,\alpha} n_\alpha = -n_\alpha \sum_{\rho=1}^{2} b_\rho \otimes w_\alpha^{(\rho)} \otimes \mu \quad \text{on } \partial D.$$

Moreover, the integral system (26.6.8) gives that the function $c_4(t)$ is the solution of the following Volterra integral equation:

$$D_{44} \otimes c_4 = -\int_D \epsilon_{3\alpha\beta} x_\alpha \left[k_{3\beta} + T_{3\beta}(\mathbf{w}^*)\right] da.$$

26.8 The Saint-Venant Principle

We now establish a spatial decay estimate that demonstrates that the solution of the Saint-Venant problem, with prescribed pointwise tractions on the ends, can be approximated by any solution of the relaxed Saint-Venant problem with equivalent global loads on the ends.

We assume that $\mathbb{G}(t)$ is positive definite and $-\mathbb{G}'(t)$ is positive semidefinite in $D \times [0, \infty)$. More precisely, it is assumed that there exist two constants $\mu_m > 0$ and $\mu_M > 0$ such that

$$\mu_m |\xi|^2 \leq \xi \cdot \mathbb{G}(t)\xi \leq \mu_M |\xi|^2 \tag{26.8.1}$$

and

$$\xi \cdot \mathbb{G}'(t)\xi \leq 0, \tag{26.8.2}$$

for every symmetric tensor ξ.

From these assumptions, we can prove for the stress tensor $\mathbf{S}$, corresponding to the strain tensor $\mathbf{E}$, the following estimate:

$$\int_0^t |\mathbf{S}(\tau)|^2 d\tau \leq 4\mu_M \int_0^t |\mathbf{E}(\tau)|^2 d\tau \quad \text{for all } t \geq 0. \tag{26.8.3}$$

In fact, it follows from the constitutive relation (26.1.1) that

$$\int_0^t |\mathbf{S}(\tau)|^2 d\tau = \int_0^t \mathbf{S}(\tau) \cdot \mathbb{G}(0)\mathbf{E}(\tau)d\tau + \int_0^t \int_0^s \mathbf{S}(s) \cdot \mathbb{G}'(s-\tau)\mathbf{E}(\tau)\, d\tau ds \tag{26.8.4}$$

and

$$\begin{aligned}
&\int_0^t \mathbf{S}(\tau) \cdot \mathbb{G}(0)\mathbf{E}(\tau)d\tau \\
&\quad = \frac{1}{2\varepsilon^2} \int_0^t \mathbf{S}(\tau) \cdot \mathbb{G}(0)\mathbf{S}(\tau)d\tau \\
&\quad + \frac{\varepsilon^2}{2} \int_0^t \mathbf{E}(\tau) \cdot \mathbb{G}(0)\mathbf{E}(\tau)d\tau \\
&\quad - \frac{1}{2} \int_0^t \left[\frac{1}{\varepsilon}\mathbf{S}(\tau) - \varepsilon\mathbf{E}(\tau)\right] \cdot \mathbb{G}(0) \left[\frac{1}{\varepsilon}\mathbf{S}(\tau) - \varepsilon\mathbf{E}(\tau)\right] d\tau, \\
&\int_0^t \int_0^s \mathbf{S}(s)\mathbb{G}'(s-\tau)\mathbf{E}(\tau)\, d\tau ds \\
&\quad = -\frac{1}{2\varepsilon^2} \int_0^t \int_0^s \mathbf{S}(s) \cdot \mathbb{G}'(s-\tau)\mathbf{S}(s)\, d\tau ds \\
&\quad - \frac{\varepsilon^2}{2} \int_0^t \int_0^s \mathbf{E}(\tau) \cdot \mathbb{G}'(s-\tau)\mathbf{E}(\tau)\, d\tau ds \\
&\quad + \frac{1}{2} \int_0^t \int_0^s \left[\frac{1}{\varepsilon}\mathbf{S}(s) + \varepsilon\mathbf{E}(\tau)\right] \cdot \mathbb{G}'(s-\tau) \left[\frac{1}{\varepsilon}\mathbf{S}(s) + \varepsilon\mathbf{E}(\tau)\right] d\tau ds,
\end{aligned} \tag{26.8.5}$$

with ε a positive parameter at our disposal. Furthermore,

$$-\int_0^t \int_0^s \mathbf{S}(s) \cdot \mathbb{G}'(s-\tau)\mathbf{S}(s)\, d\tau ds = \int_0^t \mathbf{S}(\tau) \cdot [\mathbb{G}(0) - \mathbb{G}(\tau)]\mathbf{S}(\tau) d\tau. \quad (26.8.6)$$

Interchanging the order of integration yields

$$\begin{aligned}\int_0^t \int_0^s \mathbf{E}(\tau) \cdot \mathbb{G}'(s-\tau)\mathbf{E}(\tau)\, d\tau ds &= \int_0^t \int_\tau^t \mathbf{E}(\tau) \cdot \mathbb{G}'(s-\tau)\mathbf{E}(\tau)\, ds d\tau \\ &= \int_0^t \mathbf{E}(\tau) \cdot [\mathbb{G}(t-\tau) - \mathbb{G}(0)]\mathbf{E}(\tau)\, d\tau.\end{aligned} \quad (26.8.7)$$

Finally, we substitute (26.8.5)–(26.8.7) into (26.8.4) to obtain

$$\begin{aligned}\int_0^t |\mathbf{S}(\tau)|^2 d\tau &= \frac{1}{2\varepsilon^2} \int_0^t \mathbf{S}(\tau) \cdot [2\mathbb{G}(0) - \mathbb{G}(\tau)]\mathbf{S}(\tau)\, d\tau \\ &+ \frac{\varepsilon^2}{2} \int_0^t \mathbf{E}(\tau) \cdot [2\mathbb{G}(0) - \mathbb{G}(t-\tau)]\mathbf{E}(\tau)\, d\tau \\ &- \frac{1}{2} \int_0^t \left[\frac{1}{\varepsilon}\mathbf{S}(\tau) - \varepsilon\mathbf{E}(\tau)\right] \cdot \mathbb{G}(0) \left[\frac{1}{\varepsilon}\mathbf{S}(\tau) - \varepsilon\mathbf{E}(\tau)\right] d\tau \\ &+ \frac{1}{2} \int_0^t \int_0^s \left[\frac{1}{\varepsilon}\mathbf{S}(s) + \varepsilon\mathbf{E}(\tau)\right] \cdot \mathbb{G}'(s-\tau) \left[\frac{1}{\varepsilon}\mathbf{S}(s) + \varepsilon\mathbf{E}(\tau)\right] d\tau ds.\end{aligned}$$

Using the hypotheses described by (26.8.1) and (26.8.2) and taking $\varepsilon = \sqrt{2\mu_M}$, we deduce the estimate (26.8.3).

The estimate

$$\int_0^t \mathbf{S}(\tau) \cdot \mathbf{E}(\tau) d\tau \geq \mu_m \int_0^t |\mathbf{E}(\tau)|^2 d\tau \qquad \forall t \geq 0 \quad (26.8.8)$$

also needs to be established. Relation (26.1.1) gives

$$\begin{aligned}\int_0^t \mathbf{S}(\tau) \cdot \mathbf{E}(\tau)\, d\tau &= \int_0^t \mathbf{E}(\tau) \cdot \mathbb{G}(0)\mathbf{E}(\tau) d\tau \\ &- \frac{1}{2} \int_0^t \int_0^\tau [\mathbf{E}(\tau) - \mathbf{E}(s)] \cdot \mathbb{G}'(\tau - s)[\mathbf{E}(\tau) - \mathbf{E}(s)]\, d\tau ds \\ &+ \frac{1}{2} \int_0^t \int_0^\tau [\mathbf{E}(\tau) \cdot \mathbb{G}'(\tau - s)\mathbf{E}(\tau) + \mathbf{E}(s) \cdot \mathbb{G}'(\tau - s)\mathbf{E}(s)]\, d\tau ds,\end{aligned}$$

so that by means of assumption (26.8.2), followed by some integrations by parts, we obtain

$$\int_0^t \mathbf{S}(\tau) \cdot \mathbf{E}(\tau)\, d\tau \geq \frac{1}{2} \int_0^t \mathbf{E}(s) \cdot [\mathbb{G}(s) + \mathbb{G}(t-s)]\mathbf{E}(s)\, ds. \quad (26.8.9)$$

Thus, (26.8.8) follows from (26.8.9) and (26.8.1).

In what follows, we consider a global-static equilibrium displacement vector field $\mathbf{u} \in \mathcal{S}$ for B as a solution of the Saint-Venant problem corresponding to the pointwise tractions $\mathbf{s}^{(1)}$ and $\mathbf{s}^{(2)}$ preassigned on D_0 and D_L, respectively.

Let $\tilde{\mathbf{u}}$ be a solution of the relaxed Saint-Venant problem corresponding to the global loads equivalent to

$$\mathbf{R} = \int_{D_0} \mathbf{s}^{(1)} da, \qquad \mathbf{M} = \int_{D_0} \mathbf{r} \times \mathbf{s}^{(1)} da.$$

It is a simple matter to see that the difference

$$\mathbf{v} = \mathbf{u} - \tilde{\mathbf{u}}$$

is a quasistatic equilibrium displacement on $\bar{B}$, by which we mean that it satisfies the quasistatic equilibrium equations

$$\operatorname{div} \mathbf{S}(\mathbf{v}) = \mathbf{0} \qquad \text{in } B \times [0, \infty). \tag{26.8.10}$$

Moreover, it also satisfies the lateral boundary condition

$$\mathbf{s}(\mathbf{v}) = \mathbf{0} \qquad \text{on } \pi \times [0, \infty) \tag{26.8.11}$$

and the end global boundary conditions

$$\begin{aligned} &\int_{D_0} \mathbf{s}(\mathbf{v})\, da = \mathbf{0}, \qquad \int_{D_0} \mathbf{r} \times \mathbf{s}(\mathbf{v})\, da = \mathbf{0},\\ &\int_{D_L} \mathbf{s}(\mathbf{v})\, da = \mathbf{0}, \qquad \int_{D_L} \mathbf{r} \times \mathbf{s}(\mathbf{v})\, da = \mathbf{0}. \end{aligned} \tag{26.8.12}$$

The quasistatic equilibrium equations (26.8.10) and the boundary conditions (26.8.11) with $(26.8.12)_1$ imply that

$$\int_{D_z} \mathbf{s}(\mathbf{v})\, da = \mathbf{0}, \qquad \int_{D_z} \mathbf{r} \times \mathbf{s}(\mathbf{v})\, da = \mathbf{0} \quad \forall z \in [0, L]. \tag{26.8.13}$$

Let us fix $0 < T < \infty$. Then, with the quasistatic equilibrium displacement $\mathbf{v}$ satisfying the boundary conditions (26.8.11) and (26.8.12) we associate the following measure

$$U_{\mathbf{v}}(z) = \int_0^T \int_{B_z} \mathbf{S}(\mathbf{v}(\tau)) \cdot \mathbf{E}(\mathbf{v}(\tau))\, dV d\tau, \tag{26.8.14}$$

where $B_z \equiv D \times (z, L - z)$, with $z \in [0, \frac{L}{2})$. Using the divergence theorem and the quasistatic equilibrium equations (26.8.10) together with the boundary conditions (26.8.11) and $(26.8.12)_1$, we can write

$$U_{\mathbf{v}}(z) = \int_0^T \int_{D_{L-z}} \mathbf{s}(\mathbf{v}(\tau)) \cdot \mathbf{v}(\tau)\, da d\tau - \int_0^T \int_{D_z} \mathbf{s}(\mathbf{v}(\tau)) \cdot \mathbf{v}(\tau)\, da d\tau. \tag{26.8.15}$$

Furthermore, let us add to $\mathbf{v}$ the rigid displacements $\check{\mathbf{v}}^{(1)}$ and $\check{\mathbf{v}}^{(2)}$ and set

$$\hat{\mathbf{v}}^{(\alpha)} \equiv \mathbf{v} + \check{\mathbf{v}}^{(\alpha)} \qquad (\alpha = 1, 2),$$

so that by virtue of (26.8.13), Eq. (26.8.15) yields

$$U_{\mathbf{v}}(z) = \int_0^T \int_{D_{L-z}} \mathbf{s}(\mathbf{v}(\tau)) \cdot \hat{\mathbf{v}}^{(1)}(\tau)\, da d\tau - \int_0^T \int_{D_z} \mathbf{s}(\mathbf{v}(\tau)) \cdot \hat{\mathbf{v}}^{(2)}(\tau)\, da d\tau.$$

By the Schwarz inequality and the arithmetic–geometric mean inequality, we obtain

$$\begin{aligned} U_{\mathbf{v}}(z) \le \frac{\alpha}{2} \int_0^T \int_{D_{L-z}} |\mathbf{S}(\mathbf{v}(\tau))|^2\, da d\tau + \frac{1}{2\alpha} \int_0^T \int_{D_{L-z}} \left|\hat{\mathbf{v}}^{(1)}(\tau)\right|^2 da d\tau \\ + \frac{\alpha}{2} \int_0^T \int_{D_z} |\mathbf{S}(\mathbf{v}(\tau))|^2\, da d\tau + \frac{1}{2\alpha} \int_0^T \int_{D_z} \left|\hat{\mathbf{v}}^{(2)}(\tau)\right|^2 da d\tau, \end{aligned} \tag{26.8.16}$$

with α a positive constant at our disposal. Furthermore, the use of the estimates (26.8.3) and (26.8.8) in (26.8.16) gives

$$\begin{aligned} U_{\mathbf{v}}(z) \le \frac{2\alpha\mu_M}{\mu_m} \Bigg[\int_0^T \int_{D_{L-z}} \mathbf{S}(\mathbf{v}(\tau)) \cdot \mathbf{E}(\mathbf{v}(\tau))\, da d\tau \\ + \int_0^T \int_{D_z} \mathbf{S}(\mathbf{v}(\tau)) \cdot \mathbf{E}(\mathbf{v}(\tau))\, da d\tau \Bigg] + \frac{1}{2\alpha} \int_0^T \int_{D_{L-z}} \left|\hat{\mathbf{v}}^{(1)}(\tau)\right|^2 da d\tau \\ + \frac{1}{2\alpha} \int_0^T \int_{D_z} \left|\hat{\mathbf{v}}^{(2)}(\tau)\right|^2 da d\tau. \end{aligned} \tag{26.8.17}$$

At this instant we determine the rigid displacement $\check{\mathbf{v}}^{(1)}$ and $\check{\mathbf{v}}^{(2)}$ in such a way that

$$\int_{B_1} \hat{\mathbf{v}}^{(1)} dv = \mathbf{0}, \qquad \int_{B_1} \mathbf{r} \times \hat{\mathbf{v}}^{(1)} dv = \mathbf{0},$$
$$\int_{B_2} \hat{\mathbf{v}}^{(2)} dv = \mathbf{0}, \qquad \int_{B_2} \mathbf{r} \times \hat{\mathbf{v}}^{(2)} dv = \mathbf{0},$$

where $B_1 \equiv D \times (L - z - l, L - z)$, $B_2 = D \times (z, z + l)$ and $0 < l < \frac{L}{2}$. Then, as in [187], it follows that

$$\begin{aligned} \int_{B_1} \left|\hat{\mathbf{v}}^{(1)}\right|^2 dv \le \frac{1}{\lambda_0(l)} \int_{B_1} \left|\mathbf{E}\left(\hat{\mathbf{v}}^{(1)}\right)\right|^2 dv = \frac{1}{\lambda_0(l)} \int_{B_1} |\mathbf{E}(\mathbf{v})|^2\, dv, \\ \int_{B_2} \left|\hat{\mathbf{v}}^{(2)}\right|^2 dv \le \frac{1}{\lambda_0(l)} \int_{B_2} \left|\mathbf{E}\left(\hat{\mathbf{v}}^{(2)}\right)\right|^2 dv = \frac{1}{\lambda_0(l)} \int_{B_2} |\mathbf{E}(\mathbf{v})|^2\, dv, \end{aligned} \tag{26.8.18}$$

where $\lambda_0(l)$ is the lowest nonzero characteristic value corresponding to the free vibrations of the cylindrical disk $D \times [0, l]$, made of a hypothetical elastic material having the elasticity tensor $\mathbf{C} = \mathbf{1} \otimes \mathbf{1}$, with $\mathbf{1}$ the unit tensor in Sym. Furthermore, we integrate (26.8.17) between the limits z and $z + l$, $z \in \left(0, \frac{L}{2} - l\right)$, and then use the estimates (26.8.8) and (26.8.18) to obtain

$$\begin{aligned} lQ(z,l) &\le \frac{2\alpha\mu_M}{\mu_m}\int_0^T\int_{B_*}\mathbf{S}(\mathbf{v}(\tau))\cdot\mathbf{E}(\mathbf{v}(\tau))\,dvd\tau \\ &+ \frac{1}{2\alpha\lambda_0(l)\mu_m}\int_0^T\int_{B_*}\mathbf{S}(\mathbf{v}(\tau))\cdot\mathbf{E}(\mathbf{v}(\tau))\,dvd\tau \\ &= \left[\frac{2\alpha\mu_M}{\mu_m}+\frac{1}{2\alpha\lambda_0(l)\mu_m}\right]\int_0^T\int_{B_*}\mathbf{S}(\mathbf{v}(\tau))\cdot\mathbf{E}(\mathbf{v}(\tau))\,dvd\tau \\ &\le \varkappa(l)\int_0^T\int_{B_*}\mathbf{S}(\mathbf{v}(\tau))\cdot\mathbf{E}(\mathbf{v}(\tau))\,dvd\tau, \end{aligned} \tag{26.8.19}$$

where

$$\begin{aligned} lQ(z,l) &= \int_z^{z+l} U_{\mathbf{v}}(\eta)\,d\eta, \qquad B_* = B_1\cup B_2, \\ \varkappa(l) &= \frac{2}{\mu_m}\sqrt{\frac{\mu_M}{\lambda_0(l)}}. \end{aligned} \tag{26.8.20}$$

It follows from $(26.8.20)_1$ that

$$l\frac{\partial}{\partial z}Q(z,l) = U_{\mathbf{v}}(z+l) - U_{\mathbf{v}}(z) = -\int_0^T\int_{B_*}\mathbf{S}(\mathbf{v}(\tau))\cdot\mathbf{E}(\mathbf{v}(\tau))\,dvd\tau,$$

so that from (26.8.19), we deduce

$$\varkappa(l)\frac{\partial}{\partial z}Q(z,l) + Q(z,l) \le 0,$$

or

$$\frac{\partial}{\partial z}\left[Q(z,l)e^{\varkappa(l)z}\right] \le 0.$$

The solution of this differential inequality is

$$Q(z_2,l) \le Q(z_1,l)e^{-\frac{z_2-z_1}{\varkappa(l)}} \qquad \text{for } z_2 \ge z_1.$$

Since $U_{\mathbf{v}}(z)$ is a nonincreasing function on z and $Q(z,l)$ is the mean value of $U_{\mathbf{v}}(z)$ in the interval $[z, z+l]$, we have

$$U_{\mathbf{v}}(z+l) \le Q(z,l) \le U_{\mathbf{v}}(z)$$

and hence

$$U_{\mathbf{v}}(z_2+l) \le U_{\mathbf{v}}(z_1)e^{-\frac{z_2-z_1}{\varkappa(l)}} \qquad \text{for } z_2 \ge z_1.$$

This inequality yields the following spatial decay estimate:

$$U_{\mathbf{v}}(z) \le U_{\mathbf{v}}(0)e^{-\frac{z-l}{\varkappa(l)}} \qquad \text{for } z \ge l. \tag{26.8.21}$$

The estimate (26.8.21) proves that the displacement vector $\mathbf{v}$ can be approximated by a null displacement vector (modulo a rigid displacement) at appropriately large distances of the loaded ends. Therefore, we can approximate the exact solution of the Saint-Venant problem with a solution of the relaxed Saint-Venant problem. Thus, the estimate (26.8.21) can be interpreted as representing the Saint-Venant principle in linear viscoelasticity, giving quantitative expression to the assertion that it is sufficient to give a method for finding solutions of the relaxed Saint-Venant problem in order to approximate the solution of the Saint-Venant problem.

A result similar to (26.8.21) is given in [93] in which an alternative measure to (26.8.14) is used, based on a free energy density in the integrand, which is of course manifestly nonnegative. This measure can be shown not to exceed $U_{\mathbf{v}}$. Also, a result of the same type is given for sinusoidal histories, where the energy measure and the spatial decay constant are frequency-dependent. This is of course a manifestation of an important property of viscoelastic materials, namely that their response is influenced by the rate of application of loads.

27

Exponential Decay

27.1 Differential Problem with Nonconvex Kernels

The differential problem of the dynamics of viscoelastic solids, defined, for example, by the system (24.2.31), can be represented in an abstract form by means of the following integrodifferential equation:

$$\ddot{\mathbf{u}}(t) + A\mathbf{u} + \int_{-\infty}^{t} k(t-s)A\mathbf{u}(s)ds = \mathbf{0} \quad \forall t > 0, \tag{27.1.1}$$

with which the initial conditions

$$\mathbf{u}(t) = \phi(t) \quad \forall t \leq 0, \quad \dot{\mathbf{u}}(0) = \mathbf{u}_1 \tag{27.1.2}$$

are associated.

In [24], to which we shall refer for the study of the stability of solutions, the operator $A : D(A) \to \mathcal{H}$ is supposed to be self-adjoint, positive definite, closed, and linear; its domain $D(A)$ is a dense subset of the real Hilbert space with an inner product denoted by $(\cdot, \cdot)$ and the corresponding norm $\| \cdot \|_{\mathcal{H}}$. The resolvent operator will be denoted by $R(\mu; A) = (A - \mu I)^{-1}$, which, if it exists, is assumed to be compact. Moreover, we suppose that $A\phi : (-\infty, 0] \to \mathcal{H}$ is bounded and continuous, $\mathbf{u}_1 \in \mathcal{H}$, the scalar kernel $k : [0, \infty) \to \mathbb{R}$ is such that $k \in C^1$, and both k and k' belong to $L^1(0, \infty)$. In particular, in viscoelasticity such an operator A coincides with the Laplacian Δ.

A function

$$\mathbf{u} \in C(\mathbb{R}, D(A)) \cap C^2([0, \infty), \mathcal{H}) \tag{27.1.3}$$

that satisfies (27.1.1) for all $t > 0$ and (27.1.2) is said to be a strong solution of this problem. Any solution satisfies the Volterra integrodifferential equation

$$\ddot{\mathbf{u}}(t) + A\mathbf{u}(t) + \int_{0}^{t} k(t-s)A\mathbf{u}(s)ds = \mathbf{f}(t) \quad \forall t > 0, \tag{27.1.4}$$

G. Amendola et al., *Thermodynamics of Materials with Memory*,
https://doi.org/10.1007/978-3-030-80534-0_27

with initial conditions given by

$$\mathbf{u}(0) = \mathbf{u}_0 := \phi(0), \qquad \dot{\mathbf{u}}(0) = \mathbf{u}_1, \tag{27.1.5}$$

where the function **f** is expressed in terms of the given initial history as follows:

$$\mathbf{f}(t) = -\int_{-\infty}^{0} k(t-s)A\phi(s)ds \quad \forall t \geq 0. \tag{27.1.6}$$

Observe that a forcing function could be added to (27.1.1), to allow for the presence of body forces.

Our concern here is with the stability of solutions. It will be recalled that with respect to their existence and uniqueness, Theorem 3.1 of [263] ensures that the problem (27.1.4) and (27.1.5) has a unique strong solution on $[0, \infty)$ if $\mathbf{u}_1$ is in an intermediate space between $D(A)$ and $\mathcal{H}$.

The fact that the abstract formulation expressed by (27.1.4) is related to the problems of linear viscoelasticity is shown, for example, in [79, 80, 113, 276]. The kernel $G : [0, \infty) \to \mathbb{R}$ is expressed in terms of k by means of

$$G(t) = 1 + \int_0^t k(s)ds,$$

which is the *relaxation function*, with the instantaneous elastic modulus incorporated into A. Also, linear viscoelastic materials that are inhomogeneous are allowed in this formulation. Since we are concerned with viscoelastic solids, it will be supposed that

$$G_\infty = 1 + \int_0^\infty k(s)ds > 0. \tag{27.1.7}$$

Moreover, some other conditions on the kernel must be added. We assume that

$$\frac{1}{\eta}\mathrm{Im}k_L(i\eta) > 0 \quad \forall \eta \neq 0, \quad k(0) < 0, \tag{27.1.8}$$

where

$$k_L(\lambda) := \int_0^\infty k(t)e^{-\lambda t}dt, \quad \mathrm{Re}\lambda \geq 0,$$

is the Laplace transform of k. Relations (27.1.8) are special cases of (8.1.18) and (8.1.23). Condition $(27.1.8)_2$ is more restrictive in one sense than $(27.1.8)_1$, since if k satisfies $(27.1.8)_1$ and the function $t \longmapsto tk(t)$ belongs to $L^1(0, \infty)$, then it follows that $k(0) \leq 0$ (a special case of (8.1.21)). The *loss modulus* (see after (8.1.24)) in linear viscoelasticity corresponds to the function $\eta \longmapsto \mathrm{Im}k_L(i\eta)$ and the *dynamic viscosity* to $\eta \longmapsto \mathrm{Im}k_L(i\eta)/\eta$ [133]. We note that alternatively, the monotonicity conditions

$$k(t) < 0, \quad k'(t) \geq 0 \qquad \forall t \geq 0 \tag{27.1.9}$$

can be employed. These are a special case of (10.2.2), for example. One can show that they imply (27.1.8), though the converse is not true. An example is given by the oscillatory kernel

$$k(t) = -(1 - G_\infty)\frac{d}{dt}\left(e^{-\alpha t}\cos\omega t\right) \quad \forall t \geq 0,$$

which satisfies (27.1.7) and (27.1.8) if $0 < G_\infty < 1$ and $\alpha > 0$; indeed,

$$\frac{1}{\eta}\mathrm{Im}k_L(i\eta) = \frac{(1 - G_\infty)\alpha\left(\alpha^2 + \omega^2 + \eta^2\right)}{[\alpha^2 + (\eta + \omega)^2][\alpha^2 + (\eta - \omega)^2]} > 0 \quad \forall\eta \neq 0. \tag{27.1.10}$$

Following [24], we now choose the conditions (27.1.8), instead of (27.1.9), because of their connection with the restrictions imposed by the Second Law of Thermodynamics. Finally, observe that in (27.1.1) and (27.1.4) we have only the weak damping engendered by the memory term without any damping term that depends on the current value of $\dot{\mathbf{u}}$.

It is useful to consider frequency-domain methods because of the particular forms of (27.1.4) and $(27.1.8)_1$. For the use of these methods to study the asymptotic behavior of solutions of both parabolic and hyperbolic infinite-dimensional integrodifferential equations, see [284], while for positivity conditions such as $(27.1.8)_1$, see [81, 255–258, 304].

27.1.1 Transformed Problem and Some Useful Preliminaries

In order to use the frequency domain, we consider the transformed problem by applying the Laplace transform to (27.1.4). Thus, let us denote by $\mathcal{H}_\Omega$ the complexification of the Hilbert space $\mathcal{H}$. Consider any element $\mathbf{v} \in \mathcal{H}_\Omega$ expressed by

$$\mathbf{v} = \mathbf{v}_1 + i\mathbf{v}_2,$$

where $\mathbf{v}_1,\ \mathbf{v}_2 \in \mathcal{H}$. Then one can define

$$(\mathbf{v}, \mathbf{w})_{\mathcal{H}_\Omega} = (\mathbf{v}_1, \mathbf{w}_1) + (\mathbf{v}_2, \mathbf{w}_2) + i[(\mathbf{v}_2, \mathbf{w}_1) - (\mathbf{v}_1, \mathbf{w}_2)],$$

which is a sesquilinear form induced by the inner product $(\cdot,\ \cdot)$ of $\mathcal{H}$ on $\mathcal{H}_\Omega$. The space $\mathcal{H}_\Omega$ is therefore a complex Hilbert space with norm given by $\|\mathbf{v}\|^2_{\mathcal{H}_\Omega} = \|\mathbf{v}_1\|^2_{\mathcal{H}} + \|\mathbf{v}_2\|^2_{\mathcal{H}}$. Furthermore, the closed operator A is extended by means of an operator A_Ω : $D(A_\Omega) \to \mathcal{H}_\Omega$ defined by

$$\mathbf{v} \in D(A_\Omega) \Longleftrightarrow \mathbf{v}_1,\ \mathbf{v}_2 \in D(A),$$
$$A_\Omega\mathbf{v} = A\mathbf{v}_1 + iA\mathbf{v}_2.$$

Obviously, A_Ω inherits its properties from A. We denote by $\mathcal{L}(\mathcal{H})$ the Banach space of bounded linear operators on $\mathcal{H}$. The integrals of vector-valued functions are Bochner integrals. The convolution of two functions v and $\mathbf{w}$, defined on $[0, \infty)$, is denoted by (see (C.3.4))

$$(v * \mathbf{w})(t) = \int_0^t v(t-s)\mathbf{w}(s)ds \quad \forall t \geq 0.$$

If $v \in L^1((0,\infty),\mathbb{R})$ and $\mathbf{w} \in L^p((0,\infty),\mathcal{H})$, then

$$\|v * \mathbf{w}\|_{L^p((0,\infty),\mathcal{H})} \leq \|v\|_{L^1((0,\infty),\mathbb{R})}\|\mathbf{w}\|_{L^p((0,\infty),\mathcal{H})}. \tag{27.1.11}$$

If $t \longmapsto e^{\sigma t}\mathbf{w}(t)$ is an integrable function from $(0,\infty)$ into a Banach space [201], the Laplace transform of $\mathbf{w}$, which exists for any λ such that $\mathrm{Re}\lambda \geq -\sigma$, is continuous on the complex half-plane $\{\lambda \in \Omega; \mathrm{Re}\lambda \geq -\sigma\}$ and analytic on $\{\lambda \in \Omega; \mathrm{Re}\lambda > -\sigma\}$, where Ω denotes the set of complex numbers (cf. Sect. C.2.1). We recall that functions defined on the closure of the complex half-space $\mathbf{\Pi} = \{z \in \Omega;\ \mathrm{Re}z > 0\}$ with values in a Hilbert space are shown to be the Laplace transforms of functions by establishing that they are in a Hardy space [263].

Definition 27.1.1. *The Hardy space $H^2(\mathbf{\Pi},\mathcal{H}_\Omega)$ is the class of all analytic functions $\chi : \mathbf{\Pi} \to \mathcal{H}_\Omega$ such that*

$$\sup\left\{\int_{-\infty}^{\infty} \|\chi(\xi + i\eta)\|^2_{\mathcal{H}_\Omega} d\eta,\ \xi > 0\right\} < \infty,$$

$\chi(i\eta) := \lim_{\xi\to 0^+}\chi(\xi + i\eta)$ exists almost everywhere and the function $\eta \longmapsto \chi(i\eta)$ belongs to $L^2(\mathbb{R},\mathcal{H}_\Omega)$.

We recall a generalization of a well-known Paley–Wiener result, expressed by the following theorem.

Theorem 27.1.2. *A function $\chi : \mathbf{\Pi} \to \mathcal{H}_\Omega$ is the Laplace transform of a function in $L^2((0,\infty),\mathcal{H})$ if and only if $\chi \in H^2(\mathbf{\Pi},\mathcal{H}_\Omega)$.*

From the assumptions on A it follows that there exist a family $\{\mu_n\}$ of eigenvalues such that $0 < \mu_1 \leq \mu_2 \leq \cdots$ with $\lim_{n\to\infty}\mu_n = \infty$, and a complete orthonormal set $\{\psi_n\}$ of corresponding eingenvectors of A. Moreover, the complex μ is in the spectrum $\sigma(A)$ of A if and only if $\mu = \mu_n$, in which case $A\psi_n = \mu_n\psi_n$. The resolvent set of A is denoted by $\rho(A)$.

From Lemmas 4.1 and 4.2 of [263] we have the following useful result.

Lemma 27.1.3. *Let $p, q \in L^1(0,\infty)$, $\vartheta \in \mathbb{R}$ and $\kappa > 0$. Suppose that for each $n = 1, 2, \ldots$,*

$$m_n(\lambda) := \lambda^2 + \kappa\lambda + \lambda p_L(\lambda) + q_L(\lambda) + \vartheta + \mu_n \neq 0, \quad \mathrm{Re}\lambda \geq 0. \tag{27.1.12}$$

Then the linear operators $T_L(\lambda)$ and $R_L(\lambda)$ in $\mathcal{L}(\mathcal{H}_\Omega)$, defined by

$$T_L(\lambda) = \left\{\left[\lambda^2 + \kappa\lambda + \lambda p_L(\lambda) + q_L(\lambda) + \vartheta\right]I + A_\Omega\right\}^{-1},$$
$$R_L(\lambda) = \lambda T_L(\lambda), \tag{27.1.13}$$

exist for all λ *such that* $\mathrm{Re}\lambda \geq 0$, *and*

$$\sup\left\{\int_{-\infty}^{\infty} \|T_L(\xi + i\eta)\|^2_{\mathcal{L}(\mathcal{H}_\Omega)} d\eta,\ \xi \geq 0\right\} < \infty,$$
$$\sup\left\{\int_{-\infty}^{\infty} \|R_L(\xi + i\eta)\mathbf{v}\|^2_{\mathcal{H}_\Omega} d\eta,\ \xi \geq 0\right\} < \infty, \quad \forall \mathbf{v} \in \mathcal{H}_\Omega.$$

Regarding the notation used in (27.1.13), we observe that even if for each fixed $\mathbf{v} \in \mathcal{H}_\Omega$ the functions $\lambda \longmapsto T_L(\lambda)\mathbf{v}$ and $\lambda \longmapsto R_L(\lambda)\mathbf{v}$ are not defined as Laplace transforms, they are indeed so, by virtue of Theorem 27.1.2.

27.1.2 The Resolvent of the Kernel

Consider the *resolvent* $r : [0, \infty) \to \mathbb{R}$ of k defined as the solution of

$$r(t) + (k * r)(t) = -k(t) \qquad \forall t \geq 0. \tag{27.1.14}$$

Let us put $k_\beta(t) := e^{\beta t}k(t)$, $r_\beta(t) := e^{\beta t}r(t)$ and denote their time derivatives by $k'_\beta(t)$ and $r'_\beta(t)$; we have, in particular, $k_{\beta_L}(\lambda) = k_L(\lambda - \beta)$. From (27.1.14) it follows that r_β is the resolvent of k_β; moreover, $r(0) = -k(0) > 0$ if (27.1.8) holds.

We now prove some results related to k, already shown for symmetric bounded linear operators on $\mathcal{H}$ in Proposition 5.1 of [256], the relevance of which to linear viscoelasticity can be found in [112, 113].

Proposition 27.1.4. *Suppose that k satisfies (27.1.8). Then*

$$\frac{1}{\eta}\mathrm{Im}k_L(\xi + i\eta) > 0, \quad \xi \geq 0,\ \eta \neq 0,$$
$$1 + \mathrm{Re}k_L(\xi) > 0, \quad \xi \geq 0; \tag{27.1.15}$$

furthermore, $r \in C^1[0, \infty)$, $r \in L^1(0, \infty) \cap L^2(0, \infty)$, $r' \in L^1(0, \infty)$ *and also*

$$\frac{1}{\eta}\mathrm{Im}r_L(\lambda) < 0, \quad \xi \geq 0,\ \eta \neq 0,$$
$$1 + \mathrm{Re}r_L(\xi) > 0, \quad \xi \geq 0. \tag{27.1.16}$$

Moreover, if there exists a constant $\alpha > 0$ *such that*

$$\int_0^\infty e^{\alpha t}|k(t)|dt < \infty, \qquad \int_0^\infty e^{\alpha t}|k'(t)|dt < \infty \tag{27.1.17}$$

hold, then there is a $\beta \in (0, \alpha]$ *such that* $r_\beta,\ r'_\beta \in L^1(0, \infty)$.

Proof. The resolvent r, which is continuous, belongs to $L^1(0,\infty)$ if and only if $1 + k_L(\lambda) \neq 0$ for all λ with $\mathrm{Re}\lambda \geq 0$ (see Theorem 2.4.1 of [185]). From the analyticity of the function $\lambda \longmapsto 1 + k_L(\lambda)$ it follows that its imaginary part is a harmonic function, the minimum of which is achieved on the boundary of a closed bounded region. Let

$$\{\lambda \in \Omega;\ \mathrm{Re}\lambda \geq 0,\ \mathrm{Im}\lambda \geq 0,\ |\lambda| \leq \rho\}$$

be a region defined by the parameter $\rho > 0$, which is assumed to be large; then by virtue of the Riemann–Lebesgue lemma (C.2.13) and $(27.1.8)_1$, we have $\mathrm{Im}k_L(\lambda) \geq 0$ for all λ such that $\mathrm{Re}\lambda \geq 0$ and $\mathrm{Im}\lambda \geq 0$. The open mapping theorem, since k_L is not constant, yields $\mathrm{Im}k_L(\lambda) > 0$ for all λ with $\mathrm{Re}\lambda \geq 0$ and $\mathrm{Im}\lambda > 0$. From

$$\overline{k_L(\lambda)} = k_L(\bar{\lambda}),$$

where the bar denotes the complex conjugate, we get $(27.1.15)_1$. Using the Hilbert integral representation for the Laplace transform we have

$$\mathrm{Re}k_L(\xi) = -\frac{2}{\pi}\int_0^\infty \frac{\omega}{\xi^2 + \omega^2}\mathrm{Im}k_L(i\omega)d\omega, \quad \xi > 0.$$

Observe that the function $\xi \longmapsto \left(\xi^2 + \omega^2\right)^{-1}$ is strictly decreasing on $(0,\infty)$, while $\xi \longmapsto \mathrm{Re}k_L(\xi)$ is strictly increasing on $(0,\infty)$; thus, $(27.1.15)_2$ follows, since $k_L(0) > -1$, by virtue of (27.1.7), and $\lim_{\xi\to\infty} k_L(\xi) = 0$. Both the inequalities (27.1.15) imply that $1 + k_L(\lambda) \neq 0$ for any λ with $\mathrm{Re}\lambda \geq 0$, and consequently, $r \in L^1(0,\infty)$. The assertions related to r' follow easily from the relation $r' = k(0)r - k' * r - k'$, derived from (27.1.14).

To see that $r \in L^2(0,\infty)$, note that $\lim_{t\to\infty} r(t) = 0$, since $r, r' \in L^1(0,\infty)$; thus, there exists $T > 0$ such that $r^2(t) \leq |r(t)| \forall t \geq T$.

Let us add the hypotheses expressed by (27.1.17). To show that $r_\beta \in L^1(0,\infty)$, we prove that there exists a constant $\beta \in (0,\alpha]$ such that $1 + k_{\beta_L}(\lambda) = 1 + k_L(\lambda - \beta) \neq 0$ for all λ with $\mathrm{Re}\lambda \geq 0$. It is only necessary to see that $1 + k_L(\lambda) \neq 0$ for all λ such that $-\beta \leq \mathrm{Re}\lambda \leq 0$. From the Riemann–Lebesgue lemma it follows that there exists $M_1 > 0$ such that $1 + k_L(\xi + i\eta) \neq 0\ \forall \xi \in [-\alpha, 0]$ and $|\eta| \geq M_1$. Since $k_L(0) > -1$, there exist $\beta_1 \in (0,\alpha]$ and $M_2 \in (0, M_1)$ such that $\mathrm{Re}k_L(\xi + i\eta) > -1\ \forall \xi \in [-\beta_1, 0]$ and $|\eta| \leq M_2$. Finally, there exists $\beta_2 \in (0,\alpha]$ such that $\mathrm{Im}k_L(\xi + i\eta) > 0\ \forall \xi \in [-\beta_2, 0]$ and for any η such that $|\eta| \in [M_2, M_1]$. It is enough to take $\beta = \min\{\beta_1, \beta_2\}$ to obtain the result.

It follows from (27.1.14) that

$$r_L(\lambda) = -\frac{k_L(\lambda)}{1 + k_L(\lambda)}, \quad \mathrm{Re}\lambda \geq 0,$$

whence we have

$$\mathrm{Re}[1 + r_L(\lambda)] = \frac{1 + \mathrm{Re}k_L(\lambda)}{|1 + k_L(\lambda)|^2}, \quad \mathrm{Im}r_L(\lambda) = -\frac{\mathrm{Im}k_L(\lambda)}{|1 + k_L(\lambda)|^2}.$$

Thus, using (27.1.15), the relations (27.1.16) follow. □

27.1.3 Stability Results

Some results related to the exponential decay of strong solutions of (27.1.4)–(27.1.5), derived in [24], are established by means of three theorems.

Before giving their proofs, we firstly consider the properties of the forcing function $\mathbf{f}$, defined by (27.1.6), and show its characterization by means of the inequality (27.1.10). We observe that the assumption on the initial history, expressed by the boundedness of $A\phi$, can be relaxed by constructing a suitable influence function from k and k', as done in Lemma 2.1 of [79] by Dafermos, and requiring that ϕ lie in an appropriate weighted space. On this subject we also recall that Coleman and Noll in [73, 74] introduced influence functions as part of a theory of functionals with fading memory.

Proposition 27.1.5. *The function* $\mathbf{f}$ *defined in (27.1.6) belongs to* $C^1([0,\infty),\mathcal{H})$. *If*

$$\int_0^\infty t|k(t)|dt < \infty, \qquad \int_0^\infty t|k'(t)|dt < \infty \tag{27.1.18}$$

also hold, then $\mathbf{f},\ \mathbf{f}' \in L^1((0,\infty),\mathcal{H})$.

Proof. The function $\mathbf{f}$ is continuous, because

$$\begin{aligned}
\|\mathbf{f}(t+\tau)-\mathbf{f}(t)\|_{\mathcal{H}} &\le \int_{-\infty}^0 |k(t+\tau-s)-k(t-s)|\|A\phi(s)\|_{\mathcal{H}}ds\\
&\le \sup_{s\le 0}\|A\phi(s)\|_{\mathcal{H}}\int_{-\infty}^0 |k(t+\tau-s)-k(t-s)|ds\\
&= \sup_{s\le 0}\|A\phi(s)\|_{\mathcal{H}}\int_t^\infty |k(\tau+\sigma)-k(\sigma)|d\sigma \to 0
\end{aligned}$$

as $\tau\to 0$, since $k\in L^1(0,\infty)$.

In order to show that $\mathbf{f}$ is continuously differentiable, we firstly observe that

$$\begin{aligned}
\int_{t_1}^{t_2}\left\|\int_{-\infty}^0 k'(t-s)A\phi(s)ds\right\|_{\mathcal{H}} dt &\le \sup_{s\le 0}\|A\phi(s)\|_{\mathcal{H}}\int_{t_1}^{t_2}\int_{-\infty}^0 |k'(t-s)|dsdt\\
&\le \sup_{s\le 0}\|A\phi(s)\|_{\mathcal{H}}\int_{t_1}^{t_2}\int_t^\infty |k'(\sigma)|d\sigma dt < \infty,
\end{aligned}$$

because $k'\in L^1(0,\infty)$. Then, using the Tonelli–Fubini theorem, we obtain

$$\int_{t_1}^{t_2}\int_{-\infty}^0 k'(t-s)A\phi(s)dsdt = \int_{-\infty}^0 [k(t_2-s)-k(t_1-s)]A\phi(s)ds = \mathbf{f}(t_1)-\mathbf{f}(t_2).$$

From this result and

$$\left\| \int_{-\infty}^{0} [k'(t+\tau-s)-k'(t-s)]A\phi(s)ds \right\|_{\mathcal{H}} \leq \sup_{s\leq 0} \|A\phi(s)\|_{\mathcal{H}} \int_{t}^{\infty} |k'(\tau+\sigma)-k'(\sigma)|d\sigma \to 0$$

as $\tau \to 0$, we have $\mathbf{f} \in C^1([0,\infty),\mathcal{H})$ and

$$\mathbf{f}'(t) = -\int_{-\infty}^{0} k'(t-s)A\phi(s)ds \qquad \forall t \geq 0.$$

The function $\mathbf{f}$ is integrable, since by virtue of (27.1.18),

$$\left\| \int_{0}^{\infty}\int_{-\infty}^{0} k(t-s)A\phi(s)dsdt \right\|_{\mathcal{H}} \leq \sup_{s\leq 0} \|A\phi(s)\|_{\mathcal{H}} \int_{0}^{\infty}\int_{-\infty}^{0} |k(t-s)|dsdt \leq \sup_{s\leq 0} \|A\phi(s)\|_{\mathcal{H}} \int_{0}^{\infty} \sigma|k(\sigma)|d\sigma < \infty.$$

An analogous proof shows that $\mathbf{f}' \in L^1(0,\infty)$. □

Another result, useful later on to show the stability properties, is now given by the following lemma. It is related to a particular form of $m_n(\lambda)$, already defined by (27.1.12) in Lemma 27.1.3, now expressed in terms of $r_L(\lambda)$.

Lemma 27.1.6. *Suppose that k obeys (27.1.8). Then for each $n \in \mathbf{N}$,*

$$m_n(\lambda) := \lambda^2[1+r_L(\lambda)] + \mu_n \neq 0, \quad \mathrm{Re}\lambda \geq 0. \tag{27.1.19}$$

Proof. Now $m_n(\lambda)$ has this new form (27.1.19), instead of (27.1.12). Putting $\lambda = \xi + i\eta$, with $\xi \geq 0$ and $\eta \neq 0$, the real and imaginary parts of this $m_n(\lambda)$ are

$$\mathrm{Re}m_n(\lambda) = \left(\xi^2-\eta^2\right)[1+\mathrm{Re}r_L(\lambda)] - 2\xi\eta\mathrm{Im}r_L(\lambda) + \mu_n,$$
$$\mathrm{Im}m_n(\lambda) = 2\xi\eta[1+\mathrm{Re}r_L(\lambda)] + \left(\xi^2-\eta^2\right)\mathrm{Im}r_L(\lambda).$$

Hence, if $\mathrm{Im}m_n(\lambda) = 0$, then for any $\xi > 0$ and $\eta \neq 0$, we obtain

$$2\xi\mathrm{Re}m_n(\lambda) = -\frac{\mathrm{Im}r_L(\lambda)}{\eta}\left(\xi^2+\eta^2\right)^2 + 2\xi\mu_n > 0,$$

by virtue of $(27.1.16)_1$. If $\eta = 0$ we obtain, using (27.1.16),

$$\mathrm{Re}m_n(\xi) = \xi^2\mathrm{Re}[1+r_L(\xi)] + \mu_n > 0 \quad \forall\xi \geq 0. \tag{27.1.20}$$

Moreover, if $\xi = 0$, then

$$\mathrm{Im}m_n(i\eta) = -\eta^2\mathrm{Im}r_L(i\eta) \neq 0 \quad \forall\eta \neq 0, \tag{27.1.21}$$

whence it follows that $m_n(i\eta) \neq 0 \ \forall\eta \neq 0$. □

Remark 27.1.7. This result shows that

$$-\lambda^2[1 + r_L(\lambda)] = -\frac{\lambda^2}{1 + k_L(\lambda)} \in \rho(A_\Omega)$$

for all λ such that $\mathrm{Re}\lambda \geq 0$.

A first result on the L^2-stability of the solutions of (27.1.4) and (27.1.5) is given by the following theorem.

Theorem 27.1.8. *Suppose that k obeys (27.1.8) and that (27.1.18) also hold. Then the strong solution* $\mathbf{u}$ *of (27.1.1)–(27.1.2) satisfies*

$$\int_0^\infty \left[\|\mathbf{u}(t)\|^2_{\mathcal{H}} + \|\dot{\mathbf{u}}(t)\|^2_{\mathcal{H}} \right] dt < \infty,$$

$$\|\mathbf{u}(t)\|_{\mathcal{H}} \to 0 \quad as\ t \to \infty. \tag{27.1.22}$$

Proof. This theorem is a consequence of Theorem 4.2 of [263].

Firstly, using MacCamy's trick, the problem (27.1.4) is transformed into one in which the history of $A\mathbf{u}$ is absent, as in [263]. The convolution of (27.1.4) with r gives

$$(r * \ddot{\mathbf{u}})(t) - (k * A\mathbf{u})(t) = (r * \mathbf{f})(t).$$

Hence, we can derive and substitute the expression for $k * A\mathbf{u}$ into (27.1.4); then an integration by parts gives

$$\ddot{\mathbf{u}}(t) + r(0)\dot{\mathbf{u}}(t) + (\dot{r} * \dot{\mathbf{u}})(t) + A\mathbf{u}(t) = \mathbf{g}(t) + r(t)\mathbf{u}_1 \quad \forall t \geq 0, \tag{27.1.23}$$

where

$$\mathbf{g}(t) = \mathbf{f}(t) + (r * \mathbf{f})(t).$$

Propositions 27.1.4 and 27.1.5 yield $\mathbf{g},\ \dot{\mathbf{g}} \in L^1((0,\infty),\mathcal{H})$. Therefore, by applying the Laplace transform to (27.1.23), we have

$$\begin{aligned}&\left[\lambda^2 + r(0)\lambda + \lambda\dot{r}_L(\lambda)\right]\mathbf{u}_L(\lambda) + A_\Omega \mathbf{u}_L(\lambda)\\ &\quad = \lambda\mathbf{u}(0) + [1 + r_L(\lambda)]\mathbf{u}_1 + [\dot{r}_L(\lambda) + r(0)]\mathbf{u}_0 + \mathbf{g}_L(\lambda), \quad \mathrm{Re}\lambda \geq \sigma,\end{aligned}$$

with a suitable $\sigma \geq 0$. If we introduce in $\mathcal{L}(\mathcal{H}_\Omega)$ and for all λ with $\mathrm{Re}\lambda \geq 0$ the linear operators

$$T_L(\lambda) = \left\{\lambda^2[1 + r_L(\lambda)]I + A_\Omega\right\}^{-1}, \qquad R_L(\lambda) = \lambda T_L(\lambda), \tag{27.1.24}$$

then

$$\mathbf{u}_L(\lambda) = T_L(\lambda)\{[1 + r_L(\lambda)]\mathbf{u}_1 + \dot{r}_L(\lambda)\mathbf{u}_0 + r(0)\mathbf{u}_0 + \mathbf{g}_L(\lambda)\} + R_L(\lambda)\mathbf{u}_0. \qquad (27.1.25)$$

From Lemma 27.1.3 and Theorem 27.1.2 it follows that $\mathbf{u} \in L^2((0, \infty), \mathcal{H})$ and that (27.1.25) holds for all λ with $\mathrm{Re}\lambda \geq 0$, by virtue of (27.1.19).

We now show that $\dot{\mathbf{u}}$ is bounded.

From (27.1.25) and the definition $(27.1.13)_2$ for R_L we have

$$\begin{aligned}\dot{\mathbf{u}}_L(\lambda) &= \lambda\mathbf{u}_L(\lambda) - \mathbf{u}_0 \\ &= T_L(\lambda)[\lambda r_L(\lambda)\mathbf{u}_1 + \lambda\mathbf{g}_L(\lambda)] + T_L(\lambda)\left[\lambda^2 + r(0)\lambda + \lambda\dot{r}_L(\lambda)\right]\mathbf{u}_0 + R_L(\lambda)\mathbf{u}_1 - \mathbf{u}_0,\end{aligned}$$

whence, taking into account that

$$\lambda r_L(\lambda) = \dot{r}_L(\lambda) + r(0), \quad \lambda\mathbf{g}_L(\lambda) = \dot{\mathbf{g}}_L(\lambda) + \mathbf{g}(0)$$

and that from (27.1.24),

$$T_L(\lambda)\left\{\lambda^2[1 + r_L(\lambda)]I\right\} = I - T_L(\lambda)A_\Omega,$$

we obtain

$$\dot{\mathbf{u}}_L(\lambda) = T_L(\lambda)[\dot{r}_L(\lambda)\mathbf{u}_1 + \dot{\mathbf{g}}_L(\lambda)] + T_L(\lambda)[r(0)\mathbf{u}_1 + \mathbf{g}(0) - A_\Omega\mathbf{u}_0] + R_L(\lambda)\mathbf{u}_1.$$

Hence, by virtue of the properties of $T_L(\lambda)$ and $R_L(\lambda)$, it follows that $\dot{\mathbf{u}}_L \in H^2(\mathbf{\Pi}, \mathcal{H}_\Omega)$, whence $\dot{\mathbf{u}} \in L^2((0, \infty), \mathcal{H})$.

From (27.1.22) and since $\mathbf{u} \in L^2((0, \infty), \mathcal{H})$, $\mathbf{u}$ is uniformly continuous, because

$$\|\mathbf{u}(t_2) - \mathbf{u}(t_1)\|_{\mathcal{H}} \leq \left|\int_{t_1}^{t_2} \|\dot{\mathbf{u}}(t)\|_{\mathcal{H}} dt\right| \leq |t_2 - t_1|^{1/2}\left[\int_0^\infty \|\dot{\mathbf{u}}(t)\|^2_{\mathcal{H}} dt\right]^{1/2}.$$

Thus, the proof is complete. □

We now give another result related to the forcing function $\mathbf{f}$, namely the exponential decays of $\mathbf{f}$ and $\dot{\mathbf{f}}$, as shown in the following proposition by virtue of conditions (27.1.17).

Proposition 27.1.9. *If the two inequalities (27.1.17) hold, then*

$$\int_0^\infty e^{\alpha t}\|\mathbf{f}(t)\|_{\mathcal{H}} dt < \infty, \qquad \int_0^\infty e^{\alpha t}\|\dot{\mathbf{f}}(t)\|_{\mathcal{H}} dt < \infty.$$

Proof. Using (27.1.17), we have

$$\int_0^\infty e^{\alpha t}\|\mathbf{f}(t)\|_{\mathcal{H}}dt = \int_0^\infty e^{\alpha t}\left\|\int_{-\infty}^0 k(t-s)A\phi(s)ds\right\|_{\mathcal{H}} dt$$
$$\leq \sup_{s\leq 0}\|A\phi(s)\|_{\mathcal{H}}\int_0^\infty |k(\tau)|\int_0^\tau e^{\alpha t}dtd\tau$$
$$= \frac{1}{\alpha}\sup_{s\leq 0}\|A\phi(s)ds\|_{\mathcal{H}}\int_0^\infty |k(\tau)|(e^{\alpha\tau}-1)d\tau < \infty$$

and

$$\int_0^\infty e^{\alpha t}\|\dot{\mathbf{f}}(t)\|_{\mathcal{H}}dt \leq \frac{1}{\alpha}\sup_{s\leq 0}\|A\phi(s)ds\|_{\mathcal{H}}\int_0^\infty |k'(\tau)|(e^{\alpha\tau}-1)d\tau < \infty.$$

□

Before stating the second theorem on exponential stability, we prove the following lemma, which is analogous to Lemma 27.1.6 and similar to Theorem 6 of [255].

Lemma 27.1.10. *Suppose that (27.1.8) and (27.1.17) hold. Let $\beta > 0$ be the parameter introduced in Proposition 27.1.4, while $m_n(\lambda)$ is given by (27.1.19). Then there exists $\beta_1 \in (0,\beta]$, with $2\beta_1 < r(0)$ such that $m_n(\lambda) \neq 0$ for all λ with* $\mathrm{Re}\lambda \geq -\beta_1$.

Proof. From Proposition 27.1.4 it follows that $m_n(\lambda)$ can be defined for all λ such that $\mathrm{Re}\lambda \geq -\beta$. Using (27.1.19), we must prove that $m_n(\lambda) \neq 0$ for all λ with $\mathrm{Re}\lambda \in [-\beta_1, 0]$ for some $\beta_1 \in (0,\beta)$. Since

$$m_n(\lambda) = \lambda^2 + [r(0) + \dot{r}_L(\lambda)]\lambda + \mu_n, \quad \lambda \neq 0,$$

it is easy to derive

$$\lim_{n\to\infty}\frac{1}{\eta}\mathrm{Im}m_n(\lambda) = 2\xi + r(0), \tag{27.1.26}$$

where we have put $\lambda = \xi + i\eta$. Let $\beta_2 \in (0,\beta]$ be such that $r(0) > 2\beta_2$. Thus, (27.1.26) yields that there exists a constant $M_1 > 0$ such that $\mathrm{Im}m_n(\lambda) \neq 0$ for all $\xi \in [-\beta_2, 0]$ and $|\eta| \geq M_1$. Moreover, from (27.1.20) it follows that there are constants $\beta_3 \in (0,\beta]$ and $M_2 > 0$ such that $\mathrm{Re}m_n(\xi + i\eta) > 0\ \forall\xi \in [-\beta_3, 0]$ and $|\eta| \leq M_2$. Finally, from (27.1.21) we see that there is a constant $\beta_4 \in (0,\beta]$ such that $\mathrm{Im}m_n(\xi + i\eta) > 0\ \forall\xi \in [-\beta_4, 0]$ and $M_2 \leq |\eta| \leq M_1$. Choosing $\beta_1 = \min\{\beta_2, \beta_3, \beta_4\}$, we obtain the result. □

Remark 27.1.11. The hypothesis $r(0) = -k(0) > 0$ is essential to obtain a positive limit in (27.1.26) for small ξ. It prevents there being a sequence $\{\lambda_j\}$ satisfying $m_j(\lambda_j) = 0$, with $\mathrm{Re}\lambda_j < 0\ \forall j \in \mathbf{N}$, $\lim_{j\to\infty}\mathrm{Re}\lambda_j = 0$ and $\lim_{j\to\infty}\mathrm{Im}\lambda_j = \infty$. Also Lemma 27.1.10 requires the condition $r(0) > 0$.

We can now state the second theorem, which generalizes to our class of linear hyperbolic integrodifferential equations in a Hilbert space, a theorem in [255], related to a second-order scalar functional differential equation, by requiring the exponential integrability of k and k' only.

Theorem 27.1.12. *Suppose that k obeys (27.1.8) and that there exists a constant* $\alpha > 0$ *such that (27.1.17) holds, i.e., the relations*

$$\int_0^\infty |k(t)|e^{\alpha t}dt < \infty, \qquad \int_0^\infty |k'(t)|e^{\alpha t}dt < \infty,$$

are satisfied. Then there exists $\gamma > 0$*, depending only on k, such that the strong solution of (27.1.1) and (27.1.2) obeys*

$$\int_0^\infty e^{2\gamma t}\left[\|\mathbf{u}(t)\|_{\mathcal{H}}^2 + \|\dot{\mathbf{u}}(t)\|_{\mathcal{H}}^2\right]dt < \infty.$$

Furthermore, there exists a constant $M > 0$ *such that*

$$\|\mathbf{u}(t)\|_{\mathcal{H}} \le Me^{-\gamma t} \qquad \forall t \ge 0. \tag{27.1.27}$$

Proof. We consider $\gamma = \min\{\alpha, \beta_1\}$ and the value β_1 introduced in Lemma 27.1.10. If we put

$$\mathbf{u}_\gamma(t) := e^{\gamma t}\mathbf{u}(t),$$

the first and second derivatives $\dot{\mathbf{u}}_\gamma(t)$ and $\ddot{\mathbf{u}}_\gamma(t)$ can be evaluated, which yield the following relations:

$$\begin{aligned} e^{\gamma t}\dot{\mathbf{u}}(t) &= \dot{\mathbf{u}}_\gamma(t) - \gamma\mathbf{u}_\gamma(t), \\ e^{\gamma t}\ddot{\mathbf{u}}(t) &= \ddot{\mathbf{u}}_\gamma(t) - 2\gamma\dot{\mathbf{u}}_\gamma(t) + \gamma^2\mathbf{u}_\gamma(t). \end{aligned} \tag{27.1.28}$$

In a similar way $r_\gamma(t) := e^{\gamma t}r(t)$ is introduced, and its first derivative $\dot{r}_\gamma(t)$ yields an expression analogous to $(27.1.28)_1$.

Moreover, the product of the convolution of $\dot{r}$ and $\dot{\mathbf{u}}$ by $e^{\gamma t}$ can be expressed as

$$\begin{aligned} e^{\gamma t}(\dot{r} * \dot{\mathbf{u}})(t) &= \int_0^t e^{\gamma(t-s)}\dot{r}(t-s)e^{\gamma s}\dot{\mathbf{u}}(s)ds \\ &= \left(\dot{r}_\gamma * \dot{\mathbf{u}}_\gamma\right)(t) - \gamma\left(r_\gamma * \dot{\mathbf{u}}_\gamma\right)(t) - \gamma\left(\dot{r}_\gamma * \mathbf{u}_\gamma\right)(t) + \gamma^2\left(r_\gamma * \mathbf{u}_\gamma\right)(t), \end{aligned}$$

by virtue of $(27.1.28)_1$. Multiplying (27.1.23) by $e^{\gamma t}$ and using the last three results, we have

$$\mathbf{u}_\gamma + \kappa_\gamma\dot{\mathbf{u}}_\gamma + p_\gamma * \dot{\mathbf{u}}_\gamma + q_\gamma * \mathbf{u}_\gamma + \left(A + \vartheta_\gamma I\right)\mathbf{u}_\gamma = \mathbf{g}_\gamma + r_\gamma\mathbf{u}_1, \tag{27.1.29}$$

where

$$\begin{aligned} &p_\gamma(t) := \dot{r}_\gamma(t) - \gamma r_\gamma(t), \qquad q_\gamma(t) := \gamma\left[\gamma r_\gamma(t) - \dot{r}_\gamma(t)\right], \\ &\mathbf{g}_\gamma(t) := e^{\gamma t}\mathbf{g}(t), \quad \kappa_\gamma := r(0) - 2\gamma > 0, \quad \vartheta_\gamma := \gamma[\gamma - r(0)]. \end{aligned} \tag{27.1.30}$$

By virtue of Propositions 27.1.4 and 27.1.9 the functions p_γ, q_γ, and $\mathbf{g}_\gamma$ belong to $L^1(0,\infty)$. From (27.1.19), taking into account that

$$r_L(\lambda-\gamma)=r_{\gamma L}(\lambda),\qquad \lambda r_L(\lambda)=\dot r_{\gamma L}(\lambda)+r(0)$$

and using (27.1.30), we have

$$\begin{aligned} m_n(\lambda-\gamma) &= (\lambda-\gamma)^2\left[1+r_{\gamma L}(\lambda)\right] \\ &= \lambda^2+\kappa_\gamma\lambda+\lambda p_{\gamma L}(\lambda)+q_{\gamma L}(\lambda)+\vartheta_\gamma+\mu_n,\quad \mathrm{Re}\lambda\ge 0. \end{aligned} \tag{27.1.31}$$

From Lemma 27.1.10 and the definition of the constant γ it follows that (27.1.12) is satisfied together with $\kappa_\gamma>0$. Therefore, using Lemma 27.1.3, we can state that the linear operators

$$\begin{aligned} T_{\gamma L}(\lambda) &= \left\{\left[\lambda^2+\kappa_\gamma\lambda+\lambda p_{\gamma L}(\lambda)+q_{\gamma L}(\lambda)+\vartheta_\gamma\right]I+A_\Omega\right\}^{-1}, \\ R_{\gamma L}(\lambda) &= \lambda T_{\gamma L}(\lambda), \end{aligned} \tag{27.1.32}$$

corresponding to (27.1.31), are well defined for all λ with $\mathrm{Re}\lambda\ge 0$; moreover, we have

$$\begin{aligned} &\sup\left\{\int_{-\infty}^{\infty}\|T_{\gamma L}(\xi+i\eta)\|^2_{\mathcal{L}(\mathcal{H}_\Omega)}d\eta,\ \xi\ge 0\right\}<\infty, \\ &\sup\left\{\int_{-\infty}^{\infty}\|R_{\gamma L}(\xi+i\eta)\mathbf{v}\|^2_{\mathcal{H}_\Omega}d\eta,\ \xi\ge 0\right\}<\infty,\quad \forall\mathbf{v}\in\mathcal{H}_\Omega. \end{aligned} \tag{27.1.33}$$

From (27.1.29), by applying the Laplace transform, one can derive

$$\mathbf{u}_{\gamma L}(\lambda)=T_{\gamma L}(\lambda)\left\{\left[1+r_{\gamma L}(\lambda)\right]\mathbf{u}_1+\left[p_{\gamma L}(\lambda)+\kappa_\gamma+\gamma\right]\mathbf{u}_0+\mathbf{g}_{\gamma L}(\lambda)\right\}+R_{\gamma L}(\lambda)\mathbf{u}_0, \tag{27.1.34}$$

for $\mathrm{Re}\lambda>\gamma$, where $T_{\gamma L}(\lambda)$ and $R_{\gamma L}(\lambda)$ are given by (27.1.32), and in particular, $\kappa_\gamma+\gamma=r(0)-\gamma>0$, by virtue of $(27.1.30)_4$.

Hence, it follows that the function

$$\mathbf{z}_L(\lambda)=T_{\gamma L}(\lambda)\left\{\left[1+r_{\gamma L}(\lambda)\right]\mathbf{u}_1+\left[p_{\gamma L}(\lambda)+\kappa_\gamma+\gamma\right]\mathbf{u}_0+\mathbf{g}_{\gamma L}(\lambda)\right\}+R_{\gamma L}(\lambda)\mathbf{u}_0,$$

for $\mathrm{Re}\lambda\ge 0$, is continuous in $\bar{\mathbf{\Pi}}$ and analytic in $\mathbf{\Pi}$. Moreover, since $|p_{\gamma L}(\lambda)|\le\|p_\gamma\|_{L^1}$ and $\|\mathbf{g}_{\gamma L}(\lambda)\|_{\mathcal{H}_\Omega}\le\|\mathbf{g}_\gamma\|_{L^1}$, we obtain

$$\begin{aligned} \|\mathbf{z}_L(\lambda)\|_{\mathcal{H}_\Omega} \le{}& \|T_{\gamma L}(\lambda)\|_{\mathcal{L}(\mathcal{H}_\Omega)}\Big[\left(1+\|r_\gamma\|_{L^1}\right)\|\mathbf{u}_1\|_{\mathcal{H}} \\ &+\left(\|p_\gamma\|_{L^1}+\kappa_\gamma+\gamma\right)\|\mathbf{u}_0\|_{\mathcal{H}}+\|\mathbf{g}_\gamma\|_{L^1}\Big]+\|R_{\gamma L}(\lambda)\mathbf{u}_0\|_{\mathcal{H}_\Omega}. \end{aligned}$$

Thus, relations (27.1.33) yield

$$\sup\left\{\int_{-\infty}^{\infty}\|\mathbf{z}_L(\xi+i\eta)\|^2_{\mathcal{H}_\Omega}d\eta,\ \xi\geq 0\right\}<\infty,$$

whence the function $\lambda\longmapsto \mathbf{z}_L(\lambda)$ belongs the Hardy space $H^2(\mathbf{\Pi},\mathcal{H}_\Omega)$. From Theorem 27.1.2 it follows that there exists $\mathbf{z}\in L^2((0,\infty),\mathcal{H})$ with its Laplace transform $\mathbf{z}_L(\lambda)$ for all λ with $\mathrm{Re}\lambda\geq 0$. The uniqueness of the Laplace transform yields that $\mathbf{z}(t)=\mathbf{u}_\gamma(t)$ for almost every $t>0$ and

$$\int_0^\infty e^{2\gamma t}\|\mathbf{u}(t)\|^2_{\mathcal{H}}dt=\int_0^\infty \|\mathbf{u}_\gamma(t)\|^2_{\mathcal{H}}dt<\infty.$$

It must be shown that $\dot{\mathbf{u}}_\gamma$ is bounded.

First observe that from the definition $(27.1.32)_1$ of $T_{\gamma L}$, we have

$$T_{\gamma L}(\lambda)\Big\{\big[\lambda^2+\kappa_\gamma\lambda+\lambda p_{\gamma L}(\lambda)\big]I\Big\}=I-T_{\gamma L}(\lambda)\Big\{\big[q_{\gamma L}(\lambda)+\vartheta_\gamma\big]I+A_\Omega\Big\}.$$

Using this relation, as before, to derive the form of $\dot{\mathbf{u}}_L(\lambda)$, we obtain the following expression for $\dot{\mathbf{u}}_{\gamma L}$:

$$\begin{aligned}\mathbf{u}_{\gamma L}(\lambda)&=\lambda\mathbf{u}_{\gamma L}(\lambda)-\mathbf{u}_\gamma(0)=T_{\gamma L}(\lambda)\big\{\dot{r}_\gamma(\lambda)(\mathbf{u}_1+\gamma\mathbf{u}_0)+\dot{\mathbf{g}}_{\gamma L}(\lambda)+r(0)(\mathbf{u}_1+\gamma\mathbf{u}_0)\\&\quad+\mathbf{g}_\gamma(0)-\gamma^2\left[1+r_{\gamma L}(\lambda)\right]\mathbf{u}_0-A_\Omega\mathbf{u}_0\big\}+R_{\gamma L}(\lambda)(\mathbf{u}_1+\gamma\mathbf{u}_0).\end{aligned}$$

Let us consider the function $\mathbf{y}_L:\bar{\mathbf{\Pi}}\to\mathcal{H}_\Omega$ defined by

$$\begin{aligned}\mathbf{y}_L(\lambda)&=T_{\gamma L}(\lambda)\big\{\dot{r}_\gamma(\lambda)(\mathbf{u}_1+\gamma\mathbf{u}_0)+\dot{\mathbf{g}}_{\gamma L}(\lambda)+r(0)(\mathbf{u}_1+\gamma\mathbf{u}_0)\\&\quad+\mathbf{g}_\gamma(0)-\gamma^2\left[1+r_{\gamma L}(\lambda)\right]\mathbf{u}_0-A_\Omega\mathbf{u}_0\big\}+R_{\gamma L}(\lambda)(\mathbf{u}_1+\gamma\mathbf{u}_0).\end{aligned}$$

From the properties of $T_{\gamma L}$ and $R_{\gamma L}$, it follows that $\mathbf{y}_L\in H^2(\mathbf{\Pi},\mathcal{H}_\Omega)$. Consequently, there exists a function $\mathbf{y}\in L^2((0,\infty),\mathcal{H})$ such that $\mathbf{y}_L$ is its Laplace transform. We have $\dot{\mathbf{u}}_\gamma=\mathbf{y}$ and

$$\int_0^\infty e^{2\gamma t}\|\dot{\mathbf{u}}(t)\|^2_{\mathcal{H}}dt<\infty.$$

Therefore, $\mathbf{u}_\gamma(t)\to 0$ in $\mathcal{H}$ as $t\to\infty$, since

$$\|\mathbf{u}_\gamma(t_2)-\mathbf{u}_\gamma(t_1)\|_{\mathcal{H}}\leq\left|\int_{t_1}^{t_2}\|\dot{\mathbf{u}}_\gamma(t)\|_{\mathcal{H}}dt\right|\leq|t_2-t_1|^{1/2}\left[\int_0^\infty\|\dot{\mathbf{u}}_\gamma(t)\|^2_{\mathcal{H}}dt\right]^{1/2};$$

moreover, $\mathbf{u}_\gamma$ is uniformly continuous. Thus, (27.1.27) follows and the proof of the theorem is complete. □

The dependence of $\mathbf{f}$ on the kernel k, expressed by (27.1.5), allows us to consider, instead of (27.1.4), a problem that provides the *fundamental solution* or *resolvent* associated with (27.1.4).

Thus, consider the Volterra homogeneous equation

$$\ddot{\mathbf{u}}(t) + A\mathbf{u}(t) + \int_0^t k(t-s)A\mathbf{u}(s)ds = \mathbf{0} \quad \forall t > 0, \tag{27.1.35}$$

for which the initial conditions

$$\begin{aligned} \mathbf{u}(t) &= \mathbf{0} \quad \forall t < 0, \\ \mathbf{u}(0) &= \mathbf{u}_0, \\ \dot{\mathbf{u}}(0) &= \mathbf{u}_1, \end{aligned} \tag{27.1.36}$$

where $\mathbf{u}_0 \in D(A)$ and $\mathbf{u}_1 \in \mathcal{H}$, are assumed, instead of (27.1.2). Moreover, the solution $\mathbf{u}$, the domain of which is $\mathbb{R}$, is discontinuous at $t = 0$, and the restriction of $\mathbf{u}$ to $[0, \infty)$ is in the space $C([0, \infty), D(A)) \cap C^2([0, \infty), \mathcal{H})$, instead of that specified by (27.1.3). Because of the assumed new space, the definition of strong solution is changed.

Before giving the third theorem, we recall a useful result, derived in the proof of Theorem 2 of [273], expressed by the following lemma.

Lemma 27.1.13. *Let $h : (0, \infty) \to \mathbb{R}$ be an integrable function with a single sign almost everywhere. Suppose that there is an analytic function $a : U \to \Omega$, defined on a neighborhood U of 0 in Ω such that $h_L(\lambda) = a(\lambda)\, \forall \lambda \in U$ with $\mathrm{Re}\lambda \geq 0$. Then there is an $\alpha > 0$ such that*

$$\int_0^\infty e^{\alpha s}|h(s)|ds < \infty.$$

Actually, α can be any positive number less than the radius of convergence of the power series of a about 0.

We can now consider the following third theorem.

Theorem 27.1.14. *Let $\mathbf{u}$ be the strong solution of (27.1.35) and (27.1.36). Suppose that k obeys (27.1.9) and $\mathbf{u}_1 \in D(A)$.*

(i) Suppose that there is an $\alpha > 0$ such that the inequalities (27.1.17) hold. Then there exists $\gamma > 0$ such that

$$\int_0^\infty e^{2\gamma t}\left[\|\mathbf{u}(t)\|_{\mathcal{H}}^2 + \|A\mathbf{u}(t)\|_{\mathcal{H}}^2 + \|\dot{\mathbf{u}}(t)\|_{\mathcal{H}}^2\right]dt < \infty. \tag{27.1.37}$$

Moreover, there is a constant $M > 0$such that

$$\|\mathbf{u}(t)\|_{\mathcal{H}} \leq Me^{-\gamma t}, \quad \|\dot{\mathbf{u}}(t)\|_{\mathcal{H}} \leq Me^{-\gamma t} \tag{27.1.38}$$

for any $t \geq 0$.

(ii) Suppose that for some $\mathbf{u}_1 \neq \mathbf{0}$, *there is a* $\gamma > 0$ *such that (27.1.37) holds. Then there exists* $\alpha > 0$ *for which the inequalities (27.1.17) hold and there exists* $K > 0$ *such that*

$$|k(t)| < Ke^{-\alpha t}$$

for any $t \geq 0$.

Proof. Instead of assuming (27.1.9), we initially consider the weaker conditions expressed by (27.1.8).

The proof of (*i*) follows from the same arguments already used.

We easily see that $A_\Omega T_{\gamma L} = T_{\gamma L} A_\Omega$ on $D(A_\Omega)$; hence, $A_\Omega R_{\gamma L} = R_{\gamma L} A_\Omega$ on $D(A_\Omega)$. Thus, since we are considering the homogeneous equation (27.1.35) and (27.1.34) yields

$$A_\Omega \mathbf{u}_{\gamma L}(\lambda) = T_{\gamma L}(\lambda)\Big\{\Big[1 + r_{\gamma L}(\lambda)\Big] A_\Omega \mathbf{u}_1 + \Big[p_{\gamma L}(\lambda) + \kappa_\gamma + \gamma\Big] A_\Omega \mathbf{u}_0\Big\} + R_{\gamma L}(\lambda) A_\Omega \mathbf{u}_0,$$

for $\mathrm{Re}\lambda > \gamma$. An argument similar to those used previously gives

$$\int_0^\infty e^{2\gamma t} \|A\mathbf{u}(t)\|^2_{\mathcal{H}} dt < \infty.$$

From (27.1.11) and (27.1.29) we obtain $\ddot{\mathbf{u}}_\gamma \in L^2((0,\infty),\mathcal{H})$. This yields the uniform continuity of $\dot{\mathbf{u}}_\gamma$; moreover, the pointwise estimate on $\|\dot{\mathbf{u}}(t)\|_{\mathcal{H}}$ in $(27.1.38)_2$ follows.

To prove (*ii*) we observe that from (*i*) it follows that (27.1.37) holds for some $\gamma > 0$; therefore, the Laplace transforms of $\mathbf{u}$, $\dot{\mathbf{u}}$, and $A\mathbf{u}$ are (weakly) holomorphic in $\{\lambda \in \Omega;\ \mathrm{Re}\lambda > -\gamma\}$.

From (27.1.11) and (27.1.35) it follows that $\ddot{\mathbf{u}}_\gamma \in L^2((0,\infty),\mathcal{H})$ and that the Laplace transform of $\ddot{\mathbf{u}}$ is also (weakly) holomorphic in $\{\lambda \in \Omega;\ \mathrm{Re}\lambda > -\gamma\}$. Applying the Laplace transform to (27.1.35), we have

$$\lambda^2 \mathbf{u}_L(\lambda) - \lambda \mathbf{u}_0 - \mathbf{u}_1 + [1 + k_L(\lambda)] A_\Omega \mathbf{u}_L(\lambda) = \mathbf{0}, \quad \mathrm{Re}\lambda \geq 0.$$

Observe that A is positive definite; thus, $(A_\Omega \mathbf{u}_L(0), \mathbf{u}_L(0))_{\mathcal{H}_\Omega} = 0$ if and only if $\mathbf{u}_L(0) = \mathbf{0}$. The condition $\mathbf{u}_L(0) = \mathbf{0}$ yields $\mathbf{u}_1 = \mathbf{0}$, but this possibility is excluded by hypothesis. Therefore, the function $\lambda \longmapsto (A_\Omega \mathbf{u}_L(\lambda), \mathbf{u}_L(0))_{\mathcal{H}_\Omega}$ is analytic and nonzero on a neighborhood U of 0 and

$$k_L(\lambda) = -\frac{\left(\lambda^2 \mathbf{u}_L(\lambda) - \lambda \mathbf{u}_0 - \mathbf{u}_1 + A_\Omega \mathbf{u}_L(\lambda), \mathbf{u}_L(0)\right)_{\mathcal{H}_\Omega}}{(A_\Omega \mathbf{u}_L(\lambda), \mathbf{u}_L(0))_{\mathcal{H}_\Omega}}, \quad \mathrm{Re}\lambda \geq 0.$$

From Lemma 27.1.13 it follows that

$$\int_0^\infty e^{\alpha_1 s} |k(s)| ds < \infty \quad \forall \alpha_1 \in (0,\gamma).$$

Since if k' is integrable and $k'(s) \geq 0\ \forall s \geq 0$ (see (27.1.9)), we can apply Lemma 27.1.13 to $k'_L(\lambda) = \lambda k_L(\lambda) - k(0)$ and obtain

$$\int_0^{\infty} e^{\alpha s}|k'(s)|ds < \infty \quad \forall \alpha \in (0, \alpha_1).$$

From

$$k(t)e^{\alpha t} = \int_0^t e^{\alpha s}[k'(s) + \alpha k(s)]ds - k(0),$$

it follows that $k(t)e^{\alpha t}$ tends to a limit as $t \to \infty$, and this limit must be zero.

□

28

Semigroup Theory for Abstract Equations with Memory

We consider in this chapter, in a mathematically abstract way, the evolution equations of the kind discussed in Chaps. 24 and 27, comparing the new state formulation with the traditional history approach. Relevant background to this discussion is the concept of a minimal state discussed in Part III from Sect. 7.4 onward, in particular, certain conclusions of Sect. 16.5.

28.1 Introduction

Roughly speaking, an evolution equation with memory has the formal structure

$$\partial_t w(t) = \mathbb{F}(w(t), w^t(\cdot)), \quad t > 0, \tag{28.1.1}$$

where $w^t(s) = w(t - s)$, for $s > 0$, and $\mathbb{F}$ is an operator acting on both $w(t)$ and the past values of w up to the actual time t. The function w is assumed to be known for all $t \leq 0$. Accordingly, the initial condition takes the form $w(t)|_{t \leq 0} = w_0(t)$, for a given $w_0(\cdot)$. The main difficulty in the analysis of these equations comes from their nonlocal character, due to the presence of the memory term (typically, a convolution integral). A way to circumvent the obstacle is to rephrase (28.1.1) as an ODE in some abstract space, by introducing an auxiliary variable accounting for the past history of w. This is the strategy devised by Dafermos [79], who viewed w^t as an additional variable ruled by its own differential equation, and so translated (28.1.1) into a differential system on an extended space accounting for memory. However, what one can actually measure, when dealing with (28.1.1), is the function $w(t)$ for $t \geq 0$. The practical consequences are of some relevance, since in concrete situations, the problem of assigning the initial conditions is not only of a theoretical nature. In particular, it might happen that two *different* initial past histories w_0 lead to the *same* $w(t)$ for $t \geq 0$. From the viewpoint of the dynamics, such two different initial past histories are in a fact indistinguishable (see Sect. 24.1). This observation suggests that rather than w^t, one should employ an alternative variable to describe the initial state of the system, which we call a (minimal) *state*, satisfying the natural minimality property that two

G. Amendola et al., *Thermodynamics of Materials with Memory*,
https://doi.org/10.1007/978-3-030-80534-0_28

different values of this initial state produce different evolutions $w(t)$, for $t \geq 0$. The main task is then to determine, if possible, what is a minimal state associated with (28.1.1). Unfortunately, a universal strategy is out of reach, and the correct choice depends on the particular concrete realization of (28.1.1). Nonetheless, for a large class of equations with memory, where the memory contribution enters in the form of a convolution integral with a nonincreasing positive kernel, a general approach seems available. In what follows, we discuss an abstract sample equation with memory arising from linear viscoelasticity, presenting a theoretical scheme that can be easily extended and adapted to many other differential models containing memory terms.

Let H be a separable real Hilbert space and let A be a self-adjoint strictly positive linear operator on H with dense domain $\mathbb{D}(A) \subset H$ and compact inverse. We consider, for $t > 0$, the abstract linear differential equation with memory

$$\partial_{tt}u(t) + A\left[\alpha u(t) - \int_0^\infty \mu(s)u(t-s)ds\right] = 0, \tag{28.1.2}$$

supplemented with the initial conditions

$$u(0) = u_0, \quad \partial_t u(0) = v_0, \quad u(-s)|_{s\in(0,\infty)} = \phi_0(s), \tag{28.1.3}$$

where $u_0, v_0, \phi_0(\cdot)$ are prescribed data. Here, $\alpha > 0$ and the *memory kernel* $\mu : (0,\infty) \to \mathbb{R}^+$ is a strictly positive nonincreasing summable function, locally absolutely continuous, satisfying $\lim_{s\to\infty} \mu(s) = 0$. Setting

$$M(s) = \int_s^\infty \mu(\sigma)d\sigma = \int_0^\infty \mu(s+\sigma)d\sigma,$$

we require that $M(0) < \alpha$. Without loss of generality, we may assume that $\alpha - M(0) = 1$. The dissipativity of the system is entirely contained in the convolution term, accounting for infinite delay. Defining

$$F_0(t) = \int_0^\infty \mu(t+s)\phi_0(s)ds, \tag{28.1.4}$$

Eq. (28.1.2) takes the form

$$\partial_{tt}u(t) + A\left[\alpha u(t) - \int_0^t \mu(s)u(t-s)ds - F_0(t)\right] = 0. \tag{28.1.5}$$

Introducing the Hilbert space $V = \mathbb{D}(A^{1/2})$, with $\langle \cdot, \cdot \rangle_V = \langle A^{1/2}\cdot, A^{1/2}\cdot \rangle_H$, we stipulate the following definition of a (weak) solution.

Definition 28.1.1. *Let $u_0 \in V$, $v_0 \in H$ and $\phi_0 : (0,\infty) \to V$ such that the corresponding F_0 given by (28.1.4) belongs to $L^1_{\mathrm{loc}}([0,\infty); V)$. A function*

$$u \in C([0,\infty), V) \cap C^1([0,\infty), H)$$

is a solution of (28.1.2)–(28.1.3) if $u(0) = u_0, \partial_t u(0) = v_0$ and for every $w \in V$ and a.e. $t > 0$,

$$\langle \partial_{tt} u(t), w \rangle + \alpha \langle u(t), w \rangle_V - \int_0^t \mu(s) \langle u(t-s), w \rangle_V ds - \langle F_0(t), w \rangle_V = 0.$$

A concrete realization of (28.1.2) is obtained by setting $H = [L^2(\Omega)]^3$*, where* $\Omega \subset \mathbb{R}^3$ *is a smooth bounded domain and* $A = -\Delta$ *with the Dirichlet boundary conditions. In this case, putting* $u(t) = \boldsymbol{u}(\boldsymbol{x}, t)$ *and* $M(s) = G(s) - 1$*, the equation becomes*

$$\partial_{tt} \boldsymbol{u}(\boldsymbol{x}, t) - \Delta \left[G(0) \boldsymbol{u}(\boldsymbol{x}, t) + \int_0^\infty G'(s) \boldsymbol{x}(\boldsymbol{x}, t-s) ds \right] = \boldsymbol{0}, \quad \boldsymbol{u}(\boldsymbol{x}, t)|_{\boldsymbol{x} \in \partial\Omega} = \boldsymbol{0},$$

which governs the evolution of the relative displacement field $\boldsymbol{u}$ *in a homogeneous isotropic linearly viscoelastic solid occupying a volume* Ω *at rest [292].*

28.1.0.1 Notation

The symbols $\langle \cdot, \cdot \rangle_X$ and $\| \cdot \|_X$ stand for the inner product and the norm, respectively, on a generic Hilbert space X. In addition to the spaces H and V, we consider the dual space $V^* = \mathcal{D}(A^{-1/2})$ of V, denoting by $\langle \cdot, \cdot \rangle$ the duality product. For a nonnegative measurable function ρ on $(0, \infty)$, we define the weighted L^p-space of X-valued functions

$$L^p_\rho(0, \infty; X) = \left\{ \psi : (0, \infty) \to X; \quad \int_0^\infty \rho(s) \| \psi(s) \|^p_X ds < \infty \right\},$$

endowed with the standard norm; it is a Hilbert space if $p = 2$. Finally, given a generic function $\psi : (0, \infty) \to X$, we denote by $D\psi$ its distributional derivative.

28.2 The History Formulation

An alternative way of looking at the evolution equation is to work in the so-called Dafermos's history space framework [79], by considering the *history* variable

$$\eta^t(s) = u(t) - u(t-s), \quad t \geq 0, \ s \in (0, \infty),$$

which formally obeys the relations

$$\begin{cases} \partial_t \eta^t(s) = -\partial_s \eta^t(s) + \partial_t u(t), \\ \eta^t(0) = 0, \quad \eta^0(s) = u_0 - \phi_0(s). \end{cases}$$

To place our ideas in a precise context (cf. [183, 283]), we introduce the *history space* $\mathcal{M} = L^2_\mu(0, \infty; V)$, along with the infinitesimal generator of the right-translation semigroup on $\mathcal{M}$, i.e., the operator $T\eta = -D\eta$ with domain $\mathcal{D}(T) = \{\eta \in \mathcal{M};\ D\eta \in \mathcal{M},\ \eta(0) = 0\}$. Then Eq. (28.1.1) translates into the differential system

$$\begin{cases} \partial_{tt}u(t) + A\left[u(t) + \int_0^\infty \mu(s)\eta^t(s)ds\right] = 0, \\ \partial_t\eta^t = T\eta^t + \partial_t u(t). \end{cases} \tag{28.2.1}$$

Accordingly, the initial conditions (28.1.3) become

$$u(0) = u_0, \quad \partial_t u(0) = v_0, \quad \eta^0(0) = \eta_0, \tag{28.2.2}$$

where $\eta_0(s) = u_0 - \phi_0(s)$. Defining the *extended history space*

$$\mathfrak{M} = V \times H \times \mathcal{M},$$

the problem (28.2.1)–(28.2.2) generates a contraction semigroup $\Sigma(t)$ on $\mathfrak{M}$ such that

$$\Sigma(t)(u_0, v_0, \eta_0) = (u(t), \partial_t u(t), \eta^t) \quad \forall (u_0, v_0, \eta_0) \in \mathfrak{M}.$$

Concerning the relation between (28.2.1) and (28.2.2)and the original problem (28.1.2) and (28.1.3), the following result holds [183].

Proposition 28.2.1. *Let* $(u_0, v_0, \eta_0) \in \mathfrak{M}$. *Then the first component* $u(t)$ *of* $\Sigma(t)(u_0, v_0, \eta_0)$ *solves* (28.1.2)–(28.1.3) *with*

$$F_0(t) = \int_0^\infty \mu(t+s)[u_0 - \eta_0(s)]ds.$$

28.3 The State Formulation

An essential drawback of the history approach is that for given initial data u_0 and v_0, two different initial histories may lead to the same solution $u(t)$, for $t \geq 0$. This is not surprising, since what really enters into the definition of a solution of (28.1.2)–(28.1.3) is the function F_0, defined in (28.1.4) and appearing in (28.1.5), rather than ϕ_0 (related to the initial history η_0). Thus, from the dynamical viewpoint, two initial data ϕ_{01} and ϕ_{02} should be considered fully *equivalent* when the corresponding function F_{01} and F_{02} coincide, due to the impossibility of distinguishing their effects in the future. On this basis, it seems natural to devise a scheme whereby the function F_0 appears as an initial datum accounting for the past history of u, rather than ϕ_0. In order to translate this insight into a consistent mathematical theory, it is quite helpful to see in the first instance what happens at a *formal level*. To this end, for $t \geq 0$ and $\tau \in (0, \infty)$, we introduce the *state* variable

$$\zeta^t(\tau) = \int_0^\infty \mu(\tau + s)[u(t) - u(t-s)]ds,$$

which obeys

$$\begin{cases} \partial_t\zeta^t(\tau) = \partial_\tau\zeta^t(\tau) + M(\tau)\partial_t u(t), \\ \zeta^t(\infty) = 0, \quad \zeta^0(\tau) = \zeta_0(\tau), \end{cases}$$

where $\zeta_0(\tau) = M(\tau)u_0 - F_0(\tau)$. Accordingly, Eq. (28.2.1) takes the form

$$\partial_{tt}u(t) + A\left[u(t) + \zeta^t(0)\right] = 0,$$

where

$$\zeta^t(0) = \lim_{\tau\to 0} \zeta^t(\tau) = \int_0^\infty \mu(s)[u(t) - u(t-s)]ds.$$

Rather than ζ^t, it seems more convenient to consider as a state the new variable

$$\xi^t(\tau) = -\partial_\tau \zeta^t(\tau) = -\int_0^\infty \mu'(\tau+s)[u(t) - u(t-s)]ds,$$

which, in turn, obeys

$$\begin{cases} \partial_t \xi^t(\tau) = \partial_\tau \xi^t(\tau) + \mu(\tau)\partial_t u(t), \\ \xi^0(\tau) = \xi_0(\tau), \end{cases}$$

where the initial datum ξ_0 reads

$$\xi_0(\tau) = -\int_0^\infty \mu'(\tau+s)[u_0 - \phi_0(s)]ds = \mu(\tau)u_0 + \int_0^\infty \mu'(\tau+s)\phi_0(s)ds.$$

Since $\zeta^t(\infty) = 0$, we see that

$$\int_{\tau_0}^\infty \xi^t(\tau)d\tau = \zeta^t(\tau_0), \quad \forall \tau_0 \in (0,\infty), \tag{28.3.1}$$

and in the limit $\tau_0 \to 0$,

$$\int_0^\infty \xi^t(\tau)d\tau = \zeta^t(0).$$

Therefore, (28.1.2)–(28.1.3) is (formally) translated into the system

$$\begin{cases} \partial_{tt}u(t) + A[u(t) + \int_0^\infty \xi^t(\tau)d\tau] = 0, \\ \partial_t \xi^t(\tau) = \partial_\tau \xi^t(\tau) + \mu(\tau)\partial_t u(t), \end{cases} \tag{28.3.2}$$

with initial conditions

$$u(0) = u_0, \quad \partial_t u(0) = v_0, \quad \xi^0(\tau) = \xi_0(\tau). \tag{28.3.3}$$

At this point, two major issues need to be addressed: firstly, writing (28.3.2) and (28.3.3) as a differential equation in a suitable functional space, providing an existence and uniqueness result; secondly, establishing a correspondence (not only formal) between the solutions of (28.3.2) and (28.3.3) and those of the original problem (28.1.2) and (28.1.3).

28.4 The Semigroup in the Extended State Space

The preliminary step toward setting (28.3.2) and (28.3.3) in a proper functional framework is to interpret in a correct way the derivative ∂_τ appearing in the second equation of (28.3.2). We introduce the new memory kernel $\nu(\tau) = 1/\mu(\tau) : (0,\infty) \to [0,\infty)$ and put $\nu(0) = \lim_{\tau\to 0}\nu(\tau)$. In view of the assumptions on μ, the function ν is continuous and nondecreasing on $(0,\infty)$, with nonnegative derivative $\nu'(\tau) = -\mu'(\tau)/[\mu(\tau)]^2$. Also, $\lim_{\tau\to\infty}\nu(\tau) = \infty$. Defining the *state space* $\mathcal{V} = L^2_\nu(0,\infty;V)$, the norm of which is related to the free energy functional ψ_F (see Sect. 10.2), we consider the left-translation semigroup $L(t)$ on $\mathcal{V}$, given by

$$(L(t)\xi)(\tau) = \xi(t+\tau).$$

It is a standard matter to verify that the infinitesimal generator of $L(t)$ is the linear operator $P\xi = D\xi$ with domain $\mathbb{D}(P) = \{\xi \in \mathcal{V} : D\xi \in \mathcal{V}, \xi(\infty) = 0\}$. Note that if $\xi \in \mathbb{D}(P)$, then $\|\xi\|_V \in C(0,\infty)$; moreover, the condition $\xi(\infty) = 0$ is automatically satisfied whenever $\xi, D\xi \in \mathcal{V}$.

We are now in a position to formulate (28.3.2) and (28.3.3) as an abstract evolution equation on a suitable Hilbert space. To this end, let us define the *extended state space*

$$\mathfrak{U} = V \times H \times \mathcal{V}$$

and a linear operator $\mathbb{A}$ on $\mathfrak{U}$, with domain

$$\mathcal{D}(\mathbb{A}) = \left\{(u,v,\xi) \in \mathfrak{U};\ v \in V, u + \int_0^\infty \xi^t(\tau)d\tau \in \mathcal{D}(A), \xi \in \mathcal{D}(P)\right\},$$

acting as

$$\mathbb{A}(u,v,\xi) = \left(v, -A\left[u + \int_0^\infty \xi^t(\tau)d\tau\right], P\xi + \mu v\right).$$

Introducing the 3-component vectors $Z(t) = (u(t), v(t), \xi^t)$ and $z = (u_0, v_0, \xi_0) \in \mathfrak{U}$, we view (28.3.2) and (28.3.3) as the following Cauchy problem in $\mathfrak{U}$:

$$\begin{cases} \frac{d}{dt}Z(t) = \mathbb{A}Z(t), \\ Z(0) = z. \end{cases} \tag{28.4.1}$$

The next result establishes existence and uniqueness of a solution $Z \in C([0,\infty),\mathfrak{U})$.

Theorem 28.4.1. *The problem* (28.4.1) *generates a strongly continuous semigroup* $S(t) = e^{t\mathbb{A}}$ *of linear contractions on* $\mathfrak{U}$.

Moreover, for every $z = (u_0, v_0, \xi_0) \in \mathfrak{U}$, the third component ξ^t of the solution $S(t)z$ (i.e., the state) has the explicit representation formula

$$\xi^t(\tau) = \xi_0(t+\tau) + \mu(\tau)u(t) - \mu(t+\tau)u_0 + \int_0^t \mu(\tau+s)u(t-s)ds. \qquad (28.4.2)$$

If μ satisfies the further condition (cf. [149, 239, 270])

$$\mu'(s) + \delta\mu(s) \leq 0, \qquad (28.4.3)$$

for some $\delta > 0$ and a.e. $s \in (0, \infty)$, the semigroup $S(t)$ turns out to be exponentially stable.

Theorem 28.4.2. *Let (28.4.3) hold. Then there exist* $K > 1$ *and* $\omega > 0$ *such that*

$$\|S(t)z\|_{\mathfrak{U}} \leq K\|z\|_{\mathfrak{U}}e^{-\omega t} \quad \forall z \in \mathfrak{U}.$$

The reader is referred to [106] for the proofs of Theorems 28.4.1 and 28.4.2. Here, we limit ourselves to mentioning that the exponential stability of $S(t)$ is obtained *only* via energy estimates, without appealing to linear semigroup techniques. Thus, the method can be exported to the analysis of semilinear versions of the problem (for instance, to prove the existence of absorbing sets and global attractors).

Remark 28.4.3. The state variable ξ^t is *minimal* in the following sense: if $(u(t), \partial_t u(t), \xi^t)$ is a solution to (28.4.1) with $u(t) = 0$ for every $t \geq 0$, then ξ^t is identically zero. Indeed, on account of (28.4.1) and (28.4.2),

$$\xi^t(\tau) = \xi_0(t+\tau) \quad \forall t \geq 0$$

and

$$0 = \int_0^\infty \xi^t(\tau)d\tau = \int_0^\infty \xi_0(t+\tau)d\tau = \int_t^\infty \xi_0(\tau)d\tau \quad \forall t \geq 0,$$

which implies that $\xi_0 = 0$ and, in turn, $\xi^t = 0$.

28.5 The Original Equation Revisited

Somehow, this new state approach urges us to consider the original problem from a different perspective. Indeed, the solutions of (28.1.2)–(28.1.3) are determined, by knowledge of the function F_0 (as well as u_0 and v_0, of course) and not by the particular form of the initial past history ϕ_0.Therefore, with reference to Definition 28.1.1, let us introduce the class of *admissible past history functions*

$$\mathcal{A} = \left\{\phi : (0,\infty) \to V;\ t \mapsto \int_0^\infty \mu(t+s)\phi(s)ds \in L^1_{\text{loc}}([0,\infty); V)\right\},$$

and define the linear map $\Lambda : \mathcal{A} \to L^1_{\text{loc}}([0,\infty); V)$ as

$$\phi \mapsto \Lambda\phi(t) = \int_0^\infty \mu(t+s)\phi(s)ds.$$

Accordingly, we define the class of *state functions* $\mathcal{S} = \Lambda\mathcal{A}$. Clearly (and this is really the point), the map Λ may not be injective, meaning that different $\phi \in \mathcal{A}$ may lead to the same element of $\mathcal{S}$. Coming back to Definition 28.1.1, the assumption on F_0 can now be rephrased as $F_0 = \Lambda\phi_0$ with $\phi_0 \in \mathcal{A}$. Hence, we reformulate the definition of a solution of (28.1.2) in a more convenient (and certainly more physical) way.

Definition 28.5.1. *Let* $(u_0, v_0, F_0) \in V \times H \times \mathcal{S}$ *be given. A function*

$$u \in C([0,\infty), V) \cap C^1([0,\infty), H)$$

is a solution of (28.1.2) with initial state (u_0, v_0, F_0) *if* $u(0) = u_0$, $\partial_t u(0) = v_0$, *and for every* $w \in V$ *and a.e.* $t > 0$,

$$\langle \partial_{tt} u(t), w\rangle + \alpha\langle u(t), w\rangle_V - \int_0^t \mu(s)\langle u(t-s), w\rangle_V ds - \langle F_0(t), w\rangle_V = 0.$$

In this definition, the initial datum ϕ_0 *has completely disappeared, since the state function* F_0 *contains all necessary information to capture the future dynamics. Thus, the unphysical ambiguity caused by two different initial histories leading to the same state function has been removed.*

We now investigate the properties of the space $\mathcal{S}$. *Let us begin with a lemma, which provides a precise formulation of the formal equality (28.3.1).*

Lemma 28.5.2. *Whenever* $\phi \in \mathcal{A}$, *the map* $\tau \mapsto \int_0^\infty \mu'(\tau + s)\|\phi(s)\|_V ds$ *belongs to* $L^1(t,\infty)$ *for every* $t > 0$ *and*

$$\Lambda\phi(t) = -\int_t^\infty \left[\int_0^\infty \mu'(\tau + s)\phi(s) ds\right] d\tau \quad \forall t > 0. \tag{28.5.1}$$

Moreover, if $\phi \in L^1_\mu(0,\infty; V)$, *then* $\phi \in \mathcal{A}$ *and* (28.5.1) *holds for every* $t \geq 0$.

Proof. Let $\phi \in \mathcal{A}$ be given. For every fixed $t > 0$, we have that $\Lambda\phi(t_0) \in V$, for some $t_0 \leq t$. Since μ is nonincreasing and $\Lambda\phi(t_0)$ is a Bochner integral, this means that

$$\int_0^\infty \mu(t+s)\|\phi(s)\|_V ds \leq \int_0^\infty \mu(t_0 + s)\|\phi(s)\|_V ds < \infty.$$

Exploiting the equality

$$\mu(t+s) = -\int_t^\infty \mu'(\tau + s) d\tau$$

and exchanging the order of integration, we conclude that

$$\int_0^\infty \mu(t+s)\|\phi(s)\|_V ds = -\int_t^\infty \left[\int_0^\infty \mu'(\tau + s)\|\phi(s)\|_V ds\right] d\tau < \infty.$$

Hence, $\tau \mapsto \int_0^\infty \mu'(\tau + s)\|\phi(s)\|_V ds \in L^1(t,\infty)$ and (28.5.1) follows from Fubini's theorem. Concerning the last assertion, note that $\phi \in L^1_\mu(0,\infty; V)$ if and only if $\Lambda\phi(0) \in V$. □

A straightforward consequence is that $\mathcal{S} \subset C_0([t, \infty), V)$, for every $t > 0$, where C_0 is the space of continuous functions vanishing at infinity.

Given $F \in \mathcal{S}$, it is then interesting to see what happens to $F(t)$ in the limit $t \to 0$. Three mutually disjoint situations may occur:

(i) $\lim_{t \to 0} F(t)$ exists in V;
(ii) $F \in L^\infty(\mathbb{R}^+; V)$ but $\lim_{t \to 0} F(t)$ does not exist in V;
(iii) $\|F(t)\|_V$ is unbounded in a neighborhood of $t = 0$.

As will emerge, (i) is the most interesting case in the current context. For this reason, we introduce the further space

$$\mathcal{S}_0 = \left\{ F \in \mathcal{S};\ \exists \lim_{t \to 0} F(t) \text{ in } V \right\} \subset C_0([0, \infty), V).$$

Observe that if $F = \Lambda\phi$ with $\phi \in L^1_\mu(0, \infty; V)$, then Lemma 28.5.2 ensures that $\lim_{t \to 0} F(t) = \Lambda\phi(0)$ in V, so that $F \in \mathcal{S}_0$. However, the picture can be more complicated. Indeed, it may happen that $F \in \mathcal{S}_0$ but $\Lambda\phi(0)$ is not defined for *any* $\phi \in \Lambda^{-1}F$.

28.6 Proper States

The next goal is establishing the link between (28.4.1) and the original equation (28.1.2), up to now only formal. To this end, it is worth keeping in mind the particular form of the initial datum ξ_0, obtained in a somewhat heuristic way in Sect. 28.3. This suggests that not all states are appropriate for describing the behavior of the original equation, but only certain particular states having a well-defined structure.

Definition 28.6.1. *A vector $\xi \in \mathcal{V}$ is said to be a proper state if $\xi(\tau) = DF(\tau)$ for some $F \in \mathcal{S}$. We denote by $\mathcal{P}$ the normed subspace of $\mathcal{V}$ of proper states.*

Given a kernel μ, an immediate example of a proper state is $\xi(\tau) = \mu(\tau)u$, for any $u \in V$. Indeed, $\xi = DF$ with $F(t) = -M(t)u$.

Lemma 28.6.2. *Let $\xi \in \mathcal{P}$. Then there exists a unique $F \in \mathcal{S}$ such that $\xi = DF$. Also, F belongs to $\mathcal{S}_0$. Moreover, for every $\phi \in \mathcal{A}$ such that $F = \Lambda\phi$, it follows that*

$$\xi(\tau) = \int_0^\infty \mu'(\tau + s)\phi(s)ds.$$

Conversely, if $\xi \in \mathcal{V}$ has the above representation, then $\xi \in \mathcal{P}$ and $\xi(\tau) = D\Lambda\phi(\tau)$.

Proof. We first recall a general fact: if $\xi \in \mathcal{V}$, then $\xi \in L^1(0, \infty; V)$. Indeed, using the Hölder inequality,

$$\int_0^\infty \|\xi(\tau)\|_V d\tau = \int_0^\infty \sqrt{\mu(\tau)}\sqrt{\nu(\tau)}\|\xi(\tau)\|_V d\tau \le \sqrt{M(0)}\|\xi\|_{\mathcal{V}}.$$

Let $F \in \mathcal{S}$ be such that $\xi = DF$. Due to the relation $DF \in L^1(0, \infty; V)$, we infer that

$$t \mapsto -\int_t^\infty DF(\tau)d\tau = F(t) \in C_0([0,\infty), V).$$

Hence, $F \in \mathcal{S}_0$, and it is clearly uniquely determined by DF. The remaining assertions follow from (28.5.1). □

In particular, Lemma 28.6.2 implies that the map $\Gamma : \mathcal{P} \to \mathcal{S}$ defined as

$$\Gamma\xi(t) = -\int_t^\infty \xi(\tau)d\tau$$

is injective. Since $\Gamma\mathcal{P} \subset \mathcal{S}_0$ and the inclusion $\mathcal{S}_0 \subset \mathcal{S}$ can be strict, the map Γ is not, in general, onto. In fact, the inclusion $\Gamma\mathcal{P} \subset \mathcal{S}_0$ can also be strict.

We have now all the ingredients to state the main theorem.

Theorem 28.6.3. *Let $(u_0, v_0, F_0) \in V \times H \times \mathcal{S}$. Assume in addition that $F_0 \in \Gamma\mathcal{P}$. Then a function u is a solution to* (28.1.2) *with initial state (u_0, v_0, F_0) (according to Definition 28.5.1) if and only if*

$$(u(t), \partial_t u(t), \xi^t) = S(t)(u_0, v_0, \xi_0),$$

with ξ^t as in (28.4.2) with $\xi_0(\tau) = \mu(\tau)u_0 + DF_0(\tau)$. Conversely, if u is a solution to (28.1.2) *with initial state (u_0, v_0, F_0) and $F_0 \notin \Gamma\mathcal{P}$, then there is no corresponding solution in the extended state space.*

Proof. Since $u \in C([0,\infty), V)$, arguing as in the proof of Lemma 28.5.2, the equality

$$\int_0^\infty \left[\int_0^t \mu'(\tau + s)u(t-s)ds\right]d\tau = -\int_0^t \mu(s)u(t-s)ds$$

holds for every $t > 0$. Thus, using (28.4.2), keeping in mind the particular form of ξ_0 and the fact that $F_0 \in \mathcal{S}_0$, we readily obtain

$$\int_0^\infty \xi^t(\tau)d\tau = M(0)u(t) - \int_0^t \mu(s)u(t-s)ds - F_0(t). \tag{28.6.1}$$

This equality, in light of (28.1.5) and (28.4.1), yields the first part of the theorem. To prove the converse, assume that $u(t)$ is at the same time a solution to (28.1.2) with initial state (u_0, v_0, F_0), and equal to the first component of $S(t)(u_0, v_0, \xi_0)$, for some $\xi_0 \in \mathcal{V}$. We reach the desired result by showing that $F_0 \in \Gamma\mathcal{P}$. Indeed, now referring to ξ^t as the third component of $S(t)(u_0, v_0, \xi_0)$, one obtains (28.6.1) from (28.1.5) and (28.4.1). Since by (28.4.2),

$$\int_0^\infty \xi^t(\tau)d\tau = \int_t^\infty \xi_0(\tau)d\tau + M(0)u(t) - M(t)u_0 - \int_0^t \mu(s)u(t-s)ds,$$

we conclude that

$$\int_t^\infty [\mu(\tau)u_0 - \xi_0(\tau)]d\tau = M(t)u_0 - \int_t^\infty \xi_0(\tau)d\tau = F_0(t).$$

Hence, $-\mu(\tau)u_0 + \xi_0(\tau) = DF_0(\tau)$, meaning that $\xi_0 - \mu u_0 \in \mathcal{P}$ and $F_0 = \Gamma(\xi_0 - \mu u_0)$. □

Since we have an existence and uniqueness result in the extended state space, Theorem 28.6.3 provides an existence and uniqueness result for (28.1.2), according to Definition 28.5.1, whenever we restrict the initial states so that $F_0 \in \Gamma\mathcal{P}$. However, there are situations in which the equality $\mathcal{S} = \Gamma\mathcal{P}$ holds, as in the case of an exponential kernel.

Example 28.6.4. For $a, \kappa > 0$, let $\mu(s) = ae^{-\kappa s}$. Since $\mu(t+s) = e^{-\kappa t}\mu(s)$, it is apparent that

$$\mathcal{S} = \mathcal{S}_0 = \{F(t) = e^{-\kappa t}u;\ u \in V\} \Rightarrow \mathcal{P} = \{\xi(\tau) = e^{-\kappa\tau}u;\ u \in V\} \Rightarrow \mathcal{S} = \Gamma\mathcal{P}.$$

Incidentally, the above example sheds light on another important issue: there exist states that are not proper states; in other words, the inclusion $\mathcal{P} \subset \mathcal{V}$ is strict (and not even dense). In summary, there might be state functions of the original approach that have no corresponding (proper) states. Conversely, only the proper states describe the original problem. In this respect, the state approach is a more general model, which is able to describe within the formalism of semigroups also a certain class of Volterra's equations with nonautonomous forcing terms. Nonetheless, if we start from a proper state, it is reasonable to expect that the evolution remains confined in the space of proper states. To this end, let us define the *extended proper state space*

$$\mathfrak{U}_p = V \times H \times \mathcal{P},$$

which is a normed subspace of $\mathfrak{U}$.

Proposition 28.6.5. *If $z \in \mathfrak{U}_p$, it follows that $S(t)z \in \mathfrak{U}_p$.*

Proof. Let $z = (u_0, v_0, \xi_0) \in \mathfrak{U}_p$. Then, $\xi_0 = DF$ for some $F \in \mathcal{S}$. In turn, $F = \Lambda\phi$ for some $\phi \in \mathcal{A}$. Denoting as usual $S(t)z = (u(t), v(t), \xi^t)$ and setting

$$\psi^t(s) = u(t-s)\chi_{(0,t)}(s) + u_0\chi_{(t,\infty)}(s) - u(t), \quad \phi^t(s) = \phi(s-t)\chi_{(t,\infty)}(s),$$

the representation formula (28.4.2) can be equivalently written as

$$\xi^t(\tau) = \int_0^\infty \mu'(\tau+s)\left[\psi^t(s) + \phi^t(s)\right]ds.$$

By Lemma 28.6.2, we need to show that $\psi^t + \phi^t \in \mathcal{A}$ in order to prove that $\xi^t \in \mathcal{P}$. Indeed, since $\|S(t)z\|_{\mathfrak{U}} \leq \|z\|_{\mathfrak{U}}$,

$$\int_0^\infty \mu(s)\|\psi^t(s)\|_V ds \leq 3M(0)\|z\|_{\mathfrak{U}}.$$

Therefore, $\psi^t \in L^1_\mu(0,\infty;V) \subset \mathcal{A}$. Concerning ψ^t, we have

$$\int_0^\infty \mu(s)\phi^t(s)ds = A\phi(t) \in V,$$

which yields $\phi^t \in L^1_\mu(0,\infty;V) \subset \mathcal{A}$. □

In particular, from Theorems 28.4.1 and 28.4.2, we have the following corollary.

Corollary 28.6.6. *The restriction $S_p(t)$ of $S(t)$ on the space $\mathfrak{U}_p$ is a strongly continuous semigroup of linear contractions. Assuming also the condition (28.4.3), the semigroup $S_p(t)$ is exponentially stable.*

28.7 State Versus History

Let us finally turn to the main issue that has motivated our analysis: the comparison between the past history and the state approaches. We first show that each element of $\mathcal{M}$ gives rise to a proper state, defining the linear map $\Pi : \mathcal{M} \to \mathcal{P}$ as

$$\Pi\eta(\tau) = -\int_0^\infty \mu'(\tau+s)\eta(s)ds.$$

Lemma 28.7.1. *Let $\eta \in \mathcal{M}$. Then $\Pi\eta \in \mathcal{P}$ and $\|\Pi\eta\|_\mathcal{V} \le \|\eta\|_\mathcal{M}$.*

Proof. If $\eta \in \mathcal{M}$, we have

$$\begin{aligned}\|\Pi\eta(\tau)\|_V^2 &\le \left[\int_0^\infty -\mu'(\tau+s)\|\eta(s)\|_V ds\right]^2\\ &\le \int_0^\infty -\mu'(\tau+s)ds \int_0^\infty -\mu'(\tau+s)\|\eta(s)\|_V^2 ds\\ &= \mu(\tau)\int_0^\infty -\mu'(\tau+s)\|\eta(s)\|_V^2 ds.\end{aligned}$$

Accordingly,

$$\begin{aligned}\|\Pi\eta\|_\mathcal{V}^2 &\le \int_0^\infty d\tau \int_0^\infty -\mu'(\tau+s)\|\eta(s)\|_V^2 ds\\ &= \int_0^\infty \left[\int_0^\infty -\mu'(\tau+s)d\tau\right]\|\eta(s)\|_V^2 ds = \|\eta\|_\mathcal{M}^2.\end{aligned}$$

Thus, $\Pi\eta \in \mathcal{V}$, and the norm inequality stated above holds (in fact, equality for $\eta(s) = u$, with $u \in V$). Since $\mathcal{M} \subset L^1_\mu(0,\infty;V)$, because of the straightforward estimate

$$\int_0^\infty \mu(s)\|\eta(s)\|_V ds \le \sqrt{M(0)}\|\eta\|_\mathcal{M},$$

and $L^1_\mu(0,\infty;V) \subset \mathcal{A}$, it follows from Lemma 28.6.2 that $\Pi\eta$ is a proper state. □

We now clarify the correspondence between $\eta \in \mathcal{M}$ and its related proper state $\Pi\eta$. Letting $\bar{z} = (u_0, v_0, \eta_0) \in \mathfrak{M}$, $z = (u_0, v_0, \Pi\eta_0) \in \mathfrak{U}_p$ and setting

$$\Sigma(t)\bar{z} = (\bar{u}(t), \partial_t\bar{u}(t), \bar{\eta}^t), \quad S_p(t)z = (u(t), \partial_t u(t), \xi^t),$$

we have the following result.

Proposition 28.7.2. *The equalities $u(t) = \bar{u}(t)$ and $\xi^t = \Pi\bar{\eta}^t$ hold for every $t \geq 0$.*

Proof. Introduce the function (cf. [183, 283])

$$\eta^t(s) = \begin{cases} u(t) - u(t-s), & 0 < s \leq t, \\ \eta_0(s-t) + u(t) - u_0, & s > t, \end{cases}$$

which solves the Cauchy problem in $\mathcal{M}$,

$$\begin{cases} \frac{\mathrm{d}}{\mathrm{d}t}\eta^t = T\eta^t + \partial_t u(t), \\ \eta^0 = \eta_0. \end{cases} \tag{28.7.1}$$

The representation formula (28.4.2) for ξ^t furnishes

$$\xi^t(\tau) = \Pi\eta_0(t+\tau) + \mu(\tau)u(t) - \mu(t+\tau)u_0 + \int_0^t \mu'(\tau+s)u(t-s)ds = \Pi\eta^t(\tau).$$

Thus, exploiting (28.5.1),

$$\int_0^\infty \xi^t(\tau)d\tau = \int_0^\infty \Pi\eta^t(\tau)d\tau = \Lambda\eta^t(0) = \int_0^\infty \mu(s)\eta^t(s)ds,$$

and consequently,

$$\partial_{tt}u + A\left[u + \int_0^\infty \mu(s)\eta^t(s)ds\right] = \partial_{tt}u + A\left[u + \int_0^\infty \xi^t(\tau)d\tau\right] = 0. \tag{28.7.2}$$

Since $u(0) = u_0, \partial_t u(0) = v_0$, we conclude from (28.7.1)–(28.7.2) that

$$(u(t), \partial_t u(t), \eta^t) = \Sigma(t)\bar{z}.$$

□

Nonetheless, in general, the map $\Pi : \mathcal{M} \to \mathcal{P}$ is not injective.

Example 28.7.3. Let $N \in \mathbb{N}$. Given $a_n > 0$ and $\kappa_N > \cdots > \kappa_1 > 0$, consider the kernel $\mu(s) = \sum_{n=1}^N a_n e^{-\kappa_n s}$. For $x_m \in \mathbb{R}$ to be determined later, define

$$\eta_0(s) = u, \quad \eta_N(s) = \left[\sum_{m=1}^N x_m s^m\right] u,$$

where $u \in V$ is a fixed nonzero vector. Clearly, $\eta_0, \eta_N \in \mathcal{M}$ and $\eta_0 \neq \eta_N$. Also,

$$\Pi\eta_0(\tau) = \left[\sum_{n=1}^{N} a_n e^{-\kappa_n \tau}\right] u, \quad \Pi\eta_N(\tau) = \left[\sum_{n=1}^{N} a_n J_n e^{-\kappa_n \tau}\right] u,$$

having set

$$J_n = \kappa_n \sum_{m=1}^{N} x_m \int_0^\infty s^m e^{-\kappa_n s} ds = \sum_{m=1}^{N} b_{nm} x_m,$$

with $b_{nm} = m!/\kappa_n^m$. The matrix $\mathbb{B} = \{b_{nm}\}$ is easily seen to be invertible. Hence, we choose $\boldsymbol{x} = [x_1, \ldots, x_N]^\top$ to be the (unique) solution of the linear system $\mathbb{B}\boldsymbol{x} = [1, \ldots, 1]^\top$, in which case $J_n = 1$ for all n, so that the equality $\Pi\eta_0 = \Pi\eta_N$ holds.

However, in Example 28.7.3, one can verify that Π maps $\mathcal{M}$ *onto* $\mathcal{P}$. Thus, every proper state is realized by a history from $\mathbb{M}$. On the other hand, the next example describes a situation in which the map Π is injective on $\mathcal{M}$, but $\Pi\mathcal{M}$ is strictly contained in $\mathcal{P}$, meaning that all different histories in $\mathcal{M}$ lead to different proper states, but there are proper states that do not come from histories. We need first a definition and some preliminary results.

Definition 28.7.4. *A positive sequence* $\{\kappa_n\}, n \in \mathbb{N}$, *is called a Müntz sequence if* $\kappa_n \uparrow \infty$ *and* $\sum_{n=1}^\infty 1/\kappa_n = \infty$.

Given a function $g \in L^1_{\mathrm{loc}}([0,\infty))$ such that $s \mapsto e^{-\lambda s} g(s) \in L^1(\mathbb{R}^+)$, for some $\lambda > 0$, we denote its (real) Laplace transform by

$$\mathcal{L}g(x) = \int_0^\infty e^{-xs} g(s) ds.$$

A celebrated result due to C. Müntz says that if $\{\kappa_n\}$ is a Müntz sequence belonging to the domain of $\mathcal{L}g$ and $\mathcal{L}g(\kappa_n) = 0$ for all $n \in \mathbb{N}$, then g is identically zero (see [320]).

The following lemma is standard.

Lemma 28.7.5. *Let* $\kappa_n > 0$ *be strictly increasing and let* $\beta_n \in \mathbb{R}$ *be the general term of an absolutely convergent series. Then the function* $\sum_{n=1}^\infty \beta_n e^{-\kappa_n t} : [0,\infty) \to \mathbb{R}$ *is identically zero if and only if* $\beta_n = 0$ *for every* n.

Example 28.7.6. Consider the kernel $\mu(s) = \sum_{n=1}^\infty a_n e^{-\kappa_n s}$, with $\kappa_n > 0$ strictly increasing and $a_n > 0$ such that $\sum_{n=1}^\infty a_n < \infty$. Such a μ is summable on $\mathbb{R}^+$. We first observe that if $g \in L^1_\mu(\mathbb{R}^+)$, then $g \in L^1_{\mathrm{loc}}([0,\infty))$ and $\{\kappa_n\}$ belongs to the domain of $\mathcal{L}g$. Let us extend in the obvious way the map Π to the domain

$$\mathcal{M}_\star = \left\{\eta \in L^1_\mu(0,\infty; V);\ \tau \mapsto -\int_0^\infty \mu'(\tau+s)\eta(s) ds \in \mathcal{V}\right\}$$

(we continue to call this extended map Π). Note that $\mathcal{M} \subset \mathcal{M}_\star \subset \mathcal{A}$, and from Lemma 28.6.2 we find that $\Pi\mathcal{M}_\star \subset \mathcal{P}$. Given $\eta \in \mathcal{M}_\star$ and $w \in V^*$, consider the duality product

$$g_w(s) = \langle \eta(s), w \rangle \in L^1_\mu(\mathbb{R}^+).$$

In view of (28.5.1), the relations $\Pi\eta = 0$ and $\Lambda\eta = 0$ imply each other. But $\Lambda\eta = 0$ if and only if

$$\sum_{n=1}^{\infty} \beta_n(w) e^{-\kappa_n t} = 0, \quad \forall t \geq 0, \ \forall w \in V^*,$$

having set $\beta_n(w) = a_n \mathcal{L} g_w(\kappa_n)$. Moreover,

$$\sum_{n=1}^{\infty} |\beta_n(w)| \leq \sum_{n=1}^{\infty} a_n \int_0^{\infty} e^{-\kappa_n s} |g_w(s)| ds = \|g_w\|_{L^1_\mu(\mathbb{R}^+)} < \infty.$$

Hence, from Lemma 28.7.5, the above equality is true if and only if

$$\mathcal{L} g_w(\kappa_n) = 0, \quad \forall n \in \mathbb{N}, \ \forall w \in V^*.$$

Therefore, if $\{\kappa_n\}$ is a Müntz sequence,

$$\Pi\eta = 0 \quad \Leftrightarrow \quad g_w = 0, \ \forall w \in V^* \quad \Leftrightarrow \quad \eta = 0,$$

in which case the map Π is injective on $\mathcal{M}_\star$. Accordingly, to conclude that $\Pi\mathcal{M}$ is a proper subset of $\mathcal{P}$, we have to show that the inclusion $\mathcal{M} \subset \mathcal{M}_\star$ is strict. This is obtained, for instance, by looking at the elements $\eta(s) = \mathrm{e}^{\sigma\kappa_1 s} u$, where $\sigma \in [\frac{1}{2}, 1)$ and $u \in V$ is any nonzero vector. The details are left to the reader.

29

Identification Problems for Integrodifferential Equations

This chapter is devoted to outlining some ideas involving that part of the theory of inverse problems that is usually referred to as the identification of parameters (numbers, vectors, matrices, functions) appearing in integrodifferential equations describing the evolution of fading memory materials.

We recall the celebrated definition by Hadamard of a *well-posed problem*: it requires the existence and uniqueness of the solution to the problem and its continuous dependence on data.

Yet, from certain areas of applied sciences and technology, we are increasingly aware that many important questions are modeled by mathematical problems that turn out to be *ill-posed* in Hadamard's sense. Consequently, inverse problems have become a new and fascinating field in today's mathematical research.

From a more practical point of view, we can affirm that identification problems are, usually, far more difficult than the corresponding direct ones.

This chapter will deal with some explicit identification problems involving the recovery of unknown kernels in integrodifferential evolution equations. Problems of this kind occur in applications, e.g., in describing intrinsic properties of materials with memory, and have been intensively studied during the last two decades.

Up to now, satisfactory methods have been developed for the identification of time dependent kernels in hyperbolic equations (see [178]–[182], [209]–[213], [241]–[244], [322]) both when the data are *exactly known* and when they are affected by some (known) error.

This chapter will be devoted to a detailed discussion, with a few generalizations, of the problem studied in [244]. We will deal with a special problem related to a viscoelastic material that is, however, sufficiently complex to exhibit the basic ideas and techniques related to the general case. For this purpose, let $\Omega = \omega \times (0, \ell)$ be a bounded open cylinder in $\mathbb{R}^3$, where $\omega \subset \mathbb{R}^2$ is a general smooth bounded domain in $\mathbb{R}^2$, say at least of class C^2.

G. Amendola et al., *Thermodynamics of Materials with Memory*,
https://doi.org/10.1007/978-3-030-80534-0_29

29.1 Problem Specification

We consider a viscoelastic body $\mathcal{B}$ *with a constant density* (for the sake of simplicity), related to the reference cylinder Ω, that is governed by the integrodifferential equation

$$D_t^2 u_i(t,y) = \sum_{j=1}^{3} D_{y_j}[\sigma_{i,j}(u) + h * \sigma_{i,j}(u)](t,y) + f_i(t,y),$$
$$(t,y) \in [0,T] \times \Omega,\ i = 1,2,3, \qquad (29.1.1)$$

where $*$ denotes convolution with respect to time (see (C.3.4)), i.e.,

$$g_1 * g_2(t,y) = \int_0^t g_1(t-s,y)g_2(s,y)ds, \quad (t,y) \in [0,T] \times \Omega. \qquad (29.1.2)$$

It is assumed that the *scalar* memory kernel h depends on t only and the stress tensor components, denoted in this context by $\sigma_{i,j}$, are defined by

$$\sigma_{i,j}(\mathbf{v})(y) = \mu(y')[D_{y_j} v_i(y) + D_{y_i} v_j(y)] + \lambda(y')\delta_{i,j}\nabla \cdot \mathbf{v}(y), \quad i,j = 1,2,3,$$

where $y' = (y_1, y_2)$, $\delta_{i,j}$ denotes the Kronecker tensor and

$$\lambda, \mu \in C^{1+\sigma}(\overline{\omega}), \quad \mu(x) > 0,\ x \in \overline{\omega}, \qquad (29.1.3)$$

for some constant $\sigma \in [0, +\infty)$.

We now prescribe the initial conditions

$$u_i(0,y) = u_i^0(y), \qquad D_t u_i(0,y) = u_i^1(y), \qquad y \in \Omega,\ i = 1,2,3,$$

and, e.g., either of the boundary conditions

$$u_i(t,y) = 0, \quad (t,y) \in [0,T] \times \partial\Omega,\ i = 1,2,3,$$
$$\sum_{j=1}^{3} \nu_j[\sigma_{i,j}(\mathbf{u}) + h * \sigma_{i,j}(\mathbf{u})](t,y) = 0, \qquad (29.1.4)$$
$$(t,y) \in [0,T] \times \partial\Omega,\ i = 1,2,3,$$

where ν_j denotes the jth component of the outward unit vector ν normal to the boundary $\partial\Omega$.

In this section we will assume that the memory function h is itself *unknown*. Consequently, in order to determine h, we prescribe, in correspondence with $(29.1.4)_1$ or $(29.1.4)_2$, the following additional information:

$$\int_{\overline{\Omega}}[\varphi_0(y')D_\nu u_3(t,y)+\varphi_1(y')u_3(t,y)]d\psi(y)$$
$$+\int_0^t h(t-s)ds\int_{\overline{\Omega}}[\widetilde{\varphi}_0(y')D_\nu u_3(t,y)+\widetilde{\varphi}_1(y)u_3(s,y)]d\psi(y) \tag{29.1.5}$$
$$= g(t), \quad t\in[0,T],$$

or

$$\int_{\overline{\Omega}}\varphi_0(y)u_3(t,y)d\psi(y)=g(t), \quad t\in[0,T], \tag{29.1.6}$$

where ψ is a positive Borel measure on $\overline{\Omega}, \varphi_0, \varphi_1, \widetilde{\varphi}_0, \widetilde{\varphi}_1 \in L^1(\overline{\Omega};\psi)$ and $g:[0,T]\to\mathbb{R}$ is a (smooth) function.

Finally, it is assumed that the data $f_i, u_i^0, u_i^1 (i=1,2,3)$ *are independent of* y_3.

We can now state our identification problem: *determine a pair of functions* $\mathbf{u}: [0,T]\times\Omega\to\mathbb{R}^3, h:[0,T]\to\mathbb{R}$, $\mathbf{u}$ *being independent of* y_3*, satisfying Eqs. (29.1.1), (29.1.2) and (29.1.4)*$_1$*, (29.1.5) or (29.1.4)*$_2$*, (29.1.6).*

Since $D_{y_3}u=0$ in $(0,T)\times\Omega$, simple computations show that problem (29.1.1), (29.1.2) and (29.1.4)$_1$, (29.1.5) or (29.1.4)$_2$, (29.1.6) splits up into a *scalar identification problem* involving the third component u_3 of $\mathbf{u}$ and a *direct problem* involving the pair (u_1,u_2) (once we have determined the unknown kernel h). Such problems can be explicitly stated in the following way, where $\nu(x)$ denotes now the normal outward unit vector at $x\in\partial\omega$, while D_ν stands for the normal derivative on $\partial\omega$.

The Identification Problem (I.P.): *Determine a pair of functions* $u_3:[0,T]\times\omega\to\mathbb{R}$ *and* $h:[0,T]\to\mathbb{R}$ *such that*

$$D_t^2u_3(t,x)=\sum_{j=1}^{2}D_{x_j}[\mu D_{x_j}u_3+h*\mu D_{x_j}u_3(t,x)]+f_3(t,x), \quad (t,x)\in[0,T]\times\omega,$$
$$u_3(0,x)=u_3^0(x), \qquad D_tu_3(0,x)=u_3^1(x), \qquad x\in\omega, \tag{29.1.7}$$

where $x=(x_1,x_2)$, and satisfying

$$D_\nu u_3(t,x)+h*D_\nu u_3(t,x)=0, \quad (t,x)\in[0,T]\times\partial\omega,$$
$$\int_{\overline{\omega}}\varphi_0(x)u_3(t,x)d\psi(x)=g(t), \quad t\in[0,T], \tag{29.1.8}$$

or

$$\tilde{u}_3(t,x)=0, \qquad (t,x)\in[0,T]\times\partial\omega,$$
$$\int_{\overline{\omega}}[\varphi_0(x)D_\nu u_3(t,x)+\varphi_1(x)u_3(t,x)]d\psi(x)$$
$$+\int_0^t h(t-s)ds\int_{\overline{\omega}}[\widetilde{\varphi}_0(x)D_\nu u_3(s,x)+\widetilde{\varphi}_1(x)u_3(s,x)]d\psi(x)=g(t), \quad t\in[0,T]. \tag{29.1.9}$$

The Direct Problem (D.P.): *Determine a pair of functions* $u_1, u_2 : [0,T] \times \omega \to \mathbb{R}$ *such that*

$$D_t^2 u_i(t,x) = \sum_{j=1}^{2} D_{x_j}[\tilde{\sigma}_{i,j}(u_1,u_2) + h * \tilde{\sigma}_{i,j}(u_1,u_2)](t,x) + f_i(t,x),$$
$$(t,x) \in [0,T] \times \omega, \; i = 1,2,$$
$$u_i(0,x) = u_i^0(x), \qquad D_t u_i(0,x) = u_i^1(x) \qquad x \in \omega, \; i = 1,2,$$
$$u_i(t,x) = 0, \qquad (t,x) \in [0,T] \times \partial\omega, \; i = 1,2,$$

where

$$\tilde{\sigma}_{i,j}(v)(x) = \mu(x)[D_{x_j} v_i(x) + D_{x_i} v_j(x)] + \lambda(x)\delta_{i,j}[D_{x_1} v_1(x) + D_{x_2} v_2(x)], \quad i,j = 1,2.$$

Remark 29.1.1. Note that the additional condition $(29.1.9)_2$ covers different cases:

(i) The additional condition $u(x_0)$, where $x_0 \in \partial\omega$, corresponding to a measurement at a single (boundary) point of $\partial\omega$, is a particular case of $(29.1.8)_2$ corresponding to the case $\varphi_0 = 1$ and $\psi = \delta(x_0)$, $\delta(x_0)$ denoting the Dirac measure concentrated at x_0.

(ii) The additional condition

$$\int_\Gamma \varphi_0(x) u_3(t,x) d\sigma(x) = g(t), \quad t \in [0,T],$$

where Γ and σ denote, respectively, an open set in $\partial\omega$ and the Lebesgue measure on $\partial\omega$, corresponds to the case $\psi = \sigma$ if supp $\varphi_0 \Subset \Gamma$.

(iii) The additional condition

$$\int_\omega \varphi_0(x) u_3(t,x) dx = g(t), \quad t \in [0,T],$$

where $\varphi_0 \in L^1(\omega)$ is a function with a possibly small support in $\overline{\omega}$, to be interpreted as a sensor inserted into ω, has the meaning of a mean measurement of the (vertical) displacement of the viscoelastic body. Such a condition corresponds to the case $\psi = m_2$, the two-dimensional Lebesgue measure.

(iv) Condition $(29.1.8)_1$ corresponds to the case that the third component of the *traction* vanishes on the boundary, while condition $(29.1.9)_2$, with $\widetilde{\varphi}_i = \varphi_i$, $i = 1,2$, stands for a measurement of some mean of the third component of the *traction*.

Remark 29.1.2. Assume now that Ω is a general smooth domain in $\mathbb{R}^3$ and the initial data and source satisfy the initial conditions

$$f_i(t,y) = 0, \quad (t,y) \in [0,T] \times \Omega, \quad u_i^0(y) = 0, \quad u_i^1(y) = 0, \quad y \in \Omega, \; i = 1,2.$$

Then we can assume—via a uniqueness theorem for the direct problem—that the first two components of the displacement **u** vanish everywhere for all times, so that the third component u_3 solves the identification problem I.P., where ω is replaced with Ω. Let us conclude this section by listing some consistency conditions (C.C.) related to problem I.P. depending on whether (29.1.8) or (29.1.9) hold:

C.C.I.P.1

$$D_\nu u_3^j(x) = 0, \quad x \in \partial\omega, \ j = 0, 1,$$
$$\int_{\overline{\omega}} \varphi_0(x) u_3^j(x) d\psi(x) = g^{(j)}(0), \ j = 0, 1;$$

or

C.C.I.P.2

$$u_3^j(x) = 0, \ x \in \partial\omega,$$
$$\int_{\overline{\omega}} [\varphi_0(x) D_\nu u_3^j(x) + \varphi_1(x) u_3^j(x)] d\psi(x)$$
$$+ jh(0) \int_{\overline{\omega}} [\widetilde{\varphi}_0 D_\nu u_3^0(x) + \widetilde{\varphi}_1(x) u_3^0(x)] d\psi(x) = 0, \quad j = 0, 1.$$

Remark 29.1.3. Setting $t = 0$ in condition $(29.1.8)_1$, one can immediately deduce the first boundary consistency condition $D_\nu u_3^0(x) = 0$, $x \in \partial\omega$, in C.C.I.P.1. Differentiating condition $(29.1.8)_1$ with respect to t and taking the previous consistency condition into account, we see that $D_t u_3$ satisfies the same integral boundary condition as u_3, i.e.,

$$D_\nu D_t u_3(t, x) + h * D_\nu D_t u_3(t, x) = 0, \quad (t, x) \in [0, T] \times \partial\omega.$$

Setting here $t = 0$, one deduces the consistency condition $D_\nu u_3^1(x) = 0$, $x \in \partial\omega$.

Remark 29.1.4. Setting $t = 0$ in condition $(29.1.9)_2$, we deduce the integral condition in C.C.I.P.2 corresponding to $j = 0$. Differentiating condition $(29.1.9)_2$ with respect to t and setting $t = 0$, we deduce the integral condition in C.C.I.P.2 corresponding to $j = 1$.

Finally, consistency conditions related to problem D.P. are

C.C.D.P.

$$u_i^j(x) = 0, \quad x \in \partial\omega, \quad i = 1, 2, \quad j = 0, 1.$$

Remark 29.1.5. For lack of space, the existence of the solution to problem D.P. will not be shown. We limit ourselves to indicating that the same techniques used to solve our identification problems can be applied to prove the well-posedness of problem D.P., once the kernel h has been recovered.

29.2 Solving the First Identification Problem

We can now translate our identification problem (29.1.7), (29.1.9) into an abstract framework that is more suitable for solving it. Let H be a Hilbert space and let $A : \mathcal{D}(A) \subset H \to H$ be an *invertible* linear closed operator with $A^{-1} \in \mathcal{L}(H)$, $\mathcal{L}(H)$ denoting the Banach space consisting of all linear bounded operators from H to itself. Further, let Φ be a linear continuous functional, i.e., $\Phi \in H^*$, H^* denoting the Banach space dual to H.

Our identification problem consists in *determining two functions* $u \in W^{2,\infty}((0,T);$ $H) \cap L^\infty((0,T);\mathcal{D}(A))$ *and* $h \in W^{1,p}((0,T);\mathbb{R})$ *such that*

$$\begin{aligned} &u''(t) + Au(t) + h * Au(t) = f(t), \quad t \in [0,T],\\ &u(0) = u_0, \quad u(0) = u_1, \\ &\Phi[u(t)] = g(t), \quad t \in [0,T]. \end{aligned} \tag{29.2.1}$$

Two obvious consistency conditions for a solution (u,h) to exist are the following:

$$\Phi[u_j] = g^{(j)}(0), \quad j = 0,1. \tag{29.2.2}$$

Our strategy for solving problem (29.2.1) is to derive a fixed-point equation for the unknown kernel h.

First let us apply the functional Φ to both sides of $(29.2.1)_1$ to obtain

$$\Phi[Au(t)] + h * \Phi[Au(t)] = \Phi[f(t)] - g''(t), \quad t \in [0,T].$$

Differentiating this equation yields the relation

$$\Phi[Au'(t)] + h * \Phi[Au'(t)] + h(t)\Phi[Au_0] = \Phi[f'(t)] - g'''(t), \quad t \in [0,T].$$

Now introduce the auxiliary function

$$v(t) = Au'(t), \quad t \in [0,T] \Longleftrightarrow u(t) = u_0 + \int_0^t A^{-1}v(s)ds, \quad t \in [0,T]$$

and assume

$$\chi^{-1} := \Phi[Au_0] \neq 1.$$

Then apply the operator D_tA to both sides of $(29.2.1)_1$ and take advantage of the relation

$$D_tA\{h * Au(t)\} = D_t\{h * A^2u(t)\} = h * D_tA^2u(t) + h(t)A^2u_0, \quad t \in [0,T].$$

Relation $(29.2.1)_1$ at $t = 0$ has the form $u''(0) = -Au_0 + f(0)$, implying $v'(0) = Au''(0) = -A^2u_0 + Af(0)$. Then one concludes that the pair (v, h) solves the identification problem

$$\begin{aligned} &v''(t) + Av(t) = -h * Av(t) - h(t)A^2u_0 + Af'(t), \quad t \in [0,T],\\ &v(0) = Au_1 =: v_0, \quad v'(0) = -A^2u_0 + Af(0) =: v_1, \\ &h(t) = -\Phi[v(t)] - h * \Phi[v(t)] + \Phi[f'(t)] - g'''(t), \quad t \in [0,T]. \end{aligned} \tag{29.2.3}$$

Observe that from $(29.2.3)_3$, setting $t = 0$, we can easily compute $h_0 = h(0)$ in terms of the data:

$$h_0 = -\Phi[v_0] + \Phi[f'(0)] - g'''(0).$$

Assume now that $V \hookrightarrow H \equiv H^* \hookrightarrow V^*$ is a Gelfand triplet, V being a Hilbert space such that $\mathcal{D}(A) \hookrightarrow V$. Further, assume that the linear operator A is generated by a bilinear form $a : V \times V \to \mathbb{R}$ such that $(v, Au) = a(v, u)$ for all $u, v \in V$. Finally, assume that the quadratic form related to a satisfies

$$c_1\|v\|_V^2 \le a(v,v) \le c_2\|v\|_V^2, \quad \forall v \in V, \tag{29.2.4}$$

for some constants $c_1, c_2 \in \mathbb{R}^+$, $c_1 \le c_2$.

Unfortunately, also in this case, the direct problem is *not* weakly solvable if we assume that h is not differentiable in $[0, T]$, since we need that the right-hand side in $(29.2.3)_1$ should belong to H, while it actually belongs to the wider space V^*. However, it suffices to assume that $h \in W^{1,p}((0,T);\mathbb{R})$ for some $p \in (1, +\infty]$ and look for v in the space $W^{1,p}((0,T);H) \cap L^p((0,T);V)$. To deal with our identification problem, we need the following results relating to the direct problem

$$\begin{aligned} v''(t) + Av(t) &= f(t), \quad t \in [0,T],\\ v(0) &= v_0, \quad v'(0) = v_1. \end{aligned} \tag{29.2.5}$$

Lemma 29.2.1. *Assume that $v_0 \in V$, $v_1 \in H$, and $f \in L^1((0,T);H)$. Then problem (29.2.5) admits a unique (weak) solution $v \in W^{1,\infty}((0,T);H) \cap L^\infty((0,T);V)$ satisfying the estimate*

$$\|v'(t)\|_H^2 + \|v(t)\|_V^2 \le C_1(\|v_1\|_H^2 + \|v_0\|_V^2) + C_1\Big(\int_0^t \|f(s)\|_H ds\Big)^2, \quad t \in [0,T], \tag{29.2.6}$$

for some positive constant C_1 depending only on (c_1, c_2).

Proof. We limit ourselves to sketching the proof.* Taking the scalar product of $2v'(t)$ with $(29.2.5)_1$, one (formally) deduces the following estimate for all $t \in [0,T]$:

* The reader interested in the details is referred to [237, Chapter 9].

$$D_t[\|v'(t)\|_H^2 + a(v(t), v(t))] = 2(f(t), v'(t))_H \leq \|f(t)\|_H \|v'(t)\|_H.$$

Integrating over $[0, \tau], \tau \in (0, T]$ and taking advantage of $(29.2.3)_3$, we easily obtain the following integral inequality, holding for all $\tau \in [0, T]$:

$$\begin{aligned}
\|v'(\tau)\|_H^2 + c_1\|v(\tau)\|_V^2 &\leq \|v_1\|_H^2 + c_2\|v_0\|_V^2 + 2\int_0^\tau \|f(t)\|_H \|v'(t)\|_H dt \\
&\leq \|v_1\|_H^2 + c_2\|v_0\|_V^2 + 2\int_0^\tau \|f(t)\|_H \|v'(t)\|_H dt \\
&\leq \|v_1\|_H^2 + c_2\|v_0\|_V^2]dt \\
&\quad + 2\int_0^\tau \|f(t)\|_H [\|v'(t)\|_H^2 + c_1\|v(t)\|_V^2]^{1/2} dt.
\end{aligned}$$

Taking advantage of the following Lemma 29.2.2, which is a simple generalization of Theorem 4.9 in [27], one deduces the estimate (29.2.6).

Lemma 29.2.2. *Let z be an a.e. nonnegative $L^\infty((0, T))$ function and let b, k be a.e. nonnegative $L^1((0, T))$ functions satisfying*

$$z(\tau) \leq a + \int_0^\tau b(s)z(s)ds + \int_0^\tau k(s)z(s)^p ds, \qquad \tau \in (0, T),$$

where $p \in (0, 1)$ and $a \geq 0$ are given constants. Then for all $\tau \in (0, T)$,

$$z(\tau) \leq \exp\left(\int_0^\tau b(s)ds\right)\left[a^{1-p} + (1-p)\int_0^\tau k(s)\exp\left((p-1)\int_0^s b(\sigma)d\sigma\right)ds\right]^{1/(1-p)}.$$

We conclude by observing that the existence and uniqueness of the weak solution to problem (29.2.5) can be deduced from estimate (29.2.6) by a Faedo–Galerkin procedure. □

Lemma 29.2.3. *Assume that $v_0 \in \mathcal{D}(A)$, $v_1 \in V$, and $f \in W^{1,1}((0, T) : H)$. Then problem (29.2.5) admits a unique (strong) solution*

$$v \in W^{2,\infty}((0, T); H) \cap W^{1,\infty}((0, T); V) \cap L^\infty((0, T); \mathcal{D}(A))$$

satisfying the following estimate for all $t \in [0, T]$:

$$\begin{aligned}
&\|v''(t)\|_H^2 + \|v'(t)\|_V^2 \leq C_1(\|f(0) - Av_1\|_H^2 + \|v_1\|_V^2) + C_1\left(\int_0^t \|f'(s)\|_H ds\right)^2, \\
&\|Av(t)\| \leq C_1^{1/2}(\|f(0) - Av_1\|_H^2 + \|v_1\|_V^2)^{1/2} + \|f(0)\| + (C_1^{1/2} + 1)\int_0^t \|f'(s)\|_H ds.
\end{aligned} \tag{29.2.7}$$

Proof. Also in this case we limit ourselves to sketching the proof. Observe that the function $w = v'$ is a solution of the direct problem

$$\begin{aligned} w''(t) + Aw(t) &= f'(t), \quad t \in [0, T], \\ w(0) = v_1, \quad w'(0) &= f(0) - Av_1. \end{aligned}$$

Reasoning as above, we deduce that v' satisfies the following integral inequality for all $t \in [0, T]$:

$$\|v''(t)\|_H^2 + \|v'(t)\|_V^2 \le C_1(\|f(0) - Av_1\|_H^2 + \|v_1\|_V^2) + C_1 \left(\int_0^t \|f'(s)\|_H ds \right)^2. \quad (29.2.8)$$

Moreover, from Eq. $(29.2.5)_1$,

$$\|Av(t)\|_H \le \|v''(t)\|_H + \|f(t)\|_H \le \|v''(t)\|_H + \|f(0)\|_H + \int_0^t \|f'(s)\|_H ds, \quad t \in [0, T],$$

implying $(29.2.7)_2$ via (29.2.8). □

We can choose our data in the right functional spaces, i.e.,

$$u_0 \in \mathcal{D}(A^2), \quad u_1 \in \mathcal{D}(A^2), \quad f(0) - A^2 u_0 \in V, \quad f \in W^{2,1}((0, T); \mathcal{D}(A)). \quad (29.2.9)$$

Note that assumptions (29.2.9) imply (cf. $(29.2.3)_2$) that

$$v_0 \in \mathcal{D}(A), \quad v_1 \in V.$$

Observe now that the solution v to the direct problem (29.2.5) admits the following representation:

$$v = L_0(v_0, v_1, f(0)) + L_1(f).$$

According to Lemma 29.2.3, the linear operators L_0 and L_1 belong, respectively, to $\mathcal{L}(\mathcal{D}(A) \times V; U_T^{2,\infty})$ and $\mathcal{L}(W^{1,1}((0, T); H); U_T^{2,\infty})$, with

$$U_T^{2,\infty} = W^{2,\infty}((0, T); H) \cap W^{1,\infty}((0, T); V) \cap L^\infty((0, T); \mathcal{D}(A)).$$

Moreover, L_0 and L_1 satisfy the estimates

$$\|L_0(v_0, v_1)(t)\|_{U_t^{2,\infty}} \le C_2 \left[(\|v_1\|_H^2 + \|v_0\|_V^2)^{1/2} + \|f(0) - Av_0\|_V + \|f(0)\|_H \right], \quad t \in [0, T],$$

$$\|L_1(f)(t)\|_{U_t^{2,\infty}} \le C_2 \int_0^1 \|f'(s)\|_H ds, \quad t \in [0, T],$$

C_2 being a positive constant depending only on (c_1, c_2).

Observe that the identification problem (29.2.3) can be rewritten in the operator form

$$\begin{aligned} v &= L_0(v_0, v_1, f(0)) + L_1(Af') - L_1(h * Av + hA^2u_0), \\ h &= -\Phi[v] - h * \Phi[v] + \Phi[f'] - g'''. \end{aligned}$$

Such a problem is equivalent to the following fixed-point operator system:

$$\begin{aligned} v &= \overline{v} - L_1(h * Av + hA^2u_0), \\ h &= \overline{h} + \Phi[L_1(h * Av + hA^2u_0)] - h * \Phi[v], \end{aligned}$$

where

$$\begin{aligned} \overline{v} &= L_0(v_0, v_1, f(0)) + L_1(Af'), \\ \overline{h} &= -\Phi[L_0(v_0, v_1, f(0)) + L_1(Af') - f'] - g'''. \end{aligned} \tag{29.2.10}$$

We now introduce the new unknown

$$w = Av \Longleftrightarrow v = A^{-1}w$$

and rewrite system (29.2.10) in the following equivalent form:

$$\begin{aligned} w &= A\overline{v} - AL_1(h * w + hA^2u_0), \\ h &= \overline{h} + \Phi[L_1(h * w + hA^2u_0)] - h * \Phi[A^{-1}w]. \end{aligned}$$

The solution of such a system must be sought in $L^\infty((0,T); H) \times W^{1,p}((0,T); \mathbb{R})$, $p \in (1, = \infty]$.

The following result can be proved:

Theorem 29.2.4. *Let* $\Phi \in X^*$, $u_0 \in \mathcal{D}(A^2)$, $\Phi[Au_0] \neq 0$, $u_1 \in \mathcal{D}(A)$, $Au_1 \in V$, $f \in W^{1,p}((0,T); \mathcal{D}(A))$, $g \in W^{4,p}((0,T); \mathbb{R})$, $p \in (1, +\infty]$. *Furthermore, let the consistency conditions* (29.2.2) *hold. Then the identification problem (29.2.1) admits a unique solution* $(u, h) \in U_T^{2,\infty} \times W^{1,p}((0,T); \mathbb{R})$ *such that* $u' \in W^{1,\infty}((0,T); \mathcal{D}(A^2))$.

We can now apply our general result to problem (29.1.7) and (29.1.9) with $\psi = \delta_{x_0}$, $x_0 \in \partial\omega$ and $\varphi_1 = \widetilde{\varphi}_0 = \widetilde{\varphi}_1 = 0$, so that $\Phi[w] = w(x_0)$. Moreover, we choose $H = L^2(\omega)$, $V = H^1(\omega)$, $\mathcal{D}(A) = H^2(\omega) \cap H_0^1(\omega)$, $A = -\sum_{j=1}^2 D_{x_j}[\mu D_{x_j}]$. Indeed, in this case we use the embedding $H^2(\omega) \hookrightarrow C(\overline{\omega})$. The bilinear form a associated with A, taking into account the boundary conditions $(29.1.9)_1$, can be expressed in the form

$$a(v, u) = \int_\omega \mu(x)\nabla v(x) \cdot \nabla u(x) ds, \quad \forall u, v \in H_0^1(\omega), \tag{29.2.11}$$

after a formal integration by parts. In particular, for all $v \in H_0^1(\omega)$ we obtain

$$a(v, v) = \int_\omega \mu(x)|\nabla v(x)|^2 dx.$$

So it can be immediately concluded that a satisfies conditions (29.2.4).

We obtain the following result.

Theorem 29.2.5. *Let $\partial\Omega$ be of class C^4 and let μ satisfy properties* (29.1.3) *with* $\sigma = 2$. *Let* $u_3^0, Au_3^0, u_3^1 \in H^2(\omega) \cap H_0^1(\omega)$, $Au_3^1 \in H^1(\omega)$, $Au_3^0(x_0) \neq 0$, $f_3 \in W^{1,\widetilde{p}}((0,T); H^2(\omega) \cap H_0^1(\omega))$, $g \in W^{4,\widetilde{p}}((0,T); \mathbb{R})$, $\widetilde{p} \in (p, +\infty]$, $p \in (1, +\infty]$. *Furthermore, let the consistency conditions* $u_j(x_0) = g^{(j)}(0)$, $j = 0, 1$, *hold. Then the identification problem (29.1.7), (29.1.9) admits a unique solution* (u_3, h) *such that* $u_3, Au_3 \in W^{2,\infty}((0,T); H^2(\omega) \cap H_0^1(\omega))$ *and* $h \in W^{1,p}((0,T); \mathbb{R})$.

29.3 Solving the Second Identification Problem

We now seek to solve the identification problem related to the additional information $(29.2.1)_3$ of integral type. This allows us to relax the assumptions on the "space" regularity concerning functional Φ in our Hilbert space treatment. For this purpose, it is assumed that A is the generator of a cosine function (cf. [130]), and the auxiliary function

$$v(t) = u''(t) \Longleftrightarrow u(t) = u_0 + tu_1 + \int_0^t (t-s)v(s)ds$$

is introduced. To find an integral equation for v we start from the following well-known representation for the solution u to the direct problem (29.3.1):

$$u(t) = \overline{u}(t) - \int_0^t S(t-s)h * Av(s)ds, \quad t \in [0,T],$$

where

$$\overline{u}(t) = C(t)u_0 + S(t)u_1 + \int_0^t S(s)f(t-s)ds, \quad t \in [0,T],$$

the sine operator S being defined by $S'(t) = C(t), t \in [0,T]$, and $S(0) = O$. Then, assuming $h \in W^{1,p}((0,T); \mathbb{R})$, $p \in (1, +\infty]$, we obtain the following formulas, which hold for all $t \in [0,T]$:

$$
\begin{aligned}
&D_t^2(h * u)(t) = h * u''(t) + h'(t)u(0) + h(t)u'(0),\\
&D_t^2 \int_0^t S(s)h * Au(t-s)ds\\
&\qquad = D_t \int_0^t S(s)[h(t-s)Au_0 + h * Au'(t-s)]ds = h(0)S(t)u_0 \qquad (29.3.1)\\
&\qquad + \int_0^t S(s)[h'(t-s)Au_0 + h(t-s)Au_1 + h * Au''(t-s)]ds.
\end{aligned}
$$

Consequently, the function v solves the integral equation

$$
\begin{aligned}
v(t) = \overline{v}(t) &- \int_0^t S(t-s)[h'(s)Au_0 + h(s)Au_1]ds\\
&= - \int_0^t S(t-s)h * Av(s)ds, \quad t \in [0,T],
\end{aligned}
\tag{29.3.2}
$$

where

$$
\overline{v}(t) = C(t)v_0 + S(t)v_1 + \int_0^t S(t-s)f''(s)ds, \quad t \in [0,T], \tag{29.3.3}
$$

and

$$
v_0 = f(0) - Au_0, \quad v_1 = f'(0) - Au_1 - h(0)Au_0. \tag{29.3.4}
$$

In the present case our strategy for solving problem (29.3.2) and (29.3.4) is to assume a higher regularity of the functional Φ, i.e.,

$$
\Phi[Aw] = \Psi[w], \quad \forall w \in \mathcal{D}(A), \quad \Psi \in \mathcal{L}(H). \tag{29.3.5}
$$

To derive a fixed-point equation for the unknown kernel h, we apply the functional Φ to both sides of Eq. (29.2.1)$_1$, obtaining

$$
g''(t) + \Phi[Au(t)] + h * \Phi[Au(t)] = \Phi[f(t)], \quad t \in [0,T].
$$

Differentiating this equation twice and taking advantage of (29.3.5) yields

$$
g^{(4)}(t) + \Psi[v(t)] + h * \Psi[v(t)] + h'(t)\Phi[Au_0] + h(t)\Phi[Au_1] = \Phi[f''(t)], \quad t \in [0,T]. \tag{29.3.6}
$$

Assume now

$$
\chi^{-1} := \Phi[Au_0] \neq 0. \tag{29.3.7}
$$

Then (29.3.6) can be rewritten in the fixed-point form

$$
h'(t) = \overline{h}(t) - \chi h(t)\Phi[Au_1] - \chi\Psi[v(t)] - \chi h * \Psi[v(t)], \quad t \in [0,T], \tag{29.3.8}
$$

where

$$\overline{h}(t) = \chi\Phi[f''(t)] - \chi g^{(4)}(t), \quad t \in [0, T]. \tag{29.3.9}$$

Recall that the functions C and S satisfy the following relations for some $\rho \in \mathbb{R}$:

$$\begin{aligned}
&C \in C([0, +\infty); \mathcal{L}(H)) \cap C^2([0, +\infty); \mathcal{L}(\mathcal{D}(A); H)),\\
&C(0) = I, \quad C''(t) = -AC(t), \quad \|C(t)\|_{\mathcal{L}(H)} \le Me^{t\rho}, \quad t \in [0, T],\\
&S \in C^1([0, +\infty); \mathcal{L}(H)) \cap C^2([0, +\infty); \mathcal{L}(\mathcal{D}(A); H)),\\
&S(0) = O, \quad S''(t) = -AS(t), \quad \|S(t)\|_{\mathcal{L}(H)} \le Me^{t\rho}, \quad t \in [0, T],\\
&C'(t) = -AS(t), \quad S'(t) = C(t), \quad t \in [0, T].
\end{aligned}$$

To solve the system (29.3.2) and (29.3.8), the relations

$$\begin{aligned}
\int_0^t S(t-s)h * Av(s)ds &= \int_0^t AS(t-s)h * v(s)ds\\
&= \int_0^t D_s[C(t-s)]h * v(s)ds\\
&= h * v(t) - \int_0^t C(t-s)[h' * v(s) + h(0)v(s)]ds, \quad t \in [0, T],
\end{aligned} \tag{29.3.10}$$

which hold for all $v \in C^1([0, T]; H) \cap C^1([0, T]; \mathcal{D}(A))$, are required. Consequently, the fixed-point system for (v, h) can be rewritten in the form

$$\begin{aligned}
v(t) &= \overline{v}(t) - \int_0^t S(t-s)[h'(s)Au_0 + h(s)Au_1]ds - h * v(t)\\
&\quad + \int_0^t C(t-s)[h' * v(s) + h(0)v(s)]ds\\
&= \overline{v}(t) + L(v, h)(t), \quad t \in [0, T], \qquad (29.3.11)\\
h'(t) &= \overline{h}(t) - \chi h(t)\Phi[Au_1] - \chi\Psi[v(t)] - \chi h * \Psi[v(t)]\\
&= \{\overline{h}(t) - \chi\Psi[\overline{v}(t)]\} - \chi h(t)\Phi[Au_1]\\
&\quad - \chi\Psi[L(v, h)(t)] - \chi h * \Psi[v(t)]\\
&=: \overline{h}_1(t) - \chi h(t)\Phi[Au_1] + M(v, h)(t), \quad t \in [0, T].
\end{aligned}$$

It remains to determine the initial value $h(0)$. For this purpose let us differentiate equation $(29.2.1)_1$ and then set $t = 0$ to obtain

$$u'''(0) + Au(0) + h(0)Au(0) = f(0).$$

Applying the functional Φ to both sides of $(29.3.1)_2$ and solving for $h(0)$, we obtain

$$h(0) = \chi\Phi[f(0) - Au_0] - g'''(0) =: h_0.$$

Observe that Eq. $(29.3.11)_2$ can be rewritten in the equivalent form, where $t \in [0, T]$,

$$\begin{aligned} h(t) &= \left[h_0 \exp\{-t\chi\Phi[Au_1]\} + \int_0^t \exp\{-(t-s)\chi\Phi[Au_1]\}\overline{h}_1(s)ds\right] \\ &\quad + \int_0^t \exp\{-(t-s)\chi\Phi[Au_1]\}M(v,h)(s)ds \\ &=: \overline{h}_0(t) + M_0(v,h)(t), \end{aligned} \tag{29.3.12}$$

whence one deduces that

$$h'(t) = \overline{h}_1(t) - \chi\Phi[Au_1]\overline{h}_0(t) + M_1(v,h)(t), \quad t \in [0,T],$$

where

$$M_1(v,h)(t) = -\chi\Phi[Au_1]M_0(v,h)(t) + M(v,h)(t), \quad t \in [0,T].$$

To solve the fixed-point system $(29.3.11)_1$ and (29.3.12) we introduce the following weighted complete metric spaces depending on the parameters $\lambda \geq 0$ and $r > 0$:

$$\begin{aligned} U_T^{0,\infty}(\lambda, r) &= \{v \in L^\infty((0,T);H) : \|e^{-\lambda\cdot}(v - v_0)\|_{0,\infty,H} \leq r\}, \\ H_T^{1,p}(\lambda, r) &= \{h \in W^{1,p}((0,T);\mathbb{R}) : \|e^{-\lambda\cdot}(h - h_0)\|_{0,p,\mathbb{R}} + \|e^{-\lambda\cdot}h'\|_{0,p,\mathbb{R}} \leq r\}, \end{aligned}$$

where we have set

$$\|e^{-\lambda\cdot}w\|_{0,\infty,H} = \|e^{-\lambda\cdot}w\|_{L^\infty((0,T);H)}, \quad \|e^{-\lambda\cdot}h\|_{0,p,\mathbb{R}} = \|e^{-\lambda\cdot}h\|_{L^p((0,T);\mathbb{R})}.$$

It can be checked immediately that all the metrics of the previous spaces are equivalent for all $\lambda \geq 0$.

Let us now introduce the notation $f_\lambda(t) = e^{-\lambda t}f(t)$ and observe that the convolution commutes with the exponentials $e^{-\lambda t}$ for all $\lambda > 0$, so that

$$(h * f)_\lambda = h_\lambda * f_\lambda.$$

Consequently, multiplying by $e^{-\lambda t}$ both sides in (29.3.11), we obtain the following new (equivalent) fixed-point system, where $t \in [0, T]$:

$$\begin{aligned} v_\lambda(t) &= \overline{v}(t)_\lambda - \int_0^t S_\lambda(t-s)[(h')_\lambda(s)Au_0 + h(s)Au_1]ds - h_\lambda * v_\lambda(t) \\ &\int_0^t C_\lambda(t-s)[(h')_\lambda * v_\lambda(s) + h(0)v_\lambda(s)]ds \\ &=: \overline{v}_\lambda(t) + L_\lambda(v_\lambda, h_\lambda)(t), \end{aligned}$$

$$(h')_\lambda(t) = \overline{h}_\lambda(t) - \chi\Psi[\overline{v}_\lambda(t)] - \chi h_\lambda(t)\Phi[Au_1] - \chi\Psi[L_\lambda(v_\lambda, h_\lambda)(t)] - \chi h_\lambda * \Psi[v_\lambda(t)].$$

First note the estimates

$$\|C_\lambda\|_{0,p,\mathcal{L}(H)} \le M[p(\lambda-\omega)]^{-1}, \quad \|S_\lambda\|_{0,p,\mathcal{L}(H)} \le M[p(\lambda-\omega)]^{-1}, \quad \lambda > \omega.$$

Indeed, e.g., we have

$$\left(\int_0^t e^{-p\lambda t}\|C(t)\|^p_{\mathcal{L}(H)}dt\right)^{1/p} \le M\left(\int_0^t e^{-p(\lambda-\omega)t}dt\right)^{1/p} \le M[p(\lambda-\omega)]^{-1/p}.$$

Consequently, the estimates

$$\begin{aligned}
e^{-\lambda t}\|C(t)v_0 - v_0\| &\le e^{-\lambda t}\int_0^t \|C'(s)v_0\|ds \le \int_0^t e^{-\lambda t}\|S(s)Av_0\|_{\mathcal{L}(H)}ds \\
&\le \int_0^t e^{-\lambda t}\|S(s)\|_{\mathcal{L}(H)}\|Av_0\|ds \le M(\lambda-\omega)^{-1}\|Av_0\|, \\
e^{-\lambda t}\|S(t)v_1\| &\le e^{-\lambda t}\int_0^t \|C(s)v_0\|ds \le M(\lambda-\omega)^{-1}\|v_0\|,
\end{aligned}$$

are obtained. Therefore, via Hölder's inequality, it follows that

$$\|e^{-\lambda t}(\overline{v} - v_0)\|_{0,\infty,H} \le M(\lambda-\omega)^{-1}\|v_0\|_{\mathcal{D}(A)} + M[p'(\lambda-\omega)]^{-1/p'}\|f''_\lambda\|_{0,p}.$$

Then, from Young's inequality for convolutions with $Y \in \{\mathcal{L}(H), \mathbb{R}\}$, we have

$$\begin{aligned}
\|(h * f)_\lambda\|_{0,p,Y} &\le \|h_\lambda\|_{0,1,\mathbb{R}}\|f_\lambda\|_{0,p,H}, \quad p \in [1,+\infty], \\
\|(h * f)_\lambda\|_{0,p,Y} &\le \|h_\lambda\|_{0,p,\mathbb{R}}\|f_\lambda\|_{0,1,H}, \quad p \in [1,+\infty].
\end{aligned}$$

In particular, when $h(t) \equiv h_0$ or $f(t) \equiv v_0$,

$$\begin{aligned}
\|(h_0 * f)_\lambda\|_{0,p,Y} &\le \|e^{\lambda \cdot}h_0\|_{0,1,\mathbb{R}}\|f_\lambda\|_{0,p,H} \le \lambda^{-1}|h_0|\|f_\lambda\|_{0,p,H}, \\
\|(h * v_0)_\lambda\|_{0,p,Y} &\le \|h_\lambda\|_{0,p,\mathbb{R}}\|e^{-\lambda \cdot}v_0\|_{0,1,H} \le \lambda^{-1}\|h_\lambda\|_{0,p,\mathbb{R}}\|v_0\|,
\end{aligned}$$

This implies

$$\begin{aligned}
\|(h * v)_\lambda\|_{0,\infty,H} &= \|([h - h_0 + h_0] * [v - v_0 + v_0])_\lambda\|_{0,\infty,H} \\
&\le \|(h-h_0)_\lambda\|_{0,p,\mathbb{R}}\|(v, v_0)_\lambda\|_{0,\infty,H} + \|(h-h_0)_\lambda * (v_0)_\lambda\|_{0,p,H} \\
&\quad + \|(h_0)_\lambda * (v - v_0)_\lambda\|_{0,\infty,H} + \|(h_0)_\lambda * (v_0)_\lambda\|_{0,\infty,H} \\
&\le r^2 + r\lambda^{-1}\|v_0\| + r\lambda^{-1}|h_0| + \lambda^{-1}|h_0|\|v_0\|.
\end{aligned}$$

Therefore, we obtain the estimate

$$\begin{aligned}\|e^{-\lambda\cdot}L(v,h)\|_{0,\infty,H} &= \|L_\lambda(v_\lambda,h_\lambda)\|_{0,\infty,H}\\ &\le M[p'(\lambda-\omega)]^{-1/p'}[r\|Au_0\|+(r+\lambda^{-1}|h_0|)\|Au_1\|]+r^2+r\lambda^{-1}\|v_0\|\\ &\quad+r\lambda^{-1}|h_0|+\lambda^{-1}|h_0|\|v_0\|+M(\lambda-\omega)^{-1}[r^2+r\lambda^{-1}\|v_0\|+|h_0|r].\end{aligned}$$

Summing up, the operator $(v,h)\to\overline{v}+L(v,h)$ maps $U_T^{0,\infty}(\lambda,r)\times H_T^{1,p}(\lambda,r)$ into $U_T^{0,\infty}(\lambda,r)$ if the pair (λ,r) is a solution to the inequality

$$\begin{aligned}J_0(\lambda,r) &:= M(\lambda-\omega)^{-1}\|v_0\|_{\mathcal{D}(A)}+M[p'(\lambda-\omega)]^{-1/p'}\|f_\lambda''\|_{0,p}\\ &\quad+M[p'(\lambda-\omega)]^{-1/p'}[r\|Au_0\|+(r+\lambda^{-1}|h_0|)\|Au_1\|]+r^2+r\lambda^{-1}\|v_0\|\\ &\quad+r\lambda^{-1}|h_0|+\lambda^{-1}|h_0|\|v_0\|+M(\lambda-\omega)^{-1}[r^2+r\lambda^{-1}\|v_0\|+|h_0|r]\le r.\end{aligned} \tag{29.3.13}$$

It now remains to estimate $\overline{h}_0+M_0(v,h)$. For this purpose we begin by estimating in $U_T^{0,p}(\lambda,r)$ the function defined by $\overline{h}_1$ (cf. (29.3.9) and $(29.3.11)_2$):

$$\overline{h}_1(t)=\chi\Phi[f''(t)]-\chi g^{(4)}(t)-\chi\Psi[(\overline{v}-v_0)(t)]-\chi\Psi[v_0].$$

Assume $f''\in L^{\widetilde{p}}((0,T);H)$, $g\in W^{4,\widetilde{p}}((0,T);\mathbb{R})$ for some $\widetilde{p}\in(p,+\infty]$, $p\in(1,+\infty]$ and set $q^{-1}=p^{-1}+\widetilde{p}^{-1}$. Then, using the inequality

$$\|e^{-\lambda\cdot}l\|_{0,p}\le\|e^{-\lambda\cdot}\|_{0,q}\|l\|_{0,\widetilde{p}}\le q^{-1/q}\lambda^{-1/q}\|l\|_{0,\widetilde{p}},$$

it follows that

$$\begin{aligned}\|(\overline{h}_1)_\lambda\|_{0,p,\mathbb{R}} &\le |\chi|\|\Phi\|_{X^*}\|e^{-\lambda\cdot}f''\|_{0,p,H}+|\chi|\|e^{-\lambda\cdot}g^{(4)}\|_{0,p,\mathbb{R}}\\ &\quad+|\chi|\|\Psi\|_{X^*}\{M[(\lambda-\omega)]^{-1}\|v_0\|_{\mathcal{D}(A)}+M[p'(\lambda-\omega)]^{-1/p'}\|f_\lambda''\|_{0,p,H}\}\\ &\quad+(p\lambda)^{-1/p}|\chi|\|\Psi\|_{X^*}\|v_0\|\\ &\le|\chi|q^{-1/q}\lambda^{-1/q}[\|\Phi\|_{X^*}\|f''\|_{0,\widetilde{p},H}+\|g^{(4)}\|_{0,\widetilde{p},\mathbb{R}}]\\ &\quad+|\chi|\|\Psi\|_{X^*}\{M[(\lambda-\omega)]^{-1}\|v_0\|_{\mathcal{D}(A)}+M[p'(\lambda-\omega)]^{-1/p'}\|f_\lambda''\|_{0,p,H}\}\\ &\quad+(p\lambda)^{-1/p}|\chi|\|\Psi\|_{X^*}\|v_0\|:=J_1(\lambda,r),\quad \lambda>\max\{0,\omega\}.\end{aligned} \tag{29.3.14}$$

From (29.3.12) and (29.3.14) we easily obtain the inequalities

$$\begin{aligned}\|(\overline{h}_0)_\lambda\|_{0,p,\mathbb{R}} &\le|h_0|[p(\lambda-\chi\Phi[Au_1])]^{-1/p}+(\lambda-\chi\Phi[Au_1])^{-1}\|(\overline{h}_1)_\lambda\|_{0,p,\mathbb{R}}\\ &:=J_2(\lambda,r),\quad \lambda>\max\{0,\omega,\chi\Phi[Au_1]\}.\end{aligned}$$

Consider now the identities

$$[M_0(v,h)]_\lambda = [\exp\{-\chi\Phi[Au_1]\cdot\} * M(v,h)]_\lambda,$$
$$[M_1(v,h)]_\lambda = -\chi\Phi[Au_1][\exp\{-\chi\Phi[Au_1]\cdot\} * M(v,h)]_\lambda + [M(v,h)]_\lambda, \tag{29.3.15}$$

implying, for all $\lambda > \max\{0, \omega, \chi\Phi[Au_1]\}$, the chain of estimates

$$\begin{aligned}
\|[M(v,h)]_\lambda\|_{0,p,\mathbb{R}} &\le \|\Psi\|_{X^*}[|\chi|\|L_\lambda(v_\lambda,h_\lambda)\|_{0,p,H} + T^{1/p'}\|h_\lambda * v_\lambda\|_{0,\infty,H}] \\
&\le \|\Psi\|_{X^*}[|\chi|J_0(\lambda,r) + T^{1/p'}[r^2 + r\lambda^{-1}\|v_0\| + r\lambda^{-1}|h_0| + \lambda^{-1}|h_0|\|v_0\|]] \\
&:= J_3(\lambda,r), \\
\|[M_0(v,h)]_\lambda\|_{0,p,\mathbb{R}} &\le (\lambda - \chi\Phi[Au_1])^{-1}J_3(\lambda,r), \\
\|[M_1(v,h)]_\lambda\|_{0,p,|r} &\le [|\chi|\|\Phi\|_{X^*}\|Au_1\|(\lambda - \chi\Phi[Au_1])^{-1} + 1]J_3(\lambda,r).
\end{aligned} \tag{29.3.16}$$

Summing up, the operator $(v,h) \to \overline{h}_0 + M_0(v,h)$ maps $U_T^{0,\infty}(\lambda,r) \times H_T^{1,p}(\lambda,r)$ into $H_T^{1,p}(\lambda,r)$ if the pair (λ,r) is a solution to the inequalities

$$\begin{aligned}
&\lambda > \max\{0,\omega,\chi\Phi[Au_1]\}, \\
&J_1(\lambda,r) + J_2(\lambda,r) + [(1 + |\chi|\|\Phi\|_{X^*}\|Au_1\|)(\lambda - \chi\Phi[Au_1])^{-1} + 1]J_3(\lambda,r) \le r.
\end{aligned} \tag{29.3.17}$$

Our next task is to estimate the increments of L_λ with respect to the pair (v,h). For this purpose, the identity

$$\begin{aligned}
&L_\lambda((v_2)_\lambda,(h_2)_\lambda)(t) - L_\lambda((v_1)_\lambda,(h_1)_\lambda)(t) \\
&\quad = -\int_0^t S_\lambda(t-s)[(h_2' - h_1')_\lambda(s)Au_0 + h(s)Au_1]ds \\
&\qquad - (h_2 - h_1)_\lambda * (v_2)_\lambda(t) - (h_1)_\lambda * (v_2 - v_1)_\lambda(t) \\
&\qquad + \int_0^t C_\lambda(t-s)(h_2' - h_1')_\lambda * (v_2)_\lambda(s)ds \\
&\qquad + \int_0^t C_\lambda(t-s)[(h_1')_\lambda * (v_2 - v_1)_\lambda(s) + h_0(v_2 - v_1)_\lambda(s)]ds
\end{aligned} \tag{29.3.18}$$

is needed. Consider now the chain of inequalities

$$\begin{aligned}
&\|(h_2)_\lambda * (v_2)_\lambda - (h_1)_\lambda * (v_1)_\lambda\|_{0,\infty,H} \\
&\quad \le \|(h_2 - h_1)_\lambda * [(v_2 - v_0)_\lambda + (v_0)_\lambda\|_{0,\infty,H} \\
&\qquad + \|[(h_1 - h_0)_\lambda + (h_0)_\lambda] * (v_2 - v_1)_\lambda\|_{0,\infty,H} \\
&\quad \le r\|(h_2 - h_1)_\lambda\|_{0,1,\mathbb{R}} + \lambda^{-1}\|v_0\|\|(h_2 - h_1)_\lambda\|_{0,p,\mathbb{R}} \\
&\qquad + (r + \lambda^{-1}|h_0|)\|(v_2 - v_1)_\lambda\|_{0,\infty,H}, \\
&\|(h_2')_\lambda * (v_2)_\lambda - (h_1')_\lambda * (v_1)_\lambda\|_{0,\infty,H} \qquad\qquad (29.3.19)\\
&\quad \le \|(h_2 - h_1)_\lambda' * [(v_2 - v_0)_\lambda + (v_0)_\lambda\|_{0,\infty,H} \\
&\qquad + \|(h_1)_\lambda' * (v_2 - v_1)_\lambda\|_{0,\infty,H} \\
&\quad \le rT^{1/p'}\|(h_2 - h_1)_\lambda'\|_{0,p,\mathbb{R}} + \lambda^{-1}\|v_0\|\|(h_2 - h_1)_\lambda'\|_{0,p,\mathbb{R}} \\
&\qquad + rT^{1/p'}\|(v_2 - v_1)_\lambda\|_{0,\infty,H}.
\end{aligned}$$

From (29.3.18) and $(29.3.19)_1$ we deduce the estimates

$$\begin{aligned}
&\|L_\lambda((v_2)_\lambda, (h_2)_\lambda)(t) - L_\lambda((v_1)_\lambda, (h_1)_\lambda)(t)\| \\
&\quad \le M[p'(\lambda - \omega)]^{-1/p'}\{\|(h_2' - h_1')_\lambda\|_{0,p,\mathbb{R}}\|Au_0\| + \|(h_2 - h_1)_\lambda\|_{0,p,\mathbb{R}}\|Au_1\|\} \\
&\qquad + r\|(h_2 - h_1)_\lambda\|_{0,1,\mathbb{R}} + \lambda^{-1}\|v_0\|\|(h_2 - h_1)_\lambda\|_{0,p,\mathbb{R}} \\
&\qquad + (r + \lambda^{-1}|h_0|)\|(v_2 - v_1)_\lambda\|_{0,\infty,H} \\
&\qquad + M(\lambda - \omega)^{-1}\{rT^{1/p'}\|(h_2 - h_1)_\lambda'\|_{0,p,\mathbb{R}} + \lambda^{-1}\|v_0\|\|(h_2 - h_1)_\lambda'\|_{0,p,\mathbb{R}}\} \\
&\qquad + rT^{1/p'}\|(v_2 - v_1)_\lambda\|_{0,\infty,H} + |h_0|\|(v_2 - v_1)_\lambda\|_{0,\infty} \\
&\quad \le J_4(\lambda, r)\{\|(h_2 - h_1)_\lambda\|_{1,p,\mathbb{R}} + \|(v_2 - v_1)_\lambda\|_{0,\infty,H}\},
\end{aligned}$$

where for all $\lambda > \max\{0, \omega\}$, we have

$$\begin{aligned}
J_4(\lambda, r) :=& M[p'(\lambda - \omega)]^{-1/p'} \max(\|Au_0\|, \|Au_1\|) \\
&+ \max(T^{1/p'} r + \lambda^{-1}\|v_0\|, r + |h_0|) \\
&+ M(\lambda - \omega)^{-1} \max\{rT^{1/p'} + \lambda^{-1}\|v_0\|), r + \lambda^{-1}|h_0|\}.
\end{aligned}$$

Consider now the identity

$$\begin{aligned}
M(v_2, h_2)_\lambda - M(v_1, h_1)_\lambda =& -\chi\Psi[L_\lambda((v_2)_\lambda, (h_2)_\lambda) - L_\lambda((v_1)_\lambda, (h_1)_\lambda)] \\
&-\chi\Psi[(h_2)_\lambda * (v_2)_\lambda - (h_1)_\lambda * (v_1)_\lambda],
\end{aligned}$$

implying, for all $\lambda > \max\{0, \omega\}$, the estimates

$$\begin{aligned}
&\|M(v_2, h_2)_\lambda - M(v_1, h_1)_\lambda\|_{0,p,\mathbb{R}} \\
&\quad \le |\chi|\|\Psi\|_{X^*}\|L_\lambda((v_2)_\lambda, (h_2)_\lambda) - L_\lambda((v_1)_\lambda, (h_1)_\lambda)\|_{0,p,\mathbb{R}} \\
&\qquad + |\chi|\|\Psi\|_{X^*}\|(h_2)\lambda * (v_2)_\lambda - (h_1)\lambda * (v_1)_\lambda\|_{0,p,\mathbb{R}} \qquad\qquad (29.3.20)\\
&\quad \le |\chi|\|\Psi\|_{X^*}[T^{1/p'} J_4(\lambda, r) + \max\{r + \lambda^{-1}\|v_0\|, r + \lambda^{-1}|h_0|\}] \\
&\quad =: J_5(\lambda, r)\{\|(h_2 - h_1)_\lambda\|_{1,p,\mathbb{R}} + \|(v_2 - v_1)_\lambda\|_{0,\infty,H}\}.
\end{aligned}$$

From $(29.3.15)_1$, $(29.3.16)_1$, and (29.3.20), for all $\lambda > \max\{0, \omega, \chi\Phi[Au_1]\}$, we easily deduce the estimates

$$\begin{aligned}
&\|M_0(v_2, h_2)_\lambda - M_0(v_1, h_1)_\lambda\|_{0,p,\mathbb{R}} \\
&\quad \le (\lambda - \chi\Phi[Au_1])^{-1} J_5(\lambda, r)\{\|(h_2 - h_1)_\lambda\|_{1,p,\mathbb{R}} + \|(v_2 - v_1)_\lambda\|_{0,\infty,H}\}, \\
&\|M_1(v_2, h_2)_\lambda - M_1(v_1, h_1)_\lambda\|_{0,p,\mathbb{R}} \\
&\quad \le |\chi|\|\Psi\|_{X^*}[\|\Phi\|_{X^*}\|Au_1\|(\lambda - \chi\Phi[Au_1])^{-1} + 1] \\
&\qquad \times J_5(\lambda, r)\{\|(h_2 - h_1)_\lambda\|_{1,p,\mathbb{R}} + \|(v_2 - v_1)_\lambda\|_{0,\infty,H}\}.
\end{aligned}$$

Summing up, the vector operator $(v, h) \to (\overline{v} + L(v, h), \overline{h}_0 + M_0(v, h))$ turns out to be a contracting mapping from $U_T^{0,\infty}(\lambda, r) \times H_T^{1,p}(\lambda, r)$ into itself if the pair (λ, r) is a solution of the system of inequalities (29.3.13) and

$$\begin{aligned}
&\{(\lambda - \chi\Phi[Au_1])^{-1} + |\chi|\|\Psi\|_{X^*}[\|\Phi\|_{X^*}\|Au_1\|(\lambda - \chi\Phi[Au_1])^{-1} + 1]\}J_5(\lambda, r) \\
&\quad + J_4(\lambda, r) < 1, \quad \lambda > \max\{0, \omega, \chi\Phi[Au_1]\}.
\end{aligned} \tag{29.3.21}$$

Consequently, whenever the pair (λ, r) solves system (29.3.17), (29.3.21), we conclude that the fixed-point system (29.3.11) admits a unique solution $(v, h) \in U_T^{0,\infty}(\lambda, r) \times H_T^{1,p}(\lambda, r)$. Therefore, the next task is to show that the system (29.3.17) and (29.3.21) is solvable for large λ and small r. For this purpose, it suffices to observe that as $\lambda \to +\infty$, we have

$$\begin{aligned}
&J_0(\lambda, r) \to r^2, \quad J_1(\lambda, r) \to 0, \quad J_2(\lambda, r) \to 0, \\
&J_3(\lambda, r) \to \|\Psi\|_{X^*}(|\chi| + T^{1/p'})r^2, \\
&J_4(\lambda, r) \to r\max(T^{1/p'}, 1), \\
&J_5(\lambda, r) \to r|\chi|\|\Psi\|_{X^*}[T^{1/p'}\max(T^{1/p'}, 1) + 1].
\end{aligned}$$

Choose now $\gamma_0 r^2 < r$ and $\gamma_1 r < 1$, i.e., $r < \min\{\gamma_0^{-1}, \gamma_1^{-1}\}$, where $\gamma_0 = \max\{\|\Psi\|_{X^*}(|\chi| + T^{1/p'}), 1\}$ and $\gamma_1 = \max\{T^{1/p'}, 1, |\chi|\|\Psi\|_{X^*}[T^{1/p'}\max(T^{1/p'}, 1) + 1]\}$. Then for any large enough λ, the system (29.3.17) and (29.3.21) is solvable.

Assume then that $(v, h) \in L^\infty((0, T); H) \cap W^{1,1}((0, T); \mathbb{R})$ solves problem (29.3.11). Let $u \in U_T^{2,\infty} := W^{2,\infty}((0, T); H) \cap W^{1,\infty}((0, T); V) \cap L^\infty((0, T); \mathcal{D}(A))$ be the solution to the Cauchy problem $(29.2.1)_1$ and $(29.2.1)_2$ with the previous fixed h. Then u is represented by

$$u(t) = \overline{u}(t) - \int_0^t AS(t - s)h * u(s)ds, \quad t \in [0, T],$$

where

$$\overline{u}(t) = C(t)u_0 + S(t)u_1 + \int_0^t S(s)f(t - s)ds.$$

Let us differentiate u twice to show that $u'' = v$. In this way it will be shown that $u = u_0 + tu_1 + \int_0^t (t-s)v(s)ds$ solves the Cauchy problem $(29.2.1)_1$ and $(29.2.1)_2$, as required. For this purpose we use the following formulas (cf. (29.3.10)):

$$\begin{aligned}
\overline{u}''(t) &= C''(t)u_0 + S''(t)u_1 + C(t)f(0) + S(t)f'(0) + \int_0^t S(s)f''(t-s)ds \\
&= -C(t)Au_0 - S(t)Au_1 + C(t)f(0) + S(t)f'(0) + \int_0^t S(s)f''(t-s)ds,
\end{aligned}$$

$$\begin{aligned}
D_t^2 \int_0^t AS(s)h * u(t-s)ds &= S(t)AD_t(h*u)(0) + \int_0^t AS(s)D_t^2(h*u)(t-s)ds \\
&= S(t)h(0)Au_0 + \int_0^t S(s)[h'(t-s)Au_0 + h(t-s)Au_1]ds \\
&\quad + \int_0^t AS(s)(h*u'')(t-s)ds \\
&= \int_0^t S(s)[h'(t-s)Au_0 + h(t-s)Au_1]ds + S(t)h(0)Au_0 + (h*u'')(t) \\
&\quad - \int_0^t C(s)(h*u'')(t-s)ds, \qquad t \in [0,T].
\end{aligned}$$

Observe that the difference $v - u''$ solves the following *homogeneous* convolution equation in $[0,T]$:

$$v(t) - u''(t) = -h*(v-u'')(t) + \int_0^t C(t-s)[h'*(v-u'')(s) + h(0)(v-u'')(s)]ds,$$

whence it is easily deduced that the equality $v - u'' = 0$ holds. Therefore, the pair (u,h) solves the identification problem (29.2.1).

The result proved in this section is summarized in the following theorem.

Theorem 29.3.1. *Let* $u_0, u_1 \in \mathcal{D}(A)$, $f \in W^{2,\widetilde{p}}((0,T);H)$, $g \in W^{4,\widetilde{p}}((0,T);\mathbb{R})$, $\widetilde{p} \in (p,+\infty]$, $p \in (1,+\infty]$. *Let the linear functional* Φ *satisfy (29.3.5) and (29.3.7). Then the identification problem (29.2.1) admits a unique solution* $(u,h) \in U_T^{2,\infty} \times W^{1,p}((0,T);\mathbb{R})$.

We can now apply our general result to problem (29.1.7)–(29.1.8), with $\varphi_0 \in H^2(\omega) \cap H_0^1(\omega)$ and $D_\nu(\varphi_0\mu) = 0$ on $\partial\omega$. Moreover, let us choose $H = L^2(\omega)$, $V = H^1(\omega)$, $\mathcal{D}(A) = H^2(\omega) \cap H_0^1(\omega)$, $A = -\sum_{j=1}^2 D_{x_j}[\mu D_{x_j}]$. Then the bilinear form a associated with A is given by (29.2.11) and satisfies (29.2.4). Finally, the Borel measure ψ coincides with the two-dimensional Lebesgue measure, while the functionals Φ and Ψ are defined, respectively, by $\Phi[w] = \int_\omega \varphi_0(x)w(x)ds$ and $\Psi[w] = \int_\omega \Delta(\varphi_0\mu)(x)w(x)dx$.

Note that the link (29.3.5) between the functionals Φ and Ψ can be easily shown by a twofold integration by parts.

We obtain the following result.

Theorem 29.3.2. *Let $\partial\Omega$ be of class C^2 and let μ satisfy properties* (29.1.3) *with* $\sigma = 0$. *Let $u_3^0, u_3^1 \in H^2(\omega) \cap H_0^1(\omega)$, $f_3 \in W^{2,\overline{p}}((0,T); L^2(\omega))$, $g \in W^{4,\overline{p}}((0,T); \mathbb{R})$, $\overline{p} \in (p, +\infty]$, $p \in (1, +\infty]$. Furthermore, let the consistency conditions*

$$\int_{\overline{\omega}} \varphi_0(x) u_j(x) d\psi(x) = g^{(j)}(0), \quad j = 0, 1,$$

hold. Then the identification problem (29.1.7)–(29.1.8) admits a unique solution (u_3, h) such that $u_3 \in W^{2,\infty}((0,T); L^2(\omega)) \cap W^{1,\infty}((0,T); H^1(\omega)) \cap L^\infty((0,T); H^2(\omega) \cap H_0^1(\omega))$ and $h \in W^{1,p}((0,T); \mathbb{R})$.

29.4 Solving the Third Identification Problem

Assume that the additional information $(29.2.1)_3$ is of integral type, i.e.,

$$\begin{aligned} \Phi[u(t,\cdot)] := \int_{\partial\omega} \varphi_0(x) D_\nu u(t,x) d\sigma(x) \\ + \int_0^t h(t-s) ds \int_{\partial\omega} \varphi_0(x) D_\nu u(s,x) d\sigma(x) = g(t), \quad t \in [0,T], \end{aligned} \tag{29.4.1}$$

σ denoting the Lebesgue measure on $\partial\omega$. We now rewrite condition (29.4.1) in the form

$$\Phi_0[u(t)] + h * \Phi_0[u(t)] = g(t), \quad t \in [0,T], \tag{29.4.2}$$

where

$$\Phi_0[z] = \int_{\partial\omega} \varphi_0(x) D_\nu z(x) dx.$$

Thus, our identification problem can be rewritten in the following form, in a general Hilbert space H, with a general linear operator A: *determine two functions $u : [0,T] \to H$ and $h : [0,T] \to \mathbb{R}$ such that*

$$\begin{aligned} & u''(t) + Au(t) + h * Au(t) = f(t), \quad t \in [0,T], \\ & u(0) = u_0, \quad u'(0) = u_1, \\ & \Phi_0[u(t)] + h * \Phi_0[u(t)] = g(t), \quad t \in [0,T]. \end{aligned} \tag{29.4.3}$$

To solve this identification problem, let us introduce the auxiliary function

$$v(t) = u''(t) \Longleftrightarrow u(t) = u_0 + t u_1 + \int_0^t (t-s) v(s) ds.$$

Then, assuming $h \in W^{1,p}((0,T); \mathbb{R})$, $p \in (1, +\infty]$, and differentiating equations $(29.4.3)_1$ and $(29.4.3)_3$ twice, we obtain that the pair (v, h) solves the system

$$
\begin{aligned}
&v''(t) + Av(t) + h * Av(t) + h'(t)Au_0 + h(t)Au_1 = f''(t), \quad t \in [0, T],\\
&v(0) = f(0) - Au_0 =: v_0, \quad v'(0) = f'(0) - Au_1 - h(0)Au_0 =: v_1, \qquad (29.4.4)\\
&\Phi_0[v(t)] + h * \Phi_0[v(t)] + h'(t)\Phi_0[Au_0] + h(t)\Phi_0[u_1] = g''(t), \quad t \in [0, T].
\end{aligned}
$$

Assume now

$$\chi^{-1} := \Phi_0[Au_0] \neq 0.$$

Then we observe that the solution $(v, h) \in [W^{1,\infty}((0, T); H) \cap L^\infty((0, T); V)] \times W^{1,p}((0, T); \mathbb{R})$, $p \in (1, +\infty]$, to problem (29.4.4) is equivalent to solving—in the same function space—the fixed-point system

$$
\begin{aligned}
v(t) =& \overline{v}(t) - \int_0^t S(t-s)[h'(s)Av_0 + h(s)Av_1]ds\\
&- \int_0^t S(t-s)h * Av(s)ds, \quad t \in [0, T],\\
h'(t) =& -\chi\Phi_0[Av(t)] - \chi h * \Psi_0[Av(t)] - \chi h(t)\Phi_0[Au_1] + g''(t), \ t \in [0, T], \quad (29.4.5)
\end{aligned}
$$

where the function $\overline{v}$ is defined by (29.3.3). Moreover, we can compute the initial value $h(0) = h_0$ by differentiating both sides of Eq. (29.4.2) once and then setting $t = 0$ to obtain

$$h_0 = -\chi\{g'(0) - \Phi[Au_1] + \Phi_1[u_1]\}.$$

For this purpose, further regularity properties involving cosine and sine functions are needed:

$$
\begin{aligned}
&C \in C([0, +\infty); \mathcal{L}(V)), \quad \|C(t)\|_{\mathcal{L}(V)} \le Me^{t\rho}, \quad t \in [0, +\infty),\\
&S \in C([0, +\infty); \mathcal{L}(H; V)), \quad \|S(t)\|_{\mathcal{L}(H;V)} \le Me^{t\rho}, \quad t \in [0, +\infty).
\end{aligned}
$$

Reasoning as in Sect. 29.3, we can show that the pair (w, h), where $w = Av$, solves the fixed-point system

$$
\begin{aligned}
w(t) =& A\overline{v}(t) - \int_0^t S(t-s)[h'(s)Av_0 + h(s)Av_1]ds - h * w(t)\\
&+ \int_0^t C(t-s)[h' * w(s) + h(0)w(s)]ds\\
=:& A\overline{v}(t) + L(w, h)(t), \quad t \in [0, T],\\
h'(t) =& -\chi\Phi_0[A^{-1}w(t)] - \chi h * \Phi_0[A^{-1}w(t)]\\
&- \chi h(t)\Phi_0[Au_1] + g''(t), \quad t \in [0, T].
\end{aligned}
$$

Observe that this fixed-point system is equivalent to the following:

$$\begin{aligned}
w(t) =& A\overline{v}(t) - \int_0^t S(t-s)[h'(s)Av_0 + h(s)Av_1]ds - h * w(t)\\
&+ \int_0^t C(t-s)[h' * w(s) + h(0)w(s)]ds\\
=:& A\overline{v}(t) + L(w,h)(t), \quad t \in [0,T],\\
h'(t) =& -\chi\Phi_0[\overline{v}(t)] - \chi\Phi_0[A^{-1}L(w,h)(t)] - \chi h * \Phi_0[A^{-1}w(t)]\\
&-\chi h(t)\Phi_0[Au_1] + g''(t), \quad t \in [0,T].
\end{aligned}$$

It is now assumed that

$$|\Phi_0[A^{-1}z]| \le C\|z\|_H, \quad \forall z \in H. \tag{29.4.6}$$

Using the same fixed-point procedure as in Sect. 29.3, we can show that the system (29.4.5) admits a unique solution $(w,h) \in L^\infty((0,T);V) \times W^{1,p}((0,T);\mathbb{R})$.

Let us define the function v by the formula

$$v(t) = \overline{v}(t) - \int_0^t S(t-s)[h'(s)Au_0 + h(s)Au_1]ds - \int_0^t S(t-s)h * w(s)ds, \quad t \in [0,T].$$

Reasoning as in the previous section, one can show that

$$u(t) = u_0 + tu_1 + \int_0^t (t-s)v(s)ds, \quad t \in [0,T],$$

so that $u \in U_T^{2,\infty}$.

We summarize the result proved in this section in the following theorem.

Theorem 29.4.1. *Let* $u_0, u_1, v_0 \in \mathcal{D}(A)$, $v_1 \in V$, $f \in W^{2,\widetilde{p}}((0,T);H)$, $g \in W^{2,\widetilde{p}}((0,T);\mathbb{R})$, $\widetilde{p} \in (p,+\infty]$, $p \in (1,+\infty]$. *Let the linear functional* Φ *satisfy (29.4.6). Then the identification problem (29.2.1) admits a unique solution* $(u,h) \in U_T^{2,\infty} \times W^{1,p}((0,T);\mathbb{R})$ *such that* $u'' \in U_T^{2,\infty}$.

Using the same definitions for the spaces H and $\mathcal{D}(A)$ and for the operator A as at the end of Sect. 29.3, we obtain the following result.

Theorem 29.4.2. *Let* $\partial\Omega$ *be of class* C^4 *and let* μ *satisfy properties* (29.1.3) *with* $\sigma = 2$. *Let* $u_3^0, u_3^1 \in H^2(\omega) \cap H_0^1(\omega)$, $f_3 \in W^{2,\widetilde{p}}((0,T);L^2(\omega))$, $g \in W^{4,\widetilde{p}}((0,T);\mathbb{R})$, $\widetilde{p} \in (p,+\infty]$, $p \in (1,+\infty]$. *Furthermore, let the consistency conditions*

$$\int_{\overline{\omega}} \varphi_0(x)D_\nu u_3^j(x)d\sigma(x) + jh_0\int_{\overline{\omega}} \varphi_0(x)D_\nu u_3^0(x)d\sigma = 0, \; j = 0,1,$$

hold. Then the identification problem (29.1.7)–(29.1.8) admits a unique solution (u_3, h) *such that*

$$u_3, D_t^2 u_3 \in W^{2,\infty}((0,T); L^2(\omega)) \cap W^{1,\infty}((0,T); H^1(\omega)) \cap L^\infty((0,T); H^2(\omega) \cap H_0^1(\omega))$$

and $h \in W^{1,p}((0,T); \mathbb{R})$.

30 Dynamics of Viscoelastic Fluids

30.1 Introduction

An evolution problem in a bounded domain for viscoelastic fluids of the kind considered in Chaps. 8, 10, and 13 is now presented. Our attention is confined to infinitesimal viscoelasticity for isotropic, homogeneous, and incompressible fluids. Therefore, the constitutive equation for the stress tensor $\mathbf{T}$ is expressed by the hereditary law (8.10.1), which, with an integration by parts, can be rewritten as follows:

$$\mathbf{T}(\mathbf{x}, t) = -p(\mathbf{x}, t)\mathbf{I} + \int_0^\infty \mu(s)\dot{\mathbf{E}}^t(\mathbf{x}, s)ds. \tag{30.1.1}$$

The stability problem for such fluids was examined under a variety of conditions in many articles (see [215, 216, 300, 301]). We recall, in particular, that Slemrod in [300] showed that if $\mu(s) \in C^2(\mathbb{R}^+)$, $\mu(s) \to 0$ as $s \to \infty$, $\mu(s) > 0$, $\mu'(s) < 0$, and $\mu''(s) \geq 0$, the rest state of these fluids is stable in a suitable "fading memory" norm, and the solution of the linearized boundary initial history value problem converges to the rest state weakly in this norm as $t \to \infty$. The same author proved asymptotic stability by means of the additional assumption that $\int_0^\infty s^2\mu'(s)ds < \infty$.

In [113], it was shown that there exists a strict connection between the thermodynamic restrictions on the relaxation function μ and the existence, uniqueness, and stability theorems relating to the boundary initial history value problem for viscoelastic fluids characterized by the constitutive equation (8.10.1). For this purpose, the strict inequality in condition (8.10.12) of Theorem 8.10.3, $\mu_c(\omega) > 0\ \forall\omega \in \mathbb{R}$, is important, since the asymptotic stability of the rest state fails when the weaker condition $\mu_c(\omega) \geq 0\ \forall\omega \in \mathbb{R}$ is assumed, as occurs for a particular family of relaxation functions examined also in [113].

G. Amendola et al., *Thermodynamics of Materials with Memory*,
https://doi.org/10.1007/978-3-030-80534-0_30

30.2 An Initial Boundary Value Problem for an Incompressible Viscoelastic Fluid

Let $\Omega \subset \mathbb{R}^3$ be a smooth bounded domain occupied by a linear viscoelastic incompressible fluid, for which the constitutive equation is given by (30.1.1). The linear approximation of the equations of motion on the domain $Q = \Omega \times \mathbb{R}^{++}$ for the initial boundary value problem, characterized by Dirichlet's boundary conditions, yields the following system:

$$\begin{aligned}
&\frac{\partial}{\partial t}\mathbf{v}(\mathbf{x},t) = \nabla \cdot \left[-p(\mathbf{x},t)\mathbf{I} + \int_0^\infty \mu(s)\nabla \mathbf{v}^t(\mathbf{x},s)ds \right] + \mathbf{f}(\mathbf{x},t),\\
&\nabla \cdot \mathbf{v}(\mathbf{x},t) = 0,\\
&\quad \mathbf{v}(\mathbf{x},t) = \mathbf{0} \qquad \forall \mathbf{x} \in \partial\Omega,\ \forall t \in \mathbb{R}^{++},\\
&\quad \mathbf{v}(\mathbf{x},\tau) = \mathbf{v}_0(\mathbf{x},\tau) \qquad \forall \mathbf{x} \in \Omega,\ \forall \tau \in \mathbb{R}^{-},
\end{aligned} \tag{30.2.1}$$

where the constant mass density ρ is understood and not written, $\mathbf{v}$ and p are the velocity and the pressure fields, $\mathbf{f}$ gives the body forces, and $\mathbf{v}_0$ denotes the history of the velocity up to time $t = 0$. Note that we have used the relation $\nabla \cdot \dot{\mathbf{E}}^t(\mathbf{x},t) = \frac{1}{2}\nabla \cdot [\nabla \mathbf{v}(\mathbf{x},t)]$, which follows from the constraint of incompressibility $\nabla \cdot \mathbf{v}(\mathbf{x},t) = 0$.

The relaxation function μ is assumed to be such that $\mu \in L^1(\mathbb{R}^+)$ and such that it satisfies the following thermodynamic restriction:

$$\mu_c(\omega) = \int_0^\infty \mu(s)\cos\omega s ds > 0 \quad \forall \omega \in \mathbb{R}. \tag{30.2.2}$$

We denote by $L_s^2(\Omega)$ and $H_{s0}^2(\Omega)$ the Hilbert spaces obtained by means of the completion of solenoidal vector fields $\mathbf{v} \in C_0^\infty(\Omega)$ in the $L^2(\Omega)$ and in the $H_0^1(\Omega)$ norms, respectively. Moreover, $L_\pi^2(\Omega)$ is the Hilbert space obtained by virtue of the completion of irrotational vector fields $\mathbf{v} \in C_0^\infty(\Omega)$ in the $L^2(\Omega)$ norm. Thus, we have $L^2(\Omega) = L_s^2(\Omega) \oplus L_\pi^2(\Omega)$.

Theorem 30.2.1. *If the relaxation function μ belongs to $L^1(\mathbb{R}^+)$ and satisfies (30.2.2), the supply $\mathbf{f} \in L^2(\mathbb{R}^+; L^2(\Omega))$, the initial history $\mathbf{v}_0(\cdot,s) \in H_{s0}^1(\Omega)$ for all $s \in \mathbb{R}^-$, and the function $\mathbf{V}_0(\mathbf{x},t) = \nabla \cdot \int_t^\infty \mu(s)\nabla \mathbf{v}_0(\mathbf{x},t-s)ds \in L^2(\mathbb{R}^+; L_s^2(\Omega))$, then the problem (30.2.1) has one and only one solution $\mathbf{v} \in L^2(\mathbb{R}^{++}; H_{s0}^1(\Omega))$.*

Before giving a proof of this theorem, we consider the Laplace-transformed version of the problem and then prove some relevant and useful lemmas.

30.2.1 Transformed Problem

Firstly, recall that the Laplace transform of any smooth function $\varphi : \mathbb{R}^+ \to \mathbb{R}$ is defined by $\varphi_L(z) = \int_0^\infty e^{-zs}\varphi(s)ds$, where $z \in \mathbf{C}$ is the Laplace parameter belonging to the complex plane, here denoted by $\mathbf{C}$. This generalizes the quantity given by (C.2.4), where $z \in \mathbb{R}^+$.

Applying a Laplace transform to the dynamical problem (30.2.1) gives

$$\begin{aligned} z\mathbf{v}_L(\mathbf{x},z) &= -\nabla p_L(\mathbf{x},z) + \nabla \cdot [\mu_L(z)\nabla \mathbf{v}_L(\mathbf{x},z)] + \mathbf{F}(\mathbf{x},z) \quad \forall \mathbf{x} \in \Omega, \\ \nabla \cdot \mathbf{v}_L(\mathbf{x},z) &= 0 \qquad \forall \mathbf{x} \in \Omega, \\ \mathbf{v}_L(\mathbf{x},z) &= \mathbf{0} \qquad \forall \mathbf{x} \in \partial\Omega, \end{aligned} \tag{30.2.3}$$

where we have put

$$\begin{aligned} \mathbf{F}(\mathbf{x},z) &= \mathbf{f}_L(\mathbf{x},z) + \mathbf{v}_0(\mathbf{x},0) + \int_0^\infty e^{-zs}\nabla \cdot \left[\int_s^\infty \mu(\tau)\nabla \mathbf{v}_0(\mathbf{x}, s-\tau)d\tau\right] ds \\ &= \mathbf{f}_L(\mathbf{x},z) + \mathbf{v}_0(\mathbf{x},0) + \mathbf{V}_{0_L}(\mathbf{x},z). \end{aligned}$$

From the hypotheses $\mathbf{f} \in L^2(\mathbb{R}^+; L^2(\Omega))$ and $\mathbf{V}_0 \in L^2(\mathbb{R}^+; L_s^2(\Omega))$, it follows that $\mathbf{F}$ is well defined for any complex $z \in \mathbf{C}^+ := \{z \in \mathbf{C}; \mathrm{Re} z \geq 0\}$.

Now we give the definition of a weak solution to our problem and consider the variational formulation of the linear differential system (30.2.3).

Definition 30.2.2. *A function* $\mathbf{v}_L \in H^1_{s0}(\Omega)$ *is said to be a weak solution of (30.2.3) if*

$$\int_\Omega [\mu_L(z)\nabla \mathbf{v}_L(\mathbf{x},z) \cdot \nabla \overline{\mathbf{u}}(\mathbf{x}) + z\mathbf{v}_L(\mathbf{x},\omega) \cdot \overline{\mathbf{u}}(\mathbf{x})]d\mathbf{x} = \int_\Omega \mathbf{F}(\mathbf{x},z) \cdot \overline{\mathbf{u}}(\mathbf{x})d\mathbf{x} \tag{30.2.4}$$

for every complex vector $\mathbf{u} \in H^1_{s0}(\Omega)$.

Here, as previously, the bar over a quantity indicates complex conjugate.

If $\mathbf{v}_L$ is a weak solution, then following Teman ([307], Lemma 2.1), we can prove that there exists a scalar field $p_L \in L^2(\Omega)$ such that

$$\begin{aligned} z\mathbf{v}_L(\mathbf{x},z) &= -\nabla p_L(\mathbf{x},z) + \nabla \cdot [\mu_L(z)\nabla \mathbf{v}_L(\mathbf{x},z)] + \mathbf{F}(\mathbf{x},z) \quad \forall \mathbf{x} \in \Omega, \\ \nabla \cdot \mathbf{v}_L(\mathbf{x},z)ds &= 0 \qquad \forall \mathbf{x} \in \Omega, \end{aligned}$$

in the sense of distributions, and moreover, $\mathbf{v}_L(\mathbf{x},z) = \mathbf{0}$ on $\partial\Omega$.

Lemma 30.2.3. *Under the hypotheses of Theorem 30.2.1, the problem (30.2.3) has one and only one weak solution.*

We now observe that the bilinear form

$$a(\mathbf{v},\mathbf{u};z) = \int_\Omega [\mu_L(z)\nabla \mathbf{v}(\mathbf{x}) \cdot \nabla \overline{\mathbf{u}}(\mathbf{x}) + z\mathbf{v}(\mathbf{x}) \cdot \overline{\mathbf{u}}(\mathbf{x})]d\mathbf{x} \tag{30.2.5}$$

is coercive in $H^1_{s0}(\Omega)$. Thus, by general theorems on elliptic systems (see [134, 307, 310]), the existence and uniqueness of a weak solution to (30.2.4) for every $\mathbf{F} \in L^2(\Omega)$ follow.

To prove Lemma 30.2.3, it must be shown that there exists an $\alpha > 0$, possibly dependent on z, such that

$$|a(\mathbf{v}, \mathbf{v}; z)| \geq \alpha(z)\|\mathbf{v}\|^2_{H^1_{s0}(\Omega)}.$$

We now want to demonstrate that the coerciveness of a can be proved by means of the thermodynamic restriction (30.2.2).

To prove this, recall a property of Fourier integrals.

Proposition 30.2.4. *Let $\varphi_1, \varphi_2 \in L^1(\mathbb{R})$ with $\varphi_1\varphi_2 \in L^1(\mathbb{R})$, and let φ_F denote the Fourier transform of φ. If*

$$\Phi(\omega) = \int_{-\infty}^{\infty} \varphi_{1F}(\tau)\varphi_{2F}(\omega - \tau)d\tau$$

is continuous in ω, then

$$\int_{-\infty}^{\infty} e^{-i\omega s}\varphi_1(s)\varphi_2(s)ds = \frac{1}{2\pi}\int_{-\infty}^{\infty} \varphi_{1F}(\tau)\varphi_{2F}(\omega - \tau)d\tau \quad \forall \omega \in \mathbb{R}. \tag{30.2.6}$$

This is an inverted form of the convolution theorem (C.3.2). It follows immediately from Parseval's formula ([37] and Sect. C.3).

Lemma 30.2.5. *If the relaxation function μ belongs to $L^1(\mathbb{R}^+)$ and satisfies (30.2.2), then*

$$\int_0^{\infty} e^{-\sigma s}\cos\omega s\mu(s)ds > 0 \quad \forall \omega \in \mathbb{R},\ \forall \sigma \in \mathbb{R}^+. \tag{30.2.7}$$

Proof. If $\sigma = 0$, then (30.2.7) holds, since it coincides with (30.2.2). Let $\sigma > 0$, and consider the function [146]

$$\varphi_1(s) = \mu(|s|), \quad \varphi_2(s) = \begin{cases} 0 & \forall s < 0, \\ e^{-\sigma s} & \forall s \geq 0. \end{cases}$$

Thus, $\varphi_1, \varphi_2, \varphi_1\varphi_2 \in L^1(\mathbb{R})$, and we have

$$\varphi_{1_F}(\tau) = 2\int_0^{\infty} \cos\tau s\mu(s)ds, \quad \varphi_{2_F}(\tau) = \frac{1}{\sigma + i\tau},$$

whence it follows that

$$\Phi(\omega) = \int_{-\infty}^{\infty} \frac{2}{\sigma + i(\omega - \tau)}\int_0^{\infty} \cos\tau s\mu(s)dsd\tau$$

is continuous. From (30.2.6), applied to the functions φ_1 and φ_2, we have

$$\int_0^{\infty} e^{-(\sigma+i\omega)s}\mu(s)ds = \frac{1}{\pi}\int_{-\infty}^{\infty} \frac{1}{\sigma + i(\omega - \tau)}\int_0^{\infty} \cos\tau s\mu(s)dsd\tau,$$

which has a real part given by

$$\int_0^\infty e^{-\sigma s}\cos\omega s\mu(s)ds = \frac{1}{\pi}\int_{-\infty}^\infty \frac{\sigma}{\sigma^2+(\omega-\tau)^2}\int_0^\infty \cos\tau s\mu(s)dsd\tau,$$

whence (30.2.7) follows by virtue of the conditions $\sigma > 0$ and (30.2.2). □

We can now establish the coerciveness of $a(\mathbf{v}, \mathbf{u}, z)$.

Lemma 30.2.6. *If the relaxation function μ belongs to $L^1(\mathbb{R}^+)$ and satisfies (30.2.2), then the bilinear form a is coercive for every complex number $z \in \mathbf{C}^+$.*

Proof. Since $|a(\mathbf{v}, \mathbf{v}, z)| \geq \text{Re}a(\mathbf{v}, \mathbf{v}, z)$, it is enough to show that

$$\text{Re}a(\mathbf{v}, \mathbf{v}; z) \geq \alpha(z)\|\mathbf{v}\|^2_{H^1_{s0}(\Omega)}$$

for every $z \in \mathbf{C}^+$.

The Laplace transform μ_L, putting $z = \sigma + i\omega$, where $\omega \in \mathbb{R}$ and $\sigma \in \mathbb{R}^+$, becomes

$$\mu_L(z) = \int_0^\infty e^{-\sigma s}\cos\omega s\mu(s)ds - i\int_0^\infty e^{-\sigma s}\sin\omega s\mu(s)ds,$$

which, substituted into (30.2.5), yields

$$\text{Re}a(\mathbf{v}, \mathbf{v}; z) = \int_0^\infty e^{-\sigma s}\cos\omega s\mu(s)ds\int_\Omega |\nabla\mathbf{v}(\mathbf{x})|^2 d\mathbf{x} + \sigma\int_\Omega |\mathbf{v}(\mathbf{x})|^2 d\mathbf{x}.$$

Hence, by Korn's inequality and the arbitrariness of $\sigma > 0$, we have

$$\text{Re}a(\mathbf{v}, \mathbf{v}; z) \geq C(\Omega)\int_{-\infty}^\infty e^{-\sigma s}\cos\omega s\mu(s)ds\|\mathbf{v}(\mathbf{x})\|^2_{H^1_{s0}(\Omega)},$$

where $C(\Omega)$ is a strictly positive constant that depends on the domain Ω and the integral satisfies (30.2.7). Thus, the proof is complete. □

Since Lemma 30.2.3 yields the existence and uniqueness of the solution of (30.2.4), we can study the properties of this solution $\mathbf{v}_L$.

Consider the Green tensor function $\mathbf{\Gamma} \in H^1_{s0}(\Omega)$, defined as a solution of the problem

$$\int_\Omega [\mu_L(z)\nabla_{\mathbf{x}'}\mathbf{\Gamma}(\mathbf{x}, \mathbf{x}'; z)\nabla_{\mathbf{x}'}\mathbf{u}(\mathbf{x}') + z\mathbf{\Gamma}(\mathbf{x}, \mathbf{x}'; z)\mathbf{u}(\mathbf{x}')]d\mathbf{x}' = \int_\Omega \delta(\mathbf{x}-\mathbf{x}')\mathbf{u}(\mathbf{x}')d\mathbf{x}' \tag{30.2.8}$$

for every $\mathbf{u} \in H^1_{s0}(\Omega)$, where δ is the Dirac delta function.

The solution of (30.2.4) can be written in terms of $\mathbf{\Gamma}$ as follows:

$$\mathbf{v}_L(\mathbf{x}, z) = \int_\Omega \mathbf{\Gamma}(\mathbf{x}, \mathbf{x}'; z)\mathbf{F}(\mathbf{x}', z)d\mathbf{x}'. \tag{30.2.9}$$

We now prove existence, uniqueness, and asymptotic behavior with respect to the parameter z of solutions $\mathbf{\Gamma}$ of Eq. (30.2.8).

Lemma 30.2.7. *Under the hypotheses of Theorem 30.2.1, there exists a unique solution* $\mathbf{\Gamma}$ *of (30.2.8) such that*

(i) $\mathbf{\Gamma}(x,\cdot;z) \in H^1_{s0}(\Omega)$ *for every* $z \in \mathbf{C}^+$,
(ii) $\mathbf{\Gamma}(x,x';\cdot)$ *is continuous in* $\mathbf{C}^+$, *and*
(iii) for $z \in \mathbf{C}^+$,

$$\lim_{z\to\infty} z^{1-\alpha}\mathbf{\Gamma}(\mathbf{x},\mathbf{x}';z) = \mathbf{0}, \qquad \alpha > 0,$$

$$\lim_{z\to\infty} \int_\Omega z\mathbf{\Gamma}(\mathbf{x},\mathbf{x}';z)\mathbf{u}(\mathbf{x}')d\mathbf{x}' = \mathbf{u}(\mathbf{x}), \tag{30.2.10}$$

in the sense of distributions.

Proof. The first property holds by virtue of the coerciveness of the bilinear form $a(\mathbf{v},\mathbf{v};z)$, since δ is in the dual space $H^{-1}(\Omega)$. The second property follows from the continuity of the bilinear form $a(\mathbf{v},\mathbf{u};\cdot)$ with respect to the third argument (see [310], Lemma 44.1). Finally, note that from (30.2.8), we have

$$\int_\Omega z^{1-\alpha}\mathbf{\Gamma}(\mathbf{x},\mathbf{x}';z)\left[\mathbf{u}(\mathbf{x}') - z^{-1}\mu_L(z)\nabla_{\mathbf{x}'}\cdot\nabla_{\mathbf{x}'}\mathbf{u}(\mathbf{x}')\right]d\mathbf{x}' = z^{-\alpha}\mathbf{u}(\mathbf{x})$$

for every $\alpha > 0$ and $\mathbf{u} \in C_0^\infty(\Omega)$. Hence, since μ_L is a bounded function of z, the limit as $z \to \infty$ yields (30.2.10). □

A representation for $\nabla_\mathbf{x}\mathbf{v}_L$ can be given in terms of the Green function. We denote by $\nabla_\mathbf{x}\mathbf{\Gamma}$ the third-order tensor function such that

$$\int_\Omega \{\mu_L(z)\nabla_{\mathbf{x}'}[\nabla_\mathbf{x}\mathbf{\Gamma}(\mathbf{x},\mathbf{x}';z)]\nabla_{\mathbf{x}'}\mathbf{u}(\mathbf{x}') + z[\nabla_\mathbf{x}\mathbf{\Gamma}(\mathbf{x},\mathbf{x}';z)]\mathbf{u}(\mathbf{x}')\}d\mathbf{x}'$$
$$= \int_\Omega [\delta(\mathbf{x}-\mathbf{x}')\mathbf{I}]_\mathbf{x}\mathbf{u}(\mathbf{x}')d\mathbf{x}' \tag{30.2.11}$$

for every $\mathbf{u} \in H^1_{s0}(\Omega)$. In terms of $\nabla_\mathbf{x}\mathbf{\Gamma}$,

$$\nabla_\mathbf{x}\mathbf{v}_L(\mathbf{x},z) = \int_\Omega \nabla_\mathbf{x}\mathbf{\Gamma}(\mathbf{x},\mathbf{x}';z)\mathbf{F}(\mathbf{x}',z)d\mathbf{x}'.$$

Using the proof of Lemma 30.2.7 and replacing $\mathbf{\Gamma}$ with $\nabla_\mathbf{x}\mathbf{\Gamma}$, we obtain the following result.

Lemma 30.2.8. *Under the hypotheses of Theorem 30.2.1, there exists a unique solution* $\nabla_\mathbf{x}\mathbf{\Gamma}$ *of (30.2.11) such that*

(i) $\nabla_\mathbf{x}\mathbf{\Gamma}(\mathbf{x},\cdot;z) \in L^2_s(\Omega)$ *for every* $z \in \mathbf{C}^+$ *(see [310], Lemma 23.2),*
(ii) $\nabla_\mathbf{x}\mathbf{\Gamma}(\mathbf{x},\mathbf{x}';\cdot)$ *is continuous on* $\mathbf{C}^+$,

(iii) for $\alpha > 0, z \in \mathbf{C}^+$,

$$\lim_{z\to\infty} z^{1-\alpha}\nabla_{\mathbf{x}}\boldsymbol{\Gamma}(\mathbf{x},\mathbf{x}';z) = \mathbf{0}$$

in the sense of distributions.

These results allow us to prove existence, uniqueness, and stability of the solution to (30.2.1).

A proof of Theorem 30.2.1 is now outlined.

Proof. By hypothesis, $\mathbf{f} \in L^2(\mathbb{R}^+; L^2(\Omega))$ and

$$\mathbf{V}_0(\mathbf{x},t) = \nabla\cdot\int_t^\infty \mu(s)\nabla\mathbf{v}_0(\mathbf{x},t-s)ds$$

belongs to $L^2(\mathbb{R}^+; L^2_s(\Omega))$; therefore, we obtain

$$\begin{aligned}\lim_{z\to\infty}\mathbf{F}(\mathbf{x},z) &= \lim_{z\to\infty}\left[\mathbf{f}_L(\mathbf{x},z)+\mathbf{v}_0(\mathbf{x},0)+\int_0^\infty e^{-zs}\nabla\cdot\int_s^\infty \mu(\tau)\nabla\mathbf{v}_0(\mathbf{x},s-\tau)d\tau ds\right]\\ &= \mathbf{v}(\mathbf{x},0).\end{aligned}$$

Moreover, (30.2.9) and property *(iii)* of $\boldsymbol{\Gamma}$, for $\alpha > 0$, give

$$\begin{aligned}\lim_{z\to\infty} z^{1-\alpha}\mathbf{v}_L(\mathbf{x},z) &= \lim_{z\to\infty}\int_\Omega z^{1-\alpha}\boldsymbol{\Gamma}(\mathbf{x},\mathbf{x}';z)\mathbf{F}(\mathbf{x}',z)d\mathbf{x}'\\ &= \lim_{z\to\infty}\int_\Omega z^{1-\alpha}\boldsymbol{\Gamma}(\mathbf{x},\mathbf{x}';z)\mathbf{v}(\mathbf{x}',0)d\mathbf{x}' = \mathbf{0}.\end{aligned} \tag{30.2.12}$$

Let $z = i\omega$ and $0 < \alpha < \frac{1}{2}$; from (30.2.12), it follows that $\mathbf{v}_L(\mathbf{x}, i\omega)$ is in L^2 with respect to ω; we can view it as the Fourier transform of the function

$$\check{\mathbf{v}}(\mathbf{x},t) = \begin{cases}\mathbf{0}, & t<0,\\ \mathbf{v}(\mathbf{x},t), & t\ge 0.\end{cases}$$

Using Parseval's formula, we have

$$\frac{1}{2\pi}\int_{-\infty}^\infty\int_\Omega |\mathbf{v}_L(\mathbf{x},i\omega)|^2 d\mathbf{x}d\omega = \int_{-\infty}^\infty\int_\Omega |\check{\mathbf{v}}(\mathbf{x},t)|^2 d\mathbf{x}dt = \int_0^\infty\int_\Omega |\mathbf{v}(\mathbf{x},t)|^2 d\mathbf{x}dt \tag{30.2.13}$$

and

$$\begin{aligned}\frac{1}{2\pi}\int_{-\infty}^\infty\int_\Omega |\nabla_{\mathbf{x}}\mathbf{v}_L(\mathbf{x},i\omega)|^2 d\mathbf{x}\omega &= \int_{-\infty}^\infty\int_\Omega |\nabla_{\mathbf{x}}\check{\mathbf{v}}(\mathbf{x},t)|^2 d\mathbf{x}dt\\ &= \int_0^\infty\int_\Omega |\nabla_{\mathbf{x}}\mathbf{v}(\mathbf{x},t)|^2 d\mathbf{x}dt.\end{aligned} \tag{30.2.14}$$

From (30.2.13) and (30.2.14), it follows that

$$\int_0^\infty \int_\Omega \left[|\nabla \mathbf{v}(\mathbf{x},t)|^2 + |\mathbf{v}(\mathbf{x},t)|^2 \right] d\mathbf{x}dt < \infty.$$

Hence, we obtain the required result $\mathbf{v} \in L^2(\mathbb{R}^{++}; H^1_{s0}(\Omega))$. □

30.2.2 Counterexamples to Asymptotic Stability

We consider, for the sake of simplicity, one-dimensional evolution problems.

Suppose that an incompressible viscoelastic fluid occupies the strip $0 < x < l$, $(x,y) \in \mathbb{R}^2$, between two fixed plates. Let the supply $\mathbf{f}$ and the initial history $\mathbf{v}_0$ have the form $\mathbf{f} = f(x,t)\mathbf{k}$, $\mathbf{v}_0 = v_0(x,\tau)\mathbf{k}$ $\forall \tau \in \mathbb{R}^-$; then, $\mathbf{v} = v(x,t)\mathbf{k}$ must be a solution of the scalar boundary initial history value problem

$$\begin{aligned} &v_t(x,t) = \int_0^\infty \mu(s) v_{xx}(x,t-s)ds + f(x,t) \quad \forall x \in (0,l),\ \forall t \in \mathbb{R}^{++}, \\ &v(0,t) = v(l,t) = 0 \qquad \forall t \in \mathbb{R}^{++}, \\ &v(x,\tau) = v_0(x,\tau) \qquad \forall x \in (0,l),\ \forall \tau \in \mathbb{R}^-. \end{aligned} \tag{30.2.15}$$

Since Theorem 30.2.1 can be applied to domains that are bounded in some direction (see, for example, [307], Theorem 2.1), if the relaxation function μ satisfies (30.2.2), then the problem (30.2.15) has one and only one solution $v \in L^2(\mathbb{R}^{++}; H^1_{s0}(0,l))$, for any $f \in L^2(\mathbb{R}^+; L^2(0,l))$ and v_0 such that $\int_t^\infty \mu(s)\mathbf{v}_{0_{xx}}(\mathbf{x}, t - s)ds \in L^2(\mathbb{R}^+; L^2_s(0,l))$.

We want to show that there exist relaxation functions that comply with weaker thermodynamic requirements but do not allow asymptotic stability of the solution of the problem (30.2.15) under the hypotheses of Theorem 30.2.1.

For this purpose, let us assume a nonnegative relaxation function μ with properties

(P_1) μ is a positive decreasing function belonging to $H^1(\mathbb{R}^+, \mathbb{R}^+)$ such that

$$\mu(s) < \frac{k}{(1+s)^{2+\varepsilon}}, \qquad k, \varepsilon > 0;$$

(P_2) μ satisfies the "weak formulation" of the second law of thermodynamics for isothermal processes; that is,

$$\int_0^\infty \mu(s)\cos\omega s ds \geq 0 \qquad \forall \omega \in \mathbb{R},$$

and there exists at least one frequency $\omega^* \neq 0$ such that

$$\mu_c(\omega^*) = \int_0^\infty \mu(s)\cos\omega^* s ds = 0. \tag{30.2.16}$$

Theorem 30.2.9. *Let μ be a relaxation function that has properties (P_1) and (P_2). Then there exists a critical length l^* for the strip such that for $l = l^*$, $f = 0$, and the initial history $v_0(x, \tau) = \sin \frac{\pi}{l^*}(c_1 \cos \omega^* \tau + c_2 \sin \omega^* \tau)$, the problem (30.2.15) has a unique periodic solution (not belonging to $L^2(\mathbb{R}^{++}; H^1_{s0}(0, l^*))$,*

$$v(x, t) = \sin \frac{\pi x}{l^*}(c_1 \cos \omega^* t + c_2 \sin \omega^* t).$$

Before giving the proof of this theorem, we make an observation and prove an auxiliary lemma.

Remark 30.2.10. Observe that if (P_1) holds, then (see [148]) it is possible to prove that the function

$$V_0(x, t) = \int_t^{\infty} \mu(s) v_{0_{xx}}(x, t - s) ds$$

belongs to $L^2(\mathbb{R}^+; L^2_s(0, l))$ for any $v_0 \in L^\infty(\mathbb{R}^-; H^1_{s0}(0, l))$.

Lemma 30.2.11. *Let μ be a relaxation function that has properties (P_1) and (P_2), let $\omega^* \neq 0$ be a frequency satisfying (30.2.16), and let μ_+ be the half-range Fourier transform of μ (see* (C.1.3)*). Then there exists a critical length l^* such that the problem*

$$\begin{aligned} i\omega^* v_F(x, \omega^*) - \mu_+(\omega^*) v_{xxF}(x, \omega^*) &= 0 \qquad \forall x \in (0, l), \\ v_F(0, \omega^*) = v_F(l, \omega^*) &= 0 \end{aligned} \tag{30.2.17}$$

has infinitely many complex-valued solutions for $l = l^*$.

Proof. Substituting (30.2.16) into (30.2.17), we see that both the real and the imaginary parts of the solution v of (30.2.17) must satisfy the following:

$$\begin{aligned} \omega^* u(x, \omega^*) + \mu_s(\omega^*) u_{xx}(x, \omega^*) &= 0 \qquad \forall x \in (0, l), \\ u(0, \omega^*) = u(l, \omega^*) &= 0. \end{aligned} \tag{30.2.18}$$

Integrating by parts, we obtain

$$\begin{aligned} \frac{1}{\omega^*} \mu_s(\omega^*) &= \frac{1}{(\omega^*)^2} \left[\mu(0) + \int_0^{\infty} \mu'(s) \cos \omega^* s ds \right] \\ &= \frac{1}{(\omega^*)^2} \int_0^{\infty} \mu'(s)(\cos \omega^* s - 1) ds > 0, \end{aligned}$$

because the hypothesis (P_1) yields $\mu'(s)(\cos \omega^* s - 1) \geq 0 \ \forall s \in \mathbb{R}^+$. Note that if $\mu'(s)(\cos \omega^* s - 1) = 0 \ \forall s \in \mathbb{R}^+$, then $\mu(s) = 0 \ \forall s \in \mathbb{R}^+$. Putting

$$l = l^* = \sqrt{\frac{\mu_s(\omega^*)}{\omega^*}}\pi, \tag{30.2.19}$$

then the function $u^*(x) = c \sin \frac{\pi x}{l^*}$, $c \in \mathbb{R}$, is a solution of (30.2.18), since $\frac{\omega^*}{\mu_s(\omega^*)}$ is an eigenvalue of $-\Delta$. The connection between the two problems (30.2.17) and (30.2.18) allows us to conclude that any function $v_F(x, \omega^*) = (a_1 + ia_2) sin\frac{\pi x}{l^*}$, $a_1, a_2 \in \mathbb{R}$, is a solution of (30.2.17). □

We can now give the proof of Theorem 30.2.9.

Proof. Let l^* be given by (30.2.19); then using Lemma 30.2.11, the unique solution of the problem

$$v_t(x,t) = \int_0^\infty \mu(s) v_{xx}(x, t-s) ds \quad \forall x \in (0, l^*),\ \forall t \in \mathbb{R}^{++},$$
$$v(0,t) = v(l^*, t) = 0 \qquad \forall t \in \mathbb{R}^{++},$$
$$v(x,\tau) = \sin\frac{\pi x}{l^*}(c_1 \cos\omega^*\tau + c_2 \sin\omega^*\tau) \qquad \forall x \in (0, l^*),\ \forall \tau \in \mathbb{R}^-,$$

which agrees with the prescribed initial history value, is

$$v(x,t) = \sin\frac{\pi x}{l^*}(c_1 \cos\omega^* t + c_2 \sin\omega^* t).$$

In fact, it is easy to show that the two quantities

$$v_t(x,t) = \omega^* \sin\frac{\pi x}{l^*}(-c_1 \sin\omega^* t + c_2 \cos\omega^* t)$$

and

$$\int_0^\infty \mu(s) v_{xx}(x, t-s) ds = \left(\frac{\pi}{l^*}\right)^2 \sin\frac{\pi x}{l^*}\mu_s(\omega^*)(-c_1 \sin\omega^* t + c_2 \cos\omega^* t)$$

coincide by virtue of (30.2.19). □

Finally, we exhibit a family of nonnegative relaxation functions that comply with requirements (P_1) and (P_2). Consider the function

$$\mu(s) = \left(\frac{s^2}{\beta} - \frac{\alpha - 3}{\beta^2}s + \frac{\alpha^2 - 8\alpha + 24}{8\beta^3}\right)e^{-\beta s}$$

with two parameters α and β, which are assumed to be such that $\alpha \in \left(0, 2 + \sqrt{2}\right)$, $\beta > 0$. This satisfies (P_1); furthermore, taking into account that

$$\int_0^\infty e^{-\beta s} \cos\omega s ds = \frac{\beta}{\beta^2 + \omega^2}, \quad \int_0^\infty e^{-\beta s} s \cos\omega s ds = \frac{\beta^2 - \omega^2}{(\beta^2 + \omega^2)^2},$$

and

$$\int_0^\infty e^{-\beta s} s^2 \cos \omega s ds = 2\beta \frac{\beta^2 - 3\omega^2}{(\beta^2 + \omega^2)^3},$$

its Fourier cosine transform is given by

$$\mu_c(\omega) = \int_0^\infty \mu(s) \cos \omega s ds = \frac{\alpha^2}{8\beta^3(\beta^2 + \omega^2)^3} \left(\omega^2 - \frac{8 - \alpha}{\alpha} \beta^2 \right) \geq 0.$$

Hence, for $\omega^* = \beta \sqrt{\frac{8-\alpha}{\alpha}}$, we have $\mu_c(\omega^*) = 0$, so that (P_2) also holds.

An analogous family was first introduced by Fabrizio and Morro in [119] to show that there exist relaxation functions for linear viscoelastic solids that do not allow the quasistatic problem to have a unique solution in the space of sinusoidal (in time) strain histories. Moreover, the same family was used later on by Giorgi and Lazzari [148], though with a different choice of parameters, to give counterexamples to the asymptotic stability of the rest state for initial boundary value problems relating to linear viscoelastic solid materials.

A

Conventions and Some Properties of Vector Spaces

A.1 Notation

A group of relations with a single equation number (***) will be numbered by counting "=" signs. Thus, (***)$_5$ refers to the relation with the fifth "=" sign. Minor exceptions will be clear from context.

Vectors and tensors are denoted by boldface characters, respectively, and scalars by ordinary script. The real line is denoted by $\mathbb{R}$, the nonnegative reals by $\mathbb{R}^+$, and the strictly positive reals by $\mathbb{R}^{++}$. Also, $\mathbb{R}^-$ denotes the nonpositive reals and $\mathbb{R}^{--}$ the strictly negative reals. We will be dealing with spaces of scalar quantities with values in $\mathbb{R}$ or $\mathbb{R}^+$, vector quantities in $\mathbb{R}^3$, second-order tensors and symmetric second-order tensors, the latter space being denoted by Sym. Let one of these spaces, or a composite of more than one, be denoted by $\mathcal{V}$. The space of linear transformations $\mathcal{V} \mapsto \mathcal{V}$ is denoted by $\mathrm{Lin}(\mathcal{V})$. If $\mathcal{V}$ is omitted, it is understood to be $\mathbb{R}^3$; also Lin^+ denotes the set of linear transformations with positive determinant from $\mathbb{R}^3$ to $\mathbb{R}^3$. The space of linear transformations $\mathcal{V}_1 \mapsto \mathcal{V}_2$ is $\mathrm{Lin}(\mathcal{V}_1, \mathcal{V}_2)$.

We will be considering frequency-space quantities, defined by analytic continuation from integral definitions, as functions on the complex plane Ω, where

$$\begin{aligned} \Omega^+ &= \{\omega \in \Omega \,|\, \mathrm{Im}\omega \in \mathbb{R}^+\}, \\ \Omega^{(+)} &= \{\omega \in \Omega \,|\, \mathrm{Im}\omega \in \mathbb{R}^{++}\}. \end{aligned}$$

Similarly, Ω^- and $\Omega^{(-)}$ are the lower half-planes including and excluding the real axis, respectively.

In certain contexts, for example, where the complex plane is not necessarily related to frequency, we use the above convention but with $\mathbb{C}$ replacing Ω. The latter symbol is in some chapters used to denote the spatial region occupied by the body under discussion, for consistency with usage in the literature.

G. Amendola et al., *Thermodynamics of Materials with Memory*,
https://doi.org/10.1007/978-3-030-80534-0

A.2 Finite-Dimensional Vector Spaces

We deal extensively with quadratic forms in the main text, so that it is worthwhile describing certain notation and concepts relating to finite vector spaces. More in-depth discussion can be found in, for example, [194]. Consider first the case that $\mathcal{V}$, of dimension m, is a vector space over the reals. For any two vectors $\mathbf{L}$, $\mathbf{M}$ in $\mathcal{V}$, we denote their inner (scalar or dot) product by $\mathbf{L} \cdot \mathbf{M}$. Let $\mathbf{C}_i$, $i = 1, 2, \ldots, m$, be an orthonormal basis of $\mathcal{V}$, so that

$$\mathbf{C}_i \cdot \mathbf{C}_j = \delta_{ij},\ i, j = 1, 2, \ldots, m.$$

We have

$$\mathbf{L} = \sum_{i=1}^{m} L_i \mathbf{C}_i, \qquad \mathbf{M} = \sum_{i=1}^{m} M_i \mathbf{C}_i. \tag{A.2.1}$$

Also

$$\mathbf{L} \cdot \mathbf{M} = \mathbf{L}^{\top}\mathbf{M} = \sum_{i=1}^{m} L_i M_i = \mathbf{M} \cdot \mathbf{L} = \mathbf{M}^{\top}\mathbf{L}, \tag{A.2.2}$$

and $|\mathbf{M}|^2 = \mathbf{M} \cdot \mathbf{M}$ is the squared norm of $\mathbf{M}$. A linear transformation $\mathbb{K} \in \mathrm{Lin}(\mathcal{V})$ has the representation (we keep the same notation)

$$\mathbb{K} = \sum_{i,j=1}^{m} K_{ij} \mathbf{C}_i \otimes \mathbf{C}_j,$$

where $K_{ij}, i, j = 1, 2, \ldots, m$, are its components in this basis. We will generally refer to linear transformations (with well-defined behavior under a change of reference frame) as tensors and their representations as matrices. The tensor $\mathbb{K}^{\top}$ is the transpose of $\mathbb{K}$. Now

$$(\mathbf{C}_i \otimes \mathbf{C}_j)\mathbf{C}_k = \mathbf{C}_j \cdot \mathbf{C}_k \mathbf{C}_i = \delta_{jk} \mathbf{C}_i,$$

giving

$$(\mathbb{K}\mathbf{M})_i = \mathbb{K}\mathbf{M} \cdot \mathbf{C}_i = \sum_{j=1}^{m} K_{ij} M_j.$$

We have the quadratic form

$$\mathbf{L} \cdot \mathbb{K}\mathbf{M} = \sum_{i,j=1}^{m} L_i K_{ij} M_j = \sum_{i,j=1}^{m} (K^{\top})_{ji} L_i M_j = \mathbb{K}^{\top}\mathbf{L} \cdot \mathbf{M} = \mathbf{M} \cdot \mathbb{K}^{\top}\mathbf{L}. \tag{A.2.3}$$

A commonly used scalar product on the vector space $\mathrm{Lin}(\mathcal{V})$ is

$$\mathbb{K} \cdot \mathbb{N} = \mathrm{tr}(\mathbb{K}\mathbb{N}^{\top}) = \sum_{i,j=1}^{m} K_{ij} N_{ij}, \tag{A.2.4}$$

and the associated squared norm $|\mathbb{K}|^2$ of $\mathbb{K} \in \mathrm{Lin}(\mathcal{V})$ is

$$|\mathbb{K}|^2 = \mathrm{tr}\left(\mathbb{K}\mathbb{K}^{\top}\right) = \sum_{i,j=1}^{m} K_{ij} K_{ji}. \tag{A.2.5}$$

It is frequently the case in the present work that $\mathbb{K}$ is symmetric, so that $\mathbb{K} = \mathbb{K}^{\top}$.

If $\mathcal{V}$ is over the complex numbers—which is the case in this work when we are dealing with quantities in the frequency domain—we define the complex conjugate of $\mathbf{L}$ by

$$\overline{\mathbf{L}} = \sum_{i=1}^{m} \overline{L_i}\mathbf{C}_i, \tag{A.2.6}$$

where $\overline{L_i}$ is the ordinary complex conjugate of L_i. The dot product is still defined by (A.2.2). Thus, for example,

$$\overline{\mathbf{L}} \cdot \mathbf{M} = \overline{\mathbf{L}}^{\top}\mathbf{M} = \sum_{i=1}^{m} \overline{L_i} M_i,$$

giving a real, positive squared norm

$$|\mathbf{L}|^2 = \overline{\mathbf{L}} \cdot \mathbf{L} = \sum_{i=1}^{m} |L_i|^2.$$

If $\mathbb{K} \in \mathrm{Lin}(\mathcal{V})$, now with complex components, we have

$$\overline{\mathbf{L}} \cdot \mathbb{K}\mathbf{M} = \overline{\mathbb{K}^{*}\mathbf{L}} \cdot \mathbf{M}, \tag{A.2.7}$$

where $\mathbb{K}^{*}$ is the Hermitian conjugate of $\mathbb{K}$, defined by

$$\mathbb{K}^{*} = \overline{\mathbb{K}}^{\top}, \tag{A.2.8}$$

where the overhead bar indicates taking the complex conjugate of each element. A Hermitian tensor is one with the property

$$\mathbb{K}^{*} = \mathbb{K}. \tag{A.2.9}$$

Note that

$$\overline{\overline{\mathbf{L}} \cdot \mathbb{K}\mathbf{L}} = \overline{\mathbf{L}} \cdot \mathbb{K}^{*}\mathbf{L},$$

so that $\overline{\mathbf{L}} \cdot \mathbb{K}\mathbf{L}$ is real if $\mathbb{K}$ is Hermitian. We deal largely with symmetric tensors. A Hermitian symmetric tensor is defined by $\mathbb{K}^{*} = \overline{\mathbb{K}}$.

A natural choice of orthonormal basis is the eigenvectors of a Hermitian tensor $\mathbb{K} = \mathbb{K}^{*}$. Denoting these as above by $\mathbf{C}_i, i = 1, 2, \ldots, m,$the quantities $\mathbf{C}_i \otimes \mathbf{C}_i$,

$i = 1, 2, \ldots, m$, are the projectors on the eigenspaces of this tensor. The spectral form of $\mathbb{K}$ is the representation

$$\mathbb{K} = \sum_{i=1}^{m} \lambda_i \mathbf{C}_i \otimes \mathbf{C}_i, \tag{A.2.10}$$

where $\lambda_i, i = 1, 2, \ldots, m$, are the real eigenvalues of $\mathbb{K}$. Any tensor $\mathbb{N}$ that can be given by

$$\mathbb{N} = \sum_{i=1}^{m} \mu_i \mathbf{C}_i \otimes \mathbf{C}_i, \tag{A.2.11}$$

where $\mu_i, i = 1, 2, \ldots, m$, are arbitrary complex numbers, commutes with $\mathbb{K}$.

The tensor $\mathbb{N}$ is, in general, a normal transformation, in the sense that it commutes with its Hermitian conjugate $\mathbb{N}^*$.

Note that for any $\mathbb{N}$ given by (A.2.11) and $\mathbf{L}$, $\mathbf{M}$ given by (A.2.1),

$$\overline{\mathbf{L}} \cdot \mathbb{N}\mathbf{M} = \sum_{i=1}^{m} \mu_i \overline{L_i} M_i.$$

A.2.1 Positive Definite Tensors

A Hermitian tensor $\mathbb{K}$ is positive definite if for every $\mathbf{L} \in \mathcal{V}$ (the relation for vector spaces over the reals is given in parentheses; the tensor $\mathbb{K}$ can be taken to be symmetric, since any antisymmetric portion does not contribute),

$$\overline{\mathbf{L}} \cdot \mathbb{K}\mathbf{L} \geq 0 \; (\mathbf{L} \cdot \mathbb{K}\mathbf{L} \geq 0), \tag{A.2.12}$$

where equality is true only if $\mathbf{L} = \mathbf{0}$, where $\mathbf{0}$ is the zero in $\mathcal{V}$ or Lin($\mathcal{V}$). If equality occurs for $\mathbf{L} \neq \mathbf{0}$, then $\mathbb{K}$ is positive semidefinite. The relation $\mathbb{K} > \mathbf{0}$ indicates that $\mathbb{K}$ is positive definite, while $\mathbb{K} \geq \mathbf{0}$ implies that it is positive semidefinite. The description nonnegative for a tensor is equivalent to positive semidefinite. Negative definiteness and semidefiniteness can be defined analogously. We have an ordering on Lin($\mathcal{V}$) in that, for example,

$$\mathbb{K}_1 > \mathbb{K}_2 \Longleftrightarrow \mathbb{K}_1 - \mathbb{K}_2 > \mathbf{0}.$$

A tensor is positive definite if and only if its eigenvalues are all positive. It is positive semidefinite if all eigenvalues are nonnegative.

Remark A.2.1. An example of a positive semidefinite tensor that occurs in the main text is the following. Let $\mathbf{L} \in \mathcal{V}$ be of the form $(\text{A.2.1})_1$, but with complex components, $\overline{\mathbf{L}}$ being given by (A.2.6). Then consider

$$\mathbb{K} = \overline{\mathbf{L}} \otimes \mathbf{L} = \sum_{i,j=1}^{m} \overline{L_i} L_j \mathbf{C}_i \otimes \mathbf{C}_j.$$

Clearly, $\overline{\mathbf{N}} \cdot \mathbb{K}\mathbf{N} \geq 0$ for any $\mathbf{N} \in \mathcal{V}$, as required by (A.2.12). However, any vector $\mathbf{M}$ perpendicular to $\mathbf{L}$ in $\mathcal{V}$ will obey the relation $\mathbb{K}\mathbf{M} = \mathbf{0}$, so that it is an eigenvector of $\mathbb{K}$ with eigenvalue zero. Thus, $\mathbb{K}$ is positive semidefinite.

A.2.2 Differentiation with Respect to Vector Fields

Let ϕ be a scalar depending on $\mathbf{L}$. The quantity

$$\frac{\partial \phi}{\partial \mathbf{L}} \in \mathcal{V}$$

is a vector with components $\partial\phi/\partial L_i, i = 1, 2, \ldots, m$, in a given basis. We take it to be a column vector, the transpose of which is a row vector. The quantity

$$\frac{\partial \mathbf{L}}{\partial \mathbf{L}} \in \mathrm{Lin}(\mathcal{V})$$

is a tensor with components

$$\frac{\partial L_i}{\partial L_j} = \delta_{ij}, \ i, j = 1, 2 \ldots, m,$$

so that $\partial\mathbf{L}/\partial\mathbf{L}$ is the unit tensor. We have

$$\frac{\partial}{\partial \mathbf{L}}(\mathbf{L} \cdot \mathbb{K}\mathbf{M}) = \mathbb{K}\mathbf{M}$$

and

$$\frac{\partial}{\partial \mathbf{M}}(\mathbf{L} \cdot \mathbb{K}\mathbf{M}) = \frac{\partial}{\partial \mathbf{M}}(\mathbb{K}^{\mathsf{T}}\mathbf{L} \cdot \mathbf{M}) = \mathbb{K}^{\mathsf{T}}\mathbf{L}$$

for a vector space over $\mathbb{R}$. In the complex case, we have

$$\frac{\partial}{\partial \overline{\mathbf{L}}}(\overline{\mathbf{L}} \cdot \mathbb{K}\mathbf{M}) = \mathbb{K}\mathbf{M}$$

and

$$\frac{\partial}{\partial \mathbf{M}}(\overline{\mathbf{L}} \cdot \mathbb{K}\mathbf{M}) = \frac{\partial}{\partial \mathbf{M}}(\overline{\mathbb{K}^{*}\mathbf{L}} \cdot \mathbf{M}) = \overline{\mathbb{K}^{*}\mathbf{L}}.$$

A.2.3 The Vector Space *Sym*

The space of symmetric second-order tensors acting on $\mathbb{R}^3$ is denoted by Sym := $\{\mathbf{E} \in \mathrm{Lin}(\mathbb{R}^3) \ : \ \mathbf{E} = \mathbf{E}^{\mathsf{T}}\}$. Operating on Sym is the space of fourth-order tensors Lin(Sym).

The vector space Sym is isomorphic to $\mathbb{R}^6$. In particular, for every $\mathbf{E}, \mathbf{F} \in$ Sym, if $\mathbf{C}_i, i = 1, 2, \ldots, 6$, is an orthonormal basis of Sym with respect to the inner product (A.2.4) in $\mathrm{Lin}(\mathbb{R}^3)$, namely $\mathrm{tr}(\mathbf{E}\mathbf{F}^{\mathsf{T}})$, it is clear that the representation

$$\mathbf{E} = \sum_{i=1}^{6} E_i \mathbf{C}_i, \qquad \mathbf{F} = \sum_{i=1}^{6} F_i \mathbf{C}_i, \tag{A.2.13}$$

yields that $\mathrm{tr}(\mathbf{E}\mathbf{F}^{\top}) = \sum_{i=1}^{6} E_i F_i$. Therefore, we can treat each tensor of Sym as a vector in $\mathbb{R}^6$ and denote by $\mathbf{E} \cdot \mathbf{F}$ the inner product between elements of Sym:

$$\mathbf{E} \cdot \mathbf{F} = \mathrm{tr}(\mathbf{E}\mathbf{F}^{\top}) = \mathrm{tr}(\mathbf{E}\mathbf{F}) = \sum_{i=1}^{6} E_i F_i \tag{A.2.14}$$

and $|\mathbf{F}|^2 = \mathbf{F} \cdot \mathbf{F}$. Consequently, any fourth-order tensor $\mathbb{G} \in \mathrm{Lin}(\mathrm{Sym})$ will be identified with an element of $\mathrm{Lin}(\mathbb{R}^6)$ by the representation

$$\mathbb{G} = \sum_{i,i=1}^{6} K_{ij} \mathbf{C}_i \otimes \mathbf{C}_j, \tag{A.2.15}$$

and $\mathbb{G}^{\top}$ is the transpose of $\mathbb{G}$ as an element of $\mathrm{Lin}(\mathbb{R}^6)$. The scalar product and norm in $\mathrm{Lin}(\mathbb{R}^3)$ are given by (A.2.4) and (A.2.5) for $m = 6$.

For complex-valued tensors, let $\mathrm{Sym}(\Omega)$ and $\mathrm{Lin}(\mathrm{Sym}(\Omega))$ be, respectively, the sets of tensors represented by the forms (A.2.13) and (A.2.15) with $L_i, M_i, K_{ij} \in \Omega$. Then for $\mathbf{E}, \mathbf{F} \in \mathrm{Sym}(\Omega)$, we have, instead of (A.2.14),

$$\mathbf{E} \cdot \overline{\mathbf{F}} = \mathrm{tr}(\mathbf{E}\mathbf{F}^*) = \mathrm{tr}\left(\mathbf{E}\overline{\mathbf{F}}\right) = \sum_{i=1}^{6} E_i \overline{F_i}. \tag{A.2.16}$$

In the present work, we deal with $\mathcal{V} = \Gamma^+$ defined by (5.1.10), which has dimension $m = 10$, or vector spaces contained in Γ^+. The scalar product between two elements of Γ^+ is understood to mean the sum of (A.2.14) or (A.2.16) on Sym, the standard scalar product on $\mathbb{R}^3$ and the product of quantities in $\mathbb{R}$.

B

Some Properties of Functions on the Complex Plane

B.1 Introduction

We describe briefly, for the sake of convenient reference, some properties of analytic functions that are required in various contexts, mainly in Part III. For a more complete treatment of these topics, we refer to the numerous available standard references, for example [306]. A useful now classical reference is [267].

Of all functions defined on the xy plane, there is a very special class, termed analytic functions, that have the property that they are functions only of the combination $z = x + iy$ and have a uniquely defined derivative with respect to z at each point in the region of analyticity. This latter requirement is very restrictive in that it means that the derivative is independent of the infinite number of directions from which the limit may be taken. If we write such a function $F(z)$ in the form

$$F(z) = F(x, y) = u(x, y) + iv(x, y),$$

then the uniqueness of the limit gives the Cauchy–Riemann conditions

$$\frac{\partial u}{\partial x} = \frac{\partial v}{\partial y}, \qquad \frac{\partial u}{\partial y} = -\frac{\partial v}{\partial x}.$$

These conditions are necessary consequences of the analyticity assumption. If the derivatives are continuous at (x, y), it may also be shown that they are sufficient to ensure analyticity.

Note that the Cauchy–Riemann equations imply that if the real part of a complex function is known, its imaginary part is determined to within a constant and vice versa.

B.1.1 Cauchy's Theorem and Integral Formula

If $F(z)$ is analytic on an open set O, then for any contour C in O, we have

G. Amendola et al., *Thermodynamics of Materials with Memory*,
https://doi.org/10.1007/978-3-030-80534-0

$$\oint_C F(z)dz = 0. \tag{B.1.1}$$

This is a simple statement of Cauchy's theorem. The term contour is taken to mean a closed contour. Cauchy's integral formula states that if $F(z)$ is analytic within and on a contour C, then

$$F(z) = \frac{1}{2\pi i}\oint_C \frac{F(z')}{z' - z}dz' \tag{B.1.2}$$

if the contour is taken counterclockwise, which is the conventional positive direction, and is a manifestation of a more basic convention, namely that angles are presumed to increase in a counterclockwise direction. If C is clockwise, then the integral in (B.1.2) is equal to $-F(z)$. Unless otherwise stated, a contour C may be taken to be counterclockwise. If z is outside of the contour C, this integral gives zero. An immediate consequence of (B.1.2) is

$$F^{(n)}(z) = \frac{n!}{2\pi i}\oint_C \frac{F(z')}{(z' - z)^{n+1}}dz', \tag{B.1.3}$$

where $F^{(n)}$ is the nth derivative of F.

Now let z approach the contour from the inside toward a point z_0 on C. The integral can be assigned a particular finite value as a result of a limiting process that will now be described. We distort the contour into a small semicircle around z_0 outside of C. In the limit, as this semicircle gets smaller, it can be shown that it yields a finite contribution of $F(z_0)/2$. We define the Cauchy principal value of the integral as the value obtained by means of this limiting process minus the contribution $F(z_0)/2$ of the semicircle. Therefore, for z on the contour,

$$\frac{1}{\pi i}P\oint_C \frac{F(z')}{z' - z}dz' = F(z), \tag{B.1.4}$$

where the integral is interpreted as a Cauchy principal value.

We are mainly interested in cases in which C encloses the upper or lower half-plane $\Omega^{(\pm)}$. Let $F(z)$ be analytic in the upper half-plane and let it go to zero at infinity more strongly than z^{-1}, at least in this half-plane. We take C in (B.1.4) to be the real axis and the infinite semicircle enclosing the upper half-plane. This contour is counterclockwise, so we obtain from (B.1.4), for x on the real axis,

$$\frac{1}{\pi i}P\int_{-\infty}^{\infty} \frac{F(x')}{x' - x}dx' = F(x), \tag{B.1.5}$$

where the integral is a Cauchy principal value.

If F were analytic in the lower half-plane, the sign on the right-hand side would be negative, since the contour is clockwise.

A basic property of analytic functions is that they can be expressed as infinite (or finite in the case of polynomials) power series about any point z_0 in their region of analyticity. This power series has a radius of convergence equal to the distance between z_0 and the nearest singular point.

B.1.2 Analytic Continuation

In different parts of the complex plane, an analytic function may have different representations, as power series about different points, for example. A region of analyticity defined by a circle of convergence of one power series can be extended by considering a power series about another point. Given two representations, the question arises whether they are the same complex function or distinct functions. If they represent the same function, they are said to constitute analytic continuations of each other.

A fundamental result states that if a function is analytic in a region R and zero along any continuous arc in R, then it is zero everywhere in R. It follows trivially that if two functions, analytic in a region R, are equal on a continuous arc contained in this region, they are equal over all of R.

More generally, let two functions be analytic on R and let $O \subset R$ be a nonempty open set. If the two representations are equal on O, then they are the same analytic function defined on R.

It can be shown that if two different analytic continuations are constructed from a set R into R_1, and R_2, where $R_3 = R_1 \cap R_2$ is nonempty, then the functions are equal on R_3. Note that $R \cap R_1$, $R \cap R_2$ are nonempty.

The most widespread use of this concept in the present work is a simple application of these kinds of results. If we have a formula for a function on a certain part of the complex plane, for definiteness, let us say a part or all of the real axis, given by $G(x)$, then the analytic continuation of this function into the complex plane is given by $G(z)$, for whatever values of z this quantity is meaningful. If $G(x)$ is some combination of elementary functions, for example, then $G(z)$ will exist at all values of z that are not singular points. So, for example, if

$$G(x) = \frac{1}{1 + x^2}, \quad x \in \mathbb{R},$$

its analytic continuation to the complex plane is

$$G(z) = \frac{1}{1 + z^2}, \tag{B.1.6}$$

which is valid everywhere except at the singular points $z = \pm i$. Its uniqueness everywhere except at these points is guaranteed by the above results.

B.1.3 Liouville's Theorem

A result that is fundamental to the developments of Part III is Liouville's theorem, which we state in a somewhat generalized form.

Let a function $F(z)$ be analytic at every finite point in the complex plane and let it behave like z^n as z tends to infinity. Then it must be a polynomial of degree n. In particular, if its limit at infinity is a constant, it is equal to this constant everywhere. The most important case for our purposes is where the constant is zero, and $F(z)$ is zero everywhere.

B.1.4 Singularities

What makes analytic functions interesting are their singularities, or points where they are not analytic. In fact, the content of Liouville's theorem is that if they have no singularities, they are trivial.

The simplest singularities are poles, that is to say, behaving at z_0 like $(z - z_0)^{-n}$, where n is a positive integer called the order of the pole. These are isolated singularities. A function whose only singularities are poles is known as a meromorphic function. A function behaving like z^n for large $|z|$, where n is a positive integer, is regarded as having a pole of order n at infinity.

Remark B.1.1. If a real function has simple poles on the real axis, then it must have at least one zero between each two poles, because in passing through each pole $(x - a_i)^{-1}$, moving in a positive direction, the function switches from being a large negative number to being a large positive number. Therefore, it must pass through zero on the passage to the next pole.

Poles of infinite order are referred to as essential singularities. For example, the function $\exp(1/z)$ has an essential singularity at the origin, and $\exp(iz)$ has an essential singularity at infinity. Note that on the lower half-plane $\exp(iz)$ diverges exponentially and on the upper half-plane decays exponentially. It can be treated as analytic at infinity on the latter half-plane in the sense that it is analytic at finite points and we can take infinite contours over $\Omega^{(+)}$ for integrands with $\exp(iz)$ as a factor. For simplicity, though with some imprecision, we shall refer to its behavior in this half-plane as analytic.

Functions that are analytic over the whole complex plane are said to be entire or integral functions. They must be constant everywhere or singular at infinity. If the singularity is of finite order, then they are polynomials. This is a restatement of Liouville's theorem, as given above. Nonpolynomial entire functions must have an essential singularity at infinity. Examples are e^z, $\sin z$, and $\cos z$.

We assume, except in Chap. 15, that the functions we deal with are analytic at infinity, that is to say, behave as a constant or go to zero at large $|z|$. It is further assumed that essential singularities do not occur at finite points on Ω.

B.1.5 Branch Points

If one follows an analytic function around a contour to the initial point and it does not return to the same value, then the function is multivalued. This is associated with a branch point within the contour. A branch point is a type of singularity, distinct from a pole or an essential singularity. It is not isolated because, as we shall see below, its effects are not localized at any one point. The function $(z - a)^\gamma$ is, for noninteger values of γ, a multivalued function that is of interest in Chap. 14. In the standard polar representation, it becomes

$$(z - a)^\gamma = |z - a|^\gamma e^{i\gamma\theta}, \tag{B.1.7}$$

where θ is the argument of $(z-a)$. Let γ be a real quantity. If it is rational, let us write it as p/q, where p, q have no common factors. Then if we circle the point a, say r times, where $r < q$, the function returns to different values each time. When $r = q$, the function returns to its original value. We say that $(z - a)^{\gamma}$ has a branch point at $z = a$ and has q distinct branches. If γ is irrational, the function has an infinite number of branches.

Branch points always occur in pairs. The function $(z-a)^{\gamma}$ also has a branch point at infinity, where it behaves like z^{γ}. We join the point a to infinity by some convenient line and agree that the function undergoes a discontinuous jump in crossing this line. If a is real, this line is conventionally chosen to be the x-axis from the point a to $-\infty$. This is, however, an arbitrary choice. It is perfectly possible to choose another line of discontinuity. It would not be the same function, however.

The complex plane, excluding the line of discontinuity, is sometimes referred to as the cut plane and the line itself as a branch cut or simply a cut. A multivalued function with say q distinct branches can be completely characterized by taking q cut complex planes and defining a single-valued branch on each of them.

A process of unique analytic continuation cannot go around a branch point. The branch cut represents a barrier. Thus, if we continue R into R_1 and R_2 around a branch point, then we cannot form a nonempty overlapping set $R_3 = R_1 \cap R_2$. However, it can go around an isolated singularity.

Returning to the function $F(z) = (z - a)^{\gamma}$ (where a is real) with a cut along $(-\infty, a]$, let $F^{\pm}(x)$ be the limiting values of $F(z)$ from above and below the real axis, respectively. Using (B.1.7), one can show that

$$F^{-}(x) = e^{-2\pi i\gamma} F^{+}(x).$$

Note that this applies also if γ is complex, in which case there is a real as well as an imaginary exponential factor.

Another example is

$$F(z) = (z - a)^{\gamma}(z - b)^{1-\gamma}, \tag{B.1.8}$$

where a and b are real with $b > a$. The cut for $F(z)$ given by (B.1.8) must join a and b. The simplest choice is to take the portion of the real axis $[a, b]$ as the branch cut. The function will in general have many branches, which we can represent as follows. Let $\arg(z - a) = \theta_a$ and $\arg(z - b) = \theta_b$. Then

$$F(z) = |z - a|^{\gamma}|z - b|^{1-\gamma} exp\{i[\theta_a\gamma + \theta_b(1 - \gamma) + 2\pi m\gamma + 2\pi n(1 - \gamma)]\},$$
$$m, n \text{ integers}.$$

One can show that

$$F^{-}(x) = \Gamma F^{+}(x),$$

where

$$\Gamma = \begin{cases} 1, & x \notin [a, b], \\ e^{2\pi i\gamma}, & x \in [a, b]. \end{cases}$$

Therefore this function is analytic except in the interval $[a, b]$. It will be different for different choices of m and n, though these contribute only a constant factor. Unique analytic continuation is possible around $[a, b]$, avoiding the branch cut.

Another multivalued function that has a role in our considerations is

$$F(z) = \log(z - a) = \log|z - a| + i \arg(z - a), \tag{B.1.9}$$

which has a branch point at a. We take the choice $\arg(z - a) = 0$ if $z - a$ is real as the standard branch of the logarithm. This vanishes if $|z - a|$ is unity. The function $\arg(z - a)$ has range $[-\pi, \pi]$. If $a \in \mathbb{R}$, the branch cut is conventionally taken along $(-\infty, a]$. The function

$$F(z) = \log\left(\frac{z - b}{z - a}\right) \tag{B.1.10}$$

has a branch cut on a line joining a and b, the simplest being the straight line segment $[a, b]$.

B.1.6 Evaluation of Contour Integrals

Let F be analytic on a contour C but with singularities within C. Then (B.1.1) generalizes to

$$\int_C F(z)dz = 2\pi i \sum \text{residues within } C.$$

The contour C can be deformed at will without changing the value of the integral, provided it does not cross any singularities. If F has a pole of order n at z_0, that is to say,

$$F(z) \underset{z \to z_0}{\longrightarrow} \frac{G(z)}{(z - z_0)^n},$$

where G is analytic at z_0, then the residue of that pole can be determined to be $G^{(n-1)}(z_0)/(n-1)!$, using (B.1.3).

If F has a branch cut B within C joining branch points a, b, then the residue is obtained by shrinking C tightly around the branch cut, giving

$$\frac{1}{2\pi i} \int_a^b d(u)du, \quad d(u) = F^-(u) - F^+(u), \quad F^{\pm}(u) = \lim_{z \to u^{\pm}} F(z), \quad u \in B, \tag{B.1.11}$$

the positive side of B being that along which the shrunken contour is going from b to a and the negative side being that along which it is going from a to b.

We deal in the main text with integrals of the form

$$\int_{-\infty}^{\infty} F(u)du,$$

where $F(z)$ is analytic on an open region, including the real axis, with the behavior

$$F(z) \sim \frac{A}{z^p}, \quad p \geq 1,$$

at large $|z|$, where A is a constant. Consider a contour C in Ω^+ consisting of a semicircle of radius R on a segment $[-R, R]$ or $\mathbb{R}$ that encloses all the singularities of F in $\Omega^{(+)}$. Let

$$I_R = \oint_C F(z)dz.$$

Now on the circumference of the semicircle,

$$z = Re^{i\theta}, \qquad dz = izd\theta. \tag{B.1.12}$$

If $p = 1$, then

$$\lim_{R\to\infty} I_R = 2\pi i \sum \text{residues of singularities in } \Omega^{(+)} = \int_{-\infty}^{\infty} F(u)du + i\pi A, \tag{B.1.13}$$

the rightmost term being the contribution of the circumference of the semicircle, obtained using (B.1.12). If $p > 1$, we have

$$\lim_{R\to\infty} I_R = 2\pi i \sum \text{residues of singularities in } \Omega^{(+)} = \int_{-\infty}^{\infty} F(u)du. \tag{B.1.14}$$

Closing the contour in $\Omega^{(-)}$ gives

$$\begin{aligned}\lim_{R\to\infty} I_R &= -2\pi i \sum \text{residues of singularities in } \Omega^{(-)} \\ &= \begin{cases} \int_{-\infty}^{\infty} F(u)du - i\pi A, & p = 1, \\ \int_{-\infty}^{\infty} F(u)du, & p > 1, \end{cases}\end{aligned} \tag{B.1.15}$$

where the sign changes are a result of the fact that the contour direction is now clockwise.

Remark B.1.2. Let $F(z)$ be a function analytic on Ω except at a variety of singularities. The notation $\overline{F}(z)$ indicates the complex conjugate function, leaving the variable z untouched. Then $\overline{F}(z)$ is analytic except at singularities that are a mirror image in the real axis of those of $F(z)$.

In particular, if F is analytic in $\Omega^{(+)}(\Omega^{(-)})$, then $\overline{F}$ will be analytic in $\Omega^{(-)}(\Omega^{(+)})$.

Proposition B.1.3. *Let $F(z)$ be analytic in $\Omega^{(+)}$ and $G(z)$ in $\Omega^{(-)}$. Let both go to zero as $|z|^{-p}$, $p > 1/2$ at large $|z|$. Then*

$$\int_{-\infty}^{\infty} \overline{F}(s)G(s)ds = \int_{-\infty}^{\infty} F(s)\overline{G}(s)ds = 0,$$

so that they are orthogonal in an L^2 scalar product.

Proof. This follows from Cauchy's theorem by closing the first integral on $\Omega^{(-)}$ and the second on $\Omega^{(+)}$. □

B.2 Cauchy Integrals

We consider integrals of the following type [274]:

$$F(z) = \frac{1}{2\pi i}\int_L \frac{f(u)}{u-z}du, \tag{B.2.1}$$

where L is a sectionally smooth curve in the complex plane. By this, we mean a finite number of nonintersecting smooth arcs and contours. The term contour is used to indicate a closed curve, as before, while arc refers to a curve that is not closed and therefore has endpoints. Smoothness indicates that a tangent exists at each point of L and its slope varies continuously. In other words, each arc or contour, if represented parametrically, has continuous first derivatives with respect to its parameter.

The more interesting developments around Cauchy integrals deal largely with the case in which L is finite in length. However, we are mainly interested in the cases in which L is infinite in length, given by $\mathbb{R}$, discussed below, or $\mathbb{R}^+$ for example, or infinite segments of the imaginary axis. Provided convergence issues are taken into account, there is no difficulty in dealing with L infinite in length.

We need to choose a positive direction along L. For contours, this is generally taken to be the counterclockwise direction, but for arcs, there is no set convention. For an integral along a line segment $[a, b]$ anywhere in the complex plane, written as $\int_a^b$, the positive direction is taken to be from a to b. Thus, if $[a, b]$ is on the real axis and $a < b$, the positive direction is the positive x direction. The region of the complex plane to the left, as one moves along L in the positive direction, is denoted by S^+ and the region to the right by S^-. These are the upper and lower half-planes, respectively, for L in a positive direction along the real axis.

The function f, referred to as the density function, is assumed to be bounded everywhere, except possibly at endpoints of arcs, denoted by c_k, $k = 1, 2, \ldots$, where it may have integrable singular points with

$$f(u) \sim \frac{f_0}{|u-c_k|^{\alpha}}, \quad 0 \leq \alpha < 1, \tag{B.2.2}$$

where f_0 is a constant. Furthermore, it is assumed that f is Hölder continuous at each point of L where it is not singular. This property is defined as follows: for any two points u_1, u_2, there exist positive real constants A, μ such that

$$|f(u_1) - f(u_2)| \leq A|u_1 - u_2|^{\mu}. \tag{B.2.3}$$

It is easy to show that if $\mu > 1$, the derivative of $f(u)$ is zero, so that it is a constant. This case is not of great interest, so it is always assumed that $\mu \leq 1$. For $\mu = 1$, the Hölder condition is termed the Lipschitz condition and is obeyed by any differentiable function and others not in this class. For $\mu < 1$, the condition implies continuity in the ordinary sense. The case $\mu = 0$, which is excluded, is consistent with discontinuity. A function obeying this condition at a point, or on a line, will be described as obeying the $H(\mu)$ condition on that set if μ is specified, or otherwise just the H condition.

At large $|z|$, the function F behaves like

$$F(z) \sim -\frac{A}{z}, \quad 2\pi i A = -\int_L f(u)du,$$

if the integral is nonzero. If it is zero, F falls off as some higher power of z. Consider the limiting value as z approaches a point u on L at which f is nonsingular and that is not an endpoint of an arc. We write

$$F(z) = \frac{1}{2\pi i}\int_L \frac{f(t) - f(u)}{t - z}dt + \frac{f(u)}{2\pi i}\int_L \frac{1}{t - z}dt. \tag{B.2.4}$$

The Hölder condition (B.2.3) implies that the first term approaches a well-defined integral

$$\frac{1}{2\pi i}\int_L \frac{f(t) - f(u)}{t - u}dt$$

as $z \to u$ because the behavior at the singularity is integrable. This step illustrates the importance of the Hölder property. The second term can be assigned a finite value but one that depends on the direction in which the limit is taken. Let

$$\widetilde{\psi}(z) = \frac{1}{2\pi i}\int_L \frac{dt}{t - z}$$

and denote by $\widetilde{\psi}^+(u)$, $\widetilde{\psi}^-(u)$ the limiting values of $\widetilde{\psi}(z)$ as z approaches u from S^+ and S^-, respectively. In each of these cases, we deform the contour into a small semicircle around u and consider the limit as this semicircle shrinks to zero. It is easy to show that

$$\widetilde{\psi}^+(u) = \frac{1}{2} + \frac{1}{2\pi i}P\int_L \frac{dt}{t - u}, \quad \widetilde{\psi}^-(u) = -\frac{1}{2} + \frac{1}{2\pi i}P\int_L \frac{dt}{t - u}, \tag{B.2.5}$$

where the integrals are Cauchy principal values. The more general formulas

$$F^+(u) = \frac{1}{2} f(u) + \frac{1}{2\pi i} P \int_L \frac{f(t)}{t-u} dt,$$
$$F^-(u) = -\frac{1}{2} f(u) + \frac{1}{2\pi i} P \int_L \frac{f(t)}{t-u} dt \tag{B.2.6}$$

follow from (B.2.4) and (B.2.5), since we can write the first, well-defined, integral in (B.2.4) as its Cauchy principal value, and the two integrals can then be recombined. These are the well-known Plemelj formulas which are of great importance in Part III of the present work. Another form of these relations is given by

$$F^+(u) - F^-(u) = f(u),$$
$$F^+(u) + F^-(u) = \frac{1}{\pi i} P \int_L \frac{f(t)}{t-u} dt,$$

which show clearly that F, defined by (B.2.1), is discontinuous across L at all points where f is nonzero. This implies the existence of branch points on L, resulting in branch cuts along L.

For $z \notin L$, $f(z)$, given by (B.2.1), is analytic since it is differentiable. It should be noted that this property requires no assumption on f other than Hölder continuity. In particular, no analyticity requirements need be imposed.

Consider the Cauchy integral over a single arc $[a, b]$ of finite length:

$$F(z) = \frac{1}{2\pi i} \int_a^b \frac{f(u)}{u-z} du.$$

This is a function analytic everywhere on Ω except on $[a, b]$. If f is nonzero on this segment, F has a branch cut between a and b. It is of interest to determine the behavior of $F(z)$ as z approaches the endpoints. Consider z close to a. Let $f(a)$ be finite. Then, using the same trick as in (B.2.4), we obtain

$$F(z) = \frac{f(a)}{2\pi i} \log\left(\frac{b-z}{a-z}\right) + \frac{1}{2\pi i} \int_a^b \frac{f(t)-f(a)}{t-z} dt = \frac{f(a)}{2\pi i} \log\left(\frac{1}{a-z}\right) + F_1(z), \tag{B.2.7}$$

where $F_1(z)$ has a definite, nonsingular, limit as $z \to a$. Similarly, near $z = b$,

$$F(z) = \frac{f(b)}{2\pi i} \log(b-z) + F_2(z), \tag{B.2.8}$$

where $F_2(b)$ is nonsingular. Therefore, if $f(a)$ or $f(b)$ is finite, there is a logarithmic singularity in $F(z)$ at that endpoint. If the limit is taken along the branch cut, similar formulas may be given by applying the Plemelj formulas to the singular term. Let us write

$$a - z = |a - z| e^{i\theta},$$

where $\theta = \theta_0$ gives the limit to the cut from the positive side and $\theta = \theta_0 + 2\pi$ is the limit from the negative side. Then we see that the dominant term has the form

$$\frac{1}{2}(F^+(u) + F^-(u)) \begin{cases} \underset{u\to a}{\sim} \dfrac{f(a)}{2\pi i} \log \dfrac{1}{|a-u|}, \\ \underset{u\to b}{\sim} \dfrac{f(b)}{2\pi i} \log |b-u|. \end{cases} \tag{B.2.9}$$

If the end value of $f(u)$ is zero, then $F(z)$ approaches a definite, finite, limit at that point. If $f(u)$ has a singularity at an endpoint of the type given by (B.2.2), then $F(z)$ has a singularity of the same type. This may be seen intuitively by considering the dominant term of the integral. Therefore, if we have the behavior (B.2.2) at a, then

$$F(z) \underset{z\to a}{\sim} \frac{A}{(z-a)^\alpha} \tag{B.2.10}$$

off the cut and

$$\frac{1}{2}[F^+(u) + F^-(u)] \underset{u\to a}{\sim} \frac{A_1}{(u-a)^\alpha}, \tag{B.2.11}$$

where rigorous arguments and detailed expressions for the constants A, A_1 are given by Muskhelishvili [274] and Gakhov [143], for example. Similar formulas apply for such behavior at b.

B.2.1 Cauchy Integrals on the Real Line

A most important special case of the Cauchy integral in the present context is that in which the curve L is the real axis $\mathbb{R}$, so that

$$F(z) = \frac{1}{2\pi i} \int_{-\infty}^{\infty} \frac{f(u)}{u-z} du. \tag{B.2.12}$$

If

$$I = \frac{1}{2\pi i} \int_{-\infty}^{\infty} f(u) du$$

exists, at least in the sense of a principal value infinite integral where $f(u) \sim u^{-1}$ at large values of $|u|$ and

$$I = \frac{1}{2\pi i} \lim_{L\to\infty} \int_{-L}^{L} f(u) du,$$

then

$$F(z) = -\frac{I}{z} + O\left(\frac{1}{z^2}\right) \tag{B.2.13}$$

at large z. For most examples in the present work, we have $f(u) \sim u^{-2}$ for large u, so that I exists as an ordinary integral.

In this case, the complex plane is segmented into the upper half-plane $\Omega^{(+)}$ and the lower half-plane $\Omega^{(-)}$. For $z \in \Omega^{(+)}$, $F(z)$ is analytic in $\Omega^{(+)}$, while for $z \in \Omega^{(-)}$, it is analytic in $\Omega^{(-)}$.

We write out the Plemelj formulas in this case, for reference purposes:

$$F^{+}(x) = \frac{1}{2} f(x) + \frac{1}{2\pi i} P \int_{-\infty}^{\infty} \frac{f(u)}{u - x} du,$$
$$F^{-}(x) = -\frac{1}{2} f(x) + \frac{1}{2\pi i} P \int_{-\infty}^{\infty} \frac{f(u)}{u - x} du, \tag{B.2.14}$$

or

$$F^{+}(x) + F(x^{-}) = \frac{1}{i\pi} P \int_{-\infty}^{\infty} \frac{f(u)}{u - x} du,$$
$$F^{+}(x) - F^{-}(x) = f(x), \tag{B.2.15}$$

where $F^{\pm}(x)$ are the limits of $F(z)$ as $z \to x$ from $\Omega^{(\pm)}$. Note that $F^{\pm}$ correspond to $F_{\mp}$ in the notation used in the main text and referred to after (C.2.2).

Remark B.2.1. Thus, any Hölder continuous function defined on $\mathbb{R}$ can be written as the difference between the limits of two functions, one analytic in $\Omega^{(+)}$ and the other in $\Omega^{(-)}$.

As noted earlier, there is no requirement that f be analytic. However, if f is analytic on an open set containing Ω^{+}, we see by comparing (B.1.5) and (B.2.14), that $f(x) = F^{+}(x)$ and indeed $f(z) = F(z)$, $z \in \Omega^{+}$. Similarly, if f is analytic on Ω^{-}, we have $f(z) = -F(z)$, $z \in \Omega^{-}$. Let us introduce the notation

$$F_l(z) = F(z),\ z \in \Omega^{(-)}, \qquad F_u(z) = F(z),\ z \in \Omega^{(+)},$$

where F is defined by (B.2.12) and F_l, F_u are analytic in $\Omega^{(-)}$ and $\Omega^{(+)}$, respectively. Let f be analytic on an open set containing $\mathbb{R}$ but have singularities in $\Omega^{(\pm)}$, away from the real axis. Also, let $f(z)$ behave like z^{-p}, $p > 0$, at large $|z|$. We take $z \in \Omega^{(-)}$ and close the contour in (B.1.15) on $\Omega^{(-)}$. Then by (B.1.14),

$$F_l(z) = \sum_u \text{residues of } \frac{f(u)}{u - z},$$

where the sum is over isolated singularities and integrals on branch cuts. Thus, $F_l(z)$ is analytic in a band in $\Omega^{(+)}$, $0 \le \text{Im}z < \alpha$, where α is the position of the singularity nearest to the real axis, which can be the position of an isolated singularity or a point on a branch cut. Therefore, F_l can be analytically continued into a band parallel to the real axis in $\Omega^{(+)}$ and indeed into larger regions of this half-plane, avoiding singularities. Note, however, that branch points can cause difficulties, as discussed in Sect. B.1.5.

Similarly, F_u can be analytically continued into regions of $\Omega^{(-)}$. Note, however, that the analytical continuation of F_l into $\Omega^{(+)}$ is not equal to F_u in $\Omega^{(+)}$, and vice versa.

The following observation follows from (B.2.15).

Remark B.2.2. Let f be analytic on an open set containing $\mathbb{R}$ but have singularities in $\Omega^{(\pm)}$, away from the real axis. Then it can be expressed on $\mathbb{R}$ as the difference between two functions, one analytic in an open set containing Ω^{+} and the other in an open set containing Ω^{-}.

C

Fourier Transforms

We summarize in this appendix various properties of Fourier transforms required in the main text. References include the now classical works [302, 308] and the many modern texts on the topic.

C.1 Definitions

For any function $f : \mathbb{R} \to \mathcal{V}$, where $\mathcal{V}$ is a finite-dimensional vector space, its *Fourier transform* $f_F : \mathbb{R} \mapsto \mathcal{V}$ is defined by

$$f_F(\omega) = \int_{-\infty}^{\infty} f(s) e^{-i\omega s} ds. \tag{C.1.1}$$

This formula and each of the properties noted below apply to each component of f and f_F. If $f \in L^1(\mathbb{R})$, then f_F exists on $\mathbb{R}$. The inverse transform is defined by

$$g(s) = \frac{1}{2\pi} \int_{-\infty}^{\infty} f_F(\omega) e^{i\omega s} d\omega. \tag{C.1.2}$$

If $f \in L^1(\mathbb{R})$ and its first derivatives are piecewise continuous on $\mathbb{R}$, then

$$g(s) = \frac{1}{2}[f(s+) + f(s-)], \quad s \in \mathbb{R}.$$

This is one version of Fourier's integral theorem. Another is the following. Let $f \in L^2(\mathbb{R})$. Then $f_F \in L^2(\mathbb{R})$ and $g = f$ almost everywhere on $\mathbb{R}$. The existence of the transform and inverse transform for functions in $L^2(\mathbb{R})$ is at first sight unclear. However [308], convergent forms can be given as follows:

$$f_F(\omega) = -\frac{d}{d\omega} \int_{-\infty}^{\infty} f(s) \frac{e^{-i\omega s} - 1}{is} ds,$$
$$f(s) = \frac{1}{2\pi} \frac{d}{ds} \int_{-\infty}^{\infty} f(\omega) \frac{e^{i\omega s} - 1}{i\omega} d\omega.$$

G. Amendola et al., *Thermodynamics of Materials with Memory*,
https://doi.org/10.1007/978-3-030-80534-0

Let us define

$$f_+(\omega) = \int_0^\infty f(s)e^{-i\omega s}ds, \qquad f_-(\omega) = \int_{-\infty}^0 f(s)e^{-i\omega s}ds,$$
$$f_s(\omega) = \int_0^\infty f(s)\sin\omega s\,ds, \qquad f_c(\omega) = \int_0^\infty f(s)\cos\omega s\,ds. \tag{C.1.3}$$

We have

$$f_F(\omega) = f_+(\omega) + f_-(\omega). \tag{C.1.4}$$

Assuming that f is even, we obtain from (C.1.1) and (C.1.2) the form of f_F and the inverse cosine transform, given by (taking $f = g$)

$$f_F(\omega) = 2f_c(\omega), \qquad f(s) = \frac{2}{\pi}\int_0^\infty f_c(\omega)\cos(\omega s)d\omega, \quad s \in \mathbb{R}^+. \tag{C.1.5}$$

Also, if f is odd, we have the form of f_F and the inverse sine transform,

$$f_F(\omega) = -2if_s(\omega), \qquad f(s) = \frac{2}{\pi}\int_0^\infty f_s(\omega)\sin(\omega s)d\omega, \quad s \in \mathbb{R}^+. \tag{C.1.6}$$

For these, the statements of Fourier's integral theorem also apply but modified by replacing $\mathbb{R}$ with $\mathbb{R}^+$. For functions nonzero only on $\mathbb{R}^+$ or $\mathbb{R}^-$, the properties of f may be stated on $\mathbb{R}^\pm$ as appropriate.

We shall generally assume that $f \in L^1(\mathbb{R}) \cap L^2(\mathbb{R})$ (or $f \in L^1(\mathbb{R}^\pm) \cap L^2(\mathbb{R}^\pm)$) as appropriate), ensuring the existence of the transform and the property that $f_F \in L^2(\mathbb{R})$ (or $f_F \in L^2(\mathbb{R}^\pm)$), though in certain cases discussed below, we need to consider certain functions not in this category. Membership of these function spaces imposes restrictions on the behavior at infinity of f. For example, if a piecewise continuous function $f : \mathbb{R}^+ \mapsto \mathbb{R}$ belongs to $L^1(\mathbb{R}^+) \cap L^2(\mathbb{R}^+)$, then at large positive s, we must have

$$|f(s)| \leq As^{-p}, \; p > 1,$$

where A is a constant and, since $f_+ \in L^2(\mathbb{R}^+)$,

$$|f_+(\omega)| \leq B|\omega|^{-q}, \; q > \frac{1}{2},$$

where B is a constant, for large real ω.

It is easily shown that

$$f'_F(\omega) = i\omega f_F(\omega),$$

where f' is the derivative of f.

If f is real, which is almost always the case, then

$$\overline{f}_F(\omega) = f_F(-\omega), \tag{C.1.7}$$

where the bar denotes the complex conjugate. If $h(s) = f(s+u)$, where u is a constant, then we have

$$h_F(\omega) = e^{i\omega u} f_F(\omega). \tag{C.1.8}$$

C.2 Fourier Transforms on the Complex Plane

It is central to the considerations of the present work, most particularly Part III, that we consider various quantities defined on the frequency domain over the complex plane Ω. This is effectively an analytic continuation from their definition on $\mathbb{R}$. Consider the quantity f_+ extended to the complex plane. Its integral definition allows us to carry out this extension to $\Omega^{(-)}$ without difficulty in that the quantity

$$f_+(\omega) = \int_0^\infty f(s)e^{-i\omega s}ds, \quad \omega = \omega_r - i\omega_i, \quad \omega_r \in \mathbb{R}, \quad \omega_i \in \mathbb{R}^{++}, \tag{C.2.1}$$

exists everywhere on $\Omega^{(-)}$ if it exists on $\mathbb{R}$. It is also uniquely differentiable everywhere on the open set $\Omega^{(-)}$ with respect to $\omega = \omega_r - i\omega_i$ and therefore analytic on this set. It goes to zero as $\mathrm{Im}\omega \to -\infty$.

Also, if f_+ is free of singularities in $\Omega^{(-)}$, then

$$f(s) = \frac{1}{2\pi}\int_{-\infty}^{\infty} f_+(\omega)e^{i\omega s}d\omega = 0, \quad s \in \mathbb{R}^{--},$$

by (B.1.2). Thus, we have the following result.

Proposition C.2.1. *The function f is zero on $\mathbb{R}^{--}$ if and only if f_+ is analytic on $\Omega^{(-)}$.*

A similar result holds for f zero on $\mathbb{R}^{++}$, where the analyticity of f_- is on $\Omega^{(+)}$.

As noted in Sect. B.1.4, we exclude the possibility of essential singularities in the extension of f_F to the complex plane, at finite points, and except in the context of Chap. 15, assume analytic behavior at infinity, given in fact by (C.2.16) below.

Proposition C.2.2. *The function $f_+(\omega)$ is analytic on a band in Ω^+, $0 \leq \mathrm{Im}\omega < \alpha$ (but not in a band $\mathrm{Im}\omega < \beta, \beta > \alpha$), if and only if $f(s)$ decays like $\exp(-\alpha s)$ for large s.*

Proof. If $f(s)$ decays like $\exp(-\alpha s)$ for large s, then putting $\omega = \omega_r + i\omega_i$, where ω_r, ω_i are real, we have that

$$f_+(\omega_r + i\omega_i) = \int_0^\infty f(s)e^{-i(\omega_r+i\omega_i)s}ds$$

exists and is analytic for $\omega_i < \alpha$.

Let $f_F(\omega)$ be analytic for $\mathrm{Im}\omega < \alpha$ but not in a band $\mathrm{Im}\omega < \beta, \beta > \alpha$. We can evaluate

$$f(s) = \frac{1}{2\pi}\int_{-\infty}^{\infty} f_+(\omega)e^{i\omega s}d\omega,$$

where f_+ is analytic at infinity, as a contour integral over Ω^+ with no contribution from the infinite semicircle. Then, using (B.1.14), we have

$$f(s) = 2\pi i \sum_{\omega_i} \text{residues of } f_F(\omega)e^{i\omega s} \text{ at singular points } \omega_i \text{ in } \Omega^{(+)}.$$

The position ω_n of the singularity (or more than one) nearest the real axis, whether this is an isolated singularity or a point on a branch cut, must be such that $\mathrm{Im}\omega_n = \alpha$. All other singular points will yield more strongly decaying exponentials. □

Thus, the integral definition of f_+ will typically not exist on $\Omega^{(+)}$ except perhaps on a band of finite thickness parallel to the real axis. Outside of $\Omega^{(-)}$ and this region, we must define $f_+(\omega)$ by analytic continuation from the region of analyticity, though not using the integral definition. If we can obtain an explicit formula for the transform, then the analytic continuation is very easy, as can be seen from (B.1.6). It will certainly have singularities in the region $\mathrm{Im}\omega \geq \alpha$ unless it is a constant.

Similarly, f_- is analytic in $\Omega^{(+)}$ and perhaps in certain regions of Ω^-.

Hypothesis C.2.3. *For crucial manipulations in Part III, we will always assume that the regions of analyticity of $f_\pm$ (this quantity being an independent field variable or a relaxation function derivative) is extended to include an open region containing $\mathbb{R}$.*

This assumption is, for brevity, sometimes stated in the main text as that $f_\pm$ is analytic on $\Omega^{(\mp)}$ and $\mathbb{R}$, or on $\Omega^\mp$. It is a restrictive assumption in that it means, by virtue of Proposition C.2.2, that $f(s)$ decays exponentially at large s. For relaxation functions determined by branch-cut singularities, this is particularly important in that if the cuts are allowed to touch the real axis, interesting nonexponential behaviors are possible. This raises the issue whether one can take the limit of the cut approaching the real axis after final results have been obtained, which is discussed in Chap. 14. Isolated singularities off the real axis are always associated with exponential decay, though by taking poles sufficiently close to the real axis, slow decay can be simulated.

For $f : \mathbb{R}^+ \to \mathcal{V}$ we can always extend the domain of f to $\mathbb{R}$, by considering its *causal* extension

$$f(s) = \begin{cases} f(s) & \text{for } s \geq 0, \\ 0 & \text{for } s < 0, \end{cases}$$

in which case

$$f_F(\omega) = f_+(\omega) = f_c(\omega) - i f_s(\omega). \tag{C.2.2}$$

The quantities $f_\pm$ provide an example of the notation used in Part III whereby the subscript $\pm$ indicates that the function is analytic in $\Omega^{(\mp)}$.

Let f be zero on $\mathbb{R}^{--}$ and let it diverge like $\exp(\lambda_1 s)$ at large s. Then we consider the function

$$g(s) = f(s)e^{-\lambda s}, \ \lambda > \lambda_1, \ s \in \mathbb{R}^+.$$

Its transform is given by

$$g_+(\omega) = \int_0^\infty g(s)e^{-i\omega s}ds = \int_0^\infty f(s)e^{-i(\omega s - i\lambda)}ds = f_+(\omega - i\lambda),$$

so that f_+ exists and is analytic below the line $z = -i\lambda$. Taking the inverse transform of g_+, we obtain

$$g(s) = \frac{1}{2\pi}\int_{-\infty}^{\infty} g_+(\omega)e^{i\omega s}d\omega = \frac{1}{2\pi}\int_{-\infty}^{\infty} f_+(\omega - i\lambda)e^{i\omega s}d\omega,$$

so that

$$f(s) = \frac{1}{2\pi}\int_{-\infty}^{\infty} f_+(\omega - i\lambda)e^{i(\omega - i\lambda)s}d\omega = \frac{1}{2\pi}\int_{-\infty - i\lambda}^{\infty - i\lambda} f_+(\xi)e^{i\xi s}d\xi. \qquad \text{(C.2.3)}$$

C.2.1 Laplace Transforms

Let f be zero on $\mathbb{R}^{--}$. Then

$$f_L(\alpha) = \int_0^\infty f(s)e^{-\alpha s}ds = f_+(-i\alpha), \quad \alpha \in \Omega, \qquad \text{(C.2.4)}$$

is the Laplace transform of f. It is analytic for $\mathrm{Re}\alpha > 0$. The imaginary axis is included if Hypothesis C.2.3 is introduced. If f diverges like $\exp(\alpha_0 s)$ for large s, then f_L exists and is analytic for $\mathrm{Re}\alpha > \alpha_0$. Allowing for this possibility, we use (C.2.3) to determine the (unique) inverse Laplace transform. Making a change of variable $\alpha = i\xi$, we obtain

$$f(s) = \frac{1}{2\pi i}\int_{-i\infty + \lambda}^{i\infty + \lambda} f_L(\alpha)e^{\alpha s}d\alpha, \quad \lambda > \alpha_0. \qquad \text{(C.2.5)}$$

This gives zero if $\mathrm{Re}s < 0$, allowing the contour to be closed in the right-hand half-plane.

C.2.2 The Fourier Transform of Functions with Compact Support

Proposition C.2.4. *Consider the case that $f(s) = 0, \ s \notin [0, d]$, and f is continuous for $s \in [0, d]$, $d > 0$, with $f(0)$ and $f(d)$ nonzero. Then*

$$f_+(\omega) = \int_0^d f(s)e^{-i\omega s}ds. \qquad \text{(C.2.6)}$$

The quantity f_F *is an entire function with an essential singularity at infinity and dominant behavior given by*

$$f_+(\omega) \xrightarrow[\mathrm{Im}\omega \to +\infty]{} -\frac{f(d)}{i\omega} e^{-i\omega d}. \tag{C.2.7}$$

Also, for $|\mathrm{Re}\omega|$ *large and* $\mathrm{Im}\omega$ *fixed,*

$$f_+(\omega) \sim \frac{f(0)}{i\omega}\left(1 - e^{-i\omega d}\right). \tag{C.2.8}$$

For $\mathrm{Im}\omega \to -\infty$,

$$f_+(\omega) \sim \frac{f(0)}{i\omega}. \tag{C.2.9}$$

Conversely, if f_+ *is an entire function that has an essential singularity of the form* (C.2.7), *then* $f(s) = 0,\ s > d$.

Proof. The analyticity of f_+ at all finite points in the complex plane follows from the fact that the integral exists and is uniquely differentiable everywhere. The analyticity of f_+ on $\Omega^{(-)}$ follows from the observation after (C.2.1). By taking $|\omega| \to \infty$ along the real axis and changing integration variables, we obtain (C.2.8). This formula can be analytically continued onto $\Omega^{(-)}$, so that (C.2.9) follows. Relation (C.2.8) cannot be analytically continued into $\Omega^{(+)}$ because of the presence of a divergence at infinity. By changing the integration variable in (C.2.6), we obtain

$$f_+(\omega) = e^{-i\omega d} g(\omega), \quad g(\omega) = \int_{-d}^{0} f(s+d) e^{-i\omega s} ds.$$

It follows from its definition that $g(\omega)$ is analytic on $\Omega^{(+)}$. As before, we find that

$$g(\omega) \sim -\frac{f(d)}{i\omega}\left(1 - e^{i\omega d}\right)$$

for $|\mathrm{Re}\omega|$ large and $\mathrm{Im}\omega$ fixed, giving

$$g(\omega) \xrightarrow[\mathrm{Im}\omega \to +\infty]{} -\frac{f(d)}{i\omega},$$

and (C.2.7) follows.

Remark C.2.5. If $f(d)$ vanishes and $f'(d) \neq 0$, a slightly different version of (C.2.7) emerges.

Conversely, we assume that f_+ is analytic at all finite points of Ω and diverges as indicated by (C.2.7) on $\Omega^{(+)}$. Then g is an entire function that goes to zero as $\mathrm{Im}\omega \to +\infty$. We write

$$f(s) = \frac{1}{2\pi} \int_{-\infty}^{\infty} g(\omega) e^{i\omega(s-d)} d\omega.$$

If $s > d$, the contour can be closed in $\Omega^{(+)}$ with the contribution from the infinite portion exponentially attenuated. The analyticity of $g(\omega)$ in $\Omega^{(+)}$ ensures that the result is zero. □

Proposition C.2.4 is closely related to the Paley–Wiener theorem [291].

C.2.3 Functions that Do Not Belong to $L^1 \cap L^2$

It is necessary to include cases of functions that do not belong to $L^1 \cap L^2$. Consider the case that $f(\infty) \neq 0$ but with $f_0 \in L^1(\mathbb{R}^+) \cap L^2(\mathbb{R}^+)$ defined by

$$f_0(s) = f(s) - f(\infty).$$

In this case, we write

$$\begin{aligned} f_+(\omega) &= \int_0^\infty e^{-i\omega s} f(s)ds \\ &= \int_0^\infty e^{-i\omega s} f_{0+}(s)ds + f(\infty)\int_0^\infty e^{-i\omega s}ds \\ &= f_{0+}(\omega) + \frac{f(\infty)}{i\omega^-}; \quad \omega^- = \lim_{\alpha\to 0^+}(\omega - i\alpha), \end{aligned} \tag{C.2.10}$$

where we have moved ω in the rightmost integral of the second relation into $\Omega^{(-)}$ to give a finite result. The limit in the definition of ω^- is taken after any integrations in frequency space are carried out. This is a well-known device for handling such functions—effectively as a limit of $L^1 \cap L^2$ functions—which avoids the use of distribution theory. It is, in the present context, largely redundant, since $f_+(\omega)$ will generally multiply functions that vanish at $\omega = 0$ in such a way as to cancel the pole.

Similarly, if $f(-\infty) \neq 0$ but if $f_0 \in L^1(\mathbb{R}^-) \cap L^2(\mathbb{R}^-)$, where

$$f_0(s) = f(s) - f(-\infty),$$

then

$$f_-(\omega) = \int_{-\infty}^0 e^{-i\omega s} f(s)ds = f_0(\omega) - \frac{f(-\infty)}{i\omega^+}, \ \omega^+ = \lim_{\alpha\to 0^+}(\omega + i\alpha). \tag{C.2.11}$$

Finally, we note the formal relation, referred to on occasion in the main text,

$$\int_{-\infty}^{\infty} e^{\pm i\omega s}ds = 2\pi\delta(\omega), \tag{C.2.12}$$

where δ is the singular delta function.

C.2.4 The Form of $f_\pm$ at Large Frequencies

The Riemann–Lebesgue lemma states that if $f \in L^1(\mathbb{R})$, then

$$\lim_{\omega\to\infty} \int_{-\infty}^{\infty} f(s)e^{\pm i\omega s}ds = 0. \tag{C.2.13}$$

Similar statements apply to f defined on $\mathbb{R}^\pm$. It is of interest to determine in more detail, however, the behavior of Fourier transforms at large ω. The results apply to inverse transforms with minor changes of sign. Consider the relations

$$\int_0^{\infty} e^{-i\omega s}ds = \frac{1}{i\omega^-}, \quad \int_{-\infty}^{0} e^{-i\omega s}ds = -\frac{1}{i\omega^+}, \tag{C.2.14}$$

obtained by the device introduced in (C.2.10) and (C.2.11). Differentiating n times yields

$$\int_0^{\infty} s^n e^{-i\omega s}ds = \frac{n!}{(i\omega^-)^{n+1}}, \quad \int_{-\infty}^{0} s^n e^{-i\omega s}ds = -\frac{n!}{(i\omega^+)^{n+1}}. \tag{C.2.15}$$

If $f_\pm(\omega)$ is analytic at infinity and if the first N right and left derivatives of f exist at the origin, then we obtain, by Taylor expansion and (C.2.15), the asymptotic behavior

$$f_\pm(\omega)\underset{\omega\to\infty}{\to} \pm\sum_{n=0}^{N}\frac{f^{(n)}(0\pm)}{(i\omega)^{n+1}} + O\left(\frac{1}{\omega^{N+2}}\right), \tag{C.2.16}$$

where $f^{(n)}(0+)(f^{(n)}(0-))$ is the nth right (left) derivative of f at the origin. Thus

$$f_c(\omega)\underset{\omega\to\infty}{\to} \sum_{n\,\text{odd}}^{N}\frac{f^{(n)}(0\pm)}{(i\omega)^{n+1}} + O\left(\frac{1}{\omega^{N+2}}\right),$$

$$f_s(\omega)\underset{\omega\to\infty}{\to} i\sum_{n\,\text{even}}^{N}\frac{f^{(n)}(0\pm)}{(i\omega)^{n+1}} + O\left(\frac{1}{\omega^{N+2}}\right). \tag{C.2.17}$$

If

$$f^{(n)}(0+) = f^{(n)}(0-), \; n = 0, 1, 2, \ldots, m,$$

in other words, if f is differentiable n times at the origin, then it follows from (C.2.16) that

$$f_F(\omega) \sim \omega^{-(m+2)} \tag{C.2.18}$$

at large ω.

Note that combining (C.2.12) and (C.2.14) yields

$$\frac{1}{\omega^-} - \frac{1}{\omega^+} = 2\pi i\delta(\omega). \tag{C.2.19}$$

C.2.5 Expressions for $f_\pm$ in Terms of f_F

Using the inverse transform to express f in terms of f_F, together with (C.1.3), we obtain

$$\begin{aligned} f_+(\omega) &= -\frac{1}{2\pi i}\int_{-\infty}^{\infty}\frac{f_F(\omega')}{\omega'-\omega^-}d\omega', \\ f_-(\omega) &= \frac{1}{2\pi i}\int_{-\infty}^{\infty}\frac{f_F(\omega')}{\omega'-\omega^+}d\omega', \\ \omega^\pm &= \lim_{\alpha\to 0^+}(\omega \pm i\alpha). \end{aligned} \tag{C.2.20}$$

Thus, we move ω slightly into the half-plane of analyticity of $f_\pm$ to achieve convergence in the time integration, as in (C.2.10) and (C.2.11). This also ensures that the integrals on the right-hand side of (C.2.20) have a well-defined meaning. The limit is taken after the integration is carried out. The forms of the analytic functions $f_\pm$, $\omega \in \Omega^{(\mp)}$, are given by

$$f_\pm(\omega) = \mp\frac{1}{2\pi i}\int_{-\infty}^{\infty}\frac{f_F(\omega')}{\omega'-\omega}d\omega'. \tag{C.2.21}$$

Note that (C.1.4) follows from (C.2.20) and the Plemelj formula $(B.2.15)_2$. Using (C.1.4) in (C.2.21) and (C.2.16), we see that in the formula for f_+, the contribution from f_- in the integral vanishes by Cauchy's theorem (closing the contour on $\Omega^{(+)}$). Thus, we obtain

$$f_+(\omega) = -\frac{1}{2\pi i}\int_{-\infty}^{\infty}\frac{f_+(\omega')}{\omega'-\omega}d\omega', \quad \omega \in \Omega^{(-)}, \tag{C.2.22}$$

with a similar relation for f_- (no minus sign on the integral). These relations and their limit as ω approaches the real axis provide an example of the properties noted after Remark B.2.1, where $F^\pm$ corresponds to $f_\mp$.

C.3 Parseval's Formula and the Convolution Theorem

Parseval's formula states that

$$\int_{-\infty}^{\infty}\overline{f}(u)g(u)du = \frac{1}{2\pi}\int_{-\infty}^{\infty}\overline{f}_F(\omega)g_F(\omega)d\omega, \tag{C.3.1}$$

so that the L^2 scalar products in the time and frequency domains are proportional. We have allowed the possibility of complex functions in the time domain, which occurs only rarely in the main text. The convolution, or Faltung, theorem gives that if h is the convolution product $f * g$, namely

$$h(s) = \int_{-\infty}^{\infty} f(s-u)g(u)du = \int_{-\infty}^{\infty} f(u)g(s-u)du, \tag{C.3.2}$$

then

$$h_F(\omega) = f_F(\omega)g_F(\omega). \tag{C.3.3}$$

The converse also holds. Note that if f and g are causal, then h also has this property and is given by

$$h(s) = \int_0^s f(s-u)g(u)du, \quad s \in \mathbb{R}^+. \tag{C.3.4}$$

Remark C.3.1. From (C.3.2), we see that the convolution product is commutative if f and g commute. It can also be shown to be associative.

References

1. O. P. Agrawal, Application of Fractional Derivatives in Thermal Analysis of Disk Brakes, *Nonlinear Dynamics* **38** (2004), 191–206.
2. L. Ambrosio, N. Fusco and D. Pallara, *Functions of Bounded Variation and Free Discontinuity Problems*, Oxford University Press, Oxford, 2000.
3. G. Amendola, Existence, uniqueness and asymptotic stability for a thermoelastic plate, *Nonlinear Oscillations* **2** (6) (2003), 147–165.
4. G. Amendola, Equivalent histories, states and a variational problem for a heat conductor with memory, *Int. J. Engng. Sci.* **43** (2005), 681–696.
5. G. Amendola, The minimum free energy for incompressible viscoelastic fluids, *Math. Meth. Appl. Sci.* **29** (2006), 2201–2223.
6. G. Amendola and S. Carillo, Thermal work and minimum free energy in a heat conductor with memory, *Q. J. Mech. Appl. Math.* **57** (2004), 429–446.
7. G. Amendola and M. Fabrizio, Free energies and Fichera's quasi-static problem for materials with memory, *Rend. Lincei Mat. Appl.* **17** (2006), 15–24.
8. G. Amendola and M. Fabrizio, Maximum recoverable work for incompressible viscoelastic fluids and application to a discrete spectrum model, *Diff. Int. Eq.* **20** (4) (2007), 445–466.
9. G. Amendola, M. Fabrizio and J.M. Golden, Thermodynamics of a non-simple rigid heat conductor with memory, *Quart. Appl. Math.* **69** (2011), 787–806.
10. G. Amendola, M. Fabrizio and J.M. Golden, Minimum free energy for an electromagnetic conductor with memory, *Math. Meth. Appl. Sci.* **32** (2009), 77–111.
11. G. Amendola, M. Fabrizio, J.M. Golden and B. Lazzari, Free energies and asymptotic behavior for incompressible viscoelastic fluids, *Applied Analysis* **88** (2009), 789–805.
12. G. Amendola, M. Fabrizio and J.M. Golden, Minimum free energy in the frequency domain for a rigid heat conductor with memory, *Discrete and Continuous Dynamical Systems* **B14** (2010), 793–816.
13. G. Amendola, M. Fabrizio and J.M. Golden, Free energies for a rigid heat conductor with memory, *IMA Journal of Applied Mathematics* **75** (2010), 833–856.

G. Amendola et al., *Thermodynamics of Materials with Memory*,
https://doi.org/10.1007/978-3-030-80534-0

14. G. Amendola, M. Fabrizio and J.M. Golden, Second gradient viscoelastic fluids: dissipation principle and free energies, *Meccanica* **47** (2012), 1859–1868.
15. G. Amendola, M. Fabrizio and J.M. Golden, Free energies in a general nonlocal theory of a material with memory, *Mathematical Models and Methods in Applied Sciences* **24** (2014), 1037–1090.
16. G. Amendola, M. Fabrizio and J.M. Golden, Algebraic and numerical exploration of free energies for materials with memory, *Electronic Journal of Differential Equations* **72** (2015), 1–42.
17. G. Amendola, M. Fabrizio and J.M. Golden, Free energies and minimal states for scalar linear viscoelasticity, *J. Elasticity* **123** (1) (2016), 97–123.
18. G. Amendola, M. Fabrizio and J.M. Golden, General dissipative materials for simple histories, *Quart. Appl. Math.* **76** (2018), 627–672.
19. G. Amendola, M. Fabrizio and J.M. Golden, A mathematical model of viscoferromagnetic materials, *Progress in Fractional Differentiation and Applications* **6** (4) (2020), 1–8.
20. G. Amendola, M. Fabrizio, J.M. Golden and B. Lazzari, Second order approximation for heat conduction: dissipation principle and free energies, *Proc. Roy. Soc. London A*, https://doi.org/10.1098/rspa.2015.0707.
21. G. Amendola and A. Manes, Minimum free energy for a rigid heat conductor and application to a discrete spectrum model, *Boll. Un. Mat. Ital.* **10 B** (8) (2007), 969–987.
22. G. Amendola, A. Manes and C. Vettori, Maximum recoverable work for a rigid heat conductor with memory, *Acta Applicandae Mathematicae* **110** (3) (2010), 1011–1036.
23. L. Anand and S. Govindjeem, *Continuum Mechanics of Solids*, Oxford University Press, Oxford, 2020.
24. J.A.D. Appleby, M. Fabrizio, B. Lazzari and D.W. Reynolds, On exponential asymptotic stability in linear viscoelasticity, *Math. Mod. Meth. Appl. Sci.* **16** (10) (2006), 1677–1694.
25. L. Baggett and W. Fulks, *Fourier Analysis*, Api Anjou Press, Boulder, 1979.
26. R.L. Bagley and P.J. Torvik, A Theoretical Basis for the Application of Fractional Calculus to Viscoelasticity, *Journal of Rheology* **27** (1983), 201–2011.
27. D. Bainov and P. Simeonov, *Integral Inequalities and Applications*, Kluwer Acad. Publ., Dordrecht, 1992.
28. D. Baleanu, B. Guvenc Ziya and J.A. Machado (Eds.), *New Trends in Nanotechnology and Fractional Calculus Applications*, XI, 531, ISBN Springer,1st Edition., 978-90-481-3292-8, 2010.
29. D. Baleanu and A.K. Golmankhaneh, The dual action and fractional multi time Hamilton equations, *Int. J. Phys.* **48** (2009), 2558–2569.
30. D. Baleanu, A.K. Golmankhaneh and R.R. Nigmatullin, Newtonian law with memory, *Non Linear Dynamics* **60** (1-2) (2010), 81–86.
31. C. Banfi, Su una nuova impostazione per l'analisi dei sistemi ereditari, *Ann. Univ. Ferrara* **23** (7) (1977), 29–38.
32. C. Banfi and M. Fabrizio, Sul concetto di sottocorpo nella meccanica dei continui, *Accad. Naz. Lincei Rend. Cl. Sci. Fis. Mat. Natur.* **66** (8) (1979), 136–142.

33. V. Barbu and M. Iannelli, Controllability of the heat equation with memory, *Diff. Int. Equat.* **13** (10–12) (2000), 1393–1412.
34. C. Bardos, G. Lebeau, and J. Rauch, Sharp sufficient conditions for the observation, control, and stabilization of waves from the boundary, *SIAM J. Control Optim.* **30** (5) (1992), 1024–1065.
35. V. Berti and G. Gentili, The minimum free energy for isothermal dielectrics with memory, *J. Non-Equil. Thermodyn.* 24 (1999), 154–176.
36. D. Bland, *Theory of Linear Viscoelasticity*, Pergamon, London, 1960.
37. S. Bochner, *Lectures on Fourier Integrals*, Annals of Math. Studies, Princeton Univ. Press, Princeton (NJ), 1959.
38. L. Boltzmann, Zur Theorie der elastischen Nachwirkung, *Wien. Ber.* **70** (1874), 275–306.
39. L. Boltzmann, Zur Theorie der elastischen Nachwirkung, *Wied. Ann.* **5** (1878), 430–432.
40. R. Bowen and P.J. Chen, Thermodynamic restrictions on the initial slope of the stress relaxation function, *Arch. Rational Mech. Anal.* **51** (1973), 278–284.
41. A. Bozza and G. Gentili, Inversion and quasi-static problems in linear viscoelasticity, *Meccanica* **30** (1995), 321–335.
42. S. Breuer and E.T. Onat, On recoverable work in linear viscoelasticity, *Z. Angew. Math. Phys.* **15** (1964), 13–21.
43. S. Breuer and E.T. Onat, On the determination of free energy in linear viscoelasticity, *Z. Angew. Math. Phys.* **15** (1964), 184–191.
44. J.W. Cahn and J.E. Hilliard, Free energy of a nonuniform system. I. Interfacial energy, *J. Chem. Phys.* **28** (1958), 258–267.
45. M. Caputo, Mean fractional-order-derivatives differential equations and filters, *Annali Università di Ferrara Sez. VII, Scienze Matematiche* **41** (1995), 73–83.
46. M. Caputo, *Elasticità e dissipazione*, Zanichelli, Bologna, 1965.
47. M. Caputo, Linear model of dissipation whose Q is almost frequency independent-II, *Geophys. J. R. Astr. Soc.* **13** (1967), 529–539.
48. M.G. Caputo, Vibrations of a thin viscoelastic layer with a dissipative memory, *J. Acoust. Soc. Am.* **56** (1974), 897–904.
49. M. Caputo, A model for the fatigue in elastic materials with frequency independent, *Q. J. Acoust. Soc. Am.* **66** (1979), 176–179.
50. M. Caputo, Mean fractional-order-derivatives differential equations and filters, *Annali dell'Universitá di Ferrara* **41** (1995), 73–84.
51. M. Caputo, The Green function of the diffusion of fluids in porous media with memory, *Atti Acc. Naz. Lincei Rend. Lincei Mat. Appl.* **7** (1996), 243–250.
52. M. Caputo and C. Cametti, Memory diffusion in two cases of biological interest, *Theoretical Biology* (2008). https://doi.org/10.1016/j.jbi.2008.06.021.
53. R. Caponetto, *Fractional Order systems (Modelling and Control Applications)*, World Scientific, 2010.
54. M. Caputo, *Modeling social and economic cycles, in Alternative Public Economics*, F. Forte, P. Navarra and R. Mudambi Ed.s, Elgar, Cheltenham, 2014.
55. M. Caputo and M. Fabrizio, Damage and fatigue described by a fractional derivative model, *J. Computational Physics* **293** (2015), 400–408.

56. M. Caputo and M. Fabrizio, A new definition of fractional derivative without singular kernel, *Progr. Fract. Differ. Appl.* **1** (2) 2015, 73–85.
57. B. Carbonaro, The finite speed of propagation of disturbance in an unbounded linearly viscoelastic material, *Boll. U.M.I.* **7 B** (1993), 875–891.
58. D.E. Carlson, *Linear thermoelasticity*, in Handbuch der Physik VIa/2, Springer, New York, 297–345, 1972.
59. C. Cattaneo, Sulla conduzione del calore, *Atti Sem. Mat. Fis. Univ. Modena* **3** (1948), 83–101.
60. A.L. Cauchy, Sur les équations qui expriment les conditions d'équilibre ou les lois du mouvement intérieur d'un corps solide, élastique ou non élastique, *Ex. de Math.* **3** (1928), 158–179, Ouvres **8** (2) (1828), 253–277.
61. C. Cavaterra, A. Lorenzi and M. Yamamoto, A stability result via Carleman estimates for an inverse source problem related to a hyperbolic integro-differential equation, *Comput. Appl. Math.* **25** (2-3) (2006), 229–250.
62. F. Cesarone, M. Caputo and C. Cametti, Memory formalism in the passive diffusion across a biological membrane, *J. Membrane Science* **250** (2004), 79–84.
63. S. Chirita, Saint-Venant's problem and semi-inverse solutions in linear viscoelasticity, *Acta Mechanica* **94** (1992), 221–232.
64. S. Chirita, M. Ciarletta and M. Fabrizio, Saint-Venant's principle in linear viscoelasticity, *Int. Engn. Sci.* **35** (1997), 1221–1226.
65. R.M. Christensen, *Theory of Viscoelasticity: An Introduction*, 2nd edn, Academic Press, New York, 1982.
66. V.A. Cimmelli and K. Frischmuth, Gradient generalization to the extended thermodynamic approach and diffusive-hyperbolic heat conduction, *Physica B: Condensed Matter* **400** (1-2) (2007), 257–265.
67. B.D. Coleman, Thermodynamics of materials with fading memory, *Arch. Rational Mech. Anal.* **13** (1964), 1–46.
68. B.D. Coleman, On thermodynamics, strain impulses and viscoelasticity, *Arch. Rational Mech. Anal.* **17** (1964), 230–254.
69. B.D. Coleman and E.H. Dill, Thermodynamic restrictions on the constitutive equations of electromagnetic theory, *ZAMP* **22** (1971), 691–702.
70. B.D. Coleman and E.H. Dill, On the thermodynamics of electromagnetic fields in materials with memory, *Arch. Rational Mech. Anal.* **41** (1971), 132–162.
71. B.D. Coleman and V.J. Mizel, On the general theory of fading memory, *Arch. Rational Mech. Anal.* **29** (1968), 18–31.
72. B.D. Coleman and V.J. Mizel, Norms and semi-groups in the theory of fading memory, *Arch. Rational Mech. Anal.* **23** (1966), 87–123.
73. B.D. Coleman and W. Noll, Foundations of linear viscoelasticity, *Rev. Mod. Phys.* **33** (1961), 239–249.
74. B.D. Coleman and W. Noll, An approximation theorem for functionals, with applications in continuum mechanics, *Arch. Rational Mech. Anal.* **6** (1960), 355–370.
75. B.D. Coleman and D.R. Owen, A Mathematical foundation for thermodynamics, *Arch. Rational Mech. Anal.* **54** (1974), 1–104.

76. B.D. Coleman and D.R. Owen, On thermodynamics and elastic-plastic materials, *Arch. Rational Mech. Anal.* **59** (1975), 25–51.
77. C. Corduneanu, *Integral Equations and Applications*, Cambridge University Press, Cambridge, 1991.
78. M.J. Crochet and P.M. Naghdi, A class of simple solids with fading memory, *Int. J. Engng. Sci.* **7** (1969), 1173–1198.
79. C.M. Dafermos, Asymptotic stability in viscoelasticity, *Arch. Rational Mech. Anal.* **37** (1970), 297–308.
80. C.M. Dafermos, On abstract Volterra equations with applications to linear viscoelasticity, *J. Differential Equations* **7** (1970), 554–569.
81. C.M. Dafermos and J.A. Nohel, Energy methods for nonlinear hyperbolic Volterra integro-differential equations, *Comm. Partial Differential Equations* **4** (3)(1979), 219–278.
82. P.L. Davis, On the hyperbolicity of the equations of the linear theory of heat conduction for materials with memory, *SIAM J. Appl. Math.* **30** (1976), 75–80.
83. P.L. Davis, Hyperbolic integro-differential equations, *Proc. Amer. Math. Soc.* **47** (1975), 155–160.
84. W.A. Day, Thermodynamics based on a work axiom, *Arch. Rational Mech. Anal.* **31** (1968), 1–34.
85. W.A. Day, Reversibility, recoverable work and free energy in linear viscoelasticity, *Quart. J. Mech. Appl. Math.* **23** (1970), 1–15.
86. W.A. Day, Restrictions on relaxation functions in linear viscoelasticity, *Quart. J. Mech. Appl. Math.* **24** (1971), 487–497.
87. W.A. Day, The thermodynamics of materials with memory, in *Materials with Memory*, D. Graffi ed., Liguori, Napoli, 1979.
88. G. Del Piero and L. Deseri, Monotonic, completely monotonic and exponential relaxation functions in linear viscoelasticity, *Quart. Appl. Math.* **53** (1995), 273–300.
89. G. Del Piero and L. Deseri, On the analytic expression of the free energy in linear viscoelasticity, *J. Elasticity* **43** (1996), 247–278.
90. G. Del Piero and L. Deseri, On the concepts of state and free energy in linear viscoelasticity, *Arch. Rational Mech. Anal.* **138** (1997), 1–35.
91. L. Deseri, M. Fabrizio and J.M. Golden, On the concept of a minimal state in viscoelasticity: new free energies and applications to PDEs, *Arch. Rational Mech. Anal.* **181** (1) (2006), 43–96.
92. L. Deseri, G. Gentili and J.M. Golden, An explicit formula for the minimum free energy in linear viscoelasticity, *J. Elasticity* **54** (1999), 141–185.
93. L. Deseri, G. Gentili and J.M. Golden, On the minimum free energy and the Saint-Venant principle in linear viscoelasticity, in *Mathematical Models and Methods for Smart Materials*, M. Fabrizio, B. Lazzari, and A. Morro eds., Word Scientific, vol. 62, Singapore, 2002.
94. L. Deseri and J.M. Golden, The minimum free energy for continuous spectrum materials, *SIAM J. Appl Math.* **67** (2007), 869–892.

95. K. Diethelm, *The analysis of fractional differential equations, an application oriented, exposition using differential operators of Caputo type*, Lecture notes in mathematics nr. 2004, Springer, Heidelberg, 2010.
96. E.D. Dill, *Simple materials with fading memory*, in Continuum Physics II, Academic Press, Berlin, 1972.
97. J.E. Dunn and J. Serrin, On the thermomechanics of interstitial working, *Arch. Rational Mech. Anal.* **88** (1985), 95–133.
98. G. Duvant and J. Lions, *Inequalities in Mechanics and Physics*, Springer-Verlag, Berlin, 1976.
99. M. El Shaed, *A Fractional Calculus Model of Semilunar Heart Valve Vibrations*, International Mathematica Symposium, London, 2003.
100. M. Fabrizio, G. Gentili and J.M. Golden, The minimum free energy for a class of compressible viscoelastic fluids, *Diff. Int. Eq.* **7** (2002), 319–342.
101. M. Fabrizio, G. Gentili and J.M. Golden, Free energies for linear nonisothermal materials with memory, *Mathematical and Computer Modelling* **39** (2004), 219–253.
102. M. Fabrizio, G. Gentili and D.W. Reynolds, On rigid heat conductors with memory, *Int. J. Engng. Sci.* **36** (1998), 765–782.
103. M. Fabrizio and C. Giorgi, Sulla termodinamica dei materiali semplici, *Boll. Un. Mat. Ital.* **5B** (6) (1986), 441–464.
104. M. Fabrizio, C. Giorgi and A. Morro, Free energies and dissipation properties for systems with memory, *Arch. Rational Mech. Anal.* **125** (1994), 341–373.
105. M. Fabrizio, C. Giorgi and A. Morro, Internal dissipation, relaxation property and free energy in materials with fading memory, *J. Elasticity* **40** (1995), 107–122.
106. M. Fabrizio, C. Giorgi and V. Pata, A new approach to equations with memory, *Arch. Rational Mech. Anal.* **198** (2010), 189–232.
107. M. Fabrizio and J.M. Golden, Maximum and minimum free energies and the concept of a minimal state, *Rendiconti di Matematica* **20** (2000), 131–165.
108. M. Fabrizio, B. Lazzari and R. Nibbi, Thermodynamics of non-local materials: extra fluxes and internal powers, *Continuum Mech. Thermodyn.* **23** (2011), 509–525.
109. M. Fabrizio and A. Morro, Dissipativity and irreversibility of electromagnetic systems, *Mathematical Models and Methods in Applied Sciences* **10** (2000), 217–246.
110. M. Fabrizio and J.M. Golden, Maximum and minimum free energies for a linear viscoelastic material, *Quart. Appl. Math.* **60** (2002), 341–381.
111. M. Fabrizio and J.M. Golden, Minimum free energies for materials with finite memory, *J. Elasticity* **72** (2003), 121–143.
112. M. Fabrizio and B. Lazzari, On the existence and the asymptotic stability of solutions for linearly viscoelastic solids, *Arch. Rational Mech. Anal.* **116** (1991), 139–152.
113. M. Fabrizio and B. Lazzari, On asymptotic stability for linear viscoelastic fluids, *Diff. Int. Eq.* **6** (3) (1993), 491–504.

114. M. Fabrizio and B. Lazzari, The domain of dependence inequality and asymptotic stability for a viscoelastic solid, *Nonlinear Oscillations* **1** (1998), 117–133.
115. M. Fabrizio, B. Lazzari and J.E. Muñoz Rivera, Asymptotic behavior in linear thermoelasticity, *J. Math. Anal. Appl.* **232** (1999), 138–165.
116. M. Fabrizio, B. Lazzari and J.E. Muñoz Rivera, Asymptotic behavior of a thermoelastic plate of weakly hyperbolic type, *Diff. Int. Eq.* **13** (2000), 1347–1370.
117. M. Fabrizio, B. Lazzari and R. Nibbi, Thermodynamics of non-local materials: extra fluxes and internal powers, *Continuum Mechanics and Thermodynamics* **23** (2011), 509–525.
118. M. Fabrizio and A. Morro, Thermodynamic restrictions on relaxation functions in linear viscoelasticity, *Mech. Res. Comm.* **12** (1985), 101–105.
119. M. Fabrizio and A. Morro, On uniqueness in linear viscoelasticity: a family of counterexamples, *Quart. Appl. Math.* **45** (1987), 321–325.
120. M. Fabrizio and A. Morro, Viscoelastic relaxation functions compatible with thermodynamics, *J. Elasticity* **19** (1988), 63–75.
121. M. Fabrizio and A. Morro, *Mathematical Problems in Linear Viscoelasticity*, SIAM, Philadelphia, 1992.
122. M. Fabrizio and A. Morro, *Electromagnetism of Continuous Media*, Oxford University Press, Oxford, 2003.
123. M. Fabrizio and A. Morro, Equivalent histories, minimal states and initial value problem in viscoelasticity, *Math. Meth. Appl. Sci.* **28** (2005), 233–251.
124. M. Fabrizio and A. Morro, Dissipativity and irreversibility of electromagnetic systems, *Math. Mod. Meth. Appl. Sc.* **10** (2000), 217–246.
125. M. Fabrizio and F. Franchi, Delayed thermal models. Stability and thermodynamics, *J. Therm. Stresses* **27** (2014), 160–173.
126. M. Fabrizio and B. Lazzari, Stability and Second Law of Thermodynamics in dual-phase-lag heat conduction, *Int. J. Heat Mass Transfer* **74** (2014), 484–489.
127. M. Fabrizio, B. Lazzari and R. Nibbi, Thermodynamics of non-local materials: extra fluxes and internal powers, *Continuum Mechanics and Thermodynamics* **23** (2011), 509–525.
128. M. Fabrizio, Fractional rheological models for thermomechanical systems. Dissipation and free energies, *Fractional Calculus and Applied Analysis* **17** (2014), 206–223.
129. L.H. Fatori and J.E. Muñoz Rivera, Energy decay for hyperbolic thermoelastic systems of memory type, *Quart. Appl. Math.* **59** (3) (2001), 441–458.
130. H.O. Fattorini, *Second Order Linear Differential Equations in Banach Spaces*, Elsevier Science Publishers B.V., Amsterdam, 1985.
131. W. Feller, *An Introduction to Probability Theory and Its Applications*, Vol. 1, 3rd ed. John Wiley & Sons, 1968.
132. E. Fermi, *Thermodynamics*, Dover, New York, 1956.
133. J.D. Ferry, *Viscoelastic Properties of Polymers*, 3rd edn., John Wiley, New York, 1980.
134. G. Fichera, *Existence theorems in elasticity*, in Handbuch der Physik, vol. VIa/2, C.A. Truesdell ed., Springer, Heidelberg, 1972.

135. G. Fichera, Avere una memoria tenace crea gravi problemi, *Arch. Rational Mech. Anal.* **70** (1979), 101–112.
136. G. Fichera, Sul principio della memoria evanescente, *Rend. Sem. Univ. Padova* **68** (1982), 245–259.
137. G. Fichera, *Problemi analitici nuovi nella Fisica Matematica Classica*, Scuola Tipo-Lito Istituto Anselmi, Marigliano, Napoli, 1985.
138. G. Fichera, On linear viscoelasticity, *Mech. Res. Comm.* **12** (1985), 241–242.
139. E. Fried and M.E. Gurtin, Tractions, balances and boundary conditions for non-simple materials with applications to liquid flow at small-length scales, *Arch. Rational Mech. Anal.* **182** (3) (2006), 513–554.
140. C. Friedrich, Mechanical stress relaxation in polymers: Fractional integral model versus fractional differential model, *J. Non-Newtonian Fluid Mech.* **46** (1993), 307–314.
141. J.D. Gabano and T. Poinot, Fractional modelling applied to heat conductivity and diffusivity estimation, *Phys. Scr.* **136** (2009) 014015 (6pp), https://doi.org/10.1088/0031-8949/2009/T136/014015.
142. E. Gagliardo, Proprietà di alcune classi di funzioni in più variabili, *Ricerche Mat.* **7** (1958), 102–137.
143. F.D. Gakhov, *Boundary Value Problems*, Pergamon, Oxford, 1966.
144. G. Gentili, Dissipative conditions and variational principles for the heat flux equation with memory, *Diff. Int. Eq.* **4** (5) (1991), 977–989.
145. G. Gentili, Maximum recoverable work, minimum free energy and state space in linear viscoelasticity, *Quart. Appl. Math.* **60** (2002), 153–182.
146. C. Giorgi, Alcune conseguenze delle restrizioni termodinamiche per mezzi viscoelastici lineari, *Quaderni del Dip. Matematico Univ. Cattolica S. Cuore di Brescia*, 6 (1989).
147. C. Giorgi and G. Gentili, Thermodynamic properties and stability for the heat flux equation with linear memory, *Quart. Appl. Math.* **51** (1993), 343–362.
148. C. Giorgi and B. Lazzari, *Uniqueness and stability in linear viscoelasticity: some counterexamples*, Quaderni del Seminario Matematico di Brescia, 1/90, Proceedings of the VI International Conference on Waves and Stability in Continuous Media, Sorrento (NA), October 9–14, 1989.
149. C. Giorgi, J.E. Muñoz Rivera and V. Pata, Global attractors for a semilinear hyperbolic equation in viscoelasticity, *J. Math. Anal. Appl.* **260** (2001), 83–99.
150. I.M. Gel'fand and G.E. Shilov, *Generalized functions* 1–5, Academic Press, 1968.
151. S. Glasgow and J.M. Golden, Dielectric materials with memory I: Minimum and general free energies, *Journal of Atomic and Nuclear Physics* **1** (2017), 1–15.
152. S. Glasgow and J.M. Golden, Dielectric materials with memory II: Free energies in non-magnetic materials, *Journal of Atomic and Nuclear Physics* **1** (2017), 16–29.
153. S. Glasgow, M. Meilstrup, J. Peatross and M. Ware, Real-time recoverable and irrecoverable energy in dispersive-dissipative dielectrics, *Phys. Rev.* **E75** 016616 (2007).

154. S. Glasgow and M. Ware, Real-time dissipation of optical pulses in passive dielectrics, *Phys. Rev.* **A80** 043817 (2009).
155. S. Glasgow, J. Corson and C. Verhaaren, Dispersive dielectrics and time reversal: free energies, orthogonal spectra and parity in dissipative media, *Phys. Rev.* **E82** 011115 (2010).
156. I.C. Gohberg and M.G. Kreĭn, Systems of integral equations on a half-line with kernels depending on the difference of arguments, *Am. Math. Soc. Transl.* **14** (2) (1960), 217–287.
157. J.M. Golden, A molecular theory of adhesive rubber friction, *J. Phys. A* **39** (1975), 53–70.
158. J.M. Golden, Free energies in the frequency domain: the scalar case, *Quart. Appl. Math.* **58** (2000), 127–150.
159. J.M. Golden, Consequences of nonuniqueness in the free energy of materials with memory, *Int. J. Engng. Sci.* **39** (2001), 53–70.
160. J.M. Golden, A proposal concerning the physical dissipation of materials with memory, *Quart. Appl. Math.* **63** (2005), 117–155.
161. J.M. Golden, A proposal concerning the physical dissipation of materials with memory; the nonisothermal case, *Mathematics and Mechanics of Solids* **12** (4) (2007), 403–449.
162. J.M. Golden, Constructing free energies for materials with memory, *Evol. Equat. Contr. Theor.* **3** (2014), 447–483.
163. J.M. Golden, Free energies for materials with memory in terms of state functionals, *Meccanica* **49** (2014), 2207–2235.
164. J.M. Golden, Unique characterization of materials with memory, *Quart. Appl. Math.* **74** (2016), 361–374.
165. J.M. Golden, Free energies for singleton minimal states, *Continuum Mechanics and Thermodynamics* **28** (6) (2016), 1821–1845.
166. J.M. Golden, The minimum and other free energies for non-linear materials with memory, *Quart. Appl. Math.* **74** (2016), 137–164.
167. J.M. Golden and G.A.C. Graham, *Boundary Value Problems in Linear Viscoelasticity*, Springer, Berlin, 1988.
168. I.S. Gradshteyn and I.M. Ryzhik, *Tables of Integrals, Series, and Products*, Academic Press, New York, 1965.
169. D. Graffi, Sui problemi dell'ereditarietà lineare, *Nuovo Cimento* A **5** (1928), 53–71.
170. D. Graffi, Sopra alcuni fenomeni ereditari dell'elettrologia, *Rend. Istituto Lombardo* **69** (1936), 151–166.
171. D. Graffi, Una teoria ereditaria dell'effetto Lussemburgo, *Rend. Sem. Mat. Univ. Padova* **7** (1956), 36–54.
172. D. Graffi, Sulla propagazione nei mezzi dispersivi, *Ann. Mat. Pura Appl.* **60** (1963), 173–194.
173. D. Graffi, Sull'espressione dell'energia libera nei materiali visco-elastici lineari, *Ann. Mat. Pura Appl.* **98** (1974), 273–279.
174. D. Graffi, Sull'espressione analitica di alcune grandezze termodinamiche nei materiali con memoria, *Rend. Sem. Mat. Univ. Padova* **68** (1982), 17–29.

175. D. Graffi, Ancora sull'expressione analitica dell'energia libera nei materiali con memoria, *Atti Acc. Scienze Torino* **120** (1986), 111–124.
176. D. Graffi and M. Fabrizio, Sulla nozione di stato materiali viscoelastici di tipo "rate", *Atti Accad. Naz. Lincei* **83** (1990), 201–208.
177. D. Graffi and M. Fabrizio, Non unicità dell'energia libera per i materiali viscoelastici, *Atti Acc. Naz. Lincei* **83** (1990), 209–214.
178. M. Grasselli, Identifying relaxation kernels of linearly viscoelastic bodies, *J. Inverse Ill-Posed Probl.* **4** (5) (1996), 391–407.
179. M. Grasselli, On an inverse problem for a linear hyperbolic integro-differential equation, *Forum Math.* **6** (1994), 83–110.
180. M. Grasselli, An identification problem for an abstract linear hyperbolic integro-differential equation with applications, *J. Math. Anal. Appl.* **171** (1) (1992), 27–60.
181. M. Grasselli, S.I. Kabanikhin and A. Lorenzi, An inverse integro-differential problem arising in Geophysics I, *Sibirsk. Math. J.* **33** (1992), 415–462.
182. M. Grasselli, S.I. Kabanikhin and A. Lorenzi, An inverse integro-differential problem arising in Geophysics II, *Nonlinear Anal.: Theory, Methods and Appl.* **15** (1990), 283–298.
183. M. Grasselli and V. Pata, Uniform attractors of nonautonomous systems with memory, in *Evolution Equations, Semigroups and Functional Analysis*, A Lorenzi. and B. Ruf, eds., pp.155–178, Progr. Nonlinear Differential Equations Appl. **50**, Birkhäuser, Boston, 2002.
184. G. Grioli, *Continui con memoria*, Atti Acc. Naz. Lincei, Roma, 1990.
185. G. Gripenberg, S.O. Londen and O. Staffens, *Volterra integral and functional equations*, Encyclopaedia of Mathematics and Its Applications, Cambridge University Press, Cambridge, 1990.
186. A.V. Gurevich, *Non-Linear Phenomena in the Ionosphere*, Springer, Berlin, 1994.
187. M.E. Gurtin, *The linear theory of elasticity*, in Handbuch der Physik, vol. VIa/2, C.A. Truesdell ed., Springer, Heidelberg, 1972.
188. M.E. Gurtin, *An Introduction to Continuum Mechanics*, Academic Press, Boston, 1981.
189. M.E. Gurtin and W.J. Hrusa, On energies for nonlinear viscoelastic materials of singleintegral type, *Quart. Appl. Math.* **46** (2) (1988), 381–392.
190. M.E. Gurtin and W. Hrusa, On the thermodynamics of viscoelastic materials of singleintegral type, *Quart. Appl. Math.* **49** (1) (1991), 67–85.
191. M.E. Gurtin and A.C. Pipkin, A general theory of heat conduction with finite wave speeds, *Arch. Rational Mech. Anal.* **31** (1968), 113–126.
192. R.A. Guyer and J.A. Krumhansl, Solution of the linearized phonon Boltzmann equation, *Phys. Rev.* **148** (1966), 766–778.
193. R.A. Guyer and J.A. Krumhansl, Thermal conductivity, second sound, and phonon hydrodynamic phenomena in nonmetallic crystals, *Phys. Rev.* **148** (1966), 778–788.
194. P.R. Halmos, *Finite-Dimensional Vector Spaces*, Springer-Verlag, New York, 1972.

195. P. Haupt, *Continuum Mechanics and Theory of Materials*, Springer, Berlin, 2000.
196. E. Heinsalu, M. Patriarca, I. Goychuk, G. Schmid and P. Hänggi, Fractional Fokker Plank dynamics: Numerical algorithm and simulation, *Physical Review* E 73, 046133 (2006).
197. E. Hewitt and K. Stromberg, *Real and Abstract Analysis*, Springer-Verlag, 1963.
198. R. Hilfer, *Applications of Fractional Calculus in Physics*, World Scientific Publishing Co, Singapore, 2003.
199. R. Hersch, How to classify differential polynomials, *Amer. Math. Montly* **80** (1973), 641–654.
200. N. Heymans and J.C. Bauwens, Fractal rheological models and fractional differential equations for viscoelastic behavior, *Rheol. Acta* **33** (1994), 210–219.
201. E. Hille and R.S. Phillips, *Functional Analysis and Semigroups*, Amer. Math. Soc., Coll. Publ., 1957.
202. S. Holm and S.P. Näsholm, A causal and fractional all-frequency wave equation for lossy media, *Journal of the Acoustical Society of America* **130** (2011), 2195–2201.
203. C.O. Horgan, Recent developments concerning Saint-Venant's principle: an update, *Appl. Mech. Rev.* **42** (1989), 295–303.
204. S.C. Hunter, Tentative equations for the propagation of stress, strain and temperature fields in viscoelastic solids, *J. Mech. Phys. Solids* **9** (1961), 39–51.
205. S.C. Hunter, *Mechanics of Continuous Media*, 2nd ed., Ellis Horwood, Chichester, 1983.
206. D. Ieşan, *Saint-Venant's problem*, in Lecture notes in mathematics, vol. 1279, A. Dold and B. Eckmann eds., Springer, Berlin, 1987.
207. G. Iaffaldano, M. Caputo and S. Martino, Experimental and theoretical memory diffusion of water in sand, *Hydrology and Earth System Sciences* **10** (2006), 93–100.
208. J.D. Jackson, *Classical Electrodynamics*, Wiley, New York, 1999.
209. J. Janno, Inverse problems for determining monotone weakly singular relaxation kernels in viscoelasticity, *Nonlinear Anal.: Theory, Methods and Appl.* **41** (2000), 943–962.
210. J. Janno, Identification of weakly singular relaxation kernels in three-dimensional viscoelasticity, *J. Math. Anal. Appl.* **262** (2001), 133–159.
211. J. Janno and L. v. Wolfersdorf, Inverse problems for identification of memory kernels in viscoelasticity, *Math. Meth. Appl. Sci.* **20** (1997), 291–314.
212. J. Janno and L. v. Wolfersdorf, Inverse problems for memory kernels by Laplace transform methods, *Z. Anal. Anwen.* **19** (2000), 489–510.
213. J. Janno and L. v. Wolfersdorf, An inverse problem for identification of a time- and spacedependent memory kernel in viscoelasticity, *Inverse Problems* **17** (2001), 13–24.
214. Z. Jiao, Y.Q. Chen and I. Podlubny, *Distributed–Order Dynamic Systems*, Springer, New York - London - Heidelberg, 2012.

215. D.D. Joseph, Slow motion and viscometric motion; stability and bifurcation of the rest state of a simple fluid, *Arch. Rational Mech. Anal.* **56** (1974), 99–157.
216. D.D. Joseph, *Fluid Dynamics of Viscoelastic Liquids*, Springer, New York, 1990.
217. G. Jumarie, New stochastic fractional models of the Malthusian growth, the Poissonian birth process and optimal management of populations, *Mathematical Computer Modelling* **44** (2006), 231–254.
218. A.A. Kilbas and S.A. Marzan, Nonlinear Differential Equations with Caputo Fractional Derivative in the Space of Continuously Differentiable Functions, *Differential Equations* **41** (1) (2005), 84–95.
219. A.A. Kilbas, H.M. Srivastava and J.J. Trujillo, *Theory and Applications of Fractional Differential Equations*, North-Holland Mathematical Studies, Amsterdam, 2006.
220. J.U. Kim, Control of a plate equation with large memory, *Diff. Int. Eq.* **5** (2) (1992), 261–279.
221. J.U. Kim, Control of a second-order integro-differential equation, *SIAM J. Control Optim.* **31** (1) (1993), 101–110.
222. V. Kiryakova, *Generalised Fractional Calculus and Applications*, Pitman Research Notes in Mathematics No. 301, Longman, Harlow, 1994.
223. V. V. Kulish and J.L. Lage, Fractional-Diffusion Solutions for Transient Temperature and Heat Transfer, *ASME J. Heat Transfer* **122** (2000), 372–376.
224. O.A. Ladyzhenskaya, *The Boundary Value Problem of Mathematical Physics*, Springer-Verlag, Berlin, 1985.
225. J.E. Lagnese, The reachability problem for the thermoelastic plates, *Arch. Rational Mech. Anal.* **112** (1990), 223–267.
226. J.E. Lagnese and J.L. Lions, *Modelling, Analysis and Control of Thin Plates*, Collection RMA, Masson, Paris, 1988.
227. L.D. Landau and E.M. Lifshitz, On the theory of the dispersion of magnetic permeability in ferromagnetic bodies, *Phys. Z. Sowjet.* **8** (1935), 153–166.
228. L.D. Landau, E.M. Lifshitz and L.P. Pitaevskii, *Electrodynamics of Continuous Media*, Pergamon, Oxford, 1984.
229. N. Laskin, Fractional Schrödinger equation, *Phys. Rev.* **E 66** (2002), 056108, https://doi.org/10.1103/PhysRevE.66.056108.
230. I. Lasiecka, Controllability of a viscoelastic Kirchhoff plate, *Internat. Ser. Numer. Math.* **91** (1989), 237–247.
231. I. Lasiecka and R. Triggiani, Exact controllability of the Euler-Bernoulli equation with controls in the Dirichlet and Neumann boundary conditions: A nonconservative case, *SIAM J. Control Optim.* **27** (1989), 330–373.
232. B. Lazzari, Exponential decay on the mean in linear viscoelasticity, in *Mathematical Models and Methods for Smart Materials*, M. Fabrizio, B. Lazzari and A. Morro eds., World Scientific, vol. 62, Singapore, 2002.
233. M.J. Leitman and G.M.C. Fisher, *The linear theory of viscoelasticity*, in Handbuch der Physik, vol. VIa/3, C. Truesdell ed., Springer-Verlag, Berlin, 1973.
234. G. Leugering, Exact boundary controllability of an integro-differential equation, *Appl. Math. Optim.* **15** (1987), 223–250.

235. J.L. Lions, *Contrôlabilité exacte, perturbations et stabilisation de systèmes distribués*, Tome 1, Masson, Paris, 1988. Contrôlabilité exacte. [Exact controllability], With appendices by E. Zuazua, C. Bardos, G. Lebeau, and J. Rauch.
236. J.L. Lions and E. Magenes, *Problèmes aux limites non homogènes et applications* Vol. 1–2, Dunod, Paris, 1968.
237. J.L. Lions and E. Magenes, *Non–Homogeneous Boundary Value Problems and Applications*, Vol. 1, Springer-Verlag, Berlin, 1972.
238. I-S. Liu, *Continuum Mechanics*, Springer, Berlin, 2002.
239. Z. Liu and S. Zheng, On the exponential stability of linear viscoelasticity and thermoviscoelasticity, *Quart. Appl. Math.* **54** (1996), 21–31.
240. H.A. Lorentz, *Theory of Electrons*, Dover, New York, 1952.
241. A. Lorenzi, Direct and inverse integro-differential Maxwell problems for dispersive media related to cylindrical domains, in *Inverse Problems in Diffusion Processes*, Proc. of the GAMM–SIAM Symposium, St. Wolfgang (Austria), June 27-July 1, 1994 (eds. H. Engl and W. Rundell), SIAM, 1995, 42–89.
242. A. Lorenzi, A multidimensional identification problem related to a hyperbolic integro-differential equation, *Z. Anal. Anwen.* **18** (1999), 407–435.
243. A. Lorenzi and E. Paparoni, Identification of two unknown coefficients in an integro-differential hyperbolic equation, *J. Inv. Ill-Posed Probl.* **1** (4) (1993), 331–348.
244. A. Lorenzi and V.G. Yakhno, An identification problem related to an isotropic nonhomogeneous stratified viscoelastic body, *J. Inv. Ill-Posed Probl.* **5** (1997), 29–53.
245. I. Luca, Domain of influence and uniqueness in viscothermoelasticity of integral type, *Continuum Mech. Thermodyn.* **1** (1989), 213–226.
246. Y. Luchko and R. Gorenflo, An operational method for solving fractional differential equations with the Caputo derivative, *Acta Mathematica Vietnamica* **24** (1999), 207–233.
247. R. Magin, *Fractional Calculus in Bioengineering*, Univ. of Illinois at Chicago Press, 2006.
248. F. Mainardi, *Fractional Calculus and Waves in Linear Viscoelasticity: An Introduction to Mathematical Models*, World Scientific Publishing, 2010.
249. F. Mainardi, *Fractional Calculus and Waves in Linear Viscoelasticity*, Imperial College, London, 2010.
250. F. Mainardi, An historical perspective on fractional calculus in linear viscoelasticity, *Fract. Calc. Appl. Anal.* **15** (2012), 712–717. E-print: http://arxiv.org/abs/1007.2959.
251. T. Manacorda, *Introduzione alla termodinamica dei continui*, Quaderni dell'Un. Mat. Ital. **14**, Pitagora, Bologna, 1979.
252. T. Manacorda, Teoria macroscopica della radiazione interna nei continui, *Riv. Mat. Univ. Parma* **8** (1982), 483–494.
253. A. Marzocchi and A. Musesti, The Cauchy stress theorem for bodies with finite perimeter, *Rend. Sem. Mat. Univ. Padova* **109** (2003), 1–11.
254. J.C. Maxwell, On the dynamical theory of gases, *Philos. Trans. Roy. Soc. London* Ser. A **157** (1867), 49–88.

255. R.C. MacCamy, Exponential stability for a class of functional differential equations, *Arch. Rational Mech. Anal.* **40** (1) (1970), 120–138.
256. R.C. MacCamy, Nonlinear Volterra equations on a Hilbert space, *J. Diff. Eq.* **16** (1974), 373–393.
257. R.C. MacCamy, Stability theorems for a class of functional differential equations, *SIAM J. Appl. Math.* **30** (1976), 557–576.
258. R.C. MacCamy and J.S.W. Wong, Exponential stability for a nonlinear functional differential equation, *J. Math. Anal. Appl.* **39** (1972), 699–705.
259. A.C. McBride, *Fractional Calculus and Integral Transforms of Generalized Functions*, Pitman Res. Notes in Mathem., nr. 31, Pitman, London, 1979.
260. M.F. McCarthy, Constitutive equations for thermomechanical materials with memory, *Int. J. Engng. Sci.* **8** (1970), 467–474.
261. H.R. Metzler, A. Blumen and T.F. Nonnenmacher, Generalized viscoelastic models: Their fractional equations with solutions, *J. Phys. A* **28** (1995a), 6567–6584.
262. L.A. Mihai and A. Goriely, How to characterize a nonlinear material? A review on nonlinear constitutive parameters in isotropic finite elasticity, *Proc. R. Soc.* **A473** (2017), 1–33.
263. R.K. Miller and R.L. Wheeler, Well-Posedness and stability of linear Volterra integro-differential equations in abstract spaces, *Funkcialaj Ekvacioj* **21** (1978), 279–305.
264. K.S. Miller and B. Ross, *An Introduction to Fractional Calculus and Fractional Differential Equations*, Wiley, New York, 1993.
265. A. Morro, Negative semidefiniteness of the viscoelastic attenuation tensor via the Clausius–Duhem inequality, *Arch. Mech.* **37** (1985), 255–259.
266. A. Morro and M. Vianello, Minimal and maximal free energy for materials with memory, *Boll. Un. Mat. Ital.* **4 A** (1990) 45–55.
267. P.M. Morse and H. Feshbach, *Methods of Theoretical Physics*, McGraw-Hill, New York, 1953.
268. I. Müller, On the entropy inequality, *Arch. Rational Mech. Anal.* **26** (1967), 118–141.
269. I. Müller and T. Ruggeri, *Rational Extended Thermodynamics*, 2nd ed., Springer, New York, 1998.
270. J.E. Muñoz Rivera, Asymptotic behavior in linear viscoelasticity, *Quart. Appl. Math.* **52** (1994), 629–648.
271. J.E. Muñoz Rivera and M.G. Naso, Exact boundary controllability in thermoelasticity with memory, *Advances Diff. Eq.* **8** (4) (2003), 471–490.
272. J.E. Muñoz Rivera and M.G. Naso, Exact controllability for hyperbolic thermoelastic systems with large memory, *Advances Diff. Eq.* **9** (11–12) (2004), 1369–1394.
273. S. Murakami, Exponential asymptotic stability of scalar linear Volterra equations, *Diff. Int. Eq.* **4** (3) (1991), 519–525.
274. N.I. Muskhelishvili, *Singular Integral Equations*, Noordhoff, Groningen, 1953.
275. M. Naber, Time fractional Schrödinger equation, *J. Mathematical Physik* **45** (8) (2004), 3339–3352.

276. J. Nohel, M. Renardy and W. Hrusa, *Mathematical Problems in Linear Viscoelasticity*, Pitman Monographs Surveys in Pure and Applied Mathematics, Longman, 1987.
277. W. Noll, A new mathematical theory of simple materials, *Arch. Rational Mech. Anal.* **48** (1972), 1–50.
278. W. Noll and E.G. Virga, Fit regions and functions of bounded variation, *Arch. Rational Mech. Anal.* **102** (1988), 1–21.
279. J.W. Nunziato, On heat conduction in materials with memory, *Quart. Appl. Math.* **29** (1971), 187–204.
280. P.G. Nutting, A new general law deformation, *J. Franklin Inst.* **191** (1921), 678–685.
281. R.W. Ogden, *Non-linear Elastic Deformations*, 2nd ed., Dover, New York, 1997.
282. L. Pandolfi, The controllability of the Gurtin-Pipkin equation: a cosine operator approach, *Appl. Math. Optim.* **52** (2005), 143–165.
283. V. Pata, Exponential stability in linear viscoelasticity, *Quart. Appl. Math.* **64** (2006), 499–513.
284. J. Prüss, *Evolutionary Integral Equations and Applications*, Monographs in Mathematics, Birkhäuser, 1993.
285. I. Petras, B.M. Vinagre, L. Dorcak and V. Feliu, *Fractional digital control of a heat solid: Experimental results*, International Carpathian Control - Conference, Malenovice, 2002.
286. Ivo Petras, *Fractional Calculus Nonlinear Systems, Modelling, Analysis and Simulation*, Springer, 2011.
287. I. Podlubny, *Fractional Differential Equations*, Academic Press, New York, 1999.
288. R. Quintanilla, A condition on the delay parameters in the one-dimensional dual-phase-lag thermoelastic theory, *J. Therm. Stresses* **26** (2003), 713–721.
289. R. Quintanilla and R. Racke, A note on stability in dual-phase-lag heat conduction, *Int. J. Heat Mass Transfer* **49** (2006), 1209–1213.
290. R. Quintanilla and R. Racke, Qualitative aspects in dual-phase-lag heat conduction, *Proc. Roy. Soc. London A* **463** (2007), 659–674.
291. J. Ramanathan, *Methods of Applied Fourier Analysis*, Birkhauser, Boston, 1998.
292. M. Renardy, W.J. Hrusa and J.A. Nohel, *Mathematical problems in viscoelasticity*, Harlow John Wiley and Sons, New York, 1987.
293. P.E. Rouse and K. Sittel, Viscoelastic properties of dilute polymer solutions, *J. App. Phys.* **24** (1953), 690–696.
294. J. Sabatier, O.P. Agraval and J.A. Teneiro Machado, *Advances in Fractional Calculus*, Springer, 2009.
295. S.G. Samko, A.A. Kilbas and O.I. Marichev, *Fractional Integrals and Derivatives*, Gordon and Breach Sc. Publ., 1993.
296. G. Scott-Blair, An application of the theory of quasi-properties to the treatment of anomalous stress-strain relation, *Phys. Mag.* **40** (1949), 80–94.

297. J. Serrin, An outline on thermodynamical structure, in *New Perspective in Thermodynamics*, ed. Serrin J., Springer, Berlin, 1986.
298. R.E. Showalter, *Hilbert space methods for partial differential equations*, Pitman, London, San Francisco, Melbourne, 1977.
299. D. Sierociuk, A. Dzielinski, G. Sarwas, I. Petras. I. Podlubny and T. Skovranek, Modelling heat transfer in heterogeneous media using fractional calculus, *Philos. Trans. A Math. Phys. Eng. Sci.* **371** (1990), 2013: 20120146. https://doi.org/10.1098/rsta.2012.0146.
300. M. Slemrod, A hereditary partial differential equation with applications in the theory of simple fluids, *Arch. Rational Mech. Anal.* **62** (1976), 303–332.
301. M. Slemrod, An energy stability method for simple fluids, *Arch. Rational Mech. Anal.* **68** (1978), 1–18.
302. I.N. Sneddon, *The Use of Integral Transforms*, McGraw-Hill, New York, 1972.
303. P.D. Spanos and G.I. Evangelatos, Response of a non-linear system with restoring forces governed by fractional derivatives-time domain simulation and statistical linearization solution, *Soil. Dyn. Earthquake Eng.* **30** (1991), 6426–6434.
304. O.J. Staffans, On a nonlinear hyperbolic Volterra integral equation, *SIAM J. Math. Anal.* **11** (1980), 793–812.
305. A.J. Staverman and F. Schwarzl, Thermodynamics of viscoelastic behavior, *Proc. Konink. Nederl. Akad. Wettensch.* **B55** (1952), 474–485.
306. E.M. Stein, R. Shakarchi, *Complex Analysis*, Princeton University Press, Princeton, 2003.
307. R. Teman, *Navier–Stokes Equations*, North-Holland, Amsterdam, New York, Oxford, 1984.
308. E.C. Titchmarsh, *Introduction to the Theory of Fourier Integrals*, Clarendon, Oxford, 1937.
309. D.F.M. Torres and A.B. Malinowska, *Introduction to the Fractional Calculus of Variations*, Imperial College Press, 2012.
310. F. Treves, *Basic Linear Partial Differential Equations*, Academic Press, New York, 1975.
311. C. Truesdell, *Rational Thermodynamics*, McGraw Hill, New York, 1969.
312. C. Truesdell, The meaning of viscometry in fluid dynamics, *Ann. Rev. Fluid Mech.* **6** (1974), 111–146.
313. C. Truesdell and W. Noll, *The nonlinear Field Theories of Mechanics*, in Handbuch der Physik, vol. III/3, Flügge ed., Springer, Berlin, 1964; also Ed. S. Antman, Springer, Berlin, 2004.
314. C. Truesdell, *A First Course in Rational Continuum Mechanics*, Vol. 1, Academic Press, New York, 1977.
315. D.Y. Tzou, A unified approach for heat conduction from macro to micro-scales, *ASME J. Heat Transfer* **117** (1995), 8–16.
316. V. Volterra, Sur les équations intégro-différentielles et leurs applications, *Acta Math.* **35** (1912), 295–356.
317. V. Volterra, *Leçons sur les fonctions de lignes*, Gauthier-Villars, Paris, 1913.

318. V. Volterra, *Theory of Functional and of Integral and Integro-differential Equations*, Blackie and Son Limited, London, 1930.
319. L.Wang and M. Zu, Well-posedness of dual-phase-lagging heat equation:higher dimensions, *Int. J. Heat Mass Transfer* **45** (2002), 1165–1171.
320. D.V. Widder, *The Laplace Transform*, Princeton University Press, Princeton, 1941.
321. N.S. Wilkes, Thermodynamic restrictions on viscoelastic materials, *Quart. J. Mech. Appl. Math.* **30** (1977), 209–221.
322. L. v. Wolfersdorf, On identification of memory kernels in linear viscoelasticity, *Math. Nachr.* **161** (1993), 203–217.
323. J. Zabczyk, *Mathematical Control Theory: An Introduction*, Systems & Control: Foundations and Applications, Birkhäuser Boston Inc., Boston, MA, 1992.
324. G.Z. Voyiadjis and R.K. Abu Al-Rub, Nonlocal gradient-dependent thermodynamics for modeling scale-dependent plasticity, *Int. J. Multiscale Computational Engineering* **5** (3-4) (2007), 295–323.
325. G.Z. Voyiadjis and A.H. Almasri, Variable material length scale associated with nanoindentation experiments, *J. Engrg. Mech. ASCE* **135** (3) (2009), 139–148.
326. J.H. Williamson, *Lebesgue Integration*, Holt, Rinehart and Winston Inc., New York, 1962.
327. L. Ying and Y. Chen, *Fractional Order Motion Controls*, Wiley, 2012.
328. P. Zavada, Relativistic wave equations with fractional derivatives and pseudodifferential operators, *Journal of Applied Mathematics* **2** (4) (2002), 163–197.
329. M.W. Zemansky, *Heat and Thermodynamics*, McGraw-Hill, New York, 1957.

Index

G. Amendola et al., *Thermodynamics of Materials with Memory*,
https://doi.org/10.1007/978-3-030-80534-0

R

S

Y

Printed by Printforce, the Netherlands